# Virtual Lab CD-ROM (free in the back of this book)

One of the very important changes to engineering education in the 1990s has been the more common use of computers for analysis, design, data acquisition, and control. This book and its accompanying Virtual Lab CD-ROM is designed to permit students and instructors to experiment with various computer-aided design and analysis tools. Following are the items included on the CD-ROM.

- Many examples from the text are supplemented by electronic solutions in Matlab or MathCAD that are intended to teach students how to solve typical electrical engineering problems using such computer aids, and to stimulate them to experiment in developing their own solution methods. Many of these methods will also be useful later in the curriculum.

- Some examples and figures in the text are supplemented by circuit simulation created using Electronics Workbench, a circuit analysis and simulation program that has a particularly friendly user interface, and that permits a more in-depth analysis of realistic electrical/electronic circuits and devices. Use of this feature could be limited to just running a simulated circuit to observe its behavior (with virtually no new learning required), or could be more involved and result in the design of new circuit simulations.

- Data sheets for real devices mentioned in the text.

- Instrumentation Examples, provided courtesy of Hewlett-Packard Company.

- "Find It on the Web" links from the text.

We hope these tools enrich your experience as you use Rizzoni's PRINCIPLES AND APPLICATIONS OF ELECTRICAL ENGINEERING, Third Edition.

# PRINCIPLES AND APPLICATIONS
# OF ELECTRICAL ENGINEERING

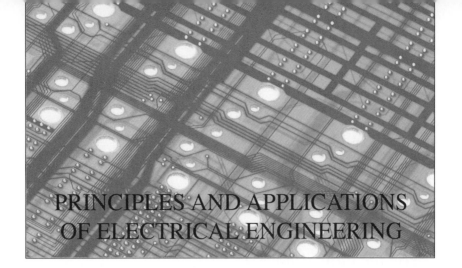

# PRINCIPLES AND APPLICATIONS OF ELECTRICAL ENGINEERING

## Third Edition

Giorgio Rizzoni
*The Ohio State University*

Boston   Burr Ridge, IL   Dubuque, IA   Madison, WI   New York   San Francisco   St. Louis
Bangkok   Bogotá   Caracas   Lisbon   London   Madrid
Mexico City   Milan   New Delhi   Seoul   Singapore   Sidney   Taipei   Toronto

**ABOUT THE COVER**

The cover photo (David H. Koether Photography) depicts The Ohio State University Smokin' Buckeye electric race car, displayed in front of Hayes Hall. Smokin' Buckeye is the winner of the 1996, 1997, and 1998 ABB University Spec Series Championships. The car is powered by a 400-volt battery pack supplying a 150-kW AC induction motor, and has a top speed of 144 miles per hour. Example 1.2 describes the design and performance characteristics of this vehicle in greater detail. For further information, visit the web site http://turbo.eng.ohio-state.edu/~lightning/.

This project was supported, in part, by the

**National Science Foundation**

Opinions expressed are those of the authors and not necessarily those of the Foundation

*McGraw-Hill Higher Education*

*A Division of The **McGraw-Hill** Companies*

domestic        2 3 4 5 6 7 8 9 0 DOW/DOW 9 0 9 8 7 6 5 4 3 2 0
international  1 2 3 4 5 6 7 8 9 0 DOW/DOW 9 0 9 8 7 6 5 4 3 2 0 9

ISBN 0-256-26116-4

Vice president/Editor-in-Chief:   *Kevin T. Kane*
Publisher:   *Thomas Casson*
Executive Editor:   *Elizabeth A. Jones*
Sponsoring editor:   *Catherine Fields*
Senior developmental editor:   *Kelly Butcher*
Senior marketing manager:   *John T. Wannemacher*
Senior project manager:   *Mary Conzachi*
Production supervisor:   *Debra R. Benson*
Designer:   *Kiera Cunningham*
Interior design: *Jamie O'Neal*
Supplement coordinator:   *Nancy Martin*
Compositor:   *Techsetters, Inc.*
Typeface:   *10/12 Times Roman*
Printer:   *R. R. Donnelley & Sons Company*

**Library of Congress Cataloging-in-Publication Data**

Rizzoni, Giorgio.
    Principles and applications of electrical engineering  / Giorgio
  Rizzoni. -- 3rd ed.
        p.      cm.
    ISBN 0-256-26116-4
    1. Electrical engineering.      I. Title.
  TK146.R473   2000
  621.3--dc21                                            99-25420

http://www.mhhe.com

Alla mia Chicca

# About the Author

Giorgio Rizzoni received the B.S., M.S., and Ph.D. degrees, all in electrical engineering, from the University of Michigan. He is currently on the faculty of the Department of Mechanical Engineering at The Ohio State University, where he teaches undergraduate courses in system dynamics, measurements, and mechatronics, and graduate courses in automotive powertrain modeling and control, hybrid vehicle modeling and control, system fault diagnosis, and digital signal processing.

Dr. Rizzoni has been involved in the development of innovative curricula and educational programs throughout his career. At the University of Michigan, where he first taught as a Lecturer, he developed a new laboratory and revamped the curriculum for the circuits and electronics engineering service course for non–electrical engineering majors. The first edition of this book was a direct result of that effort. At Ohio State, he has been involved—in collaboration with electrical and mechanical engineering colleagues—in the development of undergraduate and graduate curricula in Mechatronic Systems. Funding for this program was provided, in part, by the National Science Foundation through a curriculum development grant. The second and third editions of this book have been profoundly influenced by this interdisciplinary curriculum development.

Dr. Rizzoni and his colleagues have also developed and implemented a unique year-long graduate course sequence titled *Powertrain Modeling and Control* in collaboration with General Motors. This course sequence is offered to GM employees as a series of distance-learning courses, and is regularly taught on campus to Ohio State electrical and mechanical engineering students.

Most recently, Dr. Rizzoni has been awarded funding from the U.S. Department of Energy to establish a *Graduate Automotive Technology Education Center on Hybrid Vehicle Drivetrains and Control Systems.* This activity has resulted in the development of a graduate curriculum and of research laboratories devoted to the study of future vehicle propulsion technologies.

Dr. Rizzoni's research, in collaboration with The Ohio State University *Center for Automotive Research,* concerns the modeling, simulation, control, and diagnosis of automotive powertrains and hybrid vehicles. His work has been funded by a number of government agencies and corporations, including, among others, DOE, NASA, NSF, DaimlerChrysler, Ford, General Motors, Delphi Automotive Systems, Cummins, IBM, Motorola, and Allied Signal. He has published over 100 papers in peer-reviewed journals and conference proceedings, and has received a number of recognitions, including a 1991 NSF *Presidential Young Investigator Award.*

Dr. Rizzoni is a member of ASME, IEEE, and SAE, and has been an Associate Editor of the *ASME Journal of Dynamic Systems, Measurements, and Control* (1993–99) and of the *IEEE Transactions on Vehicular Technology* (1988–1998). He has served as Guest Editor of Special Issues of the *IEEE Transactions on Control System Technology,* of the *IEEE Control Systems Magazine,* and of *Control Engineering Practice,* and is a past Chair of the Transportation Panel of the ASME *Dynamic Systems and Control Division.*

He is The Ohio State University SAE student branch faculty advisor, and has led teams of electrical and mechanical engineering students through the development of a high-performance electric vehicle, culminating in three consecutive national championships (1996–1998). He is also an advisor of the OSU FutureCar Challenge hybrid-electric vehicle design team. OSU is one of 14 schools to have been awarded this prestigious project, sponsored by the U.S. Department of Energy, and by General Motors, Ford, and DaimlerChrysler through the United States Council for Automotive Research.

**http://rclsgi.eng.ohio-state.edu.rizzoni**

# Preface

The pervasive presence of electronic devices and instrumentation in all aspects of engineering design and analysis is one of the manifestations of the electronic revolution that has characterized the second half of the 20th century. Every aspect of engineering practice, and even of everyday life, has been affected in some way or another by electrical and electronic devices and instruments. Computers are perhaps the most obvious manifestations of this presence. However, many other areas of electrical engineering are also important to the practicing engineer, from mechanical and industrial engineering to chemical, nuclear, and materials engineering, the aerospace and astronautical disciplines, and civil engineering. Engineers today must be able to communicate effectively within the interdisciplinary teams in which they work.

## OBJECTIVES

The objectives of this book have not changed since work started on the first edition in 1987, even though engineering education and engineering professional practice have seen profound changes in the past decade. The integration of electronics and computer technologies in all engineering academic disciplines and the emergence of digital electronics and microcomputers as a central element of many engineering products and processes have become a common theme across the world.

In this context, the importance of material presented in this book has further increased, and this book is no longer aimed only at electrical engineering service courses in circuits, electronics, and electromechanics, but also at the increasing number of mechatronic systems courses and curricula that are under development in engineering schools around the world.

The basic objective of the book is to present the *principles* of electrical, electronic, and electromechanical engineering to an audience composed of non–electrical engineering majors, and ranging from sophomore students in their required introductory electrical engineering course to seniors and even first-year graduate students enrolled in more specialized courses in electronics, electromechanics, and mechatronics.

A second objective is to present the essential material in an uncomplicated fashion, focusing on the important results and applications, and presenting the students with the most appropriate *analytical and computational tools* to solve a variety of practical problems.

Finally, a third objective of the book is to illustrate, by way of examples, a number of relevant *applications* of electrical engineering principles. These examples are drawn from the author's industrial research experience, and also from ideas contributed by practicing engineers and industrial partners.

The three objectives listed above are met through the use of a number of new features, affecting the pedagogy and content of this book. The next two sections of this preface describe the organization of the book and the major changes that have been implemented in this third edition.

## ORGANIZATION AND CONTENT

The organization of the book is nearly unchanged in its basic elements: the book is divided into three sections, devoted to *Circuits, Electronics, and Electromechanics*. Two changes, both resulting from the advice of users of the second edition, have been made to the Circuits section. First, the material on AC circuit analysis and frequency response, transient response, and system analysis of linear circuits has been reorganized into three chapters (4–6) instead of two, to more clearly separate the concepts of AC circuits and sinusoidal steady-state analysis, transient analysis, and frequency response and transfer function analysis. The second change results in the material on AC power being moved to the end of the section (it is now Chapter 7), to preserve the natural continuity of the material in Chapters 2–6. Instructors who desie to introduce AC power material immediately following the presentation of AC circuit analysis can do so without any difficulty.

The organization of the book continues to be very modular. The same material can be packaged into different sequences to satisfy the needs of different audiences. Figure 1 depicts the possible sequences that have been implemented by past and current users of this book and by this author. Figure 2 depicts some suggestions for advanced courses that would suit curricula in power systems, applied electronics, or mechanatronics. Instructors will find additional suggestions on the organization of course materials at the book's web site: **http://www.mhhe.com/engcs/electrical/rizzoni**. Suggestions from users of the book are welcome!

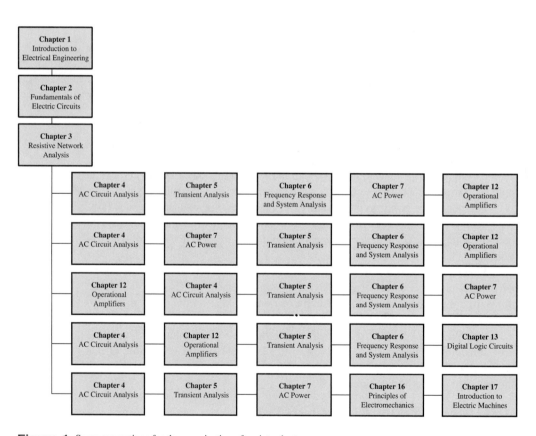

**Figure 1** Some suggestions for the organization of an introductory course

## CHANGES IN THE THIRD EDITION

While the organization of the book has not changed in a major way, the improvements to the pedagogy and to the supplements are substantial.

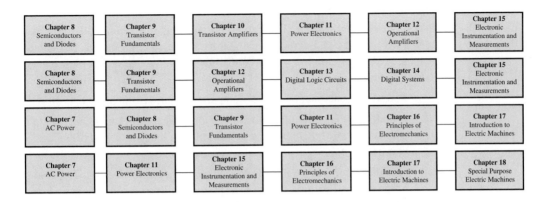

**Figure 2**  Some suggestions for the organization of an advanced course

## Pedagogy

The third edition includes a complete revision of every example in the book. The examples follow a consistent solution methodology, reinforced by the use of "Focus on Methodology" boxes that summarize important solutions methods. The emphasis on electronic instrumentation and measurements—already a feature of the first two editions—is further underscored by the introduction of "Focus on Measurements" sections, consisting of examples based on measurement and instrumentation applications of electrical engineering concepts. Further, the students and instructors can choose to take advantage of **computer-aided solutions** methods and of **Internet resources** found in the enclosed CD-ROM, and identified in the text by the icons for **Virtual Lab** and **Find It on the Web**. These resources significantly extend the material presented in the book, and provide the instructor and the student with material for further study, and for the development of special projects and laboratory and computer exercises.

## Supplements

The book includes a wealth of supplements, many available in electronic form. These include:

- A CD-ROM containing computer-aided example solutions, a list of web references for further research, device data sheets, and additional Virtual Lab material, including a demo copy of the *Electronics WorkBench*™ software package.
- A web site that will be dynamically updated to provide students and instructors with instructor notes, additional examples, suggestions for the use of the book, a forum for discussion, and other features. The URL is www.mhhe.com/engcs/electrical/rizzoni.
- A new chapter on electrical communications for schools that teach this subject in their curriculum.
- A completely revised Solutions Manual, available both in paper and electronic form.
- PowerPoint presentation slides of important figures, available on the Online Learning Center.
- Transparency masters of important figures.

## Online Learning Center

This book has an Online Learning Center, hosted on the McGraw-Hill web site at **http/www.mhhe.com**. Instructors using the text can access a special curriculum-based threaded discussion list within the Online Learning Center. Resources available are

- PowerPoint presentation slides of important figures from the text.
- A course web site builder called Page Out. Instructors can quickly build a course web site by entering basic information into McGraw-Hill's Page Out interface.
- Additional support for the CD-ROM that accompanies the text. The goal of the Online Learning Center is to provide full-service instructor and course management support for those who request it.
- An Instructor's Solutions Manual with complete solutions to all homework problems.

## Acknowledgments

This book has been critically reviewed by the following reviewers. McGraw-Hill and the author would like to thank these reviewers for their invaluable contribution to the third edition of *Principles and Applications of Electrical Engineering*.

David Cunningham, University of Missouri–Rolla
Piero Azzoni, Università di Bologna, Italy
Paul Claspy, Case Western Reserve University
Stephen A. Minnick, United States Naval Academy
Jianhua (David) Zhang, University of Illinois–Urbana
Til Glisson, North Carolina State University
Steven Bibyk, The Ohio State University
Roland Zapp, Michigan State University
M. Paul Murray, Mississippi State University
H. P. D. Lanyon, Worcester Polytechnic Institute

A number of people helped plan the third edition by completing a questionnaire about the second edition. I would like to thank these people for their feedback:

Richard S. Marleau, The University of Wisconsin
Manuel Navarro, Bradley University
Roland Zapp, Michigan State University
James D. Dilbert, Tennessee Technological University
Nelson M. Duller, Texas A&M University
In-Soo Ahn, Bradley University
Cynthia Fusse, University of Utah
Gene Stuffle, Idaho State University
Greg Bailey, San Diego State University
Gabriel Rebeiz, University of Michigan–Ann Arbor
Ron Bowman, Clemson University
Stephen Minnick, United States Naval Academy
Carl Halford, University of Memphis

The author is grateful to Professor James D. Gilbert (Tennessee Technological University) and to Professor James Kearns (York College of Pennsylvania) for their assistance in creating new homework problems. Accuracy checking and creation of the Solutions Manual have been skillfully contributed by Messrs. Fabrizio Ponti and Nicolò Cavina (Università di Bologna). Stefano Caruso and Brady Gambatese have supplied invaluable assistance in the production of the manuscript text and figures.

It is impossible for me to adequately express my love for my family, and my gratitude for my wife, Kathryn, and to my wonderful children, Alessandro, Maria, and Michael, who have continuously provided resources, inspiration, support, and good humor throughout this project. I cannot imagine a more loving and happier family.

# Contents

## PART III    ELECTROMECHANICS    766

## Chapter 16    Principles of Electromechanics    767

**Find Chapter 19 on the Web**
http://www.mhhe.com/engcs/electrical/rizzoni

# C H A P T E R

## 1

# Introduction to Electrical Engineering

The aim of this chapter is to introduce electrical engineering. The chapter is organized to provide the newcomer with a view of the different specialties making up electrical engineering and to place the intent and organization of the book into perspective. Perhaps the first question that surfaces in the mind of the student approaching the subject is, Why electrical engineering? Since this book is directed at a readership having a mix of engineering backgrounds (including electrical engineering), the question is well justified and deserves some discussion. The chapter begins by defining the various branches of electrical engineering, showing some of the interactions among them, and illustrating by means of a practical example how electrical engineering is intimately connected to many other engineering disciplines. In the second section, *mechatronic systems engineering* is introduced, with an explanation of how this book can lay the foundation for interdisciplinary mechatronic product design. This design approach is illustrated by an example. The next section introduces the Engineer-in-Training (EIT) national examination. A brief historical perspective is also provided, to outline the growth and development of this relatively young engineering specialty. Next, the fundamental physical quantities and the system of units are defined, to set the stage for the chapters that follow. Finally, the organization of the book is discussed, to give the student, as well as the teacher, a sense of continuity in the development of the different subjects covered in Chapters 2 through 18.

## 1.1  ELECTRICAL ENGINEERING

The typical curriculum of an undergraduate electrical engineering student includes the subjects listed in Table 1.1. Although the distinction between some of these subjects is not always clear-cut, the table is sufficiently representative to serve our purposes. Figure 1.1 illustrates a possible interconnection between the disciplines of Table 1.1. The aim of this book is to introduce the non-electrical engineering student to those aspects of electrical engineering that are likely to be most relevant to his or her professional career. Virtually all of the topics of Table 1.1 will be touched on in the book, with varying degrees of emphasis. The following example illustrates the pervasive presence of electrical, electronic, and electromechanical devices and systems in a very common application: the automobile. As you read through the example, it will be instructive to refer to Figure 1.1 and Table 1.1.

**Table 1.1**  Electrical engineering disciplines

Circuit analysis
Electromagnetics
Solid-state electronics
Electric machines
Electric power systems
Digital logic circuits
Computer systems
Communication systems
Electro-optics
Instrumentation systems
Control systems

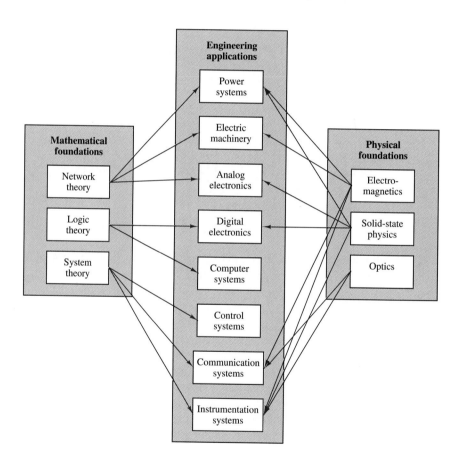

**Figure 1.1** Electrical engineering disciplines

## EXAMPLE  1.1  Electrical Systems in a Passenger Automobile

A familiar example illustrates how the seemingly disparate specialties of electrical engineering actually interact to permit the operation of a very familiar engineering system: the automobile. Figure 1.2 presents a view of electrical engineering systems in a

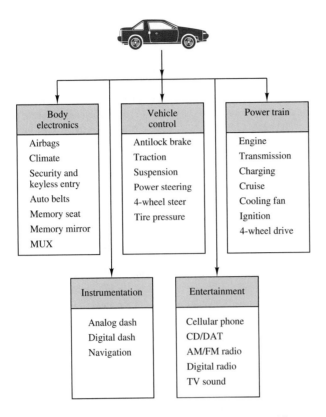

**Figure 1.2** Electrical engineering systems in the automobile

modern automobile. Even in older vehicles, the electrical system—in effect, an *electric circuit*—plays a very important part in the overall operation. An inductor coil generates a sufficiently high voltage to allow a spark to form across the spark plug gap, and to ignite the air and fuel mixture; the coil is supplied by a DC voltage provided by a lead-acid battery. In addition to providing the energy for the ignition circuits, the battery also supplies power to many other electrical components, the most obvious of which are the lights, the windshield wipers, and the radio. Electric power is carried from the battery to all of these components by means of a wire harness, which constitutes a rather elaborate electrical circuit. In recent years, the conventional electrical ignition system has been supplanted by *electronic* ignition; that is, solid-state electronic devices called *transistors* have replaced the traditional breaker points. The advantage of transistorized ignition systems over the conventional mechanical ones is their greater reliability, ease of control, and life span (mechanical breaker points are subject to wear).

Other electrical engineering disciplines are fairly obvious in the automobile. The on-board radio receives electromagnetic waves by means of the antenna, and decodes the communication signals to reproduce sounds and speech of remote origin; other common *communication systems* that exploit *electromagnetics* are CB radios and the ever more common cellular phones. But this is not all! The battery is, in effect, a self-contained 12-VDC *electric power system,* providing the energy for all of the aforementioned functions. In order for the battery to have a useful lifetime, a charging system, composed of an alternator and of power electronic devices, is present in every automobile. The alternator is an *electric machine*, as are the motors that drive the power mirrors, power windows, power seats, and other convenience features found in luxury cars. Incidentally, the loudspeakers are also electric machines!

The list does not end here, though. In fact, some of the more interesting applications of electrical engineering to the automobile have not been discussed yet. Consider *computer systems*. You are certainly aware that in the last two decades, environmental concerns related to exhaust emissions from automobiles have led to the introduction of sophisticated engine emission *control systems*. The heart of such control systems is a type of computer called a *microprocessor*. The microprocessor receives signals from devices (called *sensors*) that measure relevant variables—such as the engine speed, the concentration of oxygen in the exhaust gases, the position of the throttle valve (i.e., the driver's demand for engine power), and the amount of air aspirated by the engine—and subsequently computes the optimal amount of fuel and the correct timing of the spark to result in the cleanest combustion possible under the circumstances. The measurement of the aforementioned variables falls under the heading of *instrumentation*, and the interconnection between the sensors and the microprocessor is usually made up of *digital circuits*. Finally, as the presence of computers on board becomes more pervasive—in areas such as antilock braking, electronically controlled suspensions, four-wheel steering systems, and electronic cruise control—communications among the various on-board computers will have to occur at faster and faster rates. Some day in the not-so-distant future, these communications may occur over a fiber optic network, and *electro-optics* will replace the conventional wire harness. It should be noted that electro-optics is already present in some of the more advanced displays that are part of an automotive instrumentation system.

## 1.2    ELECTRICAL ENGINEERING AS A FOUNDATION FOR THE DESIGN OF MECHATRONIC SYSTEMS

Many of today's machines and processes, ranging from chemical plants to automobiles, require some form of electronic or computer control for proper operation. Computer control of machines and processes is common to the automotive, chemical, aerospace, manufacturing, test and instrumentation, consumer, and industrial electronics industries. The extensive use of microelectronics in manufacturing systems and in engineering products and processes has led to a new approach to the design of such engineering systems. To use a term coined in Japan and widely adopted in Europe, *mechatronic design* has surfaced as a new philosophy of design, based on the integration of existing disciplines—primarily mechanical, and electrical, electronic, and software engineering.[1]

A very important issue, often neglected in a strictly disciplinary approach to engineering education, is the integrated aspect of engineering practice, which is unavoidable in the design and analysis of large scale and/or complex systems. One aim of this book is to give engineering students of different backgrounds exposure to the integration of electrical, electronic, and software engineering into their domain. This is accomplished by making use of modern computer-aided tools and by providing relevant examples and references. Section 1.6 describes how some of these goals are accomplished.

---

[1]D. A. Bradley, D. Dawson, N. C. Burd, A. J. Loader, 1991, "Mechatronics, Electronics in Products and Processes," Chapman and Hall, London. See also ASME/IEEE *Transactions on Mechatronics*, Vol. 1, No. 1, 1996.

Example 1.2 illustrates some of the thinking behind the mechatronic system design philosophy through a practical example drawn from the design experience of undergraduate students at a number of U.S. universities.

## EXAMPLE 1.2 Mechatronic Systems—Design of a Formula Lightning Electric Race Car

The Formula Lightning electric race car competition is an interuniversity[2] competition project that has been active since 1994. This project involves the design, analysis, and testing of an electric open-wheel race car. A photo and the generic layout of the car are shown in Figures 1.3 and 1.4. The student-designed propulsion and energy storage systems have been tested in interuniversity competitions since 1994. Projects have included vehicle dynamics and race track simulation, motor and battery pack selection, battery pack and loading system design, and transmission and driveline design. This is an ongoing competition, and new projects are defined in advance of each race season. The objective of this competitive series is to demonstrate advancement in electric drive technology for propulsion applications using motorsports as a means of extending existing technology to its performance limit. This example describes some of the development that has taken place at the Ohio State University. The description given below is representative of work done at all of the participating universities.

**Figure 1.3** The Ohio State University Smokin' Buckeye

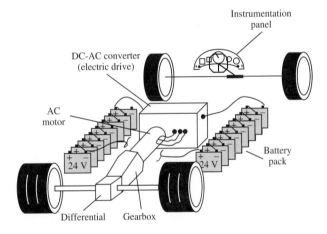

**Figure 1.4** Block diagram of electric race car

### Design Constraints:

The Formula Lightning series is based on a specification chassis; thus, extensive modifications to the frame, suspension, brakes, and body are not permitted. The focus of the competition is therefore to optimize the performance of the spec vehicle by selecting a

<hr />

[2]Universities that have participated in this competition are Arizona State University, Bowling Green State University, Case Western Reserve University, Kettering University, Georgia Institute of Technology, Indiana University—Purdue University at Indianapolis, Northern Arizona University, Notre Dame University, Ohio State University, Ohio University, Rennselaer Polytechnic Institute, University of Oklahoma, and Wright State University.

suitable combination of drivetrain and energy storage components. In addition, since the vehicle is intended to compete in a race series, issues such as energy management, quick and efficient pit stops for battery pack replacement, and the ability to adapt system performance to varying race conditions and different race tracks are also important design constraints.

### Design Solutions:[3]

Teams of undergraduate aerospace, electrical, industrial, and mechanical engineering students participate in the design of the all-electric Formula Lightning drivetrain through a special design course, made available especially for student design competitions.

In a representative course at Ohio State, the student team was divided into four groups: battery system selection, motor and controller selection, transmission and driveline design, and instrumentation and vehicle dynamics. Each of these groups was charged with the responsibility of determining the technology that would be best suited to matching the requirements of the competition and result in a highly competitive vehicle.

Figure 1.5 illustrates the interdisciplinary *mechatronics* team approach; it is apparent that, to arrive at an optimal solution, an iterative process had to be followed and that the various iterations required significant interaction between different teams.

To begin the process, a gross vehicle weight was assumed and energy storage limitations were ignored in a dynamic computer simulation of the vehicle on a simulated road course (the Cleveland Grand Prix Burke Lakefront Airport racetrack, site of the first race in the series). The simulation employed a realistic model of the vehicle and tire dynamics, but a simple model of an electric drive—energy storage limitations would be considered later.

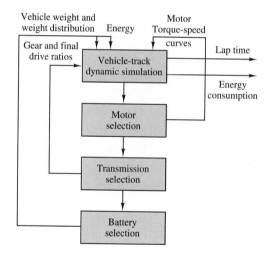

**Figure 1.5** Iterative design process for electric race car drivetrain

The simulation was exercised under various scenarios to determine the limit performance of the vehicle and the choice of a proper drivetrain design. The first round of simulations led to the conclusion that a multispeed gearbox would be a necessity for

[3]K. Grider, G. Rizzoni, "Design of the Ohio State University electric race car," SAE Technical Paper in *Proceedings*, 1996 SAE Motorsports Conference and Exposition, Dearborn, MI, Dec.10–12, 1996.

competitive performance on a road course, and also showed the need for a very high performance AC drive as the propulsion system. The motor and controller are depicted in Figure 1.6.

**Figure 1.6** Motor and controller

Once the electric drive had been selected, the results of battery tests performed by the battery team were evaluated to determine the proper battery technology, and the resulting geometry and weight distribution of the battery packs. With the preferred battery technology identified (see Figure 1.7), energy criteria was included in the simulation, and lap times and energy consumption were predicted. Finally, appropriate instrumentation was designed to permit monitoring of the most important functions in the vehicle (e.g., battery voltage and current, motor temperature, vehicle and motor speed). Figure 1.8 depicts the vehicle dashboard. Table 1.2 gives the specifications for the vehicle.

**Figure 1.7** Open side pod with battery pack and single battery

Table 1.2  Smokin' Buckeye specifications

Drive system:
    Vector controlled AC propulsion model 150
    Motor type: three-phase induction, 150 kW
    Weight: motor 100 lb, controller 75 lb
    Motor dimensions: 12-in diameter, 15-in length

Transmission/clutch:
    Webster four-speed supplied by Taylor Race Engineering
    Tilton metallic clutch

Battery system:
    Total voltage: 372 V (nominal)
    Total weight: 1440 lb
    Number of batteries: 31
    Battery: Optima spiral-wound lead-acid gel-cell battery
    Configuration: 16 battery packs, 12 or 24 V each

Instrumentation:
    Ohio Semitronics model EV1 electric vehicle monitor
    Stack model SR 800 Data Acquisition

Vehicle dimensions:
    Wheelbase: 115 in
    Total length: 163 in
    Width: 77 in
    Weight: 2690 lb

Stock components:
    Tires: Yokohama
    Chassis: 1994 Stewart Racing Formula Lightning
    Springs: Eibach
    Shocks: Penske racing coil-over shocks
    Brakes: Wilwood Dynalite II

**Figure 1.8** Dashboard

Altogether approximately 30 students from different engineering disciplines participated in the initial design process. They received credit for their effort either through the course—ME 580.04, Analysis, Design, Testing and Fabrication of Alternative Vehicles—or through a senior design project. As noted, interaction among teams and among students from different disciplines was an integral part of the design process.

**Comments:**   The example illustrates the importance of interdisciplinary thinking in the design of mechatronics systems. The aim of this book is to provide students in different engineering disciplines with the foundations of electrical/electronic engineering that are necessary to effectively participate in interdisciplinary engineering design projects. The next 17 chapters will present the foundations and vocabulary of electrical engineering.

## 1.3   FUNDAMENTALS OF ENGINEERING EXAM REVIEW

Each of the 50 states regulates the engineering profession by requiring individuals who intend to practice the profession to become registered professional engineers. To become a professional engineer, it is necessary to satisfy four requirements. The first is the completion of a B.S. degree in engineering from an accredited college or university (although it is theoretically possible to be registered without having completed a degree). The second is the successful completion of the *Fundamentals of Engineering* (FE) *Examination*. This is an eight-hour exam that covers general engineering undergraduate education. The third requirement is two to four years of engineering experience after passing the FE exam. Finally, the fourth requirement is successful completion of the *Principles and Practice of Engineering* or *Professional Engineer* (PE) *Examination*.

The FE exam is a two-part national examination given twice a year (in April and October). The exam is divided into two 4-hour sessions. The morning session consists of 140 multiple choice questions (five possible answers are given); the afternoon session consists of 70 questions. The exam is prepared by the State Board of Engineers for each state.

One of the aims of this book is to assist you in preparing for one part of the FE exam, entitled Electrical Circuits. This part of the examination consists of a total of 18 questions in the morning session and 10 questions in the afternoon session. The examination topics for the electrical circuits part are the following:

DC Circuits

AC Circuits

Three-Phase Circuits

Capacitance and Inductance

Transients

Diode Applications

Operational Amplifiers (Ideal)

Electric and Magnetic Fields

Electric Machinery

Appendix B contains a complete review of the Electrical Circuits portion of the FE examination. In Appendix B you will find a detailed listing of the topics covered in the examination, with references to the relevant material in the book. The appendix also contains a collection of sample problems similar to those found in the examination, with answers. These sample problems are arranged in two sections: The first includes worked examples with a full explanation of the solution; the second consists of a sample exam with answers supplied separately. This material is based on the author's experience in teaching the FE Electrical Circuits review course for mechanical engineering seniors at Ohio State University over several years.

## 1.4  BRIEF HISTORY OF ELECTRICAL ENGINEERING

The historical evolution of electrical engineering can be attributed, in part, to the work and discoveries of the people in the following list. You will find these scientists, mathematicians, and physicists referenced throughout the text.

**William Gilbert** (1540–1603), English physician, founder of magnetic science, published *De Magnete*, a treatise on magnetism, in 1600.

**Charles A. Coulomb** (1736–1806), French engineer and physicist, published the laws of electrostatics in seven memoirs to the French Academy of Science between 1785 and 1791. His name is associated with the unit of charge.

**James Watt** (1736–1819), English inventor, developed the steam engine. His name is used to represent the unit of power.

**Alessandro Volta** (1745–1827), Italian physicist, discovered the electric pile. The unit of electric potential and the alternate name of this quantity (voltage) are named after him.

**Hans Christian Oersted** (1777–1851), Danish physicist, discovered the connection between electricity and magnetism in 1820. The unit of magnetic field strength is named after him.

**André Marie Ampère** (1775–1836), French mathematician, chemist, and physicist, experimentally quantified the relationship between electric current and the magnetic field. His works were summarized in a treatise published in 1827. The unit of electric current is named after him.

**Georg Simon Ohm** (1789–1854), German mathematician, investigated the relationship between voltage and current and quantified the phenomenon of resistance. His first results were published in 1827. His name is used to represent the unit of resistance.

**Michael Faraday** (1791–1867), English experimenter, demonstrated electromagnetic induction in 1831. His electrical transformer and electromagnetic generator marked the beginning of the age of electric power. His name is associated with the unit of capacitance.

**Joseph Henry** (1797–1878), American physicist, discovered self-induction around 1831, and his name has been designated to represent the unit of inductance. He had also recognized the essential structure of the telegraph, which was later perfected by Samuel F. B. Morse.

**Carl Friedrich Gauss** (1777–1855), German mathematician, and **Wilhelm Eduard Weber** (1804–1891), German physicist, published a treatise in 1833 describing the measurement of the earth's magnetic field. The gauss is a unit of magnetic field strength, while the weber is a unit of magnetic flux.

**James Clerk Maxwell** (1831–1879), Scottish physicist, discovered the electromagnetic theory of light and the laws of electrodynamics. The modern theory of electromagnetics is entirely founded upon Maxwell's equations.

**Ernst Werner Siemens** (1816–1892) and **Wilhelm Siemens** (1823–1883), German inventors and engineers, contributed to the invention and development of electric machines, as well as to perfecting electrical science. The modern unit of conductance is named after them.

**Heinrich Rudolph Hertz** (1857–1894), German scientist and experimenter, discovered the nature of electromagnetic waves and published his findings in 1888. His name is associated with the unit of frequency.

**Nikola Tesla** (1856–1943), Croatian inventor, emigrated to the United States in 1884. He invented polyphase electric power systems and the induction motor and pioneered modern AC electric power systems. His name is used to represent the unit of magnetic flux density.

## 1.5  SYSTEM OF UNITS

This book employs the International System of Units (also called SI, from the French *Système International des Unités*). SI units are commonly adhered to by virtually all engineering professional societies. This section summarizes SI units and will serve as a useful reference in reading the book.

SI units are based on six fundamental quantities, listed in Table 1.3. All other units may be derived in terms of the fundamental units of Table 1.3. Since, in practice, one often needs to describe quantities that occur in large multiples or small fractions of a unit, standard prefixes are used to denote powers of 10 of SI (and derived) units. These prefixes are listed in Table 1.4. Note that, in general, engineering units are expressed in powers of 10 that are multiples of 3.

Table 1.3  SI units

| Quantity | Unit | Symbol |
| --- | --- | --- |
| Length | Meter | m |
| Mass | Kilogram | kg |
| Time | Second | s |
| Electric current | Ampere | A |
| Temperature | Kelvin | K |
| Luminous intensity | Candela | cd |

Table 1.4  Standard prefixes

| Prefix | Symbol | Power |
| --- | --- | --- |
| atto | a | $10^{-18}$ |
| femto | f | $10^{-15}$ |
| pico | p | $10^{-12}$ |
| nano | n | $10^{-9}$ |
| micro | $\mu$ | $10^{-6}$ |
| milli | m | $10^{-3}$ |
| centi | c | $10^{-2}$ |
| deci | d | $10^{-1}$ |
| deka | da | 10 |
| kilo | k | $10^{3}$ |
| mega | M | $10^{6}$ |
| giga | G | $10^{9}$ |
| tera | T | $10^{12}$ |

For example, $10^{-4}$ s would be referred to as $100 \times 10^{-6}$ s, or $100\mu s$ (or, less frequently, 0.1 ms).

## 1.6    SPECIAL FEATURES OF THIS BOOK

This book includes a number of special features designed to make learning easier and also to allow students to explore the subject matter of the book in more depth, if so desired, through the use of computer-aided tools and the Internet. The principal features of the book are described below.

### EXAMPLES

The examples in the book have also been set aside from the main text, so that they can be easily identified. All examples are solved by following the same basic methodology: A clear and simple problem statement is given, followed by a solution. The solution consists of several parts: All known quantities in the problem are summarized, and the problem statement is translated into a specific objective (e.g., "Find the equivalent resistance, $R$").

Next, the given data and assumptions are listed, and finally the analysis is presented. The analysis method is based on the following principle: All problems are solved symbolically first, to obtain more general solutions that may guide the student in solving homework problems; the numerical solution is provided at the very end of the analysis. Each problem closes with comments summarizing the findings and tying the example to other sections of the book.

The solution methodology used in this book can be used as a general guide to problem-solving techniques well beyond the material taught in the introductory electrical engineering courses. The examples contained in this book are intended to help you develop sound problem-solving habits for the remainder of your engineering career.

### Focus on Computer-Aided Tools, Virtual Lab

One of the very important changes to engineering education in the 1990s has been the ever more common use of computers for analysis, design, data acquisition, and control. This book is designed to permit students and instructors to experiment with various computer-aided design and analysis tools. Some of the tools used are generic computing tools that are likely to be in use in most engineering schools (e.g., Matlab, MathCad). Many examples are supplemented by electronic solutions that are intended to teach you how to solve typical electrical engineering problems using such computer aids, and to stimulate you to experiment in developing your own solution methods. Many of these methods will also be useful later in your curriculum.

Some examples (and also some of the figures in the main text) are supplemented by circuit simulation created using *Electronics Workbench*$^{\text{TM}}$, a circuit analysis and simulation program that has a particularly friendly user interface, and that permits a more in-depth analysis of realistic electrical/electronic circuits and devices. Use of this feature could be limited to just running a simulated circuit to observe its behavior (with virtually no new learning required), or could be more involved and result in the design of new circuit simulations. You might find it

## FOCUS ON METHODOLOGY

Each chapter, especially the early ones, includes "boxes" titled "Focus on Methodology." The content of these boxes (which are set aside from the main text) is to summarize important methods and procedures for the solution of common problems. They usually consist of step-by-step instructions, and are designed to assist you in methodically solving problems.

useful to learn how to use this tool for some of your homework and project assignments. The electronic examples supplied with the book form a veritable *Virtual Electrical and Electronic Circuits Laboratory*. The use of these computer aids is not mandatory, but you will find that the electronic supplements to the book may become a formidable partner and teaching assistant.

### Find It on the Web!

The use of the Internet as a resource for knowledge and information is becoming increasingly common. In recognition of this fact, Web site references have been included in this book to give you a starting point in the exploration of the world of electrical engineering. Typical Web references give you information on electrical engineering companies, products, and methods. Some of the sites contain tutorial material that may supplement the book's contents.

### CD-ROM Content

The inclusion of a CD-ROM in the book allows you to have a wealth of supplements. We list a few major ones: Matlab, MathCad, and *Electronics Workbench* electronic files; demo version of *Electronics Workbench*; Virtual Laboratory experiments; data sheets for common electrical/electronic circuit components; additional reference material.

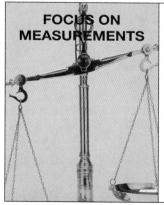

**FOCUS ON MEASUREMENTS**

As stated many times in this book, the need for measurements is a common thread to all engineering and scientific disciplines. To emphasize the great relevance of electrical engineering to the science and practice of measurements, a special set of examples focuses on measurement problems. These examples very often relate to disciplines outside electrical engineering (e.g., biomedical, mechanical, thermal, fluid system measurements). The "Focus on Measurements" sections are intended to stimulate your thinking about the many possible applications of electrical engineering to measurements in your chosen field of study. Many of these examples are a direct result of the author's work as a teacher and researcher in both mechanical and electrical engineering.

## Web Site

The list of features would not be complete without a reference to the book's Web site, **http://www.mhhe.com/engcs/electrical/rizzoni**. Create a bookmark for this site now! The site is designed to provide up-to-date additions, examples, errata, and other important information.

---

## HOMEWORK PROBLEMS

**1.1**  List five applications of electric motors in the common household.

**1.2**  By analogy with the discussion of electrical systems in the automobile, list examples of applications of the electrical engineering disciplines of Table 1.1 for each of the following engineering systems:

a.  A ship.

b.  A commercial passenger aircraft.

c.  Your household.

d.  A chemical process control plant.

**1.3**  Electric power systems provide energy in a variety of commercial and industrial settings. Make a list of systems and devices that receive electric power in:

a.  A large office building.

b.  A factory floor.

c.  A construction site.

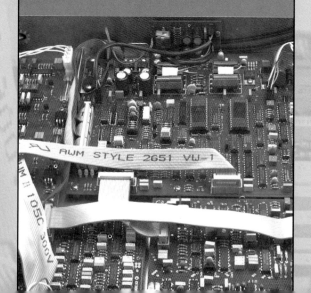

# PART I

## CIRCUITS

# CHAPTER

# 2

# Fundamentals of Electric Circuits

This chapter presents the fundamental laws of circuit analysis and serves as the foundation for the study of electrical circuits. The fundamental concepts developed in these first pages will be called upon throughout the book.

The chapter starts with definitions of charge, current, voltage, and power, and with the introduction of the basic laws of electrical circuit analysis: Kirchhoff's laws. Next, the basic circuit elements are introduced, first in their ideal form, then including the most important physical limitations. The elements discussed in the chapter include voltage and current sources, measuring instruments, and the ideal resistor. Once the basic circuit elements have been presented, the concept of an electrical circuit is introduced, and some simple circuits are analyzed using Kirchhoff's and Ohm's laws. The student should appreciate the fact that, although the material presented at this early stage is strictly introductory, it is already possible to discuss some useful applications of electric circuits to practical engineering problems. To this end, two examples are introduced which discuss simple resistive devices that can measure displacements and forces. The topics introduced in Chapter 2 form the foundations for the remainder of this book and should be mastered thoroughly. By the end of the chapter, you should have accomplished the following learning objectives:

- Application of Kirchhoff's and Ohm's laws to elementary resistive circuits.

- Power computation for a circuit element.
- Use of the passive sign convention in determining voltage and current directions.
- Solution of simple voltage and current divider circuits.
- Assigning node voltages and mesh currents in an electrical circuit.
- Writing the circuit equations for a linear resistive circuit by applying Kirchhoff's voltage law and Kirchhoff's current law.

## 2.1 CHARGE, CURRENT, AND KIRCHHOFF'S CURRENT LAW

Charles Coulomb (1736–1806). *Photo courtesy of French Embassy, Washington, D.C.*

The earliest accounts of electricity date from about 2,500 years ago, when it was discovered that static charge on a piece of amber was capable of attracting very light objects, such as feathers. The word itself—*electricity*—originated about 600 B.C.; it comes from *elektron*, which was the ancient Greek word for amber. The true nature of electricity was not understood until much later, however. Following the work of Alessandro Volta[1] and his invention of the copper-zinc battery, it was determined that static electricity and the current that flows in metal wires connected to a battery are due to the same fundamental mechanism: the atomic structure of matter, consisting of a nucleus—neutrons and protons—surrounded by electrons. The fundamental electric quantity is **charge**, and the smallest amount of charge that exists is the charge carried by an electron, equal to

$$q_e = -1.602 \times 10^{-19} \text{ C} \tag{2.1}$$

As you can see, the amount of charge associated with an electron is rather small. This, of course, has to do with the size of the unit we use to measure charge, the **coulomb (C)**, named after Charles Coulomb.[2] However, the definition of the coulomb leads to an appropriate unit when we define electric current, since current consists of the flow of very large numbers of charge particles. The other charge-carrying particle in an atom, the proton, is assigned a positive sign, and the same magnitude. The charge of a proton is

$$q_p = +1.602 \times 10^{-19} \text{ C} \tag{2.2}$$

Electrons and protons are often referred to as **elementary charges.**

**Electric current** is defined as the time rate of change of charge passing through a predetermined area. Typically, this area is the cross-sectional area of a metal wire; however, there are a number of cases we shall explore later in this book where the current-carrying material is not a conducting wire. Figure 2.1 depicts a macroscopic view of the flow of charge in a wire, where we imagine $\Delta q$ units of charge flowing through the cross-sectional area $A$ in $\Delta t$ units of time. The resulting current, $i$, is then given by

Current $i = dq/dt$ is generated by the flow of charge through the cross-sectional area $A$ in a conductor.

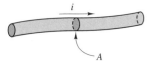

$$i = \frac{\Delta q}{\Delta t} \quad \frac{C}{s} \tag{2.3}$$

**Figure 2.1** Current flow in an electric conductor

[1] See brief biography on page 9.
[2] See brief biography on page 9.

If we consider the effect of the enormous number of elementary charges actually flowing, we can write this relationship in differential form:

$$i = \frac{dq}{dt} \quad \frac{C}{s} \tag{2.4}$$

The units of current are called **amperes (A)**, where 1 ampere = 1 coulomb/second. The name of the unit is a tribute to the French scientist André Marie Ampère.[3] The electrical engineering convention states that the positive direction of current flow is that of positive charges. In metallic conductors, however, current is carried by negative charges; these charges are the free electrons in the conduction band, which are only weakly attracted to the atomic structure in metallic elements and are therefore easily displaced in the presence of electric fields.

## EXAMPLE   2.1  Charge and Current in a Conductor

### Problem

Find the total charge in a cylindrical conductor (solid wire) and compute the current flowing in the wire.

### Solution

**Known Quantities:**  Conductor geometry, charge density, charge carrier velocity.

**Find:**  Total charge of carriers, $Q$; current in the wire, $I$.

**Schematics, Diagrams, Circuits, and Given Data:**  Conductor length: $L = 1$ m.
  Conductor diameter: $2r = 2 \times 10^{-3}$ m.
  Charge density: $n = 10^{29}$ carriers/m$^3$.
  Charge of one electron: $q_e = -1.602 \times 10^{-19}$.
  Charge carrier velocity: $u = 19.9 \times 10^{-6}$ m/s.

**Assumptions:**  None.

**Analysis:**  To compute the total charge in the conductor, we first determine the volume of the conductor:

  Volume = Length × Cross-sectional area

$$V = L \times \pi r^2 = (1 \text{ m}) \times \left[ \pi \left( \frac{2 \times 10^{-3}}{2} \right)^2 \text{m}^2 \right] = \pi \times 10^{-6} \quad \text{m}^3$$

  Next, we compute the number of carriers (electrons) in the conductor and the total charge:

  Number of carriers = Volume × Carrier density

$$N = V \times n = \left( \pi \times 10^{-6} \, \text{m}^3 \right) \times \left( 10^{29} \frac{\text{carriers}}{\text{m}^3} \right) = \pi \times 10^{23} \text{carriers}$$

  Charge = number of carriers × charge/carrier

$$Q = N \times q_e = \left( \pi \times 10^{23} \ \text{carriers} \right)$$

$$\times \left( -1.602 \times 10^{-19} \frac{\text{coulomb}}{\text{carrier}} \right) = -50.33 \times 10^3 \text{ C.}$$

---

[3] See brief biography on page 9.

To compute the current, we consider the velocity of the charge carriers, and the charge density per unit length of the conductor:

Current = Carrier charge density per unit length × Carrier velocity

$$I = \left(\frac{Q}{L}\ \frac{C}{m}\right) \times \left(u\ \frac{m}{s}\right) = \left(-50.33 \times 10^3\ \ \frac{C}{m}\right) \times \left(19.9 \times 10^{-6}\ \frac{m}{s}\right) = 1\ A$$

**Comments:**  Charge carrier density is a function of material properties. Carrier velocity is a function of the applied electric field.

---

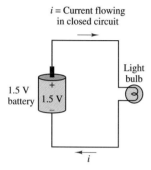

*i* = Current flowing in closed circuit

**Figure 2.2** A simple electrical circuit

In order for current to flow there must exist a closed circuit.  Figure 2.2 depicts a simple circuit, composed of a battery (e.g., a dry-cell or alkaline 1.5-V battery) and a light bulb.

Note that in the circuit of Figure 2.2, the current, $i$, flowing from the battery to the light bulb is equal to the current flowing from the light bulb to the battery. In other words, no current (and therefore no charge) is "lost" around the closed circuit.  This principle was observed by the German scientist G. R. Kirchhoff[4] and is now known as **Kirchhoff's current law (KCL).** Kirchhoff's current law states that because charge cannot be created but must be conserved, *the sum of the currents at a node must equal zero* (in an electrical circuit, a **node** is the junction of two or more conductors).  Formally:

$$\sum_{n=1}^{N} i_n = 0 \qquad \text{Kirchhoff's current law} \tag{2.5}$$

The significance of Kirchhoff's current law is illustrated in Figure 2.3, where the simple circuit of Figure 2.2 has been augmented by the addition of two light bulbs (note how the two nodes that exist in this circuit have been emphasized by the shaded areas).  In applying KCL, one usually defines currents entering a node as being negative and currents exiting the node as being positive.  Thus, the resulting expression for node 1 of the circuit of Figure 2.3 is:

$$-i + i_1 + i_2 + i_3 = 0$$

Kirchhoff's current law is one of the fundamental laws of circuit analysis, making it possible to express currents in a circuit in terms of each other; for example, one can express the current leaving a node in terms of all the other currents at the node.  The ability to write such equations is a great aid in the systematic solution of large electric circuits.  Much of the material presented in Chapter 3 will be an extension of this concept.

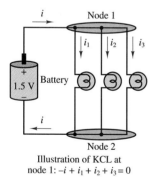

Illustration of KCL at node 1: $-i + i_1 + i_2 + i_3 = 0$

**Figure 2.3** Illustration of Kirchhoff's current law

---

[4]Gustav Robert Kirchhoff (1824–1887), a German scientist, who published the first systematic description of the laws of circuit analysis. His contribution—though not original in terms of its scientific content—forms the basis of all circuit analysis.

## EXAMPLE 2.2 Kirchhoff's Current Law Applied to an Automotive Electrical Harness

### Problem

Figure 2.4 shows an **automotive battery** connected to a variety of circuits in an automobile. The circuits include headlights, taillights, starter motor, fan, power locks, and dashboard panel. The battery must supply enough current to independently satisfy the requirements of each of the "load" circuits. Apply KCL to the automotive circuits.

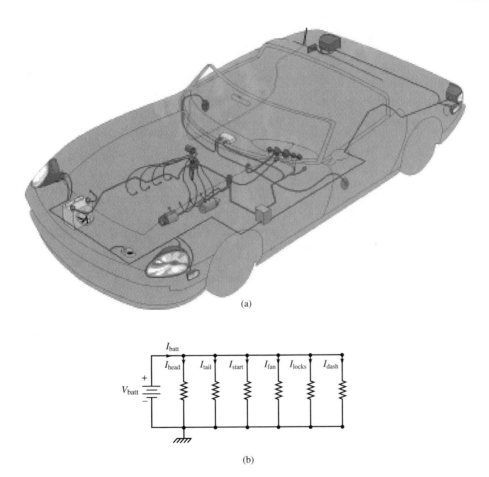

**Figure 2.4** (a) Automotive circuits (b) equivalent electrical circuit

### Solution

**Known Quantities:** Components of electrical harness: headlights, taillights, starter motor, fan, power locks, and dashboard panel.

**Find:** Expression relating battery current to load currents.

**Schematics, Diagrams, Circuits, and Given Data:** Figure 2.4.

**Assumptions:** None.

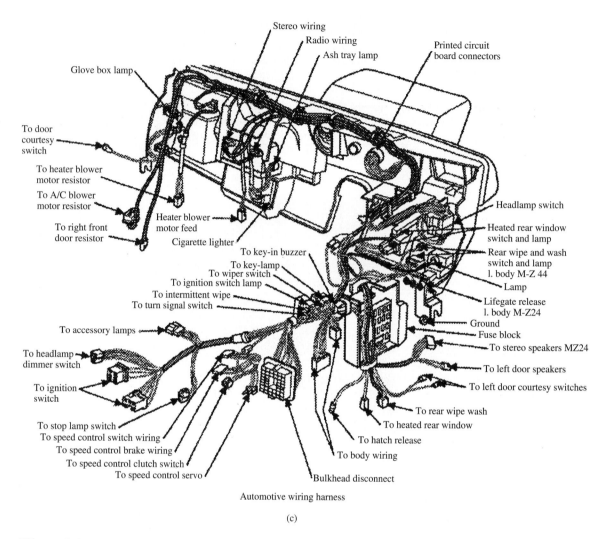

Stereo wiring

Radio wiring

Ash tray lamp

Printed circuit
board connectors

Glove box lamp

To door
courtesy
switch

To heater blower
motor resistor

To A/C blower
motor resistor

To right front
door resistor

Heater blower
motor feed

Cigarette lighter

To key-in buzzer

To key-lamp

To wiper switch

To ignition switch lamp

To intermittent wipe

To turn signal switch

To accessory lamps

To headlamp
dimmer switch

To ignition
switch

To stop lamp switch

To speed control switch wiring

To speed control brake wiring

To speed control clutch switch

To speed control servo

Bulkhead disconnect

Headlamp switch

Heated rear window
switch and lamp

Rear wipe and wash
switch and lamp
l. body M-Z 44

Lamp

Lifegate release
l. body M-Z24

Ground

Fuse block

To stereo speakers MZ24

To left door speakers

To left door courtesy switches

To rear wipe wash

To heated rear window

To hatch release

To body wiring

Automotive wiring harness

(c)

**Figure 2.4** (c) Automotive wiring harness *Copyright © 1995 by Delmar Publishers. Copyright © 1995–1997 Automotive Information Center. All rights reserved.*

**Analysis:** Figure 2.4(b) depicts the equivalent electrical circuit, illustrating how the current supplied by the battery must divide among the various circuits. The application of KCL to the equivalent circuit of Figure 2.4 requires that:

$$I_{\text{batt}} - I_{\text{head}} - I_{\text{tail}} - I_{\text{start}} - I_{\text{fan}} - I_{\text{locks}} - I_{\text{dash}} = 0$$

**Comments:** This illustration is meant to give the reader an intuitive feel for the significance of KCL; more detailed numerical examples of KCL will be presented later in this chapter, when voltage and current sources and resistors are defined more precisely. Figure 2.4(c) depicts a real automotive electrical harness—a rather complicated electrical circuit!

## 2.2   VOLTAGE AND KIRCHHOFF'S VOLTAGE LAW

Charge moving in an electric circuit gives rise to a current, as stated in the preceding section. Naturally, it must take some work, or energy, for the charge to move between two points in a circuit, say, from point *a* to point *b*. The total *work per unit charge* associated with the motion of charge between two points is called **voltage.** Thus, the units of voltage are those of energy per unit charge; they have been called **volts** in honor of Alessandro Volta:

$$1 \text{ volt} = \frac{1 \text{ joule}}{\text{coulomb}} \tag{2.6}$$

The voltage, or **potential difference**, between two points in a circuit indicates the energy required to move charge from one point to the other. As will be presently shown, the direction, or polarity, of the voltage is closely tied to whether energy is being dissipated or generated in the process. The seemingly abstract concept of work being done in moving charges can be directly applied to the analysis of electrical circuits; consider again the simple circuit consisting of a battery and a light bulb. The circuit is drawn again for convenience in Figure 2.5, with nodes defined by the letters *a* and *b*. A series of carefully conducted experimental observations regarding the nature of voltages in an electric circuit led Kirchhoff to the formulation of the second of his laws, **Kirchhoff's voltage law, or KVL.** The principle underlying KVL is that no energy is lost or created in an electric circuit; in circuit terms, the sum of all voltages associated with sources must equal the sum of the load voltages, so that *the net voltage around a closed circuit is zero.* If this were not the case, we would need to find a physical explanation for the excess (or missing) energy not accounted for in the voltages around a circuit. Kirchhoff's voltage law may be stated in a form similar to that used for KCL:

$$\sum_{n=1}^{N} v_n = 0 \qquad \text{Kirchhoff's voltage law} \tag{2.7}$$

where the $v_n$ are the individual voltages around the closed circuit. Making reference to Figure 2.5, we see that it must follow from KVL that the work generated by the battery is equal to the energy dissipated in the light bulb in order to sustain the current flow and to convert the electric energy to heat and light:

$$v_{ab} = -v_{ba}$$

or

$$v_1 = v_2$$

One may think of the work done in moving a charge from point *a* to point *b* and the work done moving it back from *b* to *a* as corresponding directly to the *voltages across individual circuit elements.* Let $Q$ be the total charge that moves around the circuit per unit time, giving rise to the current $i$. Then the work done in moving $Q$ from *b* to *a* (i.e., across the battery) is

$$W_{ba} = Q \times 1.5 \text{ V} \tag{2.8}$$

Gustav Robert Kirchhoff (1824–1887). *Photo courtesy of Deutsches Museum, Munich.*

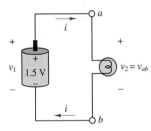

Illustration of Kirchhoff's voltage law: $v_1 = v_2$

**Figure 2.5** Voltages around a circuit

Similarly, work is done in moving $Q$ from $a$ to $b$, that is, across the light bulb. Note that the word *potential* is quite appropriate as a synonym of voltage, in that voltage represents the potential energy between two points in a circuit: if we remove the light bulb from its connections to the battery, there still exists a voltage across the (now disconnected) terminals $b$ and $a$. This is illustrated in Figure 2.6.

A moment's reflection upon the significance of voltage should suggest that it must be necessary to specify a sign for this quantity. Consider, again, the same dry-cell or alkaline battery, where, by virtue of an electrochemically induced separation of charge, a 1.5-V potential difference is generated. The potential generated by the battery may be used to move charge in a circuit. The rate at which charge is moved once a closed circuit is established (i.e., the current drawn by the circuit connected to the battery) depends now on the circuit element we choose to connect to the battery. Thus, while the voltage across the battery represents the potential for *providing energy* to a circuit, the voltage across the light bulb indicates the amount of work done in *dissipating energy*. In the first case, energy is generated; in the second, it is consumed (note that energy may also be stored, by suitable circuit elements yet to be introduced). This fundamental distinction requires attention in defining the sign (or polarity) of voltages.

We shall, in general, refer to elements that provide energy as **sources**, and to elements that dissipate energy as **loads**. Standard symbols for a generalized source-and-load circuit are shown in Figure 2.7. Formal definitions will be given in a later section.

The presence of a voltage, $v_2$, across the open terminals $a$ and $b$ indicates the potential energy that can enable the motion of charge, once a closed circuit is established to allow current to flow.

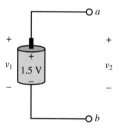

**Figure 2.6** Concept of voltage as potential difference

A symbolic representation of the battery–light bulb circuit of Figure 2.5.

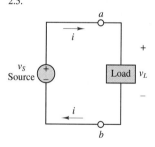

**Figure 2.7** Sources and loads in an electrical circuit

### EXAMPLE   2.3  Kirchhoff's Voltage Law—Electric Vehicle Battery Pack

**Problem**

Figure 2.8a depicts the battery pack in the Smokin' Buckeye electric race car. In this example we apply KVL to the series connection of 31 12-V batteries that make up the battery supply for the electric vehicle.

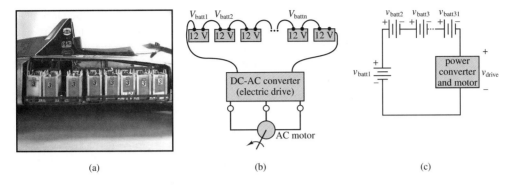

Figure 2.8 Electric vehicle battery pack: illustration of KVL

---

## Solution

**Known Quantities:**  Nominal characteristics of **Optima™ lead-acid batteries.**

**Find:**  Expression relating battery and electric motor drive voltages.

**Schematics, Diagrams, Circuits, and Given Data:** $V_{\text{batt}} = 12$ V. Figure 2.8(a), (b) and (c)

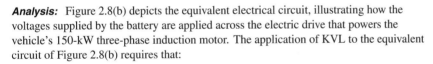

**Assumptions:**  None.

**Analysis:**  Figure 2.8(b) depicts the equivalent electrical circuit, illustrating how the voltages supplied by the battery are applied across the electric drive that powers the vehicle's 150-kW three-phase induction motor. The application of KVL to the equivalent circuit of Figure 2.8(b) requires that:

$$\sum_{n=1}^{31} V_{\text{batt}_n} - V_{\text{drive}} = 0.$$

Thus, the electric drive is nominally supplied by a $31 \times 12 = 372$-V battery pack. In reality, the voltage supplied by lead-acid batteries varies depending on the state of charge of the battery. When fully charged, the battery pack of Figure 2.8(a) is closer to supplying around 400 V (i.e., around 13 V per battery).

**Comments:**  This illustration is meant to give the reader an intuitive feel for the significance of KVL; more detailed numerical examples of KVL will be presented later in this chapter, when voltage and current sources and resistors are defined more precisely.

---

## 2.3    IDEAL VOLTAGE AND CURRENT SOURCES

In the examples presented in the preceding sections, a battery was used as a source of energy, under the unspoken assumption that the voltage provided by the battery (e.g., 1.5 volts for a dry-cell or alkaline battery, or 12 volts for an automotive lead-acid battery) is fixed. Under such an assumption, we implicitly treat the battery as an ideal source. In this section, we will formally define ideal sources. Intuitively, an ideal source is a source that can provide an arbitrary amount of energy. **Ideal sources** are divided into two types: voltage sources and current sources. Of these, you are probably more familiar with the first, since dry-cell, alkaline, and lead-acid batteries are all voltage sources (they are not ideal, of course). You might have to think harder to come up with a physical example that approximates the behavior of an ideal current source; however, reasonably good approximations of ideal current sources also exist. For instance, a voltage source connected in series with a circuit element that has a large resistance to the flow of current from the source provides a nearly constant—though small—current and therefore acts very nearly like an ideal current source.

## Ideal Voltage Sources

An **ideal voltage source** is an electrical device that will generate a prescribed voltage at its terminals. The ability of an ideal voltage source to generate its output voltage is not affected by the current it must supply to the other circuit elements. Another way to phrase the same idea is as follows:

> An ideal voltage source provides a prescribed voltage across its terminals irrespective of the current flowing through it. The amount of current supplied by the source is determined by the circuit connected to it.

Figure 2.9 depicts various symbols for voltage sources that will be employed throughout this book. Note that the output voltage of an ideal source can be a function of time. In general, the following notation will be employed in this book, unless otherwise noted. A generic voltage source will be denoted by a lowercase $v$. If it is necessary to emphasize that the source produces a time-varying voltage, then the notation $v(t)$ will be employed. Finally, a constant, or *direct current*, or *DC*, voltage source will be denoted by the uppercase character $V$. Note that by convention the direction of positive current flow out of a voltage source is *out of the positive terminal*.

The notion of an ideal voltage source is best appreciated within the context of the source-load representation of electrical circuits, which will frequently be referred to in the remainder of this book. Figure 2.10 depicts the connection of an energy source with a passive circuit (i.e., a circuit that can absorb and dissipate energy—for example, the headlights and light bulb of our earlier examples). Three different representations are shown to illustrate the conceptual, symbolic, and physical significance of this source-load idea.

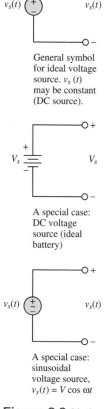

**Figure 2.9** Ideal voltage sources

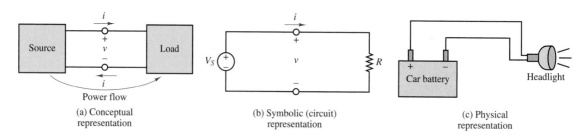

**Figure 2.10** Various representations of an electrical system.

In the analysis of electrical circuits, we choose to represent the physical reality of Figure 2.10(c) by means of the approximation provided by ideal circuit elements, as depicted in Figure 2.10(b).

## Ideal Current Sources

An **ideal current source** is a device that can generate a prescribed current independent of the circuit it is connected to. To do so, it must be able to generate an arbitrary voltage across its terminals. Figure 2.11 depicts the symbol used to represent ideal current sources. By analogy with the definition of the ideal voltage source stated in the previous section, we write:

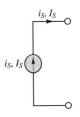

**Figure 2.11** Symbol for ideal current source

> An ideal current source provides a prescribed current to any circuit connected to it. The voltage generated by the source is determined by the circuit connected to it.

The same uppercase and lowercase convention used for voltage sources will be employed in denoting current sources.

## Dependent (Controlled) Sources

The sources described so far have the capability of generating a prescribed voltage or current independent of any other element within the circuit. Thus, they are termed *independent sources*. There exists another category of sources, however, whose output (current or voltage) is a function of some other voltage or current in a circuit. These are called **dependent** (or **controlled**) **sources**. A different symbol, in the shape of a diamond, is used to represent dependent sources and to distinguish them from independent sources. The symbols typically used to represent dependent sources are depicted in Figure 2.12; the table illustrates the relationship between the source voltage or current and the voltage or current it depends on—$v_x$ or $i_x$, respectively—which can be any voltage or current in the circuit.

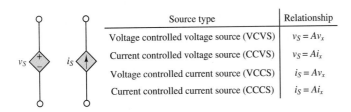

| Source type | Relationship |
|---|---|
| Voltage controlled voltage source (VCVS) | $v_S = Av_x$ |
| Current controlled voltage source (CCVS) | $v_S = Ai_x$ |
| Voltage controlled current source (VCCS) | $i_S = Av_x$ |
| Current controlled current source (CCCS) | $i_S = Ai_x$ |

**Figure 2.12** Symbols for dependent sources

Dependent sources are very useful in describing certain types of electronic circuits. You will encounter dependent sources again in Chapters 9, 10, and 12, when electronic amplifiers are discussed.

## 2.4   ELECTRIC POWER AND SIGN CONVENTION

The definition of voltage as work per unit charge lends itself very conveniently to the introduction of power. Recall that power is defined as the work done per unit time. Thus, the power, $P$, either generated or dissipated by a circuit element can be represented by the following relationship:

$$\text{Power} = \frac{\text{Work}}{\text{Time}} = \frac{\text{Work}}{\text{Charge}} \frac{\text{Charge}}{\text{Time}} = \text{Voltage} \times \text{Current} \tag{2.9}$$

Thus,

> The electrical power generated by an active element, or that dissipated or stored by a passive element, is equal to the product of the voltage across the element and the current flowing through it.

$$\boxed{P = VI} \tag{2.10}$$

It is easy to verify that the units of voltage (joules/coulomb) times current (coulombs/second) are indeed those of power (joules/second, or watts).

It is important to realize that, just like voltage, power is a signed quantity, and that it is necessary to make a distinction between *positive* and *negative power.* This distinction can be understood with reference to Figure 2.13, in which a source and a load are shown side by side. The polarity of the voltage across the source and the direction of the current through it indicate that the voltage source *is doing work in moving charge from a lower potential to a higher potential.* On the other hand, the load is dissipating energy, because the direction of the current indicates that *charge is being displaced from a higher potential to a lower potential.* To avoid confusion with regard to the sign of power, the electrical engineering community uniformly adopts the **passive sign convention**, which simply states that *the power dissipated by a load is a positive quantity* (or, conversely, that the power generated by a source is a positive quantity). Another way of phrasing the same concept is to state that if current flows from a higher to a lower voltage ($+$ to $-$), the power is dissipated and will be a positive quantity.

It is important to note also that the actual numerical values of voltages and currents do not matter: once the proper reference directions have been established and the passive sign convention has been applied consistently, the answer will be correct regardless of the reference direction chosen. The following examples illustrate this point.

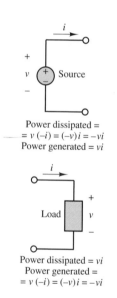

Power dissipated =
$= v\,(-i) = (-v)\,i = -vi$
Power generated $= vi$

Power dissipated $= vi$
Power generated =
$= v\,(-i) = (-v)\,i = -vi$

**Figure 2.13** The passive sign convention

## FOCUS ON METHODOLOGY

**The Passive Sign Convention**

1. Choose an arbitrary direction of current flow.
2. Label polarities of all active elements (voltage and current sources).

# FOCUS ON METHODOLOGY

3. Assign polarities to all passive elements (resistors and other loads); for passive elements, current always flows into the positive terminal.

4. Compute the power dissipated by each element according to the following rule: If positive current flows into the positive terminal of an element, then the power dissipated is positive (i.e., the element absorbs power); if the current leaves the positive terminal of an element, then the power dissipated is negative (i.e., the element delivers power).

## EXAMPLE 2.4 Use of the Passive Sign Convention

### Problem

Apply the passive sign convention to the circuit of Figure 2.14.

### Solution

**Known Quantities:** Voltages across each circuit element; current in circuit.

**Find:** Power dissipated or generated by each element.

**Schematics, Diagrams, Circuits, and Given Data:** Figure 2.15(a) and (b). The voltage drop across Load 1 is 8 V, that across Load 2 is 4 V; the current in the circuit is 0.1 A.

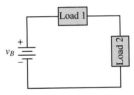

**Figure 2.14**

**Assumptions:** None.

**Analysis:** Following the passive sign convention, we first select an arbitrary direction for the current in the circuit; the example will be repeated for both possible directions of current flow to demonstrate that the methodology is sound.

1. Assume clockwise direction of current flow, as shown in Figure 2.15(a).

2. Label polarity of voltage source, as shown in Figure 2.15(a); since the arbitrarily chosen direction of the current is consistent with the true polarity of the voltage source, the source voltage will be a positive quantity.

3. Assign polarity to each passive element, as shown in Figure 2.15(a).

4. Compute the power dissipated by each element: Since current flows from − to + through the battery, the power dissipated by this element will be a negative quantity:

$$P_B = -v_B \times i = -(12 \text{ V}) \times (0.1 \text{ A}) = -1.2 \text{ W}$$

that is, the battery *generates* 1.2 W. The power dissipated by the two loads will be a positive quantity in both cases, since current flows from + to −:

$$P_1 = v_1 \times i = (8 \text{ V}) \times (0.1 \text{ A}) = 0.8 \text{ W}$$

$$P_2 = v_2 \times i = (4 \text{ V}) \times (0.1 \text{ A}) = 0.4 \text{ W}$$

Next, we repeat the analysis assuming counterclockwise current direction.

1. Assume counterclockwise direction of current flow, as shown in Figure 2.15(b).

2. Label polarity of voltage source, as shown in Figure 2.15(b); since the arbitrarily chosen direction of the current is not consistent with the true polarity of the voltage source, the source voltage will be a negative quantity.

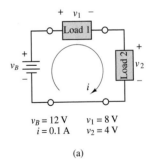

$v_B = 12 \text{ V} \qquad v_1 = 8 \text{ V}$
$i = 0.1 \text{ A} \qquad v_2 = 4 \text{ V}$

(a)

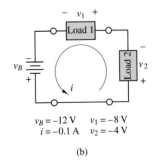

$v_B = -12 \text{ V} \qquad v_1 = -8 \text{ V}$
$i = -0.1 \text{ A} \qquad v_2 = -4 \text{ V}$

(b)

**Figure 2.15**

3. Assign polarity to each passive element, as shown in Figure 2.15(b).

4. Compute the power dissipated by each element: Since current flows from + to −
   through the battery, the power dissipated by this element will be a positive quantity;
   however, the source voltage is a negative quantity:

$$P_B = v_B \times i = (-12 \text{ V}) \times (0.1 \text{ A}) = -1.2 \text{ W}$$

that is, the battery *generates* 1.2 W, as in the previous case. The power dissipated by
the two loads will be a positive quantity in both cases, since current flows from + to
−:

$$P_1 = v_1 \times i = (8 \text{ V}) \times (0.1 \text{ A}) = 0.8 \text{ W}$$

$$P_2 = v_2 \times i = (4 \text{ V}) \times (0.1 \text{ A}) = 0.4 \text{ W}$$

**Comments:** It should be apparent that the most important step in the example is the
correct assignment of source voltage; passive elements will always result in positive power
dissipation. Note also that energy is conserved, as the sum of the power dissipated by
source and loads is zero. In other words: Power supplied always equals power dissipated.

## EXAMPLE 2.5 Another Use of the Passive Sign Convention

### Problem

Determine whether a given element is dissipating or generating power from known
voltages and currents.

### Solution

**Known Quantities:** Voltages across each circuit element; current in circuit.

**Find:** Which element dissipates power and which generates it.

**Schematics, Diagrams, Circuits, and Given Data:** Voltage across element A: 1,000 V.
Current flowing into element A: 420 A.
See Figure 2.16(a) for voltage polarity and current direction.

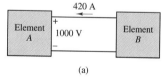

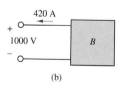

**Figure 2.16**

**Analysis:** According to the passive sign convention, an element dissipates power when
current flows from a point of higher potential to one of lower potential; thus, element A
acts as a load. Since power must be conserved, element B must be a source [Figure
2.16(b)]. Element A dissipates (1,000 V) × (420 A) = 420 kW. Element B generates the
same amount of power.

**Comments:** The procedure described in this example can be easily conducted
experimentally, by performing simple current and voltage measurements. Measuring
devices are discussed in Section 2.8.

## Check Your Understanding

**2.1** Compute the current flowing through each of the headlights of Example 2.2 if each
headlight has a power rating of 50 W. How much power is the battery providing?

**2.2**  Determine which circuit element in the illustration (below, left) is supplying power and which is dissipating power. Also determine the amount of power dissipated and supplied.

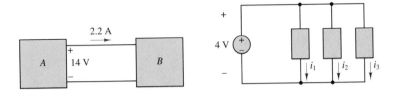

**2.3**  If the battery in the accompanying diagram (above, right) supplies a total of 10 mW to the three elements shown and $i_1 = 2$ mA and $i_2 = 1.5$ mA, what is the current $i_3$? If $i_1 = 1$ mA and $i_3 = 1.5$ mA, what is $i_2$?

## 2.5  CIRCUIT ELEMENTS AND THEIR *i-v* CHARACTERISTICS

The relationship between current and voltage at the terminals of a circuit element defines the behavior of that element within the circuit. In this section we shall introduce a graphical means of representing the terminal characteristics of circuit elements. Figure 2.17 depicts the representation that will be employed throughout the chapter to denote a generalized circuit element: the variable $i$ represents the current flowing through the element, while $v$ is the potential difference, or voltage, across the element.

Suppose now that a known voltage were imposed across a circuit element. The current that would flow as a consequence of this voltage, and the voltage itself, form a unique pair of values. If the voltage applied to the element were varied and the resulting current measured, it would be possible to construct a functional relationship between voltage and current known as the ***i-v* characteristic** (or **volt-ampere characteristic**). Such a relationship defines the circuit element, in the sense that if we impose any prescribed voltage (or current), the resulting current (or voltage) is directly obtainable from the *i-v* characteristic. A direct consequence is that the power dissipated (or generated) by the element may also be determined from the *i-v* curve.

Figure 2.18 depicts an experiment for empirically determining the *i-v* characteristic of a tungsten filament light bulb. A variable voltage source is used to apply various voltages, and the current flowing through the element is measured for each applied voltage.

We could certainly express the *i-v* characteristic of a circuit element in functional form:

$$i = f(v) \qquad v = g(i) \tag{2.11}$$

In some circumstances, however, the graphical representation is more desirable, especially if there is no simple functional form relating voltage to current. The simplest form of the *i-v* characteristic for a circuit element is a straight line, that is,

$$i = kv \tag{2.12}$$

**Figure 2.17** Generalized representation of circuit elements

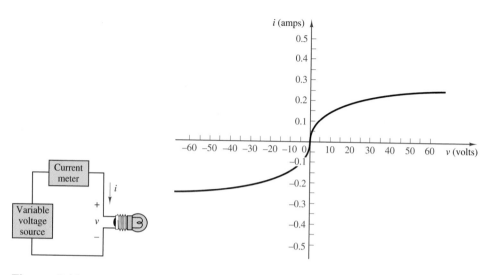

**Figure 2.18** Volt-ampere characteristic of a tungsten light bulb

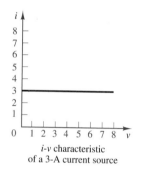

*i-v* characteristic
of a 3-A current source

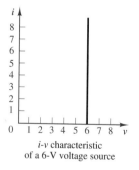

*i-v* characteristic
of a 6-V voltage source

**Figure 2.19** *i-v*
characteristics of ideal
sources

with *k* a constant. In the next section we shall see how this simple model of a circuit element is quite useful in practice and can be used to define the most common circuit elements: ideal voltage and current sources and the resistor.

We can also relate the graphical *i-v* representation of circuit elements to the power dissipated or generated by a circuit element. For example, the graphical representation of the light bulb *i-v* characteristic of Figure 2.18 illustrates that when a positive current flows through the bulb, the voltage is positive, and that, conversely, a negative current flow corresponds to a negative voltage. In both cases the power dissipated by the device is a positive quantity, as it should be, on the basis of the discussion of the preceding section, since the light bulb is a passive device. Note that the *i-v* characteristic appears in only two of the four possible quadrants in the *i-v* plane. In the other two quadrants, the product of voltage and current (i.e., power) is negative, and an *i-v* curve with a portion in either of these quadrants would therefore correspond to power generated. This is not possible for a passive load such as a light bulb; however, there are electronic devices that can operate, for example, in three of the four quadrants of the *i-v* characteristic and can therefore act as sources of energy for specific combinations of voltages and currents. An example of this dual behavior is introduced in Chapter 8, where it is shown that the photodiode can act either in a passive mode (as a light sensor) or in an active mode (as a solar cell).

The *i-v* characteristics of ideal current and voltage sources can also be useful in visually representing their behavior. An ideal voltage source generates a prescribed voltage independent of the current drawn from the load; thus, its *i-v* characteristic is a straight vertical line with a voltage axis intercept corresponding to the source voltage. Similarly, the *i-v* characteristic of an ideal current source is a horizontal line with a current axis intercept corresponding to the source current. Figure 2.19 depicts these behaviors.

## 2.6    RESISTANCE AND OHM'S LAW

When electric current flows through a metal wire or through other circuit elements, it encounters a certain amount of **resistance**, the magnitude of which depends on the electrical properties of the material. Resistance to the flow of current may be undesired—for example, in the case of lead wires and connection cable—or it may be exploited in an electrical circuit in a useful way. Nevertheless, practically all circuit elements exhibit some resistance; as a consequence, current flowing through an element will cause energy to be dissipated in the form of heat. An ideal **resistor** is a device that exhibits linear resistance properties according to **Ohm's law,** which states that

$$V = IR \qquad \text{Ohm's law} \tag{2.13}$$

that is, that the voltage across an element is directly proportional to the current flow through it. $R$ is the value of the resistance in units of **ohms ($\Omega$),** where

$$1\ \Omega = 1\ \text{V/A} \tag{2.14}$$

The resistance of a material depends on a property called **resistivity,** denoted by the symbol $\rho$; the inverse of resistivity is called **conductivity** and is denoted by the symbol $\sigma$. For a cylindrical resistance element (shown in Figure 2.20), the resistance is proportional to the length of the sample, $l$, and inversely proportional to its cross-sectional area, $A$, and conductivity, $\sigma$.

$$v = \frac{l}{\sigma A} i \tag{2.15}$$

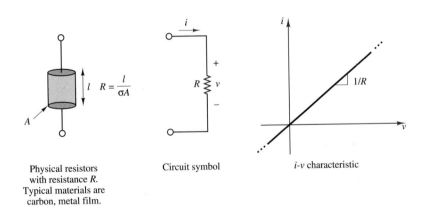

Physical resistors with resistance $R$. Typical materials are carbon, metal film.

Circuit symbol

$i$-$v$ characteristic

**Figure 2.20** The resistance element

It is often convenient to define the **conductance** of a circuit element as the inverse of its resistance. The symbol used to denote the conductance of an element is $G$, where

$$G = \frac{1}{R} \ \text{siemens (S)} \qquad \text{where} \qquad 1\ \text{S} = 1\ \text{A/V} \tag{2.16}$$

Thus, Ohm's law can be restated in terms of conductance as:

$$I = GV \tag{2.17}$$

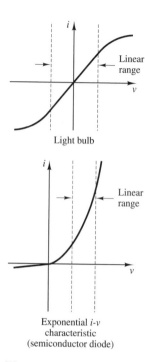

Ohm's law is an empirical relationship that finds widespread application in electrical engineering, because of its simplicity. It is, however, only an approximation of the physics of electrically conducting materials. Typically, the linear relationship between voltage and current in electrical conductors does not apply at very high voltages and currents. Further, not all electrically conducting materials exhibit linear behavior even for small voltages and currents. It is usually true, however, that for some range of voltages and currents, most elements display a linear $i$-$v$ characteristic. Figure 2.21 illustrates how the linear resistance concept may apply to elements with nonlinear $i$-$v$ characteristics, by graphically defining the linear portion of the $i$-$v$ characteristic of two common electrical devices: the light bulb, which we have already encountered, and the semiconductor diode, which we study in greater detail in Chapter 8.

The typical construction and the circuit symbol of the resistor are shown in Figure 2.20. Resistors made of cylindrical sections of carbon (with resistivity $\rho = 3.5 \times 10^{-5}$ $\Omega$-m) are very common and are commercially available in a wide range of values for several power ratings (as will be explained shortly). Another common construction technique for resistors employs metal film. A common power rating for resistors used in electronic circuits (e.g., in most consumer electronic appliances such as radios and television sets) is $\frac{1}{4}$ W. Table 2.1 lists the standard values for commonly used resistors and the color code associated with these values (i.e., the common combinations of the digits $b_1 b_2 b_3$ as defined in Figure 2.22). For example, if the first three color bands on a resistor show the colors red ($b_1 = 2$), violet ($b_2 = 7$), and yellow ($b_3 = 4$), the resistance value can be interpreted as follows:

$$R = 27 \times 10^4 = 270{,}000\ \Omega = 270\ \text{k}\Omega$$

**Figure 2.21**

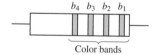

Color bands

black    0      blue    6
brown    1      violet  7
red      2      gray    8
orange   3      white   9
yellow   4      silver  10%
green    5      gold    5%

Resistor value = $(b_1\, b_2) \times 10^{b_3}$;
$b_4$ = % tolerance in actual value

**Figure 2.22** Resistor color code

Table 2.1   Common resistor values values ($\frac{1}{8}$-, $\frac{1}{4}$-, $\frac{1}{2}$-, 1-, 2-W rating)

| $\Omega$ | Code | $\Omega$ | Multiplier | k$\Omega$ | Multiplier | k$\Omega$ | Multiplier | k$\Omega$ | Multiplier |
|---|---|---|---|---|---|---|---|---|---|
| 10 | Brn-blk-blk | 100 | Brown | 1.0 | Red | 10 | Orange | 100 | Yellow |
| 12 | Brn-red-blk | 120 | Brown | 1.2 | Red | 12 | Orange | 120 | Yellow |
| 15 | Brn-grn-blk | 150 | Brown | 1.5 | Red | 15 | Orange | 150 | Yellow |
| 18 | Brn-gry-blk | 180 | Brown | 1.8 | Red | 18 | Orange | 180 | Yellow |
| 22 | Red-red-blk | 220 | Brown | 2.2 | Red | 22 | Orange | 220 | Yellow |
| 27 | Red-vlt-blk | 270 | Brown | 2.7 | Red | 27 | Orange | 270 | Yellow |
| 33 | Org-org-blk | 330 | Brown | 3.3 | Red | 33 | Orange | 330 | Yellow |
| 39 | Org-wht-blk | 390 | Brown | 3.9 | Red | 39 | Orange | 390 | Yellow |
| 47 | Ylw-vlt-blk | 470 | Brown | 4.7 | Red | 47 | Orange | 470 | Yellow |
| 56 | Grn-blu-blk | 560 | Brown | 5.6 | Red | 56 | Orange | 560 | Yellow |
| 68 | Blu-gry-blk | 680 | Brown | 6.8 | Red | 68 | Orange | 680 | Yellow |
| 82 | Gry-red-blk | 820 | Brown | 8.2 | Red | 82 | Orange | 820 | Yellow |

In Table 2.1, the leftmost column represents the complete color code; columns to the right of it only show the third color, since this is the only one that changes. For example, a 10-$\Omega$ resistor has the code brown-black-*black*, while a 100-$\Omega$ resistor has brown-black-*brown*.

In addition to the resistance in ohms, the maximum allowable power dissipation (or **power rating**) is typically specified for commercial resistors. Exceeding this power rating leads to overheating and can cause the resistor to literally burn up.  For a resistor $R$, the power dissipated can be expressed, with Ohm's Law substituted into equation 2.10, by

$$P = VI = I^2R = \frac{V^2}{R} \qquad\qquad \textbf{(2.18)}$$

That is, the power dissipated by a resistor is proportional to the square of the current flowing through it, as well as the square of the voltage across it.  The following example illustrates how one can make use of the power rating to determine whether a given resistor will be suitable for a certain application.

## EXAMPLE  2.6  Using Resistor Power Ratings

### Problem

Determine the minimum **resistor size** that can be connected to a given battery without exceeding the resistor's $\frac{1}{4}$-watt power rating.

### Solution

**Known Quantities:**  Resistor power rating $= 0.25$ W.
    Battery voltages: 1.5 and 3 V.

**Find:**  The smallest size $\frac{1}{4}$-watt resistor that can be connected to each battery.

**Schematics, Diagrams, Circuits, and Given Data:**  Figure 2.23, Figure 2.24.

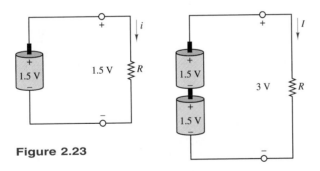

**Figure 2.23**

**Figure 2.24**

**Analysis:**  We first need to obtain an expression for resistor power dissipation as a function of its resistance. We know that $P = VI$ and that $V = IR$. Thus, the power dissipated by any resistor is:

$$P_R = V \times I = V \times \left(\frac{V}{R}\right) = \frac{V^2}{R}$$

Since the maximum allowable power dissipation is 0.25 W, we can write $V^2/R \le 0.25$, or $R \ge V^2/0.25$. Thus, for a 1.5-volt battery, the minimum size resistor will be $R = 1.5^2/0.25 = 9\Omega$. For a 3-volt battery the minimum size resistor will be $R = 3^2/0.25 = 36\Omega$.

**Comments:** Sizing resistors on the basis of power rating is very important in practice. Note how the minimum resistor size quadrupled as we doubled the voltage across it. This is because power increases as the square of the voltage. Remember that exceeding power ratings will inevitably lead to resistor failure!

---

**FOCUS ON MEASUREMENTS**

### Resistive Throttle Position Sensor

**FIND IT ON THE WEB**

**Problem:**

The aim of this example is to determine the calibration of an **automotive resistive throttle position sensor,** shown in Figure 2.25(a). Figure 2.25(b) and (c) depict the geometry of the throttle plate and the equivalent circuit of the throttle sensor. The throttle plate in a typical throttle body has a range of rotation of just under 90°, ranging from closed throttle to wide-open throttle.

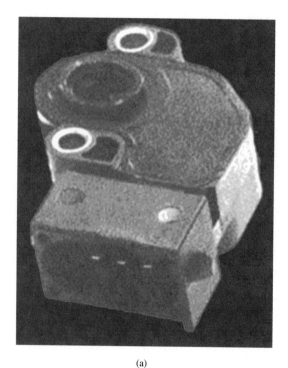

(a)

**Figure 2.25** (a) A throttle position sensor. *Photo courtesy of CTS Corporation.*

**Solution:**

**Known Quantities—** Functional specifications of throttle position sensor.

**Find—** Calibration of sensor in volts per degree of throttle plate opening.

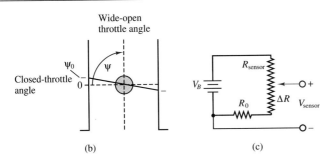

Figure 2.25 (b) Throttle blade geometry (c) Throttle position
sensor equivalent circuit

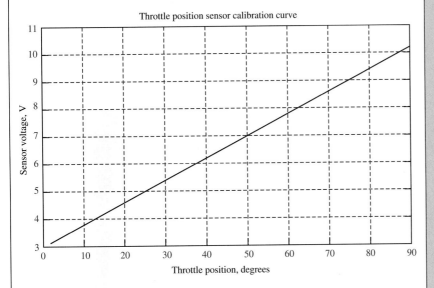

**Figure 2.25** (d) Calibration curve for throttle position sensor

## *Schematics, Diagrams, Circuits, and Given Data—*

**Functional specifications of throttle position sensor**

| Overall Resistance, $R_o + \Delta R$ | 3 to 12 kΩ |
|---|---|
| Input, $V_B$ | 5V ± 4% regulated |
| Output, $V_{\text{sensor}}$ | 5% to 95% $V_s$ |
| Current draw, $I_s$ | ≤ 20 mA |
| Recommended load, $R_L$ | ≤ 220 kΩ |
| Electrical Travel, Max. | 110 degrees |

The nominal supply voltage is 12 V and total throttle plate travel is 88°,
with a closed-throttle angle of 2° and a wide-open throttle angle of 90°.

**Analysis—** The equivalent circuit describing the resistive potentiometer that makes up the sensor is shown in Figure 2.25(b). The *wiper arm*, that is, the moving part of the potentiometer defines a voltage proportional to position. The actual construction of the potentiometer is in the shape of a circle—the figure depicts the potentiometer resistor as a straight line for simplicity. The range of the potentiometer (see specifications above) is 2 to 112° for a resistance of 3 to 12 k$\Omega$; thus, the *calibration constant* of the potentiometer is:

$$k_{pot} = \frac{112 - 0}{12 - 3} \frac{\text{degrees}}{k\Omega} = 12.22 \frac{\text{degrees}}{k\Omega}$$

The sensor voltage is proportional to the ratio $\frac{\Delta R}{R_{sensor}}$, such that

$$V_{sensor} = V_B \times \frac{R_0 + \Delta R}{R_{sensor}} = (5 \text{ V}) \left( \frac{3 + \Delta R}{12} \right)$$

$$= 1.25 + 0.417 \times \Delta R \text{ V} \quad (\Delta R \text{ in } k\Omega)$$

The calibration of the throttle position sensor is:

$$V_{sensor} = V_B \left( \frac{R_0}{R_{sensor}} + \frac{\Delta R}{R_{sensor}} \right)$$

$$= V_B \left( \frac{R_0}{R_{sensor}} + \frac{\theta}{k_{pot} \times R_{sensor}} \right) \quad (\theta \text{ in degrees})$$

The *calibration curve* for the sensor is shown in Figure 2.25(d).

So, if the throttle is closed, the sensor voltage will be:

$$V_{sensor} = V_B \left( \frac{R_0}{R_{sensor}} + \frac{\theta}{k_{pot} \times R_{sensor}} \right)$$

$$= 5 \left( \frac{3 \text{ k}\Omega}{12 \text{ k}\Omega} + \frac{2 \text{ degrees}}{12.22 \frac{\text{degrees}}{k\Omega} \times 12 \text{ k}\Omega} \right) = 1.3182 \text{ V}$$

When the throttle is wide open, the sensor voltage will be:

$$V_{sensor} = V_B \left( \frac{R_0}{R_{sensor}} + \frac{\theta}{k_{pot} \times R_{sensor}} \right)$$

$$= 5 \left( \frac{3 \text{ k}\Omega}{12 \text{ k}\Omega} + \frac{90 \text{ degrees}}{12.44 \frac{\text{degrees}}{k\Omega} \times 12 \text{ k}\Omega} \right) = 4.3187 \text{ V}$$

**Comments—** The fixed resistor $R_0$ prevents the wiper arm from shorting to ground. Note that the throttle position measurement does not use the entire range of the sensor.

**FOCUS ON MEASUREMENTS**

## Resistance Strain Gauges

FIND IT

ON THE WEB

Another common application of the resistance concept to engineering measurements is the resistance **strain gauge**. Strain gauges are devices that are bonded to the surface of an object, and whose resistance varies as a function of the surface strain experienced by the object.

Strain gauges may be used to perform measurements of strain, stress, force, torque, and pressure. Recall that the resistance of a cylindrical conductor of cross-sectional area $A$, length $L$, and conductivity $\sigma$ is given by the expression

$$R = \frac{L}{\sigma A}$$

If the conductor is compressed or elongated as a consequence of an external force, its dimensions will change, and with them its resistance. In particular, if the conductor is stretched, its cross-sectional area will decrease and the resistance will increase. If the conductor is compressed, its resistance decreases, since the length, $L$, will decrease. The relationship between change in resistance and change in length is given by the gauge factor, G, defined by

$$G = \frac{\Delta R / R}{\Delta L / L}$$

and since the strain $\epsilon$ is defined as the fractional change in length of an object, by the formula

$$\epsilon = \frac{\Delta L}{L}$$

the change in resistance due to an applied strain $\epsilon$ is given by the expression

$$\Delta R = R_0 \, G \, \epsilon$$

where $R_0$ is the resistance of the strain gauge under no strain and is called the zero strain resistance. The value of G for resistance strain gauges made of metal foil is usually about 2.

Figure 2.26 depicts a typical foil strain gauge. The maximum strain that can be measured by a foil gauge is about 0.4 to 0.5 percent; that is, $\Delta L/L = 0.004 - 0.005$. For a 120-$\Omega$ gauge, this corresponds to a change in resistance of the order of 0.96 to 1.2 $\Omega$. Although this change in resistance is very small, it can be detected by means of suitable circuitry. Resistance strain

$R_G$   Circuit symbol for the strain gauge

Metal-foil resistance strain gauge. The foil is formed by a photo-etching process and is less than 0.00002 in thick. Typical resistance values are 120, 350, and 1,000 $\Omega$. The wide areas are bonding pads for electrical connections.

**Figure 2.26** Metal-foil resistance strain gauge. The foil is formed by a photo-etching process and is less than 0.00002 in thick. Typical resistance values are 120, 350, and 1,000 $\Omega$. The wide areas are bonding pads for electrical connections.

gauges are usually connected in a circuit called the Wheatstone bridge, which we analyze later in this chapter.

*Comments*— Resistance strain gauges find application in many measurement circuits and instruments.

## EXAMPLE  2.7  Application of Kirchhoff's Laws

### Problem

Apply both KVL and KCL to each of the two circuits depicted in Figure 2.27.

### Solution

*Known Quantities:*  Current and voltage source and resistor values.

*Find:*  Obtain equations for each of the two circuits by applying KCL and KVL.

*Schematics, Diagrams, Circuits, and Given Data:*  Figure 2.27.

*Analysis:*  We start with the circuit of Figure 2.27(a). Applying KVL we write

$$V_S - V_1 - V_2 = 0$$
$$V_S = I R_1 + I R_2.$$

Applying KCL we obtain two equations, one at the top node, the other at the node between the two resistors:

$$I - \frac{V_1}{R_1} = 0 \quad \text{and} \quad \frac{V_1}{R_1} - \frac{V_2}{R_2} = 0$$

With reference to the circuit of Figure 2.27(b), we apply KVL to two equations (one for each loop):

$$V = I_2 R_2; \; V = I_1 R_1$$

Applying KCL we obtain a single equation at the top node:

$$I_S - I_1 - I_2 = 0 \quad \text{or} \quad I_S - \frac{V_1}{R_1} - \frac{V_2}{R_2} = 0$$

*Comments:*  Note that in each circuit one of Kirchhoff's laws results in a single equation, while the other results in two equations. In Chapter 3 we shall develop methods for systematically writing the smallest possible number of equations sufficient to solve a circuit.

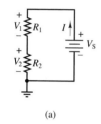

(a)

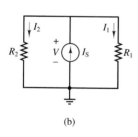

(b)

**Figure 2.27**

## Open and Short Circuits

Two convenient idealizations of the resistance element are provided by the limiting cases of Ohm's law as the resistance of a circuit element approaches zero or infinity. A circuit element with resistance approaching zero is called a **short circuit**. Intuitively, one would expect a short circuit to allow for unimpeded flow of current. In fact, metallic conductors (e.g., short wires of large diameter) approximate the behavior of a short circuit. Formally, a short circuit is defined as a circuit element across which the voltage is zero, regardless of the current flowing through it. Figure 2.28 depicts the circuit symbol for an ideal short circuit.

Physically, any wire or other metallic conductor will exhibit some resistance, though small. For practical purposes, however, many elements approximate a short circuit quite accurately under certain conditions. For example, a large-diameter copper pipe is effectively a short circuit in the context of a residential electrical power supply, while in a low-power microelectronic circuit (e.g., an FM radio) a short length of 24 gauge wire (refer to Table 2.2 for the resistance of 24 gauge wire) is a more than adequate short circuit.

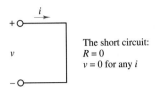

The short circuit:
$R = 0$
$v = 0$ for any $i$

**Figure 2.28** The short circuit

Table 2.2   Resistance of copper wire

| AWG size | Number of strands | Diameter per strand | Resistance per 1,000 ft ($\Omega$) |
|---|---|---|---|
| 24 | Solid | 0.0201 | 28.4 |
| 24 | 7 | 0.0080 | 28.4 |
| 22 | Solid | 0.0254 | 18.0 |
| 22 | 7 | 0.0100 | 19.0 |
| 20 | Solid | 0.0320 | 11.3 |
| 20 | 7 | 0.0126 | 11.9 |
| 18 | Solid | 0.0403 | 7.2 |
| 18 | 7 | 0.0159 | 7.5 |
| 16 | Solid | 0.0508 | 4.5 |
| 16 | 19 | 0.0113 | 4.7 |

A circuit element whose resistance approaches infinity is called an **open circuit**. Intuitively, one would expect no current to flow through an open circuit, since it offers infinite resistance to any current. In an open circuit, we would expect to see zero current regardless of the externally applied voltage. Figure 2.29 illustrates this idea.

In practice, it is not too difficult to approximate an open circuit: any break in continuity in a conducting path amounts to an open circuit. The idealization of the open circuit, as defined in Figure 2.29, does not hold, however, for very high voltages. The insulating material between two insulated terminals will break down at a sufficiently high voltage. If the insulator is air, ionized particles in the neighborhood of the two conducting elements may lead to the phenomenon of arcing; in other words, a pulse of current may be generated that momentarily jumps a gap between conductors (thanks to this principle, we are able to ignite the air-fuel mixture in a spark-ignition internal combustion engine by means of spark plugs). The ideal open and short circuits are useful concepts and find extensive use in circuit analysis.

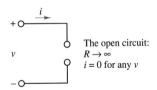

The open circuit:
$R \to \infty$
$i = 0$ for any $v$

**Figure 2.29** The open circuit

## Series Resistors and the Voltage Divider Rule

Although electrical circuits can take rather complicated forms, even the most involved circuits can be reduced to combinations of circuit elements *in parallel* and

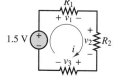

The current $i$ flows through each of the four series elements. Thus, by KVL,

$$1.5 = v_1 + v_2 + v_3$$

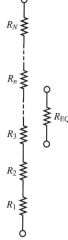

$N$ series resistors are equivalent to a single resistor equal to the sum of the individual resistances.

**Figure 2.30**

*in series.* Thus, it is important that you become acquainted with parallel and series circuits as early as possible, even before formally approaching the topic of network analysis. Parallel and series circuits have a direct relationship with Kirchhoff's laws. The objective of this section and the next is to illustrate two common circuits based on series and parallel combinations of resistors: the voltage and current dividers. These circuits form the basis of all network analysis; it is therefore important to master these topics as early as possible.

For an example of a series circuit, refer to the circuit of Figure 2.30, where a battery has been connected to resistors $R_1$, $R_2$, and $R_3$. The following definition applies:

**Definition**

Two or more circuit elements are said to be **in series** if the identical current flows through each of the elements.

By applying KVL, you can verify that the sum of the voltages across the three resistors equals the voltage externally provided by the battery:

$$1.5 \text{ V} = v_1 + v_2 + v_3$$

and since, according to Ohm's law, the separate voltages can be expressed by the relations

$$v_1 = iR_1 \qquad v_2 = iR_2 \qquad v_3 = iR_3$$

we can therefore write

$$1.5 \text{ V} = i(R_1 + R_2 + R_3)$$

This simple result illustrates a very important principle: To the battery, the three series resistors appear as a single equivalent resistance of value $R_{\text{EQ}}$, where

$$R_{\text{EQ}} = R_1 + R_2 + R_3$$

The three resistors could thus be replaced by a single resistor of value $R_{\text{EQ}}$ without changing the amount of current required of the battery. From this result we may extrapolate to the more general relationship defining the equivalent resistance of $N$ series resistors:

$$R_{\text{EQ}} = \sum_{n=1}^{N} R_n \tag{2.19}$$

which is also illustrated in Figure 2.30. A concept very closely tied to series resistors is that of the **voltage divider**. This terminology originates from the observation that the source voltage in the circuit of Figure 2.30 divides among the three resistors according to KVL. If we now observe that the series current, $i$, is given by

$$i = \frac{1.5 \text{ V}}{R_{\text{EQ}}} = \frac{1.5 \text{ V}}{R_1 + R_2 + R_3}$$

we can write each of the voltages across the resistors as:

$$v_1 = i R_1 = \frac{R_1}{R_{EQ}}(1.5\ \text{V})$$

$$v_2 = i R_2 = \frac{R_2}{R_{EQ}}(1.5\ \text{V})$$

$$v_3 = i R_3 = \frac{R_3}{R_{EQ}}(1.5\ \text{V})$$

That is:

> The voltage across each resistor in a series circuit is directly proportional to the ratio of its resistance to the total series resistance of the circuit.

An instructive exercise consists of verifying that KVL is still satisfied, by adding the voltage drops around the circuit and equating their sum to the source voltage:

$$v_1 + v_2 + v_3 = \frac{R_1}{R_{EQ}}(1.5\ \text{V}) + \frac{R_2}{R_{EQ}}(1.5\ \text{V}) + \frac{R_3}{R_{EQ}}(1.5\ \text{V}) = 1.5\ \text{V}$$

since

$$R_{EQ} = R_1 + R_2 + R_3$$

Therefore, since KVL is satisfied, we are certain that the voltage divider rule is consistent with Kirchhoff's laws. By virtue of the voltage divider rule, then, we can always determine the proportion in which voltage drops are distributed around a circuit. This result will be useful in reducing complicated circuits to simpler forms. The general form of the voltage divider rule for a circuit with $N$ series resistors and a voltage source is:

$$\boxed{v_n = \frac{R_n}{R_1 + R_2 + \cdots + R_n + \cdots + R_N}\, v_S \quad \text{Voltage divider}} \qquad \textbf{(2.20)}$$

## EXAMPLE  2.8  Voltage Divider

**Problem**

Determine the voltage $v_3$ in the circuit of Figure 2.31.

**Solution**

***Known Quantities:***  Source voltage, resistance values

***Find:***  Unknown voltage $v_3$.

***Schematics, Diagrams, Circuits, and Given Data:***  $R_1 = 10\,\Omega$; $R_2 = 6\,\Omega$; $R_3 = 8\,\Omega$; $V_S = 3$ V. Figure 2.31.

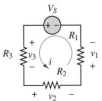

**Figure 2.31**

**Analysis:** Figure 2.31 indicates a reference direction for the current (dictated by the polarity of the voltage source). Following the passive sign convention, we label the polarities of the three resistors, and apply KVL to determine that

$$V_S - v_1 - v_2 - v_3 = 0$$

The voltage divider rule tells us that

$$v_3 = V_S \times \frac{R_3}{R_1 + R_2 + R_3} = 3 \times \frac{8}{10 + 6 + 8} = 1 \text{ V}$$

**Comments:** Application of the voltage divider rule to a series circuit is very straightforward. The difficulty usually arises in determining whether a circuit is in fact a series circuit. This point is explored later in this section, and in Example 2.10.

VIRTUAL LAB

**Focus on Computer-Aided Tools:** The simple voltagedivider circuit introduced in this example provides an excellent introduction to the capabilities of the *Electronics Workbench*, or EWB™, a computer-aided tool for solving electrical and electronic circuits. You will find the EWB™ version of the circuit of Figure 2.31 in the electronic files that accompany this book in CD-ROM format. This simple example may serve as a workbench to practice your own skills in constructing circuits using *Electronics Workbench*.

## Parallel Resistors and the Current Divider Rule

A concept analogous to that of the voltage divider may be developed by applying Kirchhoff's current law to a circuit containing only parallel resistances.

---

**Definition**

Two or more circuit elements are said to be **in parallel** if the identical voltage appears across each of the elements.

---

Figure 2.32 illustrates the notion of parallel resistors connected to an ideal current source. Kirchhoff's current law requires that the sum of the currents into, say, the top node of the circuit be zero:

$$i_S = i_1 + i_2 + i_3$$

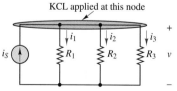

The voltage $v$ appears across each parallel element; by KCL, $i_S = i_1 + i_2 + i_3$

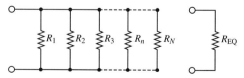

$N$ resistors in parallel are equivalent to a single equivalent resistor with resistance equal to the inverse of the sum of the inverse resistances.

**Figure 2.32** Parallel circuits

But by virtue of Ohm's law we may express each current as follows:

$$i_1 = \frac{v}{R_1} \qquad i_2 = \frac{v}{R_2} \qquad i_3 = \frac{v}{R_3}$$

since, by definition, the same voltage, $v$, appears across each element. Kirchhoff's current law may then be restated as follows:

$$i_S = v \left( \frac{1}{R_1} + \frac{1}{R_2} + \frac{1}{R_3} \right)$$

Note that this equation can be also written in terms of a single equivalent resistance:

$$i_S = v \frac{1}{R_{EQ}}$$

where

$$\frac{1}{R_{EQ}} = \frac{1}{R_1} + \frac{1}{R_2} + \frac{1}{R_3}$$

As illustrated in Figure 2.32, one can generalize this result to an arbitrary number of resistors connected in parallel by stating that $N$ resistors in parallel act as a single equivalent resistance, $R_{EQ}$, given by the expression

$$\frac{1}{R_{EQ}} = \frac{1}{R_1} + \frac{1}{R_2} + \cdots + \frac{1}{R_N} \qquad \textbf{(2.21)}$$

or

$$R_{EQ} = \frac{1}{1/R_1 + 1/R_2 + \cdots + 1/R_N} \qquad \textbf{(2.22)}$$

Very often in the remainder of this book we shall refer to the parallel combination of two or more resistors with the following notation:

$$R_1 \parallel R_2 \parallel \cdots$$

where the symbol $\parallel$ signifies "in parallel with."

From the results shown in equations 2.21 and 2.22, which were obtained directly from KCL, the **current divider** rule can be easily derived. Consider, again, the three-resistor circuit of Figure 2.32. From the expressions already derived from each of the currents, $i_1$, $i_2$, and $i_3$, we can write:

$$i_1 = \frac{v}{R_1} \qquad i_2 = \frac{v}{R_2} \qquad i_3 = \frac{v}{R_3}$$

and since $v = R_{EQ} i_S$, these currents may be expressed by:

$$i_1 = \frac{R_{EQ}}{R_1} i_S = \frac{1/R_1}{1/R_{EQ}} i_S = \frac{1/R_1}{1/R_1 + 1/R_2 + 1/R_3} i_S$$

$$i_2 = \frac{1/R_2}{1/R_1 + 1/R_2 + 1/R_3} i_S$$

$$i_3 = \frac{1/R_3}{1/R_1 + 1/R_2 + 1/R_3} i_S$$

One can easily see that the current in a parallel circuit divides in inverse proportion to the resistances of the individual parallel elements. The general expression for the current divider for a circuit with $N$ parallel resistors is the following:

$$i_n = \frac{1/R_n}{1/R_1 + 1/R_2 + \cdots + 1/R_n + \cdots + 1/R_N} i_S \quad \text{Current divider} \qquad \textbf{(2.23)}$$

Example 2.9 illustrates the application of the current divider rule.

### EXAMPLE 2.9 Current Divider

**Problem**

Determine the current $i_1$ in the circuit of Figure 2.33.

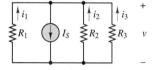

**Figure 2.33**

**Solution**

**Known Quantities:** Source current, resistance values.

**Find:** Unknown current $i_1$.

**Schematics, Diagrams, Circuits, and Given Data:**
$R_1 = 10\Omega$; $R_2 = 2\Omega$; $R_3 = 20\Omega$; $I_S = 4$ A. Figure 2.33.

**Analysis:** Application of the current divider rule yields:

$$i_1 = I_S \times \frac{\frac{1}{R_1}}{\frac{1}{R_1} + \frac{1}{R_2} + \frac{1}{R_3}} = 4 \times \frac{\frac{1}{10}}{\frac{1}{10} + \frac{1}{2} + \frac{1}{20}} = 0.6154 \text{ A}$$

**Comments:** While application of the current divider rule to a parallel circuit is very straightforward, it is sometimes not so obvious whether two or more resistors are actually in parallel. A method for ensuring that circuit elements are connected in parallel is explored later in this section, and in Example 2.10.

**Focus on Computer-Aided Tools:** You will find the EWB™ version of the circuit of Figure 2.33 in the electronic files that accompany this book in CD-ROM format. This simple example may serve as a workbench to practice your own skills in constructing circuits using *Electronics Workbench*.

VIRTUAL LAB

Much of the resistive network analysis that will be introduced in Chapter 3 is based on the simple principles of the voltage and current dividers introduced in this section. Unfortunately, practical circuits are rarely composed only of parallel or only of series elements. The following examples and Check Your Understanding exercises illustrate some simple and slightly more advanced circuits that combine parallel and series elements.

## EXAMPLE 2.10 Series-Parallel Circuit

**Problem**

Determine the voltage $v$ in the circuit of Figure 2.34.

**Solution**

***Known Quantities:*** Source voltage, resistance values.

***Find:*** Unknown voltage $v$.

***Schematics, Diagrams, Circuits, and Given Data:*** See Figures 2.34, 2.35.

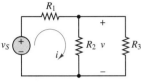

**Figure 2.34**

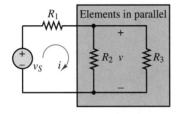
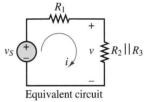

**Figure 2.35**

***Analysis:*** The circuit of Figure 2.34 is neither a series nor a parallel circuit because the following two conditions do not apply:

1. The current through all resistors is the same (series circuit condition)
2. The voltage across all resistors is the same (parallel circuit condition)

The circuit takes a much simplier appearance once it becomes evident that the same voltage appears across both $R_2$ and $R_3$ and, therefore, that these elements are in parallel. If these two resistors are replaced by a single equivalent resistor according to the procedures described in this section, the circuit of Figure 2.35 is obtained. Note that now the equivalent circuit is a simple series circuit and the voltage divider rule can be applied to determine that:

VIRTUAL LAB

$$v = \frac{R_2 \| R_3}{R_1 + R_2 \| R_3} v_S$$

while the current is found to be

$$i = \frac{v_S}{R_1 + R_2 \| R_3}$$

***Comments:*** Systematic methods for analyzing arbitrary circuit configurations are explored in Chapter 3.

## EXAMPLE 2.11 The Wheatstone Bridge

### Problem

The **Wheatstone bridge** is a resistive circuit that is frequently encountered in a variety of measurement circuits. The general form of the bridge circuit is shown in Figure 2.36(a), where $R_1$, $R_2$, and $R_3$ are known while $R_x$ is an unknown resistance, to be determined. The circuit may also be redrawn as shown in Figure 2.36(b). The latter circuit will be used to demonstrate the use of the voltage divider rule in a mixed series-parallel circuit. The objective is to determine the unknown resistance, $R_x$.

1. Find the value of the voltage $v_{ab} = v_{ad} - v_{bd}$ in terms of the four resistances and the source voltage, $v_S$. Note that since the reference point $d$ is the same for both voltages, we can also write $v_{ab} = v_a - v_b$.

2. If $R_1 = R_2 = R_3 = 1$ k$\Omega$, $v_S = 12$ V, and $v_{ab} = 12$ mV, what is the value of $R_x$?

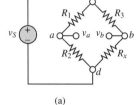

(a)

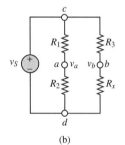

(b)

**Figure 2.36** Wheatstone bridge circuits

### Solution

***Known Quantities:*** Source voltage, resistance values, bridge voltage.

***Find:*** Unknown resistance $R_x$.

***Schematics, Diagrams, Circuits, and Given Data:*** See Figure 2.36.
$$R_1 = R_2 = R_3 = 1 \text{ k}\Omega; v_S = 12 \text{ V}; v_{ab} = 12 \text{ mV}.$$

***Analysis:***

1. First, we observe that the circuit consists of the parallel combination of three subcircuits: the voltage source, the series combination of $R_1$ and $R_2$, and the series combination of $R_3$ and $R_x$. Since these three subcircuits are in parallel, the same voltage will appear across each of them, namely, the source voltage, $v_S$. Thus, the source voltage divides between each resistor pair, $R_1 - R_2$ and $R_3 - R_x$, according to the voltage divider rule: $v_a$ is the fraction of the source voltage appearing across $R_2$, while $v_b$ is the voltage appearing across $R_x$:

$$v_a = v_S \frac{R_2}{R_1 + R_2} \quad \text{and} \quad v_b = v_S \frac{R_x}{R_3 + R_x}$$

Finally, the voltage difference between points $a$ and $b$ is given by:

$$v_{ab} = v_a - v_b = v_S \left( \frac{R_2}{R_1 + R_2} - \frac{R_x}{R_3 + R_x} \right)$$

This result is very useful and quite general.

2. In order to solve for the unknown resistance, we substitute the numerical values in the preceding equation to obtain

$$0.012 = 12 \left( \frac{1,000}{2,000} - \frac{R_x}{1,000 + R_x} \right)$$

which may be solved for $R_x$ to yield

$$R_x = 996 \ \Omega$$

**Comments:** The Wheatstone bridge finds application in many measurement circuits and instruments.

**Focus on Computer-Aided Tools: Virtual Lab**  You will find a Virtual Lab version of the circuit of Figure 2.36 in the electronic files that accompany this book.  If you have practiced building some simple circuits using *Electronics Workbench*, you should by now be convinced that this is an invaluable tool in validating numerical solutions to problems, and in exploring more advanced concepts.

VIRTUAL LAB

---

## The Wheatstone Bridge and Force Measurements

FOCUS ON MEASUREMENTS

Strain gauges, which were introduced in a Focus on Measurements section earlier in this chapter, are frequently employed in the measurement of force. One of the simplest applications of strain gauges is in the measurement of the force applied to a cantilever beam, as illustrated in Figure 2.37.  Four strain gauges are employed in this case, of which two are bonded to the upper surface of the beam at a distance $L$ from the point where the external force, $F$, is applied and two are bonded on the lower surface, also at a distance $L$.  Under the influence of the external force, the beam deforms and causes the upper gauges to extend and the lower gauges to compress.  Thus, the resistance of the upper gauges will increase by an amount $\Delta R$, and that of the lower gauges will decrease by an equal amount, assuming that the gauges are symmetrically placed.  Let $R_1$ and $R_4$ be the upper gauges and $R_2$ and $R_3$ the lower gauges.  Thus, under the influence of the external force, we have:

$$R_1 = R_4 = R_0 + \Delta R$$
$$R_2 = R_3 = R_0 - \Delta R$$

where $R_0$ is the zero strain resistance of the gauges.  It can be shown from elementary statics that the relationship between the strain $\epsilon$ and a force $F$

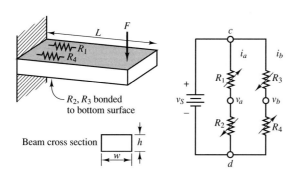

A force-measuring instrument

applied at a distance $L$ for a cantilever beam is:

$$\epsilon = \frac{6LF}{wh^2Y}$$

where $h$ and $w$ are as defined in Figure 2.37 and $Y$ is the beam's modulus of elasticity.

In the circuit of Figure 2.37, the currents $i_a$ and $i_b$ are given by

$$i_a = \frac{v_S}{R_1 + R_2} \quad \text{and} \quad i_b = \frac{v_S}{R_3 + R_4}$$

The bridge output voltage is defined by $v_o = v_b - v_a$ and may be found from the following expression:

$$v_o = i_b R_4 - i_a R_2 = \frac{v_S R_4}{R_3 + R_4} - \frac{v_S R_2}{R_1 + R_2}$$

$$= v_S \frac{R_0 + \Delta R}{R_0 + \Delta R + R_0 - \Delta R} - v_S \frac{R_0 - \Delta R}{R_0 + \Delta R + R_0 - \Delta R}$$

$$= v_S \frac{\Delta R}{R_0} = v_S \, G\epsilon$$

where the expression for $\Delta R/R_0$ was obtained in "Focus on Measurements: Resistance Strain Gauges" section. Thus, it is possible to obtain a relationship between the output voltage of the bridge circuit and the force, $F$, as follows:

$$v_o = v_S \, G\epsilon = v_S \, G\frac{6LF}{wh^2Y} = \frac{6v_S GL}{wh^2Y} F = kF$$

where $k$ is the calibration constant for this force transducer.

*Comments*— **Strain gauge bridges** are commonly used in mechanical, chemical, aerospace, biomedical, and civil engineering applications (and wherever measurements of force, pressure, torque, stress, or strain are sought).

## Check Your Understanding

**2.4**   Repeat Example 2.8 by reversing the reference direction of the current, to show that the same result is obtained.

**2.5**   The circuit in the accompanying illustration contains a battery, a resistor, and an unknown circuit element.

1.   If the voltage $V_{battery}$ is 1.45 V and $i = 5$ mA, find power supplied to or by the battery.
2.   Repeat part 1 if $i = -2$ mA.

**2.6**   The battery in the accompanying circuit supplies power to the resistors $R_1$, $R_2$, and $R_3$. Use KCL to determine the current $i_B$, and find the power supplied by the battery if $V_{battery} = 3$ V.

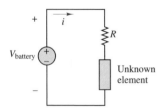

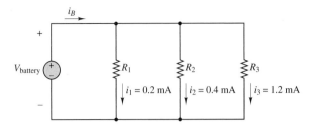

**2.7**   Use the results of part 1 of Example 2.11 to find the condition for which the voltage $v_{ab} = v_a - v_b$ is equal to zero (this is called the balanced condition for the bridge). Does this result necessarily require that all four resistors be identical? Why?

**2.8**   Verify that KCL is satisfied by the current divider rule and that the source current $i_S$ divides in inverse proportion to the parallel resistors $R_1$, $R_2$, and $R_3$ in the circuit of Figure 2.33. (This should not be a surprise, since we would expect to see more current flow through the smaller resistance.)

**2.9**   Compute the full-scale (i.e., largest) output voltage for the force-measuring apparatus of "Focus on Measurements: The Wheatstone Bridge and Force Measurements." Assume that the strain gauge bridge is to measure forces ranging from 0 to 500 N, $L = 0.3$ m, $w = 0.05$ m, $h = 0.01$ m, $G = 2$, and the modulus of elasticity for the beam is $69 \times 10^9$ N/m$^2$ (aluminum). The source voltage is 12 V. What is the calibration constant of this force transducer?

**2.10**   Repeat the derivation of the current divider law by using conductance elements—that is, by replacing each resistance with its equivalent conductance, $G = 1/R$.

---

## 2.7   PRACTICAL VOLTAGE AND CURRENT SOURCES

The idealized models of voltage and current sources we discussed in Section 2.3 fail to consider the internal resistance of practical voltage and current sources. The objective of this section is to extend the ideal models to models that are capable of describing the physical limitations of the voltage and current sources used in practice. Consider, for example, the model of an ideal voltage source shown in Figure 2.9. As the load resistance ($R$) decreases, the source is required to provide increasing amounts of current to maintain the voltage $v_S(t)$ across its terminals:

$$i(t) = \frac{v_S(t)}{R} \qquad\qquad \textbf{(2.24)}$$

This circuit suggests that the ideal voltage source is required to provide an infinite amount of current to the load, in the limit as the load resistance approaches zero. Naturally, you can see that this is impossible; for example, think about the ratings of a conventional car battery: 12 V, 450 A-h (ampere-hours). This implies that there is a limit (albeit a large one) to the amount of current a practical source can deliver to a load. Fortunately, it will not be necessary to delve too deeply into the physical nature of each type of source in order to describe the behavior of a practical voltage source: The limitations of practical sources can be approximated quite simply by exploiting the notion of the internal resistance of a source. Although the models

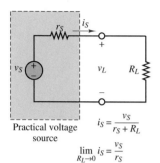

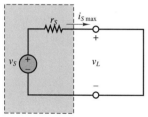

$$i_S = \frac{v_S}{r_S + R_L}$$

$$\lim_{R_L \to 0} i_S = \frac{v_S}{r_S}$$

The maximum (short circuit) current which can be supplied by a practical voltage source is

$$i_{S\,max} = \frac{v_S}{r_S}$$

**Figure 2.38** Practical voltage source

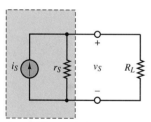

A model for practical current sources consists of an ideal source in *parallel* with an internal resistance.

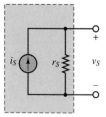

Maximum output voltage for practical current source with open-circuit load:

$$v_{S\,max} = i_S r_S$$

**Figure 2.39** Practical current source

described in this section are only approximations of the actual behavior of energy sources, they will provide good insight into the limitations of practical voltage and current sources. Figure 2.38 depicts a model for a practical voltage source, composed of an ideal voltage source, $v_S$, in series with a resistance, $r_S$. The resistance $r_S$ in effect poses a limit to the maximum current the voltage source can provide:

$$i_{S\,max} = \frac{v_S}{r_S} \tag{2.25}$$

Typically, $r_S$ is small. Note, however, that its presence affects the voltage across the load resistance: Now this voltage is no longer equal to the source voltage. Since the current provided by the source is

$$i_S = \frac{v_S}{r_S + R_L} \tag{2.26}$$

the load voltage can be determined to be

$$v_L = i_S R_L = v_S \frac{R_L}{r_S + R_L} \tag{2.27}$$

Thus, in the limit as the source internal resistance, $r_S$, approaches zero, the load voltage, $v_L$, becomes exactly equal to the source voltage. It should be apparent that a desirable feature of an ideal voltage source is a very small internal resistance, so that the current requirements of an arbitrary load may be satisfied. Often, the effective internal resistance of a voltage source is quoted in the technical specifications for the source, so that the user may take this parameter into account.

A similar modification of the ideal current source model is useful to describe the behavior of a practical current source. The circuit illustrated in Figure 2.39 depicts a simple representation of a practical current source, consisting of an ideal source in parallel with a resistor. Note that as the load resistance approaches infinity (i.e., an open circuit), the output voltage of the current source approaches its limit,

$$v_{S\,max} = i_S r_S \tag{2.28}$$

A good current source should be able to approximate the behavior of an ideal current source. Therefore, a desirable characteristic for the internal resistance of a current source is that it be as large as possible.

## 2.8    MEASURING DEVICES

In this section, you should gain a basic understanding of the desirable properties of practical devices for the measurement of electrical parameters. The measurements most often of interest are those of current, voltage, power, and resistance. In analogy with the models we have just developed to describe the nonideal behavior of voltage and current sources, we shall similarly present circuit models for practical measuring instruments suitable for describing the nonideal properties of these devices.

### The Ohmmeter

The **ohmmeter** is a device that, when connected across a circuit element, can measure the resistance of the element. Figure 2.40 depicts the circuit connection of an ohmmeter to a resistor. One important rule needs to be remembered:

The resistance of an element can be measured only when the element is disconnected from any other circuit.

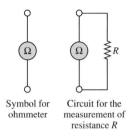

Symbol for     Circuit for the
ohmmeter      measurement of
                    resistance $R$

**Figure 2.40** Ohmmeter and measurement of resistance

## The Ammeter

The **ammeter** is a device that, when connected in series with a circuit element, can measure the current flowing through the element. Figure 2.41 illustrates this idea. From Figure 2.41, two requirements are evident for obtaining a correct measurement of current:

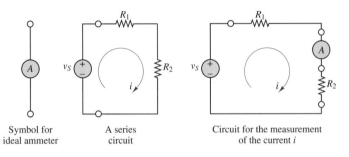

Symbol for          A series          Circuit for the measurement
ideal ammeter       circuit           of the current $i$

**Figure 2.41** Measurement of current

1.  The ammeter must be placed in series with the element whose current is to be measured (e.g., resistor $R_2$).
2.  The ammeter should not restrict the flow of current (i.e., cause a voltage drop), or else it will not be measuring the true current flowing in the circuit. *An ideal ammeter has zero internal resistance.*

## The Voltmeter

The **voltmeter** is a device that can measure the voltage across a circuit element. Since voltage is the difference in potential between two points in a circuit, the voltmeter needs to be connected across the element whose voltage we wish to measure. A voltmeter must also fulfill two requirements:

1.  The voltmeter must be placed in parallel with the element whose voltage it is measuring.
2.  The voltmeter should draw no current away from the element whose voltage it is measuring, or else it will not be measuring the true voltage across that element. Thus, *an ideal voltmeter has infinite internal resistance.*

Figure 2.42 illustrates these two points.

Once again, the definitions just stated for the ideal voltmeter and ammeter need to be augmented by considering the practical limitations of the devices. A practical ammeter will contribute some series resistance to the circuit in which it is measuring current; a practical voltmeter will not act as an ideal open circuit but will always draw some current from the measured circuit. The homework problems verify that these practical restrictions do not necessarily pose a limit to the accuracy of the measurements obtainable with practical measuring devices, as long as the internal resistance of the measuring devices is known. Figure 2.43 depicts the circuit models for the practical ammeter and voltmeter.

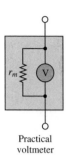

Practical
voltmeter

Practical
ammeter

**Figure 2.43** Models for
practical ammeter and voltmeter

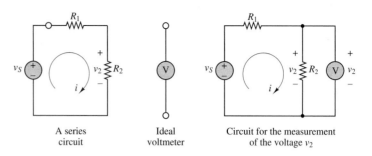

A series          Ideal          Circuit for the measurement
circuit          voltmeter          of the voltage $v_2$

**Figure 2.42** Measurement of voltage

All of the considerations that pertain to practical ammeters and voltmeters can be applied to the operation of a **wattmeter**, a measuring instrument that provides a measurement of the power dissipated by a circuit element, since the wattmeter is in effect made up of a combination of a voltmeter and an ammeter. Figure 2.44 depicts the typical connection of a wattmeter in the same series circuit used in the preceding paragraphs. In effect, the wattmeter measures the current flowing through the load and, simultaneously, the voltage across it and multiplies the two to provide a reading of the power dissipated by the load. The internal power consumption of a practical wattmeter is explored in the homework problems.

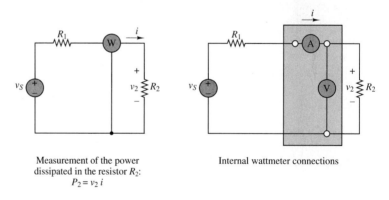

Measurement of the power
dissipated in the resistor $R_2$:
$P_2 = v_2 i$

Internal wattmeter connections

**Figure 2.44** Measurement of power

## 2.9    ELECTRICAL NETWORKS

In the previous sections we have outlined models for the basic circuit elements: sources, resistors, and measuring instruments. We have assembled all the tools and parts we need in order to define an **electrical network**. It is appropriate at this stage to formally define the elements of the electrical circuit; the definitions that follow are part of standard electrical engineering terminology.

### Branch

A **branch** is any portion of a circuit with two terminals connected to it. A branch may consist of one or more circuit elements (Figure 2.45). In practice, any circuit element with two terminals connected to it is a branch.

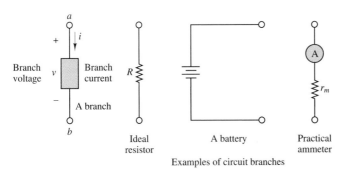

**Figure 2.45** Definition of a branch

# DC Measurements with the Digital MultiMeter
## (Courtesy: Hewlett-Packard)

**FOCUS ON
MEASUREMENTS**

Digital multimeters (DMMs) are the workhorse of all measurement laboratories. Figure 2.46 depicts the front panel of a typical benchtop DMM. Tables 2.3 and 2.4 list the features and specifications of the multimeter.

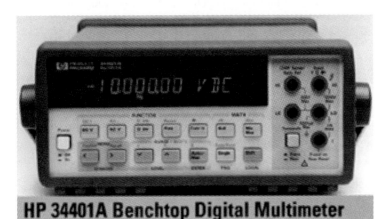

**Figure 2.46** Hewlett-Packard 34401A 6.5-digit multimeter.

Table 2.3  Features of the 34401A multimeter

- 6.5 digit resolution uncovers the details that hide from other DMMs
- Accuracy you can count on: 0.0015% for dc, 0.06% for ac
- Perfect for your bench - more than a dozen functions one or two key presses away
- True RMS AC volts and current
- Perfect for your system - 1000 rdgs/sec in ASCII format across the HP-IB bus
- RS-232 and HP-IB Standard

The *Measurements* section in the accompanying CD-ROM contains interactive programs that illustrate the use of the DMM and of other common measuring instruments.

Table 2.4  Specifications for the 34401A multimeter

**DC Voltage Accuracy specs**

| Range dc voltage | 6.5 Digits Resolution | Accuracy: 1 year (%reading + %range) | Input resistance |
|---|---|---|---|
| 100mV | 100nV | 0.0050 + 0.0035 | 10 MΩ or >10 GΩ |
| 1V | 1μV | 0.0040 + 0.0007 | 10 MΩ or >10 GΩ |
| 10V | 10μV | 0.0035 + 0.0005 | 10 MΩ or >10 GΩ |
| 100V | 100μV | 0.0045 + 0.0006 | 10 MΩ |
| 1000V | 1mV | 0.0045 + 0.0010 | 10 MΩ |

**True RMS AC Voltage Accuracy specs**

| | Frequency | Accuracy: 1 year (%reading + %range) |
|---|---|---|
| 100 mV range | 3 Hz–5 Hz | 1.00 + 0.04 |
| | 5 Hz–10 Hz | 0.35 + 0.04 |
| | 10 Hz–20 kHz | 0.06 + 0.04 |
| | 20 kHz–50 kHz | 0.12 + 0.04 |
| | 50 kHz–100 kHz | 0.60 + 0.08 |
| | 100 kHz–300 kHz | 4.00 + 0.50 |
| 1 V–750 V ranges | 3 Hz–5 Hz | 1.00 + 0.03 |
| | 5 Hz–10 Hz | 0.35 + 0.03 |
| | 10 Hz–20 kHz | 0.06 + 0.03 |
| | 20 kHz–50 kHz | 0.12 + 0.05 |
| | 50 kHz–100 kHz | 0.60 + 0.08 |
| | 100 kHz–300 kHz | 400 + 0.50 |

**Resistance Accuracy specs**

| Range | Resolution | Accuracy: 1 year (%reading + %range) | Current Source |
|---|---|---|---|
| 100 ohm | 100 Ω | 0.010 + 0.004 | 1 mA |
| 1 kΩ | 1 mΩ | 0.010 + 0.001 | 1 mA |
| 10 kΩ | 10 mΩ | 0.010 + 0.001 | 100 μA |
| 100 kohm | 100 mΩ | 0.010 + 0.001 | 10 μA |
| 1 MΩ | 1 Ω | 0.010 + 0.001 | 5 μA |
| 10 MΩ | 10 Ω | 0.040 + 0.001 | 500 nA |
| 100 Mohm | 100 Ω | 0.800 + 0.010 | 500 nA |

**Other Accuracy specs (basic 1 year accuracy)**

| | |
|---|---|
| dc current accuracy: (10 mA to 3 A ranges) | 0.05% of reading + 0.005% of range |
| ac current accuracy: (1 A to 3 A ranges) | 0.1% of reading + 0.04% of range |
| Frequency (and Period): (3 Hz to 300 kHz, 0.333 sec to 3.33 μsec) | 0.01% of reading |
| Continuity: (1000 Ω range, 1 mA test current) | 0.01% of reading + 0.02% of range |
| Diode test: 1 V range, 1 mA test current | 0.01% of reading = 0.02% of range |

## Node

A node is the junction of two or more branches (one often refers to the junction of only two branches as a *trivial node*). Figure 2.47 illustrates the concept. In effect, any connection that can be accomplished by soldering various terminals together is a node. It is very important to identify nodes properly in the analysis of electrical networks.

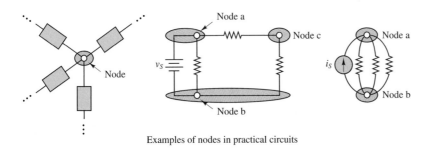

Examples of nodes in practical circuits

**Figure 2.47** Definition of a node

## Loop

A **loop** is any closed connection of branches. Various loop configurations are illustrated in Figure 2.48.

Note how two different loops in the same circuit may include some of the same elements or branches.

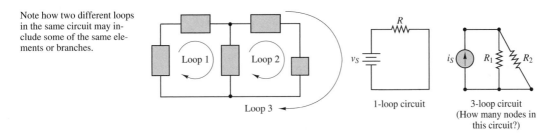

**Figure 2.48** Definition of a loop

## Mesh

A **mesh** is a loop that does not contain other loops. Meshes are an important aid to certain analysis methods. In Figure 2.48, the circuit with loops 1, 2, and 3 consists of two meshes: loops 1 and 2 are meshes, but loop 3 is not a mesh, because it encircles both loops 1 and 2. The one-loop circuit of Figure 2.48 is also a one-mesh circuit. Figure 2.49 illustrates how meshes are simpler to visualize in complex networks than loops are.

## Network Analysis

The analysis of an electrical network consists of determining each of the unknown branch currents and node voltages. It is therefore important to define all of the

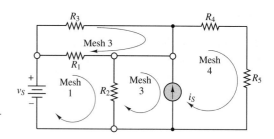

How many loops can you identify in this four-mesh circuit? (Answer: 14)

**Figure 2.49** Definition of a mesh

relevant variables as clearly as possible, and in systematic fashion. Once the known and unknown variables have been identified, a set of equations relating these variables is constructed, and these are solved by means of suitable techniques. The analysis of electrical circuits consists of writing the smallest set of equations sufficient to solve for all of the unknown variables. The procedures required to write these equations are the subject of Chapter 3 and are very well documented and codified in the form of simple rules. The analysis of electrical circuits is greatly simplified if some standard conventions are followed. The objective of this section is precisely to outline the preliminary procedures that will render the task of analyzing an electrical circuit manageable.

## Circuit Variables

The first observation to be made is that the relevant variables in network analysis are the node voltages and the branch currents. This fact is really nothing more than a consequence of Ohm's law. Consider the branch depicted in Figure 2.50, consisting of a single resistor. Here, once a voltage $v_R$ is defined across the resistor $R$, a current $i_R$ will flow through the resistor, according to $v_R = i_R R$. But the voltage $v_R$, which causes the current to flow, is really the difference in electric potential between nodes $a$ and $b$:

$$v_R = v_a - v_b \tag{2.29}$$

What meaning do we assign to the variables $v_a$ and $v_b$? Was it not stated that voltage is a potential difference? Is it then legitimate to define the voltage at a single point (node) in a circuit? Whenever we reference the voltage at a node in a circuit, we imply an assumption that the voltage at that node is the potential difference between the node itself and a reference node called **ground**, which is located somewhere else in the circuit and which for convenience has been assigned a potential of zero volts. Thus, in Figure 2.50, the expression

$$v_R = v_a - v_b$$

really signifies that $v_R$ is the difference between the voltage differences $v_a - v_c$ and $v_b - v_c$, where $v_c$ is the (arbitrary) ground potential. Note that the equation $v_R = v_a - v_b$ would hold even if the reference node, $c$, were not assigned a potential of zero volts, since

$$v_R = v_a - v_b = (v_a - v_c) - (v_b - v_c) \tag{2.30}$$

What, then, is this ground or reference voltage?

**Figure 2.50** Variables in a network analysis problem

## Ground

The choice of the word *ground* is not arbitrary. This point can be illustrated by a simple analogy with the physics of fluid motion. Consider a tank of water, as shown in Figure 2.51, located at a certain height above the ground. The potential energy due to gravity will cause water to flow out of the pipe at a certain flow rate. The pressure that forces water out of the pipe is directly related to the head, $(h_1 - h_2)$, in such a way that this pressure is zero when $h_2 = h_1$. Now the point $h_3$, corresponding to the ground level, is defined as having zero potential energy. It should be apparent that the pressure acting on the fluid in the pipe is really caused by the difference in potential energy, $(h_1 - h_3) - (h_2 - h_3)$. It can be seen, then, that it is not necessary to assign a precise energy level to the height $h_3$; in fact, it would be extremely cumbersome to do so, since the equations describing the flow of water would then be different, say, in Denver ($h_3 = 1,600$ m above sea level) from those that would apply in Miami ($h_3 = 0$ m above sea level). You see, then, that it is the relative difference in potential energy that matters in the water tank problem.

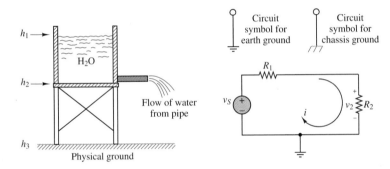

**Figure 2.51** Analogy between electrical and earth ground

In analogous fashion, in every circuit a point can be defined that is recognized as "ground" and is assigned the electric potential of zero volts *for convenience*. Note that, unless they are purposely connected together, the grounds in two completely separate circuits are not necessarily at the same potential. This last statement may seem puzzling, but Example 2.12 should clarify the idea.

It is a useful exercise at this point to put the concepts illustrated in this chapter into practice by identifying the relevant variables in a few examples of electrical circuits. In the following example, we shall illustrate how it is possible to define unknown voltages and currents in a circuit in terms of the source voltages and currents and of the resistances in the circuit.

## EXAMPLE 2.12

Identify the branch and node voltages and the loop and mesh currents in the circuit of Figure 2.52.

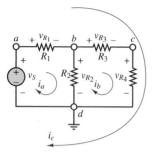

**Figure 2.52**

### Solution

The following node voltages may be identified:

| Node voltages | Branch voltages |
|---|---|
| $v_a = v_S$ (source voltage) | $v_S = v_a - v_d = v_a$ |
| $v_b = v_{R2}$ | $v_{R1} = v_a - v_b$ |
| $v_c = v_{R4}$ | $v_{R2} = v_b - v_d = v_b$ |
| $v_d = 0$ (ground) | $v_{R3} = v_b - v_c$ |
| | $v_{R4} = v_c - v_d = v_c$ |

***Comments:*** Currents $i_a$, $i_b$, and $i_c$ are loop currents, but only $i_a$ and $i_b$ are mesh currents.

---

It should be clear at this stage that some method is needed to organize the wealth of information that can be generated simply by applying Ohm's law at each branch in a circuit. What would be desirable at this point is a means of reducing the number of equations needed to solve a circuit to the minimum necessary, that is, a method for obtaining $N$ equations in $N$ unknowns. The next chapter is devoted to the development of systematic circuit analysis methods that will greatly simplify the solution of electrical network problems.

## Check Your Understanding

**2.11**   Write expressions for the voltage across each resistor in Example 2.12 in terms of the mesh currents.

**2.12**   Write expressions for the current through each resistor in Example 2.12 in terms of the node voltages.

---

## Conclusion

The objective of this chapter was to introduce the background needed in the following chapters for the analysis of linear resistive networks. The fundamental laws of circuit analysis, Kirchhoff's current law, Kirchhoff's voltage law, and Ohm's law, were introduced, along with the basic circuit elements, and all were used to analyze the most basic circuits: voltage and current dividers. Measuring devices and a few other practical circuits employed in common engineering measurements were also introduced to provide a flavor of the applicability of these basic ideas to practical engineering problems. The remainder of the book draws on the concepts developed in this chapter. Mastery of the principles exposed in these first pages is therefore of fundamental importance.

## CHECK YOUR UNDERSTANDING ANSWERS

**CYU 2.1**       $I_P = I_D = 4.17$ A; 100 W

**CYU 2.2**       $A$, supplying 30.8 W; $B$, dissipating 30.8 W

**CYU 2.3**       $i_3 = -1$ mA; $i_2 = 0$ mA

**CYU 2.5**       $P_1 = 7.25 \times 10^{-3}$ W (supplied by); $P_2 = 2.9 \times 10^{-3}$ W (supplied to)

| CYU 2.6 | $i_B = 1.8 \text{ mA}$    $P_B = 5.4 \text{ mW}$ |
|---|---|
| CYU 2.7 | $R_1 R_x = R_2 R_3$ |
| CYU 2.9 | $v_o \text{ (full scale)} = 62.6 \text{ mV}; k = 0.125 \text{ mV/N}$ |
| CYU 2.11 | $v_{R1} = i_a R_1; v_{R2} = (i_a - i_b) R_2; v_{R3} = i_b R_3; v_{R4} = i_b R_4$ |
| CYU 2.12 | $i_1 = \dfrac{v_a - v_b}{R_1}; i_2 = \dfrac{v_b - v_d}{R_2}; i_3 = \dfrac{v_b - v_c}{R_3}; i_4 = \dfrac{v_c - v_d}{R_4}$ |

# HOMEWORK PROBLEMS

## Section 1: Charge and Kirchhoff's Laws; Voltages and Currents

**2.1**  An isolated free electron is traveling through an electric field from some initial point where its Coulombic potential energy per unit charge (*voltage*) is 17 kJ/C and velocity = 93 Mm/s to some final point where its Coulombic potential energy per unit charge is 6 kJ/C. Determine the change in velocity of the electron. Neglect gravitational forces.

**2.2**  The unit used for voltage is the volt, for current the ampere, and for resistance the ohm. Using the definitions of voltage, current, and resistance, express each quantity in fundamental MKS units.

**2.3**  Suppose the current flowing through a wire is given by the curve shown in Figure P2.3.

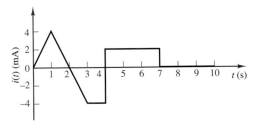

**Figure P2.3**

a. Find the amount of charge, $q$, that flows through the wire between $t_1 = 0$ and $t_2 = 1$ s.

b. Repeat part a for $t_2 = 2, 3, 4, 5, 6, 7, 8, 9,$ and 10 s.

c. Sketch $q(t)$ for $0 \leq t \leq 10$ s.

**2.4**  The capacity of a car battery is usually specified in ampere-hours. A battery rated at, say, 100 A-h should be able to supply 100 A for 1 hour, 50 A for 2 hours, 25 A for 4 hours, 1 A for 100 hours, or any other combination yielding a product of 100 A-h.

a. How many coulombs of charge should we be able to draw from a fully charged 100 A-h battery?

b. How many electrons does your answer to part a require?

**2.5**  The current in a semiconductor device results from the motion of two different kinds of charge carriers: electrons and holes. The holes and electrons have charge of equal magnitude but opposite sign. In a particular device, suppose the electron density is $2 \times 10^{19}$ electrons/m$^3$, and the hole density is $5 \times 10^{18}$ holes/m$^3$. This device has a cross-sectional area of 50 nm$^2$. If the electrons are moving to the left at a velocity of 0.5 mm/s, and the holes are moving to the right at a velocity of 0.2 mm/s, what are:

a. The direction of the current in the semiconductor.

b. The magnitude of the current in the device.

**2.6**  The charge cycle shown in Figure P2.6 is an example of a *two-rate charge.* The current is held constant at 50 mA for 5 h. Then it is switched to 20 mA for the next 5 h. Find:

a. The total charge transferred to the battery.

b. The energy transferred to the battery.

Hint: Recall that energy, $w$, is the integral of power, or $P = dw/dt$.

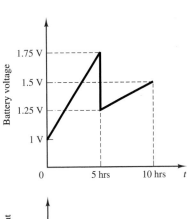

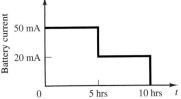

**Figure P2.6**

**2.7**   Batteries (e.g., lead-acid batteries) store chemical energy and convert it to electrical energy on demand. Batteries do not store electrical charge or charge carriers. Charge carriers (electrons) enter one terminal of the battery, acquire electrical potential energy and exit from the other terminal at a lower voltage. Remember the electron has a negative charge! It is convenient to think of positive carriers flowing in the opposite direction, i.e., conventional current, and exiting at a higher voltage. All currents in this course, unless otherwise stated, will be conventional current. (Benjamin Franklin caused this mess!) For a battery with a rated voltage = 12 V and a rated capacity = 350 ampere-hours (A-h), determine:

a. The rated chemical energy stored in the battery.

b. The total charge that can be supplied at the rated voltage.

**2.8**   What determines:

a. How much current is supplied (at a constant voltage) by an ideal voltage source?

b. How much voltage is supplied (at a constant current) by an ideal current source?

**2.9**   Determine the current through $R_3$ in Figure P2.9 for:
$$V_S = 12V \qquad R_S = 1 \text{ k}\Omega$$
$$R_1 = 2 \text{ k}\Omega \qquad R_2 = 4 \text{ k}\Omega \qquad R_3 = 6 \text{ k}\Omega$$

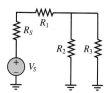

**Figure P2.9**

## Section 2: Electric Power

**2.10**   In the block diagram in Figure P2.10:

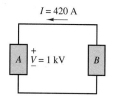

**Figure P2.10**

a. Which component *must* be a voltage or current source?

b. What could the other component be? Include all possible answers.

**2.11**   If an electric heater requires 23 A at 110 V, determine:

a. The power it dissipates as heat or other losses.

b. The energy dissipated by the heater in a 24-hour period.

c. The cost of the energy if the power company charges at the rate 6 cents/kW-h.

**2.12**   Determine which elements in the circuit of Figure P2.12 are supplying power and which are dissipating power. Also determine the amount of power dissipated and supplied.

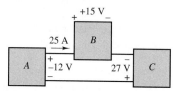

**Figure P2.12**

**2.13**   In the circuit shown in Figure P2.13, determine the terminal voltage of the source, the power supplied to the circuit (or load), and the efficiency of the circuit. Assume that the only loss is due to the internal resistance of the source. Efficiency is defined as the ratio of load power to source power.

$$V_S = 12 \text{ V} \qquad R_S = 5 \text{ k}\Omega \qquad R_L = 7 \text{ k}\Omega$$

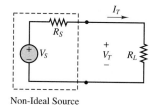

Non-Ideal Source

**Figure P2.13**

**2.14**   For the circuit shown in Figure P2.14:

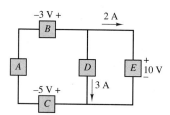

**Figure P2.14**

a. Determine which components are absorbing power and which are delivering power.

b. Is conservation of power satisfied? Explain your answer.

**2.15**  Suppose one of the two headlights in Example 2.2 has been replaced with the wrong part and the 12-V battery is now connected to a 75-W and a 50-W headlight. What is the resistance of each headlight, and what is the total resistance seen by the battery?

**2.16**  What is the equivalent resistance seen by the battery of Example 2.2 if two 10-W taillights are added to the 50-W (each) headlights?

**2.17**  For the circuit shown in Figure P2.17, determine the power absorbed by the 5 Ω resistor.

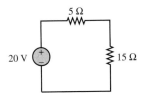

**Figure P2.17**

**2.18**  With reference to Figure P2.18, determine:

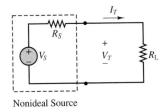

Nonideal Source

**Figure P2.18**

a. The total power supplied by the ideal source.
b. The power dissipated and lost within the nonideal source.
c. The power supplied by the source to the circuit as modeled by the load resistance.
d. Plot the terminal voltage and power supplied to the circuit as a function of current.

Calculate for $I_T = 0, 5, 10, 20, 30$ A.

$$V_S = 12 \text{ V} \qquad R_S = 0.3 \ \Omega$$

**2.19**  In the circuit of Figure P2.19, if $v_1 = v/8$ and the power delivered by the source is 8 mW, find $R, v, v_1$, and $i$.

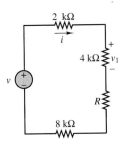

**Figure P2.19**

**2.20**  A GE SoftWhite Longlife light bulb is rated as follows:

$$P_R = \text{Rated power} = 60 \text{ W}$$
$$P_{OR} = \textit{Rated optical power} = 820 \text{ lumens (average)}$$
$$\text{Operating life} = 1500 \text{ h (average)}$$
$$V_R = \text{Rated operating voltage} = 115 \text{ V}$$

The resistance of the filament of the bulb, measured with a standard multimeter, is 16.7 Ω. When the bulb is connected into a circuit and is operating at the rated values given above, determine:

a. The resistance of the filament.
b. The efficiency of the bulb.

**2.21**  An incandescent light bulb rated at 100 W will dissipate 100 W as heat and light when connected across a 110-V ideal voltage source. If three of these bulbs are connected in series across the same source, determine the power each bulb will dissipate.

**2.22**  An incandescent light bulb rated at 60 W will dissipate 60 W as heat and light when connected across a 100-V ideal voltage source. A 100-W bulb will dissipate 100 W when connected across the same source. If the bulbs are connected in series across the same source, determine the power that either one of the two bulbs will dissipate.

## Section 3:  Resistance Calculations

**2.23**  Use Kirchhoff's current law to determine the current in each of the 30-Ω resistors in the circuit of Figure P2.23.

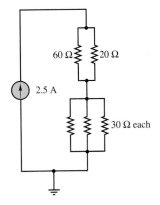

**Figure P2.23**

**2.24**  Cheap resistors are fabricated by depositing a thin layer of carbon onto a nonconducting cylindrical substrate (see Figure P2.24). If such a cylinder has radius $a$ and length $d$, determine the thickness of the film required for a resistance $R$ if:

$$a = 1 \text{ mm} \qquad\qquad R = 33 \text{ k}\Omega$$

$$\sigma = \frac{1}{\rho} = 2.9 \text{ M}\frac{S}{m} \quad d = 9 \text{ mm}$$

Neglect the end surfaces of the cylinder and assume that the thickness is much smaller than the radius.

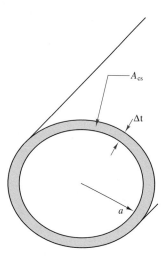

**Figure P2.24**

**2.25** The resistive elements of fuses, light bulbs, heaters, etc., are significantly nonlinear, i.e., the resistance is dependent on the current through the element. Assume the resistance of a fuse (Figure P2.25) is given by the expression: $R = R_0[1 + A(T - T_0)]$ with $T - T_0 = kP$; $T_0 = 25°C$; $A = 0.7[°C]^{-1}$; $k = 0.35\frac{°C}{W}$; $R_0 = 0.11 \ \Omega$; and P is the power dissipated in the resistive element of the fuse. Determine:

a. The rated current at which the circuit will melt and open, i.e., "blow." *Hint:* The fuse blows when $R$ becomes infinite.

b. The temperature of the element at which this occurs.

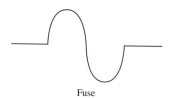

Fuse

**Figure P2.25**

**2.26** The voltage divider network of Figure P2.26 is expected to provide 2.5 V at the output. The resistors,

however, may not be exactly the same; that is, their tolerances are such that the resistances may not be exactly 10 k$\Omega$.

a. If the resistors have ±10 percent tolerance, find the worst-case output voltages.

b. Find these voltages for tolerances of ±5 percent.

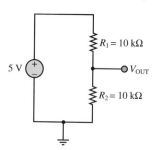

**Figure P2.26**

**2.27** For the circuits of Figure P2.27, determine the resistor values (including the power rating) necessary to achieve the indicated voltages.
Resistors are available in $\frac{1}{8}$-, $\frac{1}{4}$-, $\frac{1}{2}$-, and 1-W ratings.

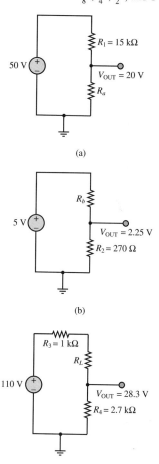

**Figure P2.27**

**2.28**  For the circuit shown in Figure P2.28, find

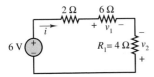

**Figure P2.28**

a. The equivalent resistance seen by the source.
b. The current, $i$.
c. The power delivered by the source.
d. The voltages, $v_1$, $v_2$.
e. The minimum power rating required for $R_1$.

**2.29**  Find the equivalent resistance of the circuit of Figure P2.29 by combining resistors in series and in parallel.

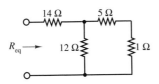

**Figure P2.29**

**2.30**  Find the equivalent resistance seen by the source and the current $i$ in the circuit of Figure P2.30.

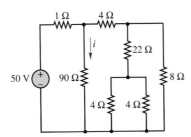

**Figure P2.30**

**2.31**  In the circuit of Figure P2.31, the power absorbed by the 15-Ω resistor is 15 W. Find $R$.

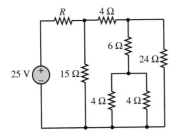

**Figure P2.31**

**2.32**  Find the equivalent resistance between terminals $a$ and $b$ in the circuit of Figure P2.32.

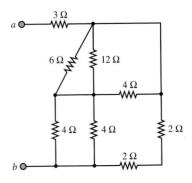

**Figure P2.32**

**2.33**  For the circuit shown in Figure P2.33:

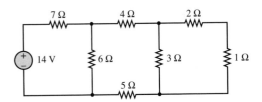

**Figure P2.33**

a. Find the equivalent resistance seen by the source.
b. How much power is delivered by the source?

**2.34**  In the circuit of Figure P2.34, find the equivalent resistance looking in at terminals $a$ and $b$ if terminals $c$ and $d$ are open and again if terminals $c$ and $d$ are shorted together. Also, find the equivalent resistance looking in at terminals $c$ and $d$ if terminals $a$ and $b$ are open and if terminals $a$ and $b$ are shorted together.

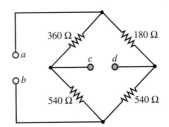

**Figure P2.34**

**2.35**  Find the currents $i_1$ and $i_2$, the power delivered by the 2-A current source and by the 10-V voltage source, and the total power dissipated by the circuit of Figure P2.35. $R_1 = 32\ \Omega$, $R_2 = R_3 = 6\ \Omega$, and $R_4 = 50\ \Omega$.

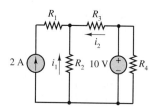

**Figure P2.35**

**2.36**  Determine the power delivered by the dependent source in the circuit of Figure P2.36.

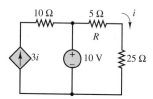

**Figure P2.36**

**2.37**  Consider the circuit shown in Figure P2.37.

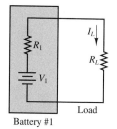

**Figure P2.37**

a. If $V_1 = 10.0$ V, $R_1 = 0.05$ Ω, and $R_L = 0.45$ Ω, find the load current $I_L$ and the power dissipated by the load.

b. If we connect a second battery in parallel with battery 1 that has voltage $V_2 = 10$ V and $R_2 = 0.1$ Ω, will the load current $I_L$ increase or decrease? Will the power dissipated by the load increase or decrease? By how much?

**2.38**  With no load attached, the voltage at the terminals of a particular power supply is 25.5 V. When a 5 Ω

load is attached, the voltage drops to 25 V.

a. Determine $v_S$ and $R_S$ for this nonideal source.

b. What voltage would be measured at the terminals in the presence of a 10-Ω load resistor?

c. How much current could be drawn from this power supply under short-circuit conditions?

**2.39**  A 120-V electric heater has two heating coils which can be switched such that either can be used independently, or the two can be connected in series or parallel, yielding a total of four possible configurations. If the warmest setting corresponds to 1500-W power dissipation and the coolest corresponds to 200 W, determine:

a. The resistance of each of the two coils.

b. The power dissipation for each of the other two possible arrangements.

**2.40**  At an engineering site which you are supervising, a 1-horsepower motor must be sited a distance $d$ from a portable generator (Figure P2.40). Assume the generator can be modeled as an ideal source with the voltage given. The nameplate on the motor gives the following rated voltages and the corresponding full-load current:

$$V_G = 110 \text{ V}$$
$$V_{M \text{ min}} = 105 \text{ V} \rightarrow I_{M \text{ FL}} = 7.10 \text{ A}$$
$$V_{M \text{ max}} = 117 \text{ V} \rightarrow I_{M \text{ FL}} = 6.37 \text{ A}$$

If $d = 150$ m and the motor must deliver its full rated power, determine the minimum AWG conductors which must be used in a rubber insulated cable. Assume that the only losses in the circuit occur in the wires.

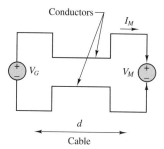

**Figure P2.40**

**2.41**  A building has been added to your plant to house an additional production line. The total electrical load in the building is 23 kW. The nameplates on the various loads give the minimum and maximum voltages below with the related full-load current:

$$V_S = 450 \text{ V}$$
$$V_{L \text{ min}} = 446 \text{ V} \rightarrow I_{L \text{ FL}} = 51.5 \text{ A}$$
$$V_{L \text{ max}} = 463 \text{ V} \rightarrow I_{L \text{ FL}} = 49.6 \text{ A}$$

The building is sited a distance $d$ from the transformer bank which can be modeled as an ideal source (see Figure P2.41). If $d = 85$ m, determine the AWG of the smallest conductors which can be used in a rubber-insulated cable used to supply the load.

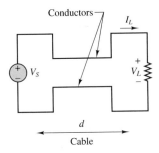

**Figure P2.41**

**2.42**    At an engineering site which you are supervising, a 1-horsepower motor must be sited a distance $d$ from a portable generator (Figure P2.42). Assume the generator can be modeled as an ideal source with the voltage given. The nameplate on the motor gives the rated voltages and the corresponding full load current:

$$V_G = 110 \text{ V}$$
$$V_{M \, min} = 105 \text{ V} \rightarrow I_{M \, FL} = 7.10 \text{ A}$$
$$V_{M \, max} = 117 \text{ V} \rightarrow I_{M \, FL} = 6.37 \text{ A}$$

The cable must have AWG #14 or larger conductors to carry a current of 7.103 A without overheating. Determine the maximum length of a rubber insulated cable with AWG #14 conductors which can be used to connect the motor and generator.

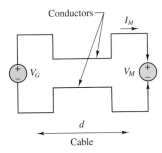

**Figure P2.42**

**2.43**    An additional building has been added to your plant to house a production line. The total electrical load in the building is 23 kW. The nameplates on the loads give the minimum and maximum voltages with the related full load current:

$$V_S = 450 \text{ V}$$
$$V_{L \, min} = 446 \text{ V} \rightarrow I_{L \, FL} = 51.57 \text{ A}$$
$$V_{L \, max} = 463 \text{ V} \rightarrow I_{L \, FL} = 49.68 \text{ A}$$

The building is sited a distance $d$ from the transformer bank which can be modeled as an ideal source (Figure P2.43). The cable must have AWG 4 or larger conductors to carry a current of 51.57 A without overheating. Determine the maximum length $d$ of a rubber-insulated cable with AWG 4 conductors which can be used to connect the source to the load.

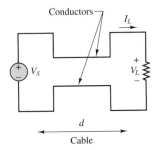

**Figure P2.43**

**2.44**    In the bridge circuit in Figure P2.44, if nodes (or terminals) $C$ and $D$ are shorted, and:

$$R_1 = 2.2 \text{ k}\Omega \quad R_2 = 18 \text{ k}\Omega$$
$$R_3 = 4.7 \text{ k}\Omega \quad R_4 = 3.3 \text{ k}\Omega$$

determine the equivalent resistance between the nodes or terminals $A$ and $B$.

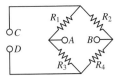

**Figure P2.44**

**2.45**    Determine the voltage between nodes $A$ and $B$ in the circuit shown in Figure P2.45.

$$V_S = 12 \text{ V}$$
$$R_1 = 11 \text{ k}\Omega \quad R_3 = 6.8 \text{ k}\Omega$$
$$R_2 = 220 \text{ k}\Omega \quad R_4 = 0.22 \text{ m}\Omega$$

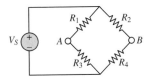

**Figure P2.45**

**2.46**    Determine the voltage between the nodes $A$ and $B$ in the circuit shown in Figure P2.45.

$$V_S = 5 \text{ V}$$
$$R_1 = 2.2 \text{ k}\Omega \quad R_2 = 18 \text{ k}\Omega$$
$$R_3 = 4.7 \text{ k}\Omega \quad R_4 = 3.3 \text{ k}\Omega$$

**2.47**  Determine the voltage across $R_3$ in Figure P2.47.

$V_S = 12$ V    $R_1 = 1.7$ mΩ
$R_2 = 3$ kΩ    $R_3 = 10$ kΩ

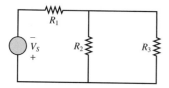

**Figure P2.47**

## Section 4: Measuring Devices

**2.48**  A *thermistor* is a device whose terminal resistance changes with the temperature of its surroundings. Its resistance is an exponential relationship:

$$R_{th}(T) = R_A e^{-\beta T}$$

where $R_A$ is the terminal resistance at $T = 0°$C and $\beta$ is a material parameter with units $[°C]^{-1}$.

a. If $R_A = 100$ Ω and $\beta = 0.10/$C°, plot $R_{th}(T)$ versus $T$ for $0 \leq T \leq 100°$C.
b. The thermistor is placed in parallel with a resistor whose value is 100 Ω.
    i. Find an expression for the equivalent resistance.
    ii. Plot $R_{eq}(T)$ on the same plot you made in part a.

**2.49**  A certain resistor has the following nonlinear characteristic:

$$R(x) = 100e^x$$

where $x$ is a normalized displacement. The nonlinear resistor is to be used to measure the displacement $x$ in the circuit of Figure P2.49.

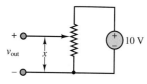

**Figure P2.49**

a. If the total length of the resistor is 10 cm, find an expression for $v_{out}(x)$.
b. If $v_{out} = 4$ V, what is the distance, $x$?

**2.50**  A moving coil meter movement has a meter resistance $r_m = 200$ Ω and full-scale deflection is caused by a meter current $I_m = 10\mu$A. The movement must be used to indicate pressure measured by the

sensor up to a maximum of 100 kPa. See Figure P2.50.

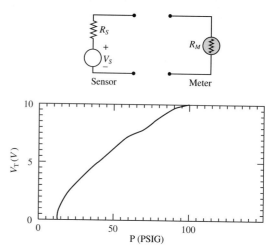

**Figure P2.50**

a. Draw a circuit required to do this showing all appropriate connections between the terminals of the sensor and meter movement.
b. Determine the value of each component in the circuit.
c. What is the linear range, i.e., the minimum and maximum pressure that can accurately be measured?

**2.51**  A moving coil meter and pressure transducer are used to monitor the pressure at a critical point in a system. The meter movement is rated at 1.8 kΩ and 50 $\mu$A (full scale). A new transducer must be installed with the pressure-voltage characteristic shown in Figure P2.51 (different from the previous transducer). The maximum pressure that must be measured by the monitoring system is 100 kPa.

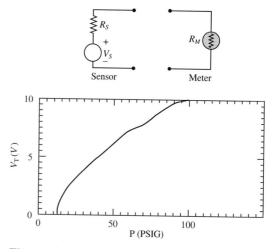

**Figure P2.51**

a. Redesign the meter circuit required for these specifications and draw the circuit between the terminals of the sensor and meter showing all appropriate connections.

b. Determine the value for each component in your circuit.

c. What is the linear range (i.e., the minimum and maximum pressure that can accurately be measured) of this system?

**2.52**    In the circuit shown in Figure P2.52 the temperature sensor and moving coil meter movement are used to monitor the temperature in a chemical process. The sensor has malfunctioned and must be replaced with another sensor with the current-temperature characteristic shown (not the same as the previous sensor). Temperatures up to a maximum of 400°C must be measured. The meter is rated at 2.5 k$\Omega$ and 250 mV (full scale). Redesign the meter circuit for these specifications.

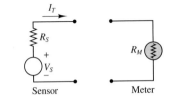

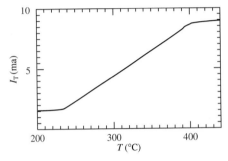

**Figure P2.53**

a. Draw the circuit between the terminals of the sensor and meter showing all appropriate connections.

b. Determine the value of each component in the circuit.

c. What is the minimum temperature that can accurately be measured?

**2.54**    The circuit of Figure P2.54 is used to measure the internal impedance of a battery. The battery being tested is a zinc-carbon dry cell.

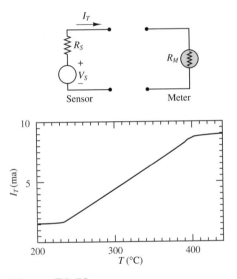

**Figure P2.52**

a. Draw the circuit between the terminals of the sensor and meter showing all appropriate connections.

b. Determine the value of each component in the circuit.

c. What is the linear range (i.e., the minimum and maximum temperature that can accurately be measured) of the system?

**2.53**    In the circuit in Figure P2.53, a temperature sensor with the current-temperature characteristic shown and a Triplett Electric Manufacturing Company Model 321L moving coil meter will be used to monitor the condenser temperature in a steam power plant. Temperatures up to a maximum of 350°C must be measured. The meter is rated at 1 k$\Omega$ and 100 $\mu$A (full scale). Design a circuit for these specifications.

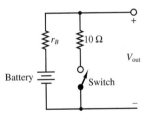

**Figure P2.54**

a. A fresh battery is being tested, and it is found that the voltage, $V_{out}$, is 1.64 V with the switch open and 1.63 V with the switch closed. Find the internal resistance of the battery.

b. The same battery is tested one year later, and $V_{out}$ is found to be 1.6 V with the switch open but 0.17 V with the switch closed. Find the internal resistance of the battery.

**2.55**    Consider the practical ammeter, diagrammed in Figure P2.55, consisting of an ideal ammeter in series with a 2-k$\Omega$ resistor. The meter sees a full-scale deflection when the current through it is 50$\mu$A. If we wished to construct a multirange ammeter reading full-scale values of 1 mA, 10 mA, or 100 mA, depending on the setting of a rotary switch, what should $R_1$, $R_2$, and $R_3$ be?

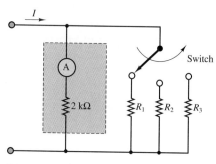

**Figure P2.55**

**2.56**  A circuit that measures the internal resistance of a practical ammeter is shown in Figure P2.56, where $R_S = 10,000\ \Omega$, $V_S = 10$ V, and $R_p$ is a variable resistor that can be adjusted at will.

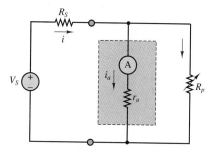

**Figure P2.56**

a. Assume that $r_a \ll 10,000\ \Omega$. Estimate the current $i$.

b. If the meter displays a current of 0.43 mA when $R_p = 7\ \Omega$, find the internal resistance of the meter, $r_a$.

**2.57**  A practical voltmeter has an internal resistance $r_m$. What is the value of $r_m$ if the meter reads 9.89 V when connected as shown in Figure P2.57.

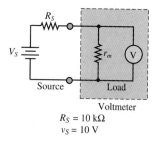

$R_S = 10\ \text{k}\Omega$
$v_S = 10$ V

**Figure P2.57**

**2.58**  Using the circuit of Figure P2.57, find the voltage that the meter reads if $V_S = 10$ V and $R_S$ has the following values:
$R_S = 0.1r_m, 0.3r_m, 0.5r_m, r_m, 3r_m, 5r_m,$ and $10r_m$.
How large (or small) should the internal resistance of the meter be relative to $R_S$?

**2.59**  A voltmeter is used to determine the voltage across a resistive element in the circuit of Figure P2.59. The instrument is modeled by an ideal voltmeter in parallel with a 97-k$\Omega$ resistor, as shown. The meter is placed to measure the voltage across $R_3$. Let $R_1 = 10$ k$\Omega$, $R_S = 100$ k$\Omega$, $R_2 = 40$ k$\Omega$, and $I_S = 90$ mA. Find the voltage across $R_3$ with and without the voltmeter in the circuit for the following values:

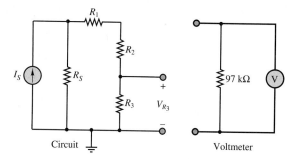

**Figure P2.59**

a.  $R_3 = 100\ \Omega$
b.  $R_3 = 1\ \text{k}\Omega$
c.  $R_3 = 10\ \text{k}\Omega$
d.  $R_3 = 100\ \text{k}\Omega$

**2.60**  An ammeter is used as shown in Figure P2.60. The ammeter model consists of an ideal ammeter in series with a resistance. The ammeter model is placed in the branch as shown in the figure. Find the current through $R_3$ both with and without the ammeter in the circuit for the following values, assuming that $V_S = 10$ V, $R_S = 10\ \Omega$, $R_1 = 1$ k$\Omega$, and $R_2 = 100\ \Omega$: (a) $R_3 = 1$ k$\Omega$, (b) $R_3 = 100\ \Omega$, (c) $R_3 = 10\ \Omega$, (d) $R_3 = 1\ \Omega$.

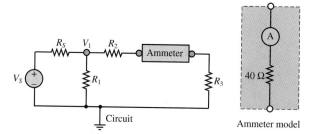

**Figure P2.60**

**2.61** Shown in Figure P2.61 is an aluminum cantilevered beam loaded by the force $F$. Strain gauges $R_1$, $R_2$, $R_3$, and $R_4$ are attached to the beam as shown in Figure P2.61 and connected into the circuit shown. The force causes a tension stress on the top of the beam that causes the length (and therefore the resistance) of $R_1$ and $R_4$ to increase and a compression stress on the bottom of the beam that causes the length [and therefore the resistance] of $R_2$ and $R_3$ to decrease. This causes a voltage 50 mV at node $B$ with respect to node $A$. Determine the force if:

$$R_o = 1 \text{ k}\Omega \qquad V_S = 12 \text{ V} \qquad L = 0.3 \text{ m}$$
$$w = 25 \text{ mm} \qquad h = 100 \text{ mm} \qquad Y = 69 \text{ GN/m}^2$$

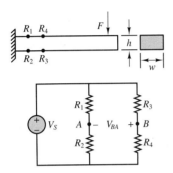

**Figure P2.61**

**2.62** Shown in Figure P2.62 is a structural steel cantilevered beam loaded by a force $F$. Strain gauges $R_1$, $R_2$, $R_3$, and $R_4$ are attached to the beam as shown and connected into the circuit shown. The force causes a tension stress on the top of the beam that causes the length (and therefore the resistance) of $R_1$ and $R_4$ to increase and a compression stress on the bottom of the beam that causes the length (and therefore the resistance) of $R_2$ and $R_3$ to decrease. This generates a voltage between nodes $B$ and $A$. Determine this voltage if $F = 1.3 \text{ MN}$ and:

$$R_o = 1 \text{ k}\Omega \qquad V_S = 24 \text{ V} \qquad L = 1.7 \text{ m}$$
$$w = 3 \text{ cm} \qquad h = 7 \text{ cm} \qquad Y = 200 \text{ GN/m}^2$$

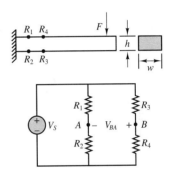

**Figure P2.62**

# CHAPTER

# 3

# Resistive Network Analysis

T his chapter will illustrate the fundamental techniques for the analysis of resistive circuits. The methods introduced are based on the circuit laws presented in Chapter 2: Kirchhoff's and Ohm's laws. The main thrust of the chapter is to introduce and illustrate various methods of circuit analysis that will be applied throughout the book.

The first topic is the analysis of resistive circuits by the methods of mesh currents and node voltages; these are fundamental techniques, which you should master as early as possible. The second topic is a brief introduction to the principle of superposition. Section 3.5 introduces another fundamental concept in the analysis of electrical circuits: the reduction of an arbitrary circuit to equivalent circuit forms (Thévenin and Norton equivalent circuits). In this section it will be shown that it is generally possible to reduce all linear circuits to one of two equivalent forms, and that any linear circuit analysis problem can be reduced to a simple voltage or current divider problem. The Thévenin and Norton equivalent representations of electrical circuits naturally lead to the description of electrical circuits in terms of sources and loads. This notion, in turn, leads to the analysis of the transfer of power between a source and a load, and of the phenomenon of source loading. Finally, some graphical and numerical techniques are introduced for the analysis of nonlinear circuit elements.

Upon completing this chapter, you should have developed confidence in your ability to compute numerical solutions for a wide range of resistive circuits. Good familiarity with the techniques illustrated in this chapter will greatly simplify the study of AC circuits in Chapter 4. The objective of the chapter is to develop a solid understanding of the following topics:

- Node voltage and mesh current analysis.
- The principle of superposition.
- Thévenin and Norton equivalent circuits.
- Numerical and graphical (load-line) analysis of nonlinear circuit elements.

## 3.1    THE NODE VOLTAGE METHOD

Chapter 2 introduced the essential elements of network analysis, paving the way for a systematic treatment of the analysis methods that will be introduced in this chapter. You are by now familiar with the application of the three fundamental laws of network analysis: KCL, KVL, and Ohm's law; these will be employed to develop a variety of solution methods that can be applied to linear resistive circuits. The material presented in the following sections presumes good understanding of Chapter 2. You can resolve many of the doubts and questions that may occasionally arise by reviewing the material presented in the preceding chapter.

Node voltage analysis is the most general method for the analysis of electrical circuits. In this section, its application to linear resistive circuits will be illustrated. The **node voltage method** is based on defining the voltage at each node as an independent variable. One of the nodes is selected as a **reference node** (usually— but not necessarily—ground), and each of the other node voltages is referenced to this node. Once each node voltage is defined, Ohm's law may be applied between any two adjacent nodes in order to determine the current flowing in each branch. In the node voltage method, *each branch current is expressed in terms of one or more node voltages*; thus, currents do not explicitly enter into the equations. Figure 3.1 illustrates how one defines branch currents in this method. You may recall a similar description given in Chapter 2.

Once each branch current is defined in terms of the node voltages, Kirchhoff's current law is applied at each node:

$$\sum i = 0 \qquad \text{(3.1)}$$

Figure 3.2 illustrates this procedure.

The systematic application of this method to a circuit with $n$ nodes would lead to writing $n$ linear equations. However, one of the node voltages is the reference voltage and is therefore already known, since it is usually assumed to be zero (recall that the choice of reference voltage is dictated mostly by convenience, as explained in Chapter 2). Thus, we can write $n - 1$ *independent linear equations* in the $n - 1$ independent variables (the node voltages). Nodal analysis provides the minimum number of equations required to solve the circuit, since any branch voltage or current may be determined from knowledge of nodal voltages. At this stage, you might wish to review Example 2.12, to verify that, indeed, knowledge of the node voltages is sufficient to solve for any other current or voltage in the circuit.

The nodal analysis method may also be defined as a sequence of steps, as outlined in the following box:

In the node voltage method, we assign the node voltages $v_a$ and $v_b$; the branch current flowing from $a$ to $b$ is then expressed in terms of these node voltages.

$$i = \frac{v_a - v_b}{R}$$

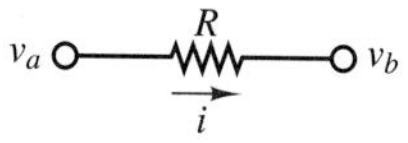

$v_a \circ\!\!-\!\!\!\wedge\!\!\wedge\!\!\wedge\!\!-\!\!\circ v_b$
$\overset{R}{\underset{i}{}}$

**Figure 3.1** Branch current formulation in nodal analysis

By KCL: $i_1 - i_2 - i_3 = 0$. In the node voltage method, we express KCL by

$$\frac{v_a - v_b}{R_1} - \frac{v_b - v_c}{R_2} - \frac{v_b - v_d}{R_3} = 0$$

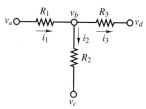

**Figure 3.2** Use of KCL in nodal analysis

**Node Voltage Analysis Method**

1. Select a reference node (usually ground). All other node voltages will be referenced to this node.
2. Define the remaining $n - 1$ node voltages as the independent variables.
3. Apply KCL at each of the $n - 1$ nodes, expressing each current in terms of the adjacent node voltages.
4. Solve the linear system of $n - 1$ equations in $n - 1$ unknowns.

Following the procedure outlined in the box guarantees that the correct solution to a given circuit will be found, provided that the nodes are properly identified and KCL is applied consistently. As an illustration of the method, consider the circuit shown in Figure 3.3. The circuit is shown in two different forms to illustrate equivalent graphical representations of the same circuit. The bottom circuit leaves no question where the nodes are. The direction of current flow is selected arbitrarily (assuming that $i_S$ is a positive current). Application of KCL at node $a$ yields:

$$i_S - i_1 - i_2 = 0 \tag{3.2}$$

whereas, at node $b$,

$$i_2 - i_3 = 0 \tag{3.3}$$

It is instructive to verify (at least the first time the method is applied) that it is not necessary to apply KCL at the reference node. The equation obtained at node $c$,

$$i_1 - i_3 - i_S = 0 \tag{3.4}$$

is not independent of equations 3.2 and 3.3; in fact, it may be obtained by adding the equations obtained at nodes $a$ and $b$ (verify this, as an exercise). This observation confirms the statement made earlier:

In a circuit containing $n$ nodes, we can write at most $n - 1$ independent equations.

Now, in applying the node voltage method, the currents $i_1$, $i_2$, and $i_3$ are expressed as functions of $v_a$, $v_b$, and $v_c$, the independent variables. Ohm's law requires that $i_1$, for example, be given by

$$i_1 = \frac{v_a - v_c}{R_1} \tag{3.5}$$

since it is the potential difference, $v_a - v_c$, across $R_1$ that causes the current $i_1$ to flow from node $a$ to node $c$. Similarly,

$$i_2 = \frac{v_a - v_b}{R_2}$$

$$\tag{3.6}$$

$$i_3 = \frac{v_b - v_c}{R_3}$$

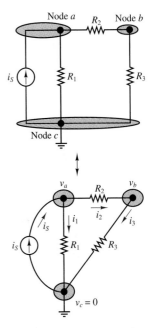

**Figure 3.3** Illustration of nodal analysis

Substituting the expression for the three currents in the nodal equations (equations 3.2 and 3.3), we obtain the following relationships:

$$i_S - \frac{v_a}{R_1} - \frac{v_a - v_b}{R_2} = 0 \qquad (3.7)$$

$$\frac{v_a - v_b}{R_2} - \frac{v_b}{R_3} = 0 \qquad (3.8)$$

Equations 3.7 and 3.8 may be obtained directly from the circuit, with a little practice. Note that these equations may be solved for $v_a$ and $v_b$, assuming that $i_S$, $R_1$, $R_2$, and $R_3$ are known. The same equations may be reformulated as follows:

$$\left(\frac{1}{R_1} + \frac{1}{R_2}\right)v_a + \left(-\frac{1}{R_2}\right)v_b = i_S$$
$$\left(-\frac{1}{R_2}\right)v_a + \left(\frac{1}{R_2} + \frac{1}{R_3}\right)v_b = 0 \qquad (3.9)$$

The following examples further illustrate the application of the method.

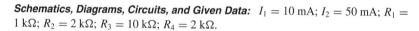

### EXAMPLE 3.1 Nodal Analysis

**Problem**

Solve for all unknown currents and voltages in the circuit of Figure 3.4.

VIRTUAL LAB

**Solution**

**Known Quantities:** Source currents, resistor values.

**Find:** All node voltages and branch currents.

**Schematics, Diagrams, Circuits, and Given Data:** $I_1 = 10$ mA; $I_2 = 50$ mA; $R_1 = 1$ kΩ; $R_2 = 2$ kΩ; $R_3 = 10$ kΩ; $R_4 = 2$ kΩ.

**Assumptions:** The reference (ground) node is chosen to be the node at the bottom of the circuit.

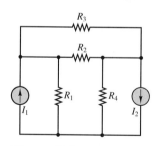

**Figure 3.4**

**Analysis:** The circuit of Figure 3.4 is shown again in Figure 3.5, with a graphical indication of how KCL will be applied to determine the nodal equations. Note that we have selected to ground the lower part of the circuit, resulting in a reference voltage of zero at that node. Applying KCL at nodes 1 and 2 we obtain

$$I_1 - \frac{v_1 - 0}{R_1} - \frac{v_1 - v_2}{R_2} - \frac{v_1 - v_2}{R_3} = 0 \qquad \text{(node 1)}$$

$$\frac{v_1 - v_2}{R_2} + \frac{v_1 - v_2}{R_3} - \frac{v_2 - 0}{R4} - I_2 = 0 \qquad \text{(node 2)}$$

Now we can write the same equations more systematically as a function of the unknown node voltages, as was done in equation 3.9.

$$\left(\frac{1}{R_1} + \frac{1}{R_2} + \frac{1}{R_3}\right)v_1 + \left(-\frac{1}{R_2} - \frac{1}{R_3}\right)v_2 = I_1 \qquad \text{(node 1)}$$

$$\left(-\frac{1}{R_2} - \frac{1}{R_3}\right)v_1 + \left(\frac{1}{R_2} + \frac{1}{R_3} + \frac{1}{R_4}\right)v_2 = -I_2 \qquad \text{(node 2)}$$

With some manipulation, the equations finally lead to the following form:

$$1.6v_1 - 0.6v_2 = 10$$
$$-0.6v_1 + 1.1v_2 = -50$$

These equations may be solved simultaneously to obtain

$$v_1 = -13.57 \text{ V}$$
$$v_2 = -52.86 \text{ V}$$

Knowing the node voltages, we can determine each of the branch currents and voltages in the circuit. For example, the current through the 10-kΩ resistor is given by:

$$i_{10 \text{ k}\Omega} = \frac{v_1 - v_2}{10,000} = 3.93 \text{ mA}$$

indicating that the initial (arbitrary) choice of direction for this current was the same as the actual direction of current flow. As another example, consider the current through the 1-kΩ resistor:

$$i_{1 \text{ k}\Omega} = \frac{v_1}{1,000} = -13.57 \text{ mA}$$

In this case, the current is negative, indicating that current actually flows from ground to node 1, as it should, since the voltage at node 1 is negative with respect to ground. You may continue the branch-by-branch analysis started in this example to verify that the solution obtained in the example is indeed correct.

**Comments:**  Note that we have chose to assign a positive sign to currents entering a node, and a negative sign to currents exiting a node; this choice is arbitrary (one could use the opposite convention), but we shall use it consistently in this book.

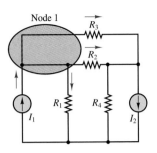

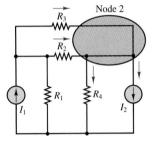

**Figure 3.5**

---

# EXAMPLE  3.2  Nodal Analysis

## Problem

Write the nodal equations and solve for the node voltages in the circuit of Figure 3.6.

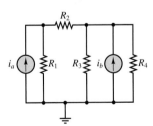

---

## Solution

**Known Quantities:**  Source currents, resistor values.

**Figure 3.6**

**Find:**  All node voltages and branch currents.

**Schematics, Diagrams, Circuits, and Given Data:**  $i_a = 1$ mA; $i_b = 2$ mA; $R_1 = 1$ kΩ; $R_2 = 500$ Ω; $R_3 = 2.2$ kΩ; $R_4 = 4.7$ kΩ.

**Assumptions:**  The reference (ground) node is chosen to be the node at the bottom of the circuit.

**Analysis:**  To write the node equations, we start by selecting the reference node (step 1). Figure 3.7 illustrates that two nodes remain after the selection of the reference node. Let us label these $a$ and $b$ and define voltages $v_a$ and $v_b$ (step 2).

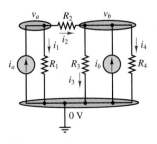

**Figure 3.7**

Next, we apply KCL at each of the nodes, $a$ and $b$ (step 3):

$$i_a - \frac{v_a}{R_1} - \frac{v_a - v_b}{R_2} = 0 \qquad \text{(node } a\text{)}$$

$$\frac{v_a - v_b}{R_2} + i_b - \frac{v_b}{R_3} - \frac{v_b}{R_4} = 0 \qquad \text{(node } b\text{)}$$

and rewrite the equations to obtain a linear system:

$$\left( \frac{1}{R_1} + \frac{1}{R_2} \right) v_a + \left( -\frac{1}{R_2} \right) v_b = i_a$$

$$\left( -\frac{1}{R_2} \right) v_a + \left( \frac{1}{R_2} + \frac{1}{R_3} + \frac{1}{R_4} \right) v_b = i_b$$

Substituting the numerical values in these equations, we get

$$3 \times 10^{-3} v_a - 2 \times 10^{-3} v_b = 1 \times 10^{-3}$$
$$-2 \times 10^{-3} v_a + 2.67 \times 10^{-3} v_b = 2 \times 10^{-3}$$

or

$$3v_a - 2v_b = 1$$
$$-2v_a + 2.67v_b = 2$$

The solution $v_a = 1.667$ V, $v_b = 2$ V may then be obtained by solving the system of equations.

---

## EXAMPLE  3.3  Solution of Linear System of Equations Using Cramer's Rule

### Problem

Solve the circuit equations obtained in Example 3.2 using Cramer's rule (see Appendix A).

---

### Solution

**Known Quantities:**  Linear system of equations.

**Find:**  Node voltages.

**Analysis:**  The system of equations generated in Example 3.2 may also be solved by using linear algebra methods, by recognizing that the system of equations can be written as:

$$\begin{bmatrix} 3 & -2 \\ -2 & 2.67 \end{bmatrix} \begin{bmatrix} v_a \\ v_b \end{bmatrix} = \begin{bmatrix} 1 \text{ V} \\ 2 \text{ V} \end{bmatrix}$$

By using Cramer's rule (see Appendix A), the solution for the two unknown variables, $v_a$ and $v_b$, can be written as follows:

$$v_a = \frac{\begin{vmatrix} 1 & -2 \\ 2 & 2.67 \end{vmatrix}}{\begin{vmatrix} 3 & -2 \\ -2 & 2.67 \end{vmatrix}} = \frac{(1)(2.67) - (-2)(2)}{(3)(2.67) - (-2)(-2)} = \frac{6.67}{4} = 1.667 \text{ V}$$

$$v_b = \frac{\begin{vmatrix} 3 & 1 \\ -2 & 2 \end{vmatrix}}{\begin{vmatrix} 3 & -2 \\ -2 & 2.67 \end{vmatrix}} = \frac{(3)(2) - (-2)(1)}{(3)(2.67) - (-2)(-2)} = \frac{8}{4} = 2 \text{ V}$$

The result is the same as in Example 3.2.

***Comments:*** While Cramer's rule is an efficient solution method for simple circuits (e.g., two nodes), it is customary to use computer-aided methods for larger circuits. Once the nodal equations have been set in the general form presented in equation 3.9, a variety of computer aids may be employed to compute the solution. You will find the solution to the same example computed using MathCad in the electronic files that accompany this book.

## Nodal Analysis with Voltage Sources

It would appear from the examples just shown that the node voltage method is very easily applied when current sources are present in a circuit. This is, in fact, the case, since current sources are directly accounted for by KCL. Some confusion occasionally arises, however, when voltage sources are present in a circuit analyzed by the node voltage method. In fact, the presence of voltage sources actually simplifies the calculations. To further illustrate this point, consider the circuit of Figure 3.8. Note immediately that one of the node voltages is known already! The voltage at node $a$ is forced to be equal to that of the voltage source; that is, $v_a = v_S$. Thus, only two nodal equations will be needed, at nodes $b$ and $c$:

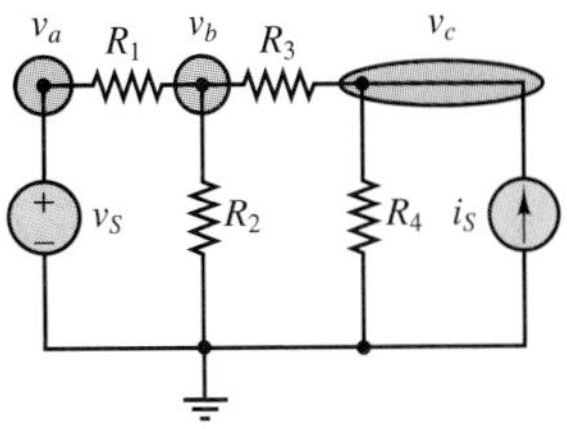

$$\frac{v_S - v_b}{R_1} - \frac{v_b}{R_2} - \frac{v_b - v_c}{R_3} = 0 \quad \text{(node } b\text{)}$$

$$\frac{v_b - v_c}{R_3} + i_S - \frac{v_c}{R_4} = 0 \quad \text{(node } c\text{)}$$

(3.10)

Rewriting these equations, we obtain:

$$\left(\frac{1}{R_1} + \frac{1}{R_2} + \frac{1}{R_3}\right) v_b + \left(-\frac{1}{R_3}\right) v_c = \frac{v_S}{R_1}$$

$$\left(-\frac{1}{R_3}\right) v_b + \left(\frac{1}{R_3} + \frac{1}{R_4}\right) v_c = i_S$$

(3.11)

Note how the term $v_S/R_1$ on the right-hand side of the first equation is really a current, as is dimensionally required by the nature of the node equations.

**Figure 3.8** Nodal analysis with voltage sources

---

## EXAMPLE   3.4  Nodal Analysis with Voltage Sources

### Problem

Find the node voltages in the circuit of Figure 3.9.

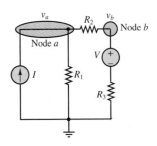

### Solution

***Known Quantities:*** Source current and voltage; resistor values.

***Find:*** Node voltages.

***Schematics, Diagrams, Circuits, and Given data:*** $I = -2$ mA; $V = 3$ V; $R_1 = 1$ kΩ; $R_2 = 2$ kΩ; $R_3 = 3$ kΩ.

**Figure 3.9**

***Assumptions:*** Place the reference node at the bottom of the circuit.

**Analysis:** Apply KCL at nodes $a$ and $b$:

$$I - \frac{v_a - 0}{R_1} - \frac{v_a - v_b}{R_2} = 0$$

$$\frac{v_a - v_b}{R_2} - \frac{v_b - 3}{R_3} = 0$$

Reformulating the last two equations, we derive the following system:

$$1.5v_a - 0.5v_b = -2$$

$$-0.5v_a + 0.833v_b = 1$$

Solving the last set of equations, we obtain the following values:

$$v_a = -1.167 \text{ V} \qquad \text{and} \qquad v_b = 0.5 \text{ V}$$

**Comments:** To compute the current flowing through resistor $R_3$ we noted that the voltage immediately above resistor $R_3$ (at the negative terminal of the voltage source) must be 3 volts lower than $v_b$; thus, the current through $R_3$ is equal to $(v_b - 3)/R_3$.

## Check Your Understanding

**3.1** Find the current $i_L$ in the circuit shown on the left, using the node voltage method.

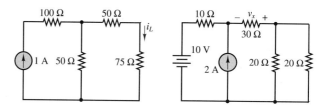

**3.2** Find the voltage $v_x$ by the node voltage method for the circuit shown on the right.

**3.3** Show that the answer to Example 3.2 is correct by applying KCL at one or more nodes.

## 3.2    THE MESH CURRENT METHOD

The second method of circuit analysis discussed in this chapter, which is in many respects analogous to the method of node voltages, employs **mesh currents** as the independent variables. The idea is to write the appropriate number of independent equations, using mesh currents as the independent variables. Analysis by mesh currents consists of defining the currents around the individual meshes as the independent variables. Subsequent application of Kirchhoff's voltage law around each mesh provides the desired system of equations.

In the mesh current method, we observe that a current flowing through a resistor in a specified direction defines the polarity of the voltage across the resistor, as illustrated in Figure 3.10, and that the sum of the voltages around a closed circuit

The current $i$, defined as flowing from left to right, establishes the polarity of the voltage across $R$.

**Figure 3.10** Basic principle of mesh analysis

must equal zero, by KVL. Once a convention is established regarding the direction of current flow around a mesh, simple application of KVL provides the desired equation. Figure 3.11 illustrates this point.

The number of equations one obtains by this technique is equal to the number of meshes in the circuit. All branch currents and voltages may subsequently be obtained from the mesh currents, as will presently be shown. Since meshes are easily identified in a circuit, this method provides a very efficient and systematic procedure for the analysis of electrical circuits. The following box outlines the procedure used in applying the mesh current method to a linear circuit.

Once the direction of current flow has been selected, KVL requires that $v_1 - v_2 - v_3 = 0$.

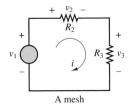

A mesh

**Figure 3.11** Use of KVL in mesh analysis

## FOCUS ON METHODOLOGY

**Mesh Current Analysis Method**

1. Define each mesh current consistently. We shall always define mesh currents clockwise, for convenience.
2. Apply KVL around each mesh, expressing each voltage in terms of one or more mesh currents.
3. Solve the resulting linear system of equations with mesh currents as the independent variables.

In mesh analysis, it is important to be consistent in choosing the direction of current flow. To avoid confusion in writing the circuit equations, mesh currents will be defined exclusively clockwise when we are using this method. To illustrate the mesh current method, consider the simple two-mesh circuit shown in Figure 3.12. This circuit will be used to generate two equations in the two unknowns, the mesh currents $i_1$ and $i_2$. It is instructive to first consider each mesh by itself. Beginning with mesh 1, note that the voltages around the mesh have been assigned in Figure 3.13 according to the direction of the mesh current, $i_1$. Recall that as long as signs are assigned consistently, an arbitrary direction may be assumed for any current in a circuit; if the resulting numerical answer for the current is negative, then the chosen reference direction is opposite to the direction of actual current flow. Thus, one need not be concerned about the actual direction of current flow in mesh analysis, once the directions of the mesh currents have been assigned. The correct solution will result, eventually.

According to the sign convention, then, the voltages $v_1$ and $v_2$ are defined as shown in Figure 3.13. Now, it is important to observe that while mesh current $i_1$ is equal to the current flowing through resistor $R_1$ (and is therefore also the branch current through $R_1$), it is not equal to the current through $R_2$. The branch current through $R_2$ is the difference between the two mesh currents, $i_1 - i_2$. Thus, since the polarity of the voltage $v_2$ has already been assigned, according to the convention discussed in the previous paragraph, it follows that the voltage $v_2$ is given by:

$$v_2 = (i_1 - i_2)R_2 \tag{3.12}$$

Finally, the complete expression for mesh 1 is

$$v_S - i_1 R_1 - (i_1 - i_2)R_2 = 0 \tag{3.13}$$

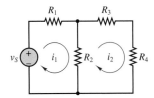

**Figure 3.12** A two-mesh circuit

Mesh 1: KVL requires that $v_S - v_1 - v_2 = 0$, where $v_1 = i_1 R_1$, $v_2 = (i_1 - i_2)R_1$.

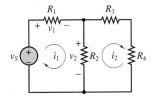

**Figure 3.13** Assignment of currents and voltages around mesh 1

Mesh 2: KVL requires that

$$v_2 + v_3 + v_4 = 0$$

where

$$v_2 = (i_2 - i_1)R_2,$$

$$v_3 = i_2 R_3,$$

$$v_4 = i_2 R_4$$

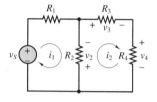

**Figure 3.14** Assignment of currents and voltages around mesh 2

The same line of reasoning applies to the second mesh. Figure 3.14 depicts the voltage assignment around the second mesh, following the clockwise direction of mesh current $i_2$. The mesh current $i_2$ is also the branch current through resistors $R_3$ and $R_4$; however, the current through the resistor that is shared by the two meshes, $R_2$, is now equal to $(i_2 - i_1)$, and the voltage across this resistor is

$$v_2 = (i_2 - i_1)R_2 \tag{3.14}$$

and the complete expression for mesh 2 is

$$(i_2 - i_1)R_2 + i_2 R_3 + i_2 R_4 = 0 \tag{3.15}$$

Why is the expression for $v_2$ obtained in equation 3.14 different from equation 3.12? The reason for this apparent discrepancy is that the voltage assignment for each mesh was dictated by the (clockwise) mesh current. Thus, since the mesh currents flow through $R_2$ in opposing directions, the voltage assignments for $v_2$ in the two meshes will also be opposite. This is perhaps a potential source of confusion in applying the mesh current method; you should be very careful to carry out the assignment of the voltages around each mesh separately.

Combining the equations for the two meshes, we obtain the following system of equations:

$$\begin{aligned} (R_1 + R_2)i_1 - R_2 i_2 &= v_S \\ -R_2 i_1 + (R_2 + R_3 + R_4)i_2 &= 0 \end{aligned} \tag{3.16}$$

These equations may be solved simultaneously to obtain the desired solution, namely, the mesh currents, $i_1$ and $i_2$. You should verify that knowledge of the mesh currents permits determination of all the other voltages and currents in the circuit. The following examples further illustrate some of the details of this method.

### EXAMPLE 3.5 Mesh Analysis

**Problem**

Find the mesh currents in the circuit of Figure 3.15.

**Solution**

**Known Quantities:** Source voltages; resistor values.

**Find:** Mesh currents.

**Schematics, Diagrams, Circuits, and Given Data:** $V_1 = 10$ V; $V_2 = 9$ V; $V_3 = 1$ V; $R_1 = 5\ \Omega$; $R_2 = 10\ \Omega$; $R_3 = 5\ \Omega$; $R_4 = 5\ \Omega$.

**Assumptions:** Assume clockwise mesh currents $i_1$ and $i_2$.

**Analysis:** The circuit of Figure 3.15 will yield two equations in two unknowns, $i_1$ and $i_2$. It is instructive to consider each mesh separately in writing the mesh equations; to this end, Figure 3.16 depicts the appropriate voltage assignments around the two meshes,

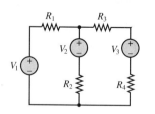

**Figure 3.15**

based on the assumed directions of the mesh currents. From Figure 3.16, we write the mesh equations:

$$V_1 - R_1 i_1 - V_2 - R_2(i_1 - i_2) = 0$$
$$R_2(i_2 - i_1) + V_2 - R_3 i_2 - V_3 - R_4 i_2 = 0$$

Rearranging the linear system of the equation, we obtain

$$15i_1 - 10i_2 = 1$$
$$-10i_1 + 20i_2 = 8$$

which can be solved to obtain $i_1$ and $i_2$:

$$i_1 = 0.5 \text{ A} \qquad \text{and} \qquad i_2 = 0.65 \text{ A}$$

**Comments:** Note how the voltage $v_2$ across resistor $R_2$ has different polarity in Figure 3.16, depending on whether we are working in mesh 1 or mesh 2.

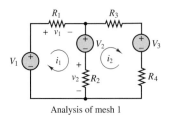

Analysis of mesh 1

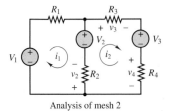

Analysis of mesh 2

**Figure 3.16**

## EXAMPLE 3.6 Mesh Analysis

### Problem

Write the mesh current equations for the circuit of Figure 3.17.

### Solution

**Known Quantities:** Source voltages; resistor values.

**Find:** Mesh current equations.

**Schematics, Diagrams, Circuits, and Given Data:** $V_1 = 12$ V; $V_2 = 6$ V; $R_1 = 3 \, \Omega$; $R_2 = 8 \, \Omega$; $R_3 = 6 \, \Omega$; $R_4 = 4 \, \Omega$.

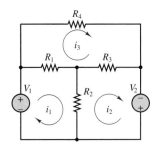

**Figure 3.17**

**Assumptions:** Assume clockwise mesh currents $i_1$, $i_2$, and $i_3$.

**Analysis:** Starting from mesh 1 we apply KVL to obtain

$$V_1 - R_1(i_1 - i_3) - R_2(i_1 - i_2) = 0.$$

KVL applied to mesh 2 yields

$$-R_2(i_2 - i_1) - R_3(i_2 - i_3) + V_2 = 0$$

while in mesh 3 we find

$$-R_1(i_3 - i_1) - R_4 i_3 - R_3(i_3 - i_2) = 0.$$

These equations can be rearranged in standard form to obtain

$$(3 + 8)i_1 - 8i_2 - 3i_3 = 12$$
$$-8i_1 + (6 + 8)i_2 - 6i_3 = 6$$
$$-3i_1 - 6i_2 + (3 + 6 + 4)i_3 = 0$$

You may verify that KVL holds around any one of the meshes, as a test to check that the answer is indeed correct.

*Comments:* The solution of the mesh current equations with computer-aided tools (MathCad) may be found in the electronic files that accompany this book.

A comparison of this result with the analogous result obtained by the node voltage method reveals that we are using Ohm's law in conjunction with KVL (in contrast with the use of KCL in the node voltage method) to determine the minimum set of equations required to solve the circuit.

## Mesh Analysis with Current Sources

Mesh analysis is particularly effective when applied to circuits containing voltage sources exclusively; however, it may be applied to mixed circuits, containing both voltage and current sources, if care is taken in identifying the proper current in each mesh. The method is illustrated by solving the circuit shown in Figure 3.18. The first observation in analyzing this circuit is that the presence of the current source requires that the following relationship hold true:

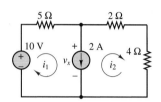

**Figure 3.18** Mesh analysis with current sources

$$i_1 - i_2 = 2 \text{ A} \tag{3.17}$$

If the unknown voltage across the current source is labeled $v_x$, application of KVL around mesh 1 yields:

$$10 - 5i_1 - v_x = 0 \tag{3.18}$$

while KVL around mesh 2 dictates that

$$v_x - 2i_2 - 4i_2 = 0 \tag{3.19}$$

Substituting equation 3.19 in equation 3.18, and using equation 3.17, we can then obtain the system of equations

$$
\begin{aligned}
5i_1 + 6i_2 &= 10 \\
-i_1 + i_2 &= -2
\end{aligned}
\tag{3.20}
$$

which we can solve to obtain

$$
\begin{aligned}
i_1 &= 2 \text{ A} \\
i_2 &= 0 \text{ A}
\end{aligned}
\tag{3.21}
$$

Note also that the voltage across the current source may be found by using either equation 3.18 or equation 3.19; for example, using equation 3.19,

$$v_x = 6i_2 = 0 \text{ V} \tag{3.22}$$

The following example further illustrates the solution of this type of circuit.

### EXAMPLE   3.7  Mesh Analysis with Current Sources

**Problem**

Find the mesh currents in the circuit of Figure 3.19.

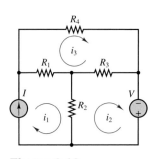

**Solution**

**Known Quantities:** Source current and voltage; resistor values.

**Find:** Mesh currents.

**Schematics, Diagrams, Circuits, and Given Data:** $I = 0.5$ A; $V = 6$ V; $R_1 = 3\ \Omega$; $R_2 = 8\ \Omega$; $R_3 = 6\ \Omega$; $R_4 = 4\ \Omega$.

**Assumptions:** Assume clockwise mesh currents $i_1$, $i_2$, and $i_3$.

**Figure 3.19**

**Analysis:** Starting from mesh 1, we see immediately that the current source forces the mesh current to be equal to $I$:

$$i_1 = I$$

There is no need to write any further equations around mesh 1, since we already know the value of the mesh current. Now we turn to meshes 2 and 3 to obtain:

$$-R_2(i_2 - i_1) - R_3(i_2 - i_3) + V = 0 \qquad \text{mesh 2}$$
$$-R_1(i_3 - i_1) - R_4 i_3 - R_3(i_3 - i_2) = 0 \qquad \text{mesh 3}$$

Rearranging the equations and substituting the known value of $i_1$, we obtain a system of two equations in two unknowns:

$$14 i_2 - 6 i_3 = 10$$
$$-6 i_2 + 13 i_3 = 1.5$$

which can be solved to obtain

$$i_2 = 0.95\ \text{A} \qquad i_3 = 0.55\ \text{A}$$

As usual, you should verify that the solution is correct by applying KVL.

**Comments:** Note that the current source has actually simplified the problem by constraining a mesh current to a fixed value.

---

## Check Your Understanding

**3.4** Find the unknown voltage, $v_x$, by mesh current analysis in the circuit of Figure 3.20.

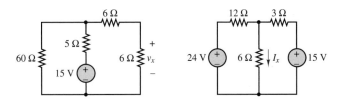

**Figure 3.20**                 **Figure 3.21**

**3.5** Find the unknown current, $I_x$, using mesh current methods in the circuit of Figure 3.21.

**3.6**  Show that the equations given in Example 3.6 are correct, by applying KCL at each node.

---

## 3.3  NODAL AND MESH ANALYSIS WITH CONTROLLED SOURCES

The methods just described also apply, with relatively minor modifications, in the presence of dependent (controlled) sources. Solution methods that allow for the presence of controlled sources will be particularly useful in the study of *transistor amplifiers* in Chapter 8. Recall from the discussion in Section 2.3 that a dependent source is a source that generates a voltage or current that depends on the value of another voltage or current in the circuit. When a dependent source is present in a circuit to be analyzed by node or mesh analysis, one can initially treat it as an ideal source and write the node or mesh equations accordingly. In addition to the equation obtained in this fashion, there will also be an equation relating the dependent source to one of the circuit voltages or currents. This **constraint equation** can then be substituted in the set of equations obtained by the techniques of nodal and mesh analysis, and the equations can subsequently be solved for the unknowns.

It is important to remark that once the constraint equation has been substituted in the initial system of equations, the number of unknowns remains unchanged. Consider, for example, the circuit of Figure 3.22, which is a simplified model of a bipolar transistor amplifier (transistors will be introduced in Chapter 8). In the circuit of Figure 3.22, two nodes are easily recognized, and therefore nodal analysis is chosen as the preferred method. Applying KCL at node 1, we obtain the following equation:

$$i_S = v_1 \left( \frac{1}{R_S} + \frac{1}{R_b} \right) \tag{3.23}$$

KCL applied at the second node yields:

$$\beta i_b + \frac{v_2}{R_C} = 0 \tag{3.24}$$

Next, it should be observed that the current $i_b$ can be determined by means of a simple current divider:

$$i_b = i_S \frac{1/R_b}{1/R_b + 1/R_S} = i_S \frac{R_S}{R_b + R_S} \tag{3.25}$$

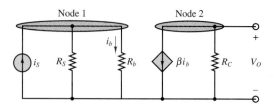

**Figure 3.22** Circuit with dependent source

which, when inserted in equation 3.24, yields a system of two equations:

$$i_S = v_1 \left( \frac{1}{R_S} + \frac{1}{R_b} \right)$$

$$-\beta i_S \frac{R_S}{R_b + R_S} = \frac{v_2}{R_C}$$

(3.26)

which can be used to solve for $v_1$ and $v_2$. Note that, in this particular case, the two equations are independent of each other. The following example illustrates a case in which the resulting equations are not independent.

## EXAMPLE 3.8 Analysis with Dependent Sources

### Problem

Find the node voltages in the circuit of Figure 3.23.

### Solution

VIRTUAL LAB

**Known Quantities:** Source current; resistor values; dependent voltage source relationship.

**Find:** Unknown node voltage $v$.

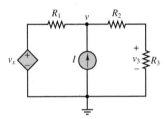

**Schematics, Diagrams, Circuits, and Given Data:** $I = 0.5$ A; $R_1 = 5 \, \Omega$; $R_2 = 2 \, \Omega$; $R_3 = 4 \, \Omega$. Dependent source relationship: $v_x = 2 \times v_3$.

**Assumptions:** Assume reference node is at the bottom of the circuit.

**Figure 3.23**

**Analysis:** Applying KCL to node $v$ we find that

$$\frac{v_x - v}{R_1} + I - \frac{v - v_3}{R_2} = 0$$

Applying KCL to node $v_3$ we find

$$\frac{v - v_3}{R_2} - \frac{v_3}{R_3} = 0$$

If we substitute the dependent source relationship into the first equation, we obtain a system of equations in the two unknowns $v$ and $v_3$:

$$\left( \frac{1}{R_1} + \frac{1}{R_2} \right) v + \left( -\frac{2}{R_1} - \frac{1}{R_2} \right) v_3 = I$$

$$\left( -\frac{1}{R_2} \right) v + \left( \frac{1}{R_2} + \frac{1}{R_3} \right) v_3 = 0$$

Substituting numerical values, we obtain:

$$0.7v - 0.9v_3 = 0.5$$
$$-0.5v + 0.75v_3 = 0$$

Solution of the above equations yields $v = 5$ V; $v_3 = 3.33$ V.

**Comments:** You will find the solution to the same example computed using MathCad in the electronic files that accompany this book.

## Remarks on Node Voltage and Mesh Current Methods

The techniques presented in this section and the two preceding sections find use more generally than just in the analysis of resistive circuits. These methods should be viewed as general techniques for the analysis of any linear circuit; they provide systematic and effective means of obtaining the minimum number of equations necessary to solve a network problem. Since these methods are based on the fundamental laws of circuit analysis, KVL and KCL, they also apply to any electrical circuit, even circuits containing nonlinear circuit elements, such as those to be introduced later in this chapter.

You should master both methods as early as possible. Proficiency in these circuit analysis techniques will greatly simplify the learning process for more advanced concepts.

---

## Check Your Understanding

**3.7**  The current source $i_x$ is related to the voltage $v_x$ in Figure 3.24 by the relation

$$i_x = \frac{v_x}{3}$$

Find the voltage across the 8-$\Omega$ resistor by nodal analysis.

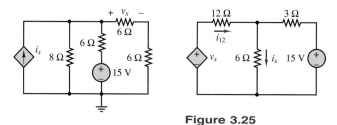

**Figure 3.25**

**Figure 3.24**

**3.8**  Find the unknown current $i_x$ in Figure 3.25 using the mesh current method. The dependent voltage source is related to the current $i_{12}$ through the 12-$\Omega$ resistor by $v_x = 2i_{12}$.

---

## 3.4    THE PRINCIPLE OF SUPERPOSITION

This brief section discusses a concept that is frequently called upon in the analysis of linear circuits. Rather than a precise analysis technique, like the mesh current and node voltage methods, the principle of superposition is a conceptual aid that can be very useful in visualizing the behavior of a circuit containing multiple sources. The principle of superposition applies to any linear system and for a linear circuit may be stated as follows:

> In a linear circuit containing $N$ sources, each branch voltage and current is the sum of $N$ voltages and currents each of which may be computed by setting all but one source equal to zero and solving the circuit containing that single source.

An elementary illustration of the concept may easily be obtained by simply considering a circuit with two sources connected in series, as shown in Figure 3.26.

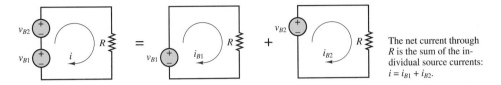

The net current through $R$ is the sum of the individual source currents: $i = i_{B1} + i_{B2}$.

**Figure 3.26** The principle of superposition

The circuit of Figure 3.26 is more formally analyzed as follows. The current, $i$, flowing in the circuit on the left-hand side of Figure 3.26 may be expressed as:

$$i = \frac{v_{B1} + v_{B2}}{R} = \frac{v_{B1}}{R} + \frac{v_{B2}}{R} = i_{B1} + i_{B2} \qquad \textbf{(3.27)}$$

Figure 3.26 also depicts the circuit as being equivalent to the combined effects of two circuits, each containing a single source. In each of the two subcircuits, a short circuit has been substituted for the missing battery. This should appear as a sensible procedure, since a short circuit—by definition—will always "see" zero voltage across itself, and therefore this procedure is equivalent to "zeroing" the output of one of the voltage sources.

If, on the other hand, one wished to cancel the effects of a current source, it would stand to reason that an open circuit could be substituted for the current source, since an open circuit is by definition a circuit element through which no current can flow (and which will therefore generate zero current). These basic principles are used frequently in the analysis of circuits, and are summarized in Figure 3.27.

The principle of superposition can easily be applied to circuits containing multiple sources and is sometimes an effective solution technique. More often,

1. In order to set a voltage source equal to zero, we replace it with a short circuit.

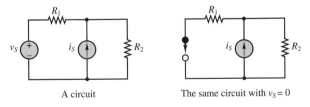

A circuit                    The same circuit with $v_S = 0$

2. In order to set a current source equal to zero, we replace it with an open circuit.

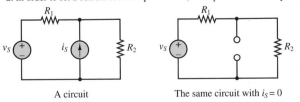

A circuit                    The same circuit with $i_S = 0$

**Figure 3.27** Zeroing voltage and current sources

however, other methods result in a more efficient solution. Example 3.9 further illustrates the use of superposition to analyze a simple network. The Check Your Understanding exercises at the end of the section illustrate the fact that superposition is often a cumbersome solution method.

---

### EXAMPLE  3.9  Principle of Superposition

#### Problem

Determine the current $i_2$ in the circuit of Figure 3.18 using the principle of superposition.

---

#### Solution

**Known Quantities:**  Source voltage and current values. Resistor values.

**Find:**  Unknown current $i_2$.

**Given Data**  Figure 3.18.

**Assumptions:**  Assume reference node is at the bottom of the circuit.

**Analysis:**  *Part 1:* Zero the current source. Once the current source has been set to zero (replaced by an open circuit), the resulting circuit is a simple series circuit; the current flowing in this circuit, $i_{2-V}$, is the current we seek. Since the total series resistance is $5 + 2 + 4 = 11\ \Omega$, we find that $i_{2-V} = 10/11 = 0.909$ A.

*Part 2:* Zero the voltage source. After zeroing of the voltage source by replacing it with a short circuit, the resulting circuit consists of three parallel branches: On the left we have a single 5-$\Omega$ resistor; in the center we have a $-2$-A current source (negative because the source current is shown to flow into the ground node); on the right we have a total resistance of $2 + 4 = 6\ \Omega$. Using the current divider rule, we find that the current flowing in the right branch, $i_{2-I}$, is given by:

$$i_{2-I} = \frac{\dfrac{1}{6}}{\dfrac{1}{5} + \dfrac{1}{6}}(-2) = -0.909 \text{ A}$$

And, finally, the unknown current $i_2$ is found to be

$$i_2 = i_{2\text{-}V} + i_{2-I} = 0 \text{ A}.$$

The result is, of course, identical to that obtained by mesh analysis.

**Comments:**  Superposition may appear to be a very efficient tool. However, beginners may find it preferable to rely on more systematic methods, such as nodal analysis, to solve circuits. Eventually, experience will suggest the preferred method for any given circuit.

---

## Check Your Understanding

**3.9**    Find the voltages $v_a$ and $v_b$ for the circuits of Example 3.4 by superposition.

**3.10**   Repeat Check Your Understanding Exercise 3.2, using superposition. This exercise illustrates that superposition is not necessarily a computationally efficient solution method.

**3.11**  Solve Example 3.5, using superposition.

**3.12**  Solve Example 3.7, using superposition.

---

# 3.5  ONE-PORT NETWORKS
# AND EQUIVALENT CIRCUITS

You may recall that, in the discussion of ideal sources in Chapter 2, the flow of energy from a source to a load was described in a very general form, by showing the connection of two "black boxes" labeled source and load (see Figure 2.10). In the same figure, two other descriptions were shown: a symbolic one, depicting an ideal voltage source and an ideal resistor; and a physical representation, in which the load was represented by a headlight and the source by an automotive battery. Whatever the form chosen for source-load representation, each block—source or load—may be viewed as a two-terminal device, described by an $i$-$v$ characteristic. This general circuit representation is shown in Figure 3.28. This configuration is called a **one-port network** and is particularly useful for introducing the notion of equivalent circuits. Note that the network of Figure 3.28 is completely described by its $i$-$v$ characteristic; this point is best illustrated by the next example.

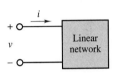

**Figure 3.28** One-port network

---

## EXAMPLE  3.10  Equivalent Resistance Calculation

### Problem

Determine the source (load) current $i$ in the circuit of Figure 3.29 using equivalent resistance ideas.

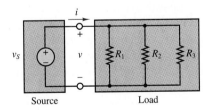

**Figure 3.29** Illustration of equivalent-circuit concept

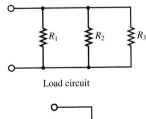

Load circuit

Equivalent load circuit

**Figure 3.30** Equivalent load resistance concept

### Solution

**Known Quantities:**  Source voltage, resistor values.

**Find:**  Source current.

**Given Data:**  Figures 3.29, 3.30.

**Assumptions:**  Assume reference node is at the bottom of the circuit.

**Analysis:** *Insofar as the source is concerned*, the three parallel resistors appear identical to a single equivalent resistance of value

$$R_{EQ} = \cfrac{1}{\cfrac{1}{R_1} + \cfrac{1}{R_2} + \cfrac{1}{R_3}}$$

Thus, we can replace the three load resistors with the single equivalent resistor $R_{EQ}$, as shown in Figure 3.30, and calculate

$$i = \frac{v_S}{R_{EQ}}$$

**Comments:** Similarly, *insofar as the load is concerned*, it would not matter whether the source consisted, say, of a single 6-V battery or of four 1.5-V batteries connected in series.

For the remainder of this chapter, we shall focus on developing techniques for computing equivalent representations of linear networks. Such representations will be useful in deriving some simple—yet general—results for linear circuits, as well as analyzing simple nonlinear circuits.

## Thévenin and Norton Equivalent Circuits

This section discusses one of the most important topics in the analysis of electrical circuits: the concept of an **equivalent circuit**. It will be shown that it is always possible to view even a very complicated circuit in terms of much simpler *equivalent* source and load circuits, and that the transformations leading to equivalent circuits are easily managed, with a little practice. In studying node voltage and mesh current analysis, you may have observed that there is a certain correspondence (called **duality**) between current sources and voltage sources, on the one hand, and parallel and series circuits, on the other. This duality appears again very clearly in the analysis of equivalent circuits: it will shortly be shown that equivalent circuits fall into one of two classes, involving either voltage or current sources and (respectively) either series or parallel resistors, reflecting this same principle of duality. The discussion of equivalent circuits begins with the statement of two very important theorems, summarized in Figures 3.31 and 3.32.

**Figure 3.31** Illustration of Thévenin theorem

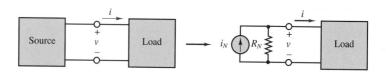

**Figure 3.32** Illustration of Norton theorem

**The Thévenin Theorem**

As far as a load is concerned, any network composed of ideal voltage and current sources, and of linear resistors, may be represented by an equivalent circuit consisting of an ideal voltage source, $v_T$, in series with an equivalent resistance, $R_T$.

**The Norton Theorem**

As far as a load is concerned, any network composed of ideal voltage and current sources, and of linear resistors, may be represented by an equivalent circuit consisting of an ideal current source, $i_N$, in parallel with an equivalent resistance, $R_N$.

The first obvious question to arise is, how are these equivalent source voltages, currents, and resistances computed? The next few sections illustrate the computation of these equivalent circuit parameters, mostly through examples. A substantial number of Check Your Understanding exercises are also provided, with the following caution: The only way to master the computation of Thévenin and Norton equivalent circuits is by patient repetition.

## Determination of Norton or Thévenin Equivalent Resistance

The first step in computing a Thévenin or Norton equivalent circuit consists of finding the equivalent resistance presented by the circuit at its terminals. This is done by setting all sources in the circuit equal to zero and computing the effective resistance between terminals. The voltage and current sources present in the circuit are set to zero by the same technique used with the principle of superposition: voltage sources are replaced by short circuits, current sources by open circuits. To illustrate the procedure, consider the simple circuit of Figure 3.33; the objective is to compute the equivalent resistance the load $R_L$ "sees" at port $a$-$b$.

In order to compute the equivalent resistance, we remove the load resistance from the circuit and replace the voltage source, $v_S$, by a short circuit. At this point—seen from the load terminals—the circuit appears as shown in Figure 3.34. You can see that $R_1$ and $R_2$ are in parallel, since they are connected between the same two nodes. If the total resistance between terminals $a$ and $b$ is denoted by $R_T$, its value can be determined as follows:

$$R_T = R_3 + R_1 \parallel R_2 \tag{3.28}$$

An alternative way of viewing $R_T$ is depicted in Figure 3.35, where a hypothetical 1-A current source has been connected to the terminals $a$ and $b$. The voltage $v_x$ appearing across the $a$-$b$ pair will then be numerically equal to $R_T$ (only because $i_S = 1$ A!). With the 1-A source current flowing in the circuit, it should be apparent that the source current encounters $R_3$ as a resistor in series with the parallel combination of $R_1$ and $R_2$, prior to completing the loop.

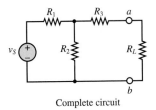

Complete circuit

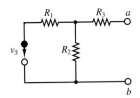

Circuit with load removed for computation of $R_T$. The voltage source is replaced by a short circuit.

**Figure 3.33** Computation of Thévenin resistance

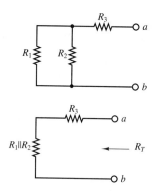

What is the total resistance the current $i_S$ will encounter in flowing around the circuit?

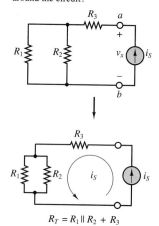

**Figure 3.34** Equivalent resistance seen by the load

$$R_T = R_1 \parallel R_2 + R_3$$

**Figure 3.35** An alternative method of determining the Thévenin resistance

Summarizing the procedure, we can produce a set of simple rules as an aid in the computation of the Thévenin (or Norton) equivalent resistance for a linear resistive circuit:

## FOCUS ON METHODOLOGY

**Computation of Equivalent Resistance of a One-Port Network**

1. Remove the load.
2. Zero all independent voltage and current sources.
3. Compute the total resistance between load terminals, *with the load removed.* This resistance is equivalent to that which would be encountered by a current source connected to the circuit in place of the load.

We note immediately that this procedure yields a result that is independent of the load. This is a very desirable feature, since once the equivalent resistance has been identified for a source circuit, the equivalent circuit remains unchanged if we connect a different load. The following examples further illustrate the procedure.

## EXAMPLE 3.11 Thévenin Equivalent Resistance

### Problem

Find the Thévenin equivalent resistance seen by the load $R_L$ in the circuit of Figure 3.36.

### Solution

***Known Quantities:*** Resistor and current source values.

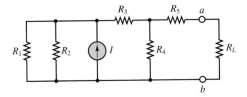

**Figure 3.36**

**Find:** Thévenin equivalent resistance $R_T$.

**Schematics, Diagrams, Circuits, and Given Data:**  $R_1 = 20\,\Omega$; $R_2 = 20\,\Omega$; $I = 5$ A; $R_3 = 10\,\Omega$; $R_4 = 20\,\Omega$; $R_5 = 10\,\Omega$.

**Assumptions:**  Assume reference node is at the bottom of the circuit.

**Analysis:**  Following the methodology box introduced in the present section, we first set the current source equal to zero, by replacing it with an open circuit. The resulting circuit is depicted in Figure 3.37. Looking into terminal $a$-$b$ we recognize that, starting from the left (away from the load) and moving to the right (toward the load) the equivalent resistance is given by the expression

$$R_T = [((R_1 || R_2) + R_3) || R_4] + R_5$$

$$= [((20||20) + 10) ||20] + 10 = 20\,\Omega$$

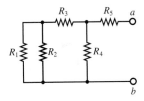

**Figure 3.37**

**Comments:**  Note that the reduction of the circuit started at the farthest point away from the load.

---

## EXAMPLE  3.12  Thévenin Equivalent Resistance

### Problem

Compute the Thévenin equivalent resistance seen by the load in the circuit of Figure 3.38.

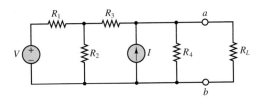

**Figure 3.38**

---

### Solution

**Known Quantities:**  Resistor values.

**Find:**  Thévenin equivalent resistance $R_T$.

**Schematics, Diagrams, Circuits, and Given Data:** $V = 5$ V; $R_1 = 2\,\Omega$; $R_2 = 2\,\Omega$; $R_3 = 1\,\Omega$; $I = 1$ A, $R_4 = 2\,\Omega$.

**Assumptions:** Assume reference node is at the bottom of the circuit.

**Analysis:** Following the Thévenin equivalent resistance methodology box, we first set the current source equal to zero, by replacing it with an open circuit, then set the voltage source equal to zero by replacing it with a short circuit. The resulting circuit is depicted in Figure 3.39. Looking into terminal $a$-$b$ we recognize that, starting from the left (away from the load) and moving to the right (toward the load), the equivalent resistance is given by the expression

$$R_T = ((R_1 || R_2) + R_3) \, || R_4$$

$$= ((2||2) + 1) \, ||2 = 1 \; \Omega$$

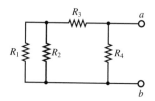

**Figure 3.39**

**Comments:** Note that the reduction of the circuit started at the farthest point away from the load.

---

As a final note, it should be remarked that the Thévenin and Norton equivalent resistances are one and the same quantity:

$$R_T = R_N \tag{3.29}$$

Therefore, the preceding discussion holds whether we wish to compute a Norton or a Thévenin equivalent circuit. From here on we shall use the notation $R_T$ exclusively, for both Thévenin and Norton equivalents. Check Your Understanding Exercise 3.13 will give you an opportunity to explain why the two equivalent resistances are one and the same.

## Check Your Understanding

**3.13** Apply the methods described in this section to show that $R_T = R_N$ in the circuits of Figure 3.40.

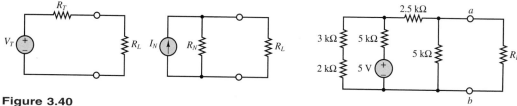

**Figure 3.40**

**Figure 3.41**

**3.14** Find the Thévenin equivalent resistance of the circuit of Figure 3.41 seen by the load resistor, $R_L$.

**3.15** Find the Thévenin equivalent resistance seen by the load resistor, $R_L$, in the circuit of Figure 3.42.

**3.16** For the circuit of Figure 3.43, find the Thévenin equivalent resistance seen by the load resistor, $R_L$.

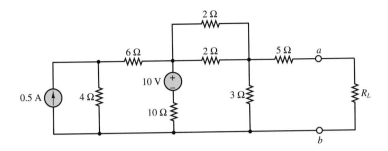

**Figure 3.42**

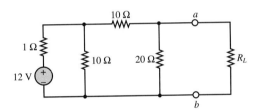

**Figure 3.43**

**3.17**   For the circuit of Figure 3.44, find the Thévenin equivalent resistance seen by the load resistor, $R_L$.

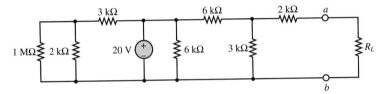

**Figure 3.44**

## Computing the Thévenin Voltage

This section describes the computation of the Thévenin equivalent voltage, $v_T$, for an arbitrary linear resistive circuit. The Thévenin equivalent voltage is defined as follows:

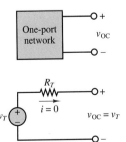

> The equivalent (Thévenin) source voltage is equal to the **open-circuit voltage** present at the load terminals (with the load removed).

This states that in order to compute $v_T$, it is sufficient to remove the load and to compute the open-circuit voltage at the one-port terminals. Figure 3.45 illustrates that the open-circuit voltage, $v_{OC}$, and the Thévenin voltage, $v_T$, must

**Figure 3.45** Equivalence of open-circuit and Thévenin voltage

be the same if the Thévenin theorem is to hold. This is true because in the circuit consisting of $v_T$ and $R_T$, the voltage $v_{OC}$ must equal $v_T$, since no current flows through $R_T$ and therefore the voltage across $R_T$ is zero. Kirchhoff's voltage law confirms that

$$v_T = R_T(0) + v_{OC} = v_{OC} \tag{3.30}$$

## FOCUS ON METHODOLOGY

### Computing the Thévenin Voltage

1. Remove the load, leaving the load terminals open-circuited.
2. Define the open-circuit voltage $v_{OC}$ across the open load terminals
3. Apply any preferred method (e.g., nodal analysis) to solve for $v_{OC}$.
4. The Thévenin voltage is $v_T = v_{OC}$.

The actual computation of the open-circuit voltage is best illustrated by examples; there is no substitute for practice in becoming familiar with these computations. To summarize the main points in the computation of open-circuit voltages, consider the circuit of Figure 3.33, shown again in Figure 3.46 for convenience. Recall that the equivalent resistance of this circuit was given by $R_T = R_3 + R_1 \parallel R_2$. To compute $v_{OC}$, we disconnect the load, as shown in Figure 3.47, and immediately observe that no current flows through $R_3$, since there is no closed circuit connection at that branch. Therefore, $v_{OC}$ must be equal to the voltage across $R_2$, as illustrated in Figure 3.48. Since the only closed circuit is the mesh consisting of $v_S$, $R_1$, and $R_2$, the answer we are seeking may be obtained by means of a simple voltage divider:

$$v_{OC} = v_{R2} = v_S \frac{R_2}{R_1 + R_2}$$

It is instructive to review the basic concepts outlined in the example by considering the original circuit and its Thévenin equivalent side by side, as shown in Figure 3.49. The two circuits of Figure 3.49 are equivalent in the sense that the current drawn by the load, $i_L$, is the same in both circuits, that current being given by:

$$i_L = v_S \cdot \frac{R_2}{R_1 + R_2} \cdot \frac{1}{(R_3 + R_1 \parallel R_2) + R_L} = \frac{v_T}{R_T + R_L} \tag{3.31}$$

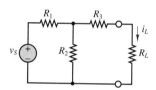

**Figure 3.46**

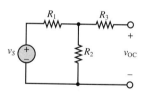

**Figure 3.47**

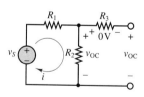

**Figure 3.48**

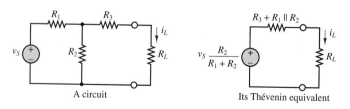

**Figure 3.49** A circuit and its Thévenin equivalent

The computation of Thévenin equivalent circuits is further illustrated in the following examples.

## EXAMPLE 3.13 Thévenin Equivalent Voltage (Open-Circuit Load Voltage)

**Problem**

Compute the open-circuit voltage, $v_{OC}$, in the circuit of Figure 3.50.

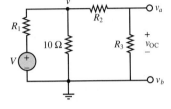

**Figure 3.50**

**Solution**

**Known Quantities:** Source voltage, resistor values.

**Find:** Open-circuit voltage, $v_{OC}$.

**Schematics, Diagrams, Circuits, and Given Data:** $V = 12$ V; $R_1 = 1\,\Omega$; $R_2 = 10\,\Omega$; $R_3 = 20\,\Omega$.

**Assumptions:** Assume reference node is at the bottom of the circuit.

**Analysis:** Following the Thévenin voltage methodology box, we first remove the load and label the open-circuit voltage, $v_{OC}$. Next, we observe that, since $v_b$ is equal to the reference voltage, (i.e., zero), the node voltage $v_a$ will be equal, numerically, to the open-circuit voltage. If we define the other node voltage to be $v$, nodal analysis will be the natural technique for arriving at the solution. Figure 3.50 depicts the original circuit ready for nodal analysis. Applying KCL at the two nodes, we obtain the following two equations:

$$\frac{12 - v}{1} - \frac{v}{10} - \frac{v - v_a}{10} = 0$$

$$\frac{v - v_a}{10} - \frac{v_a}{20} = 0$$

In matrix form we can write:

$$\begin{bmatrix} 1.2 & -0.1 \\ -0.1 & 0.15 \end{bmatrix} \begin{bmatrix} v \\ v_a \end{bmatrix} = \begin{bmatrix} 12 \\ 0 \end{bmatrix}$$

Solving the above matrix equations yields: $v = 10.588$ V; $v_a = 7.059$ V.

**Comments:** Note that the determination of the Thévenin voltage is nothing more than the careful application of the basic circuit analysis methods presented in earlier sections. The only difference is that we first need to properly identify and define the open-circuit load voltage. You will find the solution to the same example computed by MathCad in the electronic files that accompany this book.

## EXAMPLE 3.14 Load Current Calculation by Thévenin Equivalent Method

**Problem**

Compute the load current, $i$, by the Thévenin equivalent method in the circuit of Figure 3.51.

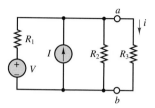

**Figure 3.51**

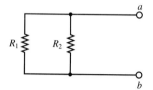

**Figure 3.52**

## Solution

**Known Quantities:**  Source voltage, resistor values.

**Find:**  Load current, $i$.

**Schematics, Diagrams, Circuits, and Given Data:**  $V = 24$ V; $I = 3$ A; $R_1 = 4\ \Omega$; $R_2 = 12\ \Omega$; $R_1 = 6\ \Omega$.

**Assumptions:**  Assume reference node is at the bottom of the circuit.

**Analysis:**  We first compute the Thévenin equivalent resistance. According to the method proposed earlier, we zero the two sources by shorting the voltage source and opening the current source. The resulting circuit is shown in Figure 3.52. We can clearly see that $R_T = R_1 \| R_2 = 4 \| 12 = 3\ \Omega$.

Following the Thévenin voltage methodology box, we first remove the load and label the open-circuit voltage, $v_{OC}$. The circuit is shown in Figure 3.53. Next, we observe that, since $v_b$ is equal to the reference voltage (i.e., zero) the node voltage $v_a$ will be equal, numerically, to the open-circuit voltage. In this circuit, a single nodal equation is required to arrive at the solution:

$$\frac{V - v_a}{R_1} + I - \frac{v_a}{R_2} = 0$$

Substituting numerical values, we find that $v_a = v_{OC} = v_T = 27$ V.

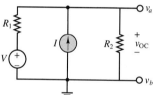

**Figure 3.53**

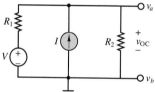

**Figure 3.54** Thévenin equivalent

Finally, we assemble the Thévenin equivalent circuit, shown in Figure 3.54, and reconnect the load resistor. Now the load current can be easily computed to be:

$$i = \frac{v_T}{R_T + R_L} = \frac{27}{3 + 6} = 3 \text{ A}$$

**Comments:**  It may appear that the calculation of load current by the Thévenin equivalent method leads to more complex calculations than, say, node voltage analysis (you might wish to try solving the same circuit by nodal analysis to verify this). However, there is one major advantage to equivalent circuit analysis: Should the load change (as is often the case in many practical engineering situations), the equivalent circuit calculations still hold, and only the (trivial) last step in the above example needs to be repeated. Thus, knowing the Thévenin equivalent of a particular circuit can be very useful whenever we need to perform computations pertaining to any load quantity.

## Check Your Understanding

**3.18**  With reference to Figure 3.46, find the load current, $i_L$, by mesh analysis, if $v_S = 10$ V, $R_1 = R_3 = 50\ \Omega$, $R_2 = 100\ \Omega$, $R_L = 150\ \Omega$.

**3.19**  Find the Thévenin equivalent circuit seen by the load resistor, $R_L$, for the circuit of Figure 3.55.

**3.20**  Find the Thévenin equivalent circuit for the circuit of Figure 3.56.

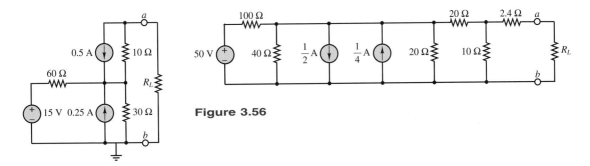

**Figure 3.56**

**Figure 3.55**

## Computing the Norton Current

The computation of the Norton equivalent current is very similar in concept to that of the Thévenin voltage. The following definition will serve as a starting point:

> **Definition**
>
> The Norton equivalent current is equal to the **short-circuit current** that would flow were the load replaced by a short circuit.

An explanation for the definition of the Norton current is easily found by considering, again, an arbitrary one-port network, as shown in Figure 3.57, where the one-port network is shown together with its Norton equivalent circuit.

It should be clear that the current, $i_{SC}$, flowing through the short circuit replacing the load is exactly the Norton current, $i_N$, since all of the source current in the circuit of Figure 3.57 must flow through the short circuit. Consider the circuit of Figure 3.58, shown with a short circuit in place of the load resistance. Any of the techniques presented in this chapter could be employed to determine the current $i_{SC}$. In this particular case, mesh analysis is a convenient tool, once it is recognized that the short-circuit current is a mesh current. Let $i_1$ and $i_2 = i_{SC}$ be the mesh currents in the circuit of Figure 3.58. Then, the following mesh equations can be derived and solved for the short-circuit current:

$$(R_1 + R_2)i_1 - R_2 i_{SC} = v_S$$
$$-R_2 i_1 + (R_2 + R_3)i_{SC} = 0$$

An alternative formulation would employ nodal analysis to derive the equation

$$\frac{v_S - v}{R_1} = \frac{v}{R_2} + \frac{v}{R_3}$$

leading to

$$v = v_S \frac{R_2 R_3}{R_1 R_3 + R_2 R_3 + R_1 R_2}$$

**Figure 3.57** Illustration of Norton equivalent circuit

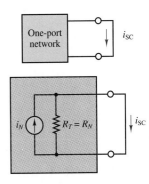

**Figure 3.58** Computation of Norton current

Recognizing that $i_{SC} = v/R_3$, we can determine the Norton current to be:

$$i_N = \frac{v}{R_3} = \frac{v_S R_2}{R_1 R_3 + R_2 R_3 + R_1 R_2}$$

Thus, conceptually, the computation of the Norton current simply requires identifying the appropriate short-circuit current. The following example further illustrates this idea.

### FOCUS ON METHODOLOGY

**Computing the Norton Current**

1. Replace the load with a short circuit.
2. Define the short circuit current, $i_{SC}$, to be the Norton equivalent current.
3. Apply any preferred method (e.g., nodal analysis) to solve for $i_{SC}$.
4. The Norton current is $i_N = i_{SC}$.

## EXAMPLE 3.15 Norton Equivalent Circuit

**Problem**

Determine the Norton current and the Norton equivalent for the circuit of Figure 3.59.

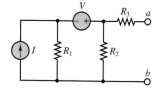

**Figure 3.59**

**Solution**

**Known Quantities:** Source voltage and current, resistor values.

**Find:** Equivalent resistance, $R_T$. Norton current, $i_N = i_{SC}$.

**Schematics, Diagrams, Circuits, and Given Data:** $V = 6$ V; $I = 2$ A; $R_1 = 6\ \Omega$; $R_2 = 3\ \Omega$; $R_3 = 2\ \Omega$.

**Assumptions:** Assume reference node is at the bottom of the circuit.

**Analysis:** We first compute the Thévenin equivalent resistance. We zero the two sources by shorting the voltage source and opening the current source. The resulting circuit is shown in Figure 3.60. We can clearly see that $R_T = R_1 \| R_2 + R_3 = 6 \| 3 + 2 = 4\ \Omega$.

Next we compute the Norton current. Following the Norton current methodology box, we first replace the load with a short circuit, and label the short-circuit current, $i_{SC}$. The circuit is shown in Figure 3.61 ready for node voltage analysis. Note that we have identified two node voltages, $v_1$ and $v_2$, and that the voltage source requires that $v_2 - v_1 = V$. The unknown current flowing through the voltage source is labeled $i$. Applying KCL at nodes 1 and 2, we obtain the following set of equations:

$$I - \frac{v_1}{R_1} - i = 0 \qquad \text{node 1}$$

$$i - \frac{v_2}{R_2} - \frac{v_2}{R_3} = 0 \qquad \text{node 2}$$

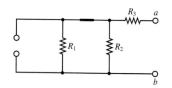

**Figure 3.60**

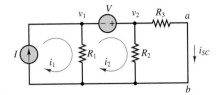

**Figure 3.61**

To eliminate one of the three unknowns, we substitute $v_2 - V = v_1$ in the first equation:

$$I - \frac{v_2 - V}{R_1} - i = 0 \qquad \text{node 1}$$

and we rewrite the equations, recognizing that the unknowns are $i$ and $v_2$. Note that the short-circuit current is $i_{SC} = v_2/R_3$.

$$(1)\, i + \left(\frac{1}{R_1}\right) v_2 = I + \left(\frac{1}{R_1}\right) V$$

$$(-1)i + \left(\frac{1}{R_2} + \frac{1}{R_3}\right) v_2 = 0$$

Substituting numerical values we obtain

$$\begin{bmatrix} 1 & 0.1667 \\ -1 & 0.8333 \end{bmatrix} \begin{bmatrix} i \\ v_2 \end{bmatrix} = \begin{bmatrix} 3 \\ 0 \end{bmatrix}$$

and can numerically solve for the two unknowns to find that $i = 2.5$ A and $v_2 = 3$ V. Finally, the Norton or short-circuit current is $i_N = i_{SC} = v_2/R_3 = 1.5$ A.

**Comments:** In this example it was not obvious whether nodal analysis, mesh analysis, or superposition might be the quickest method to arrive at the answer. It would be a very good exercise to try the other two methods and compare the complexity of the three solutions. The complete Norton equivalent circuit is shown in Figure 3.62.

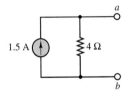

**Figure 3.62** Norton equivalent circuit

## Source Transformations

This section illustrates **source transformations,** a procedure that may be very useful in the computation of equivalent circuits, permitting, in some circumstances, replacement of current sources with voltage sources, and vice versa. The Norton and Thévenin theorems state that any one-port network can be represented by a voltage source in series with a resistance, or by a current source in parallel with a resistance, and that either of these representations is equivalent to the original circuit, as illustrated in Figure 3.63.

An extension of this result is that any circuit in Thévenin equivalent form may be replaced by a circuit in Norton equivalent form, provided that we use the following relationship:

$$v_T = R_T i_N \tag{3.32}$$

Thus, the subcircuit to the left of the dashed line in Figure 3.64 may be replaced by its Norton equivalent, as shown in the figure. Then, the computation of $i_{SC}$

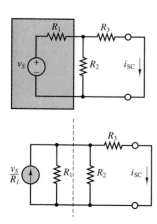

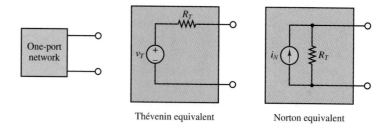

Figure 3.63 Equivalence of Thévenin and Norton representations

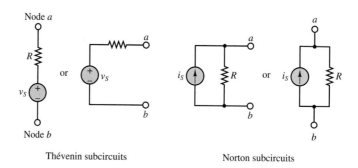

**Figure 3.64** Effect of source
transformation

becomes very straightforward, since the three resistors are in parallel with the
current source and therefore a simple current divider may be used to compute the
short-circuit current. Observe that the short-circuit current is the current flowing
through $R_3$; therefore,

$$i_{SC} = i_N = \frac{1/R_3}{1/R_1 + 1/R_2 + 1/R_3}\frac{v_S}{R_1} = \frac{v_S R_2}{R_1 R_3 + R_2 R_3 + R_1 R_2} \tag{3.33}$$

which is the identical result obtained for the same circuit in the preceding section,
as you may easily verify. This source transformation method can be very useful,
if employed correctly. Figure 3.65 shows how one can recognize subcircuits
amenable to such source transformations. Example 3.16 is a numerical example
illustrating the procedure.

Figure 3.65 Subcircuits amenable to source transformation

---

## EXAMPLE  3.16  Source Transformations

### Problem

Compute the Norton equivalent of the circuit of Fig. 3.66 using source transformations.

---

### Solution

**Known Quantities:**  Source voltages and current, resistor values.

**Find:**  Equivalent resistance, $R_T$; Norton current, $i_N = i_{SC}$.

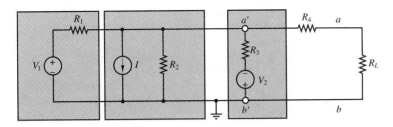

**Figure 3.66**

***Schematics, Diagrams, Circuits, and Given Data:*** $V_1 = 50$ V; $I = 0.5$ A; $V_2 = 5$ V; $R_1 = 100\ \Omega$; $R_2 = 100\ \Omega$; $R_3 = 200\ \Omega$; $R_4 = 160\ \Omega$.

***Assumptions:*** Assume reference node is at the bottom of the circuit.

***Analysis:*** First, we sketch the circuit again, to take advantage of the source transformation technique; we emphasize the location of the nodes for this purpose, as shown in Figure 3.67. Nodes $a'$ and $b'$ have been purposely separated from nodes $a''$ and $b''$ even though these are the same pairs of nodes. We can now replace the branch consisting of $V_1$ and $R_1$, which appears between nodes $a''$ and $b''$, with an equivalent Norton circuit with Norton current source $V_1/R_1$ and equivalent resistance $R_1$. Similarly, the series branch between nodes $a'$ and $b'$ is replaced by an equivalent Norton circuit with Norton current source $V_2/R_3$ and equivalent resistance $R_3$. The result of these manipulations is shown in Figure 3.68. The same circuit is now depicted in Figure 3.69 with numerical values substituted for each component. Note how easy it is to visualize the equivalent resistance: if each current source is replaced by an open circuit, we find:

$$R_T = R_1||R_2||R_3|| + R_4 = 200||100||100 + 160 = 200\ \Omega$$

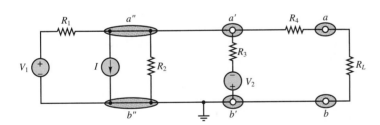

**Figure 3.67**

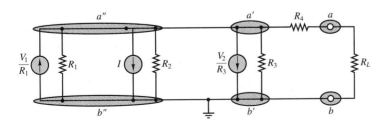

**Figure 3.68**

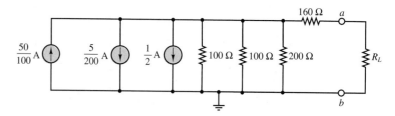

**Figure 3.69**

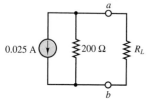

**Figure 3.70**

The calculation of the Norton current is similarly straightforward, since it simply involves summing the currents:

$$i_N = 0.5 - 0.025 - 0.5 = -0.025 \text{ A}$$

Figure 3.70 depicts the complete Norton equivalent circuit connected to the load.

**Comments:** It is not always possible to reduce a circuit as easily as was shown in this example by means of source transformations. However, it may be advantageous to use source transformation as a means of converting parts of a circuit to a different form, perhaps more naturally suited to a particular solution method (e.g., nodal analysis).

## Experimental Determination of Thévenin and Norton Equivalents

The idea of equivalent circuits as a means of representing complex and sometimes unknown networks is useful not only analytically, but in practical engineering applications as well. It is very useful to have a measure, for example, of the equivalent internal resistance of an instrument, so as to have an idea of its power requirements and limitations. Fortunately, Thévenin and Norton equivalent circuits can also be evaluated experimentally by means of very simple techniques. The basic idea is that the Thévenin voltage is an open-circuit voltage and the Norton current is a short-circuit current. It should therefore be possible to conduct appropriate measurements to determine these quantities. Once $v_T$ and $i_N$ are known, we can determine the Thévenin resistance of the circuit being analyzed according to the relationship

$$R_T = \frac{v_T}{i_N} \tag{3.34}$$

How are $v_T$ and $i_N$ measured, then?

Figure 3.71 illustrates the measurement of the open-circuit voltage and short-circuit current for an arbitrary network connected to any load and also illustrates that the procedure requires some special attention, because of the nonideal nature of any practical measuring instrument. The figure clearly illustrates that in the presence of finite meter resistance, $r_m$, one must take this quantity into account in the computation of the short-circuit current and open-circuit voltage; $v_{OC}$ and $i_{SC}$ appear between quotation marks in the figure specifically to illustrate that the measured "open-circuit voltage" and "short-circuit current" are in fact affected by the internal resistance of the measuring instrument and are not the true quantities.

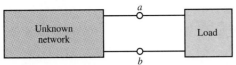

An unknown network connected to a load

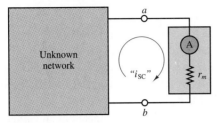

Network connected for measurement of short-circuit current

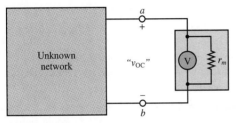

Network connected for measurement of open-circuit voltage

**Figure 3.71** Measurement of open-circuit voltage
and short-circuit current

You should verify that the following expressions for the true short-circuit
current and open-circuit voltage apply (see the material on nonideal measuring
instruments in Section 2.8):

$$i_N = \text{``}i_{\text{SC}}\text{''} \left( 1 + \frac{r_m}{R_T} \right)$$

$$v_T = \text{``}v_{\text{OC}}\text{''} \left( 1 + \frac{R_T}{r_m} \right)$$

(3.35)

where $i_N$ is the ideal Norton current, $v_T$ the Thévenin voltage, and $R_T$ the true
Thévenin resistance. If you recall the earlier discussion of the properties of ideal
ammeters and voltmeters, you will recall that for an ideal ammeter, $r_m$ should
approach zero, while in an ideal voltmeter, the internal resistance should approach
an open circuit (infinity); thus, the two expressions just given permit the determination of the true Thévenin and Norton equivalent sources from an (imperfect)
measurement of the open-circuit voltage and short-circuit current, provided that
the internal meter resistance, $r_m$, is known. Note also that, in practice, the internal resistance of voltmeters is sufficiently high to be considered infinite relative
to the equivalent resistance of most practical circuits; on the other hand, it is
impossible to construct an ammeter that has zero internal resistance. If the internal ammeter resistance is known, however, a reasonably accurate measurement
of short-circuit current may be obtained. The following example illustrates the
point.

**FOCUS ON
MEASUREMENTS**

### Experimental Determination of Thévenin Equivalent Circuit

**Problem:**

Determine the Thévenin equivalent of an unknown circuit from measurements of open-circuit voltage and short-circuit current.

**Solution:**

**Known Quantities—** Measurement of short-circuit current and open-circuit voltage. Internal resistance of measuring instrument.

**Find—** Equivalent resistance, $R_T$; Thévenin voltage, $v_T = v_{OC}$.

**Schematics, Diagrams, Circuits, and Given Data—** Measured $v_{OC} = 6.5$ V; Measured $i_{SC} = 3.75$ mA; $r_m = 15\ \Omega$.

**Assumptions—** The unknown circuit is a linear circuit containing ideal sources and resistors only.

**Analysis—** The unknown circuit, shown on the top left in Figure 3.72, is replaced by its Thévenin equivalent, and is connected to an ammeter for a measurement of the short-circuit current (Figure 3.72, top right), and then to a voltmeter for the measurement of the open-circuit voltage (Figure 3.72, bottom). The open-circuit voltage measurement yields the Thévenin voltage:

$$v_{OC} = v_T = 6.5 \text{ V}$$

To determine the equivalent resistance, we observe in the figure depicting the voltage measurement that, according to the circuit diagram,

$$\frac{v_{OC}}{i_{SC}} = R_T + r_m$$

Thus,

$$R_T = \frac{v_{OC}}{i_{SC}} - r_m = 1{,}733 - 15 = 1{,}718\ \Omega$$

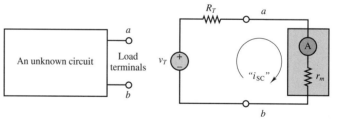

Network connected for measurement of
short-circuit current (practical ammeter)

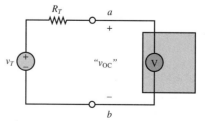

Network connected for measurement of
open-circuit voltage (ideal voltmeter)

**Figure 3.72**

One last comment is in order concerning the practical measurement of the internal resistance of a network. In most cases, it is not advisable to actually short-circuit a network by inserting a series ammeter as shown in Figure 3.71; permanent damage to the circuit or to the ammeter may be a consequence. For example, imagine that you wanted to estimate the internal resistance of an automotive battery; connecting a laboratory ammeter between the battery terminals would surely result in immediate loss of the instrument. Most ammeters are not designed to withstand currents of such magnitude. Thus, the experimenter should pay attention to the capabilities of the ammeters and voltmeters used in measurements of this type, as well as to the (approximate) power ratings of any sources present. However, there are established techniques especially designed to measure large currents.

## 3.6    MAXIMUM POWER TRANSFER

The reduction of any linear resistive circuit to its Thévenin or Norton equivalent form is a very convenient conceptualization, as far as the computation of load-related quantities is concerned. One such computation is that of the power absorbed by the load. The Thévenin and Norton models imply that some of the power generated by the source will necessarily be dissipated by the internal circuits within the source. Given this unavoidable power loss, a logical question to ask is, how much power can be transferred to the load from the source under the most ideal conditions? Or, alternatively, what is the value of the load resistance that will absorb maximum power from the source? The answer to these questions is contained in the **maximum power transfer theorem,** which is the subject of the present section.

The model employed in the discussion of power transfer is illustrated in Figure 3.73, where a practical source is represented by means of its Thévenin equivalent circuit. The maximum power transfer problem is easily formulated if we consider that the power absorbed by the load, $P_L$, is given by the expression

$$P_L = i_L^2 R_L \qquad (3.36)$$

and that the load current is given by the familiar expression

$$i_L = \frac{v_T}{R_L + R_T} \qquad (3.37)$$

Combining the two expressions, we can compute the load power as

$$P_L = \frac{v_T^2}{(R_L + R_T)^2} R_L \qquad (3.38)$$

To find the value of $R_L$ that maximizes the expression for $P_L$ (assuming that $V_T$ and $R_T$ are fixed), the simple maximization problem

$$\frac{dP_L}{dR_L} = 0 \qquad (3.39)$$

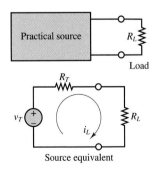

Given $v_T$ and $R_T$, what value of $R_L$ will allow for maximum power transfer?

**Figure 3.73** Power transfer between source and load

must be solved. Computing the derivative, we obtain the following expression:

$$\frac{dP_L}{dR_L} = \frac{v_T^2(R_L + R_T)^2 - 2v_T^2 R_L(R_L + R_T)}{(R_L + R_T)^4} \tag{3.40}$$

which leads to the expression

$$(R_L + R_T)^2 - 2R_L(R_L + R_T) = 0 \tag{3.41}$$

It is easy to verify that the solution of this equation is:

$$R_L = R_T \tag{3.42}$$

Thus, in order to transfer maximum power to a load, the equivalent source and load resistances must be **matched,** that is, equal to each other.

This analysis shows that in order to transfer maximum power to a load, given a fixed equivalent source resistance, the load resistance must match the equivalent source resistance. What if we reversed the problem statement and required that the load resistance be fixed? What would then be the value of source resistance that maximizes the power transfer in this case? The answer to this question can be easily obtained by solving Check Your Understanding Exercise 3.23.

A problem related to power transfer is that of **source loading.** This phenomenon, which is illustrated in Figure 3.74, may be explained as follows: when a practical voltage source is connected to a load, the current that flows from the source to the load will cause a voltage drop across the internal source resistance, $v_{int}$; as a consequence, the voltage actually seen by the load will be somewhat lower than the *open-circuit voltage* of the source. As stated earlier, the open-circuit voltage is equal to the Thévenin voltage. The extent of the internal voltage drop within the source depends on the amount of current drawn by the load. With reference to Figure 3.75, this internal drop is equal to $i R_T$, and therefore the load voltage will be:

$$v_L = v_T - i R_T \tag{3.43}$$

It should be apparent that it is desirable to have as small an internal resistance as possible in a practical voltage source.

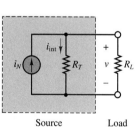

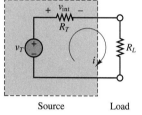

**Figure 3.74** Source loading effects

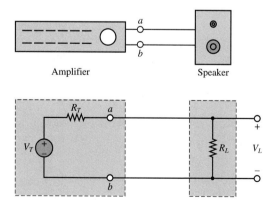

**Figure 3.75** A simplified model of an audio system

In the case of a current source, the internal resistance will draw some current away from the load because of the presence of the internal source resistance; this current is denoted by $i_{\text{int}}$ in Figure 3.74. Thus the load will receive only part of the *short-circuit current* available from the source (the Norton current):

$$i_L = i_N - \frac{v}{R_T} \tag{3.44}$$

It is therefore desirable to have a very large internal resistance in a practical current source. You may wish to refer back to the discussion of practical sources to verify that the earlier interpretation of practical sources can be expanded in light of the more recent discussion of equivalent circuits.

## EXAMPLE 3.17 Maximum Power Transfer

### Problem

Use the maximum power transfer theorem to determine the increase in power delivered to a loudspeaker resulting from matching the speaker load resistance to the amplifier equivalent source resistance.

### Solution

**Known Quantities:** Source equivalent resistance, $R_T$; unmatched speaker load resistance, $R_{\text{LU}}$; matched loudspeaker load resistance, $R_{\text{LM}}$.

**Find:** Difference between power delivered to loudspeaker with unmatched and matched loads, and corresponding percent increase.

**Schematics, Diagrams, Circuits, and Given Data:** $R_T = 8\ \Omega$; $R_{\text{LU}} = 16\ \Omega$; $R_{\text{LM}} = 8\ \Omega$.

**Assumptions:** The amplifier can be modeled as a linear resistive circuit, for the purposes of this analysis.

**Analysis:** Imagine that we have unknowingly connected an 8-$\Omega$ amplifier to a 16-$\Omega$ speaker. We can compute the power delivered to the speaker as follows. The load voltage is found by using the voltage divider rule:

$$v_{\text{LU}} = \frac{R_{\text{LU}}}{R_{\text{LU}} + R_T} v_T = \frac{2}{3} v_T$$

and the load power is then computed to be:

$$P_{\text{LU}} = \frac{v_L^2}{R_{\text{LU}}} = \frac{4}{9} \frac{v_T^2}{R_{\text{LU}}} = 0.0278 v_T^2$$

Let us now repeat the calculation for the case of a matched 8-$\Omega$, speaker resistance, $R_{\text{LM}}$. Let the new load voltage be $v_{\text{LM}}$ and the corresponding load power be $P_{\text{LM}}$. Then,

$$v_{\text{LM}} = \frac{1}{2} v_T$$

and

$$P_{\text{LM}} = \frac{v_{\text{LM}}^2}{R_{\text{LM}}} = \frac{1}{4} \frac{v_T^2}{R_{\text{LM}}} = 0.03125 v_T^2$$

The increase in load power is therefore

$$\Delta P = \frac{0.03125 - 0.0278}{0.0278} \times 100 = 12.5\%$$

**Comments:** In practice, an audio amplifier and a speaker are not well represented by the simple resistive Thévenin equivalent models used in the present example. Circuits that are appropriate to model amplifiers and loudspeakers are presented in later chapters. The audiophile can find further information concerning hi-fi circuits in Chapters 7 and 16.

**Focus on Computer-Aided Tools:** A very nice illustration of the **maximum power transfer theorem** based on MathCad may be found in the Web references.

---

## Check Your Understanding

**3.21**  A practical voltage source has an internal resistance of 1.2 Ω and generates a 30-V output under open-circuit conditions. What is the smallest load resistance we can connect to the source if we do not wish the load voltage to drop by more than 2 percent with respect to the source open-circuit voltage?

**3.22**  A practical current source has an internal resistance of 12 kΩ and generates a 200-mA output under short-circuit conditions. What percent drop in load current will be experienced (with respect to the short-circuit condition) if a 200-Ω load is connected to the current source?

**3.23**  Repeat the derivation leading to equation 3.42 for the case where the load resistance is fixed and the source resistance is variable. That is, differentiate the expression for the load power, $P_L$, with respect to $R_S$ instead of $R_L$. What is the value of $R_S$ that results in maximum power transfer to the load?

---

## 3.7    NONLINEAR CIRCUIT ELEMENTS

Until now the focus of this chapter has been on linear circuits, containing ideal voltage and current sources, and linear resistors. In effect, one reason for the simplicity of some of the techniques illustrated in the earlier part of this chapter is the ability to utilize Ohm's law as a simple, linear description of the *i-v* characteristic of an ideal resistor. In many practical instances, however, the engineer is faced with elements exhibiting a nonlinear *i-v* characteristic. This section explores two methods for analyzing nonlinear circuit elements.

### Description of Nonlinear Elements

There are a number of useful cases in which a simple functional relationship exists between voltage and current in a nonlinear circuit element. For example, Figure 3.76 depicts an element with an exponential *i-v* characteristic, described by the following equations:

$$i = I_0 e^{\alpha v} \qquad v > 0$$
$$i = -I_0 \qquad v \le 0$$

(3.45)

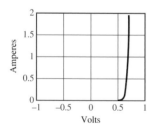

**Figure 3.76** *i-v* characteristic of exponential resistor

There exists, in fact, a circuit element (the semiconductor diode) that very nearly satisfies this simple relationship. The difficulty in the *i-v* relationship of equation 3.45 is that it is not possible, in general, to obtain a closed-form analytical solution, even for a very simple circuit.

With the knowledge of equivalent circuits you have just acquired, one approach to analyzing a circuit containing a nonlinear element might be to treat the nonlinear element as a load, and to compute the Thévenin equivalent of the remaining circuit, as shown in Figure 3.77. Applying KVL, the following equation may then be obtained:

$$v_T = R_T i_x + v_x \tag{3.46}$$

To obtain the second equation needed to solve for both the unknown voltage, $v_x$, and the unknown current, $i_x$, it is necessary to resort to the $i$-$v$ description of the nonlinear element, namely, equation 3.45. If, for the moment, only positive voltages are considered, the circuit is completely described by the following system:

$$i_x = I_0 e^{\alpha v_x} \qquad v_x > 0$$
$$v_T = R_T i_x + v_x \tag{3.47}$$

The two parts of equation 3.47 represent a system of two equations in two unknowns; however, one of these equations is nonlinear. If we solve for the load voltage and current, for example, by substituting the expression for $i_x$ in the linear equation, we obtain the following expression:

$$v_T = R_T I_0 e^{\alpha v_x} + v_x \tag{3.48}$$

or

$$v_x = v_T - R_T I_0 e^{\alpha v_x} \tag{3.49}$$

Equations 3.48 and 3.49 do not have a closed-form solution; that is, they are *transcendental equations*. How can $v_x$ be found? One possibility is to generate a solution numerically, by guessing an initial value (e.g., $v_x = 0$) and iterating until a sufficiently precise solution is found. This solution is explored further in the homework problems. Another method is based on a graphical analysis of the circuit and is described in the following section.

## Graphical (Load-Line) Analysis of Nonlinear Circuits

The nonlinear system of equations of the previous section may be analyzed in a different light, by considering the graphical representation of equation 3.46, which may also be written as follows:

$$i_x = -\frac{1}{R_T} v_x + \frac{v_T}{R_T} \tag{3.50}$$

We notice first that equation 3.50 describes the behavior of any load, linear or nonlinear, since we have made no assumptions regarding the nature of the load voltage and current. Second, it is the equation of a line in the $i_x$-$v_x$ plane, with slope $-1/R_T$ and $i_x$ intercept $V_T/R_T$. This equation is referred to as the **load-line equation;** its graphical interpretation is very useful and is shown in Figure 3.78.

The load-line equation is but one of two $i$-$v$ characteristics we have available, the other being the nonlinear-device characteristic of equation 3.45. The intersection of the two curves yields the solution of our nonlinear system of equations. This result is depicted in Figure 3.79.

Finally, another important point should be emphasized: the linear network reduction methods introduced in the preceding sections can always be employed to

Nonlinear element as a load. We wish to solve for $v_x$ and $i_x$.

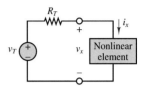

**Figure 3.77** Representation of nonlinear element in a linear circuit

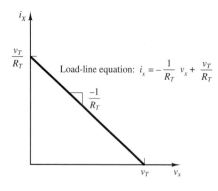

Figure 3.78 Load line

Figure 3.79 Graphical solution equations 3.48 and 3.49

reduce any circuit containing a single nonlinear element to the Thévenin equivalent form, as illustrated in Figure 3.80. The key is to identify the nonlinear element and to treat it as a load. Thus, the equivalent-circuit solution methods developed earlier can be very useful in simplifying problems in which a nonlinear load is present. Example 3.19 illustrates this point.

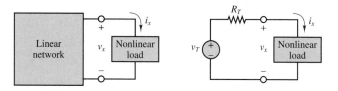

Figure 3.80 Transformation of nonlinear circuit of Thévenin equivalent

### EXAMPLE 3.18 Nonlinear Load Power Dissipation

**Problem**

A linear generator is connected to a nonlinear load in the configuration of Figure 3.80. Determine the power dissipated by the load.

**Solution**

**Known Quantities:** Generator Thévenin equivalent circuit; load $i$-$v$ characteristic and load line.

**Find:** Power dissipated by load, $P_x$.

**Schematics, Diagrams, Circuits, and Given Data:** $R_T = 30\ \Omega$; $v_T = 15$ V.

**Assumptions:** None.

**Analysis:** We can model the circuit as shown in Figure 3.80. The objective is to determine the voltage $v_x$ and the current $i_x$ using graphical methods. The load-line equation for the circuit is given by the expression

$$i_x = -\frac{1}{R_T}v_x + \frac{v_T}{R_T}$$

or

$$i_x = -\frac{1}{30}v_x + \frac{15}{30}$$

This equation represents a line in the $i_x$-$v_x$ plane, with $i_x$ intercept at 0.5 A and $v_x$ intercept at 15 V. In order to determine the operating point of the circuit, we superimpose the load line on the device $i$-$v$ characteristic, as shown in Figure 3.81, and determine the solution by finding the intersection of the load line with the device curve. Inspection of the graph reveals that the intersection point is given approximately by

$$i_x = 0.14 \text{ A} \qquad v_x = 11 \text{ V}$$

and therefore the power dissipated by the nonlinear load is

$$P_x = 0.14 \times 11 = 1.54 \text{ W}$$

It is important to observe that the result obtained in this example is, in essence, a description of experimental procedures, indicating that the analytical concepts developed in this chapter also apply to practical measurements.

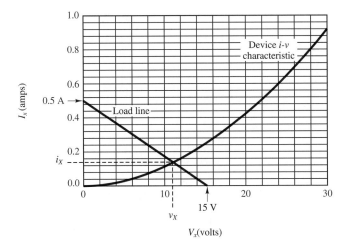

**Figure 3.81**

# CONCLUSION

The objective of this chapter was to provide a practical introduction to the analysis of linear resistive circuits. The emphasis on examples is important at this stage, since we believe that familiarity with the basic circuit analysis techniques will greatly ease the task of learning more advanced ideas in circuits and electronics. In particular, your goal at this point should be to have mastered four analysis methods, summarized as follows:

1. *Node voltage and mesh current analysis.* These methods are analogous in concept; the choice of a preferred method depends on the specific circuit. They are generally applicable to the circuits we will analyze in this book and are amenable to solution by matrix methods.

2. *The principle of superposition.* This is primarily a conceptual aid that may simplify the solution of circuits containing multiple sources. It is usually not an efficient method.

3. *Thévenin and Norton equivalents.* The notion of equivalent circuits is at the heart of circuit analysis. Complete mastery of the reduction of linear resistive circuits to either equivalent form is a must.

4. *Numerical and graphical analysis.* These methods apply in the case of nonlinear circuit elements. The load-line analysis method is intuitively appealing and will be employed again in this book to analyze electronic devices.

The material covered in this chapter will be essential to the development of more advanced techniques throughout the remainder of the book.

## CHECK YOUR UNDERSTANDING ANSWERS

| | | | |
|---|---|---|---|
| CYU 3.1 | 0.2857 A | CYU 3.16 | $R_T = 4.0 \text{ k}\Omega$ |
| CYU 3.2 | $-18$ V | CYU 3.17 | $R_T = 7.06 \ \Omega$ |
| CYU 3.4 | 5 V | CYU 3.18 | $i_L = 0.02857$ A |
| CYU 3.5 | 2 A | CYU 3.19 | $R_T = 30 \ \Omega; v_{OC} = v_T = 5\text{V}$ |
| CYU 3.7 | 12 V | CYU 3.20 | $R_T = 10 \ \Omega; v_{OC} = v_T = 0.704$ V |
| CYU 3.8 | 1.39 A | CYU 3.21 | 58.8 $\Omega$ |
| CYU 3.14 | $R_T = 2.5 \text{ k}\Omega$ | CYU 3.22 | 1.64% |
| CYU 3.15 | $R_T = 7 \ \Omega$ | CYU 3.23 | $R_S = 0$ for maximum power transfer to the load |

## HOMEWORK PROBLEMS

### Section 1: Node/Mesh Analysis

**3.1**  In the circuit shown in Figure P3.1, the mesh currents are:

$$I_1 = 5 \text{ A} \quad I_2 = 3 \text{ A} \quad I_3 = 7 \text{ A}$$

Determine the branch currents through:
a. $R_1$. b. $R_2$. c. $R_3$.

**3.2**  In the circuit shown in Figure P3.2, the source and node voltages are:

$$V_{S1} = V_{S2} = 110 \text{ V}$$
$$V_A = 103 \text{ V} \qquad V_B = -107 \text{ V}$$

Determine the voltage across each of the five resistors.

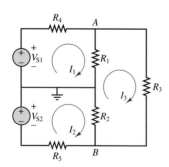

**Figure P3.1**

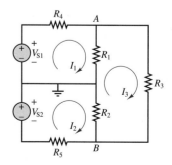

**Figure P3.2**

**3.3** Using node voltage analysis in the circuit of Figure P3.3, find the currents $i_1$ and $i_2$.

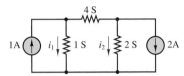

**Figure P3.3**

**3.4** Using node voltage analysis in the circuit of Figure P3.4, find the voltage, $v$, across the 4-siemens conductance.

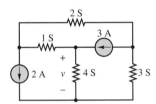

**Figure P3.4**

**3.5** Using node voltage analysis in the circuit of Figure P3.5, find the current, $i$, through the voltage source.

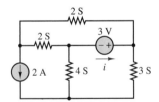

**Figure P3.5**

**3.6** Using node voltage analysis in the circuit of Figure P3.6, find the three indicated node voltages.

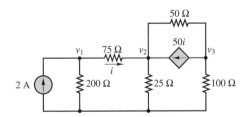

**Figure P3.6**

**3.7** Using node voltage analysis in the circuit of Figure P3.7, find the current, $i$, drawn from the independent voltage source.

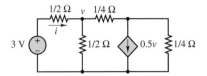

**Figure P3.7**

**3.8** The circuit shown in Figure P3.8 is a Wheatstone bridge circuit. Use node voltage analysis to determine $V_a$ and $V_b$, and thus determine $V_a - V_b$.

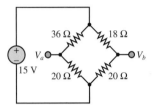

**Figure P3.8**

**3.9** In the circuit in Figure P3.9, assume the source voltage and source current and all resistances are known.

a. Write the node equations required to determine the node voltages.

b. Write the matrix solution for each node voltage in terms of the known parameters.

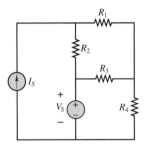

**Figure P3.9**

**3.10** For the circuit of Figure P3.10 determine:

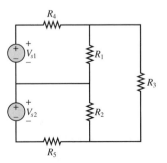

**Figure P3.10**

a. The most efficient way to solve for the voltage across $R_3$. Prove your case.

b. The voltage across $R_3$.

$$V_{S1} = V_{S2} = 110 \text{ V}$$
$$R_1 = 500 \text{ m}\Omega \qquad R_2 = 167 \text{ m}\Omega$$
$$R_3 = 700 \text{ m}\Omega$$
$$R_4 = 200 \text{ m}\Omega \qquad R_5 = 333 \text{ m}\Omega$$

**3.11**  In the circuit shown in Figure P3.11, $V_{S2}$ and $R_s$ model a temperature sensor, i.e.,

$$V_{S2} = kT \qquad k = 10 \text{ V/}^\circ\text{C}$$
$$V_{S1} = 24 \text{ V} \qquad R_s = R_1 = 12 \text{ k}\Omega$$
$$R_2 = 3 \text{ k}\Omega \qquad R_3 = 10 \text{ k}\Omega$$
$$R_4 = 24 \text{ k}\Omega \qquad V_{R3} = -2.524 \text{ V}$$

The voltage across $R_3$, which is given, indicates the temperature. Determine the temperature.

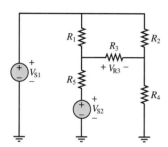

**Figure P3.11**

**3.12**  Using KCL, perform a node analysis on the circuit shown in Figure P3.12 and determine the voltage across $R_4$. Note that one source is a controlled voltage source!

$$V_S = 5 \text{ V} \qquad A_V = 70 \qquad R_1 = 2.2 \text{ k}\Omega$$
$$R_2 = 1.8 \text{ k}\Omega \qquad R_3 = 6.8 \text{ k}\Omega \qquad R_4 = 220 \text{ }\Omega$$

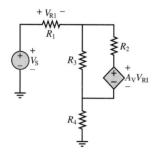

**Figure P3.12**

**3.13**  Using mesh current analysis, find the voltage, $v$, across the 3-$\Omega$ resistor in the circuit of Figure P3.13.

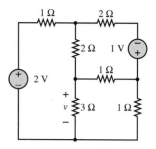

**Figure P3.13**

**3.14**  Using mesh current analysis, find the current, $i$, through the 2-$\Omega$ resistor on the right in the circuit of Figure P3.14.

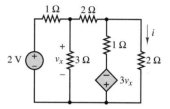

**Figure P3.14**

**3.15**  The circuit shown in Figure P3.10 is a simplified DC model of a 3-wire distribution service to residential and commercial buildings. The two ideal sources, $R_4$ and $R_5$, are the Thévenin equivalent circuit of the distribution system. $R_1$ and $R_2$ represent 110-V lighting and utility loads of about 800 W and 300 W respectively. $R_3$ represents a 220-V heating load of about 3 kW. The numbers above are not actual values *rated* (or nominal) values, that is, the typical values for which the circuit has been designed. Determine the actual voltages across the three loads.

$$V_{S1} = V_{S2} = 110 \text{ V} \qquad R_4 = R_5 = 1.3 \text{ }\Omega$$
$$R_1 = 15 \text{ }\Omega \qquad R_2 = 40 \text{ }\Omega \qquad R_3 = 16 \text{ }\Omega$$

**3.16**  Using mesh current analysis, find the voltage, $v$, across the current source in the circuit of Figure P3.16.

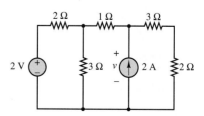

**Figure P3.16**

**3.17**  Using mesh current analysis, find the current, $i$, through the voltage source in the circuit of Figure P3.5.

**3.18**  Using mesh current analysis, find the current, $i$, in the circuit of Figure P3.6.

**3.19**  Using mesh current analysis, find the equivalent resistance, $R = v/i$, seen by the source of the circuit in Figure P3.19.

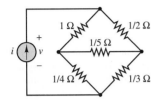

**Figure P3.19**

**3.20**  Using mesh current analysis, find the voltage gain, $A_v = v_2/v_1$, in the circuit of Figure P3.20.

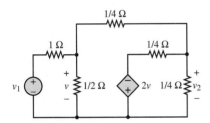

**Figure P3.20**

**3.21**  In the circuit shown in Figure P3.21:

$$V_{S1} = V_{S2} = 450 \text{ V}$$
$$R_4 = R_5 = 0.25 \text{ } \Omega$$
$$R_1 = 8 \text{ } \Omega \qquad R_2 = 5 \text{ } \Omega$$
$$R_3 = 32 \text{ } \Omega$$

Determine, using KCL and a node analysis, the voltage across $R_1$, $R_2$, and $R_3$.

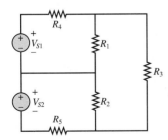

**Figure P3.21**

**3.22**  $F_1$ and $F_2$ in the circuit shown in Figure P3.22 are fuses. Under normal conditions they are modeled as a

short circuit. However, if excess current flows through a fuse, its element melts and the fuse "blows," i.e., it becomes an open circuit.

$$V_{S1} = V_{S2} = 115 \text{ V}$$
$$R_1 = R_2 = 5 \text{ } \Omega \qquad R_3 = 10 \text{ } \Omega$$
$$R_4 = R_5 = 200 \text{ m } \Omega$$

Normally, the voltages across $R_1$, $R_2$, and $R_3$ are 106.5 V, $-106.5$ V, and 213.0 V. If $F_1$ now blows, or opens, determine, using KCL and a node analysis, the new voltages across $R_1$, $R_2$, and $R_3$.

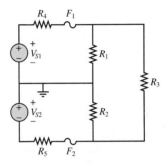

**Figure P3.22**

**3.23**  $F_1$ and $F_2$ in the circuit shown in Figure P3.22 are fuses. Under normal conditions they are modeled as a short circuit. However, if excess current flows through a fuse, it "blows" and the fuse becomes an open circuit.

$$V_{S1} = V_{S2} = 120 \text{ V}$$
$$R_1 = R_2 = 2 \text{ } \Omega \qquad R_3 = 8 \text{ } \Omega$$
$$R_4 = R_5 = 250 \text{ m } \Omega$$

If $F_1$ blows, or opens, determine, using KCL and a node analysis, the voltages across $R_1$, $R_2$, $R_3$, and $F_1$.

**3.24**  The circuit shown in Figure P3.24 is a simplified DC version of an AC three-phase Y-Y electrical distribution system commonly used to supply industrial loads, particularly rotating machines.

$$V_{S1} = V_{S2} = V_{S3} = 170 \text{ V}$$
$$R_{W1} = R_{W2} = R_{W3} = 0.7 \text{ } \Omega$$
$$R_1 = 1.9 \text{ } \Omega \qquad R_2 = 2.3 \text{ } \Omega$$
$$R_3 = 11 \text{ } \Omega$$

Determine:

a. The number of unknown node voltages and mesh currents.

b. Node voltages.

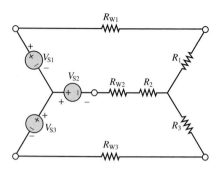

**Figure P3.24**

**3.25**  The circuit shown in Figure P3.24 is a simpled DC version of an AC three-phase Y-Y electrical distribution system commonly used to supply industrial loads, particularly rotating machines.

$$V_{S1} = V_{S2} = V_{S3} = 170 \text{ V}$$
$$R_{W1} = R_{W2} = R_{W3} = 0.7 \text{ Ω}$$
$$R_1 = 1.9 \text{ Ω} \qquad R_2 = 2.3 \text{ Ω}$$
$$R_3 = 11 \text{ Ω}$$

A node analysis with KCL and a ground at the terminal common to the three sources gives the only unknown node voltage $V_N = 28.94$ V. If the node voltages in a circuit are known, all other voltages and currents in the circuit can be determined. Determine the current through and voltage across $R_1$.

**3.26**  The circuit shown in Figure P3.24 is a simplified DC version of a typical 3-wire, 3-phase AC Y-Y distribution system. Write the mesh (or loop) equations and any additional equations required to determine the current through $R_1$ in the circuit shown.

**3.27**  Determine the branch currents using KVL and loop analysis in the circuit of Figure P3.24.

$$V_{S2} = V_{S3} = 110 \text{ V} \qquad V_{S1} = 90 \text{ V}$$
$$R_1 = 7.9 \text{ Ω} \qquad R_2 = R_3 = 3.7 \text{ Ω}$$
$$R_{W1} = R_{W2} = R_{W3} = 1.3 \text{ Ω}$$

**3.28**  $F_1$ and $F_2$ in the circuit shown in Figure P3.22 are fuses. Under normal conditions they are modeled as a short circuit. However, if excess current flows through a fuse, its element melts and the fuse "blows"; i.e., it becomes an open circuit.

$$V_{S1} = V_{S2} = 115 \text{ V}$$
$$R_1 = R_2 = 5 \text{ Ω} \qquad R_3 = 10 \text{ Ω}$$
$$R_4 = R_5 = 200 \text{ mΩ}$$

Determine, using KVL and a mesh analysis, the voltages across $R_1$, $R_2$, and $R_3$ under normal conditions, i.e., no blown fuses.

**3.29**  Using KVL and a mesh analysis only, determine the voltage across $R_1$ in the 2-phase, 3-wire power distribution system shown in Figure P3.22. $R_1$ and $R_2$

represent the 110-V loads. A light bulb rated at 100 W and 110 V has a resistance of about 100 Ω. $R_3$ represents the 220-V loads. A microwave oven rated at 750 W and 220 V has a resistance of about 65 Ω. $R_4$ and $R_5$ represent losses in the distribution system (normally much, much smaller than the values given below). Fuses are normally connected in the path containing these resistances to protect against current overloads.

$$V_{S1} = V_{S2} = 110 \text{ V} \qquad R_4 = R_5 = 13 \text{ Ω}$$
$$R_1 = 100 \text{ Ω} \qquad R_2 = 22 \text{ Ω} \qquad R_3 = 70 \text{ Ω}$$

**3.30**  $F_1$ and $F_2$ in the circuit shown in Figure P3.22 are fuses. Under normal conditions they are modeled as short circuits, in which case the voltages across $R_1$ and $R_2$ are 106.5 V and that across $R_3$ is 213.0 V. However, if excess current flows through a fuse, its element melts and the fuse "blows"; i.e., it becomes an open circuit.

$$V_{S1} = V_{S2} = 115 \text{ V} \qquad R_4 = R_5 = 200 \text{ mΩ}$$
$$R_1 = R_2 = 5 \text{ Ω} \qquad R_3 = 10 \text{ Ω}$$

If $F_1$ "blows" or opens, determine, using KVL and a mesh analysis, the voltages across $R_1$, $R_2$, and $R_3$ and across the open fuse.

**3.31**  $F_1$ and $F_2$ in the circuit shown in Figure P3.22 are fuses. Under normal conditions they are modeled as short circuits. Because of the voltage drops across the distribution losses, modeled here as $R_4$ and $R_5$, the voltages across $R_1$ and $R_2$ (the 110-V loads) are somewhat less than the source voltages and across $R_3$ [the 220-V loads] somewhat less than twice one of the source voltages. If excess current flows through a fuse, its element melts and the fuse "blows"; i.e., it becomes an open circuit.

$$V_{S1} = V_{S2} = 115 \text{ V} \qquad R_4 = R_5 = 1 \text{ Ω}$$
$$R_1 = 4 \text{ Ω} \quad R_2 = 7.5 \text{ Ω} \qquad R_3 = 12.5 \text{ Ω}$$

If $F_1$ blows, or opens, determine, using KVL and a mesh analysis, the voltages across $R_1$, $R_2$, and $R_3$ and across the open fuse.

## Section 2: Equivalent Circuits

**3.32**  Find the Thévenin equivalent circuit as seen by the 3-Ω resistor for the circuit of Figure P3.32.

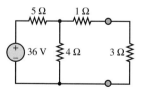

**Figure P3.32**

**3.33**  Find the voltage, $v$, across the 3-Ω resistor in the circuit of Figure P3.33 by replacing the remainder of the circuit with its Thévenin equivalent.

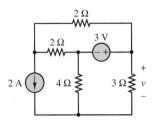

**Figure P3.33**

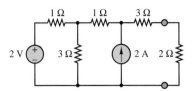

**Figure P3.38**

**3.34**  Find the Thévenin equivalent for the circuit of Figure P3.34.

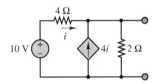

**Figure P3.34**

**3.39**  Find the Norton equivalent to the left of terminals *a* and *b* of the circuit shown in Figure P3.39.

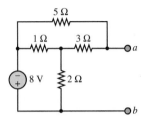

**Figure P3.39**

**3.35**  Find the Thévenin equivalent for the circuit of Figure P3.35.

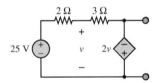

**Figure P3.35**

**3.40**  In the circuit shown in Figure P3.40, $V_S$ models the voltage produced by the generator in a power plant, and $R_s$ models the losses in the generator, distribution wire, and transformers. The three resistances model the various loads connected to the system by a customer. How much does the voltage across the total load change when the customer connects the third load $R_3$ in parallel with the other two loads?

$$V_S = 110 \text{ V} \qquad R_s = 19 \text{ m}\Omega$$
$$R_1 = R_2 = 930 \text{ m}\Omega \qquad R_3 = 100 \text{ m}\Omega$$

**3.36**  Find the Norton equivalent of the circuit of Figure P3.34.

**3.37**  Find the Norton equivalent of the circuit of Figure P3.37.

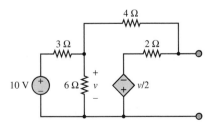

**Figure P3.37**

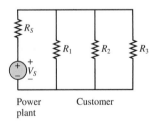

**Figure P3.40**

**3.41**  In the circuit shown in Figure P3.41, $V_S$ models the voltage produced by the generator in a power plant, and $R_s$ models the losses in the generator, distribution wire, and transformers. $R_1$, $R_2$, and $R_3$ model the various loads connected by a customer. How much does the voltage across the total load change when the customer closes switch $S_3$ and connects the third load $R_3$ in parallel with the other two loads?

$$V_S = 450 \text{ V} \qquad R_s = 19 \text{ m}\Omega$$
$$R_1 = R_2 = 1.3 \ \Omega \qquad R_3 = 500 \text{ m}\Omega$$

**3.38**  Find the Norton equivalent of the circuit to the left of the 2-Ω resistor in Figure P3.38.

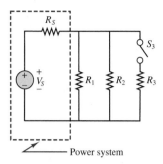

Figure P3.41

**3.42** A nonideal voltage source is modeled in Figure P3.42 as an ideal source in series with a resistance that models the internal losses; i.e., dissipates the same power as the internal losses. In the circuit shown in Figure P3.42, with the load resistor removed so that the current is zero (i.e., no load), the terminal voltage of the source is measured and is 20 V. Then, with $R_L = 2.7$ kΩ, the terminal voltage is again measured and is now 18 V. Determine the internal resistance and the voltage of the ideal source.

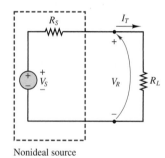

Figure P3.42

**3.43** The circuit of Figure P3.43 is part of the DC biasing network in many transistor amplifier stages. Determining its Thévenin equivalent circuit considerably simplifies analysis of the amplifier. Determine the Thévenin equivalent circuit with respect to the port shown.

$$R_1 = 1.3 \text{ M}\Omega \qquad R_2 = 220 \text{ k}\Omega \qquad V_{CC} = 20 \text{ V}$$

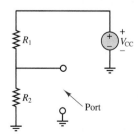

Figure P3.43

**3.44** The circuit of Figure P3.44 shows a battery in parallel with a mechanical generator supplying a load.

$$V_B = 11 \text{ V} \qquad V_G = 12 \text{ V}$$
$$R_B = 0.7 \text{ }\Omega \qquad R_G = 0.3 \text{ }\Omega \qquad R_L = 7 \text{ }\Omega.$$

Determine:
a. The Thévenin equivalent of the circuit to the right of the terminal pair or port $X$-$X'$.
b. The terminal voltage of the battery, i.e., the voltage between $X$ and $X'$.

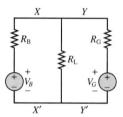

Figure P3.44

**3.45** The circuit of Figure P3.45 shows a battery in parallel with a mechanical generator supplying a load.

$$V_B = 11 \text{ V} \qquad V_G = 12 \text{ V}$$
$$R_B = 0.7 \text{ }\Omega \qquad R_G = 0.3 \text{ }\Omega \qquad R_L = 7.2 \text{ }\Omega$$

Determine:
a. The Thévenin equivalent of the circuit to the left of the terminal pair or port $Y$-$Y'$.
b. The terminal voltage of the battery, i.e., the voltage between $Y$ and $Y'$.

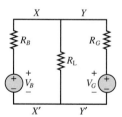

Figure P3.45

**3.46** Find the Norton equivalent resistance of the circuit in Figure P3.46 by applying a voltage source $v_o$ and calculating the resulting current $i_o$.

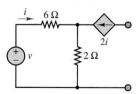

Figure P3.46

**3.47**  The circuit shown in Figure P3.47 is in the form of what is known as a *differential amplifier.* Find an expression for $v_o$ in terms of $v_1$ and $v_2$ using Thévenin's or Norton's theorem.

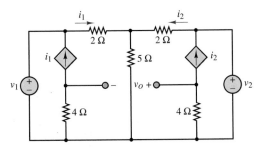

**Figure P3.47**

**3.48**  Refer to the circuit of Figure P3.35. Assume the Thévenin voltage is known to be 2 V, positive at the bottom terminal. Find the new source voltage.

## Section 3: Superposition

**3.49**  With reference to Figure P3.49, determine the current through $R_1$ due only to the source $V_{S2}$.

$$V_{S1} = 110 \text{ V} \qquad V_{S2} = 90 \text{ V}$$
$$R_1 = 560 \text{ }\Omega \qquad R_2 = 3.5 \text{ k}\Omega$$
$$R_3 = 810 \text{ }\Omega$$

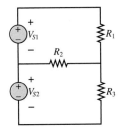

**Figure P3.49**

**3.50**  Determine, using superposition, the voltage across $R$ in the circuit of Figure P3.50.

$$I_B = 12 \text{ A} \qquad R_B = 1 \text{ }\Omega$$
$$V_G = 12 \text{ V} \qquad R_G = 0.3 \text{ }\Omega$$
$$R = 0.23 \text{ }\Omega$$

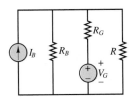

**Figure P3.50**

**3.51**  Using superposition, determine the voltage across $R_2$ in the circuit of Figure P3.51.

$$V_{S1} = V_{S2} = 12 \text{ V}$$
$$R_1 = R_2 = R_3 = 1 \text{ k}\Omega$$

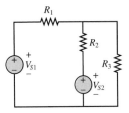

**Figure P3.51**

**3.52**  With reference to Figure P3.52, using superposition, determine the component of the current through $R_3$ that is due to $V_{S2}$.

$$V_{S1} = V_{S2} = 450 \text{ V}$$
$$R_1 = 7 \text{ }\Omega \qquad R_2 = 5 \text{ }\Omega$$
$$R_3 = 10 \text{ }\Omega \qquad R_4 = R_5 = 1 \text{ }\Omega$$

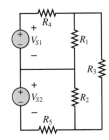

**Figure P3.52**

**3.53**  The circuit shown in Figure P3.24 is a simplified DC version of an AC three-phase electrical distribution system.

$$V_{S1} = V_{S2} = V_{S3} = 170 \text{ V}$$
$$R_{W1} = R_{W2} = R_{W3} = 0.7 \text{ }\Omega$$
$$R_1 = 1.9 \text{ }\Omega \qquad R_2 = 2.3 \text{ }\Omega$$
$$R_3 = 11 \text{ }\Omega$$

To prove how cumbersome and inefficient (although sometimes necessary) the method is, determine, using superposition, the current through $R_1$.

## Section 4: Maximum Power Transfer

**3.54**  The equivalent circuit of Figure P3.54 has:

$$V_{\text{TH}} = 12 \text{ V} \qquad R_{\text{eq}} = 8 \text{ }\Omega$$

If the conditions for maximum power transfer exist, determine:

a. The value of $R_L$.

b. The power developed in $R_L$.

c. The efficiency of the circuit, that is, the ratio of power absorbed by the load to power supplied by the source.

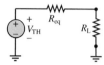

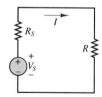

**Figure P3.54**

**3.55**   The equivalent circuit of Figure P3.54 has:

$$V_{TH} = 35 \text{ V} \qquad R_{eq} = 600 \text{ }\Omega$$

If the conditions for maximum power transfer exist, determine:

a. The value of $R_L$.

b. The power developed in $R_L$.

c. The efficiency of the circuit.

**3.56**   A nonideal voltage source can be modeled as an ideal voltage source in series with a resistance representing the internal losses of the source as shown in Figure P3.56. A load is connected across the terminals of the nonideal source.

$$V_S = 12 \text{ V} \qquad R_S = 0.3 \text{ }\Omega$$

a. Plot the power dissipated in the load as a function of the load resistance. What can you conclude from your plot?

b. Prove, analytically, that your conclusion is valid in all cases.

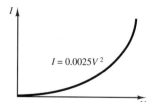

**Figure P3.56**

## Section 5: Nonlinear Circuit Elements

**3.57**   Write the node voltage equations in terms of $v_1$ and $v_2$ for the circuit of Figure P3.57. The two nonlinear resistors are characterized by

$$i_a = 2v_a^3$$
$$i_b = v_b^3 + 10v_b$$

*Do not solve* the resulting equations.

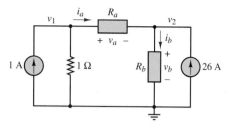

**Figure P3.57**

**3.58**   We have seen that some devices do not have a linear current-voltage characteristic for all $i$ and $v$—that is, $R$ is not constant for all values of current and voltage. For many devices, however, we can estimate the characteristics by piecewise linear approximation. For a portion of the characteristic curve around an operating point, the slope of the curve is relatively constant. The inverse of this slope at the operating point is defined as "incremental resistance," $R_{inc}$:

$$R_{inc} = \frac{dV}{dI}\bigg|_{[V_0, I_0]} \approx \frac{\Delta V}{\Delta I}\bigg|_{[V_0, I_0]}$$

where $[V_0, I_0]$ is the operating point of the circuit.

a. For the circuit of Figure P3.58, find the operating point of the element that has the characteristic curve shown.

b. Find the incremental resistance of the nonlinear element at the operating point of part a.

c. If $V_T$ were increased to 20 V, find the new operating point and the new incremental resistance.

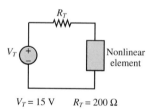

$$V_T = 15 \text{ V} \qquad R_T = 200 \text{ }\Omega$$

$$I = 0.0025V^2$$

**Figure P3.58**

**3.59**   The device in the circuit in Figure P3.59 is a temperature sensor with the nonlinear $i$-$v$ characteristic shown. The remainder of the circuit in

which the device is connected has been reduced to a
Thévenin equivalent circuit with:

$$V_{TH} = 2.4 \text{ V} \qquad R_{eq} = 19.2 \ \Omega$$

Determine the current through the nonlinear device.

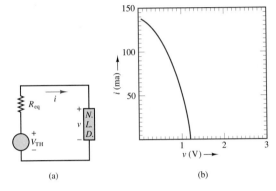

(a)    (b)

**Figure P3.59**

**3.60**   The device in the circuit in Figure P3.60 is an
induction motor with the nonlinear $i$-$v$ characteristic
shown. Determine the current through and the voltage
across the nonlinear device.

$$V_S = 450 \text{ V} \qquad R = 9 \ \Omega$$

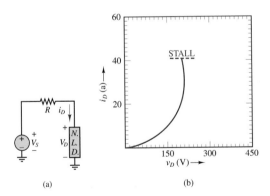

(a)    (b)

**Figure P3.60**

**3.61**   The nonlinear device in the circuit shown in Figure
P3.61 has the $i$-$v$ characteristic given.

$$V_S = V_{TH} = 1.5 \text{ V} \qquad R = R_{eq} = 60 \ \Omega$$

Determine the voltage across and the current through
the nonlinear device.

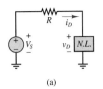

(a)

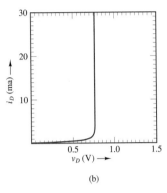

(b)

**Figure P3.61**

**3.62**   The resistance of the nonlinear device in the circuit
in Figure P3.62 is a nonlinear function of pressure.
The $i$-$v$ characteristic of the device is shown as a
family of curves for various pressures. Construct the
DC load line. Plot the voltage across the device as a
function of pressure. Determine the current through
the device when $P = 30$ psig.

$$V_S = V_{TH} = 2.5 \text{ V} \qquad R = R_{eq} = 125 \ \Omega$$

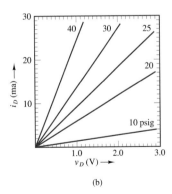

(a)

(b)

**Figure P3.62**

**3.63**   The resistance of the nonlinear device in the circuits shown in Figure P3.63 is a nonlinear function of pressure. The $i$-$v$ characteristic of the device is shown as a family of curves for various pressures. Construct the DC load line and determine the current through the device when $P = 40$ kPa.

$$V_S = V_{TH} = 2.5 \text{ V} \qquad R = R_{eq} = 125 \ \Omega$$

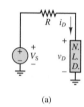

(a)

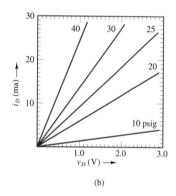

(b)

**Figure P3.63**

**3.64**   The nonlinear device in the circuit shown in Figure P3.64 has the $i$-$v$ characteristic:

$$i_D = I_o e^{v_D/V_T}$$
$$I_o = 10^{-15} \text{ A} \qquad V_T = 26 \text{ mV}$$
$$V_S = V_{TH} = 1.5 \text{ V}$$
$$R = R_{eq} = 60 \ \Omega$$

Determine an expression for the DC load line. Then use an iterative technique to determine the voltage across and current through the nonlinear device.

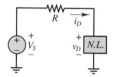

**Figure P3.64**

# CHAPTER

4

# AC Network Analysis

I n this chapter we introduce energy-storage elements and the analysis of circuits excited by sinusoidal voltages and currents. Sinusoidal (or AC) signals constitute the most important class of signals in the analysis of electrical circuits. The simplest reason is that virtually all of the electric power used in households and industries comes in the form of sinusoidal voltages and currents.

The chapter is arranged as follows. First, energy-storage elements are introduced, and time-dependent signal sources and the concepts of average and root-mean-square (rms) values are discussed. Next, we analyze the circuit equations that arise when time-dependent signal sources excite circuits containing energy-storage elements; in the course of this discussion, it will become apparent that differential equations are needed to describe the dynamic behavior of these circuits. The remainder of the chapter is devoted to the development of circuit analysis techniques that greatly simplify the solution of dynamic circuits for the special case of sinusoidal signal excitation; the more general analysis of these circuits will be completed in Chapter 5.

By the end of the chapter, you should have mastered a number of concepts that will be used routinely in the remainder of the book; these are summarized as follows:

- Definition of the *i-v* relationship for inductors and capacitors.
- Computation of rms values for periodic waveforms.

· Representation of sinusoidal signals by complex phasors.
· Impedance of common circuit elements.
· AC circuit analysis by Kirchhoff's laws and equivalent-circuit methods.

## 4.1   ENERGY-STORAGE (DYNAMIC) CIRCUIT ELEMENTS

The ideal resistor was introduced through Ohm's law in Chapter 2 as a useful idealization of many practical electrical devices. However, in addition to resistance to the flow of electric current, which is purely a dissipative (i.e., an energy-loss) phenomenon, electric devices may also exhibit energy-storage properties, much in the same way a spring or a flywheel can store mechanical energy. Two distinct mechanisms for energy storage exist in electric circuits: **capacitance** and **inductance,** both of which lead to the storage of energy in an electromagnetic field. For the purpose of this discussion, it will not be necessary to enter into a detailed electromagnetic analysis of these devices. Rather, two ideal circuit elements will be introduced to represent the ideal properties of capacitive and inductive energy storage: the **ideal capacitor** and the **ideal inductor.** It should be stated clearly that ideal capacitors and inductors do not exist, strictly speaking; however, just like the ideal resistor, these "ideal" elements are very useful for understanding the behavior of physical circuits. In practice, any component of an electric circuit will exhibit some resistance, some inductance, and some capacitance—that is, some energy dissipation and some energy storage.

### The Ideal Capacitor

A physical capacitor is a device that can store energy in the form of a charge separation when appropriately polarized by an electric field (i.e., a voltage). The simplest capacitor configuration consists of two parallel conducting plates of cross-sectional area $A$, separated by air (or another **dielectric**[1] material, such as mica or Teflon). Figure 4.1 depicts a typical configuration and the circuit symbol for a capacitor.

The presence of an insulating material between the conducting plates does not allow for the flow of DC current; thus, *a capacitor acts as an open circuit in the presence of DC currents.* However, if the voltage present at the capacitor terminals changes as a function of time, so will the charge that has accumulated at the two capacitor plates, since the degree of polarization is a function of the applied electric field, which is time-varying. In a capacitor, the charge separation caused by the polarization of the dielectric is proportional to the external voltage, that is, to the applied electric field:

$$Q = CV \tag{4.1}$$

where the parameter $C$ is called the *capacitance* of the element and is a measure of the ability of the device to accumulate, or store, charge. The unit of capacitance is the coulomb/volt and is called the **farad (F).** The farad is an unpractically large unit; therefore it is common to use microfarads (1 $\mu$F $= 10^{-6}$ F) or picofarads (1 pF $= 10^{-12}$ F). From equation 4.1 it becomes apparent that if the external

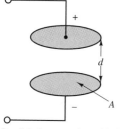

Parallel-plate capacitor with air gap $d$ (air is the dielectric)

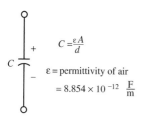

$$C = \frac{\varepsilon A}{d}$$

$\varepsilon =$ permittivity of air

$= 8.854 \times 10^{-12} \ \frac{F}{m}$

Circuit symbol

**Figure 4.1** Structure of parallel-plate capacitor

---

[1]A dielectric material is a material that is not an electrical conductor but contains a large number of electric dipoles, which become polarized in the presence of an electric field.

voltage applied to the capacitor plates changes in time, so will the charge that is internally stored by the capacitor:

$$q(t) = Cv(t) \tag{4.2}$$

Thus, although no current can flow through a capacitor if the voltage across it is constant, a time-varying voltage will cause charge to vary in time.

The change with time in the stored charge is analogous to a current. You can easily see this by recalling the definition of current given in Chapter 2, where it was stated that

$$i(t) = \frac{dq(t)}{dt} \tag{4.3}$$

that is, that electric current corresponds to the time rate of change of charge. Differentiating equation 4.2, one can obtain a relationship between the current and voltage in a capacitor:

$$\boxed{i(t) = C\frac{dv(t)}{dt}} \tag{4.4}$$

Equation 4.4 is the defining circuit law for a capacitor. If the differential equation that defines the $i$-$v$ relationship for a capacitor is integrated, one can obtain the following relationship for the voltage across a capacitor:

$$v_C(t) = \frac{1}{C}\int_{-\infty}^{t} i_C\, dt' \tag{4.5}$$

Equation 4.5 indicates that the capacitor voltage depends on the past current through the capacitor, up until the present time, $t$. Of course, one does not usually have precise information regarding the flow of capacitor current for all past time, and so it is useful to define the initial voltage (or *initial condition*) for the capacitor according to the following, where $t_0$ is an arbitrary initial time:

$$V_0 = v_C(t = t_0) = \frac{1}{C}\int_{-\infty}^{t_0} i_C\, dt' \tag{4.6}$$

The capacitor voltage is now given by the expression

$$v_C(t) = \frac{1}{C}\int_{t_0}^{t} i_C\, dt' + V_0 \qquad t \geq t_0 \tag{4.7}$$

The significance of the initial voltage, $V_0$, is simply that at time $t_0$ some charge is stored in the capacitor, giving rise to a voltage, $v_C(t_0)$, according to the relationship $Q = CV$. Knowledge of this initial condition is sufficient to account for the entire past history of the capacitor current.

Capacitors connected in series and parallel can be combined to yield a single equivalent capacitance. The rule of thumb, which is illustrated in Figure 4.2, is the following:

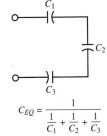

$$C_{EQ} = \cfrac{1}{\dfrac{1}{C_1} + \dfrac{1}{C_2} + \dfrac{1}{C_3}}$$

Capacitances in series combine like resistors in parallel

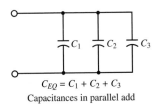

$$C_{EQ} = C_1 + C_2 + C_3$$

Capacitances in parallel add

**Figure 4.2** Combining capacitors in a circuit

Capacitors in parallel add. Capacitors in series combine according to the same rules used for resistors connected in parallel.

## EXAMPLE 4.1 Calculating Capacitor Current from Voltage

### Problem

Calculate the current through a capacitor from knowledge of its terminal voltage.

---

### Solution

**Known Quantities:** Capacitor terminal voltage; capacitance value.

**Find:** Capacitor current.

**Schematics, Diagrams, Circuits, and Given Data:** $v(t) = 5\left(e^{-t/10^{-6}}\right)$ V $t \geq 0$ s; $C = 0.1\ \mu\text{F}$. The terminal voltage is plotted in Figure 4.3.

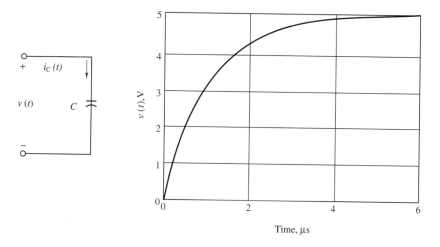

**Figure 4.3**

**Assumptions:** The capacitor is initially discharged: $v(t = 0) = 0$.

**Analysis:** Using the defining differential relationship for the capacitor, we may obtain the current by differentiating the voltage:

$$i_C(t) = C\frac{dv(t)}{dt} = 10^{-7}\frac{5}{10^{-6}}\left(e^{-t/10^{-6}}\right) = 0.5e^{-t/10^{-6}} \quad \text{A} \qquad t \geq 0$$

A plot of the capacitor current is shown in Figure 4.4. Note how the current jumps to 0.5 A instantaneously as the voltage rises exponentially: The ability of a capacitor's current to change instantaneously is an important property of capacitors.

**Comments:** As the voltage approaches the constant value 5 V, the capacitor reaches its maximum charge-storage capability for that voltage (since $Q = CV$) and no more current flows through the capacitor. The total charge stored is $Q = 0.5 \times 10^{-6}$ C. This is a fairly small amount of charge, but it can produce a substantial amount of current for a brief period of time. For example, the fully charged capacitor could provide 100 mA of current

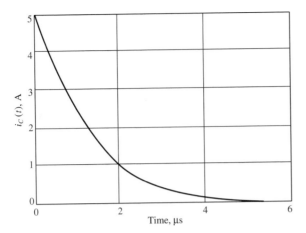

**Figure 4.4**

for a period of time equal to 5 $\mu$s:

$$I = \frac{\Delta Q}{\Delta t} = \frac{0.5 \times 10^{-6}}{5 \times 10^{-6}} = 0.1 \text{ A}$$

There are many useful applications of this energy-storage property of capacitors in practical circuits.

***Focus on Computer-Aided Tools:*** The Matlab™ *m-files* used to generate the plots of Figures 4.3 and 4.4 may be found in the CD-ROM that accompanies this book.

VIRTUAL LAB

# EXAMPLE 4.2 Calculating Capacitor Voltage from Current and Initial Conditions

## Problem

Calculate the voltage across a capacitor from knowledge of its current and initial state of charge.

## Solution

***Known Quantities:*** Capacitor current; initial capacitor voltage; capacitance value.

***Find:*** Capacitor voltage.

***Schematics, Diagrams, Circuits, and Given Data:***

$$i_C(t) = \begin{cases} 0 & t < 0 \text{ s} \\ I = 10 \text{ mA} & 0 \leq t \leq 1 \text{ s} \\ 0 & t > 1 \text{ s} \end{cases}$$

$v_C(t = 0) = 2 \text{ V}; C = 1{,}000 \ \mu\text{F}.$

The capacitor current is plotted in Figure 4.5(a).

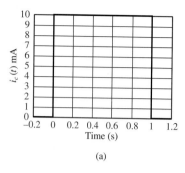

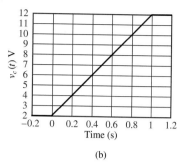

(a)                                                          (b)

**Figure 4.5**

**Assumptions:** The capacitor is initially charged such that $v_C(t = t_0 = 0) = 2$ V.

**Analysis:** Using the defining integral relationship for the capacitor, we may obtain the voltage by integrating the current:

$$v_C(t) = \frac{1}{C} \int_{t_0}^{t} i_C(t')\, dt' + v_C(t_0) \qquad t \geq t_0$$

$$v_C(t) = \begin{cases} \dfrac{1}{C} \displaystyle\int_0^1 I\, dt' + V_0 = \dfrac{I}{C}t + V_0 = 10t + 2\ V & 0 \leq t \leq 1\ \text{s} \\[2mm] 12\ V & t > 1\ \text{s} \end{cases}$$

**Comments:** Once the current stops, at $t = 1$ s, the capacitor voltage cannot develop any further but remains now at the maximum value it reached at $t = 1$ s: $v_C(t = 1) = 12$ V. The final value of the capacitor voltage after the current source has stopped charging the capacitor depends on two factors: (1) the initial value of the capacitor voltage, and (2) the history of the capacitor current. Figure 4.5(a) and (b) depicts the two waveforms.

**Focus on Computer-Aided Tools:** The Matlab™ *m-files* used to generate the plots of Figures 4.5(a) and (b) may be found in the CD-ROM that accompanies this book.

**Physical capacitors** are rarely constructed of two parallel plates separated by air, because this configuration yields very low values of capacitance, unless one is willing to tolerate very large plate areas. In order to increase the capacitance (i.e., the ability to store energy), physical capacitors are often made of tightly rolled sheets of metal film, with a dielectric (paper or Mylar) sandwiched in between. Table 4.1 illustrates typical values, materials, maximum voltage ratings, and useful frequency ranges for various types of capacitors. The voltage rating is particularly important, because any insulator will break down if a sufficiently high voltage is applied across it.

## Energy Storage in Capacitors

You may recall that the capacitor was described earlier in this section as an energy-storage element. An expression for the energy stored in the capacitor, $W_C(t)$, may be derived easily if we recall that energy is the integral of power, and that the

Table 4.1  Capacitors

| Material | Capacitance range | Maximum voltage (V) | Frequency range (Hz) |
|----------|------------------|---------------------|----------------------|
| Mica | 1 pF to 0.1 $\mu$F | 100–600 | $10^3$–$10^{10}$ |
| Ceramic | 10 pF to 1 $\mu$F | 50–1,000 | $10^3$–$10^{10}$ |
| Mylar | 0.001 $\mu$F to 10 $\mu$F | 50–500 | $10^2$–$10^8$ |
| Paper | 1,000 pF to 50 $\mu$F | 100–105 | $10^2$–$10^8$ |
| Electrolytic | 0.1 $\mu$F to 0.2 F | 3–600 | $10$–$10^4$ |

instantaneous power in a circuit element is equal to the product of voltage and current:

$$W_C(t) = \int P_C(t')\,dt'$$

$$= \int v_C(t')i_C(t')\,dt' \qquad (4.8)$$

$$= \int v_C(t')C\frac{dv_C(t')}{dt'}\,dt'$$

$$W_C(t) = \frac{1}{2}Cv_C^2(t) \qquad \text{Energy stored in a capacitor (J)}$$

Example 4.3 illustrates the calculation of the energy stored in a capacitor.

## EXAMPLE  4.3  Energy Stored in a Capacitor

### Problem

Calculate the energy stored in a capacitor.

### Solution

**Known Quantities:**  Capacitor voltage; capacitance value.

**Find:**  Energy stored in capacitor.

**Schematics, Diagrams, Circuits, and Given Data:**  $v_C(t = 0) = 12$ V; $C = 10\ \mu$F.

**Analysis:**

$$Q = Cv_C = 10^{-5} \times 12 = 120\ \mu C$$

$$W_C = \frac{1}{2}Cv_C^2 = \frac{1}{2} \times 10^{-5} \times 144 = 720 \times 10^{-6} = 720\ \mu J$$

## Capacitive Displacement Transducer and Microphone

As shown in Figure 4.1, the capacitance of a parallel-plate capacitor is given by the expression

$$C = \frac{\varepsilon A}{d}$$

where $\varepsilon$ is the **permittivity** of the dielectric material, $A$ the area of each of the plates, and $d$ their separation. The permittivity of air is $\varepsilon_0 = 8.854 \times 10^{-12}$ F/m, so that two parallel plates of area 1 m$^2$, separated by a distance of 1 mm, would give rise to a capacitance of $8.854 \times 10^{-3}$ $\mu$F, a very small value for a very large plate area. This relative inefficiency makes parallel-plate capacitors impractical for use in electronic circuits. On the other hand, parallel-plate capacitors find application as *motion transducers*, that is, as devices that can measure the motion or displacement of an object. In a capacitive motion transducer, the air gap between the plates is designed to be variable, typically by fixing one plate and connecting the other to an object in motion. Using the capacitance value just derived for a parallel-plate capacitor, one can obtain the expression

$$C = \frac{8.854 \times 10^{-3} A}{x}$$

where $C$ is the capacitance in pF, $A$ is the area of the plates in mm$^2$, and $x$ is the (variable) distance in mm. It is important to observe that the change in capacitance caused by the displacement of one of the plates is nonlinear, since the capacitance varies as the inverse of the displacement. For small displacements, however, the capacitance varies approximately in a linear fashion.

The *sensitivity*, $S$, of this motion transducer is defined as the slope of the change in capacitance per change in displacement, $x$, according to the relation

$$S = \frac{dC}{dx} = -\frac{8.854 \times 10^{-3} A}{2x^2} \frac{\text{pF}}{\text{mm}}$$

Thus, the sensitivity increases for small displacements. This behavior can be verified by plotting the capacitance as a function of $x$ and noting that as $x$ approaches zero, the slope of the nonlinear $C(x)$ curve becomes steeper (thus the greater sensitivity). Figure 4.6 depicts this behavior for a transducer with area equal to 10 mm$^2$.

This simple capacitive displacement transducer actually finds use in the popular *capacitive (or condenser) microphone*, in which the sound pressure waves act to displace one of the capacitor plates. The change in capacitance can then be converted into a change in voltage or current by means of a suitable circuit. An extension of this concept that permits measurement of differential pressures is shown in simplified form in Figure 4.7. In the figure, a three-terminal variable capacitor is shown to be made up of two fixed surfaces (typically, spherical depressions ground into glass disks and coated

**FIND IT**

**ON THE WEB**

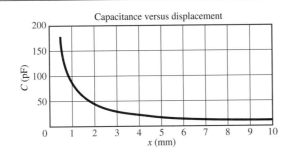

**Figure 4.6** Response of a capacitive displacement transducer

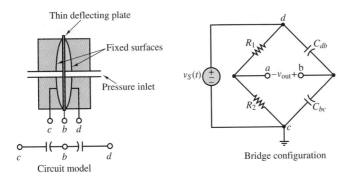

**Figure 4.7** Capacitive pressure transducer, and related bridge circuit

with a conducting material) and of a deflecting plate (typically made of steel) sandwiched between the glass disks. Pressure inlet orifices are provided, so that the deflecting plate can come into contact with the fluid whose pressure it is measuring. When the pressure on both sides of the deflecting plate is the same, the capacitance between terminals $b$ and $d$, $C_{bd}$, will be equal to that between terminals $b$ and $c$, $C_{bc}$. If any pressure differential exists, the two capacitances will change, with an increase on the side where the deflecting plate has come closer to the fixed surface and a corresponding decrease on the other side.

This behavior is ideally suited for the application of a bridge circuit, similar to the Wheatstone bridge circuit illustrated in Example 2.12, and also shown in Figure 4.7. In the bridge circuit, the output voltage, $v_{\text{out}}$, is precisely balanced when the differential pressure across the transducer is zero, but it will deviate from zero whenever the two capacitances are not identical because of a pressure differential across the transducer. We shall analyze the bridge circuit later.

## The Ideal Inductor

The ideal inductor is an element that has the ability to store energy in a magnetic field. Inductors are typically made by winding a coil of wire around a **core,** which can be an insulator or a ferromagnetic material, as shown in Figure 4.8. When a

Magnetic flux
lines

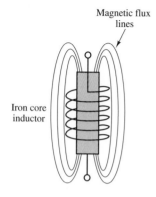

Iron core
inductor

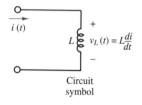

$$i(t) \quad \begin{array}{c} + \\ L \lessgtr v_L(t) = L\frac{di}{dt} \\ - \end{array}$$

Circuit
symbol

**Figure 4.8** Iron-core
inductor

FIND IT

ON THE WEB

current flows through the coil, a magnetic field is established, as you may recall from early physics experiments with electromagnets.[2] In an ideal inductor, the resistance of the wire is zero, so that a constant current through the inductor will flow freely without causing a voltage drop. In other words, *the ideal inductor acts as a short circuit in the presence of DC currents.* If a time-varying voltage is established across the inductor, a corresponding current will result, according to the following relationship:

$$v_L(t) = L\frac{di_L}{dt} \tag{4.9}$$

where $L$ is called the *inductance* of the coil and is measured in **henrys (H),** where

$$1 \text{ H} = 1 \text{ V-s/A} \tag{4.10}$$

Henrys are reasonable units for **practical inductors;** millihenrys (mH) and microhenrys ($\mu$H) are also used.

It is instructive to compare equation 4.9, which defines the behavior of an ideal inductor, with the expression relating capacitor current and voltage:

$$i_C(t) = C\frac{dv_C}{dt} \tag{4.11}$$

We note that the roles of voltage and current are reversed in the two elements, but that both are described by a differential equation of the same form. This *duality* between inductors and capacitors can be exploited to derive the same basic results for the inductor that we already have for the capacitor simply by replacing the capacitance parameter, $C$, with the inductance, $L$, and voltage with current (and vice versa) in the equations we derived for the capacitor. Thus, the inductor current is found by integrating the voltage across the inductor:

$$i_L(t) = \frac{1}{L}\int_{-\infty}^{t} v_L \, dt' \tag{4.12}$$

If the current flowing through the inductor at time $t = t_0$ is known to be $I_0$, with

$$I_0 = i_L(t = t_0) = \frac{1}{L}\int_{-\infty}^{t_0} v_L \, dt' \tag{4.13}$$

then the inductor current can be found according to the equation

$$i_L(t) = \frac{1}{L}\int_{t_0}^{t} v_L \, dt' + I_0 \qquad t \geq t_0 \tag{4.14}$$

Series and parallel combinations of inductors behave like resistors, as illustrated in Figure 4.9, and stated as follows:

Inductors in series add. Inductors in parallel combine according to the same rules used for resistors connected in parallel.

---

[2]See also Chapter 15.

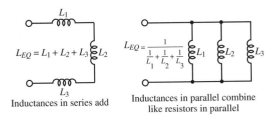

**Figure 4.9** Combining inductors in a circuit

## EXAMPLE 4.4 Calculating Inductor Voltage from Current

### Problem

Calculate the voltage across the inductor from knowledge of its current.

### Solution

**Known Quantities:**  Inductor current; inductance value.

**Find:**  Inductor voltage.

**Schematics, Diagrams, Circuits, and Given Data:**

$$
i_L(t) = \begin{cases}
0 & t < 1\,\text{ms} \\[2mm]
-\dfrac{0.1}{4} + \dfrac{0.1}{4}t & 1 \le t \le 5\,\text{ms} \\[2mm]
0.1 & 5 \le t \le 9\,\text{ms} \\[2mm]
13 \times \dfrac{0.1}{4} - \dfrac{0.1}{4}t & 9 \le t \le 13\,\text{ms} \\[2mm]
0 & t > 13\,\text{ms}
\end{cases}
$$

$L = 10\,\text{H}.$

The inductor current is plotted in Figure 4.10.

**Assumptions:**  $i_L(t = 0) \le 0$.

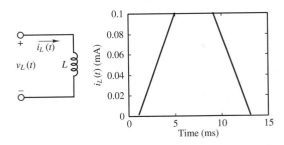

**Figure 4.10**

**Analysis:** Using the defining differential relationship for the inductor, we may obtain the voltage by differentiating the current:

$$v_L(t) = L\frac{di_L(t)}{dt}$$

Piecewise differentiating the expression for the inductor current, we obtain:

$$v_L(t) = \begin{cases} 0\text{ V} & t < 1\text{ ms} \\ 0.25\text{ V} & 1 < t \le 5\text{ ms} \\ 0\text{ V} & 5 < t \le 9\text{ ms} \\ -0.25\text{ V} & 9 < t \le 13\text{ ms} \\ 0\text{ V} & t > 13\text{ ms} \end{cases}$$

The inductor voltage is plotted in Figure 4.11.

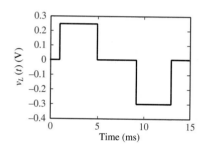

**Figure 4.11**

**VIRTUAL LAB**

**Comments:** Note how the inductor voltage has the ability to change instantaneously!

**Focus on Computer-Aided Tools:** The Matlab™ *m-files* used to generate the plots of Figures 4.10 and 4.11 may be found in the CD-ROM that accompanies this book.

## EXAMPLE 4.5 Calculating Inductor Current from Voltage

### Problem

Calculate the current through the inductor from knowledge of the terminal voltage and of the initial current.

### Solution

**Known Quantities:** Inductor voltage; initial condition (current at $t = 0$); inductance value.

**Find:** Inductor current.

**Schematics, Diagrams, Circuits, and Given Data:**

$$v(t) = \begin{cases} 0\text{ V} & t < 0\text{ s} \\ -10\text{ mV} & 0 < t \le 1\text{ s} \\ 0\text{ V} & t > 1\text{ s} \end{cases}$$

$$L = 10\text{ mH; } i_L(t = 0) = I_0 = 0\text{ A}.$$

The terminal voltage is plotted in Figure 4.12(a).

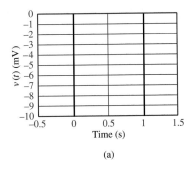

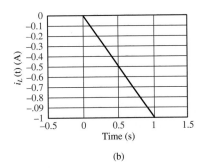

(a)                                                   (b)

**Figure 4.12**

**Assumptions:** $i_L(t = 0) = I_0 = 0$.

**Analysis:** Using the defining integral relationship for the inductor, we may obtain the voltage by integrating the current:

$$i_L(t) = \frac{1}{L} \int_{t_0}^{t} v(t) \, dt' + i_L(t_0) \qquad t \geq t_0$$

$$i_L(t) = \begin{cases} \dfrac{1}{L} \displaystyle\int_0^{t'} (-10 \times 10^{-3}) \, dt' + I_0 = \dfrac{-10^{-2}}{10^{-2}} t + 0 = -t \text{ A} & 0 \leq t \leq 1 \text{ s} \\[3mm] -1 \text{ A} & t > 1 \text{ s} \end{cases}$$

The inductor current is plotted in Figure 4.12b.

**Comments:** Note how the inductor voltage has the ability to change instantaneously!

**Focus on Computer-Aided Tools:** The Matlab™ *m-files* used to generate the plots of Figures 4.12(a) and (b) may be found in the CD-ROM that accompanies this book.

VIRTUAL LAB

## Energy Storage in Inductors

The magnetic energy stored in an ideal inductor may be found from a power calculation by following the same procedure employed for the ideal capacitor. The instantaneous power in the inductor is given by

$$P_L(t) = i_L(t) v_L(t) = i_L(t) L \frac{d i_L(t)}{dt} = \frac{d}{dt} \left[ \frac{1}{2} L i_L^2(t) \right] \qquad \textbf{(4.15)}$$

Integrating the power, we obtain the total energy stored in the inductor, as shown in the following equation:

$$W_L(t) = \int P_L(t') \, dt' = \int \frac{d}{dt'} \left[ \frac{1}{2} L i_L^2(t') \right] dt' \qquad \textbf{(4.16)}$$

$$W_L(t) = \frac{1}{2}Li_L^2(t) \qquad \text{Energy stored in an inductor (J)}$$

Note, once again, the duality with the expression for the energy stored in a capacitor, in equation 4.8.

## EXAMPLE 4.6 Energy Storage in an Ignition Coil

### Problem

Determine the energy stored in an automotive ignition coil.

### Solution

**Known Quantities:** Inductor current initial condition (current at $t = 0$); inductance value.

**Find:** Energy stored in inductor.

**Schematics, Diagrams, Circuits, and Given Data:** $L = 10$ mH; $i_L = I_0 = 8$ A.

**Analysis:**

$$W_L = \frac{1}{2}Li_L^2 = \frac{1}{2} \times 10^{-2} \times 64 = 32 \times 10^{-2} = 320 \text{ mJ}$$

**Comments:** A more detailed analysis of an automotive ignition coil is presented in Chapter 5 to accompany the discussion of transient voltages and currents.

**FOCUS ON MEASUREMENTS**

## Analogy between Electrical and Hydraulic Circuits

A useful analogy can be made between the flow of electrical current through electrical components and the flow of incompressible fluids (e.g., water, oil) through hydraulic components. The analogy starts with the observation that the volume flow rate of a fluid in a pipe is analogous to current flow in a conductor. Similarly, the pressure drop across the pipe is analogous to the voltage drop across a resistor. Figure 4.13 depicts this relationship graphically. The **fluid resistance** presented by the pipe to the fluid flow is analogous to an electrical resistance: The pressure difference between the two ends of the pipe, $(P_1 - P_2)$, causes fluid flow, $q_f$, much like a potential difference across a resistor forces a current flow:

$$q_f = \frac{1}{R_f}(p_1 - p_2)$$

$$i = \frac{1}{R}(v_1 - v_2)$$

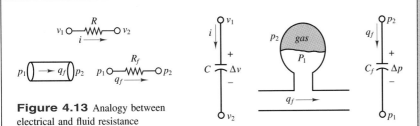

**Figure 4.13** Analogy between electrical and fluid resistance

**Figure 4.14** Analogy between fluid capacitance and electrical capacitance

The analogy between electrical and hydraulic circuits can also be extended to include energy storage effects corresponding to capacitance and inductance. If the fluid enters a vessel that has some elasticity (compressibility), energy can be stored in the expansion and contraction of the vessel walls (if this reminds you of a mechanical spring, you are absolutely right!). This phenomenon gives rise to a **fluid capacitance** effect very similar to the electrical capacitance phenomenon we have just introduced. Energy is stored in the compression and expansion of the gas; this form of energy storage is of the *potential energy* type. Figure 4.14 depicts a so-called gas bag accumulator, which consists of a two-chamber arrangement that permits fluid to displace a membrane separating the incompressible fluid from a compressible fluid (e.g., air). If, for a moment, we imagine that the reference pressure, $p_2$, is zero (think of this as a ground or reference pressure), and that the voltage is the reference or ground voltage, we can create an analogy between an electrical capacitor and a fluid capacitor (the gas-bag accumulator) as shown in Figure 4.14.

$$q_f = C_f \frac{d\Delta p}{dt} = C_f \frac{dp_1}{dt}$$

$$i = C \frac{d\Delta v}{dt} = C \frac{dv_1}{dt}$$

The final element in the analogy is the so-called **fluid inertance** parameter, which is analogous to inductance in the electrical circuit. Fluid inertance, as the name suggests, is caused by the inertial properties, i.e., the mass, of the fluid in motion. As you know from physics, a particle in motion has kinetic energy associated with it; fluid in motion consists of a collection of particles, and it also therefore must have kinetic energy storage properties. If you wish to experience the kinetic energy contained in a fluid in motion, all you have to do is hold a fire hose and experience the reaction force caused by the fluid in motion on your body! Figure 4.15 depicts the analogy between electrical inductance and fluid inertance. These analogies and the energy equations are summarized in Table 4.2.

$$\Delta p = p_1 - p_2 = I_f \frac{dq_f}{dt}$$

$$\Delta v = v_1 - v_2 = L \frac{di}{dt}$$

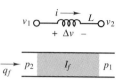

**Figure 4.15** Analogy
between fluid inertance and
electrical inertance

Table 4.2  Analogy between electrical and fluid circuits

| Property | Electrical element or equation | Hydraulic or fluid analogy |
|---|---|---|
| Potential variable | Voltage or potential difference, $v_1 - v_2$ | Pressure difference, $P_1 - P_2$ |
| Flow variable | Current flow, $i$ | Fluid volume flow rate, $q_f$ |
| Resistance | Resistor, $R$ | Fluid resistor, $R_f$ |
| Capacitance | Capacitor, $C$ | Fluid capacitor, $C_f$ |
| Inductance | Inductor, $L$ | Fluid inertor, $I_f$ |
| Power dissipation | $P = i^2 R$ | $P_f = q_f^2 R_f$ |
| Potential energy storage | $W_p = \frac{1}{2} C v^2$ | $W_p = \frac{1}{2} C_f p^2$ |
| Kinetic energy storage | $W_k = \frac{1}{2} L i^2$ | $W_k = \frac{1}{2} I_f q_f^2$ |

## Check Your Understanding

**4.1**   The current waveform shown in Figure 4.16 flows through a 50-mH inductor. Plot
the inductor voltage, $v_L(t)$.

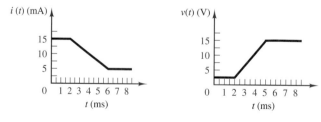

**Figure 4.16**              **Figure 4.17**

**4.2**   The voltage waveform of Figure 4.17 appears across a 1,000-$\mu$F capacitor. Plot the
capacitor current, $i_C(t)$.

**4.3**   Calculate the energy stored in the inductor (in joules) at $t = 3$ ms by the waveform
of Exercise 4.1. Assume $i(-\infty) = 0$.

**4.4**   Perform the calculation of Exercise 4.3 for the capacitor if $v_C(-\infty) = 0$ V.

**4.5**  Compute and plot the inductor energy (in joules) and power (in watts) for the case of Exercise 4.1.

---

## 4.2    TIME-DEPENDENT SIGNAL SOURCES

In Chapter 2, the general concept of an ideal energy source was introduced. In the present chapter, it will be useful to specifically consider sources that generate time-varying voltages and currents and, in particular, sinusoidal sources. Figure 4.18 illustrates the convention that will be employed to denote time-dependent signal sources.

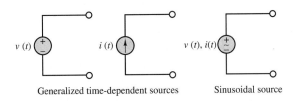

Generalized time-dependent sources        Sinusoidal source

**Figure 4.18** Time-dependent signal sources

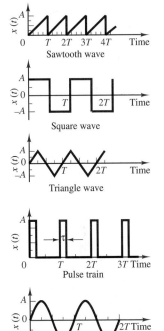

**Figure 4.19** Periodic signal waveforms

One of the most important classes of time-dependent signals is that of **periodic signals.** These signals appear frequently in practical applications and are a useful approximation of many physical phenomena. A periodic signal $x(t)$ is a signal that satisfies the following equation:

$$x(t) = x(t + nT) \qquad n = 1, 2, 3, \ldots \tag{4.17}$$

where $T$ is the **period** of $x(t)$. Figure 4.19 illustrates a number of the periodic waveforms that are typically encountered in the study of electrical circuits. Waveforms such as the sine, triangle, square, pulse, and sawtooth waves are provided in the form of voltages (or, less frequently, currents) by commercially available **signal** (or **waveform**) **generators.** Such instruments allow for selection of the waveform peak amplitude, and of its period.

As stated in the introduction, sinusoidal waveforms constitute by far the most important class of time-dependent signals. Figure 4.20 depicts the relevant parameters of a sinusoidal waveform. A generalized sinusoid is defined as follows:

$$x(t) = A \cos(\omega t + \phi) \tag{4.18}$$

where $A$ is the **amplitude,** $\omega$ the **radian frequency,** and $\phi$ the **phase.** Figure 4.20 summarizes the definitions of $A$, $\omega$, and $\phi$ for the waveforms

$$x_1(t) = A \cos(\omega t) \qquad \text{and} \qquad x_2(t) = A \cos(\omega t + \phi)$$

where

$$f = \text{natural frequency} = \frac{1}{T} \text{ (cycles/s, or Hz)}$$

$$\omega = \text{radian frequency} = 2\pi f \text{ (radians/s)} \tag{4.19}$$

$$\phi = 2\pi \frac{\Delta t}{T} \text{(radians)} = 360 \frac{\Delta t}{T} \text{ (degrees)}$$

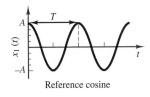

Reference cosine

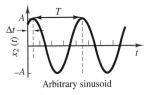

Arbitrary sinusoid

**Figure 4.20** Sinusoidal waveforms

The phase shift, $\phi$, permits the representation of an arbitrary sinusoidal signal. Thus, the choice of the reference cosine function to represent sinusoidal signals—arbitrary as it may appear at first—does not restrict the ability to represent all sinusoids. For example, one can represent a sine wave in terms of a cosine wave simply by introducing a phase shift of $\pi/2$ radians:

$$A \sin(\omega t) = A \cos\left(\omega t - \frac{\pi}{2}\right) \tag{4.20}$$

Although one usually employs the variable $\omega$ (in units of radians per second) to denote sinusoidal frequency, it is common to refer to natural frequency, $f$, in units of cycles per second, or **hertz (Hz).** The reader with some training in music theory knows that a sinusoid represents what in music is called a *pure tone;* an A-440, for example, is a tone at a frequency of 440 Hz. It is important to be aware of the factor of $2\pi$ that differentiates radian frequency (in units of rad/s) from natural frequency (in units of Hz). The distinction between the two units of frequency—which are otherwise completely equivalent—is whether one chooses to define frequency in terms of revolutions around a trigonometric circle (in which case the resulting units are rad/s), or to interpret frequency as a repetition rate (cycles/second), in which case the units are Hz. The relationship between the two is the following:

$$\boxed{\omega = 2\pi f} \tag{4.21}$$

## Why Sinusoids?

You should by now have developed a healthy curiosity about why so much attention is being devoted to sinusoidal signals. Perhaps the simplest explanation is that the electric power used for industrial and household applications worldwide is generated and delivered in the form of either 50- or 60-Hz sinusoidal voltages and currents. Chapter 7 will provide more detail regarding the analysis of electric power circuits. The more ambitious reader may explore the box "Fourier Analysis" in Chapter 6 to obtain a more comprehensive explanation of the importance of sinusoidal signals. It should be remarked that the methods developed in this section and the subsequent sections apply to many engineering systems, not just to electrical circuits, and will be encountered again in the study of dynamic-system modeling and of control systems.

## Average and RMS Values

Now that a number of different signal waveforms have been defined, it is appropriate to define suitable measurements for quantifying the strength of a time-varying electrical signal. The most common types of measurements are the **average** (or **DC) value** of a signal waveform—which corresponds to just measuring the mean voltage or current over a period of time—and the **root-mean-square** (or **rms) value,** which takes into account the fluctuations of the signal about its average value. Formally, the operation of computing the average value of a signal corresponds to integrating the signal waveform over some (presumably, suitably chosen) period of time. We define the time-averaged value of a signal $x(t)$ as

$$\langle x(t) \rangle = \frac{1}{T} \int_0^T x(t') \, dt' \tag{4.22}$$

where $T$ is the period of integration. Figure 4.21 illustrates how this process does, in fact, correspond to computing the average amplitude of $x(t)$ over a period of $T$ seconds.

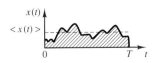

**Figure 4.21** Averaging a signal waveform

## EXAMPLE 4.7 Average Value of Sinusoidal Waveform

### Problem

Compute the average value of the signal $x(t) = 10 \cos(100t)$.

### Solution

**Known Quantities:** Functional form of the periodic signal $x(t)$.

**Find:** Average value of $x(t)$.

**Analysis:** The signal is periodic with period $T = 2\pi/\omega = 2\pi/100$, thus we need to integrate over only one period to compute the average value:

$$\langle x(t) \rangle = \frac{1}{T} \int_0^T x(t') \, dt' = \frac{100}{2\pi} \int_0^{2\pi/100} 10 \cos(100t) dt$$

$$= \frac{10}{2\pi} \langle \sin(2\pi) - \sin(0) \rangle = 0$$

**Comments:** The average value of a sinusoidal signal is zero, independent of its amplitude and frequency.

The result of Example 4.7 can be generalized to state that

$$\langle A \cos(\omega t + \phi) \rangle = 0 \qquad \textbf{(4.23)}$$

a result that might be perplexing at first: If any sinusoidal voltage or current has zero average value, is its average power equal to zero? Clearly, the answer must be no. Otherwise, it would be impossible to illuminate households and streets and power industrial machinery with 60-Hz sinusoidal current! There must be another way, then, of quantifying the strength of an AC signal.

Very conveniently, a useful measure of the voltage of an AC waveform is the root-mean-square, or rms, value of the signal, $x(t)$, defined as follows:

$$x_{\text{rms}} = \sqrt{\frac{1}{T} \int_0^T x^2(t') \, dt'} \qquad \textbf{(4.24)}$$

Note immediately that if $x(t)$ is a voltage, the resulting $x_{\text{rms}}$ will also have units of volts. If you analyze equation 4.24, you can see that, in effect, the rms value consists of the square *root* of the average (or *mean*) of the *square* of the signal. Thus, the notation *rms* indicates exactly the operations performed on $x(t)$ in order to obtain its rms value.

### EXAMPLE 4.8 Rms Value of Sinusoidal Waveform

**Problem**

Compute the rms value of the sinusoidal current $i(t) = I \cos(\omega t)$.

---

**Solution**

**Known Quantities:** Functional form of the periodic signal $i(t)$.

**Find:** Rms value of $i(t)$.

**Analysis:** Applying the definition of rms value in equation 4.24, we compute:

$$i_{rms} = \sqrt{\frac{1}{T} \int_0^T i^2(t')dt'} = \sqrt{\frac{\omega}{2\pi} \int_0^{2\pi/\omega} I^2 \cos^2(\omega t')dt'}$$

$$= \sqrt{\frac{\omega}{2\pi} \int_0^{2\pi/\omega} I^2 \left(\frac{1}{2} + \cos(2\omega t')\right) dt'}$$

$$= \sqrt{\frac{1}{2}I^2 + \frac{\omega}{2\pi} \int_0^{2\pi/\omega} \frac{I^2}{2} \cos(2\omega t')dt'}$$

At this point, we recognize that the integral under the square root sign is equal to zero (see Example 4.7), because we are integrating a sinusoidal waveform over two periods. Hence:

$$i_{rms} = \frac{I}{\sqrt{2}} = 0.707I$$

where $I$ is the *peak value* of the waveform $i(t)$.

**Comments:** The rms value of a sinusoidal signal is equal to 0.707 times the peak value, independent of its amplitude and frequency.

---

The preceding example illustrates how the rms value of a sinusoid is proportional to its peak amplitude. The factor of $0.707 = 1/\sqrt{2}$ is a useful number to remember, since it applies to any sinusoidal signal. It is not, however, generally applicable to signal waveforms other than sinusoids, as the Check Your Understanding exercises will illustrate.

---

## Check Your Understanding

**4.6** Express the voltage $v(t) = 155.6 \sin(377t + 60°)$ in cosine form. You should note that the radian frequency $\omega = 377$ will recur very often, since $377 = 2\pi 60$; that is, 377 is the radian equivalent of the natural frequency of 60 cycles/second, which is the frequency of the electric power generated in North America.

**4.7** Compute the average value of the sawtooth waveform shown in Figure 4.22.

**4.8** Compute the average value of the shifted triangle wave shown in Figure 4.23.

Figure 4.22

Figure 4.23

**4.9**  Find the rms value of the sawtooth wave of Exercise 4.7.

**4.10**  Find the rms value of the half cosine wave shown in Figure 4.24.

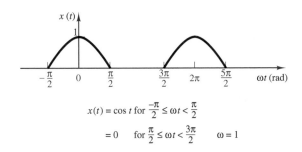

$$x(t) = \cos t \text{ for } \frac{-\pi}{2} \le \omega t < \frac{\pi}{2}$$

$$= 0 \quad \text{for } \frac{\pi}{2} \le \omega t < \frac{3\pi}{2} \qquad \omega = 1$$

Figure 4.24

## 4.3    SOLUTION OF CIRCUITS CONTAINING DYNAMIC ELEMENTS

The first two sections of this chapter introduced energy-storage elements and time-dependent signal sources. The logical next task is to analyze the behavior of circuits containing such elements. The major difference between the analysis of the resistive circuits studied in Chapters 2 and 3 and the circuits we will explore in the remainder of this chapter is that now the equations that result from applying Kirchhoff's laws are differential equations, as opposed to the algebraic equations obtained in solving resistive circuits. Consider, for example, the circuit of Figure 4.25, which consists of the series connection of a voltage source, a resistor, and a capacitor. Applying KVL around the loop, we may obtain the following equation:

$$v_S(t) = v_R(t) + v_C(t) \tag{4.25}$$

Observing that $i_R = i_C$, equation 4.25 may be combined with the defining equation for the capacitor (equation 4.5) to obtain

$$v_S(t) = Ri_C(t) + \frac{1}{C} \int_{-\infty}^{t} i_C \, dt' \tag{4.26}$$

Equation 4.26 is an integral equation, which may be converted to the more familiar form of a differential equation by differentiating both sides of the equation, and recalling that

$$\frac{d}{dt} \left( \int_{-\infty}^{t} i_C(t') \, dt' \right) = i_C(t) \tag{4.27}$$

A circuit containing energy-storage elements is described by a differential equation. The differential equation describing the series $RC$ circuit shown is

$$\frac{di_C}{dt} + \frac{1}{RC} i_C = \frac{dv_S}{dt}$$

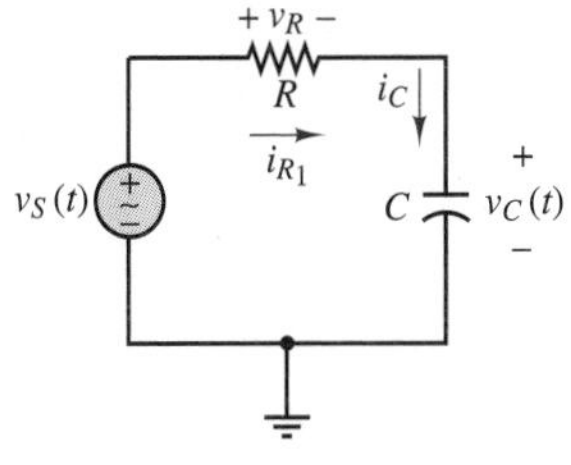

**Figure 4.25**  Circuit containing energy-storage element

to obtain the following differential equation:

$$\frac{di_C}{dt} + \frac{1}{RC}i_C = \frac{1}{R}\frac{dv_S}{dt} \tag{4.28}$$

where the argument $(t)$ has been dropped for ease of notation.

Observe that in equation 4.28, the independent variable is the series current flowing in the circuit, and that this is not the only equation that describes the series $RC$ circuit. If, instead of applying KVL, for example, we had applied KCL at the node connecting the resistor to the capacitor, we would have obtained the following relationship:

$$i_R = \frac{v_S - v_C}{R} = i_C = C\frac{dv_C}{dt} \tag{4.29}$$

or

$$\frac{dv_C}{dt} + \frac{1}{RC}v_C = \frac{1}{RC}v_S \tag{4.30}$$

Note the similarity between equations 4.28 and 4.30. The left-hand side of both equations is identical, except for the independent variable, while the right-hand side takes a slightly different form. The solution of either equation is sufficient, however, to determine all voltages and currents in the circuit.

## Forced Response of Circuits Excited by Sinusoidal Sources

Consider again the circuit of Figure 4.25, where now the external source produces a sinusoidal voltage, described by the expression

$$v_S(t) = V\cos(\omega t) \tag{4.31}$$

Substituting the expression $V\cos(\omega t)$ in place of the source voltage, $v_S(t)$, in the differential equation obtained earlier (equation 4.30), we obtain the following differential equation:

$$\frac{d}{dt}v_C + \frac{1}{RC}v_C = \frac{1}{RC}V\cos\omega t \tag{4.32}$$

Since the forcing function is a sinusoid, the solution may also be assumed to be of the same form. An expression for $v_C(t)$ is then the following:

$$v_C(t) = A\sin\omega t + B\cos\omega t \tag{4.33}$$

which is equivalent to

$$v_C(t) = C\cos(\omega t + \phi) \tag{4.34}$$

Substituting equation 4.33 in the differential equation for $v_C(t)$ and solving for the coefficients $A$ and $B$ yields the expression

$$A\omega\cos\omega t - B\omega\sin\omega t + \frac{1}{RC}(A\sin\omega t + B\cos\omega t)$$

$$= \frac{1}{RC}V\cos\omega t \tag{4.35}$$

and if the coefficients of like terms are grouped, the following equation is obtained:

$$\left(\frac{A}{RC} - B\omega\right)\sin\omega t + \left(A\omega + \frac{B}{RC} - \frac{V}{RC}\right)\cos\omega t = 0 \qquad \textbf{(4.36)}$$

The coefficients of $\sin\omega t$ and $\cos\omega t$ must both be identically zero in order for equation 4.36 to hold. Thus,

$$\frac{A}{RC} - B\omega = 0$$

and                                                                                      **(4.37)**

$$A\omega + \frac{B}{RC} - \frac{V}{RC} = 0$$

The unknown coefficients, $A$ and $B$, may now be determined by solving equation 4.37, to obtain:

$$A = \frac{V\omega RC}{1 + \omega^2(RC)^2}$$

$$B = \frac{V}{1 + \omega^2(RC)^2} \qquad \textbf{(4.38)}$$

Thus, the solution for $v_C(t)$ may be written as follows:

$$v_C(t) = \frac{V\omega RC}{1 + \omega^2(RC)^2}\sin\omega t + \frac{V}{1 + \omega^2(RC)^2}\cos\omega t \qquad \textbf{(4.39)}$$

This response is plotted in Figure 4.26.

The solution method outlined in the previous paragraphs can become quite complicated for circuits containing a large number of elements; in particular, one may need to solve higher-order differential equations if more than one energy-storage element is present in the circuit. A simpler and preferred method for the solution of AC circuits will be presented in the next section. This brief section has provided a simple, but complete, illustration of the key elements of AC circuit analysis. These can be summarized in the following statement:

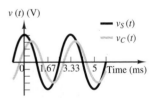

**Figure 4.26** Waveforms for the AC circuit of Figure 4.25

> In a sinusoidally excited linear circuit, all branch voltages and currents are *sinusoids* at the *same frequency* as the excitation signal. The amplitudes of these voltages and currents are a *scaled* version of the excitation *amplitude,* and the voltages and currents may be *shifted in phase* with respect to the excitation signal.

These observations indicate that three parameters uniquely define a sinusoid: *frequency, amplitude,* and *phase.* But if this is the case, is it necessary to carry the "excess luggage," that is, the sinusoidal functions? Might it be possible to simply keep track of the three parameters just mentioned? Fortunately, the answers to these two questions are no and yes, respectively. The next section will describe the use of a notation that, with the aid of complex algebra, eliminates the need for the sinusoidal functions of time, and for the formulation and solution of differential equations, permitting the use of simpler algebraic methods.

## Check Your Understanding

**4.11** Show that the solution to either equation 4.28 or equation 4.30 is sufficient to compute all of the currents and voltages in the circuit of Figure 4.25.

**4.12** Show that the equality

$$A \sin \omega t + B \cos \omega t = C \cos(\omega t + \phi)$$

holds if

$$A = -C \sin \phi$$
$$B = C \cos \phi$$

or, conversely, if

$$C = \sqrt{A^2 + B^2}$$

$$\phi = \tan^{-1}\left(\frac{-A}{B}\right)$$

**4.13** Use the result of Exercise 4.12 to compute $C$ and $\phi$ as functions of $V$, $\omega$, $R$, and $C$ in equation 4.39.

## 4.4    PHASORS AND IMPEDANCE

In this section, we introduce an efficient notation to make it possible to represent sinusoidal signals as *complex numbers,* and to eliminate the need for solving differential equations. The student who needs a brief review of complex algebra will find a reasonably complete treatment in Appendix A, including solved examples and Check Your Understanding exercises. For the remainder of the chapter, it will be assumed that you are familiar with both the rectangular and the polar forms of complex number coordinates, with the conversion between these two forms, and with the basic operations of addition, subtraction, multiplication, and division of complex numbers.

### Euler's Identity

Named after the Swiss mathematician Leonhard Euler (the last name is pronounced "Oiler"), Euler's identity forms the basis of phasor notation. Simply stated, the identity defines the **complex exponential** $e^{j\theta}$ as a point in the complex plane, which may be represented by real and imaginary components:

Leonhard Euler (1707–1783). *Photo courtesy of Deutsches Museum, Munich.*

$$e^{j\theta} = \cos \theta + j \sin \theta \tag{4.40}$$

Figure 4.27 illustrates how the complex exponential may be visualized as a point (or vector, if referenced to the origin) in the complex plane. Note immediately that the magnitude of $e^{j\theta}$ is equal to 1:

$$|e^{j\theta}| = 1 \tag{4.41}$$

since

$$|\cos \theta + j \sin \theta| = \sqrt{\cos^2 \theta + \sin^2 \theta} = 1 \tag{4.42}$$

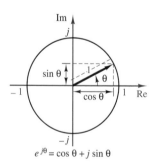

$$e^{j\theta} = \cos \theta + j \sin \theta$$

**Figure 4.27** Euler's identity

and note also that writing Euler's identity corresponds to equating the polar form of a complex number to its rectangular form. For example, consider a vector of length $A$ making an angle $\theta$ with the real axis. The following equation illustrates the relationship between the rectangular and polar forms:

$$Ae^{j\theta} = A \cos \theta + jA \sin \theta = A\angle\theta \tag{4.43}$$

In effect, Euler's identity is simply a trigonometric relationship in the complex plane.

## Phasors

To see how complex numbers can be used to represent sinusoidal signals, rewrite the expression for a generalized sinusoid in light of Euler's equation:

$$A \cos(\omega t + \phi) = \text{Re}\,[Ae^{j(\omega t + \phi)}] \tag{4.44}$$

This equality is easily verified by expanding the right-hand side, as follows:

$$\text{Re}\,[Ae^{j(\omega t + \phi)}] = \text{Re}\,[A \cos(\omega t + \phi) + jA \sin(\omega t + \phi)]$$

$$= A \cos(\omega t + \phi)$$

We see, then, that *it is possible to express a generalized sinusoid as the real part of a complex vector* whose **argument,** or **angle,** is given by $(\omega t + \phi)$ and whose length, or **magnitude,** is equal to the peak amplitude of the sinusoid. The **complex phasor** corresponding to the sinusoidal signal $A \cos(\omega t + \phi)$ is therefore defined to be the complex number $Ae^{j\phi}$:

$$Ae^{j\phi} = \text{complex phasor notation for } A \cos(\omega t + \phi) = A\angle\theta \tag{4.45}$$

It is important to explicitly point out that this is a *definition.* Phasor notation arises from equation 4.44; however, this expression is simplified (for convenience, as will be promptly shown) by removing the "real part of" operator (Re) and factoring out and deleting the term $e^{j\omega t}$. The next equation illustrates the simplification:

$$A \cos(\omega t + \phi) = \text{Re}\,[Ae^{j(\omega t + \phi)}] = \text{Re}\,[Ae^{j\phi}e^{j\omega t}] \tag{4.46}$$

The reason for this simplification is simply mathematical convenience, as will become apparent in the following examples; you will have to remember that the $e^{j\omega t}$ term that was removed from the complex form of the sinusoid is really still present, indicating the specific frequency of the sinusoidal signal, $\omega$. With these caveats, you should now be prepared to use the newly found phasor to analyze AC circuits. The following comments summarize the important points developed thus far in the section.

## FOCUS ON METHODOLOGY

1. Any sinusoidal signal may be mathematically represented in one of two ways: a **time-domain form,**

$$v(t) = A \cos(\omega t + \phi)$$

and a **frequency-domain** (or **phasor**) **form,**

$$\mathbf{V}(j\omega) = Ae^{j\phi} = A\angle\theta$$

Note the $j\omega$ in the notation $\mathbf{V}(j\omega)$, indicating the $e^{j\omega t}$ dependence of the phasor. In the remainder of this chapter, bold uppercase quantities will be employed to indicate phasor voltages or currents.

*(Continued)*

> (*Concluded*)
>
> 2. A phasor is a complex number, expressed in polar form, consisting of a *magnitude* equal to the peak amplitude of the sinusoidal signal and a *phase angle* equal to the phase shift of the sinusoidal signal *referenced to a cosine signal.*
>
> 3. When using phasor notation, it is important to make a note of the specific frequency, $\omega$, of the sinusoidal signal, since this is not explicitly apparent in the phasor expression.

### EXAMPLE 4.9 Addition of Two Sinusoidal Sources in Phasor Notation

#### Problem

Compute the phasor voltage resulting from the series connection of two sinusoidal voltage sources (Figure 4.28).

#### Solution

**Known Quantities:**

$$v_1(t) = 15 \, \cos(377t + \pi/4) \text{ V}$$
$$v_2(t) = 15 \, \cos(377t + \pi/12) \text{ V}$$

**Find:** Equivalent phasor voltage $v_S(t)$.

**Analysis:** Write the two voltages in phasor form:

$$\mathbf{V}_1(j\omega) = 15\angle\pi/4 \text{ V}$$
$$\mathbf{V}_2(j\omega) = 15e^{j\pi/12} = 15\angle\pi/12 \text{ V}$$

Convert the phasor voltages from polar to rectangular form:

$$\mathbf{V}_1(j\omega) = 10.61 + j10.61 \text{ V}$$
$$\mathbf{V}_2(j\omega) = 14.49 + j3.88$$

Then

$$\mathbf{V}_S(j\omega) = \mathbf{V}_1(j\omega) + \mathbf{V}_2(j\omega) = 25.10 + j14.49 = 28.98e^{j\pi/6} = 28.98\angle\pi/6 \text{ V}$$

Now we can convert $\mathbf{V}_S(j\omega)$ to its time-domain form:

$$v_S(t) = 28.98 \, \cos(377t + \pi/6) \text{ V}.$$

**Comments:** Note that we could have obtained the same result by adding the two sinusoids in the time domain, using trigonometric identities:

$$v_1(t) = 15 \, \cos(377t + \pi/4) = 15 \, \cos(\pi/4) \, \cos(377t) - 15 \, \sin(\pi/4) \, \sin(377t) \text{ V}$$
$$v_2(t) = 15 \, \cos(377t + \pi/12) = 15 \, \cos(\pi/12) \, \cos(377t) - 15 \, \sin(\pi/12) \, \sin(377t) \text{ V}.$$

Combining like terms, we obtain

$$v_1(t) + v_2(t) = 15[\cos(\pi/4) + \cos(\pi/12)] \, \cos(377t) - 15[\sin(\pi/4) + \sin(\pi/12)] \, \sin(377t)$$

$$= 15(1.673 \, \cos(377t) - 0.966 \, \sin(377t))$$

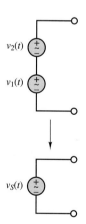

**Figure 4.28**

$v_2(t)$

$v_1(t)$

$v_S(t)$

$$= 15\sqrt{(1.673)^2 + (0.966)^2} \times \cos\left[377t + \arctan\left(\frac{0.966}{1.673}\right)\right]$$

$$= 15(1.932\cos(377t + \pi/6) = 28.98 \ \cos(377t + \pi/6) \ \text{V}.$$

The above expression is, of course, identical to the one obtained by using phasor notation, but it required a greater amount of computation. In general, phasor analysis greatly simplifies calculations related to sinusoidal voltages and currents.

It should be apparent by now that phasor notation can be a very efficient technique to solve AC circuit problems. The following sections will continue developing this new method to build your confidence in using it.

## Superposition of AC Signals

Example 4.9 explored the combined effect of two sinusoidal sources of different phase and amplitude, but of the same frequency. It is important to realize that the simple answer obtained there does not apply to the superposition of two (or more) sinusoidal sources that *are not at the same frequency*. In this subsection, the case of two sinusoidal sources oscillating at different frequencies will be used to illustrate how phasor analysis can deal with this more general case.

The circuit shown in Figure 4.29 depicts a source excited by two current sources connected in parallel, where

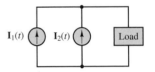

**Figure 4.29** Superposition of AC currents

$$i_1(t) = A_1\cos(\omega_1 t)$$
$$i_2(t) = A_2\cos(\omega_2 t)$$

(4.47)

The load current is equal to the sum of the two source currents; that is,

$$i_L(t) = i_1(t) + i_2(t)$$

(4.48)

or, in phasor form,

$$\mathbf{I}_L = \mathbf{I}_1 + \mathbf{I}_2$$

(4.49)

At this point, you might be tempted to write $\mathbf{I}_1$ and $\mathbf{I}_2$ in a more explicit phasor form as

$$\mathbf{I}_1 = A_1 e^{j0}$$
$$\mathbf{I}_2 = A_2 e^{j0}$$

(4.50)

and to add the two phasors using the familiar techniques of complex algebra. However, this approach *would be incorrect*. Whenever a sinusoidal signal is expressed in phasor notation, the term $e^{j\omega t}$ is implicitly present, where $\omega$ is the actual radian frequency of the signal. In our example, the two frequencies are not the same, as can be verified by writing the phasor currents in the form of equation 4.46:

$$\mathbf{I}_1 = \text{Re}\,[A_1 e^{j0} e^{j\omega_1 t}]$$
$$\mathbf{I}_2 = \text{Re}\,[A_2 e^{j0} e^{j\omega_2 t}]$$

(4.51)

Since phasor notation does not *explicitly* include the $e^{j\omega t}$ factor, this can lead to serious errors if you are not careful! The two phasors of equation 4.50 cannot be added, but must be kept separate; thus, the only unambiguous expression for the load current in this case is equation 4.48. In order to complete the analysis of any circuit with multiple sinusoidal sources at different frequencies using phasors, it is necessary to solve the circuit separately for each signal and then add the individual answers obtained for the different excitation sources. Example 4.10 illustrates the response of a circuit with two separate AC excitations using AC superposition.

### EXAMPLE   4.10  Example of AC Superposition

#### Problem

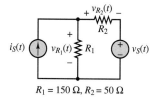

$R_1 = 150\ \Omega,\ R_2 = 50\ \Omega$

**Figure 4.30**

Compute the voltages $v_{R1}(t)$ and $v_{R2}(t)$ in the circuit of Figure 4.30.

#### Solution

**Known Quantities:**

$$i_S(t) = 0.5\ \cos(2\pi\,100t)\ \text{A}$$
$$v_S(t) = 20\ \cos(2\pi\,1{,}000t)\ \text{V}$$

**Find:** $v_{R1}(t)$ and $v_{R2}(t)$.

**Analysis:** Since the two sources are at different frequencies, we must compute a separate solution for each. Consider the current source first, with the voltage source set to zero (short circuit) as shown in Figure 4.31. The circuit thus obtained is a simple current divider. Write the source current in phasor notation:

$$I_S(j\omega) = 0.5e^{j0} = 0.5\angle 0\ \text{A} \qquad \omega = 2\pi\,100\angle\text{rad/s}$$

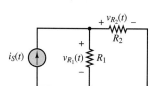

**Figure 4.31**

Then,

$$\mathbf{V}_{R1}(\mathbf{I}_S) = \mathbf{I}_S\,\frac{R_2}{R_1 + R_2}\,R_1 = 0.5\ \angle 0\left(\frac{50}{150 + 50}\right)150 = 18.75\ \angle 0\ \text{V}$$

$$\omega = 2\pi\,100\ \text{rad/s}$$

$$\mathbf{V}_{R2}(\mathbf{I}_S) = \mathbf{I}_S\,\frac{R_1}{R_1 + R_2}\,R_2 = 0.5\ \angle 0\left(\frac{150}{150 + 50}\right)50 = 18.75\ \angle 0\ \text{V}$$

$$\omega = 2\pi\,100\ \text{rad/s}$$

Next, we consider the voltage source, with the current source set to zero (open circuit), as shown in Figure 4.32. We first write the source voltage in phasor notation:

$$\mathbf{V}_S(j\omega) = 20e^{j0} = 20\angle 0\ \text{V} \qquad \omega = 2\pi\,1{,}000\ \text{rad/s}$$

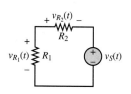

**Figure 4.32**

Then we apply the voltage divider law to obtain

$$\mathbf{V}_{R1}(\mathbf{V}_S) = \mathbf{V}_S\,\frac{R_1}{R_1 + R_2} = 20\angle 0\left(\frac{150}{150 + 50}\right) = 15\ \angle 0\ \text{V}$$

$$\omega = 2\pi\,1{,}000\ \text{rad/s}$$

$$\mathbf{V}_{R2}(\mathbf{V}_S) = -\mathbf{V}_S\,\frac{R_2}{R_1 + R_2} = -20\angle 0\left(\frac{50}{150 + 50}\right) = -5\angle 0 = 5\angle\pi\ \text{V}$$

$$\omega = 2\pi\,1{,}000\ \text{rad/s}$$

Now we can determine the voltage across each resistor by adding the contributions from each source and converting the phasor form to time-domain representation:

$$\mathbf{V}_{R1} = \mathbf{V}_{R1}(\mathbf{I}_S) + \mathbf{V}_{R1}(\mathbf{V}_S)$$
$$v_{R1}(t) = 18.75\ \cos(2\pi\,100t) + 15\ \cos(2\pi\,1{,}000t)\ \text{V}$$

and

$$\mathbf{V}_{R2} = \mathbf{V}_{R2}(\mathbf{I}_S) + \mathbf{V}_{R2}(\mathbf{V}_S)$$
$$v_{R2}(t) = 18.75 \ \cos(2\pi 100t) + 5 \ \cos(2\pi 1{,}000t + \pi) \ \text{V}.$$

**Comments:**  Note that it is impossible to simplify the final expression any further, because the two components of each voltage are at different frequencies.

---

## Impedance

We now analyze the *i-v* relationship of the three ideal circuit elements in light of the new phasor notation. The result will be a new formulation in which resistors, capacitors, and inductors will be described in the same notation. A direct consequence of this result will be that the circuit theorems of Chapter 3 will be extended to AC circuits. In the context of AC circuits, any one of the three ideal circuit elements defined so far will be described by a parameter called **impedance,** which may be viewed as a *complex resistance*. The impedance concept is equivalent to stating that capacitors and inductors act as *frequency-dependent resistors*, that is, as resistors whose resistance is a function of the frequency of the sinusoidal excitation. Figure 4.33 depicts the same circuit represented in conventional form (top) and in phasor-impedance form (bottom); the latter representation explicitly shows phasor voltages and currents and treats the circuit element as a generalized "impedance." It will presently be shown that each of the three ideal circuit elements may be represented by one such impedance element.

Let the source voltage in the circuit of Figure 4.33 be defined by

$$v_S(t) = A \cos \omega t \qquad \text{or} \qquad \mathbf{V}_S(j\omega) = Ae^{j0^\circ} = A\angle 0 \tag{4.52}$$

without loss of generality. Then the current $i(t)$ is defined by the *i-v* relationship for each circuit element. Let us examine the frequency-dependent properties of the resistor, inductor, and capacitor, one at a time.

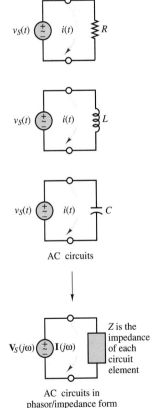

AC circuits

Z is the impedance of each circuit element

AC circuits in phasor/impedance form

**Figure 4.33** The impedance element

### The Resistor

Ohm's law dictates the well-known relationship $v = iR$. In the case of sinusoidal sources, then, the current flowing through the resistor of Figure 4.33 may be expressed as

$$i(t) = \frac{v_S(t)}{R} = \frac{A}{R} \cos(\omega t) \tag{4.53}$$

Converting the voltage $v_S(t)$ and the current $i(t)$ to phasor notation, we obtain the following expressions:

$$\mathbf{V}_S(j\omega) = A\angle 0$$
$$\mathbf{I}(j\omega) = \frac{A}{R}\angle 0 \tag{4.54}$$

Finally, the *impedance* of the resistor is defined as the ratio of the phasor voltage across the resistor to the phasor current flowing through it, and the symbol $Z_R$ is used to denote it:

$$Z_R(j\omega) = \frac{\mathbf{V}_S(j\omega)}{\mathbf{I}(j\omega)} = R \qquad \text{Impedance of a resistor} \tag{4.55}$$

Equation 4.55 corresponds to Ohm's law in phasor form, and the result should be intuitively appealing: Ohm's law applies to a resistor independent of the particular form of the voltages and currents (whether AC or DC, for instance). The ratio of phasor voltage to phasor current has a very simple form in the case of the resistor. In general, however, the impedance of an element is a complex function of frequency, as it must be, since it is the ratio of two phasor quantities, which are frequency-dependent. This property will become apparent when the impedances of the inductor and capacitor are defined.

### The Inductor

Recall the defining relationships for the ideal inductor (equations 4.9 and 4.12), repeated here for convenience:

$$v_L(t) = L\frac{di_L(t)}{dt}$$

$$i_L(t) = \frac{1}{L}\int v_L(t') \tag{4.56}$$

Let $v_L(t) = v_S(t)$ and $i_L(t) = i(t)$ in the circuit of Figure 4.33. Then the following expression may be derived for the inductor current:

$$i_L(t) = i(t) = \frac{1}{L}\int v_S(t')\, dt'$$

$$i_L(t) = \frac{1}{L}\int A\cos\omega t'\, dt' \tag{4.57}$$

$$= \frac{A}{\omega L}\sin\omega t$$

Note how a dependence on the radian frequency of the source is clearly present in the expression for the inductor current. Further, the inductor current is shifted in phase (by 90°) with respect to the voltage. This fact can be seen by writing the inductor voltage and current in time-domain form:

$$v_S(t) = v_L(t) = A\cos\omega t$$

$$i(t) = i_L(t) = \frac{A}{\omega L}\cos\left(\omega t - \frac{\pi}{2}\right) \tag{4.58}$$

It is evident that the current is not just a scaled version of the source voltage, as it was for the resistor. Its magnitude depends on the frequency, $\omega$, and it is shifted (delayed) in phase by $\pi/2$ radians, or 90°. Using phasor notation, equation 4.58 becomes

$$\mathbf{V}_S(j\omega) = A\angle 0$$

$$\mathbf{I}(j\omega) = \frac{A}{\omega L}\angle{-\pi/2} \tag{4.59}$$

Thus, the impedance of the inductor is defined as follows:

$$Z_L(j\omega) = \frac{\mathbf{V}_S(j\omega)}{\mathbf{I}(j\omega)} = \omega L \angle \pi/2 = j\omega L \qquad \text{Impedance of an inductor} \qquad \textbf{(4.60)}$$

Note that the inductor now appears to behave like a *complex frequency-dependent resistor,* and that the magnitude of this complex resistor, $\omega L$, is proportional to the signal frequency, $\omega$. Thus, an inductor will "impede" current flow in proportion to the sinusoidal frequency of the source signal. This means that at low signal frequencies, an inductor acts somewhat like a short circuit, while at high frequencies it tends to behave more as an open circuit.

## The Capacitor

An analogous procedure may be followed to derive the equivalent result for a capacitor. Beginning with the defining relationships for the ideal capacitor,

$$i_C(t) = C \frac{dv_C(t)}{dt}$$

$$\qquad \qquad \qquad \qquad \qquad \qquad \qquad \qquad \textbf{(4.61)}$$

$$v_C(t) = \frac{1}{C} \int i_C(t')\, dt'$$

with $i_C = i$ and $v_C = v_S$ in Figure 4.33, the capacitor current may be expressed as:

$$i_C(t) = C \frac{dv_C(t)}{dt}$$

$$= C \frac{d}{dt}(A \cos \omega t) \qquad \qquad \qquad \textbf{(4.62)}$$

$$= -C(A\omega \sin \omega t)$$

$$= \omega C A \cos(\omega t + \pi/2)$$

so that, in phasor form,

$$\mathbf{V}_S(j\omega) = A\angle 0$$

$$\qquad \qquad \qquad \qquad \qquad \qquad \qquad \textbf{(4.63)}$$

$$\mathbf{I}(j\omega) = \omega C A \angle \pi/2$$

The impedance of the ideal capacitor, $Z_C(j\omega)$, is therefore defined as follows:

$$Z_C(j\omega) = \frac{\mathbf{V}_S(j\omega)}{\mathbf{I}(j\omega)} = \frac{1}{\omega C}\angle -\pi/2$$

$$= \frac{-j}{\omega C} = \frac{1}{j\omega C} \qquad \text{Impedance of a capacitor} \qquad \textbf{(4.64)}$$

where we have used the fact that $1/j = e^{-j\pi/2} = -j$. Thus, the impedance of a capacitor is also a frequency-dependent complex quantity, with the impedance of the capacitor varying as an inverse function of frequency; and so a capacitor acts

156

Chapter 4 AC Network Analysis

like a short circuit at high frequencies, whereas it behaves more like an open circuit at low frequencies. Figure 4.34 depicts $Z_C(j\omega)$ in the complex plane, alongside $Z_R(j\omega)$ and $Z_L(j\omega)$.

The impedance parameter defined in this section is extremely useful in solving AC circuit analysis problems, because it will make it possible to take advantage of most of the network theorems developed for DC circuits by replacing resistances with complex-valued impedances. The examples that follow illustrate how branches containing series and parallel elements may be reduced to a single equivalent impedance, much in the same way resistive circuits were reduced to equivalent forms. It is important to emphasize that although the impedance of simple circuit elements is either purely real (for resistors) or purely imaginary (for capacitors and inductors), the general definition of impedance for an arbitrary circuit must allow for the possibility of having both a real and an imaginary part, since practical circuits are made up of more or less complex interconnections of different circuit elements. In its most general form, the impedance of a circuit element is defined as the sum of a real part and an imaginary part:

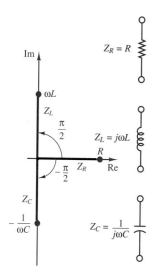

**Figure 4.34** Impedances of $R$, $L$, and $C$ in the complex plane

$$Z(j\omega) = R(j\omega) + jX(j\omega) \tag{4.65}$$

where $R$ is called the **AC resistance** and $X$ is called the **reactance.** The frequency dependence of $R$ and $X$ has been indicated explicitly, since it is possible for a circuit to have a frequency-dependent resistance. Note that the reactances of equations 4.60 and 4.64 have units of ohms, and that **inductive reactance** is always positive, while **capacitive reactance** is always negative. The following examples illustrate how a complex impedance containing both real and imaginary parts arises in a circuit.

### EXAMPLE 4.11 Impedance of a Practical Capacitor

#### Problem

A practical capacitor can be modeled by an ideal capacitor in parallel with a resistor (Figure 4.35). The parallel resistance represents leakage losses in the capacitor and is usually quite large. Find the impedance of a practical capacitor at the radian frequency $\omega = 377$ rad/s. How will the impedance change if the capacitor is used at a much higher frequency, say 800 MHz?

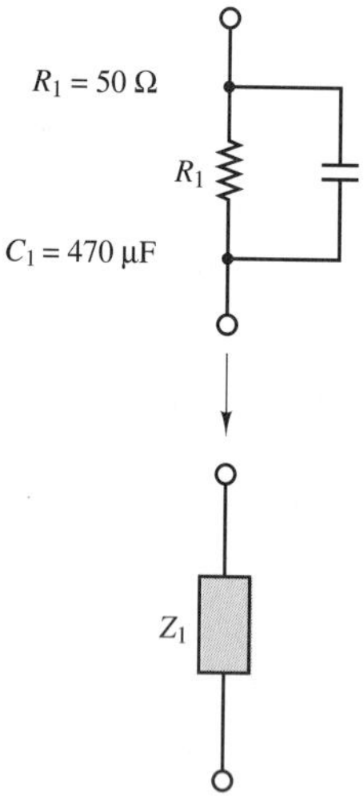

$R_1 = 50\ \Omega$

$C_1 = 470\ \mu F$

**Figure 4.35**

#### Solution

**Known Quantities:** $C_1 = 0.1\ \mu F = 0.1 \times 10^{-6}$ F; $R_1 = 1$ MΩ.

**Find:** The equivalent impedance of the parallel circuit, $Z_1$.

**Analysis:** To determine the equivalent impedance we combine the two impedances in parallel:

$$Z_1 = R_1 \left\| \frac{1}{j\omega C_1} = \frac{R_1 \frac{1}{j\omega C_1}}{R_1 + \frac{1}{j\omega C_1}} = \frac{R_1}{1 + j\omega C_1 R_1} \right.$$

Substituting numerical values, we find

$$Z_1(\omega = 377) = \frac{10^6}{1 + j377 \times 10^6 \times 0.1 \times 10^{-6}} = \frac{10^6}{1 + j37.7}$$

$$= 2.6516 \times 10^4 \angle - 1.5443 \; \Omega$$

The impedance of the capacitor alone at this frequency would be:

$$Z_{C_1}(\omega = 377) = \frac{1}{j377 \times 0.1 \times 10^{-6}} = 26.53 \times 10^3 \angle -\pi/2 \; \Omega$$

If the frequency is increased to 800 MHz, or $1600\pi \times 10^6$ rad/s—a radio frequency in the AM range—we can recompute the impedance to be:

$$Z_1(\omega = 1600\pi \times 10^6) = \frac{10^6}{1 + j1600\pi \times 10^6 \times 0.1 \times 10^{-6} \times 10^6}$$

$$= \frac{10^6}{1 + j160\pi \times 10^6} = 0.002 \angle -1.5708 \; \Omega$$

The impedance of the capacitor alone at this frequency would be:

$$Z_{C_1}(\omega = 1600\pi \times 10^6) = \frac{1}{j1600\pi \times 10^6 \times 0.1 \times 10^{-6}} = 0.002 \; \angle -\pi/2 \; \Omega$$

**Comments:**  Note that the effect of the parallel resistance at the lower frequency (corresponding to the well-known 60-Hz AC power frequency) is significant: The effective impedance of the practical capacitor is substantially different from that of the ideal capacitor. On the other hand, at much higher frequency, the parallel resistance has an impedance so much larger than that of the capacitor that it effectively acts as an open circuit, and there is no difference between the ideal and practical capacitor impedances. This example suggests that the behavior of a circuit element depends very much in the frequency of the voltages and currents in the circuit. We should also note that the inductance of the wires may become significant at high frequencies.

---

## EXAMPLE  4.12  Impedance of a Practical Inductor

### Problem

A practical inductor can be modeled by an ideal inductor in series with a resistor. Figure 4.36 shows a *toroidal* (doughnut-shaped) inductor. The series resistance represents the resistance of the coil wire and is usually small. Find the range of frequencies over which the impedance of this practical inductor is largely *inductive* (i.e., due to the inductance in the circuit). We shall consider the impedance to be inductive if the impedance of the inductor in the circuit of Figure 4.37 is at least 10 times as large as that of the resistor.

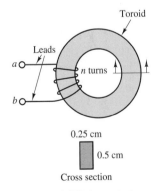

**Figure 4.36** A practical inductor

### Solution

**Known Quantities:**  $L = 0.098$ H; lead length $= l_c = 2 \times 10$ cm; $n = 250$ turns; wire is 30 gauge. Resistance of 30 gauge wire $= 0.344 \; \Omega/m$.

**Find:**  The range of frequencies over which the practical inductor acts nearly like an ideal inductor.

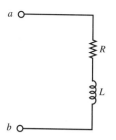

**Figure 4.37**

**Analysis:** We first determine the equivalent resistance of the wire used in the practical inductor using the cross section as an indication of the wire length, $l_w$, used in the coil:

$$l_w = 250 \times (2 \times 0.25 + 2 \times 0.5) = 375 \text{ cm}$$

$$l = \text{ Total length} = l_w + l_c = 375 + 20 = 395 \text{ cm}$$

The total resistance is therefore

$$R = 0.344 \text{ }\Omega/\text{m } \times 0.395 \text{ m} = 0.136 \text{ }\Omega$$

Thus, we wish to determine the range of radian frequencies, $\omega$, over which the magnitude of $j\omega L$ is greater than $10 \times 0.136$ $\Omega$:

$$\omega L > 1.36, \text{ or } \omega > 1.36/L = 1.36/0.098 = 1.39 \text{ rad/s}.$$

Alternatively, the range is $f = \omega/2\pi > 0.22$ Hz.

**Comments:** Note how the resistance of the coil wire is relatively insignificant. This is true because the inductor is rather large; wire resistance can become significant for very small inductance values. At high frequencies, a capacitance should be added to the model because of the effect of the insulator separating the coil wires.

---

### EXAMPLE 4.13 Impedance of a More Complex Circuit

**Problem**

Find the equivalent impedance of the circuit shown in Figure 4.38.

---

**Solution**

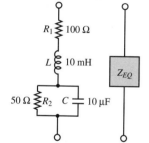

**Figure 4.38**

**Known Quantities:** $\omega = 10^4$ rad/s; $R_1 = 100$ $\Omega$; $L = 10$ mH; $R_2 = 50$ $\Omega$, $C = 10$ $\mu$F.

**Find:** The equivalent impedance of the series-parallel circuit.

**Analysis:** We determine first the parallel impedance of the $R_2$-$C$ circuit, $Z_{\|}$.

$$Z_{\|} = R_2 \left\| \frac{1}{j\omega C} = \frac{R_2 \frac{1}{j\omega C}}{R_2 + \frac{1}{j\omega C}} = \frac{R_2}{1 + j\omega C R_2} \right.$$

$$= \frac{50}{1 + j10^4 \times 10 \times 10^{-6} \times 50} = \frac{50}{1 + j5} = 1.92 - j9.62$$

$$= 9.81\angle -1.3734 \text{ }\Omega$$

Next, we determine the equivalent impedance, $Z_{eq}$:

$$Z_{eq} = R_1 + j\omega L + Z_{\|} = 100 + j10^4 \times 10^{-2} + 1.92 - j9.62$$

$$= 101.92 + j90.38 = 136.2\angle 0.723 \text{ }\Omega$$

Is this impedance inductive or capacitive in nature?

**Comments:** At the frequency used in this example, the circuit has an inductive impedance, since the reactance is positive (or, alternatively, the phase angle is positive).

## Capacitive Displacement Transducer

Earlier, we introduced the idea of a capacitive displacement transducer when
we considered a parallel-plate capacitor composed of a fixed plate and a
movable plate. The capacitance of this variable capacitor was shown to be a
*nonlinear* function of the position of the movable plate, $x$ (see Figure 4.6).
In this example, we show that under certain conditions the impedance of the
capacitor varies as a *linear* function of displacement—that is, the
movable-plate capacitor can serve as a linear transducer.

Recall the expression derived earlier:

$$C = \frac{8.854 \times 10^{-3} A}{x} \text{ pF}$$

where $C$ is the capacitance in pF, $A$ is the area of the plates in mm$^2$, and $x$ is
the (variable) distance in mm. If the capacitor is placed in an AC circuit, its
impedance will be determined by the expression

$$Z_C = \frac{1}{j\omega C}$$

so that

$$Z_C = \frac{x}{j\omega 8.854 A} \ \Omega$$

Thus, at a fixed frequency $\omega$, the impedance of the capacitor will vary
linearly with displacement. This property may be exploited in the bridge
circuit of Figure 4.7, where a differential pressure transducer was shown as
being made of two movable-plate capacitors, such that if the capacitance of
one increased as a consequence of a pressure differential across the
transducer, the capacitance of the other had to decrease by a corresponding
amount (at least for small displacements). The circuit is shown again in
Figure 4.39, where two resistors have been connected in the bridge along
with the variable capacitors (denoted by $C(x)$). The bridge is excited by a
sinusoidal source.

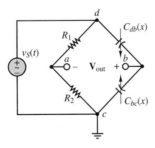

**Figure 4.39** Bridge circuit
for capacitive displacement
transducer

Using phasor notation, we can express the output voltage as follows:

$$\mathbf{V}_{out}(j\omega) = \mathbf{V}_S(j\omega) \left( \frac{Z_{C_{bc}(x)}}{Z_{C_{db}(x)} + Z_{C_{bc}(x)}} - \frac{R_2}{R_1 + R_2} \right)$$

If the nominal capacitance of each movable-plate capacitor with the diaphragm in the center position is given by

$$C = \frac{\varepsilon A}{d}$$

where $d$ is the nominal (undisplaced) separation between the diaphragm and the fixed surfaces of the capacitors (in mm), the capacitors will see a change in capacitance given by

$$C_{db} = \frac{\varepsilon A}{d - x} \qquad \text{and} \qquad C_{bc} = \frac{\varepsilon A}{d + x}$$

when a pressure differential exists across the transducer, so that the impedances of the variable capacitors change according to the displacement:

$$Z_{C_{db}} = \frac{d - x}{j\omega 8.854A} \qquad \text{and} \qquad Z_{C_{bc}} = \frac{d + x}{j\omega 8.854A}$$

and we obtain the following expression for the phasor output voltage:

$$\mathbf{V}_{\text{out}}(j\omega) = \mathbf{V}_S(j\omega) \left( \frac{\dfrac{d + x}{j\omega 8.854A}}{\dfrac{d - x}{j\omega 8.854A} + \dfrac{d + x}{j\omega 8.854A}} - \frac{R_2}{R_1 + R_2} \right)$$

$$= \mathbf{V}_S(j\omega) \left( \frac{1}{2} + \frac{x}{2d} - \frac{R_2}{R_1 + R_2} \right)$$

$$= \mathbf{V}_S(j\omega) \frac{x}{2d}$$

if we choose $R_1 = R_2$. Thus, the output voltage will vary as a scaled version of the input voltage in proportion to the displacement. A typical $v_{\text{out}}(t)$ is displayed in Figure 4.40 for a 0.05-mm "triangular" diaphragm displacement, with $d = 0.5$ mm and $\mathbf{V}_S$ a 25-Hz sinusoid with 1-V amplitude.

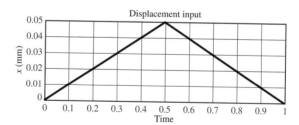

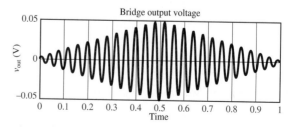

**Figure 4.40** Displacement input and bridge output voltage for capacititve displacement transducer

## Admittance

In Chapter 3, it was suggested that the solution of certain circuit analysis problems was handled more easily in terms of conductances than resistances. In AC circuit analysis, an analogous quantity may be defined, the reciprocal of complex impedance. Just as the conductance, $G$, of a resistive element was defined as the inverse of the resistance, the **admittance** of a branch is defined as follows:

$$Y = \frac{1}{Z} \text{ S} \tag{4.66}$$

Note immediately that whenever $Z$ is purely real—that is, when $Z = R + j0$—the admittance $Y$ is identical to the conductance $G$. In general, however, $Y$ is the complex number

$$Y = G + jB \tag{4.67}$$

where $G$ is called the **AC conductance** and $B$ is called the **susceptance;** the latter plays a role analogous to that of reactance in the definition of impedance. Clearly, $G$ and $B$ are related to $R$ and $X$. However, this relationship is not as simple as an inverse. Let $Z = R + jX$ be an arbitrary impedance. Then, the corresponding admittance is

$$Y = \frac{1}{Z} = \frac{1}{R + jX} \tag{4.68}$$

In order to express $Y$ in the form $Y = G + jB$, we multiply numerator and denominator by $R - jX$:

$$Y = \frac{1}{R + jX} \frac{R - jX}{R - jX} = \frac{R - jX}{R^2 + X^2}$$
$$= \frac{R}{R^2 + X^2} - j\frac{X}{R^2 + X^2} \tag{4.69}$$

and conclude that

$$G = \frac{R}{R^2 + X^2}$$
$$B = \frac{-X}{R^2 + X^2} \tag{4.70}$$

Notice in particular that $G$ *is not the reciprocal of* $R$ in the general case!

The following example illustrates the determination of $Y$ for some common circuits.

## EXAMPLE  4.14  Admittance

### Problem

Find the equivalent admittance of the two circuits shown in Figure 4.41.

### Solution

**Known Quantities:** $\omega = 2\pi \times 10^3$ rad/s; $R_1 = 150 \ \Omega$; $L = 16$ mH; $R_2 = 100 \ \Omega$, $C = 3 \ \mu F$.

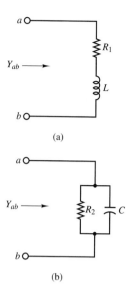

(a)

$Y_{ab}$ ⟶

$Y_{ab}$ ⟶

(b)

**Figure 4.41**

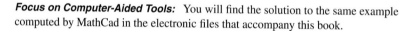

VIRTUAL LAB

**Find:** The equivalent admittance of the two circuits.

**Analysis:** Circuit (a): First, determine the equivalent impedance of the circuit:

$$Z_{ab} = R_1 + j\omega L$$

Then compute the inverse of $Z_{ab}$ to obtain the admittance:

$$Y_{ab} = \frac{1}{R_1 + j\omega L} = \frac{R_1 - j\omega L}{R_1^2 + (\omega L)^2}$$

Substituting numerical values gives

$$Y_{ab} = \frac{1}{50 + j2\pi \times 10^3} = \frac{1}{50 + j100.5} = 3.968 \times 10^{-3} - j7.976 \times 10^{-3} \text{ S}$$

Circuit (b): First, determine the equivalent impedance of the circuit:

$$Z_{ab} = R_2 \left\| \frac{1}{j\omega C} = \frac{R_2}{1 + j\omega R_2 C} \right.$$

Then compute the inverse of $Z_{ab}$ to obtain the admittance:

$$Y_{ab} = \frac{1 + j\omega R_2 C}{R_2} = \frac{1}{R_2} + j\omega C = 0.01 + j0.019 \text{ S}$$

**Comments:** Note that the units of admittance are siemens, that is, the same as the units of conductance.

**Focus on Computer-Aided Tools:** You will find the solution to the same example computed by MathCad in the electronic files that accompany this book.

## Check Your Understanding

**4.14**  Add the sinusoidal voltages $v_1(t) = A \cos(\omega t + \phi)$ and $v_2(t) = B \cos(\omega t + \theta)$ using phasor notation, and then convert back to time-domain form, for:

    a. $A = 1.5$ V, $\phi = 10°$; $B = 3.2$ V, $\theta = 25°$.
    b. $A = 50$ V, $\phi = -60°$; $B = 24$, $\theta = 15°$.

**4.15**  Add the sinusoidal currents $i_1(t) = A \cos(\omega t + \phi)$ and $i_2(t) = B \cos(\omega t + \theta)$ for:

    a. $A = 0.09$ A, $\phi = 72°$; $B = 0.12$ A, $\theta = 20°$.
    b. $A = 0.82$ A, $\phi = -30°$; $B = 0.5$ A, $\theta = -36°$.

**4.16**  Compute the equivalent impedance of the circuit of Example 4.13 for $\omega = 1,000$ and 100,000 rad/s.

**4.17**  Compute the equivalent admittance of the circuit of Example 4.13.

**4.18**  Calculate the equivalent series capacitance of the parallel $R_2$-$C$ circuit of Example 4.13 at the frequency $\omega = 10$ rad/s.

## 4.5    AC CIRCUIT ANALYSIS METHODS

This section will illustrate how the use of phasors and impedance facilitates the solution of AC circuits by making it possible to use the same solution methods developed in Chapter 3 for DC circuits. The AC circuit analysis problem of interest in this section consists of determining the unknown voltage (or currents) in a circuit containing linear passive circuit elements ($R$, $L$, $C$) and excited by a sinusoidal source. Figure 4.42 depicts one such circuit, represented in both conventional time-domain form and phasor-impedance form.

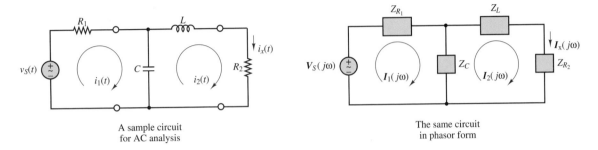

A sample circuit
for AC analysis

The same circuit
in phasor form

**Figure 4.42** An AC circuit

The first step in the analysis of an AC circuit is to note the frequency of the sinusoidal excitation. Next, all sources are converted to phasor form, and each circuit element to impedance form. This is illustrated in the phasor circuit of Figure 4.42. At this point, if the excitation frequency, $\omega$, is known numerically, it will be possible to express each impedance in terms of a known amplitude and phase, and a numerical answer to the problem will be found. It does often happen, however, that one is interested in a more general circuit solution, valid for an arbitrary excitation frequency. In this latter case, the solution becomes a function of $\omega$. This point will be developed further in Chapter 6, where the concept of sinusoidal frequency response is discussed.

With the problem formulated in phasor notation, the resulting solution will be in phasor form and will need to be converted to time-domain form. In effect, the use of phasor notation is but an intermediate step that greatly facilitates the computation of the final answer. In summary, here is the procedure that will be followed to solve an AC circuit analysis problem. Example 4.15 illustrates the various aspects of this method.

## FOCUS ON METHODOLOGY

### AC Circuit Analysis

1. Identify the sinusoidal source(s) and note the excitation frequency.
2. Convert the source(s) to phasor form.
3. Represent each circuit element by its impedance.
4. Solve the resulting phasor circuit, using appropriate network analysis tools.
5. Convert the (phasor-form) answer to its time-domain equivalent, using equation 4.46.

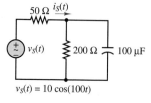

$v_S(t) = 10 \cos(100t)$

**Figure 4.43**

### EXAMPLE 4.15 Phasor Analysis of AC Circuit

**Problem**

Apply the phasor analysis method just described to the circuit of Figure 4.43 to determine the source current.

---

**Solution**

**Known Quantities:** $\omega = 100$ rad/s; $R_1 = 50\ \Omega$; $R_2 = 200\ \Omega$, $C = 100\ \mu$F.

**Find:** The source current $i_S(t)$.

**Analysis:** Define the voltage $v$ at the top node and use nodal analysis to determine $v$. Then observe that

$$i_S(t) = \frac{v_S(t) - v(t)}{R_1}$$

Next, we follow the steps outlined in the Methodology Box: "AC Circuit Analysis."

Step 1: $v_S(t) = 10 \cos(100t)$ V; $\omega = 100$ rad/s.

Step 2: $\mathbf{V}_S(j\omega) = 10\angle 0$ V.

Step 3: $Z_{R1} = 50\ \Omega$, $Z_{R2} = 200\ \Omega$, $Z_C = 1/(j100 \times 10^{-4}) = -j100\ \Omega$. The resulting phasor circuit is shown in Figure 4.44.

Step 4: Next, we solve for the source current using nodal analysis. First we find $\mathbf{V}$:

$$\frac{\mathbf{V}_S - \mathbf{V}}{Z_{R1}} = \frac{\mathbf{V}}{Z_{R2}||Z_C}$$

$$\frac{\mathbf{V}_S}{Z_{R1}} = \mathbf{V}\left(\frac{1}{Z_{R2}||Z_C} + \frac{1}{Z_{R1}}\right)$$

$$\mathbf{V} = \left(\frac{1}{Z_{R2}||Z_C} + \frac{1}{Z_{R1}}\right)^{-1}\frac{\mathbf{V}_S}{Z_{R1}} = \left(\frac{1}{40 - j80} + \frac{1}{50}\right)^{-1}\frac{\mathbf{V}_S}{50}$$

$$= 7.428\ \angle -0.381\ \text{V}$$

Then we compute $\mathbf{I}_S$:

$$\mathbf{I}_S = \frac{\mathbf{V}_S - \mathbf{V}}{Z_{R1}} = 0.083\angle 0.727\ \text{A}$$

Step 5: Finally, we convert the phasor answer to time domain notation:
$i_s(t) = 0.083\ \cos(100t + 0.727)$ A.

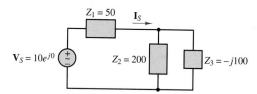

**Figure 4.44**

VIRTUAL LAB

**Focus on Computer-Aided Tools:** You will find the solution to the same example computed by MathCad in the electronic files that accompany this book. An EWB solution is also enclosed.

## EXAMPLE 4.16  AC Circuit Solution for Arbitrary Sinusoidal Input

### Problem

Determine the general solution of Example 4.15 for any sinusoidal source, $A \cos(\omega t + \phi)$.

### Solution

**Known Quantities:** $R_1 = 50\ \Omega$; $R_2 = 200\ \Omega$, $C = 100\ \mu\text{F}$.

**Find:** The phasor source current $\mathbf{I}_S(j\omega)$.

**Analysis:** Since the radian frequency is arbitrary, it will be impossible to determine a numerical answer. The answer will be a function of $\omega$. The source in phasor form is represented by the expression $\mathbf{V}_S(j\omega) = A\angle\phi$. The impedances will be $Z_{R1} = 50\ \Omega$; $Z_{R2} = 200\ \Omega$; $Z_C = -j10^4/\omega\ \Omega$. Note that the impedance of the capacitor is a function of $\omega$.

Taking a different approach from Example 4.15, we observe that the source current is given by the expression

$$\mathbf{I}_S = \frac{\mathbf{V}_S}{Z_{R1} + Z_{R2}||Z_C}$$

The parallel impedance $Z_{R2}||Z_C$ is given by the expression

$$Z_{R2}||Z_C = \frac{Z_{R2} \times Z_C}{Z_{R2} + Z_C} = \frac{200 \times 10^4/j\omega}{200 + 10^4/j\omega} = \frac{2 \times 10^6}{10^4 + j\omega200}\ \Omega$$

Thus, the total series impedance is

$$Z_{R1} + Z_{R2}||Z_C = 50 + \frac{2 \times 10^6}{10^4 + j\omega200} = \frac{2.5 \times 10^6 + j\omega10^4}{10^4 + j\omega200}\ \Omega$$

and the phasor source current is

$$\mathbf{I}_S = \frac{\mathbf{V}_S}{Z_{R1} + Z_{R2}||Z_C} = A\angle\phi\frac{10^4 + j\omega200}{2.5 \times 10^6 + j\omega10^4}\ \text{A}$$

**Comments:** The expression obtained in this example can be evaluated for an arbitrary sinusoidal excitation, by substituting numerical values for $A$, $\phi$, and $\omega$ in the above expression. The answer can then be computed as the product of two complex numbers. As an example, you might wish to substitute the values used in Example 4.15 ($A = 10$ V, $\phi = 0$ rad, $\omega = 100$ rad/s) to verify that the same answer is obtained.

**Focus on Computer-Aided Tools:** An EWB file simulating this circuit for an arbitrary sinusoidal input is enclosed in the accompanying CD-ROM.

VIRTUAL LAB

By now it should be apparent that the laws of network analysis introduced in Chapter 3 are also applicable to phasor voltages and currents. This fact suggests that it may be possible to extend the node and mesh analysis methods developed earlier to circuits containing phasor sources and impedances, although the resulting simultaneous complex equations are difficult to solve without the aid of a computer, even for relatively simple circuits. On the other hand, it is very useful to extend the concept of equivalent circuits to the AC case, and to define complex Thévenin (or Norton) equivalent impedances. The fundamental difference between resistive and AC equivalent circuits is that the AC Thévenin (or Norton) equivalent circuits will be frequency-dependent and complex-valued. In general, then, one may think of the resistive circuit analysis of Chapter 3 as a special case of AC analysis in which all impedances are real.

## AC Equivalent Circuits

In Chapter 3, we demonstrated that it was convenient to compute equivalent circuits, especially in solving for load-related variables. Figure 4.45 depicts the two representations analogous to those developed in Chapter 3. Figure 4.45(a) shows an *equivalent load*, as viewed by the source, while Figure 4.45(b) shows an *equivalent source* circuit, from the perspective of the load.

In the case of linear resistive circuits, the equivalent load circuit can always be expressed by a single equivalent resistor, while the equivalent source circuit may take the form of a Norton or a Thévenin equivalent. This section extends these concepts to AC circuits and demonstrates that the notion of equivalent circuits applies to phasor sources and impedances as well. The techniques described in this section are all analogous to those used for resistive circuits, with resistances replaced by impedances, and arbitrary sources replaced by phasor sources. The principal difference between resistive and AC equivalent circuits will be that the latter are frequency-dependent. Figure 4.46 summarizes the fundamental principles used in computing an AC equivalent circuit. Note the definite analogy between impedance and resistance elements, and between conductance and admittance elements.

The computation of an equivalent impedance is carried out in the same way as that of equivalent resistance in the case of resistive circuits:

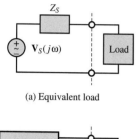

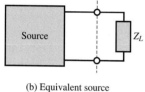

(a) Equivalent load

(b) Equivalent source

**Figure 4.45** AC equivalent circuits

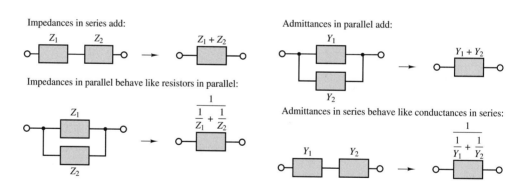

**Figure 4.46** Rules for impedance and admittance reduction

1.  Short-circuit all voltage sources, and open-circuit all current sources.
2.  Compute the equivalent impedance between load terminals, with the load disconnected.

In order to compute the Thévenin or Norton equivalent form, we recognize that the Thévenin equivalent voltage source is the open-circuit voltage at the load terminals and the Norton equivalent current source is the short-circuit current (the current with the load replaced by a short circuit). Figure 4.47 illustrates these points by outlining the steps in the computation of an equivalent circuit. The remainder of the section will consist of examples aimed at clarifying some of the finer points in the calculation of such equivalent circuits. Note how the initial circuit reduction

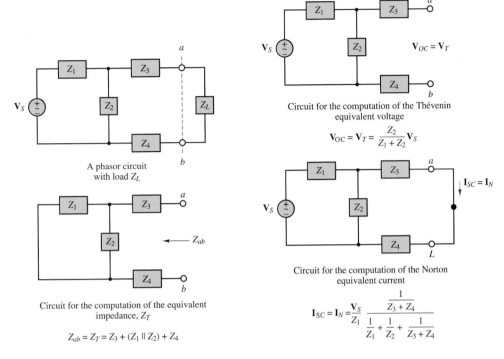

**Figure 4.47** Reduction of AC circuit to equivalent form

proceeds exactly as in the case of a resistive circuit; the details of the complex algebra required in the calculations are explored in the examples.

## EXAMPLE 4.17 Solution of AC Circuit by Nodal Analysis

### Problem

The electrical characteristics of electric motors (which are described in greater detail in the last three chapters of this book) can be approximately represented by means of a series R-L circuit. In this problem we analyze the currents drawn by two different motors connected to the same AC voltage supply (Figure 4.48).

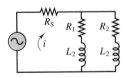

**Figure 4.48**

### Solution

**Known Quantities:** $R_S = 0.5\ \Omega$; $R_1 = 2\ \Omega$; $R_2 = 0.2\ \Omega$, $L_1 = 0.1$ H; $L_2 = 20$ mH. $v_S(t) = 155\ \cos(377t)$ V.

**Find:** The motor load currents, $i_1(t)$ and $i_2(t)$.

**Analysis:** First, we calculate the impedances of the source and of each motor:

$$Z_S = 0.5\ \Omega$$

$$Z_1 = 2 + j377 \times 0.1 = 2 + j37.7 = 37.75\angle 1.52\ \Omega$$

$$Z_2 = 0.2 + j377 \times 0.02 = 0.2 + j7.54 = 7.54\angle 1.54\ \Omega$$

The source voltage is $\mathbf{V}_S = 155\angle 0$ V.

Next, we apply KCL at the top node, with the aim of solving for the node voltage $\mathbf{V}$:

$$\frac{\mathbf{V}_S - \mathbf{V}}{Z_S} = \frac{\mathbf{V}}{Z_1} + \frac{\mathbf{V}}{Z_2}$$

$$\frac{\mathbf{V}_S}{Z_S} = \frac{\mathbf{V}}{Z_S} + \frac{\mathbf{V}}{Z_1} + \frac{\mathbf{V}}{Z_2} = \mathbf{V}\left(\frac{1}{Z_S} + \frac{1}{Z_1} + \frac{1}{Z_2}\right)$$

$$\mathbf{V} = \left(\frac{1}{Z_S} + \frac{1}{Z_1} + \frac{1}{Z_2}\right)^{-1}\frac{\mathbf{V}_S}{Z_S} = \left(\frac{1}{0.5} + \frac{1}{2+j37.7} + \frac{1}{0.2+j7.54}\right)^{-1}\frac{\mathbf{V}_S}{0.5}$$

$$= 154.1\angle 0.079 \text{ V}$$

Having computed the phasor node voltage, $\mathbf{V}$, we can now easily determine the phasor motor currents, $\mathbf{I}_1$ and $\mathbf{I}_2$:

$$\mathbf{I}_1 = \frac{\mathbf{V}}{Z_1} = \frac{82\angle -0.305}{2+j37.7} = 4.083\angle -1.439$$

$$\mathbf{I}_2 = \frac{\mathbf{V}}{Z_2} = \frac{82.05\angle -0.305}{0.2+j7.54} = 20.44 \angle -1.465.$$

Finally, we can write the time-domain expressions for the currents:

$$i_1(t) = 4.083 \cos(377t - 1.439) \text{ A}$$

$$i_2(t) = 20.44 \cos(377t - 1.465) \text{ A}$$

Figure 4.49 depicts the source voltage (scaled down by a factor of 10) and the two motor currents.

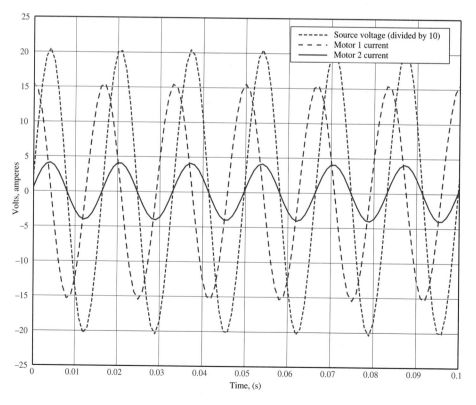

**Figure 4.49** Plot of source voltage and motor currents for Example 4.17

**Comments:** Note the phase shift between the source voltage and the two motor currents. A Matlab-generated computer-aided solution of this problem, including plotting of the graph of Figure 4.49, may be found in the CD that accompanies this book. An EWB solution is also included.

VIRTUAL LAB

---

## EXAMPLE 4.18 Thévenin Equivalent of AC Circuit

### Problem

Compute the Thévenin equivalent of the circuit of Figure 4.50.

---

### Solution

**Known Quantities:** $Z_1 = 5\ \Omega$; $Z_2 = j20\ \Omega$. $v_S(t) = 110\ \cos(377t)$ V.

**Find:** Thévenin equivalent circuit.

**Analysis:** First compute the equivalent impedance seen by the (arbitrary) load, $Z_L$. As illustrated in Figure 4.47, we remove the load, short-circuit the voltage source, and compute the equivalent impedance seen by the load; this calculation is illustrated in Figure 4.51.

$$Z_T = Z_1 || Z_2 = \frac{Z_1 \times Z_2}{Z_1 + Z_2} = \frac{5 \times j20}{5 + j20} = 4.71 + j1.176\ \Omega$$

Next, we compute the open-circuit voltage, between terminals $a$ and $b$:

$$V_T = \frac{Z_2}{Z_1 + Z_2}V_S = \frac{j20}{5 + j20}110\angle 0 = \frac{20\angle \pi/2}{20.6\angle 1.326}110\angle 0 = 106.7\angle 0.245\ \text{V}.$$

The complete Thévenin equivalent circuit is shown in Figure 4.52.

$V_S = 110\angle 0°\quad Z_1 = 5\ \Omega\quad Z_2 = j20\ \Omega$

**Figure 4.50**

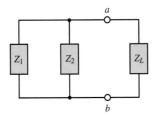

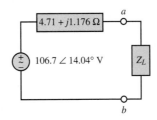

**Figure 4.51**          **Figure 4.52**

**Comments:** Note that the procedure followed for the computation of the equivalent circuit is completely analogous to that used in the case of resistive circuits (Section 3.5), the only difference being in the use of complex impedances in place of resistances. Thus, other than the use of complex quantities, there is no difference between the analysis leading to DC and AC equivalent circuits.

---

## Check Your Understanding

**4.19** Compute the magnitude of the current $I_S(j\omega)$ of Example 4.16 if $A = 1$ and $\phi = 0$, for $\omega = 10, 10^2, 10^3, 10^4$, and $10^5$ rad/s. Can you explain these results intuitively?

[*Hint:* Evaluate the impedance of the capacitor relative to that of the two resistors at each frequency.]

**4.20**   Find the voltage across the capacitor in Example 4.15.

**4.21**   Determine the Norton current in Example 4.17.

## CONCLUSION

In this chapter we have introduced concepts and tools useful in the analysis of AC circuits. The importance of AC circuit analysis cannot be overemphasized, for a number of reasons. First, circuits made up of resistors, inductors, and capacitors constitute reasonable models for more complex devices, such as transformers, electric motors, and electronic amplifiers. Second, sinusoidal signals are ever present in the analysis of many physical systems, not just circuits. The skills developed in Chapter 4 will be called upon in the remainder of the book. In particular, they form the basis of Chapters 5 and 6.

- In addition to elements that dissipate electric power, there are also electric energy-storage elements. The ideal inductor and capacitor are ideal elements that represent the energy-storage properties of electric circuits.

- Since the *i*-*v* relationship for the ideal capacitor and the ideal inductor consists of a differential equation, application of the fundamental circuit laws in the presence of such dynamic circuit elements leads to the formulation of differential equations.

- For the very special case of sinusoidal sources, the differential equations describing circuits containing dynamic elements can be converted into algebraic equations and solved using techniques similar to those employed in Chapter 3 for resistive circuits.

- Sinusoidal voltages and currents can be represented by means of complex phasors, which explicitly indicate the amplitude and phase of the sinusoidal signal and implicitly denote the sinusoidal frequency dependence.

- Circuit elements can be represented in terms of their impedance, which may be conceptualized as a frequency-dependent resistance. The rules of circuit analysis developed in Chapters 2 and 3 can then be employed to analyze AC circuits by using impedance elements as complex resistors. Thus, the only difference between the analysis of AC and resistive circuits lies in the use of complex algebra instead of real algebra.

## CHECK YOUR UNDERSTANDING ANSWERS

**CYU 4.1**        Plot for Check Your Understanding 4.1

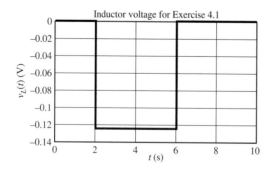

**CYU 4.2**        Plot for Check Your Understanding Exercise 4.2

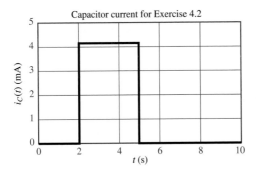

Capacitor current for Exercise 4.2

**CYU 4.3**        $w(t = 3 \text{ ms}) = 3.9 \ \mu\text{J}$

**CYU 4.4**        $w(t = 3 \text{ ms}) = 22.22 \text{ mJ}$

**CYU 4.5**
$$w(t) = \begin{cases} 5.625 \times 10^{-6} \text{ J} & 0 \le t < 2 \text{ ms} \\ 0.156 \times 10^{-6}t^2 - 2.5 \times 10^{-6}t \\ +10^{-5} & 2 \le t < 6 \text{ ms} \\ 0.625 \times 10^{-6} & t \ge 6 \text{ ms} \end{cases}$$

$$p(t) = \begin{cases} (20 \times 10^{-3} - 2.5t) \times (-0.125) \text{ W} & 2 \le t < 6 \text{ ms} \\ 0 & \text{otherwise} \end{cases}$$

**CYU 4.6**        $v(t) = 155.6 \ \cos(377t - \frac{\pi}{6})$

**CYU 4.7**        $\langle v(t) \rangle = 2.5 \text{ V}$

**CYU 4.8**        $\langle v(t) \rangle = 1.5 \text{ V}$

**CYU 4.9**        2.89 V

**CYU 4.10**       0.5 V

**CYU 4.13**

$$C = \frac{V}{\sqrt{1 + (\omega RC)^2}}$$

$$\phi = \tan^{-1}(-\omega RC)$$

**CYU 4.14**       (a) $v_1 + v_2 = 4.67 \cos(\omega t + 0.3526°)$; (b) $v_1 + v_2 = 60.8 \cos(\omega t - 0.6562°)$

**CYU 4.15**       (a) $i_1 + i_2 = 0.19 \cos(\omega t + 0.733°)$; (b) $i_1 + i_2 = 1.32 \cos(\omega t - 0.5637°)$

**CYU 4.16**       $Z(1,000) = 140 - j10$; $Z(100,000) = 100 + j999$

**CYU 4.17**       $Y_{\text{EQ}} = 5.492 \times 10^{-3} - j4.871 \times 10^{-3}$

**CYU 4.18**       $X_{\|} = 0.25$; $C = 0.4 \text{ F}$

**CYU 4.19**       $|\mathbf{I}_S| = 0.0041 \text{ A}$; 0.0083 A; 0.0194 A; 0.02 A; 0.02 A

**CYU 4.20**       $7.424e^{-j0.381}$

**CYU 4.21**       $22e^{j0} \text{ A}$

# HOMEWORK PROBLEMS

## Section 1: Energy Storage Elements

**4.1**  The current through a 0.5-H inductor is given by $i_L = 2\cos(377t + \pi/6)$. Write the expression for the voltage across the inductor.

**4.2**  The voltage across a 100-$\mu$F capacitor takes the following values. Calculate the expression for the current through the capacitor in each case.

  a.  $v_C(t) = 40\cos(20t - \pi/2)$ V

  b.  $v_C(t) = 20\sin 100t$ V

  c.  $v_C(t) = -60\sin(80t + \pi/6)$ V

  d.  $v_C(t) = 30\cos(100t + \pi/4)$ V

**4.3**  The current through a 250-mH inductor takes the following values. Calculate the expression for the voltage across the inductor in each case.

  a.  $i_L(t) = 5\sin 25t$ A

  b.  $i_L(t) = -10\cos 50t$ A

  c.  $i_L(t) = 25\cos(100t + \pi/3)$ A

  d.  $i_L(t) = 20\sin(10t - \pi/12)$ A

**4.4**  In the circuit shown in Figure P4.4, let

$$
\begin{aligned}
i(t) &= 0 & \text{for } -\infty < t < 0 \\
&= t & \text{for } 0 \le t < 1 \text{ s} \\
&= -(t-2) & \text{for } 1 \text{ s} \le t < 2 \text{ s} \\
&= 0 & \text{for } 2 \text{ s} \le t < \infty
\end{aligned}
$$

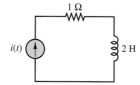

**Figure P4.4**

Find the energy stored in the inductor for all time.

**4.5**  In the circuit shown in Figure P4.5, let

$$
\begin{aligned}
v(t) &= 0 & \text{for } -\infty < t < 0 \\
&= 2t & \text{for } 0 \le t < 1 \text{ s} \\
&= -(2t-4) & \text{for } 1 \text{ s} \le t < 2 \text{ s} \\
&= 0 & \text{for } 2 \text{ s} \le t < \infty
\end{aligned}
$$

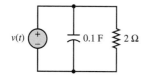

**Figure P4.5**

Find the energy stored in the capacitor for all time.

**4.6**  Find the energy stored in each capacitor and inductor, under steady-state conditions, in the circuit shown in Figure P4.6.

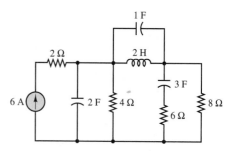

**Figure P4.6**

**4.7**  Find the energy stored in each capacitor and inductor, under steady-state conditions, in the circuit shown in Figure P4.7.

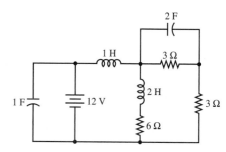

**Figure P4.7**

**4.8**  The plot of time-dependent voltage is shown in Figure P4.8. The waveform is piecewise continuous. If this is the voltage across a capacitor and $C = 80\ \mu$F, determine the current through the capacitor. How can current flow "through" a capacitor?

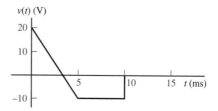

**Figure P4.8**

**4.9**  The plot of a time-dependent voltage is shown in Figure P4.8. The waveform is piecewise continuous. If this is the voltage across an inductor $L = 35$ mH, determine the current through the inductor. Assume the initial current is $i_L(0) = 0$.

**4.10**  The voltage across an inductor plotted as a function of time is shown in Figure P4.10. If $L = 0.75$ mH, determine the current through the inductor at $t = 15 \ \mu$s.

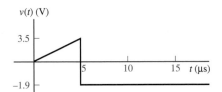

**Figure P4.10**

**4.11**  If the waveform shown in Figure P4.11 is the voltage across a capacitor plotted as a function of time with:

$$v_{PK} = 20 \text{ V} \quad T = 40 \ \mu\text{s} \quad C = 680 \text{ nF}$$

determine and plot the waveform for the current through the capacitor as a function of time.

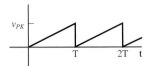

**Figure P4.11**

**4.12**  If the current through a 16 $\mu$h inductor is zero at $t = 0$ and the voltage across the inductor (shown in Figure P4.12) is:

$$\begin{aligned} v_L(f) &= 0 & t < 0 \\ &= 3t^2 & 0 < t < 20 \ \mu\text{s} \\ &= 1.2 \text{ nV} & t > 20 \ \mu\text{s} \end{aligned}$$

determine the current through the inductor at $t = 30 \ \mu$s.

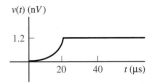

**Figure P4.12**

**4.13**  Determine and plot as a function of time the current through a component if the voltage across it has the waveform shown in Figure P4.13 and the component is a:

a.  Resistor $R = 7 \ \Omega$.
b.  Capacitor $C = 0.5 \ \mu$F.
c.  Inductor $L = 7$ mH.

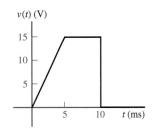

**Figure P4.13**

**4.14**  If the plots shown in Figure P4.14 are the voltage across and the current through an ideal capacitor, determine the capacitance.

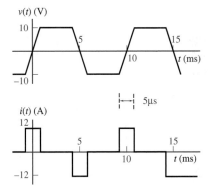

**Figure P4.14**

**4.15**  If the plots shown in Figure P4.15 are the voltage across and the current through an ideal inductor, determine the inductance.

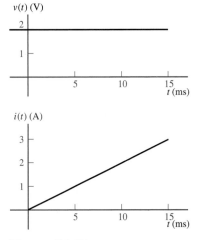

**Figure P4.15**

**4.16** The voltage across and the current through a capacitor are shown in Figure 4.16. Determine the value of the capacitance.

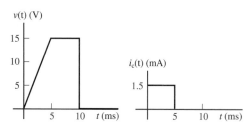

**Figure P4.16**

**4.17** The voltage across and the current through a capacitor are shown in Figure P4.17. Determine the value of the capacitance.

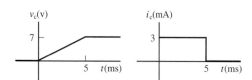

**Figure P4.17**

## Section 2: Time-Dependent Waveforms

**4.18** Find the rms value of $x(t)$ if $x(t)$ is a sinusoid that is offset by a DC value:

$$x(t) = 2 \sin(\omega t) + 2.5$$

**4.19** For the waveform of Figure P4.19:

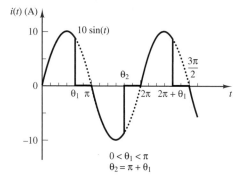

**Figure P4.19**

a. Find the rms current.

b. If $\theta_1$ is $\pi/2$, what is the rms current of this waveform?

**4.20** Find the rms value of the waveform of Figure P4.20.

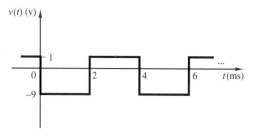

**Figure P4.20**

**4.21** Find the rms value of the waveform of Figure P4.21.

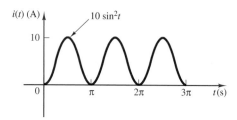

**Figure P4.21**

**4.22** Find the rms voltage of the waveform of Figure P4.22.

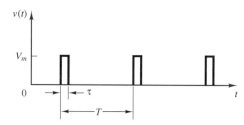

**Figure P4.22**

**4.23** Find the rms value of the waveform shown in Figure P4.23.

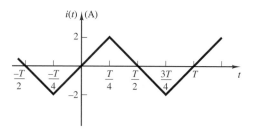

**Figure P4.23**

**4.24** Determine the rms (or effective) value of:

$$v(t) = V_{DC} + v_{AC} = 50 + 70.7 \cos(377t) \text{ V}$$

# Section 3: Phasor Analysis

**4.25**  If the current through and the voltage across a component in an electrical circuit are:

$$i(t) = 17 \cos[\omega t - \tfrac{\pi}{12}] \text{ mA}$$
$$v(t) = 3.5 \cos[\omega t + 1.309] \text{ V}$$

where $\omega = 628.3$ rad/s, determine:

a.  Whether the component is a resistor, capacitor, or inductor.
b.  The value of the component in ohms, farads, or henrys.

**4.26**  Describe the sinusoidal waveform shown in Figure P4.26 using time-dependent and phasor notation.

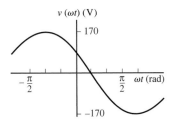

**Figure P4.26**

**4.27**  Describe the sinusoidal waveform shown in Figure P4.27 using time-dependent and phasor notation.

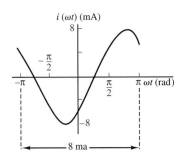

**Figure P4.27**

**4.28**  Describe the sinusoidal waveform shown in Figure P4.28 using time-dependent and phasor notation.

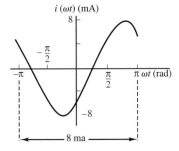

**Figure P4.28**

**4.29**  If the current through and the voltage across an electrical component are:

$$i(t) = I_o \cos(\omega t + \tfrac{\pi}{4}) \quad v(t) = V_o \cos \omega t$$

where:

$$I_o = 3 \text{ mA} \quad V_o = 700 \text{ mV} \quad \omega = 6.283 \text{ rad/s}$$

a.  Is the component inductive or capacitive?
b.  Plot the instantaneous power $p(t)$ as a function of $\omega t$ over the range $0 < \omega t < 2\pi$.
c.  Determine the average power dissipated as heat in the component.
d.  Repeat parts (b) and (c) if the phase angle of the current is changed to zero degrees.

**4.30**  Determine the equivalent impedance in the circuit shown in Figure P4.30:

$$v_s(t) = 7 \cos\left(3{,}000t + \tfrac{\pi}{6}\right) \text{ V}$$
$$R_1 = 2.3 \text{ k}\Omega \qquad R_2 = 1.1 \text{ k}\Omega$$
$$L = 190 \text{ mH} \qquad C = 55 \text{ nF}$$

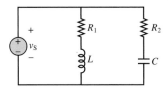

**Figure P4.30**

**4.31**  Determine the equivalent impedance in the circuit shown in Figure P4.30:

$$v_s(t) = 636 \cos\left(3{,}000t + \tfrac{\pi}{12}\right) \text{ V}$$
$$R_1 = 3.3 \text{ k}\Omega \qquad R_2 = 22 \text{ k}\Omega$$
$$L = 1.90 \text{ H} \qquad C = 6.8 \text{ nF}$$

**4.32**  In the circuit of Figure P4.32,

$$i_s(t) = I_o \cos\left(\omega t + \tfrac{\pi}{6}\right)$$
$$I_o = 13 \text{ mA} \qquad \omega = 1{,}000 \text{ rad/s}$$
$$C = 0.5 \text{ }\mu\text{F}$$

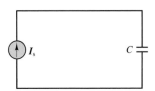

**Figure P4.32**

a.  State, using phasor notation, the source current.
b.  Determine the impedance of the capacitor.

c. Using phasor notation only and showing all work, determine the voltage across the capacitor, including its polarity.

**4.33**  Determine $i_3(t)$ in the circuit shown in Figure P4.33, if:

$$i_1(t) = 141.4 \cos(\omega t + 2.356) \text{ mA}$$
$$i_2(t) = 50 \sin(\omega t - 0.927) \text{ mA}$$
$$\omega = 377 \text{ rad/s}$$

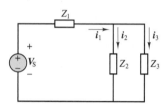

**Figure P4.33**

**4.34**  Determine the current through $Z_3$ in the circuit of Figure P4.34.

$$V_{s1} = v_{s2} = 170 \cos(377t) \text{ V}$$
$$Z_1 = 5.9\angle 0.122 \ \Omega$$
$$Z_2 = 2.3\angle 0 \ \Omega$$
$$Z_3 = 17\angle 0.192 \ \Omega$$

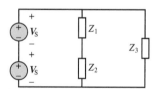

**Figure P4.34**

**4.35**  Determine the frequency so that the current $I_i$ and the voltage $V_o$ in the circuit of of Figure P4.35 are in phase.

$$Z_s = 13,000 + j\omega 3 \ \Omega$$
$$R = 120 \ \Omega$$
$$L = 19 \text{ mH} \qquad C = 220 \text{ pF}$$

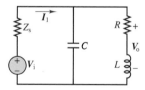

**Figure P4.35**

**4.36**  In the circuit of Figure P4.35, determine the frequency $\omega_r$ at which $\mathbf{I}_i$ and $\mathbf{V}_o$ are in phase.

**4.37**  The coil resistor in series with $L$ models the internal losses of an inductor in the circuit of Figure P4.37. Determine the current supplied by the source if:

$$v_s(t) = V_o \cos(\omega t + 0)$$
$$V_o = 10 \text{ V} \qquad \omega = 6 \text{ Mrad/s} \qquad R_s = 50 \ \Omega$$
$$R_c = 40 \ \Omega \qquad L = 20 \ \mu\text{H} \qquad C = 1.25 \text{ nF}$$

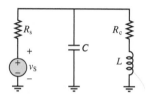

**Figure P4.37**

**4.38**  Using phasor techniques, solve for the current in the circuit shown in Figure P4.38.

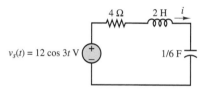

**Figure P4.38**

**4.39**  Using phasor techniques, solve for the voltage, $v$, in the circuit shown in Figure P4.39.

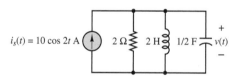

**Figure P4.39**

**4.40**  Solve for $\mathbf{I}_1$ in the circuit shown in Figure P4.40.

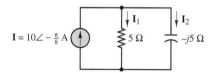

**Figure P4.40**

**4.41**  Solve for $\mathbf{V}_2$ in the circuit shown in Figure P4.41. Assume $\omega = 2$.

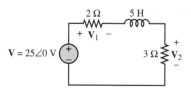

**Figure P4.41**

**4.42**  Find the current through the resistor in the circuit shown in Figure P4.42.

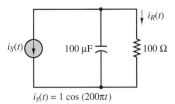

$$i_S(t) = 1\cos(200\pi t)$$

**Figure P4.42**

**4.43**  Find $v_{out}(t)$ for the circuit shown in Figure P4.43.

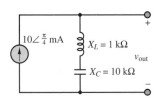

**Figure P4.43**

**4.44**  For the circuit shown in Figure P4.44, find the impedance $Z$, given $\omega = 4$ rad/s.

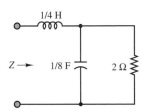

**Figure P4.44**

**4.45**  Find the admittance, $Y$, for the circuit shown in Figure P4.45, when $\omega = 5$ rad/s.

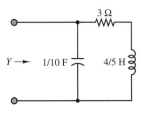

**Figure P4.45**

## Section 4: AC Circuit Analysis

**4.46**  Using phasor techniques, solve for $v$ in the circuit shown in Figure P4.46.

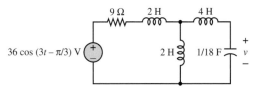

**Figure P4.46**

**4.47**  Using phasor techniques, solve for $i$ in the circuit shown in Figure P4.47.

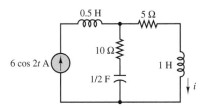

**Figure P4.47**

**4.48**  Determine the Thévenin equivalent circuit as seen by the load shown in Figure P4.48 if

a.  $v_S(t) = 10\cos(1{,}000t)$.

b.  $v_S(t) = 10\cos(1{,}000{,}000t)$.

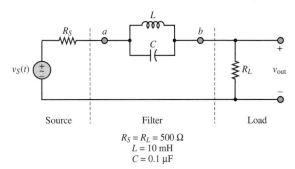

$$R_S = R_L = 500\ \Omega$$
$$L = 10\ \text{mH}$$
$$C = 0.1\ \mu\text{F}$$

**Figure P4.48**

**4.49**  Find the Thévenin equivalent of the circuit shown in Figure P4.49 as seen by the load resistor.

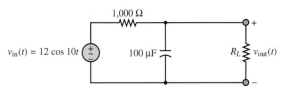

**Figure P4.49**

**4.50**  Solve for $i(t)$ in the circuit of Figure P4.50, using phasor techniques, if $v_S(t) = 2\cos(2t)$, $R_1 = 4\ \Omega$, $R_2 = 4\ \Omega$, $L = 2$ H, and $C = \frac{1}{4}$ F.

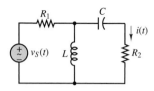

**Figure P4.50**

**4.51**  Using mesh current analysis, determine the currents $i_1(t)$ and $i_2(t)$ in the circuit shown in Figure P4.51.

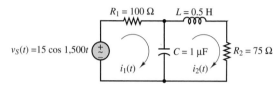

**Figure P4.51**

**4.52**  Using node voltage methods, determine the voltages $v_1(t)$ and $v_2(t)$ in the circuit shown in Figure P4.52.

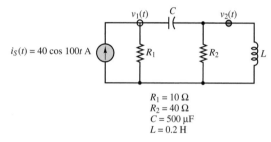

**Figure P4.52**

**4.53**  The circuit shown in Figure P4.53 is a Wheatstone bridge that will allow you to determine the reactance of an inductor or a capacitor. The circuit is adjusted by changing $R_1$ and $R_2$ until $v_{ab}$ is zero.

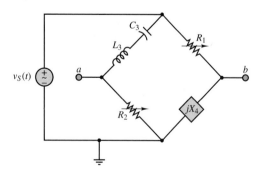

**Figure P4.53**

a.  Assuming that the circuit is balanced, that is, that $v_{ab} = 0$, determine $X_4$ in terms of the circuit elements.

b.  If $C_3 = 4.7\ \mu$ F, $L_3 = 0.098$ H, $R_1 = 100\ \Omega$, $R_2 = 1\ \Omega$, $v_S(t) = 24\sin(2,000t)$, and $v_{ab} = 0$, what is the reactance of the unknown circuit element? Is it a capacitor or an inductor? What is its value?

c.  What frequency should be avoided by the source in this circuit, and why?

**4.54**  Compute the Thévenin impedance seen by resistor $R_2$ in Problem 4.50.

**4.55**  Compute the Thévenin voltage seen by the inductance, $L$, in Problem 4.52.

**4.56**  Find the Thévenin equivalent circuit as seen from terminals $a$-$b$ for the circuit shown in Figure 4.56.

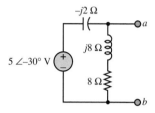

**Figure P4.56**

**4.57**  Compute the Thévenin voltage seen by resistor $R_2$ in Problem 4.50.

**4.58**  Find the Norton equivalent circuit seen by resistor $R_2$ in Problem 4.50.

**4.59**  Write the two loop equations required to solve for the loop currents in the circuit of Figure P4.59 in:

a.  Integral-differential form.

b.  Phasor form.

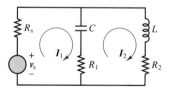

**Figure P4.59**

**4.60**  Write the node equations required to solve for all voltages and currents in the circuit of Figure P4.59. Assume all impedances and the two source voltages are known.

**4.61**  In the circuit shown in Figure P4.61:

$$v_{s1} = 450\ \cos\omega t \text{ V} \qquad v_{s2} = 450\ \cos\omega t \text{ V}$$

A solution of the circuit with the ground at node $e$ as shown gives:

$$\mathbf{V}_a = 450\angle 0 \text{ V} \qquad \mathbf{V}_b = 440\angle\tfrac{\pi}{6} \text{ V}$$
$$\mathbf{V}_c = 420\angle -3.49 \text{ V}$$
$$\mathbf{V}_{bc} = 779.5\angle 0.098 \text{ V} \qquad \mathbf{V}_{cd} = 153.9\angle 1.2 \text{ V}$$
$$\mathbf{V}_{ba} = 230.6\angle 1.875 \text{ V}$$

If the ground is now moved from node $e$ to node $d$, determine $\mathbf{V}_b$ and $\mathbf{V}_{bc}$.

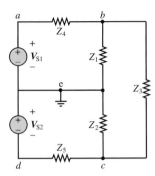

**Figure P4.61**

**4.62**  Determine $V_o$ in the circuit of Figure P4.62 if:

$$v_i = 4 \cos(1{,}000t + \tfrac{\pi}{6}) \text{ V}$$
$$L = 60 \text{ mH} \qquad C = 12.5 \ \mu\text{F}$$
$$R_L = 120 \ \Omega$$

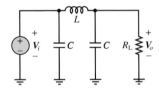

**Figure P4.62**

**4.63**  The mesh currents and node voltages in the circuit shown in Figure P4.63 are:

$$i_1(t) = 3.127 \ \cos(\omega t - 0.825) \text{ A}$$
$$i_2(t) = 3.914 \ \cos(\omega t - 1.78) \text{ A}$$
$$i_3(t) = 1.900 \ \cos(\omega t + 0.655) \text{ A}$$
$$v_1(t) = 130.0 \ \cos(\omega t + 0.176) \text{ V}$$
$$v_2(t) = 130.0 \ \cos(\omega t - 0.436) \text{ V}$$

where $\omega = 377.0$ rad/s.  Determine one of the following $L_1$, $C_2$, $R_3$, or $L_3$.

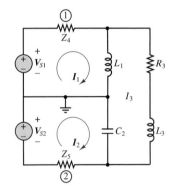

**Figure P4.63**

# C H A P T E R

# 5

# Transient Analysis

## 5.1 INTRODUCTION

The aim of this chapter is to explore the solution of circuits that contain resistances, inductances, capacitances, voltage and current sources, and switches. The response of a circuit to the sudden application of a voltage or current is called *transient response*. The most common instance of a transient response in a circuit occurs when a switch is turned on or off—a rather common event in electrical circuits. Although there are many possible types of transients that can be introduced in a circuit, in the present chapter we shall focus exclusively on the transient response of circuits in which a switch activates or deactivates a DC source. Further, we shall restrict our analysis, for the sake of simplicity, to *first-* and *second-order transients*, that is to circuits that have only one or two energy storage elements.

The graphs of Figure 5.1 illustrate the result of the sudden appearance of a voltage across a hypothetical load [a DC voltage in Figure 5.1(a), an AC voltage in Figure 5.1(b)]. In the figure, the source voltage is turned on at time $t = 0.2$ s. The voltage waveforms of Figure 5.1 can be subdivided into three regions: a *steady-state* region, for $0 \leq t \leq 0.2$ s; a *transient* region for $0.2 \leq t \leq 2$ s (approximately); and a new steady-state region for $t > 2$ s, where the voltage reaches a steady DC or AC condition. The objective of **transient analysis** is to describe the behavior of a voltage or a current during the transition that takes place between two distinct steady-state conditions.

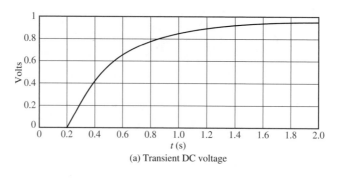

(a) Transient DC voltage

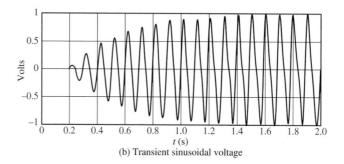

(b) Transient sinusoidal voltage

**Figure 5.1** Examples of transient response

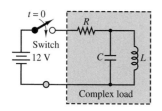

**Figure 5.2** Circuit with switched DC excitation

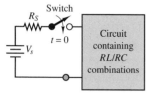

**Figure 5.3** A general model of the transient analysis problem

You already know how to analyze circuits in a sinusoidal steady state by means of phasors. The material presented in the remainder of this chapter will provide the tools necessary to describe the *transient response* of circuits containing resistors, inductors, and capacitors. A general example of the type of circuit that will be discussed in this section is shown in Figure 5.2. The switch indicates that we turn the battery power on at time $t = 0$. Transient behavior may be expected whenever a source of electrical energy is switched on or off, whether it be AC or DC. A typical example of the transient response to a switched DC voltage would be what occurs when the ignition circuits in an automobile are turned on, so that a 12-V battery is suddenly connected to a large number of electrical circuits. The degree of complexity in transient analysis depends on the number of energy-storage elements in the circuit; the analysis can became quite involved for high-order circuits. In this chapter, we shall analyze only first- and second-order circuits—that is, circuits containing one or two energy-storage elements, respectively. In electrical engineering practice, we would typically resort to computer-aided analysis for higher-order circuits.

A convenient starting point in approaching the transient response of electrical circuits is to consider the general model shown in Figure 5.3, where the circuits in the box consist of a combination of resistors connected to a *single energy-storage element*, either an inductor or a capacitor. Regardless of how many resistors the circuit contains, it is a **first-order circuit**. In general, the response of a first-order circuit to a switched DC source will appear in one of the two forms shown in Figure 5.4, which represent, in order, a **decaying exponential** and a **rising exponential** waveform. In the next sections, we will systematically analyze these responses by recognizing that they are exponential in nature and can be computed very easily once we have the proper form of the differential equation describing the circuit.

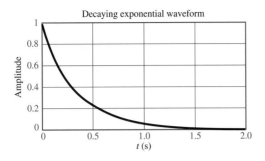

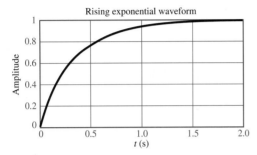

**Figure 5.4** Decaying and rising exponential responses

## 5.2    SOLUTION OF CIRCUITS CONTAINING DYNAMIC ELEMENTS

The major difference between the analysis of the resistive circuits studied in Chapters 2 and 3 and the circuits we will explore in the remainder of this chapter is that now the equations that result from applying Kirchhoff's laws are differential equations, as opposed to the algebraic equations obtained in solving resistive circuits. Consider, for example, the circuit of Figure 5.5, which consists of the series connection of a voltage source, a resistor, and a capacitor. Applying KVL around the loop, we may obtain the following equation:

$$v_S(t) - v_R(t) - v_C(t) = 0 \tag{5.1}$$

Observing that $i_R = i_C$, we may combine equation 5.1 with the defining equation for the capacitor (equation 4.6) to obtain

$$v_S(t) - R i_C(t) - \frac{1}{C} \int_{-\infty}^{t} i_C \, dt' = 0 \tag{5.2}$$

Equation 5.2 is an integral equation, which may be converted to the more familiar form of a differential equation by differentiating both sides of the equation, and recalling that

$$\frac{d}{dt} \left( \int_{-\infty}^{t} i_C(t') \, dt' \right) = i_C(t) \tag{5.3}$$

to obtain the following differential equation:

$$\frac{d i_C}{dt} + \frac{1}{RC} i_C = \frac{1}{R} \frac{d v_S}{dt} \tag{5.4}$$

where the argument $(t)$ has been dropped for ease of notation.

A circuit containing energy-storage elements is described by a differential equation. The differential equation describing the series *RC* circuit shown is

$$\frac{d i_C}{dt} + \frac{1}{RC} i_C = \frac{d v_S}{dt}$$

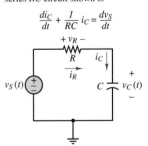

**Figure 5.5** Circuit containing energy-storage element

Observe that in equation 5.4, the independent variable is the series current flowing in the circuit, and that this is not the only equation that describes the series *RC* circuit. If, instead of applying KVL, for example, we had applied KCL at the node connecting the resistor to the capacitor, we would have obtained the following relationship:

$$i_R = \frac{v_S - v_C}{R} = i_C = C\frac{dv_C}{dt} \tag{5.5}$$

or

$$\frac{dv_C}{dt} + \frac{1}{RC}v_C = \frac{1}{RC}v_S \tag{5.6}$$

Note the similarity between equations 5.4 and 5.6. The left-hand side of both equations is identical, except for the variable, while the right-hand side takes a slightly different form. The solution of either equation is sufficient, however, to determine all voltages and currents in the circuit. The following example illustrates the derivation of the differential equation for another simple circuit containing an energy-storage element.

### EXAMPLE   5.1   Writing the Differential Equation of an *RL* Circuit

**Problem**

Derive the differential equation of the circuit shown in Figure 5.6.

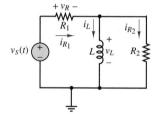

**Figure 5.6**

**Solution**

**Known Quantities:** $R_1 = 10\ \Omega$; $R_2 = 5\ \Omega$; $L = 0.4$ H.

**Find:** The differential equation in $i_L(t)$.

**Assumptions:** None.

**Analysis:** Apply KCL at the top node (nodal analysis) to write the circuit equation. Note that the top node voltage is the inductor voltage, $v_L$.

$$i_{R1} - i_L - i_{R2} = 0$$

$$\frac{v_S - v_L}{R_1} - i_L - \frac{v_L}{R_2} = 0$$

Next, use the definition of inductor voltage to eliminate the variable $v_L$ from the nodal equation.

$$\frac{v_S}{R_1} - \frac{L}{R_1}\frac{di_L}{dt} - i_L - \frac{L}{R_2}\frac{di_L}{dt} = 0$$

$$\frac{di_L}{dt} + \frac{R_1 R_2}{L(R_1 + R_2)}i_L = \frac{R_2}{L(R_1 + R_2)}v_S$$

Substituting numerical values, we obtain the following differential equation:

$$\frac{di_L}{dt} + 8.33i_L = 0.833v_S$$

**Comments:** Deriving differential equations for dynamic circuits requires the same basic circuit analysis skills that were developed in Chapter 3. The only difference is the introduction of integral or derivative terms originating from the defining relations for capacitors and inductors.

We can generalize the results presented in the preceding pages by observing that any circuit containing a single energy-storage element can be described by a differential equation of the form

$$a_1 \frac{dx(t)}{dx} + a_0 x(t) = f(t) \tag{5.7}$$

where $x(t)$ represents the capacitor voltage in the circuit of Figure 5.5 and the inductor current in the circuit of Figure 5.6, and where the constants $a_0$ and $a_1$ consist of combinations of circuit element parameters. Equation 5.7 is a **first-order ordinary differential equation** with constant coefficients. The equation is said to be of first order because the highest derivative present is of first order; it is said to be ordinary because the derivative that appears in it is an ordinary derivative (in contrast to a *partial* derivative); and the coefficients of the differential equation are constant in that they depend only on the values of resistors, capacitors, or inductors in the circuit, and not, for example, on time, voltage, or current.

Consider now a circuit that contains two energy-storage elements, such as that shown in Figure 5.7. Application of KVL results in the following equation:

$$Ri(t) - L\frac{di(t)}{dt} - \frac{1}{C}\int_{-\infty}^{t} i(t')\, dt' - v_S(t) = 0 \tag{5.8}$$

Equation 5.8 is called an integro-differential equation, because it contains both an integral and a derivative. This equation can be converted into a differential equation by differentiating both sides, to obtain:

$$R\frac{di(t)}{dt} + L\frac{d^2 i(t)}{dt^2} + \frac{1}{C}i(t) = \frac{dv_S(t)}{dt} \tag{5.9}$$

or, equivalently, by observing that the current flowing in the series circuit is related to the capacitor voltage by $i(t) = Cdv_C/dt$, and that equation 5.8 can be rewritten as:

$$RC\frac{dv_C}{dt} + LC\frac{d^2 v_C(t)}{dt^2} + v_C(t) = v_S(t) \tag{5.10}$$

Note that, although different variables appear in the preceding differential equations, both equations 5.9 and 5.10 can be rearranged to appear in the same general form, as follows:

$$a_2 \frac{d^2 x(t)}{dt^2} + a_1 \frac{dx(t)}{dt} + a_0 x(t) = F(t) \tag{5.11}$$

where the general variable $x(t)$ represents either the series current of the circuit of Figure 5.7 or the capacitor voltage. By analogy with equation 5.7, we call equation 5.11 a **second-order ordinary differential equation** with constant coefficients. As the number of energy-storage elements in a circuit increases, one can therefore

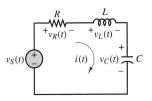

**Figure 5.7** Second-order circuit

expect that higher-order differential equations will result. Computer aids are often employed to solve differential equations of higher order; some of these software packages are specifically targeted at the solution of the equations that result from the analysis of electrical circuits (e.g., *Electronics Workbench*$^{TM}$).

---

### EXAMPLE 5.2 Writing the Differential Equation of an *RLC* Circuit

**Problem**

Derive the differential equation of the circuit shown in Figure 5.8.

---

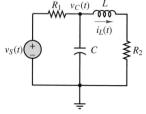

**Figure 5.8** Second-order circuit of Example 5.2

**Solution**

**Known Quantities:** $R_1 = 10 \text{ k}\Omega$; $R_2 = 50\ \Omega$; $L = 10 \text{ mH}$; $C = 0.1\ \mu\text{F}$.

**Find:** The differential equation in $i_L(t)$.

**Assumptions:** None.

**Analysis:** Apply KCL at the top node (nodal analysis) to write the first circuit equation. Note that the top node voltage is the capacitor voltage, $v_C$.

$$\frac{v_S - v_C}{R_1} - C\frac{dv_C}{dt} - i_L = 0$$

Now, we need a second equation to complete the description of the circuit, since the circuit contains two energy storage elements (second-order circuit). We can obtain a second equation in the capacitor voltage, $v_C$, by applying KVL to the mesh on the right-hand side:

$$v_C - L\frac{di_L}{dt} - R_2 i_L = 0$$

$$v_C = L\frac{di_L}{dt} + R_2 i_L$$

Next, we can substitute the above expression for $v_C$ into the first equation, to obtain a *second-order* differential equation, shown below.

$$\frac{v_S}{R_1} - \frac{L}{R_1}\frac{di_L}{dt} - \frac{R_2}{R_1}i_L - C\frac{d}{dt}\left(L\frac{di_L}{dt} + R_2 i_L\right) - i_L = 0$$

Rearranging the equation we can obtain the standard form similar to equation 5.11:

$$R_1 C L\frac{d^2 i_L}{dt^2} + (R_1 R_2 C + L)\frac{di_L}{dt} + (R_1 + R_2)i_L = v_S$$

**Comments:** Note that we could have derived an analogous equation using the capacitor voltage as an independent variable; either energy storage variable is an acceptable choice. You might wish to try obtaining a second-order equation in $v_C$ as an exercise. In this case, you would want to substitute an expression for $i_L$ in the first equation into the second equation in $v_C$.

---

## 5.3  TRANSIENT RESPONSE OF FIRST-ORDER CIRCUITS

**First-order systems** occur very frequently in nature: any system that has the ability to store energy in one form and to dissipate the energy stored is a first-order system. In electrical circuits, we recognize that any circuit containing a single energy storage element (inductor or capacitor) and a combination of sources and resistors (and possibly switches) is a first-order system. In other domains, we also encounter first-order systems. For example, a mechanical system that has mass and damping (e.g., friction), but not elasticity, will be a first-order system. A fluid system with fluid resistance and fluid capacitance (fluid storage) will also be of first order; an example of a first-order fluid system is a storage tank with a valve. In thermal systems, we also encounter first-order systems quite frequently: The ability to store heat (heat capacity) and to dissipate it leads to a first-order thermal system; heating and cooling of bodies is, at its simplest level, described by first-order behavior.

In the present section we analyze the transient response of first-order circuits. In what follows, we shall explain that the **initial condition**, the **steady-state solution**, and the **time constant** of the first-order system are the three quantities that uniquely determine its response.

## Natural Response of First-Order Circuits

Figure 5.9 compares an $RL$ circuit with the general form of the series $RC$ circuit, showing the corresponding differential equation. From Figure 5.9, it is clear that equation 5.12 is in the general form of the equation for any first-order circuit:

$$a_1 \frac{dx(t)}{dt} + a_2 x(t) = f(t) \tag{5.12}$$

where $f$ is the **forcing function** and $x(t)$ represents either $v_C(t)$ or $i_L(t)$. The constant $a = a_2/a_1$ is the inverse of the parameter $\tau$, called the **time constant** of the system: $a = 1/\tau$.

To gain some insight into the solution of this equation, consider first the **natural solution**, or **natural response**, of the equation,[1] which is obtained by setting the forcing function equal to zero. This solution, in effect, describes the response of the circuit in the absence of a source and is therefore characteristic of all $RL$ and $RC$ circuits, regardless of the nature of the excitation. Thus, we are interested in the solution of the equation

$$\frac{dx_N(t)}{dt} + \frac{1}{\tau} x_N(t) = 0 \tag{5.13}$$

or

$$\frac{dx_N(t)}{dt} = -\frac{1}{\tau} x_N(t) \tag{5.14}$$

where the subscript $N$ has been chosen to denote the natural solution. One can easily verify by substitution that the general form of the solution of the homogeneous equation for a first-order circuit must be exponential in nature, that is, that

$$x_N(t) = K e^{-at} = K e^{-t/\tau} \tag{5.15}$$

To evaluate the constant $K$, we need to know the **initial condition**. The initial condition is related to the energy stored in the capacitor or inductor, as will be further explained shortly. Knowing the value of the capacitor voltage or inductor current at $t = 0$ allows for the computation of the constant $K$, as follows:

$$x_N(t = 0) = K e^{-0} = K = x_0 \tag{5.16}$$

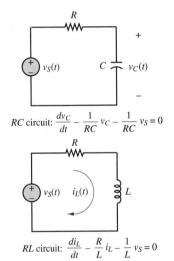

$RC$ circuit: $\dfrac{dv_C}{dt} - \dfrac{1}{RC} v_C - \dfrac{1}{RC} v_S = 0$

$RL$ circuit: $\dfrac{di_L}{dt} - \dfrac{R}{L} i_L - \dfrac{1}{L} v_S = 0$

**Figure 5.9** Differential equations of first-order circuits

---

[1]Mathematicians usually refer to the unforced solution as the **homogeneous solution**.

Thus, the natural solution, which depends on the initial condition of the circuit at $t = 0$, is given by the expression

$$x_N(t) = x_0 e^{-t/\tau} \tag{5.17}$$

where, once again, $x_N(t)$ represents either the capacitor voltage or the inductor current and $x_0$ is the initial condition (i.e., the value of the capacitor voltage or inductor current at $t = 0$).

## Energy Storage in Capacitors and Inductors

Before delving into the complete solution of the differential equation describing the response of first-order circuits, it will be helpful to review some basic results pertaining to the response of energy-storage elements to DC sources. This knowledge will later greatly simplify the complete solution of the differential equation describing a circuit. Consider, first, a capacitor, which accumulates charge according to the relationship $Q = CV$. The charge accumulated in the capacitor leads to the storage of energy according to the following equation:

$$W_C = \frac{1}{2} C v_C^2(t) \tag{5.18}$$

To understand the role of stored energy, consider, as an illustration, the simple circuit of Figure 5.10, where a capacitor is shown to have been connected to a battery, $V_B$, for a long time. The capacitor voltage is therefore equal to the battery voltage: $v_C(t) = V_B$. The charge stored in the capacitor (and the corresponding energy) can be directly determined using equation 5.18. Suppose, next, that at $t = 0$ the capacitor is disconnected from the battery and connected to a resistor, as shown by the action of the switches in Figure 5.10. The resulting circuit would be governed by the $RC$ differential equation described earlier, subject to the initial condition $v_C(t = 0) = V_B$. Thus, according to the results of the preceding section, the capacitor voltage would decay exponentially according to the following equation:

$$v_C(t) = V_B e^{-t/RC} \tag{5.19}$$

Physically, this exponential decay signifies that the energy stored in the capacitor at $t = 0$ is dissipated by the resistor at a rate determined by the time constant of the circuit, $\tau = RC$. Intuitively, the existence of a closed circuit path allows for the flow of a current, thus draining the capacitor of its charge. All of the energy initially stored in the capacitor is eventually dissipated by the resistor.

A very analogous reasoning process explains the behavior of an inductor. Recall that an inductor stores energy according to the expression

$$W_L = \frac{1}{2} L i_L^2(t) \tag{5.20}$$

Thus, in an inductor, energy storage is associated with the flow of a current (note the dual relationship between $i_L$ and $v_C$). Consider the circuit of Figure 5.11, which is similar to that of Figure 5.10 except that the battery has been replaced with a current source and the capacitor with an inductor. For $t < 0$, the source current, $I_B$, flows through the inductor, and energy is thus stored; at $t = 0$, the inductor current is equal to $I_B$. At this point, the current source is disconnected by means of the left-hand switch and a resistor is simultaneously connected to the inductor,

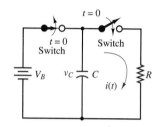

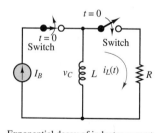

**Figure 5.10** Decay through a resistor of energy stored in a capacitor

Exponential decay of capacitor current

**Figure 5.11** Decay through a resistor of energy stored in an inductor

Exponential decay of inductor current

to form a closed circuit.[2] The inductor current will now continue to flow through the resistor, which dissipates the energy stored in the inductor. By the reasoning in the preceding discussion, the inductor current will decay exponentially:

$$i_L(t) = I_B e^{-tR/L} \qquad (5.21)$$

That is, the inductor current will decay exponentially from its initial condition, with a time constant $\tau = L/R$. Example 5.3 further illustrates the significance of the time constant in a first-order circuit.

## EXAMPLE 5.3 First-Order Systems and Time Constants

### Problem

Create a table illustrating the exponential decay of a voltage or current in a first-order circuit versus the number of time constants.

### Solution

**Known Quantities:** Exponential decay equation.

**Find:** Amplitude of voltage or current, $x(t)$, at $t = 0, \tau, 2\tau, 3\tau, 4\tau, 5\tau$.

**Assumptions:** The initial condition at $t = 0$ is $x(0) = X_0$.

**Analysis:** We know that the exponential decay of $x(t)$ is governed by the equation:

$$x(t) = X_0 e^{-t/\tau}$$

Thus, we can create the following table for the ratio $x(t)/X_0 = e^{-n\tau/\tau}$, $n = 0, 1, 2, \ldots$, at each value of $t$:

| $\dfrac{x(t)}{X_0}$ | n |
|---|---|
| 1 | 0 |
| 0.3679 | 1 |
| 0.1353 | 2 |
| 0.0498 | 3 |
| 0.0183 | 4 |
| 0.0067 | 5 |

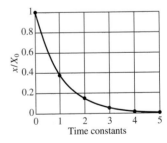

**Figure 5.12** First-order exponential decay and time constants

Figure 5.12 depicts the five points on the exponential decay curve.

**Comments:** Note that after three time constants, $x$ has decayed to approximately 5 percent of the initial value, and after five time constants to less than 1 percent.

---

[2]Note that in theory an ideal current source cannot be connected in series with a switch. For the purpose of this hypothetical illustration, imagine that upon opening the right-hand-side switch, the current source is instantaneously connected to another load, not shown.

## EXAMPLE 5.4 Charging a Camera Flash—Time Constants

### Problem

A capacitor is used to store energy in a camera flash light. The camera operates on a 6-V battery. Determine the time required for the energy stored to reach 90 percent of the maximum. Compute this time in seconds, and as a multiple of the time constant. The equivalent circuit is shown in Figure 5.13.

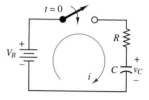

**Figure 5.13** Equivalent circuit of camera flash charging circuit

### Solution

**Known Quantities:** Battery voltage; capacitor and resistor values.

**Find:** Time required to reach 90 percent of the total energy storage.

**Schematics, Diagrams, Circuits, and Given Data:** $V_B = 6$ V; $C = 1,000\ \mu$F; $R = 1$ kΩ.

**Assumptions:** Charging starts at $t = 0$, when the flash switch is turned on. The capacitor is completely discharged at the start.

**Analysis:** First, we compute the total energy that can be stored in the capacitor:

$$E_{total} = \tfrac{1}{2}Cv_C^2 = \tfrac{1}{2}CV_B^2 = 18 \times 10^{-3}\quad \text{J}$$

Thus, 90 percent of the total energy will be reached when $E_{total} = 0.9 \times 18 \times 10^{-3} = 16.2 \times 10^{-3}$ J. This corresponds to a voltage calculated from

$$\tfrac{1}{2}Cv_C^2 = 16.2 \times 10^{-3}$$

$$v_C = \sqrt{\frac{2 \times 16.2 \times 10^{-3}}{C}} = 5.692\quad \text{V}$$

Next, we determine the time constant of the circuit: $\tau = RC = 10^{-3} \times 10^3 = 1$ s; and we observe that the capacitor will charge exponentially according to the expression

$$v_C = 6\left(1 - e^{-t/\tau}\right) = 6\left(1 - e^{-t}\right)$$

To compute the time required to reach 90 percent of the energy, we must therefore solve for $t$ in the equation

$$v_{C\text{-}90\%} = 5.692 = 6\left(1 - e^{-t}\right)$$

$$0.949 = 1 - e^{-t}$$

$$0.051 = e^{-t}$$

$$t = -\log_e(0.051) = 2.97\quad \text{s}$$

The result corresponds to a charging time of approximately 3 time constants.

**Comments:** This example demonstrates the physical connection between the time constant of a first-order circuit and a practical device. If you wish to practice some of the calculations related to time constants, you might calculate the number of time constants required to reach 95 percent and 99 percent of the total energy stored in a capacitor.

# Forced and Complete Response of First-Order Circuits

In the preceding section, the natural response of a first-order circuit was found by setting the forcing function equal to zero and considering the energy initially stored in the circuit as the driving force. The **forced response**, $x_F(t)$, of the **inhomogeneous equation**

$$\frac{dx_F(t)}{dt} + \frac{1}{\tau}x_F(t) = f(t) \tag{5.22}$$

is defined as the response to a particular forcing function $f(t)$, *without regard for the initial conditions.*[3] Thus, the forced response depends exclusively on the nature of the forcing function. The distinction between natural and forced response is particularly useful because it clarifies the nature of the transient response of a first-order circuit: the voltages and currents in the circuit are due to the superposition of two effects, the presence of *stored energy* (which can either decay, or further accumulate if a source is present) and the action of *external sources* (forcing functions). The natural response considers only the former, while the forced response describes the latter. The sum of these two responses forms the **complete response** of the circuit:

$$x(t) = x_N(t) + x_F(t) \tag{5.23}$$

The forced response depends, in general, on the form of the forcing function, $f(t)$. For the purpose of the present discussion, it will be assumed that $f(t)$ is a constant, applied at $t = 0$, that is, that

$$f(t) = F \qquad t \geq 0 \tag{5.24}$$

(Note that this is equivalent to turning a switch on or off.) In this case, the differential equation describing the circuit may be written as follows:

$$\frac{dx_F}{dt} = -\frac{x_F}{\tau} + F \qquad t \geq 0 \tag{5.25}$$

For the case of a DC forcing function, the form of the forced solution is also a constant. Substituting $x_F(t) = X_F = $ constant in the inhomogeneous differential equation, we obtain

$$0 = -\frac{X_F}{\tau} + F \tag{5.26}$$

or

$$X_F = \tau F$$

Thus, the complete solution of the original differential equation subject to initial condition $x(t = 0) = x_0$ and to a DC forcing function $F$ for $t \geq 0$ is

$$x(t) = x_N(t) + x_F(t) \tag{5.27}$$

or

$$x(t) = Ke^{-t/\tau} + \tau F$$

---

[3]Mathematicians call this solution the **particular solution.**

where the constant $K$ can be determined from the initial condition $x(t = 0) = x_0$:

$$x_0 = K + \tau F$$
$$K = x_0 - \tau F \tag{5.28}$$

Electrical engineers often classify this response as the sum of a **transient response** and a **steady-state response,** rather than a sum of a natural response and a forced response. The transient response is the response of the circuit following the switching action *before the exponential decay terms have died out*; that is, the transient response is the sum of the natural and forced responses during the transient readjustment period we have just described. The steady-state response is the response of the circuit *after all of the exponential terms have died out.* Equation 5.27 could therefore be rewritten as

$$x(t) = x_T(t) + x_{SS} \tag{5.29}$$

where

$$x_T(t) = (x_0 - \tau F)e^{-t/\tau} \tag{5.30}$$

and in the case of a DC excitation, $F$,

$$x_{SS}(t) = \tau F = x_\infty$$

Note that the transient response is not equal to the natural response, but it includes part of the forced response. The representation in equations 5.30 is particularly convenient, because it allows for solution of the differential equation that results from describing the circuit *by inspection.* The key to solving first-order circuits subject to DC transients by inspection is in considering *two separate circuits*: the circuit prior to the switching action, to determine the initial condition, $x_0$; and the circuit following the switching action, to determine the time constant of the circuit, $\tau$, and the steady-state (final) condition, $x_\infty$. Having determined these three values, you can write the solution directly in the form of equation 5.29, and you can then evaluate it using the initial condition to determine the constant $K$.

To summarize, the transient behavior of a circuit can be characterized in three stages. Prior to the switching action, the circuit is in a steady-state condition (the initial condition, determined by $x_0$). For a period of time following the switching action, the circuit sees a transient readjustment, which is the sum of the effects of the natural response and of the forced response. Finally, after a suitably long time (which depends on the time constant of the system), the natural response decays to zero (i.e., the term $e^{-t/\tau} \to 0$ as $t \to \infty$) and the new steady-state condition of the circuit is equal to the forced response: as $t \to \infty$, $x(t) \to x_F(t)$. You may recall that this is exactly the sequence of events described in the introductory paragraphs of Section 5.3. Analysis of the circuit differential equation has formalized our understanding of the transient behavior of a circuit.

## Continuity of Capacitor Voltages and Inductor Currents

As has already been stated, the primary variables employed in the analysis of circuits containing energy-storage elements are *capacitor voltages* and *inductor currents*. This choice stems from the fact that the energy-storage process in capacitors and inductors is closely related to these respective variables. The amount of charge stored in a capacitor is directly related to the voltage present across the capacitor, while the energy stored in an inductor is related to the current flowing through it. A fundamental property of inductor currents and capacitor voltages makes it easy to identify the initial condition and final value for the differential equation describing a circuit: *capacitor voltages and inductor currents cannot change instantaneously.* An instantaneous change in either of these variables would require an infinite amount of power. Since power equals energy per unit time, it follows that a truly instantaneous change in energy (i.e., a finite change in energy in zero time) would require infinite power.

Another approach to illustrating the same principle is as follows. Consider the defining equation for the capacitor:

$$i_C(t) = C \frac{dv_C(t)}{dt}$$

and assume that the capacitor voltage, $v_C(t)$, can change instantaneously, say, from 0 to $V$ volts, as shown in Figure 5.14. The value of $dv_C/dt$ at $t = 0$ is simply the slope of the voltage, $v_C(t)$, at $t = 0$. Since the slope is infinite at that point, because of the instantaneous transition, it would require an infinite amount of current for the voltage across a capacitor to change instantaneously. But this is equivalent to requiring an infinite amount of power, since power is the product of voltage and current. A similar argument holds if we assume a "step" change in inductor current from, say, 0 to $I$ amperes: an infinite voltage would be required to cause an instantaneous change in inductor current. This simple fact is extremely useful in determining the response of a circuit. Its immediate consequence is that *the value of an inductor current or a capacitor voltage just prior to the closing (or opening) of a switch is equal to the value just after the switch has been closed (or opened).* Formally,

**Figure 5.14** Abrupt change in capacitor voltage

$$v_C(0^+) = v_C(0^-) \tag{5.31}$$

$$i_L(0^+) = i_L(0^-) \tag{5.32}$$

where the notation $0^+$ signifies "just after $t = 0$" and $0^-$ means "just before $t = 0$."

---

### EXAMPLE 5.5 Continuity of Inductor Current

**Problem**

Find the initial condition and final value of the inductor current in the circuit of Figure 5.15.

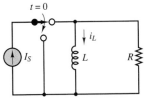

**Figure 5.15**

**Solution**

**Known Quantities:** Source current, $I_S$; inductor and resistor values.

**Find:** Inductor current at $t = 0^+$ and as $t \to \infty$.

**Schematics, Diagrams, Circuits, and Given Data:** $I_S = 10$ mA.

**Assumptions:** The current source has been connected to the circuit for a very long time.

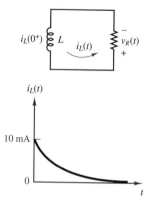

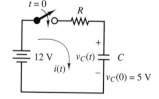

**Figure 5.16**

**Analysis:** At $t = 0^-$, since the current source has been connected to the circuit for a very long time, the inductor acts as a short circuit, and $i_L(0^-) = I_S$. Since all the current flows through the inductor, the voltage across the resistor must be zero. At $t = 0^+$, the switch opens and we can state that

$$i_L(0^+) = i_L(0^-) = I_S$$

because of the continuity of inductor current.

The circuit for $t \geq 0$ is shown in Figure 5.16, where the presence of the current $i_L(0^+)$ denotes the initial condition for the circuit. A qualitative sketch of the current as a function of time is also shown in Figure 5.16, indicating that the inductor current eventually becomes zero as $t \to \infty$.

**Comments:** Note that the direction of the current in the circuit of Figure 5.16 is dictated by the initial condition, since the inductor current cannot change instantaneously. Thus, the current will flow counterclockwise, and the voltage across the resistor will therefore have the polarity shown in the figure.

## Complete Solution of First-Order Circuits

In this section, we illustrate the application of the principles put forth in the preceding sections by presenting a number of examples. The first example summarizes the complete solution of a simple $RC$ circuit.

### EXAMPLE 5.6 Complete Solution of First-Order Circuit

**Problem**

Determine an expression for the capacitor voltage in the circuit of Figure 5.17.

**Figure 5.17**

**Solution**

**Known Quantities:** Initial capacitor voltage; battery voltage, resistor and capacitor values.

**Find:** Capacitor voltage as a function of time, $v_C(t)$, for all $t$.

**Schematics, Diagrams, Circuits, and Given Data:** $v_C(t = 0^-) = 5$ V; $R = 1$ k$\Omega$; $C = 470$ $\mu$F; $V_B = 12$ V.

**Assumptions:** None.

**Analysis:** We first observe that the capacitor had previously been charged to an initial voltage of 5 V. Thus,

$$v_C(t) = 5 \text{ V} \qquad t < 0$$

At $t = 0$ the switch closes, and the circuit is described by the following differential equation, obtained by application of KVL:

$$V_B - RC\frac{dv_C(t)}{dt} - v_C(t) = 0 \qquad t > 0$$

$$\frac{dv_C(t)}{dt} + \frac{1}{RC}v_C(t) = \frac{1}{RC}V_B \qquad t > 0$$

In the above equation we recognize the following variables, with reference to equation 5.22:

$$x = v_C \qquad \tau = RC \qquad f(t) = \frac{1}{RC}V_B \qquad t > 0 \text{ s}$$

The natural response of the circuit is therefore of the form:

$$x_N(t) = v_{CN}(t) = Ke^{-t/\tau} = Ke^{-t/RC} \qquad t > 0 \text{ s},$$

while the forced response is of the form:

$$x_F(t) = v_{CF}(t) = \tau f(t) = V_B \qquad t > 0 \text{ s}.$$

Thus, the complete response of the circuit is given by the expression

$$x(t) = v_C(t) = v_{CN}(t) + v_{CF}(t) = Ke^{-t/RC} + V_B \qquad t > 0 \text{ s}$$

Now that we have the complete response, we can apply the initial condition to determine the value of the constant $K$. At time $t = 0$,

$$v_C(0) = 5 = Ke^{-0/RC} + V_B$$

$$K = 5 - 12 = -7 \text{ V}$$

We can finally write the complete response with numerical values:

$$v_C(t) = -7e^{-t/0.47} + 12 \text{ V} \qquad t > 0 \text{ s}$$

$$= v_{CT}(t) + v_{CSS}(t)$$

$$= 12\left(1 - e^{-t/0.47}\right) + 5e^{-t/0.47} \text{ V} \qquad t > 0 \text{ s}$$

$$= v_{CF}(t) + v_{CN}(t)$$

The complete response described by the above equations is shown graphically in Figure 5.18 (a) and (b).

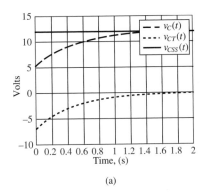

(a)

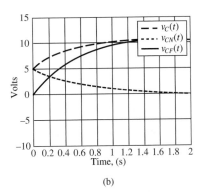

(b)

**Figure 5.18** (a) Complete, transient, and steady-state responses of the circuit of Figure 5.17. (b) Complete, natural, and forced responses of the circuit of Figure 5.17.

**Comments:** Note how in Figure 5.18(a) the *steady-state response* $v_{CSS}(t)$ is simply equal to the battery voltage, while the *transient response*, $v_{CT}(t)$, rises from $-7$ V to $0$ V exponentially. In Figure 5.18(b), on the other hand, we can see that the energy initially stored in the capacitor decays to zero via its *natural response*, $v_{CN}(t)$, while the external forcing function causes the capacitor voltage to eventually rise exponentially to 12 V, as shown in the forced response, $V_{CF}(t)$. The example just completed, though based on a very simple circuit, illustrates all the steps required to complete the solution of a first-order circuit. The methodology applied in the example is summarized in a box, next.

**Focus on Computer-Aided Tools:** An electronic file generated using Matlab to create the graphs of Figure 5.18 may be found in the accompanying CD-ROM. An EWB solution is also enclosed.

VIRTUAL LAB

## F O C U S   O N   M E T H O D O L O G Y

### Solution of First-Order Circuits

1. Determine the initial condition of the energy storage element.
2. Write the differential equation for the circuit for $t > 0$.
3. Determine the time constant of the circuit for $t > 0$.
4. Write the complete solution as the sum of the natural and forced responses.
5. Apply the initial condition to the *complete solution*, to determine the constant $K$.

## EXAMPLE 5.7  Starting Transient of DC Motor

### Problem

An approximate circuit representation of a DC motor consists of series $RL$ circuit, shown in Figure 5.19. Apply the first-order circuit solution methodology just described to this approximate DC motor equivalent circuit to determine the transient current.

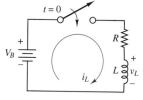

**Figure 5.19**

### Solution

**Known Quantities:** Initial motor current; battery voltage, resistor and inductor values.

**Find:** Inductor current as a function of time, $i_L(t)$, for all $t$.

**Schematics, Diagrams, Circuits, and Given Data:** $i_L(t = 0^-) = 0$ A; $R = 4\ \Omega$; $L = 0.1$ H; $V_B = 50$ V.

**Assumptions:** None.

**Analysis:** At $t = 0$ the switch closes, and the circuit is described by the following differential equation, obtained by application of KVL:

$$V_B - Ri_L - L\frac{di_L(t)}{dt} = 0 \qquad t > 0$$

$$\frac{di_L(t)}{dt} + \frac{R}{L}i_L(t) = \frac{1}{L}V_B \qquad t > 0$$

In the above equation we recognize the following variables, with reference to equation 5.22:

$$x = i_L \qquad \tau = \frac{L}{R} \qquad f(t) = \frac{1}{L}V_B \qquad t > 0$$

The natural response of the circuit is therefore of the form:

$$x_N(t) = i_{LN}(t) = Ke^{-t/\tau} = Ke^{-Rt/L} \qquad t > 0$$

while the forced response is of the form:

$$x_F(t) = i_{LF}(t) = \tau f(t) = \frac{V_B}{R} \qquad t > 0.$$

Thus, the complete response of the circuit is given by the expression

$$x(t) = i_L(t) = i_{LN}(t) + i_{LF}(t) = Ke^{-Rt/L} + \frac{1}{R}V_B \qquad t > 0$$

Now that we have the complete response, we can apply the initial condition to determine the value of the constant $K$. At time $t = 0$,

$$i_L(0) = 0 = Ke^{-0} + \frac{1}{R}V_B$$

$$K = -\frac{1}{R}V_B$$

We can finally write the complete response with numerical values:

$$i_L(t) = \frac{V_B}{R}(1 - e^{-t/\tau}) \qquad t > 0$$
$$= 12.5(1 - e^{-t/0.025}) \qquad t > 0$$

The complete response described by the above equations is shown graphically in Figure 5.20.

**Comments:**  Note that in practice it is not a good idea to place a switch in series with an inductor. As the switch opens, the inductor current is forced to change instantaneously, with the result that $di_L/dt$, and therefore $v_L(t)$ approaches infinity. The large voltage transient resulting from this *inductive kick* can damage circuit components. A practical solution to this problem, the free-wheeling diode, is presented in Section 11.5.

**Focus on Computer-Aided Tools:**  An electronic file generated using Matlab to create the graph of Figure 5.20 may be found in the accompanying CD-ROM.

VIRTUAL LAB

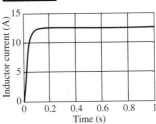

**Figure 5.20** Complete response of the circuit of Fig. 5.19

---

In the preceding examples we have seen how to systematically determine the solution of first-order circuits. The solution methodology was applied to two simple cases, but it applies in general to any first-order circuit, providing that one is careful to identify a Thévenin (or Norton) equivalent circuit, determined with respect to the energy storage element (i.e., treating the energy storage element as the load). Thus the equivalent circuit methodology for resistive circuits presented in Chapter 3 applies to transient circuits as well. Figure 5.21 depicts the general appearance of a first-order circuit once the resistive part of the circuit has been reduced to Thévenin equivalent form.

An important comment must be made before demonstrating the equivalent circuit approach to more complex circuit topologies. Since the circuits that are the subject of the present discussion usually contain a switch, one must be careful to determine the equivalent circuits *before and after the switch changes position.* In other words, it is possible that the equivalent circuit seen by the load before activating the switch is different from the circuit seen after the switch changes position.

To illustrate the procedure, consider the *RC* circuit of Figure 5.22. The objective is to determine the capacitor voltage for all time. The switch closes at $t = 0$. For $t < 0$, we recognize that the capacitor has been connected to the battery $V_2$ through resistor $R_2$. This circuit is already in Thévenin equivalent form, and we know that the capacitor must have charged to the battery voltage, $V_2$, provided

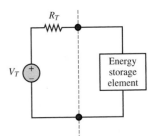

**Figure 5.21** Equivalent-circuit representation of first-order circuits

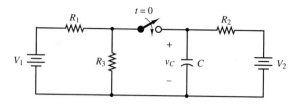

**Figure 5.22** A more involved $RC$ circuit

that the switch has been closed for a sufficient time (we shall assume so). Thus:

$$v_C(t) = V_2 \qquad t \leq 0$$
$$V_C(0) = V_2 \qquad \qquad (5.33)$$

After the switch closes, the circuit on the left-hand side of Figure 5.22 must be accounted for. Figure 5.23 depicts the new arrangement, in which we have moved the capacitor to the far right-hand side, in preparation for the evaluation of the equivalent circuit. Using the Thévenin-to-Norton source transformation technique (introduced in Chapter 3), we next obtain the circuit at the top of Figure 5.24, which can be easily reduced by adding the two current sources and computing the equivalent parallel resistance of $R_1$, $R_2$, and $R_3$. The last step illustrated in the figure is the conversion to Thévenin form. Figure 5.25 depicts the final appearance of the equivalent circuit for $t \geq 0$.

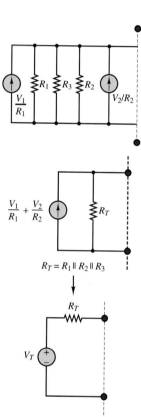

**Figure 5.24** Reduction of the circuit of Figure 5.23 to Thévenin equivalent form

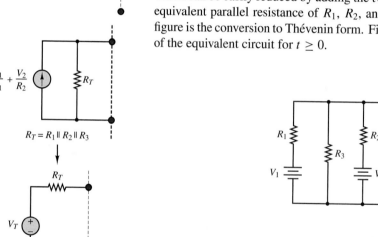

**Figure 5.23** The circuit of Figure 6.45 for $t \geq 0$

Now we are ready to write the differential equation for the equivalent circuit:

$$\frac{dv_C}{dt} + \frac{1}{R_T C} v_C = \frac{1}{R_T C} V_T \qquad t \geq 0 \qquad (5.34)$$

$$\tau = R_T C \qquad v_C(0) = V_2$$

The complete solution is then computed following the usual procedure, as shown below.

$$v_C(t) = K e^{-t/\tau} + \tau f(t)$$
$$v_C(0) = K e^0 + V_T$$
$$K = v_C(0) - V_T = V_2 - V_T \qquad (5.35)$$
$$v_C(t) = (V_2 - V_T)e^{-t/R_T C} + V_T$$

The method illlustrated above is now applied to two examples.

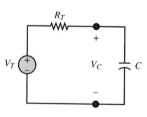

**Figure 5.25** The circuit of Figure 5.22 in equivalent form for $t \geq 0$

## EXAMPLE 5.8 Turn-Off Transient of DC Motor

### Problem

Determine the motor voltage for all time in the simplified electric motor circuit model shown in Figure 5.26. The motor is represented by the series $RL$ circuit in the shaded box.

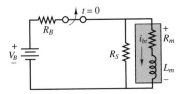

**Figure 5.26**

### Solution

**Known Quantities:** Battery voltage, resistor, and inductor values.

**Find:** The voltage across the motor as a function of time.

**Schematics, Diagrams, Circuits, and Given Data:** $R_B = 2\ \Omega$; $R_S = 20\ \Omega$; $R_m = 0.8\ \Omega$; $L = 3$ H; $V_B = 100$ V.

**Assumptions:** The switch has been closed for a long time.

**Analysis:** With the switch closed for a long time, the inductor in the circuit of Figure 5.26 behaves like a short circuit. The current through the motor can then be calculated by the current divider rule in the modified circuit of Figure 5.27, where the inductor has been replaced with a short circuit and the Thévenin circuit on the left has been replaced by its Norton equivalent:

$$i_m = \frac{\dfrac{1}{R_m}}{\dfrac{1}{R_B} + \dfrac{1}{R_s} + \dfrac{1}{R_m}}\frac{V_B}{R_B} = \frac{\dfrac{1}{0.8}}{\dfrac{1}{2} + \dfrac{1}{20} + \dfrac{1}{0.8}}\frac{100}{2} = 34.72 \quad \text{A}$$

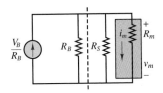

**Figure 5.27**

This current is the initial condition for the inductor current: $i_L(0) = 34.72$ A. Since the motor inductance is effectively a short circuit, the motor voltage for $t < 0$ is equal to

$$v_m(t) = i_m R_m = 27.8 \quad \text{V} \qquad t < 0$$

When the switch opens and the motor voltage supply is turned off, the motor sees only the *shunt* (parallel) resistance $R_s$, as depicted in Figure 5.28. Remember now that the inductor current cannot change instantaneously; thus, the motor (inductor) current, $i_m$, must continue to flow in the same direction. Since all that is left is a series $RL$ circuit, with resistance $R = R_s + R_m = 20.8\ \Omega$, the inductor current will decay exponentially with time constant $\tau = L/R = 0.1442$ s:

$$i_L(t) = i_m(t) = i_L(o)e^{-t/\tau} = 34.7e^{-t/0.1442} \qquad t > 0$$

**Figure 5.28**

The motor voltage is then computed by adding the voltage drop across the motor resistance and inductance:

$$v_m(t) = R_m i_L(t) + L\frac{di_L(t)}{dt}$$

$$= 0.8 \times 34.7e^{-t/0.1442} + 3 \times \left(-\frac{34.7}{0.1442}\right)e^{-t/0.1442} \qquad t > 0$$

$$= -694.1e^{-t/0.1442} \qquad t > 0$$

The motor voltage is plotted in Figure 5.29.

**Comments:** Notice how the motor voltage rapidly changes from the steady-state value of 27.8 V for $t < 0$ to a large negative value due to the turn-off transient. This *inductive kick* is typical of $RL$ circuits, and results from the fact that, although the inductor current cannot change instantaneously, the inductor voltage can and does, as it is proportional to

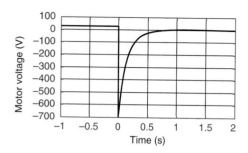

**Figure 5.29** Motor voltage transient response

the derivative of $i_L$. This example is based on a simplified representation of an electric motor, but illustrates effectively the need for special starting and stopping circuits in electric motors. Some of these ideas are explored in Chapters 11 ("Power Electronics"), 17 ("Introduction to Electric Machines") and 18 ("Special-Purpose Electric Machines").

VIRTUAL LAB

***Focus on Computer-Aided Tools:*** The Matlab *m-file* containing the numerical analysis and plotting commands for this example may be found in the CD that accompanies this book.

---

## FOCUS ON MEASUREMENTS

## Coaxial Cable Pulse Response

FIND IT

ON THE WEB

**Problem:**

A problem of great practical importance is the transmission of *pulses* along cables. Short voltage pulses are used to represent the two-level binary signals that are characteristic of digital computers; it is often necessary to transmit such voltage pulses over long distances through **coaxial cables,** which are characterized by a finite resistance per unit length and by a certain capacitance per unit length, usually expressed in pF/m. A simplified model of a long coaxial cable is shown in Figure 5.30. If a 10-m cable has a capacitance of 1,000 pF/m and a series resistance of 0.2 Ω/m, what will the output of the pulse look like after traveling the length of the cable?

**Solution:**

***Known Quantities—*** Cable length, resistance, and capacitance; voltage pulse amplitude and time duration.

***Find—*** The cable voltage as a function of time.

***Schematics, Diagrams, Circuits, and Given Data—*** $r_1 = 0.2$ Ω/m; $R_L = 150$ Ω; $c = 1,000$ pF/m; $l = 10$ m; pulse duration $= 1$ μs.

***Assumptions—*** The short voltage pulse is applied to the cable at $t = 0$. Assume zero initial conditions.

***Analysis—*** The voltage pulse can be modeled by a 5-V battery connected to a switch; the switch will then close at $t = 0$ and open again at $t = 1$ μs. The solution strategy will therefore proceed as follows. First, we determine the initial condition; next, we solve the transient problem for $t > 0$; finally, we compute the value of the capacitor voltage at $t = 1$ μs—that is, when the switch opens again—and solve a different transient problem. Intuitively, we

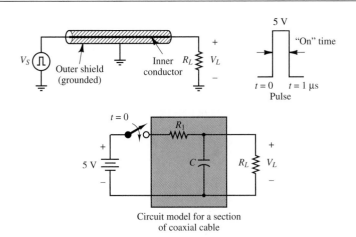

**Figure 5.30** Pulse transmission in a coaxial cable

know the equivalent capacitor will charge for 1 $\mu$s, and the voltage will reach a certain value. This value will be the initial condition for the capacitor voltage when the switch is opened; the capacitor voltage will then decay to zero, since the voltage source has been disconnected. Note that the circuit will be characterized by two different time constants during the two transient stages of the problem. The initial condition for this problem is zero, assuming that the switch has been open for a long time.

The differential equation for $0 < t < 1$ $\mu$s is obtained by computing the Thévenin equivalent circuit relative to the capacitor when the switch is closed:

$$V_T = \frac{R_L}{R_1 + R_L} V_B \qquad R_T = R_1 \| R_L \qquad \tau = R_T C \qquad 0 < t < 1 \ \mu s$$

As we have already seen, the differential equation is given by the expression

$$\frac{dv_C}{dt} + \frac{1}{R_T C} v_C = \frac{1}{R_T C} V_T \qquad 0 < t < 1 \ \mu s$$

and the solution is of the form

$$v_C(t) = K e^{-t/\tau_{\text{on}}} + \tau f(t) = K e^{-t/\tau_{\text{on}}} + V_T$$

$$v_C(0) = K e^0 + V_T$$

$$K = v_C(0) - V_T = -V_T$$

$$v_C(t) = -V_T e^{-t/R_T C} + V_T = V_T \left(1 - e^{-t/R_T C}\right) \qquad 0 < t < 1 \ \mu s$$

We can assign numerical values to the solution by calculating the effective resistance and capacitance of the cable:

$$R_1 = r_1 \times l = 0.2 \times 10 = 2 \ \Omega \quad C = c \times l$$
$$= 1{,}000 \times 10 = 10{,}000 \ \text{pF}$$

$$R_T = 2 \| 150 = 1.97 \ \Omega \qquad V_T = \frac{150}{152} V_B = 4.93 \ V$$

$$\tau_{\text{on}} = R_T C = 19.74 \times 10^{-9} \ \text{s}$$

so that

$$v_C(t) = 4.93 \left(1 - e^{-t/19.74 \times 10^{-9}}\right) \qquad 0 < t < 1\ \mu s$$

At the time when the switch opens again, $t = 1\ \mu s$, the capacitor voltage can be found to be $v_C(t = 1\ \mu s) = 4.93$ V.

   When the switch opens again, the capacitor will discharge through the load resistor, $R_L$; this discharge is described by the natural response of the circuit consisting of $C$ and $R_L$ and is governed by the following values: $v_C(t = 1\ \mu s) = 4.93$ V, $\tau_{\text{off}} = R_L C = 1.5\ \mu s$. We can directly write the natural solution as follows:

$$v_C(t) = v_C(t = 1 \times 10^{-6}) \times e^{-(t-1 \times 10^{-6}/\tau_{\text{off}}}$$
$$= 4.93 \times e^{-(t-1 \times 10^{-6})/1.5 \times 10^{-6}} \qquad t \geq 1\ \mu s$$

Figure 5.31 shows a plot of the solution for $t > 0$, along with the voltage pulse.

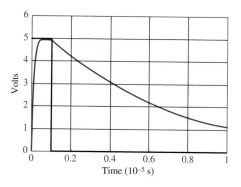

**Figure 5.31** Coaxial cable pulse response

***Comments—*** Note that the voltage response shown in Figure 5.31 rapidly reaches the desired value, near 5 volts, thanks to the very short charging time constant, $\tau_{\text{on}}$. On the other hand, the discharging time constant, $\tau_{\text{off}}$, is significantly slower. As the length of the cable is increased, however, $\tau_{\text{on}}$ will increase, to the point that the voltage pulse may not rise sufficiently close to the desired 5-V value in the desired time. While the numbers used in this example are somewhat unrealistic, you should remember that cable length limitations may exist in some applications because of the cable intrinsic capacitance and resistance.

***Focus on Computer-Aided Tools—*** The Matlab *m-file* containing the numerical analysis and plotting commands for this example may be found in the CD that accompanies this book. An EWB solution is also enclosed.

VIRTUAL LAB

---

## Check Your Understanding

**5.1**  Write the differential equation for the circuit shown in Figure 5.32.

**5.2**  Write the differential equation for the circuit shown in Figure 5.33.

**5.3**  Write the differential equation for the circuit shown in Figure 5.34.

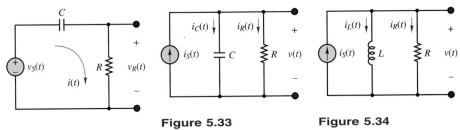

**Figure 5.33**          **Figure 5.34**

**Figure 5.32**

**5.4**  It is instructive to repeat the analysis of Example 5.5 for a capacitive circuit. For the circuit shown in Figure 5.35, compute the quantities $v_C(0^-)$ and $i_R(0^+)$, and sketch the response of the circuit, that is, $v_C(t)$, if the switch opens at $t = 0$.

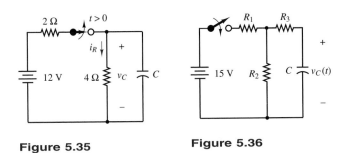

**Figure 5.35**          **Figure 5.36**

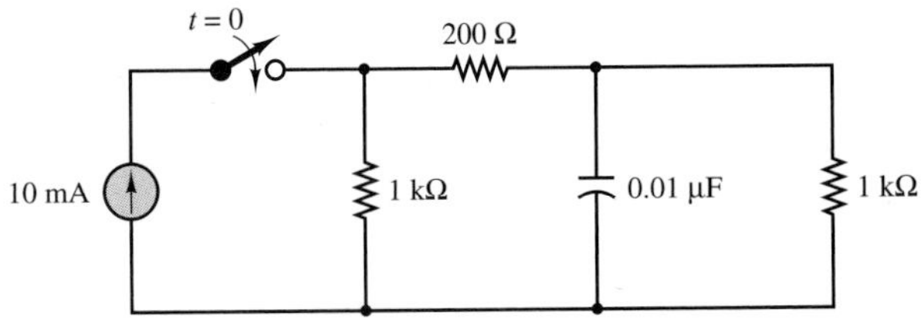

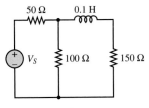

**Figure 5.37**

**5.5**  The circuit of Figure 5.36 has a switch that can be used to connect and disconnect a battery. The switch has been open for a very long time. At $t = 0$, the switch closes, and then at $t = 50$ ms, the switch opens again. Assume that $R_1 = R_2 = 1{,}000\ \Omega$, $R_3 = 500\ \Omega$, and $C = 25\ \mu F$.

   a.  Determine the capacitor voltage as a function of time.

   b.  Plot the capacitor voltage from $t = 0$ to $t = 100$ ms.

**5.6**  If the 10-mA current source is switched on at $t = 0$ in the circuit of Figure 5.37, how long will it take for the capacitor to charge to 90 percent of its final voltage?

**5.7**  Find the time constant for the circuit shown in Figure 5.38.

**5.8**  Repeat the calculations of Example 5.9 if the load resistance is 1,000 $\Omega$. What is the effect of this change?

**Figure 5.38**

## 5.4  TRANSIENT RESPONSE OF SECOND-ORDER CIRCUITS

In many practical applications, understanding the behavior of first- and second-order systems is often all that is needed to describe the response of a physical system to external excitation. In this section, we discuss the solution of the second-order differential equations that characterize second-order circuits.

## Deriving the Differential Equations for Second-Order Circuits

A simple way of introducing second-order circuits consists of replacing the box labeled "Circuit containing $RL/RC$ combinations" in Figure 5.3 with a combination of two energy-storage elements, as shown in Figure 5.39. Note that two different cases are considered, depending on whether the energy-storage elements are connected in series or in parallel.

Consider the parallel case first, which has been redrawn in Figure 5.40 for clarity. Practice and experience will eventually suggest the best method for writing the circuit equations. At this point, the most sensible procedure consists of applying the basic circuit laws to the circuit of Figure 5.40. Start with KVL around the left-hand loop:

$$v_T(t) - R_T i_S(t) - v_C(t) = 0 \tag{5.36}$$

Then apply KCL to the top node, to obtain

$$i_S(t) - i_C(t) - i_L(t) = 0 \tag{5.37}$$

Further, KVL applied to the right-hand loop yields

$$v_C(t) = v_L(t) \tag{5.38}$$

It should be apparent that we have all the equations we need (in fact, more). Using the defining relationships for capacitor and inductor, we can express equation 5.37 as

$$\frac{v_T(t) - v_C(t)}{R_T} - C\frac{dv_C}{dt} - i_L(t) = 0 \tag{5.39}$$

and equation 5.38 becomes

$$v_C(t) = L\frac{di_L}{dt} \tag{5.40}$$

Substituting equation 5.40 in equation 5.39, we can obtain a differential circuit equation in terms of the variable $i_L(t)$:

$$\frac{1}{R_T}v_T(t) - \frac{L}{R_T}\frac{di_L}{dt} = LC\frac{d^2 i_L}{dt^2} + i_L(t) \tag{5.41}$$

or

$$\frac{d^2 i_L}{dt^2} + \frac{1}{R_T C}\frac{di_L}{dt} + \frac{1}{LC}i_L = \frac{v_T(t)}{R_T LC} \tag{5.42}$$

The solution to this differential equation (which depends, as in the case of first-order circuits, on the initial conditions and on the forcing function) completely determines the behavior of the circuit. By now, two questions should have appeared in your mind:

1. Why is the differential equation expressed in terms of $i_L(t)$? (Why not $v_C(t)$?)
2. Why did we not use equation 5.36 in deriving equation 5.42?

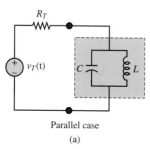

$R_T$

$v_T(t)$

$C$     $L$

Parallel case

(a)

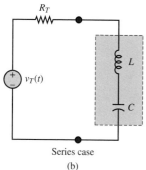

$R_T$

$v_T(t)$

$L$

$C$

Series case

(b)

**Figure 5.39** Second-order circuits

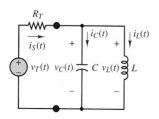

$R_T$

$i_S(t)$     $i_C(t)$     $i_L(t)$

$v_T(t)$     $v_C(t)$     $C$     $v_L(t)$     $L$

**Figure 5.40** Parallel case

In response to the first question, it is instructive to note that, knowing $i_L(t)$, we can certainly derive any one of the voltages and currents in the circuit. For example,

$$v_C(t) = v_L(t) = L\frac{di_L}{dt} \tag{5.43}$$

$$i_C(t) = C\frac{dv_C}{dt} = LC\frac{d^2i_L}{dt^2} \tag{5.44}$$

To answer the second question, note that equation 5.42 is not the only form the differential circuit equation can take. By using equation 5.36 in conjunction with equation 5.37, one could obtain the following equation:

$$v_T(t) = R_T[i_C(t) + i_L(t)] + v_C(t) \tag{5.45}$$

Upon differentiating both sides of the equation and appropriately substituting from equation 5.39, the following second-order differential equation in $v_C$ would be obtained:

$$\frac{d^2v_C}{dt^2} + \frac{1}{R_TC}\frac{dv_C}{dt} + \frac{1}{LC}v_C = \frac{1}{R_TC}\frac{dv_T(t)}{dt} \tag{5.46}$$

Note that the left-hand side of the equation is identical to equation 5.42, except that $v_C$ has been substituted for $i_L$. The right-hand side, however, differs substantially from equation 5.42, because the forcing function is the derivative of the equivalent voltage.

Since all of the desired circuit variables may be obtained either as a function of $i_L$ or as a function of $v_C$, the choice of the preferred differential equation depends on the specific circuit application, and we conclude that there is no unique method to arrive at the final equation. As a case in point, consider the two circuits depicted in Figure 5.41. If the objective of the analysis were to determine the output voltage, $v_{out}$, then for the circuit in Figure 5.41(a), one would choose to write the differential equation in $v_C$, since $v_C = v_{out}$. In the case of Figure 5.41(b), however, the inductor current would be a better choice, since $v_{out} = R_T i_{out}$.

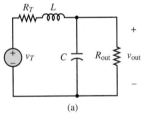

(a)

## Natural Response of Second-Order Circuits

From the previous discussion, we can derive a general form for the governing equation of a second-order circuit:

$$a_2\frac{d^2x(t)}{dt^2} + a_1\frac{dx(t)}{dt} + a_0x(t) = f(t) \tag{5.47}$$

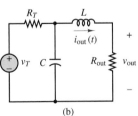

(b)

**Figure 5.41** Two second-order circuits

It is now appropriate to derive a general form for the solution. The same classification used for first-order circuits is also valid for second-order circuits. Therefore, the complete solution of the second-order equation is the sum of the natural and forced responses:

$$x(t) = \underset{\text{Natural response}}{x_N(t)} + \underset{\text{Forced response}}{x_F(t)} \tag{5.48}$$

where the natural response is the solution of the homogeneous equation without regard for the forcing function (i.e., with $f(t) = 0$) and the forced response is the solution of the forced equation with no consideration of the effects of the initial conditions. Once the general form of the complete response is found, the unknown constants are evaluated subject to the initial conditions, and the solution can then be divided into transient and steady-state parts, with

$$x(t) = \underbrace{x_T(t)}_{\text{Transient part}} + \underbrace{x_{SS}(t)}_{\text{Steady-state part}} \tag{5.49}$$

The aim of this section is to determine the natural response, which satisfies the homogeneous equation:

$$\frac{d^2x_N(t)}{dt^2} + b\frac{dx_N(t)}{dt} + cx_N(t) = 0 \tag{5.50}$$

where $b = a_1/a_2$ and $c = a_0/a_2$. Just as in the case of first-order circuits, $x_N(t)$ takes on an exponential form:

$$x_N(t) = Ke^{st} \tag{5.51}$$

This is easily verifiable by direct substitution in the differential equation:

$$s^2Ke^{st} + bsKe^{st} + cKe^{st} = 0 \tag{5.52}$$

and since it is possible to divide both sides by $Ke^{st}$, the natural response of the differential equation is, in effect, determined by the solution of the quadratic equation

$$s^2 + bs + c = 0 \tag{5.53}$$

This polynomial in the variable $s$ is called the **characteristic polynomial** of the differential equation. Thus, the natural response, $x_N(t)$, is of the form

$$x_N(t) = K_1e^{s_1t} + K_2e^{s_2t} \tag{5.54}$$

where the exponents $s_1$ and $s_2$ are found by applying the quadratic formula to the characteristic polynomial:

$$s_{1,2} = -\frac{b}{2} \pm \frac{1}{2}\sqrt{b^2 - 4c} \tag{5.55}$$

The exponential solution in terms of the exponents $s_{1,2}$ can take different forms depending on whether the roots of the quadratic equation are real or complex. As an example, consider the parallel circuit of Figure 5.40, and the governing differential equation, 5.42. The natural response for $i_L(t)$ in this case is the solution of the following equation:

$$\frac{d^2i_L(t)}{dt^2} + \frac{1}{RC}\frac{di_L(t)}{dt} + \frac{1}{LC}i_L(t) = 0 \tag{5.56}$$

where $R = R_T$ in Figure 5.40. The solution of equation 5.56 is determined by solving the quadratic equation

$$s^2 + \frac{1}{RC}s + \frac{1}{LC} = 0 \tag{5.57}$$

The roots are

$$s_{1,2} = -\frac{1}{2RC} \pm \frac{1}{2}\sqrt{\left(\frac{1}{RC}\right)^2 - \frac{4}{LC}} \tag{5.58}$$

where

$$s_1 = -\frac{1}{2RC} + \frac{1}{2}\sqrt{\left(\frac{1}{RC}\right)^2 - \frac{4}{LC}} \qquad\qquad \textbf{(5.59a)}$$

$$s_2 = -\frac{1}{2RC} - \frac{1}{2}\sqrt{\left(\frac{1}{RC}\right)^2 - \frac{4}{LC}} \qquad\qquad \textbf{(5.59b)}$$

The key to interpreting this solution is to analyze the term under the square root sign; we can readily identify three cases:

- Case I:

$$\left(\frac{1}{RC}\right)^2 > \frac{4}{LC} \qquad\qquad \textbf{(5.60)}$$

  $s_1$ and $s_2$ are real and distinct roots: $s_1 = \alpha_1$ and $s_2 = \alpha_2$.

- Case II:

$$\left(\frac{1}{RC}\right)^2 = \frac{4}{LC} \qquad\qquad \textbf{(5.61)}$$

  $s_1$ and $s_2$ are real, repeated roots: $s_1 = s_2 = \alpha$.

- Case III:

$$\left(\frac{1}{RC}\right)^2 < \frac{4}{LC} \qquad\qquad \textbf{(5.62)}$$

  $s_1$, $s_2$ are complex conjugate roots: $s_1 = s_2^* = \alpha + j\beta$.

It should be remarked that a special case of the solution (5.62) arises when the value of $R$ is identically zero. This is known as the **resonance** condition; we shall return to it later in this section. For each of these three cases, as we shall see, the solution of the differential equation takes a different form. The remainder of this section will explore the three different cases that can arise.

---

## EXAMPLE 5.9 Natural Response of Second-Order Circuit

### Problem

Find the natural response of $i_L(t)$ in the circuit of Figure 5.42.

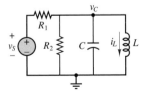

---

### Solution

**Known Quantities:** Resistor, capacitor, inductor values.

**Find:** The inductor current as a function of time.

**Figure 5.42**

**Schematics, Diagrams, Circuits, and Given Data:** $R_1 = 8\ \text{k}\Omega$; $R_2 = 8\ \text{k}\Omega$; $C = 10\ \mu\text{F}$; $L = 1\ \text{H}$.

**Assumptions:** None.

**Analysis:** To determine the natural response of the circuit, we set the arbitrary voltage source equal to zero by replacing it with a short circuit. Next, we observe that the two resistors can be replaced by a single resistor, $R = R_1||R_2$, and that we now are faced with a parallel $RLC$ circuit. Applying KCL at the top node, we write:

$$\frac{v_C}{R} + C\frac{dv_C}{dt} + i_L = 0$$

We recognize that the top node voltage is also equal to the inductor voltage, and that

$$v_C = v_L = L\frac{di_L}{dt}$$

Next, we substitute the expression for $v_C$ in the first equation to obtain

$$\frac{d^2i_L}{dt^2} + \frac{1}{RC}\frac{di_L}{dt} + \frac{1}{LC}i_L = 0$$

The characteristic equation corresponding to this differential equation is:

$$s^2 + \frac{1}{RC}s + \frac{1}{LC} = 0$$

with roots

$$s_{1,2} = -\frac{1}{2RC} \pm \frac{1}{2}\sqrt{\left(\frac{1}{RC}\right)^2 - \frac{4}{LC}}$$

$$= -12.5 \pm j316$$

Finally, the natural response is of the form

$$i_L(t) = K_1 e^{s_1 t} + K_2 e^{s_2 t}$$

$$= K_1 e^{(-12.5+j316)t} + K_2 e^{(-12.5-j316)t}$$

The constants $K_1$ and $K_2$ in the above expression can be determined once the complete solution is known, that is, once the forced response to the source $v_S(t)$ is found. The constants $K_1$ and $K_2$ will have to be complex conjugates to assure that the solution is real.

Although the previous example dealt with a specific circuit, one can extend the result by stating that the natural response of any second-order system can be described by one of the following three expressions:

- Case I. Real, distinct roots: $s_1 = \alpha_1$, $s_2 = \alpha_2$.

$$x_N(t) = K_1 e^{\alpha_1 t} + K_2 e^{\alpha_2 t} \tag{5.63}$$

- Case II. Real, repeated roots: $s_1 = s_2 = \alpha$.

$$x_N(t) = K_1 e^{\alpha t} + K_2 t e^{\alpha t} \tag{5.64}$$

- Case III. Complex conjugate roots: $s_1 = \alpha + j\beta$, $s_2 = \alpha - j\beta$.

$$x_N(t) = K_1 e^{(\alpha+j\beta)t} + K_2 e^{(\alpha-j\beta)t} \tag{5.65}$$

The solution of the homogeneous second-order differential equation will now be discussed for each of the three cases.

## Overdamped Solution

The case of real and distinct roots yields the so-called **overdamped solution**, which consists of a sum of real exponentials. An overdamped system naturally decays to zero in the absence of a forcing function, according to the expression

$$x_N(t) = K_1 e^{-\alpha_1 t} + K_2 e^{-\alpha_2 t} \tag{5.66}$$

where $\alpha_1$ and $\alpha_2$ are now assumed to be positive constants. Note that $\alpha_1$ and $\alpha_2$ are the reciprocals of two time constants:

$$\tau_1 = \frac{1}{\alpha_1} \qquad \tau_2 = \frac{1}{\alpha_2} \tag{5.67}$$

so that the behavior of an overdamped system may be portrayed, for example, as in Figure 5.43 ($K_1 = K_2 = 1$, $\alpha_1 = 5$, and $\alpha_2 = 2$ in the figure).

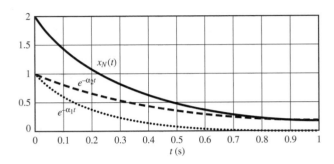

**Figure 5.43** Response of overdamped second-order circuit

## Critically Damped Solution

When the roots are real and repeated, the natural solution is said to be **critically damped**, and is of the form

$$x_N(t) = K_1 e^{-\alpha t} + K_2 t e^{-\alpha t} \tag{5.68}$$

The first term, $K_1 e^{-\alpha t}$, is the familiar exponential decay term. The term $K_2 t e^{-\alpha t}$, on the other hand, has a behavior that differs from a decaying exponential: for small $t$, the function $t$ grows faster than $e^{-\alpha t}$ decays, so that the function initially increases, reaches a maximum at $t = 1/\alpha$, and finally decays to zero. Figure 5.44 depicts the critically damped solution for $K_1 = K_2 = 1$, $\alpha = 5$.

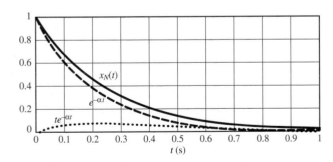

**Figure 5.44** Response of critically damped second-order circuit

## Underdamped Solution

A slightly more involved form of the natural response of a second-order circuit occurs when the roots of the characteristic polynomial form a complex conjugate pair, that is, $s_1 = s_2^*$. In this case, the solution is said to be **underdamped**. The solution for $x_N(t)$, then, is of the form

$$x_N(t) = K_1 e^{s_1 t} + K_2 e^{s_2 t} \qquad (5.69)$$

or

$$x_N(t) = K_1 e^{\alpha t} e^{j\beta t} + K_2 e^{\alpha t} e^{-j\beta t} \qquad (5.70)$$

where $s_1 = \alpha + j\beta$ and $s_2 = \alpha - j\beta$. What is the significance of the complex exponential in the case of underdamped natural response? Recall Euler's identity, which was introduced in Chapter 4:

$$e^{j\theta} = \cos\theta + j\sin\theta \qquad (5.71)$$

If we assume for the moment that $K_1 = K_2 = K$, then the natural response takes the form

$$\begin{aligned}
x_N(t) &= K e^{\alpha t}(e^{j\beta t} + e^{-j\beta t}) \\
&= K e^{\alpha t}(2\cos\beta t)
\end{aligned} \qquad (5.72)$$

Thus, in the case of complex roots, the natural response of a second-order circuit can have oscillatory behavior! The function $2K e^{\alpha t}\cos\beta t$ is a **damped sinusoid**; it is depicted in Figure 5.45 for $\alpha = -5$, $\beta = 50$, and $K = 0.5$. Note that $K_1$ and $K_2$ will be complex conjugates; nonetheless, the underdamped response will still display damped sinusoidal oscillations.

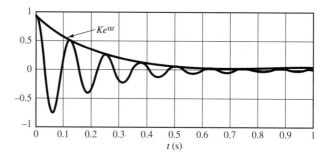

**Figure 5.45** Response of underdamped second-order circuit

As a final note, we return to the special situation in Case III when $R$ is identically zero. We defined this earlier as the **resonance** condition. The resonant solution is not characterized by an exponential decay (damping), and gives rise to a pure sinusoidal waveform, oscillating at the natural frequency, $\beta$. **Resonant circuits** find application in **filters**, which are presented in Chapter 6. We shall not discuss this case any longer in the present chapter.

## Forced and Complete Response of Second-Order Circuits

Once we obtain the natural response using the techniques described in the preceding section, we may find the forced response using the same method employed for first-order circuits. Once again, we shall limit our analysis to a switched DC forcing function, for the sake of simplicity (the form of the forced response when the forcing function is a switched sinusoid is explored in the homework problems). The form of the forced differential equation is

$$a\frac{d^2x(t)}{dt^2} + b\frac{dx(t)}{dt} + cx(t) = F \qquad (5.73)$$

where $F$ is a constant. Therefore we assume a solution of the form $x_F(t) = X_F =$ constant, and we substitute in the forced equation to find that

$$X_F = \frac{F}{c} \qquad (5.74)$$

Finally, in order to compute the complete solution, we sum the natural and forced responses, to obtain

$$x(t) = x_N(t) + x_F(t) = K_1 e^{s_1 t} + K_2 e^{s_2 t} + \frac{F}{c} \qquad (5.75)$$

For a second-order differential equation, we need two initial conditions to solve for the constants $K_1$ and $K_2$. These are the values of $x(t)$ at $t = 0$ and of the derivative of $x(t)$, $dx/dt$, at $t = 0$. To complete the solution, we therefore need to solve the two equations

$$x(t = 0) = x_0 = K_1 + K_2 + \frac{F}{c} \qquad (5.76)$$

and

$$\frac{dx}{dt}(t = 0) = \dot{x}_0 = s_1 K_1 + s_2 K_2 \qquad (5.77)$$

To summarize, we must follow the steps in the accompanying methodology box to obtain the complete solution of a second-order circuit excited by a switched DC source.

## FOCUS ON METHODOLOGY

### Solution of Second-Order Circuits

1. Write the differential equation for the circuit.
2. Find the roots of the characteristic polynomial, and determine the natural response.
3. Find the forced response.
4. Write the complete solution as the sum of natural and forced responses.
5. Determine the initial conditions for inductor currents and capacitor voltages.
6. Apply the initial conditions *to the complete solution* to determine the constants $K_1$ and $K_2$.

Although these steps are straightforward, the successful application of this technique will require some practice, especially the determination of the initial conditions and the computation of the constants. There is no substitute for practice in gaining familiarity with these techniques! The following examples should be of help in illustrating the methods just described.

### EXAMPLE   5.10   Complete Response of Overdamped Second-Order Circuit

#### Problem

Determine the complete response of the circuit of Figure 5.46.

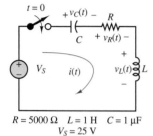

$R = 5000\ \Omega$   $L = 1$ H   $C = 1\ \mu$F
$V_S = 25$ V

**Figure 5.46**

#### Solution

**Known Quantities:**  Resistor, capacitor, inductor values; source voltage.

**Find:**  The capacitor voltage as a function of time.

**Schematics, Diagrams, Circuits, and Given Data:**  $R = 5$ k$\Omega$; $C = 1\ \mu$F; $L = 1$ H; $V_s = 25$ V.

**Assumptions:**  The capacitor has been charged (through a separate circuit, not shown) prior to the switch closing, such that $v_C(0) = 5$ V.

**Analysis:**

1. Apply KVL to determine the circuit differential equation:

$$V_S - v_C(t) - v_R(t) - v_L(t) = 0$$

$$V_S - \frac{1}{C}\int_{-\infty}^{t} i\,dt - iR - L\frac{di}{dt} = 0$$

$$\frac{d^2 i}{dt^2} + \frac{R}{L}\frac{di}{dt} + \frac{1}{LC}i = \frac{1}{L}\frac{dV_S}{dt} = 0 \qquad t > 0$$

We note that the above equations that we have chosen the series (inductor) current as the variable in the differential equation; we also observe that the DC forcing function is zero, because the capacitor acts as an open circuit in the steady state, and the current will therefore be zero as $t \to \infty$.

2. We determine the characteristic polynomial by substituting $s$ for $d/dt$:

$$s^2 + \frac{R}{L}s + \frac{1}{LC} = 0$$

$$s_{1,2} = -\frac{L}{2R} \pm \frac{1}{2}\sqrt{\left(\frac{R}{L}\right)^2 - \frac{4}{LC}}$$

$$= -2{,}500 \pm \sqrt{(5{,}000^2 - 4 \times 10^6}$$

$$s_1 = -208.7; \quad s_2 = -4{,}791.3$$

These are real, distinct roots, therefore we have an overdamped circuit with natural response given by equation 5.63:

$$i_N(t) = K_1 e^{-208.7t} + K_2 e^{-4{,}791.3t}$$

3. The forced response is zero, as stated earlier, because of the behavior of the capacitor as $t \to \infty$: $F = 0$.

4. The complete solution is therefore equal to the natural response:

$$i(t) = i_N(t) = K_1 e^{-208.7t} + K_2 e^{-4{,}791.3t}$$

5. The initial conditions for the energy storage elements are $v_C(0^+) = 5$ V; $i_L(0^+) = 0$ A.

6. To evaluate the coefficients $K_1$ and $K_2$, we consider the initial conditions $i_L(0^+)$ and $di_L(0^+)/dt$. The first of these is given by $i_L(0^+) = 0$, as stated above. Thus,

$$i(0^+) = 0 = K_1e^0 + K_2e^0$$
$$K_1 + K_2 = 0$$
$$K_1 = -K_2$$

To use the second initial condition, we observe that

$$\frac{di}{dt}(0^+) = \frac{di_L}{dt}(0^+) = \frac{1}{L}v_L(0^+)$$

and we note that the inductor voltage *can* change instantaneously; i.e., $v_L(0^-) \neq v_L(0^+)$. To determine $v_L(0^+)$ we need to apply KVL once again at $t = 0^+$:

$$V_S - v_C(0^+) - v_R(0^+) - v_L(0^+) = 0$$
$$v_R(0^+) = i(0^+)R = 0$$
$$v_C(0^+) = 5$$

Therefore

$$v_L(0^+) = V_S - v_C(0^+) - v_R(0^+) = 25 - 5 - 0 = 20 \text{ V}$$

and we conclude that

$$\frac{di}{dt}(0^+) = \frac{1}{L}v_L(0^+) = 20$$

Now we can obtain a second equation in $K_1$ and $K_2$,

$$\frac{di}{dt}(0^+) = 20 = -208.7K_1e^0 - 4.791.3K_2e^0$$

and since

$$K_1 = -K_2$$
$$20 = 208.7K_2 - 4.791.3K_2$$
$$K_1 = 4.36 \times 10^3$$
$$K_2 = -4.36 \times 10^{-3}$$

Finally, the complete solution is:

$$i(t) = 4.36 \times 10^{-3}e^{-208.7t} - 4.36 \times 10^{-3}e^{-4,791.3t} \text{ A}$$

To compute the desired quantity, that is, $v_C(t)$, we can now simply integrate the result above, remembering that the capacitor initial voltage was equal to 5 V:

$$v_C(t) = \frac{1}{C}\int_0^t i(t)dt + v_C(0)$$

$$\frac{1}{C}\int_0^t i(t)dt = 10^6 \left( \int_0^t \left( 4.36 \times 10^{-3}e^{-208.7t} - 4.36 \times 10^{-3}e^{-4,791.3t} \right) dt \right)$$

$$= \frac{10^6 \times 4.36 \times 10^{-3}}{(-208.7)} \left[ e^{-208.7t} - 1 \right]$$

$$- \frac{10^6 \times 4.36 \times 10^{-3}}{(-4,791.3)} \left[ e^{-4,791.3t} - 1 \right]$$

$$= -20.9e^{-208.7t} + 20.9 + 0.9e^{-4,791.3t} - 0.9 \quad t > 0$$

$$= 20 - 20.9e^{-208.7t} + 0.9e^{-4,791.3t}$$

$$v_C(t) = 25 - 20.9e^{-208.7t} + 0.9e^{-4,791.3t} \text{ V}$$

The capacitor voltage is plotted in Figure 5.47.

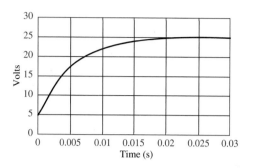

**Figure 5.47** Overdamped circuit capacitor voltage response

**Focus on Computer-Aided Tools:** The Matlab *m-file* containing the numerical analysis and plotting commands for this example may be found in the CD that accompanies this book. An EWB circuit simulation is also included in the CD.

## EXAMPLE 5.11  Complete Response of Critically Damped Second-Order Circuit

### Problem

Determine the complete response of $v(t)$ in the circuit of Figure 5.48.

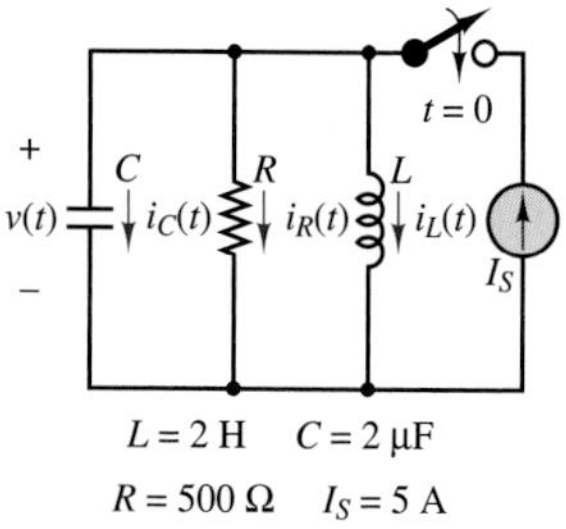

$L = 2$ H    $C = 2$ μF

$R = 500$ Ω    $I_S = 5$ A

**Figure 5.48**

### Solution

**Known Quantities:** Resistor, capacitor, inductor values.

**Find:** The capacitor voltage as a function of time.

**Schematics, Diagrams, Circuits, and Given Data:** $I_S = 5$ A; $R = 500$ Ω; $C = 2$ μF; $L = 2$ H.

**Assumptions:** The capacitor voltage and inductor current are equal to zero at $t = 0^+$.

**Analysis:**

1. Apply KCL to determine the circuit differential equation:

$$I_S - i_L(t) - i_R(t) - i_C(t) = 0$$

$$I_S - \frac{1}{L}\int_{-\infty}^{t} v(t)dt - \frac{v(t)}{R} - C\frac{dv(t)}{dt} = 0$$

$$\frac{d^2v}{dt^2} + \frac{1}{RC}\frac{dv}{dt} + \frac{1}{LC}v = \frac{1}{C}\frac{dI_S}{dt} = 0 \qquad t > 0$$

We note that the DC forcing function is zero, because the inductor acts as a short circuit in the steady state, and the voltage across the inductor (and therefore across the parallel circuit) will be zero as $t \to \infty$.

2. We determine the characteristic polynomial by substituting $s$ for $d/dt$:

$$s^2 + \frac{1}{RC}s + \frac{1}{LC} = 0$$

$$s_{1,2} = -\frac{1}{2RC} \pm \frac{1}{2}\sqrt{\left(\frac{1}{RC}\right)^2 - \frac{4}{LC}}$$

$$= -500 \pm \frac{1}{2}\sqrt{(1,000)^2 - 10^6}$$

$$s_1 = -500 \qquad s_2 = -500$$

These are real, repeated roots, therefore we have a critically damped circuit with natural response given by equation 5.64:

$$v_N(t) = K_1 e^{-500t} + K_2 t e^{-500t}$$

3. The forced response is zero, as stated earlier, because of the behavior of the inductor as $t \to \infty$: $F = 0$.

4. The complete solution is therefore equal to the natural response:

$$v(t) = v_N(t) = K_1 e^{-500t} + K_2 t e^{-500t}$$

5. The initial conditions for the energy storage elements are: $v_C(0^+) = 0$ V; $i_L(0^+) = 0$ A.

6. To evaluate the coefficients $K_1$ and $K_2$, we consider the initial conditions $v_C(0^+)$ and $dv_C(0^+)/dt$. The first of these is given by $v_C(0^+) = 0$, as stated above. Thus,

$$v(0^+) = 0 = K_1 e^0 + K_2 \times 0 e^0$$

$$K_1 = 0 \text{ V}$$

To use the second initial condition, we observe that

$$\frac{dv}{dt}(0^+) = \frac{dv_C(0^+)}{dt} = \frac{1}{C} i_C(0^+)$$

and we note that the capacitor current *can* change instantaneously; i.e., $i_C(0^-) \neq i_C(0^+)$. To determine $i_C(0^+)$ we need to apply KCL once again at $t = 0^+$:

$$I_S - i_L(0^+) - i_R(0^+) - i_C(0^+) = 0$$

$$i_L(0^+) = 0; \quad i_R(0^+) = \frac{v(0^+)}{R} = 0;$$

Therefore

$$i_C(0^+) = I_S - 0 - 0 - 0 = 5 \text{ A}$$

and we conclude that

$$\frac{dv}{dt}(0^+) = \frac{1}{C} i_C(0^+) = \frac{5 \text{ A}}{2\mu \text{ F}} = 2.5 \times 10^6 \frac{\text{V}}{s}$$

Now we can obtain a second equation in $K_1$ and $K_2$,

$$i_C(t) = C\frac{dv}{dt} = C\left[K_1(-500) e^{-500t} + K_2 e^{-500t} + K_2(-500) t e^{-500t}\right]$$

$$i_C(0^+) = C\left[K_1(-500) e^0 + K_2 e^0 + K_2(-500)(0) e^0\right]$$

$$5 = C[K_1(-500) + K_2]$$

$$K_2 = \frac{i_C(0^+)}{C} = \frac{5 \text{ A}}{2\mu \text{ F}} = 2.5 \times 10^6 \frac{\text{V}}{s}$$

Finally, the complete solution is:

$$v(t) = 2.5 \times 10^6 t e^{-500t} \text{ V}$$

A plot of the voltage response of this critically damped circuit is shown in Figure 5.49.

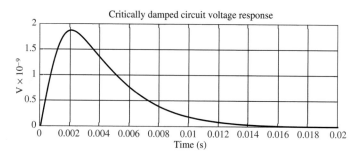

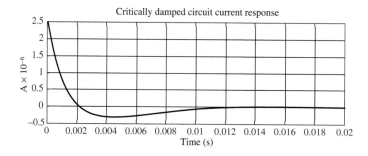

**Figure 5.49**

*Focus on Computer-Aided Tools:*  The Matlab *m-file* containing the numerical analysis and plotting commands for this example may be found in the CD that accompanies this book.

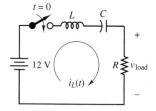

VIRTUAL LAB

---

## EXAMPLE   5.12   Complete Response of Underdamped Second-Order Circuit

### Problem

Determine the complete response of the circuit of Figure 5.50.

---

### Solution

*Known Quantities:*  Source voltage, resistor, capacitor, inductor values.

*Find:*  The load voltage as a function of time.

*Schematics, Diagrams, Circuits, and Given Data:*  $R = 10 \ \Omega$; $C = 10 \ \mu\text{F}$; $L = 5 \text{ mH}$.

*Assumptions:*  No energy is stored in the capacitor and inductor before the switch closes; i.e., $v_C(0^-) = 0 \text{ V}$; $i_L(0^-) = 0 \text{ A}$.

*Analysis:*  Since the load voltage is given by the expression $v_{\text{load}} = R i_L(t)$, we shall solve for the inductor current.

**Figure 5.50**

1. Apply KVL to determine the circuit differential equation:

$$V_B - v_L(t) - v_C(t) - v_R(t) = 0$$

$$V_B - L\frac{di_L}{dt} - \frac{1}{C}\int_{-\infty}^{t} i_L dt - i_L R = 0$$

$$\frac{d^2 i_L}{dt^2} + \frac{R}{L}\frac{di_L}{dt} + \frac{1}{LC}i_L = \frac{1}{L}\frac{dV_B}{dt} = 0 \qquad t > 0$$

2. We determine the characteristic polynomial by substituting $s$ for $d/dt$:

$$s^2 + \frac{R}{L}s + \frac{1}{LC} = 0$$

$$s_{1,2} = -\frac{L}{2R} \pm \frac{1}{2}\sqrt{\left(\frac{R}{L}\right)^2 - \frac{4}{LC}}$$

$$= -1{,}000 \pm j4359$$

These are complex conjugate roots, therefore we have an underdamped circuit with natural response given by equation 5.65:

$$i_{LN}(t) = K_1 e^{(-1,000+j4,359)t} + K_2 e^{(-1,000-j4,359)t}$$

3. The forced response is zero, as stated earlier, because of the behavior of the capacitor as $t \to \infty$: $F = 0$.

4. The complete solution is therefore equal to the natural response:

$$i_L(t) = i_{LN}(t) = K_1 e^{(-1,000+j4,359)t} + K_2 e^{(-1,000-j4,359)t}$$

5. The initial conditions for the energy storage elements are $v_C(0^+) = 0$ V; $i_L(0^+) = 0$ A.

6. To evaluate the coefficients $K_1$ and $K_2$, we consider the initial conditions $i_L(0^+)$ and $di_L(0^+)/dt$. The first of these is given by $i_L(0^+) = 0$, as stated above. Thus,

$$i(0^+) = 0 = K_1 e^0 + K_2 e^0$$

$$K_1 + K_2 = 0$$

$$K_1 = -K_2$$

To use the second initial condition, we observe that

$$\frac{di_L}{dt}(0^+) = \frac{1}{L}v_L(0^+)$$

and we note that the inductor voltage *can* change instantaneously; i.e., $v_L(0^-) \neq v_L(0^+)$. To determine $v_L(0^+)$ we need to apply KVL once again at $t = 0^+$:

$$V_S - v_C(0^+) - v_R(0^+) - v_L(0^+) = 0$$

$$v_R(0^+) = i_L(0^+)R = 0; \quad v_C(0^+) = 0$$

Therefore

$$v_L(0^+) = V_S - 0 - 0 = 12 \quad V$$

and

$$\frac{di_L}{dt}(0^+) = \frac{v_L(0^+)}{L} = 2{,}400$$

Now we can obtain a second equation in $K_1$ and $K_2$,

$$\frac{di}{dt}(0^+) = (-1{,}000 + j4{,}359)K_1 e^0 - (-1{,}000 - j4{,}359)K_2 e^0$$

and since

$$K_1 = -K_2$$

$$2{,}400 = K_1 \left[ (-1{,}000 + j4{,}359) - (-1{,}000 - j4{,}359) \right]$$

$$K_1 = \frac{2{,}400}{j8{,}718} = -j0.2753$$

$$K_2 = -K_1 = j0.2753$$

Note that $K_1$ and $K_2$ are complex conjugates. Finally, the complete solution is:

$$v_{\text{Load}}(t) = Ri_L(t) = 10 \left( -j0.2753 e^{(-1{,}000+j4{,}359)t} + j0.2753 e^{(-1{,}000-j4{,}359)t} \right)$$

$$= 2.753 e^{-1{,}000t} \left( -j e^{j4{,}359t} + j e^{-j4{,}359t} \right)$$

$$= 5.506 e^{-1{,}000t} \sin(4{,}359t) \text{ V}$$

The output voltage of the circuit is plotted in Figure 5.51.

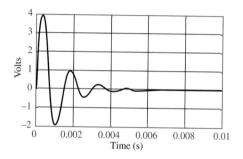

**Figure 5.51** Underdamped circuit voltage response

*Focus on Computer-Aided Tools:* The Matlab *m-file* containing the numerical analysis and plotting commands for this example may be found in the CD that accompanies this book. An EWB simulation is also enclosed.

---

### EXAMPLE 5.13 Transient Response of Automotive Ignition Circuit

#### Problem

The circuit shown in Figure 5.52 is a simplified but realistic representation of an automotive ignition system. The circuit includes an **automotive battery,** a transformer[4] (**ignition coil**), a capacitor (known as *condenser* in old-fashioned automotive parlance) and a switch. The switch is usually an electronic switch (e.g., a transistor—see Chapter 9), and can be treated as an ideal switch. The circuit on the left represents the ignition circuit immediately after the electronic switch has closed, following a spark discharge. Thus, one can assume that no energy is stored in the inductor prior to the switch closing, say at $t = 0$. Furthermore, no energy is stored in the capacitor, as the short

---

[4]Transformers are discussed more formally in Chapters 7 amd 17; the operation of the transformer in an ignition coil will be explained *ad hoc* in this example.

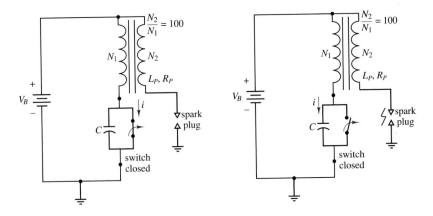

**Figure 5.52**

circuit (closed switch) across it would have dissipated any charge in the capacitor. The primary winding of the ignition coil (left-hand-side inductor) is then given a suitable length of time to build up stored energy, and then the switch opens, say at $t = \Delta t$, leading to a rapid voltage buildup across the secondary winding of the coil (right-hand-side inductor). The voltage rises to a very high value because of two effects: the *inductive voltage kick* described in Examples 5.3 and 5.8, and the voltage multiplying effect of the transformer. The result is a very short high-voltage transient (reaching thousands of volts), which causes a spark to be generated across the spark plug.

---

**Solution**

**Known Quantities:**  Battery voltage, resistor, capacitor, inductor values.

**Find:**  The ignition coil current, $i(t)$, and the open circuit voltage across the spark plug, $v_{OC}(t)$.

**Schematics, Diagrams, Circuits, and Given Data:**  $V_B = 12$ V; $R_P = 2\ \Omega$; $C = 10\ \mu F$; $L_p = 5$ mH.

**Assumptions:**  The switch has been open for a long time, and closes at $t = 0$. The switch opens again at $t = \Delta t$.

**Analysis:**  With no energy stored in either the inductor or the capacitor, the action of closing the switch will create a closed circuit comprising the battery, $V_B$, the coil primary inductance, $L_P$, and the coil primary resistance, $R_P$. The inductor current will therefore rise exponentially to a final value equal to $V_B/R_P$, as described in the following equation:

$$i_L(t) = \frac{V_B}{R_P}\left(1 - e^{-t/\tau}\right) = \frac{V_B}{R_P}\left(1 - e^{-R_P t/L}\right) = 6\left(1 - e^{-t/2.5\times10^{-3}}\right) \quad 0 < t < \Delta t$$

We know from Example 5.4 that the energy storage element will acquire approximately 90 percent of its energy in 3 time constants; let's assume that the switch remains closed for 5 time constants; i.e., $\Delta t = 12.5$ ms. Thus, at $t = \Delta t$, the inductor current will be equal to

$$i_L(\Delta t) = \frac{V_B}{R_P}\left(1 - e^{-5\tau/\tau}\right) = 6\left(1 - e^{-5}\right) = 5.96 \quad \text{A}$$

that is, the current reaches 99 percent of its final value in 5 time constants.

Now, when the switch opens at $t = \Delta t$, we are faced with a series $RLC$ circuit similar to that of Example 5.13. The inductor current at this time is 5.96 A, and the capacitor voltage is zero, because a short circuit (the closed switch) had been placed across the capacitor. The differential equation describing the circuit for $t > \Delta t$ is given below.

$$V_B - v_L(t) - v_R(t) - v_C(t) = 0$$

$$V_B - L\frac{di_L}{dt} - i_L R - \frac{1}{C}\int_{-\infty}^{t} i_L dt = 0$$

$$\frac{d^2 i_L}{dt^2} + \frac{R}{L}\frac{di_L}{dt} + \frac{1}{LC}i_L = \frac{1}{L}\frac{dV_B}{dt} = 0 \qquad t > \Delta t$$

Next, we solve for the roots of the characteristic polynomial:

$$s^2 + \frac{R}{L}s + \frac{1}{LC} = 0$$

$$s_{1,2} = -\frac{L}{2R} \pm \frac{1}{2}\sqrt{\left(\frac{R}{L}\right)^2 - \frac{4}{LC}} = -200 \pm j4{,}468$$

These are complex conjugate roots, therefore we have an underdamped circuit with natural response given by equation 5.65. By analogy with Example 5.13, the complete solution is given by:

$$i_L(t) = i_{LN}(t) = K_1 e^{(-200+j4,468)(t-\Delta t)} + K_2 e^{(-200-j4,468)(t-\Delta t)} \qquad t > \Delta t$$

The initial conditions for the energy storage elements are: $v_C(\Delta t^+) = 0$ V; $i_L(\Delta t^+) = 5.96$ A. Thus,

$$i_L(\Delta t^+) = 5.96 = K_1 e^0 + K_2 e^0$$

$$K_1 + K_2 = 5.96$$

$$K_1 = 5.96 - K_2$$

To use the second initial condition, we observe that

$$\frac{di_L}{dt}(\Delta t^+) = \frac{1}{L_P}v_L(\Delta t^+)$$

and we note that the inductor voltage *can* change instantaneously; i.e., $v_L(0^-) \neq v_L(0^+)$. To determine $v_L(0^+)$ we need to apply KVL once again at $t = 0^+$:

$$V_B - v_C(\Delta t^+) - v_R(\Delta t^+) - v_L(\Delta t^+) = 0$$

$$v_R(\Delta t^+) = i_L(\Delta t^+)R = 5.96 \times 2 = 11.92$$

$$v_C(\Delta t^+) = 0$$

Therefore

$$v_L(\Delta t^+) = V_B - 11.92 = 12 - 11.92 = 0.08 \quad \text{V}$$

and

$$\frac{di_L}{dt}(\Delta t^+) = \frac{v_L(\Delta t^+)}{L_P} = \frac{0.08}{5 \times 10^{-3}} = 16$$

Now we can obtain a second equation in $K_1$ and $K_2$,

$$\frac{di_L}{dt}(\Delta t^+) = (-200 + j4, 468)K_1 e^0 + (-200 - j4, 468)K_2 e^0$$

and since

$$K_1 = 5.96 - K_2$$

$$16 = (-200 + j4, 468)(5.96 - K_2) + (-200 - j4, 468)K_2$$

$$= -1192 + j26, 629 - (-200 + j4, 468)K_2 + (-200 - j4, 468)K_2$$

$$= -1192 + j26, 629 - j8.936K_2$$

$$K_2 = \frac{1}{-j8,936}(1208 - j26,629) = 2.98 + j0.1352$$

$$K_1 = 5.96 - K_2 = 2.98 - j0.1352$$

Note, again, that $K_1$ and $K_2$ are complex conjugates, as suggested earlier.

Finally, the complete solution is:

$$i_L(t) = (2.98 - j0.1352)e^{(-200+j4,468)(t-\Delta t)}$$

$$+ (2.98 + j0.1352)e^{(-200-j4,468)(t-\Delta t)} \qquad t > \Delta t$$

$$= 2.98e^{-200(t-\Delta t)} \left( e^{j4,468(t-\Delta t)} + e^{-j4,468(t-\Delta t)} \right)$$

$$- j0.1352e^{-200(t-\Delta t)} \left( e^{j4,468(t-\Delta t)} - e^{-j4,468(t-\Delta t)} \right)$$

$$= 2 \times 2.98e^{-200(t-\Delta t)} \cos(4,468(t-\Delta t))$$

$$-2 \times 0.1352e^{-200(t-\Delta t)} \sin(4,468(t-\Delta t)) \text{ A}$$

The coil primary current is plotted in Figure 5.53.

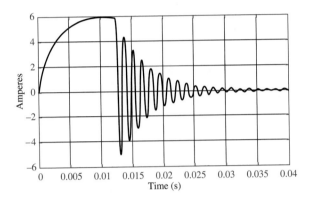

**Figure 5.53** Ignition circuit primary current response

To compute the primary voltage, we simply differentiate the inductor current and multiply by $L_P$; to determine the secondary voltage, which is that applied to the spark plug, we simply remark that a 1:100 transformer steps up the voltage by a factor of 100, so that the secondary voltage is 100 times larger than the primary voltage. Thus, the expression for the secondary voltage is:

$$v_{\text{spark plug}} = 100 \times L_P \frac{di_L}{dt} = 0.5 \times \frac{d}{dt} \left[ (2 \times 2.98e^{-200t} \cos(4,468t) \right.$$

$$\left. - 2 \times 0.1352e^{-200t} \sin(4,468t) \right]$$

$$= 0.5 \times [-200 \times 5.96 \times e^{-200t} \cos(4,468t) - 4,468 \times 5.96$$

$$\times e^{-200t} \sin(4,468t)]$$

$$- 0.5 \times [-200 \times 0.1352 \times e^{-200t} \sin(4,468t)$$

$$+ 4,468 \times 0.1352 \times e^{-200t} \cos(4,468t)]$$

where we have "reset" time to $t = 0$ for simplicity. We are actually interested in the value of this voltage at $t = 0$, since this is what will generate the spark; evaluating the above expression at $t = 0$, we obtain:

$$v_{\text{spark plug}}(t = 0) = 0.5 \times [-200 \times 5.96] - 0.5 \times [4,468 \times 0.1352]$$

$$v_{\text{spark plug}}(t = 0) = -596 - 302 = -898 \quad \text{V}$$

One can clearly see that the result of the switching is a very large (negative) voltage spike, capable of generating a spark across the plug gap. A plot of the secondary voltage starting at the time when the switch is opened is shown in Figure 5.54, showing that approximately 0.3 ms after the switching transient, the secondary voltage reaches approximately −12,500 volts! This value is typical of the voltages required to generate a spark across an **automotive spark plug.**

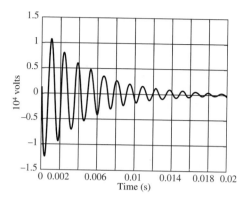

**Figure 5.54** Secondary ignition voltage response

VIRTUAL LAB

**Focus on Computer-Aided Tools:** The Matlab *m-file* containing the numerical analysis and plotting commands for this example may be found in the CD that accompanies this book. An EWB simulation is also enclosed.

## Check Your Understanding

**5.9** Derive the differential equation for the series circuit of Figure 5.39(b). Show that one can write the equation either as

$$\frac{d^2 v_C}{dt^2} + \frac{R_T}{L}\frac{dv_C}{dt} + \frac{1}{LC}v_C = \frac{1}{LC}v_T(t)$$

or as

$$\frac{d^2 i_L}{dt^2} + \frac{R_T}{L}\frac{di_L}{dt} + \frac{1}{LC}i_L = \frac{1}{L}\frac{dv_T(t)}{dt}$$

**5.10** Determine the roots of the characteristic equation of the series $RLC$ circuit of Figure 5.39(b) with $R = 100\ \Omega$, $C = 10\ \mu F$, and $L = 1$ H.

**5.11** For the series $RLC$ circuit of Figure 5.39(b), with $L = 1$ H and $C = 10\ \mu F$, find the ranges of values of $R$ for which the circuit response is overdamped and underdamped, respectively.

## CHECK YOUR UNDERSTANDING ANSWERS

**CYU 5.1** $\dfrac{dv_C}{dt} + \dfrac{1}{RC}v_C = \dfrac{1}{RC}v_S$

**CYU 5.2** $\dfrac{dv}{dt} + \dfrac{1}{RC}v = \dfrac{1}{C}i_S$

**CYU 5.3** $\dfrac{di_L}{dt} + \dfrac{R}{L}i_L = \dfrac{R}{L}i_S$

**CYU 5.4** $v_C(0^-) = 8$ V and $i_R(0^+) = 2$ A

**CYU 5.5** $v_C = 7.5 - 7.5e^{-t/0.025}$ V, $0 \le t < 0.05$ s; $v_C = 6.485e^{-(t-0.05)/0.0375}$ V, $t \ge 0.05$ s

**CYU 5.6** $t_{90\%} = 12.5\ \mu s$

**CYU 5.7** $545\ \mu s$

**CYU 5.8** The output pulse has a higher peak.

**CYU 5.9** $-50 \pm j312.25$

**CYU 5.10** Overdamped: $R > 632.46\ \Omega$; underdamped: $R < 632.46\ \Omega$

# HOMEWORK PROBLEMS

## Section 1: First-Order Transients

**5.1**  Just before the switch is opened at $t = 0$, the current through the inductor is 1.70 mA in the direction shown in Figure P5.1. Did steady-state conditions exist just before the switch was opened?

$$L = 0.9 \text{ mH} \qquad V_S = 12 \text{ V}$$
$$R_1 = 6 \text{ k}\Omega \qquad R_2 = 6 \text{ k}\Omega$$
$$R_3 = 3 \text{ k}\Omega$$

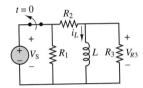

**Figure P5.1**

**5.2**  At $t < 0$, the circuit shown in Figure P5.2 is at steady state. The switch is changed as shown at $t = 0$.

$$V_{S1} = 35 \text{ V} \qquad V_{S2} = 130 \text{ V}$$
$$C = 11 \ \mu\text{F} \qquad R_1 = 17 \text{ k}\Omega$$
$$R_2 = 7 \text{ k}\Omega \qquad R_3 = 23 \text{ k}\Omega$$

Determine at $t = 0^+$ the initial current through $R_3$ just after the switch is changed.

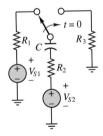

**Figure P5.2**

**5.3**  Determine the current through the capacitor just before and just after the switch is closed in Figure P5.3. Assume steady-state conditions for $t < 0$.

$$V_1 = 12 \text{ V} \qquad C = 0.5 \ \mu\text{F}$$
$$R_1 = 0.68 \text{ k}\Omega \qquad R_2 = 1.8 \text{ k}\Omega$$

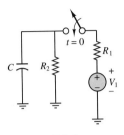

**Figure P5.3**

**5.4**  Determine the current through the capacitor just before and just after the switch is closed in Figure P5.3. Assume steady-state conditions for $t < 0$.

$$V_1 = 12 \text{ V} \qquad C = 150 \ \mu\text{F}$$
$$R_1 = 400 \text{ m}\Omega \qquad R_2 = 2.2 \text{ k}\Omega$$

**5.5**  Just before the switch is opened at $t = 0$ in Figure P5.1, the current through the inductor is 1.70 mA in the direction shown. Determine the voltage across $R_3$ just after the switch is opened.

$$V_S = 12 \text{ V} \qquad L = 0.9 \text{ mH}$$
$$R_1 = 6 \text{ k}\Omega \qquad R_2 = 6 \text{ k}\Omega$$
$$R_3 = 3 \text{ k}\Omega$$

**5.6**  Determine the voltage across the inductor just before and just after the switch is changed in Figure P5.6. Assume steady-state conditions exist for $t < 0$.

$$V_S = 12 \text{ V} \qquad R_s = 0.7 \ \Omega$$
$$R_1 = 22 \text{ k}\Omega \qquad L = 100 \text{ mH}$$

**Figure P5.6**

**5.7**  Steady-state conditions exist in the circuit shown in Figure P5.7 at $t < 0$. The switch is closed at $t = 0$.

$$V_1 = 12 \text{ V} \qquad R_1 = 0.68 \text{ k}\Omega$$
$$R_2 = 2.2 \text{ k}\Omega \qquad R_3 = 1.8 \text{ k}\Omega$$
$$C = 0.47 \ \mu\text{F}$$

Determine the current through the capacitor at $t = 0^+$, just after the switch is closed.

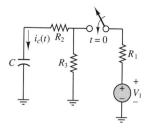

**Figure P5.7**

**5.8**  At $t > 0$, the circuit shown in Figure P5.2 is at steady state. The switch is changed as shown at $t = 0$.

$$V_{S1} = 35 \text{ V} \qquad V_{S2} = 130 \text{ V}$$
$$C = 11 \ \mu\text{F} \qquad R_1 = 17 \text{ k}\Omega$$
$$R_2 = 7 \text{ k}\Omega \qquad R_3 = 23 \text{ k}\Omega$$

Determine the time constant of the circuit for $t > 0$.

**5.9**  At $t < 0$, the circuit shown in Figure P5.9 is at steady state. The switch is changed as shown at $t = 0$.

$$V_{S1} = 13 \text{ V} \qquad V_{S2} = 13 \text{ V}$$
$$L = 170 \text{ mH} \qquad R_1 = 2.7 \ \Omega$$
$$R_2 = 4.3 \text{ k}\Omega \qquad R_3 = 29 \text{ k}\Omega$$

Determine the time constant of the circuit for $t > 0$.

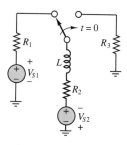

**Figure P5.9**

**5.10**  Steady-state conditions exist in the circuit shown in Figure P5.7 for $t < 0$. The switch is closed at $t = 0$.

$$V_1 = 12 \text{ V} \qquad C = 0.47 \ \mu\text{F}$$
$$R_1 = 680 \ \Omega \qquad R_2 = 2.2 \text{ k}\Omega$$
$$R_3 = 1.8 \text{ k}\Omega$$

Determine the time constant of the circuit for $t > 0$.

**5.11**  Just before the switch is opened at $t = 0$ in Figure P5.1, the current through the inductor is 1.70 mA in the direction shown.

$$V_S = 12 \text{ V} \qquad L = 0.9 \text{ mH}$$
$$R_1 = 6 \text{ k}\Omega \qquad R_2 = 6 \text{ k}\Omega$$
$$R_3 = 3 \text{ k}\Omega$$

Determine the time constant of the circuit for $t > 0$.

**5.12**  Determine $v_C(t)$ for $t > 0$. The voltage across the capacitor in Figure P5.12 just before the switch is changed is given below.

$$v_C(0^-) = -7 \text{ V} \qquad I_o = 17 \text{ mA} \qquad C = 0.55 \ \mu\text{F}$$
$$R_1 = 7 \text{ k}\Omega \qquad R_2 = 3.3 \text{ k}\Omega$$

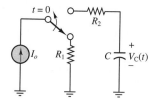

**Figure P5.12**

**5.13**  Determine $i_{R_3}(t)$ for $t > 0$ in Figure P5.9.

$$V_{S1} = 23 \text{ V} \qquad V_{S2} = 20 \text{ V}$$
$$L = 23 \text{ mH} \qquad R_1 = 0.7 \ \Omega$$
$$R_2 = 13 \ \Omega \qquad R_3 = 330 \text{ k}\Omega$$

**5.14**  Assume DC steady-state conditions exist in the circuit shown in Figure P5.14 for $t < 0$. The switch is changed at $t = 0$ as shown.

$$V_{S1} = 17 \text{ V} \qquad V_{S2} = 11 \text{ V}$$
$$R_1 = 14 \text{ k}\Omega \qquad R_2 = 13 \text{ k}\Omega$$
$$R_3 = 14 \text{ k}\Omega \qquad C = 70 \text{ nF}$$

Determine:

a.  $v(t)$ for $t > 0$.

b.  The time required, after the switch is operated, for $V(t)$ to change by 98 percent of its total change in voltage.

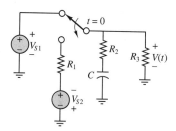

**Figure P5.14**

**5.15**  The circuit of Figure P5.15 is a simple model of an automotive ignition system. The switch models the "points" that switch electrical power to the cylinder when the fuel-air mixture is compressed. $R$ is the resistance between the electrodes (i.e., the "gap") of the spark plug.

$$V_G = 12 \text{ V} \qquad R_G = 0.37 \ \Omega$$
$$R = 1.7 \text{ k}\Omega$$

Determine the value of $L$ and $R_1$ so that the voltage across the spark plug gap just after the switch is changed is 23 kV and so that this voltage will change exponentially with a time constant $\tau = 13$ ms.

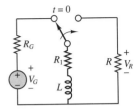

**Figure P5.15**

**5.16**  The inductor $L$ in the circuit shown in Figure P5.16 is the coil of a relay. When the current through the coil is equal to or greater than +2 mA the relay functions. Assume steady-state conditions at $t < 0$. If:

$V_S = 12$ V

$L = 10.9$ mH $\qquad R_1 = 3.1$ k$\Omega$

determine $R_2$ so that the relay functions at $t = 2.3$ s.

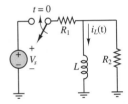

**Figure P5.16**

**5.17**  Determine the current through the capacitor just before and just after the switch is closed in Figure P5.17. Assume steady-state conditions for $t < 0$.

$V_1 = 12$ V $\qquad C = 150$ $\mu$F

$R_1 = 400$ m$\Omega$ $\qquad R_2 = 2.2$ k$\Omega$

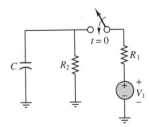

**Figure P5.17**

**5.18**  Determine the voltage across the inductor just before and just after the switch is changed in Figure P5.18. Assume steady-state conditions exist for $t < 0$.

$V_S = 12$ V $\qquad R_S = 0.24$ $\Omega$

$R_1 = 33$ k$\Omega$ $\qquad L = 100$ mH

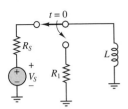

**Figure P5.18**

**5.19**  Steady-state conditions exist in the circuit shown in Figure P5.7 for $t < 0$. The switch is closed at $t = 0$.

$V_1 = 12$ V $\qquad C = 150$ $\mu$F

$R_1 = 4$ M$\Omega$ $\qquad R_2 = 80$ M$\Omega$

$R_3 = 6$ M$\Omega$

Determine the time constant of the circuit for $t > 0$.

**5.20**  Just before the switch is opened at $t = 0$ in Figure P5.1, the current through the inductor is 1.70 mA in the direction shown.

$V_S = 12$ V $\qquad L = 100$ mH

$R_1 = 400$ $\Omega$ $\qquad R_2 = 400$ $\Omega$

$R_3 = 600$ $\Omega$

Determine the time constant of the circuit for $t > 0$.

**5.21**  For the circuit shown in Figure P5.21, assume that switch $S_1$ is always open and that switch $S_2$ closes at $t = 0$.
a. Find the capacitor voltage, $v_C(t)$, at $t = 0^+$.
b. Find the time constant, $\tau$, for $t \geq 0$.
c. Find an expression for $v_C(t)$ and sketch the function.
d. Find $v_C(t)$ for each of the following values of $t$: $0, \tau, 2\tau, 5\tau, 10\tau$.

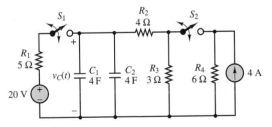

**Figure P5.21**

**5.22**  For the circuit shown in Figure P5.21, assume that switch $S_1$ is always open; switch $S_2$ has been closed for a long time, and opens at $t = 0$.
a. Find the capacitor voltage, $v_C(t)$, at $t = 0^+$.
b. Find the time constant, $\tau$, for $t \geq 0$.
c. Find an expression for $v_C(t)$ and sketch the function.
d. Find $v_C(t)$ for each of the following values of $t$: $0, \tau, 2\tau, 5\tau, 10\tau$.

**5.23** For the circuit of Figure P5.21, assume that switch $S_2$ is always open, and that switch $S_1$ has been closed for a long time and opens at $t = 0$. At $t = t_1 = 3\tau$, switch $S_1$ closes again.

a. Find the capacitor voltage, $v_C(t)$, at $t = 0^+$.

b. Find an expression for $v_C(t)$ for $t > 0$ and sketch the function.

**5.24** Assume both switches $S_1$ and $S_2$ in Figure P5.21 close at $t = 0$.

a. Find the capacitor voltage, $v_C(t)$, at $t = 0^+$.

b. Find the time constant, $\tau$, for $t \geq 0$.

c. Find an expression for $v_C(t)$ and sketch the function.

d. Find $v_C(t)$ for each of the following values of $t$: $0, \tau, 2\tau, 5\tau, 10\tau$.

**5.25** Assume both switches $S_1$ and $S_2$ in Figure P5.21 have been closed for a long time and switch $S_2$ opens at $t = 0^+$.

a. Find the capacitor voltage, $v_C(t)$, at $t = 0^+$.

b. Find an expression for $v_C(t)$ and sketch the function.

c. Find $v_C(t)$ for each of the following values of $t$: $0, \tau, 2\tau, 5\tau, 10\tau$.

**5.26** For the circuit of Figure P5.26, determine the time constants $\tau$ and $\tau'$ before and after the switch opens, respectively. $R_S = 4 \text{ k}\Omega$, $R_1 = 2 \text{ k}\Omega$, $R_2 = R_3 = 6 \text{ k}\Omega$, and $C = 1 \ \mu\text{F}$.

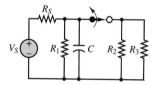

**Figure P5.26**

**5.27** For the circuit of Figure P5.27, find the initial current through the inductor, the final current through the inductor, and the expression for $i_L(t)$ for $t \geq 0$.

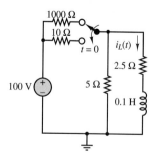

**Figure P5.27**

**5.28** At $t = 0$, the switch in the circuit of Figure P5.28 opens. At $t = 10$ s, the switch closes.

a. What is the time constant for $0 < t < 10$ s?

b. What is the time constant for $t > 10$ s?

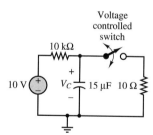

**Figure P5.28**

**5.29** The circuit of Figure P5.29 includes a model of a voltage-controlled switch. When the voltage across the capacitor reaches 7 V, the switch is closed. When the capacitor voltage reaches 0.5 V, the switch opens. Assume that the capacitor voltage is initially $V_C = 0.5$ V and that the switch has just opened.

a. Sketch the capacitor voltage versus time, showing explicitly the periods when the switch is open and when the switch is closed.

b. What is the period of the voltage waveform across the 10-$\Omega$ resistor?

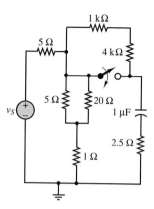

**Figure P5.29**

**5.30** At $t = 0$, the switch in the circuit of Figure P5.30 closes. Assume that $i_L(0) = 0$ A. For $t \geq 0$,

a. Find $i_L(t)$.

b. Find $v_{L_1}(t)$.

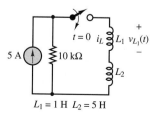

**Figure P5.30**

## Section 2: **Second-Order Transients**

**5.31**  In the circuit shown in Figure P5.31:

$V_{S1} = 15$ V          $V_{S2} = 9$ V
$R_{S1} = 130$ Ω         $R_{S2} = 290$ Ω
$R_1 = 1.1$ kΩ           $R_2 = 700$ Ω
$L = 17$ mH             $C = 0.35$ μF

Assume that DC steady-state conditions exist for $t < 0$. Determine the voltage across the capacitor and the current through the inductor and $R_{S2}$ as $t$ approaches infinity.

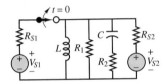

**Figure P5.31**

**5.32**  In the circuit shown in Figure P5.31:

$V_{S1} = 12$ V          $V_{S2} = 12$ V
$R_{S1} = 50$ Ω          $R_{S2} = 50$ Ω
$R_1 = 2.2$ kΩ           $R_2 = 600$ Ω
$L = 7.8$ mH            $C = 68$ μF

Assume that DC steady-state conditions exist at $t < 0$. Determine the voltage across the capacitor and the current through the inductor as $t$ approaches infinity. Remember to specify the polarity of the voltage and the direction of the current that you assume for your solution.

**5.33**  If the switch in the circuit shown in Figure P5.33 is closed at $t = 0$ and:

$V_S = 170$ V           $R_S = 7$ kΩ
$R_1 = 2.3$ kΩ          $R_2 = 7$ kΩ
$L = 30$ mH            $C = 130$ μF

determine, after the circuit has returned to a steady state, the current through the inductor and the voltage across the capacitor and $R_1$.

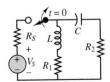

**Figure P5.33**

**5.34**  If the switch in the circuit shown in Figure P5.34 is closed at $t = 0$ and:

$V_S = 12$ V           $C = 130$ μF
$R_1 = 2.3$ kΩ         $R_2 = 7$ kΩ
$L = 30$ mH

Determine the current through the inductor and the voltage across the capacitor and across $R_1$ after the circuit has returned to a steady state.

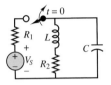

**Figure P5.34**

**5.35**  If the switch in the circuit shown in Figure P5.35 is closed at $t = 0$ and:

$V_S = 12$ V           $C = 0.5$ μF
$R_1 = 31$ kΩ          $R_2 = 22$ kΩ
$L = 0.9$ mH

Determine the current through the inductor and the voltage across the capacitor after the circuit has returned to a steady state.

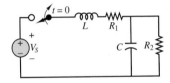

**Figure P5.35**

**5.36**  At $t < 0$, the circuit shown in Figure P5.36 is at steady state and the voltage across the capacitor is $+7$ V. The switch is changed as shown at $t = 0$ and:

$V_S = 12$ V           $C = 3300$ μF
$R_1 = 9.1$ kΩ         $R_2 = 4.3$ kΩ
$R_3 = 4.3$ kΩ         $L = 16$ mH

Determine the initial voltage across $R_2$ just after the switch is changed.

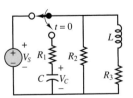

**Figure P5.36**

**5.37**  In the circuit shown in Figure P5.37, assume that DC steady-state conditions exist for $t < 0$. Determine at $t = 0^+$, just after the switch is opened, the current through and voltage across the inductor and the capacitor and the current through $R_{S2}$.

$V_{S1} = 15$ V        $V_{S2} = 9$ V
$R_{S1} = 130$ Ω        $R_{S2} = 290$ Ω
$R_1 = 1.1$ kΩ        $R_2 = 700$ Ω
$L = 17$ mH        $C = 0.35$ μF

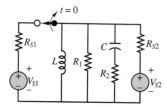

**Figure P5.37**

**5.38**   In the circuit shown in Figure P5.37:

$V_{S1} = 12$ V        $V_{S2} = 12$ V
$R_{S1} = 50$ Ω        $R_{S2} = 50$ Ω
$R_1 = 2.2$ kΩ        $R_2 = 600$ Ω
$L = 7.8$ mH        $C = 68$ μF

Assume that DC steady-state conditions exist for $t < 0$. Determine the voltage across the capacitor and the current through the inductor as $t$ approaches infinity. Remember to specify the polarity of the voltage and the direction of the current that you assume for your solution.

**5.39**   Assume the switch in the circuit of Figure P5.39 has been closed for a very long time. It is suddenly opened at $t = 0$, and then reclosed at $t = 5$ s. Determine an expression for the inductor current for $t \geq 0$.

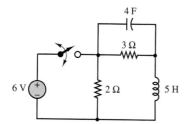

**Figure P5.39**

**5.40**   Assume the circuit of Figure P5.40 initially stores no energy. The switch is closed at $t = 0$, and then reopened at $t = 50\,\mu$s. Determine an expression for the capacitor voltage for $t \geq 0$.

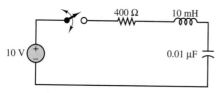

**Figure P5.40**

**5.41**   Assume the circuit of Figure P5.41 initially stores no energy. Switch $S_1$ is open, and $S_2$ is closed. Switch $S_1$ is closed at $t = 0$, and switch $S_2$ is opened at $t = 5$ s. Determine an expression for the capacitor voltage for $t \geq 0$.

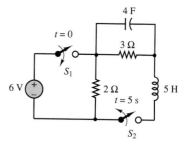

**Figure P5.41**

**5.42**   Assume that the circuit shown in Figure P5.42 is underdamped and that the circuit initially has no energy stored. It has been observed that, after the switch is closed at $t = 0$, the capacitor voltage reaches an initial peak value of 70 V when $t = 5\pi/3$ μs, a second peak value of 53.2 V when $t = 5\pi$ μs, and eventually approaches a steady-state value of 50 V. If $C = 1.6$ nF, what are the values of $R$ and $L$?

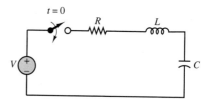

**Figure P5.42**

**5.43**   Given the information provided in Problem 5.42, explain how to modify the circuit so that the first two peaks occur at $5\pi$ μs and $15\pi$ μs. Assume that $C$ cannot be changed.

**5.44**   Find $i$ for $t > 0$ in the circuit of Figure P5.44 if $i(0) = 4$ A and $v(0) = 6$ V.

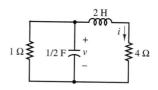

**Figure P5.44**

**5.45**   Find $v$ for $t > 0$ in the circuit of Figure P5.45 if the circuit is in steady state at $t = 0^-$.

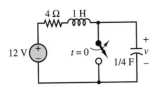

**Figure P5.45**

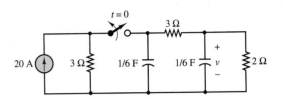

**Figure P5.48**

**5.46** Find $i$ for $t > 0$ in the circuit of Figure P5.46 if the circuit is in steady state at $t = 0^-$.

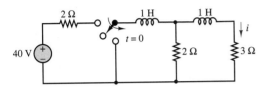

**Figure P5.46**

**5.47** Find $i$ for $t > 0$ in the circuit of Figure P5.47 if the circuit is in steady state at $t = 0^-$.

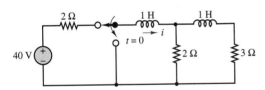

**Figure P5.47**

**5.48** Find $v$ for $t > 0$ in the circuit of Figure P5.48 if the circuit is in steady state at $t = 0^-$.

**5.49** The circuit of Figure P5.49 is in steady state at $t = 0^-$. Find $v$ for $t > 0$ if $L$ is (a) 2.4 H, (b) 3 H, and (c) 4 H.

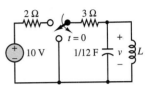

**Figure P5.49**

**5.50** Find $v$ for $t > 0$ in the circuit of Figure P5.50 if the circuit is in steady state at $t = 0^-$.

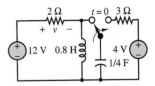

**Figure P5.50**

# CHAPTER

# 6

# Frequency Response
# and System Concepts

hapter 4 introduced the notions of energy-storage elements and dynamic circuit equations and developed appropriate tools (complex algebra and phasors) for the solution of AC circuits. In Chapter 5, we explored the solution of first- and second-order circuits subject to switching transients. The aim of the present chapter is to exploit AC circuit analysis methods to study the frequency response of electric circuits.

It is common, in engineering problems, to encounter phenomena that are frequency-dependent. For example, structures vibrate at a characteristic frequency when excited by wind forces (some high-rise buildings experience perceptible oscillation!). The propeller on a ship excites the shaft at a vibration frequency related to the engine's speed of rotation and to the number of blades on the propeller. An internal combustion engine is excited periodically by the combustion events in the individual cylinder, at a frequency determined by the firing of the cylinders. Wind blowing across a pipe excites a resonant vibration that is perceived as sound (wind instruments operate on this principle). Electrical circuits are no different from other dynamic systems in this respect, and a large body of knowledge has been developed for understanding the frequency response of electrical circuits, mostly based on the ideas behind phasors and impedance. These ideas, and the concept of filtering, will be explored in this chapter.

The ideas developed in this chapter will also be applied, by analogy, to the analysis of other physical systems (e.g., mechanical systems), to illustrate the generality of the concepts. By the end of the chapter, you should be able to:

- Compute the frequency response function for an arbitrary circuit.
- Use knowledge of the frequency response to determine the output of a circuit.
- Recognize the analogy between electrical circuits and other dynamic systems.

## 6.1   SINUSOIDAL FREQUENCY RESPONSE

The **sinusoidal frequency response** (or, simply, **frequency response**) of a circuit provides a measure of how the circuit responds to sinusoidal inputs of arbitrary frequency. In other words, given the input signal amplitude, phase, and frequency, knowledge of the frequency response of a circuit permits the computation of the output signal. The box "Fourier Analysis" provides further explanation of the importance of sinusoidal signals. Suppose, for example, that you wanted to determine how the load voltage or current varied in response to different excitation signal frequencies in the circuit of Figure 6.1. An analogy could be made, for example, with how a speaker (the load) responds to the audio signal generated by a CD player (the source) when an amplifier (the circuit) is placed between the two.[1] In the circuit of Figure 6.1, the signal source circuitry is represented by its Thévenin equivalent. Recall that the impedance $Z_S$ presented by the source to the remainder of the circuit is a function of the frequency of the source signal (Section 4.4). For the purpose of illustration, the amplifier circuit is represented by the idealized connection of two impedances, $Z_1$ and $Z_2$, and the load is represented by an additional impedance, $Z_L$. What, then, is the frequency response of this circuit? The following is a fairly general definition:

> The frequency response of a circuit is a measure of the variation of a load-related voltage or current as a function of the frequency of the excitation signal.

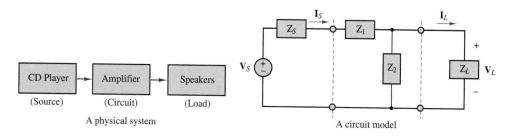

A physical system                              A circuit model

**Figure 6.1** A circuit model

---

[1] In reality, the circuitry in a hi-fi stereo system is far more complex than the circuits that will be discussed in this chapter and in the homework problems. However, from the standpoint of intuition and everyday experience, the audio analogy provides a useful example; it allows you to build a quick feeling for the idea of frequency response. Practically everyone has an intuitive idea of bass, mid range, and treble as coarsely defined frequency regions in the audio spectrum. The material presented in the next few sections should give you a more rigorous understanding of these concepts.

According to this definition, frequency response could be defined in a variety of ways. For example, we might be interested in determining how the load voltage varies as a function of the source voltage. Then, analysis of the circuit of Figure 6.1 might proceed as follows.

To express the frequency response of a circuit in terms of variation in output voltage as a function of source voltage, we use the general formula

$$H_V(j\omega) = \frac{\mathbf{V}_L(j\omega)}{\mathbf{V}_S(j\omega)} \tag{6.1}$$

One method that allows for representation of the load voltage as a function of the source voltage (this is, in effect, what the frequency response of a circuit implies) is to describe the source and attached circuit by means of the Thévenin equivalent circuit. (This is not the only useful technique; the node voltage or mesh current equations for the circuit could also be employed.) Figure 6.2 depicts the original circuit of Figure 6.1 with the load removed, ready for the computation of the Thévenin equivalent.

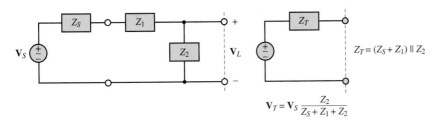

**Figure 6.2** Thévenin equivalent source circuit

## Fourier Analysis

In this brief introduction to **Fourier theory,** we shall explain in an intuitive manner how it is possible to represent many signals by means of the superposition of various sinusoidal signals of different amplitude, phase, and frequency. Any periodic finite-energy signal may be expressed by means of an infinite sum of sinusoids, as illustrated in the following paragraphs.

Consider a periodic waveform, $x(t)$. Its Fourier series representation is defined below by the infinite summation of sinusoids at the frequencies $n\omega_0$ (integer multiples of the **fundamental frequency,** $\omega_0$), with amplitudes $A_n$ and phases $\phi_n$.

$$x(t) = x(t + T_0) \qquad T_0 = \text{period} \tag{6.2}$$

$$x(t) = \sum_{n=0}^{\infty} A_n \cos\left(\frac{2\pi nt}{T_0} + \phi_n\right) \tag{6.3}$$

One could also write the term $2\pi n/T_0$ as $n\omega_0$, where

$$\omega_0 = \frac{2\pi}{T_0} = 2\pi f_0 \tag{6.4}$$

is the fundamental (radian) frequency and the frequencies $2\omega_0$, $3\omega_0$, $4\omega_0$, and so on, are called its **harmonics.**

The notion that a signal may be represented by sinusoidal components is particularly useful, and not only in the study of electrical circuits—in the sense that we need only understand the response of a circuit to an arbitrary sinusoidal excitation in order to be able to infer the circuit's response to more complex signals. In fact, the frequently employed *sinusoidal frequency response* discussed in this chapter is a function that enables us to explain how a circuit would respond to a signal made up of a superposition

(*continued*)

of sinusoidal components at various frequencies. These sinusoidal components form the **spectrum** of the signal, that is, its frequency composition; the amplitude and phase of each of the sinusoids contribute to the overall "character" of the signal, in the same sense as the timbre of a musical instrument is made up of the different harmonics that are generated when a note is played (the timbre is what differentiates, for example, a viola from a cello or a violin). An example of the amplitude spectrum of a "square-wave" signal is shown in Figure 6.3. In order to further illustrate how the superposition of sinusoids can give rise to a signal that at first might appear substantially different from a sinusoid, the evolution of a sine wave

into a square wave is displayed in Figure 6.4, as more Fourier components are added. The first picture represents the fundamental component, that is, the sinusoid that has the same frequency as the square wave. Then one harmonic at a time is added, up to the fifth nonzero component (the ninth frequency component; see Figure 6.3), illustrating how, little by little, the rounded peaks of the sinusoid transform into the flat top of the square wave!

Although this book will not deal with the mathematical aspects of Fourier series, it is important to recognize that this analysis tool provides excellent motivation for the study of sinusoidal signals, and of the sinusoidal frequency response of electric circuits.

Jean Baptiste Joseph Fourier (1768–1830), French mathematician and physicist who formulated the Fourier series. *Photo courtesy of Deutsches Museum, Munich.*

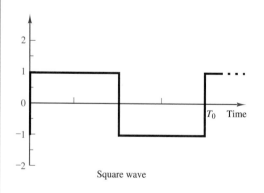

Square wave

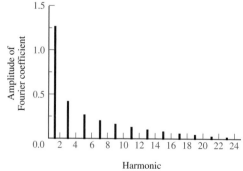

Fourier spectrum of square wave

**Figure 6.3** Amlitude spectrum of square wave

*(continued)*

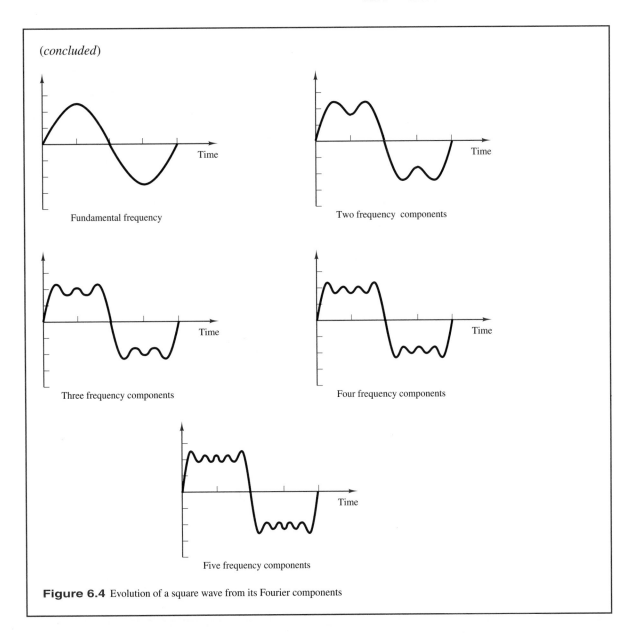

(*concluded*)

Fundamental frequency

Two frequency components

Three frequency components

Four frequency components

Five frequency components

**Figure 6.4** Evolution of a square wave from its Fourier components

Next, an expression for the load voltage, $\mathbf{V}_L$, may be found by connecting the load to the Thévenin equivalent source circuit and by computing the result of a simple voltage divider, as illustrated in Figure 6.5 and by the following equation:

$$
\begin{aligned}
\mathbf{V}_L &= \frac{Z_L}{Z_L + Z_T}\mathbf{V}_T \\
&= \frac{Z_L}{Z_L + \dfrac{(Z_S + Z_1)Z_2}{Z_S + Z_1 + Z_2}} \cdot \frac{Z_2}{Z_S + Z_1 + Z_2}\mathbf{V}_S \\
&= \frac{Z_L Z_2}{Z_L(Z_S + Z_1 + Z_2) + (Z_S + Z_1)Z_2}\mathbf{V}_S
\end{aligned}
$$

**(6.5)**

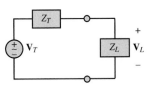

**Figure 6.5** Complete equivalent circuit

Thus, the frequency response of the circuit, as defined in equation 6.4, is given by the expression

$$\frac{\mathbf{V}_L}{\mathbf{V}_S}(j\omega) = H_V(j\omega) = \frac{Z_L Z_2}{Z_L(Z_S + Z_1 + Z_2) + (Z_S + Z_1)Z_2} \tag{6.6}$$

The expression for $H_V(j\omega)$ is therefore known if the impedances of the circuit elements are known. Note that $H_V(j\omega)$ is a complex quantity (dimensionless, because it is the ratio of two voltages), and that it therefore follows that

$\mathbf{V}_L(j\omega)$ is a phase-shifted and amplitude-scaled version of $\mathbf{V}_S(j\omega)$.

If the phasor source voltage and the frequency response of the circuit are known, the phasor load voltage can be computed as follows:

$$\mathbf{V}_L(j\omega) = H_V(j\omega) \cdot \mathbf{V}_S(j\omega) \tag{6.7}$$

$$V_L e^{j\phi_L} = |H_V| e^{j\phi_H} \cdot V_S e^{j\phi_S} \tag{6.8}$$

or

$$V_L e^{j\phi_L} = |H_V| V_S e^{j(\phi_H + \phi_S)} \tag{6.9}$$

where

$$V_L = |H_V| \cdot V_S$$

and

$$\phi_L = \phi_H + \phi_S \tag{6.10}$$

Thus, the effect of inserting a linear circuit between a source and a load is best understood by considering that, at any given frequency, $\omega$, the load voltage is a sinusoid at the same frequency as the source voltage, with amplitude given by $V_L = |H_V| \cdot V_S$ and phase equal to $\phi_L = \phi_H + \phi_S$, where $|H_V|$ is the magnitude of the frequency response and $\phi_H$ its phase angle. Both $|H_V|$ and $\phi_H$ are functions of frequency.

---

### EXAMPLE 6.1 Computing the Frequency Response of a Circuit Using Equivalent Circuit Ideas

**Problem**

Compute the frequency response $H_V(j\omega)$ for the circuit of Figure 6.6.

---

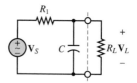

**Figure 6.6**

**Solution**

**Known Quantities:** $R_1 = 1$ k$\Omega$; $C = 10$ $\mu$F; $R_L = 10$ k$\Omega$.

**Find:** The frequency response $H_V(j\omega) = \mathbf{V}_L(j\omega)/\mathbf{V}_S(j\omega)$.

**Assumptions:** None.

***Analysis:*** To solve this problem we use an equivalent circuit approach. Recognizing that $R_L$ is the load resistance, we determine the equivalent circuit representation of the circuit to the left of the load, using the techniques perfected in Chapters 3 and 4. The Thévenin equivalent circuit is shown in Figure 6.7. Using the voltage divider rule and the equivalent circuit shown in the figure, we obtain the following expression

$$\mathbf{V}_L = \frac{Z_L}{Z_T + Z_L}\mathbf{V}_T = \frac{Z_L}{\dfrac{Z_1 Z_2}{Z_1 + Z_2} + Z_L}\frac{Z_2}{Z_1 + Z_2}\mathbf{V}_S = H_V \mathbf{V}_S$$

and

$$\frac{\mathbf{V}_L}{\mathbf{V}_S}(j\omega) = H_V(j\omega) = \frac{Z_L Z_2}{Z_L(Z_1 + Z_2) + Z_1 Z_2}$$

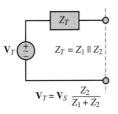

**Figure 6.7**

The impedances of the circuit elements are: $Z_1 = 10^3 \ \Omega$; $Z_2 = \frac{1}{j\omega \times 10^{-5}} \ \Omega$; $Z_L = 10^4 \ \Omega$. The resulting frequency response can be calculated to be:

$$H_V(j\omega) = \frac{\dfrac{10^4}{j\omega \times 10^{-5}}}{10^4\left(10^3 + \dfrac{1}{j\omega \times 10^{-5}}\right) + \dfrac{10^3}{j\omega \times 10^{-5}}} = \frac{100}{110 + j\omega}$$

$$= \frac{100}{\sqrt{110^2 + \omega^2}e^{j \arctan\left(\frac{\omega}{110}\right)}} = \frac{100}{\sqrt{110^2 + \omega^2}}\angle - \arctan\left(\frac{\omega}{110}\right)$$

***Comments:*** The use of equivalent circuit ideas is often helpful in deriving frequency response functions, because it naturally forces us to identify source and load quantities. However, it is certainly not the only method of solution. For example, nodal analysis would have yielded the same results just as easily, by recognizing that the top node voltage is equal to the load voltage, and by solving directly for $\mathbf{V}_L$ as a function of $\mathbf{V}_S$, without going through the intermediate step of computing the Thévenin equivalent source circuit.

***Focus on Computer-Aided Tools:*** A computer-generated solution of this problem may be found in the CD-ROM that accompanies this book.

VIRTUAL LAB

---

The importance and usefulness of the frequency response concept lies in its ability to summarize the response of a circuit in a single function of frequency, $H(j\omega)$, which can predict the load voltage or current at any frequency, given the input. Note that the frequency response of a circuit can be defined in four different ways:

$$\boxed{\begin{array}{cc} H_V(j\omega) = \dfrac{\mathbf{V}_L(j\omega)}{\mathbf{V}_S(j\omega)} & H_I(j\omega) = \dfrac{\mathbf{I}_L(j\omega)}{\mathbf{I}_S(j\omega)} \\[2ex] H_Z(j\omega) = \dfrac{\mathbf{V}_L(j\omega)}{\mathbf{I}_S(j\omega)} & H_Y(j\omega) = \dfrac{\mathbf{I}_L(j\omega)}{\mathbf{V}_S(j\omega)} \end{array}}$$

$$(6.11)$$

If $H_V(j\omega)$ and $H_I(j\omega)$ are known, one can directly derive the other two expressions:

$$H_Z(j\omega) = \frac{\mathbf{V}_L(j\omega)}{\mathbf{I}_S(j\omega)} = Z_L(j\omega)\frac{\mathbf{I}_L(j\omega)}{\mathbf{I}_S(j\omega)} = Z_L(j\omega)H_I(j\omega) \qquad (6.12)$$

$$H_Y(j\omega) = \frac{\mathbf{I}_L(j\omega)}{\mathbf{V}_S(j\omega)} = \frac{1}{Z_L(j\omega)}\frac{\mathbf{V}_L(j\omega)}{\mathbf{V}_S(j\omega)} = \frac{1}{Z_L(j\omega)}H_V(j\omega) \qquad \textbf{(6.13)}$$

With these definitions in hand, it is now possible to introduce one of the central concepts of electrical circuit analysis: **filters.** The concept of filtering an electrical signal will be discussed in the next section.

---

### EXAMPLE 6.2 Computing the Frequency Response of a Circuit

**Problem**

Compute the frequency response $H_Z(j\omega)$ for the circuit of Figure 6.8.

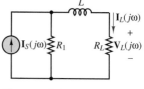

---

**Solution**

**Known Quantities:** $R_1 = 1\ \text{k}\Omega$; $L = 2\ \text{mH}$; $R_L = 4\ \text{k}\Omega$.

**Find:** The frequency response $H_Z(j\omega) = \mathbf{V}_L(j\omega)/\mathbf{I}_S(j\omega)$.

**Assumptions:** None.

**Analysis:** To determine expressions for the load voltage, we recognize that the load current can be obtained simply by using a current divider between the two branches connected to the current source, and that the load voltage is simply the product of the load current times $R_L$.

Using the current divider rule, we obtain the following expression for $\mathbf{I}_L$:

$$\mathbf{I}_L = \frac{\dfrac{1}{R_L + j\omega L}}{\dfrac{1}{R_L + j\omega L} + \dfrac{1}{R_1}}\mathbf{I}_S = \frac{1}{1 + \dfrac{R_L}{R_1} + j\dfrac{\omega L}{R_1}}\mathbf{I}_S$$

and

$$\frac{\mathbf{V}_L}{\mathbf{I}_S}(j\omega) = H_Z(j\omega) = \frac{I_L R_L}{I_S} = \frac{R_L}{1 + \dfrac{R_L}{R_1} + j\dfrac{\omega L}{R_1}}$$

Substituting numerical values, we obtain:

$$H_Z(j\omega) = \frac{4 \times 10^3}{1 + 4 + j\dfrac{2 \times 10^{-3}\omega}{10^3}} = \frac{0.8 \times 10^3}{1 + j0.4 \times 10^{-6}\omega}$$

**Comments:** You should verify that the untis of the expression for $H_Z(j\omega)$ are indeed ohms, as they should be from the definition of $H_Z$.

**Focus on Computer-Aided Tools:** A computer-generated solution of this problem may be found in the CD-ROM that accompanies this book.

---

## 6.2    FILTERS

There are many practical, everyday applications that involve filters of one kind or another. Just to mention two, filtration systems are used to eliminate impurities

**Figure 6.8**

from drinking water, and sunglasses are used to filter out eye-damaging ultraviolet radiation and to reduce the intensity of sunlight reaching the eyes. An analogous concept applies to electrical circuits: it is possible to *attenuate* (i.e., reduce in amplitude) or altogether eliminate signals of unwanted frequencies, such as those that may be caused by electrical noise or other forms of interference. This section will be devoted to the analysis of electrical filters.

## Low-Pass Filters

Figure 6.9 depicts a simple **RC filter** and denotes its input and output voltages by $V_i$ and $V_o$. The frequency response for the filter may be obtained by considering the function

$$H(j\omega) = \frac{V_o}{V_i}(j\omega) \qquad (6.14)$$

and noting that the output voltage may be expressed as a function of the input voltage by means of a voltage divider, as follows:

$$V_o(j\omega) = V_i(j\omega)\frac{1/j\omega C}{R + 1/j\omega C} = V_i(j\omega)\frac{1}{1 + j\omega RC} \qquad (6.15)$$

Thus, the frequency response of the *RC* filter is

$$\frac{V_o}{V_i}(j\omega) = \frac{1}{1 + j\omega CR} \qquad (6.16)$$

*RC* low-pass filter. The circuit preserves lower frequencies while attenuating the frequencies above the cutoff frequency, $\omega_0 = 1/RC$. The voltages $V_i$ and $V_o$ are the filter input and output voltages, respectively.

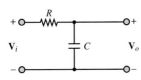

**Figure 6.9** A simple *RC* filter

An immediate observation upon studying this frequency response is that if the signal frequency, $\omega$, is zero, the value of the frequency response function is 1. That is, the filter is passing all of the input. Why? To answer this question, we note that at $\omega = 0$, the impedance of the capacitor, $1/j\omega C$, becomes infinite. Thus, the capacitor acts as an open circuit, and the output voltage equals the input:

$$V_o(j\omega = 0) = V_i(j\omega = 0) \qquad (6.17)$$

Since a signal at sinusoidal frequency equal to zero is a DC signal, this filter circuit does not in any way affect DC voltages and currents. As the signal frequency increases, the magnitude of the frequency response decreases, since the denominator increases with $\omega$. More precisely, equations 6.18 to 6.21 describe the magnitude and phase of the frequency response of the *RC* filter:

$$H(j\omega) = \frac{V_o}{V_i}(j\omega) = \frac{1}{1 + j\omega CR}$$

$$= \frac{1}{\sqrt{1 + (\omega CR)^2}}\frac{e^{j0°}}{e^{j\arctan(\omega CR/1)}} \qquad (6.18)$$

$$= \frac{1}{\sqrt{1 + (\omega CR)^2}} \cdot e^{-j\arctan(\omega CR)}$$

or

$$H(j\omega) = |H(j\omega)|e^{j\phi_H(j\omega)} \qquad (6.19)$$

with

$$|H(j\omega)| = \frac{1}{\sqrt{1 + (\omega CR)^2}} = \frac{1}{\sqrt{1 + (\omega/\omega_0)^2}} \qquad (6.20)$$

and

$$\phi_H(j\omega) = -\arctan(\omega CR) = -\arctan\left(\frac{\omega}{\omega_0}\right) \tag{6.21}$$

with

$$\omega_0 = \frac{1}{RC} \tag{6.22}$$

The simplest way to envision the effect of the filter is to think of the phasor voltage $\mathbf{V}_i = V_i e^{j\phi_i}$ scaled by a factor of $|H|$ and shifted by a phase angle $\phi_H$ by the filter *at each frequency*, so that the resultant output is given by the phasor $V_o e^{j\phi_o}$, with

$$V_o = |H| \cdot V_i$$
$$\phi_o = \phi_H + \phi_i \tag{6.23}$$

and where $|H|$ and $\phi_H$ are functions of frequency. The frequency $\omega_0$ is called the **cutoff frequency** of the filter and, as will presently be shown, gives an indication of the filtering characteristics of the circuit.

It is customary to represent $H(j\omega)$ in two separate plots, representing $|H|$ and $\phi_H$ as functions of $\omega$. These are shown in Figure 6.10 in normalized form—that is, with $|H|$ and $\phi_H$ plotted versus $\omega/\omega_0$, corresponding to a cutoff frequency $\omega_0 = 1$ rad/s. Note that, in the plot, the frequency axis has been scaled logarithmically. This is a common practice in electrical engineering, because it allows viewing a

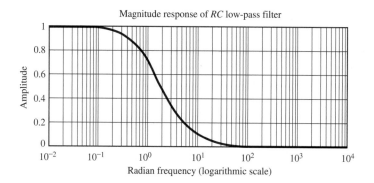

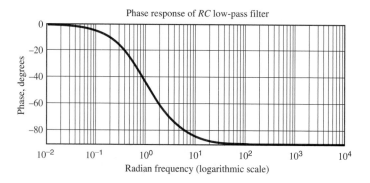

**Figure 6.10** Magnitude and phase response plots for *RC* filter

very broad range of frequencies on the same plot without excessively compressing the low-frequency end of the plot. The frequency response plots of Figure 6.10 are commonly employed to describe the frequency response of a circuit, since they can provide a clear idea at a glance of the effect of a filter on an excitation signal. For example, the *RC* filter of Figure 6.9 has the property of "passing" signals at low frequencies ($\omega \ll 1/RC$) and of filtering out signals at high frequencies ($\omega \gg 1/RC$). This type of filter is called a **low-pass filter.** The cutoff frequency $\omega = 1/RC$ has a special significance in that it represents—approximately—the point where the filter begins to filter out the higher-frequency signals. The value of $H(j\omega)$ at the cutoff frequency is $1/\sqrt{2} = 0.707$. Note how the cutoff frequency depends exclusively on the values of $R$ and $C$. Therefore, one can adjust the filter response as desired simply by selecting appropriate values for $C$ and $R$, and therefore choose the desired filtering characteristics.

### EXAMPLE 6.3 Frequency Response of *RC* Filter

#### Problem

Compute the response of the *RC* filter of Figure 6.9 to sinusoidal inputs at the frequencies of 60 and 10,000 Hz.

#### Solution

**Known Quantities:**  $R = 1\ \text{k}\Omega; C = 0.47\ \mu\text{F}; v_i(t) = 5\ \cos(\omega t)$ V.

**Find:**  The output voltage, $v_o(t)$, at each frequency.

**Assumptions:**  None.

**Analysis:**  In this problem, we know the input signal voltage and the frequency response of the circuit (equation 6.18), and we need to find the output voltage at two different frequencies. If we represent the voltages in phasor form, we can use the frequency response to calculate the desired quantities:

$$\frac{\mathbf{V}_o}{\mathbf{V}_i}(j\omega) = H_V(j\omega) = \frac{1}{1 + j\omega CR}$$

$$\mathbf{V}_o(j\omega) = H_V(j\omega)\mathbf{V}_i(j\omega) = \frac{1}{1 + j\omega CR}\mathbf{V}_i(j\omega)$$

If we recognize that the cutoff frequency of the filter is $\omega_0 = 1/RC = 2,128$ rad/s, we can write the expression for the frequency response in the form of equations 6.20 and 6.21:

$$H_V(j\omega) = \frac{1}{1 + \dfrac{j\omega}{\omega_0}} \qquad |H_V(j\omega)| = \frac{1}{\sqrt{1 + \left(\dfrac{\omega}{\omega_0}\right)^2}} \qquad \phi_H(j\omega) = -\arctan\left(\frac{\omega}{\omega_0}\right)$$

Next, we recognize that at $\omega = 120\pi$ rad/s, the ratio $\omega/\omega_0 = 0.177$, and at $\omega = 20,000\pi$, $\omega/\omega_0 = 29.5$. Thus we compute the output voltage at each frequency as follows:

$$\mathbf{V}_o(\omega = 2\pi 60) = \frac{1}{1 + j0.177}\mathbf{V}_i(\omega = 2\pi 60) = 0.985 \times 5\angle{-0.175}\ \text{V}$$

$$\mathbf{V}_o(\omega = 2\pi 10,000) = \frac{1}{1 + j29.5}\mathbf{V}_i(\omega = 2\pi 10,000) = 0.0345 \times 5\angle{-1.537}\ \text{V}$$

And finally write the time-domain response for each frequency:

$$v_o(t) = 4.923 \, \cos(2\pi 60t - 0.175) \text{ V} \qquad \text{at } \omega = 2\pi 60 \text{ rad/s}$$
$$v_o(t) = 0.169 \, \cos(2\pi 10,000t - 1.537) \text{ V} \qquad \text{at } \omega = 2\pi 10,000 \text{ rad/s}$$

The magnitude and phase responses of the filter are plotted in Figure 6.11. It should be evident from these plots that only the low-frequency components of the signal are passed by the filter. This low-pass filter would pass only the *bass range* of the audio spectrum.

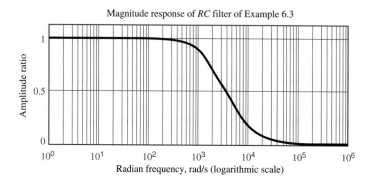

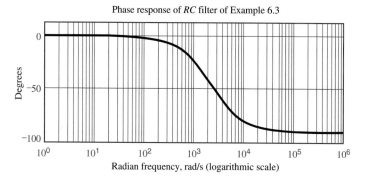

**Figure 6.11** Response of *RC* filter of Example 6.3

**Comments:** Can you think of a very quick, approximate way of obtaining the answer to this problem from the magnitude and phase plots of Figure 6.11? Try to multiply the input voltage amplitude by the magnitude response at each frequency, and determine the phase shift at each frequency. Your answer should be pretty close to the one computed analytically.

VIRTUAL LAB

**Focus on Computer-Aided Tools:** A computer-generated solution of this problem generated by MathCad may be found in the CD-ROM that accompanies this book.

### EXAMPLE 6.4 Frequency Response of *RC* Low-Pass Filter in a More Realistic Circuit

**Problem**

Compute the response of the *RC* filter in the circuit of Figure 6.12.

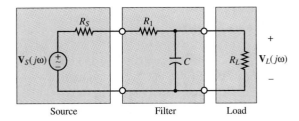

**Figure 6.12** *RC* filter inserted in a circuit

## Solution

**Known Quantities:** $R_S = 50\ \Omega$; $R_1 = 200\ \Omega$; $R_L = 500\ \Omega$; $C = 10\ \mu F$.

**Find:** The output voltage, $v_o(t)$, at each frequency.

**Assumptions:** None.

**Analysis:** The circuit shown in this problem is a more realistic representation of a filtering problem, in that we have inserted the *RC* filter circuit between source and load circuits (where the source and load are simply represented in equivalent form). To determine the response of the circuit, we compute the Thévenin equivalent representation of the circuit with respect to the load, as shown in Figure 6.13. Let $R' = R_S + R_1$ and

$$Z' = R_L \| \frac{1}{j\omega C} = \frac{R_L}{1 + j\omega C R_L}$$

Then the circuit response may be computed as follows:

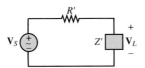

**Figure 6.13** Equivalent-circuit representation of Figure 6.12

$$\frac{\mathbf{V}_L}{\mathbf{V}_S}(j\omega) = \frac{Z'}{R' + Z'} = \frac{\dfrac{R_L}{1 + j\omega C R_L}}{R_S + R_1 + \dfrac{R_L}{1 + j\omega C R_L}}$$

$$= \frac{R_L}{R_L + R_S + R_1 + j\omega C R_L(R_S + R_1)} = \frac{\dfrac{R_L}{R_L + R'}}{1 + j\omega C R_L \| R'}$$

The above expression can be written as follows:

$$H(j\omega) = \frac{\dfrac{R_L}{R_L + R'}}{1 + j\omega C R_L \| R'} = \frac{K}{1 + j\omega C R_{EQ}} = \frac{0.667}{1 + j\dfrac{\omega}{600}}$$

**Comments:** Note the similarity and difference between the above expression and equation 6.16: The numerator is different than 1, because of the voltage divider effect resulting from the source and load resistances, and the cutoff frequency is given by the expression

$$\omega_0 = \frac{1}{C R_{EQ}}$$

**FOCUS ON
MEASUREMENTS**

## Wheatstone Bridge Filter

FIND IT

ON THE WEB

The Wheatstone bridge circuit of Examples 2.10 and Focus on Measurements: Wheatstone Bridge in Chapter 2 is used in a number of instrumentation applications, including the **measurement of force** (see Example 2.13, describing the strain gauge bridge). Figure 6.14 depicts the appearance of the bridge circuit. When undesired noise and interference are present in a measurement, it is often appropriate to use a low-pass filter to reduce the effect of the noise. The capacitor that is connected to the output terminals of the bridge in Figure 6.14 constitutes an effective and simple low-pass filter, in conjunction with the bridge resistance. Assume that the average resistance of each leg of the bridge is 350 Ω (a standard value for strain gauges) and that we desire to measure a sinusoidal force at a frequency of 30 Hz. From prior measurements, it has been determined that a filter with a cutoff frequency of 300 Hz is sufficient to reduce the effects of noise. Choose a capacitor that matches this filtering requirement.

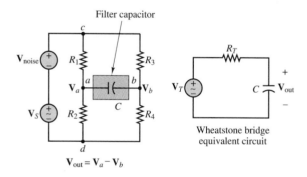

**Figure 6.14** Wheatstone bridge with equivalent circuit and simple capacitive filter

### Solution:

By evaluating the Thévenin equivalent circuit for the Wheatstone bridge, calculating the desired value for the filter capacitor becomes relatively simple, as illustrated at the bottom of Figure 6.14. The Thévenin resistance for the bridge circuit may be computed by short-circuiting the two voltage sources and removing the capacitor placed across the load terminals:

$$R_T = R_1 \parallel R_2 + R_3 \parallel R_4 = 350 \parallel 350 + 350 \parallel 350 = 350 \ \Omega$$

Since the required cutoff frequency is 300 Hz, the capacitor value can be computed from the expression

$$\omega_0 = \frac{1}{R_T C} = 2\pi \times 300$$

or

$$C = \frac{1}{R_T \omega_0} = \frac{1}{350 \times 2\pi \times 300} = 1.51 \ \mu F$$

The frequency response of the bridge circuit is of the same form as equation 6.16:

$$\frac{\mathbf{V}_{\text{out}}}{\mathbf{V}_T}(j\omega) = \frac{1}{1 + j\omega C R_T}$$

This response can be evaluated at the frequency of 30 Hz to verify that the attenuation and phase shift at the desired signal frequency are minimal:

$$\frac{\mathbf{V}_{\text{out}}}{\mathbf{V}_T}(j\omega = j2\pi \times 30) = \frac{1}{1 + j2\pi \times 30 \times 1.51 \times 10^{-6} \times 350}$$
$$= 0.9951\angle{-5.7°}$$

Figure 6.15 depicts the appearance of a 30-Hz sinusoidal signal before and after the addition of the capacitor to the circuit.

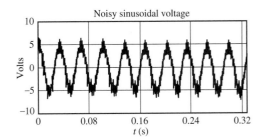

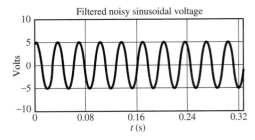

**Figure 6.15** Unfiltered and filtered bridge output

*Focus on Computer-Aided Tools—* An EWB simulation of this circuit may be found in the accompanying CD-ROM.

Much more complex low-pass filters than the simple *RC* combinations shown so far can be designed by using appropriate combinations of various circuit elements. The synthesis of such advanced filter networks is beyond the scope of this book; however, we shall discuss the practical implementation of some commonly used filters in Chapters 12 and 15, in connection with the discussion of the operational amplifier. The next two sections extend the basic ideas introduced in the preceding pages to high- and band-pass filters—that is, to filters that emphasize the higher frequencies or a band of frequencies, respectively.

## High-Pass Filters

Just as you can construct a simple filter that preserves low frequencies and attenuates higher frequencies, you can easily construct a **high-pass filter** that passes mainly those frequencies *above a certain cutoff frequency*. The analysis of a simple high-pass filter can be conducted by analogy with the preceding discussion of the low-pass filter. Consider the circuit shown in Figure 6.16. The frequency response for the high-pass filter,

$$H(j\omega) = \frac{\mathbf{V}_o}{\mathbf{V}_i}(j\omega)$$

may be obtained by noting that

$$\mathbf{V}_o(j\omega) = \mathbf{V}_i(j\omega)\frac{R}{R + 1/j\omega C} = \mathbf{V}_i(j\omega)\frac{j\omega C R}{1 + j\omega C R} \qquad (6.24)$$

Thus, the frequency response of the filter is:

$$\frac{\mathbf{V}_o}{\mathbf{V}_i}(j\omega) = \frac{j\omega C R}{1 + j\omega C R} \qquad (6.25)$$

which can be expressed in magnitude-and-phase form by

$$H(j\omega) = \frac{\mathbf{V}_o}{\mathbf{V}_i}(j\omega) = \frac{j\omega C R}{1 + j\omega C R} = \frac{\omega C R e^{j90°}}{\sqrt{1 + (\omega C R)^2} e^{j\,\arctan(\omega C R/1)}}$$
$$= \frac{\omega C R}{\sqrt{1 + (\omega C R)^2}} \cdot e^{j(90° - \arctan(\omega C R))} \qquad (6.26)$$

or

$$H(j\omega) = |H|e^{j\phi_H}$$

with

$$H(j\omega) = \frac{\omega C R}{\sqrt{1 + (\omega C R)^2}} \qquad (6.27)$$
$$\phi_H(j\omega) = 90° - \arctan(\omega C R)$$

You can verify by inspection that the amplitude response of the high-pass filter will be zero at $\omega = 0$ and will asymptotically approach 1 as $\omega$ approaches infinity, while the phase shift is 90° at $\omega = 0$ and tends to zero for increasing $\omega$. Amplitude-and-phase response curves for the high-pass filter are shown in Figure 6.17. These plots have been normalized to have the filter cutoff frequency $\omega_0 = 1$ rad/s. Note that, once again, it is possible to define a cutoff frequency at $\omega_0 = 1/RC$ in the same way as was done for the low-pass filter.

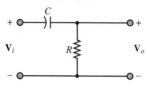

*RC* high-pass filter. The circuit preserves higher frequencies while attenuating the frequencies below the cutoff frequency, $\omega_0 = 1/RC$.

**Figure 6.16** High-pass filter

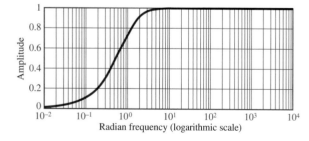
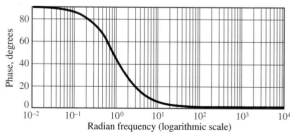

**Figure 6.17** Frequency response of a high-pass filter

## EXAMPLE  6.5  Frequency Response of *RC* High-Pass Filter

### Problem

Compute the response of the *RC* filter in the circuit of Figure 6.16. Evaluate the response
of the filter at $\omega = 2\pi \times 100$ and $2\pi \times 10{,}000$ rad/s.

### Solution

**Known Quantities:**  $R = 200\ \Omega$; $C = 0.199\ \mu$F.

**Find:**  The frequency response, $H_V(j\omega)$.

**Assumptions:**  None.

**Analysis:**  We first recognize that the cutoff frequency of the high-pass filter is
$\omega_0 = 1/RC = 2\pi \times 4{,}000$ rad/s. Next, we write the frequency response as in equation
6.25:

$$H_V(j\omega) = \frac{\mathbf{V}_o}{\mathbf{V}_i}(j\omega) = \frac{j\omega CR}{1 + j\omega CR} = \frac{\dfrac{\omega}{\omega_0}}{\sqrt{1 + \left(\dfrac{\omega}{\omega_0}\right)^2}} \angle \left[\frac{\pi}{2} - \arctan\left(\frac{\omega}{\omega_0}\right)\right]$$

We can now evaluate the response at the two frequencies:

$$H_V(\omega = 2\pi \times 100) = \frac{\dfrac{100}{4000}}{\sqrt{1 + \left(\dfrac{100}{4000}\right)^2}} \angle \left[\frac{\pi}{2} - \arctan\left(\frac{100}{4000}\right)\right] = 0.025\angle 1.546$$

$$H_V(\omega = 2\pi \times 10{,}000) = \frac{\dfrac{10{,}000}{4000}}{\sqrt{1 + \left(\dfrac{10{,}000}{4000}\right)^2}} \angle \left[\frac{\pi}{2} - \arctan\left(\frac{10{,}000}{4000}\right)\right]$$

$$= 0.929\angle 0.38$$

The frequency response plots are shown in Figure 6.18.

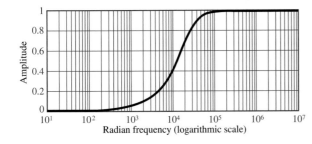
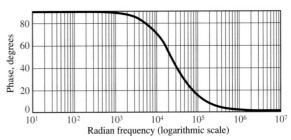

**Figure 6.18** Response of high-pass filter of Example 6.5

**Comments:**  The effect of this high-pass filter is to preserve the amplitude of the input signal at frequencies substantially greater than $\omega_0$, while signals at frequencies below $\omega_0$ would be strongly attenuated. With $\omega_0 = 2\pi \times 4,000$ (i.e., 4,000 Hz), this filter would pass only the *treble range* of the audio frequency spectrum.

## Band-Pass Filters

Building on the principles developed in the preceding sections, we can also construct a circuit that acts as a **band-pass filter,** passing mainly those frequencies *within a certain frequency range.* The analysis of a simple *second-order* band-pass filter (i.e., a filter with two energy-storage elements) can be conducted by analogy with the preceding discussions of the low-pass and high-pass filters. Consider the circuit shown in Figure 6.19, and the related frequency response function for the filter

*RLC* band-pass filter. The circuit preserves frequencies within a band.

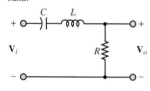

**Figure 6.19** *RLC* band-pass filter

$$H(j\omega) = \frac{\mathbf{V}_o}{\mathbf{V}_i}(j\omega)$$

Noting that

$$
\begin{aligned}
\mathbf{V}_o(j\omega) &= \mathbf{V}_i(j\omega) \cdot \frac{R}{R + 1/j\omega C + j\omega L} \\
&= \mathbf{V}_i(j\omega)\frac{j\omega C R}{1 + j\omega C R + (j\omega)^2 LC}
\end{aligned}
\tag{6.28}
$$

we may write the frequency response of the filter as

$$\frac{\mathbf{V}_o}{\mathbf{V}_i}(j\omega) = \frac{j\omega C R}{1 + j\omega C R + (j\omega)^2 LC} \tag{6.29}$$

Equation 6.29 can often be factored into the following form:

$$\frac{\mathbf{V}_o}{\mathbf{V}_i}(j\omega) = \frac{j A\omega}{(j\omega/\omega_1 + 1)(j\omega/\omega_2 + 1)} \tag{6.30}$$

where $\omega_1$ and $\omega_2$ are the two frequencies that determine the **pass-band** (or **bandwidth**) of the filter—that is, the frequency range over which the filter "passes" the input signal—and $A$ is a constant that results from the factoring. An immediate observation we can make is that if the signal frequency, $\omega$, is zero, the response of the filter is equal to zero, since at $\omega = 0$ the impedance of the capacitor, $1/j\omega C$, becomes infinite. Thus, the capacitor acts as an open circuit, and the output voltage equals zero. Further, we note that the filter output in response to an input signal at sinusoidal frequency approaching infinity is again equal to zero. This result can be verified by considering that as $\omega$ approaches infinity, the impedance of the inductor becomes infinite, that is, an open circuit. Thus, the filter cannot pass signals at very high frequencies. In an intermediate band of frequencies, the band-pass filter circuit will provide a variable attenuation of the input signal, dependent on the frequency of the excitation. This may be verified by taking a closer look at equation 6.30:

$$H(j\omega) = \frac{\mathbf{V}_o}{\mathbf{V}_i}(j\omega) = \frac{jA\omega}{(j\omega/\omega_1 + 1)(j\omega/\omega_2 + 1)}$$

$$= \frac{A\omega e^{j90°}}{\sqrt{1 + \left(\frac{\omega}{\omega_1}\right)^2}\sqrt{1 + \left(\frac{\omega}{\omega_2}\right)^2}\, e^{j\,\arctan(\omega/\omega_1)} e^{j\,\arctan(\omega/\omega_2)}} \qquad \textbf{(6.31)}$$

$$= \frac{A\omega}{\sqrt{\left[1 + \left(\frac{\omega}{\omega_1}\right)^2\right]\left[1 + \left(\frac{\omega}{\omega_2}\right)^2\right]}} \cdot e^{j[90° - \arctan(\omega/\omega_1) - \arctan(\omega/\omega_2)]}$$

Equation 6.31 is of the form $H(j\omega) = |H|e^{j\phi_H}$, with

$$|H(j\omega)| = \frac{A\omega}{\sqrt{\left[1 + \left(\frac{\omega}{\omega_1}\right)^2\right]\left[1 + \left(\frac{\omega}{\omega_2}\right)^2\right]}} \qquad \textbf{(6.32)}$$

and

$$\phi_H(j\omega) = 90° - \arctan\left(\frac{\omega}{\omega_1}\right) - \arctan\left(\frac{\omega}{\omega_2}\right)$$

The magnitude and phase plots for the frequency response of the band-pass filter of Figure 6.19 are shown in Figure 6.20. These plots have been normalized to have the filter pass-band centered at the frequency $\omega = 1$ rad/s.

The frequency response plots of Figure 6.20 suggest that, in some sense, the band-pass filter acts as a combination of a high-pass and a low-pass filter. As illustrated in the previous cases, it should be evident that one can adjust the filter response as desired simply by selecting appropriate values for $L$, $C$, and $R$.

The expression for the frequency response of a second-order band-pass filter (equation 6.29) can also be rearranged to illustrate two important features of this circuit: the **quality factor, $Q$,** and the **resonant frequency, $\omega_0$.** Let

$$\omega_0 = \frac{1}{\sqrt{LC}} \qquad \text{and} \qquad Q = \frac{1}{\omega_0 C R} = \frac{\omega_0 L}{R} \qquad \textbf{(6.33)}$$

Then we can write

$$\omega C R = \omega_0 C R \frac{\omega}{\omega_0} = \frac{1}{Q}\frac{\omega}{\omega_0}; \quad \frac{\omega L}{R} = \frac{\omega_0 L}{R}\frac{\omega}{\omega_0} = Q\frac{\omega}{\omega_0}$$

and rearrange equation 6.29 as follows:

$$\frac{\mathbf{V}_o}{\mathbf{V}_i}(j\omega) = \frac{j\omega C R}{1 + j\omega C R + (j\omega)^2 LC} = \frac{1}{\frac{1}{j\omega C R} + 1 + \frac{j\omega L}{R}} = \frac{1}{1 + jQ\left(\frac{\omega}{\omega_0} - \frac{\omega_0}{\omega}\right)}$$

$$\textbf{(6.34)}$$

In equation 6.34, the resonant frequency, $\omega_0$, corresponds to the center frequency of the filter, while $Q$, the quality factor, indicates the *sharpness* of the resonance,

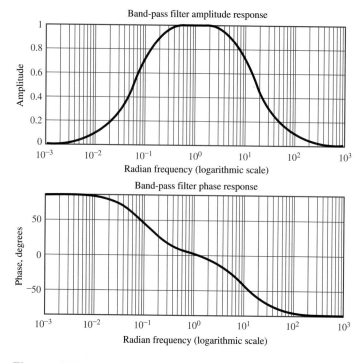

**Figure 6.20** Frequency response of *RLC* band-pass filter

that is, how narrow or wide the shape of the pass-band of the filter is. The width of the pass-band is also referred to as the *bandwidth*, and it can easily be shown that the bandwidth of the filter is given by the expression

$$B = \frac{\omega_0}{Q} \tag{6.35}$$

Thus, a high-$Q$ filter has a narrow bandwidth, while a low-$Q$ filter has a large bandwidth and is therefore less selective. The quality factor of a filter provides an immediate indication of the nature of the filter. The following examples illustrate the significance of these parameters in the response of various $RLC$ filters.

### EXAMPLE 6.6 Frequency Response of Band-Pass Filter

**Problem**

Compute the frequency response of the band-pass filter of Figure 6.19 for two sets of component values.

**Solution**

***Known Quantities:***

(a)  $R = 1 \text{ k}\Omega; C = 10 \ \mu\text{F}; L = 5 \text{ mH}.$

(b)  $R = 10 \ \Omega; C = 10 \ \mu\text{F}; L = 5 \text{ mH}.$

***Find:*** The frequency response, $H_V(j\omega)$.

**Assumptions:** None.

**Analysis:** We write the frequency response of the band-pass filter as in equation 6.29:

$$H_V(j\omega) = \frac{\mathbf{V}_o}{\mathbf{V}_i}(j\omega) = \frac{j\omega CR}{1 + j\omega CR + (j\omega)^2 LC}$$

$$= \frac{\omega CR}{\sqrt{(1 - \omega^2 LC)^2 + (\omega CR)^2}} \angle \left[\frac{\pi}{2} - \arctan\left(\frac{\omega CR}{1 - \omega^2 LC}\right)\right]$$

We can now evaluate the response for two different values of the series resistance. The frequency response plots for case $a$ (large series resistance) are shown in Figure 6.21. Those for case $b$ (small series resistance) are shown in Figure 6.22. Let us calculate some quantities for each case. Since $L$ and $C$ are the same in both cases, the *resonant frequency* of the two circuits will be the same:

$$\omega_0 = \frac{1}{\sqrt{LC}} = 4.47 \times 10^3 \text{ rad/s}$$

On the other hand, the *quality factor*, $Q$, will be substantially different:

$$Q_a = \omega_0 CR \approx 0.45 \qquad \text{case } a$$
$$Q_b = \omega_0 CR \approx 45 \qquad \text{case } b$$

From these values of $Q$ we can calculate the approximate bandwidth of the two filters:

$$B_a = \frac{\omega_0}{Q_a} \approx 10{,}000 \text{ rad/s} \qquad \text{case } a$$
$$B_b = \frac{\omega_0}{Q_b} \approx 100 \text{ rad/s} \qquad \text{case } b$$

The frequency response plots in Figures 6.21 and 6.22 confirm these observations.

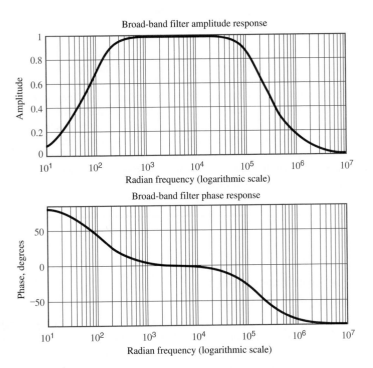

**Figure 6.21** Frequency response of broad-band band-pass filter of Example 6.6

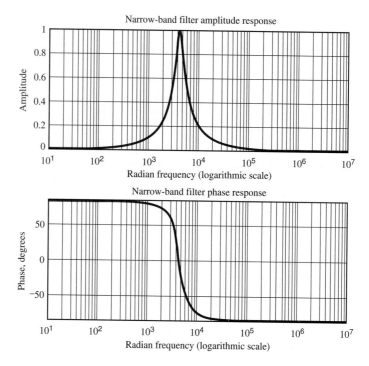

Figure 6.22 Frequency response of narrow-band band-pass filter of Example 6.6

**Comments:** It should be apparent that, while at the higher and lower frequencies most of the amplitude of the input signal is filtered from the output, at the mid-band frequency (4,500 rad/s) most of the input signal amplitude passes through the filter. The first band-pass filter analyzed in this example would "pass" the *mid-band range* of the audio spectrum, while the second would pass only a very narrow band of frequencies around the **center frequency** of 4,500 rad/s. Such narrow-band filters find application in **tuning circuits,** such as those employed in conventional AM radios (although at frequencies much higher than that of the present example). In a tuning circuit, a narrow-band filter is used to tune in a frequency associated with the **carrier** of a radio station (for example, for a station found at a setting of "AM 820," the carrier wave transmitted by the radio station is at a frequency of 820 kHz). By using a variable capacitor, it is possible to tune in a range of carrier frequencies and therefore select the preferred station. Other circuits are then used to decode the actual speech or music signal modulated on the carrier wave; some of these will be discussed in Chapter 8.

**FOCUS ON MEASUREMENTS**

## AC Line Interference Filter

**FIND IT ON THE WEB**

Problem:

One application of narrow-band filters is in rejecting interference due to AC line power. Any undesired 60-Hz signal originating in the AC line power can cause serious interference in sensitive instruments. In medical instruments such as the **electrocardiograph,** 60-Hz notch filters are often provided to reduce the effect of this interference[2] on cardiac measurements. Figure 6.23 depicts a circuit in which the effect of 60-Hz noise is represented by way of a 60-Hz sinusoidal generator connected in series with a signal source ($\mathbf{V}_S$), representing the desired signal. In this example we design a 60-Hz narrow-band (or *notch*) filter to remove the unwanted 60-Hz noise.

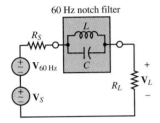

**Figure 6.23** 60-Hz notch filter

**Solution:**

***Known Quantities***— $R_S = 50\ \Omega$.

***Find***— Appropriate values of $L$ and $C$ for the notch filter.

***Assumptions***— None.

***Analysis***— To determine the appropriate capacitor and inductor values, we write the expression for the notch filter impedance:

$$Z_{\parallel} = Z_L \| Z_C = \frac{\dfrac{j\omega L}{j\omega C}}{j\omega L + \dfrac{1}{j\omega C}} = \frac{j\omega L}{1 - \omega^2 LC}.$$

Note that when $\omega^2 LC = 1$, the impedance of the circuit is infinite! The frequency

$$\omega_0 = \frac{1}{\sqrt{LC}}$$

is the resonant frequency of the $LC$ circuit. If this resonant frequency were selected to be equal to 60 Hz, then the series circuit would show an infinite impedance to 60-Hz currents, and would therefore block the interference signal, while passing most of the other frequency components. We thus select values of $L$ and $C$ that result in $\omega_0 = 2\pi \times 60$. Let $L = 100$ mH. Then

$$C = \frac{1}{\omega_0^2 L} = 70.36\ \mu F$$

The frequency response of the complete circuit is given below:

$$H_V(j\omega) = \frac{\mathbf{V}_o(j\omega)}{\mathbf{V}_i(j\omega)} = \frac{R_L}{R_S + R_L + Z_{\parallel}} = \frac{R_L}{R_S + R_L + \dfrac{j\omega L}{1 - \omega^2 LC}}$$

and is plotted in Figure 6.24.

[2] See Example 13.3 and Section 15.2 for further information on electrocardiograms and line noise, respectively.

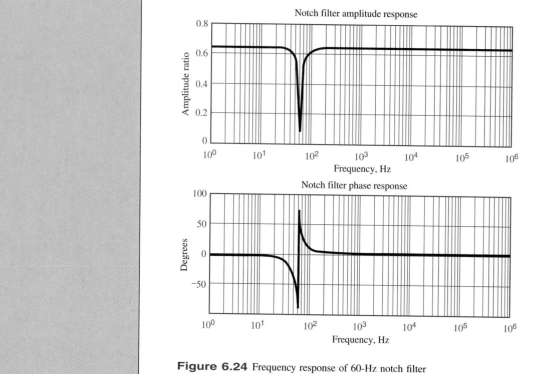

**Figure 6.24** Frequency response of 60-Hz notch filter

*Comments*— It would be instructive for you to calculate the response of the notch filter at frequencies in the immediate neighborhood of 60 Hz, to verify the attenuation effect of the notch filter.

### FOCUS ON MEASUREMENTS

### Seismic Transducer

This example illustrates the application of the frequency response idea to a practical displacement transducer. The frequency response of a **seismic displacement transducer** is analyzed, and it is shown that there is an analogy between the equations describing the mechanical transducer and those that describe a second-order electrical circuit.

The configuration of the transducer is shown in Figure 6.25. The transducer is housed in a case rigidly affixed to the surface of a body whose motion is to be measured. Thus, the case will experience the same displacement as the body, $x_i$. Inside the case, a small mass, $M$, rests on a spring characterized by stiffness $K$, placed in parallel with a damper, $B$. The wiper arm of a potentiometer is connected to the floating mass, $M$; the potentiometer is attached to the transducer case, so that the voltage $V_o$ is proportional to the *relative displacement of the mass with respect to the case, $x_o$.*

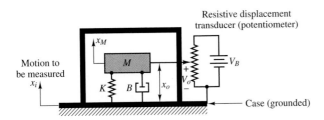

**Figure 6.25** Seismic displacement transducer

The equation of motion for the mass-spring-damper system may be obtained by summing all the forces acting on the mass $M$:

$$Kx_o + B\frac{dx_o}{dt} = M\frac{d^2x_M}{dt^2} = M\left(\frac{d^2x_i}{dt^2} - \frac{d^2x_o}{dt^2}\right)$$

where we have noted that the motion of the mass is equal to the difference between the motion of the case and the motion of the mass relative to the case itself; that is,

$$x_M = x_i - x_o$$

If we assume that the motion of the mass is sinusoidal, we may use phasor analysis to obtain the frequency response of the transducer by defining the phasor quantities

$$\mathbf{X}_i(j\omega) = |X_i|e^{j\phi_i} \qquad \text{and} \qquad \mathbf{X}_o(j\omega) = |X_o|e^{j\phi_o}$$

The assumption of a sinusoidal motion may be justified in light of the discussion of Fourier analysis in Section 6.1. If we then recall (from Chapter 4) that taking the derivative of a phasor corresponds to multiplying the phasor by $j\omega$, we can rewrite the second-order differential equation as follows:

$$M(j\omega)^2\mathbf{X}_o + B(j\omega)\mathbf{X}_o + K\mathbf{X}_o = M(j\omega)^2\mathbf{X}_i$$
$$(-\omega^2 M + j\omega B + K)\mathbf{X}_o = -\omega^2 M\mathbf{X}_i$$

and we can write an expression for the frequency response:

$$\frac{\mathbf{X}_o(j\omega)}{\mathbf{X}_i(j\omega)} = H(j\omega) = \frac{-\omega^2 M}{-\omega^2 M + j\omega B + K}$$

The frequency response of the transducer is plotted in Figure 6.26 for the component values $M = 0.005$ kg and $K = 1,000$ N/m and for three values of $B$:

$\quad B = 10$ N·s/m    (dotted line)

$\quad B = 2$ N·s/m    (dashed line)

and

$\quad B = 1$ N·s/m    (solid line)

The transducer clearly displays a high-pass response, indicating that for a sufficiently high input signal frequency, the measured displacement

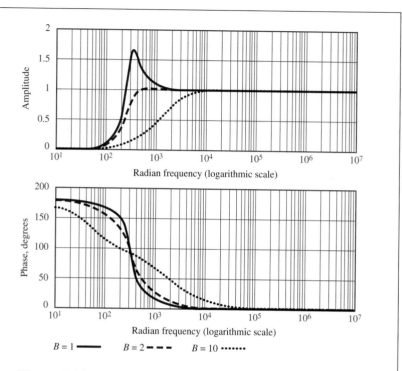

**Figure 6.26** Frequency response of seismic transducer

(proportional to the voltage $V_o$) is equal to the input displacement, $x_i$, which is the desired quantity. Note how sensitive the frequency response of the transducer is to changes in damping: as $B$ changes from 2 to 1, a sharp **resonant peak** appears around the frequency $\omega = 316$ rad/s (approximately 50 Hz). As $B$ increases to a value of 10, the amplitude response curve shifts to the right. Thus, this transducer, with the preferred damping given by $B = 2$, would be capable of correctly measuring displacements at frequencies above a minimum value, about 1,000 rad/s (or 159 Hz). The choice of $B = 2$ as the preferred design may be explained by observing that, ideally, we would like to obtain a constant amplitude response at all frequencies. The magnitude response that most closely approximates the ideal case in Figure 6.26 corresponds to $B = 2$. This concept is commonly applied to a variety of **vibration measurements.**

**FIND IT**

**ON THE WEB**

We now illustrate how a second-order electrical circuit will exhibit the same type of response as the seismic transducer. Consider the circuit shown in Figure 6.27. The frequency response for the circuit may be obtained by using the principles developed in the preceding sections:

$$\frac{\mathbf{V}_o}{\mathbf{V}_i}(j\omega) = \frac{j\omega L}{R + 1/j\omega C + j\omega L} = \frac{(j\omega L)(j\omega C)}{j\omega C R + 1 + (j\omega L)(j\omega C)}$$

$$= \frac{-\omega^2 L}{-\omega^2 L + j\omega R + 1/C}$$

Comparing this expression with the frequency response of the seismic

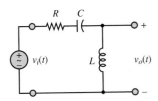

**Figure 6.27** Electrical
circuit analog of the seismic
transducer

transducer,

$$\frac{\mathbf{X}_o(j\omega)}{\mathbf{X}_i(j\omega)} = H(j\omega) = \frac{-\omega^2 M}{-\omega^2 M + j\omega B + K}$$

we find that there is a definite resemblance between the two. In fact, it is possible to draw an analogy between input and output motions and input and output voltages. Note also that the mass, $M$, plays a role analogous to that of the inductance, $L$. The damper, $B$, acts in analogy with the resistor, $R$; and the spring, $K$, is analogous to the inverse of the capacitance, $C$. This analogy between the mechanical system and the electrical circuit derives simply from the fact that the equations describing the two systems have the same form. Engineers often use such analogies to construct electrical *models*, or *analogs*, of physical systems. For example, to study the behavior of a large mechanical system, it might be easier and less costly to start by modeling the mechanical system with an inexpensive electrical circuit and testing the model, rather than the full-scale mechanical system.

## Decibel (dB) or Bode Plots

Frequency response plots are often displayed in the form of logarithmic plots, where the horizontal axis represents frequency on a logarithmic scale (to base 10) and the vertical axis represents the amplitude of the frequency response, in units of **decibels (dB).** In a dB plot, the ratio $|\mathbf{V}_{out}/\mathbf{V}_{in}|$ is given in units of decibels (dB), where

$$\left|\frac{\mathbf{V}_{out}}{\mathbf{V}_{in}}\right|_{dB} = 20\log_{10}\left|\frac{\mathbf{V}_{out}}{\mathbf{V}_{in}}\right| \qquad\qquad \textbf{(6.36)}$$

and this is plotted as a function of frequency on a $\log_{10}$ scale. Note that the use of decibels implies that one is measuring a *ratio*. Decibel plots are usually displayed on semilogarithmic paper, with decibels on the linear axis and frequency on the logarithmic axis.

    **Bode plots** are named after Hendrik W. Bode, a research mathematician who is among the pioneers in modern electrical network analysis and feedback amplifier design.

    Let us examine the appearance of dB plots for typical low-pass and high-pass filter circuits. From Figure 6.28, we can see that both plots have a very simple appearance: either the low-frequency part of the plot (for a low-pass filter) or the

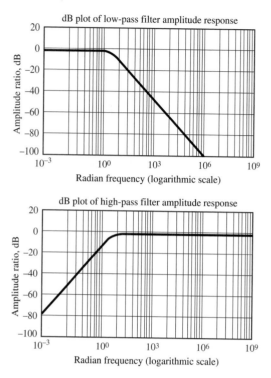

**Figure 6.28** dB magnitude plots of low- and high-pass filters

high-frequency part (for the high-pass filter) is well approximated by a flat line, indicating that for some range of frequencies, the filter has a constant amplitude response, equal to 1. Further, the filter cutoff frequency, $\omega_0$, appears quite clearly as the approximate frequency where the filter response starts to fall. The response of the circuit decreases (or increases) with a constant slope with respect to $\omega$ (on a logarithmic scale). For the high-pass and low-pass filters described earlier, this slope is equal to $\pm 20$ dB/decade ($-$ for the low-pass filter, $+$ for the high-pass), where a **decade** is a range of frequencies $f_1$ to $f_2$ such that

$$\frac{f_2}{f_1} = 10 \tag{6.37}$$

What kind of decrease in gain is $-20$ dB/decade? The expression

$$|H(j\omega)|_{dB} = -20 \text{ dB} \tag{6.38}$$

means that

$$-20 = 20\log_{10}|H(j\omega)|$$

or

$$|H(j\omega)| = 0.1 \tag{6.39}$$

That is, the gain decreases by a factor of 10 for every increase in frequency by a factor of 10. You see how natural these units are. Further, if $\omega_0$ is known, a plot of $|H(j\omega)|_{dB}$ versus $\omega$ (on a logarithmic scale) may be readily sketched using the asymptotic approximations of two straight lines, one of slope zero and the other with slope equal to $-20$ dB/decade, and with intersection at $\omega_0$. The homework problems and exercises provide a good number of practical examples of this technique.

## Check Your Understanding

**6.1**  Derive an expression for $H_I(j\omega) = \frac{I_L}{I_S}(j\omega)$ for the circuit of Figure 6.1.

**6.2**  Use the method of node voltages to derive $H_Y(j\omega)$ for the circuit of Figure 6.1.

**6.3**  Use the method of mesh currents to derive $H_V(j\omega)$ for the circuit of Figure 6.1.

**6.4**  Connect the filter of Example 6.3 to a 1-V sinusoidal source with internal resistance of 50 $\Omega$ to form a circuit similar to that of Figure 6.12. Determine the circuit cutoff frequency, $\omega_0$.

**6.5**  Determine the cutoff frequency for each of the four "prototype" filters shown in Figure 6.29. Which are high-pass and which are low-pass?

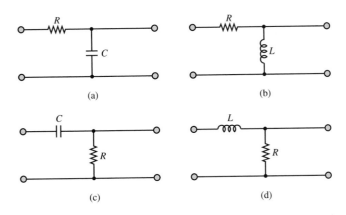

(a)          (b)

(c)          (d)

**Figure 6.29**

**6.6**  Show that it is possible to obtain a high-pass filter response simply by substituting an inductor for the capacitor in the circuit of Figure 6.9. Derive the frequency response for the circuit.

**6.7**  Determine the cutoff frequency for the high-pass $RC$ filter shown in Figure 6.30. [*Hint:* First find the frequency response in the form $j\omega a/(1 + j\omega b)$, where $a$ and $b$ are constants related to $R_1$, $R_2$, and $C_1$, and then solve numerically.] Sketch the amplitude and frequency responses.

**6.8**  A simple $RC$ low-pass filter is constructed using a 10-$\mu$F capacitor and a 2.2-k$\Omega$ resistor. Over what range of frequencies will the output of the filter be within 1 percent of the input signal amplitude (i.e., when will $V_L \geq 0.99V_S$)?

**6.9**  Compute the frequency at which the phase shift introduced by the circuit of Example 6.3 is equal to $-10°$.

**6.10**  Compute the frequency at which the output of the circuit of Example 6.3 is attenuated by 10 percent (i.e., $V_L = 0.9V_S$).

**6.11**  Compute the frequency at which the output of the circuit of Example 6.6 is attenuated by 10 percent (i.e., $V_L = 0.9V_S$).

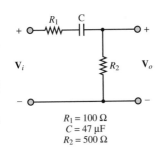

$R_1 = 100\ \Omega$
$C = 47\ \mu F$
$R_2 = 500\ \Omega$

**Figure 6.30**

**6.12** Compute the frequency at which the phase shift introduced by the circuit of Example 6.6 is equal to $20°$.

**6.13** Compute the frequencies $\omega_1$ and $\omega_2$ for the band-pass filter of Example 6.7 (with $R = 1 \text{ k}\Omega$) for equating the magnitude of the band-pass filter frequency response to $1/\sqrt{2}$ (this will result in a quadratic equation in $\omega$, which can be solved for the two frequencies).

## 6.3   COMPLEX FREQUENCY AND THE LAPLACE TRANSFORM

The transient analysis methods illustrated in the preceding chapter for first- and second-order circuits can become rather cumbersome when applied to higher-order circuits. Moreover, solving the differential equations directly does not reveal the strong connection that exists between the transient response and the frequency response of a circuit. The aim of this section is to introduce an alternate solution method based on the notions of complex frequency and of the **Laplace transform.** The concepts presented in this section will demonstrate that the frequency response of linear circuits is but a special case of the general transient response of the circuit, when analyzed by means of Laplace methods. In addition, the use of the Laplace transform method allows the introduction of *systems* concepts, such as poles, zeros, and transfer functions, that cannot be otherwise recognized.

### Complex Frequency

In Chapter 4, we considered circuits with sinusoidal excitations such as

$$v(t) = A \cos(\omega t + \phi) \tag{6.40}$$

which we also wrote in the equivalent phasor form

$$\mathbf{V}(j\omega) = Ae^{j\phi} = A\angle\phi \tag{6.41}$$

The two expressions just given are related by

$$v(t) = \text{Re}(\mathbf{V}e^{j\omega t}) \tag{6.42}$$

As was shown in Chapter 4, phasor notation is extremely useful in solving AC steady-state circuits, in which the voltages and currents are *steady-state sinusoids* of the form of equation 6.40. We now consider a different class of waveforms, useful in the transient analysis of circuits, namely, *damped sinusoids*. The most general form of a damped sinusoid is

$$v(t) = Ae^{\sigma t} \cos(\omega t + \phi) \tag{6.43}$$

As one can see, a damped sinusoid is a sinusoid multiplied by a real exponential, $e^{\sigma t}$. The constant $\sigma$ is real and is usually zero or negative in most practical circuits. Figures 6.31(a) and (b) depict the case of a damped sinusoid with negative $\sigma$ and with positive $\sigma$, respectively. Note that the case $\sigma = 0$ corresponds exactly to a sinusoidal waveform. The definition of phasor voltages and currents given in Chapter 4 can easily be extended to account for the case of damped sinusoidal waveforms by defining a new variable, $s$, called the *complex frequency*:

$$s = \sigma + j\omega \tag{6.44}$$

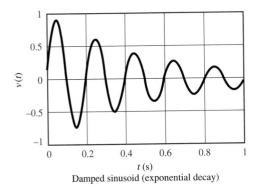

**Figure 6.31(a)** Damped sinusoid: negative $\sigma$

**Figure 6.31(b)** Damped sinusoid: positive $\sigma$

You may wish to compare this expression with the term $\alpha \pm j\beta$ in equation 5.65. Note that the special case $\sigma = 0$ corresponds to $s = j\omega$, that is, the familiar steady-state sinusoidal (phasor) case. We shall now refer to the complex variable $\mathbf{V}(s)$ as the **complex frequency domain** representation of $v(t)$. It should be observed that from the viewpoint of circuit analysis, the use of the Laplace transform is analogous to phasor analysis; that is, substituting the variable $s$ wherever $j\omega$ was used is the only step required to describe a circuit using the new notation.

## Check Your Understanding

**6.14**  Find the complex frequencies that are associated with

    a. $5e^{-4t}$

    b. $\cos 2\omega t$

    c. $\sin(\omega t + 2\theta)$

    d. $4e^{-2t}\sin(3t - 50°)$

    e. $e^{-3t}(2 + \cos 4t)$

**6.15**  Find $s$ and $\mathbf{V}(s)$ if $v(t)$ is given by

    a. $5e^{-2t}$

    b. $5e^{-2t}\cos(4t + 10°)$

    c. $4\cos(2t - 20°)$

**6.16**  Find $v(t)$ if

    a. $s = -2, \mathbf{V} = 2\angle 0°$

    b. $s = j2, \mathbf{V} = 12\angle -30°$

    c. $s = -4 + j3, \mathbf{V} = 6\angle 10°$

All the concepts and rules used in AC network analysis (see Chapter 4), such as impedance, admittance, KVL, KCL, and Thévenin's and Norton's theorems, carry over to the damped sinusoid case exactly. In the complex frequency domain, the current $\mathbf{I}(s)$ and voltage $\mathbf{V}(s)$ are related by the expression

$$\mathbf{V}(s) = Z(s)\mathbf{I}(s) \qquad\qquad \textbf{(6.45)}$$

where $Z(s)$ is the familiar impedance, with $s$ replacing $j\omega$. We may obtain $Z(s)$ from $Z(j\omega)$ by simply replacing $j\omega$ by $s$. For a resistance, $R$, the impedance is

$$Z_R(s) = R \tag{6.46}$$

For an inductance, $L$, the impedance is

$$Z_L(s) = sL \tag{6.47}$$

For a capacitance, $C$, it is

$$Z_C(s) = \frac{1}{sC} \tag{6.48}$$

Impedances in series or parallel are combined in exactly the same way as in the AC steady-state case, since we only replace $j\omega$ by $s$.

---

### EXAMPLE 6.7 Complex Frequency Notation

#### Problem

Use complex impedance ideas to determine the response of a series $RL$ circuit to a damped exponential voltage.

---

#### Solution

**Known Quantities:** Source voltage, resistor, inductor values.

**Find:** The time-domain expression for the series current, $i_L(t)$.

**Schematics, Diagrams, Circuits, and Given Data:** $v_s(t) = 10e^{-2t} \cos(5t)$ V; $R = 4\ \Omega$; $L = 2$ H.

**Assumptions:** None.

**Analysis:** The input voltage phasor can be represented by the expression

$$\mathbf{V}(s) = 10\angle 0 \text{ V}$$

The impedance seen by the voltage source is

$$Z(s) = R + sL = 4 + 2s$$

Thus, the series current is:

$$\mathbf{I}(s) = \frac{\mathbf{V}(s)}{Z(s)} = \frac{10}{4 + 2s} = \frac{10}{4 + 2(-2 + j5)} = \frac{10}{j10} = j1 = 1\angle\left(-\frac{\pi}{2}\right)$$

Finally, the time-domain expression for the current is:

$$i_L(t) = e^{-2t} \cos(5t - \pi/2) \text{ A}$$

**Comments:** The phasor analysis method illustrated here is completely analogous to the method introduced in Chapter 4, with the complex frequency $j\omega$ (steady-state sinusoidal frequency) related by $s$ (damped sinusoidal frequency).

Just as frequency response functions $H(j\omega)$ were defined in this chapter, it is possible to define a **transfer function,** $H(s)$. This can be a ratio of a voltage to a current, a ratio of a voltage to a voltage, a ratio of a current to a current, or a ratio of a current to a voltage. The transfer function $H(s)$ is a function of network elements and their interconnections. Using the transfer function and knowing the input (voltage or current) to a circuit, we can find an expression for the output either in the complex frequency domain or in the time domain. As an example, suppose $\mathbf{V}_i(s)$ and $\mathbf{V}_o(s)$ are the input and output voltages to a circuit, respectively, in complex frequency notation. Then

$$H(s) = \frac{\mathbf{V}_o(s)}{\mathbf{V}_i(s)} \tag{6.49}$$

from which we can obtain the output in the complex frequency domain by computing

$$\mathbf{V}_o(s) = H(s)\mathbf{V}_i(s) \tag{6.50}$$

If $\mathbf{V}_i(s)$ is a known damped sinusoid, we can then proceed to determine $v_o(t)$ by means of the method illustrated earlier in this section.

## Check Your Understanding

**6.17**  Given the transfer function $H(s) = 3(s+2)/(s^2+2s+3)$ and the input $\mathbf{V}_i(s) = 4\angle 0°$, find the forced response $v_o(t)$ if

    a. $s = -1$

    b. $s = -1 + j1$

    c. $s = -2 + j1$

**6.18**  Given the transfer function $H(s) = 2(s+4)/(s^2+4s+5)$ and the input $\mathbf{V}_i(s) = 6\angle 30°$, find the forced response $v_o(t)$ if

    a. $s = -4 + j1$

    b. $s = -2 + j2$

## The Laplace Transform

The Laplace transform, named after the French mathematician and astronomer Pierre Simon de Laplace, is defined by

$$\mathcal{L}[f(t)] = F(s) = \int_0^\infty f(t)e^{-st}\,dt \tag{6.51}$$

The function $F(s)$ is the Laplace transform of $f(t)$ and is a function of the complex frequency, $s = \sigma + j\omega$, considered earlier in this section. Note that the function $f(t)$ is defined only for $t \geq 0$. This definition of the Laplace transform applies to what is known as the **one-sided** or **unilateral Laplace transform,** since $f(t)$ is evaluated only for positive $t$. In order to conveniently express arbitrary functions only for positive time, we introduce a special function called the **unit step function,** $u(t)$, defined by the expression

$$u(t) = \begin{cases} 0 & t < 0 \\ 1 & t > 0 \end{cases} \tag{6.52}$$

## EXAMPLE 6.8 Computing a Laplace Transform

### Problem

Find the Laplace transform of $f(t) = e^{-at}u(t)$.

### Solution

**Known Quantities:** Function to be Laplace-transformed.

**Find:** $F(s) = \mathcal{L}\{f(t)\}$.

**Schematics, Diagrams, Circuits, and Given Data:** $f(t) = e^{-at}u(t)$.

**Assumptions:** None.

**Analysis:** From equation 6.51:

$$F(s) = \int_0^\infty e^{-at}e^{-st}\,dt = \int_0^\infty e^{-(s+a)t}\,dt = \frac{1}{s+a}e^{-(s+a)t}\Big|_0^\infty = \frac{1}{s+a}$$

**Comments:** Table 6.1 contains a list of common Laplace transform pairs.

## EXAMPLE 6.9 Computing a Laplace Transform

### Problem

Find the Laplace transform of $f(t) = \cos(\omega t)u(t)$.

### Solution

**Known Quantities:** Function to be Laplace-transformed.

**Find:** $F(s) = \mathcal{L}\{f(t)\}$.

**Schematics, Diagrams, Circuits, and Given Data:** $f(t) = \cos(\omega t)u(t)$.

**Assumptions:** None.

**Analysis:** Using equation 6.51 and applying Euler's identity to $\cos(\omega t)$ gives:

$$F(s) = \int_0^\infty \frac{1}{2}\left(e^{j\omega t} + e^{-j\omega t}\right)e^{-st}\,dt = \frac{1}{2}\int_0^\infty \left(e^{(-s+j\omega)t} + e^{(-s-j\omega)t}\right)dt$$

$$= \frac{1}{-s+j\omega}e^{-(s+j\omega)t}\Big|_0^\infty + \frac{1}{-s-j\omega}e^{-(s-j\omega)t}\Big|_0^\infty$$

$$= \frac{1}{-s+j\omega} + \frac{1}{-s-j\omega} = \frac{s}{s^2+\omega^2}$$

**Comments:** Table 6.1 contains a list of common Laplace transform pairs.

# Check Your Understanding

**6.19**  Find the Laplace transform of the following functions:

    a.  $u(t)$

    b.  $\sin(\omega t)u(t)$

    c.  $tu(t)$

**6.20**  Find the Laplace transform of the following functions:

    a.  $e^{-at}\sin \omega t\, u(t)$

    b.  $e^{-at}\cos \omega t\, u(t)$

Table 6.1  Laplace transform pairs

| $f(t)$ | $F(s)$ |
|---|---|
| $\delta(t)$ (unit impulse) | $1$ |
| $u(t)$ (unit step) | $\dfrac{1}{s}$ |
| $e^{-at}u(t)$ | $\dfrac{1}{s+a}$ |
| $\sin \omega t\, u(t)$ | $\dfrac{\omega}{s^2+\omega^2}$ |
| $\cos \omega t\, u(t)$ | $\dfrac{s}{s^2+\omega^2}$ |
| $e^{-at}\sin \omega t\, u(t)$ | $\dfrac{\omega}{(s+a)^2+\omega^2}$ |
| $e^{-at}\cos \omega t\, u(t)$ | $\dfrac{s+a}{(s+a)^2+\omega^2}$ |
| $tu(t)$ | $\dfrac{1}{s^2}$ |

From what has been said so far about the Laplace transform, it is obvious that we may compile a lengthy table of functions and their Laplace transforms by repeated application of equation 6.51 for various functions of time, $f(t)$. Then, we could obtain a wide variety of inverse transforms by matching entries in the table. Table 6.1 lists some of the more common **Laplace transform pairs.** The computation of the **inverse Laplace transform** is in general rather complex if one wishes to consider arbitrary functions of $s$. In many practical cases, however, it is possible to use combinations of known transform pairs to obtain the desired result.

## EXAMPLE  6.10  Computing an Inverse Laplace Transform

### Problem

Find the inverse Laplace transform of

$$F(s) = \frac{2}{s+3} + \frac{4}{s^2+4} + \frac{4}{s}$$

### Solution

***Known Quantities:***  Function to be inverse Laplace–transformed.

***Find:***  $f(t) = \mathcal{L}^{-1}\{F(s)\}$.

***Schematics, Diagrams, Circuits, and Given Data:***

$$F(s) = \frac{2}{s+3} + \frac{4}{s^2+4} + \frac{4}{s} = F_1(s) + F_2(s) + F_3(s)$$

***Assumptions:***  None.

***Analysis:***  Using Table 6.1, we can individually inverse-transform each of the elements of $F(s)$:

$$f_1(t) = 2\mathcal{L}^{-1}\left(\frac{1}{s+3}\right) = 2e^{-3t}u(t)$$

$$f_2(t) = 2\mathcal{L}^{-1}\left(\frac{2}{s^2+2^2}\right) = 2\sin(2t)u(t)$$

$$f_3(t) = 4\mathcal{L}^{-1}\left(\frac{1}{s}\right) = 4u(t)$$

Thus

$$f(t) = f_1(t) + f_2(t) + f_3(t) = \left(2e^{-3t} + 2\,\sin(2t) + 4\right)u(t).$$

---

## EXAMPLE 6.11 Computing an Inverse Laplace Transform

### Problem

Find the inverse Laplace transform of

$$F(s) = \frac{2s + 5}{s^2 + 5s + 6}$$

---

### Solution

***Known Quantities:*** Function to be inverse Laplace–transformed.

***Find:*** $f(t) = \mathcal{L}^{-1}\{F(s)\}$.

***Assumptions:*** None.

***Analysis:*** A direct entry for the function cannot be found in Table 6.1. In such cases, one must compute a *partial fraction expansion* of the function $F(s)$, and then individually transform each term in the expansion. A partial fraction expansion is the inverse operation of obtaining a common denominator, and is illustrated below.

$$F(s) = \frac{2s + 5}{s^2 + 5s + 6} = \frac{A}{s + 2} + \frac{B}{s + 3}$$

To obtain the constants $A$ and $B$, we multiply the above expression by each of the denominator terms:

$$(s + 2)F(s) = A + \frac{(s + 2)B}{s + 3}$$

$$(s + 3)F(s) = \frac{(s + 3)A}{s + 2} + B$$

From the above two expressions, we can compute $A$ and $B$ as follows:

$$A = (s + 2)F(s)|_{s=-2} = \left.\frac{2s + 5}{s + 3}\right|_{s=-2} = 1$$

$$B = (s + 3)F(s)|_{s=-3} = \left.\frac{2s + 5}{s + 2}\right|_{s=-3} = 1$$

Finally,

$$F(s) = \frac{2s + 5}{s^2 + 5s + 6} = \frac{1}{s + 2} + \frac{1}{s + 3}$$

and using Table 6.1, we compute

$$f(t) = \left(e^{-2t} + e^{-3t}\right)u(t)$$

## Check Your Understanding

**6.21**   Find the inverse Laplace transform of each of the following functions:

a. $F(s) = \dfrac{1}{s^2 + 5s + 6}$

b. $F(s) = \dfrac{s - 1}{s(s + 2)}$

c. $F(s) = \dfrac{3s}{(s^2 + 1)(s^2 + 4)}$

d. $F(s) = \dfrac{1}{(s + 2)(s + 1)^2}$

## Transfer Functions, Poles, and Zeros

It should be clear that the Laplace transform can be quite a convenient tool for analyzing the transient response of a circuit. The Laplace variable, $s$, is an extension of the steady-state frequency response variable $j\omega$ already encountered in this chapter. Thus, it is possible to describe the input-output behavior of a circuit using Laplace transform ideas in the same way in which we used frequency response ideas earlier. Now, we can define voltages and currents in the complex frequency domain as $\mathbf{V}(s)$ and $\mathbf{I}(s)$, and denote impedances by the notation $Z(s)$, where $s$ replaces the familiar $j\omega$. We define an extension of the frequency response of a circuit, called the transfer function, as the ratio of any input variable to any output variable, i.e.:

$$H_1(s) = \frac{\mathbf{V}_o(s)}{\mathbf{V}_i(s)} \qquad \text{or} \qquad H_2(s) = \frac{\mathbf{I}_o(s)}{\mathbf{V}_i(s)} \qquad \text{etc.} \qquad \textbf{(6.53)}$$

As an example, consider the circuit of Figure 6.32. We can analyze it using a method analogous to phasor analysis by defining impedances

$$Z_1 = R_1 \qquad Z_C = \frac{1}{sC} \qquad Z_L = sL \qquad Z_2 = R_2 \qquad \textbf{(6.54)}$$

Then, we can use mesh analysis methods to determine that

$$\mathbf{I}_o(s) = \mathbf{V}_i(s) \frac{Z_C}{(Z_L + Z_2)Z_C + (Z_L + Z_2)Z_1 + Z_1 Z_C} \qquad \textbf{(6.55)}$$

or, upon simplifying and substituting the relationships of equation 6.54,

$$H_2(s) = \frac{\mathbf{I}_o(s)}{\mathbf{V}_i(s)} = \frac{1}{R_1 L C s^2 + (R_1 R_2 C + L)s + R_1 + R_2} \qquad \textbf{(6.56)}$$

If we were interested in the relationship between the input voltages and, say, the capacitor voltage, we could similarly calculate

$$H_1(s) = \frac{\mathbf{V}_C(s)}{\mathbf{V}_i(s)} = \frac{sL + R_2}{R_1 L C s^2 + (R_1 R_2 C + L)s + R_1 + R_2} \qquad \textbf{(6.57)}$$

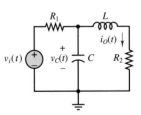

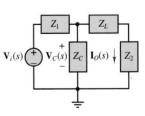

**Figure 6.32** A circuit and its Laplace transform domain equivalent

Note that a transfer function consists of a *ratio of polynomials;* this ratio can also be expressed in factored form, leading to the discovery of additional important properties of the circuit. Let us, for the sake of simplicity, choose numerical values for the components of the circuit of Figure 6.32. For example, let $R_1 = 0.5 \ \Omega$, $C = \frac{1}{4}$ F, $L = 0.5$ H, and $R_2 = 2 \ \Omega$. Then we can substitute these values into equation 6.57 to obtain

$$H_1(s) = \frac{0.5s + 2}{0.0625s^2 + 0.375s + 2.5} = 8 \left( \frac{s + 4}{s^2 + 6s + 40} \right) \qquad (6.58)$$

Equation 6.58 can be factored into products of first-order terms as follows:

$$H_1(s) = 8 \left[ \frac{s + 4}{s - 3.0000 + j5.5678)(s - 3.0000 - j5.5678)} \right] \qquad (6.59)$$

where it is apparent that the response of the circuit has very special characteristics for three values of $s$: $s = -4$; $s = +3.0000 + j5.5678$; and $s = +3.0000 - j5.5678$. In the first case, at the complex frequency $s = -4$, the numerator of the transfer function becomes zero, and the response of the circuit is zero, regardless of how large the input voltage is. We call this particular value of $s$ a **zero** of the transfer function. In the latter two cases, for $s = +3.0000 \pm j5.5678$, the response of the circuit becomes infinite, and we refer to these values of $s$ as **poles** of the transfer function.

It is customary to represent the response of electric circuits in terms of poles and zeros, since knowledge of the location of these poles and zeros is equivalent to knowing the transfer function and provides complete information regarding the response of the circuit. Further, if the poles and zeros of the transfer function of a circuit are plotted in the complex plane, it is possible to visualize the response of the circuit very effectively. Figure 6.33 depicts the pole–zero plot of the circuit of Figure 6.32; in plots of this type it is customary to denote zeros by a small circle, and poles by an "×."

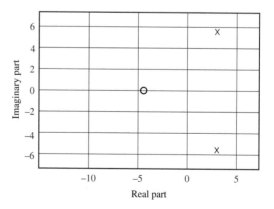

**Figure 6.33** Zero–pole plot for the circuit of Figure 6.32

The poles of a transfer function have a special significance, in that they are equal to the roots of the natural response of the system. They are also called the **natural frequencies** of the circuit. Example 6.13 illustrates this point.

## EXAMPLE 6.12 Poles of a Second-Order Circuit

### Problem

Determine the poles of the circuit of Example 5.11.

### Solution

**Known Quantities:** Values of resistor, inductor, and capacitor.

**Find:** Poles of the circuit.

**Assumptions:** None.

**Analysis:** The differential equation describing the circuit of Example 5.11 was found to be

$$\frac{d^2 i}{dt^2} + \frac{R}{L}\frac{di}{dt} + \frac{1}{LC}i = 0$$

with characteristic equation given by

$$s^2 + \frac{R}{L}s + \frac{1}{LC} = 0$$

Now, let us determine the transfer function of the circuit, say $\mathbf{V}_L(s)/\mathbf{V}_S(s)$. Applying the voltage divider rule, we can write

$$\frac{\mathbf{V}_L(s)}{\mathbf{V}_S(s)} = \frac{sL}{\dfrac{1}{sC} + R + sL} = \frac{s^2}{s^2 + \dfrac{R}{L}s + \dfrac{1}{LC}}$$

The denominator of this function, which determines the poles of the circuit, is identical to the characteristic equation of the circuit: The poles of the transfer function are identical to the roots of the characteristic equation!

$$s_{1,2} = -\frac{L}{2R} \pm \frac{1}{2}\sqrt{\left(\frac{R}{L}\right)^2 - \frac{4}{LC}}$$

**Comments:** Describing a circuit by means of its transfer function is completely equivalent to representing it by means of its differential equation. However, it is often much easier to derive a transfer function by basic circuit analysis than it is to obtain the differential equation of a circuit.

## CONCLUSION

- In many practical applications it is important to analyze the *frequency response* of a circuit, that is, the response of the circuit to sinusoidal signals of different frequencies. This can be accomplished quite effectively by means of the phasor analysis methods developed in Chapter 4, where the radian frequency, $\omega$, is now a variable. The frequency response of a circuit is then defined as the ratio of an output phasor quantity (voltage or current) to an input phasor quantity (voltage or current), as a function of frequency.

- One of the primary applications of frequency analysis is in the study of electrical *filters,* that is, circuits that can selectively attenuate signals in certain frequency regions. Filters can be designed, using standard resistors, inductors, and capacitors, to have one of four types of characteristics: low-pass, high-pass, band-pass, and band-reject. Such filters find widespread application in many practical engineering applications that involve signal conditioning. Filters will be studied in more depth in Chapters 12 and 15.

• Although the analysis of electrical circuits by means of phasors—that is, steady-state sinusoidal voltages and currents—is quite useful in many applications, there are situations where these methods are not appropriate. In particular, when a circuit (or another system) is subjected to an abrupt change in input voltage or current, different analysis methods must be employed to determine the *transient response* of the circuit. In this chapter, we have studied the analysis methods that are required to determine the transient response of first- and second-order circuits (that is, circuits containing one or two energy-storage elements, respectively). One method involves identifying the differential equation that describes the circuit during the transient period and recognizing important parameters, such as the time constant of a first-order circuit and the damping ratio and natural frequency of a second-order circuit. A second method exploits the idea of complex frequency and the Laplace transform.

# CHECK YOUR UNDERSTANDING ANSWERS

**CYU 6.1** $\quad H_I(j\omega) = \dfrac{Z_2}{Z_L + Z_2}$

**CYU 6.4** $\quad \omega_0 = 2{,}026.3$ rad/s

**CYU 6.5** $\quad$ (a) $\omega_0 = \dfrac{1}{RC}$ (low); (b) $\omega_0 = \dfrac{R}{L}$ (high);

$\qquad$ (c) $\omega_0 = \dfrac{1}{RC}$ (high); (d) $\omega_0 = \dfrac{R}{L}$ (low)

**CYU 6.6** $\quad H(j\omega) = \dfrac{\omega L/R}{\sqrt{1 + (\omega L/R)^2}}$

$\qquad \phi(j\omega) = 90° + \arctan \dfrac{-\omega L}{R}$

**CYU 6.7** $\quad \omega_0 = 35.46$ rad/s

**CYU 6.8** $\quad 0 \le \omega \le 6.48$ rad/s

**CYU 6.9** $\quad \omega = 375.17$ rad/s

**CYU 6.10** $\quad \omega = 1{,}030.49$ rad/s

**CYU 6.11** $\quad \omega = 51{,}878$ rad/s

**CYU 6.12** $\quad \omega = 69{,}032$ rad/s

**CYU 6.13** $\quad \omega_1 = 99.95$ rad/s; $\omega_2 = 200.1$ krad/s

**CYU 6.14** $\quad$ a. $-4$; b. $\pm j2\omega$; c. $\pm j\omega$; d. $-2 \pm j3$; e. $-3$ and $-3 \pm j4$

**CYU 6.15** $\quad$ a. $-2, 5\angle 0°$; b. $-2 + j4, 5\angle 10°$; c. $j2, 4\angle -20°$

**CYU 6.16** $\quad$ a. $2e^{-2t}$; b. $12\cos(2t - 30°)$; c. $6e^{-4t}\cos(3t + 10°)$

**CYU 6.17** $\quad$ a. $6e^{-t}$; b. $12\sqrt{2}e^{-t}\cos(t + 45°)$; c. $6e^{-2t}\cos(t + 135°)$

**CYU 6.18** $\quad$ a. $3e^{-4t}\cos(t + 165°)$; b. $8\sqrt{2}e^{-2t}\cos(2t - 105°)$

**CYU 6.19** $\quad$ a. $\dfrac{1}{s}$; b. $\dfrac{\omega}{s^2 + \omega^2}$; c. $\dfrac{1}{s^2}$

**CYU 6.20** $\quad$ a. $\dfrac{\omega}{(s + a)^2 + \omega^2}$; b. $\dfrac{(s + a)}{(s + a)^2 + \omega^2}$

**CYU 6.21**    a. $f(t) = (e^{-2t} - e^{-3t})u(t)$; b. $f(t) = \left(\frac{3}{2}e^{-2t} - \frac{1}{2}\right)u(t)$; c. $f(t) = (\cos t - \cos 2t)u(t)$; d. $f(t) = (e^{-2t} + te^{-t} - e^{-t})u(t)$

# HOMEWORK PROBLEMS

## Section 1: Frequency Response

### 6.1

a. Determine the frequency response $\mathbf{V}_{out}(j\omega)/\mathbf{V}_{in}(j\omega)$ for the circuit of Figure P6.1.
b. Plot the magnitude and phase of the circuit for frequencies between 1 and 100 rad/s on graph paper, with a linear scale for frequency.
c. Repeat part b, using semilog paper. (Place the frequency on the logarithmic axis.)
d. Plot the magnitude response on semilog paper with magnitude in dB.

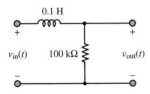

**Figure P6.1**

### 6.2    Repeat Problem 6.1 for the circuit of Figure P6.2.

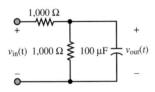

**Figure P6.2**

### 6.3    Repeat Problem 6.1 for the circuit of Figure P6.3.

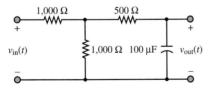

**Figure P6.3**

### 6.4    Assume in a certain frequency range that the ratio of output amplitude to input amplitude is proportional to $1/\omega^2$. What is the slope of the Bode plot in this frequency range, expressed in dB per decade?

### 6.5    Assume that the output amplitude of a circuit depends on frequency according to:

$$V = \frac{A\omega}{\sqrt{B + C\omega^2}}$$

Find:
a. The break frequency.
b. The slope of the Bode plot (in dB per decade) above the break frequency.
c. The slope of the Bode plot below the break frequency.
d. The high-frequency limit of V.

### 6.6    The function of a loudspeaker *crossover network* is to channel frequencies higher than a given crossover frequency, $f_c$, into the high-frequency speaker (tweeter) and frequencies below $f_c$ into the low-frequency speaker (woofer). Figure P6.6 shows an approximate equivalent circuit where the amplifier is represented as a voltage source with zero internal resistance and each speaker acts as an 8 Ω resistance. If the crossover frequency is chosen to be 1200 Hz, evaluate C and L. Hint: The break frequency would be a reasonable value to set as the crossover frequency.

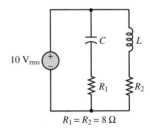

$R_1 = R_2 = 8\ \Omega$

**Figure P6.6**

### 6.7    Consider the circuit shown in Figure P6.7. Determine the resonance frequency and the bandwidth for the circuit.

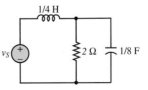

**Figure P6.7**

**6.8**  Repeat Problem 6.7 for the circuit of Figure P6.8.

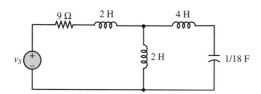

**Figure P6.8**

**6.9**

a. What is the equivalent impedance, $Z_{ab}$, of the filter of Figure P6.9?
b. At what frequency does the magnitude of the impedance go to infinity?

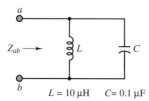

$L = 10 \, \mu\text{H}$    $C = 0.1 \, \mu\text{F}$

**Figure P6.9**

**6.10**  In the circuit shown in Figure P6.10:

a. Determine how the driving point impedance

$$Z(j\omega) = \frac{\mathbf{V}_i(j\omega)}{\mathbf{I}_i(j\omega)}$$

behaves at extremely high or low frequencies.
b. Find an expression for the driving point impedance.
c. Show that this expression can be manipulated into the form:

$$Z(j\omega) = Z_o(1 \pm jf(\omega))$$

where

$$Z_o = R \qquad f(\omega) = \frac{1}{\omega RC}$$

$$C = 0.5 \, \mu\text{F} \qquad R = 2 \, \text{k}\Omega$$

d. Determine the cutoff frequency $\omega = \omega_c$ at which $f(\omega_c) = 1$.
e. Determine the magnitude and angle of $Z(\omega)$ at $\omega = 100$ rad/s, 1 krad/s, and 10 krad/s.
f. Predict (without computing it) the magnitude and angle of $Z(j\omega)$ at $\omega = 10$ rad/s and 100 krad/s. Construct the Bode plot for the magnitude of the impedance [in dB!] as a function of the log of the frequency.

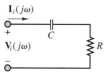

**Figure P6.10**

**6.11**  In the circuit shown in Figure P6.11:

a. Determine how the driving point impedance:

$$Z(j\omega) = \frac{\mathbf{V}_i(j\omega)}{\mathbf{I}_i(j\omega)}$$

behaves at extremely high or low frequencies.
b. Find an expression for the driving point impedance.
c. Show that this expression can be manipulated into the form:

$$Z(j\omega) = Z_o(1 + jf(\omega))$$

where

$$Z_o = R \qquad f(\omega) = \frac{\omega L}{R}$$

$$L = 2 \, \text{mH} \qquad R = 2 \, \text{k}\Omega$$

d. Determine the frequency $\omega = \omega_c$ at which $f(\omega_c) = 1$.
e. Determine the magnitude and angle of $Z(\omega)$ at $\omega = 100$ krad/s, 1 Mrad/s, and 10 Mrad/s.
f. Predict (without computing it) the magnitude and angle of $Z(j\omega)$ at $\omega = 10$ k rad/s and 100 M rad/s. Construct the Bode plot for the magnitude of the impedance [in dB] as a function of the log of the frequency.

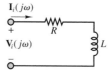

**Figure P6.11**

**6.12**  In the circuit shown in Figure P6.12, if:

$$L = 190 \, \text{mH} \qquad R_1 = 2.3 \, \text{k}\Omega$$
$$C = 55 \, \text{nF} \qquad R_2 = 1.1 \, \text{k}\Omega$$

a. Determine how the driving point or input impedance behaves at extremely high or low frequencies.
b. Find an expression for the driving point impedance in the form:

$$Z(j\omega) = Z_o\left[\frac{1 + jf_1[\omega]}{1 + jf_2[\omega]}\right]$$

$$Z_o = R_1 + \frac{L}{R_2 C}$$

$$f_1(\omega) = \frac{\omega^2 R_1 LC - R_1 - R_2}{\omega[R_1 R_2 C + L]}$$

$$f_2(\omega) = \frac{\omega^2 LC - 1}{\omega C R_2}$$

c. Determine the four cutoff frequencies at which $f_1(\omega) = +1$ or $-1$ and $f_2(\omega) = +1$ or $-1$.
d. Determine the resonant frequency of the circuit.

e. Plot the magnitude of the impedance (in dB) as a function of the log of the frequency, i.e., a Bode plot.

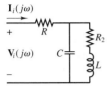

**Figure P6.12**

**6.13** Determine an expression for the circuit of Figure P6.13(a) for the equivalent impedance in standard form. Choose the Bode plot from Figure P6.13(b) that best describes the behavior of the impedance as a function of frequency and describe how (a simple one-line statement with no analysis is sufficient) you would obtain the resonant and cutoff frequencies and the magnitude of the impedance where it is constant over some frequency range. Label the Bode plot to indicate which feature you are discussing.

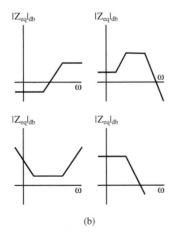

(a)

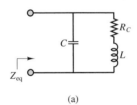

(b)

**Figure P6.13**

**6.14** In the circuit of Figure P6.14:

$R_1 = 1.3 \text{ k}\Omega$    $R_2 = 1.9 \text{ k}\Omega$

$C = 0.5182 \ \mu\text{F}$

Determine:

a. How the voltage transfer function:

$$H_V(j\omega) = \frac{\mathbf{V}_0(j\omega)}{\mathbf{V}_i(j\omega)}$$

behaves at extremes of high and low frequencies.

b. An expression for the voltage transfer function and show that it can be manipulated into the form:

$$H_v(j\omega) = \frac{H_o}{1 + jf(\omega)}$$

where

$$H_o = \frac{R_2}{R_1 + R_2} \qquad f(\omega) = \frac{\omega R_1 R_2 C}{R_1 + R_2}$$

c. The cutoff frequency at which $f(\omega) = 1$ and the value of $H_o$ in dB.

d. The value of the voltage transfer function at the cutoff frequency and at $\omega = 25$ rad/s, 250 rad/s, 25 krads, and 250 krad/s.

e. How the magnitude (in dB) and the angle of the transfer function behave at low frequencies, the cutoff frequency, and high frequencies.

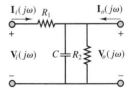

**Figure P6.14**

**6.15** The circuit shown in Figure P6.15 is not a filter but illustrates the undesirable effects of capacitances (and sometimes inductances) in a circuit. The circuit is a simple model of an amplifier state. Capacitors are often necessary for the proper operation of such circuits, or they may be unwanted but inherent in one of the circuit components. At high or low frequencies these capacitors adversely affect the proper operation of the circuit. The input impedance is used to demonstrate. Determine:

a. An expression, in the form:

$$Z_i(j\omega) = \frac{\mathbf{V}_i(j\omega)}{\mathbf{I}_i(j\omega)} = Z_o \left( \frac{1 + jf_1(\omega)}{1 + jf_2(\omega)} \right)$$

for the input impedance. Note the output current $= 0$.

b. The cutoff frequencies at which $f_1(\omega) = 1$ and $f_2(\omega) = 1$ if:

$R_1 = 1.3 \text{ k}\Omega$        $R_2 = 5.6 \text{ k}\Omega$

$C = 0.5 \ \mu\text{F}$        $g_m = 35 \text{ mS}$

c. The limiting value of $Z_i$ as $\omega$ increases toward infinity. As $\omega$ decreases toward zero.

d. The Bode plot for the input impedance.

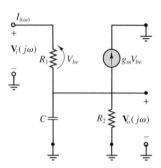

**Figure P6.15**

**6.16**   The circuit shown in Figure P6.16 is a very simplified model of a transistor amplifier stage. The capacitance $C$ is an internal effect of the transistor. It causes the transfer function:

$$H_v(j\omega) = \frac{V_o(j\omega)}{V_i(j\omega)}$$

where

$$R = 100 \text{ k}\Omega \qquad R_\pi = 750 \ \Omega$$
$$C = 0.125 \text{ nF} \qquad g_m = 7.5 \text{ mS}$$

to decrease at high frequencies as shown in the Bode plot. Determine:

a. The two cutoff frequencies.
b. The magnitude of the transfer function at very low and very high frequencies.

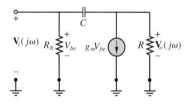

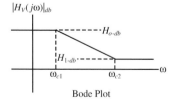

Bode Plot

**Figure P6.16**

**6.17**   The circuit shown in Figure P6.17 is a second-order filter because it has two reactive components ($L$ and $C$). A complete solution will not be attempted. However, determine:

a. The behavior of the voltage transfer function or gain at extremely high and low frequencies.
b. The output voltage $V_o$ if the input voltage has a

frequency where:

$$V_i = 7.07\angle\tfrac{\pi}{8} \text{ V} \qquad R_1 = 2.2 \text{ k}\Omega$$
$$R_2 = 3.8 \text{ k}\Omega \qquad X_c = 5 \text{ k}\Omega \qquad X_L = 1.25 \text{ k}\Omega$$

c. The output voltage if the frequency of the input voltage doubles so that:

$$X_C = 2.5 \text{ k}\Omega \qquad X_L = 2.5 \text{ k}\Omega$$

d. The output voltage if the frequency of the input voltage again doubles so that:

$$X_C = 1.25 \text{ k}\Omega \qquad X_L = 5 \text{ k}\Omega$$

e. The possible type of filter this might be, considering how the output voltage changes with frequency.

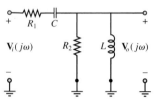

**Figure P6.17**

**6.18**   Are the filters shown in Figure P6.18 low-pass, high-pass, band-pass, or band-stop (notch) filters?

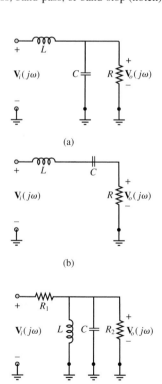

**Figure P6.18**

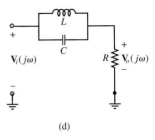

(d)

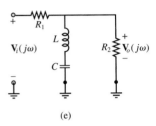

(e)

**Figure P6.18**

**6.19**  Determine if each of the circuits shown in Figure P6.19 is a low-pass, high-pass, band-pass, or band-stop (notch) filter.

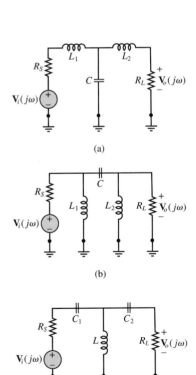

(a)

(b)

(c)

**Figure P6.19**

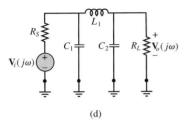

(d)

**Figure P6.19**

**6.20**  In the circuit shown in Figure P6.20, determine:

  a.  The voltage transfer function in the form

$$H_v(j\omega) = \frac{\mathbf{V}_o[j\omega]}{\mathbf{V}_i[j\omega]} = \frac{H_{vo}}{1 \pm jf(\omega)}$$

  b.  The gain or insertion loss in the pass-band in dB if

$$R_1 = R_2 = 16\ \Omega \qquad C = 0.47\ \mu\text{F}$$

  c.  The cutoff frequency.

  d.  The Bode plot, i.e., a semilog plot where the magnitude [in dB!] of the transfer function is plotted on a linear scale as a function of frequency on a log scale.

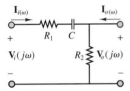

**Figure P6.20**

**6.21**  The circuit shown in Figure P6.21 is a high-pass filter in which

$$R_1 = 100\ \Omega \qquad R_L = 100\ \Omega$$
$$R_2 = 50\ \Omega \qquad C = 80\ \text{nF}$$

Determine:

  a.  The magnitude of the voltage transfer function, i.e., the gain or insertion loss, at very low and at very high frequencies.

  b.  The two cutoff frequencies.

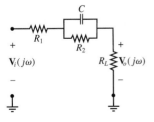

**Figure P6.21**

**6.22** Determine, for the filter circuit shown in Figure P6.22:

a. If this is a low-pass, high-pass, band-pass, or band-stop filter.

b. The magnitude (in dB!) of the voltage transfer function gain (or gain or insertion loss) in the pass-band if:

$$L = 11 \text{ mH} \qquad C = 0.47 \text{ nF}$$
$$R_1 = 2.2 \text{ k}\Omega \qquad R_2 = 3.8 \text{ k}\Omega$$

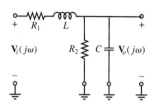

**Figure P6.22**

**6.23** In the filter circuit shown in Figure P6.23:

$$R_S = 100 \ \Omega \qquad R_L = 5 \text{ k}\Omega$$
$$R_c = 400 \ \Omega \qquad L = 1 \text{ mH}$$
$$C = 0.5 \text{ nF}$$

Determine the magnitude, in dB, of the voltage transfer function or gain at:

a. Very high and very low frequencies.

b. The resonant frequency.

c. What type of filter is this?

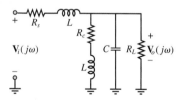

**Figure P6.23**

**6.24** In the filter circuit shown in Figure P6.23:

$$R_S = 100 \ \Omega \qquad R_L = 5 \text{ k}\Omega$$
$$R_c = 4 \ \Omega \qquad L = 1 \text{ mH}$$
$$C = 0.5 \text{ nF}$$

Determine the magnitude, in dB, of the voltage transfer function or gain at:

a. High frequencies.

b. Low frequencies.

c. The resonant frequency.

d. What type of filter is this?

**6.25** In the filter circuit shown in Figure P6.25:

$$R_S = 5 \text{ k}\Omega \qquad C = 56 \text{ nF}$$
$$R_L = 5 \text{ k}\Omega \qquad L = 9 \ \mu\text{H}$$

Determine:

a. An expression for the voltage transfer function:

$$H_v(j\omega) = \frac{\mathbf{V}_o(j\omega)}{\mathbf{V}_i(j\omega)}$$

b. The resonant frequency.

c. The cutoff frequencies.

d. The magnitude of the voltage transfer function (gain) at the two cutoff frequencies and the resonant frequency.

e. The bandwidth and $Q$.

f. The magnitude of the voltage transfer function at high, resonant, and low frequencies without using the expression above.

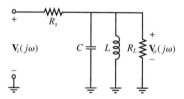

**Figure P6.25**

**6.26** In the filter circuit shown in Figure P6.25:

$$R_S = 5 \text{ k}\Omega \qquad C = 0.5 \text{ nF}$$
$$R_L = 5 \text{ k}\Omega \qquad L = 1 \text{ mH}$$

Determine:

a. An expression for the voltage transfer function:

$$H_v(j\omega) = \frac{\mathbf{V}_o(j\omega)}{\mathbf{V}_i(j\omega)}$$

b. The resonant frequency.

c. The cutoff frequencies.

d. The magnitude of the voltage transfer function (gain) at the two cutoff frequencies and the resonant frequency.

e. The bandwidth and $Q$.

f. The magnitude of the voltage transfer function at high, resonant, and low frequencies without using the expression above.

**6.27** In the filter circuit shown in Figure P6.27:

$$R_S = 500 \ \Omega \qquad R_L = 5 \text{ k}\Omega$$
$$R_c = 4 \text{ k}\Omega \qquad L = 1 \text{ mH}$$
$$C = 5 \text{ pF}$$

Determine the magnitude, in dB, of the voltage transfer

function or gain at:

$$H(j\omega) = \frac{\mathbf{V}_o(j\omega)}{\mathbf{V}_i(j\omega)}$$

a.  High frequencies.
b.  Low frequencies.
c.  The resonant frequency.
d.  What type of filter is this?

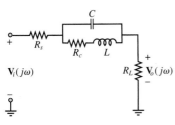

**Figure P6.27**

**6.28**   In the filter circuit shown in Figure P6.28, derive the equation for the voltage transfer function in standard form. Then, if

$R_S = 500\ \Omega$       $R_L = 5\ k\Omega$
$C = 5\ pF$           $L = 1\ mH$

determine the:

a.  Magnitude, in dB, of the voltage transfer function or gain at:

$$H(j\omega) = \frac{\mathbf{V}_o(j\omega)}{\mathbf{V}_i(j\omega)}$$

at high and low frequencies and at the resonant frequency.
b.  Resonant and cutoff frequencies.

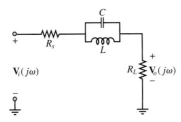

**Figure P6.28**

**6.29**   In the filter circuit shown in Figure P6.28, derive the equation for the voltage transfer function in standard form. Then, if

$R_s = 500\ \Omega$                $R_L = 5\ k\Omega$
$\omega_r = 12.1278\ Mrad/s$       $C = 68\ nF$
$L = 0.1\ \mu H$

determine the cutoff frequencies, bandwidth, and $Q$.

**6.30**   In the filter circuit shown in Figure P6.28, derive the equation for the voltage transfer function in standard form. Then, if

$R_s = 4.4\ k\Omega$      $R_L = 600\ \Omega$      $\omega_r = 25\ Mrad/s$
$C = 0.8\ nF$          $L = 2\ \mu H$

determine the cutoff frequencies, bandwidth, and $Q$.

**6.31**   In the bandstop (notch) filter shown in Figure P6.31:

$L = 0.4\ mH$        $R_c = 100\ \Omega$
$C = 1\ pF$         $R_s = R_L = 3.8\ k\Omega$

Determine:

a.  An expression for the voltage transfer function or gain in the form:

$$H_v(j\omega) = \frac{\mathbf{V}_o(j\omega)}{\mathbf{V}_i(j\omega)} = H_o \frac{1 + jf_1(\omega)}{1 + jf_2(\omega)}$$

b.  The magnitude of the voltage transfer function or gain at high and low frequencies and at the resonant frequency.
c.  The resonant frequency.
d.  The four cutoff frequencies.

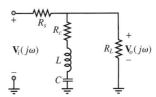

**Figure P6.31**

**6.32**   In the filter circuit shown in Figure P6.25:

$R_S = 5\ k\Omega$       $C = 5\ nF$
$R_L = 5\ k\Omega$       $L = 2\ mH$

Determine:

a.  An expression for the voltage transfer function:

$$H_V(j\omega) = \frac{\mathbf{V}_o(j\omega)}{\mathbf{V}_i(j\omega)}$$

b.  The resonant frequency.
c.  The cutoff frequencies.
d.  The magnitude of the voltage transfer function (gain) at the two cutoff frequencies and the resonant frequency.
e.  The bandwidth and $Q$.
f.  The magnitude of the voltage transfer function at high, resonant, and low frequencies without using the expression above.

**6.33**   In the filter circuit shown in Figure P6.28, derive the equation for the voltage transfer function in standard form. Then, if

$R_s = 500 \, \Omega$          $R_L = 1 \, k\Omega$

$\omega_r = 12.1278 \, \text{Mrad/s}$      $C = 470 \, \text{nF}$

$L = 0.1 \, \mu H$

determine the cutoff frequencies, bandwidth, and $Q$.

**6.34**  In the filter circuit shown in Figure P6.28, derive the equation for the voltage transfer function in standard form. Then, if

$R_s = 2.2 \, k\Omega$      $R_L = 600 \, \Omega$      $\omega_r = 25 \, \text{Mrad/s}$
$C = 2 \, \text{nF}$        $L = 2 \, \mu H$

determine the cutoff frequencies, bandwidth, and $Q$.

**6.35**  A 60 Hz notch filter was discussed in Focus on Measurements: AC Line Interference Filter. If the inductor has a 0.2-$\Omega$ series resistance, and the capacitor has a 10-M$\Omega$ parallel resistance,

a. What is the impedance of the *nonideal* notch filter at 60 Hz?

b. How much of the 60-Hz interference signal will appear at $V_L$?

**6.36**  It is very common to see interference caused by the power lines, at a frequency of 60 Hz. This problem outlines the design of a notch filter, shown in Figure P6.36, to reject a band of frequencies around 60 Hz.

a. Write the impedance function for the filter of Figure P6.36 (the resistor $r_L$ represents the internal resistance of a practical inductor).

b. For what value of $C$ will the center frequency of the filter equal 60 Hz if $L = 100 \, \text{mH}$ and $r_L = 5 \, \Omega$?

c. Would the "sharpness," or selectivity, of the filter increase or decrease if $r_L$ were to increase?

d. Assume that the filter is used to eliminate the 60-Hz noise from a signal generator with output frequency of 1 kHz. Evaluate the frequency response $\mathbf{V}_L(j\omega)/\mathbf{V}_{in}(j\omega)$ at both frequencies if:

$v_g(t) = \sin(2\pi 1{,}000t) \, \text{V}$      $r_g = 50 \, \Omega$
$v_n(t) = 3\sin(2\pi 60t)$          $R_L = 300 \, \Omega$

and if $L$ and $C$ are as in part b.

e. Plot the magnitude frequency response $|\mathbf{V}_L(j\omega)/\mathbf{V}_{in}(j\omega)|$ in dB versus $\log(j\omega)$, and indicate the value of $|\mathbf{V}_L(j\omega)/\mathbf{V}_{in}(j\omega)|_{dB}$ at the frequencies 60 Hz and 1,000 Hz on your plot.

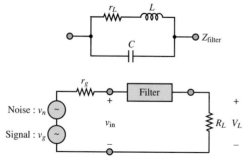

**Figure P6.36**

**6.37**  The circuit of Figure P6.37 is representative of an amplifier–speaker connection. The crossover circuit (filter) is a low-pass filter that is connected to a woofer. The filter's topography is known as a $\pi$ network.

a. Find the frequency response $\mathbf{V}_o(j\omega)/\mathbf{V}_S(j\omega)$.

b. If $C_1 = C_2 = C$, $R_S = R_L = 600 \, \Omega$, and $1/\sqrt{LC} = R/L = 1/RC = 2{,}000\pi$, plot $|\mathbf{V}_o(j\omega)/\mathbf{V}_S(j\omega)|$ in dB versus frequency (logarithmic scale) in the range $100 \, \text{Hz} \leq f \leq 10{,}000 \, \text{Hz}$.

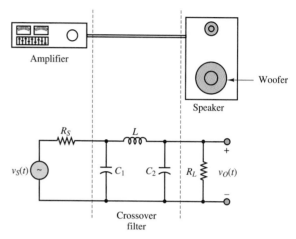

**Figure P6.37**

**6.38**  The $\pi$ filter of the circuit of Figure P6.38 is a high-pass filter that may be used to pass signals to the tweeter portion of a speaker.

a. Find the frequency response $\mathbf{V}_o(j\omega)/\mathbf{V}_S(j\omega)$.

b. If $L_1 = L_2 = L$, $R_S = R_L = 600 \, \Omega$, and $1/\sqrt{LC} = R/L = 1/RC = 2{,}000\pi$, plot $|\mathbf{V}_o(j\omega)/\mathbf{V}_S(j\omega)|$ in dB versus frequency (logarithmic scale) in the range $100 \, \text{Hz} \leq f \leq 10{,}000 \, \text{Hz}$.

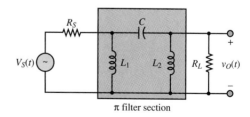

**Figure P6.38**

**6.39**  The circuit of Figure P6.39 is representative of an amplifier–speaker connection (see the left side of Figure P6.39). The crossover circuit (filter) is a high-pass filter that is connected to a tweeter. The filter's topography is known as a T network.

a. Find the frequency response $\mathbf{V}_o(j\omega)/\mathbf{V}_S(j\omega)$.

b. If $C_1 = C_2 = C$, $R_L = R_S = 600\ \Omega$, and $1/\sqrt{LC} = R/L = 1/RC = 2,000\pi$, plot $|\mathbf{V}_o(j\omega)/\mathbf{V}_S(j\omega)|$ in dB versus frequency (logarithmic scale) in the range 100 Hz $\le f \le 10,000$ Hz.

**6.40**   The T filter of the circuit of Figure P6.40 is a low-pass filter that may be used to pass signals to the woofer portion of a speaker.

a. Find the frequency response $\mathbf{V}_o(j\omega)/\mathbf{V}_S(j\omega)$.

b. If $L_1 = L_2 = L$, $R_S = R_L = 600\ \Omega$, and $1/\sqrt{LC} = R/L = 1/RC = 2,000\pi$, plot $|\mathbf{V}_o(j\omega)/\mathbf{V}_S(j\omega)|$ in dB versus frequency (logarithmic scale) in the range 100 Hz $\le f \le 10,000$ Hz.

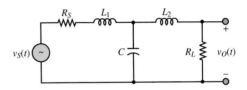

**Figure P6.40**

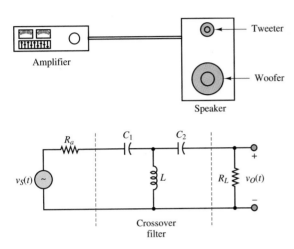

**Figure P6.39**

# C H A P T E R

# 7

# AC Power

The aim of this chapter is to introduce the student to simple AC power calculations, and to the generation and distribution of electric power. The chapter builds on the material developed in Chapter 4—namely, phasors and complex impedance—and paves the way for the material on electric machines in Chapters 16, 17, and 18.

The chapter starts with the definition of AC average and complex power and illustrates the computation of the power absorbed by a complex load; special attention is paid to the calculation of the power factor, and to power factor correction. The next subject is a brief discussion of ideal transformers and of maximum power transfer. This is followed by an introduction to three-phase power. The chapter ends with a discussion of electric power generation and distribution.

Upon completing this chapter, you should have mastered the following basic concepts:

- Calculation of real and reactive power for a complex load.
- Operation of ideal transformers.
- Impedance matching and maximum power transfer.
- Basic notions of residential circuit wiring, including grounding and safety.
- Configuration of electric power distribution networks.

## 7.1  POWER IN AC CIRCUITS

The objective of this section is to introduce the notion of AC power. As already mentioned in Chapter 4, 50- or 60-Hz AC power constitutes the most common form of electric power; in this section, the phasor notation developed in Chapter 4 will be employed to analyze the power absorbed by both resistive and complex loads.

### Instantaneous and Average Power

From Chapter 4, you already know that when a linear electric circuit is excited by a sinusoidal source, all voltages and currents in the circuit are also sinusoids of the same frequency as that of the excitation source. Figure 7.1 depicts the general form of a linear AC circuit. The most general expressions for the voltage and current delivered to an arbitrary load are as follows:

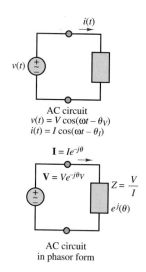

$$v(t) = V \cos(\omega t - \theta_V)$$
$$i(t) = I \cos(\omega t - \theta_I) \tag{7.1}$$

where $V$ and $I$ are the peak amplitudes of the sinusoidal voltage and current, respectively, and $\theta_V$ and $\theta_I$ are their phase angles. Two such waveforms are plotted in Figure 7.2, with unit amplitude and with phase angles $\theta_V = \pi/6$ and $\theta_I = \pi/3$. From here on, let us assume that the reference phase angle of the voltage source, $\theta_V$, is zero, and let $\theta_I = \theta$.

**AC circuit**
$v(t) = V \cos(\omega t - \theta_V)$
$i(t) = I \cos(\omega t - \theta_I)$

**AC circuit in phasor form**

**Figure 7.1** Circuit for illustration of AC power

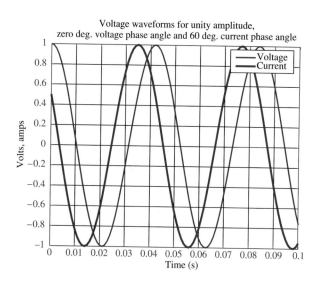

**Figure 7.2** Current and voltage waveforms for illustration of AC power

Since the **instantaneous power** dissipated by a circuit element is given by the product of the instantaneous voltage and current, it is possible to obtain a general expression for the power dissipated by an AC circuit element:

$$p(t) = v(t)i(t)$$
$$= VI \cos(\omega t) \cos(\omega t - \theta) \tag{7.2}$$

Equation 7.2 can be further simplified with the aid of trigonometric identities to yield

$$p(t) = \frac{VI}{2}\cos(\theta) + \frac{VI}{2}\cos(2\omega t - \theta) \qquad (7.3)$$

where $\theta$ is the difference in phase between voltage and current. Equation 7.3 illustrates how the instantaneous power dissipated by an AC circuit element is equal to the sum of an average component, $\frac{1}{2}VI\cos(\theta)$, plus a sinusoidal component, $\frac{1}{2}VI\cos(2\omega t - \theta)$, oscillating at a frequency double that of the original source frequency.

The instantaneous and average power are plotted in Figure 7.3 for the signals of Figure 7.2. The **average power** corresponding to the voltage and current signals of equation 7.1 can be obtained by integrating the instantaneous power over one cycle of the sinusoidal signal. Let $T = 2\pi/\omega$ represent one cycle of the sinusoidal signals. Then the average power, $P_{av}$, is given by the integral of the instantaneous power, $p(t)$, over one cycle:

$$P_{av} = \frac{1}{T}\int_0^T p(t)\, dt$$

$$= \frac{1}{T}\int_0^T \frac{VI}{2}\cos(\theta)\, dt + \frac{1}{T}\int_0^T \frac{VI}{2}\cos(2\omega t - \theta)\, dt \qquad (7.4)$$

$$P_{av} = \frac{VI}{2}\cos(\theta) \qquad \text{Average power} \qquad (7.5)$$

since the second integral is equal to zero and $\cos(\theta)$ is a constant.

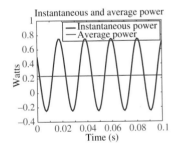

**Figure 7.3** Instantaneous and average power dissipation corresponding to the signals plotted in Figure 7.2.

As shown in Figure 7.1, the same analysis carried out in equations 7.1 to 7.3 can also be repeated using phasor analysis. In phasor notation, the current and voltage of equation 7.1 are given by

$$\mathbf{V}(j\omega) = Ve^{j0}$$
$$\mathbf{I}(j\omega) = Ie^{-j\theta} \qquad (7.6)$$

Note further that the impedance of the circuit element shown in Figure 7.1 is defined by the phasor voltage and current of equation 7.6 to be

$$Z = \frac{V}{I}e^{j(\theta)} = |Z|e^{j\theta_z} \qquad (7.7)$$

and therefore that the phase angle of the impedance is

$$\theta_Z = \theta \tag{7.8}$$

The expression for the average power obtained in equation 7.4 can therefore also be represented using phasor notation, as follows:

$$P_{\text{av}} = \frac{1}{2}\frac{V^2}{|Z|}\cos\theta = \frac{1}{2}I^2|Z|\cos\theta \tag{7.9}$$

## AC Power Notation

It has already been noted that AC power systems operate at a fixed frequency; in North America, this frequency is 60 cycles per second (Hz), corresponding to a radian frequency

$$\omega = 2\pi \cdot 60 = 377 \text{ rad/s} \qquad \text{AC power frequency} \tag{7.10}$$

In Europe and most other parts of the world, AC power is generated at a frequency of 50 Hz (this is the reason why some appliances will not operate under one of the two systems).

It will therefore be understood that for the remainder of this chapter the radian frequency, $\omega$, is fixed at 377 rad/s.

With knowledge of the radian frequency of all voltages and currents, it will always be possible to compute the exact magnitude and phase of any impedance in a circuit.

A second point concerning notation is related to the factor $\frac{1}{2}$ in equation 7.9. It is customary in AC power analysis to employ the rms value of the AC voltages and currents in the circuit (see Section 4.2). Use of the rms value eliminates the factor $\frac{1}{2}$ in power expressions and leads to considerable simplification. Thus, the following expressions will be used in this chapter:

$$V_{\text{rms}} = \frac{V}{\sqrt{2}} = \tilde{V} \tag{7.11}$$

$$I_{\text{rms}} = \frac{I}{\sqrt{2}} = \tilde{I} \tag{7.12}$$

$$P_{\text{av}} = \frac{1}{2}\frac{V^2}{|Z|}\cos\theta = \frac{\tilde{V}^2}{|Z|}\cos\theta \tag{7.13}$$

$$= \frac{1}{2}I^2|Z|\cos\theta = \tilde{I}^2|Z|\cos\theta = \tilde{V}\tilde{I}\cos\theta$$

Figure 7.4 illustrates the so-called **impedance triangle,** which provides a convenient graphical interpretation of impedance as a vector in the complex plane. From the figure, it is simple to verify that

$$R = |Z| \cos \theta \tag{7.14}$$

$$X = |Z| \sin \theta \tag{7.15}$$

Finally, the amplitudes of phasor voltages and currents will be denoted throughout this chapter by means of the rms amplitude. We therefore introduce a slight modification in the phasor notation of Chapter 4 by defining the following **rms phasor** quantities:

$$\tilde{\mathbf{V}} = V_{\text{rms}} e^{j\theta_V} = \tilde{V} e^{j\theta_V} = \tilde{V} \angle \theta_V \tag{7.16}$$

and

$$\tilde{\mathbf{I}} = I_{\text{rms}} e^{j\theta_I} = \tilde{I} e^{j\theta_I} = \tilde{I} \angle \theta_I \tag{7.17}$$

In other words,

> throughout the remainder of this chapter the symbols $\tilde{V}$ and $\tilde{I}$ will denote the rms value of a voltage or a current, and the symbols $\tilde{\mathbf{V}}$ and $\tilde{\mathbf{I}}$ will denote rms phasor voltages and currents.

Also recall the use of the symbol $\angle$ to represent the complex exponential. Thus, the sinusoidal waveform corresponding to the phasor current $\tilde{\mathbf{I}} = \tilde{I} \angle \theta_I$ corresponds to the time-domain waveform

$$i(t) = \sqrt{2} \tilde{I} \cos(\omega t + \theta_I) \tag{7.18}$$

and the sinusoidal form of the phasor voltage $\mathbf{V} = \tilde{V} \angle \theta_V$ is

$$v(t) = \sqrt{2} \tilde{V} \cos(\omega t + \theta_V) \tag{7.19}$$

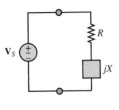

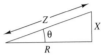

**Figure 7.4** Impedance triangle

## EXAMPLE 7.1   Computing Average and Instantaneous AC Power

### Problem

Compute the average and instantaneous power dissipated by the load of Figure 7.5.

### Solution

**Known Quantities:** Source voltage and frequency, load resistance and inductance values.

**Find:** $P_{\text{av}}$ and $p(t)$ for the $RL$ load.

**Schematics, Diagrams, Circuits, and Given Data:** $v(t) = 14.14 \sin(377t)$ V; $R = 4 \, \Omega$; $L = 8$ mH.

**Assumptions:** Use rms values for all phasor quantities in the problem.

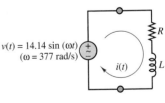

$v(t) = 14.14 \sin(\omega t)$
$(\omega = 377 \text{ rad/s})$

**Figure 7.5**

**Analysis:** First, we define the phasors and impedances at the frequency of interest in the problem, $\omega = 377$ rad/s:

$$\tilde{V} = 10\angle\left(-\frac{\pi}{2}\right) \qquad Z = R + j\omega L = 4 + j3 = 5\angle(0.644)$$

$$\tilde{I} = \frac{\tilde{V}}{Z} = \frac{10\angle\left(-\frac{\pi}{2}\right)}{5\angle(0.644)} = 2\angle(-2.215)$$

The average power can be computed from the phasor quantities:

$$P_{av} = \tilde{V}\tilde{I}\cos(\theta) = 10 \times 2 \times \cos(0.644) = 16 \text{ W}$$

The instantaneous power is given by the expression:

$$p(t) = v(t) \times i(t) = \sqrt{2} \times 10 \ \sin(377t) \times \sqrt{2} \times 2 \ \cos(377t - 2.215) \text{ W}$$

The instantaneous voltage and current waveforms and the instantaneous and average power are plotted in Figure 7.6.

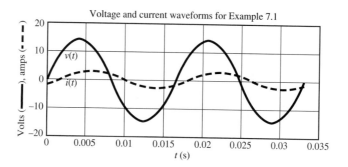

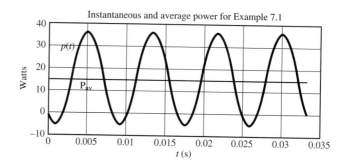

**Figure 7.6**

**Comments:** Please pay attention to the use of rms values in this example: It is very important to remember that we have defined phasors to have rms amplitude in power calculation. This is a standard procedure in electrical engineering practice.

Note that the instantaneous power can be negative for brief periods of time, even though the average power is positive.

# EXAMPLE  7.2  Computing Average AC Power

## Problem

Compute the average power dissipated by the load of Figure 7.7.

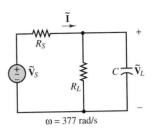

**Figure 7.7**

$\omega = 377$ rad/s

## Solution

**Known Quantities:** Source voltage, internal resistance and frequency, load resistance and inductance values.

**Find:** $P_{av}$ for the $RC$ load.

**Schematics, Diagrams, Circuits, and Given Data:** $\tilde{\mathbf{V}}_s = 110\angle 0$; $R_S = 2\ \Omega$; $R_L = 16\ \Omega$; $C = 100\ \mu F$.

**Assumptions:** Use rms values for all phasor quantities in the problem.

**Analysis:** First, we compute the load impedance at the frequency of interest in the problem, $\omega = 377$ rad/s:

$$Z_L = R\|\frac{1}{j\omega C} = \frac{R_L}{1 + j\omega C R_L} = \frac{16}{1 + j0.6032} = 13.7\angle(-0.543)\ \Omega$$

Next, we compute the load voltage, using the voltage divider rule:

$$\tilde{\mathbf{V}}_L = \frac{Z_L}{R_S + Z_L}\tilde{\mathbf{V}}_S = \frac{13.7\angle(-0.543)}{2 + 13.7\angle(-0.543)}110\angle(0) = 97.6\angle(-0.067)\ V$$

Knowing the load voltage, we can compute the average power according to:

$$P_{av} = \frac{|\tilde{\mathbf{V}}_L|^2}{|Z_L|}\cos(\theta) = \frac{97.6^2}{13.7}\cos(-0.543) = 595\ W$$

or, alternatively, we can compute the load current and calculate average power according to the equation below:

$$\tilde{\mathbf{I}}_L = \frac{\tilde{\mathbf{V}}_L}{Z_L} = 7.1\angle(0.476)\ A$$

$$P_{av} = |\tilde{\mathbf{I}}_L|^2|Z_L|\cos(\theta) = 7.1^2 \times 13.7 \times \cos(-0.543) = 595\ W$$

**Comments:** Please observe that it is very important to determine *load* current and/or voltage before proceeding to the computation of power; the internal source resistance in this problem causes the source and load voltages to be different.

**Focus on Computer-Aided Tools:** A file containing the computer-generated solution to this problem may be found in the CD-ROM that accompanies this book.

VIRTUAL LAB

---

## EXAMPLE 7.3 Computing Average AC Power

### Problem

Compute the average power dissipated by the load of Figure 7.8.

An AC circuit

---

### Solution

**Known Quantities:** Source voltage, internal resistance and frequency, load resistance, capacitance and inductance values.

**Find:** $P_{av}$ for the complex load.

**Schematics, Diagrams, Circuits, and Given Data:** $\tilde{\mathbf{V}}_s = 110\angle 0$ V; $R = 10\ \Omega$; $L = 0.05$ H; $C = 470\ \mu F$.

**Assumptions:** Use rms values for all phasor quantities in the problem.

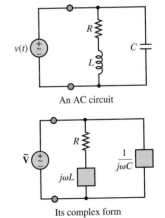

Its complex form

**Figure 7.8**

**Analysis:** First, we compute the load impedance at the frequency of interest in the problem, $\omega = 377$ rad/s:

$$Z_L = (R + j\omega L) \| \frac{1}{j\omega C} = \frac{\dfrac{(R + j\omega L)}{j\omega C}}{R + j\omega L + \dfrac{1}{j\omega C}}$$

$$= \frac{R + j\omega L}{-\omega^2 LC + j\omega CR} = 1.16 - j7.18$$

$$= 7.27\angle(-1.41)\ \Omega$$

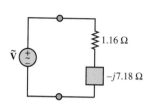

1.16 Ω

−j7.18 Ω

**Figure 7.9**

VIRTUAL LAB

Note that the equivalent load impedance consists of a capacitive load at this frequency, as shown in Figure 7.9. Knowing that the load voltage is equal to the source voltage, we can compute the average power according to:

$$P_{av} = \frac{|\tilde{\mathbf{V}}_L|^2}{|Z_L|} \cos(\theta) = \frac{110^2}{7.27} \cos(-1.41) = 266 \text{ W}$$

**Focus on Computer-Aided Tools:** A file containing the computer-generated solution to this problem may be found in the CD-ROM that accompanies this book.

## Power Factor

The phase angle of the load impedance plays a very important role in the absorption of power by a load impedance. As illustrated in equation 7.13 and in the preceding examples, the average power dissipated by an AC load is dependent on the cosine of the angle of the impedance. To recognize the importance of this factor in AC power computations, the term $\cos(\theta)$ is referred to as the **power factor (pf).** Note that the power factor is equal to 0 for a purely inductive or capacitive load and equal to 1 for a purely resistive load; in every other case,

$$0 < \text{pf} < 1 \tag{7.20}$$

Two equivalent expressions for the power factor are given in the following:

$$\text{pf} = \cos(\theta) = \frac{P_{av}}{\tilde{V}\tilde{I}} \qquad \text{Power factor} \tag{7.21}$$

where $\tilde{V}$ and $\tilde{I}$ are the rms values of the load voltage and current.

## Check Your Understanding

**7.1**  Show that the equalities in equation 7.9 hold when phasor notation is used.

**7.2**  Consider the circuit shown in Figure 7.10. Find the load impedance of the circuit, and compute the average power dissipated by the load.

**7.3**  Use the expression $P_{av} = \tilde{I}^2 |Z| \cos\theta$ to compute the average power dissipated by the load of Example 7.2.

**7.4**  Compute the power dissipated by the internal source resistance in Example 7.2.

**7.5**  Compute the power factor for an inductive load with $L = 100$ mH and $R = 0.4$ Ω. Assume $\omega = 377$ rad/s.

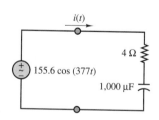

i(t)

4 Ω

155.6 cos (377t)

1,000 μF

**Figure 7.10**

## 7.2    Complex Power

The expression for the instantaneous power given in equation 7.3 may be further expanded to provide further insight into AC power. Using trigonometric identities, we obtain the following expressions:

$$p(t) = \frac{\tilde{V}^2}{|Z|}[\cos\theta + \cos\theta\cos(2\omega t) + \sin\theta\sin(2\omega t)]$$
$$= \tilde{I}^2|Z|[\cos\theta + \cos\theta\cos(2\omega t) + \sin\theta\sin(2\omega t)] \qquad \textbf{(7.22)}$$
$$= \tilde{I}^2|Z|\cos\theta(1 + \cos(2\omega t)) + \tilde{I}^2|Z|\sin\theta\sin(2\omega t)$$

Recalling the geometric interpretation of the impedance $Z$ of Figure 7.4, you may recognize that

$$|Z|\cos\theta = R$$

and $\qquad\qquad\qquad\qquad\qquad\qquad\qquad\qquad\qquad\qquad\qquad\qquad\qquad$ **(7.23)**

$$|Z|\sin\theta = X$$

are the resistive and reactive components of the load impedance, respectively. On the basis of this fact, it becomes possible to write the instantaneous power as:

$$p(t) = \tilde{I}^2 R(1 + \cos(2\omega t)) + \tilde{I}^2 X\sin(2\omega t)$$
$$= \tilde{I}^2 R + \tilde{I}^2 R\cos(2\omega t) + \tilde{I}^2 X\sin(2\omega t) \qquad \textbf{(7.24)}$$

The physical interpretation of this expression for the instantaneous power should be intuitively appealing at this point. As equation 7.23 suggests, the instantaneous power dissipated by a complex load consists of the following three components:

1.  An average component, which is constant; this is called the *average power* and is denoted by the symbol $P_{av}$:

    $$P_{av} = \tilde{I}^2 R \qquad \textbf{(7.25)}$$

    where $R = \text{Re}\,(Z)$.

2.  A time-varying (sinusoidal) component with zero average value that is contributed by the power fluctuations in the resistive component of the load and is denoted by $p_R(t)$:

    $$p_R(t) = \tilde{I}^2 R\cos 2\omega t \qquad \textbf{(7.26)}$$
    $$= P_{av}\cos 2\omega t$$

3.  A time-varying (sinusoidal) component with zero average value, due to the power fluctuation in the reactive component of the load and denoted by $p_X(t)$:

    $$p_X(t) = +\tilde{I}^2 X\sin(2\omega t) \qquad \textbf{(7.27)}$$
    $$= Q\sin 2\omega t$$

    where $X = \text{Im}\,(Z)$ and $Q$ is called the **reactive power.** *Note that since reactive elements can only store energy and not dissipate it, there is no net average power absorbed by X.*

Since $P_{av}$ corresponds to the power absorbed by the load resistance, it is also called the **real power,** measured in units of watts (W). On the other hand, $Q$ takes the

**Table 7.1** Real and reactive power

| Real power, $P_{av}$ | Reactive power, $Q$ |
|---|---|
| $\tilde{V}\tilde{I}\cos(\theta)$ | $\tilde{V}\tilde{I}\sin(\theta)$ |
| $\tilde{I}^2 R$ | $\tilde{I}^2 X$ |

name of *reactive power*, since it is associated with the load reactance. Table 7.1 shows the general methods of calculating $P$ and $Q$.

The units of $Q$ are **volt-amperes reactive,** or **VAR.** Note that $Q$ represents an exchange of energy between the source and the reactive part of the load; thus, no net power is gained or lost in the process, since the average reactive power is zero. In general, it is desirable to minimize the reactive power in a load. Example 7.5 will explain the reason for this statement.

The computation of AC power is greatly simplified by defining a fictitious but very useful quantity called the **complex power,** $S$:

$$S = \tilde{\mathbf{V}}\tilde{\mathbf{I}}^* \qquad \text{Complex power} \tag{7.28}$$

where the asterisk denotes the complex conjugate (see Appendix A). You may easily verify that this definition leads to the convenient expression

$$S = \tilde{V}\tilde{I}\cos\theta + j\tilde{V}\tilde{I}\sin\theta = \tilde{I}^2 R + j\tilde{I}^2 X = \tilde{I}^2 Z$$

or $\tag{7.29}$

$$S = P_{av} + jQ$$

The complex power $S$ may be interpreted graphically as a vector in the complex plane, as shown in Figure 7.11.

The magnitude of $S$, $|S|$, is measured in units of **volt-amperes (VA)** and is called **apparent power,** because this is the quantity one would compute by measuring the rms load voltage and currents without regard for the phase angle of the load. Note that the right triangle of Figure 7.11 is similar to the right triangle of Figure 7.4, since $\theta$ is the load impedance angle. The complex power may also be expressed by the product of the square of the rms current through the load and the complex load impedance:

$$S = \tilde{I}^2 Z$$

or $\tag{7.30}$

$$\tilde{I}^2 R + j\tilde{I}^2 X = \tilde{I}^2 Z$$

or, equivalently, by the ratio of the square of the rms voltage across the load to the complex conjugate of the load impedance:

$$S = \frac{\tilde{V}^2}{Z^*} \tag{7.31}$$

The power triangle and complex power greatly simplify load power calculations, as illustrated in the following examples.

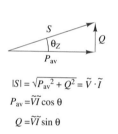

$$|S| = \sqrt{P_{av}^2 + Q^2} = \tilde{V} \cdot \tilde{I}$$

$$P_{av} = \tilde{V}\tilde{I}\cos\theta$$

$$Q = \tilde{V}\tilde{I}\sin\theta$$

**Figure 7.11** The complex power triangle

### EXAMPLE 7.4 Complex Power Calculations

#### Problem

Use the definition of complex power to calculate real and reactive power for the load of Figure 7.12.

## Solution

**Known Quantities:**  Source, load voltage and current.

**Find:**  $S = P_{av} + jQ$ for the complex load.

**Schematics, Diagrams, Circuits, and Given Data:**  $v(t) = 100 \, \cos(\omega t + 0.262)$ V; $i(t) = 2 \, \cos(\omega t - 0.262)$ A.

**Assumptions:**  Use rms values for all phasor quantities in the problem.

**Analysis:**  First, we convert the voltage and current into phasor quantities:

$$\tilde{\mathbf{V}} = \frac{100}{\sqrt{2}}\angle(0.262) \text{ V} \qquad \tilde{\mathbf{I}} = \frac{2}{\sqrt{2}}\angle(-0.262) \text{ A}$$

Next, we compute real and reactive power using the definitions of equation 7.13:

$$P_{av} = |\tilde{\mathbf{V}}||\tilde{\mathbf{I}}| \, \cos(\theta) = \frac{200}{2} \, \cos(0.524) = 86.6 \text{ W}$$

$$Q = |\tilde{\mathbf{V}}||\tilde{\mathbf{I}}| \, \sin(\theta) = \frac{200}{2} \, \sin(0.524) = 50 \text{ VAR}$$

Now we apply the definition of complex power (equation 7.28) to repeat the same calculation:

$$S = \tilde{\mathbf{V}}\tilde{\mathbf{I}}^* = \frac{100}{\sqrt{2}}\angle(0.262) \times \frac{2}{\sqrt{2}}\angle -(-0.262) = 100\angle(0.524)$$

$$= 86.6 + j50 \text{ W}$$

Therefore

$$P_{av} = 86.6 \text{ W} \qquad Q = 50 \text{ VAR}$$

**Comments:**  Note how the definition of complex power yields both quantities at one time.

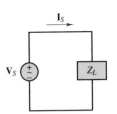

**Figure 7.12**

---

# EXAMPLE  7.5  Real and Reactive Power Calculations

## Problem

Use the definition of complex power to calculate real and reactive power for the load of Figure 7.13.

---

## Solution

**Known Quantities:**  Source voltage and resistance; load impedance.

**Find:**  $S = P_{av} + jQ$ for the complex load.

**Schematics, Diagrams, Circuits, and Given Data:**  $\tilde{\mathbf{V}}_S = 110\angle0$ V; $R_S = 2 \, \Omega$; $R_L = 5 \, \Omega$; $C = 2{,}000 \, \mu\text{F}$.

**Assumptions:**  Use rms values for all phasor quantities in the problem.

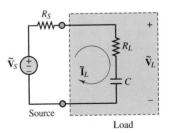

**Figure 7.13**

**Analysis:** Define the load impedance:

$$Z_L = R_L + \frac{1}{j\omega C} = 5 - j1.326 = 5.173\angle(-0.259) \ \Omega$$

Next, compute the load voltage and current:

$$\tilde{\mathbf{V}}_L = \frac{Z_L}{R_S + Z_L}\tilde{\mathbf{V}}_S = \frac{5 - j1.326}{7 - j1.326} \times 110 = 79.9\angle(-0.072) \text{ V}$$

$$\tilde{\mathbf{I}}_L = \frac{\tilde{\mathbf{V}}_L}{Z_L} = \frac{79.9\angle(-0.072)}{5\angle(-0.259)} = 15.44\angle(0.187) \text{ A}$$

Finally, we compute the complex power, as defined in equation 7.28:

$$S = \tilde{\mathbf{V}}_L\tilde{\mathbf{I}}_L^* = 79.9\angle(-0.072) \times 15.44\angle(-0.187) = 1,233\angle(-0.259)$$

$$= 1,192 - j316 \text{ W}$$

Therefore

$$P_{av} = 1,192 \text{ W} \qquad Q = -316 \text{ VAR}$$

**Comments:** Is the reactive power capacitive or inductive?

**Focus on Computer-Aided Tools:** A file containing the computer-generated solution to this problem may be found in the CD-ROM that accompanies this book.

VIRTUAL LAB

---

Although the reactive power does not contribute to any average power dissipation in the load, it may have an adverse effect on power consumption, because it increases the overall rms current flowing in the circuit. Recall from Example 7.2 that the presence of any source resistance (typically, the resistance of the line wires in AC power circuits) will cause a loss of power; the power loss due to this line resistance is unrecoverable and constitutes a net loss for the electric company, since the user never receives this power. The following example illustrates quantitatively the effect of such **line losses** in an AC circuit.

---

### EXAMPLE  7.6  Real Power Transfer for Complex Loads

#### Problem

Use the definition of complex power to calculate real and reactive power for the load of Figure 7.14. Repeat the calculation when the inductor is removed from the load, and compare the real power transfer between source and load for the two cases.

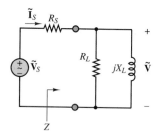

**Figure 7.14**

#### Solution

**Known Quantities:** Source voltage and resistance; load impedance.

**Find:**

1. $S_a = P_{ava} + jQ_a$ for the complex load.
2. $S_b = P_{avb} + jQ_b$ for the real load.
3. Compare $P_{av}/P_S$ for the two cases.

**Schematics, Diagrams, Circuits, and Given Data:** $\tilde{\mathbf{V}}_S = 110\angle(0)$ V; $R_S = 4\,\Omega$; $R_L = 10\,\Omega$; $jX_L = j6\,\Omega$.

**Assumptions:** Use rms values for all phasor quantities in the problem.

**Analysis:**

1. The inductor is part of the load. Define the load impedance.

$$Z_L = R_L \| j\omega L = \frac{10 \times j6}{10 + j6} = 5.145\angle(1.03)\,\Omega$$

Next, compute the load voltage and current:

$$\tilde{\mathbf{V}}_L = \frac{Z_L}{R_S + Z_L}\tilde{\mathbf{V}}_S = \frac{5.145\angle(1.03)}{4 + 5.145\angle(1.03)} \times 110 = 70.9\angle(0.444)\text{ V}$$

$$\tilde{\mathbf{I}}_L = \frac{\tilde{\mathbf{V}}_L}{Z_L}\frac{70.9\angle(0.444)}{5.145\angle(1.03)} = 13.8\angle(-0.586)\text{ A}$$

Finally, we compute the complex power, as defined in equation 7.28:

$$S_a = \tilde{\mathbf{V}}_L\tilde{\mathbf{I}}_L^* = 70.9\angle(0.444) \times 13.8\angle(0.586) = 978\angle(1.03)$$

$$= 503 + j839\text{ W}$$

Therefore

$$P_{\text{ava}} = 503\text{ W} \qquad Q_a = +839\text{ VAR}$$

2. The inductor is removed from the load (Figure 7.15). Define the load impedance:

$$Z_L = R_L = 10$$

Next, compute the load voltage and current:

$$\tilde{\mathbf{V}}_L = \frac{Z_L}{R_S + Z_L}\tilde{\mathbf{V}}_S = \frac{10}{4 + 10} \times 110 = 78.6\angle(0)\text{ V}$$

$$\tilde{\mathbf{I}}_L = \frac{\tilde{\mathbf{V}}_L}{Z_L} = \frac{78.6\angle(0)}{10} = 7.86\angle(0)\text{ A}$$

Finally, we compute the complex power, as defined in equation 7.28:

$$S_b = \tilde{\mathbf{V}}_L\tilde{\mathbf{I}}_L^* = 78.6\angle(0) \times 7.86\angle(0) = 617\angle(0) = 617\text{ W}$$

Therefore

$$P_{\text{avb}} = 617\text{ W} \qquad Q_b = 0\text{ VAR}$$

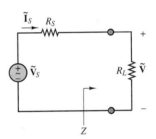

**Figure 7.15**

3. Compute the percent power transfer in each case. To compute the power transfer we must first compute the power delivered by the source in each case, $S_S = \tilde{\mathbf{V}}_S\tilde{\mathbf{I}}_S^*$. For Case 1:

$$\tilde{\mathbf{I}}_S = \frac{\tilde{\mathbf{V}}_S}{Z_{\text{total}}} = \frac{\tilde{\mathbf{V}}_S}{R_S + Z_L} = \frac{110}{4 + 5.145\angle(1.03)} = 13.8\angle(-0.586)\text{ A}$$

$$S_{Sa} = \tilde{\mathbf{V}}_S\tilde{\mathbf{I}}_S^* = 110 \times 13.8\angle - (-0.586) = 1{,}264 + j838\text{ W } = P_{Sa} + jQ_{Sa}$$

and the percent real power transfer is:

$$100 \times \frac{P_a}{P_{Sa}} = \frac{503}{1{,}264} = 39.8\%$$

For Case 2:

$$\tilde{\mathbf{I}}_S = \frac{\tilde{\mathbf{V}}_S}{Z_{\text{total}}} = \frac{\tilde{\mathbf{V}}}{R_S + R_L} = \frac{110}{4 + 10} = 7.86\angle(0) \text{ A}$$

$$S_{Sb} = \tilde{\mathbf{V}}_S \tilde{\mathbf{I}}_S^* = 110 \times 7.86 = 864 + j0 \text{ W} = P_{Sb} + jQ_{Sb}$$

and the percent real power transfer is:

$$100 \times \frac{P_b}{P_{Sb}} = \frac{617}{864} = 71.4\%$$

**Comments:** You can see that if it were possible to eliminate the reactive part of the impedance, the percentage of real power transferred from the source to the load would be significantly increased! A procedure that accomplishes this goal, called *power factor correction*, is discussed next.

**Focus on Computer-Aided Tools:** A file containing the computer-generated solution to this problem may be found in the CD-ROM that accompanies this book.

## Power Factor, Revisited

The power factor, defined earlier as the cosine of the angle of the load impedance, plays a very important role in AC power. A power factor close to unity signifies an efficient transfer of energy from the AC source to the load, while a small power factor corresponds to inefficient use of energy, as illustrated in Example 7.6. It should be apparent that if a load requires a fixed amount of real power, $P$, the source will be providing the smallest amount of current when the power factor is the greatest, that is, when $\cos\theta = 1$. If the power factor is less than unity, some additional current will be drawn from the source, lowering the efficiency of power transfer from the source to the load. However, it will be shown shortly that it is possible to correct the power factor of a load by adding an appropriate reactive component to the load itself.

Since the reactive power, $Q$, is related to the reactive part of the load, its sign depends on whether the load reactance is inductive or capacitive. This leads to the following important statement:

> If the load has an inductive reactance, then $\theta$ is positive and the current *lags* (or *follows*) the voltage. Thus, when $\theta$ and $Q$ are positive, the corresponding power factor is termed *lagging*. Conversely, a capacitive load will have a negative $Q$, and hence a negative $\theta$. This corresponds to a *leading* power factor, meaning that the load current *leads* the load voltage.

Table 7.2 illustrates the concept and summarizes all of the important points so far. In the table, the phasor voltage $\tilde{\mathbf{V}}$ has a zero phase angle and the current phasor is referenced to the phase of $\tilde{\mathbf{V}}$.

The following examples illustrate the computation of complex power for a simple circuit.

Table 7.2  Important facts related to complex power

|  | **Resistive load** | **Capacitive load** | **Inductive load** |
|---|---|---|---|
| Ohm's law | $\tilde{\mathbf{V}}_L = Z_L\tilde{\mathbf{I}}_L$ | $\tilde{\mathbf{V}}_L = Z_L\tilde{\mathbf{I}}_L$ | $\tilde{\mathbf{V}}_L = Z_L\tilde{\mathbf{I}}_L$ |
| Complex impedance | $Z_L = R_L$ | $Z_L = R_L - jX_L$ | $Z_L = R_L + jX_L$ |
| Phase angle | $\theta = \theta$ | $\theta < \theta$ | $\theta > \theta$ |
| Complex plane sketch | Im, $\theta=0$, $\tilde{\mathbf{I}}$ $\tilde{\mathbf{V}}$, Re | Im, $\tilde{\mathbf{I}}$, $\theta$ $\tilde{\mathbf{V}}$, Re | Im, $\tilde{\mathbf{V}}$, $\theta$, Re, $\tilde{\mathbf{I}}$ |
| Explanation | The current is in phase with the voltage. | The current "leads" the voltage. | The current "lags" the voltage. |
| Power factor | Unity | Leading, $< 1$ | Lagging, $< 1$ |
| Reactive power | 0 | Negative | Positive |

## EXAMPLE 7.7  Complex Power and Power Triangle

### Problem

Find the reactive and real power for the load of Figure 7.16. Draw the associated power triangle.

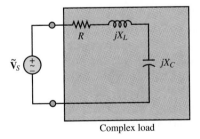

Complex load

**Figure 7.16**

### Solution

**Known Quantities:** Source voltage; load impedance.

**Find:** $S = P_{av} + jQ$ for the complex load.

**Schematics, Diagrams, Circuits, and Given Data:** $\tilde{\mathbf{V}}_S = 60\angle(0)$ V; $R = 3$ Ω; $jX_L = j9$ Ω; $jX_C = -j5$ Ω.

**Assumptions:** Use rms values for all phasor quantities in the problem.

**Analysis:** First, we compute the load current:

$$\tilde{I}_L = \frac{\tilde{V}_L}{Z_L} = \frac{60\angle(0)}{3+j9-j5} = \frac{60\angle(0)}{5\angle(0.644)} = 12\angle(-0.644)\text{ A}$$

Next, we compute the complex power, as defined in equation 7.28:

$$S = \tilde{V}_L\tilde{I}_L^* = 60\angle(0) \times 12\angle(0.644) = 720\angle(0.644) = 432 + j576\text{ W}$$

Therefore

$$P_{av} = 432\text{ W} \qquad Q = 576\text{ VAR}$$

If we observe that the total reactive power must be the sum of the reactive powers in each of the elements, we can write $Q = Q_C + Q_L$, and compute each of the two quantities as follows:

$$Q_C = |\tilde{I}_L|^2 \times X_C = (144)(-5) = -720\text{ VAR}$$

$$Q_L = |\tilde{I}_L|^2 \times X_L = (144)(9) = 1{,}296\text{ VAR}$$

and

$$Q = Q_L + Q_C = 576\text{ VAR}$$

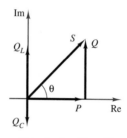

Note: $S = P_R + jQ_C + jQ_L$

**Figure 7.17**

**Comments:** The power triangle corresponding to this circuit is drawn in Figure 7.17. The vector diagram shows how the complex power, $S$, results from the vector addition of the three components, $P$, $Q_C$, and $Q_L$.

The distinction between leading and lagging power factors made in Table 7.2 is important, because it corresponds to opposite signs of the reactive power: $Q$ is positive if the load is inductive ($\theta > 0$) and the power factor is lagging; $Q$ is negative if the load is capacitive and the power factor is leading ($\theta < 0$). It is therefore possible to improve the power factor of a load according to a procedure called **power factor correction**—that is, by placing a suitable reactance in parallel with the load so that the reactive power component generated by the additional reactance is of opposite sign to the original load reactive power. Most often the need is to improve the power factor of an inductive load, because many common industrial loads consist of electric motors, which are predominantly inductive loads. This improvement may be accomplished by placing a capacitance in parallel with the load. The following example illustrates a typical power factor correction for an industrial load.

### EXAMPLE 7.8 Power Factor Correction

#### Problem

Calculate the complex power for the circuit of Figure 7.18 and correct the power factor to unity by connecting a parallel reactance to the load.

**Solution**

***Known Quantities:*** Source voltage; load impedance.

***Find:***

1. $S = P_{av} + jQ$ for the complex load.
2. Value of parallel reactance required for power factor correction resulting in pf $= 1$.

***Schematics, Diagrams, Circuits, and Given Data:*** $\tilde{\mathbf{V}}_S = 117\angle(0)$ V; $R_L = 50\ \Omega$; $jX_L = j86.7\ \Omega$.

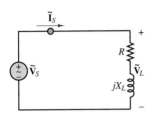

**Figure 7.18**

***Assumptions:*** Use rms values for all phasor quantities in the problem.

***Analysis:***

1. First, we compute the load impedance:

$$Z_L = R + jX_L = 50 + j86.7 = 100\angle(1.047)\ \Omega$$

Next, we compute the load current:

$$\tilde{\mathbf{I}}_L = \frac{\tilde{\mathbf{V}}_L}{Z_L} = \frac{117\angle(0)}{50 + j86.6} = \frac{117\angle(0)}{100\angle(1.047)} = 1.17\angle(-1.047)\ \text{A}$$

and the complex power, as defined in equation 7.28:

$$S = \tilde{\mathbf{V}}_L\tilde{\mathbf{I}}_L^* = 117\angle(0) \times 1.17\angle(1.047) = 137\angle(1.047) = 68.4 + j118.5\ \text{W}$$

Therefore

$$P_{av} = 68.4\ \text{W} \qquad Q = 118.5\ \text{VAR}$$

The power triangle corresponding to this circuit is drawn in Figure 7.19. The vector diagram shows how the complex power, $S$, results from the vector addition of the two components, $P$ and $Q_L$. To eliminate the reactive power due to the inductance, we will need to add an equal and opposite reactive power component, $-Q_L$, as described below.

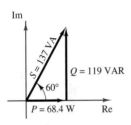

**Figure 7.19**

2. To compute the reactance needed for the power factor correction, we observe that we need to contribute a negative reactive power equal to $-118.5$ VAR. This requires a negative reactance, and therefore a capacitor with $Q_C = -118.5$ VAR. The reactance of such a capacitor is given by the expression:

$$X_C = \frac{|\tilde{\mathbf{V}}_L|^2}{Q_C} = -\frac{(117)^2}{118.5} = -115\ \Omega$$

and, since

$$C = -\frac{1}{\omega X_C}$$

we have

$$C = -\frac{1}{\omega X_C} = -\frac{1}{377 \times (-115)} = 23.1\ \mu\text{F}$$

***Comments:*** The power factor correction is illustrated in Figure 7.20. You can see that it is possible to eliminate the reactive part of the impedance, thus significantly increasing the percentage of real power transferred from the source to the load. Power factor correction is a very common procedure in electrical power systems.

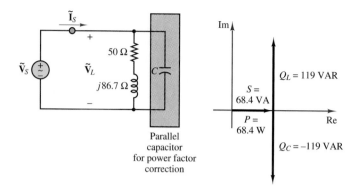

Figure 7.20 Power factor correction

**Focus on Computer-Aided Tools:** A file containing the computer-generated solution to this problem may be found in the CD-ROM that accompanies this book.

## EXAMPLE 7.9 Can a Series Capacitor Be Used for Power Factor Correction?

### Problem

The circuit of Figure 7.21 proposes the use of a series capacitor to perform power factor correction. Show why this is *not* a feasible alternative to the parallel capacitor approach demonstrated in Example 7.8.

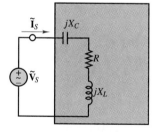

**Figure 7.21**

### Solution

**Known Quantities:** Source voltage; load impedance.

**Find:** Load (source) current.

**Schematics, Diagrams, Circuits, and Given Data:** $\tilde{\mathbf{V}}_S = 117\angle(0)$ V; $R_L = 50\ \Omega$; $jX_L = j86.7\ \Omega$; $jX_C = -j86.7\ \Omega$.

**Assumptions:** Use rms values for all phasor quantities in the problem.

**Analysis:** To determine the feasibility of the approach, we compute the load current and voltage, to observe any differences between the circuit of Figure 7.21 and that of Figure 7.20. First, we compute the load impedance:

$$Z_L = R + jX_L - jX_C = 50 + j86.7 - j86.7 = 50\ \Omega$$

Next, we compute the load (source) current:

$$\tilde{\mathbf{I}}_L = \tilde{\mathbf{I}}_S = \frac{\tilde{\mathbf{V}}_L}{Z_L} = \frac{117\angle(0)}{50} = 2.34\ \text{A}$$

**Comments:** Note that a twofold increase in the series current results from the addition of the series capacitor. This would result in a doubling of the power required by the generator, with respect to the solution found in Example 7.8. Further, in practice the parallel connection is much easier to accomplish, since a parallel element can be added externally, without the need for breaking the circuit.

## The Wattmeter

**FOCUS ON MEASUREMENTS**

The instrument used to measure power is called a *wattmeter.* The external part of a wattmeter consists of four connections and a metering mechanism that displays the amount of real power dissipated by a circuit. The external and internal appearance of a wattmeter are depicted in Figure 7.22. Inside the wattmeter are two coils: a current-sensing coil, and a voltage-sensing coil. In this example, we assume for simplicity that the impedance of the current-sensing coil, $Z_I$, is zero and the impedance of the voltage-sensing coil, $Z_V$, is infinite. In practice, this will not necessarily be true; some correction mechanism will be required to account for the impedance of the sensing coils.

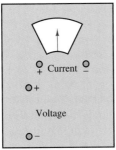
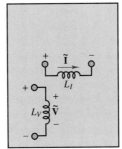

External connections          Wattmeter coils (inside)

**Figure 7.22**

A wattmeter should be connected as shown in Figure 7.23, to provide both current and voltage measurements. We see that the current-sensing coil is placed in series with the load and the voltage-sensing coil is placed in parallel with the load. In this manner, the wattmeter is seeing the current through and the voltage across the load. Remember that the power dissipated by a circuit element is related to these two quantities. The wattmeter, then, is constructed to provide a readout of the product of the rms values of the load current and the voltage, which is the real power absorbed by the load: $P = \text{Re}(S) = \text{Re}(\mathbf{VI}^*)$.

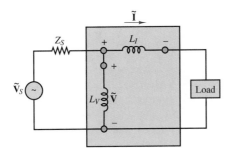

**Figure 7.23**

1.  For the circuit shown in Figure 7.24, show the connections of the wattmeter, and find the power dissipated by the load.

2.  Show the connections that will determine the power dissipated by $R_2$. What should the meter read?

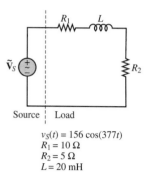

Source ┆ Load

$v_S(t) = 156 \cos(377t)$
$R_1 = 10\ \Omega$
$R_2 = 5\ \Omega$
$L = 20$ mH

**Figure 7.24**

**Solution:**

1.  To measure the power dissipated by the load, we must know the current through and the voltage across the entire load circuit. This means that the wattmeter must be connected as shown in Figure 7.25. The wattmeter should read:

$$P = \text{Re}\,(\tilde{\mathbf{V}}_S \tilde{\mathbf{I}}^*) = \text{Re}\left\{\left(\frac{156}{\sqrt{2}}\angle 0\right)\left(\frac{\frac{156}{\sqrt{2}}\angle 0}{R_1 + R_2 + j\omega L}\right)^*\right\}$$

$$= \text{Re}\left\{110\angle 0°\left(\frac{110\angle 0}{15 + j7.54}\right)^*\right\}$$

$$\text{Re}\left\{110\angle 0°\left(\frac{110\angle 0}{16.79\angle 0.466}\right)^*\right\} = \text{Re}\left(\frac{110^2}{16.79\angle -0.466}\right)$$

$$= \text{Re}\,(720.67\angle 0.466)$$

$$= 643.88\text{ W}$$

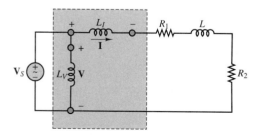

**Figure 7.25**

2. To measure the power dissipated by $R_2$ alone, we must measure the current through $R_2$ and the voltage across $R_2$ *alone*. The connection is shown in Figure 7.26. The meter will read

$$P = \tilde{I}^2 R_2 = \left( \frac{110}{(15^2 + 7.54^2)^{1/2}} \right)^2 \times 5 = \left( \frac{110^2}{15^2 + 7.54^2} \right) \times 5$$

$$= 215 \text{ W}$$

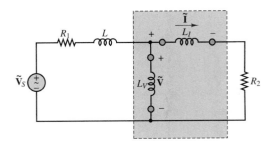

**Figure 7.26**

## How Hall-Effect Current Transducers Work[1]

In 1879, E. H. Hall noticed that if a conducting material is placed in a magnetic field perpendicular to a current flow, a voltage perpendicular to both the initial current flow and the magnetic field is developed. This voltage is called the *Hall voltage* and is directly proportional to both the strength of the magnetic field and the current. It results from the deflection of the moving charge carriers from their normal path by the magnetic field and its resulting transverse electric field.

To illustrate the physics involved, consider a confined stream of free particles each having a charge $e$ and an initial velocity $u_x$. A magnetic field in the $Z$ direction will produce a deflection in the $y$ direction. Therefore, a charge imbalance is created; this results in an electric field $E_y$. This electric field, the Hall field, will build up until the force it exerts on a charged particle counterbalances the force resulting from the magnetic field. Now subsequent particles of the same charge and velocity are no longer deflected. A steady state exists. Figure 7.27 depicts this effect.

The Hall effect occurs in any conductor. In most conductors the Hall voltage is very small and is difficult to measure. Dr. Warren E. Bulman, working with others, developed semiconductor compounds in the early 1950s that made the Hall effect practical for measuring magnetic fields.

The choice of materials for the active Hall element of most Hall probes is indium arsenide (InAs). This semiconductor compound is manufactured from highly refined elemental arsenic and indium. From an ingot of the semiconductor compound, thin slices are taken. These slices are then diced

[1]Courtesy Ohio Semitronics, Inc., Columbus, Ohio.

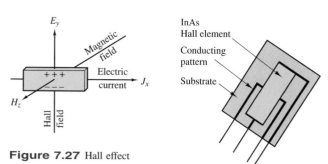

**Figure 7.27** Hall effect

**Figure 7.28** Hall-effect
probe

by an ultrasonic cutter into small, rectangular chips. These chips of indium
arsenide are then placed on a thin ceramic substrate and soldered to a
conducting pattern on the substrate, as shown in Figure 7.28.

Once made, the Hall probe is normally coated with epoxy to protect the
semiconductor compound and other components.

To use the probe, an electric current is passed through the length of the
InAs chip, as shown in Figure 7.29. Note that the contact areas for passing
current through the Hall element are made larger than the ones for detecting
the Hall voltage. Typically, current on the order of $10^{-1}$ A is passed through
the Hall element. This is known as the control current. Care must be taken
when using a Hall probe never to put the control current through the output.
Because the solder contacts for the voltage sensing are very small, the
control current can melt these solder joints. This may destroy or damage the
Hall probe.

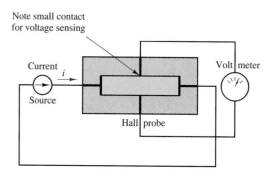

**Figure 7.29** Hall-effect probe circuit

A wire carrying a current will have a closed magnetic field around it, as
depicted in Figures 7.30 and 7.31. If a Hall probe is placed perpendicular to
the magnetic flux lines around a current-carrying conductor, then the Hall
probe will have a voltage output proportional to that magnetic field and the
control current through the Hall probe. Since the magnetic field is directly
proportional to the current, $I$, the output of the Hall probe is directly

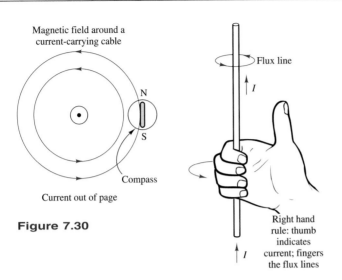

Magnetic field around a
current-carrying cable

N

S

Compass

Current out of page

**Figure 7.30**

Flux line

$I$

$I$

Right hand
rule: thumb
indicates
current; fingers
the flux lines

**Figure 7.31** Right-hand
rule

proportional to the current, $I$, and to the control current. We have a current
transducer.

Unfortunately, this method will provide adequate output only if the
current being measured, $I$, is of the order of $10^4$ amperes. Also, the strength
of the magnetic field is proportional to the inverse of the distance from the
center of the conductor.

**A practical current transducer** (Figure 7.32) can be made
by using a magnetic field concentrator with a Hall probe placed
in a gap. Typically, a laminated iron core with very low magnetic
retention is utilized. This arrangement makes a simple but very effective
current transducer.

Unfortunately, a Hall probe is temperature-sensitive. Hence, the voltage
output of the current transducer as described will be dependent upon the

FIND IT

ON THE WEB

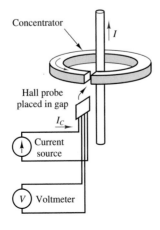

Concentrator

$I$

Hall probe
placed in gap

$I_C$

Current
source

$V$ Voltmeter

**Figure 7.32**

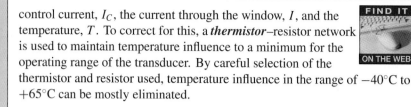

control current, $I_C$, the current through the window, $I$, and the temperature, $T$. To correct for this, a ***thermistor***–resistor network is used to maintain temperature influence to a minimum for the operating range of the transducer. By careful selection of the thermistor and resistor used, temperature influence in the range of $-40°C$ to $+65°C$ can be mostly eliminated.

The measurement and correction of the power factor for the load are an extremely important aspect of any engineering application in industry that requires the use of substantial quantities of electric power. In particular, industrial plants, construction sites, heavy machinery, and other heavy users of electric power must be aware of the power factor their loads present to the electric utility company. As was already observed, a low power factor results in greater current draw from the electric utility, and in greater line losses. Thus, computations related to the power factor of complex loads are of great practical utility to any practicing engineer. To provide you with deeper insight into calculations related to power factor, a few more advanced examples are given in the remainder of the section.

**FOCUS ON MEASUREMENTS**

## Power Factor

A capacitor is being used to correct the power factor to unity. The circuit is shown in Figure 7.33. The capacitor value is varied, and measurements of the total current are taken. Explain how it is possible to "zero in" on the capacitance value necessary to bring the power factor to unity just by monitoring the current $\tilde{\mathbf{I}}_S$.

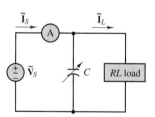

**Figure 7.33**

**Solution:**
The current through the load is

$$\tilde{\mathbf{I}}_L = \frac{\tilde{V}_S \angle 0°}{R + j\omega L} = \frac{\tilde{V}_S}{R^2 + \omega^2 L^2}(R - j\omega L)$$

$$= \frac{\tilde{V}_S R}{R^2 + \omega^2 L^2} - j\frac{\tilde{V}_S \omega L}{R^2 + \omega^2 L^2}$$

The current through the capacitor is

$$\tilde{\mathbf{I}}_C = \frac{\tilde{V}_S \angle 0°}{1/j\omega C} = j\tilde{V}_S \omega C$$

The source current to be measured is

$$\tilde{\mathbf{I}}_S = \tilde{\mathbf{I}}_L + \tilde{\mathbf{I}}_C = \frac{\tilde{V}_S R}{R^2 + \omega^2 L^2} + j \left( \tilde{V}_S \omega C - \frac{\tilde{V}_S \omega L}{R^2 + \omega^2 L^2} \right)$$

The magnitude of the source current is

$$\tilde{I}_S = \sqrt{ \left( \frac{\tilde{V}_S R}{R^2 + \omega^2 L^2} \right)^2 + \left( \tilde{V}_S \omega C - \frac{\tilde{V}_S \omega L}{R^2 + \omega^2 L^2} \right)^2 }$$

We know that when the load is a pure resistance, the current and voltage are in phase, the power factor is 1, and all the power delivered by the source is dissipated by the load as real power. This corresponds to equating the imaginary part of the expression for the source current to zero, or, equivalently, to the following expression:

$$\frac{\tilde{V}_S \omega L}{R^2 + \omega^2 L^2} = \tilde{V}_S \omega C$$

in the expression for $\tilde{I}_S$. Thus, the magnitude of the source current is actually a minimum when the power factor is unity! It is therefore possible to "tune" a load to a unity pf by observing the readout of the ammeter while changing the value of capacitor and selecting the capacitor value that corresponds to the lowest source current value.

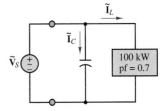

VIRTUAL LAB

---

## EXAMPLE  7.10  Power Factor Correction

### Problem

A capacitor is used to correct the power factor of the load of Figure 7.34. Determine the reactive power when the capacitor is not in the circuit, and compute the required value of capacitance for perfect pf correction.

---

### Solution

**Known Quantities:**  Source voltage; load power and power factor.

**Find:**

1. $Q$ when the capacitor is not in the circuit.

2. Value of capacitor required for power factor correction resulting in pf $= 1$.

**Schematics, Diagrams, Circuits, and Given Data:**  $\tilde{\mathbf{V}}_S = 480\angle(0)$; $P = 10^5$ W; pf $= 0.7$ lagging.

**Assumptions:**  Use rms values for all phasor quantities in the problem.

**Analysis:**

1. With reference to the power triangle of Figure 7.11, we can compute the reactive power of the load from knowledge of the real power and of the power factor, as

**Figure 7.34**

shown below:

$$|S| = \frac{P}{\cos(\theta)} = \frac{P}{\text{pf}} = \frac{10^5}{0.7} = 1.429 \times 10^5 \text{ VA}$$

Since the power factor is lagging, we know that the reactive power is positive (see Table 7.2), and we can calculate $Q$ as shown below:

$$Q = |S| \, \sin(\theta) \qquad \theta = \arccos(pf) = 0.795$$

$$Q = 1.429 \times 10^5 \times \sin(0.795) = 102 \text{ kVAR}$$

2. To compute the reactance needed for the power factor correction, we observe that we need to contribute a negative reactive power equal to $-102$ kVAR. This requires a negative reactance, and therefore a capacitor with $Q_C = -102$ kVAR. The reactance of such a capacitor is given by the expression:

$$X_C = \frac{|\tilde{\mathbf{V}}_L|^2}{Q_C} = \frac{(480)^2}{-102 \times 10^5} = -2.258$$

and, since

$$C = -\frac{1}{\omega X_C}$$

we have

$$C = -\frac{1}{\omega X_C} = -\frac{1}{377 \times -2.258} = 1,175 \ \mu\text{F}.$$

**VIRTUAL LAB**

**Comments:**  Note that it is not necessary to know the load impedance to perform power factor correction; it is sufficient to know the *apparent* power and the power factor.

**Focus on Computer-Aided Tools:**  A file containing the computer-generated solution to this problem may be found in the CD-ROM that accompanies this book.

---

## EXAMPLE  7.11  Power Factor Correction

### Problem

A second load is added to the circuit of Figure 7.34, as shown in Figure 7.35. Determine the required value of capacitance for perfect pf correction after the second load is added. Draw the phasor diagram showing the relationship between the two load currents and the capacitor current.

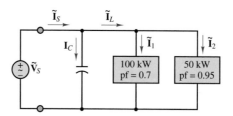

**Figure 7.35**

---

### Solution

**Known Quantities:**  Source voltage; load power and power factor.

**Find:**

1. Power factor correction capacitor

2. Phasor diagram

**Schematics, Diagrams, Circuits, and Given Data:** $\tilde{\mathbf{V}}_S = 480\angle(0)$ V; $P_1 = 10^5$ W; $pf_1 = 0.7$ lagging; $P_2 = 5 \times 10^4$ W; $pf_2 = 0.95$ leading.

**Assumptions:** Use rms values for all phasor quantities in the problem.

**Analysis:**

1. We first compute the two load currents, using the relationships given in equations 7.28 and 7.29:

$$P = |\tilde{\mathbf{V}}_S||\tilde{\mathbf{I}}_1^*|\cos(\theta_1);$$

$$|\tilde{\mathbf{I}}_1^*| = \frac{P_1}{|\tilde{\mathbf{V}}_S|\cos(\theta_1)};$$

$$\tilde{\mathbf{I}}_1 = \frac{P_1}{|\tilde{\mathbf{V}}_S|pf_1}\angle\left(\arccos(pf_1)\right) = \frac{10^5}{480 \times 0.7}\angle\left(\arccos(0.7)\right)$$
$$= 298\angle(0.795) \text{ A}$$

and, similarly

$$\tilde{\mathbf{I}}_2 = \frac{P_2}{|\tilde{\mathbf{V}}_S|pf_2}\angle-\left(\arccos(pf_2)\right) = \frac{5 \times 10^4}{480 \times 0.95}\angle-\left(\arccos(0.95)\right)$$
$$= 360\angle(-0.318) \text{ A}$$

where we have selected the positive value of arccos $(pf_1)$ because $pf_1$ is lagging, and the negative value of arccos $(pf_2)$ because $pf_2$ is leading. Now we compute the reactive power at each load:

$$|S_1| = \frac{P}{pf_1} = \frac{P}{\cos(\theta_1)} = \frac{10^5}{0.7} = 1.429 \times 10^5 \text{ VA}$$

$$|S_2| = \frac{P}{pf_2} = \frac{P}{\cos(\theta_2)} = \frac{5 \times 10^4}{0.95} = 1.634 \times 10^4 \text{ VA}$$

and from these values we can calculate $Q$ as shown below:

$$Q_1 = |S_1| \sin(\theta_1) \qquad \theta_1 = \arccos(pf_1) = 0.795$$

$$Q_1 = 1.429 \times 10^5 \times \sin(0.795) = 102 \text{ kVAR}$$

$$Q_2 = |S_2| \sin(\theta_2) \qquad \theta_2 = -\arccos(pf_2) = -0.318$$

$$Q_2 = 5.263 \times 10^4 \times \sin(-0.318) = -16.43 \text{ kVAR}$$

where, once again, $\theta_1$ is positive because $pf_1$ is lagging, $\theta_2$ is negative because $pf_2$ is leading (see Table 7.2).

The total reactive power is therefore $Q = Q_1 + Q_2 = 85.6 \text{ kVAR}$.

To compute the reactance needed for the power factor correction, we observe that we need to contribute a negative reactive power equal to $-85.6$ kVAR. This requires a negative reactance, and therefore a capacitor with $Q_C = -85.6$ kVAR. The reactance of such a capacitor is given by the expression:

$$X_C = \frac{|\tilde{\mathbf{V}}_S|^2}{Q_C} = \frac{(480)^2}{-85.6 \times 10^5} = -2.692$$

and, since

$$C = -\frac{1}{\omega X_C}$$

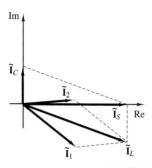

**Figure 7.36**

VIRTUAL LAB

we have

$$C = \frac{1}{\omega X_C} = -\frac{1}{377 \times (-2.692)} = 985.3 \ \mu F$$

2. To draw the phasor diagram, we need only to compute the capacitor current, since we have already computed the other two:

$$Z_C = jX_C = j2.692 \ \Omega$$

$$\tilde{\mathbf{I}}_C = \frac{\tilde{\mathbf{V}}_S}{Z_C} = 178.3\angle(1.571) \ A$$

The total current is $\tilde{\mathbf{I}}_S = \tilde{\mathbf{I}}_1 + \tilde{\mathbf{I}}_2 + \tilde{\mathbf{I}}_C = 312.5\angle 0° \ A$. The phasor diagram corresponding to these three currents is shown in Figure 7.36.

**Focus on Computer-Aided Tools:** A file containing the computer-generated solution to this problem may be found in the CD-ROM that accompanies this book.

## Check Your Understanding

**7.6**   Compute the power factor for the load of Example 7.6 with and without the inductance in the circuit.

**7.7**   Show that one can also express the instantaneous power for an arbitrary complex load $Z = |Z|\angle\theta$ as

$$p(t) = \tilde{I}^2|Z|\cos\theta + \tilde{I}^2|Z|\cos(2\omega t + \theta)$$

**7.8**   Determine the power factor for the load in the circuit of Figure 7.37, and state whether it is leading or lagging for the following conditions:

   a. $v_S(t) = 540\cos(\omega t + 15°) \ V$     $i(t) = 2\cos(\omega t + 47°) \ A$
   b. $v_S(t) = 155\cos(\omega t - 15°) \ V$     $i(t) = 2\cos(\omega t - 22°) \ A$

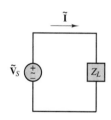

**Figure 7.37**

**7.9**   Determine whether the load is capacitive or inductive for the circuit of Figure 7.37 if

   a. pf = 0.87 (leading)
   b. pf = 0.42 (leading)
   c. $v_S(t) = 42\cos(\omega t)$     $i(t) = 4.2\sin(\omega t)$
   d. $v_S(t) = 10.4\cos(\omega t - 12°)$     $i(t) = 0.4\cos(\omega t - 12°)$

**7.10**   Prove that the power factor is indeed 1 after the addition of the parallel capacitor in Example 7.8.

**7.11**   Compute the magnitude of the current drawn from the source after the power factor correction in the circuit of Example 7.8.

## 7.3   TRANSFORMERS

AC circuits are very commonly connected to each other by means of **transformers.** A transformer is a device that couples two AC circuits magnetically rather than through any direct conductive connection and permits a "transformation" of the voltage and current between one circuit and the other (for example, by matching a high-voltage, low-current AC output to a circuit requiring a low-voltage, high-current source). Transformers play a major role in electric power engineering and are a necessary part of the electric power distribution network. The objective of this section is to introduce the ideal transformer and the concepts of impedance reflection and impedance matching. The physical operations of practical transformers, and more advanced models, will be discussed in Chapter 16.

## The Ideal Transformer

The ideal transformer consists of two coils that are coupled to each other by some magnetic medium. There is no electrical connection between the coils. The coil on the input side is termed the **primary,** and that on the output side the **secondary.** The primary coil is wound so that it has $n_1$ turns, while the secondary has $n_2$ turns. We define the **turns ratio $N$** as

$$N = \frac{n_2}{n_1} \tag{7.32}$$

Figure 7.38 illustrates the convention by which voltages and currents are usually assigned at a transformer. The dots in Figure 7.38 are related to the polarity of the coil voltage: coil terminals marked with a dot have the same polarity.

Since an ideal inductor acts as a short circuit in the presence of DC currents, transformers do not perform any useful function when the primary voltage is DC. However, when a time-varying current flows in the primary winding, a corresponding time-varying voltage is generated in the secondary because of the magnetic coupling between the two coils. This behavior is due to Faraday's law, as will be explained in Chapter 16. The relationship between primary and secondary current in an ideal transformer is very simply stated as follows:

$$\tilde{\mathbf{V}}_2 = N\tilde{\mathbf{V}}_1$$
$$\tilde{\mathbf{I}}_2 = \frac{\tilde{\mathbf{I}}_1}{N} \tag{7.33}$$

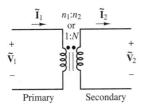

**Figure 7.38** Ideal transformer

> An ideal transformer multiplies a sinusoidal input voltage by a factor of $N$ and divides a sinusoidal input current by a factor of $N$.

If $N$ is greater than 1, the output voltage is greater than the input voltage and the transformer is called a **step-up transformer.** If $N$ is less than 1, then the transformer is called a **step-down transformer,** since $\tilde{\mathbf{V}}_2$ is now smaller than $\tilde{\mathbf{V}}_1$. An ideal transformer can be used in either direction (i.e., either of its coils may be viewed as the input side or primary). Finally, a transformer with $N = 1$ is called an **isolation transformer** and may perform a very useful function if one needs to electrically isolate two circuits from each other; note that any DC currents at the primary will not appear at the secondary coil. An important property of ideal transformers is conservation of power; one can easily verify that an ideal transformer conserves power, since

$$S_1 = \tilde{\mathbf{I}}_1^* \tilde{\mathbf{V}}_1 = N\tilde{\mathbf{I}}_2^* \frac{\tilde{\mathbf{V}}_2}{N} = \tilde{\mathbf{I}}_2^* \tilde{\mathbf{V}}_2 = S_2 \tag{7.34}$$

That is, the power on the primary side equals that on the secondary.

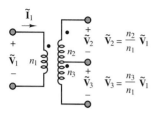

**Figure 7.39** Center-tapped transformer

In many practical circuits, the secondary is tapped at two different points, giving rise to two separate output circuits, as shown in Figure 7.39. The most common configuration is the **center-tapped transformer,** which splits the secondary voltage into two equal voltages. The most common occurrence of this type of transformer is found at the entry of a power line into a household, where a high-voltage primary (see Figure 7.64) is transformed to 240 V, and split into two 120-V lines. Thus, $\tilde{\mathbf{V}}_2$ and $\tilde{\mathbf{V}}_3$ in Figure 7.39 are both 120-V lines, and a 240-V line ($\tilde{\mathbf{V}}_2 + \tilde{\mathbf{V}}_3$) is also available.

---

## EXAMPLE  7.12  Ideal Transformer Turns Ratio

### Problem

We require a transformer to deliver 500 mA at 24 V from a 120-V rms line source. How many turns are required in the secondary? What is the primary current?

---

### Solution

**Known Quantities:**  Primary and secondary voltages; secondary current. Number of turns in the primary coil.

**Find:**  $n_2$ and $\tilde{\mathbf{I}}_1$.

**Schematics, Diagrams, Circuits, and Given Data:**  $\tilde{V}_1 = 120$ V; $\tilde{V}_2 = 24$ V; $\tilde{I}_2 = 500$ mA; $n_1 = 3{,}000$ turns.

**Assumptions:**  Use rms values for all phasor quantities in the problem.

**Analysis:**  Using equation 7.33 we compute the number of turns in the secondary coil as follows:

$$\frac{\tilde{V}_1}{n_1} = \frac{\tilde{V}_2}{n_2} \qquad n_2 = n_1 \frac{\tilde{V}_2}{\tilde{V}_1} = 3{,}000 \times \frac{24}{120} = 600 \text{ turns}$$

Knowing the number of turns, we can now compute the primary current, also from equation 7.33:

$$n_1 \tilde{I}_1 = n_2 \tilde{I}_2 \qquad \tilde{I}_1 = \frac{n_2}{n_1} \tilde{I}_2 = \frac{600}{3{,}000} \times 500 = 100 \text{ mA}$$

**Comments:**  Note that since the transformer does not affect the phase of the voltages and currents, we could solve the problem using simply the rms amplitudes.

---

## EXAMPLE  7.13  Center-Tapped Transformer

### Problem

A center-tapped power transformer has a primary voltage of 4,800 V and two 120-V secondaries (see Figure 7.39). Three loads (all resistive, i.e., with unity power factor) are connected to the transformer. The first load, $R_1$, is connected across the 240-V line (the two outside taps in Figure 7.39). The second and third loads, $R_2$ and $R_3$, are connected across each of the 120-V lines. Compute the current in the primary if the power absorbed by the three loads is known.

## Solution

**Known Quantities:** Primary and secondary voltages; load power ratings.

**Find:** $\tilde{I}_{\text{primary}}$

**Schematics, Diagrams, Circuits, and Given Data:** $\tilde{V}_1 = 4{,}800$ V; $\tilde{V}_2 = 120$ V; $\tilde{V}_3 = 120$ V. $P_1 = 5{,}000$ W; $P_2 = 1{,}000$ W; $P_3 = 1{,}500$ W.

**Assumptions:** Use rms values for all phasor quantities in the problem.

**Analysis:** Since we have no information about the number of windings, nor about the secondary current, we cannot solve this problem using equation 7.33. An alternative approach is to apply conservation of power (equation 7.34). Since the loads all have unity power factor, the voltages and currents will all be in phase, and we can use the rms amplitudes in our calculations:

$$\left| S_{\text{primary}} \right| = \left| S_{\text{secondary}} \right|$$

or

$$\tilde{V}_{\text{primary}} \times \tilde{I}_{\text{primary}} = P_{\text{secondary}} = P_1 + P_2 + P_3.$$

Thus,

$$4{,}800 \times \tilde{I}_{\text{primary}} = 5{,}000 + 1{,}000 + 1{,}500 = 7{,}500 \text{ W}$$

$$\tilde{I}_{\text{primary}} = \frac{7{,}500 \text{ W}}{4{,}800 \text{ A}} = 1.5625 \text{ A}$$

## Impedance Reflection and Power Transfer

As stated in the preceding paragraphs, transformers are commonly used to couple one AC circuit to another. A very common and rather general situation is that depicted in Figure 7.40, where an AC source, represented by its Thévenin equivalent, is connected to an equivalent load impedance by means of a transformer.

It should be apparent that expressing the circuit in phasor form does not alter the basic properties of the ideal transformer, as illustrated in the following equation:

$$\tilde{\mathbf{V}}_1 = \frac{\tilde{\mathbf{V}}_2}{N} \qquad \tilde{\mathbf{I}}_1 = N\tilde{\mathbf{I}}_2$$

$$\tilde{\mathbf{V}}_2 = N\tilde{\mathbf{V}}_1 \qquad \tilde{\mathbf{I}}_2 = \frac{\tilde{\mathbf{I}}_1}{N} \tag{7.35}$$

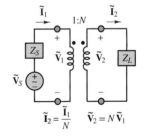

**Figure 7.40** Operation of an ideal transformer

These expressions are very useful in determining the equivalent impedance seen by the source and by the load, on opposite sides of the transformer. At the primary connection, the equivalent impedance seen by the source must equal the ratio of $\tilde{\mathbf{V}}_1$ to $\tilde{\mathbf{I}}_1$:

$$Z' = \frac{\tilde{\mathbf{V}}_1}{\tilde{\mathbf{I}}_1} \tag{7.36}$$

which can be written as:

$$Z' = \frac{\tilde{\mathbf{V}}_2/N}{N\tilde{\mathbf{I}}_2} = \frac{1}{N^2}\frac{\tilde{\mathbf{V}}_2}{\tilde{\mathbf{I}}_2} \tag{7.37}$$

But the ratio $\tilde{\mathbf{V}}_2/\tilde{\mathbf{I}}_2$ is by definition the load impedance, $Z_L$. Thus,

$$Z' = \frac{1}{N^2} Z_L \tag{7.38}$$

That is, the AC source "sees" the load impedance reduced by a factor of $1/N^2$.

The load impedance also sees an equivalent source. The open-circuit voltage is given by the expression

$$\tilde{\mathbf{V}}_{OC} = N\tilde{\mathbf{V}}_1 = N\tilde{\mathbf{V}}_S \tag{7.39}$$

since there is no voltage drop across the source impedance in the circuit of Figure 7.40. The short-circuit current is given by the expression

$$\tilde{\mathbf{I}}_{SC} = \frac{\tilde{\mathbf{V}}_S}{Z_S} \frac{1}{N} \tag{7.40}$$

and the load sees a Thévenin impedance equal to

$$Z'' = \frac{\tilde{\mathbf{V}}_{OC}}{\tilde{\mathbf{I}}_{SC}} = \frac{N\tilde{\mathbf{V}}_S}{\dfrac{\tilde{\mathbf{V}}_S}{Z_S} \dfrac{1}{N}} = N^2 Z_S \tag{7.41}$$

Thus the load sees the source impedance multiplied by a factor of $N^2$. Figure 7.41 illustrates this **impedance reflection** across a transformer. It is very important to note that an ideal transformer changes the magnitude of the load impedance seen by the source by a factor of $1/N^2$. This property naturally leads to the discussion of power transfer, which we consider next.

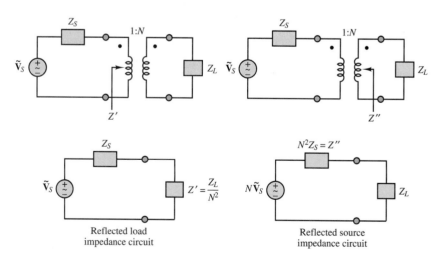

**Figure 7.41** Impedance reflection across a transformer

Recall that in DC circuits, given a fixed equivalent source, maximum power is transferred to a resistive load when the latter is equal to the internal resistance of the source; achieving an analogous maximum power transfer condition in an AC circuit is referred to as **impedance matching.** Consider the general form of an AC circuit, shown in Figure 7.42, and assume that the source impedance, $Z_S$, is given by

$$Z_S = R_S + jX_S \tag{7.42}$$

The problem of interest is often that of selecting the load resistance and reactance that will maximize the real (average) power absorbed by the load. Note that the requirement is to maximize the real power absorbed by the load. Thus, the problem can be restated by expressing the real load power in terms of the impedance of the source and load. The real power absorbed by the load is

$$P_L = \tilde{V}_L \tilde{I}_L \cos\theta = \text{Re}\,(\tilde{\mathbf{V}}_L \tilde{\mathbf{I}}_L^*) \tag{7.43}$$

where

$$\tilde{\mathbf{V}}_L = \frac{Z_L}{Z_S + Z_L}\tilde{\mathbf{V}}_S \tag{7.44}$$

and

$$\tilde{\mathbf{I}}_L^* = \left(\frac{\tilde{\mathbf{V}}_S}{Z_S + Z_L}\right)^* = \frac{\tilde{\mathbf{V}}_S^*}{(Z_S + Z_L)^*} \tag{7.45}$$

Thus, the complex load power is given by

$$S_L = \tilde{\mathbf{V}}_L \tilde{\mathbf{I}}_L^* = \frac{Z_L \tilde{\mathbf{V}}_S}{Z_S + Z_L} \times \frac{\tilde{\mathbf{V}}_S^*}{(Z_S + Z_L)^*} = \frac{\tilde{V}_S^2}{|Z_S + Z_L|^2} Z_L \tag{7.46}$$

and the average (real) power by

$$\begin{aligned}
P_L &= \text{Re}\,(\tilde{\mathbf{V}}_L \tilde{\mathbf{I}}_L^*) = \text{Re}\left(\frac{\tilde{V}_S^2}{|Z_S + Z_L|^2}\right)\text{Re}\,(Z_L) \\
&= \frac{\tilde{V}_S^2}{(R_S + R_L)^2 + (X_S + X_L)^2}\,\text{Re}\,(Z_L) \\
&= \frac{\tilde{V}_S^2 R_L}{(R_S + R_L)^2 + (X_S + X_L)^2}
\end{aligned} \tag{7.47}$$

The expression for $P_L$ is maximized by selecting appropriate values of $R_L$ and $X_L$; it can be shown that the average power is greatest when $R_L = R_S$ and $X_L = -X_S$, that is, when the load impedance is equal to the complex conjugate of the source impedance, as shown in the following equation:

$$Z_L = Z_S^* \quad \text{i.e.,} \quad R_L = R_S \quad X_L = -X_S \tag{7.48}$$

When the load impedance is equal to the complex conjugate of the source impedance, the load and source impedances are matched and maximum power is transferred to the load.

In many cases, it may not be possible to select a matched load impedance, because of physical limitations in the selection of appropriate components. In these situations, it is possible to use the impedance reflection properties of a transformer to maximize the transfer of AC power to the load. The circuit of Figure 7.43

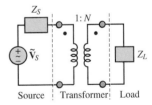

**Figure 7.42** The maximum power transfer problem in AC circuits

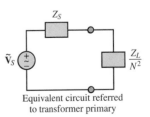

**Figure 7.43** Maximum power transfer in an AC circuit with a transformer

illustrates how the reflected load impedance, as seen by the source, is equal to $Z_L/N^2$, so that maximum power transfer occurs when

$$\frac{Z_L}{N^2} = Z_S^*$$

$$R_L = N^2 R_S$$

$$X_L = -N^2 X_S$$

(7.49)

### EXAMPLE 7.14 Maximum Power Transfer Through a Transformer

#### Problem

Find the transformer turns ratio and load reactance that results in maximum power transfer in the circuit of Figure 7.44.

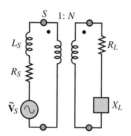

**Figure 7.44**

#### Solution

**Known Quantities:**  Source voltage, frequency and impedance; load resistance.

**Find:**  Transformer turns ratio and load reactance.

**Schematics, Diagrams, Circuits, and Given Data:**  $\tilde{V}_S = 240\angle(0)$ V; $R_S = 10\ \Omega$; $L_S = 0.1$ H; $R_L = 400\ \Omega$; $\omega = 377$ rad/s.

**Assumptions:**  Use rms values for all phasor quantities in the problem.

**Analysis:**  For maximum power transfer, we require that $R_L = N^2 R_S$ (equation 7.48). Thus,

$$N^2 = \frac{R_L}{R_S} = \frac{400}{10} = 40 \qquad N = \sqrt{40} = 6.325$$

Further, to cancel the reactive power we require that $X_L = -N^2 X_S$, i.e.,

$$X_S = \omega \times 0.1 = 37.7$$

and

$$X_L = -40 \times 37.7 = -1{,}508$$

Thus, the load reactance should be a capacitor with value

$$C = -\frac{1}{X_L \omega} = -\frac{1}{(-1{,}508) \times 377} = 1.76\ \mu F$$

## Check Your Understanding

**7.12**  If the transformer shown in Figure 7.45 is ideal, find the turns ratio, $N$, that will ensure maximum power transfer to the load. Assume that $Z_S = 1{,}800\ \Omega$ and $Z_L = 8\ \Omega$.

**7.13**  If the circuit of Exercise 7.12 has $Z_L = (2 + j10)\ \Omega$ and the turns ratio of the transformer is $N = 5.4$, what should $Z_S$ be in order to have maximum power transfer?

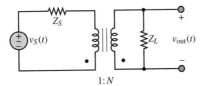

**Figure 7.45**

## 7.4 THREE-PHASE POWER

The material presented so far in this chapter has dealt exclusively with **single-phase AC power,** that is, with single sinusoidal sources. In fact, most of the AC power used today is generated and distributed as **three-phase power,** by means of an arrangement in which three sinusoidal voltages are generated out of phase with each other. The primary reason is efficiency: The weight of the conductors and other components in a three-phase system is much lower than in a single-phase system delivering the same amount of power. Further, while the power produced by a single-phase system has a pulsating nature (recall the results of Section 7.1), a three-phase system can deliver a steady, constant supply of power. For example, later in this section it will be shown that a three-phase generator producing three **balanced voltages**—that is, voltages of equal amplitude and frequency displaced in phase by 120°—has the property of delivering constant instantaneous power.

Another important advantage of three-phase power is that, as will be explained in Chapter 17, three-phase motors have a nonzero starting torque, unlike their single-phase counterpart. The change to three-phase AC power systems from the early DC system proposed by Edison was therefore due to a number of reasons: the efficiency resulting from transforming voltages up and down to minimize transmission losses over long distances; the ability to deliver constant power (an ability not shared by single- and two-phase AC systems); a more efficient use of conductors; and the ability to provide starting torque for industrial motors.

To begin the discussion of three-phase power, consider a three-phase source connected in the **wye** (or **Y**) **configuration,** as shown in Figure 7.46. Each of the three voltages is 120° out of phase with the others, so that, using phasor notation,

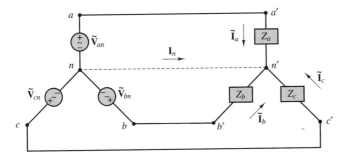

**Figure 7.46** Balanced three-phase AC circuit

we may write:

$$\tilde{\mathbf{V}}_{an} = \tilde{V}_{an}\angle 0°$$

$$\tilde{\mathbf{V}}_{bn} = \tilde{V}_{bn}\angle -120°$$

$$\tilde{\mathbf{V}}_{cn} = \tilde{V}_{cn}\angle -240° = \tilde{V}_{cn}\angle 120°$$

(7.50)

where the quantities $\tilde{V}_{an}$, $\tilde{V}_{bn}$, and $\tilde{V}_{cn}$ are rms values and are equal to each other. To simplify the notation, it will be assumed from here on that

$$\tilde{V}_{an} = \tilde{V}_{bn} = \tilde{V}_{cn} = \tilde{V}$$

(7.51)

Chapter 17 will discuss how three-phase AC electric generators may be constructed to provide such balanced voltages. In the circuit of Figure 7.46, the resistive loads are also wye-connected and balanced (i.e., equal). The three AC sources are all connected together at a node called the *neutral node,* denoted by $n$. The voltages $\tilde{\mathbf{V}}_{an}$, $\tilde{\mathbf{V}}_{bn}$, and $\tilde{\mathbf{V}}_{cn}$ are called the **phase voltages** and form a balanced set in the sense that

$$\tilde{\mathbf{V}}_{an} + \tilde{\mathbf{V}}_{bn} + \tilde{\mathbf{V}}_{cn} = 0$$

(7.52)

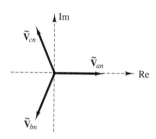

**Figure 7.47** Positive, or *abc*, sequence for balanced three-phase voltages

This last statement is easily verified by sketching the phasor diagram. The sequence of phasor voltages shown in Figure 7.47 is usually referred to as the **positive (or *abc*) sequence.**

Consider now the "lines" connecting each source to the load and observe that it is possible to also define **line voltages** (also called *line-to-line voltages*) by considering the voltages between the lines $aa'$ and $bb'$, $aa'$ and $cc'$, and $bb'$ and $cc'$. Since the line voltage, say, between $aa'$ and $bb'$ is given by

$$\tilde{\mathbf{V}}_{ab} = \tilde{\mathbf{V}}_{an} + \tilde{\mathbf{V}}_{nb} = \tilde{\mathbf{V}}_{an} - \tilde{\mathbf{V}}_{bn}$$

(7.53)

the line voltages may be computed relative to the phase voltages as follows:

$$\tilde{\mathbf{V}}_{ab} = \tilde{V}\angle 0° - \tilde{V}\angle -120° = \sqrt{3}\tilde{V}\angle 30°$$

$$\tilde{\mathbf{V}}_{bc} = \tilde{V}\angle -120° - \tilde{V}\angle 120° = \sqrt{3}\tilde{V}\angle -90°$$

$$\tilde{\mathbf{V}}_{ca} = \tilde{V}\angle 120° - \tilde{V}\angle 0° = \sqrt{3}\tilde{V}\angle 150°$$

(7.54)

It can be seen, then, that the magnitude of the line voltages is equal to $\sqrt{3}$ times the magnitude of the phase voltages. It is instructive, at least once, to point out that the circuit of Figure 7.46 can be redrawn to have the appearance of the circuit of Figure 7.48.

One of the important features of a balanced three-phase system is that it does not require a fourth wire (the neutral connection), since the current $\tilde{\mathbf{I}}_n$ is identically zero (for balanced load $Z_a = Z_b = Z_c = Z$). This can be shown by applying KCL at the neutral node $n$:

$$\tilde{\mathbf{I}}_n = (\tilde{\mathbf{I}}_a + \tilde{\mathbf{I}}_b + \tilde{\mathbf{I}}_c)$$

$$= \frac{1}{Z}(\tilde{\mathbf{V}}_{an} + \tilde{\mathbf{V}}_{bn} + \tilde{\mathbf{V}}_{cn})$$

$$= 0$$

(7.55)

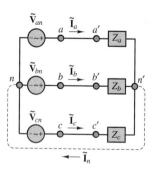

**Figure 7.48** Balanced three-phase AC circuit (redrawn)

Another, more important characteristic of a balanced three-phase power system may be illustrated by simplifying the circuits of Figures 7.46 and 7.48 by replacing the balanced load impedances with three equal resistances, $R$. With this simplified configuration, one can show that the total power delivered to the balanced load by the three-phase generator is constant. This is an extremely important result, for a very practical reason: delivering power in a smooth fashion (as opposed to the pulsating nature of single-phase power) reduces the wear and stress on the generating equipment. Although we have not yet discussed the nature of the machines used to generate power, a useful analogy here is that of a single-cylinder engine versus a perfectly balanced V-8 engine. To show that the total power delivered by the three sources to a balanced resistive load is constant, consider the instantaneous power delivered by each source:

$$p_a(t) = \frac{\tilde{V}^2}{R}(1 + \cos 2\omega t)$$

$$p_b(t) = \frac{\tilde{V}^2}{R}[1 + \cos(2\omega t - 120°)]$$   **(7.56)**

$$p_c(t) = \frac{\tilde{V}^2}{R}[1 + \cos(2\omega t + 120°)]$$

The total instantaneous load power is then given by the sum of the three contributions:

$$p(t) = p_a(t) + p_b(t) + p_c(t)$$

$$= \frac{3\tilde{V}^2}{R} + \frac{\tilde{V}^2}{R}[\cos 2\omega t + \cos(2\omega t - 120°)$$
$$+ \cos(2\omega t + 120°)]$$   **(7.57)**

$$= \frac{3\tilde{V}^2}{R} = \text{a constant!}$$

You may wish to verify that the sum of the trigonometric terms inside the brackets is identically zero.

It is also possible to connect the three AC sources in a three-phase system in a so-called **delta (or Δ) connection**, although in practice this configuration is rarely used. Figure 7.49 depicts a set of three delta-connected generators.

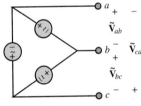

A delta-connected three-phase generator with line voltages $V_{ab}$, $V_{bc}$, $V_{ca}$

**Figure 7.49** Delta-connected generators

---

## EXAMPLE 7.15 Per-Phase Solution of Balanced Wye-Wye Circuit

### Problem

Compute the power delivered to the load by the three-phase generator in the circuit shown in Figure 7.50.

---

### Solution

**Known Quantities:** Source voltage, line resistance, load impedance.

**Find:** Power delivered to the load, $P_L$.

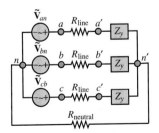

**Figure 7.50**

**Schematics, Diagrams, Circuits, and Given Data:** $\tilde{\mathbf{V}}_{an} = 480\angle(0)$ V; $\tilde{\mathbf{V}}_{bn} = 480\angle(-2\pi/3)$ V; $\tilde{\mathbf{V}}_{cn} = 480\angle(2\pi/3)$ V; $Z_y = 2 + j4 = 4.47\angle(1.107)$ $\Omega$; $R_{\text{line}} = 2$ $\Omega$; $R_{\text{neutral}} = 10$ $\Omega$.

**Assumptions:** Use rms values for all phasor quantities in the problem.

**Analysis:** Since the circuit is balanced, we can use per-phase analysis, and the current through the neutral line is zero, i.e., $\tilde{\mathbf{V}}_{n-n'} = 0$. The resulting per-phase circuit is shown in Figure 7.51. Using phase $a$ for the calculations, we look for the quantity

$$P_a = |\tilde{\mathbf{I}}|^2 R_L$$

where

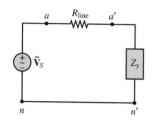

**Figure 7.51** One phase of the three-phase circuit

$$|\tilde{\mathbf{I}}| = \left|\frac{\tilde{\mathbf{V}}_a}{Z_y + R_{\text{line}}}\right| = \left|\frac{480\angle 0}{2 + j4 + 2}\right| = \left|\frac{480\angle 0}{5.66\angle\left(\frac{\pi}{4}\right)}\right| = 84.85 \text{ A}$$

and $P_a = (84.85)^2 \times 2 = 14.4$ kW. Since the circuit is balanced, the results for phases $b$ and $c$ are identical, and we have:

$$P_L = 3P_a = 43.2 \text{ kW}$$

**Comments:** Note that, since the circuit is balanced, there is zero voltage across neutrals. This fact is shown explicitly in Figure 7.51, where $n$ and $n'$ are connected to each other directly. Per-phase analysis for balanced circuits turn three-phase power calculations into a very simple exercise.

## Balanced Wye Loads

In the previous section we performed some power computations for a purely resistive balanced wye load. We shall now generalize those results for an arbitrary balanced complex load. Consider again the circuit of Figure 7.46, where now the balanced load consists of the three complex impedances

$$Z_a = Z_b = Z_c = Z_y = |Z_y|\angle\theta \tag{7.58}$$

From the diagram of Figure 7.46, it can be verified that each impedance sees the corresponding phase voltage across itself; thus, since the currents $\tilde{\mathbf{I}}_a$, $\tilde{\mathbf{I}}_b$, and $\tilde{\mathbf{I}}_c$ have the same rms value, $\tilde{I}$, the phase angles of the currents will differ by $\pm 120°$. It is therefore possible to compute the power for each phase by considering the phase voltage (equal to the load voltage) for each impedance, and the associated line current. Let us denote the complex power for each phase by $S$:

$$S = \tilde{\mathbf{V}} \cdot \tilde{\mathbf{I}}^* \tag{7.59}$$

so that

$$\begin{aligned} S &= P + jQ \\ &= \tilde{V}\tilde{I}\cos\theta + j\tilde{V}\tilde{I}\sin\theta \end{aligned} \tag{7.60}$$

where $\tilde{V}$ and $\tilde{I}$ denote, once again, the rms values of each phase voltage and line current. Consequently, the total real power delivered to the balanced wye load is $3P$, and the total reactive power is $3Q$. Thus, the total complex power, $S_T$, is given by

$$\begin{aligned} S_T &= P_T + jQ_T = 3P + j3Q \\ &= \sqrt{(3P)^2 + (3Q)^2}\angle\theta \end{aligned} \tag{7.61}$$

and the apparent power is

$$|S_T| = 3\sqrt{(VI)^2 \cos^2 \theta + (VI)^2 \sin^2 \theta}$$
$$= 3VI$$

and the total real and reactive power may be expressed in terms of the apparent power:

$$P_T = |S_T| \cos \theta \tag{7.62}$$
$$Q_T = |S_T| \sin \theta$$

## Balanced Delta Loads

In addition to a wye connection, it is also possible to connect a balanced load in the delta configuration. A wye-connected generator and a delta-connected load are shown in Figure 7.52.

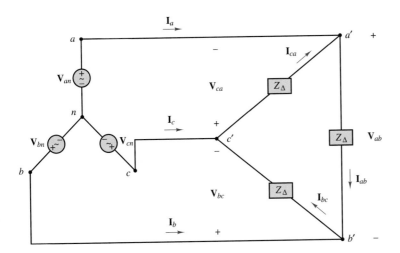

**Figure 7.52** Balanced wye generators with balanced delta load

It should be noted immediately that now the corresponding line voltage (not phase voltage) appears across each impedance. For example, the voltage across $Z_{c'a'}$ is $\tilde{\mathbf{V}}_{ca}$. Thus, the three load currents are given by the following expressions:

$$\tilde{\mathbf{I}}_{ab} = \frac{\tilde{\mathbf{V}}_{ab}}{Z_\Delta} = \frac{\sqrt{3}V\angle(\pi/6)}{|Z_\Delta|\angle\theta}$$

$$\tilde{\mathbf{I}}_{bc} = \frac{\tilde{\mathbf{V}}_{bc}}{Z_\Delta} = \frac{\sqrt{3}V\angle(-\pi/4)}{|Z_\Delta|\angle\theta} \tag{7.63}$$

$$\tilde{\mathbf{I}}_{ca} = \frac{\tilde{\mathbf{V}}_{ca}}{Z_\Delta} = \frac{\sqrt{3}V\angle(5\pi/6)}{|Z_\Delta|\angle\theta}$$

To understand the relationship between delta-connected and wye-connected loads, it is reasonable to ask the question, For what value of $Z_\Delta$ would a delta-connected load draw the same amount of current as a wye-connected load with impedance $Z_y$ for a given source voltage? This is equivalent to asking what value of $Z_\Delta$ would make the line currents the same in both circuits (compare Figure 7.48 with Figure 7.52).

The line current drawn, say, in phase $a$ by a wye-connected load is

$$(\tilde{\mathbf{I}}_{an})_y = \frac{\tilde{\mathbf{V}}_{an}}{Z} = \frac{\tilde{V}}{|Z_y|}\angle-\theta \tag{7.64}$$

while that drawn by the delta-connected load is

$$
\begin{aligned}
(\tilde{\mathbf{I}}_a)_\Delta &= \tilde{\mathbf{I}}_{ab} - \tilde{\mathbf{I}}_{ca} \\[2mm]
&= \frac{\tilde{\mathbf{V}}_{ab}}{Z_\Delta} - \frac{\tilde{\mathbf{V}}_{ca}}{Z_\Delta} \\[2mm]
&= \frac{1}{Z_\Delta}(\tilde{\mathbf{V}}_{an} - \tilde{\mathbf{V}}_{bn} - \tilde{\mathbf{V}}_{cn} + \tilde{\mathbf{V}}_{an}) \\[2mm]
&= \frac{1}{Z_\Delta}(2\tilde{\mathbf{V}}_{an} - \tilde{\mathbf{V}}_{bn} - \tilde{\mathbf{V}}_{cn}) \\[2mm]
&= \frac{3\tilde{\mathbf{V}}_{an}}{Z_\Delta} = \frac{3\tilde{V}}{|Z_\Delta|}\angle-\theta
\end{aligned}
\tag{7.65}
$$

One can readily verify that the two currents $(\tilde{\mathbf{I}}_a)_\Delta$ and $(\tilde{\mathbf{I}}_a)_y$ will be equal if the magnitude of the delta-connected impedance is 3 times larger than $Z_y$:

$$Z_\Delta = 3Z_y \tag{7.66}$$

This result also implies that a delta load will necessarily draw 3 times as much current (and therefore absorb 3 times as much power) as a wye load with the same branch impedance.

---

### EXAMPLE 7.16 Parallel Wye-Delta Load Circuit

**Problem**

Compute the power delivered to the wye-delta load by the three-phase generator in the circuit shown in Figure 7.53.

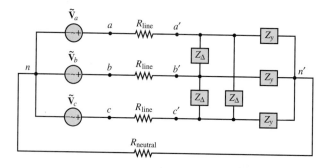

**Figure 7.53** AC circuit with delta and wye loads

## Solution

**Known Quantities:** Source voltage, line resistance, load impedance.

**Find:** Power delivered to the load, $P_L$.

**Schematics, Diagrams, Circuits, and Given Data:** $\tilde{\mathbf{V}}_{an} = 480\angle(0)$ V;
$\tilde{\mathbf{V}}_{bn} = 480\angle(-2\pi/3)$ V; $\tilde{\mathbf{V}}_{cn} = 480\angle(2\pi/3)$ V; $Z_y = 2 + j4 = 4.47\angle(1.107)$ Ω;
$Z_\Delta = 5 - j2 = 5.4\angle(-0.381)$ Ω; $R_{line} = 2$ Ω; $R_{neutral} = 10$ Ω.

**Assumptions:** Use rms values for all phasor quantities in the problem.

**Analysis:** We first convert the balanced delta load to an equivalent wye load, according to equation 7.66. Figure 7.54 illustrates the effect of this conversion.

$$Z_{\Delta-y} = \frac{Z_\Delta}{3} = 1.667 - j0.667 = 1.8\angle(-0.381) \ \Omega.$$

Since the circuit is balanced, we can use per-phase analysis, and the current through the neutral line is zero, i.e., $\tilde{\mathbf{V}}_{n-n'} = 0$. The resulting per-phase circuit is shown in Figure 7.55. Using phase $a$ for the calculations, we look for the quantity

$$P_a = |\tilde{\mathbf{I}}|^2 R_L$$

where

$$Z_L = Z_y \| Z_{\Delta-y} = \frac{Z_y \times Z_{\Delta-y}}{Z_y + Z_{\Delta-y}} = 1.62 - j0.018 = 1.62\angle(-0.011) \ \Omega$$

and the load current is given by:

$$|\tilde{\mathbf{I}}| = \left| \frac{\tilde{\mathbf{V}}_a}{Z_L + R_{line}} \right| = \left| \frac{480\angle0}{1.62 + j0.018 + 2} \right| = 132.6 \ \text{A}$$

and $P_a = (132.6)^2 \times \text{Re}(Z_L) = 28.5$ kW. Since the circuit is balanced, the results for phase $b$ and $c$ are identical, and we have:

$$P_L = 3P_a = 85.5 \ \text{kW}$$

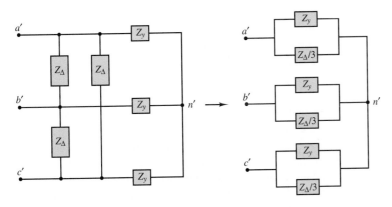

**Figure 7.54** Conversion of delta load to equivalent wye load

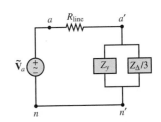

**Figure 7.55** Per-phase circuit

**Comments:** Note that per-phase analysis for balanced circuits turns three-phase power calculations into a very simple exercise.

**Focus on Computer-Aided Tools:**  A computer-generated solution of this example may be found in the accompanying CD-ROM.

---

## Check Your Understanding

**7.14**  Find the power lost in the lines in the circuit of Example 7.15.

**7.15**  Draw the phasor diagram and power triangle for a single phase and compute the power delivered to the balanced load of Example 7.15 if the lines have zero resistance and $Z_L = 1 + j3 \ \Omega$.

**7.16**  Show that the voltage across each branch of the balanced wye load in Exercise 7.15 is equal to the corresponding phase voltage (e.g., the voltage across $Z_a$ is $\tilde{\mathbf{V}}_a$).

**7.17**  Prove that the sum of the instantaneous powers absorbed by the three branches in a balanced wye-connected load is constant and equal to $3\tilde{V}\tilde{I}\cos\theta$.

**7.18**  Derive an expression for the rms line current of a delta load in terms of the rms line current for a wye load with the same branch impedances (i.e., $Z_y = Z_\Delta$) and same source voltage. Assume $Z_S = 0$.

**7.19**  The equivalent wye load of Example 7.16 is connected in a delta configuration. Compute the line currents.

---

## 7.5  RESIDENTIAL WIRING; GROUNDING AND SAFETY

Common residential electric power service consists of a three-wire AC system supplied by the local power company. The three wires originate from a utility pole and consist of a neutral wire, which is connected to earth ground, and two "hot" wires. Each of the hot lines supplies 120 V rms to the residential circuits; the two lines are 180° out of phase, for reasons that will become apparent during the course of this discussion. The phasor line voltages, shown in Figure 7.56, are usually referred to by means of a subscript convention derived from the color of the insulation on the different wires: $W$ for white (neutral), $B$ for black (hot), and $R$ for red (hot). This convention is adhered to uniformly.

The voltages across the hot lines are given by:

$$\tilde{\mathbf{V}}_B - \tilde{\mathbf{V}}_R = \tilde{\mathbf{V}}_{BR} = \tilde{\mathbf{V}}_B - (-\tilde{\mathbf{V}}_B) = 2\tilde{\mathbf{V}}_B = 240\angle 0° \qquad \textbf{(7.67)}$$

Thus, the voltage between the hot wires is actually 240 V rms. Appliances such as electric stoves, air conditioners, and heaters are powered by the 240-V rms arrangement. On the other hand, lighting and all of the electric outlets in the house used for small appliances are powered by a single 120-V rms line.

The use of 240-V rms service for appliances that require a substantial amount of power to operate is dictated by power transfer considerations. Consider the two circuits shown in Figure 7.57. In delivering the necessary power to a load, a lower line loss will be incurred with the 240-V rms wiring, since the power loss in the lines (the $I^2R$ **loss,** as it is commonly referred to) is directly related to the current required by the load. In an effort to minimize line losses, the size of the wires is increased for the lower-voltage case. This typically reduces the wire resistance by a factor of 2. In the top circuit, assuming $R_S/2 = 0.01 \ \Omega$, the current required

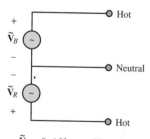

$\tilde{\mathbf{V}}_W = 0\angle 0°$    (Neutral)
$\tilde{\mathbf{V}}_B = 120\angle 0°$    (Hot)
$\tilde{\mathbf{V}}_R = 120\angle 180°$    (Hot)
or $\tilde{\mathbf{V}}_R = -\tilde{\mathbf{V}}_B$

**Figure 7.56** Line voltage convention for residential circuits

by the 10-kW load is approximately 83.3 A, while in the bottom circuit, with $R_S = 0.02 \ \Omega$, it is approximately half as much (41.7 A). (You should be able to verify that the approximate $I^2R$ losses are 69.4 W in the top circuit and 34.7 W in the bottom circuit.) Limiting the $I^2R$ losses is important from the viewpoint of efficiency, besides reducing the amount of heat generated in the wiring for safety considerations. Figure 7.58 shows some typical wiring configurations for a home. Note that several circuits are wired and fused separately.

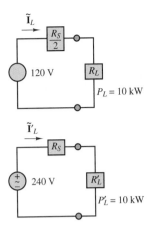

**Figure 7.57** Line losses in 120-VAC and 240-VAC circuits

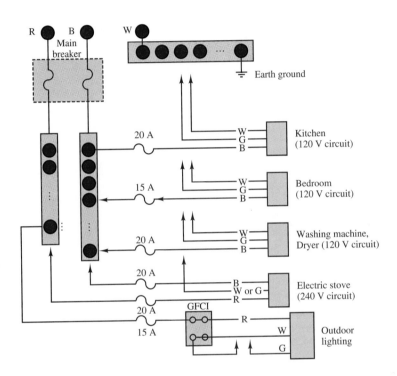

**Figure 7.58** A typical residential wiring arrangement

Today, most homes have three wire connections to their outlets. The outlets appear as sketched in Figure 7.59. Then why are both the ground and neutral connections needed in an outlet? The answer to this question is *safety:* the ground connection is used to connect the chassis of the appliance to earth ground. Without this provision, the appliance chassis could be at any potential with respect to ground, possibly even at the hot wire's potential if a segment of the hot wire were to lose some insulation and come in contact with the inside of the chassis! Poorly grounded appliances can thus be a significant hazard. Figure 7.60 illustrates schematically how, even though the chassis is intended to be insulated from the electric circuit, an unintended connection (represented by the dashed line) may occur, for example, because of corrosion or a loose mechanical connection. A path to ground might be provided by the body of a person touching the chassis with a hand. In the figure, such an undesired ground loop current is indicated by $I_G$. In this case, the ground current $I_G$ would flow directly through the body to ground and could be harmful.

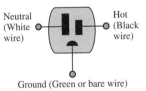

**Figure 7.59** A three-wire outlet

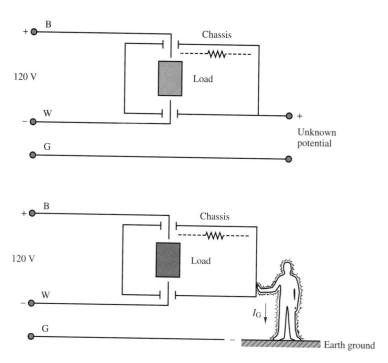

**Figure 7.60** Unintended connection

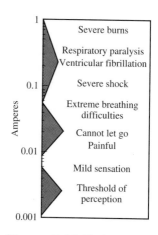

**Figure 7.61** Physiological effects of electric currents

In some cases the danger posed by such undesired ground loops can be great, leading to death by electric shock. Figure 7.61 describes the effects of electric currents on an average male when the point of contact is dry skin. Particularly hazardous conditions are liable to occur whenever the natural resistance to current flow provided by the skin breaks down, as would happen in the presence of water. The **ground fault circuit interrupter,** labeled **GFCI** in Figure 7.58, is a special safety circuit used primarily with outdoor circuits and in bathrooms, where the risk of death by electric shock is greatest. Its application is best described by an example.

Consider the case of an outdoor pool surrounded by a metal fence, which uses an existing light pole for a post, as shown in Figure 7.62. The light pole and the metal fence can be considered as forming a chassis. If the fence were not properly grounded all the way around the pool and if the light fixture were poorly insulated from the pole, a path to ground could easily be created by an unaware swimmer reaching, say, for the metal gate. A GFCI provides protection from potentially lethal ground loops, such as this one, by sensing both the hot-wire (B) and the neutral (W) currents. If the difference between the hot-wire current, $I_B$, and the neutral current, $I_W$, is more than a few milliamperes, then the GFCI disconnects the circuit nearly instantaneously. Any significant difference between the hot and neutral (return-path) currents means that a second path to ground has been created (by the unfortunate swimmer, in this example) and a potentially dangerous condition has arisen. Figure 7.63 illustrates the idea. GFCIs are typically resettable circuit breakers, so that one does not need to replace a fuse every time the GFCI circuit is enabled.

**Figure 7.62** Outdoor pool

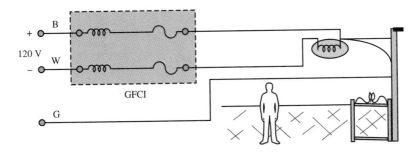

**Figure 7.63** Use of a GFCI in a potentially hazardous setting

## Check Your Understanding

**7.20**    Use the circuit of Figure 7.57 to show that the $I^2R$ losses will be higher for a 120-V service appliance than a 240-V service appliance if both have the same power usage rating.

## 7.6    GENERATION AND DISTRIBUTION OF AC POWER

We now conclude the discussion of power systems with a brief description of the various elements of a power system. Electric power originates from a variety of sources; in Chapter 17, electric generators will be introduced as a means of producing electric power from a variety of energy-conversion processes. In general, electric power may be obtained from hydroelectric, thermoelectric, geothermal, wind, solar, and nuclear sources. The choice of a given source is typically dictated by the power requirement for the given application, and by economic and environmental factors. In this section, the structure of an AC power network, from the power-generating station to the residential circuits discussed in the previous section, is briefly outlined.

A typical generator will produce electric power at 18 kV, as shown in the diagram of Figure 7.64. To minimize losses along the conductors, the output of the generators is processed through a step-up transformer to achieve line voltages of hundreds of kilovolts (345 kV, in Figure 7.64). Without this transformation, the majority of the power generated would be lost in the **transmission lines** that carry the electric current from the power station.

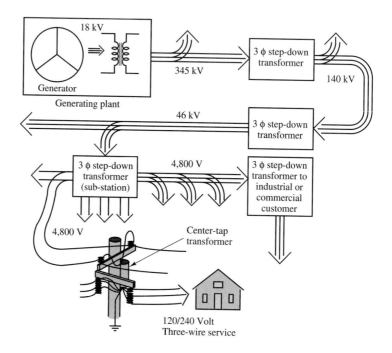

**Figure 7.64** Structure of an AC power distribution network

The local electric company operates a power-generating plant that is capable of supplying several hundred megavolt-amperes (MVA) on a three-phase basis. For this reason, the power company uses a three-phase step-up transformer at the generation plant to increase the line voltage to around 345 kV. One can immediately see that at the rated power of the generator (in MVA) there will be a significant reduction of current beyond the step-up transformer.

Beyond the generation plant, an electric power network distributes energy to several **substations.** This network is usually referred to as the **power grid.** At the substations, the voltage is stepped down to a lower level (10 to 150 kV, typically). Some very large loads (for example, an industrial plant) may be served directly from the power grid, although most loads are supplied by individual substations in the power grid. At the local substations (one of which you may have seen in your own neighborhood), the voltage is stepped down further by a three-phase step-down transformer to 4,800 V. These substations distribute the energy to residential and industrial customers. To further reduce the line voltage to levels that are safe for residential use, step-down transformers are mounted on utility poles. These drop the voltage to the 120/240-V three-wire single-phase residential service discussed in the previous section. Industrial and commercial customers receive 460- and/or 208-V three-phase service.

## CONCLUSION

This chapter introduced the essential elements leading to the analysis of AC power systems. Single-phase AC power, ideal transformers, and three-phase power were discussed. A brief review of residential circuit wiring and safety, and a description of an electric distribution network, were also given to underscore the importance of these concepts in electric power.

- The power dissipated by a load in an AC circuit consists of the sum of an average and a fluctuating component. In practice, the average power is the quantity of interest.
- AC power can best be analyzed with the aid of complex notation. Complex power is defined as the product of the phasor load voltage and the complex conjugate of the phasor load current. Complex power consists of the sum of a real component (the average, or real, power) and an imaginary component (reactive power). Real power corresponds to the electric power for which a user is billed by a utility company; reactive power corresponds to energy storage and cannot be directly used.
- Although reactive power is of no practical use, it does cause an undesirable increase in the current that must be generated by the electric company, resulting in additional line losses. Thus, it is customary to try to reduce reactive power. A measure of the presence of reactive power at a load is the power factor, equal to the cosine of the angle of the load impedance. By adding a suitable reactance to the load, it is possible to attain power factors close to ideal (unity). This procedure is called *power factor correction.*
- Electric power is most commonly generated in three-phase form, for reasons of efficiency. Three-phase power entails the generation of three 120° out-of-phase AC voltages of equal amplitude, so that the instantaneous power is actually constant. Three-phase sources and loads can be configured in either wye or delta configurations; of these, the wye form is more common. The calculation of currents, voltages, and power in three-phase circuits is greatly simplified if one uses per-phase calculations.

## CHECK YOUR UNDERSTANDING ANSWERS

| | |
|---|---|
| **CYU 7.2** | $Z = 4.8e^{-j33.5°}$ Ω; $P_{av} = 2,103.4$ W |
| **CYU 7.3** | See Example 7.2. |
| **CYU 7.4** | 101.46 W |
| **CYU 7.5** | pf $= \cos 89.36° = 0.0105$ |
| **CYU 7.6** | 0.514 lagging, 1 |
| **CYU 7.8** | (a) 0.848, leading; (b) 0.9925, lagging |
| **CYU 7.9** | (a) capacitive; (b) capacitive; (c) inductive; (d) neither (resistive) |
| **CYU 7.11** | 0.584 A |
| **CYU 7.12** | $N = 0.0667$ |
| **CYU 7.13** | $Z_S = 0.0686 - j0.3429$ Ω |
| **CYU 7.14** | $P_{loss} = 43.2$ kW |
| **CYU 7.15** | $\mathbf{V}_a = 480\angle 0°$ V; $\mathbf{I}_a = 151.8\angle -71.6°$ A; $S_T = 69.12$ W $+ j207.4 \times 10^3$ VA |

**CYU 7.18**   $I_\Delta = 3I_y$

**CYU 7.19**   $\mathbf{I}_a = 189\angle 0°$ A; $\mathbf{I}_b = 189\angle -120°$ A; $\mathbf{I}_c = 189\angle 120°$ A

**CYU 7.20**   Losses for a 120-V circuit are approximately double the losses for a 240-V circuit of the same power rating.

# HOMEWORK PROBLEMS

## Section 1:
## Basic AC Power Calculations

**7.1**  The heating element in a soldering iron has a resistance of 391 $\Omega$. Find the average power dissipated in the soldering iron if it is connected to a voltage source of 117 V rms.

**7.2**  The heating element in an electric heater has a resistance of 10 $\Omega$. Find the power dissipated in the heater when it is connected to a voltage source of 240 V rms.

**7.3**  A current source $i(t)$ is connected to a 100-$\Omega$ resistor. Find the average power delivered to the resistor, given that $i(t)$ is:
   a. $4\cos 100t$ A
   b. $4\cos(100t - 0.873)$ A
   c. $4\cos 100t - 3\cos(100t - 0.873)$ A
   d. $4\cos 100t - 3$ A

**7.4**  Find the rms value of each of the following periodic currents:
   a. $\cos 377t + \cos 377t$
   b. $\cos 2t + \sin 2t$
   c. $\cos 377t + 1$
   d. $\cos 2t + \cos(2t + 3\pi/4)$
   e. $\cos 2t + \cos 3t$

**7.5**  A current of 10 A rms flows when a single-phase circuit is placed across a 220-V rms source. The current lags the voltage by $\pi/3$ rad. Find the power dissipated by the circuit and the power factor.

**7.6**  A single-phase circuit is placed across a 120-V rms, 60-Hz source, with an ammeter, a voltmeter, and a wattmeter connected. The instruments indicate 12 A, 120 V, and 800 W, respectively. Find
   a. The power factor.
   b. The phase angle.
   c. The impedance.
   d. The resistance.

**7.7**  The nameplate on a single-phase induction machine reads 2 horsepower (output), 110 V rms, 60 Hz, and 24 A rms. Find the power factor of the machine if the efficiency at rated output is 80 percent. [*Note:* 1 horsepower = 0.746 kW.]

**7.8**  Given the waveform of a voltage source shown in Figure P7.8, find:
   a. the average and rms values of the voltage.
   b. the average power supplied to a 10-$\Omega$ resistor connected across the voltage source.

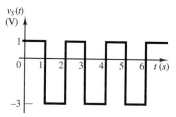

**Figure P7.8**

## Section 2:  Complex Power

**7.9**  For the following numerical values, determine the average power, $P$, the reactive power, $Q$, and the complex power, $S$, of the circuit shown in Figure P7.9. Note: phasor quantities are rms.
   a. $v_S(t) = 650\cos(377t)$ V
      $i_L(t) = 20\cos(377t - 0.175)$ A
   b. $\tilde{\mathbf{V}}_S = 460\angle 0$ V
      $\tilde{\mathbf{I}}_L = 14.14\angle -\pi/4$ A
   c. $\tilde{\mathbf{V}}_S = 100\angle 0$ V
      $\tilde{\mathbf{I}}_L = 8.6\angle -1.5$ A
   d. $\tilde{\mathbf{V}}_S = 208\angle -\pi/6$ V
      $\tilde{\mathbf{I}}_L = 2.3\angle -1.1$ A

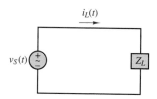

**Figure P7.9**

**7.10**  For the circuit of Figure P7.9, determine the power factor for the load and state whether it is leading or lagging for the following conditions:

a. $v_S(t) = 540\cos(\omega t + \pi/12)$ V
   $i_L(t) = 20\cos(\omega t + 0.82)$ A
b. $v_S(t) = 155\cos(\omega t - \pi/12)$ V
   $i_L(t) = 20\cos(\omega t - 0.384)$ A
c. $v_S(t) = 208\cos(\omega t)$ V
   $i_L(t) = 1.7\sin(\omega t + 3.054)$ A
d. $Z_L = (48 + j16)$ Ω

**7.11**  For the circuit of Figure P7.9, determine whether the load is capacitive or inductive for the circuit shown if

a. pf = 0.87 (leading)
b. pf = 0.42 (leading)
c. $v_S(t) = 42\cos(\omega t)$
   $i_L(t) = 4.2\sin(\omega t)$
d. $v_S(t) = 10.4\cos(\omega t - \pi/15)$
   $i_L(t) = 0.4\cos(\omega t - \pi/15)$

**7.12**  Find the real and reactive power supplied by the source in the circuit shown in Figure P7.12.

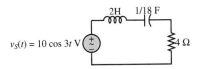

Figure P7.12

**7.13**  For the circuit shown in Figure P7.13, find the real and reactive power supplied by each source. The sources are $\tilde{\mathbf{V}}_{s1} = 36\angle -\pi/3$ V and $\tilde{\mathbf{V}}_{s2} = 24\angle 0.644$ V.

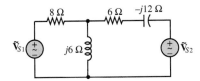

Figure P7.13

**7.14**  The load $Z_L$ in the circuit of Figure P7.14 consists of a 25-Ω resistor in parallel with a 100-μF capacitor. Assume $\omega = 377$ rad/s. Calculate

a. The apparent power delivered to the load.
b. The apparent power supplied by the source.
c. The power factor of the load.

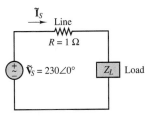

Figure P7.14

**7.15**  Calculate the apparent power, real power, and reactive power for the circuit shown in Figure P7.15. Draw the power triangle.

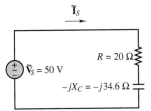

Figure P7.15

**7.16**  A single-phase motor draws 220 W at a power factor of 80 percent (lagging) when connected across a 200-V, 60-Hz source. A capacitor is connected in parallel with the load to give a unity power factor, as shown in Figure P7.16. Find the required capacitance.

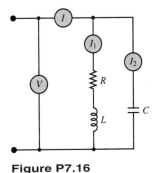

Figure P7.16

**7.17**  Suppose that the electricity in your home has gone out and the power company will not be able to have you hooked up again for several days. The freezer in the basement contains several hundred dollars' worth of food that you cannot afford to let spoil. You have also been experiencing very hot, humid weather and would like to keep one room air-conditioned with a window air conditioner, as well as run the refrigerator in your kitchen. When the appliances are on, they draw the following currents (all values are rms):

Air conditioner:   9.6 A @ 120 V
                   pf = 0.90 (lagging)
Freezer:           4.2 A @ 120 V
                   pf = 0.87 (lagging)
Refrigerator:      3.5 A @ 120 V
                   pf = 0.80 (lagging)

In the worst-case scenario, how much power must an emergency generator supply?

**7.18**  The load on a single-phase three-wire system in a home is generally not balanced. For the system shown in Figure P7.18, let $\tilde{\mathbf{V}}_{s1} = 115\angle 0\ \text{V}_{\text{rms}}$ and $\tilde{\mathbf{V}}_{s2} = 115\angle 0\ \text{V}_{\text{rms}}$. Determine:

a. The total average power delivered to the connected loads: $Z_{L1}$, $Z_{L2}$, and $Z_{L3}$.

b. The total average power lost in the lines: $Z_{g1}$, $Z_{g2}$, and $Z_n$.

c. The average power supplied by each source.

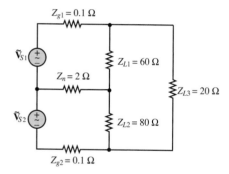

**Figure P7.18**

**7.19**  A large consumer of electricity requires 10 kW of power at 230 $\text{V}_{\text{rms}}$ at a pf angle of $\pi/3$ rad lagging. The transmission line between the electric utility and the consumer has a resistance of 0.1 Ω. If the consumer can increase the pf from 0.5 to 0.9 lagging, determine the change in transmission line losses and load current.

**7.20**  A 1000-W electric motor is connected to a source of 120 $\text{V}_{\text{rms}}$, 60 Hz, and the result is a lagging pf of 0.8. To correct the pf to 0.95 lagging, a capacitor is placed in parallel with the motor. Calculate the current drawn from the source with and without the capacitor connected. Determine the value of the capacitor required to make the correction.

**7.21**  If the voltage and current given below are supplied by a source to a circuit or load, determine:

a. The power supplied by the source which is dissipated as heat or work in the circuit (load).

b. The power stored in reactive components in the circuit (load).

c. The power factor angle and the power factor.

$$\tilde{\mathbf{V}}_s = 7\angle 0.873\ \text{V} \qquad \tilde{\mathbf{I}}_s = 13\ \angle -0.349\ \text{A}$$

**7.22**  Determine $C$ so that the plant power factor of Figure P7.22 is corrected to 1; i.e., $\tilde{\mathbf{I}}_s$ is minimized and in phase with $\tilde{\mathbf{V}}_o$.

$$v_s(t) = 450\ \cos(\omega t)\ \text{V} \qquad \omega = 377\ \text{rad/s}$$
$$Z = 7 + j1\ \Omega$$
$$Z_G = 3 + j0.11\ \text{m}\Omega$$

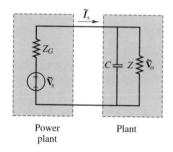

**Figure P7.22**

**7.23**  Determine $C$ so that the plant power factor of Figure P7.22 is corrected to 1 (or the power factor angle to zero) so that $\tilde{\mathbf{I}}_s$ is minimized and in phase with $\tilde{\mathbf{V}}_o$.

$$v_s(t) = 450\ \cos(\omega t)\ \text{V} \qquad \omega = 377\ \text{rad/s}$$
$$Z = 7\angle 0.175\ \Omega$$

**7.24**  Without the capacitor connected into the circuit of Figure P7.22,

$$\tilde{\mathbf{V}}_o = 450\angle 0\ \text{V} \qquad \tilde{\mathbf{I}}_s = 17\angle -0.175\ \text{A}$$
$$f = 60\ \text{Hz} \qquad C = 17.40\ \mu\text{F}$$

The value of $C$ is that which will correct the power factor angle to zero, i.e., reduces $\tilde{\mathbf{I}}_s$ to a minimum value in phase with $\tilde{\mathbf{V}}_o$. Determine the reduction of current which resulted from connecting the capacitor into the circuit.

**7.25**  Without the capacitor connected into the circuit:

$$v_o(t) = 170\ \cos \omega t\ \text{V}$$
$$i_s(t) = 130\ \cos(\omega t - 0.192)\ \text{A}$$
$$f = 60\ \text{Hz} \qquad C = 387\ \mu\text{F}$$

The value of $C$ given is that which will correct the power factor angle to zero, i.e., reduces $\tilde{\mathbf{I}}_s$ to a minimum value in phase with $\tilde{\mathbf{V}}_o$. Determine by how much the current supplied to the plant is reduced by connecting the capacitor.

**7.26**  Determine the time-averaged total power, the real power dissipated, and the reactive power stored in each of the impedances in the circuit shown in Figure P7.26 if:

$$\tilde{V}_{s1} = 170\angle 0 \text{ V}$$

$$\tilde{V}_{s2} = 170 \text{ V}\angle(\pi/2) \text{ V}$$

$$\omega = 377 \text{ rad/s}$$

$$Z_1 = 0.7\angle(\pi/6) \ \Omega$$

$$Z_2 = 1.5\angle 0.105 \ \Omega$$

$$Z_3 = 0.3 + j0.4 \ \Omega$$

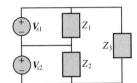

**Figure P7.26**

**7.27**  If the voltage and current supplied to a circuit or load by a source are:

$$\tilde{V}_s = 170\angle -0.157° \text{ V} \qquad \tilde{I}_s = 13\angle 0.28° \text{ A}$$

Determine:

a. The power supplied by the source which is dissipated as heat or work in the circuit (load).

b. The power stored in reactive components in the circuit (load).

c. The power factor angle and power factor.

## Section 3: Transformers

**7.28**  A center-tap transformer has the schematic representation shown in Figure P7.28. The primary-side voltage is stepped down to a secondary-side voltage, $\tilde{V}_{sec}$, by a ratio of $n : 1$. On the secondary side, $\tilde{V}_{sec1} = \tilde{V}_{sec2} = \frac{1}{2}\tilde{V}_{sec}$.

a. If $\tilde{V}_{prim} = 120\angle 32°$ V and $n = 9$, find $\tilde{V}_{sec}$, $\tilde{V}_{sec1}$, and $\tilde{V}_{sec2}$.

b. What must $n$ be if $\tilde{V}_{prim} = 208\angle 0.175$ V and we desire $|\tilde{V}_{sec2}|$ to be 8.7 V?

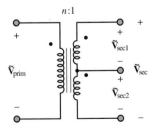

**Figure P7.28**

**7.29**  For the circuit shown in Figure P7.29, find:

a. The total resistance seen by the voltage source.

b. The voltage gain $v_2/v_g$.

c. The value to which the 16-$\Omega$ load resistance should be changed so it will absorb maximum power from the given source.

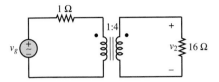

**Figure P7.29**

**7.30**  An ideal transformer is rated to deliver 400 kVA at 460 V to a customer as shown in Figure P7.30.

a. How much current can the transformer supply to the customer?

b. If the customer's load is purely resistive (i.e., if pf = 1), what is the maximum power that the customer can receive?

c. If the customer's power factor is 0.8 (lagging), what is the maximum usable power the customer can receive?

d. What is the maximum power if the pf is 0.7 (lagging)?

e. If the customer requires 300 kW to operate, what is the minimum power factor with the given size transformer?

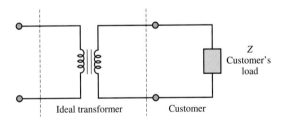

**Figure P7.30**

**7.31**  For the ideal transformer shown in Figure P7.31, find $v_o(t)$ if $v_S(t)$ is $294 \cos 377t$.

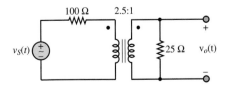

**Figure P7.31**

**7.32**  If the transformer shown in Figure P7.32 is ideal, find the turns ratio $N = 1/n$ that will provide maximum power transfer to the load.

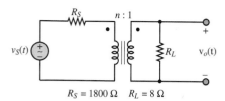

$R_S = 1800\ \Omega$   $R_L = 8\ \Omega$

**Figure P7.32**

**7.33**  Assume the 8-$\Omega$ resistor is the load in the circuit
shown in Figure P7.33. Assume a turns ratio of $1 : n$.
What value of $n$ will result in the load resistor
absorbing maximum power from the source?

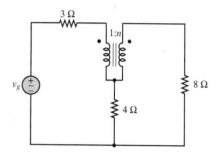

**Figure P7.33**

**7.34**  If we knew that the transformer shown in Figure
P7.34 was to deliver 50 A at 110 V rms with a certain
resistive load, what rms phasor voltage source, $\mathbf{V}_S$,
would provide this voltage and current?

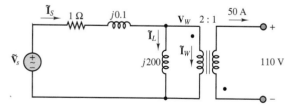

**Figure P7.34**

**7.35**  A method for determining the equivalent circuit of
a transformer consists of two tests: the open-circuit test
and the short-circuit test. The open-circuit test, shown
in Figure P7.35(a), is usually done by applying rated
voltage to the primary side of the transformer while

leaving the secondary side open. The current into the
primary side is measured, as is the power dissipated.

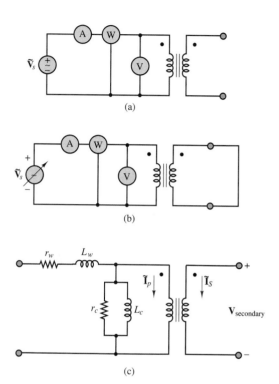

**Figure P7.35**

The short-circuit test, shown in Figure P7.35(b),
is performed by increasing the primary voltage until
rated current is going into the transformer while the
secondary side is short-circuited. The current into the
transformer, the applied voltage, and the power
dissipated are measured.

The equivalent circuit of a transformer is shown
in Figure P7.35(c), where $r_w$ and $L_w$ represent the
winding resistance and inductance, respectively, and $r_c$
and $L_c$ represent the losses in the core of the
transformer and the inductance of the core. The ideal
transformer is also included in the model.

With the open-circuit test, we may assume that
$\tilde{\mathbf{I}}_P = \tilde{\mathbf{I}}_S = 0$. Then all of the current that is measured
is directed through the parallel combination of $r_c$ and
$L_c$. We also assume that $|r_c||j\omega L_c|$ is much greater
than $r_w + j\omega L_w$. Using these assumptions and the
open-circuit test data, we can find the resistance $r_c$ and
the inductance $L_c$.

In the short-circuit test, we assume that $\tilde{\mathbf{V}}_{\text{secondary}}$
is zero, so that the voltage on the primary side of the
ideal transformer is also zero, causing no current flow
through the $r_c - L_c$ parallel combination. Using this

assumption with the short-circuit test data, we are able to find the resistance $r_w$ and inductance $L_w$.

Using the following test data, find the equivalent circuit of the transformer:

Open-circuit test:    $\tilde{V} = 241$ V
$\phantom{Open-circuit test:}$ $\tilde{I} = 0.95$ A
$\phantom{Open-circuit test:}$ $P = 32$ W

Short-circuit test:    $\tilde{V} = 5$ V
$\phantom{Short-circuit test:}$ $\tilde{I} = 5.25$ A
$\phantom{Short-circuit test:}$ $P = 26$ W

Both tests were made at $\omega = 377$ rad/s.

**7.36**  Using the methods of Problem 7.35 and the following data, find the equivalent circuit of the transformer tested:

Open-circuit test:    $\tilde{V}_P = 4{,}600$ V
$\phantom{Open-circuit test:}$ $\tilde{I}_{OC} = 0.7$ A
$\phantom{Open-circuit test:}$ $P = 200$ W

Short-circuit test:    $P = 50$ W
$\phantom{Short-circuit test:}$ $\tilde{V}_P = 5.2$ V

The transformer is a 460-kVA transformer, and the tests are performed at 60 Hz.

## Section 4: Three-Phase Power

**7.37**  The magnitude of the phase voltage of a balanced three-phase wye system is 100 V. Express each phase and line voltage in both polar and rectangular coordinates.

**7.38**  The phase currents in a four-wire wye-connected load are as follows:

$$\tilde{I}_{an} = 10\angle 0, \tilde{I}_{bn} = 12\angle 5\pi/6, \tilde{I}_{cn} = 8\angle 2.88$$

Determine the current in the neutral wire.

**7.39**  For the circuit shown in Figure P7.39, we see that each voltage source has a phase difference of $2\pi/3$ in relation to the others.

a. Find $\tilde{V}_{RW}$, $\tilde{V}_{WB}$, and $\tilde{V}_{BR}$, where
$\tilde{V}_{RW} = \tilde{V}_R - \tilde{V}_W$, $\tilde{V}_{WB} = \tilde{V}_W - \tilde{V}_B$, and
$\tilde{V}_{BR} = \tilde{V}_B - \tilde{V}_R$.

b. Repeat part a, using the calculations

$$\tilde{V}_{RW} = \tilde{V}_R\sqrt{3}\angle -\pi/6$$

$$\mathbf{V}_{WB} = \mathbf{V}_W\sqrt{3}\angle -\pi/6$$

$$\mathbf{V}_{BR} = \mathbf{V}_B\sqrt{3}\angle -\pi/6$$

c. Compare the results of part a with the results of part b.

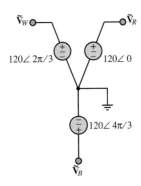

**Figure P7.39**

**7.40**  For the three-phase circuit shown in Figure P7.40, find the currents $\tilde{I}_W$, $\tilde{I}_B$, $\tilde{I}_R$, and $\tilde{I}_N$.

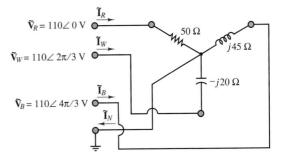

**Figure P7.40**

**7.41**  For the circuit shown in Figure P7.41, find the currents $\tilde{I}_R$, $\tilde{I}_W$, $\tilde{I}_B$, and $\tilde{I}_N$.

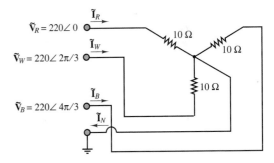

**Figure P7.41**

**7.42**  In the circuit of Figure P7.42:

$$v_{s1} = 170\,\cos(\omega t)\text{ V}$$
$$v_{s2} = 170\,\cos(\omega t + 2\pi/3)\text{ V}$$
$$v_{s3} = 170\,\cos(\omega t - 2\pi/3)\text{ V}$$

$$f = 60\text{ Hz}\qquad Z_1 = 0.5\angle 20°\ \Omega$$
$$Z_2 = 0.35\angle 0°\ \Omega \qquad Z_3 = 1.7\angle -90°\ \Omega$$

Determine the current through $Z_1$ using:

a. Loop/mesh analysis.

b. Node analysis.

c. Superposition.

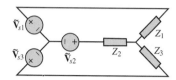

**Figure P7.42**

**7.43** Determine the current through $R$ in the circuit of Figure P7.43:

$$v_1 = 170\cos(\omega t) \text{ V}$$
$$v_2 = 170\cos(\omega t - 2\pi/3) \text{ V}$$
$$v_3 = 170\cos(\omega t + 2\pi/3) \text{ V}$$
$$f = 400 \text{ Hz} \qquad R = 100 \text{ }\Omega$$
$$C = 0.47 \text{ }\mu\text{F} \qquad L = 100 \text{ mH}$$

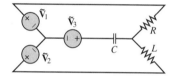

**Figure P7.43**

**7.44** The three sources in the circuit of Figure P7.44 are connected in wye configuration and the loads in a delta configuration. Determine the current through each impedance.

$$v_{s1} = 170\cos(\omega t) \text{ V}$$
$$v_{s2} = 170\cos(\omega t + 2\pi/3) \text{ V}$$
$$v_{s3} = 170\cos(\omega t - 2\pi/3) \text{ V}$$

$$f = 60 \text{ Hz} \qquad Z_1 = 3\angle 0 \text{ }\Omega$$
$$Z_2 = 7\angle \pi/2 \text{ }\Omega \qquad Z_3 = 0 - j11 \text{ }\Omega$$

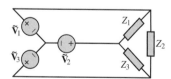

**Figure P7.44**

**7.45** If we model each winding of a three-phase motor like the circuit shown in Figure P7.45(a) and connect the windings as shown in Figure P7.45(b), we have the three-phase circuit shown in Figure P7.45(c). The motor can be constructed so that $R_1 = R_2 = R_3$ and $L_1 = L_2 = L_3$, as is the usual case. If we connect the motor as shown in Figure P7.45(c), find the currents $\tilde{\mathbf{I}}_R, \tilde{\mathbf{I}}_W, \tilde{\mathbf{I}}_B$, and $\tilde{\mathbf{I}}_N$, assuming that the resistances are 40 $\Omega$ each and each inductance is 5 mH. The frequency of each of the sources is 60 Hz.

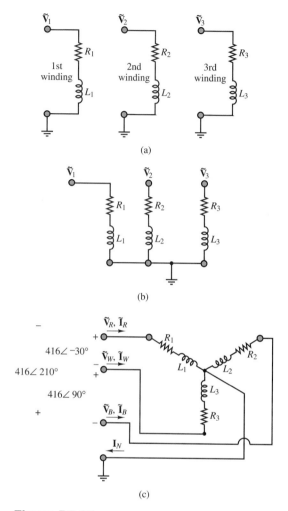

**Figure P7.45**

**7.46** With reference to the motor of Problem 7.44,

a. How much power (in watts) is delivered to the motor?

b. What is the motor's power factor?

c. Why is it common in industrial practice *not* to connect the ground lead to motors of this type?

**7.47** Find the apparent power and the real power delivered to the load in the Y-$\Delta$ circuit shown in Figure P7.47. What is the power factor? Assume rms values.

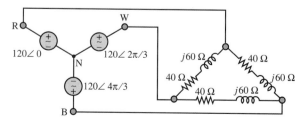

**Figure P7.47**

**7.48** The electric power company is concerned with the loading of its transformers. Since it is responsible to a large number of customers, it must be certain that it can supply the demands of *all* customers. The power company's transformers will deliver rated kVA to the secondary load. However, if the demand were to increase to a point where greater than rated current were required, the secondary voltage would have to drop below rated value. Also, the current would increase, and with it the $I^2R$ losses (due to winding resistance), possibly causing the transformer to overheat. Unreasonable current demand could be caused, for example, by excessively low power factors at the load.

The customer, on the other hand, is not greatly concerned with an inefficient power factor, provided that sufficient power reaches the load. To make the customer more aware of power factor considerations, the power company may install a penalty on the customer's bill. A typical penalty-power factor chart is shown in Table 7.3. Power factors below 0.7 are not permitted. A 25 percent penalty will be applied to any billing after two consecutive months in which the customer's power factor has remained below 0.7.

Table 7.3

| Power factor | Penalty |
| --- | --- |
| 0.850 and higher | None |
| 0.8 to 0.849 | 1% |
| 0.75 to 0.799 | 2% |
| 0.7 to 0.749 | 3% |

Courtesy of Detroit Edison.

The Y-Y circuit shown in Figure P7.48 is representative of a three-phase motor load. Assume rms values.

a. Find the total power supplied to the motor.
b. Find the power converted to mechanical energy if the motor is 80 percent efficient.
c. Find the power factor.
d. Does the company risk facing a power factor penalty on its next bill if all the motors in the factory are similar to this one?

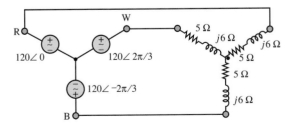

**Figure P7.48**

**7.49** A balanced, three-phase Y-connected source with 230-$V_{rms}$ line voltages has a balanced Y-connected load of $3+j4$ $\Omega$ per phase. For the case that the lines have zero impedance, find all three line currents and the total real power absorbed by the load.

**7.50** The circuit shown in Figure P7.50 is a Y-$\Delta$-Y connected three-phase circuit. The primaries of the transformers are wye-connected, the secondaries are delta-connected, and the load is wye-connected. Find the currents $\tilde{I}_{RP}$, $\tilde{I}_{WP}$, $\tilde{I}_{BP}$, $\tilde{I}_A$, $\tilde{I}_B$, and $\tilde{I}_C$.

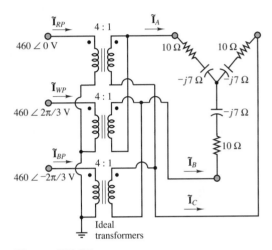

**Figure P7.50**

**7.51** For the circuit shown in Figure P7.51, find the currents $\tilde{I}_A$, $\tilde{I}_B$, $\tilde{I}_C$, and $\tilde{I}_N$, and the real power dissipated by the load.

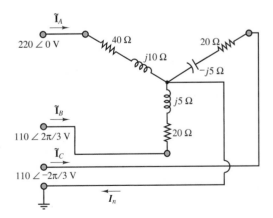

**Figure P7.51**

# PART II
## ELECTRONICS

# C H A P T E R

## 8

# Semiconductors and Diodes

his chapter introduces semiconductor-based electronic devices, and in so doing, it provides a transition between the fundamentals of electrical circuit analysis and the study of electronic circuits. Although the theme of this chapter may seem somewhat different from the circuit analysis of the first seven chapters, the analysis of electrical circuits is still at the core of the material. For example, the operation of diodes will be explained in part using linear circuit models containing resistors and voltage and current sources. In fact, the primary emphasis in this and the next two chapters will be the use of linear circuit models for understanding and analyzing the behavior of more complex nonlinear electronic devices; we show how it is possible to construct models of devices having nonlinear $i$-$v$ characteristics by means of linear circuits. The alternative to this approach would be to conduct an in-depth study of the physics of each class of device: diodes, bipolar transistors, field-effect devices, and other types of semiconductors. Such an approach is neither practical nor fruitful from the viewpoint of this book, since it would entail lengthy explanations and require a significant background in semiconductor physics. Thus, the approach here will be first to provide a qualitative understanding of the physics of each family of devices, and then to describe the devices in terms of their $i$-$v$ characteristics and simple circuit models, illustrating their analysis and applications.

The chapter starts with a discussion of semiconductors and of the $pn$ junction and the semiconductor diode. The second part of this chapter is devoted to a study of diode circuit models, and numerous practical applications. By the end of Chapter 8, you should have accomplished the following objectives:

- A qualitative understanding of electrical conduction in semiconductor materials.
- The ability to explain the $i$-$v$ characteristic of a semiconductor diode (or of a $pn$ junction).
- The ability to use the ideal, offset, and piecewise linear diode models in simple circuits.
- The ability to analyze diode rectifier, peak limiter, peak detector, and regulator circuits and the behavior of LEDs and photocells.

## 8.1  ELECTRICAL CONDUCTION IN SEMICONDUCTOR DEVICES

This section briefly introduces the mechanism of conduction in a class of materials called **semiconductors.** Semiconductors are materials consisting of elements from group IV of the periodic table and having electrical properties falling somewhere between those of conducting and of insulating materials. As an example, consider the conductivity of three common materials. Copper, a good conductor, has a conductivity of $0.59 \times 10^6$ S/cm; glass, a common insulator, may range between $10^{-16}$ and $10^{-13}$ S/cm; while silicon, a semiconductor, has a conductivity that varies from $10^{-8}$ to $10^{-1}$ S/cm. You see, then, that the name *semiconductor* is an appropriate one.

A conducting material is characterized by a large number of conduction-band electrons, which have a very weak bond with the basic structure of the material. Thus, an electric field easily imparts energy to the outer electrons in a conductor and enables the flow of electric current. In a semiconductor, on the other hand, one needs to consider the lattice structure of the material, which in this case is characterized by **covalent bonding.** Figure 8.1 depicts the lattice arrangement for silicon (Si), one of the more common semiconductors. At sufficiently high temperatures, thermal energy causes the atoms in the lattice to vibrate; when sufficient kinetic energy is present, some of the valence electrons break their bonds with the lattice structure and become available as conduction electrons. These **free electrons** enable current flow in the semiconductor. It should be noted that in a conductor valence electrons have a very loose bond with the nucleus and are therefore available for conduction to a much greater extent than valence electrons in a semiconductor. One important aspect of this type of conduction is that the number of charge carriers depends on the amount of thermal energy present in the structure. Thus, many semiconductor properties are a function of temperature.

The free valence electrons are not the only mechanism of conduction in a semiconductor, however. Whenever a free electron leaves the lattice structure, it creates a corresponding positive charge within the lattice. Figure 8.2 depicts the situation in which a covalent bond is missing because of the departure of a free electron from the structure. The vacancy caused by the departure of a free electron is called a **hole.** Note that whenever a hole is present, we have, in effect, a positive charge. The positive charges also contribute to the conduction process, in the sense that if a valence-band electron "jumps" to fill a neighboring hole, thereby neutralizing a positive charge, it correspondingly creates a new hole at a different location. Thus, the effect is equivalent to that of a positive charge moving to the

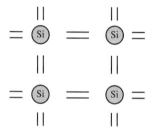

**Figure 8.1** Lattice structure of silicon, with four valence electrons

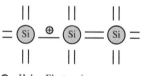

$\oplus$ = Hole  Electron jumps to fill hole

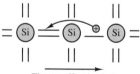

The net effect is a hole moving to the right

A vacancy (or hole) is created whenever a free electron leaves the structure.
This "hole" can move around the lattice if other electrons replace the free electron.

**Figure 8.2** Free electrons and "holes" in the lattice structure

right, in the sketch of Figure 8.2. This phenomenon becomes relevant when an external electric field is applied to the material. It is important to point out here that the **mobility**—that is, the ease with which charge carriers move across the lattice—differs greatly for the two types of carriers. Free electrons can move far more easily around the lattice than holes. To appreciate this, consider the fact that a free electron has already broken the covalent bond, whereas for a hole to travel through the structure, an electron must overcome the covalent bond each time the hole jumps to a new position.

According to this relatively simplified view of semiconductor materials, we can envision a semiconductor as having two types of charge carriers—holes and free electrons—which travel in opposite directions when the semiconductor is subjected to an external electric field, giving rise to a net flow of current in the direction of the electric field. Figure 8.3 illustrates the concept.

An additional phenomenon, called **recombination,** reduces the number of charge carriers in a semiconductor. Occasionally, a free electron traveling in the immediate neighborhood of a hole will recombine with the hole, to form a covalent bond. Whenever this phenomenon takes place, two charge carriers are lost. However, in spite of recombination, the net balance is such that a number of free electrons always exist at a given temperature. These electrons are therefore available for conduction. The number of free electrons available for a given material is called the **intrinsic concentration,** $n_i$. For example, at room temperature, silicon has

$$n_i = 1.5 \times 10^{16} \text{ electrons/m}^3 \tag{8.1}$$

Note that there must be an equivalent number of holes present as well.

Semiconductor technology rarely employs pure, or intrinsic, semiconductors. To control the number of charge carriers in a semiconductor, the process of **doping** is usually employed. Doping consists of adding impurities to the crystalline structure of the semiconductor. The amount of these impurities is controlled, and the impurities can be of one of two types. If the dopant is an element from the fifth column of the periodic table (e.g., arsenic), the end result is that wherever an impurity is present, an additional free electron is available for conduction. Figure 8.4 illustrates the concept. The elements providing the impurities are called **donors** in the case of group V elements, since they "donate" an additional free electron to the lattice structure. An equivalent situation arises when group III elements (e.g., indium) are used to dope silicon. In this case, however, an additional hole is created by the doping element, which is called an **acceptor,** since it accepts a free electron from the structure and generates a hole in doing so.

Semiconductors doped with donor elements conduct current predominantly by means of free electrons and are therefore called **$n$-type semiconductors.** When an acceptor element is used as the dopant, holes constitute the most common carrier, and the resulting semiconductor is said to be a **$p$-type semiconductor.** Doping usually takes place at such levels that the concentration of carriers due to the dopant is significantly greater than the intrinsic concentration of the original semiconductor. If $n$ is the total number of free electrons and $p$ that of holes, then in an $n$-type doped semiconductor, we have

$$n \gg n_i \tag{8.2}$$

and

$$p \ll p_i \tag{8.3}$$

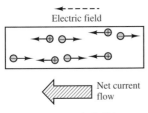

Electric field

Net current flow

An external electric field forces holes to migrate to the left and free electrons to the right. The net current flow is to the left.

**Figure 8.3** Current flow in a semiconductor

An additional free electron is created when Si is "doped" with a group V element.

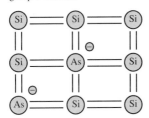

**Figure 8.4** Doped semiconductor

Thus, free electrons are the **majority carriers** in an *n*-type material, while holes are the **minority carriers.** In a *p*-type material, the majority and minority carriers are reversed.

Doping is a standard practice for a number of reasons. Among these are the ability to control the concentration of charge carriers, and the increase in the conductivity of the material that results from doping.

## 8.2   THE *pn* JUNCTION
## AND THE SEMICONDUCTOR DIODE

A simple section of semiconductor material does not in and of itself possess properties that make it useful for the construction of electronic circuits. However, when a section of *p*-type material and a section of *n*-type material are brought in contact to form a *pn* **junction,** a number of interesting properties arise. The *pn* junction forms the basis of the **semiconductor diode**, a widely used circuit element.

Figure 8.5 depicts an idealized *pn* junction, where on the *p* side, we see a dominance of positive charge carriers, or holes, and on the *n* side, the free electrons dominate. Now, in the neighborhood of the junction, in a small section called the **depletion region**, the mobile charge carriers (holes and free electrons) come into contact with each other and recombine, thus leaving virtually no charge carriers at the junction. What is left in the depletion region, in the absence of the charge carriers, is the lattice structure of the *n*-type material on the right, and of the *p*-type material on the left. But the *n*-type material, deprived of the free electrons, which have recombined with holes in the neighborhood of the junction, is now positively ionized. Similarly, the *p*-type material at the junction is negatively ionized, because holes have been lost to recombination. The net effect is that, while most of the material (*p*- or *n*-type) is charge-neutral because the lattice structure and the charge carriers neutralize each other (on average), the depletion region sees a separation of charge, giving rise to an electric field pointing from the *n* side to the *p* side. The charge separation therefore causes a **contact potential** to exist at the junction. This potential is typically on the order of a few tenths of a volt and depends on the material (about 0.6 to 0.7 V for silicon). The contact potential is also called the *offset voltage, $V_\gamma$*.

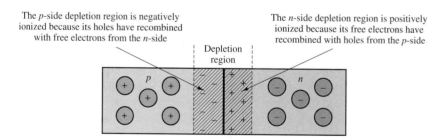

**Figure 8.5** A *pn* junction

In effect, then, if one were to connect the two terminals of the *pn* junction to each other, to form a closed circuit, two currents would be present. First, a small current, called **reverse saturation current,** $I_0$, exists because of the presence of the contact potential and the associated electric field. In addition, it also happens that holes and free electrons with sufficient thermal energy can cross the junction. This current across the junction flows opposite to the reverse saturation current and is called **diffusion current,** $I_d$. Of course, if a hole from the $p$ side enters the $n$ side, it is quite likely that it will quickly recombine with one of the $n$-type carriers on the $n$ side. One way to explain diffusion current is to visualize the diffusion of a gas in a room: gas molecules naturally tend to diffuse from a region of higher concentration to one of lower concentration. Similarly, the $p$-type material—for example—has a much greater concentration of holes than the $n$-type material. Thus, some holes will tend to diffuse into the $n$-type material across the junction, although only those that have sufficient (thermal) energy to do so will succeed. Figure 8.6 illustrates this process.

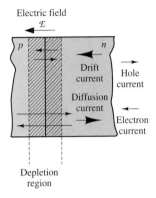

**Figure 8.6** Drift and diffusion currents in a *pn* junction

The phenomena of drift and diffusion help explain how a $pn$ junction behaves when it is connected to an external energy source. Consider the diagrams of Figure 8.7, where a battery has been connected to a $pn$ junction in the **reverse-biased** direction (Figure 8.7(a)), and in the **forward-biased** direction (Figure 8.7(b)). We assume that some suitable form of contact between the battery wires and the semiconductor material can be established (this is called an **ohmic contact**). The effect of a reverse bias is to increase the contact potential at the junction. Now, the majority carriers trying to diffuse across the junction need to overcome a greater barrier (a larger potential) and a wider depletion region. Thus, the diffusion current becomes negligible. The only current that flows under reverse bias is the very small reverse saturation current, so that the diode current, $i_D$ (defined in the figure), is

$$i_D = -I_0 \tag{8.4}$$

When the $pn$ junction is forward-biased, the contact potential across the junction is lowered (note that $V_B$ acts in opposition to the contact potential). Now, the diffusion of majority carriers is aided by the external voltage source; in fact, the diffusion current increases as a function of the applied voltage, according to equation 8.5

$$I_d = I_0 e^{qv_D/kT} \tag{8.5}$$

where $v_D$ is the voltage across the $pn$ junction, $k = 1.381 \times 10^{-23}$ J/K is Boltzmann's constant, $q$ the charge of one electron, and $T$ the temperature of the material in kelvins (K). The quantity $kT/q$ is constant at a given temperature and is approximately equal to 25 mV at room temperature. The net diode current under forward bias is given by equation 8.6

$$\boxed{i_D = I_d - I_0 = I_0(e^{qv_D/kT} - 1)} \qquad \text{Diode equation} \qquad \textbf{(8.6)}$$

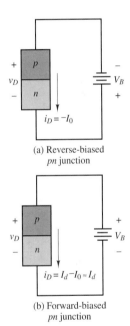

(a) Reverse-biased *pn* junction

(b) Forward-biased *pn* junction

**Figure 8.7** Forward- and reverse-biased *pn* junctions

which is known as the **diode equation.** Figure 8.8 depicts the diode $i$-$v$ characteristic described by the diode equation for a fairly typical silicon diode for positive diode voltages. Since the reverse saturation current, $I_0$, is typically very small ($10^{-9}$ to $10^{-15}$ A), equation 8.7:

$$i_D = I_0 e^{qv_D/kT} \tag{8.7}$$

is a good approximation if the diode voltage, $v_D$, is greater than a few tenths of a volt.

The ability of the $pn$ junction to essentially conduct current in only one direction—that is, to conduct only when the junction is forward-biased—makes it

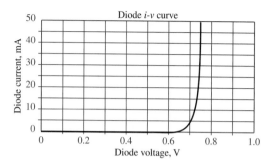

**Figure 8.8** Semiconductor diode $i$-$v$ characteristic

valuable in circuit applications. A device having a single *pn* junction and ohmic contacts at its terminals, as described in the preceding paragraphs, is called a *semiconductor diode*, or simply *diode*. As will be shown later in this chapter, it finds use in many practical circuits. The circuit symbol for the diode is shown in Figure 8.9, along with a sketch of the *pn* junction.

Figure 8.10 summarizes the behavior of the semiconductor diode by means of its *i*-*v* characteristic; it will become apparent later that this *i*-*v* characteristic plays an important role in constructing circuit models for the diode. Note that a third region appears in the diode *i*-*v* curve that has not been discussed yet. The **reverse-breakdown** region to the far left of the curve represents the behavior of the diode when a sufficiently high reverse bias is applied. Under such a large reverse bias (greater in magnitude than the voltage $V_Z$, a quantity that will be explained shortly), the diode conducts current again, this time *in the reverse direction*. To explain the mechanism of reverse conduction, one needs to visualize the phenomenon of *avalanche breakdown*. When a very large negative bias is applied to the *pn* junction, sufficient energy is imparted to charge carriers that reverse current can flow, well beyond the normal reverse saturation current. In addition, because of the large electric field, electrons are energized to such levels that if they collide with other charge carriers at a lower energy level, some of their energy is transferred to the carriers with lower energy, and these can now contribute to the reverse conduction process, as well. This process is called *impact ionization*. Now, these new carriers may also have enough energy to energize other low-energy electrons by impact ionization, so that once a sufficiently high reverse bias is provided, this process of conduction takes place very much like an avalanche: a single electron can ionize several others.

The phenomenon of **Zener breakdown** is related to avalanche breakdown. It is usually achieved by means of heavily doped regions in the neighborhood of the metal-semiconductor junction (the ohmic contact). The high density of charge carriers provides the means for a substantial reverse breakdown current to be sustained, at a nearly constant reverse bias, the **Zener voltage, $V_Z$.** This phenomenon is very useful in applications where one would like to hold some load voltage constant—for example, in **voltage regulators,** which are discussed in a later section.

To summarize the behavior of the semiconductor diode, it is useful to refer to the sketch of Figure 8.10, observing that when the voltage across the diode, $v_D$, is greater than the offset voltage, $V_\gamma$, the diode is said to be forward-biased and

The arrow in the circuit symbol for the diode indicates the direction of current flow when the diode is forward-biased.

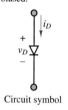

Circuit symbol

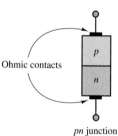

*pn* junction

**Figure 8.9** Semiconductor diode circuit symbol

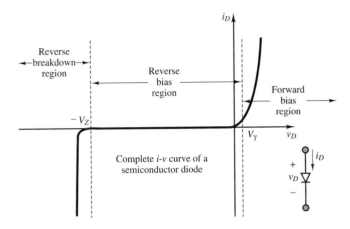

**Figure 8.10** $i$-$v$ characteristic of semiconductor diode

acts nearly as a short circuit, readily conducting current. When $v_D$ is between $V_\gamma$ and the Zener breakdown voltage, $-V_Z$, the diode acts very much like an open circuit, conducting a small reverse current, $I_0$, of the order of only nanoamperes (nA). Finally, if the voltage $v_D$ is more negative than the Zener voltage, $-V_Z$, the diode conducts again, this time in the reverse direction.

# 8.3    CIRCUIT MODELS FOR THE SEMICONDUCTOR DIODE

From the viewpoint of a *user* of electronic circuits (as opposed to a *designer*), it is often sufficient to characterize a device in terms of its $i$-$v$ characteristic, using either load-line analysis or appropriate circuit models to determine the operating currents and voltages. This section shows how it is possible to use the $i$-$v$ characteristics of the semiconductor diode to construct simple yet useful *circuit models*. Depending on the desired level of detail, it is possible to construct *large-signal models* of the diode, which describe the gross behavior of the device in the presence of relatively large voltages and currents; or *small-signal models*, which are capable of describing the behavior of the diode in finer detail and, in particular, the response of the diode to small changes in the average diode voltage and current. From the user's standpoint, these circuit models greatly simplify the analysis of diode circuits and make it possible to effectively analyze relatively "difficult" circuits simply by using the familiar circuit analysis tools of Chapter 3. The first two major divisions of this section will describe different diode models and the assumptions under which they are obtained, to provide the knowledge you will need to select and use the appropriate model for a given application.

## Large-Signal Diode Models

### *Ideal Diode Model*

Our first large-signal model treats the diode as a simple on-off device (much like a check valve in hydraulic circuits—see box, "Hydraulic Check Valves").

## Hydraulic Check Valves

**FIND IT ON THE WEB**

To understand the operation of the semiconductor diode intuitively, we make reference to a very common hydraulic device that finds application whenever one wishes to restrict the flow of a fluid to a single direction, and prevent (check) reverse flow. Hydraulic **check valves** perform this task in a number of ways. We illustrate a few examples in this box.

Figure 1 depicts a *swing check valve*. In this design, flow from left to right is permitted, as the greater fluid pressure on the right side of the valve forces the swing "door" to open. If flow were to reverse, the reversal of fluid pressure (greater pressure on the right) would cause the swing door to shut.

Figure 2 depicts a *flapper check valve*. The principle is similar to that described above for the swing check valve. In Figure 2, fluid flow is permitted from left to right, and not in the reverse direction. The response of the valve of Figure 2 is faster (due to the shorter travel distance of the flapper) than that of Figure 1.

You will find the analysis of the diode circuits in this chapter much easier to understand intuitively if you visualize the behavior of the diode to be similar to that of the check valves shown here, with the pressure difference across the valve orifice being analogous to the voltage across the diode, and the fluid flow rate being analogous to the current through the diode. Figure 3 depicts the diode circuit symbol. Current flows only from left to right whenever the voltage across the diode is positive, and no current flows when the diode voltage is reversed. The circuit element of Figure 3 is functionally analogous to the two check valves of Figures 1 and 2.

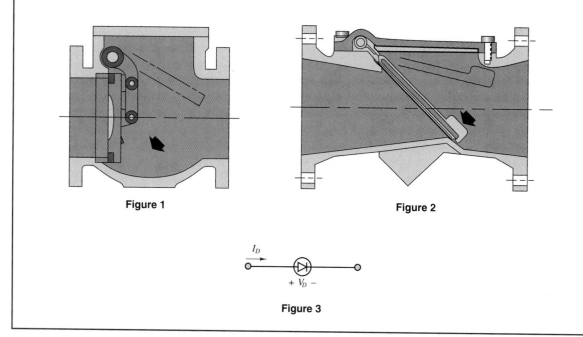

**Figure 1**

**Figure 2**

**Figure 3**

Figure 8.11 illustrates how, on a large scale, the *i-v* characteristic of a typical diode may be approximated by an open circuit when $v_D < 0$ and by a short circuit when $v_D \geq 0$ (recall the *i-v* curves of the ideal short and open circuits presented in Chapter 2). The analysis of a circuit containing a diode may be greatly simplified by using the short-circuit–open-circuit model. From here on, this diode model will be known as the **ideal diode model.** In spite of its simplicity, the ideal diode model (indicated by the symbol shown in Figure 8.11) can be very useful in analyzing diode circuits.

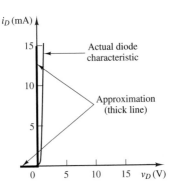

> In the remainder of the chapter, ideal diodes will always be represented by the filled (black) triangle symbol shown in Figure 8.11.

Consider the circuit shown in Figure 8.12, which contains a 1.5-V battery, an ideal diode, and a 1-k$\Omega$ resistor. A technique will now be developed to determine whether the diode is conducting or not, with the aid of the ideal diode model.

Assume first that the diode is conducting (or, equivalently, that $v_D \geq 0$). This enables us to substitute a short circuit in place of the diode, as shown in Figure 8.13, since the diode is now represented by a short circuit, $v_D = 0$. This is consistent with the initial assumption (i.e., diode "on"), since the diode is assumed to conduct for $v_D \geq 0$ and since $v_D = 0$ does not contradict the assumption. The series current in the circuit (and through the diode) is $i_D = 1.5/1,000 = 1.5$ mA. To summarize, the assumption that the diode is on in the circuit of Figure 8.13 allows us to assume a positive (clockwise) current in the circuit. Since the direction of the current and the diode voltage are consistent with the assumption that the diode is on ($v_D \geq 0$, $i_D > 0$), it must be concluded that the diode is indeed conducting.

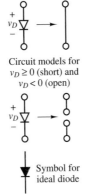

**Figure 8.11** Large-signal on-off diode model

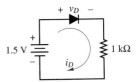

**Figure 8.12** Circuit containing ideal diode

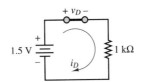

**Figure 8.13** Circuit of Figure 8.12, assuming that the ideal diode conducts

Suppose, now, that the diode had been assumed to be off. In this case, the diode would be represented by an open circuit, as shown in Figure 8.14. Applying KVL to the circuit of Figure 8.14 reveals that the voltage $v_D$ must equal the battery voltage, or $v_D = 1.5$ V, since the diode is assumed to be an open circuit and no current flows through the circuit. Equation 8.8 must then apply.

$$1.5 = v_D + 1,000i_D = v_D \tag{8.8}$$

But the result $v_D = 1.5$ V is contrary to the initial assumption (i.e., $v_D < 0$). Thus, assuming that the diode is off leads to an inconsistent answer. Clearly, the assumption must be incorrect, and therefore the diode must be conducting.

This method can be very useful in more involved circuits, where it is not quite so obvious whether a diode is seeing a positive or a negative bias. The method is particularly effective in these cases, since one can make an educated guess whether the diode is on or off and solve the resulting circuit to verify the

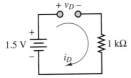

**Figure 8.14** Circuit of Figure 8.12, assuming that the ideal diode does not conduct

correctness of the initial assumption. Some solved examples are perhaps the best way to illustrate the concept.

---

**FOCUS ON METHODOLOGY**

**Determining the Conduction State of an Ideal Diode**

1. Assume a diode conduction state (on or off).
2. Substitute ideal circuit model into circuit (short circuit if on, open circuit if off)
3. Solve for diode current and voltage using linear circuit analysis techniques.
4. If the solution is consistent with the assumption, then the initial assumption was correct; if not, the diode conduction state is opposite to that initially assumed. For example, if the diode has been assumed to be off but the diode voltage computed after replacing the diode with an open circuit is a forward bias, then it must be true that the actual state of the diode is on.

---

## EXAMPLE 8.1 Determining the Conduction State of an Ideal Diode

### Problem

Determine whether the ideal diode of Figure 8.15 is conducting.

---

### Solution

**Known Quantities:** $V_S = 12$ V; $V_B = 11$ V; $R_1 = 5$ $\Omega$; $R_2 = 10$ $\Omega$; $R_3 = 10$ $\Omega$.

**Find:** The conduction state of the diode.

**Assumptions:** Use the ideal diode model.

**Analysis:** Assume initially that the ideal diode does not conduct and replace it with an open circuit, as shown in Figure 8.16. The voltage across $R_2$ can then be computed using the voltage divider rule:

$$v_1 = \frac{R_2}{R_1 + R_2} V_S = \frac{10}{5 + 10} 12 = 8 \text{ V}$$

Applying KVL to the right-hand-side mesh (and observing that no current flows in the circuit since the diode is assumed off), we obtain:

$$v_1 = v_D + V_B \text{ or } v_D = 8 - 11 = -3 \text{ V}$$

The result indicates that the diode is reverse-biased, and confirms the initial assumption. Thus, the diode is not conducting.

As further illustration, let us make the opposite assumption, and assume that the diode conducts. In this case, we should replace the diode with a short circuit, as shown in Figure 8.17. The resulting circuit is solved by nodal analysis, noting that $v_1 = v_2$ since the

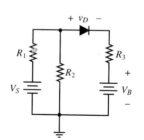

**Figure 8.15**

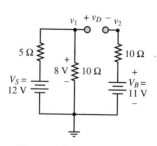

**Figure 8.16**

diode is assumed to act as a short circuit.

$$\frac{V_S - v_1}{R_1} = \frac{v_1}{R_2} + \frac{v_1 - V_B}{R_3}$$

$$\frac{V_S}{R_1} + \frac{V_B}{R_3} = \frac{v_1}{R_1} + \frac{v_1}{R_2} + \frac{v_1}{R_3}$$

$$\frac{12}{5} + \frac{11}{10} = \left(\frac{1}{5} + \frac{1}{10} + \frac{1}{10}\right) v_1$$

$$v_1 = 2.5(2.4 + 1.1) = 8.75V$$

Since $v_1 = v_2 < V_B = 11$ V, we must conclude that current is flowing in the reverse direction (from $V_B$ to node $v_2/v_1$) through the diode. This observation is inconsistent with the initial assumption, since if the diode were conducting, we can see current flow only in the forward direction. Thus, the initial assumption was incorrect, and we must conclude that the diode is not conducting.

**Comments:** The formulation of diode problems illustrated in this example is based on making an initial assumption. The assumption results in replacing the ideal diode with either a short or an open circuit. Once this step is completed, the resulting circuit is a linear circuit and can be solved by known methods to verify the consistency of the initial assumption.

**Focus on Computer-Aided Solution:** The circuit of Figure 8.15 is simulated by *Electronics Workbench*™ in the CD that accompanies the book. Try changing the values of resistors in the simulation circuit to see if it is possible to cause the diode to conduct (*Hint:* Use a very large value for $R_2$). Note that the computer simulation employs an ideal diode model, but could also use a physically correct model of the diode (that of equation 8.6) (click on the diode symbol to see the list of options).

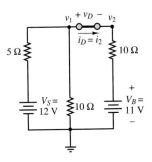

**Figure 8.17**

VIRTUAL LAB

---

# EXAMPLE 8.2 Determining the Conduction State of an Ideal Diode

## Problem

Determine whether the ideal diode of Figure 8.18 is conducting.

---

## Solution

**Known Quantities:** $V_S = 12$ V; $V_B = 11$ V; $R_1 = 5\ \Omega$; $R_2 = 4\ \Omega$.

**Find:** The conduction state of the diode.

**Assumptions:** Use the ideal diode model.

**Analysis:** Assume initially that the ideal diode does not conduct and replace it with an open circuit, as shown in Figure 8.19. The current flowing in the resulting series circuit (shown in Figure 8.19) is:

$$i = \frac{V_S - V_B}{R_1 + R_2} = \frac{1}{9}\ \text{A}$$

The voltage at node $v_1$ is:

$$\frac{12 - v_1}{5} = \frac{v_1 - 11}{4}$$

$$v_1 = 11.44\ \text{V}$$

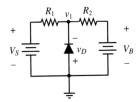

**Figure 8.18**

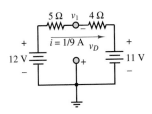

**Figure 8.19**

The result indicates that the diode is strongly reverse-biased, since $v_D = 0 - v_1 = -11.44$ V, and confirms the initial assumption. Thus, the diode is not conducting.

**Focus on Computer-Aided Solution:**  The circuit of Figure 8.18 is simulated by *Electronics Workbench*™ in the CD that accompanies the book. Try changing the values of resistors in the simulation circuit to see if it is possible to cause the diode to conduct.

One of the important applications of the semiconductor diode is **rectification** of AC signals, that is, the ability to convert an AC signal with zero average (DC) value to a signal with a nonzero DC value. The application of the semiconductor diode as a rectifier is very useful in obtaining DC voltage supplies from the readily available AC line voltage. Here, we illustrate the basic principle of rectification, using an ideal diode—for simplicity, and also because the large-signal model is appropriate when the diode is used in applications involving large AC voltage and current levels.

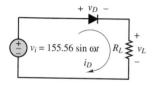

Consider the circuit of Figure 8.20, where an AC source, $v_i = 155.56 \sin \omega t$, is connected to a load by means of a series ideal diode. From the analysis of Example 8.1, it should be apparent that the diode will conduct only during the positive half-cycle of the sinusoidal voltage—that is, that the condition $v_D \geq 0$ will be satisfied only when the AC source voltage is positive—and that it will act as an open circuit during the negative half-cycle of the sinusoid ($v_D < 0$). Thus, the appearance of the load voltage will be as shown in Figure 8.21, with the negative portion of the sinusoidal waveform cut off. The rectified waveform clearly has a nonzero DC (average) voltage, whereas the average input waveform voltage was zero. When the diode is conducting, or $v_D \geq 0$, the unknowns $v_L$ and $i_D$ can be found by using the following equations:

$$i_D = \frac{v_i}{R_L} \quad \text{when} \quad v_i > 0 \tag{8.9}$$

and

$$v_L = i_D R_L \tag{8.10}$$

**Figure 8.20**

The load voltage, $v_L$, and the input voltage, $v_i$, are sketched in Figure 8.21. From equation 8.10, it is obvious that the current waveform has the same shape as the load voltage. The average value of the load voltage is obtained by integrating the load voltage over one period and dividing by the period:

$$v_{\text{load, DC}} = \frac{\omega}{2\pi} \int_0^{\frac{\pi}{\omega}} 155.56 \sin \omega t \, dt = \frac{155.56}{\pi} = 49.52 \text{ V} \tag{8.11}$$

The circuit of Figure 8.20 is called a **half-wave rectifier,** since it preserves only half of the waveform. This is not usually a very efficient way of rectifying an AC signal, since half the energy in the AC signal is not recovered. It will be shown in a later section that it is possible to recover also the negative half of the AC waveform by means of a *full-wave rectifier.*

### Offset Diode Model

While the ideal diode model is useful in approximating the large-scale characteristics of a physical diode, it does not account for the presence of an offset voltage,

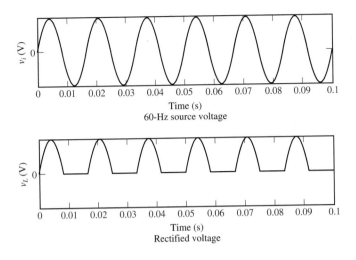

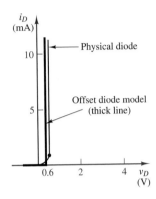

**Figure 8.21** Ideal diode rectifier input and output voltages

which is an unavoidable component in semiconductor diodes (recall the discussion of the contact potential in Section 8.2). The **offset diode model** consists of an ideal diode in series with a battery of strength equal to the offset voltage (we shall use the value $V_\gamma = 0.6$ V for silicon diodes, unless otherwise indicated). The effect of the battery is to shift the characteristic of the ideal diode to the right on the voltage axis, as shown in Figure 8.22. This model is a better approximation of the large-signal behavior of a semiconductor diode than the ideal diode model.

According to the offset diode model, the diode of Figure 8.22 acts as an open circuit for $v_D < 0.6$ V, and it behaves like a 0.6-V battery for $v_D \geq 0.6$ V. The equations describing the offset diode model are as follows:

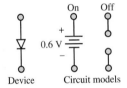

**Figure 8.22**

$$v_D \geq 0.6 \text{ V} \qquad \text{Diode} \rightarrow 0.6\text{-V battery}$$
$$v_D < 0.6 \text{ V} \qquad \text{Diode} \rightarrow \text{Open circuit}$$

(8.12)

The diode offset model may be represented by an ideal diode in series with a 0.6-V ideal battery, as shown in Figure 8.23. Use of the offset diode model is best described by means of examples.

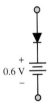

**Figure 8.23** Offset diode as an extension of ideal diode model

---

## EXAMPLE  8.3  Using the Offset Diode Model in a Half-Wave Rectifier

### Problem

Compute and plot the rectified load voltage, $v_R$, in the circuit of Figure 8.24.

---

### Solution

**Known Quantities:**  $v_S(t) = 3 \cos(\omega t)$; $V_\gamma = 0.6$ V.

**Find:**  An analytical expression for the load voltage.

**Assumptions:**  Use the offset diode model.

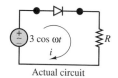

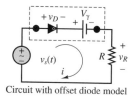

Actual circuit

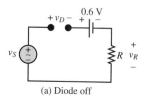

Circuit with offset diode model

**Figure 8.24**

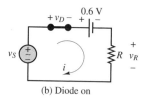

(a) Diode off

(b) Diode on

**Figure 8.25**

*Analysis:* We start by replacing the diode with the offset diode model, as shown in the lower half of Figure 8.24. Now we can use the method developed earlier for ideal diode analysis, that is, we can focus on determining whether the voltage $v_D$ across the ideal diode is positive (diode on) or negative (diode off).

Assume first that the diode is off. The resulting circuit is shown in Figure 8.25(a). Since no current flows in the circuit, we obtain the following expression for $v_D$:

$$v_D = v_S - 0.6$$

To be consistent with the assumption that the diode is off, we require that $v_D$ be negative, which in turns corresponds to

$$v_S < 0.6 \text{ V} \qquad \text{Diode off condition}$$

With the diode off, the current in the circuit is zero, and the load voltage is also zero. If the source voltage is greater than 0.6 V, the diode conducts, and the current flowing in the circuit and resulting load voltage are given by the expressions:

$$i = \frac{v_S - 0.6}{R} \qquad v_R = iR = v_S - 0.6$$

We summarize these results as follows:

$$v_R = 0 \qquad \text{for } v_S < 0.6 \text{ V}$$
$$v_R = v_S - 0.6 \qquad \text{for } v_S \geq 0.6 \text{ V}$$

The resulting waveform is plotted with $v_S$ in Figure 8.26.

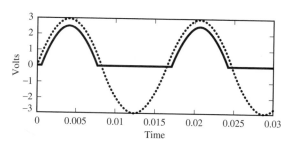

**Figure 8.26** Source voltage (dotted curve) and rectified voltage (solid curve) for the circuit of Figure 8.24.

*Comments:* Note that use of the offset diode model leads to problems that are very similar to ideal diode problems, with the addition of a voltage source in the circuit.

Also observe that the load voltage waveform is shifted downward by an amount equal to the offset voltage, $V_\gamma$. The shift is visible in the case of this example because $V_\gamma$ is a substantial fraction of the source voltage. If the source voltage had peak values of tens or hundreds of volts, such a shift would be negligible, and an ideal diode model would serve just as well.

*Focus on Computer-Aided Solution:* The half-wave rectifier of Figure 8.20 is simulated using *Electronics Workbench*™ in the CD that accompanies the book. The circuit is simulated by using an ideal diode model. Replace the ideal diode with any of the other available options (physical diodes), and observe any differences in the result (*Hint:* The differences will be more dramatic for small peak source voltage values, say 5 volts).

# EXAMPLE   8.4  Using the Offset Diode Model

## Problem

Use the offset diode model to determine the value of $v_1$ for which diode $D_1$ first conducts in the circuit of Figure 8.27.

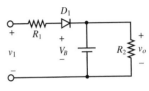

**Figure 8.27**

## Solution

**Known Quantities:** $V_B = 2$ V; $R_1 = 1$ k$\Omega$; $R_2 = 500$ $\Omega$; $V_\gamma = 0.6$ V.

**Find:** The lowest value of $v_1$ for which diode $D_1$ conducts.

**Assumptions:** Use the offset diode model.

**Analysis:** We start by replacing the diode with the offset diode model, as shown in Figure 8.28. Based on our experience with previous examples, we can state immediately that if $v_1$ is negative, the diode will certainly be off. To determine the point at which the diode turns on as $v_1$ is increased, we write the circuit equation assuming that the diode is off. If you were conducting a laboratory experiment, you might monitor $v_1$ and progressively increase it until the diode conducts; the equation below is an analytical version of this experiment. With the diode off, no current flows through $R_1$, and

$$v_1 = v_{D1} + V_\gamma + V_B$$

According to this equation

$$v_{D1} = v_1 - 2.6$$

and the condition required for the diode to conduct is:

$$v_1 > 2.6 \text{ V} \qquad \text{Diode on condition}$$

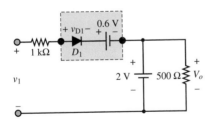

**Figure 8.28**

**Comments:** Once again, the offset diode model permits using the same analysis method that was developed for the ideal diode model.

## Small-Signal Diode Models

As one examines the diode $i$-$v$ characteristic more closely, it becomes apparent that the short-circuit approximation is not adequate to represent the *small-signal behavior* of the diode. The term *small-signal behavior* usually signifies the response of the diode to small time-varying signals that may be superimposed on the average diode current and voltage. Figure 8.8 depicts a close-up view of a silicon diode $i$-$v$ curve. From this figure, it should be apparent that the short-circuit approximation is not very accurate when a diode's behavior is viewed on an expanded scale. To a first-order approximation, however, the $i$-$v$ characteristic resembles that of a resistor (i.e., is linear) for voltages greater than the offset voltage. Thus, it may be reasonable to model the diode as a resistor (instead of a short circuit) *once it is conducting*, to account for the slope of its $i$-$v$ curve. In the following discussion, the method of load-line analysis (which was introduced in Chapter 3) will be exploited to determine the **small-signal resistance** of a diode.

Consider the circuit of Figure 8.29, which represents the Thévenin equivalent circuit of an arbitrary linear resistive circuit connected to a diode. Equations 8.13 and 8.14 describe the operation of the circuit:

$$v_T = i_D R_T + v_D \tag{8.13}$$

arises from application of KVL, and

$$i_D = I_0(e^{qv_D/kT} - 1) \tag{8.14}$$

is the diode equation (8.6).

Although we have two equations in two unknowns, these cannot be solved analytically, since one of the equations contains $v_D$ in exponential form. As discussed in Chapter 3, two methods exist for the solution of *transcendental equations* of this type: graphical and numerical. In the present case, only the graphical solution shall be considered. The graphical solution is best understood if we associate a curve in the $i_D$-$v_D$ plane with each of the two preceding equations. The diode equation gives rise to the familiar curve of Figure 8.8. The *load-line equation*, obtained by KVL, is the equation of a line with slope $-1/R$ and ordinate intercept given by $V_T/R_T$.

$$i_D = -\frac{1}{R_T}v_D + \frac{1}{R_T}V_T \qquad \text{Load line equation} \tag{8.15}$$

The superposition of these two curves gives rise to the plot of Figure 8.30, where the solution to the two equations is graphically found to be the pair of values $(I_Q, V_Q)$. The intersection of the two curves is called the **quiescent (operating) point, or $Q$ point.** The voltage $v_D = V_Q$ and the current $i_D = I_Q$ are the actual diode voltage and current when the diode is connected as in the circuit of Figure 8.29. Note that this method is also useful for circuits containing a larger number of elements, provided that we can represent these circuits by their Thévenin equivalents, with the diode appearing as the load.

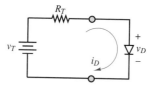

**Figure 8.29** Diode circuit for illustration of load-line analysis

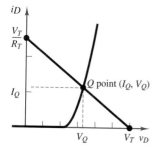

**Figure 8.30** Graphical solution of equations 8.13 and 8.14

# FOCUS ON METHODOLOGY

**Determining the Operating Point of a Diode**

1. Reduce the circuit to a Thévenin or Norton equivalent circuit with the diode as the load.
2. Write the load line equation (8.15).
3. Solve numerically two simultaneous equations in two unknowns (the load line equations and the diode equation) for the diode current and voltage.

   Or

4. Solve graphically by finding the intersection of the diode curve (e.g., from a data sheet) with the load line curve. The intersection of the two curves is the diode operating point.

# FOCUS ON METHODOLOGY

**Using Device Data Sheets**

One of the most important design tools available to engineers is the **device data sheet**. In this box we illustrate the use of a device data sheet for the 1N400X diode. This is a *general-purpose rectifier* diode, designed to conduct average currents in the 1.0-A range. Excerpts from the data sheet are shown below, with some words of explanation. The complete data sheets can be found in the accompanying CD-ROM.

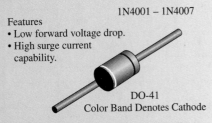

1N4001 – 1N4007

Features
• Low forward voltage drop.
• High surge current capability.

DO-41
Color Band Denotes Cathode

1.0 Ampere General Purpose Rectifiers

## ABSOLUTE MAXIMUM RATINGS:

This table summarizes the limitations of the device. For example, in the first column one can find the maximum allowable average current (1 A), and the maximum *surge current*, that is the maximum short-time burst current the diode can sustain without being destroyed. Also mentioned are the **power rating** and operating temperatures. Note that in the entry for the total device power dissipation, **derating** information is also given. Derating implies that the device power dissipation will change as a function of temperature, in this case at the rate of 20 mW/°C. For example, if we expect to operate the diode at a temperature of 100°C, we would calculate a derated power of:

$$P = 2.5 \text{ W} - (75°C \times 0.02 \text{ mW/}°C) = 1.0 \text{ W}$$

Thus, the diode operated at a higher temperature can dissipate only 1 W.

*(Continued)*

**Absolute Maximum Ratings***    $T = 25°C$ unless otherwise noted

| Symbol | Parameter | Value | Units |
|---|---|---|---|
| $I_0$ | Average Rectified Current<br>.375″ lead length @ $T_A = 75°C$ | 1.0 | A |
| $i_{t(surge)}$ | Peak Forward Surge Current<br>8.3 ms single half-sine-wave<br>Superimposed on rated load (JEDEC method) | 30 | A |
| $P_D$ | Total Device Dissipation<br>Derate above 25°C | 2.5<br>20 | W<br>mW/°C |
| $R_{8JA}$ | Thermal Resistance, Junction to Ambient | 50 | °C/W |
| $T_{stg}$ | Storage Temperature Range | −55 to +175 | °C |
| $T_J$ | Operating Junction Temperature | −55 to +150 | °C |

*These ratings are limiting values above which the serviceability of any semiconductor device may be impaired.

# ELECTRICAL CHARACTERISTICS:

The section on electrical characteristics summarizes some of the important voltage and current specifications of the diode. For example, the maximum DC reverse voltage is listed for each diode in the 1N400X family. Similarly, you will find information on the maximum forward voltage, reverse current, and typical junction capacitance.

**Electrical Characteristics**    $T = 25°C$ unless otherwise noted

| Parameter | Device | | | | | | | Units |
|---|---|---|---|---|---|---|---|---|
| | **4001** | **4002** | **4003** | **4004** | **4005** | **4006** | **4007** | |
| Peak Repetitive Reverse Voltage | 50 | 100 | 200 | 400 | 600 | 800 | 1000 | V |
| Maximum RMS Voltage | 35 | 70 | 140 | 280 | 420 | 560 | 700 | V |
| DC Reverse Voltage    (Rated $V_R$) | 50 | 100 | 200 | 400 | 600 | 800 | 1000 | V |
| Maximum Reverse Current<br>    @ rated $V_R$    $T_A = 25°$<br>                $T_A = 100°$ | 5.0<br>500 | | | | | | | $\mu$A<br>$\mu$A |
| Maximum Forward Voltage @ 1.0 A | 1.1 | | | | | | | V |
| Maximum Full Load Reverse Current,<br>Full Cycle    $T_A = 75°$ | 30 | | | | | | | $\mu$A |
| Typical Junction Capacitance<br>    $V_R = 4.0$ V, $f = 1.0$ MHz | 15 | | | | | | | pF |

# TYPICAL CHARACTERISTIC CURVES:

Device data sheets always include characteristic curves that may be useful to a designer. In this example, we include the forward-current derating curve, in which the maximum forward current is derated as a function of temperature. To illustrate this curve, we point out that at a temperature of 100°C the maximum diode current

*(Concluded)*

is around 0.65 A (down from 1 A). A second curve is related to the diode forward current versus forward voltage (note that this curve was obtained for a very particular type of input, consisting of a pulse of width equal to 300 $\mu$s and 2 percent duty cycle.

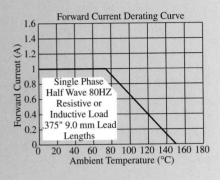

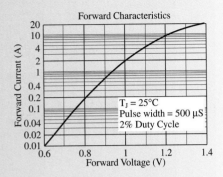

---

# EXAMPLE 8.5  Using Load Line Analysis and Diode Curves to Determine the Operating Point of a Diode

## Problem

Determine the operating point of the 1N941 diode in the circuit of Figure 8.31 and compute the total power output of the 12-V battery.

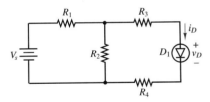

**Figure 8.31**

---

## Solution

**Known Quantities:**  $V_S = 12$ V; $R_1 = 50$ $\Omega$; $R_2 = 10$ $\Omega$; $R_3 = 20$ $\Omega$; $R_4 = 20$ $\Omega$.

**Find:**  The diode operating voltage and current and the power supplied by the battery.

**Assumptions:**  Use the diode nonlinear model, as described by its $i$-$v$ curve (Figure 8.32).

**Analysis:**  We first compute the Thévenin equivalent representation of the circuit of Figure 8.31 to reduce it to prepare the circuit for load-line analysis (see Figures 8.29 and 8.30).

$$R_T = R_3 + R_4 + (R_1 \| R_2) = 20 + 20 + (10 \| 50) = 48.33 \ \Omega;$$

$$V_T = \frac{R_2}{R_1 + R_2} V_S = \frac{10}{60} 12 = 2 \text{ V}$$

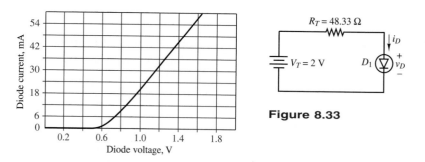

**Figure 8.33**

**Figure 8.32** 1N914 diode $i$-$v$ curve

The equivalent circuit is shown in Figure 8.33. Next we plot the load line (see Figure 8.30), with $y$ intercept $V_T/R_T = 41$ mA, and with $x$ intercept $V_T = 2$ V; the diode curve and load line are shown in Figure 8.34. The intersection of the two curves is the *quiescent* ($Q$) or operating point of the diode, which is given by the values $V_Q = 0.67$ V, $I_Q = 27.5$ mA.

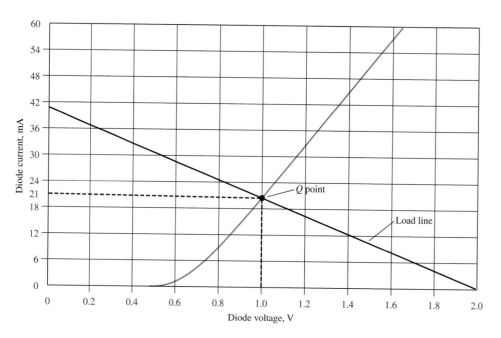

**Figure 8.34** Superposition of load line and diode $i$-$v$ curve

To determine the battery power output, we observe that the power supplied by the battery is $P_B = 12 \times I_B$ and that $I_B$ is equal to current through $R_1$. Upon further inspection, we see that the battery current must, by KCL, be equal to the sum of the currents through $R_2$ and through the diode. We already know the current through the diode, $I_Q$. To determine the current through $R_2$, we observe that the voltage across $R_2$ is equal to the sum of the voltages across $R_3$, $R_4$ and $D_1$:

$$V_{R2} = I_Q(R_3 + R_4) + V_Q = 0.021 \times 40 + 1 = 1.84 \text{ V}$$

and therefore the current through $R_2$ is $I_{R2} = V_{R2}/R_2 = 0.184$ A.

Finally,

$$P_B = 12 \times I_B = 12 \times (0.021 + 0.184) = 12 \times 0.205 = 2.46 \text{ W}$$

**Comments:** Graphical solutions are not the only means of solving the nonlinear equations that result from using a nonlinear model for a diode. The same equations could be solved numerically by using a nonlinear equation solver. The code in *Electronics Workbench*™ accomplishes exactly this task.

## Piecewise Linear Diode Model

The graphical solution of diode circuits can be somewhat tedious, and its accuracy is limited by the resolution of the graph; it does, however, provide insight into the **piecewise linear diode model.** In the piecewise linear model, the diode is treated as an open circuit in its off state, and as a linear resistor in series with $V_\gamma$ in the on state. Figure 8.35 illustrates the graphical appearance of this model. Note that the straight line that approximates the "on" part of the diode characteristic is tangent to the $Q$ point. Thus, in the neighborhood of the $Q$ point, the diode does act as a linear small-signal resistance, with slope given by $1/r_D$, where

$$\frac{1}{r_D} = \left.\frac{\partial i_D}{\partial v_D}\right|_{(I_Q,V_Q)} \tag{8.16}$$

That is, it acts as a linear resistance whose $i$-$v$ characteristic is the tangent to the diode curve at the operating point. The tangent is extended to meet the voltage axis, thus defining the intersection as the diode offset voltage. Thus, rather than represent the diode by a short circuit in its forward-biased state, we treat it as a linear resistor, with resistance $r_D$. The piecewise linear model offers the convenience of a linear representation once the state of the diode is established, and of a more accurate model than either the ideal or the offset diode model. This model is very useful in illustrating the performance of diodes in real-world applications.

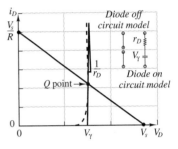

**Figure 8.35** Piecewise linear diode model

---

## EXAMPLE 8.6 Computing the Incremental (Small-Signal) Resistance of a Diode

### Problem

Determine the incremental resistance of a diode using the diode equation.

---

### Solution

**Known Quantities:** $I_0 = 10^{-14}$ A; $kT/q = 0.025$ V (at $T = 300$ K); $I_Q = 50$ mA.

**Find:** The diode small signal resistance, $r_D$.

**Assumptions:** Use the approximate diode equation (equation 8.7).

**Analysis:** The approximate diode equation relates diode voltage and current according to:

$$i_D = I_0 e^{q v_D / kT}$$

From the above expression we can compute the incremental resistance using equation 8.16:

$$\frac{1}{r_D} = \left.\frac{\partial i_D}{\partial v_D}\right|_{(I_Q,V_Q)} = \frac{q I_0}{kT} e^{q v_D / kT}$$

To calculate the numerical value of the above expression, we must first compute the quiescent diode voltage corresponding to the quiescent current $I_Q = 50$ mA:

$$V_Q = \frac{kT}{q} \log_e \frac{I_Q}{I_0} = 0.731 \text{ V}$$

Substituting the numerical value of $V_Q$ in the expression for $r_D$ we obtain:

$$\frac{1}{r_D} = \frac{10^{-14}}{0.025} e^{0.731/0.025} = 2 \text{ S} \qquad \text{or} \qquad r_D = 0.5 \text{ }\Omega$$

**Comments:** It is important to understand that, while one can calculate the linearized incremental resistance of a diode at an operating point, this does not mean that the diode can be treated simply as a resistor. The linearized small-signal resistance of the diode is used in the piecewise linear diode model to account for the fact that there is a dependence between diode voltage and current (i.e., the diode $i$-$v$ curve is not exactly a vertical line for voltages above the offset voltage—see Figure 8.35).

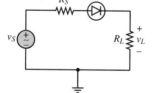

**Figure 8.36**

## EXAMPLE 8.7 Using the Piecewise Linear Diode Model

### Problem

Determine the load voltage in the rectifier of Figure 8.36 using a piecewise linear approximation.

### Solution

**Known Quantities:** $v_S(t) = 10 \cos(\omega t)$; $V_\gamma = 0.6$ V; $r_D = 0.5$ $\Omega$; $R_S = 1$ $\Omega$; $R_L = 10$ $\Omega$.

**Find:** The load voltage, $v_L$.

**Assumptions:** Use the piecewise linear diode model (Figure 8.35).

**Analysis:** We replace the diode in the circuit of Figure 8.36 with the piecewise linear model, as shown in Figure 8.37. Next, we determine the conduction condition for the ideal diode by applying KVL to the circuit of Figure 8.37:

$$v_S = v_1 + v_2 + v_D + 0.6 + v_L$$

$$v_D = v_S - v_1 - v_2 - 0.6 - v_L$$

We use the above equation as was done in Example 8.4—that is, to determine the source voltage value for which the diode first conducts. Observe first that the diode will be off for

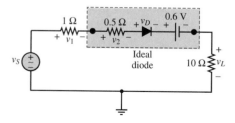

**Figure 8.37**

negative values of $v_S$. With the diode off, that is, an open circuit, the voltages $v_1$, $v_2$, and $v_L$ are zero and

$$v_D = v_S - 0.6$$

Thus, the condition for the ideal diode to conduct ($v_D > 0$) corresponds to:

$$v_S \geq 0.6 \text{ V} \qquad \text{Diode on condition}$$

Once the diode conducts, we replace the ideal diode with a short circuit, and compute the load voltage using the voltage divider rule. The resulting load equations are:

$$v_L = 0 \qquad\qquad\qquad\qquad\qquad\qquad\qquad v_S < 0.6 \text{ V}$$

$$v_L = \frac{R_L}{R_S + r_D + R_L}\left(v_S - V_\gamma\right) = 8.17\,\cos(\omega t) - 0.52 \qquad v_S \geq 0.6 \text{ V}$$

The source and load voltage are plotted in Figure 8.38(a).

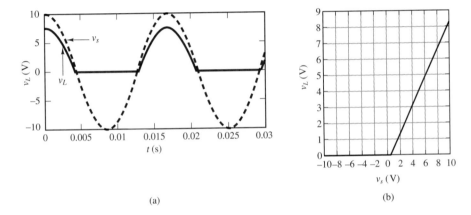

(a)                                              (b)

**Figure 8.38** (a) Source voltage and rectified load voltage; (b) Voltage transfer characteristic

It is instructive to compute the *transfer characteristic* of the diode circuit by generating a plot of $v_L$ versus $v_S$. This is done with reference to the equation for $v_L$ given above; the result is plotted in Figure 8.38(b).

**Comments:** The methods developed in this example will be very useful in analyzing some practical diode circuits in the next section.

**Focus on Computer-Aided Tools:** The Matlab™ code used to generate the plot of Figure 8.38(b) may be found in the CD-ROM that accompanies this book.

VIRTUAL LAB

## Check Your Understanding

**8.1** Repeat Example 8.2, assuming that the diode is conducting, and show that this assumption leads to an inconsistent result.

**8.2** Compute the DC value of the rectified waveform for the circuit of Figure 8.20 for $v_i = 52\cos\omega t$ V.

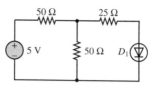

**Figure 8.39**

**8.3**  Use load-line analysis to determine the operating point ($Q$ point) of the diode in the circuit of Figure 8.39. The diode has the characteristic curve shown in Figure 8.32.

**8.4**  Compute the incremental resistance of the diode of Example 8.6 if the current through the diode is 250 mA.

**8.5**  Consider a half-wave rectifier similar to that of Figure 8.20, with $v_i(t) = 18\cos(t)$, and a 4-$\Omega$ load resistor. Sketch the output waveform if the piecewise linear diode model is used to describe the operation of the diode, with $V_\gamma = 0.6$ V and $r_D = 1$ $\Omega$. What is the peak value of the rectifier waveform?

**8.6**  Determine which of the diodes in the circuit of Figure 8.40 conducts. Each diode has an offset voltage of 0.6 V.

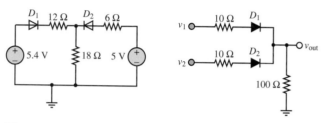

**Figure 8.40**

**Figure 8.41**

**8.7**  Determine which of the diodes in Figure 8.41 conducts for the following voltages (in V): (a) $v_1 = 0$, $v_2 = 0$; (b) $v_1 = 5$, $v_2 = 5$; (c) $v_1 = 0$, $v_2 = 5$; (d) $v_1 = 5$, $v_2 = 0$. Treat the diodes as ideal.

## 8.4    PRACTICAL DIODE CIRCUITS

This section illustrates some of the applications of diodes to practical engineering circuits. The nonlinear behavior of diodes, especially the rectification property, makes these devices valuable in a number of applications. In this section, more advanced rectifier circuits (the **full-wave rectifier** and the **bridge rectifier**) will be explored, as well as **limiter** and *peak detector* circuits. These circuits will be analyzed by making use of the circuit models developed in the preceding sections; as stated earlier, these models are more than adequate to develop an understanding of the operation of diode circuits.

In addition to the operation of diodes as rectifiers and limiters, there is another useful class of applications that takes advantage of the reverse-breakdown characteristic of the semiconductor diode discussed in the opening section. The phenomenon of Zener breakdown is exploited in a class of devices called **Zener diodes,** which enjoy the property of a sharp reverse-bias breakdown with relatively constant breakdown voltage. These devices are used as voltage regulators, that is, to provide a nearly constant output (DC) voltage from a voltage source whose output might ordinarily fluctuate substantially (for example, a rectified sinusoid).

### The Full-Wave Rectifier

The half-wave rectifier discussed earlier is one simple method of converting AC energy to DC energy. The need for converting one form of electrical energy into the other arises frequently in practice. The most readily available form of electric power is AC (the standard 110- or 220-V rms AC line power), but one frequently needs a DC power supply, for applications ranging from the control of certain types of electric motors to the operation of electronic circuits such as those discussed in Chapters 8 through 14. You will have noticed that most consumer electronic circuits, from CD players to personal computers, require AC-DC power adapters.

The half-wave rectifier, however, is not a very efficient AC-DC conversion circuit, because it fails to utilize half the energy available in the AC waveform, by not conducting current during the negative half-cycle of the AC waveform. The full-wave rectifier shown in Figure 8.42 offers a substantial improvement in efficiency over the half-wave rectifier. The first section of the full-wave rectifier circuit includes an AC source and a center-tapped transformer (see Chapter 7) with $1:2N$ turns ratio. The purpose of the transformer is to obtain the desired voltage amplitude prior to rectification. Thus, if the peak amplitude of the AC source voltage is $v_S$, the amplitude of the voltage across each half of the output side of the transformer will be $Nv_S$; this scheme permits scaling the source voltage up or down (depending on whether $N$ is greater or less than 1), according to the specific requirements of the application. In addition to scaling the source voltage, the transformer also isolates the rectifier circuit from the AC source voltage, since there is no direct electrical connection between the input and output of a transformer (see Chapter 16).

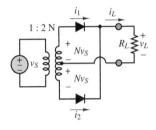

**Figure 8.42** Full-wave rectifier

In the analysis of the full-wave rectifier, the diodes will be treated as ideal, since in most cases the source voltage is the AC line voltage (110 V rms, 60 Hz) and therefore the offset voltage is negligible in comparison. The key to the operation of the full-wave rectifier is to note that during the positive half-cycle of $v_S$, the top diode is forward-biased while the bottom diode is reverse-biased; therefore, the load current during the positive half-cycle is

$$i_L = i_1 = \frac{Nv_S}{R_L} \qquad v_S \geq 0 \tag{8.17}$$

while during the negative half-cycle, the bottom diode conducts and the top diode is off, and the load current is given by

$$i_L = i_2 = \frac{-Nv_S}{R_L} \qquad v_S < 0 \tag{8.18}$$

Note that the direction of $i_L$ is always positive, because of the manner of connecting the diodes (when the top diode is off, $i_2$ is forced to flow from $+$ to $-$ across $R_L$).

The source voltage, the load voltage, and the currents $i_1$ and $i_2$ are shown in Figure 8.43 for a load resistance $R_L = 1\ \Omega$ and $N = 1$. The full-wave

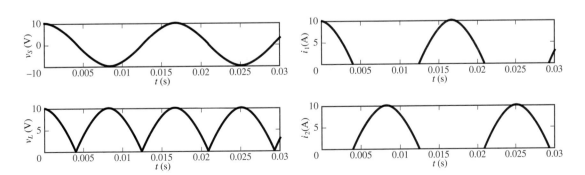

**Figure 8.43** Full-wave rectifier current and voltage waveforms ($R_L = 1\ \Omega$)

rectifier results in a twofold improvement in efficiency over the half-wave rectifier introduced earlier.

## The Bridge Rectifier

Another rectifier circuit commonly available "off the shelf" as a single *integrated circuit package*[1] is the *bridge rectifier*, which employs four diodes in a bridge configuration, similar to the Wheatstone bridge already explored in Chapter 2. Figure 8.44 depicts the bridge rectifier, along with the associated integrated circuit (IC) package.

The analysis of the bridge rectifier is simple to understand by visualizing the operation of the rectifier for the two half-cycles of the AC waveform separately. The key is that, as illustrated in Figure 8.45, diodes $D_1$ and $D_3$ conduct during the positive half-cycle, while diodes $D_2$ and $D_4$ conduct during the negative half-cycle. Because of the structure of the bridge, the flow of current through the load resistor is in the same direction (from $c$ to $d$) during both halves of the cycle; hence, the full-wave rectification of the waveform. The original and rectified waveforms are shown in Figure 8.46(a) for the case of ideal diodes and a 30-V peak AC source. Figure 8.46(b) depicts the rectified waveform if we assume diodes with a 0.6-V offset voltage. Note that the waveform of Figure 8.46(b) is not a pure rectified sinusoid any longer: The effect of the offset voltage is to shift the waveform downward by twice the offset voltage. This is most easily understood by considering that the load seen by the source during either half-cycle consists of two diodes in series with the load resistor.

Although the conventional and bridge full-wave rectifier circuits effectively convert AC signals that have zero average, or DC, value to a signal with a nonzero average voltage, either rectifier's output is still an oscillating waveform. Rather than provide a smooth, constant voltage, the full-wave rectifier generates a sequence of sinusoidal pulses at a frequency double that of the original AC signal. The **ripple**—that is, the fluctuation about the mean voltage that is characteristic of these rectifier circuits—is undesirable if one desires a true DC supply. A simple yet effective means of eliminating most of the ripple (i.e., AC component) associated with the output of a rectifier is to take advantage of the energy-storage properties of capacitors to filter out the ripple component of the load voltage. A low-pass filter that preserves the DC component of the rectified voltage while filtering out components at frequencies at or above twice the AC signal frequency would be an appropriate choice to remove the ripple component from the rectified voltage. In most practical applications of rectifier circuits, the signal waveform to be rectified is the 60-Hz, 110-V rms line voltage. The ripple frequency is, therefore, $f_{ripple} = 120$ Hz, or $\omega_{ripple} = 2\pi \cdot 120$ rad/s. A low-pass filter is required for which

$$\omega_0 \ll \omega_{ripple} \tag{8.19}$$

For example, the filter could be characterized by

$$\omega_0 = 2\pi \cdot 2 \text{ rad/s}$$

A simple low-pass filter circuit similar to those studied in Chapter 6 that accomplishes this task is shown in Figure 8.47.

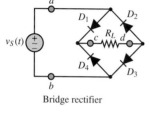

Bridge rectifier

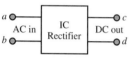

Corresponding IC package

**Figure 8.44** Full-wave bridge rectifier

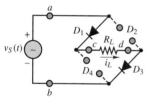

During the positive half-cycle of $v_S(t)$, $D_1$ and $D_3$ are forward-biased and $i_L = v_S(t)/R_L$ (ideal diodes).

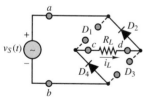

During the negative half-cycle of $v_S(t)$, $D_2$ and $D_4$ are forward-biased and $i_L = v_S(t)/R_L$ (ideal diodes).

**Figure 8.45** Operation of bridge rectifier

---

[1] An integrated circuit is a collection of electronic devices interconnected on a single silicon chip.

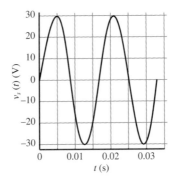
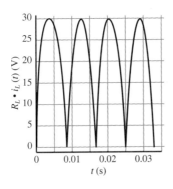
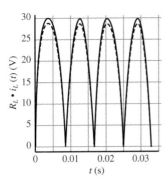

**Figure 8.46** (a) Unrectified source voltage; (b) Rectified load voltage (ideal diodes); (c) Rectified load voltage (ideal and offset diodes)

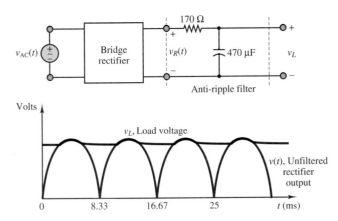

**Figure 8.47** Bridge rectifier with filter circuit

## Diode Thermometer

**FOCUS ON MEASUREMENTS**

### Problem:

An interesting application of a diode, based on the diode equation, is an electronic thermometer. The concept is based on the empirical observation that if the current through a diode is nearly constant, the offset voltage is nearly a linear function of temperature, as shown in Figure 8.48(a).

1. Show that $i_D$ in the circuit of Figure 8.48(b) is nearly constant in the face of variations in the diode voltage, $v_D$. This can be done by computing the percent change in $i_D$ for a given percent change in $v_D$. Assume that $v_D$ changes by 10 percent, from 0.6 to 0.66 V.

2. On the basis of the graph of Figure 8.48(a), write an equation for $v_D(T°)$ of the form

$$v_D = \alpha T° + \beta$$

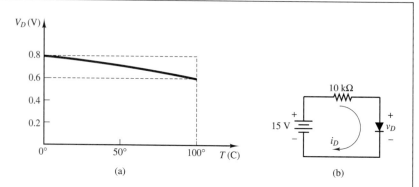

**Figure 8.48**

Solution:

1. With reference to the circuit of Figure 8.48(a), the current $i_D$ is

$$i_D = \frac{15 - v_D}{10} \text{ mA}$$

For

$$v_D = 0.8 \text{ V}(0°), i_D = 1.42 \text{ mA} \quad v_D = 0.7 \text{ V}(50°), i_D = 1.43 \text{ mA}$$
$$v_D = 0.6 \text{ V}(100°), i_D = 1.44 \text{ mA}$$

The percent change in $v_D$ over the full scale of the thermometer (assuming the midrange temperature of 50° to be the reference value) is:

$$\Delta v_D\% = \pm\frac{0.1 \text{ V}}{0.7 \text{ V}} \times 100 = \pm 14.3\%$$

The corresponding percent change in $i_D$ is:

$$\Delta i_D\% = \pm\frac{0.01 \text{ mA}}{1.43 \text{ mA}} \times 100 = \pm 0.7\%$$

Thus, $i_D$ is nearly constant over the range of operation of the diode thermometer.

2. The diode voltage versus temperature equation can be extracted from the graph of Figure 8.48(a):

$$v_D(T) = \frac{(0.8 - 0.6) \text{ V}}{(0 - 100)°\text{C}}T + 0.8 \text{ V} = -0.02T + 0.8 \text{ V}$$

*Comments*—The graph of Figure 8.48(a) was obtained experimentally by calibrating a commercial diode in both hot water and an ice bath. The circuit of Figure 8.48(b) is rather simple, and one could fairly easily design a better constant-current source; however, this example illustrates than an inexpensive diode can serve quite well as the sensing element in an **electronic thermometer.**

FIND IT

ON THE WEB

## DC Power Supplies, Zener Diodes, and Voltage Regulation

The principal application of rectifier circuits is in the conversion of AC to DC power. A circuit that accomplishes this conversion is usually called a **DC power**

**supply.** In power supply applications, transformers are employed to obtain an AC voltage that is reasonably close to the desired DC supply voltage. DC power supplies are very useful in practice: Many familiar electrical and electronic appliances (e.g., radios, personal computers, TVs) require DC power to operate. For most applications, it is desirable that the DC supply be as steady and ripple-free as possible. To ensure that the DC voltage generated by a DC supply is constant, DC supplies contain voltage regulators, that is, devices that can hold a DC load voltage relatively constant in spite of possible fluctuations in the DC supply. This section describes the fundamentals of voltage regulators.

A typical DC power supply is made up of the components shown in Figure 8.49. In the figure, a transformer is shown connecting the AC source to the rectifier circuit to permit scaling of the AC voltage to the desired level. For example, one might wish to step the 110-V rms line voltage down to 24 V rms by means of a transformer prior to rectification and filtering, to eventually obtain a 12-VDC regulated supply (*regulated* here means that the output voltage is a DC voltage that is constant and independent of load and supply variations). Following the step-down transformer are a bridge rectifier, a filter capacitor, a voltage regulator, and, finally, the load.

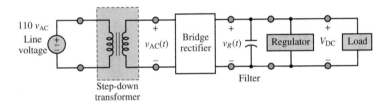

**Figure 8.49** DC power supply

The most common device employed in voltage regulation schemes is the Zener diode. Zener diodes function on the basis of the reverse portion of the $i$-$v$ characteristic of the diode discussed in Section 8.2. Figure 8.10 in Section 8.2 illustrates the general characteristic of a diode, with forward offset voltage $V_\gamma$ and **reverse Zener voltage** $V_Z$. Note how steep the $i$-$v$ characteristic is at the Zener breakdown voltage, indicating that in the Zener breakdown region the diode can hold a very nearly constant voltage for a large range of currents. This property makes it possible to use the Zener diode as a voltage regulator.

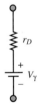

**Figure 8.50** Zener diode model for forward bias

The operation of the Zener diode may be analyzed by considering three modes of operation:

1. For $v_D \geq V_\gamma$, the device acts as a conventional forward-biased diode (Figure 8.50).
2. For $V_Z < v_D < V_\gamma$, the diode is reverse-biased but Zener breakdown has not taken place yet. Thus, it acts as an open circuit.
3. For $v_D \leq V_Z$, Zener breakdown occurs and the device holds a nearly constant voltage, $-V_Z$ (Figure 8.51).

The combined effect of forward and reverse bias may be lumped into a single model with the aid of ideal diodes, as shown in Figure 8.52.

To illustrate the operation of a Zener diode as a voltage regulator, consider the circuit of Figure 8.53(a), where the unregulated DC source, $V_S$, is regulated to the value of the Zener voltage $V_Z$. Note how the diode must be connected "upside

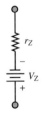

**Figure 8.51** Zener diode model for reverse bias

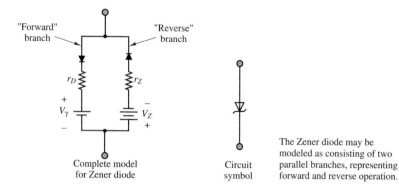

The Zener diode may be modeled as consisting of two parallel branches, representing forward and reverse operation.

**Figure 8.52** Complete model for Zener diode

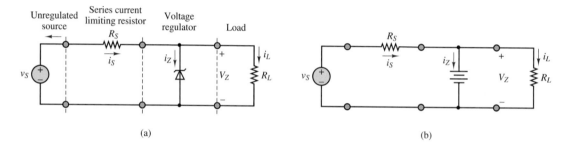

**Figure 8.53** (a) A Zener diode voltage regulator; (b) Simplified circuit for Zener regulator

down" to obtain a positive regulated voltage. Note also that if $v_S$ is greater than $V_Z$, it follows that the Zener diode is in its reverse-breakdown mode. Thus, one need not worry whether the diode is conducting or not in simple voltage regulator problems, provided that the unregulated supply voltage is guaranteed to stay above $V_Z$ (a problem arises, however, if the unregulated supply can drop below the Zener voltage). Assuming that the resistance $r_Z$ is negligible with respect to $R_S$ and $R_L$, we replace the Zener diode with the simplified circuit model of Figure 8.53(b), consisting of a battery of strength $V_Z$ (the effects of the nonzero Zener resistance are explored in the examples and homework problems).

Three simple observations are sufficient to explain the operation of this voltage regulator:

1. The load voltage must equal $V_Z$, as long as the Zener diode is in the reverse-breakdown mode. Then,

$$i_L = \frac{V_Z}{R_L} \tag{8.20}$$

2. The load current (which should be constant if the load voltage is to be regulated to sustain $V_Z$) is the difference between the unregulated supply current, $i_S$, and the diode current, $i_Z$:

$$i_L = i_S - i_Z \tag{8.21}$$

This second point explains intuitively how a Zener diode operates: Any current in excess of that required to keep the load at the constant voltage $V_Z$ is "dumped" to ground through the diode. Thus, the Zener diode acts as a sink to the undesired source current.

3.    The source current is given by

$$i_S = \frac{v_S - V_Z}{R_S} \tag{8.22}$$

In the ideal case, the operation of a Zener voltage regulator can be explained very simply on the basis of this model. The examples and exercises will illustrate the effects of the practical limitations that arise in the design of a practical voltage regulator; the general principles will be discussed in the following paragraphs.

The Zener diode is usually rated in terms of its maximum allowable power dissipation. The power dissipated by the diode, $P_Z$, may be computed from

$$P_Z = i_Z V_Z \tag{8.23}$$

Thus, one needs to worry about the possibility that $i_Z$ will become too large. This may occur either if the supply current is very large (perhaps because of an unexpected upward fluctuation of the unregulated supply), or if the load is suddenly removed and all of the supply current sinks through the diode. The latter case, of an open-circuit load, is an important design consideration.

Another significant limitation occurs when the load resistance is small, thus requiring large amounts of current from the unregulated supply. In this case, the Zener diode is hardly taxed at all in terms of power dissipation, but the unregulated supply may not be able to provide the current required to sustain the load voltage. In this case, regulation fails to take place. Thus, in practice, the range of load resistances for which load voltage regulation may be attained is constrained to a finite interval:

$$R_{L\,min} \leq R_L \leq R_{L\,max} \tag{8.24}$$

where $R_{L\,max}$ is typically limited by the Zener diode power dissipation and $R_{L\,min}$ by the maximum supply current. The following examples illustrate these concepts.

---

### EXAMPLE  8.8  Determining the Power Rating of a Zener Diode

#### Problem

We wish to design a regulator similar to the one depicted in Figure 8.53(a). Determine the minimum acceptable power rating of the Zener diode.

---

#### Solution

**Known Quantities:**  $v_S = 24$ V; $V_Z = 12$ V; $R_S = 50\ \Omega$; $R_L = 250\ \Omega$.

**Find:**  The maximum power dissipated by the Zener diode under worst-case conditions.

**Assumptions:**  Use the piecewise linear Zener diode model (Fig. 8.52) with $r_Z = 0$.

**Analysis:**  When the regulator operates according to the intended design specifications, i.e., with a 250-$\Omega$ load, the source and load currents may be computed as follows:

$$i_S = \frac{v_S - V_Z}{R_S} = \frac{12}{50} = 0.24 \text{ A}$$

$$i_L = \frac{V_Z}{R_L} = \frac{12}{250} = 0.048 \text{ A}$$

Thus, the Zener current would be:

$$i_Z = i_S - i_L = 0.192 \text{ A}$$

corresponding to a nominal power dissipation

$$P_Z = i_Z V_Z = 0.192 \times 12 = 2.304 \text{ W}$$

However, if the load were accidentally (or intentionally) disconnected from the circuit, all of the load current would be diverted to flow through the Zener diode. Thus, the *worst-case* Zener current is actually equal to the source current, since the Zener diode would sink all of the source current for an open-circuit load:

$$i_{Z\,\max} = i_S = \frac{v_S - V_Z}{R_S} = \frac{12}{50} = 0.24 \text{ A}$$

Therefore the maximum power dissipation that the Zener diode must sustain is:

$$P_{Z\,\max} = i_{Z\,\max} V_Z = 2.88 \text{ W}$$

**Comments:**  A safe design would exceed the value of $P_{Z\,\max}$ computed above. For example, one might select a 3-W Zener diode.

---

### EXAMPLE  8.9  Calculation of Allowed Load Resistances for a Given Zener Regulator

#### Problem

Calculate the allowable range of load resistances for the Zener regulator of Figure 8.54 such that the diode power rating is not exceeded.

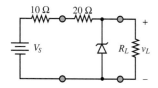

**Figure 8.54**

#### Solution

**Known Quantities:**  $V_S = 50$ V; $V_Z = 14$ V; $P_Z = 5$ W.

**Find:**  The smallest and largest values of $R_L$ for which load voltage regulation to 14 V is achieved, and which do not cause the diode power rating to be exceeded.

**Assumptions:**  Use the piecewise linear Zener diode model (Fig. 8.52) with $r_Z = 0$.

**Analysis:**

1. *Determining the minimum acceptable load resistance.* To determine the minimum acceptable load, we observe that the regulator can at most supply the load with the amount of current that can be provided by the source. Thus, the minimum theoretical resistance can be computed by assuming that all the source current goes to the load, and that the load voltage is regulated at the nominal value:

$$R_{L\,\min} = \frac{V_Z}{i_S} = \frac{V_Z}{\dfrac{V_S - V_Z}{30}} = \frac{14}{\dfrac{36}{30}} = 11.7 \text{ }\Omega$$

If the load required any more current, the source would not be able to supply it. Note that for this value of the load, the Zener diode dissipates zero power, because the Zener current is zero.

2. *Determining the maximum acceptable load resistance.* The second constraint we need to invoke is the power rating of the diode. For the stated 5-W rating, the maximum Zener current is:

$$i_{Z\,max} = \frac{P_Z}{V_Z} = \frac{5}{14} = 0.357\ \text{A}$$

Since the source can generate

$$i_{S\,max} = \frac{V_S - V_Z}{30} = \frac{50 - 14}{30} = 1.2\ \text{A}$$

the load must not require any less than $1.2 - 0.357 = 0.843$ A; if it required any less current (i.e., if the resistance were too large), the Zener diode would be forced to sink more current than its power rating permits. From this requirement we can compute the maximum allowable load resistance:

$$R_{L\,min} = \frac{V_Z}{i_{S\,max} - i_{Z\,max}} = \frac{14}{0.843} = 16.6\ \Omega$$

Finally, the range of allowable load resistance is $11.7\ \Omega \le R_L \le 16.6\ \Omega$.

**Comments:**  Note that this regulator *cannot* operate with an open-circuit load!

---

## EXAMPLE  8.10  Effect of Nonzero Zener Resistance in a Regulator

### Problem

Calculate the amplitude of the ripple present in the output voltage of the regulator of Figure 8.55. The unregulated supply voltage is depicted in Figure 8.56.

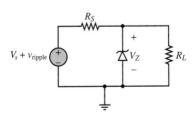

**Figure 8.55**

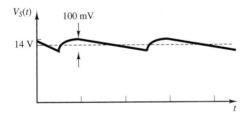

**Figure 8.56**

---

### Solution

**Known Quantities:**  $v_S = 14$ V; $v_{ripple} = 100$ mV; $V_Z = 8$ V; $r_Z = 10\ \Omega$; $R_S = 50\ \Omega$; $R_L = 150\ \Omega$.

**Find:**  Amplitude of ripple component in load voltage.

**Assumptions:**  Use the piecewise linear Zener diode model (Fig. 8.52).

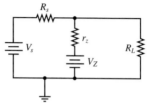

DC equivalent circuit

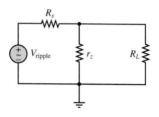

AC equivalent circuit

**Figure 8.57**

**Analysis:** To analyze the circuit, we consider the DC and AC equivalent circuits of Figure 8.57 separately.

1. *DC equivalent circuit.* The DC equivalent circuit reveals that the load voltage consists of two contributions; that due to the unregulated DC supply and that due to the Zener diode ($V_Z$). Applying superposition and the voltage divider rule, we obtain:

$$V_L = V_S \left( \frac{r_Z \| R_L}{r_Z \| R_L + R_S} \right) + V_Z \left( \frac{R_S \| R_L}{R_S \| R_L + R_S} \right) = 2.21 + 6.32 = 8.53 \text{ V}$$

2. *AC equivalent circuit.* The AC equivalent circuit allows us to compute the AC component of the load voltage as follows:

$$v_L = v_{\text{ripple}} \left( \frac{r_Z \| R_L}{r_Z \| R_L + R_S} \right) = 0.016 \text{ V}$$

that is, 16 mV of ripple is present in the load voltage, or approximately one-sixth the source ripple.

**Comments:** Note that the DC load voltage is affected by the unregulated source voltage; if the unregulated supply were to fluctuate significantly, the regulated voltage would also change. Thus, one of the effects of the Zener resistance is to cause imperfect regulation. If the Zener resistance is significantly smaller than both $R_S$ and $R_L$, its effects would not be as pronounced (see Check Your Understanding Exercise 8.10).

## Check Your Understanding

**8.8** Show that the DC voltage output of the full-wave rectifier of Figure 8.42 is $2Nv_{S\max}/\pi$.

**8.9** Compute the peak voltage output of the bridge rectifier of Figure 8.44, assuming diodes with 0.6-V offset voltage and a 110-V rms AC supply.

**8.10** Compute the actual DC load voltage and the percentage of the ripple reaching the load (relative to the initial 100-mV ripple) for the circuit of Example 8.10 if $r_Z = 1\ \Omega$.

## Signal-Processing Applications

Among the numerous applications of diodes, there are a number of interesting signal-conditioning or signal-processing applications that are made possible by the nonlinear nature of the device. We explore three such applications here: the **diode limiter, or clipper; the diode clamp; and the peak detector.** Other applications are left for the homework problems.

### *The Diode Clipper (Limiter)*

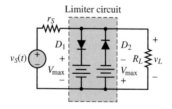

**Figure 8.58** Two-sided doide clipper

The *diode clipper* is a relatively simple diode circuit that is often employed to protect loads against excessive voltages. The objective of the clipper circuit is to keep the load voltage within a range—say, $-V_{\max} \le v_L(t) \le V_{\max}$—so that the maximum allowable load voltage (or power) is never exceeded. The circuit of Figure 8.58 accomplishes this goal.

The circuit of Figure 8.58 is most easily analyzed by first considering just the branch containing $D_1$. This corresponds to clipping only the positive peak voltages; the analysis of the negative voltage limiter is left as a drill exercise. The circuit containing the $D_1$ branch is sketched in Figure 8.59; note that we have exchanged the location of the $D_1$ branch and that of the load branch for convenience. Further, the circuit is reduced to Thévenin equivalent form. Having reduced the circuit to a simpler form, we can now analyze its operation for two distinct cases: the ideal diode and the piecewise linear diode.

**1. Ideal diode model**    For the ideal diode case, we see immediately that $D_1$ conducts if

$$\frac{R_L}{r_S + R_L} v_S(t) \geq V_{max} \tag{8.25}$$

and that if this condition occurs, then ($D_1$ being a short circuit) the load voltage, $v_L$, becomes equal to $V_{max}$. The equivalent circuit for the "on" condition is shown in Figure 8.60.

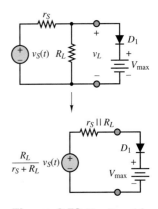

**Figure 8.59** Circuit model for the diode clipper

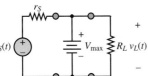

**Figure 8.60** Equivalent circuit for the one-sided limiter (diode on)

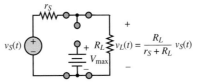

**Figure 8.61** Equivalent circuit for the one-sided limiter (doide off)

If, on the other hand, the source voltage is such that

$$\frac{R_L}{r_S + R_L} v_S(t) < V_{max} \tag{8.26}$$

then $D_1$ is an open circuit and the load voltage is simply

$$v_L(t) = \frac{R_L}{r_S + R_L} v_S(t) \tag{8.27}$$

The equivalent circuit for this case is depicted in Figure 8.61.

The analysis for the negative branch of the circuit of Figure 8.58 can be conducted by analogy with the preceding derivation, resulting in the waveform for the two-sided clipper shown in Figure 8.62. Note how the load voltage is drastically "clipped" by the limiter in the waveform of Figure 8.62. In reality, such hard clipping does not occur, because the actual diode characteristic does not have the sharp on-off breakpoint the ideal diode model implies. One can develop a reasonable representation of the operation of a physical diode limiter by using the piecewise linear model.

**2. Piecewise linear diode model**    To avoid unnecessary complexity in the analysis, assume that $V_{max}$ is much greater than the diode offset voltage, and therefore assume that $V_\gamma \approx 0$. We do, however, consider the finite diode resistance $r_D$. The circuit of Figure 8.59 still applies, and thus the determination of the diode

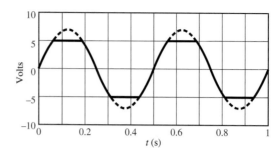

**Figure 8.62** Two-sided (ideal diode) clipper input and output voltages

The effect of finite diode resistance on the limiter circuit.

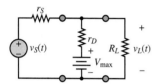

**Figure 8.63** Circuit model for the diode clipper (piecewise linear diode model)

on-off state is still based on whether $[R_L/(r_S + R_L)]v_S(t)$ is greater or less than $V_{\max}$. When $D_1$ is open, the load voltage is still given by

$$v_L(t) = \frac{R_L}{r_S + R_L} v_S(t) \tag{8.28}$$

When $D_1$ is conducting, however, the corresponding circuit is as shown in Figure 8.63.

The primary effect the diode resistance has on the load waveform is that some of the source voltage will reach the load even when the diode is conducting. This is most easily verified by applying superposition; it can be readily shown that the load voltage is now composed of two parts, one due to the voltage $V_{\max}$, and one proportional to $v_S(t)$:

$$v_L(t) = \frac{R_L \parallel r_S}{r_D + (R_L \parallel r_S)} V_{\max} + \frac{r_D \parallel R_L}{r_S + (r_D \parallel R_L)} v_S(t) \tag{8.29}$$

It may easily be verified that as $r_D \to 0$, the expression for $v_L(t)$ is the same as for the ideal diode case. The effect of the diode resistance on the limiter circuit is depicted in Figure 8.64. Note how the clipping has a softer, more rounded appearance.

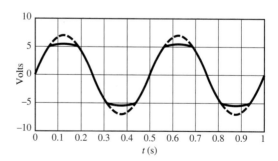

**Figure 8.64** Voltages for the diode clipper (piecewise linear diode model)

### The Diode Peak Detector

Another common application of semiconductor diodes, the *peak detector*, is very similar in appearance to the half-wave rectifier with capacitive filtering described in an earlier section. One of its more classic applications is in the demodulation of amplitude-modulated (AM) signals. We study this circuit in the following, "Focus on Measurements" box.

## Peak Detector Circuit for Capacitive Displacement Transducer

In Chapter 4, a capacitive displacement transducer was introduced in Focus on Measurements: Capacitive Displacement Transducer and Microphone. It took the form of a parallel-plate capacitor composed of a fixed plate and a movable plate. The capacitance of this variable capacitor was shown to be a function of displacement, that is, it was shown that a movable-plate capacitor can serve as a linear transducer. Recall the expression derived in Chapter 4

$$C = \frac{8.854 \times 10^{-3} A}{x}$$

where $C$ is the capacitance in pF, $A$ is the area of the plates in mm$^2$, and $x$ is the (variable) distance in mm. If the capacitor is placed in an AC circuit, its impedance will be determined by the expression

$$Z_C = \frac{1}{j\omega C}$$

so that

$$Z_C = \frac{x}{j\omega 8.854 \times 10^{-3} A}$$

Thus, at a fixed frequency $\omega$, the impedance of the capacitor will vary linearly with displacement. This property may be exploited in the bridge circuit of Figure 8.65, where a differential-pressure transducer is shown made of two movable-plate capacitors. If the capacitance of one of these capacitors increases as a consequence of a pressure difference across the transducer, the capacitance of the other must decrease by a corresponding amount, at least for small displacements (you may wish to refer to Focus on Measurements: Capacitive Displacement Transducer and Microphone, Chapter 4, p. 133, for a picture of this transducer). The bridge is excited by a sinusoidal source.

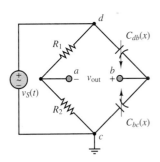

**Figure 8.65** Bridge circuit for displacement transducer

Using phasor notation, in Chapter 4 we showed that the output voltage of the bridge circuit is given by

$$\mathbf{V}_{out}(j\omega) = \mathbf{V}_S(j\omega)\frac{x}{2d}$$

provided that $R_1 = R_2$. Thus, the output voltage will vary as a scaled version of the input voltage in proportion to the displacement. A typical $v_{\text{out}}(t)$ is displayed in Figure 8.66 for a 0.05-mm "triangle" diaphragm displacement, with $d = 0.5$ mm and $\mathbf{V}_S$ a 50-Hz sinusoid with 1-V amplitude. Clearly, although the output voltage is a function of the displacement, $x$, it is not in a convenient form, since the displacement is proportional to the amplitude of the sinusoidal peaks.

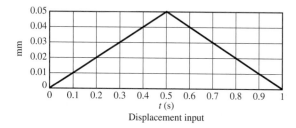

Displacement input

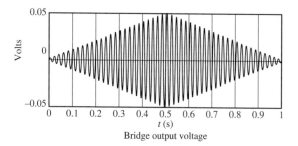

Bridge output voltage

**Figure 8.66** Displacement and bridge output voltage waveforms

The diode peak detector is a circuit capable of tracking the sinusoidal peaks without exhibiting the oscillations of the bridge output voltage. The peak detector operates by rectifying and filtering the bridge output in a manner similar to that of the circuit of Figure 8.47. The ideal peak detector circuit is shown in Figure 8.67, and the response of a practical peak detector is shown in Figure 8.68. Its operation is based on the rectification property of the diode, coupled with the filtering effect of the shunt capacitor, which acts as a low-pass filter.

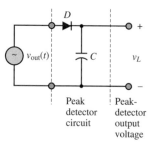

Peak detector circuit     Peak-detector output voltage

**Figure 8.67** Peak detector circuit

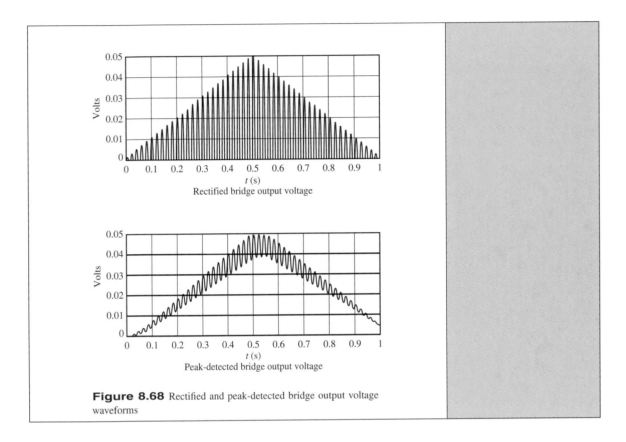

**Figure 8.68** Rectified and peak-detected bridge output voltage waveforms

### The Diode Clamp

Another circuit that finds common application is the *diode clamp*, which permits "clamping" a waveform to a fixed DC value. Figure 8.69 depicts two different types of clamp circuits.

The operation of the simple clamp circuit is based on the notion that the diode will conduct current only in the forward direction, and that therefore the capacitor will charge during the positive half-cycle of $v_S(t)$ but will not discharge during the negative half-cycle. Thus, the capacitor will eventually charge up to the peak voltage of $v_S(t)$, $V_{peak}$. The DC voltage across the capacitor has the effect of shifting the source waveform down by $V_{peak}$, so that, after the initial transient period, the output voltage is

$$v_{out}(t) = v_S(t) - V_{peak} \tag{8.30}$$

and the positive peaks of $v_S(t)$ are now clamped at 0 V. For equation 8.30 to be accurate, it is important that the $RC$ time constant be greater than the period, $T$, of $v_S(t)$:

$$RC \gg T \tag{8.31}$$

Figure 8.70 depicts the behavior of the diode clamp for a sinusoidal input waveform, where the dashed line is the source voltage and the solid line represents the clamped voltage.

The clamp circuit can also work with the diode in the reverse direction; the capacitor will charge to $-V_{peak}$ with the output voltage given by

$$v_{out}(t) = v_S(t) + V_{peak} \tag{8.32}$$

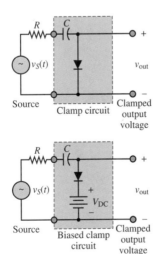

**Figure 8.69** Diode clamp circuits

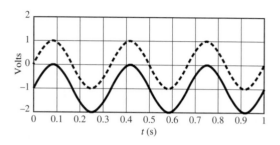

**Figure 8.70** Ideal diode clamp input and output voltages

Now the output voltage has its negative peaks clamped to zero, since the entire waveform is shifted upward by $V_{peak}$ volts. Note that in either case, the diode clamp has the effect of introducing a DC component in a waveform that does not originally have one. It is also possible to shift the input waveform by a voltage different from $V_{peak}$ by connecting a battery, $V_{DC}$, in series with the diode, provided that

$$V_{DC} < V_{peak} \tag{8.33}$$

The resulting circuit is called a *biased diode clamp*; it is discussed in Example 8.11.

### EXAMPLE 8.11    Biased Diode Clamp

**Problem**

Design a biased diode clamp to shift the DC level of the signal $v_S(t)$ up by 3 V.

**Solution**

***Known Quantities:*** $v_S(t) = 5 \cos(\omega t)$.

***Find:*** The value of $V_{DC}$ in the circuit in the lower half of Figure 8.69.

***Assumptions:*** Use the ideal diode model.

***Analysis:*** With reference to the circuit in the lower half of Figure 8.69, we observe that once the capacitor has charged to $V_{peak} - V_{DC}$, the output voltage will be given by:

$$v_{out} = v_S - V_{peak} + V_{DC}$$

Since $V_{DC}$ must be smaller than $V_{peak}$ (otherwise the diode would never conduct!), this circuit would never permit raising the DC level of $v_{out}$. To resolve this problem, we must invert both the diode and the battery, as shown in the circuit of Figure 8.71. Now the output voltage is given by:

$$v_{out} = v_S + V_{peak} - V_{DC}$$

To have a DC level of 3 V, we choose $V_{DC} = 2$ V. The resulting waveforms are shown in Figure 8.72.

***Focus on Computer-Aided Tools:*** A simulation of the circuit of Figure 8.71 generated by *Electronics Workbench*™ may be found in the accompanying CD-ROM.

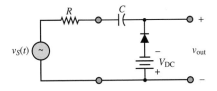

**Figure 8.71**

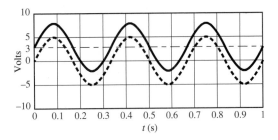

**Figure 8.72**

## Photodiodes

Another property of semiconductor materials that finds common application in measurement systems is their response to light energy. In appropriately fabricated diodes, called **photodiodes,** when light reaches the depletion region of a *pn* junction, photons cause hole-electron pairs to be generated by a process called *photo-ionization*. This effect can be achieved by using a surface material that is transparent to light. As a consequence, the reverse saturation current depends on the light intensity (i.e., on the number of incident photons), in addition to the other factors mentioned earlier, in Section 8.2. In a photodiode, the reverse current is given by $-(I_0 + I_p)$, where $I_p$ is the additional current generated by photo-ionization. The result is depicted in the family of curves of Figure 8.73, where the diode characteristic is shifted downward by an amount related to the additional current generated by photo-ionization. Figure 8.73 depicts the appearance of the *i-v* characteristic of a photodiode for various values of $I_p$, where the *i-v* curve is shifted to lower values for progressively larger values of $I_p$. The circuit symbol is depicted in Figure 8.74.

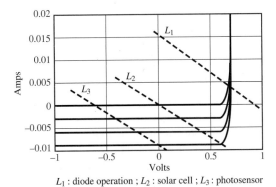

$L_1$ : diode operation ; $L_2$ : solar cell ; $L_3$ : photosensor

**Figure 8.73** Photodiode *i-v* curves

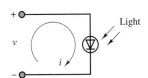

**Figure 8.74** Photodiode circuit symbol

Also displayed in Figure 8.73 are three load lines, which depict the three modes of operation of a photodiode. Curve $L_1$ represents normal diode operation, under forward bias. Note that the operating point of the device is in the positive *i*, positive *v* (first) quadrant of the *i-v* plane; thus, the diode dissipates positive power in this mode, and is therefore a passive device, as we already know. On the other hand, load line $L_2$ represents operation of the photodiode as a **solar cell;** in this mode, the operating point is in the negative $i$, positive $v$, or fourth, quadrant, and therefore the power dissipated by the diode is *negative*. In other words, the photodiode is generating power by converting light energy to electrical energy. Note further that the load line intersects the voltage axis at zero, meaning that no supply voltage is required to bias the photodiode in the solar-cell mode. Finally, load line $L_3$ represents the operation of the diode as a light sensor: when the diode is reverse-biased, the current flowing through the diode is determined by the light intensity; thus, the diode current changes in response to changes in the incident light intensity.

The operation of the photodiode can also be reversed by forward-biasing the diode and causing a significant level of recombination to take place in the depletion region. Some of the energy released is converted to light energy by emission of photons. Thus, a diode operating in this mode emits light when forward-biased. Photodiodes used in this way are called **light-emitting diodes (LEDs);** they exhibit a forward (offset) voltage of 1 to 2 volts. The circuit symbol for the LED is shown in Figure 8.75.

Gallium arsenide (GaAs) is one of the more popular substrates for creating LEDs; gallium phosphide (GaP) and the alloy $GaAs_{1-x}P_x$ are also quite common. Table 8.1 lists combinations of materials and dopants used for common LEDs and the colors they emit. The dopants are used to create the necessary *pn* junction.

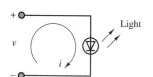

**Figure 8.75** Light-emitting diode (LED) circuit symbol

**Table 8.1**  LED materials and wavelengths

| Material | Dopant | Wavelength | Color |
|----------|--------|------------|-------|
| GaAs | Zn | 900 nm | Infrared |
| GaAs | Si | 910–1,020 nm | Infrared |
| GaP | N | 570 nm | Green |
| GaP | N | 590 nm | Yellow |
| GaP | Zn, O | 700 nm | Red |
| $GaAs_{0.6}P_{0.4}$ | | 650 nm | Red |
| $GaAs_{0.35}P_{0.65}$ | N | 632 nm | Orange |
| $GaAs_{0.15}P_{0.85}$ | N | 589 nm | Yellow |

The construction of a typical LED is shown in Figure 8.76, along with the schematic representation for an LED. A shallow *pn* junction is created with electrical contacts made to both *p* and *n* regions. As much of the upper surface of the *p* material is uncovered as possible, so that light can leave the device unimpeded. It is important to note that, actually, only a relatively small fraction of the emitted light leaves the device; the majority stays inside the semiconductor. A photon that stays inside the device will eventually collide with an electron in the valence band, and the collision will force the electron into the conduction band, emitting an electron-hole pair and absorbing the photon. To minimize the probability that a photon will be absorbed before it has an opportunity to leave the LED, the depth of the *p*-doped region is left very thin. Also, it is advantageous

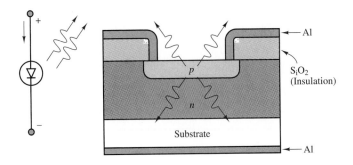

**Figure 8.76** Light-emitting diode (LED)

to have most of the recombinations that emit photons occur as close to the surface of the diode as possible. This is made possible by various doping schemes, but even so, of all of the carriers going through the diode, only a small fraction emit photons that are able to leave the semiconductor.

A simple LED drive circuit is shown in Figure 8.77. From the standpoint of circuit analysis, LED characteristics are very similar to those of the silicon diode, except that the offset voltage is usually quite a bit larger. Typical values of $V_\gamma$ can be in the range of 1.2 to 2 volts, and operating currents can range from 20 mA to 100 mA. Manufacturers usually specify an LED's characteristics by giving the rated operating-point current and voltage.

## EXAMPLE 8.12 Analysis of Light-Emitting Diode

### Problem

For the circuit of Figure 8.77, determine: (1) the LED power consumption; (2) the resistance $R_S$; (3) the power required by the voltage source.

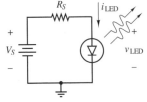

### Solution

**Known Quantities:** Diode operating point: $V_{LED} = 1.7$ V; $I_{LED} = 40$ mA; $V_S = 5$ V.

**Find:** $P_{LED}$; $R_S$; $P_S$.

**Assumptions:** Use the ideal diode model.

**Analysis:**

1. The power consumption of the LED is determined directly from the specification of the operating point:

   $$P_{LED} = V_{LED} \times I_{LED} = 68 \text{ mW}$$

2. To determine the required value of $R_S$ to achieve the desired operating point, we apply KVL around the circuit of Figure 8.77:

   $$V_S = I_{LED} R_S + V_{LED}$$

   $$R_S = \frac{V_S - V_{LED}}{I_{LED}} = \frac{5 - 1.7}{40 \times 10^{-3}} = 82.5 \text{ } \Omega$$

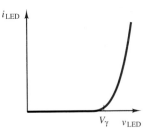

**Figure 8.77** LED drive circuit and $i\text{-}v$ characteristic

3.  To satisfy the power requirement of the circuit, the battery must be able to supply 40 mA to the diode. Thus,

$$P_S = V_S \times I_{\text{LED}} = 200 \text{ mW}$$

**Comments:**  A more practical LED biasing circuit may be found in Chapter 9 (Example 9.7).

**FOCUS ON MEASUREMENTS**

## Opto-Isolators

One of the common applications of photodiodes and LEDs is the **opto-coupler,** or **opto-isolator.** This device, which is usually enclosed in a sealed package, uses the light-to-current and current-to-light conversion property of photodiodes and LEDs to provide signal connection between two circuits without any need for electrical connections. Figure 8.78 depicts the circuit symbol for the opto-isolator.

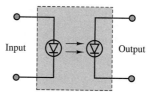

Input          Output

**Figure 8.78**  Opto-isolator

Because diodes are nonlinear devices, the opto-isolator is not used in transmitting analog signals: the signals would be distorted because of the nonlinear diode $i$-$v$ characteristic. However, opto-isolators find a very important application when on-off signals need to be transmitted from high-power machinery to delicate computer control circuitry. The optical interface ensures that potentially damaging large currents cannot reach delicate instrumentation and computer circuits.

## Check Your Understanding

**8.11**    Repeat the analysis of the diode clipper of Figure 8.58 for the branch containing $D_2$.

**8.12**    For the one-sided diode clipper of Figure 8.59, find the percentage of the source voltage that reaches the load if $R_L = 150 \ \Omega$, $r_S = 50 \ \Omega$, and $r_D = 5 \ \Omega$. Assume that the diode is conducting, and use the circuit model of Figure 8.63.

**8.13**    How would the diode clipper output waveform change if we used the offset diode model instead of the piecewise linear model in the analysis? [*Hint:* Compare Figures 8.26 and 8.38(a).]

## CONCLUSION

- Semiconductor materials have conductive properties that fall between those of conductors and insulators. These properties make such materials useful in the construction of many electronic devices that exhibit nonlinear $i$-$v$ characteristics. Of these devices, the diode is one of the most commonly employed.

- The semiconductor diode acts like a one-way current valve, permitting the flow of current only when biased in the forward direction. Although the behavior of the diode is described by an exponential equation, it is possible to approximate the operation of the diode by means of simple circuit models. The simplest model treats the diode as either a short circuit or an open circuit (the on-off, or ideal, model). The ideal model can be extended to include an offset voltage (usually 0.2 to 0.7 V), which represents the contact potential at the diode junction. A slightly more realistic model, the piecewise linear diode model, accounts for the effects of the diode forward resistance. With the aid of these circuit models it becomes possible to analyze diode circuits using the DC and AC circuit analysis techniques developed in earlier chapters.

- One of the most important properties of the semiconductor diode is that of rectification, which permits the conversion of AC voltages and currents to DC voltages and currents. Diode rectifiers can be of the half-wave type, or they can be full-wave. Full-wave rectifiers can be constructed in a conventional two-diode configuration, or in a bridge configuration. Diode rectifiers are an essential part of DC power supplies and are usually employed in conjunction with filter capacitors to obtain a relatively smooth DC voltage waveform. In addition to rectification and smoothing, it is also necessary to regulate the output of a DC power supply; Zener diodes accomplish this task by holding a constant voltage when reverse-biased above the Zener voltage.

- In addition to power supply applications, diodes find use in many signal-processing and signal-conditioning circuits. Of these, the diode limiter, peak detector, and clamp have been explored in the chapter. Further, since semiconductor material properties are also affected by light intensity, certain types of diodes, known as *photodiodes*, find application as light detectors, solar cells, or light-emitting diodes.

# CHECK YOUR UNDERSTANDING ANSWERS

| | |
|---|---|
| **CYU 8.2** | 16.55 V |
| **CYU 8.3** | $V_Q = 0.65$ V; $I_Q = 37$ mA |
| **CYU 8.4** | 0.1 $\Omega$ |
| **CYU 8.5** | 13.92 V |
| **CYU 8.6** | Both diodes conduct. |
| **CYU 8.7** | (a) Neither conducts. (b) Both conduct. (c) Only $D_2$ conducts. (d) Only $D_1$ conducts. |
| **CYU 8.9** | 154.36 V |
| **CYU 8.10** | 8.06 V; 2% |
| **CYU 8.12** | 8.8% |

# HOMEWORK PROBLEMS

## Section 1: Semiconductors

**8.1**  In a semiconductor material, the net charge is zero. This requires the density of positive charges to be equal to the density of negative charges. Both charge carriers (free electrons and holes) and ionized dopant atoms have a charge equal to the magnitude of one electronic charge. Therefore the charge neutrality equation (CNE) is:

$$p_o + N_d^+ - n_o - N_a^- = 0$$

where

$n_o$ = Equilibrium negative carrier density

$p_o$ = Equilibrium positive carrier density

$N_a^- = $ Ionized acceptor density

$N_d^+ = $ Ionized donor density

The carrier product equation (CPE) states that as a semiconductor is doped the product of the charge carrier densities remains constant:

$$n_o p_o = \text{Constant}$$

For intrinsic silicon at $T = 300$ K:

$$\text{Constant} = n_{io} p_{io} = n_{io}^2 = p_{io}^2$$

$$= \left(1.5 \times 10^{16} \frac{1}{\text{m}^3}\right)^2 = 2.25 \times 10^{32} \frac{1}{\text{m}^2}$$

The semiconductor material is $n$- or $p$-type depending on whether donor or acceptor doping is greater. Almost all dopant atoms are ionized at room temperature. If intrinsic silicon is doped:

$$N_A \approx N_a^- = 10^{17} \frac{1}{\text{m}^3} \qquad N_d = 0$$

Determine:

a. If this is an $n$- or $p$-type extrinsic semiconductor.

b. Which are the major and which the minority charge carriers.

c. The density of majority and minority carriers.

**8.2**   If intrinsic silicon is doped:

$$N_a \approx N_a^- = 10^{17} \frac{1}{\text{m}^3} \qquad N_d \approx N_d^+ = 5 \times 10^{18} \frac{1}{\text{m}^3}$$

Determine:

a. If this is an $n$- or $p$-type extrinsic semiconductor.

b. Which are the majority and which the minority charge carriers.

c. The density of majority and minority carriers.

**8.3**   Describe the microscopic structure of semiconductor materials. What are the three most commonly used semiconductor materials?

**8.4**   Describe the thermal production of charge carriers in a semiconductor and how this process limits the operation of a semiconductor device.

**8.5**   Describe the properties of donor and acceptor dopant atoms and how they affect the densities of charge carriers in a semiconductor material.

**8.6**   Physically describe the behavior of the charge carriers and ionized dopant atoms in the vicinity of a semiconductor $pn$ junction that causes the potential (energy) barrier that tends to prevent charge carriers from crossing the junction.

## Section 2: Diode Circuit Models

**8.7**   Find voltage $v_L$ in the circuit of Figure P8.7, where $D$ is an ideal diode. Use values of $v_S < $ and $ > 0$.

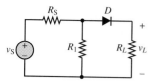

**Figure P8.7**

**8.8**   In the circuit of Figure P8.7, $v_S = 6$ V and $R_1 = R_S = R_L = 500$ Ω. Determine $i_D$ and $v_D$ graphically, using the diode characteristic of the 1N461A.

**8.9**   Assume that the diode in Figure P8.9 requires a minimum current of 1 mA to be above the knee of its $i$-$v$ characteristic.

a. What should be the value of $R$ to establish 5 mA in the circuit?

b. With the value of $R$ determined in part a, what is the minimum value to which the voltage $E$ could be reduced and still maintain diode current above the knee? Use $V_\gamma = 0.7$ V.

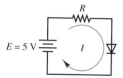

**Figure P8.9**

**8.10**   In Figure P8.10, a sinusoidal source of 50 V rms drives the circuit. Use the offset diode model for a silicon diode.

a. What is the maximum forward current?

b. What is the peak inverse voltage across the diode?

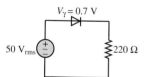

**Figure P8.10**

**8.11**   Determine which diodes are forward biased and which are reverse biased in each of the configurations shown in Figure P8.11.

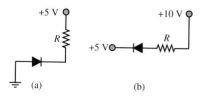

**Figure P8.11**

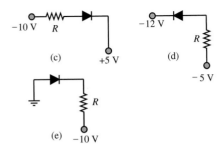

(c)

(d)

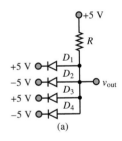

(e)  −10 V

**Figure P8.11** Continued

**8.12**  In the circuit of Figure P8.12, find the range of $V_{\text{in}}$ for which $D_1$ is forward-biased. Assume ideal diodes.

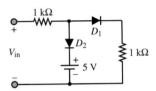

**Figure P8.12**

**8.13**  Determine which diodes are forward biased and which are reverse-biased in the configurations shown in Figure P8.13. Assuming a 0.7-V drop across each forward-biased diode, determine the output voltage.

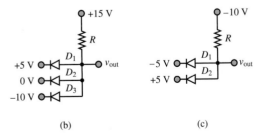

**Figure P8.13**

**8.14**  Sketch the output waveform and the voltage transfer characteristic for the circuit of Figure P8.14. Assume ideal diode characteristics, $v_S(t) = 10\sin(2{,}000\pi t)$.

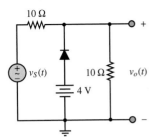

**Figure P8.14**

**8.15**  The diode in the circuit shown in Figure P8.15 is fabricated from silicon and:

$$i_D = I_o(e^{v_D/V_T} - 1)$$

where at $T = 300$ K:

$$I_o = 250 \times 10^{-12} \text{ A} \qquad V_T = \frac{kT}{q} \approx 26 \text{ mV}$$

$$v_S = 4.2 \text{ V} + 110\cos(\omega t) \text{ mV}$$

$$\omega = 377 \text{ rad/s} \qquad R = 7 \text{ k}\Omega$$

Determine, using superposition, the DC or $Q$ point current through the diode:

a. Using the DC offset model for the diode.

b. By iteratively solving the circuit characteristic (i.e., the DC load line equation) and the device characteristic (i.e., the diode equation).

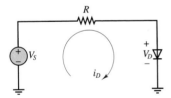

**Figure P8.15**

**8.16**  If the diode in the circuit shown in Figure P8.15 is fabricated from silicon and:

$$i_D = I_o[e^{v_D/V_T} - 1]$$

where at $T = 300$ K:

$$I_o = 2.030 \times 10^{-15} \text{ A} \qquad V_T = \frac{kT}{q} \approx 26 \text{ mV}$$

$$v_S = 5.3 \text{ V} + 7\cos(\omega t) \text{ mV}$$

$$\omega = 377 \text{ rad/s} \qquad R = 4.6 \text{ k}\Omega$$

Determine, using superposition and the offset (or threshold) voltage model for the diode, the DC or $Q$ point current through the diode.

**8.17**  If the diode in the circuit shown in Figure P8.15 is fabricated from silicon and:

$$i_D = I_o[e^{v_D/V_T} - 1]$$

where at $T = 300$ K:

$$I_o = 250 \times 10^{-12} \text{ A} \qquad V_T = \frac{kT}{q} \approx 26 \text{ mV}$$

$$v_S = 4.2 \text{ V} + 110 \cos(\omega t) \text{ mV}$$

$$\omega = 377 \text{ rad/s} \qquad R = 7 \text{ k}\Omega$$

and the DC operating point or quiescent point ($Q$ point) is:

$$I_{DQ} = 0.5458 \text{ mA} \qquad V_{DQ} = 379.5 \text{ mV}$$

determine the equivalent small-signal AC resistance of the diode at room temperature at the $Q$ point given.

**8.18**    If the diode in the circuit shown in Figure P8.15 is fabricated from silicon and:

$$i_D = I_o[e^{v_D/V_T} - 1]$$

where at $T = 300$ K:

$$I_o = 2.030 \times 10^{-15} \text{ A} \qquad V_T = \frac{kT}{q} \approx 26 \text{ mV}$$

$$v_S = 5.3 \text{ V} + 70 \cos(\omega t) \text{ mV}$$

$$\omega = 377 \text{ rad/s} \qquad R = 4.6 \text{ k}\Omega$$

and the DC operating point or quiescent point ($Q$ point) is:

$$I_{DQ} = 1.000 \text{ mA} \qquad V_{DQ} = 0.700 \text{ V}$$

determine the equivalent small-signal AC resistance of the diode at room temperature at the $Q$ point given.

**8.19**    If the diode in the circuit shown in Figure P8.15 is fabricated from silicon and:

$$i_D = I_o[e^{v_D/V_T} - 1]$$

where at $T = 300$ K:

$$I_o = 250 \times 10^{-12} \text{ A} \qquad V_T = \frac{kT}{q} \approx 26 \text{ mV}$$

$$v_S = V_S + v_s = 4.2 \text{ V} + 110 \cos(\omega t) \text{ mV}$$

$$\omega = 377 \text{ rad/s} \qquad R = 7 \text{ k}\Omega$$

The DC operating point or quiescent point ($Q$ point) and the AC small signal equivalent resistance at this $Q$ point are:

$$I_{DQ} = 0.548 \text{ mA} \qquad V_{DQ} = 0.365 \text{ V} \qquad r_d = 47.45 \text{ }\Omega$$

Determine, using superposition, the AC voltage across the diode and the AC current through it.

**8.20**    The diode in the circuit shown in Figure P8.20 is fabricated from silicon and:

$$R = 2.2 \text{ k}\Omega \qquad V_{S2} = 3 \text{ V}$$

Determine the minimum value of $V_{S1}$ at and above which the diode will conduct with a significant current.

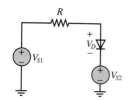

**Figure P8.20**

## Section 3: Rectifiers and Voltage Supplies

**8.21**    Find the average value of the output voltage for the circuit of Figure P8.21 if the input voltage is sinusoidal with an amplitude of 5 V. Let $V_\gamma = 0.7$ V.

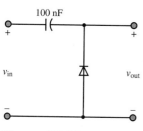

**Figure P8.21**

**8.22**    In the rectifier circuit shown in Figure P8.22, $v(t) = A \sin (2\pi 100)t$ V. Assume a forward voltage drop of 0.7 V across the diode when it is conducting. If conduction must begin during each positive half-cycle at an angle no greater than 5°, what is the minimum peak value, $A$, that the AC source must produce?

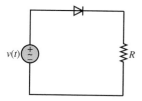

**Figure P8.22**

**8.23**    A half-wave rectifier is to provide an average voltage of 50 V at its output.

a.  Draw a schematic diagram of the circuit.
b.  Sketch the output voltage waveshape.
c.  Determine the peak value of the output voltage.
d.  Sketch the input voltage waveshape.
e.  What is the rms voltage at the input?

**8.24**    You have been asked to design a full-wave bridge rectifier for a power supply. A step-down transformer has already been chosen. It will supply 12 V rms to your rectifier. The full-wave rectifier is shown in the circuit of Figure P8.24.

a. If the diodes have an offset voltage of 0.6 V, sketch the input source voltage, $v_S(t)$, and the output voltage, $v_L(t)$, and state which diodes are on and which are off in the appropriate cycles of $v_S(t)$. The frequency of the source is 60 Hz.

b. If $R_L = 1,000 \, \Omega$ and a capacitor, placed across $R_L$ to provide some filtering, has a value of 8 $\mu$F, sketch the output voltage, $v_L(t)$.

c. Repeat part b, with the capacitance equal to 100 $\mu$F.

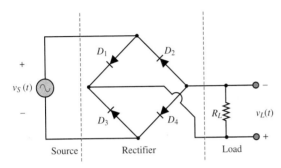

**Figure P8.24**

**8.25** In the full-wave power supply shown in Figure P8.25 the diodes are 1N4001 with a rated peak reverse voltage (also called peak inverse voltage) of 25 V. They are fabricated from silicon.

$$n = 0.05883$$
$$C = 80 \, \mu\text{F} \qquad R_L = 1 \, \text{k}\Omega$$
$$V_{\text{line}} = 170 \cos(377t) \, \text{V}$$

a. Determine the actual peak reverse voltage across each diode.

b. Explain why these diodes are or are not suitable for the specifications given.

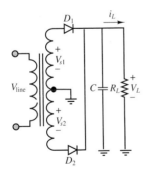

**Figure P8.25**

**8.26** In the full-wave power supply shown in Figure P8.25:

$$n = 0.1$$
$$C = 80 \, \mu\text{F} \qquad R_L = 1 \, \text{k}\Omega$$
$$V_{\text{line}} = 170 \cos(377t) \, \text{V}$$

The diodes are 1N914 switching diodes (but used here

for AC-DC conversion), fabricated from silicon, with the following rated performance:

$$P_{\text{max}} = 500 \, \text{mW} \qquad \text{at } T = 25°C$$
$$V_{\text{pk–rev}} = 30 \, \text{V}$$

The derating factor is 3 mW/°C for $25°\text{C} < T \leq 125°\text{C}$ and 4 mW/°C for $125°\text{C} < T \leq 175°\text{C}$.

a. Determine the actual peak reverse voltage across each diode.

b. Explain why these diodes are or are not suitable for the specifications given.

**8.27** The diodes in the full-wave DC power supply shown in Figure P8.25 are silicon. The load voltage waveform is shown in Figure P8.27. If:

$$I_L = 60 \, \text{mA} \quad V_L = 5 \, \text{V} \quad V_{\text{ripple}} = (0.05) \, V_L$$
$$V_{\text{line}} = 170 \cos(\omega t) \, \text{V} \qquad \omega = 377 \, \text{rad/s}$$

determine the value of:

a. The turns ratio, $n$.

b. The value of the capacitor, $C$.

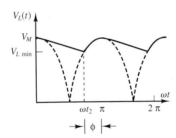

**Figure P8.27**

**8.28** The diodes in the full-wave DC power supply shown in Figure P8.25 are silicon. If:

$$I_L = 600 \, \text{mA} \qquad V_L = 50 \, \text{V}$$
$$V_r = 8\% = 4 \, \text{V}$$
$$V_{\text{line}} = 170 \cos(\omega t) \, \text{V} \qquad \omega = 377 \, \text{rad/s}$$

determine the value of:

a. The turns ratio, $n$.

b. The value of the capacitor, $C$.

**8.29** The diodes in the full-wave DC power supply shown in Figure P8.25 are silicon. If:

$$I_L = 5 \, \text{mA} \qquad V_L = 10 \, \text{V}$$
$$V_r = 20\% = 2 \, \text{V}$$
$$V_{\text{line}} = 170 \cos(\omega t) \, \text{V} \qquad \omega = 377 \, \text{rad/s}$$

determine the:

a. Turns ratio, $n$.

b. The value of the capacitor, $C$.

**8.30** In the circuit shown in Figure 8.15:

$$I_L = 600 \text{ mA} \qquad V_L = 50 \text{ V}$$

$$V_R = 4 \text{ V} \qquad C = 1000 \ \mu\text{F}$$

$$v_{s1}(t) = v_{s2}(t) = V_{s0} \cos(\omega t) \qquad \omega = 377 \text{ rad/s}$$

The diodes are silicon. If the power rating of one of the diodes is exceeded and it burns out or opens, determine the new values of the DC output or load voltage and the ripple voltage.

**8.31** In the full-wave power supply shown in Figure P8.31 the diodes are 1N4001 with a rated peak reverse voltage (also called peak inverse voltage) of 50 V. They are fabricated from silicon.

$$V_{\text{line}} = 170 \cos(377t) \text{V}$$

$$n = 0.2941$$

$$C = 700 \ \mu\text{F} \qquad R_L = 2.5 \text{ k}\Omega$$

a. Determine the actual peak reverse voltage across each diode.

b. Explain why these diodes are or are not suitable for the specifications given.

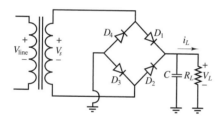

**Figure P8.31**

**8.32** In the full-wave power supply shown in Figure P8.31 the diodes are 1N4001 general-purpose silicon diodes with a rated peak reverse voltage of 10 V and:

$$V_{\text{line}} = 156 \cos(377t) \text{V}$$

$$n = 0.04231 \qquad V_r = 0.2 \text{ V}$$

$$I_L = 2.5 \text{ mA} \qquad V_L = 5.1 \text{ V}$$

a. Determine the actual peak reverse voltage across the diodes.

b. Explain why these diodes are or are not suitable for the specifications given.

**8.33** The diodes in the full-wave DC power supply shown in Figure P8.31 are silicon. If:

$$I_L = 650 \text{ mA} \qquad V_L = 10 \text{ V}$$

$$V_r = 1 \text{ V} \qquad \omega = 377 \text{ rad/s}$$

$$V_{\text{line}} = 170 \cos(\omega t) \text{ V} \qquad \phi = 23.66°$$

determine the value of the average and peak current through each diode.

**8.34** The diodes in the full-wave DC power supply shown in Figure P8.31 are silicon. If:

$$I_L = 85 \text{ mA} \qquad V_L = 5.3 \text{ V}$$

$$V_r = 0.6 \text{ V} \qquad \omega = 377 \text{ rad/s}$$

$$V_{\text{line}} = 156 \cos(\omega t) \text{ V}$$

Determine the value of:

a. The turns ratio, $n$.

b. The capacitor, $C$.

**8.35** The diodes in the full-wave DC power supply shown are silicon. If:

$$I_L = 250 \text{ mA} \qquad V_L = 10 \text{ V}$$

$$V_r = 2.4 \text{ V} \qquad \omega = 377 \text{ rad/s}$$

$$V_{\text{line}} = 156 \cos(\omega t) \text{ V}$$

Determine the value of:

a. The turns ratio, $n$.

b. The capacitor, $C$.

## Section 4: Zener Diodes and Voltage Regulation

**8.36** The diode shown in Figure P8.36 has a piecewise linear characteristic that passes through the points $(-10 \text{ V}, -5 \ \mu\text{A})$, $(0, 0)$, $(0.5 \text{ V}, 5 \text{ mA})$, and $(1 \text{ V}, 50 \text{ mA})$. Determine the piecewise linear model, and, using that model, solve for $i$ and $v$.

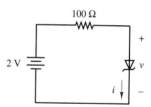

**Figure P8.36**

**8.37** Find the minimum value of $R_L$ in the circuit shown in Figure P8.37 for which the output voltage remains at just 5.6 V.

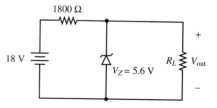

**Figure P8.37**

**8.38** Determine the minimum value and the maximum value that the series resistor may have in a regulator circuit whose output voltage is to be 25 V, whose input voltage varies from 35 to 40 V, and whose maximum load current is 75 mA. The Zener diode used in this circuit has a maximum current rating of 250 mA.

**8.39** The $i$-$v$ characteristic of a semiconductor diode designed to operate in the Zener breakdown region is shown in Figure P8.39. The Zener or breakdown region extends from a minimum current at the knee of the curve, equal here to about $-5$ mA (from the graph) and the maximum rated current equal $-90$ mA (from the specification sheet). Determine the Zener resistance and Zener voltage of the diode.

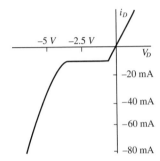

**Figure P8.39**

**8.40** The Zener diode in the simple voltage regulator circuit shown in Figure P8.40 is a 1N5231B. The source voltage is obtained from a DC power supply. It has a DC and a ripple component:

$$v_S = V_S + V_r$$

where:

$$V_S = 20 \text{ V} \qquad V_r = 250 \text{ mV}$$
$$R = 220 \ \Omega \qquad I_L = 65 \text{ mA} \qquad V_L = 5.1 \text{ V}$$
$$V_z = 5.1 \text{ V} \qquad r_z = 17 \ \Omega \qquad P_{\text{Rated}} = 0.5 \text{ W}$$
$$i_{z\min} = 10 \text{ mA}$$

Determine the maximum rated current the diode can handle without exceeding its power limitation.

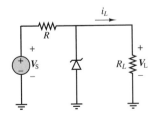

**Figure P8.40**

**8.41** The 1N963 Zener diode in the simple voltage regulator circuit shown in Figure P8.40 has the

specifications:

$$V_z = 12 \text{ V} \qquad r_z = 11.5 \ \Omega \qquad P_{\text{Rated}} = 400 \text{ mW}$$

At knee of curve:

$$i_{zk} = 0.25 \text{ mA} \qquad r_{zk} = 700 \ \Omega$$

Determine the maximum rated current the diode can handle without exceeding its power limitation.

**8.42** In the simple voltage regulator circuit shown in Figure P8.40, $R$ must maintain the Zener diode current within its specified limits for all values of the source voltage, load current, and Zener diode voltage. Determine the minimum and maximum values of $R$ which can be used.

$$V_z = 5 \text{ V} \pm 10\% \qquad r_z = 15 \ \Omega$$
$$i_{z\min} = 3.5 \text{ mA} \qquad i_{z\max} = 65 \text{ mA}$$
$$V_S = 12 \text{ V} \pm 3 \text{ V} \qquad I_L = 70 \pm 20 \text{ mA}$$

**8.43** In the simple voltage regulator circuit shown in Figure P8.40, $R$ must maintain the Zener diode current within its specified limits for all values of the source voltage, load current, and Zener diode voltage. If:

$$V_z = 12 \text{ V} \pm 10\% \qquad r_z = 9 \ \Omega$$
$$i_{z\min} = 3.25 \text{ mA} \qquad i_{z\max} = 80 \text{ mA}$$
$$V_S = 25 \pm 1.5 \text{ V}$$
$$I_L = 31.5 \pm 21.5 \text{ mA}$$

determine the minimum and maximum values of $R$ which can be used.

**8.44** In the simple voltage regulator circuit shown in Figure P8.40, the Zener diode is a 1N4740A.

$$V_z = 10 \text{ V} \pm 5\% \qquad r_z = 7 \ \Omega \qquad i_{z\min} = 10 \text{ mA}$$
$$P_{\text{rated}} = 1 \text{ W} \qquad i_{z\max} = 91 \text{ mA}$$
$$V_S = 14 \pm 2 \text{ V} \qquad R = 19.8 \ \Omega$$

Determine the minimum and maximum load current for which the diode current remains within its specified values.

**8.45** In the simple voltage regulator circuit shown in Figure P8.40, the Zener diode is a 1N963. Determine the minimum and maximum load current for which the diode current remains within its specified values.

$$V_z = 12 \text{ V} \pm 10\% \qquad r_z = 11.5 \ \Omega$$
$$i_{z\min} = 2.5 \text{ mA} \qquad i_{z\max} = 32.6 \text{ mA}$$
$$P_R = 400 \text{ mW}$$
$$V_S = 25 \pm 2 \text{ V} \qquad R = 470 \ \Omega$$

**8.46** In the simple voltage regulator circuit shown in Figure P8.40, the Zener diode is a 1N4740A. Determine the minimum and maximum source voltage for which the diode current remains within its specified values.

$$V_z = 10 \text{ V} \pm 5\% \qquad r_z = 7 \, \Omega$$

$$i_{z\,min} = 10 \text{ mA} \qquad i_{z\,max} = 91 \text{ mA}$$

$$P_{Rated} = 1 \text{ W} \qquad R = 80 \, \Omega \pm 5\%$$

$$I_L = 35 \pm 10 \text{ mA}$$

**8.47**  The Zener diode in the simple voltage regulator circuit shown in Figure P8.40 is a 1N4740. The source voltage is obtained from a DC power supply. It has a DC and a ripple component:

$$v_S = V_S + V_r$$

where:

$$V_S = 16 \text{ V} \qquad V_r = 2 \text{ V}$$

$$I_L = 35 \text{ mA} \qquad V_L = 10 \text{ V}$$

$$V_z = 10 \text{ V} \qquad r_z = 7 \, \Omega$$

$$i_{z\,max} = 91 \text{ mA} \qquad i_{z\,min} = 10 \text{ mA}$$

Determine the ripple voltage across the load.

**8.48**  The Zener diode in the simple voltage regulator circuit shown in Figure P8.40 is a 1N5231B. The source voltage is obtained from a DC power supply. It has a DC and a ripple component:

$$v_S = V_S + V_r$$

where:

$$V_S = 20 \text{ V} \qquad V_r = 250 \text{ mV}$$

$$R = 220 \, \Omega$$

$$I_L = 65 \text{ mA} \qquad V_L = 5.1 \text{ V}$$

$$V_z = 5.1 \text{ V} \qquad r_z = 17 \, \Omega$$

$$P_{rated} = 0.5 \text{ W}$$

$$i_{z\,min} = 10 \text{ mA}$$

Determine the ripple voltage across the load.

**8.49**  The Zener diode in the simple voltage regulator circuit shown in Figure P8.40 is a 1N970. The source voltage is obtained from a DC power supply. It has a DC and a ripple component:

$$v_S = V_S + V_r$$

where:

$$V_S = 30 \text{ V} \qquad V_r = 3 \text{ V} \qquad I_L = 8 \text{ A}$$

$$V_z = 24 \text{ V} \qquad r_z = 33 \, \Omega \qquad V_L = 24 \text{ V}$$

$$i_{z\,max} = 15 \text{ A} \qquad i_{z\,min} = 1.5 \text{ A} \qquad R = 1 \, \Omega$$

Determine the ripple voltage across the load.

## Section 5: Other Diode Circuits

**8.50**  Assuming that the diodes are ideal, determine and sketch the $i$-$v$ characteristics for the circuit of Figure P8.50. Consider the range $10 \geq v \geq 0$.

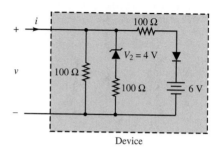

**Figure P8.50**

**8.51**  Given the input voltage waveform and the circuit shown in Figure P8.51, sketch the output voltage.

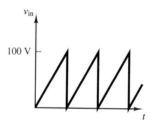

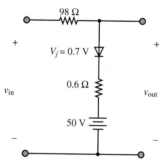

**Figure P8.51**

**8.52**  We are using a voltage source to charge an automotive battery as shown in the circuit of Figure P8.52(a). At $t = t_1$, the protective circuitry of the source causes switch $S_1$ to close, and the source voltage goes to zero. Find the currents, $I_S$, $I_B$, and $I_{SW}$, for the following conditions:

a.  $t = t_1^-$

b.  $t = t_1^+$

c.  What will happen to the battery after the switch closes?

Now we are going to charge the battery, using the circuit of Figure P8.52(b). Repeat parts a and b if the diode has an offset voltage of 0.6 V.

**8.53** Find the output voltage of the peak detector shown in Figure P8.53. Use sinusoidal input voltages with amplitude 6, 1.5, and 0.4 V and zero average value. Let $V_\gamma = 0.7$ V.

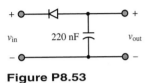

**Figure P8.53**

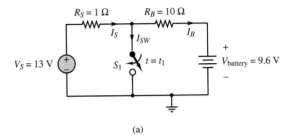

(a)

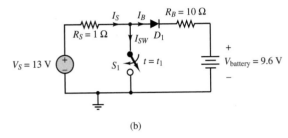

(b)

**Figure P8.52**

# CHAPTER

# 9

# Transistor Fundamentals

C hapter 9 continues the discussion of electronic devices that began in Chapter 8 with the semiconductor diode. This chapter describes the operating characteristics of the two major families of electronic devices: bipolar and field-effect transistors. The first half of Chapter 9 is devoted to a brief, qualitative discussion of the physics and operation of the bipolar junction transistor (BJT), which naturally follows the discussion of the *pn* junction in Chapter 8. The $i$-$v$ characteristics of bipolar transistors and their operating states are presented. Large-signal circuit models for the BJT are then introduced, to illustrate how one can analyze transistor circuits using basic circuit analysis methods. A few practical examples are discussed to illustrate the use of the circuit models. The second half of the chapter focuses on field-effect transistors; the basic operation and $i$-$v$ characteristics of enhancement- and depletion-mode MOS transistors and of junction field-effect transistors are presented. Universal curves for each of these devices and large-signal circuit models are also discussed. By the end of Chapter 9, you should be able to:

- Describe the basic operation of bipolar junction transistors.
- Interpret BJT characteristic curves and extract large-signal model parameters from these curves.
- Identify the operating state of a BJT from measured data and determine its operating point.

- Analyze simple large-signal BJT amplifiers.
- Describe the basic operation of enhancement- and depletion-mode metal-oxide-semiconductor field-effect transistors (MOSFETs) and of junction field-effect transistors (JFETs).
- Interpret the universal curves for these devices and extract linear (small-signal) models for simple amplifiers from device curves and data sheets.
- Identify the operating state of a field-effect transistor from measured data and determine its operating point.
- Analyze simple large-signal FET amplifiers.

## 9.1  TRANSISTORS AS AMPLIFIERS AND SWITCHES

A transistor is a three-terminal semiconductor device that can perform two functions that are fundamental to the design of electronic circuits: **amplification** and **switching.** Put simply, amplification consists of magnifying a signal by transferring energy to it from an external source; whereas a transistor switch is a device for controlling a relatively large current between or voltage across two terminals by means of a small control current or voltage applied at a third terminal. In this chapter, we provide an introduction to the two major families of transistors: *bipolar junction transistors*, or *BJTs*; and *field-effect transistors*, or *FETs*.

The operation of the transistor as a linear amplifier can be explained qualitatively by the sketch of Figure 9.1, in which the four possible modes of operation of a transistor are illustrated by means of circuit models employing controlled sources (you may wish to review the section on controlled sources in Chapter 2). In Figure 9.1, controlled voltage and current sources are shown to generate an output proportional to an input current or voltage; the proportionality constant, $\mu$, is called the internal *gain* of the transistor. As will be shown, the BJT acts essentially as a current-controlled device, while the FET behaves as a voltage-controlled device.

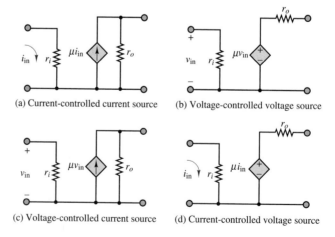

(a) Current-controlled current source          (b) Voltage-controlled voltage source

(c) Voltage-controlled current source          (d) Current-controlled voltage source

**Figure 9.1** Controlled-source models of linear amplifier transistor operation

Transistors can also act in a nonlinear mode, as voltage- or current-controlled switches. When a transistor operates as a switch, a small voltage or current is used to control the flow of current between two of the transistor terminals in an on-off fashion. Figure 9.2 depicts the idealized operation of the transistor as a switch, suggesting that the switch is closed (on) whenever a control voltage or current is greater than zero and is open (off) otherwise. It will later become apparent that the conditions for the switch to be on or off need not necessarily be those depicted in Figure 9.2. The operation of transistors as switches will be discussed in more detail in Chapter 10.

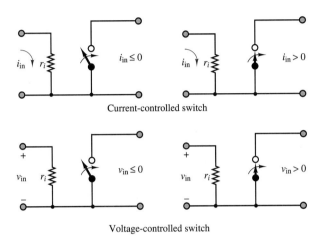

**Figure 9.2** Models of ideal transistor switches

## EXAMPLE  9.1  Model of Linear Amplifier

### Problem

Determine the voltage gain of the amplifier circuit model shown in Figure 9.3.

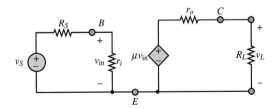

**Figure 9.3**

### Solution

**Known Quantities:**  Amplifier internal input and output resistances, $r_i$ and $r_o$; amplifier internal gain, $\mu$; source and load resistances, $R_S$ and $R_L$.

**Find:**  $A_V = \dfrac{v_L}{v_S}$

**Analysis:**  First determine the input voltage, $v_{\text{in}}$, using the voltage divider rule:

$$v_{\text{in}} = \frac{r_i}{r_i + R_S} v_S$$

Then, the output of the controlled voltage source is:

$$\mu v_{\text{in}} = \mu \frac{r_i}{r_i + R_S} v_S$$

and the output voltage can be found by the voltage divider rule:

$$v_L = \mu \frac{r_i}{r_i + R_S} v_S \times \frac{R_L}{r_o + R_L}$$

Finally, the amplifier voltage gain can be computed:

$$A_V = \frac{v_L}{v_S} = \mu \frac{r_i}{r_i + R_S} \times \frac{R_L}{r_o + R_L}$$

**Comments:**  Note that the voltage gain computed above is always less than the transistor internal voltage gain, $\mu$. One can easily show that if the conditions $r_i \gg R_S$ and $r_o \ll R_L$ hold, then the gain of the amplifier becomes approximately equal to the gain of the transistor. One can therefore conclude that the actual gain of an amplifier always depends on the relative values of source and input resistance, and of output and load resistance.

## Check Your Understanding

**9.1**  Repeat the analysis of Example 9.1 for the current-controlled voltage source model of Figure 9.1(d). What is the amplifier voltage gain? Under what conditions would the gain $A$ be equal to $\mu/R_S$?

**9.2**  Repeat the analysis of Example 9.1 for the current-controlled current source model of Figure 9.1(a). What is the amplifier voltage gain?

**9.3**  Repeat the analysis of Example 9.1 for the voltage-controlled current source model of Figure 9.1(c). What is the amplifier voltage gain?

## 9.2    THE BIPOLAR JUNCTION TRANSISTOR (BJT)

The *pn* junction studied in Chapter 8 forms the basis of a large number of semiconductor devices. The semiconductor diode, a two-terminal device, is the most direct application of the *pn* junction. In this section, we introduce the **bipolar junction transistor (BJT)**. As we did in analyzing the diode, we will introduce the physics of transistor devices as intuitively as possible, resorting to an analysis of their $i$-$v$ characteristics to discover important properties and applications.

A BJT is formed by joining three sections of semiconductor material, each with a different doping concentration. The three sections can be either a thin $n$ region sandwiched between $p^+$ and $p$ layers, or a $p$ region between $n$ and $n^+$ layers, where the superscript "plus" indicates more heavily doped material. The resulting BJTs are called *pnp* and *npn* transistors, respectively; we shall discuss only the latter in this chapter. Figure 9.4 illustrates the approximate construction, symbols, and nomenclature for the two types of BJTs.

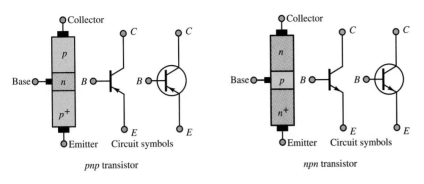

**Figure 9.4** Bipolar junction transistors

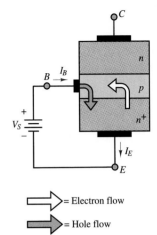

= Electron flow

= Hole flow

The *BE* junction acts very much as an ordinary diode when the collector is open. In this case, $I_B = I_E$.

**Figure 9.5** Current flow in an *npn* BJT

The operation of the *npn* BJT may be explained by considering the transistor as consisting of two back-to-back *pn* junctions. The **base-emitter (*BE*) junction** acts very much like a diode when it is forward-biased; thus, one can picture the corresponding flow of hole and electron currents from base to emitter when the collector is open and the *BE* junction is forward-biased, as depicted in Figure 9.5. Note that the electron current has been shown larger than the hole current, because of the heavier doping of the *n* side of the junction. Some of the electron-hole pairs in the base will recombine; the remaining charge carriers will give rise to a net flow of current from base to emitter. It is also important to observe that the base is much narrower than the emitter section of the transistor.

Imagine, now, reverse-biasing the **base-collector (*BC*) junction**. In this case, an interesting phenomenon takes place: the electrons "emitted" by the emitter with the *BE* junction forward-biased reach the very narrow base region, and after a few are lost to recombination in the base, most of these electrons are "collected" by the collector. Figure 9.6 illustrates how the reverse bias across the *BC* junction is in such a direction as to sweep the electrons from the emitter into the collector. This phenomenon can take place because the base region is kept particularly narrow. Since the base is narrow, there is a high probability that the electrons will have gathered enough momentum from the electric field to cross the reverse-biased collector-base junction and make it into the collector. The result is that there is a net flow of current from collector to emitter (opposite in direction to the flow of electrons), in addition to the hole current from base to emitter. The electron current flowing into the collector through the base is substantially larger than that which flows into the base from the external circuit. One can see from Figure 9.6 that if KCL is to be satisfied, we must have

$$I_E = I_B + I_C \tag{9.1}$$

The most important property of the bipolar transistor is that the small base current controls the amount of the much larger collector current

$$I_C = \beta I_B \tag{9.2}$$

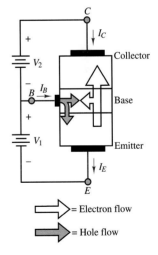

= Electron flow

= Hole flow

When the *BC* junction is reverse-biased, the electrons from the emitter region are swept across the base into the collector.

**Figure 9.6** Flow of emitter electrons into the collector in an *npn* BJT

where $\beta$ is a current amplification factor dependent on the physical properties of the transistor. Typical values of $\beta$ range from 20 to 200. The operation of a *pnp* transistor is completely analogous to that of the *npn* device, with the roles of the charge carriers (and therefore the signs of the currents) reversed. The symbol for a *pnp* transistor was shown in Figure 9.4.

The exact operation of bipolar transistors can be explained by resorting to a detailed physical analysis of the *npn* or *pnp* structure of these devices. The reader interested in such a discussion of transistors is referred to any one of a number of excellent books on semiconductor electronics. The aim of this book, on the other hand, is to provide an introduction to the basic principles of transistor operation by means of simple linear circuit models based on the device *i-v* characteristic. Although it is certainly useful for the non-electrical engineer to understand the basic principles of operation of electronic devices, it is unlikely that most readers will engage in the design of high-performance electronic circuits or will need a detailed understanding of the operation of each device. The present chapter will therefore serve as a compendium of the basic ideas, enabling an engineer to read and understand electronic circuit diagrams and to specify the requirements of electronic instrumentation systems. The focus of this section will be on the analysis of the *i-v* characteristic of the *npn* BJT, based on the circuit notation defined in Figure 9.7. The device *i-v* characteristics will be presented qualitatively, without deriving the underlying equations, and will be utilized in constructing circuit models for the device.

The number of independent variables required to uniquely define the operation of the transistor may be determined by applying KVL and KCL to the circuit of Figure 9.7. Two voltages and two currents are sufficient to specify the operation of the device. Note that, since the BJT is a three-terminal device, it will not be sufficient to deal with a single *i-v* characteristic; two such characteristics are required to explain the operation of this device. One of these characteristics relates the base current, $i_B$, to the base-emitter voltage, $v_{BE}$; the other relates the collector current, $i_C$, to the collector-emitter voltage, $v_{CE}$. The latter characteristic actually consists of a *family* of curves. To determine these *i-v* characteristics, consider the *i-v* curves of Figures 9.8 and 9.9, using the circuit notation of Figure 9.7. In Figure 9.8, the collector is open and the *BE* junction is shown to be very similar to a diode. The ideal current source, $I_{BB}$, injects a base current, which causes the junction to be forward-biased. By varying $I_{BB}$, one can obtain the open-collector *BE* junction *i-v* curve shown in the figure.

If a voltage source were now to be connected to the collector circuit, the voltage $v_{CE}$ and, therefore, the collector current, $i_C$, could be varied, in addition to the base current, $i_B$. The resulting circuit is depicted in Figure 9.9(a). By varying both the base current and the collector-emitter voltage, one could then generate a plot of the device **collector characteristic.** This is also shown in Figure 9.9(b). Note that this figure depicts not just a single $i_C$-$v_{CE}$ curve, but an entire family, since for each value of the base current, $i_B$, an $i_C$-$v_{CE}$ curve can be generated. Four regions are identified in the collector characteristic:

1. The **cutoff region,** where both junctions are reverse-biased, the base current is very small, and essentially no collector current flows.

2. The **active linear region,** in which the transistor can act as a linear amplifier, where the *BE* junction is forward-biased and the *CB* junction is reverse-biased.

The operation of the BJT is defined in terms of two currents and two voltages: $i_B$, $i_C$, $v_{CE}$, and $v_{BE}$.

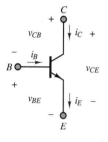

KCL: $i_E = i_B + i_C$
KVL: $v_{CE} = v_{CB} + v_{BE}$

**Figure 9.7** Definition of BJT voltages and currents

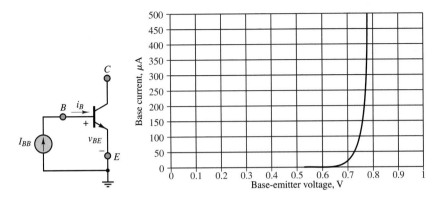

**Figure 9.8** BE junction open-collector curve

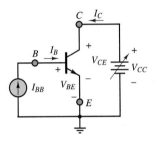

**Figure 9.9(a)** Ideal test circuit to determine the $i$-$v$ characteristic of a BJT

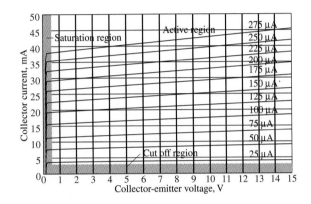

**Figure 9.9(b)** The collector-emitter output characteristics of a BJT

3. The **saturation region,** in which both junctions are forward-biased.
4. The **breakdown region,** which determines the physical limit of operation of the device.

From the curves of Figure 9.9(b), we note that as $v_{CE}$ is increased, the collector current increases rapidly, until it reaches a nearly constant value; this condition holds until the collector junction breakdown voltage, $BV_{CEO}$, is reached (for the purposes of this book, we shall not concern ourselves with the phenomenon of breakdown, except in noting that there are maximum allowable voltages and currents in a transistor). If we were to repeat the same measurement for a set of different values of $i_B$, the corresponding value of $i_C$ would change accordingly; hence, the family of collector characteristic curves.

## Determining the Operating Region of a BJT

Before we discuss common circuit models for the BJT, it will be useful to consider the problem of determining the operating region of the transistor. A few simple

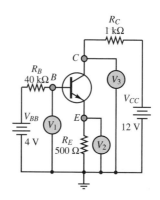

**Figure 9.10** Determination of the operation region of a BJT

voltage measurements permit a quick determination of the state of a transistor placed in a circuit. Consider, for example, the BJT described by the curves of Figure 9.9 when it is placed in the circuit of Figure 9.10. In this figure, voltmeters are used to measure the value of the collector, emitter, and base voltages. Can these simple measurements identify the operating region of the transistor? Assume that the measurements reveal the following conditions:

$$V_B = V_1 = 2 \text{ V} \qquad V_E = V_2 = 1.3 \text{ V} \qquad V_C = V_3 = 8 \text{ V}$$

What can be said about the operating region of the transistor?

The first observation is that knowing $V_B$ and $V_E$ permits determination of $V_{BE} : V_B - V_E = 0.7$ V. Thus, we know that the $BE$ junction is forward-biased. Another quick calculation permits determination of the relationship between base and collector current: the base current is equal to

$$I_B = \frac{V_{BB} - V_B}{R_B} = \frac{4 - 2}{40,000} = 50 \text{ } \mu\text{A}$$

while the collector current is

$$I_C = \frac{V_{CC} - V_C}{R_C} = \frac{12 - 8}{1,000} = 4 \text{ mA}$$

Thus, the current amplification (or gain) factor for the transistor is

$$\frac{I_C}{I_B} = \beta = 80$$

Such a value for the current gain suggests that the transistor is in the linear active region, because substantial current amplification is taking place (typical values of current gain range from 20 to 200). Finally, the collector-to-emitter voltage, $V_{CE}$, is found to be: $V_{CE} = V_C - V_E = 8 - 1.3 = 6.7$ V.

At this point, you should be able to locate the operating point of the transistor on the curves of Figures 9.8 and 9.9. The currents $I_B$ and $I_C$ and the voltage $V_{CE}$ uniquely determine the state of the transistor in the $I_C$-$V_{CE}$ and $I_B$-$V_{BE}$ characteristic curves. What would happen if the transistor were not in the linear active region? The following examples answer this question and provide further insight into the operation of the bipolar transistor.

---

### EXAMPLE  9.2  Determining the Operating Region of a BJT

**Problem**

Determine the operating region of the BJT in the circuit of Figure 9.10 when the base voltage source, $V_{BB}$, is short-circuited.

---

**Solution**

**Known Quantities:**  Base and collector supply voltages; base, emitter, and collector resistance values.

**Find:**  Operating region of the transistor.

**Schematics, Diagrams, Circuits, and Given Data:** $V_{BB} = 0$; $V_{CC} = 12$ V; $R_B = 40$ kΩ; $R_C = 1$ kΩ; $R_E = 500$ Ω.

**Analysis:** Since $V_{BB} = 0$, the base will be at zero volts, and therefore the base-emitter junction is reverse-biased and the base current is zero. Thus the emitter current will also be nearly zero. From equation 9.1 we conclude that the collector current must also be zero. Checking these observations against Figure 9.9(b) leads to the conclusion that the transistor is in the cutoff state. In these cases the three voltmeters of Figure 9.10 will read zero for $V_B$ and $V_E$ and +12 V for $V_C$, since there is no voltage drop across $R_C$.

**Comments:** In general, if the base supply voltage is not sufficient to forward-bias the base-emitter junction, the transistor will be in the cutoff region.

VIRTUAL LAB

# EXAMPLE  9.3  Determining the Operating Region of a BJT

## Problem

Determine the operating region of the BJT in the circuit of Figure 9.11.

## Solution

**Known Quantities:** Base, collector, and emitter voltages with respect to ground.

**Find:** Operating region of the transistor.

**Schematics, Diagrams, Circuits, and Given Data:** $V_1 = V_B = 2.7$ V; $V_2 = V_E = 2$ V; $V_3 = V_C = 2.3$ V.

**Analysis:** To determine the region of the transistor we shall compute $V_{BE}$ and $V_{BC}$ to determine whether the $BE$ and $BC$ junctions are forward or reverse-biased. Operation in the *saturation region* corresponds to forward bias at both junctions (and very small voltage drops); operation in the *active region* is characterized by a forward-biased $BE$ junction and a reverse-biased $BC$ junction.

From the available measurements, we compute:

$$V_{BE} = V_B - V_E = 0.7 \text{ V}$$
$$V_{BC} = V_B - V_C = 0.4 \text{ V}$$

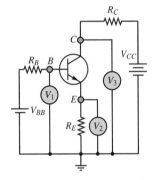

**Figure 9.11**

Since both junctions are forward-biased, the transistor is operating in the saturation region. The value of $V_{CE} = V_C - V_E = 0.3$ V is also very small. This is usually a good indication that the BJT is operating in saturation.

**Comments:** Try to locate the operating point of this transistor in Figure 9.9(b), assuming that

$$I_C = \frac{V_{CC} - V_3}{R_C} = \frac{12 - 2.3}{1,000} = 9.7 \text{ mA}$$

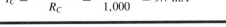

# Selecting an Operating Point for a BJT

The family of curves shown for the collector $i$-$v$ characteristic in Figure 9.9(b) reflects the dependence of the collector current on the base current. For each value of the base current, $i_B$, there exists a corresponding $i_C$-$v_{CE}$ curve. Thus, by appropriately selecting the base current and collector current (or collector-emitter

By appropriate choice of $I_{BB}$, $R_C$ and $V_{CC}$, the desired $Q$ point may be selected.

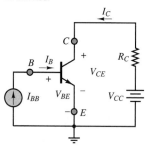

**Figure 9.12** A simplified bias circuit for a BJT amplifier

voltage), we can determine the operating point, or **$Q$ point,** of the transistor. The $Q$ point of a device is defined in terms of the **quiescent** (or **idle**) **currents** and **voltages** that are present at the terminals of the device when DC supplies are connected to it. The circuit of Figure 9.12 illustrates an ideal **DC bias circuit,** used to set the $Q$ point of the BJT in the approximate center of the collector characteristic. The circuit shown in Figure 9.12 is not a practical DC bias circuit for a BJT amplifier, but it is very useful for the purpose of introducing the relevant concepts. A practical bias circuit is discussed later in this section.

Applying KVL around the base-emitter and collector circuits, we obtain the following equations:

$$I_B = I_{BB} \tag{9.3}$$

and

$$V_{CE} = V_{CC} - I_C R_C \tag{9.4}$$

which can be rewritten as

$$I_C = \frac{V_{CC}}{R_C} - \frac{V_{CE}}{R_C} \tag{9.5}$$

Note the similarity of equation 9.5 to the load-line curves of Chapters 3 and 8. Equation 9.5 represents a line that intersects the $I_C$ axis at $I_C = V_{CC}/R_C$ and the $V_{CE}$ axis at $V_{CE} = V_{CC}$. The slope of the load line is $-1/R_C$. Since the base current, $I_B$, is equal to the source current, $I_{BB}$, the operating point may be determined by noting that the load line intersects the entire collector family of curves. The intersection point at the curve that corresponds to the base current $I_B = I_{BB}$ constitutes the operating, or $Q$, point. The load line corresponding to the circuit of Figure 9.12 is shown in Figure 9.13, superimposed on the collector curves for the 2N3904 transistor (data sheets for the 2N3904 transistor are included in the CD-ROM that accompanies the book). In Figure 9.13, $V_{CC} = 15$ V, $V_{CC}/R_C = 40$ mA, and $I_{BB} = 150$ $\mu$A; thus, the $Q$ point is determined by the intersection of the load line with the $I_C$-$V_{CE}$ curve corresponding to a base current of 150 $\mu$A.

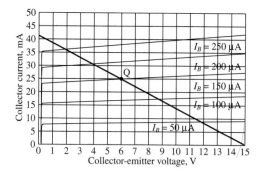

**Figure 9.13** Load-line analysis of a simplified BJT amplifier

Once an operating point is established and DC currents $I_{CQ}$ and $I_{BQ}$ are flowing into the collector and base, respectively, the BJT can serve as a linear amplifier, as will be explained in Section 9.3. Example 9.4 serves as an illustration of the DC biasing procedures just described.

# EXAMPLE 9.4 Calculation of DC Operating Point for BJT Amplifier

## Problem

Determine the DC operating point of the BJT amplifier in the circuit of Figure 9.14.

## Solution

**Known Quantities:** Base, collector, and emitter resistances; base and collector supply voltages; collector characteristic curves; $BE$ junction offset voltage.

**Find:** DC (quiescent) base and collector currents, $I_{BQ}$ and $I_{CQ}$, and collector-emitter voltage, $V_{CEQ}$.

**Schematics, Diagrams, Circuits, and Given Data:** $R_B = 62.7$ k$\Omega$; $R_C = 375$ $\Omega$; $V_{BB} = 10$ V; $V_{CC} = 15$ V; $V_\gamma = 0.6$ V. The collector characteristic curves are shown in Figure 9.13.

**Assumptions:** The transistor is in the active state.

**Analysis:** Write the load line equation for the collector circuit:

$$V_{CE} = V_{CC} - R_C I_C = 15 - 375 I_C$$

The load line is shown in Figure 9.13; to determine the $Q$ point, we need to determine which of the collector curves intersects the load line; that is, we need to know the base current. Applying KVL around the base circuit, and assuming that the BE junction is forward-biased (this results from the assumption that the transistor is in the active region),

$$I_B = \frac{V_{BB} - V_{BE}}{R_B} = \frac{V_{BB} - V_\gamma}{R_B} = \frac{10 - 0.6}{62,700} = 150 \ \mu A$$

The intersection of the load line with the 150 $\mu$A base curve is the DC operating or quiescent point of the transistor amplifier, defined below by the three values:

$$V_{CEQ} = 7 \text{ V} \qquad I_{CQ} = 22 \text{ mA} \qquad I_{BQ} = 150 \ \mu A$$

**Comments:** The base circuit consists of a battery in series with a resistance; we shall soon see that it is not necessary to employ two different voltage supplies for base and collector circuits, but that a single collector supply is sufficient to bias the transistor. Note that even in the absence of an external input to be amplified (AC source), the transistor dissipates power; most of the power is dissipated by the collector circuit:
$P_{CQ} = V_{CEQ} \times I_{CQ} = 154$ mW.

**Focus on Computer-Aided Tools:** An *Electronics Workbench*™ simulation of the circuit analyzed in this example is available in the CD-ROM that accompanies the book. If you run the simulation, you will note that the values of the bias currents and voltage are slightly different from the ones computed in the example. Double-click on the BJT icon to access the transistor model parameters, and compare the value of the $BE$ junction voltage used by *Electronics Workbench* with that used in the example. Does this discrepancy explain the observed differences?

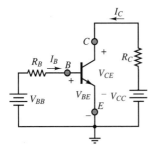

**Figure 9.14**

How can a transistor amplify a signal, then, given the $V_{BE}$-$I_B$ and $V_{CE}$-$I_C$ curves discussed in this section? The small-signal amplifier properties of the transistor are best illustrated by analyzing the effect of a small sinusoidal current

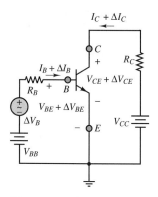

**Figure 9.15** Circuit illustrating the amplification effect in a BJT

superimposed on the DC current flowing into the base. The circuit of Figure 9.15 illustrates the idea, by including a small-signal AC source, of strength $\Delta V_B$, in series with the base circuit. The effect of this AC source is to cause sinusoidal oscillations $\Delta I_B$ about the $Q$ point, that is, around $I_{BQ}$. A study of the collector characteristic indicates that for a sinusoidal oscillation in $I_B$, a corresponding, but larger, oscillation will take place in the collector current. Figure 9.16 illustrates the concept. Note that the base current oscillates between 110 and 190 $\mu$A, causing the collector current to correspondingly fluctuate between 15.3 and 28.6 mA. The notation that will be used to differentiate between DC and AC (or fluctuating) components of transistor voltages and currents is as follows: DC (or quiescent) currents and voltages will be denoted by uppercase symbols; for example: $I_B$, $I_C$, $V_{BE}$, $V_{CE}$. AC components will be preceded by a "$\Delta$": $\Delta I_B(t)$, $\Delta I_C(t)$, $\Delta V_{BE}(t)$, $\Delta V_{CE}(t)$. The complete expression for one of these quantities will therefore include both a DC term and a time-varying, or AC, term. For example, the collector current may be expressed by $i_C(t) = I_C + \Delta I_C(t)$.

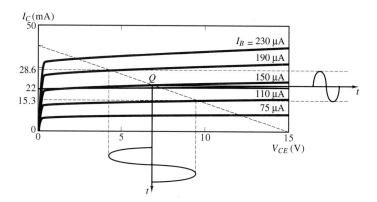

**Figure 9.16** Amplification of sinusoidal oscillations in a BJT

The $i$-$v$ characteristic of Figure 9.16 illustrates how an increase in collector current follows the same sinusoidal pattern of the base current but is greatly amplified. Thus, the BJT acts as a *current amplifier*, in the sense that any oscillations in the base current appear amplified in the collector current. Since the voltage across the collector resistance, $R_C$, is proportional to the collector current, one can see how the collector voltage is also affected by the amplification process. Example 9.5 illustrates numerically the effective amplification of the small AC signal that takes place in the circuit of Figure 9.15.

### EXAMPLE   9.5  A BJT Small-Signal Amplifier

**Problem**

With reference to the BJT amplifier of Figure 9.17 and to the collector characteristic curves of Figure 9.13, determine: (1) the DC operating point of the BJT; (2) the nominal current gain, $\beta$, at the operating point; (3) the AC voltage gain $A_V = \Delta V_o / \Delta V_B$.

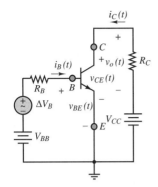

## Solution

**Known Quantities:**  Base, collector, and emitter resistances; base and collector supply voltages; collector characteristic curves; $BE$ junction offset voltage.

**Find:**  (1) DC (quiescent) base and collector currents, $I_{BQ}$ and $I_{CQ}$, and collector-emitter voltage, $V_{CEQ}$; (2) $\beta = \Delta I_C / \Delta I_B$; (3) $A_V = \Delta V_o / \Delta V_B$.

**Schematics, Diagrams, Circuits, and Given Data:**  $R_B = 10\ \text{k}\Omega$; $R_C = 375\ \Omega$; $V_{BB} = 2.1\ \text{V}$; $V_{CC} = 15\ \text{V}$; $V_\gamma = 0.6\ \text{V}$. The collector characteristic curves are shown in Figure 9.13.

**Assumptions:**  Assume that the $BE$ junction resistance is negligible when compared to the base resistance. Assume that each voltage and current can be represented by the superposition of a DC (quiescent) value and an AC component. For example: $v_0(t) = V_{0Q} + \Delta V_0(t)$.

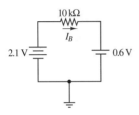

**Figure 9.17**

## Analysis:

1. *DC operating point.* On the assumption the BE junction resistance is much smaller than $R_B$, we can state that the junction voltage is constant: $v_{BE}(t) = V_{BEQ} = V_\gamma$, and plays a role only in the DC circuit. The DC equivalent circuit for the base is shown in Figure 9.18 and described by the equation

$$V_{BB} = R_B I_{BQ} + V_{BEQ}$$

from which we compute the quiescent base current:

$$I_{BQ} = \frac{V_{BB} - V_{BEQ}}{R_B} = \frac{V_{BB} - V_\gamma}{R_B} = \frac{2.1 - 0.6}{10{,}000} = 150\ \mu\text{A}$$

To determine the DC operating point, we write the load line equation for the collector circuit:

$$V_{CE} = V_{CC} - R_C I_C = 15 - 375 I_C$$

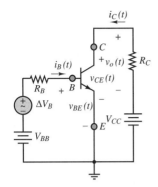

The load line is shown in Figure 9.19. The intersection of the load line with the 150 $\mu$A base curve is the DC operating or quiescent point of the transistor amplifier, defined below by the three values:  $V_{CEQ} = 7.2\ \text{V}$; $I_{CQ} = 22\ \text{mA}$; $I_{BQ} = 150\ \mu\text{A}$.

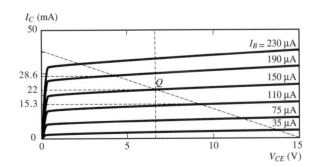

**Figure 9.19** Operating point on the characteristic curve

2. *AC current gain.* To determine the current gain, we resort, again, to the collector curves. Figure 9.19 indicates that if we consider the values corresponding to base currents of 190 and 110 $\mu$A, the collector will see currents of 28.6 and 15.3 mA,

respectively. We can think of these collector current excursions, $\Delta I_C$, from the $Q$ point as corresponding to the effects of an oscillation $\Delta I_B$ in the base current, and calculate the current gain of the BJT amplifier according to:

$$\beta = \frac{\Delta I_C}{\Delta I_B} = \frac{28.6 \times 10^{-3} - 15.3 \times 10^{-3}}{190 \times 10^{-6} - 110 \times 10^{-6}} = 166.25$$

Thus, the nominal current gain of the transistor is approximately $\beta = 166$.

3. *AC voltage gain.* To determine the AC voltage gain, $A_V = \Delta V_o/\Delta V_B$, we need to express $\Delta V_o$ as a function of $\Delta V_B$. Observe that $v_o(t) = R_C i_C(t) = R_C I_{CQ} + R_C \Delta I_C(t)$. Thus we can write:

$$\Delta V_o(t) = -R_C \Delta I_C(t) = -R_C \beta \, \Delta I_B(t).$$

Using the principle of superposition in considering the base circuit, we observe that $\Delta I_B(t)$ can be computed from the KVL base equation

$$\Delta V_B(t) = R_B \Delta I_B(t) + \Delta V_{BE}(t)$$

but we had stated in part 1 that, since the $BE$ junction resistance is negligible relative to $R_B$, $\Delta V_{BE}(t)$ is also negligible. Thus,

$$\Delta I_B = \frac{\Delta V_B}{R_B}$$

Substituting this result into the expression for $\Delta V_o(t)$, we can write

$$\Delta V_o(t) = -R_C \beta \Delta I_B(t) = -\frac{R_C \beta \Delta V_B(t)}{R_B}$$

or

$$\frac{\Delta V_o(t)}{\Delta V_B} = A_V = -\frac{R_C}{R_B}\beta = -6.23$$

**Comments:**  The circuit examined in this example is not quite a practical transistor amplifier yet, but it demonstrates most of the essential features of BJT amplifiers. We summarize them as follows.

- Transistor amplifier analysis is greatly simplified by considering the DC bias circuit and the AC equivalent circuits separately. This is an application of the principle of superposition.
- Once the bias point (or DC operating or quiescent point) has been determined, the current gain of the transistor can be determined from the collector characteristic curves. This gain is somewhat dependent on the location of the operating point.
- The AC voltage gain of the amplifier is strongly dependent on the base and collector resistance values. Note that the AC voltage gain is negative! This corresponds to a 180° phase inversion if the signal to be amplified is a sinusoid.

Many issues remain to be considered before we can think of designing and analyzing a practical transistor amplifier. It is extremely important that you master this example before studying the remainder of the section.

**Focus on Computer-Aided Tools:**  An *Electronics Workbench*™ simulation of the circuit analyzed in this example is available in the CD-ROM that accompanies the book. Run the simulation to see the effect of the negative voltage gain on the output signal waveform.

---

In discussing the DC biasing procedure for the BJT, we pointed out that the simple circuit of Figure 9.12 would not be a practical one to use in an application circuit. In fact, the more realistic circuit of Example 9.4 is also not a practical biasing circuit. The reasons for this statement are that two different supplies are required ($V_{CC}$ and $V_{BB}$)—a requirement that is not very practical—and that the resulting DC bias (operating) point is not very stable. This latter point may be made clearer by pointing out that the location of the operating point could vary significantly if, say, the current gain of the transistor, $\beta$, were to vary from device to device. A circuit that provides great improvement on both counts is shown in Figure 9.20. Observe, first, that the voltage supply, $V_{CC}$, appears across the pair of resistors $R_1$ and $R_2$, and that therefore the base terminal for the transistor will see the Thévenin equivalent circuit composed of the equivalent voltage source,

$$V_{BB} = \frac{R_2}{R_1 + R_2} V_{CC} \tag{9.6}$$

and of the equivalent resistance,

$$R_B = R_1 \| R_2 \tag{9.7}$$

Figure 9.21(b) shows a redrawn DC bias circuit that makes this observation more evident. The circuit to the left of the dashed line in Figure 9.21(a) is represented in Figure 9.21(b) by the equivalent circuit composed of $V_{BB}$ and $R_B$.

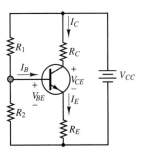

**Figure 9.20** Practical BJT self-bias DC circuit

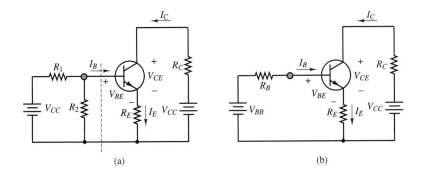

(a)                        (b)

**Figure 9.21** DC self-bias circuit represented in equivalent-circuit form

Recalling that the $BE$ junction acts much as a diode, the following equations describe the DC operating point of the self-bias circuit. Around the base-emitter circuit,

$$V_{BB} = I_B R_B + V_{BE} + I_E R_E = [R_B + (\beta + 1)R_E]I_B + V_{BE} \tag{9.8}$$

where $V_{BE}$ is the $BE$ junction voltage (diode forward voltage) and $I_E = (\beta+1)I_B$. Around the collector circuit, on the other hand, the following equation applies:

$$V_{CC} = I_C R_C + V_{CE} + I_E R_E = I_C \left( R_C + \frac{\beta + 1}{\beta} R_E \right) + V_{CE} \tag{9.9}$$

since

$$I_E = I_B + I_C = (\frac{1}{\beta} + 1)I_C$$

These two equations may be solved to obtain: (1) an expression for the base current,

$$I_B = \frac{V_{BB} - V_{BE}}{R_B + (\beta + 1)R_E} \tag{9.10}$$

from which the collector current can be determined as $I_C = \beta I_B$; and (2) an expression for the collector-emitter voltage:

$$V_{CE} = V_{CC} - I_C \left( R_C + \frac{\beta + 1}{\beta} R_E \right) \tag{9.11}$$

This last equation is the load-line equation for the bias circuit. Note that the effective load resistance seen by the DC collector circuit is no longer just $R_C$, but is now given by

$$R_C + \frac{\beta + 1}{\beta} R_E \approx R_C + R_E$$

The following example provides a numerical illustration of the analysis of a DC self-bias circuit for a BJT.

### EXAMPLE  9.6  Practical BJT bias circuit

**Problem**

Determine the DC bias point of the transistor in the circuit of Figure 9.20.

---

**Solution**

**_Known Quantities:_**  Base, collector, and emitter resistances; collector supply voltage; nominal transistor current gain; $BE$ junction offset voltage.

**_Find:_**  DC (quiescent) base and collector currents, $I_{BQ}$ and $I_{CQ}$, and collector-emitter voltage, $V_{CEQ}$.

**_Schematics, Diagrams, Circuits, and Given Data:_**  $R_1 = 100$ k$\Omega$; $R_2 = 50$ k$\Omega$; $R_C = 5$ k$\Omega$; $R_E = 3$ k$\Omega$; $V_{CC} = 15$ V; $V_\gamma = 0.7$ V.

**_Analysis:_**  We first determine the equivalent base voltage from equation 9.6:

$$V_{BB} = \frac{R_1}{R_1 + R_2} V_{CC} = \frac{100}{100 + 50} 15 = 5 \text{ V}$$

and the equivalent base resistance from equation 9.7:

$$R_B = R_1 \| R_2 = 33.3 \text{ k}\Omega$$

Now we can compute the base current from equation 9.10:

$$I_B = \frac{V_{BB} - V_{BE}}{R_B + (\beta + 1) R_E} = \frac{V_{BB} - V_\gamma}{R_B + (\beta + 1) R_E} = \frac{5 - 0.7}{33,000 + 101 \times 3000} = 128 \ \mu\text{A}$$

and knowing the current gain of the transistor, $\beta$, we can determine the collector current:

$$I_C = \beta I_B = 1.28 \text{ mA}$$

Finally, the collector-emitter junction voltage can be computed with reference to equation 9.11:

$$V_{CE} = V_{CC} - I_C \left( R_C + \frac{\beta + 1}{\beta} R_E \right)$$

$$= 15 - 1.28 \times 10^{-3} \left( 5 \times 10^{-3} + \frac{101}{100} \times 3 \times 10^{-3} \right) = 4.72 \text{ V}$$

Thus, the $Q$ point of the transistor is given by:

$$V_{CEQ} = 4.72 \text{ V} \qquad I_{CQ} = 1.28 \text{ mA} \qquad I_{BQ} = 128 \, \mu\text{A}$$

**Focus on Computer-Aided Tools:** An *Electronics Workbench*™ simulation of the circuit analyzed in this example is available in the CD-ROM that accompanies the book.

---

The material presented in this section has illustrated the basic principles that underlie the operation of a BJT and the determination of its $Q$ point. In the next section and later, in Chapter 10, we shall develop some simple circuit models for the BJT that will enable us to analyze the transistor amplifier in the linear active region using the familiar tools of linear circuit analysis.

## Check Your Understanding

**9.4**  Describe the operation of a *pnp* transistor in the active region, by analogy with that of the *npn* transistor.

**9.5**  For the circuit given in Figure 9.11, the readings are $V_1 = 3$ V, $V_2 = 2.4$ V, and $V_3 = 2.7$ V. Determine the operating region of the transistor.

**9.6**  For the circuit given in Figure 9.21, find the value of $V_{BB}$ that yields a collector current $I_C = 6.3$ mA. What is the corresponding collector-emitter voltage (assume that $V_{BE} = 0.6$ V and that the transistor is in the active region)? Assume that $R_B = 50$ k$\Omega$, $R_E = 200$ $\Omega$, $R_C = 1$ k$\Omega$, $\beta = 100$, and $V_{CC} = 14$ V.

**9.7**  What percent change in collector current would result if $\beta$ were changed to 150 in Example 9.6? Why does the collector current increase by less than 50 percent?

---

## 9.3    BJT LARGE-SIGNAL MODEL

The $i$-$v$ characteristics and the simple circuits of the previous sections indicate that the BJT acts very much as a current-controlled current source: A small amount of current injected into the base can cause a much larger current to flow into the collector. This conceptual model, although somewhat idealized, is useful in describing a **large-signal model** for the BJT, that is, a model that describes the behavior of the BJT in the presence of relatively large base and collector currents, close to the limit of operation of the device. A more careful analysis of the collector curves in Chapter 10 will reveal that it is also possible to generate a small-signal model, a model that describes the operation of the transistor as a linear amplifier of small AC signals. These models are certainly not a complete description of the properties of the BJT, nor do they accurately depict all of the effects that characterize the operation of such devices (for example, temperature effects, saturation, and cutoff); however, they are adequate for the intended objectives of this book, in that they provide a good qualitative feel for the important features of transistor amplifiers.

### Large-Signal Model of the *npn* BJT

The large-signal model for the BJT recognizes three basic operating modes of the transistor. When the *BE* junction is reverse-biased, no base current (and therefore no forward collector current) flows, and the transistor acts virtually as an open circuit; the transistor is said to be in the *cutoff region*. In practice, there is always a leakage current flowing through the collector, even when $V_{BE} = 0$ and $I_B = 0$. This leakage current is denoted by $I_{CEO}$. When the $BE$ junction becomes forward-biased, the transistor is said to be in the *active region,* and the base current is amplified by a factor of $\beta$ at the collector:

$$I_C = \beta I_B \qquad\qquad\qquad\qquad\qquad\qquad\qquad \textbf{(9.12)}$$

Since the collector current is controlled by the base current, the controlled-source symbol is used to represent the collector current. Finally, when the base current becomes sufficiently large, the collector-emitter voltage, $V_{CE}$, reaches its saturation limit, and the collector current is no longer proportional to the base current; this is called the *saturation region*. The three conditions are described in Figure 9.22 in terms of simple circuit models. The corresponding collector curves are shown in Figure 9.23.

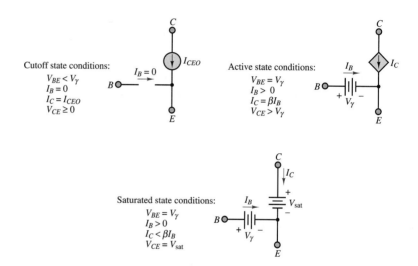

**Figure 9.22** *npn* BJT large-signal model

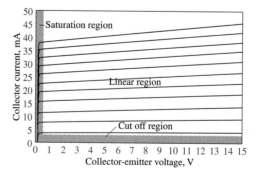

**Figure 9.23** BJT collector characteristic

Example 9.7 illustrates the application of this large-signal model in a practical circuit and illustrates how to determine which of the three states is applicable, using relatively simple analysis.

# FOCUS ON METHODOLOGY

## Using Device Data Sheets

One of the most important design tools available to engineers is the **device data sheet.** In this box we illustrate the use of a device data sheet for the 2N3904 bipolar transistor. This is an *npn general-purpose amplifier* transistor. Excerpts from the data sheet are shown below, with some words of explanation. The complete data sheet can be found in the accompanying CD-ROM.

2N3904

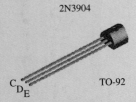

C
D
E          TO-92

**NPN General Purpose Amplifier**
This device is designed as a general purpose amplifier and switch. The useful dynamic range extends to 100 mA as a switch and to 100 MHz as an amplifier.

# ELECTRICAL CHARACTERISTICS:

The section on electrical characteristics summarizes some of the important voltage and current specifications of the transistor. For example, you will find breakdown voltages (not to be exceeded), and cutoff currents. In this section you also find important modeling information, related to the *large-signal model* described in this chapter. The large-signal current gain of the transistor, $h_{FE}$, or $\beta$, is given as a function of collector current. Note that this parameter varies significantly (from 30 to 100) as the DC collector current varies. Also important are the $CE$ and $BE$ junction saturation voltages (the batteries in the large-signal model of Figure 9.22).

**Electrical Characteristics**    TA = 25°C unless otherwise noted

| Symbol | Parameter | Test Conditions | Min | Max | Units |
|--------|-----------|-----------------|-----|-----|-------|
| **OFF CHARACTERISTICS** | | | | | |
| $V_{(BR)CEO}$ | Collector-Emitter Breakdown Voltage | $I_C = 1.0$ mA, $I_B = 0$ | 40 | | V |
| $V_{(BR)CBO}$ | Collector-Base Breakdown Voltage | $I_C = 10\ \mu$A, $I_E = 0$ | 60 | | V |
| $V_{(BR)EBO}$ | Emitter-Base Breakdown Voltage | $I_E = 10\ \mu$A, $I_C = 0$ | 6.0 | | V |
| $I_{BL}$ | Base Cutoff Current | $V_{CE} = 30$ V, $V_{EB} = 0$ | | 50 | nA |
| $I_{CEX}$ | Collector Cutoff Current | $V_{CE} = 30$ V, $V_{EB} = 0$ | | 50 | nA |
| **ON CHARACTERISTICS\*** | | | | | |
| $h_{FE}$ | DC Current Gain | $I_C = 0.1$ mA, $V_{CE} = 1.0$ V<br>$I_C = 1.0$ mA, $V_{CE} = 1.0$ V<br>$I_C = 10$ mA, $V_{CE} = 1.0$ V<br>$I_C = 50$ mA, $V_{CE} = 1.0$ V<br>$I_C = 100$ mA, $V_{CE} = 1.0$ V | 40<br>70<br>100<br>60<br>30 | 300 | |
| $V_{CE(sat)}$ | Collector-Emitter Saturation Voltage | $I_C = 10$ mA, $I_B = 1.0$ mA<br>$I_C = 50$ mA, $I_B = 5.0$ mA | | 0.2<br>0.3 | V<br>V |
| $V_{BE(sat)}$ | Base-Emitter Saturation Voltage | $I_C = 10$ mA, $I_B = 1.0$ mA<br>$I_C = 50$ mA, $I_B = 5.0$ mA | 0.065 | 0.85<br>0.95 | V<br>V |

*(Continued)*

*(Concluded)*

# THERMAL CHARACTERISTICS:

This table summarizes the thermal limitations of the device. For example, one can find the **power rating**, listed at 625 mW at 25°C. Note that in the entry for the total device power dissipation, **derating** information is also given. Derating implies that the device power dissipation will change as a function of temperature, in this case at the rate of 5 mW/°C. For example, if we expect to operate the diode at a temperature of 100°C, we would calculate a derated power of:

$$P = 625 \text{ mW} - 75°C \times 0.005 \text{ mW/}°C = 250 \text{ mW}$$

Thus, the diode operated at a higher temperature can dissipate only 250 mW.

**Thermal Characteristics**      TA = 25°C unless otherwise noted

| Symbol | Characteristic | Max | | Units |
|--------|----------------|-----|-----|-------|
| | | 2N3904 | *PZT3904 | |
| $P_D$ | Total Device Dissipation<br>Derate above 25°C | 625<br>5.0 | 1,000<br>8.0 | mW<br>mW/°C |
| $R_{0JC}$ | Thermal Resistance, Junction to Case | 83.3 | | °C/W |
| $R_{0JA}$ | Thermal Resistance, Junction to Ambient | 200 | 125 | °C/W |

# TYPICAL CHARACTERISTIC CURVES:

Device data sheets always include characteristic curves that may be useful to a designer. In this example, we include the base-emitter "on" voltage as a function of collector current, for three device temperatures. We also show the power dissipation versus ambient temperature derating curve for three different device *packages*. The transistor's ability to dissipate power is determined by its heat transfer properties; the package shown above is the TO-92 package; the SOT-223 and SOT-23 packages have different heat transfer characteristics, leading to different power dissipation capabilities.

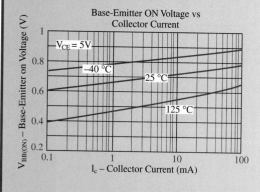

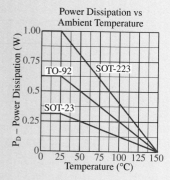

## EXAMPLE 9.7 LED Driver

### Problem

Design a transistor amplifier to supply a LED. The LED is required to turn on and off
following the on-off signal from a digital output port of a microcomputer. The circuit is
shown in Figure 9.24.

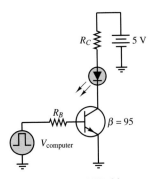

**Figure 9.24** LED driver
circuit

### Solution

**Known Quantities:**  Microcomputer output resistance and output signal voltage and
current levels; LED offset voltage, required current, and power rating; BJT current gain
and base-emitter junction offset voltage.

**Find:**  Collector resistance, $R_C$, such that the transistor is in the saturation region when
the computer outputs 5 V; power dissipated by LED.

**Schematics, Diagrams, Circuits, and Given Data:**
    Microcomputer: output resistance $= R_B = 1$ k$\Omega$; $V_{on} = 5$ V; $V_{OFF} = 0$ V; $I = 5$ mA.
Transistor: $V_{CC} = 5$ V; $V_\gamma = 0.7$ V; $\beta = 95$.
LED: $V_{\gamma\text{LED}} = 1.4$ V; $I_{\text{LED}} > 15$ mA; $P_{\text{max}} = 100$ mW.

**Assumptions:**  Use the large-signal model of Figure 9.22.

**Analysis:**  When the computer output voltage is zero, the BJT is clearly in the cutoff
region, since no base current can flow. When the computer output voltage is $V_{\text{ON}} = 5$ V,
we wish to drive the transistor into the saturation region. Recall that operation in
saturation corresponds to small values of collector-emitter voltages, with typical values of
$V_{CE}$ around 0.2 V. Figure 9.25(a) depicts the equivalent base-emitter circuit when the
computer output voltage is $V_{\text{ON}} = 5$ V. Figure 9.25(b) depicts the collector circuit, and
Figure 9.25(c), the same collector circuit with the large-signal model for the transistor (the
battery $V_{CE\text{sat}}$) in place of the BJT. From this saturation model we write:

$$V_{CC} = R_C I_C + V_{\gamma\text{LED}} + V_{CE\text{sat}}$$

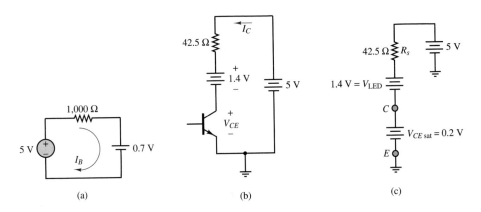

(a)                                    (b)                          (c)

**Figure 9.25** (a) $BE$ circuit for LED driver; (b) Equivalent collector circuit of LED driver, assuming
that the BJT is in the linear active region; (c) LED driver equivalent collector circuit, assuming that the
BJT is in the saturation region

or

$$R_C = \frac{V_{CC} - V_{\gamma \text{LED}} - V_{CE\text{sat}}}{I_C} = \frac{3.4}{I_C}$$

We know that the LED requires at least 15 mA to be on; let us suppose that 30 mA is a reasonable LED current to ensure good brightness, then the value of collector resistance that would complete our design is, approximately, $R_C = 113\ \Omega$.

With the above design, the BJT LED driver will clearly operate as intended to turn the LED on and off. But how do we know that the BJT is in fact in the saturation region? Recall that the major difference between operation in the active and saturation regions is that in the active region the transistor displays a nearly constant current gain, $\beta$, while in the saturation region the current gain is much smaller. Since we know that the nominal $\beta$ for the transistor is 95, we can calculate the base current using the equivalent base circuit of Figure 9.25(a) and determine the ratio of base to collector current:

$$I_B = \frac{V_{\text{ON}} - V_\gamma}{R_B} = \frac{4.3}{1,000} = 4.3\ \text{mA}$$

The actual large-signal current gain is therefore equal to $30/4.3 = 6.7 \ll \beta$. Thus, it can be reasonably assumed that the BJT is operating in saturation.

We finally compute the LED power dissipation:

$$P_{\text{LED}} = V_{\gamma \text{LED}} I_C = 1.4 \times 0.3 = 42\ \text{mW} < 100\ \text{mW}$$

Since the power rating of the LED has not been exceeded, the design is complete.

**Comments:**  Using the large-signal model of the BJT is quite easy, since the model simply substitutes voltage sources in place of the $BE$ and $CE$ junctions. To be sure that the correct model (e.g., saturation versus active region) has been employed, it is necessary to verify either the current gain or the value of the $CE$ junction voltage. Current gains near the nominal $\beta$ indicate active region operation, while small $CE$ junction voltages denote operation in saturation.

**Focus on Computer-Aided Tools:**  An *Electronics Workbench*™ simulation of the circuit analyzed in this example is available in the CD-ROM that accompanies the book. Try changing the value of the collector resistance and see the resulting changes in collector current and collector-emitter voltage. For what approximate value of $R_C$ does the BJT go back into the active region?

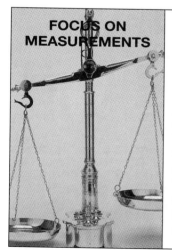

**FOCUS ON MEASUREMENTS**

## Large-Signal Amplifier for Diode Thermometer

**Problem:**

In Chapter 8 we explored the use of a diode as the sensing element in an electronic thermometer (see "Focus on Measurements: Diode Thermometer"). In the present example, we illustrate the design of a transistor amplifier for such a diode thermometer. The circuit is shown in Figure 9.26.

**Solution:**

*Known Quantities*— Diode and transistor amplifier bias circuits; diode voltage versus temperature response.

*Find*— Collector resistance and transistor output voltage versus temperature.

*Schematics, Diagrams, Circuits, and Given Data*— $V_{CC} = 12$ V; large signal $\beta = 188.5$; $V_{BE} = 0.75$ V; $R_S = 500\ \Omega$; $R_B = 10$ k$\Omega$.

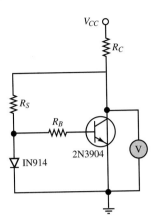

**Figure 9.26** Large signal
amplifier for diode thermometer

*Assumptions—* Use a 1N914 diode and a 2N3904 transistor.

*Analysis—* With reference to the circuit of Figure 9.26 and to the diode
temperature response characteristic of Figure 9.27(a), we observe that the
midrange diode thermometer output voltage is approximately 1.1 V. Thus,
we should design the transistor amplifier so that when $v_D = 1.1$ V the
transistor output is in the center of the collector characteristic for minimum
distortion. Since the collector supply is 12 V, we choose to have the $Q$ point
at $V_{CEQ} = 6$ V.

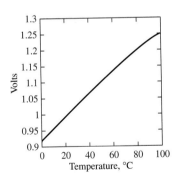

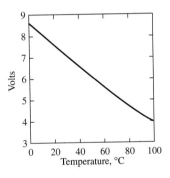

**Figure 9.27(a)** Diode voltage
temperature dependence

**Figure 9.27(b)** Amplifier
output

Knowing that the diode output voltage at the quiescent point is 1.1 V,
we compute the quiescent base current

$$v_D - I_{BQ}R_B - V_{BEQ} = 0$$

$$I_{BQ} = \frac{v_D - V_{BEQ}}{R_B} = \frac{1.1 - 0.75}{10{,}000} = 35 \; \mu A$$

Knowing $\beta$, we can compute the collector current:

$$I_{CQ} = \beta I_{BQ} = 188.5 \times 35\ \mu A\ = 6.6\ mA$$

Now we can write the collector equation and solve for the desired collector resistance:

$$V_{CC} - I_{CQ}R_C - V_{CEQ} = 0$$

$$R_C = \frac{V_{CC} - V_{CEQ}}{I_{CQ}} = \frac{12\ V\ - 6\ V}{6.6\ mA} = 0.909\ k\Omega$$

Once the circuit is designed according to these specifications, the output voltage can be determined by computing the base current as a function of the diode voltage (which is a function of temperature); from the base current, we can compute the collector current and use the collector equation to determine the output voltage, $v_{out} = v_{CE}$. The result is plotted in Figure 9.27(b).

**Comments—** Note that the transistor amplifies the slope of the temperature by a factor of approximately 6. Observe also that the common emitter amplifier used in this example causes a sign inversion in the output (the output voltage now decreases for increasing temperatures, while the diode voltage increases). Finally, we note that the design shown in this example assumes that the impedance of the voltmeter is infinite. This is a good assumption in the circuit shown in this example, because a practical voltmeter will have a very large input resistance relative to the transistor output resistance. Should the thermometer be connected to another circuit, one would have to pay close attention to the input resistance of the second circuit to ensure that loading does not occur.

**Focus on Computer-Aided Tools—** An *Electronics Workbench*$^{TM}$ version of this example is available in the accompanying CD. If you wish to verify the results obtained here, you may change the diode temperature by opening the 1N914 diode template, clicking on the Edit button, and changing the parameter TNOM (temperature in degrees Celsius) in sheet 2 of the template. You may also wish to look at the parameters of the 2N3904 transistor.

The large-signal model of the BJT presented in this section treats the *BE* junction as an offset diode and assumes that the BJT in the linear active region acts as an ideal controlled current source. In reality, the *BE* junction is better modeled by considering the forward resistance of the *pn* junction; further, the BJT does not act quite like an ideal current-controlled current source. These phenomena will be partially taken into account in the small-signal model introduced in Chapter 10.

## Check Your Understanding

**9.8**   Repeat the analysis of Example 9.7 for $R_S = 400\ \Omega$. Which region is the transistor in? What is the collector current?

**9.9**   What is the power dissipated by the LED of Example 9.7 if $R_S = 30\ \Omega$?

## 9.4    FIELD-EFFECT TRANSISTORS

The second transistor family discussed in this chapter operates on the basis of a principle that is quite different from that of the *pn* junction devices. The concept that forms the basis of the operation of the **field-effect transistor,** or **FET,** is that the width of a conducting channel in a semiconductor may be varied by the external application of an electric field. Thus, FETs behave as *voltage-controlled resistors.* This family of electronic devices can be subdivided into three groups, all of which will be introduced in the remainder of this chapter. Figure 9.28 depicts the classification of field-effect transistors, as well as the more commonly used symbols for these devices. These devices can be grouped into three major categories. The first two categories are both types of **metal-oxide-semiconductor field-effect transistors,** or **MOSFETs: enhancement-mode MOSFETs** and **depletion-mode MOSFETs.** The third category consists of **junction field-effect transistors,** or **JFETs.** In addition, each of these devices can be fabricated either as an ***n*-channel** device or as a ***p*-channel** device, where the *n* or *p* designation indicates the nature of the doping in the semiconductor channel. All these transistors behave in a very similar fashion, and we shall predominantly discuss enhancement MOSFETs in this chapter, although some discussion of depletion devices and JFETs will also be included.

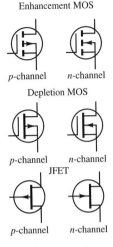

**Figure 9.28** Classification of field-effect transistors

## 9.5    OVERVIEW OF ENHANCEMENT-MODE MOSFETS

Figure 9.29 depicts the circuit symbol and the approximate construction of a typical *n*-channel enhancement-mode MOSFET. The device has three terminals: the **gate** (analogous to the base in a BJT); the **drain** (analogous to the collector); and the **source** (analogous to the emitter). The **bulk** or **substrate** of the device is shown to be electrically connected to the source, and therefore does not appear in the electrical circuit diagram as a separate terminal. The gate consists of a metal film layer, separated from the *p*-type bulk by a thin oxide layer (hence the terminology *metal-oxide-semiconductor*). The drain and source are both constructed of $n^+$ material.

Imagine now that the drain is connected to a positive voltage supply, $V_{DD}$, and the source is connected to ground. Since the *p*-type bulk is connected to the source, and hence to ground, the drain-bulk $n^+ p$ junction is strongly reverse-biased. The junction voltage for the $pn^+$ junction formed by the bulk and the source is zero, since both are connected to ground. Thus, the path between drain and source consists of two reverse-biased *p-n* junctions, and no current can flow. This situation is depicted in Figure 9.30(a): in the absence of a gate voltage, the *n*-channel enhancement-mode MOSFET acts as an open circuit. Thus, enhancement-mode devices are *normally off.*

Suppose now that a positive voltage is applied to the gate; this voltage will create an electric field in the direction shown in Figure 9.30(b). The effect of the electric field is to repel positive charge carriers away from the surface of the *p*-type bulk, and to form a narrow **channel** near the surface of the bulk in which negative charge carriers dominate, and are available for conduction. For a fixed drain bias, the greater the strength of the externally applied electric field (that is, the higher the gate voltage), the deeper the channel. This behavior explains the terminology *enhancement-mode*, because the application of an external electric field *enhances* the conduction in the channel by creating *n*-type charge carriers. It should also be

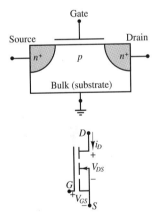

**Figure 9.29** *n*-channel enhancement MOSFET construction and circuit symbol

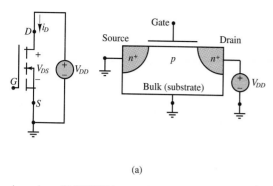

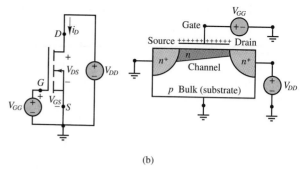

(a)

An *n*-channel MOSFET is normally off in the absence of an external electric field

(b)

When a gate voltage is applied, a conducting *n*-type channel is formed near the surface of the substrate; now current can flow from drain to source

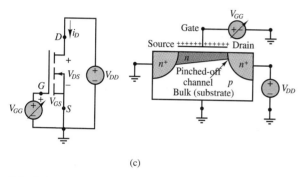

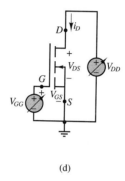

(c)

If the drain-source voltage is kept fixed and the gate supply voltage is varied, the MOSFET will behave as a voltage-controlled resistor until the pinch-off condition is reached (see Figure 9.31(a))

(d)

If the drain and gate supply voltages are both varied, a family of curves (shown in Figure 9.31(b)) can be generated, illustrating the MOSFET cutoff, ohmic, saturation, and breakdown regions

**Figure 9.30** Operation of *n*-channel enchancement MOSFET

clear why these devices are called *field-effect*, since it is an external electric field that determines the conduction properties of the transistor.

It is also possible to create *depletion-mode* devices in which an externally applied field depletes the channel of charge carriers by reducing the effective channel width (see Section 9.6). Depletion-mode MOSFETs are normally on, and can be turned off by application of an external electric field. We shall discuss depletion-mode devices in the next section.

To complete this brief summary of the operation of MOS transistors we note that, in analogy with *pnp* bipolar transistors, it is also possible to construct *p*-channel MOSFETS. In these transistors, conduction occurs in a channel formed in *n*-type bulk material via positive charge carriers.

## Operation of the *n*-Channel Enhancement-Mode MOSFET

To explain the operation of this family of transistors, we will mimic a laboratory experiment designed to generate the characteristic curves of the transistor, much as was done for the bipolar transistor. The experiment is depicted in Figure 9.30(a) to (d), where each pair of figures depicts the test circuit and a corresponding qualitative sketch of the electric field and channel for a particular set of voltages. In the test circuit, we have gate and drain supply voltages and the source is connected to ground.

We have already seen that when no gate voltage is applied, both *pn* junctions are reverse-biased and the transistor cannot conduct current (Figure 9.30(a)). This is the **cutoff region** of the *n*-channel enhancement MOSFET. When a gate voltage is applied, an electric field is generated and a *channel* is formed when the gate-source voltage, $v_{GS}$, exceeds a **threshold voltage,** $V_T$ (Figure 9.30(b)). The threshold voltage is a physical characteristic of a given transistor, and can be treated as a parameter (e.g., like the junction voltage in a diode or BJT). Note that the channel is narrower near the drain and wider near the source. This is because the electric field is spatially distributed, and is stronger near the source (which is at zero volts) than near the drain (which is at $V_{DD}$ volts). When the channel has formed, conduction can take place between drain and source. Note that, as a result of the nature of the *n*-type channel, charge carriers are exclusively of the negative type; thus MOSFETs are *unipolar* devices (as opposed to *bipolar* transistors). Conduction occurs through electrons injected from the $n^+$ source region into the channel (thus the terminology *source*, analogous to the emitter in a BJT). These electrons are then swept into the $n^+$ *drain* (analogous to the collector in a BJT). Thus, the direction of current flow is from drain to source, and we shall refer to this current as the **drain current,** $i_D$.

Suppose now that the drain-source voltage is fixed at some value (Figure 9.30(c)). Then, as the gate-source voltage is raised above the threshold voltage (so that a channel is formed), the width of the conduction channel increases because of the increasing strength of the electric field. In this mode of operation, the MOSFET acts as a **voltage controlled resistor**: as the gate voltage is further increased, the resistance of the channel decreases because the channel width increases. This phenomenon, however, can take place only up to a certain gate-source voltage: when the difference between gate-source voltage and threshold voltage, $v_{GS} - V_T$, equals the (fixed for the moment) drain-source voltage, $v_{DS}$, the width of the channel reaches a minimum in the vicinity of the drain. The channel width is thus reduced near the drain, because the field strength is near zero at the drain end of the channel (recall that the field strength is always at a minimum near the drain). This condition is called **pinch-off,** and the channel is said to be pinched off. Once pinch-off occurs, increasing the gate-source voltage will not cause an appreciable increase in drain current since the channel width is fixed at the drain, and the MOSFET behaves much like a **constant-current source,** with the drain current limited to a *saturation* value by the pinched-off channel width. Figure 9.31(a) depicts the relationship between drain current and gate-source voltage for fixed drain-source voltage of an *n*-channel enhancement MOSFET.

The pinch-off condition allows us to divide the operation of the *n*-channel enhancement MOSFET into two major operating regions: the **ohmic** or **triode region** (before pinch-off occurs), and the **saturation** region (after pinch-off has occurred). In the ohmic region, the MOSFET acts as a *voltage-controlled resistor*; in the saturation region it acts as a *current source*.

If we finally allow both the drain and the gate supply to be varied (Figure 9.30(d)), we can generate a family of curves. These curves, depicted in Figure 9.31(b), are called the **drain characteristic curves**. They represent the behavior of the MOSFET in terms of drain current versus drain-source voltage curves; each curve in the drain characteristic corresponds to a different value of gate-source

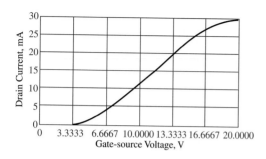

**Figure 9.31(a)** *n*-channel enhancement MOSFET drain curves

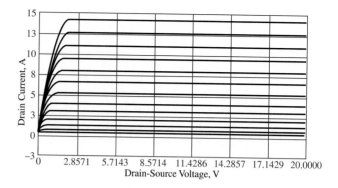

**Figure 9.31(b)** *n*-channel enhancement MOSFET as a controlled current source

voltage. There is an obvious analogy between the MOSFET drain characteristic and the BJT collector characteristic, but we also note the important differences: In the BJT collector curves, the family of curves is indexed as a function of base current, while the drain characteristic is a function of gate voltage. Thus, we can think of the BJT as a *current-controlled device*, while the MOSFET is intrinsically a *voltage-controlled device*. Note also that to completely describe the behavior of a BJT we needed to also define a separate base junction curve in terms of the base current variation versus the base-emitter voltage, while MOSFETs do not require an additional gate characteristic curve, because the gate is insulated and no gate current flows. Finally, we observe that if the drain-source voltage is increased above a **breakdown** value, $V_B$, the drain current rapidly increases, eventually leading to device destruction by thermally induced damage. This condition defines the last region of operation of the MOSFET, namely, the **breakdown region**. The four regions of operation are summarized in Table 9.1. The equations describing the ohmic and saturation regions are also given in the table. Note that in these equations we have introduced another important MOSFET physical parameter, $I_{DSS}$. It is also important to note that the equations describing MOSFET operation are nonlinear.

Table 9.1 Regions of operation and equations of *n*-channel enhancement MOSFET

---

**Cutoff region:**

$v_{GS} < V_T$

**Ohmic or triode region:** $v_{DS} < 0.25(v_{GS} - V_T), v_{GS} > V_T$

$$R_{DS} = \frac{V_T^2}{2I_{DSS}(V_{GS} - V_T)} \qquad \text{(equivalent drain-to-source resistance)}$$

$$i_D \approx \frac{v_{DS}}{R_{DS}}$$

**Saturation region:**

$$v_{DS} \geq v_{GS} - V_T, v_{GS} > V_T$$

$$i_D = \frac{I_{DSS}}{V_T^2}(v_{GS} - V_T)^2 = k(v_{GS} - V_T)^2$$

**Breakdown region:**

$v_{DS} > V_B$

Examples 9.8 to 9.10 illustrate the use of the MOSFET drain curves of Figure 9.31(b) in establishing the $Q$-point of a MOSFET amplifier.

## EXAMPLE 9.8 MOSFET Q-Point Graphical Determination

### Problem

Determine the $Q$ point for the MOSFET in the circuit of Figure 9.32.

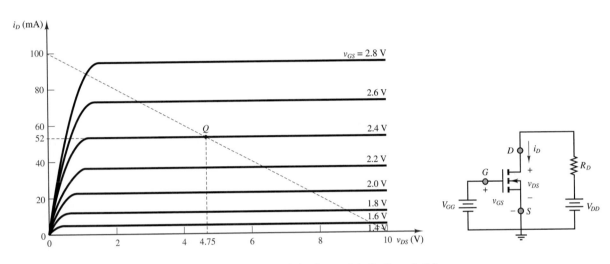

**Figure 9.32** $n$-channel enhancement MOSFET circuit and drain characteristic for Example 9.8

### Solution

**Known Quantities:** MOSFET drain resistance; drain and gate supply voltages; MOSFET drain curves.

**Find:** MOSFET quiescent drain current, $i_{DQ}$, and quiescent drain-source voltage, $v_{DSQ}$.

**Schematics, Diagrams, Circuits, and Given Data:** $V_{GG} = 2.4$ V; $V_{DD} = 10$ V; $R_D = 100\ \Omega$.

**Assumptions:** Use the drain curves of Figure 9.32.

**Analysis:** To determine the $Q$ point we write the drain circuit equation, applying KVL:

$$V_{DD} = R_D i_D + v_{DS}$$

$$10 = 100 i_D + v_{DS}$$

The resulting curve is plotted as a dashed line on the drain curves of Figure 9.32 by noting that the drain current axis intercept is equal to $V_{DD}/R_D = 100$ mA and that the drain-source voltage axis intercept is equal to $V_{DD} = 10$ V. The $Q$ point is then given by the intersection of the load line with the $V_{GG} = 2.4$ V curve. Thus, $i_{DQ} = 52$ mA and $v_{DSQ} = 4.75$ V.

VIRTUAL LAB

**Comments:** Note that the $Q$ point determination for a MOSFET is easier than for a BJT, since there is no need to consider the gate circuit, because gate current flow is essentially zero. In the case of the BJT, we also needed to consider the base circuit.

## EXAMPLE 9.9 MOSFET Q-Point Calculation

### Problem

Determine the $Q$ point for the MOSFET in the circuit of Figure 9.32.

### Solution

**Known Quantities:** MOSFET drain resistance; drain and gate supply voltages; MOSFET universal equations.

**Find:** MOSFET quiescent drain current, $i_{DQ}$, and quiescent drain-source voltage, $v_{DSQ}$.

**Schematics, Diagrams, Circuits, and Given Data:** $V_{GG} = 2.4$ V; $V_{DD} = 10$ V; $R_D = 100\ \Omega$.

**Assumptions:** Use the MOSFET universal equations of Table 9.1.

**Analysis:** We determine the threshold voltage by observing (in the curves of Figure 9.32) that the smallest gate voltage for which the drain current is nonzero is 1.4 V. Thus, $V_T = 1.4$ V. From the same curves, the drain current corresponding to $2V_T$ is approximately 95 mA. Thus $I_{DSS} = 95$ mA. Knowing these two parameters and the gate voltage, we apply the appropriate equation in Table 9.1. Since $v_{GS} = v_{GG} > V_T$, we write $i_{DQ} = I_{DSS}(\frac{v_{GS}}{V_T} - 1)^2 = 95(\frac{2.4}{1.4} - 1) = 48.5$ mA and $v_{DSQ} = V_{DD} - R_D i_{DQ} = 10 - 100 \times 48.5 \times 10^{-3} = 5.15$ V.

**Comments:** Note that the results differ slightly from the (approximate) graphical analysis in Example 9.8.

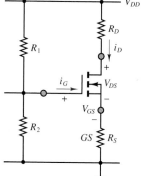

**Figure 9.33(a)** Self-bias circuit for Example 9.10

## EXAMPLE 9.10 MOSFET Self-Bias Circuit

### Problem

Determine the $Q$ point for the MOSFET self-bias circuit of Figure 9.33(a). Choose $R_S$ such that $v_{DSQ} = 8$ V.

### Solution

**Known Quantities:** MOSFET drain and gate resistances; drain supply voltage; MOSFET parameters $V_T$ and $I_{DSS}$.

**Find:** MOSFET quiescent gate-source voltage, $v_{GSQ}$; quiescent drain current, $i_{DQ}$, and $R_S$ such that the quiescent drain-source voltage, $v_{DSQ}$, is 8 V.

**Schematics, Diagrams, Circuits, and Given Data:** $V_{DD} = 30$ V; $R_D = 10$ kΩ; $R_1 = R_2 = 1.2$ MΩ; $R_D = 1.2$ MΩ; $V_T = 4$ V; $I_{DSS} = 7.2$ mA.

**Assumptions:** Operation is in the active region.

**Analysis:** Let all currents be expressed in mA and all resistances in kΩ. Applying KVL around the equivalent gate circuit of Figure 9.33(b) yields:

$$V_{GG} = v_{GSQ} + i_{GQ}R_G + i_{DQ}R_S = v_{GSQ} + i_{DQ}R_S$$

where $V_{GG} = V_{DD}/2$ and $R_G = R_1 \| R_2$. Note that $i_{GQ} = 0$ because of the infinite input resistance of the MOSFET.

The drain circuit equation is

$$V_{DD} = i_{DQ}R_D + v_{DSQ} + i_{DQ}R_S \qquad \textbf{(b)}$$

and the MOSFET universal equation (chosen from table 9.1 for operation in the active region) is:

$$i_{DQ} = I_{DSS}\left(\frac{v_{GS}}{V_T} - 1\right)^2. \tag{c}$$

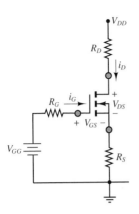

The above three equations will now be used to solve for the three unknowns: $v_{GSQ}, i_{DQ}$, and $v_{DSQ}$. From equation (a) we write:

$$i_{DQ}R_S = V_{GG} - v_{GSQ} == V_{DD}/2 - v_{GSQ}$$

and substitute the result into equation (b):

$$V_{DD} = i_{DQ}R_D + v_{DSQ} + V_{DD}/2 - v_{GSQ} \quad \text{or} \quad i_{DQ} = \frac{1}{R_D}\left(\frac{V_{DD}}{2} - v_{DSQ} + v_{GSQ}\right).$$

Substituting the above equation for $i_{DQ}$ into (c), we finally obtain a quadratic equation that can be solved for $v_{GSQ}$ since we know the desired value of $v_{DSQ}$:

$$\frac{1}{R_D}\left(\frac{V_{DD}}{2} - v_{DSQ} + v_{GSQ}\right) = I_{DSS}\left(\frac{v_{GSQ}}{V_T} - 1\right)^2$$

$$\frac{1}{10}(7 + v_{GSQ}) = 7.2\left(\frac{v_{GSQ}}{4} - 1\right)^2 \qquad 0.45v_{GSQ}^2 - 3.7v_{GSQ} + 6.5 = 0$$

**Figure 9.33(b)** Equivalent circuit for Fig. 9.33(a)

The two solutions for the above quadratic equation are:

$$v_{GSQ} = 5.68 \ V \text{ and } v_{GSQ} = 2.54 \text{ V}.$$

Only the first of these two values is acceptable for operation in the active region, since the second is lower than the threshold voltage ($V_T = 4$ V). Substituting the first value into the universal equation (c), we can compute the quiescent drain current:

$$i_{DQ} = 1.268 \text{ mA}.$$

Using this value in the drain equation (b), we compute the solution for the source resistance:

$$R_S = 7.35 \text{ k}\Omega.$$

**Comments:** Why are there two solutions to the problem posed in this example? Mathematically, we know that this should be the case because the drain universal equation is a quadratic equation. We then use the physical constraints of the problem to select the appropriate solution.

**Focus on computer-aided tools:** The above example is available as a EWB simulation in the CD that accompanies the book. Verify that if $R_S$ is chosen according to the second solution for $v_{GSQ}$, the transistor will be in the cutoff region.

## p-Channel MOSFETs and CMOS Devices

As the designation indicates, a p-channel MOSFET is characterized by p-type doping; the construction and symbol are shown in Figure 9.34. Note the opposite direction of the arrow to indicate that the pn junction formed by the channel and substrate is now in the opposite direction. The direction of drain current is opposite; therefore $v_{DS}$ and $v_{GS}$ are now negative. A more convenient reference is obtained if voltages are defined in the direction opposite to that for the n-channel device: if one defines $v_{SD} = -v_{DS}$ and $v_{SG} = -v_{GS}$, then these voltages will be positive for the drain current direction indicated in Figure 9.34. The carriers are holes in this device, since the channel, when formed, is p-type. Aside from the nature of the charge carriers and the direction of current and polarity of the voltages, the p-channel and n-channel transistors behave in conceptually the same way; however, since holes are in general less mobile (recall the discussion of carrier *mobility* in Section 8.1), p-channel MOSFETs are not used very much by themselves. They do find widespread application in **complementary metal-oxide-semiconductor (CMOS) devices.** CMOS devices take advantage of the complementary symmetry of p- and n-channel transistors built on the same integrated circuit. "Focus on Measurements: MOSFET Bidirectional Analog Gate" illustrates one application of CMOS technology.

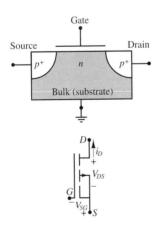

**Figure 9.34** p-channel enhancement MOSFET construction and circuit symbol

**FOCUS ON MEASUREMENTS**

## MOSFET Bidirectional Analog Gate

The variable-resistor feature of MOSFETs in the ohmic state finds application in the **analog transmission gate.** The circuit shown in Figure 9.35 depicts a circuit constructed using CMOS technology. The circuit operates on the basis of a control voltage, $v$, that can be either "low" (say, 0 V), or "high" ($v > V_T$), where $V_T$ is the threshold voltage for the $n$-channel MOSFET and $-V_T$ is the threshold voltage for the $p$-channel MOSFET. The circuit operates in one of two modes. When the gate of $Q_1$ is connected to the high voltage and the gate of $Q_2$ is connected to the low voltage, the path between $v_{\text{in}}$ and $v_{\text{out}}$ is a relatively small resistance, and the transmission gate conducts. When the gate of $Q_1$ is connected to the low voltage and the gate of $Q_2$ is connected to the high voltage, the transmission gate acts like a very large resistance and is an open circuit for all practical purposes. A more precise analysis follows.

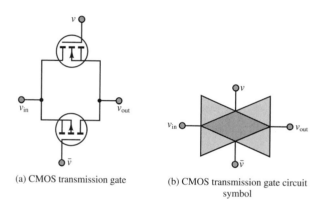

(a) CMOS transmission gate

(b) CMOS transmission gate circuit symbol

**Figure 9.35** Analog transmission gate

Let $v = V > V_T$ and $\bar{v} = 0$. Assume that the input voltage, $v_{\text{in}}$, is in the range $0 \leq v_{\text{in}} \leq V$. To determine the state of the transmission gate, we shall consider only the extreme cases $v_{\text{in}} = 0$ and $v_{\text{in}} = V$. When $v_{\text{in}} = 0$, $v_{GS1} = v - v_{\text{in}} = V - 0 = V > V_T$. Since $V$ is above the threshold voltage, MOSFET $Q_1$ conducts (in the ohmic region). Further, $v_{GS2} = \bar{v} - v_{\text{in}} = 0 > -V_T$. Since the gate-source voltage is not more negative than the threshold voltage, $Q_2$ is in cutoff and does not conduct. Since one of the two possible paths between $v_{\text{in}}$ and $v_{\text{out}}$ is conducting, the transmission gate is on. Now consider the other extreme, where $v_{\text{in}} = V$. By reversing the previous argument, we can see that $Q_1$ is now off, since $v_{GS1} = 0 < V_T$. However, now $Q_2$ is in the ohmic state, because $v_{GS2} = \bar{v} - v_{\text{in}} = 0 - V < -V_T$. In this case, then, it is $Q_2$ that provides a conducting path between the input and the output of the transmission gate, and the transmission gate is also on. We have therefore concluded that when $v = V$ and $\bar{v} = 0$, the transmission gate conducts and provides a near-zero-resistance (typically tens of ohms) connection between the input and the output of the transmission gate, for values of the input ranging from 0 to $V$.

Let us now reverse the control voltages and set $v = 0$ and $\overline{v} = V > V_T$. It is very straightforward to show that in this case, regardless of the value of $v_{\text{in}}$, both $Q_1$ and $Q_2$ are always off; therefore, the transmission gate is essentially an open circuit.

The analog transmission gate finds common application in *analog multiplexers* and *sample-and-hold* circuits, to be discussed in Chapter 15.

## Check Your Understanding

**9.10**  Determine the operating region of the MOSFET of Example 9.10 when $v_{GS} = 3.5$ V.

**9.11**  Determine the appropriate value of $R_S$ if we wish to move the operating point of the MOSFET of Example 9.10 to $v_{DSQ} = 12$ V. Also find the values of $v_{GSQ}$ and $i_{DQ}$. Are these values unique?

**9.12**  Show that the CMOS bidirectional gate described in the "Focus on Measurements: MOSFET Bidirectional Analog Gate" box is off for all values of $v_{\text{in}}$ between 0 and $V$ whenever $v = 0$ and $\overline{v} = V > V_T$.

**9.13**  Find the lowest value of $R_D$ for the MOSFET of Example 9.9 that will place the MOSFET in the ohmic region.

## 9.6    DEPLETION MOSFETs AND JFETs

To complete this brief discussion of field-effect transistors, we summarize the characteristics of depletion-mode MOSFETs and of JFETs. While the construction details of these two families of devices differ, their operation is actually quite similar, and we shall develop one set of equations describing the operation of both.

### Depletion MOSFETs

The construction of a depletion-mode MOSFET and its circuit symbol are shown schematically in Figure 9.36. We note that the only difference with respect to the enhancement type devices is the addition of a lightly dopes $n$-type region between the oxide layer and the $p$-type substrate. The presence of this $n$ region results in the presence of conducting channel in the absence of an externally applied electric field, as shown in Figure 9.37(a). Thus, depletion MOSFETs are normally on or normally conducting devices.

Since a channel already exists for $v_{GS} = 0$, increasing the gate-source voltage will further enhance conductivity by drawing additional electrons to the channel, to reduce channel resistance. If, on the other hand, $v_{GS}$ is made negative, the channel will be *depleted* of charge carriers, and channel resistance will decrease. When $v_{GS}$ is sufficiently negative (less than a **threshold voltage,** $V_t$), the channel electrons are all repelled into the substrate, and the channel ceases to conduct. This corresponds to the **cutoff region**, depicted in Figure 9.37(b). It is important to note that *the threshold voltage is negative in a depletion-mode device*. If we now repeat the qualitative analysis illustrated in Figure 9.30 for an enhancement-mode device for a depletion-mode MOSFET, we see that for a given drain-source voltage,

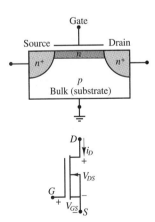

**Figure 9.36** $n$-channel depletion MOSFET construction and circuit symbol

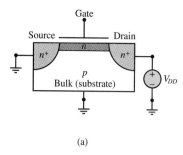

(a)

When the gate voltage is zero, the *n*-type channel permits drain current flow

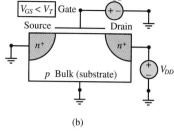

(b)

When the gate voltage is below the threshold voltage, the *n*-type channel has been depleted of charge carriers, and the MOSFET is in the **cutoff region**

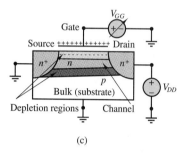

(c)

When the gate source voltage is increased above the threshold voltage for small values of drain-source voltage, the MOSFET is in the **ohmic region** and acts as a voltage-controlled resistor

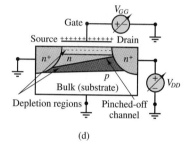

(d)

As the drain-source voltage is increased, the channel is eventually pinched off, and the transistor is in the **saturation region**

**Figure 9.37** Operation of *n*-channel depletion MOSFET

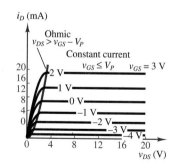

**Figure 9.38** Depletion MOSFET characteristic curves

as we increase the gate-source voltage, the transistor acts as a variable resistor, dependent on the gate-source voltage. This corresponds to the **ohmic region** of operation of the depletion-mode MOSFET, depicted in Figure 9.37(c). As the drain-source voltage is increased for a fixed gate-source voltage, the relative strength of the field will decrease near the drain end of the device, until the channel is *pinched off*. When the pinch-off condition is reached, the transistor acts as a constant-current source, and is in the **saturation region**. This condition is depicted in Figure 9.37(d).

In effect, a depletion-mode MOSFET acts very much like an enhancement mode device with a negative threshold voltage. Unlike an enhancement MOSFET, however, the depletion MOSFET

- Allows negative as well as positive gate voltages
- Can be in the saturation region for $v_{GS} = 0$

Figure 9.38 depicts the characteristic curves of a depletion-mode MOSFET. Once again, note the similarity to the enhancement-MOSFET curves: The primary difference results from the fact that, since a channel exists even for zero gate voltage, the device can operate for both positive and negative gate-source voltages.

## Junction Field-Effect Transistors

The last member of the FET family that will be discussed in this chapter is the junction field-effect transistor (JFET). The construction of a JFET and its circuit symbol are shown schematically in Figure 9.39. The *n*-type JFET consists of an *n*-type bulk element, with heavily doped *p*-type regions forming the gate. Metal contacts at the gates and at the ends of the *n*-type material provide external circuit connections. If the material composition is reversed, a *p*-type JFET can be similarly constructed.

To understand the operation of a JFET we refer to Figure 9.40. When the gate-source and drain-source voltages are both zero, the transistor will not conduct any current, and is said to be in the **cutoff region** (see Figure 9.40(a)). The JFET will remain in cutoff until the gate-source voltage exceeds a threshold or pinch-off voltage, $-V_P$. We shall give a definition of this voltage shortly. As the gate-source voltage is increased above $-V_P$ for small values of the drain-source voltage, the *pn* junction between gate and channel becomes more reverse biased, and the width of the depletion region increases, thus narrowing the channel. This has the effect of increasing the channel resistance, resulting in the *voltage-controlled resistor* behavior that is characteristics of the **ohmic region**. As shown in Figure 9.40(b), the channel will become narrower towards the drain end of the device, because the reverse bias of the *pn* junctions is larger near the drain (because the drain is at higher voltage than the source). If we now increase the drain-source voltage, the reverse bias will increase to the point where the channel is *pinched off*. This condition is shown in Figure 9.40(c). Now any further increase in $v_{DS}$ will not result in an appreciable increase in drain current, leading to operation in the **saturation region**. This corresponds to a flattening of the curve of drain current versus drain-source voltage. When the drain-source voltage is increased above a **breakdown voltage**, $V_B$, the drain current will increase very rapidly due to avalanche conduction, leading to excessive heat generation and device destruction. This is the **breakdown region** of the JFET.

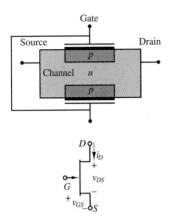

**Figure 9.39** JFET construction and circuit symbol

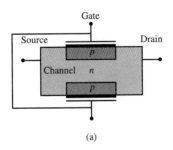

(a)

When the gate-source voltage is lower than $-V_P$, no current flows. This is the **cutoff region**

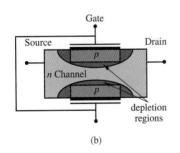

(b)

For small values of drain-source voltage, depletion regions form around the gate sections. As the gate voltage is increased, the depletion regions widen, and the channel width (i.e., the resistance) is controlled by the gate-source voltage. This is the **ohmic region** of the JFET

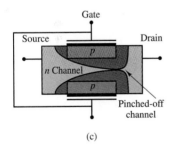

(c)

As the drain-source voltage is increased, the depletion regions further widen near the drain end, eventually pinching off the channel. This corresponds to the **saturation region**

**Figure 9.40** JFET operation

The behavior just described can also be visualized in the JFET characteristic curves of Figure 9.41. Note the similarity between the curves of Figure 9.41

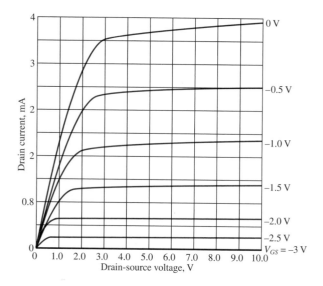

**Figure 9.41** JFET characteristic curves

and those of Figure 9.38, which describe the depletion MOSFET. Because of this similarity, the two families of devices can be described by the same set of equations, given in the following subsection. The operation of *p*-channel JFET is identical to that of an *n*-channel JFET, with the exception that the polarities of the voltages and the directions of the currents are all reversed.

## Depletion MOSFET and JFET Equations

The equations describing the operation of depletion MOSFETs and JFETs are summarized in the Table 9.2. We note that the equations describing depletion MOSFETs and JFETs are identical if we recognize that the depletion MOSFET threshold voltage and the JFET pinch-off voltage play the same role. *In Table 9.2 we use the symbol $V_P$ to represent both the depletion MOSFET threshold voltage and the JFET pinch-off voltage.* The equations of Table 9.2 are also valid for *p*-channel devices if one substitutes $v_{SG}$ for $v_{GS}$ and $v_{SD}$ for $v_{DS}$. The following examples illustrate analysis and biasing methods and an application of JFETs.

**Table 9.2** Regions of operation and equations of *n*-channel depletion MOSFET and of JFET

---

**Cutoff region:**

$v_{GS} < -V_P$

**Ohmic or triode region:** $v_{DS} < 0.25(v_{GS} + V_P), v_{GS} > -V_P$

$$R_{DS} = \frac{V_P^2}{2I_{DSS}(v_{GS} + V_P)} \qquad \text{(equivalent drain-to-source resistance)}$$

$$i_D \approx \frac{v_{DS}}{R_{DS}}$$

**Saturation region:**

$v_{DS} \geq v_{GS} + V_P, v_{GS} > -V_P$

$$i_D = \frac{I_{DSS}}{V_P^2}(v_{GS} + V_P)^2$$

**Breakdown region:**

$v_{DS} > V_B$

---

## EXAMPLE  9.11  Determining the Operating Region of a JFET

### Problem

Determine the operating region of each of the JFETs in Figure 9.42.

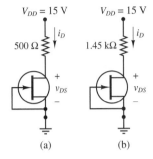

### Solution

**Known Quantities:**  Drain resistance; JFET parameters $I_{DSS}$ and $V_P$.

**Find:**  Operating region of each JFET.

**Schematics, Diagrams, Circuits, and Given Data:**  $V_{DD} = 15$ V; $R_{D(a)} = 500\ \Omega$; $R_{D(b)} = 1.45$ k$\Omega$; $V_P = 4$ V; $I_{DSS} = 10$ mA.

**Assumptions:**  Use the JFET equations of Table 9.2

**Figure 9.42**

### Analysis:

1. *Circuit (a).* Since $v_{GS} = 0$, we know (by definition) that $i_D = I_{DSS} = 10$ mA. We can therefore write the drain circuit equation:

$$V_{DD} = R_D i_D + v_{DS}$$

and calculate

$$v_{DS} = V_{DD} - R_D i_D = 15 - 500 \times 10^{-2} = 10\ \text{V}$$

From Table 9.2, we see that the condition for operation in the saturation region is $v_{DS} > v_{GS} + V_P$. Since $10 > 0 + 4$, we conclude that the JFET of circuit (a) is operating in the active region.

2. *Circuit (b).* The drain circuit equation is:

$$V_{DD} = R_D i_D + v_{DS}$$

and

$$v_{DS} = V_{DD} - R_D i_D = 15 - 1.45 \times 10^3 \times 10^{-2} = 0.5\ \text{V}$$

From Table 9.2, we see that the condition for operation in the saturation region is $v_{DS} > v_{GS} + V_P$, but in this case, $0.5 < 0 + 4$, and we therefore conclude that the JFET is in the ohmic region. We can also directly check the condition for ohmic operation given in Table 9.2:

$$|v_{DS}| < 0.25(v_{GS} + V_P)$$

and confirm that, indeed, the JFET operates in the ohmic region since $0.5 < 1$.

## EXAMPLE  9.12  Biasing a JFET

### Problem

Design the JFET bias circuit of Figure 9.43 to operate in the saturation region with a drain current of 4 mA and a drain-source voltage of 10 V.

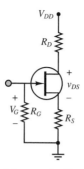

$V_{DD}$

$R_D$

$+$
$v_{DS}$
$-$

$+$
$V_G$ $R_G$
$-$

$R_S$

**Figure 9.43**

## Solution

**Known Quantities:**  Drain resistance; JFET parameters $I_{DSS}$ and $V_P$; JFET breakdown voltage.

**Find:**  Drain, gate and source resistances, drain supply voltage.

**Schematics, Diagrams, Circuits, and Given Data:**  $V_P = 3$ V; $I_{DSS} = 6$ mA; $i_{DQ} = 4$ mA; $V_B = 30$ V.

**Assumptions:**  Use the JFET equations of Table 9.2.

**Analysis:**  The information regarding the breakdown voltage, $V_B$, is useful to select the drain supply voltage. The drain supply must be less than the breakdown voltage to prevent device failure; $V_{DD} = 24$ V is a reasonable choice.

The resistance $R_G$ serves the purpose of tying the gate to ground. This is usually accomplished with a large resistance. Any leakage current would be only of the order of nanoamperes, so the gate would be at most a few mV above ground. Let's choose $R_G = 1$ MΩ.

Using the universal equation for the saturation region from Table 9.2, we write:

$$i_D = \frac{I_{DSS}}{V_P^2}(v_{GS} + V_P)^2 = I_{DSS}\left(\frac{v_{GS}}{V_P} + 1\right)^2$$

Knowing the desired drain current, we can expand the above equation to obtain:

$$\frac{I_{DSS}}{i_{DQ}V_P^2}v_{GS}^2 + 2\frac{I_{DSS}}{i_{DQ}V_P}v_{GS} + \left(\frac{I_{DSS}}{i_{DQ}} - 1\right) = 0$$

$$0.166v_{GS}^2 + v_{GS} + 0.5 = 0$$

with roots

$$v_{GS} = -5.45V \qquad v_{GS} = -0.55V$$

Since the JFET is in the active region only if $v_{GS} > -V_P$, the only acceptable solution is $v_{GSQ} = -0.55$ V.

To obtain the desired value of $v_{GSQ}$ we must choose an appropriate value of $R_S$. Remember that we know the value of the quiescent drain current (we desire 4 mA). Thus, we desire $v_{GSQ} = v_G - v_S = -0.55$ V, and since $v_G = 0$:

$$v_S = 0.55 = i_{DQ}R_S \qquad \text{or} \qquad R_S = \frac{0.55}{0.004} = 137.5\ \Omega$$

Now we can write the drain circuit equation to determine the appropriate value of $R_D$ to ensure that $v_{DS} = 10$ V:

$$V_{DD} = R_D i_D + v_{DS} + R_S i_D$$

or

$$24 = 0.004R_D + 10 + 0.55$$

from which equation $R_D$ can be computed to be approximately 3.36 kΩ.

**Comments:**  The value of source and drain resistance computed in the above example are not standard resistor values (see Chapter 2, Table 2.1); engineering practice would require that the nearest available standard values be used: $R_S = 150\ \Omega$, $R_D = 3.33\ \text{k}\Omega$.

---

## EXAMPLE 9.13 JFET Current Source

### Problem

Show that the circuit of Figure 9.44 can act as a current source.

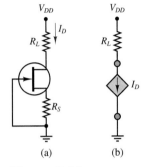

**Figure 9.44**

---

### Solution

**Known Quantities:**  Drain and source resistance; JFET parameters $I_{DSS}$ and $V_P$; Drain supply voltage.

**Find:**  Explain why the circuit of Figure 9.44(a) can be modeled by the circuit of Figure 9.44(b).

**Schematics, Diagrams, Circuits, and Given Data:**  $V_P = 3\ \text{V}$; $I_{DSS} = 6\ \text{mA}$; $R_S = 1\ \text{k}\Omega$.

**Assumptions:**  Use the JFET equations of Table 9.2.

**Analysis:**  We have already seen in Example 9.12 that a JFET can be biased in the saturation region with $v_G = 0$. The simple circuit shown in Figure 9.44(a) can therefore be biased in the saturation region provided that $v_{DS} > v_{GS} + V_P$ (see Table 9.2).

Note that if the source resistor, $R_S$, were set equal to zero, then we would have $v_{GS} = 0$, and the saturation region equation from Table 9.2 would simplify to $i_D = I_{DSS}$. Thus, *any JFET can be operated as a current source with* $i_D = I_{DSS}$ simply by tying the gate and source to ground. Unfortunately, because of the uncertainties in the fabrication process, the parameter $I_{DSS}$ can vary significantly even among nominally identical transistors, fabricated in the same batch. The addition of the source resistor permits adjusting the current source output, as shown below.

With a source resistor in the circuit, as shown in Figure 9.44(a), KVL applied around the gate-source circuit yields $v_{GS} = -i_D R_{GS}$. Applying the drain current equation for the saturation region, we have:

$$i_D = \frac{I_{DSS}}{V_P^2}(v_{GS} + V_P)^2 = I_{DSS}\left(\frac{v_{GS}}{V_P} + 1\right)^2 = I_{DSS}\left(\frac{-i_D R_S}{V_P} + 1\right)^2$$

We can expand the above equation to obtain:

$$\frac{I_{DSS}}{V_P^2}R_S^2 i_D^2 - \left(1 + \frac{2I_{DSS}R_S}{V_P}\right)i_D + I_{DSS} = 0$$

and solve for the given values to obtain $i_D = 6\ \text{mA}$ and $i_D = 1.5\ \text{mA}$. These lead to values of $v_{GS} = -6\ \text{V}$ and $v_{GS} = -1.5\ \text{V}$. Since the condition $v_{GS} = -V_P$ is satisfied only for the second solution, we conclude that the circuit of Figure 9.44(a) indeed operates as a current source in the saturation region, as indicated in Figure 9.44(b). The strength of the current source is $i_D = 1.5\ \text{mA}$. Note that this value is different from the parameter $I_{DSS}$, which is the nominal current source strength when $R_S = 0$.

**Comments:**  The source resistor can be adjusted to yield the desired source current in spite of variations in $I_{DSS}$.

## Check Your Understanding

**9.14**  What is $R_{DS}$ for the circuit of Figure 9.44(b)?

**9.15**  What is the drain current for the circuit of Figure 9.44(b)?

**9.16**  Determine the actual operating point of the JFET of Example 9.12, given the choice of standard resistor values.

**9.17**  Repeat the design of Example 9.12 (i.e., calculate $R_S$ and $R_D$) if the required drain current is halved.

## CONCLUSION

- Transistors are three-terminal electronic semiconductor devices that can serve as linear amplifiers or switches.
- The bipolar junction transistor (BJT) acts as a current-controlled current source, amplifying a small base current by a factor ranging from 20 to 200. The operation of the BJT can be explained in terms of the device base-emitter and collector $i$-$v$ characteristics. Large-signal linear circuit models for the BJT can be obtained by treating the transistor as a controlled current source.
- Field-effect transistors (FETs) can be grouped into three major families: enhancement MOSFETs, depletion MOSFETs, and JFETs. All FETs behave like voltage-controlled current sources. FET $i$-$v$ characteristics are intrinsically nonlinear, characterized by a quadratic dependence of the drain current on gate voltage. The nonlinear equations that describe FET drain characteristics can be summarized in a set of universal curves for each family.

## CHECK YOUR UNDERSTANDING ANSWERS

| | |
|---|---|
| **CYU 9.1** | $A = \mu \frac{1}{(r_i + R_S)} \frac{R_L}{(r_o + R_L)}; r_i \to 0, r_o \to 0$ |
| **CYU 9.2** | $A = \frac{1}{(r_i + R_S)} \frac{r_o R_L}{(r_o + R_L)} \mu$ |
| **CYU 9.3** | $A = \mu \frac{r_i}{(r_i + R_S)} \frac{r_o R_L}{(r_o + R_L)}$ |
| **CYU 9.5** | Saturation |
| **CYU 9.6** | $V_{BB} = 5$ V; $V_{CE} = 6.44$ V |
| **CYU 9.7** | 3.74%; because $R_E$ provides a negative feedback action that will keep $I_C$ and $I_E$ almost constant. |
| **CYU 9.8** | Saturation; 8.5 mA |
| **CYU 9.9** | 159 mW |
| **CYU 9.10** | The MOSFET is in the ohmic region. |
| **CYU 9.11** | Choosing the smaller value of $v_{GS}$, $R_S = 20.7$ k$\Omega$, $v_{GS} = 2.86$ V, $i_D = 0.586$ mA. The answer is not unique: Selecting the larger gate voltage, we find $R_S = 11.5$ k$\Omega$. |
| **CYU 9.13** | Approximately 400 $\Omega$ |
| **CYU 9.14** | 200 $\Omega$ |
| **CYU 9.15** | Approximately 2.5 mA |
| **CYU 9.16** | $i_D = 3.89$ mA; $v_{DS} = 10.57$ V; $v_{GS} = -0.58$ V |
| **CYU 9.17** | $R_S = 634$ $\Omega$; $R_D = 6.366$ k$\Omega$ |

# HOMEWORK PROBLEMS

## Section 1: Bipolar Transistors

**9.1**  For each transistor shown in Figure P9.1, determine whether the $BE$ and $BC$ junctions are forward- or reverse-biased, and determine the operating region.

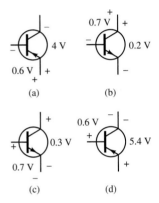

(a)                    (b)

(c)                    (d)

**Figure P9.1**

**9.2**  Determine the region of operation for the following transistors:

a. $npn$, $V_{BE} = 0.8$ V, $V_{CE} = 0.4$ V
b. $npn$, $V_{CB} = 1.4$ V, $V_{CE} = 2.1$ V
c. $pnp$, $V_{CB} = 0.9$ V, $V_{CE} = 0.4$ V
d. $npn$, $V_{BE} = -1.2$ V, $V_{CB} = 0.6$ V

**9.3**  Given the circuit of Figure P9.3, determine the operating point of the transistor. Assume the BJT is a silicon device with $\beta = 100$. In what region is the transistor?

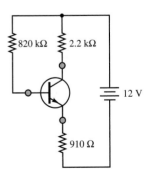

**Figure P9.3**

**9.4**  The magnitudes of a $pnp$ transistor's emitter and base currents are 6 mA and 0.1 mA, respectively. The magnitudes of the voltages across the emitter-base and collector-base junctions are 0.65 and 7.3 V. Find

a. $V_{CE}$.
b. $I_C$.
c. The total power dissipated in the transistor, defined here as $P = V_{CE}I_C + V_{BE}I_B$.

**9.5**  Given the circuit of Figure P9.5, determine the emitter current and the collector-base voltage. Assume the BJT has $V_\gamma = 0.6$ V.

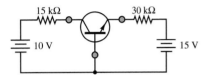

**Figure P9.5**

**9.6**  Given the circuit of Figure P9.6, determine the operating point of the transistor. Assume a 0.6-V offset voltage and $\beta = 150$. In what region is the transistor?

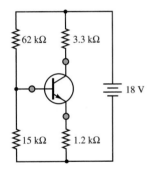

**Figure P9.6**

**9.7**  Given the circuit of Figure P9.7, determine the emitter current and the collector-base voltage. Assume the BJT has a 0.6-V offset voltage at the $BE$ junction.

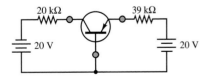

**Figure P9.7**

**9.8**  If the emitter resistor in Problem 9.7 (Figure P9.7) is changed to 22 k$\Omega$, how does the operating point of the BJT change?

**9.9**  The collector characteristics for a certain transistor are shown in Figure P9.9.

a.  Find the ratio $I_C/I_B$ for $V_{CE} = 10$ V and $I_B = 100$ $\mu$A, 200 $\mu$A, and 600 $\mu$A.

b.  The maximum allowable collector power dissipation is 0.5 W for $I_B = 500$ $\mu$A. Find $V_{CE}$.

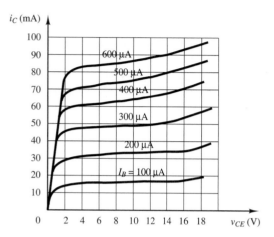

**Figure P9.9**

[*Hint:* A reasonable approximation for the power dissipated at the collector is the product of the collector voltage and current: $P = I_C V_{CE}$
where   $P$ is the permissible power dissipation,
$I_C$ is the quiescent collector current,
$V_{CE}$ is the operating point collector-emitter voltage.]

**9.10**   Given the circuit of Figure P9.10, assume both transistors are silicon-based with $\beta = 100$. Determine:

a.  $I_{C1}$, $V_{C1}$, $V_{CE1}$
b.  $I_{C2}$, $V_{C2}$, $V_{CE2}$

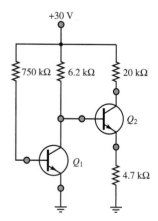

**Figure P9.10**

**9.11**   Use the collector characteristics of the 2N3904 *npn* transistor to determine the operating point ($I_{CQ}$, $V_{CEQ}$) of the transistor in Figure P9.11. What is the value of $\beta$ at this point?

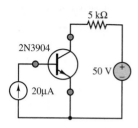

**Figure P9.11**

**9.12**   For the circuit given in Figure P9.12, verify that the transistor operates in the saturation region by computing the ratio of collector current to base current. (*Hint*: With reference to Figure 9.22, $V_\gamma = 0.6$ V, $V_{sat} = 0.2$ V.)

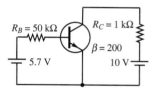

**Figure P9.12**

**9.13**   It has been found that $V_E$ in the circuit of Figure P9.13 is 1 V. If the transistor has $V_\gamma = 0.6$ V, determine:

a.  $V_B$
b.  $I_B$
c.  $I_E$
d.  $I_C$
e.  $\beta$
f.  $\alpha$

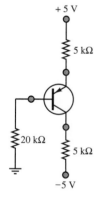

**Figure P9.13**

**9.14**   The circuit shown in Figure P9.14 is a common-emitter amplifier stage. Determine the Thévenin equivalent of the part of the circuit containing $R_1$, $R_2$, and $V_{CC}$ with respect to the terminals of $R_2$. Redraw the schematic using the Thévenin equivalent.

$$V_{CC} = 20 \text{ V} \quad \beta = 130$$
$$R_1 = 1.8 \text{ M}\Omega \quad R_2 = 300 \text{ k}\Omega$$
$$R_C = 3 \text{ k}\Omega \quad R_E = 1 \text{ k}\Omega$$
$$R_L = 1 \text{ k}\Omega \quad R_S = 0.6 \text{ k}\Omega$$
$$v_S = 1 \ \cos(6.28 \times 10^3 t) \text{ mV}$$

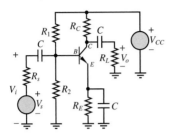

**Figure P9.14**

**9.15** Shown in Figure P9.14 is a common-emitter amplifier stage implemented with an *npn* silicon transistor. Determine $V_{CEQ}$ and the region of operation.

$$V_{CC} = 15 \text{ V} \quad \beta = 100$$
$$R_1 = 68 \text{ k}\Omega \quad R_2 = 11.7 \text{ k}\Omega$$
$$R_C = 200 \ \Omega \quad R_E = 200 \ \Omega$$
$$R_L = 1.5 \text{ k}\Omega \quad R_S = 0.9 \text{ k}\Omega$$
$$v_S = 1 \ \cos(6.28 \times 10^3 t) \text{ mV}$$

**9.16** Shown in Figure P9.14 is a common-emitter amplifier stage implemented with an *npn* silicon transistor. Determine $V_{CEQ}$ and the region of operation.

$$V_{CC} = 15 \text{ V} \quad \beta = 100$$
$$R_1 = 68 \text{ k}\Omega \quad R_2 = 11.7 \text{ k}\Omega$$
$$R_C = 4 \text{ k}\Omega \quad R_E = 200 \ \Omega$$
$$R_L = 1.5 \text{ k}\Omega \quad R_S = 0.9 \text{ k}\Omega$$
$$v_S = 1 \ \cos(6.28 \times 10^3 t) \text{ mV}$$

**9.17** The circuit shown in Figure P9.17 is a common-collector (also called an emitter follower) amplifier stage implemented with an *npn* silicon transistor. Determine $V_{CEQ}$ at the DC operating or $Q$ point.

$$V_{CC} = 12 \text{ V} \quad \beta = 130$$
$$R_1 = 82 \text{ k}\Omega \quad R_2 = 22 \text{ k}\Omega$$
$$R_S = 0.7 \text{ k}\Omega \quad R_E = 0.5 \text{ k}\Omega$$
$$R_L = 16 \ \Omega$$

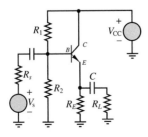

**Figure P9.17**

**9.18** Shown in Figure P9.18 is a common-emitter amplifier stage implemented with an *npn* silicon transistor and two DC supply voltages (one positive and one negative) instead of one. The DC bias circuit connected to the base consists of a single resistor. Determine $V_{CEQ}$ and the region of operation.

$$V_{CC} = 12 \text{ V} \quad V_{EE} = 4 \text{ V}$$
$$\beta = 100 \quad R_B = 100 \text{ k}\Omega$$
$$R_C = 3 \text{ k}\Omega \quad R_E = 3 \text{ k}\Omega$$
$$R_L = 6 \text{ k}\Omega \quad R_S = 0.6 \text{ k}\Omega$$
$$v_S = 1 \ \cos(6.28 \times 10^3 t) \text{ mV}$$

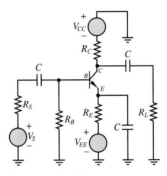

**Figure P9.18**

**9.19** Shown in Figure P9.19 is a common-emitter amplifier stage implemented with an *npn* silicon transistor. The DC bias circuit connected to the base consists of a single resistor; however, it is connected directly between base and collector. Determine $V_{CEQ}$ and the region of operation.

$$V_{CC} = 12 \text{ V}$$
$$\beta = 130 \quad R_B = 325 \text{ k}\Omega$$
$$R_C = 1.9 \text{ k}\Omega \quad R_E = 2.3 \text{ k}\Omega$$
$$R_L = 10 \text{ k}\Omega \quad R_S = 0.5 \text{ k}\Omega$$
$$v_S = 1 \ \cos(6.28 \times 10^3 t) \text{ mV}$$

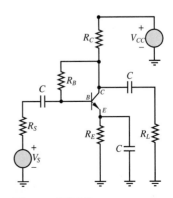

**Figure P9.19**

**9.20** Shown in Figure P9.19 is a common-emitter amplifier stage implemented with an *npn* silicon transistor. Determine $V_{CEQ}$ and the region of operation.

$$V_{CC} = 15 \text{ V} \qquad C = 0.5 \text{ } \mu\text{F}$$
$$\beta = 170 \qquad R_B = 22 \text{ k}\Omega$$
$$R_C = 3.3 \text{ k}\Omega \qquad R_E = 3.3 \text{ k}\Omega$$
$$R_L = 1.7 \text{ k}\Omega \qquad R_S = 70 \text{ }\Omega$$
$$v_S = 1 \text{ } \cos(6.28 \times 10^3 t) \text{ mV}$$

**9.21** Shown in Figure P9.14 is a common-emitter amplifier stage with:

$$V_{CC} = 15 \text{ V} \qquad C = 0.47 \text{ } \mu\text{F}$$
$$R_1 = 220 \text{ k}\Omega \qquad R_2 = 55 \text{ k}\Omega$$
$$R_C = 3 \text{ k}\Omega \qquad R_E = 710 \text{ }\Omega$$
$$R_L = 3 \text{ k}\Omega \qquad R_S = 0.6 \text{ k}\Omega$$
$$v_i = V_{io} \text{ } \sin(\omega t) \qquad V_{io} = 10 \text{ mV}$$

A DC analysis gives the DC operating point or $Q$ point:

$$I_{BQ} = 19.9 \text{ } \mu\text{A} \qquad V_{CEQ} = 7.61 \text{ V}$$

The transistor is an *npn* silicon transistor with a transfer characteristic and beta:

$$i_C \approx I_S e^{v_{BE}/V_T} = I_S e^{V_{BEQ}+v_{be}/V_T} \qquad \beta = 100$$

The device $i$-$v$ characteristic is plotted in Figure P9.21.

a. Determine the no-load large signal gain $(v_o/v_i)$.
b. Sketch the waveform of the output voltage as a function of time.
c. Discuss how the output voltage is distorted compared to the input waveform.

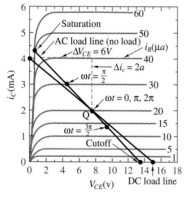

**Figure P9.21**

## Section 2: Field-Effect Transistors

**9.22** Shown in Figure P9.22 are the circuit symbols for a depletion-mode and an enhancement-mode MOSFET.

For the enhancement-mode MOSFET:

$$I_{DSS} = 7 \text{ mA} \qquad V_T = +5 \text{ V}$$

For the depletion-mode MOSFET:

$$I_{DSS} = 7 \text{ mA} \qquad V_P = -5 \text{ V}$$

a. Are these *n*- or *p*-channel devices?
b. Which is the depletion-mode device? The enhancement-mode device?
c. For each device, state the conditions for the operation in the active region in terms of the voltages shown for the device and the threshold or pinch-off voltages given above.

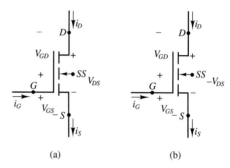

(a)                    (b)

**Figure P9.22**

**9.23** The transistors shown in Figure P9.23 have $|V_T| = 3$ V. Determine the operating region.

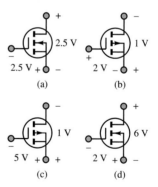

**Figure P9.23**

**9.24** The three terminals of an *n*-channel enhancement-mode MOSFET are at potentials of 4 V, 5 V, and 10 V with respect to ground. Draw the circuit symbol, with the appropriate voltages at each terminal, if the device is operating

a. In the ohmic region.
b. In the active region.

**9.25** An enhancement-type NMOS transistor with $V_T = 2$ V has its source grounded and a 3-VDC source connected to the gate. Determine the operating state if

a. $v_D = 0.5$ V

b. $v_D = 1$ V

c. $v_D = 5$ V

**9.26**  In the circuit shown in Figure P9.26, the $p$-channel transistor has $V_T = 2$ V and $k = 10$ mA/$V^2$. Find $R$ and $v_D$ for $i_D = 0.4$ mA.

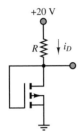

**Figure P9.26**

**9.27**  An enhancement-type NMOS transistor has $V_T = 2$ V and $i_D = 1$ mA when $v_{GS} = v_{DS} = 3$ V. Find the value of $i_D$ for $v_{GS} = 4$ V.

**9.28**  An $n$-channel enhancement-mode MOSFET is operated in the ohmic region, with $v_{DS} = 0.4$ V and $V_T = 3.2$ V. The effective resistance of the channel is given by $R_{DS} = 500/(V_{GS} - 3.2)$ $\Omega$. Find $i_D$ when $v_{GS} = 5$ V, $R_{DS} = 500$ $\Omega$, and $v_{GD} = 4$ V.

**9.29**  An $n$-channel JFET has $V_p = -2.8$ V. It is operating in the ohmic region, with $v_{GS} = -1$ V, $v_{DS} = 0.05$ V, and $i_D = 0.3$ mA. Find $i_D$ if

a. $v_{GS} = -1$ V, $v_{DS} = 0.08$ V

b. $v_{GS} = 0$ V, $v_{DS} = 0.1$ V

c. $v_{GS} = -3.2$ V, $v_{DS} = 0.06$ V

**9.30**  An enhancement-type NMOS transistor with $V_T = 2.5$ V has its source grounded and a 4-VDC source connected to the gate. Find the operating region of the device if

a. $v_D = 0.5$ V

b. $v_D = 1.5$ V

**9.31**  An enhancement-type NMOS transistor has $V_T = 4$ V, $i_D = 1$ mA when $v_{GS} = v_{DS} = 6$ V. Neglecting the dependence of $i_D$ on $v_{DS}$ in saturation, find the value of $i_D$ for $v_{GS} = 5$ V.

**9.32**  The NMOS transistor shown in Figure P9.32 has $V_T = 1.5$ V, $k = 0.4$ mA/$V^2$. Now if $v_G$ is a pulse with 0 V to 5 V, find the voltage levels of the pulse signal at the drain output.

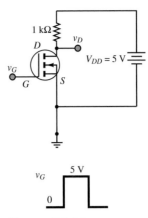

**Figure P9.32**

**9.33**  A JFET having $V_p = -2$ V and $I_{DSS} = 8$ mA is operating at $v_{GS} = -1$ V and a very small $v_{DS}$. Find

a. $r_{DS}$

b. $v_{GS}$ at which $r_{DS}$ is half of its value in (a)

**9.34**  Shown in Figure P9.34 is the schematic for a common-source amplifier stage. Determine the DC operating point and verify that the device is operating in the saturation region.

$$I_{DSS} = 1.125 \text{ mA} \qquad V_T = 1.5 \text{ V}$$
$$R_1 = 1.32 \text{ M}\Omega \qquad R_2 = 2.2 \text{ M}\Omega$$
$$R_D = 4 \text{ k}\Omega \qquad R_S = 4 \text{ k}\Omega$$
$$R_L = 1.3 \text{ k}\Omega \qquad C = 0.47 \text{ }\mu\text{F}$$
$$R_{SS} = 0.7 \text{ k}\Omega \qquad V_{DD} = 12 \text{ V}$$

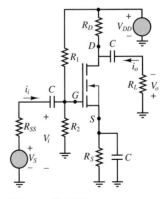

**Figure P9.34**

**9.35**  Shown in Figure P9.35 is the simplified common-source amplifier stage where the DC bias network in the gate circuit has already been replaced by its Thévenin equivalent. Determine the DC operating or $Q$ point and the region of operation.

$$I_{DSS} = 7 \text{ mA} \qquad V_P = 2.65 \text{ V}$$
$$R_G = 330 \text{ k}\Omega \qquad V_{GG} = 4.7 \text{ V}$$
$$R_D = 3.3 \text{ k}\Omega \qquad R_S = 3.3 \text{ k}\Omega$$
$$R_L = 1.7 \text{ k}\Omega \qquad C = 0.5 \text{ }\mu\text{F}$$
$$R_{SS} = 70 \text{ k}\Omega \qquad V_{DD} = 15 \text{ V}$$

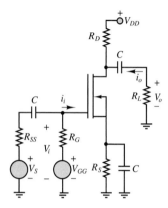

**Figure P9.35**

**9.36**   In the circuit of Figure P9.35, a common-source amplifier stage, $R_G$ is an actual component in the circuit and $V_{GG}$ has been eliminated (i.e., made equal to zero). A partial solution for the DC operating or $Q$ point gives:

$$I_{DQ} = 1.324 \text{ mA} \qquad V_{GSQ} = -4.368 \text{ V}$$
$$I_{DSS} = 18 \text{ mA} \qquad V_T = 6 \text{ V}$$
$$R_G = 1.7 \text{ M}\Omega \qquad R_S = 3.3 \text{ k}\Omega$$
$$R_L = 3 \text{ k}\Omega \qquad C = 0.5 \text{ }\mu\text{F}$$
$$V_{DD} = 20 \text{ V} \qquad V_{GG} = 0.0 \text{ V}$$
$$v_i[t] = 1 \ \cos(6.28 \times 10^3 t) \text{ V}$$

Determine $R_D$ so that $V_{DSQ} = 6$ V.

**9.37**   In the circuit shown in Figure P9.35, a common-source amplifier stage, $R_G$ is an actual component in the circuit and $V_{GG}$ has been eliminated (i.e., shorted). A partial solution for the $Q$ point gives:

$$I_{DQ} = 2.97 \text{ mA} \qquad V_{GSQ} = -3.56 \text{ V}$$

Determine $R_D$ so that $V_{DSQ} = 6$ V.

$$I_{DSS} = 18 \text{ mA} \qquad V_P = 6 \text{ V}$$
$$R_G = 1.7 \text{ M}\Omega \qquad R_S = 1.2 \text{ k}\Omega$$
$$R_L = 3 \text{ k}\Omega \qquad C = 0.5 \text{ }\mu\text{F}$$
$$V_{DD} = 20 \text{ V} \qquad V_{GG} = 0.0 \text{ V}$$
$$v_i(t) = 1 \ \cos(6.28 \times 10^3 t) \text{ V}$$

**9.38**   Shown in Figure P9.38(a) is a common-source amplifier stage implemented with an $n$-channel depletion-mode MOSFET with the static $i$-$v$ characteristics shown in Figure P9.38(b). The $Q$ point and component values are:

$$V_{DSQ} = 13.6 \text{ V} \qquad V_{GSQ} = 1.5 \text{ V}$$
$$R_D = 1.0 \text{ k}\Omega \qquad R_S = 400 \text{ }\Omega$$
$$R_L = 3.2 \text{ k}\Omega \qquad R_{SS} = 600 \text{ }\Omega$$
$$v_i[t] = 2 \ \sin(\omega t) \text{ V}$$

Determine the DC supply voltage required for the $Q$ point and component values specified.

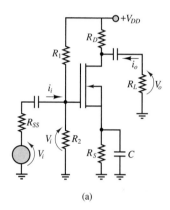

(a)

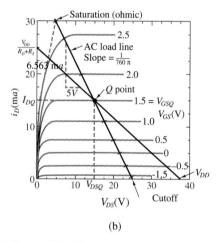

(b)

**Figure P9.38**

# C H A P T E R

# 10

# Transistor Amplifiers and Switches

T
he aim of this chapter is to describe the application of transistors as amplifiers and switches. Small-signal transistor amplifiers can be analyzed by means of linear small-signal models that make it possible, through the use of linear circuit analysis techniques, to determine an amplifier's input and output impedance and current and voltage gain. Small-signal models of transistor amplifier circuits also permit analysis of multistage amplifiers.

The chapter begins with the analysis of the BJT $h$ parameters; these are linear approximations that are valid in the neighborhood of an operating point and are directly derived from the base and collector characteristic curves. Subsequently, the common-emitter BJT amplifier is discussed in some detail, and the common-base and emitter-follower amplifiers are briefly introduced. Next, a similar analysis is conducted for MOSFET amplifiers. The material on amplifiers closes with a general discussion of multistage transistor amplifiers and of amplifier frequency response.

In addition to serving as the essential component of electronic amplifiers, transistors find common application in switching circuits and logic gates. The last section of the chapter describes BJT and MOSFET inverters and gates and introduces the two major families of logic devices, TTL and CMOS.

Upon completing this chapter, you should be able to:

- Use small signal models of bipolar and field-effect transistors to construct small signal amplifier models, from which voltage and current gain and input and output resistance can be computed.
- Qualitatively evaluate the frequency response characteristics of a transistor amplifier and understand the major mechanisms limiting an amplifier's frequency response.
- Understand the major requirements in the design of multistage amplifiers.
- Understand the switching characteristics of BJTs and MOSFETs and be able to analyze the fundamental behavior of TTL and CMOS logic gates.

## 10.1    SMALL-SIGNAL MODELS OF THE BJT

**Small-signal models** for the BJT take advantage of the relative linearity of the base and collector curves in the vicinity of an operating point. These linear circuit models work very effectively provided that the transistor voltages and currents remain within some region around the operating point. This condition is usually satisfied in small-signal amplifiers used to magnify low-level signals (e.g., sensor signals). For the purpose of our discussion, we use the **hybrid-parameter (h-parameter) small-signal model** of the BJT, to be discussed presently.

Note that a small-signal model assumes that the DC bias point of the transistor has been established. As was done in Chapter 9, the following convention will be used: each voltage and current is assumed to be the superposition of a DC component (the quiescent voltage or current) and a small-signal AC component. The former is denoted by an uppercase letter, and the latter by an uppercase letter preceded by the symbol $\Delta$. Thus,

$$i_B = I_{BQ} + \Delta I_B$$
$$i_C = I_{CQ} + \Delta I_C$$
$$v_{CE} = V_{CEQ} + \Delta V_{CE}$$

Figure 10.1 depicts the appearance of the collector current $i_C(t)$ when $I_{CQ} = 5 \times 10^{-3}$ A and $\Delta I_C(t) = 0.5 \times 10^{-3} \sin \omega t$ A.

Imagine the collector curves of Figure 9.19 magnified about the $Q$ point. Figure 10.2 graphically depicts the interpretation of each of the $h$ parameters relative to the operating point of the BJT.

The parameter $h_{ie}$ is approximately equal to the ratio $\Delta V_{BE}/\Delta I_B$ in the neighborhood of the $Q$ point; Figure 10.2(a) illustrates how this parameter is equal to the reciprocal of the slope of the $I_B$-$V_{BE}$ curve at the operating point. Physically, this parameter represents the forward resistance of the $BE$ junction.

The parameter $h_{re}$ is representative of the fact that the $I_B$-$V_{BE}$ curve is slightly dependent on the actual value of the collector-emitter voltage, $V_{CE}$. However, this effect is virtually negligible in any applications of interest to us. Thus, we shall assume that $h_{re} \approx 0$. Figure 10.2(b) depicts the shift in the $I_B$-$V_{BE}$ curves represented by $h_{re}$. A typical value of $h_{re}$ for $V_{CE} \geq 1$ V is around $10^{-2}$.

The parameter $h_{fe}$ is approximated in Figure 10.2(c) by the current ratio $\Delta I_C/\Delta I_B$. This parameter represents the current gain of the transistor and is approximately equivalent to the parameter $\beta$ introduced earlier. For the purpose of our discussion, $\beta$ and $h_{fe}$ will be interchangeable, although they are not exactly identical.

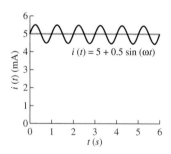

**Figure 10.1** Superposition of AC and DC signals

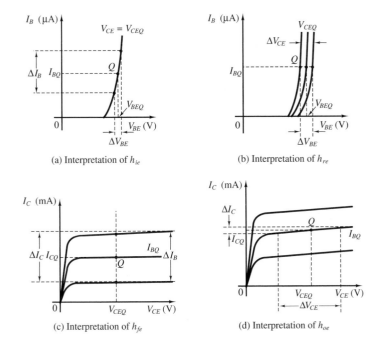

**Figure 10.2** Graphical interpretation of $h$ parameters

The parameter $h_{oe}$ may be calculated as $h_{oe} = \Delta I_C / \Delta V_{CE}$ from the collector characteristic curves, as shown in Figure 10.2(d). This parameter is a physical indication of the fact that the $I_C$-$V_{CE}$ curves in the linear active region are not exactly flat; $h_{oe}$ represents the upward slope of these curves and therefore has units of conductance (S). Typical values of $h_{oe}$ are around $10^{-5}$ S. We shall often assume that this effect is negligible.

To be more precise, the $h$ parameters are defined by the following set of equations:

$$h_{ie} = \left.\frac{\partial v_{BE}}{\partial i_B}\right|_{I_{BQ}} \qquad (\Omega) \qquad\qquad (10.1)$$

$$h_{oe} = \left.\frac{\partial i_C}{\partial v_{CE}}\right|_{V_{CEQ}} \qquad (S) \qquad\qquad (10.2)$$

$$h_{fe} = \left.\frac{\partial i_C}{\partial i_B}\right|_{I_{BQ}} \qquad \left(\frac{A}{A}\right) \qquad\qquad (10.3)$$

$$h_{re} = \left.\frac{\partial v_{BE}}{\partial v_{CE}}\right|_{V_{CEQ}} \qquad \left(\frac{V}{V}\right) \qquad\qquad (10.4)$$

The circuit of Figure 10.3 illustrates the small-signal model side by side with the BJT circuit symbol. Representative parameters for a small-signal transistor are listed in Table 10.1.

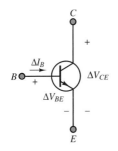

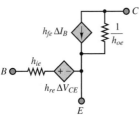

**Figure 10.3** $h$-parameter small-signal model for BJT

Table 10.1 $h$ parameters for the
2N2222A BJT

| Parameter | Minimum | Maximum |
|---|---|---|
| $h_{ie}$ (k$\Omega$) | 2 | 4 |
| $h_{re}(\times 10^{-4})$ | | 8 |
| $h_{fe}$ | 50 | 300 |
| $h_{oe}$ ($\mu$S) | 5 | 35 |

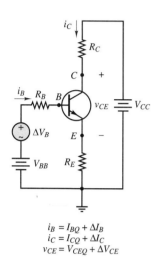

$$i_B = I_{BQ} + \Delta I_B$$
$$i_C = I_{CQ} + \Delta I_C$$
$$v_{CE} = V_{CEQ} + \Delta V_{CE}$$

**Figure 10.4** BJT amplifier

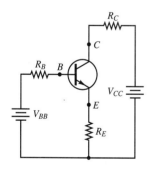

**Figure 10.5** DC
equivalent circuit for the BJT
amplifier of Figure 10.4

To illustrate the application of the $h$-parameter small-signal model of Figure 10.3, consider the transistor amplifier circuit shown in Figure 10.4. This circuit is most readily analyzed if DC and AC equivalent circuits are treated separately. To obtain the DC circuit, the AC source is replaced by a short circuit. The resulting DC circuit is shown in Figure 10.5. The DC circuit may be employed to carry out a $Q$-point analysis similar to that of Examples 9.4 and 9.6. Since our objective at present is to illustrate the AC circuit model for the transistor amplifier, we shall assume that the DC analysis (i.e., selection of the appropriate $Q$ point) has already been carried out, and that a suitable operating point has been established.

Replacing the DC voltage sources with short circuits, we obtain the AC equivalent circuit of Figure 10.6. The transistor may now be replaced by its $h$-parameter small-signal model, also shown in Figure 10.6. We may simplify the model by observing that $h_{oe}^{-1}$ is a very large resistance and that if the load resistance $R_L$ (in parallel with $h_{oe}^{-1}$) is small (i.e., if $R_L h_{oe} \leq 0.1$), the resistor $h_{oe}^{-1}$ in the model may be ignored. The linear AC equivalent circuit makes it possible to take advantage of the circuit analysis techniques developed in Chapters 2 and 3 to analyze the operation of the amplifier. For example, application of KVL around the base circuit loop yields the following equation:

$$\Delta V_B = \Delta I_B R_B + \Delta I_B h_{ie} + (h_{fe} + 1)\Delta I_B R_E \tag{10.5}$$

which can be solved to obtain

$$\frac{\Delta I_B}{\Delta V_B} = \frac{1}{R_B + h_{ie} + (h_{fe} + 1)R_E} \tag{10.6}$$

and

$$\Delta V_C = -\Delta I_C R_C$$
$$= -h_{fe}\Delta I_B R_C = \frac{-h_{fe}R_C}{R_B + h_{ie} + (h_{fe} + 1)R_E}\Delta V_B \tag{10.7}$$

Then, the **AC/open-loop voltage gain** of the amplifier is given by the expression

$$\mu = \frac{\Delta V_C}{\Delta V_B} = \frac{-h_{fe}R_C}{R_B + h_{ie} + (h_{fe} + 1)R_E} \tag{10.8}$$

You may recall that the open-loop voltage gain $\mu$ was introduced in Section 9.1 and Example 9.1. The small-signal model for the BJT will be further explored in the next section.

## Transconductance

In addition to the $h$ parameters described above, another useful small-signal transistor parameter is the **transfer conductance,** or **transconductance**. The transconductance of a bipolar transistor is defined as the local slope of the collector current base-emitter voltage curve:

$$g_m = \frac{\partial i_C}{\partial v_{BE}} \qquad\qquad (10.9)$$

and it can be expressed in terms of the $h$ parameters if we observe that we can write

$$g_m = \frac{\partial i_C}{\partial v_{BE}} = \frac{\partial i_C}{\partial i_B} \frac{\partial i_B}{\partial v_{BE}} = \frac{h_{fe}}{h_{ie}} \qquad\qquad (10.10)$$

It can be shown that the expression for the transconductance can be approximated by

$$g_m \cong \frac{I_{CQ}}{\frac{kT}{q}} = \frac{I_{CQ}}{0.026} = 39 I_{CQ} \frac{\text{mA}}{\text{V}} \qquad\qquad (10.11)$$

at room temperature, where $k$ is Boltzmann's constant, $T$ is the temperature in degrees Kelvin, and $q$ the electron charge (see Chapter 8 for a review of the $pn$ junction equation). Now, the transconductance is an important measure of the voltage amplification of a BJT amplifier, because it relates small oscillations in the base-emitter junction voltage to the corresponding oscillations in the collector current. We shall see in the next sections that this parameter can be related to the voltage gain of the transistor. Note that the approximation of equation 10.11 suggests that the transconductance parameter is dependent on the operating point.

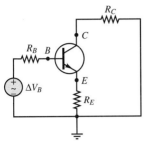

AC equivalent circuit

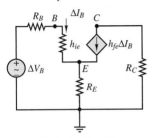

Small-signal model

**Figure 10.6** AC equivalent circuit and small-signal model for the amplifier of Figure 10.4

---

## EXAMPLE  10.1  Determining the AC Open-Loop Voltage Gain of a Common-Emitter Amplifier

### Problem

Determine the $Q$ point and AC open-loop voltage gain of the amplifier of Figure 10.4; the amplifier employs a 2N5088 *npn* transistor.

---

### Solution

**Known Quantities:**  Amplifier supply voltages; base, collector, and emitter resistances; $h$ parameters; AC circuit model of Figure 10.3.

**Find:**  Quiescent values of $I_B$, $I_C$, and $V_{CE}$; open-loop AC voltage gain, $\mu$.

**Schematics, Diagrams, Circuits, and Given Data:**  $V_{BB} = 6$ V; $V_{CC} = 12$ V; $R_B = 100$ k$\Omega$; $R_C = 500$ $\Omega$; $R_E = 100$ $\Omega$; $V_\gamma = 0.6$ V. $h_{fe} = 350$; $h_{ie} = 1.4$ k$\Omega$; $h_{oe} = 150$ $\mu$S.

**Assumptions:**  Use the linear small-signal $h$-parameter model of the BJT.

*Analysis:*

1. *Q-point calculation* (also see Example 9.4). We first write the collector circuit equation by applying KVL:

$$V_{CC} = V_{CE} + R_C I_C + R_E I_E = V_{CE} + R_C I_C + R_E (I_B + I_C)$$
$$\approx V_{CE} + (R_C + R_E) I_C$$

where the emitter current has been approximately set equal to the collector current since the current gain is large and $I_C \gg I_B$.

Next, we write the base circuit equation, also via KVL:

$$V_{BB} = R_B I_B + V_{BE} + R_E I_E = (R_B + (\beta + 1) R_E) I_B + V_\gamma$$

The above equation can be solved numerically (with $h_{fe} = \beta$) to obtain:

$$I_B = \frac{V_{BB} - V_\gamma}{R_B + (\beta + 1) R_E} = \frac{6 - 0.6}{100 \times 10^3 + 351 \times 100} = 40 \, \mu A$$

Then,

$$I_C = \beta I_B = 350 \times 40 \times 10^{-6} = 14 \text{ mA}$$

and

$$V_{CE} = V_{CC} - (R_C + R_E) I_C = 12 - 600 \times 14 \times 10^{-3} = 3.6 \text{ V}$$

Thus, the $Q$ point for the amplifier is:

$$I_{BQ} = 40 \, \mu A \qquad I_{CQ} = 14 \text{ mA} \qquad V_{CEQ} = 3.6 \text{ V}$$

confirming that the transistor is indeed in the active region.

2. *AC open-loop gain calculation.* The AC open-loop gain for the amplifier of Figure 9.4 can be computed by using equation 10.7:

$$\mu = \frac{-h_{fe} R_C}{R_B + h_{ie} + (h_{fe} + 1) R_E} = \frac{-350 \times 500 \times 10^3}{100 \times 10^3 + 1.4 \times 10^3 + 351 \times 100}$$
$$= -1.28 \, \frac{V}{V}$$

**Comments:** You may wish to examine the data sheets for the 2N5088 *npn* transistor. They are available in electronic form in the accompanying CD-ROM. Look for the values of the parameters used in this example.

Note that if the parameter $h_{ie}$ were neglected in the expression for $\mu$, the answer would not change significantly. This is true in this particular case because of the large values of the base resistor and of $h_{fe}$, but is not true in the general case.

## Check Your Understanding

**10.1**  Determine the AC/open-loop voltage gain of a 2N2222A transistor, using the results of Example 10.1 and Table 10.1. Use maximum values of the $h$ parameters, and assume that $I_{CQ} = 50$ mA, $R_C = 1$ k$\Omega$, $R_B = 100$ k$\Omega$, and $R_E = 100$ $\Omega$.

## 10.2    BJT SMALL-SIGNAL AMPLIFIERS

The *h*-parameter model developed in the previous section is very useful in the small-signal analysis of various configurations of BJT amplifiers. In this section, a set of techniques will be developed to enable you to first establish the $Q$ point of a transistor amplifier, then construct the small-signal model, and finally use the small-signal model for analyzing the small-signal behavior of the amplifier. A major portion of this section will be devoted to the analysis of the **common-emitter amplifier,** using this circuit as a case study to illustrate the analysis methods. At the end of the section, we will look briefly at two other common amplifier circuits: the **voltage follower,** or **common-collector amplifier;** and the **common-base amplifier.** The discussion of the common-emitter amplifier will also provide an occasion to introduce, albeit qualitatively, the important issue of transistor amplifier frequency response. A detailed treatment of this last topic is beyond the intended scope of this book.

A complete common-emitter amplifier circuit is shown in Figure 10.7. The circuit may appear to be significantly different from the simple examples studied in the previous sections; however, it will soon become apparent that all the machinery necessary to understand the operation of a complete transistor amplifier is already available.

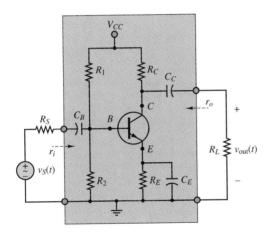

**Figure 10.7** The BJT common-emitter amplifier

We shall create a small-signal linear AC equivalent-circuit model for the amplifier based on a **two-port**[1] **equivalent circuit**; this equivalent circuit can then be used in connection with equivalent circuits for the load and source to determine the actual gains of the amplifier as a function of the load and source impedances.

Figure 10.8 depicts the appearance of this simplified representation for a transistor amplifier, where $r_i$ and $r_o$ represent the input and output resistance of

---

[1]A two-port circuit is a circuit that has an input and an output port, in contrast with the one-port circuits studied in Chapter 3, which had only a single port connecting the source to a load. The amplifier configuration shown in Figure 10.7 is representative of a general two-port circuit.

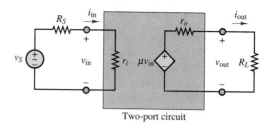

**Figure 10.8** Equivalent-circuit model of voltage amplifier

the amplifier, respectively, and $\mu$ is the open-loop voltage gain of the amplifier. Throughout this section, it will be shown how such a model can be obtained for a variety of amplifier configurations. You may wish to compare this model with that shown in Figure 9.3 in Section 9.1. Note that the model of Figure 10.8 makes use of the simple Thévenin equivalent-circuit model developed in Chapter 3 in representing the input to the amplifier as a single equivalent resistance, $r_i$. Similarly, the circuit seen by the load consists of a Thévenin equivalent circuit. In the remainder of this section, it will be shown how the values of $r_i$, $r_o$, and $\mu$ may be computed, given a transistor amplifier design. It is also useful to define the overall voltage and current gains for the amplifier model of Figure 10.8 as follows. The amplifier voltage gain is defined as

$$A_V = \frac{v_{\text{out}}}{v_S} = \frac{R_L}{R_L + r_o} \mu \cdot \frac{r_i}{R_S + r_i} \tag{10.12}$$

while the current gain is

$$A_I = \frac{i_{\text{out}}}{i_{\text{in}}} = \frac{v_{\text{out}}/R_L}{v_s/(R_S + r_i)} = \frac{R_S + r_i}{R_L} \cdot A_V \tag{10.13}$$

The first observation that can be made regarding the circuit of Figure 10.7 is that the AC input signal, $v_S(t)$, has been "coupled" to the remainder of the circuit through a capacitor, $C_B$ (called a **coupling capacitor**). Similarly, the load resistance, $R_L$, has been connected to the circuit by means of an identical coupling capacitor. The reason for the use of coupling capacitors is that they provide separate paths for DC and AC currents in the circuit. In particular, *the quiescent DC currents cannot reach the source or the load.* This is especially useful, since the aim of the circuit is to amplify the AC input signal only, and it would be undesirable to have DC currents flowing through the load. In fact, the presence of DC currents would cause unnecessary and undesired power consumption at the load. The operation of the coupling capacitors is best explained by observing that a capacitor acts as an open circuit to DC currents, while—if the capacitance is sufficiently large—it will act as a short circuit at the frequency of the input signal. Thus, in general, one wishes to make $C_C$ as large as possible, within reason. The **emitter bypass capacitor,** $C_E$, serves a similar purpose, by "bypassing" the emitter resistance $R_E$ *insofar as AC currents are concerned,* since the capacitor acts as a short circuit at the signal frequency. On the other hand, $C_E$ is an open circuit to DC currents, and therefore the quiescent current will flow through the emitter resistor, $R_E$. Thus, the emitter resistor can be chosen to select a given $Q$ point, but it will not appear in the calculations of the AC gains. This dual role served by coupling and bypass capacitors in transistor amplifiers is of fundamental importance in their practical operation.

Figure 10.9 depicts the path taken by AC and DC currents in the circuit of Figure 10.7. Example 10.2 further explains the use of coupling capacitors.

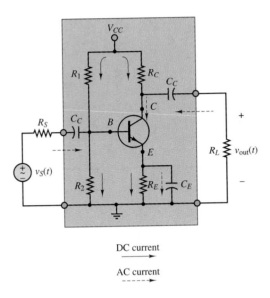

DC current

AC current

**Figure 10.9** Effect of coupling capacitors on DC and AC current paths

---

**EXAMPLE  10.2  Computing the Value of the AC Coupling Capacitor for Audio-Range Amplifier Operation**

### Problem

Determine the value of the coupling capacitor $C_C$ in Figure 10.9 that will permit amplifier operation in the audio range.

---

### Solution

**Known Quantities:**  Audio-frequency range.

**Find:**  Value of $C_C$ such that the series impedance of the capacitors is high at low frequencies and low at frequencies in the audio range.

**Schematics, Diagrams, Circuits, and Given Data:**  Audio-frequency range: $40\pi \leq \omega \leq 40,000\pi$ (20 to 20,000 Hz).

**Assumptions:**  The input resistance of the amplifier is in the range of 1 k$\Omega$.

**Analysis:**  If the input resistance of the amplifier is expected to be around 1 k$\Omega$, the impedance of a series capacitor to be used for AC coupling in the amplifier of Figure 10.9 should be significantly smaller in the frequency range of interest, say, 10 $\Omega$. The expression for the impedance of the capacitor is:

$$Z_C = \frac{1}{j\omega C}$$

thus, the capacitor impedance will be larger at the lower frequencies, and we should require the magnitude of the above impedance to be less than or equal to the desired value of 10 Ω at the frequency $\omega = 40\pi$. We therefore require that

$$|Z_C| = \frac{1}{\omega C} = 10 \qquad \text{at} \qquad \omega = 40\pi$$

or

$$C = \frac{1}{\omega |Z_C|} = \frac{1}{400\pi} = 2,190 \; \mu F$$

**Comments:** The coupling capacitor offers infinite resistance to DC currents (i.e., for $\omega = 0$). The capacitor value calculated in this example is fairly large; in practice it would be reasonable to select a somewhat smaller value. A standard value would be 470 $\mu F$.

---

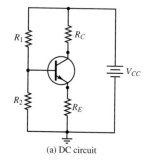

(a) DC circuit

The amplifier of Figure 10.7 employs a single DC supply, $V_{CC}$, as does the DC self-bias circuit of Example 9.6. The resistors $R_1$ and $R_2$, in effect, act as a voltage divider that provides a suitable bias for the BE junction. This effect is most readily understood if separate DC and AC equivalent circuits for the common-emitter amplifier are portrayed as in Figure 10.10.

To properly interpret the DC and AC equivalent circuits of Figure 10.10, a few comments are in order. Consider, first, the DC circuit. As far as DC currents are concerned, the two coupling capacitors and the emitter bypass capacitor are open circuits. Further, note that the supply voltage, $V_{CC}$, appears across two branches, the first consisting of the emitter and collector resistors and of the CE "junction," the second of the base resistors. This DC equivalent circuit is used in determining the $Q$ point of the amplifier—that is, the quantities $I_{BQ}$, $V_{CEQ}$, and $I_{CQ}$. In drawing the AC equivalent circuit, each of the capacitors has been replaced by a short circuit, as has the DC supply. The effect of the latter substitution (which applies only to AC signals) is to create a direct path to ground for the resistors $R_1$ and $R_C$. Thus, $R_1$ appears in parallel with $R_2$. Note, also, that in the AC equivalent circuit, the collector resistance $R_C$ appears in parallel with the load. This will have an important effect on the overall gain of the amplifier.

## DC Analysis of the Common-Emitter Amplifier

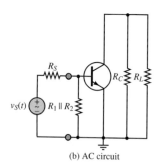

(b) AC circuit

**Figure 10.10** DC and AC circuits for the common-emitter amplifier

We redraw the DC circuit of Figure 10.10 in a slightly different form, recognizing that the Thévenin equivalent circuit seen by the base consists of the equivalent voltage

$$V_{BB} = \frac{R_2}{R_1 + R_2} V_{CC} \tag{10.14}$$

and of the equivalent resistance

$$R_B = R_1 \parallel R_2 \tag{10.15}$$

The resulting circuit is sketched in Figure 10.11. Application of KVL around the base and collector-emitter circuits yields the following equations, the solution of which determines the $Q$ point of the transistor amplifier:

$$V_{CEQ} = V_{CC} - I_{CQ}R_C - \frac{\beta + 1}{\beta}I_{CQ}R_E \qquad (10.16)$$

$$V_{BEQ} = V_{BB} - I_{BQ}R_B - \frac{\beta + 1}{\beta}I_{CQ}R_E \qquad (10.17)$$

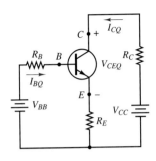

**Figure 10.11** DC bias circuit for the common-emitter amplifier

In these equations, the quiescent emitter current, $I_{EQ}$, has been expressed in terms of the collector current, $I_{CQ}$, according to the relation

$$I_{EQ} = \frac{\beta + 1}{\beta}I_{CQ} \qquad (10.18)$$

The next two examples illustrate a number of practical issues in the choice of DC bias point, and in the determination of other important features of the common-emitter transistor amplifier.

---

## EXAMPLE 10.3 Analysis of Common-Emitter Amplifier Operating Point

### Problem

Determine which of the two amplifiers designs, design A and design B, offers the better choice of operating point for the common-emitter amplifier of Figure 10.12, and explain why one is superior to the other.

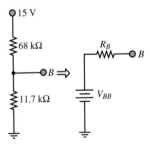

**Figure 10.12** Common-emitter amplifier for Example 10.3

---

### Solution

**Known Quantities:** Amplifier supply voltages; base, collector, and emitter resistances; transistor parameters.

**Find:** Quiescent values of $I_B$, $I_C$, and $V_{CE}$ for each design.

**Schematics, Diagrams, Circuits, and Given Data:** $V_\gamma = 0.7$ V; $\beta = 100$; $V_{CC} = 15$ V.
Design A: $R_1 = 68$ k$\Omega$; $R_2 = 11.7$ k$\Omega$; $R_C = 200$ $\Omega$; $R_E = 200$ $\Omega$.
Design B: $R_1 = 23.7$ k$\Omega$; $R_2 = 17.3$ k$\Omega$; $R_C = 200$ $\Omega$; $R_E = 200$ $\Omega$.

**Assumptions:** Use the linear small-signal $h$-parameter model of the BJT.

### Analysis:

1. *Design A.* The DC equivalent supply seen by the transistor is shown in Figure 10.13. We can compute the equivalent base supply and resistance as follows (also see Example 9.6):

$$V_{BB} = \frac{R_1}{R_1 + R_2}V_{CC} = \frac{11.7}{11.7 + 68}15 = 2.2 \text{ V}$$

and the equivalent base resistance from equation 9.7:

$$R_B = R_1 \| R_2 = \frac{68 \times 11.7}{68 + 11.7} \approx 10 \text{ k}\Omega$$

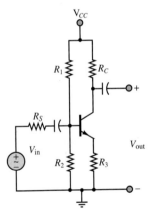

**Figure 10.13** Equivalent base supply circuit of design A

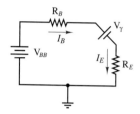

**Figure 10.14** Equivalent base-emitter circuit of design A

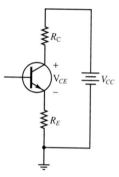

**Figure 10.15** Collector-emitter circuit of design A

Next, we consider the equivalent base-emitter circuit, shown in Figure 10.14. Applying KVL we compute the base and collector currents as follows:

$$V_{BB} = I_B R_B + V_\gamma + I_E R_E = I_B R_B + V_\gamma + (\beta + 1)I_B R_E$$

$$I_B = \frac{V_{BB} - V_\gamma}{R_B + (\beta + 1)R_E} = \frac{2.2 - 0.7}{10,000 + 101 \times 200} = 50\ \mu A$$

$$I_C = \beta I_B = 5\ mA$$

Next, we turn to the equivalent collector-emitter circuit (see Figure 10.15), and apply KVL to determine the collector-emitter voltage:

$$V_{CC} = I_C R_C + V_{CE} + I_E R_E = I_C R_C + V_{CE} + \frac{(\beta + 1)}{\beta} I_C R_E$$

$$V_{CE} = V_{CC} - I_C \left( R_C + \frac{(\beta + 1)}{\beta} R_E \right) = 15 - 5 \times 10^{-3} \left( 200 + \frac{101}{100} \times 200 \right)$$

$$= 13\ V$$

Thus, the $Q$ point for the amplifier of design A is:

$$\boxed{I_{BO} = 50\ \mu A \qquad I_{CO} = 5\ mA \qquad V_{CEO} = 13\ V}$$

2. *Design B.* We repeat the calculations of part 1 for the amplifier of design B:

$$V_{BB} = \frac{R_1}{R_1 + R_2} V_{CC} = \frac{17.3}{23.7 + 17.3} 15 = 6.33\ V$$

$$R_B = R_1 || R_2 = \frac{17.3 \times 23.7}{17.3 + 23.7} \approx 10\ k\Omega$$

$$I_B = \frac{V_{BB} - V_\gamma}{R_B + (\beta + 1)R_E} = \frac{6.33 - 0.7}{10,000 + 101 \times 200} = 186\ \mu A$$

$$I_C = \beta I_B = 18.6\ mA$$

$$V_{CE} = V_{CC} - I_C \left( R_C + \frac{(\beta + 1)}{\beta} R_E \right)$$

$$= 15 - 18.6 \times 10^{-3} \left( 200 + \frac{101}{100} \times 200 \right) = 7.5\ V$$

Thus, the $Q$ point for the amplifier of design B is:

$$\boxed{I_{BO} = 186\ \mu A \qquad I_{CO} = 18.6\ mA \qquad V_{CEO} = 7.5\ V}$$

To compare the two designs, we plot the load line in the $I_C - V_{CE}$ plane, and observe that the maximum collector current swing that can be achieved by design B is far greater than that permitted by design A (Figure 10.16). The reason is quite simply that the $Q$ point of design B is much closer to the center of the active region of the transistor, while the $Q$ point of design A will cause the transistor to move into the cutoff region if the collector current swing is to be more than a couple of milliamperes.

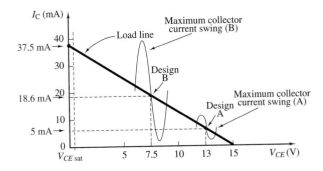

**Figure 10.16** Operating points for designs A and B

**Comments:**  A simple rule that can be gleaned from this example is that, for linear amplifier designs, it is desirable to place the $Q$ point near the center of the collector characteristic active region.

**Focus on Computer-Aided Tools:**  This example is available as an *Electronics Workbench*™ simulation in the accompanying CD-ROM. Both circuits are simulated, and you can verify that design A leads to cutoff distortion by applying an AC input signal larger than a few millivolts to the amplifier of design A. Another virtual experiment that you can run is to increase the amplitude of the input signal for the amplifier of design B and determine the maximum sinusoidal input voltage amplitude that can be amplified without distortion. What causes distortion first: cutoff or saturation?

---

## EXAMPLE 10.4 Compensating for Variation in $\beta$ in a Common-Emitter Amplifier

### Problem

Current gain variability from transistor to transistor is a practical problem that complicates amplifier design. In particular, it is desirable to obtain a stable operating point, relatively independent of variation in $\beta$ (which can be as much as $\pm 50$ percent). A common rule of thumb to reduce operating point variability is to require that

$$R_B = \frac{\beta_{\min} R_E}{10}$$

1. Find the operating point of the design B amplifier of Example 10.3 if $75 \leq \beta \leq 150$.
2. Using the above design rule, design a new amplifier (call it design C) with the same quiescent collector current as design B.
3. Demonstrate that the operating point of design C is more stable than that of design B.

---

### Solution

**Known Quantities:**  Amplifier supply voltages; base, collector, and emitter resistances; transistor parameters.

**Find:**

1. Quiescent values of $I_B$, $I_C$, and $V_{CE}$ for design B amplifier for extreme values of $\beta$.
2. Quiescent values of $I_B$, $I_C$, and $V_{CE}$ for design C amplifier.
3. Variation in $Q$ point of design C versus design B.

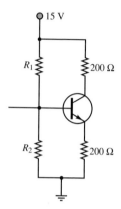

15 V

$R_1$     200 Ω

$R_2$     200 Ω

**Figure 10.17** DC
circuit for Example 10.4

**Schematics, Diagrams, Circuits, and Given Data:** $V_\gamma = 0.7$ V; $\beta_{min} = 75$; $\beta_{max} = 150$; $V_{CC} = 15$ V.

Design B (Figure 10.17): $R_1 = 23.7$ kΩ; $R_2 = 17.3$ kΩ; $R_C = 200$ Ω; $R_E = 200$ Ω.

**Analysis:**

1. *Design B.* We compute the $Q$ point for design B for each of the extreme values of $\beta$. The calculations are identical to those already carried out in Example 10.3. For $\beta = \beta_{min} = 75$:

$$I_B = \frac{V_{BB} - V_\gamma}{R_B + (\beta_{min} + 1)R_E} = \frac{6.33 - 0.7}{10{,}000 + 76 \times 200} = 233 \ \mu A$$

$$I_C = \beta_{min} I_B = 16.7 \text{ mA}$$

$$V_{CE} = V_{CC} - I_C \left( R_C + \frac{\beta_{min} + 1}{\beta_{min}} R_E \right)$$

$$= 15 - 16.7 \times 10^{-3} \left( 200 + \frac{76}{75} \times 200 \right) = 8.3 \text{ V}$$

Thus, the $Q$ point for $\beta = \beta_{min} = 75$ is:

$$I_{BO} = 233 \ \mu A \qquad I_{CO} = 16.7 \text{ mA} \qquad V_{CEO} = 8.3 \text{ V}$$

For $\beta = \beta_{max} = 150$:

$$I_B = \frac{V_{BB} - V_\gamma}{R_B + (\beta_{max} + 1)R_E} = \frac{6.33 - 0.7}{10{,}000 + 151 \times 200} = 140 \ \mu A$$

$$I_C = \beta_{max} I_B = 21 \text{ mA}$$

$$V_{CE} = V_{CC} - I_C \left( R_C + \frac{\beta_{max} + 1}{\beta_{max}} R_E \right)$$

$$= 15 - 21 \times 10^{-3} \left( 200 + \frac{151}{150} \times 200 \right) = 6.57 \text{ V}$$

Thus, the $Q$ point for $\beta = \beta_{max} = 150$ is:

$$I_{BO} = 140 \ \mu A \qquad I_{CO} = 21 \text{ mA} \qquad V_{CEO} = 6.57 \text{ V}$$

The change in quiescent base current, relative to the nominal value of design B (for $\beta = 100$) is 50 percent; the changes in quiescent collector current and collector-emitter voltage (relative to the same quantities for the $\beta = 100$ design) are around 23 percent. Figure 10.18 depicts the location of the two extreme $Q$ points.

2. *Design C.* Using the design rule stated in the problem statement we compute:

$$R_B = \frac{\beta_{min} R_E}{10} = \frac{75 \times 200}{10} = 1.5 \text{ kΩ}$$

$R_B = R_1 \| R_2$ is an equivalent resistance; further, the value of $V_{BB}$ depends on $R_1$ and $R_2$. Thus, we need to select these components in such a way as to satisfy the requirement that $I_C = 18.6$ mA. We write the base circuit equation in terms of the

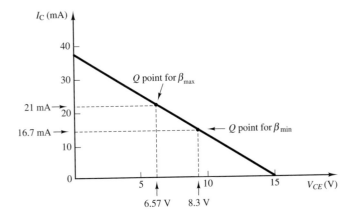

**Figure 10.18** Variability in $Q$ point due to change in $\beta$

collector current and compute the desired equivalent base supply voltage. Note that in the calculation below we have used the nominal design value of $\beta = 100$.

$$V_{BB} = I_B R_B + V_\gamma + I_E R_E = \frac{I_C}{\beta} R_B + V_\gamma + \frac{(\beta + 1)}{\beta} I_C R_E = 4.74 \text{ V}$$

Using the following relationships, we can calculate the values of the two resistors $R_1$ and $R_2$:

$$V_{BB} = \frac{R_1}{R_1 + R_2} V_{CC}$$

$$4.74 \, (R_1 + R_2) = 15 R_1$$

$$4.74 R_2 = 10.26 R_1$$

$$R_2 = 2.16 R_1$$

and

$$R_B = \frac{R_1 \times R_2}{R_1 + R_2} = 1,500 \, \Omega$$

$$1,500 = \frac{2.16 R_1^2}{3.16 R_1} = 0.68 R_1$$

$$R_1 = 2,194 \, \Omega \approx 2.2 \text{ k}\Omega$$

$$R_2 = 4,740 \, \Omega \approx 4.7 \text{ k}\Omega$$

Note that we have selected the closest 5 percent tolerance resistor standard values (see Table 2.1); resistor tolerance is another source of variability in transistor amplifier design. With the stated value we can now proceed to complete the $Q$ point determination for the nominal design C. The DC bias circuit is shown in Figure 10.19.

$$I_B = \frac{V_{BB} - V_\gamma}{R_B + (\beta + 1)R_E} = \frac{4.74 - 0.7}{1,500 + 101 \times 200} = 186 \, \mu\text{A}$$

$$I_C = \beta I_B = 18.6 \text{ mA}$$

$$V_{CE} = V_{CC} - I_C \left( R_C + \frac{(\beta + 1)}{\beta} R_E \right)$$

$$= 15 - 18.6 \times 10^{-3} \left( 200 + \frac{101}{100} \times 200 \right) = 7.5 \text{ V}$$

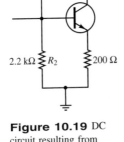

**Figure 10.19** DC circuit resulting from application of rule of thumb

Thus, the $Q$ point for the amplifier of design C is identical to that of design B:

$$I_{BO} = 186 \ \mu A \qquad I_{CO} = 18.6 \ \text{mA} \qquad V_{CEO} = 7.5 \ \text{V}$$

3. *Q point variability.* Now we are ready to determine the $Q$ point variability for design C amplifier. For $\beta = \beta_{\min} = 75$:

$$I_B = \frac{V_{BB} - V_\gamma}{R_B + (\beta_{\min} + 1)R_E} = \frac{4.74 - 0.7}{1,500 + 76 \times 200} = 242 \ \mu A$$

$$I_C = \beta_{\min} I_B = 18.1 \ \text{mA}$$

$$V_{CE} = V_{CC} - I_C \left( R_C + \frac{\beta_{\min} + 1}{\beta_{\min}} R_E \right)$$

$$= 15 - 18.1 \times 10^{-3} \left( 200 + \frac{76}{75} \times 200 \right) = 7.7 \ \text{V}$$

Thus, the $Q$ point for $\beta = \beta_{\min} = 75$ is:

$$I_{BO} = 242 \ \mu A \qquad I_{CO} = 18.1 \ \text{mA} \qquad V_{CEO} = 7.7 \ \text{V}$$

For $\beta = \beta_{\max} = 150$:

$$I_B = \frac{V_{BB} - V_\gamma}{R_B + (\beta_{\max} + 1)R_E} = \frac{6.33 - 0.7}{10,000 + 151 \times 200} = 127 \ \mu A$$

$$I_C = \beta_{\max} I_B = 19.1 \ \text{mA}$$

$$V_{CE} = V_{CC} - I_C \left( R_C + \frac{(\beta_{\max} + 1)}{\beta_{\max}} R_E \right)$$

$$= 15 - 19.1 \times 10^{-3} \left( 200 + \frac{151}{150} \times 200 \right) = 7.3 \ \text{V}$$

Thus, the $Q$ point for $\beta = \beta_{\max} = 150$ is:

$$I_{BO} = 127 \ \mu A \qquad I_{CO} = 19.1 \ \text{mA} \qquad V_{CEO} = 7.3 \ \text{V}$$

The change in quiescent base current, relative to the nominal value of design C (for $\beta = 100$) has actually increased to 62 percent; however, the changes in quiescent collector current and collector-emitter voltage (relative to the same quantities for the $\beta = 100$ design) have decreased, to approximately 5 percent. This is a substantial improvement (nearly by a factor of 5!). You may wish to approximately locate the new $Q$ points on the load line of Figure 10.18.

**Comments:** This example may have been repetitive, but it presents some very important points about the importance of biasing and the effect of component variability in transistor amplifier design.

## Check Your Understanding

**10.2**  Find the operating point of the circuit of design B in Example 10.3 if (a) $\beta = 90$ and (b) $\beta = 120$.

**10.3**  Verify that the operating point in design 2 of Example 10.4 is more stable than that of design 1.

**10.4**  Repeat all parts of Example 10.4 if $\beta = \beta_{\min} = 60$, $\beta = \beta_{\max} = 200$, and $I_C = 10$ mA.

## AC Analysis of the Common-Emitter Amplifier

To analyze the AC circuit of the common-emitter amplifier, we substitute the hybrid-parameter small-signal model of Figure 10.3 in the AC circuit of Figure 10.10, obtaining the linear AC equivalent circuit of Figure 10.20. Note how the emitter resistance is bypassed by the emitter capacitor at AC frequencies; this, in turn, implies that the base-to-emitter (*BE*) junction appears in parallel with the equivalent resistance $R_B = R_1 \parallel R_2$. Similarly, the collector-to-emitter (*CE*) junction is replaced by the parallel combination of the controlled current source, $h_{fe} \Delta I_B$, with the resistance $1/h_{oe}$. Once again, since the DC supply provides a direct path to ground for the AC currents, the collector resistance appears in parallel with the load. It is important to understand why we have defined input and output voltages, $v_{in}$ and $v_{out}$, rather than use the complete circuit containing the source, $v_S$, its internal resistance, $R_S$, and the load resistance, $R_L$. The AC circuit model is most useful if an equivalent input resistance and the equivalent output resistance and amplifier gain are defined as quantities independent of the signal source and load properties. This approach permits the computation of the parameters $r_i$, $r_o$, and $\mu$ shown in Figure 10.8, and therefore provides a circuit model that may be called upon for any source-and-load configuration. The equivalent circuit shown in Figure 10.20 allows viewing the transistor amplifier as a single equivalent circuit either from the source or from the load end, as will presently be illustrated.

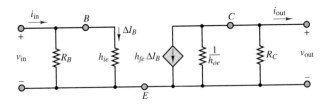

**Figure 10.20** AC equivalent-circuit model for the common-emitter amplifier

First, compute the AC input current,

$$i_{\text{in}} = \frac{v_{\text{in}}}{h_{ie} \parallel R_B} \tag{10.19}$$

and the AC base current,

$$\Delta I_B = \frac{v_{\text{in}}}{h_{ie}} \tag{10.20}$$

noting, further, that the input resistance of the circuit consists of the parallel combination of $h_{ie}$ and $R_B$:

$$r_i = h_{ie} \parallel R_B \tag{10.21}$$

Next, observe that the AC base current is amplified by a factor of $h_{fe}$ and that it flows through the parallel combination of the collector resistance, $R_C$, and the *CE* junction resistance, $1/h_{oe}$. The latter term is often, but not always, negligible with respect to load and collector resistances, since the slope of the collector $i$-$v$ curves is very shallow. Thus, the AC short-circuit output current $i_{\text{out}}$ is given by the expression

$$i_{\text{out}} = -h_{fe}\,\Delta I_B = -h_{fe}\frac{v_{\text{in}}}{h_{ie}} = -g_m v_{\text{in}} \tag{10.22}$$

and the AC open-circuit output voltage is given by the expression

$$v_{\text{out}} = -h_{fe}\,\Delta I_B\,R_C \parallel \frac{1}{h_{oe}} = -h_{fe}\frac{v_{\text{in}}}{h_{ie}}R_C \parallel \frac{1}{h_{oe}}$$
$$= -g_m v_{\text{in}} R_C \parallel \frac{1}{h_{oe}} \tag{10.23}$$

Note the negative sign, due to the direction of the controlled current source. This sign reversal is a typical characteristic of the common-emitter amplifier. Knowing the open-circuit output voltage and the short-circuit output current, we can find the output resistance of the amplifier as the ratio of these two quantities:

$$r_o = \frac{v_{\text{out}}}{i_{\text{out}}} = R_C \parallel \frac{1}{h_{oe}} \tag{10.24}$$

Next, if we define the **AC open-circuit voltage gain** of the amplifier, $\mu$, by the expression

$$\mu = \frac{v_{\text{out}}}{v_{\text{in}}} \tag{10.25}$$

we find that

$$\mu = -h_{fe}\frac{R_C \parallel \frac{1}{h_{oe}}}{h_{ie}} = -g_m R_C \parallel \frac{1}{h_{oe}} \tag{10.26}$$

At this point, it is possible to take advantage of the equivalent-circuit representation of Figure 10.8, with the expressions for $\mu$, $r_o$, and $r_i$ just given. Figure 10.21 illustrates the equivalent circuit for the common-emitter amplifier. This circuit replaces the transistor amplifier and enables us to calculate the actual voltage and current gain of the amplifier for any given load-and-source pair. Referring to

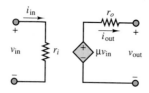

**Figure 10.21** Simplified equivalent circuit for the common-emitter amplifier

the circuit of Figure 10.8 and to equations 10.12 and 10.13, we can compute these gains to be

$$A_V = \frac{v_L}{v_S} = -\mu \frac{R_L}{r_o + R_L} \frac{r_i}{R_S + r_i}$$

$$= -h_{fe} \frac{R_C \| h_{oe}}{h_{ie}} \frac{R_L}{r_o + R_L} \frac{r_i}{R_S + r_i} \qquad \textbf{(10.27)}$$

$$= -g_m \frac{R_L}{r_o + R_L} \frac{r_i}{R_S + r_i}$$

and

$$A_I = \frac{i_{\text{out}}}{i_{\text{in}}} = \frac{v_{\text{out}}/R_L}{v_S/(r_i + R_S)} = \frac{r_i + R_S}{R_L} A_V \qquad \textbf{(10.28)}$$

Example 10.5 provides numerical values for a typical small-signal common-emitter amplifier.

---

## EXAMPLE 10.5 Common-Emitter Amplifier Analysis

### Problem

Compute the input and output resistances and the voltage, current and power gains of the common-emitter amplifier of Figure 10.9.

---

### Solution

**Known Quantities:** Amplifier supply voltages; base, collector, and emitter resistances; source and load resistances; transistor parameters.

**Find:** $r_i$; $r_o$; $\mu$;

$$A_V = \frac{v_{\text{out}}}{v_S} \qquad A_I = \frac{i_{\text{out}}}{i_S} \qquad A_P = \frac{P_{\text{out}}}{P_S}$$

**Schematics, Diagrams, Circuits, and Given Data:** $h_{ie} = 1,400\ \Omega$; $h_{fe} = 100$; $h_{oe} = 125\ \mu\text{S}$. $R_1 = 20\ \text{k}\Omega$; $R_2 = 5\ \text{k}\Omega$; $R_C = 4\ \text{k}\Omega$; $R_E = 1\ \text{k}\Omega$; $R_L = 500\ \Omega$; $R_S = 50\ \Omega$.

**Assumptions:** A suitable $Q$ point has already been established. The coupling capacitors have appropriately been selected to separate the AC circuit from the DC bias circuit.

**Analysis:** We replace the BJT in the circuit of Figure 10.9 with the BJT $h$-parameter circuit model of Figure 10.20. The resulting AC equivalent circuit is shown in Figure 10.22, where $R_B = R_1 \| R_2 = 4\ \text{k}\Omega$.

The input resistance is given by the parallel combination of $h_{ie}$ and $R_B$, as stated in equation 10.21:

$$r_i = h_{ie} \| R_B = 4,000 \| 1,400 = 1.04\ \text{k}\Omega$$

To determine the output resistance, we short circuit the voltage source, $v_S$, leading to $v_{\text{in}} = 0$ as well. Thus,

$$r_o = R_C \| \frac{1}{h_{oe}} = \frac{4 \times 8}{4 + 8} = 2.67\ \text{k}\Omega$$

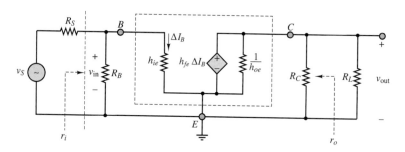

**Figure 10.22**

The open-circuit voltage gain of the amplifier is found by using equation 10.26:

$$\mu = -h_{fe}\frac{R_C}{h_{ie}} = -100 \times \frac{4,000}{1,400} = -286\;\frac{V}{V}$$

With the above three parameters in place, we can build the equivalent AC circuit model of the amplifier shown in Figure 10.23. Then we can use equations 10.27 and 10.28 to compute the actual amplifier gains:

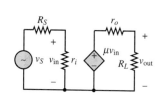

**Figure 10.23**

$$A_V = \frac{v_{\text{out}}}{v_S} = -\mu\frac{R_L}{r_o + R_L}\frac{r_i}{R_S + r_i} = -286\frac{0.5}{2.67 + 0.5}\frac{1.04}{0.05 + 1.04} = -43\;\frac{V}{V}$$

$$A_I = \frac{i_{\text{out}}}{i_{\text{in}}} = \frac{\frac{v_{\text{out}}}{R_L}}{\frac{v_S}{R_S+r_i}} = \frac{R_S + r_i}{R_L}A_V = -94\;\frac{A}{A}$$

$$A_P = \frac{P_{\text{out}}}{P_S} = \frac{i_{\text{out}}v_{\text{out}}}{i_{in}v_S} = A_I A_V = 4,042\;\frac{W}{W}$$

**Comments:** Although the open-loop voltage gain, $\mu$, is very large, the actual gain of the amplifier is significantly smaller. You may have already noted that it is the small value of load resistance as compared to the AC output resistance of the amplifier that causes $A_V$ to be smaller than $\mu$. The current gain is actually quite large because of the relatively large input resistance of the amplifier. A word of warning is appropriate with regard to the power gain. While such a large power gain may seem a desirable feature, one always has to contend with the allowable power dissipation of the transistor. For example, the data sheets for the 2N3904 general purpose *npn* transistor show a maximum power dissipation of 350 mW at room temperature for the TO-236 package, and of 600 mW for the TO-92 package. Thus, a power gain of 4,000 can be meaningful only if the input signal power is very low (less than 150 $\mu$W), or else the transistor will exceed its thermal rating and will be destroyed.

**VIRTUAL LAB**

***Focus on Computer-Aided Solutions:*** An *Electronics Workbench*™ simulation of the circuit illustrated in this example has been included in the accompanying CD-ROM. You may wish to compare run the simulation using the "ideal" transistor model, as well as the models for the 2N3904 and 2N2222 transistors, to observe the difference in voltage gains.

## Check Your Understanding

**10.5** Compute the actual voltage gain, $v_{\text{out}}/v_S$, of the amplifier of Example 10.5 for the following source-load pairs:

a. $R_S = 50\;\Omega$, $R_L = 150\;\Omega$

b.  $R_S = 50\ \Omega,\ R_L = 1,500\ \Omega$

c.  $R_S = 500\ \Omega,\ R_L = 150\ \Omega$

What conclusions can you draw from these results?

**10.6**  Calculate the current gains for the amplifier parameters of Check Your Understanding 10.5.

**10.7**  Repeat Example 10.5 for $h_{ie} = 2\ k\Omega$ and $h_{fe} = 60$.

---

## Other BJT Amplifier Circuits

The common-emitter amplifier is a commonly employed configuration but is by no means the only type of BJT amplifier. Other amplifier configurations are also used, depending on the specific application requirements. Each type of amplifier is classified in terms of properties such as input and output resistance, and voltage and current gain. Rather than duplicate the detailed analysis just conducted for the common-emitter amplifier, we summarize the properties of the three more common BJT amplifier circuits in Table 10.2, which depicts the amplifier circuits and summarizes their properties. The methodology employed to derive these results is completely analogous to that surveyed in the previous section. Examples and analysis of these amplifiers may be found in the homework problems.

Table 10.2  BJT amplifier configurations

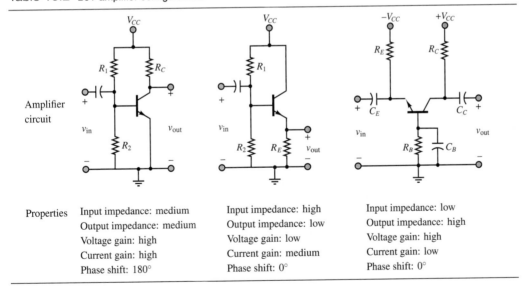

| | Amplifier circuit | | |
|---|---|---|---|
| Properties | Input impedance: medium<br>Output impedance: medium<br>Voltage gain: high<br>Current gain: high<br>Phase shift: 180° | Input impedance: high<br>Output impedance: low<br>Voltage gain: low<br>Current gain: medium<br>Phase shift: 0° | Input impedance: low<br>Output impedance: high<br>Voltage gain: high<br>Current gain: low<br>Phase shift: 0° |

## 10.3    FET SMALL-SIGNAL AMPLIFIERS

The discussion of FETs as amplifiers is analogous to that of BJT amplifiers. In particular, the **common-source amplifier** circuit is equivalent in structure to the common-emitter amplifier circuit studied earlier, and the **common-drain amplifier (source follower)** is analogous to the common-collector amplifier (emitter follower). In this section, we discuss the general features of FET amplifiers; to simplify the discussion, we have selected the $n$-channel enhancement MOSFET

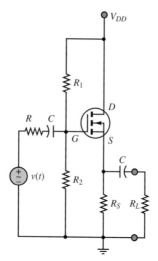

(a) Common-drain amplifier

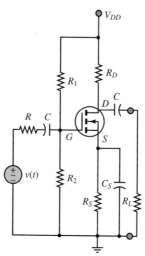

(b) Common-source amplifier

**Figure 10.24** Typical common-drain and common-source MOSFET amplifiers

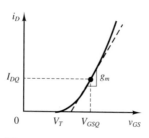

**Figure 10.25** MOSFET transconductance parameter

to represent the FET family. Although some of the details differ depending on the specific device, the discussion that follows applies in general to all FET amplifiers. A summary of FET symbols was given in Figure 9.30 in Chapter 9; reviewing it will help you recognize a specific device in a circuit diagram.

Figure 10.24 depicts typical common-drain and common-source amplifiers, including coupling and bypass capacitors. One of the great features of FETs, and especially MOSFETs, is the high input impedance that can be achieved because the gate is effectively insulated from the substrate material. We shall illustrate this property in analyzing the source-follower circuit. Before proceeding with the analysis of FET amplifiers, though, we shall discuss how one can construct a small-signal model analogous to the one that was obtained for the BJT.

In the case of MOSFETs, we can make use of the analytic relation between drain current and gate-source voltage,

$$i_D = k(v_{GS} - V_T)^2 \qquad k = I_{DSS}/V_T^2 \tag{10.29}$$

to establish the $Q$ point for the transistor, which is defined by the quiescent voltages, $V_{GSQ}$ and $V_{DSQ}$, and by the quiescent current, $I_{DQ}$. The quadratic relationship allows us to determine the drain current, $I_{DQ}$, that will flow, given that $v_{GS} = V_{GSQ}$; the parameters $k$ and $V_T$ are a property of any given device. Thus,

$$I_{DQ} = k(V_{GSQ} - V_T)^2 \tag{10.30}$$

and it is now possible to determine the quiescent drain-to-source voltage from the load-line equation for the drain circuit:

$$V_{DD} = V_{DSQ} + R_D I_{DQ} \tag{10.31}$$

In a MOSFET, it is possible to approximate the small-signal behavior of the device as a linear relationship by using the transconductance parameter $g_m$, where $g_m$ is the slope of the $i_D$-$v_{GS}$ curve at the $Q$ point, as shown in Figure 10.25. Formally, we can write

$$g_m = \left.\frac{\partial i_D}{\partial v_{GS}}\right|_{I_{DQ}, V_{GSQ}} \tag{10.32}$$

Since an analytical expression for the drain current is known (equation 10.26), we can actually determine $g_m$ as given by

$$g_m = \frac{\partial}{\partial v_{GS}}[k(v_{GS} - V_T)^2] \tag{10.33}$$

at the operating point $(I_{DQ}, V_{CSQ})$. Thus, evaluating the expression for the transconductance parameter, we obtain the expression

$$g_m = 2k\,(v_{GS} - V_T)|_{I_{DQ}, V_{GSQ}}$$

$$= 2\sqrt{kI_{DQ}} = 2\frac{\sqrt{I_{DSS}I_{DQ}}}{V_T} \tag{10.34}$$

where $g_m$ is a function of the quiescent drain current, as expected, since the tangent to the $i_D$-$v_{GS}$ curve of Figure 10.25 has a slope that is dependent on the value of $I_{DQ}$. The next two examples illustrate the calculation and significance of the transconductor parameter in MOSFETs.

## EXAMPLE 10.6 MOSFET Transconductance Calculation

### Problem

Compute the value of the transconductance parameter for a MOSFET.

### Solution

**Known Quantities:** MOSFET quiescent voltage and current values; MOSFET parameters.

**Find:** $g_m$.

**Schematics, Diagrams, Circuits, and Given Data:** $V_{DSQ} = 3.5$ V; $V_{GSQ} = 2.4$ V; $V_T = 1.0$ V; $k = 0.125$ mA/V$^2$.

**Assumptions:** Use the MOSFET model of equation 10.29.

**Analysis:** First, we compute the quiescent drain current:

$$I_{DQ} = k \left(V_{GSQ} - V_T\right)^2 = 0.125 \, (2.4 - 1)^2 = 0.245 \text{ mA}$$

Next, we evaluate the transconductance parameter.

$$g_m = 2\sqrt{kI_{DQ}} = 2\sqrt{0.125 \times 0.245 \times 10^{-3}} = 0.35 \, \frac{\text{mA}}{\text{V}}$$

**Comments:** The transconductance parameter tells us that for every volt increase in gate-source voltage, the drain current will increase by 0.35 mA. The MOSFET clearly acts as a *voltage-controlled current source*.

## EXAMPLE 10.7 Analysis of MOSFET Amplifier

### Problem

Determine the gate and drain-source voltage and the drain current for the MOSFET amplifier of Figure 10.26.

### Solution

**Known Quantities:** Drain, source, and gate resistors; drain supply voltage; MOSFET parameters.

**Find:** $v_{GS}$; $v_{DS}$; $i_D$.

**Schematics, Diagrams, Circuits, and Given Data:** $R_1 = R_2 = 1 \text{ M}\Omega$; $R_D = 6 \text{ k}\Omega$; $R_S = 6 \text{ k}\Omega$; $V_{DD} = 10$ V. $V_T = 1$ V; $k = 0.5$ mA/V$^2$.

**Assumptions:** The MOSFET is operating in the saturation region. All currents are expressed in mA and all resistors in k$\Omega$.

**Analysis:** The gate voltage is computed by applying the voltage divider rule between resistors $R_1$ and $R_2$ (remember that no current flows into the transistor):

$$v_G = \frac{R_2}{R_1 + R_2} V_{DD} = \frac{1}{2} V_{DD} = 5 \text{ V}$$

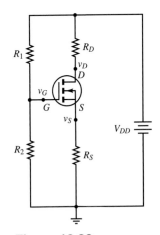

**Figure 10.26**

Assuming saturation region operation, we write:

$$v_{GS} = v_G - v_S = v_G - R_S i_D = 5 - 6i_D$$

The drain current can be computed from equation 10.29:

$$i_D = k(v_{GS} - V_T)^2 = 0.5(5 - 6i_D - 1)^2$$

leading to

$$36i_D^2 - 50i_D + 16 = 0$$

with solutions:

$$i_D = 0.89 \text{ mA and } i_D = 0.5 \text{ mA}$$

To determine which of the two solutions should be chosen, we compute the gate-source voltage for each. For $i_D = 0.89$ mA, $v_{GS} = 5 - 6i_D = -0.34$ V. For $i_D = 0.5$ mA, $v_{GS} = 5 - 6i_D = 2$ V. Since $v_{GS}$ must be greater than $V_T$ for the MOSFET to be in the saturation region, we select the solution:

$$i_D = 0.5 \text{ mA} \qquad v_{GS} = 2 \text{ V}$$

The corresponding drain voltage is therefore found to be:

$$v_D = v_{DD} - R_D i_D = 10 - 6i_D = 7 \text{ V}$$

And therefore

$$v_{DS} = v_D - v_S = 7 - 3 = 4 \text{ V}$$

**Comments:** Now that we have computed the desired voltages and current, we can verify that the condition for operation in the saturation region is indeed satisfied: $v_{DS} > v_{GS} - V_T$ leads to $4 > 2 - 1$; since the inequality is satisfied, the MOSFET is indeed operating in the saturation region.

---

The transconductance parameter allows us to define a very simple model for the MOSFET during small-signal operation: we replace the input circuit (gate) by an open circuit, since no current can flow into the insulated gate, and model the drain-to-source circuit by a controlled current source, $g_m \Delta V_{GS}$. This small-signal model is depicted in Figure 10.27.

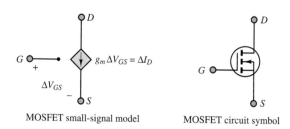

MOSFET small-signal model              MOSFET circuit symbol

**Figure 10.27** MOSFET small-signal model

It is important to appreciate the fact that the transconductance, $g_m$, is dependent on the quiescent value of the drain current, and therefore any MOSFET amplifier design is going to be very strongly dependent on the operating point.

## The MOSFET Common-Source Amplifier

It is useful at this stage to compare the performance of the common-source amplifier of Figure 10.24(b) with that of the BJT common-emitter amplifier. The DC
equivalent circuit is shown in Figure 10.28; note the remarkable similarity with
the BJT common-emitter amplifier DC circuit.

The equations for the DC equivalent circuit are obtained most easily by
reducing the gate circuit to a Thévenin equivalent with $V_{GG} = R_2/R_1 + R_2 V_{DD}$
and $R_G = R_1 \parallel R_2$. Then the gate circuit is described by the equation

$$V_{GG} = V_{GSQ} + I_{DQ} R_S \tag{10.35}$$

and the drain circuit by

$$V_{DD} = V_{DSQ} + I_{DQ}(R_D + R_S) \tag{10.36}$$

Note that, given $V_{DD}$, any value for $V_{GG}$ may be achieved by appropriate selection
of $R_1$ and $R_2$. Example 10.8 illustrates the computation of the $Q$ point for a
MOSFET amplifier.

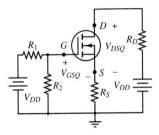

**Figure 10.28** DC circuit for
the common-source amplifier

## EXAMPLE 10.8 Analysis of MOSFET Common-Source Amplifier

### Problem

Design a common-source MOSFET amplifier (Figure 10.24(b)) to operate at a specified
$Q$ point.

### Solution

**Known Quantities:** Drain supply voltage; MOSFET threshold voltage and $k$; desired
gate-source and drain-source voltages.

**Find:** $R_1, R_2, R_D, R_S$.

**Schematics, Diagrams, Circuits, and Given Data:** $V_{GSQ} = 2.4$ V; $V_{DSQ} = 4.5$ V;
$V_{DD} = 10$ V. $V_T = 1.4$ V; $k = 95$ mA/V$^2$.

**Assumptions:** All currents are expressed in mA and all resistors in k$\Omega$.

**Analysis:** First, we compute the quiescent drain current for operation in the saturation
region; we know that the MOSFET is operating in the saturation region from the
equations of Table 9.1, since

$$v_{DS} > v_{GS} - V_T \text{ and } v_{GS} > V_T.$$
$$I_{DQ} = k \left(V_{GSQ} - V_T\right)^2 = 95 \times (2.4 - 1.4)^2 = 95 \text{ mA}$$

Applying KVL around the gate loop requires that:

$$V_{GG} = V_{GSQ} + I_{DQ} R_S = 2.4 - 95 R_S$$

while the drain circuit imposes the condition:

$$V_{DD} = V_{DSQ} + I_{DQ}(R_D + R_S) = 4.5 + 95(R_D + R_S)$$

or

$$10 = 4.5 + 95(R_D + R_S)$$
$$(R_D + R_S) = 0.058 \text{ k}\Omega$$

Since the choice of drain and source resistors is arbitrary at this point, we select $R_D = 11\ \Omega$ and $R_S = 47\ \Omega$. Then we can solve for $V_{GG}$ in the first equation:

$$V_{GG} = 6.865\ \text{V}$$

To achieve the desired value of $V_{GG}$ we need to select resistors $R_1$ and $R_2$ such that

$$V_{GG} = \frac{R_2}{R_1 + R_2} V_{DD} = 6.865\ \text{V}$$

To formulate a problem with a unique solution, we can also (arbitrarily) impose the condition that $R_1 \| R_2 = 100\ \text{k}\Omega$. Then we can solve the two equations to obtain $R_1 = 319\ \text{k}\Omega$, $R_2 = 146\ \text{k}\Omega$. The nearest standard 5 percent resistor values will be: $R_1 = 333\ \text{k}\Omega$, $R_2 = 150\ \text{k}\Omega$, resulting in $R_1 \| R_2 = 103.4\ \text{k}\Omega$.

**Comments:** Since the role of the resistors $R_1$ and $R_2$ is simply to serve as a voltage divider, you should not be overly concerned with the arbitrariness of the choice made in the example. In general, one selects rather large values to reduce current flow and therefore power consumption. On the other hand, the choice of the drain and source resistances may be more delicate, as it affects the output resistance and gain of the amplifier.

**Focus on Computer-Aided Solutions:** The calculations carried out in the present example may also be found in electronic form in a Mathcad™ file in the accompanying CD-ROM.

VIRTUAL LAB

Substituting the small-signal model in the common-source amplifier circuit of Figure 10.24(b), we obtain the small-signal AC equivalent circuit of Figure 10.29, where we have assumed the coupling and bypass capacitors to be short circuits at the frequency of the input signal $v(t)$. The circuit is analyzed as follows. The load voltage, $v_L(t)$, is given by the expression

$$v_L(t) = -\Delta I_D \cdot (R_D \| R_L) \qquad \text{(10.37)}$$

where

$$\Delta I_D = g_m \Delta V_{GS} \qquad \text{(10.38)}$$

Thus, we need to determine $\Delta V_{GS}$ to write expressions for the voltage and current gains of the amplifier. Since the gate circuit is equivalent to an open circuit, we

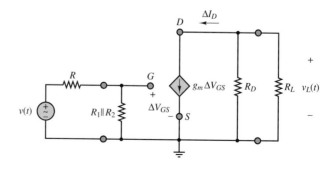

**Figure 10.29** AC circuit for the common-source MOSFET amplifier

have:

$$\Delta V_{GS}(t) = \frac{(R_1 \parallel R_2)}{R + (R_1 \parallel R_2)} v(t) \qquad \textbf{(10.39)}$$

and

$$\Delta V_{GS}(t) \approx v(t) \qquad \textbf{(10.40)}$$

because we have purposely selected $R_1$ and $R_2$ to be quite large in the design of the DC bias circuit and therefore $R_1 \parallel R_2 \gg R$. Thus, we can express the load (output) voltage in terms of the input voltage $v(t)$ as

$$v_L(t) = -g_m \cdot (R_D \parallel R_L) \cdot v(t) \qquad \textbf{(10.41)}$$

This expression corresponds to a voltage gain of

$$A_V = \frac{v_L(t)}{v(t)} = -g_m(R_D \parallel R_L) \qquad \textbf{(10.42)}$$

Note that we have an upper bound on the voltage gain for this amplifier, $A_{V_{max}}$, that is independent of the value of the load resistance:

$$|A_{V_{max}}| \le g_m R_D \qquad \textbf{(10.43)}$$

This upper bound is achieved, of course, only when the load is an open circuit. This open-circuit voltage gain plays the same role in the analysis of the MOSFET amplifier as the parameter $\mu$ we defined in equation 10.26 for BJT amplifiers:

$$\mu = -g_m R_D \qquad \textbf{(10.44)}$$

If we compute the value of $g_m$ for the MOSFET amplifier of Example 10.8 ($g_m = 0.1033$ A/V), we find that the open-loop voltage gain of this amplifier is $\mu = -g_m R_D = -10.33$. If we computed a typical value of transconductance, a comparable BJT design could yield a voltage gain around $-200$. A further disadvantage of the MOSFET amplifier is, of course, the much more nonlinear drain characteristic. Why then use MOSFETs as amplifiers if a BJT can supply both higher gain and improved linearity? The answer lies in the large input impedance of FET amplifiers (theoretically infinite in MOSFETs).

In the MOSFET common-source amplifier, the current drawn by the input is given by the expression

$$i_{in} = \frac{v(t)}{R + R_G} \approx \frac{v(t)}{R_G} \qquad \textbf{(10.45)}$$

while the output current is given by

$$i_L(t) = \frac{v_L(t)}{R_L} = \frac{A_V \cdot v(t)}{R_L} \qquad \textbf{(10.46)}$$

Thus, the current gain is given by the following expression:

$$A_I = \frac{i_L}{i_{in}} = \frac{A_V \cdot R_G}{R_L} = -g_m \frac{R_D R_L}{R_D + R_L} \frac{R_G}{R_L}$$

$$= -g_m \frac{R_D R_G}{R_D + R_L} \qquad \textbf{(10.47)}$$

Note that we have purposely made $R_G$ large (100 k$\Omega$), to limit the current required of the signal source, $v(t)$. Thus, the current gain for a common-source amplifier can be significant. For a 100-$\Omega$ load, we find that the effective voltage gain for the common-source amplifier is

$$A_V = -g_m R_D \parallel R_L \approx -5$$

and the corresponding current gain is

$$A_I = -g_m \frac{R_D R_G}{R_D + R_L} = -5{,}000$$

for the design values quoted in Example 10.8. Thus, the net **power gain** of the transistor, $A_P$, is quite significant.

$$A_P = A_V \cdot A_I = 25{,}000$$

This brief discussion of the common-source amplifier has pointed to some important features of MOSFETs. We first noted that the transconductance of a MOSFET is highly variable. However, MOSFETs make up for this drawback by their inherently high input impedance, which requires very little current of a signal source and thus affords substantial current gains. Finally, the output resistance of this amplifier can be made quite small by design. Example 10.9 provides an illustration of the amplification characteristics of a power MOSFET.

---

### EXAMPLE 10.9 Analysis of a Power MOSFET Common-Source Amplifier

**Problem**

Design a common-source power MOSFET amplifier (Figure 10.24(b)) using a BS170 transistor to operate at a specified $Q$ point. Compute the voltage and current gain of the amplifier.

---

**Solution**

**Known Quantities:** Drain supply voltage; load resistance; MOSFET transconductance and threshold voltage; desired drain current and drain-source voltage.

**Find:** $R_1$, $R_2$, $R_D$, $R_S$, $A_V$, and $A_I$.

**Schematics, Diagrams, Circuits, and Given Data:** $V_{DD} = 25$ V; $R_L = 80$ $\Omega$; $g_m = 200$ mA/V for $V_{DS} = 10$ V and $I_D = 250$ mA; $V_T = 0.8$ V. The device data sheets for the BS170 transistor may be found in the accompanying CD-ROM.

**Assumptions:** All currents are expressed in mA and all resistors in k$\Omega$.

**Analysis:** Knowing the transconductance and threshold voltage of the MOSFET allows us to compute the values of $k$ and $I_{DSS}$ (see equations in Table 9.1 and in the following paragraph):

$$k = \frac{1}{4} \frac{\left( g_m \frac{\text{mA}}{\text{V}} \right)^2}{I_{DQ} \text{ mA}} = \frac{1}{4} \frac{200^2}{250} = 40 \frac{\text{mA}}{\text{V}^2}$$

$$I_{DSS} = k \times V_T^2 = 40\,\frac{mA}{V^2} \times (0.8)^2\ V^2 = 25.6\ mA$$

Next, we calculate $V_{GSQ}$ using the equation for operation in the saturation region in Table 9.1:

$$V_{GSQ} = \sqrt{V_T^2 \times \frac{I_{DQ}}{I_{DSS}}} + V_T = 3.3\ V$$

Since the conditions for saturation region operation are satisfied ($v_{DS} > v_{GS} - V_T$; $v_{GS} > V_T$), we can use the above result to calculate the desired resistances. Applying KVL around the gate loop requires that:

$$V_{GG} = V_{GSQ} + I_{DQ}R_S = 3.3 + 250R_S$$

while the drain circuit imposes the condition:

$$V_{DD} = V_{DSQ} + I_{DQ}(R_D + R_S) = 10 + 250(R_D + R_S)$$

or

$$25 = 10 + 250(R_D + R_S)$$

$$(R_D + R_S) = 0.06\ k\Omega$$

We select $R_D = 22\ \Omega$ and $R_S = 39\ \Omega$ (note that these are both standard 5 percent resistor values—they add up to 61 $\Omega$, a small error). Then we can solve for $V_{GG}$ in the first equation: $V_{GG} = 10\ V$.

To achieve the desired value of $V_{GG}$ we need to select resistors $R_1$ and $R_2$ such that

$$V_{GG} = \frac{R_2}{R_1 + R_2}V_{DD} = 13.05\ V$$

To formulate a problem with a unique solution, we can also (arbitrarily) impose the condition that $R_G = R_1||R_2 = 100\ k\Omega$. Then we can solve the two equations to obtain: $R_1 = 196\ k\Omega$; $R_2 = 209\ k\Omega$. The nearest standard 5 percent resistor values are 180 k$\Omega$ and 220 k$\Omega$, leading to an equivalent resistance $R_G = R_1||R_2 = 99\ k\Omega$.

Now we can compute the amplifier voltage and current gains from equations 10.42 and 10.47:

$$A_V = -g_m(R_D||R_L) = -200(0.022||0.08) \approx -3.5\,\frac{V}{V}$$

$$A_I = -g_m\left(\frac{R_D R_G}{R_D + R_L}\right) = -200\left(\frac{0.022 \times 99}{0.022 + 0.08}\right) \approx -4300\,\frac{A}{A}$$

**Comments:** Note that we assumed saturation region operation in calculating the quiescent gate-source voltage. If the resulting calculation had yielded a value of $V_{GSQ}$ that did not match the conditions for active region operation (see Table 9.1), we would have had to recompute $V_{GSQ}$ using the ohmic region equations.

**Focus on Computer-Aided Solutions:** The calculations carried out in the present example may also be found in electronic form in a Mathcad™ file in the accompanying CD-ROM.

## The MOSFET Source Follower

FET source followers are commonly used as input stages to many common instruments, because of their very high input impedance. A commonly used source follower is shown in Figure 10.30. The desirable feature of this amplifier configuration is that, as we shall soon verify, it provides a high input impedance and a

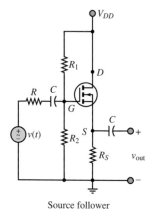

Source follower

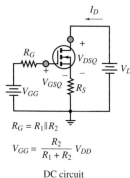

$R_G = R_1 \| R_2$

$V_{GG} = \dfrac{R_2}{R_1 + R_2} V_{DD}$

DC circuit

**Figure 10.30** MOSFET
source follower and DC circuit

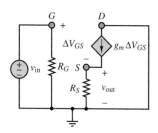

**Figure 10.31** AC
equivalent circuit for the
MOSFET source follower

low output impedance. These features are both very useful and compensate for the fact that the voltage gain of the device is at most unity.

As usual, the $Q$ point of the amplifier is found by writing KVL around the drain circuit:

$$V_{DSQ} = V_{DD} - I_{DQ} R_S \tag{10.48}$$

and by assuming that no current flows into the gate. The voltage $V_{GSQ}$, which controls the gate-to-source bias, is then given by the expression

$$V_{GSQ} = V_{GG} - I_{DQ} R_S \tag{10.49}$$

so that the desired $Q$ point may be established by selecting appropriate values of $R_1$, $R_2$, and $R_S$ (Example 10.10 will illustrate a typical design). The AC equivalent circuit is shown in Figure 10.31.

Since it is possible to select $R_1$ and $R_2$ arbitrarily, we see that the input resistance, $R_G$, of the source follower can be made quite large. Another observation is that, for this type of amplifier, the voltage gain is always less than 1. This fact may be verified by considering that

$$\Delta V_{GS} = v_{in}(t) - g_m \Delta V_{GS} R_S \tag{10.50}$$

which implies

$$\frac{\Delta V_{GS}}{v_{in}(t)} = \frac{1}{1 + g_m R_S} \tag{10.51}$$

Since the AC output voltage is related to $\Delta V_{GS}$ by

$$v_{out} = g_m \Delta V_{GS} R_S \tag{10.52}$$

it follows that

$$\frac{v_{out}}{\Delta V_{GS}} = g_m R_S \tag{10.53}$$

Thus, the open-circuit voltage gain, $\mu$, may be obtained as follows:

$$\frac{v_{out}}{v_{in}} = \frac{v_{out}}{\Delta V_{GS}} = (g_m R_S) \left( \frac{1}{1 + g_m R_S} \right)$$
$$= \frac{g_m R_S}{g_m R_S + 1} < 1 \tag{10.54}$$

We can see that the open-circuit voltage gain will always be less than 1. The output resistance can be found by inspection from the circuit of Figure 10.31 to be equal to $R_S$.

In summary, the source-follower amplifier has an input resistance, $r_i = R_G = R_1 \| R_2$, which can be made arbitrarily large; a voltage gain less than unity; and an output resistance, $r_o = R_S$, which can be made small by appropriate design.

---

### EXAMPLE 10.10 Analysis of MOSFET Source Follower

#### Problem

Find the small-signal voltage gain and the input resistance of the enhancement MOSFET source follower amplifier of Figure 10.32.

## Solution

**Known Quantities:** Drain, source and gate resistors; drain supply voltage; MOSFET $k$ and threshold voltage parameters.

**Find:** $V_{GSQ}$; $V_{DSQ}$; $I_{DQ}$. $A_V$ and $r_i$.

**Schematics, Diagrams, Circuits, and Given Data:** $R_G = 10 \ \mathrm{M\Omega}$; $R_D = 10 \ \mathrm{k\Omega}$; $R_L = 10 \ \mathrm{k\Omega}$; $V_{DD} = 15 \ \mathrm{V}$; $V_T = 1.5 \ \mathrm{V}$; $k = 0.125 \ \mathrm{mA/V^2}$.

**Assumptions:** All currents are expressed in mA and all resistors in $\mathrm{k\Omega}$.

**Analysis:** To determine the $Q$ point of the amplifier we use the DC equivalent circuit shown in Figure 10.33. We immediately observe that, since the input resistance of the transistor is infinite, the current $I_{GQ}$ must be zero. Thus, there is no voltage drop across resistor $R_G$. If there is no voltage drop across $R_G$, then the gate and drain must be at the same potential, and $V_{GSQ} = V_{DQ} = V_{DSQ}$, since the source is grounded. The drain current equation for the MOSFET in the active region is:

$$I_{DQ} = k \left( V_{GSQ} - V_T \right)^2 = 0.125 \left( V_{GSQ} - 1.5 \right)^2 = 0.125 \left( V_{DSQ} - 1.5 \right)^2$$

The drain circuit imposes the condition:

$$V_{DD} = V_{DSQ} + R_D I_{DQ}$$

or

$$15 = V_{DSQ} + 10 I_{DQ}$$

Solving the two above equations we obtain the quiescent operating point of the amplifier: $I_{DQ} = 1.06 \ \mathrm{mA}$; $V_{DSQ} = 4.4 \ \mathrm{V}$. Note that for these quiescent values, the MOSFET is, indeed, operating in the saturation region, since $v_{DS} > v_{GS} - V_T$ and $v_{GS} > V_T$. Next, we compute the transconductance parameter from equation 10.34:

$$g_m = 2k \left( V_{GSQ} - V_T \right) = 2 \times 0.125 \times (4.4 - 1.5) = 0.725 \ \frac{\mathrm{mA}}{\mathrm{V}}$$

With the above parameter we can finally construct the AC equivalent circuit, shown in Figure 10.34, and derive an expression for the output voltage:

$$v_{\mathrm{out}} = \Delta V_{DS} = -g_m \Delta V_{GS} \left( R_D || R_L \right)$$

and since $\Delta V_{GS} = v_{\mathrm{in}}$, the AC small signal voltage gain of the amplifier is:

$$A_V = \frac{v_{\mathrm{out}}}{v_{\mathrm{in}}} = \frac{\Delta V_{DS}}{\Delta V_{GS}} = -g_m \left( R_D || R_L \right) = -0.725 \times 5 = -3.625 \ \frac{\mathrm{V}}{\mathrm{V}}$$

To determine the input resistance of this amplifier, we determine the input current:

$$i_{\mathrm{in}} = \frac{v_{\mathrm{in}} - v_{\mathrm{out}}}{R_G} = \frac{v_{\mathrm{in}}}{R_G} \left( 1 - \frac{v_{\mathrm{out}}}{v_{\mathrm{in}}} \right) = \frac{v_{\mathrm{in}}}{R_G} \left( 1 - A_V \right)$$

Since the input resistance is defined as the ratio of input voltage to input current, we have:

$$R_i = \frac{v_{\mathrm{in}}}{i_{\mathrm{in}}} = \frac{v_{\mathrm{in}}}{\dfrac{v_{\mathrm{in}}}{R_G} \left( 1 - A_V \right)} = \frac{R_G}{\left( 1 - A_V \right)} = \frac{10^6}{4.625} = 2.16 \ \mathrm{M\Omega}$$

**Comments:** By selecting a very large value for $R_G$ (with no effect on the $Q$ point) it is possible to design MOSFET amplifiers that have remarkably high input impedance, though their voltage gain is modest. This is a very useful feature in designing input stages for multistage amplifiers, as we shall see in the next section.

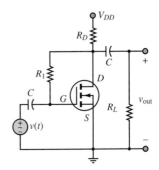

**Figure 10.32**

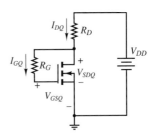

**Figure 10.33** DC equivalent circuit for Example 10.10

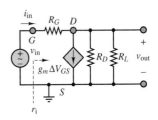

**Figure 10.34**

## Check Your Understanding

**10.8**  Compute the quiescent drain current and the small-signal transconductance parameter for a MOSFET with $V_T = 1.4$ V, $k = 2$ mA/V$^2$, for the following values of $V_{GS}$: 1.8 V, 2.0 V, 2.2 V, 2.4 V, 2.6 V, 2.8 V, 3.0 V.

**10.9**  Find the open-circuit AC voltage gain, $\mu$, for the amplifier design of Example 10.8. What is the effective voltage gain of the amplifier if the load resistance is 1 k$\Omega$?

**10.10**  Repeat Exercise 10.9 for the values of $g_m$ found in Exercise 10.8.

## 10.4  TRANSISTOR AMPLIFIERS

The design of a practical transistor amplifier is a more complex process than what has been described in the preceding sections. A number of issues must be addressed in order to produce a useful amplifier design; such a detailed discussion is beyond the scope of this book, primarily because the intended audience of this text is not likely to be involved in the design of advanced amplifier circuits. However, it is important to briefly mention two topics: frequency response limitations of transistor amplifiers; and the need for **multistage amplifiers** consisting of several transistor amplifier stages—that is, several individual amplifiers like the ones described in the preceding sections. Other issues that will not be addressed in greater detail in this book include: amplifier input and output impedance; impedance matching between amplifier stages, and between source, amplifier and load; *direct-coupled* and *transformer-coupled amplifiers; differential amplifiers;* and *feedback.* However, some of these issues will be indirectly addressed in Chapter 12, devoted to *operational amplifiers,* where it will be shown that for many instrumentation and signal-conditioning needs, a non–electrical engineer can be satisfied with the use of *integrated circuit amplifiers,* consisting of multistage transistor amplifiers completely assembled into an integrated circuit "chip."

### Frequency Response of Small-Signal Amplifiers

When the idea of a coupling capacitor was introduced earlier in this chapter, the observation was made that the input and output coupling capacitors (see Figure 10.7, for example) acted very nearly like a short circuit at AC frequencies. In fact, the presence of a series capacitor in the AC small-signal equivalent input circuit has very much the effect of a high-pass filter. Consider the small-signal equivalent circuit of Figure 10.35, which corresponds to a BJT common-emitter amplifier. In this circuit, a single coupling capacitor has been placed at the input, for simplicity. Let us compute the frequency response of the small-signal equivalent circuit of Figure 10.34. If we assume that $R_B \gg R_S$ and $R_B \gg h_{ie}$, then we can approximate the base current by

$$\Delta i_B = \frac{v_{\text{in}}}{h_{ie}} \tag{10.55}$$

and

$$v_{\text{in}} = \frac{h_{ie}}{h_{ie} + R_S + \dfrac{1}{j\omega C_B}} v_S \tag{10.56}$$

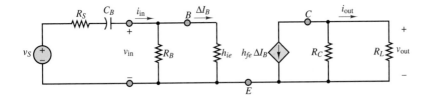

**Figure 10.35** Small-signal equivalent circuit

so that the output voltage is given by

$$v_{\text{out}} = -R_C \parallel R_L h_{fe} \Delta i_B = R_C \parallel R_L \frac{v_{\text{in}}}{h_{ie}} \tag{10.57}$$

or

$$\frac{v_{\text{out}}}{v_S} = R_C \parallel R_L \frac{h_{fe}}{h_{ie} + R_S + \dfrac{1}{j\omega C_B}} \tag{10.58}$$

This expression for $v_{\text{out}}$ is clearly a function of frequency, approaching a constant as $\omega$ tends to infinity. Note that, in effect, the cutoff frequency is determined at the input circuit by the effective $RC$ combination of equation 10.59:

$$\omega_B = \frac{1}{(R_S + h_{ie})C_B} \tag{10.59}$$

Clearly, if we wished to design an *audio amplifier* (that is, an amplifier that can provide an undistorted frequency response over the range of frequencies audible to the human ear), we would select a coupling capacitor such that the lower cutoff frequency in the circuit of Figure 10.35 would be greater than $2\pi \times 20$ rad/s. A similar effect can be attributed to the output coupling capacitor, $C_C$, in Figure 10.7, and to the emitter bypass capacitor, $C_E$, resulting in the cutoff frequencies

$$\omega_C = \frac{1}{(R_C + R_L)C_C} \tag{10.60}$$

and

$$\omega_E = \frac{1}{r_o C_E} \tag{10.61}$$

where $r_o$ is the output impedance of the amplifier. The highest of the three frequencies given in equations 10.59 through 10.61 is often $\omega_E$, and this is therefore the lower cutoff frequency of the amplifier. Figure 10.36 illustrates the general effect of a nonideal coupling capacitor on the frequency response of the common-emitter amplifier.

The frequency response of a bipolar transistor amplifier is generally limited at the high-frequency end by the so-called **parasitic capacitance** present between pairs of terminals at the transistor (don't forget that any two conductors separated by a dielectric—air, for example—form a capacitor). Figure 10.37 depicts a high-frequency small-signal model (called the **hybrid-pi model**) of the BJT, which includes the effect of two capacitances, $C_{BE}$ and $C_{CB}$. These two capacitances can be safely ignored (by treating them as open circuits) at low and mid frequencies.

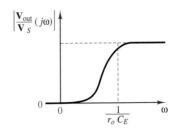

**Figure 10.36** Approximate low-end frequency response of a common-emitter amplifier, assuming finite coupling capacitors

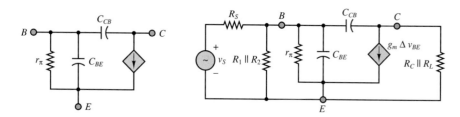

**Figure 10.37** High-frequency BJT model and equivalent circuit

It can be shown that at high frequency both capacitances contribute to a low-pass response with cutoff frequency given by:

$$\omega_T = \frac{1}{R_T C_T} \tag{10.62}$$

where $R_T = R_S \parallel R_1 \parallel R_2 \parallel r_\pi$ and $C_T$ is an equivalent capacitance related to $C_{CB}$ and $C_{BE}$. The calculation of $C_T$ requires making use of Miller's theorem, which is well beyond the scope of this book. The effect of these capacitances on the frequency response of the amplifier is depicted in Figure 10.38. Figure 10.39 summarizes the discussion by illustrating that a practical BJT common-emitter amplifier stage will be characterized by a band-pass frequency response with low cutoff frequency $\omega_{\text{low}}$ and high cutoff frequency $\omega_{\text{high}}$.

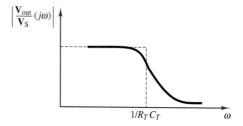

**Figure 10.38** Low-pass filter effect of parasitic capacitance in a common-emitter amplifier

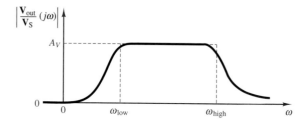

**Figure 10.39** Frequency response of a common-emitter amplifier

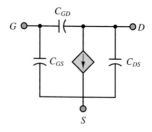

**Figure 10.40** High-frequency FET model

A completely analogous discussion could be made for FET amplifiers, where the effect of the input and output coupling capacitors and of the source bypass capacitor is to reduce the amplifier frequency response at the low end, creating a high-pass effect. Similarly, a high-frequency small-signal model valid for all FETs will include the effect of parasitic capacitances between gate, drain, and source terminals. This model is shown in Figure 10.40.

## Multistage Amplifiers

The design of a practical amplifier involves a variety of issues, as mentioned earlier in this chapter. To resolve the various design trade-offs and to obtain acceptable performance characteristics, it is usually necessary (except in the simplest applications) to design an amplifier in various stages. In general, three stages are needed to address three important issues:

1. Choosing an appropriate input impedance for the amplifier, so as not to load the small-signal source. The input impedance should, in general, be large.
2. Providing suitable gain.
3. Matching the output impedance of the amplifier to the load. This usually requires choosing a low output impedance.

Each of these tasks can be accomplished in a different manner and with more than one amplifier stage, depending on the intended application. Although the task of designing a multistage amplifier is very advanced, and beyond the scope of this book, the minimum necessary tools to understand such a design process have already been introduced. Each amplifier stage can, through the use of small-signal models, be represented in the form of a two-port circuit and characterized by a gain and an input and an output impedance; the overall response of the amplifier can then be obtained by cascading the individual two-port blocks, as shown in Figure 10.41.

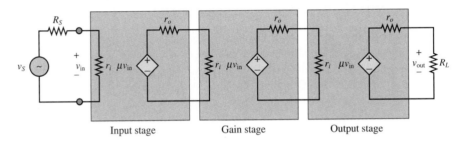

**Figure 10.41** Block diagram of a multistage amplifier

A three-stage amplifier is shown in Figure 10.42. The input stage consists of a MOSFET amplifier; the choice of a MOSFET for the input stage is quite

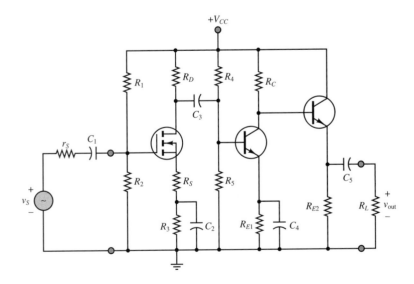

**Figure 10.42** Three-stage amplifier

natural, because of the high input impedance of this device. Note that the first stage is AC-coupled; thus, the amplifier will have a band-pass characteristic, as discussed in the preceding section. The choice of a MOSFET as the input stage of a linear amplifier is acceptable in spite of the nonlinearity of the MOSFET transfer characteristic, because the role of the first stage is to amplify a very low-amplitude signal, and therefore the MOSFET is required to amplify in only a relatively small operating region. Thus, the linear transconductance approximation will be valid, and relatively little distortion should be expected. The second stage consists of a BJT common-emitter stage, which is also AC-coupled to the first stage and provides most of the gain. The output stage is a BJT emitter follower, which produces no additional gain but has a relatively low output impedance, needed to match the load. Note that this last stage is DC-coupled to the preceding stage but is AC-coupled to the load. The amplifier of Figure 10.42 will be further explored in the homework problems.

## 10.5   TRANSISTOR GATES AND SWITCHES

In describing the properties of transistors in Chapter 9, it was suggested that, in addition to serving as amplifiers, three-terminal devices can be used as electronic switches in which one terminal controls the flow of current between the other two. It had also been hinted in Chapter 8 that diodes can act as on-off devices as well. In this section, we discuss the operation of diodes and transistors as electronic switches, illustrating the use of these electronic devices as the switching circuits that are at the heart of **analog** and **digital gates.** Transistor switching circuits form the basis of digital logic circuits, which will be discussed in more detail in the next chapter. The objective of this section is to discuss the internal operation of these circuits and to provide the reader interested in the internal workings of digital circuits with an adequate understanding of the basic principles.

An **electronic gate** is a device that, on the basis of one or more input signals, produces one of two or more prescribed outputs; as will be seen shortly, one can construct both digital and analog gates. A word of explanation is required, first, regarding the meaning of the words *analog* and *digital*. An analog voltage or current—or, more generally, an analog signal—is one that varies in a continuous fashion over time, in *analogy* (hence the expression *analog*) with a physical quantity. An example of an analog signal is a sensor voltage corresponding to ambient temperature on any given day, which may fluctuate between, say, 30° and 50°F. A digital signal, on the other hand, is a signal that can take only a finite number of values; in particular, a commonly encountered class of digital signals consists of **binary signals,** which can take only one of two values (for example, 1 and 0). A typical example of a binary signal would be the control signal for the furnace in a home heating system controlled by a conventional thermostat, where one can think of this signal as being "on" (or 1) if the temperature of the house has dropped below the thermostat setting (desired value), or "off" (or 0) if the house temperature is greater than or equal to the set temperature (say, 68°F). Figure 10.43 illustrates the appearance of the analog and digital signals in this furnace example.

The discussion of digital signals will be continued and expanded in Chapters 13, 14, and 15. Digital circuits are an especially important topic, because a large part of today's industrial and consumer electronics is realized in digital form.

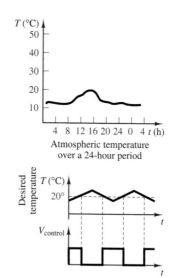

Atmospheric temperature over a 24-hour period

Average temperature in a house and related digital control voltage

**Figure 10.43** Illustration of analog and digital signals

## Analog Gates

A common form of analog gate employs an FET and takes advantage of the fact that current can flow in either direction in an FET biased in the ohmic region. Recall that the drain characteristic of the MOSFET discussed in Chapter 9 consists of three regions: ohmic, active, and breakdown. A MOSFET amplifier is operated in the active region, where the drain current is nearly constant for any given value of $v_{GS}$. On the other hand, a MOSFET biased in the ohmic state acts very much as a linear resistor. For example, for an $n$-channel enhancement MOSFET, the conditions for the transistor to be in the ohmic region are:

$$v_{GS} > V_T \quad \text{and} \quad |v_{DS}| \leq \frac{1}{4}(v_{GS} - V_T) \tag{10.63}$$

As long as the FET is biased within these conditions, it acts simply as a linear resistor, and it can conduct current in either direction (provided that $v_{DS}$ does not exceed the limits stated in equation 10.63). In particular, the resistance of the channel in the ohmic region is found to be

$$R_{DS} = \frac{V_T^2}{2I_{DSS}(v_{GS} - V_T)} \tag{10.64}$$

so that the drain current is equal to

$$i_D \approx \frac{v_{DS}}{r_{DS}} \quad \text{for} \quad |v_{DS}| \leq \frac{1}{4}(v_{GS} - V_T) \quad \text{and} \quad v_{GS} > V_T \tag{10.65}$$

The most important feature of the MOSFET operating in the ohmic region, then, is that it acts as a voltage-controlled resistor, with the gate-source voltage, $v_{GS}$, controlling the channel resistance, $R_{DS}$. The use of the MOSFET as a switch in the ohmic region, then, consists of providing a gate-source voltage that can either hold the MOSFET in the cutoff region ($v_{GS} \leq V_T$) or bring it into the ohmic region. In this fashion, $v_{GS}$ acts as a control voltage for the transistor.

Consider the circuit shown in Figure 10.44, where we presume that $v_C$ can be varied externally and that $v_{in}$ is some analog signal source that we may wish to connect to the load $R_L$ at some appropriate time. The operation of the switch is as follows. When $v_C \leq V_T$, the FET is in the cutoff region and acts as an open circuit. When $v_C > V_T$ (with a value of $v_{GS}$ such that the MOSFET is in the ohmic region), the transistor acts as a linear resistance, $R_{DS}$. If $R_{DS} \ll R_L$, then $v_{out} \approx v_{in}$. By using a pair of MOSFETs, it is possible to improve the dynamic range of signals one can transmit through this analog gate.

MOSFET analog switches are usually produced in integrated circuit (IC) form and denoted by the symbol shown in Figure 10.45. A CMOS gate is described in Chapter 9 in a "Focus on Measurements" section.

## Digital Gates

In this section, we explore the operation of diodes, BJTs, and FETs as digital gates. Digital circuits are simpler to analyze than analog circuits, because one is only interested in determining whether a given device is conducting or not.

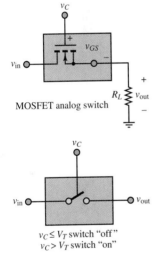

MOSFET analog switch

$v_C \leq V_T$ switch "off"
$v_C > V_T$ switch "on"

Functional model

**Figure 10.44** MOSFET analog switch

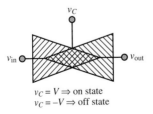

$v_C = V \Rightarrow$ on state
$v_C = -V \Rightarrow$ off state

**Figure 10.45** Symbol for bilateral FET analog gate

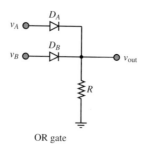

OR gate

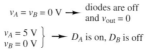

OR gate operation

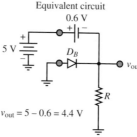

$v_A = v_B = 0$ V $\longrightarrow$ diodes are off and $v_{out} = 0$

$\left. \begin{array}{l} v_A = 5\text{ V} \\ v_B = 0\text{ V} \end{array} \right\} \longrightarrow D_A$ is on, $D_B$ is off

Equivalent circuit

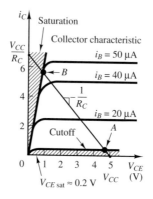

$v_{out} = 5 - 0.6 = 4.4$ V

**Figure 10.46** Diode OR gate

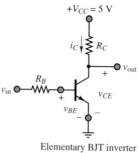

Elementary BJT inverter

**Figure 10.47** BJT switching characteristic

### Diode Gates

You will recall that a diode conducts current when it is forward-biased and otherwise acts very much as an open circuit. Thus, the diode can serve as a switch if properly employed. The circuit of Figure 10.46 is called an **OR gate;** it operates as follows. Let voltage levels greater than, say, 2 V correspond to a "logic 1" and voltages less than 2 V represent a "logic 0." Suppose, then, that the input voltages $v_A$ and $v_B$ can be equal to either 0 V or 5 V. If $v_A = 5$ V, diode $D_A$ will conduct; if $v_A = 0$ V, $D_A$ will act as an open circuit. The same argument holds for $D_B$. It should be apparent, then, that the voltage across the resistor $R$ will be 0 V, or logic 0, if both $v_A$ and $v_B$ are 0. If either $v_A$ or $v_B$ is equal to 5 V, though, the corresponding diode will conduct, and—assuming an offset model for the diode with $V_\gamma = 0.6$ V—we find that $v_{out} = 4.4$ V, or logic 1. Similar analysis yields an equivalent result if both $v_A$ and $v_B$ are equal to 5 V.

This type of gate is called an OR gate, because $v_{out}$ is equal to logic 1 (or "high") if either $v_A$ *or* $v_B$ is on, while it is logic 0 (or "low") if neither $v_A$ nor $v_B$ is on. Other functions can also be implemented; however, the discussion of diode gates will be limited to this simple introduction, because diode gate circuits, such as the one of Figure 10.46, are rarely if ever employed in practice. Most modern digital circuits employ transistors to implement switching and gate functions.

### BJT Gates

In discussing large-signal models for the BJT, we observed that the $i$-$v$ characteristic of this family of devices includes a *cutoff* region, where virtually no current flows through the transistor. On the other hand, when a sufficient amount of current is injected into the base of the transistor, a bipolar transistor will reach *saturation,* and a substantial amount of collector current will flow. This behavior is quite well suited to the design of electronic gates and switches and can be visualized by superimposing a load line on the collector characteristic, as shown in Figure 10.47.

The operation of the simple **BJT switch** is illustrated in Figure 10.47, by means of load-line analysis. Writing the load-line equation at the collector circuit, we have

$$v_{CE} = V_{CC} - i_C R_C \tag{10.66}$$

and

$$v_{out} = v_{CE} \tag{10.67}$$

Thus, when the input voltage, $v_{in}$, is low (say, 0 V, for example) the transistor is in the cutoff region and little or no current flows, and

$$v_{out} = v_{CE} = V_{CC} \tag{10.68}$$

so that the output is "logic high."

When $v_{in}$ is large enough to drive the transistor into the saturation region, a substantial amount of collector current will flow and the collector-emitter voltage will be reduced to the small saturation value, $V_{CE\ sat}$, which is typically a fraction of a volt. This corresponds to the point labeled $B$ on the load line. For the input voltage $v_{in}$ to drive the BJT of Figure 10.47 into saturation, a base current of approximately 50 $\mu$A will be required. Suppose, then, that the voltage $v_{in}$

could take the values 0 V or 5 V. Then, if $v_{in} = 0$ V, $v_{out}$ will be nearly equal to $V_{CC}$, or, again, 5 V. If, on the other hand, $v_{in} = 5$ V and $R_B$ is, say, equal to 89 k$\Omega$ (so that the base current required for saturation flows into the base: $i_B = (v_{in} - V_\gamma)/R_B = (5 - .06)/89,000 \approx 50\ \mu A$), we have the BJT in saturation, and $v_{out} = V_{CE\ sat} \approx 0.2$ V.

Thus, you see that whenever $v_{in}$ corresponds to a logic high (or logic 1), $v_{out}$ takes a value close to 0 V, or logic low (or 0); conversely, $v_{in} = $ "0" (logic "low") leads to $v_{out} = $ "1." The values of 5 V and 0 V for the two logic levels 1 and 0 are quite common in practice and are the standard values used in a family of logic circuits denoted by the acronym **TTL,** which stands for **transistor-transistor logic.**[2] One of the more common TTL blocks is the **inverter** shown in Figure 10.47, so called because it "inverts" the input by providing a low output for a high input, and vice versa. This type of inverting, or "negative," logic behavior is quite typical of BJT gates (and of transistor gates in general).

In the following paragraphs, we introduce some elementary BJT logic gates, similar to the diode gates described previously; the theory and application of digital logic circuits is discussed in Chapter 13. Example 10.11 illustrates the operation of a **NAND gate,** that is, a logic gate that acts as an inverted AND gate (thus the prefix N in NAND, which stands for NOT).

## EXAMPLE 10.11 TTL NAND Gate

### Problem

Complete the table below to determine the logic gate operation of the TTL NAND gate of Figure 10.48.

| $v_1$ | $v_2$ | State of $Q_1$ | State of $Q_2$ | $v_{out}$ |
|-------|-------|----------------|----------------|-----------|
| 0 V   | 0 V   |                |                |           |
| 0 V   | 5 V   |                |                |           |
| 5 V   | 0 V   |                |                |           |
| 5 V   | 5 V   |                |                |           |

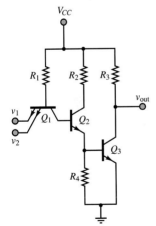

**Figure 10.48** TTL NAND gate

### Solution

**Known Quantities:** Resistor values; $V_{BE\ on}$ and $V_{CE\ sat}$ for each transistor.

**Find:** $v_{out}$ for each of the four combinations of $v_1$ and $v_2$.

**Schematics, Diagrams, Circuits, and Given Data:** $R_1 = 5.7$ k$\Omega$; $R_2 = 2.2$ k$\Omega$; $R_3 = 2.2$ k$\Omega$; $R_4 = 1.8$ k$\Omega$; $V_{CC} = 5$ V; $V_{BE\ on} = V_\gamma = 0.7$ V; $V_{CE\ sat} = 0.2$ V.

**Assumptions:** Treat the *BE* and *BC* junctions of $Q_1$ as offset diodes. Assume that the transistors are in saturation when conducting.

---

[2]TTL logic values are actually quite flexible, with $v_{HIGH}$ as low as 2.4 V and $v_{LOW}$ as high as 0.4 V.

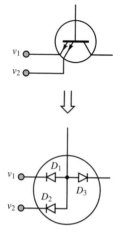

**Figure 10.49**

**Analysis:** The inputs to the TTL gate, $v_1$ and $v_2$, are applied to the emitter of transistor $Q_1$. The transistor is designed so as to have two emitter circuits in parallel. $Q_1$ is modeled by the offset diode model, as shown in Figure 10.49. We shall now consider each of the four cases.

1.  $v_1 = v_2 = 0$ V. With the emitters of $Q_1$ connected to ground and the base of $Q_1$ at 5 V, the *BE* junction will clearly be forward biased and $Q_1$ is on. This result means that the base current of $Q_2$ (equal to the collector current of $Q_1$) is negative, and therefore $Q_2$ must be off. If $Q_2$ is off, its emitter current must be zero, and therefore no base current can flow into $Q_3$, which is in turn also off. With $Q_3$ off, no current flows through $R_3$, and therefore $v_{out} = 5 - v_{R3} = 5$ V.

2.  $v_1 = 5$ V; $v_2 = 0$ V. Now, with reference to Figure 10.49, we see that diode $D_1$ is still forward-biased, but $D_2$ is now reverse-biased because of the 5-V potential at $v_2$. Since one of the two emitter branches is capable of conducting, base current will flow and $Q_1$ will be on. The remainder of the analysis is the same as in case (1), and $Q_2$ and $Q_3$ will both be off, leading to $v_{out} = 5$ V.

3.  $v_1 = 0$ V; $v_2 = 5$ V. By symmetry with case (2), we conclude that, again, one emitter branch is conducting, and therefore $Q_1$ will be on, $Q_2$ and $Q_3$ will both be off, and $v_{out} = 5$ V.

4.  $v_1 = 5$ V; $v_2 = 5$ V. When both $v_1$ and $v_2$ are at 5 V, diodes $D_1$ and $D_2$ are both strongly reverse-biased, and therefore no emitter current can flow. Thus, $Q_1$ must be off. Note, however, that while $D_1$ and $D_2$ are reverse-biased, $D_3$ is forward-biased, and therefore a current will flow into the base of $Q_2$; thus, $Q_2$ is on and since the emitter of $Q_2$ is connected to the base of $Q_3$, $Q_3$ will also see a positive base current and will be on. To determine the output voltage, we assume that $Q_3$ is operating in saturation. Then, applying KVL to the collector circuit we have:

$$V_{CC} = I_{C_3} R_3 + V_{CE_3}$$

or

$$I_{C3} = \frac{V_{CC} - V_{CE_3}}{R_C} = \frac{V_{CC} - V_{CE \text{ sat}}}{R_C} = \frac{5 - 0.2}{2{,}200} = 2.2 \text{ mA}$$

and

$$v_{out} = V_{CC} - I_C R_3 = 5 - 2.2 \times 10^{-3} \times 2.2 \times 10^{-3} = 5 - 4.84 = 0.16 \text{ V}$$

These results are summarized in the table below. The output values are consistent with TTL logic; the output voltage for case (4) is sufficiently close to zero to be considered zero for logic purposes.

| $v1$ | $v2$ | State of $Q_2$ | State of $Q_3$ | $v_{out}$ |
| --- | --- | --- | --- | --- |
| 0 V | 0 V | Off | Off | 5 V |
| 0 V | 5 V | Off | Off | 5 V |
| 5 V | 0 V | Off | Off | 5 V |
| 5 V | 5 V | On | On | 0.16 V |

**Comments:** While exact analysis of TTL logic gate circuits could be tedious and involved, the method demonstrated in this example—to determine whether transistors are on or off—leads to very simple analysis. Since in logic devices one is interested primarily in logic levels and not in exact values, this approximate analysis method is very appropriate.

**Focus on Computer-Aided Solutions:**  An *Electronics Workbench*™ simulation of the TTL NAND gate may be found in the accompanying CD-ROM. You may wish to validate the saturation assumption for transistors $Q_2$ and $Q_3$ by "measuring" $V_{CE2}$ and $V_{CE3}$ in the simulation.

---

The analysis method employed in Example 10.11 can be used to analyze any TTL gate. With a little practice, the calculations of this example will become familiar. The Check Your Understanding exercises and homework problems will reinforce the concepts developed in this section.

### MOSFET Logic Gates

Having discussed the BJT as a switching element, we might suspect that FETs may similarly serve as logic gates. In fact, in some respects, FETs are better suited to be employed as logic gates than BJTs. The *n*-channel enhancement MOSFET, discussed in Chapter 9, serves as an excellent illustration: because of its physical construction, it is normally off (that is, it is off until a sufficient gate voltage is provided), and therefore it does not require much current from the input signal source. Further, MOS devices offer the additional advantage of easy fabrication into integrated circuit form, making production economical in large volume. On the other hand, MOS devices cannot provide as much current as BJTs, and their switching speeds are not quite as fast—although these last statements may not hold true for long, because great improvements are taking place in MOS technology. Overall, it is certainly true that in recent years it has become increasingly common to design logic circuits based on MOS technology. In particular, a successful family of logic gates called *CMOS* (for *complementary metal-oxide-semiconductor*) takes advantage of both *p*- and *n*-channel enhancement-mode MOSFETs to exploit the best features of both types of transistors. CMOS logic gates (and many other types of digital circuits constructed by using the same technology) consume very little supply power, and have become the mainstay in pocket calculators, wristwatches, portable computers, and many other consumer electronics products. Without delving into the details of CMOS technology (a brief introduction is provided in Chapter 9), we shall briefly illustrate the properties of MOSFET logic gates and of CMOS gates in the remainder of this section.

Figure 10.50 depicts a MOSFET switch with its drain *i-v* characteristic. Note the general similarity with the switching characteristic of the BJT shown in the previous section. When the input voltage, $v_{in}$, is zero, the MOSFET conducts virtually no current, and the output voltage, $v_{out}$, is equal to $V_{DD}$. When $v_{in}$ is equal to 5 V, the MOSFET $Q$ point moves from point $A$ to point $B$ along the load line, with $v_{DS} = 0.5$ V. Thus, the circuit acts as an inverter. Much as in the case of the BJT, the inverter forms the basis of all MOS logic gates.

An elementary CMOS inverter is shown in Figure 10.51. Note first the simplicity of this configuration, which simply employs two enhancement-mode MOSFETs: *p*-channel at the top, denoted by the symbol $Q_p$, and *n*-channel at the bottom, denoted by $Q_n$. Recall from Chapter 9 that when $v_{in}$ is low, transistor $Q_n$

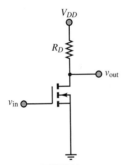

MOSFET inverter

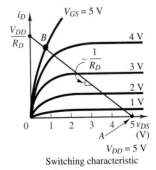

Switching characteristic

**Figure 10.50** MOSFET switching characteristic

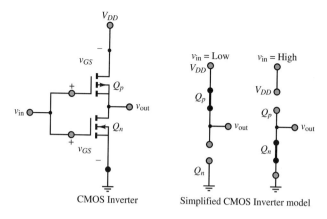

**Figure 10.51** CMOS Inverter and circuit model

is off. However, transistor $Q_p$ sees a gate-to-source voltage $v_{GS} = v_{in} - V_{DD} = -V_{DD}$; in a $p$-channel device, this condition is the counterpart of having $v_{GS} = V_{DD}$ for an $n$-channel MOSFET. Thus, when $Q_n$ is off, $Q_p$ is on and acts very much as a small resistance. In summary, when $v_{in}$ is low, the output is $v_{out} \approx V_{DD}$. When $v_{in}$ is high, the situation is reversed: $Q_n$ is now on and acts nearly as a short circuit, while $Q_p$ is open (since $v_{GS} = 0$ for $Q_p$). Thus, $v_{out} \approx 0$. The complementary MOS operation is depicted in Figure 10.51 in simplified form by showing each transistor as either a short or an open circuit, depending on its state. This simplified analysis is sufficient for the purpose of a qualitative analysis. The following examples illustrate methods for analyzing MOS switches and gates.

## EXAMPLE 10.12 MOSFET Switch

### Problem

Determine the operating points of the MOSFET switch of Figure 10.52 when the signal source outputs zero volts and 2.5 volts, respectively.

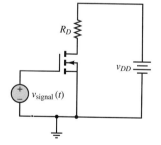

**Figure 10.52**

### Solution

**Known Quantities:** Drain resistor; $V_{DD}$; signal source output voltage as a function of time.

**Find:** $Q$ point for each value of the signal source output voltage.

**Schematics, Diagrams, Circuits, and Given Data:** $R_D = 125 \ \Omega$; $V_{DD} = 10$ V; $v_{signal}(t) = 0$ V for $t < 0$; $v_{signal}(t) = 2.5$ V for $t = 0$.

**Assumptions:** Use the drain characteristic curves for the MOSFET (Figure 10.53).

**Analysis:** We first draw the load line using the drain circuit equation:

$$V_{DD} = R_D I_D + V_{DS} \qquad 10 = 125 I_D + V_{DS}$$

recognizing a $V_{DS}$ axis intercept at 10 V and an $I_D$ axis intercept at $10/125 = 80$ mA.

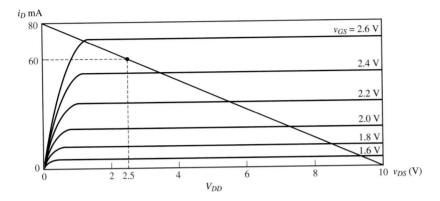

**Figure 10.53** Drain curves for MOSFET of Figure 10.52

1. $t < 0$ s. When the signal source output is zero, the gate voltage is zero and the MOSFET is in the cutoff region. The $Q$ point is:

$$V_{GSQ} = 0 \text{ V} \qquad I_{DQ} = 0 \text{ mA} \qquad V_{DSQ} = 10 \text{ V}$$

2. $t \geq 0$ s. When the signal source output is 2.5 V, the gate voltage is 2.5 V and the MOSFET is in the saturation region. The $Q$ point is:

$$V_{GSQ} = 0 \text{ V} \qquad I_{DQ} = 60 \text{ mA} \qquad V_{DSQ} = 2.5 \text{ V}$$

This result satisfies the drain equation, since $R_D I_D = 0.06 \times 125 = 7.5$ V.

**Comments:**  The simple MOSFET configuration shown can quite effectively serve as a switch, conducting 60 mA when the gate voltage is switched to 2.5 V.

---

## EXAMPLE  10.13  CMOS Gate

### Problem

Determine the logic function implemented by the CMOS gate of Figure 10.54. Use the table below to summarize the behavior of the circuit.

| $v_1$ | $v_2$ | State of $M_1$ | State of $M_2$ | State of $M_3$ | State of $M_4$ | $v_{out}$ |
|-------|-------|----------------|----------------|----------------|----------------|-----------|
| 0 V   | 0 V   |                |                |                |                |           |
| 0 V   | 5 V   |                |                |                |                |           |
| 5 V   | 0 V   |                |                |                |                |           |
| 5 V   | 5 V   |                |                |                |                |           |

---

### Solution

**Find:**  $v_{out}$ for each of the four combinations of $v_1$ and $v_2$.

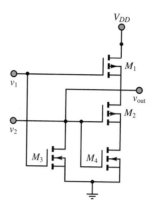

The transistors in this circuit show the substrate for each transistor connected to its respective gate. In a true CMOS IC, the substrates for the $p$-channel transistors are connected to 5 V and the substrates of the $n$-channel transistors are connected to ground.

**Figure 10.54**

**Schematics, Diagrams, Circuits, and Given Data:**  $V_T = 1.7$ V; $V_{DD} = 5$ V.

**Assumptions:**  Treat the MOSFETs as open circuits when off and as linear resistors when on.

**Analysis:**

1. $v_1 = v_2 = 0$ V. With both input voltages equal to zero, neither $M_3$ nor $M_4$ can conduct, since the gate voltage is less than the threshold voltage for both transistors. $M_1$ and $M_2$ will similarly be off, and no current will flow through the drain-source circuits of $M_1$ and $M_2$. Thus, $v_{out} = V_{DD} = 5$ V. This condition is depicted in Figure 10.55.

2. $v_1 = 5$ V; $v_2 = 0$ V. Now $M_2$ and $M_4$ are off because of the zero gate voltage, while $M_1$ and $M_3$ are on. Figure 10.56 depicts this condition.

3. $v_1 = 0$ V; $v_2 = 5$ V. By symmetry with case (2), we conclude that, again, one emitter branch is conducting, and therefore $Q_1$ will be on, $Q_2$ and $Q_3$ will both be off, and $v_{out} = 5$ V. See Figure 10.57.

4. $v_1 = 5$ V; $v_2 = 5$ V. When both $v_1$ and $v_2$ are at 5 V, diodes $D_1$ and $D_2$ are both strongly reverse-biased, and therefore no emitter current can flow. Thus, $Q_1$ must be off. Note, however, that while $D_1$ and $D_2$ are reverse-biased, $D_3$ is forward-biased, and therefore a current will flow into the base of $Q_2$; thus, $Q_2$ is on, and, since the emitter of $Q_2$ is connected to the base of $Q_3$, $Q_3$ will also see a positive base current

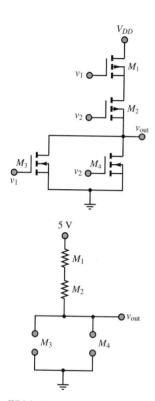

With both $v_1$ and $v_2$ at zero volts, $M_3$ and $M_4$ will be turned off (in cutoff), since $v_{GS}$ is less than $V_T$ (0 V < 1.7 V). $M_1$ and $M_2$ will be turned on, since the gate-to-source voltages will be greater than $V_T$.

**Figure 10.55**

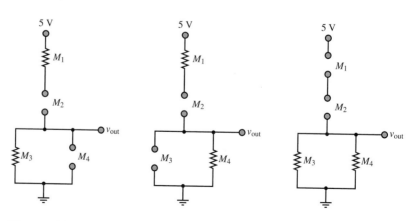

**Figure 10.56**          **Figure 10.57**          **Figure 10.58**

and will be on. See Figure 10.58. To determine the output voltage, we assume that $Q_3$ is operating in saturation. Then, applying KVL to the collector circuit we have:

$$V_{CC} = I_{C_3} R_3 + V_{CE_3}$$

or

$$I_{C_3} = \frac{V_{CC} - V_{CE_3}}{R_C} = \frac{V_{CC} - V_{CE\ \text{sat}}}{R_C} = \frac{5 - 0.2}{2,200} = 2.2 \text{ mA}$$

and

$$v_{\text{out}} = V_{CC} - I_C R_3 = 5 - 2.2 \times 10^{-3} \times 2.2 \times 10^{-3} = 5 - 4.84 = 0.16 \text{ V}$$

These results are summarized in the table below. The output values are consistent with TTL logic; the output voltage for case (4) is sufficiently close to zero to be considered zero for logic purposes.

| $v_1$ | $v_2$ | State of $Q_2$ | State of $Q_3$ | $v_{\text{out}}$ |
|-------|-------|----------------|----------------|------------------|
| 0 V | 0 V | Off | Off | 5 V |
| 0 V | 5 V | Off | Off | 5 V |
| 5 V | 0 V | Off | Off | 5 V |
| 5 V | 5 V | On | On | 0.16 V |

**Comments:**  While exact analysis of TTL logic gate circuits could be tedious and involved, the method demonstrated in this example—to determine whether transistors are on or off—leads to very simple analysis. Since in logic devices one is interested primarily in logic levels and not in exact values, this approximate analysis method is very appropriate.

**Focus on Computer-Aided Solutions:**  An *Electronics Workbench*™ simulation of the TTL NAND gate may be found in the accompanying CD-ROM. You may wish to validate the saturation assumption for transistors $Q_2$ and $Q_3$ by "measuring" $V_{CE_2}$ and $V_{CE_3}$ in the simulation.

## Check Your Understanding

**10.11**   Show that both $v_1$ and $v_2$ must be high for the AND gate circuit shown in Figure 10.59 to give a logic high output.

**10.12**   Show that the circuit in Figure 10.60 acts as an AND gate, and construct a truth table as in Example 10.11.

**10.13**   What value of $R_D$ would ensure a drain-to-source voltage, $v_{DS}$, of 5 V in the circuit of Example 10.12?

**10.14**   Analyze the CMOS gate of Figure 10.61 and find the output voltages for the following conditions: (a) $v_1 = 0$, $v_2 = 0$; (b) $v_1 = 5$ V, $v_2 = 0$; (c) $v_1 = 0$, $v_2 = 5$ V; (d) $v_1 = 5$ V, $v_2 = 5$ V. Identify the logic function accomplished by the circuit.

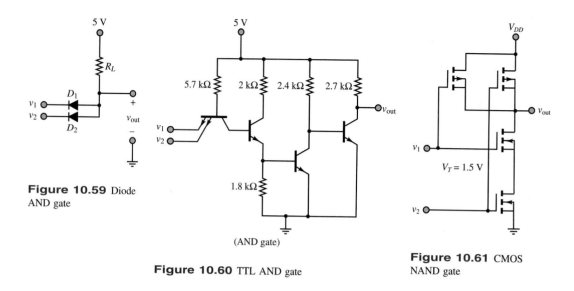

**Figure 10.59** Diode AND gate

**Figure 10.60** TTL AND gate

**Figure 10.61** CMOS NAND gate

# CONCLUSION

- Small-signal models of transistors permit linear circuit analysis of amplifier circuits, using the well-established techniques of Chapters 3 and 4.

- Small-signal, or $h$-parameter, BJT models take into account the base and collector $i$-$v$ characteristics in terms of linearized resistance parameters and controlled sources. These models can be used to analyze the operation of the BJT as a linear amplifier. Various amplifier configurations exist, each of which can be characterized by an equivalent input and output resistance and by an open-circuit voltage gain.

- Field-effect transistors can also be modeled by means of small-signal parameters and controlled sources. Small-signal FET amplifiers can be designed on the basis of linear circuit models in a manner similar to BJT amplifiers.

- BJTs are characterized by a more linear transfer characteristic than FETs and provide, in general, greater current output. However, the input impedance of FETs is significantly larger. In general, the favorable features of each family of transistors can be exploited when multistage amplifiers are designed.

- All transistor amplifiers are limited in their frequency response by the presence of coupling capacitors and by internal transistor parasitic capacitances.

- Transistors form the basis of many switching circuits. Transistor switching circuits can employ either BJT or FET circuits, giving rise to two very large families of digital logic circuits: TTL and CMOS. Each family is characterized by certain advantages; in particular, TTL circuits are faster and can provide greater load currents, while CMOS circuits are characterized by extremely low power consumption and are more easily fabricated. Transistor switching circuits can be analyzed more easily than linear amplifier circuits, since one is usually concerned only with whether the device is on or off.

# CHECK YOUR UNDERSTANDING ANSWERS

**CYU 10.1**    $\mu = 2.24$

**CYU 10.2**    (a) $I_{BQ} = 199.6\,\mu\text{A}$, $I_{CQ} = 17.96\,\text{mA}$, $V_{CEQ} = 7.77\,\text{V}$

(b) $I_{BQ} = 164.6\,\mu\text{A}$, $I_{CQ} = 19.75\,\text{mA}$, $V_{CEQ} = 7.065\,\text{V}$

| **CYU 10.3** | Design A: $\Delta V_{CEQ} = 1.73$ V, $\Delta I_{CQ} = 4.25$ mA |
| | Design B: $\Delta V_{CER} = 0.37$ V, $\Delta I_{CQ} = 0.99$ mA |

**CYU 10.4**   For $\beta = \beta_{min}$: $I_{BQ} = 254\ \mu$A, $I_{CQ} = 15.2$ mA, $V_{CEQ} = 8.87$ V
For $\beta = \beta_{max}$: $I_{BQ} = 112\ \mu$ A, $I_{CQ} = 22$ mA, $V_{CEQ} = 6.01$ V
For $I_C = 10$ mA: $R_B = 1.2$ k$\Omega$, $R_1 = 6.14$ k$\Omega$, $R_2 = 1.49$ k$\Omega$

**CYU 10.5**   (a) $-14.5$; (b) $-98$; (c) $-10.25$. The amplifier behaves more closely to the ideal model if the source resistance is small and the load resistance is large.

**CYU 10.6**   (a) $-105$; (b) $-71$; (c) $-105$

**CYU 10.7**   $A_V = -51.39$; $A_I = -34.17$

**CYU 10.8**   $I_{DQ} = 0.32, 0.72, 1.28, 2, 2.88, 3.92, 5.12$ mA; $g_m = 1.6, 2.4, 3.2, 4, 4.8, 5.6, 6.4$ mA/V

**CYU 10.9**   $\mu = -10.3$; $A_V = -9.36$

**CYU 10.10**   $\mu = -0.16, -0.24, -0.32, -0.4, -0.48, -0.56, -0.64$. $A_V = -0.146, -0.218, -0.291, -0.364, -0.436,$
$-0.509, -0.582$

**CYU 10.13**   62.5 $\Omega$

**CYU 10.14**

| $v_1$ | $v_2$ | $v_{\text{out}}$ |
|-------|-------|------------------|
| 0 V   | 0 V   | 5 V              |
| 5 V   | 0 V   | 5 V              |
| 0 V   | 5 V   | 5 V              |
| 5 V   | 5 V   | 0 V              |

NAND gate

---

# HOMEWORK PROBLEMS

## Section 1: Bipolar Transistor Amplifiers

**10.1**   The circuit shown in Figure P10.1 is a simplified *common-base circuit.*

   a. Determine the operating point of the transistor circuit.

   b. Draw the $h$-parameter model for this circuit.

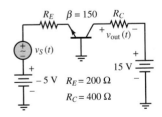

**Figure P10.1**

**10.2**   Consider a common-emitter amplifier circuit with the following parameter values (see Figure 10.7):
$V_{CC} = 10$ V, $R_1 = 3000\ \Omega$, $R_2 = 3000\ \Omega$, $R_S = 50\ \Omega$,
$R_C = R_E = 100\ \Omega$, $R_L = 150\ \Omega$, $\beta = 100$.

   a. Find the operating point of the transistor.

   b. Draw the AC equivalent circuit of the amplifier.

   c. If $h_{oe} = 10^{-5}$ S and $h_{fe} = \beta$, determine the voltage gain as defined by $v_{\text{out}}/v_{\text{in}}$.

   d. Find the input resistance, $r_i$.

   e. Find the output resistance, $r_o$.

**10.3**   The transistor shown in the amplifier circuit in Figure P10.3 has $h_{ie} = 1.3$ k$\Omega$, $h_{fe} = 90$, and $h_{oe} = 120\ \mu$S. Find the voltage gain.

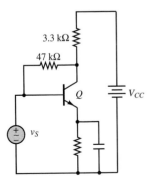

**Figure P10.3**

**10.4**   The circuit shown in Figure P10.4 is a variation of the common-base amplifier.

a. Find the $Q$ point of the transistor.

b. Draw the AC equivalent circuit, using $h$ parameters.

c. Find the voltage gain $v_L/v_{in}$.

d. Find the input resistance, $r_i$.

e. Find the output resistance, $r_o$.

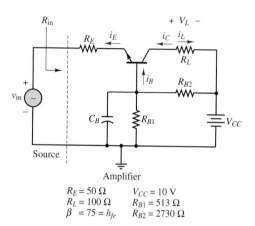

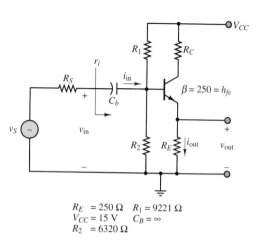

d. Find the input resistance, $r_i$.

e. Find the output resistance, $r_o$.

$$R_E = 50\,\Omega \qquad V_{CC} = 10\text{ V}$$
$$R_L = 100\,\Omega \qquad R_{B1} = 513\,\Omega$$
$$\beta = 75 = h_{fe} \qquad R_{B2} = 2730\,\Omega$$

**Figure P10.4**

$$R_E = 250\,\Omega \quad R_1 = 9221\,\Omega$$
$$V_{CC} = 15\text{ V} \quad C_B = \infty$$
$$R_2 = 6320\,\Omega$$

**Figure P10.6**

**10.5**  For the circuit shown in Figure P10.5, $v_S$ is a small sine wave signal with average value of 3 V. If $\beta = 100$ and $R_B = 60$ k$\Omega$,

a. Find the value of $R_E$ so that $I_E$ is 1 mA.

b. Find $R_C$ so that $V_C$ is 5 V.

c. For $R_L = 5$ k$\Omega$, find the small-signal equivalent circuit of the amplifier.

d. Find the voltage gain.

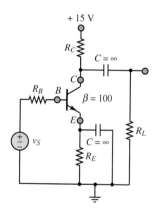

**Figure P10.5**

**10.6**  The circuit in Figure P10.6 is in a common-collector configuration. Using Table 10.2 and assuming $R_C = 200\,\Omega$:

a. Find the operating point of the transistor.

b. Draw the AC equivalent circuit for $h_{oe} = 10^{-4}$ S.

c. If the voltage gain is defined as $v_{out}/v_{in}$, find the voltage gain. If the current gain is defined as $i_{out}/i_{in}$, find the current gain.

**10.7**  The circuit that supplies energy to an automobile's fuel injector is shown in Figure P10.7(a). The internal circuitry of the injector can be modeled as shown in Figure P10.7(b). The injector will inject gasoline into the intake manifold when $I_{inj} \geq 0.1$ A. The voltage $V_{signal}$ is a pulse train whose shape is as shown in Figure P10.7(c). If the engine is cold and under start-up conditions, the signal duration, $\tau$, is determined by the equation

$$\tau = \text{BIT} \times K_C + \text{VCIT}$$

where

$$\text{BIT} = \text{Basic injection time} = 1\text{ ms}$$
$$K_C = \text{Compensation constant of temperature}$$
$$\text{of coolant } (T_C)$$
$$\text{VCIT} = \text{Voltage-compensated injection time}$$

The characteristics of VCIT and $K_C$ are shown in Figure P10.7(d).

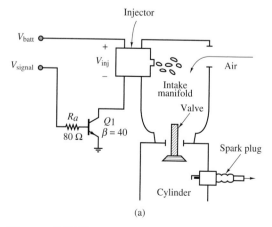

(a)

**Figure P10.7**

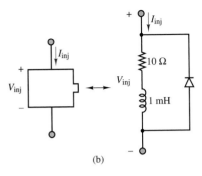

(b)

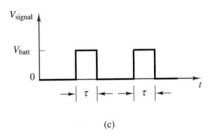

(c)

dissipate 0.5 W when $V_R = 10$ V. If $R_E = 40\ \Omega$, $R_B = 450\ \Omega$, $R_S = 20\ \Omega$, $L = 100$ mH, $R_{w1} = R_{w2} = 100\ \Omega$, $V_{CE\ \text{sat}} = 0.3$ V, $V_\gamma = 0.75$ V, and $V_S(t)$ is a square wave that ranges from 0 volts to 4.8 volts and has a maximum current of 20 mA.

a. Determine the maximum power dissipated by this circuit.

b. Determine the time it takes for the relay to switch from the closed to the open position if $V_S$ has been at 4.8 V for a very long time.

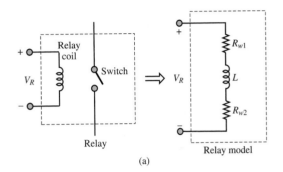

(a)

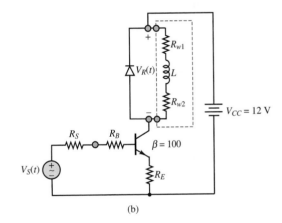

(b)

**Figure P10.8**

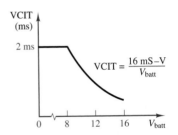

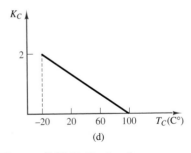

(d)

**Figure P10.7** (Continued)

If the transistor, $Q_1$, saturates at $V_{CE} = 0.3$ V and $V_{BE} = 0.9$ V, find the duration of the fuel injector pulse if

a. $V_{\text{batt}} = 13$ V, $T_C = 100°$ C
b. $V_{\text{batt}} = 8.6$ V, $T_C = 20°$ C

**10.8** A DC relay coil can be modeled as an inductor with series resistance due to the windings, as shown in Figure P10.8(a). The relay in Figure P10.8(b) is being driven by a transistor circuit. When $V_R$ is greater than 7.2 VDC, the relay switch will close if it is in the open state. If the switch is closed, it will open when $V_R$ is less than 2.4 VDC. The diode on resistance is 10 $\Omega$, and its off resistance is $\infty$. The relay is rated to

**10.9** The circuit shown in Figure P10.9 is used to switch a relay that turns a light off and on under the control of a computer. The relay dissipates 0.5 W at 5 VDC. It switches on at 3 VDC and off at 1.0 VDC. What is the maximum frequency with which the light can be switched? The inductance of the relay is 5 mH, and the transistor saturates at 0.2 V, $V_\gamma = 0.8$ V.

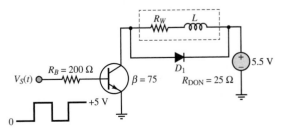

**Figure P10.9**

**10.10**  A Darlington pair of transistors is connected as shown in Figure P10.10. The transistor parameters for small-signal operation are $Q_1$: $h_{ie} = 1.5\ k\Omega$, $h_{re} = 4 \times 10^{-4}$, $h_{oe} = 110\ \mu A/V$, and $h_{fe} = 130$; $Q_2$: $h_{ie} = 200\ \Omega$, $h_{re} = 10^{-3}$, $h_{oe} = 500\ \mu A/V$, and $h_{fe} = 70$. Calculate:

a. The overall current gain.

b. The input impedance.

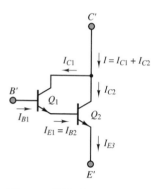

**Figure P10.10**

**10.11**  Given the common-emitter amplifier shown in Figure P10.11, where the transistor has the following $h$ parameters:

|          | **Maximum**         | **Minimum**              |
| -------- | ------------------- | ------------------------ |
| $h_{ie}$ | 15 k$\Omega$        | 1 k$\Omega$              |
| $h_{fe}$ | 500                 | 40                       |
| $h_{re}$ | $8 \times 10^{-4}$  | $0.1 \times 10^{-4}$     |
| $h_{oe}$ | 30 $\mu$S           | $1 \times 10^{-6}\ \mu$S |

Determine maximum and minimum values for:

a. The open-circuit voltage gain $A_V$.

b. The open-circuit current gain $A_I$.

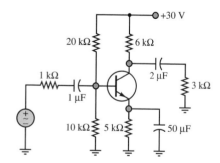

**Figure P10.11**

**10.12**  The transistor shown in Figure P10.12 has $V_x = 0.6$ V. Determine values for $R_1$ and $R_2$ such that

a. The quiescent collector-emitter voltage, $V_{CEQ}$, is 5V.

b. The quiescent collector current, $I_{CQ}$, will vary no more than 10 percent as $\beta$ varies from 20 to 50.

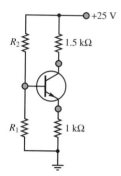

**Figure P10.12**

**10.13**  Consider again the amplifier of Figure P10.12. Determine values of $R_1$ and $R_2$ which will permit maximum symmetrical swing in the collector current. Assume $\delta = 100$.

**10.14**  In the small-signal analysis of circuits with transistors, the transistor is modeled as a small-signal AC circuit. The model shown in Figure P10.14 (the "hybrid *pi*" model) is valid for any BJT, *npn* or *pnp*.

a. The capacitors are internal to the device. Their capacitance can be determined only from the transistor specification sheet. In a mid-frequency analysis, they are modeled as open circuits. Why can this be done?

b. State the definitions of and determine expressions for the transconductance and $r_\pi$.

c. Illustrate graphically the definition of the transconductance and $r_\pi$ and the significance of the $Q$ point in determining their values.

d. The small signal output resistance of a transistor is determined either by a graphical analysis or from the transistor specification sheet. It is normally large and is often modeled as an open circuit. State its definition and determine its value using the $i$-$v$ characteristic for the 2N3904 transistor for a $Q$ point at a base current of 30 $\mu$A and a collector-emitter voltage of 8 V.

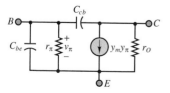

**Figure P10.14**

**10.15** The circuit shown in Figure P10.15 is a common-collector (also called an emitter follower) amplifier stage. The transistor is fabricated from silicon. A DC analysis gives the $Q$ or DC operating point:

$$V_{CEQ} = 10.21 \text{ V} \qquad I_C = 358 \ \mu\text{A} \qquad \beta = 130$$

At mid-frequency, determine the small signal AC model of the transistor and draw the AC small-signal equivalent circuit.

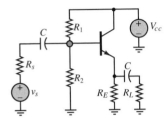

**Figure P10.15**

**10.16** Using the AC small-signal mid-frequency equivalent circuit shown in Figure P10.16 to specify the voltages, currents, etc. used in your definitions and conditions:

a. Define the input resistance, output resistance and no-load voltage gain of an amplifier stage. Include the conditions required to determine each.

b. Draw a simplified model using these three circuit parameters. Include the signal source and gain and derive an expression for the overall gain.

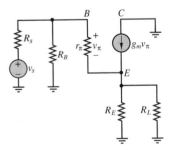

**Figure P10.16**

**10.17** The circuit shown in Figure P10.15 is a common-collector (or emitter follower) amplifier stage. The transistor is fabricated from silicon. A solution for the DC operating ($Q$ or bias) point gives:

$$
\begin{aligned}
I_{CQ} &= 2.5 \text{ mA} & \beta &= 70 \\
R_1 &= 330 \text{ k}\Omega & R_2 &= 100 \text{ k}\Omega \\
R_L &= 16 \ \Omega & R_S &= 0.6 \text{ k}\Omega \\
R_E &= 1.7 \text{ k}\Omega & C &= 0.47 \ \mu\text{F}
\end{aligned}
$$

Determine, at mid-frequency, the AC small-signal mid-frequency equivalent circuit.

**10.18** The AC small-signal mid-frequency equivalent circuit for a common collector [or emitter follower] amplifier stage is shown in Figure P10.18. The DC bias point is:

$$
\begin{aligned}
I_{CQ} &= 717 \ \mu\text{A} & \beta &= 70 \\
V_{CC} &= 20 \text{ V} & V_{BB} &= 6.5 \text{ V} \\
R_B &= 1.7 \text{ k}\Omega & R_E &= 1.9 \text{ k}\Omega \\
R_L &= 3 \text{ k}\Omega & R_S &= 0.6 \text{ k}\Omega \\
C &= 0.47 \ \mu\text{F}
\end{aligned}
$$

Determine, the no-load voltage gain $v_o/v_i$; i.e., do not include the loading effects of the source and load resistance.

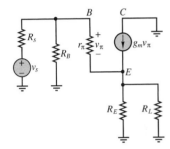

**Figure P10.18**

**10.19** The circuit shown in Figure P10.18 is the small-signal equivalent circuit of a common-collector (or emitter follower) amplifier stage. The transistor is fabricated from silicon.

$$
\begin{aligned}
g_m &= 96.15 \text{ mS} & \beta &= 70 \\
r_\pi &= 728.0 \ \Omega & R_B &= 74.76 \text{ k}\Omega \\
R_L &= 16 \ \Omega & R_S &= 0.6 \text{ k}\Omega \\
R_E &= 1.7 \text{ k}\Omega
\end{aligned}
$$

Determine, at mid-frequency, the output resistance.

**10.20** Shown in Figure P10.18 is the AC small-signal mid-frequency equivalent circuit of a common collector [or emitter follower] amplifier stage. The $Q$ point and the parameters for the small-signal AC model of the transistor are:

$$
\begin{aligned}
g_m &= 14.3 \text{ mS} & r_\pi &= 9.09 \text{ k}\Omega \\
\beta &= 130 & R_S &= 6.8 \text{ k}\Omega \\
R_B &= 17.35 \text{ k}\Omega & R_E &= 5 \text{ k}\Omega \\
R_L &= 16 \ \Omega & C &= 0.47 \ \mu\text{F}
\end{aligned}
$$

The input resistance (you may wish to verify it), including the effect of the load resistor, is 6.80 k$\Omega$. Determine, at mid-frequency, the power gain $P_o/P_i$ (in dB).

**10.21** The circuit shown in Figure P10.18 is the mid-frequency, AC small-signal equivalent circuit for a common-collector (or emitter follower) amplifier stage.

The small-signal AC parameters of the transistor are:

$$g_m = 40 \text{ mS} \qquad r_\Pi = 1.25 \text{ k}\Omega$$
$$R_B = 50 \text{ k}\Omega \qquad R_E = 1.7 \text{ k}\Omega$$
$$R_L = 2 \text{ k}\Omega \qquad R_S = 7 \text{ k}\Omega$$
$$C = 0.47 \text{ } \mu\text{F}$$

Determine input resistance.

**10.22**    Shown in Figure P10.22 is the schematic of a common-emitter amplifier stage, the most common BJT amplifier state. The DC operating ($Q$ or bias) point is:

$$I_{CQ} = 2.5 \text{ mA} \qquad \beta = 70$$
$$R_1 = 330 \text{ k}\Omega \qquad R_2 = 100 \text{ k}\Omega$$
$$R_C = 3.3 \text{ k}\Omega \qquad R_E = 1.7 \text{ k}\Omega$$
$$R_L = 16 \text{ }\Omega \qquad R_S = 0.6 \text{ k}\Omega$$
$$C = 0.47 \text{ } \mu\text{F}$$

Determine the mid-frequency

a. AC small-signal equivalent circuit.

b. Input resistance.

c. Output resistance.

d. No load voltage gain $v_o/v_i$.

e. Power gain in dB.

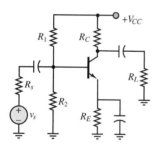

**Figure P10.22**

**10.23**    Shown in Figure P10.23 is the small-signal high-frequency AC equivalent circuit for a common-emitter BJT amplifier stage. Coupling and bypass capacitors have been modeled as short circuits and are not shown. For simplicity only one of the two internal BJT capacitances $C$ ($= C_{cb}$) is included. Recall that, at high frequencies, the impedance of these capacitances decreases and this affects the input and output resistance and the gain. The no-load gain at mid-frequency, the small signal AC transistor parameters, and component values are:

$$A_{vo} = \frac{v_o}{v_i} = -120 \qquad g_m = 40 \text{ mS} \qquad r_\pi = 2.5 \text{ k}\Omega$$
$$C = C_{cb} = 3 \text{ pF} \qquad R_B = 44 \text{ k}\Omega \qquad R_C = 3 \text{ k}\Omega$$

Determine the frequency at which the magnitude of the no-load gain decreases to 0.707 of its value at

mid-frequency, or 3 dB below the mid-frequency gain. This frequency is the same as the cutoff frequency in a filter.

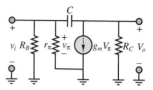

**Figure P10.23**

## Section 2:
## Field-Effect Transistor Amplifiers

**10.24**    The $i$-$v$ characteristic of a depletion MOSFET is shown in Figure P10.24(a) and an amplifier circuit based on the MOSFET is shown in Figure P10.24(b). Determine the quiescent operating point if $V_{DD} = 30$ V and $R_D = 500 \text{ }\Omega$.

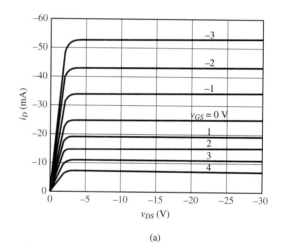

(a)

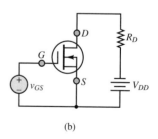

(b)

**Figure P10.24**

**10.25**    Consider again the amplifier of Figure P10.24. Determine the quiescent operating point if $V_{DD} = 15$ V and $R_D = 330 \text{ }\Omega$.

**10.26**  Again consider the amplifier of Figure P10.24. Determine the quiescent operating point if $V_{DD} = 50$ V and $R_D = 1.5$ kΩ.

**10.27**  Consider the amplifier of Figure P10.24. Let $R_D = 1$ kΩ. If the quiescent operating point is $I_{DQ} = -25$ mA, $V_{DSQ} = -12.5$ V, determine the value of $V_{DD}$.

**10.28**  Consider the amplifier of Figure P10.24. Let $R_D = 2$ kΩ. If the quiescent operating point is $I_{DQ} = -25$ mA, $V_{DSQ} = -12.5$ V, determine the value of $V_{DD}$.

**10.29**  Consider again the amplifier of Figure P10.24 with component values as specified in Problem 10.24. Let $v_{GS} = 1 \sin \omega t$ V. Sketch the output voltage and current, and estimate the voltage amplification.

**10.30**  Consider again the amplifier of Figure P10.24 with component values as specified in Problem 10.24. Let $v_{GS} = 3 \sin \omega t$ V. Sketch the output voltage and current, and estimate the voltage amplification.

**10.31**  Consider again the amplifier of Figure P10.24 with component values as specified in Problem 10.25. Let $v_{GS} = 1 \sin \omega t$ V. Sketch the output voltage and current, and estimate the voltage amplification.

**10.32**  Consider again the amplifier of Figure P10.24 with component values as specified in Problem 10.25. Let $v_{GS} = 3 \sin \omega t$ V. Sketch the output voltage and current, and estimate the voltage amplification.

**10.33**  Consider again the amplifier of Figure P10.24 with component values as specified in Problem 10.24. Estimate $\mu$ and $g_m$ at the quiescent operating point.

**10.34**  Shown in Figure P10.34 is a common-source amplifier stage implemented with an $n$-channel depletion mode MOSFET with the static $i$-$v$ characteristics shown in Chapter 9. The $Q$ point and component values are:

$$V_{DSQ} = 13.6 \text{ V} \qquad V_{GSQ} = 1.5 \text{ V}$$
$$R_D = 1.0 \text{ k}\Omega \qquad R_S = 400 \text{ }\Omega$$
$$R_L = 3.2 \text{ k}\Omega \qquad R_{SS} = 600 \text{ }\Omega$$
$$V_{DD} = 35 \text{ V} \qquad v_i(t) = 2 \sin(\omega t) \text{ V}$$

a. Construct the DC load line.
b. Construct the AC load line.
c. Using values along the AC load line, plot (do not sketch) the transfer characteristic, i.e., the drain current as a function of the gate-source voltage.
d. Using the transfer function, sketch the output voltage as a function of $\omega t$ for $0 < \omega t < 2\pi$. Recall: $v_I = V_{IQ} + v_i$.
e. Determine the loaded gain of the stage.

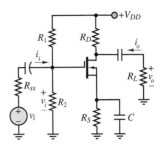

**Figure P10.34**

**10.35**  Shown in Figure P10.34 is a common-source amplifier stage implemented with an $n$-channel depletion MOSFET with the static $i$-$v$ characteristics shown in Chapter 9. Assume that the $Q$ point and component values are:

$$V_{GSQ} = 1.5 \text{ V} \qquad V_{DSQ} = 13.6 \text{ V}$$
$$I_{DQ} = 15.3 \text{ mA}$$
$$R_D = 1 \text{ k}\Omega \qquad R_S = 400 \text{ }\Omega$$
$$R_L = 3.2 \text{ k}\Omega \qquad R_{SS} = 600 \text{ }\Omega$$
$$V_{DD} = 35 \text{ V} \qquad v_i(t) = 1 \sin(\omega t) \text{ V}$$

a. For loaded conditions (include the effects of the load resistance), construct the AC load line, plot the output voltage as a function of $\omega t$, and determine the large-signal voltage gain.
b. Repeat without including the loading effect of the load resistance.

**10.36**  In the small-signal analysis of circuits with transistors, the transistor is replaced with a small-signal AC model or equivalent circuit. A simplified version of this model is shown in Figure P10.36. It is valid for any FET, $n$- or $p$-channel, junction, depletion, or enhancement. Assume here it is for a depletion MOSFET.

a. In a mid-frequency analysis the capacitors which are internal to the device are modeled as open circuits. Explain why this can be done.
b. State the definition of and determine an expression for the transconductance.
c. Illustrate graphically the definition of the transconductance and the significance of the $Q$ point in determining its value.
d. The small-signal output resistance of a transistor normally very large and in a very simple model can be approximated as an open circuit. State its definition and determine its value using the $i$-$v$ characteristic (given in Chapter 9 for depletion MOSFETs) at the $Q$ point:

$$V_{GSQ} = 0.5 \text{ V} \qquad V_{DSQ} = 20\text{V}$$

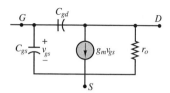

**Figure P10.36**

**10.37**  An *n*-channel enhancement-mode MOSFET has the parameters

$$I_{DSS} = 7 \text{ mA} \qquad V_T = 5 \text{ V}$$

It is operated at the $Q$ point:

$$V_{GSQ} = 7 \text{ V} \qquad I_{DQ} = 1.120 \text{ mA}$$

a. Draw the symbol for the transistor.
b. Determine numerically the parameters for the simplest small-signal AC model of the transistor.
c. Draw the small-signal AC model.

**10.38**  In a small-signal AC equivalent circuit of a circuit containing FETs, AC coupling or DC blocking capacitors, bypass capacitors, internal transistor capacitance, and ideal DC sources, how are the following components in the circuit modeled in the mid-frequency range:

a. Transistors?
b. AC coupling or DC blocking capacitors?
c. Bypass capacitors?
d. Internal capacitances in the transistor? There is a small internal capacitance between the gate and drain terminals and another between the gate and source terminals.
e. Ideal DC source?

**10.39**  Given below are the transistor parameters, a hypothetical $Q$ point for the transistor, and the component values in the circuit shown in Figure P10.39.

$$I_{DSS} = 0.5 \text{ mA} \qquad V_T = 1 \text{ V}$$
$$I_{DQ} = 0.5 \text{ mA} \qquad V_{GSQ} = 2 \text{ V}$$
$$R_1 = 10 \text{ M}\Omega \qquad R_2 = 10 \text{ M}\Omega$$
$$R_D = 6 \text{ k}\Omega \qquad R_S = 6 \text{ k}\Omega$$
$$R_{SS} = 50 \text{ }\Omega \qquad R_L = 3 \text{ k}\Omega$$

a. Determine the AC model for the transistor including numerical values.
b. Draw the small-signal AC equivalent circuit.

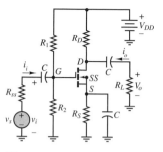

**Figure P10.39**

**10.40**  Given below are the transistor parameters, a hypothetical $Q$ point for the transistor, and the component values in the circuit shown in Figure P10.40.

$$I_{DSS} = 0.5 \text{ mA} \qquad V_T = 1 \text{ V}$$
$$I_{DQ} = 0.5 \text{ mA} \qquad V_{GSQ} = 2 \text{ V}$$
$$R_1 = 10 \text{ M}\Omega$$
$$R_D = 6 \text{ k}\Omega \qquad R_S = 6 \text{ k}\Omega$$
$$R_{SS} = 50 \text{ }\Omega \qquad R_L = 3 \text{ k}\Omega$$

a. Determine the AC model for the transistor including numerical values.
b. Draw the small-signal AC equivalent circuit.

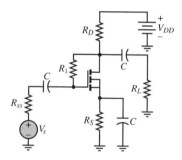

**Figure P10.40**

**10.41**  Define the input resistance, output resistance, and no-load voltage gain of an amplifier stage. Include the conditions required for determining each. Use the AC small-signal equivalent circuit shown in Figure P10.41 to specify the voltages, currents, etc. used in your definitions and conditions. Draw a simplified model using these three circuit parameters. Include the signal source and load and derive an expression for the overall gain.

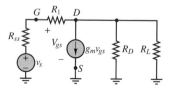

**Figure P10.41**

**10.42**  The circuit shown in Figure P10.39 is a common-source stage. The transistor parameters, $Q$ point, and component values in the circuit are:

$$I_{DSS} = 1.125 \text{ mA} \qquad V_T = 1.5 \text{ V}$$
$$I_{DQ} = 1.125 \text{ mA} \qquad V_{GSQ} = 3 \text{ V}$$
$$R_1 = 1.32 \text{ M}\Omega \qquad R_2 = 2.2 \text{ M}\Omega$$
$$R_D = 4 \text{ k}\Omega \qquad R_S = 4 \text{ k}\Omega$$
$$R_L = 1.3 \text{ k}\Omega \qquad R_{SS} = 700 \text{ }\Omega$$
$$V_{DD} = 12 \text{ V}$$

a. Determine all the component values for and draw the AC small-signal equivalent circuit (for mid-frequencies, of course).

b. Derive expressions for and determine the values of the input resistance $R_i$, the output resistance $R_o$, and the no-load voltage gain $A_{vo}$.

c. Determine the overall voltage gain $v_o/v_s$ (which accounts for the effects of the load and signal source resistance) in dB.

**10.43**  The circuit shown in Figure P10.43 is a common-drain (or source follower) amplifier stage. The transistor parameters, the $Q$ point, and component values are:

$$I_{DSS} = 1.125 \text{ mA} \qquad V_T = 1.5 \text{ V}$$
$$I_{DQ} = 1.125 \text{ mA} \qquad V_{GSQ} = 3 \text{ V}$$
$$R_1 = 1.32 \text{ M}\Omega \qquad R_2 = 2.2 \text{ M}\Omega$$
$$R_S = 4 \text{ k}\Omega$$
$$R_L = 1.3 \text{ k}\Omega \qquad R_{SS} = 700 \text{ }\Omega$$
$$V_{DD} = 12 \text{ V}$$

a. Draw the AC small-signal equivalent circuit.

b. Derive expressions for and determine the values of the input resistance $R_i$, the output resistance $R_o$, and the no-load voltage gain $A_{vo}$.

c. Determine the overall voltage gain $v_o/v_s$ (which accounts for the effects of the load and signal source resistance) in dB.

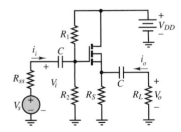

**Figure P10.43**

## Section 3: Gates and Switches

**10.44**  Show that the circuit of Figure P10.44 functions as an OR gate if the output is taken at $v_{o1}$.

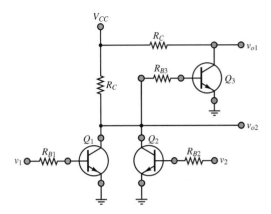

**Figure P10.44**

**10.45**  Show that the circuit of Figure P10.44 functions as a NOR gate if the output is taken at $v_{o2}$.

**10.46**  Show that the circuit of Figure P10.46 functions as an AND gate if the output is taken at $v_{o1}$.

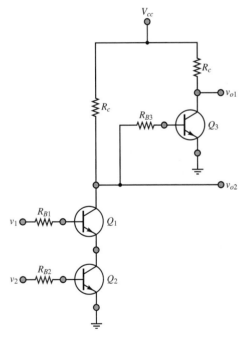

**Figure P10.46**

**10.47**  Show that the circuit of Figure P10.46 functions as a NAND gate if the output is taken at $v_{o2}$.

**10.48**  In Figure P10.48, the minimum value of $v_{in}$ for a high input is 2.0 V. Assume that transistor $Q_1$ has a $\beta$

of at least 10. Find the range for resistor $R_B$ that can guarantee that the transistor $Q_1$ is on.

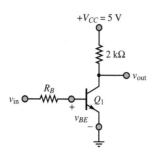

**Figure P10.48**

**10.49**   Figure P10.49 shows a circuit with two transistor inverters connected in series, where $R_{1C} = R_{2C} = 10$ kΩ and $R_{1B} = R_{2B} = 27$ kΩ.

a. Find $v_B$, $v_{out}$, and the state of transistor $Q_1$ when $v_{in}$ is low.

b. Find $v_B$, $v_{out}$, and the state of transistor $Q_1$ when $v_{in}$ is high.

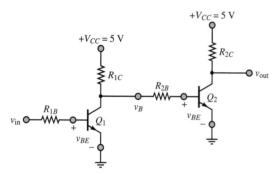

**Figure P10.49**

**10.50**   For the inverter of Figure P10.50, $R_B = 5$ kΩ and $R_{C1} = R_{C2} = 2$ kΩ. Find the minimum values of $\beta_1$ and $\beta_2$ to ensure that $Q_1$ and $Q_2$ saturate when $v_{in}$ is high.

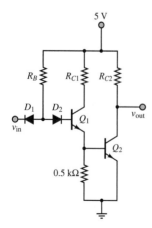

**Figure P10.50**

**10.51**   For the inverter of Figure P10.50, $R_B = 4$ kΩ, $R_{C1} = 2.5$ kΩ, and $\beta_1 = \beta_2 = 4$. Show that $Q_1$ saturates when $v_{in}$ is high. Find a condition for $R_{C2}$ to ensure that $Q_2$ also saturates.

**10.52**   The basic circuit of a TTL gate is shown in the circuit of Figure P10.52. Determine the logic function performed by this circuit.

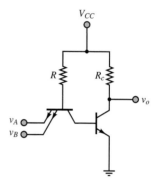

**Figure P10.52**

**10.53**   Figure P10.53 is a circuit diagram for a three-input TTL NAND gate. Assuming that all the input voltages are high, find $v_{B1}$, $v_{B2}$, $v_{B3}$, $v_{C2}$, and $v_{out}$. Also, indicate the operating region of each transistor.

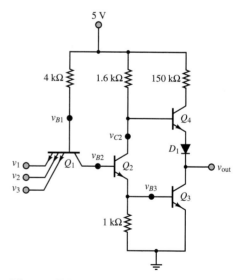

**Figure P10.53**

**10.54**   Show that when two or more emitter-follower outputs are connected to a common load, as shown in the circuit of Figure P10.54, the OR operation results; that is, $v_o = v_1$ OR $v_2$.

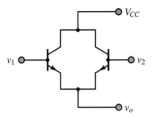

**Figure P10.54**

**10.55** For the CMOS NAND gate of Check Your Understanding Exercise 10.14 identify the state of each transistor for $v_1 = v_2 = 5$ V.

**10.56** Repeat Problem 10.55 for $v_1 = 5$ V and $v_2 = 0$ V.

**10.57** Draw the schematic diagram of a two-input CMOS OR gate.

**10.58** Draw the schematic diagram of a two-input CMOS AND gate.

**10.59** Draw the schematic diagram of a two-input TTL OR gate.

**10.60** Draw the schematic diagram of a two-input TTL AND gate.

**10.61** Show that the circuit of Figure P10.61 functions as a logic inverter.

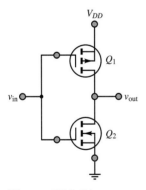

**Figure P10.61**

**10.62** Show that the circuit of Figure P10.62 functions as a NOR gate.

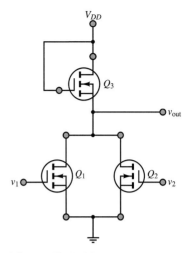

**Figure P10.62**

**10.63** Show that the circuit of Figure P10.63 functions as a NAND gate.

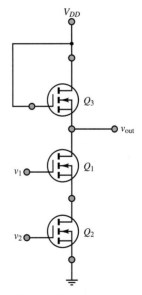

**Figure P10.63**

# C H A P T E R
# 11

# Power Electronics

T he objective of this chapter is to present a survey of power electronic devices and systems. Power electronic devices form the "muscle" of many
electromechanical systems. For example, one finds such devices in many
appliances, in industrial machinery, and virtually wherever an electric motor is found, since one of the foremost applications of power electronic devices is to
supply and control the currents and voltages required to power electric machines,
such as those introduced in Part III of this book.

Power electronic devices are specially designed diodes and transistors that
have the ability to carry large currents and sustain large voltages; thus, the basis
for this chapter is the material on diodes and transistors introduced in Chapters 8
through 10. A detailed understanding of diode and transistor small-signal models
is not necessary for acquiring an essential knowledge of power semiconductor
devices.

This chapter will describe the basic properties of each type of power electronic device, and it will illustrate the application of a selected few, especially in
electric motor power supplies. After completing the chapter, you should be able
to recognize the symbols for the major power semiconductor devices and understand their principles of operation. You should also understand the operation of
the principal electronic power supplies for DC and AC motors.

Upon completing this chapter, you should be able to:

• Provide a classification of power electronic devices and circuits.
• Understand the operation of voltage regulators, transistor power amplifiers, and power switches.
• Analyze rectifier and controlled rectifier circuits.
• Understand the basic principles behind DC and AC electric motor drives.

## 11.1  CLASSIFICATION OF POWER ELECTRONIC DEVICES

Power semiconductors can be broadly subdivided into five groups: (1) power diodes, (2) thyristors, (3) power bipolar junction transistors (BJTs), (4) insulated-gate bipolar transistors (IGBTs), and (5) static induction transistors (SITs). Figure 11.1 depicts the symbols for the most common power electronic devices.

**Power diodes** are functionally identical to the diodes introduced in Chapter 8, except for their ability to carry much larger currents. You will recall that a diode conducts in the forward-biased mode when the anode voltage ($V_A$) is higher than the cathode voltage ($V_K$). Three types of power diodes exist: *general-purpose, high-speed (fast-recovery),* and *Schottky.* Typical ranges of voltage and current are 3,000 V and 3,500 A for general-purpose diodes and 3,000 V and 1,000 A for high-speed devices. The latter have switching times as low as a fraction of a microsecond. Schottky diodes can switch much faster (in the nanosecond range) but are limited to around 100 V and 300 A. The forward voltage drop of power diodes is not much higher than that of low-power diodes, being between 0.5 and 1.2 V. Since power diodes are used with rather large voltages, the forward bias voltage is usually considered negligible relative to other voltages in the circuit, and the switching characteristics of power diodes may be considered near ideal. The principal consideration in choosing power diodes is their power rating.

**Thyristors** function like power diodes with an additional gate terminal that controls the time when the device begins conducting; a thyristor starts to conduct when a small gate current is injected into the gate terminal, provided that the anode voltage is greater than the cathode voltage (or $V_{AK} > 0$ V). The forward voltage drop of a thyristor is of the order of 0.5 to 2 V. Once conduction is initiated, the gate current has no further control. To stop conduction, the device must be reverse-biased; that is, one must ensure that $V_{AK} \leq 0$ V. Thyristors can be rated at up to 6,000 V and 3,500 A. The **turn-off time** is an important characteristic of thyristors; it represents the time required for the device current to return to zero after external switching of $V_{AK}$. The fastest turn-off times available are in the range of 10 $\mu$s; however, such turn-off times are achieved only in devices with slightly lower power ratings (1,200 V, 1,000 A). Thyristors can be subclassified into the following groups: force-commutated and line-commutated thyristors, gate turn-off thyristors (GTOs), reverse-conducting thyristors (RCTs), static induction thyristors (SITs), gate-assisted turn-off thyristors (GATTs), light-activated silicon controlled rectifiers (LASCRs), and MOS controlled thyristors (MCTs). It is beyond the scope of this chapter to go into a detailed description of each of these types of devices; their operation is typically a slight modification of the basic operation of the thyristor. The reader who wishes to gain greater insight into this topic may refer to one of a number of excellent books specifically devoted to the subject of power electronics.

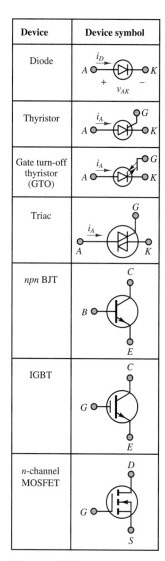

| Device | Device symbol |
|---|---|
| Diode | |
| Thyristor | |
| Gate turn-off thyristor (GTO) | |
| Triac | |
| *npn* BJT | |
| IGBT | |
| *n*-channel MOSFET | |

**Figure 11.1** Classification of power electronic devices

Two types of thyristor-based device deserve some more attention. The **triac,** as can be seen in Figure 11.1, consists of a pair of thyristors connected back to back, with a single gate; this allows for current control in either direction. Thus, a triac may be thought of as a bidirectional thyristor. The gate turn-off thyristor (GTO), on the other hand, can be turned on by applying a short positive pulse to the gate, like a thyristor, and can also be turned off by application of a short negative pulse. Thus, GTOs are very convenient in that they do not require separate commutation circuits to be turned on and off.

**Power BJTs** can reach ratings up to 1,200 V and 400 A, and they operate in much the same way as a conventional BJT. Power BJTs are used in power converter applications at frequencies up to around 10 kHz. **Power MOSFETs** can operate at somewhat higher frequencies (a few to several tens of kHz), but are limited in power (typically up to 1,000 V, 50 A). **Insulated-gate bipolar transistors (IGBTs)** are voltage-controlled (because of their insulated gate, reminiscent of insulated-gate FETs) power transistors that offer superior speed with respect to BJTs but are not quite as fast as power MOSFETs.

## 11.2    CLASSIFICATION OF POWER ELECTRONIC CIRCUITS

The devices that will be discussed in the present chapter find application in a variety of **power electronic circuits.** This section will briefly summarize the principal types of power electronic circuits and will qualitatively describe their operation. The following sections will describe the devices and their operation in these circuits in more detail.

One possible classification of power electronic circuits is given in Table 11.1. Many of the types of circuits are similar to circuits that were introduced in earlier chapters. Voltage regulators were introduced in Chapter 8 (see Fig. 8.52); this chapter will present a more detailed discussion of practical regulators. Power electronic switches function exactly like the transistor switches described in Chapter 10 (see Figures 10.47 and 10.55); their function is to act as voltage- or

Table 11.1  Power electronic circuits

| Circuit type | Essential features |
| --- | --- |
| Voltage regulators | Regulate a DC supply to a fixed voltage output |
| Power amplifiers | Large-signal amplification of voltages and currents |
| Switches | Electronic switches (for example, transistor switches) |
| Diode rectifier | Converts fixed AC voltage (single- or multiphase) to fixed DC voltage |
| AC-DC converter (controlled rectifier) | Converts fixed AC voltage (single- or multiphase) to variable DC voltage |
| AC-AC converter (AC voltage controller) | Converts fixed AC voltage to variable AC voltage (single- or multiphase) |
| DC-DC converter (chopper) | Converts fixed DC voltage to variable DC voltage |
| DC-AC converter (inverter) | Converts fixed DC voltage to variable AC voltage (single- or multiphase) |

current-controlled switches to turn AC or DC supplies on and off. Transistor power amplifiers are the high-power version of the BJT and MOSFET amplifiers studied in Chapters 9 and 10; it is important to consider power limitations and signal distortion more carefully in power amplifiers than in the small-signal amplifiers described in Chapter 10.

Diode rectifiers were discussed in Chapter 8 in their single-phase form (see Figures 8.20, 8.42, and 8.44); similar rectifiers can also be designed to operate with three-phase sources. The operation of a single-phase full-wave rectifier was summarized in Figure 8.43. AC-DC converters are also rectifiers, but they take advantage of the controlled properties of thyristors. The thyristor gate current can be timed to "fire" conduction at variable times, resulting in a variable DC output, as illustrated in Figure 11.2, which shows the circuit and behavior of a single-phase AC-DC converter. This type of converter is very commonly used as a supply for DC electric motors. In Figure 11.2, $\alpha$ is the firing angle of thyristor $T_1$, where the device starts to conduct.

AC-AC converters are used to obtain a variable AC voltage from a fixed AC source. Figure 11.3 shows a triac-based AC-AC converter, which takes advantage of the bidirectional capability of triacs to control the rms value of an alternating voltage. Note in particular that the resulting AC waveform is no longer a pure sinusoid even though its fundamental period (frequency) is unchanged. A **DC-DC converter,** also known as a *chopper,* or *switching regulator,* permits conversion of a fixed DC source to a variable DC supply. Figure 11.4 shows how such an effect may be obtained by controlling the base-emitter voltage of a bipolar transistor, enabling conduction at the desired time. This results in the conversion of the DC input voltage to a variable–duty-cycle output voltage, whose average value can be controlled by selecting the "on" time of the transistor. DC-DC converters find application as variable voltage supplies for DC electric motors used in electric vehicles.

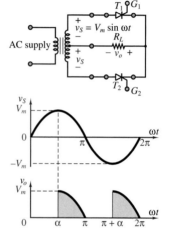

**Figure 11.2** AC-DC converter circuit and waveform

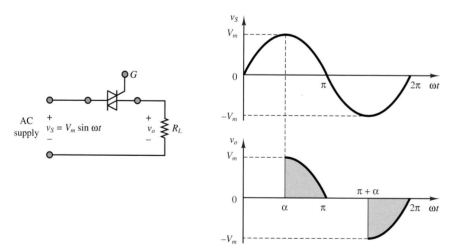

**Figure 11.3** AC-AC converter circuit and waveform

Finally, **DC-AC converters, or inverters,** are used to convert a fixed DC supply to a variable AC supply; they find application in AC motor control. The operation of these circuits is rather complex; it is illustrated conceptually in the wave-

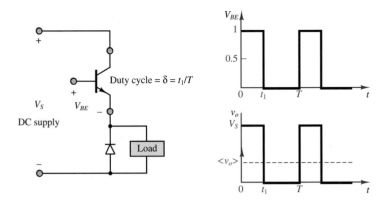

**Figure 11.4** DC-DC converter circuit and waveform

forms of Figure 11.5, where it is shown that by appropriately switching two pairs of transistors it is possible to generate an alternating current waveform (square wave).

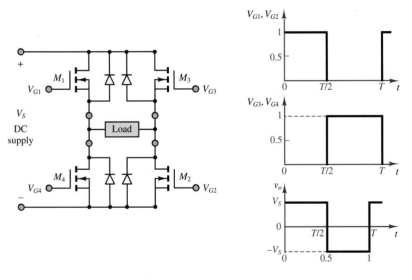

**Figure 11.5** DC-AC converter circuit and waveform

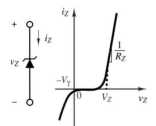

Each of the circuits of Table 11.1 will be analyzed in greater detail later in this chapter.

## 11.3    VOLTAGE REGULATORS

You will recall the discussion of the Zener diode as a voltage regulator in Chapter 8, where we introduced a voltage regulator as a three-terminal device that acts nearly as an ideal battery. Figure 11.6 depicts the appearance of a Zener diode $i$-$v$ characteristic and shows a block diagram of a three-terminal regulator.

A simple Zener diode is often inadequate for practical voltage regulation. In some cases, the Zener resistance alone might cause excessive power dissipation

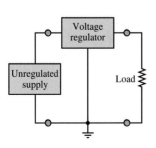

**Figure 11.6** Zener diode characteristic and voltage regulator circuit

in the Zener diode (especially when little current is required by the load). A more practical—and often-used—circuit is a regulator that includes a series pass transistor, shown in Figure 11.7. The operation of this voltage regulator is as follows. If the unregulated supply voltage, $v_S$, exceeds the Zener voltage, $v_Z$, by an amount sufficient to maintain the BJT in the active region, then $v_{BE} \approx V_\gamma$ and $v_Z \approx V_Z$, the Zener voltage. Thus, the load voltage is equal to

$$v_L = V_Z - V_\gamma = \text{ Constant} \qquad (11.1)$$

and is relatively independent of fluctuations in the unregulated source voltage, or in the required load current. The difference between the unregulated source voltage and the load voltage will appear across the *CE* "junction." Thus, the required power rating of the BJT may be determined by considering the largest unregulated voltage, $V_{S\text{max}}$:

$$P_{\text{BJT}} = (V_{S\text{max}} - V_L)i_C$$
$$\approx (V_{S\text{max}} - V_L)i_L \qquad (11.2)$$

The operation of a practical voltage regulator is discussed in more detail in Example 11.1.

---

### EXAMPLE  11.1  Analysis of Voltage Regulator

#### Problem

Determine the maximum allowable load current and the required Zener diode rating for the Zener regulator of Figure 11.7.

---

#### Solution

**Known Quantities:**  Transistor parameters; Zener voltage; unregulated source voltage; BJT base and load resistors.

**Find:**  $I_{L\,\text{max}}$ and $P_Z$.

**Schematics, Diagrams, Circuits, and Given Data:**  $V_S = 20$ V; $V_Z = 12.7$ V; $R_B = 47\ \Omega$; $R_L = 10\ \Omega$. Transistor data: TIP31 (see Table 11.2).

**Assumptions:**  Use the large-signal model of the BJT. Assume that the BJT is in the active region and the Zener diode is on and therefore regulating to the nominal voltage.

**Analysis:**  Figure 11.8 depicts the equivalent load circuit. Applying KVL we obtain:

$$V_Z = V_{BE} + R_L I$$

From which we can compute the load current:

$$\boxed{I = \frac{V_Z - V_\gamma}{R_L} = \frac{12.7 - 1.3}{10} = 1.14 \text{ A}}$$

We then note that $I$ is also the BJT emitter current, $I_E$.

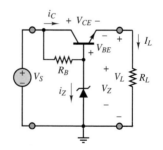

**Figure 11.7** Practical voltage regulator

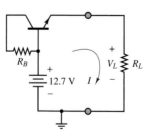

**Figure 11.8**

Applying KVL to the base circuit, shown in Figure 11.9, we compute the current through the base resistor:

$$I_{RB} = \frac{V_S - V_Z}{R_B} = \frac{20 - 12.7}{47} = 0.155 \text{ A}$$

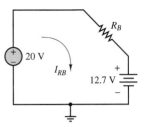

**Figure 11.9**

Knowing the base and emitter currents, we can determine whether the transistor is indeed is the active region, as assumed. With reference to Figure 11.10, we do so by computing $V_{CB}$ and $V_{BE}$ to determine the value of $V_{CE}$:

The base voltage is fixed by the presence of the Zener diode:

$$V_B = V_Z = 12.7 \text{ V}$$

$$V_{CB} = I_{RB} R_B = 0.155 \times 47 = 7.3 \text{ V}$$

$$V_E = I_L R_L = 1.14 \times 10 = 11.4 \text{ V}$$

Thus,

$$V_{CE} = V_{CB} + V_{BE} = V_{CB} + (V_B - V_E) = 7.3 + (12.7 - 11.4) = 8.6 \text{ V}$$

This value of the collector-emitter voltage indicates that the BJT is in the active region. Thus, we can use the large-signal model and compute the base current and subsequently the Zener current:

$$I_B = \frac{I_E}{\beta + 1} = \frac{1.14}{10 + 1} = 103.6 \text{ mA}$$

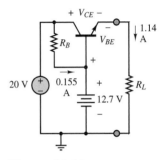

**Figure 11.10**

Applying KVL at the base junction (see Figure 11.11), we find

$$I_{RB} - I_B - I_Z = 0$$

$$I_Z = I_{RB} - I_B = 0.155 - 0.1036 = 51.4 \text{ mA}$$

and the power dissipated by the Zener diode is:

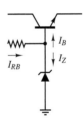

**Figure 11.11**

$$\boxed{P_Z = I_Z \times V_Z = 0.0514 \times 12.7 = 0.652 \text{ W}}$$

**Comments:** It will be instructive to compare these results with the Zener regulator examples of Chapter 8 (Examples 8.8, 8.9, and 8.10). Note that the Zener current is kept at a reasonably low level by the presence of the BJT (the load current is an amplified version of the base current). Further, disconnecting the load would result in cutting off the BJT, thus resulting in no Zener power dissipation. This is a significant advantage over the designs of Chapter 8.

**Focus on Computer-Aided Solutions:** An *Electronics Workbench*™ simulation of the practical Zener regulator analyzed in this example is included in the accompanying CD-ROM. You may wish to disconnect the load and verify that the Zener diode consumes no power.

VIRTUAL LAB

Three-terminal voltage regulators are available in packaged form to include all the necessary circuitry (often including protection against excess heat dissipation). Regulators are rated in terms of the regulated voltage and power dissipation.

Some types provide a variable regulated voltage by means of an external adjustment. Because of their requirement for relatively large power (and therefore heat) dissipation, voltage regulators often need to be attached to a **heat sink,** a thermally conductive assembly that aids in the cooling process. Figure 11.12 depicts the appearance of typical heat sinks. Heat sinking is a common procedure with many power electronic devices.

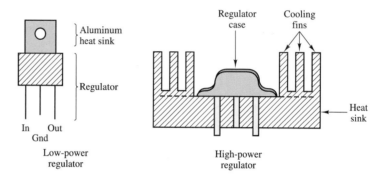

**Figure 11.12** Heat-sink construction for voltage regulators

## Check Your Understanding

**11.1**    Repeat Example 10.1 using the TIP31 transistor (see Table 11.2).

## 11.4    POWER AMPLIFIERS AND TRANSISTOR SWITCHES

So far, we have primarily considered low-power electronic devices, either in the form of small-signal linear amplifiers, or as switches and digital logic gates (the latter will be discussed in more detail in Chapter 13). There are many applications, however, in which it is desirable to provide a substantial amount of power to a load. Among the most common applications are loudspeakers (these can draw several amperes); electric motors and electromechanical actuators, which will be considered in greater detail in Chapters 16 through 18, and DC power supplies, which have already been analyzed to some extent in Chapter 8. In addition to such applications, the usage and control of electric power in industry requires electronic devices that can carry currents as high as hundreds of amperes, and voltages up to thousands of volts. Examples are readily found in the control of large motors and heavy industrial machinery.

   The aim of the present section is to discuss some of the more relevant issues in the design of *power amplifiers,* such as distortion and heat dissipation, and to introduce power switching transistors.

### Power Amplifiers

The brief discussion of power amplifiers in this section makes reference exclusively to the BJTs; this family of devices has traditionally dominated the field of **power**

**amplifiers,** although in recent years semiconductor technology has made power MOSFETs competitive with the performance of bipolar devices.

You may recall the notion of breakdown from the introductory discussion of BJTs. In practice, a bipolar transistor is limited in its operation by three factors: the maximum collector current, the maximum collector-emitter voltage, and the maximum power dissipation, which is the product of $I_C$ and $V_{CE}$. Figure 11.13 illustrates graphically the power limitation of a BJT by showing the regions where the maximum capabilities of the transistor are exceeded:

1. Exceeding the maximum allowable current $I_{C\,max}$ on a continuous basis will result in melting the wires that bond the device to the package terminals.
2. Maximum power dissipation is the locus of points for which $V_{CE}I_C = P_{max}$ at a case temperature of 25°C. The average power dissipation should not exceed $P_{max}$.
3. The instantaneous value of $v_{CE}$ should not exceed $V_{CEmax}$; otherwise, avalanche breakdown of the collector-base junction may occur.

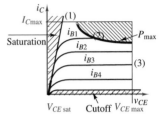

**Figure 11.13** Limitations of a BJT amplifier

It is important to note that the linear operation of the transistor as an amplifier is also limited by the saturation and cutoff limits.

The operation of a BJT as a linear amplifier is rather severely limited by these factors. Consider, first, the effect of driving an amplifier beyond the limits of the linear active region, into saturation or cutoff. The result will be signal distortion. For example, a sinusoid amplified by a transistor amplifier that is forced into saturation, either by a large input or by an excessive gain, will be compressed around the peaks, because of the decreasing device gain in the extreme regions. Thus, to satisfy these limitations—and to fully take advantage of the relatively distortion-free linear active region of operation for a BJT—the $Q$ point should be placed in the center of the device characteristic to obtain the **maximum symmetrical swing.** This point has already been discussed in Chapter 10 (see Example 10.3 and Figure 10.16, in particular).

The maximum power dissipation of the device, of course, presents a more drastic limitation on the performance of the amplifier, in that the transistor can be irreparably damaged if its power rating is exceeded. Values of the maximum allowable collector current, $I_{C\,max}$, of the maximum allowable transistor power dissipation, $P_{max}$, and of other relevant power BJT parameters are given in Table 11.2 for a few typical devices. Because of their large geometry and high operating

Table 11.2  Typical parameters for representative power BJTs

|  | **MJE3055T** | **TIP31** | **MJE170** |
|---|---|---|---|
| Type | *npn* | *npn* | *pnp* |
| Maximum $I_C$ (continuous) | 10 A | 3 A | −3 A |
| $V_{CEO}$ | 60 V | 40 V | −40 A |
| Power rating | 75 W | 40 W | 12.5 W |
| $\beta$ | 20@ $I_C = 4$ A | 10 @ $I_C = 3$ A | 30 @ $I_C = -0.5$ A |
| $V_{CE\,sat}$ | 1.1 V | 1.3 V | −1.7 V |
| $V_{BE\,on}$ | 8 V | 1.8 V | −2 V |

currents, power transistors have typical parameters quite different from those of small-signal transistors.

From Table 11.2, we can find some of these differences:

1. $\beta$ is low. It can be as low as 5; the typical value is 20 to 80.
2. $I_{C\ max}$ is typically in the ampere range; it can be as high as 100 A.
3. $V_{CEO}$ is usually 40 to 100 V, but it can reach 500 V.

---

### EXAMPLE   11.2  Power Amplifier Limitations

#### Problem

Verify the limitations imposed by the power dissipation, saturation, and cutoff limits of the TIP31 power BJT on a linear power amplifier design. The amplifier is in the (DC-coupled) common-emitter configuration.

---

#### Solution

**Known Quantities:**  Amplifier component values.

**Find:**  Approximate values of maximum gain for distortionless operation.

**Schematics, Diagrams, Circuits, and Given Data:**  Data sheets for the 2N6306 amplifier may be found in the accompanying CD-ROM.

**Comments:**  You may wish to progressively increase the input signal amplitude to see the effects of cutoff and saturation. Monitor transistor power dissipation to ensure that the limit is not exceeded.

---

## BJT Switching Characteristics

In addition to their application in power amplifiers, power BJTs can also serve as controlled power switches, taking advantage of the switching characteristic described in Chapter 10 (see Figure 10.47). In addition to the properties already discussed, it is important to understand the phenomena that limit the switching speed of bipolar devices. The parasitic capacitances $C_{CB}$ and $C_{BE}$ that exist at the $CB$ and $BE$ junctions have the effect of imposing a charging time constant; since the transistor is also characterized by an internal resistance, you see that it is impossible for the transistor to switch from the cutoff to the saturation region instantaneously, because the inherent $RC$ circuits physically present inside the transistor must first be charged. Figure 11.14 illustrates the behavior of the base and collector currents in response to a step change in base voltage. If a step voltage up to amplitude $V_1$ is applied to the base of the transistor and a base current begins to flow, the collector current will not begin to flow until after a delay, because the base capacitance needs to charge up before the $BE$ junction voltage reaches

$V_\gamma$; this **delay time,** $t_d$, is an important parameter. After the *BE* junction finally becomes forward-biased, the collector current will rise to the final value in a finite time, called **rise time,** $t_r$. An analogous process (though the physics are different) takes place when the base voltage is reversed to drive the BJT into cutoff. Now the excess charge that had been accumulated in the base must be discharged before the *BE* junction can be reverse-biased. This discharge takes place over a **storage time,** $t_s$. To accelerate this process, the base voltage is usually driven to negative values ($-V_2$), so that the negative base current can accelerate the discharge of the charge stored in the base. Finally, the reverse-biased *BE* junction capacitance must now be charged to the negative base voltage value before the switching transient is complete; this process takes place during the **fall time,** $t_f$. In the figure, $I_{CS}$ represents the collector saturation current. Thus, the turn-on time of the BJT is given by:

$$t_{\text{on}} = t_d + t_r \tag{11.3}$$

and the turn-off time by

$$t_{\text{off}} = t_s + t_f \tag{11.4}$$

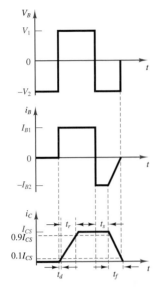

**Figure 11.14** BJT switching waveforms

## EXAMPLE 11.3 Power BJT Switching Characteristics

### Problem

The MJE170 power BJT is now used as a switch. Determine the delay time, rise time, and turn-off time for this transistor switch.

### Solution

**Known Quantities:** BJT switch component values.

**Find:** $t_d, t_r, t_s$, and $t_f$, as defined in Figure 11.14.

**Schematics, Diagrams, Circuits, and Given Data:** Data sheets for the MJE170 amplifier may be found in the accompanying CD-ROM.

**Focus on Computer-Aided Solutions:** The analysis of this design has been conducted in simulation, using *Electronics Workbench*™. The simulation of this circuit may be found in the accompanying CD-ROM.

**Comments:** You may wish to substitute a power MOSFET or an IGBT in the same circuit to explore differences in switching behavior between devices.

## Power MOSFETs

MOSFETs can also be used as power switches, like BJTs. The preferred mode of operation of a power MOSFET when operated as a switch is in the ohmic region, where substantial drain current can flow for relatively low drain voltages (see Table 9.1 and Figure 10.55). Thus, a MOSFET switch is driven from cutoff to the ohmic

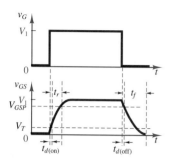

**Figure 11.15** MOSFET switching waveforms

state by the gate voltage. In an enhancement MOSFET, positive gate voltages are required to turn the transistor on; in depletion MOSFETs, either positive or negative voltages can be used.

To understand the switching behavior of MOSFETs, recall once again the parasitic capacitances that exist between pairs of terminals: $C_{GS}$, $C_{GD}$, and $C_{DS}$. As a consequence of these capacitances, the transistor experiences a **turn-on delay,** $t_{d(\text{on})}$, corresponding to the time required to charge the equivalent input capacitance to the threshold voltage, $V_T$. As shown in Figure 11.15, the rise time, $t_r$, is defined as the time it takes to charge the gate from the threshold voltage to the gate voltage required to have the MOSFET in the ohmic state, $V_{GSP}$. The **turn-off delay time,** $t_{d(\text{off})}$, is the time required for the input capacitance to discharge, so that the gate voltage can drop and $v_{DS}$ can begin to rise. As $v_{GS}$ continues to decrease, we define the **fall time,** $t_f$, which is the time required for $v_{GS}$ to drop below the threshold voltage and turn the transistor off.

# FOCUS ON METHODOLOGY

## Using Device Data Sheets

One of the most important design tools available to engineers is the **device data sheet.** In this box we illustrate the use of a device data sheet for the NDS8410 power MOSFET transistor. Excerpts from the data sheet are shown below, with some words of explanation. The complete data sheet can be found in the accompanying CD-ROM.

### NDS8410
### Single N-Channel Enhancement Mode Field Effect Transistor

### General Description
These N-Channel enhancement mode power field effect transistors are produced using Fairchild's proprietary, high cell density, DMOS technology. This very high density process is especially tailored to minimize on-state resistance and provide superior switching performance. These devices are particularly suited for low voltage applications such as notebook computer power management and other battery powered circuits where fast switching, low in-line power loss, and resistance to transients are needed.

### Features

- 10A, 30V. $R_{DS(ON)} = 0.015\Omega$ @ $V_{GS} = 10V$.
  $R_{DS(ON)} = 0.020\Omega$ @ $V_{GS} = 4.5V$.
- High density cell design for extremely low $R_{DS(ON)}$.
- High power and current handling capability in a widely used surface mount package.

## ABSOLUTE MAXIMUM RATINGS

This table summarizes the limitations of the device. For example, one can find the maximum allowable gate-source and drain source voltages, and the **power rating**.

**Absolute Maximum Ratings**     $T_A = 25°C$ unless otherwise noted

| Symbol | Parameter | NDS8410 | Units |
|--------|-----------|---------|-------|
| $V_{DSS}$ | Drain-Source Voltage | 30 | V |
| $V_{GSS}$ | Gate-Source Voltage | 20 | V |
| $I_D$ | Drain Current-Continuous | ±10 | A |
|  | - Pulsed | ± 50 |  |
| $P_D$ | Maximum Power Dissipation | 2.5 | W |
|  |  | 1.2 |  |
|  |  | 1 |  |
| $T_J, T_{STG}$ | Operating and Storage Temperature Range | −55 to 150 | °C |

# ELECTRICAL CHARACTERISTICS:

The table summarizing electrical characteristics is divided into various sections, including "on" characteristics, "off" characteristics, dynamic characteristics, and switching characteristics. We focus on the last of these, and make reference to Figure 11.15. Note how all of the relevant parameters shown in this figure are listed in the data sheet.

**ELECTRICAL CHARACTERISTICS**     $T_A = 25°C$ unless otherwise noted

| Symbol | Parameter | Conditions | Min | Typ | Max | Units |
|--------|-----------|------------|-----|-----|-----|-------|
| **OFF CHARACTERISTICS** | | | | | | |
| $BV_{DSS}$ | Drain-Source Breakdown Voltage | $V_{GS} = 0$ V, $I_D = 250$ $\mu$A | 30 | | | V |
| $I_{DSS}$ | Zero Gate Voltage Drain Current | $V_{DS} = 24$ V, $V_{GS} = 0$ V | | | 1 | $\mu$A |
|  |  | $T_j = 55°C$ | | | 10 | $\mu$A |
| $I_{GSSF}$ | Gate-Body Leakage, Forward | $V_{GS} = 20$ V, $V_{DS} = 0$ V | | | 100 | nA |
| $I_{GSSR}$ | Gate-Body Leakage, Reverse | $V_{GS} = -20$ V, $V_{DS} = 0$ V | | | −100 | nA |
| **ON CHARACTERISTICS**   (Note 2) | | | | | | |
| $V_{GS(ON)}$ | Gate-Threshold Voltage | $V_{DS} = V_{GS}$, $I_D = 250$ $\mu$A | 1 | 1.5 | | V |
| $R_{DS(ON)}$ | Static Drain-Source On-Resistance | $V_{GS} = 10$ V, $I_D = 10$ A | | 0.013 | 0.015 | $\Omega$ |
|  |  | $V_{GS} = 4.5$ V, $I_D = 9$ A | | 0.018 | 0.02 | |
| $I_{D(ON)}$ | On-State Drain Current | $V_{GS} = 10$ V, $V_{DS} = 5$ V | 20 | | | A |
| $g_{FS}$ | Forward Transconductance | $V_{DS} = 10$ V, $I_D = 10$ A | | 22 | | S |
| **DYNAMIC CHARACTERISTICS** | | | | | | |
| $C_{ISS}$ | Input Capacitance | $V_{DS} = 15$ V, $V_{GS} = 0$ V, | | 1350 | | pF |
| $C_{DSS}$ | Output Capacitance | $f= 1.0$ MHz | | 800 | | pF |
| $C_{ISS}$ | Reverse Transfer Capacitance | | | 300 | | pf |

(*Continued*)

*(Concluded)*

**ELECTRICAL CHARACTERISTICS** *(Continued)*     $T_A = 25°C$ unless otherwise noted

| Symbol | Parameter | Conditions | Min | Typ | Max | Units |
|--------|-----------|------------|-----|-----|-----|-------|
| **SWITCHING CHARACTERISTICS** (Note 2) | | | | | | |
| $t_{D(on)}$ | Turn-On Delay Time | $V_{DD} = 10$ V, $I_D = 1$ A, | | 14 | 30 | ns |
| $t_T$ | Turn-On Rise Time | $V_{GEN} = 10$ V, $R_{GEN} = 6 \ \Omega$ | | 20 | 25 | ns |
| $t_{D(off)}$ | Turn-Off Delay Time | | | 56 | 100 | ns |
| $t_F$ | Turn-Off Fall Time | | | 31 | 80 | ns |
| $Q_g$ | Total Gate Charge | $V_{DS} = 15$ V, | | 46 | 60 | nC |
| $Q_{gs}$ | Gate-Source Charge | $I_D = 10$ A, $V_{GS} = 10$ V | | 5.6 | | nC |
| $Q_{gd}$ | Gate-Drain Charge | | | 14 | | nC |

## Insulated-Gate Bipolar Transistors (IGBTs)

The insulated-gate bipolar transistor, or IGBT, is a hybrid device, combining features of both field-effect and bipolar devices. The circuit symbol of the IGBT is shown in Figure 11.1; a simplified equivalent circuit is shown in Figure 11.16. The IGBT is a voltage-controlled device, like a MOSFET, but its performance is closer to that of a BJT. The switching and conduction losses of the IGBT are lower than those of a MOSFET, and the switching speed is greater than that of a BJT (but somewhat lower than that of a MOSFET); the convenience of a MOSFET-like gate drive is an advantage over BJTs.

IGBTs can be rated up to 400 A and 1,200 V, and can have switching frequencies as high as 20 kHz. These devices have in recent years found increasing application in medium-power applications, such as AC and DC motor drives.

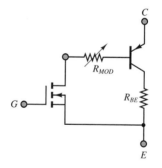

**Figure 11.16** IGBT simplified equivalent circuit

## 11.5     RECTIFIERS AND CONTROLLED RECTIFIERS (AC-DC CONVERTERS)

As explained in Chapter 8, one of the most immediate applications of the semiconductor diode is rectification of AC voltages and currents, to convert AC waveforms to DC. Rectification can be achieved both with conventional diodes and with controlled diodes, such as thyristors. A simple diode rectifier can provide only a fixed DC voltage level; however, variable DC supplies can be easily obtained with the aid of thyristors. The aim of this section is to illustrate the basic features of diode rectifiers, and to introduce thyristor-based controlled rectifiers.

The basic diode half-wave rectifier and also full-wave and bridge rectifiers were discussed in Sections 8.3 and 8.4. In addition to the considerations noted in Chapter 8, one often has to take into account the nature of the load seen by such DC supplies.

    In practice, loads are not always resistive, as will be seen in Chapters 16
through 18, where circuit models for electromechanical actuators and electric
motors are introduced. A very common occurrence consists of a DC voltage supply
providing current to a *DC motor*. For the purpose of the present discussion, it will
suffice to state that a DC motor presents an inductive impedance to the voltage
supply and requires a constant current from the supply to operate at a constant
speed. The circuit of Figure 11.17 illustrates, as an example, a simple half-wave
rectifier connected to an $RL$ load.

    The circuit on top in Figure 11.17, assuming an ideal diode, would present
a serious problem during the negative half-cycle of the source voltage, since the
requirement for continuity of current in the inductor (recall the discussion on
continuity of inductor currents and capacitor voltages in Chapter 5) would be
violated with $D_1$ off. Whenever the current flow through the inductor is interrupted
(during the negative half-cycles of $v_{AC}$), the inductor attempts to build a **flyback
voltage** proportional to $di_L/dt$. Since the rectifier does not provide any current
during the negative half-cycle of the source voltage, the instantaneous inductor
voltage could be very large and could lead to serious damage to either the motor
or the rectifier.

    The circuit shown on the bottom in Figure 11.17 contains a so-called **free-
wheeling diode,** $D_2$. The role of $D_2$ is to provide continuity of current when $D_1$
is in the off state. $D_2$ is off during the positive half-cycle but turns on when $D_1$
ceases to conduct, because of the flyback voltage, $Ldi_L/dt$. Rather than build up
a large voltage, the inductor now has a path for current to flow, through $D_2$, when
$D_1$ is off. Thus, the energy stored by the inductor during the positive half-cycle of
$v_{AC}$ is utilized to preserve a continuous current through the inductor during the off
period. Figure 11.18 depicts the load current for the circuit including the diode.
Note that $D_2$ allows the energy-storage properties of the inductor to be utilized to
smooth the pulselike supply current and to produce a nearly constant load current.

    Analyzing the circuit on the bottom of Figure 11.17, with

$$v_{AC}(t) = A \sin(\omega t) \tag{11.5}$$

(and assuming that both $D_1$ and $D_2$ are ideal), we conclude that the DC component
of the load voltage, $V_L$, must appear across the load resistor, $R$ (no steady-state
DC voltage can appear across the inductor, since $v_L = Ldi_L/dt$). Thus, the
approximate DC current flowing through the load is given by

$$I_L = \frac{A}{\pi R} \tag{11.6}$$

since the average output voltage of a half-wave rectifier is $A/\pi$ V for an AC source
of peak amplitude $A$ (see Chapter 8). The AC component of the load current (or
"ripple" current) is not as simple to compute, since it is due to the AC component
of $v_L$, which is not a pure sinusoid. The exact analysis would require the use of a
Fourier series expansion. For the purposes of this discussion, it is not unreasonable
to assume that most of the energy is at a frequency equal to that of the AC source:

$$i_L(t) \approx I_L + I_{AC} \cos(\omega t + \theta) \tag{11.7}$$

where $I_L$ is the average load current, $I_{AC}$ is the peak value of the ripple current,
and $\theta$ is its phase. An acceptable approximation from which the amplitude of $I_{AC}$

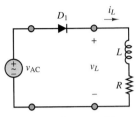

Simple half-wave rectifier

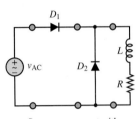

Same arrangement with
free-wheeling diode

**Figure 11.17** Rectifier
connected to an inductive load

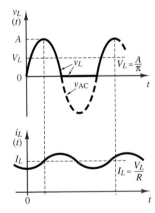

**Figure 11.18** Operation of
a free-wheeling diode

may be computed is

$$v_L(t) \approx \frac{A}{2\pi} + \frac{A}{2\pi} \sin \omega t \qquad (11.8)$$

Figure 11.19 graphically illustrates the extent of this approximation.

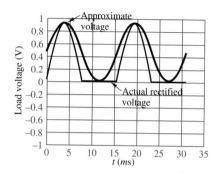

**Figure 11.19** Approximation of ripple voltage for a half-wave rectifier

A common alternative to the half-wave rectifier is the full-wave rectifier, which was discussed in Chapter 8.

## EXAMPLE   11.4  Rectifiers and Inductive Loads

### Problem

Analyze the circuit depicted in the lower half of Figure 11.17 to determine the rms amplitude of the ripple in the load voltage.

### Solution

**Known Quantities:**  Source voltage; load resistance and inductance.

**Find:**  $\tilde{V}_{\text{ripple}}$.

**Schematics, Diagrams, Circuits, and Given Data:**  $v_{\text{AC}} = 100$ V rms, 60 Hz. The load is a DC motor with $R_a = 1.1$ $\Omega$ and $L_a = 0.001$ H. The motor specifications may be found in the DC motor template in *Electronics Workbench*™.

**Assumptions:**  Ignore the DC motor mechanical load.

**Focus on Computer-Aided Solutions:**  The analysis of this design has been conducted in simulation, using *Electronics Workbench*™. The simulation of this circuit may be found in the accompanying CD-ROM.

VIRTUAL LAB

**Comments:**  What happens if the freewheeling diode is not in the circuit? You may try to run a simulation without diode $D_2$ in the circuit.

## Three-Phase Rectifiers

It is important to realize that the same type of circuit that can be used for single-phase rectifiers can also be employed to design multiphase rectifiers. Recall the analysis of three-phase AC power systems in Section 7.4. In many high-power applications, three-phase voltages need to be rectified to give rise to a single DC supply; such rectification can be achieved by means of an extension of the bridge rectifier. Consider the balanced three-phase circuit shown in Figure 11.20. The three-phase wye-connected source is connected to a resistive load by means of a three-phase transformer, with a delta-connected primary and a wye-connected secondary. The circuit could also operate without the transformer. The three secondary currents, $i_a$, $i_b$, and $i_c$, flow through pairs of diodes $D_1$ to $D_6$ in a manner very similar to the single-phase bridge rectifier described in Figure 8.45. The diodes will conduct in pairs depending on the relative line voltages, according to the following sequence: $D_1$-$D_2$, $D_2$-$D_3$, $D_3$-$D_4$, $D_4$-$D_5$, $D_5$-$D_6$, and $D_6$-$D_1$. Recall from the analysis of Section 7.4, equation 7.54, that the line-to-line voltage is $\sqrt{3}$ times the phase voltage in a three-phase wye-connected source. The instantaneous source voltages and the related diode conduction periods, as well as the load voltage, are shown in Figure 11.21.

It can be shown that the average output voltage is given by the expression:

$$V_L = \frac{2}{2\pi/6} \int_0^{\pi/6} \sqrt{3} V_m \cos \omega t \, d(\omega t) = \frac{3\sqrt{3}}{\pi} V_m = 1.654 V_m \qquad (11.9)$$

where $V_m$ is the peak phase voltage. The rms output voltage can be calculated to be

$$V_{\text{rms}} = \sqrt{\frac{2}{2\pi/6} \int_0^{\pi/6} \sqrt{3} V_m^2 \cos^2 \omega t \, d(\omega t)} = \left(\frac{3}{2} + \frac{9\sqrt{3}}{4\pi}\right) \qquad (11.10)$$

$$V_m = 1.6554 V_m$$

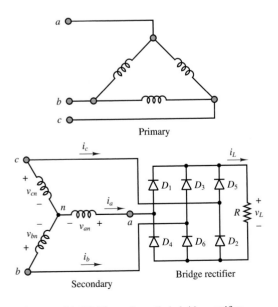

**Figure 11.20** Three-phase diode bridge rectifier

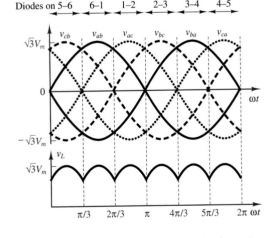

**Figure 11.21** Waveforms and conduction times of three-phase bridge rectifier

### EXAMPLE 11.5 Three-phase Bridge Rectifier

**Problem**

Simulate the three-phase bridge rectifier of Figure 11.20 and verify numerically that the average and rms output voltages are given by equations 11.9 and 11.10, respectively.

---

**Solution**

**Known Quantities:** Source voltage; load resistance.

**Find:** $\tilde{V}_{\text{ripple}}$

**Schematics, Diagrams, Circuits, and Given Data:** The load resistance is $R = 1.2\ \Omega$; the source is 208 V, 3 phase.

**Comments:** What is the effect of an inductive load on the load current and voltage waveforms? Try adding a 0.01-H series inductance to the load circuit.

---

## Thyristors and Controlled Rectifiers

In a number of applications, it is useful to be able to externally control the amount of current flowing from an AC source to the load. A family of power semiconductor devices called **controlled rectifiers** allows for control of the rectifier state by means of a third input, called the **gate.** Figure 11.22 depicts the appearance of a **thyristor, or silicon controlled rectifier (SCR),** illustrating how the physical structure of this device consists of four layers, alternating $p$-type and $n$-type material. Note that the circuit symbol for the thyristor suggests that this device acts as a diode, with provision for an additional external control signal.

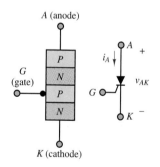

**Figure 11.22** Thyristor structure and circuit symbol

The operation of the thyristor can be explained in an intuitive fashion as follows. When the voltage $v_{AK}$ is negative (i.e., providing reverse bias), the thyristor acts just like a conventional $pn$ junction in the off state. When $v_{AK}$ is forward-biased *and* a small amount of current is injected into the gate, the thyristor conducts forward current. The thyristor then continues to conduct (even in the absence of gate current), provided that $v_{AK}$ remains positive. Figure 11.23 depicts the $i$-$v$ curve for the thyristor. Note that the thyristor has two stable states, determined by the bias $v_{AK}$ and by the gate current. In summary, the thyristor acts as a diode with a control gate that determines the time when conduction begins.

A somewhat more accurate description of thyristor operation may be provided if we realize that the four-layer $pnpn$ device can be modeled as a $pnp$ transistor connected to an $npn$ transistor. Figure 11.24 clearly shows that, physically, this is a realistic representation. Note that the anode current, $i_A$, is equal to the emitter current of the $pnp$ transistor (labeled $Q_p$) and the base current of $Q_p$ is equal to the collector current of the $npn$ transistor, $Q_n$. Likewise, the base current of $Q_n$ is the sum of the gate current and the collector current of $Q_p$. The

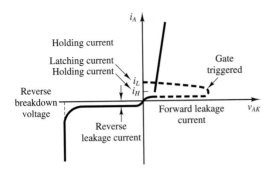

**Figure 11.23** Thyristor $i$-$v$ characteristic

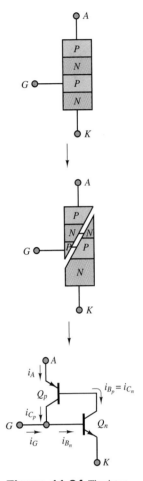

**Figure 11.24** Thyristor two-transistor model

behavior of this transistor model is explained as follows. Suppose, initially, $i_G$ and $i_{B_n}$ are both zero. Then it follows that $Q_n$ is in cutoff, and therefore $i_{C_n} = 0$. But if $i_{C_n} = 0$, then the base current going into $Q_p$ is also zero and $Q_p$ is also in cutoff, and $i_{C_p} = 0$, consistent with our initial assumption. Thus, this is a stable state, in the sense that unless an external condition perturbs the thyristor, it will remain off.

Now, suppose a small pulse of current is injected at the gate. Then $i_{B_n} > 0$, and $Q_n$ starts to conduct, provided, of course, that $v_{AK} > 0$. At this point, $i_{C_n}$, and therefore $i_{B_p}$, must be greater than zero, so that $Q_p$ conducts. It is important to note that once the gate current has turned $Q_n$ on, $Q_p$ also conducts, so that $i_{C_p} > 0$. Thus, even though $i_G$ may cease, once this "on" state is reached, $i_{C_p} = i_{B_n}$ continues to drive $Q_n$ so that the on state is also self-sustaining. The only condition that will cause the thyristor to revert to the off state is the condition in which $v_{AK}$ becomes negative; in this case, both transistors return to the cutoff state.

In a typical controlled rectifier application, the device is used as a half-wave rectifier that conducts only after a trigger pulse is applied to the gate. Without concerning ourselves with how the trigger pulse is generated, we can analyze the general waveforms for the circuit of Figure 11.25 as follows. Let the voltage $v_{\text{trigger}}$ be applied to the gate of the thyristor at $t = \tau$. The voltage $v_{\text{trigger}}$ can be a short pulse, provided by a suitable trigger-timing circuit (Chapter 13 will discuss timing and switching circuits). At $t = \tau$, the thyristor begins to conduct, and it continues to do so until the AC source enters its negative cycle. Figure 11.26 depicts the relevant waveforms.

Note how the DC load voltage is controlled by the firing time $\tau$, according to the following expression:

$$\langle v_L \rangle = V_L = \frac{1}{T} \int_{\tau}^{T/2} v_{\text{AC}}(t)\, dt \qquad \textbf{(11.11)}$$

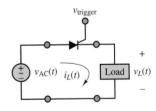

**Figure 11.25** Controlled rectifier circuit

where $T$ is the period of $v_{\text{AC}}(t)$. Now, if we let

$$v_{\text{AC}}(t) = A \sin \omega t \qquad \textbf{(11.12)}$$

we can express the average (DC) value of the load voltage

$$V_L = \frac{1}{T} \int_{\tau}^{T/2} A \sin \omega t\, dt = (1 + \cos \omega t)\frac{A}{2\pi} \qquad \textbf{(11.13)}$$

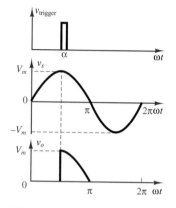

**Figure 11.26** Half-wave
controlled rectifier waveforms

in terms of the **firing angle**, $\alpha$, defined as

$$\alpha = \omega \tau \tag{11.14}$$

By evaluating the integral of equation 10.13, we can see that the (DC) load voltage amplitude depends on the firing angle, $\alpha$:

$$V_L = (1 + \cos \alpha) \frac{A}{2\pi} \tag{11.15}$$

The following examples illustrate applications of thyristor circuits.

### EXAMPLE   11.6  Thyristor-based Variable Voltage Supply

#### Problem

Analyze the thyristor-based variable voltage supply shown in Figure 11.27. Determine: (1) the rms load voltage as a function of the firing angle and (2) the power supplied to the resistive load at zero firing angle and at firing angles equal to $\pi/2$ and $\pi$.

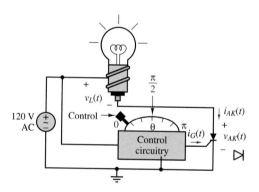

**Figure 11.27**

#### Solution

**Known Quantities:** Load resistance.

**Find:** $\tilde{V}_L$, $P_L|_{\alpha=0}$, $P_L|_{\alpha=\pi/2}$, $P_L|_{\alpha=\pi}$.

**Schematics, Diagrams, Circuits, and Given Data:** $V_{AK \text{ on}} = 0$ V; $R_L = 240 \ \Omega$. The pulsed gate current, $i_G(t)$, is timed as shown in Figure 11.28.

**Assumptions:** The thyristor acts as an ideal diode when on ($V_{AK} > 0$).

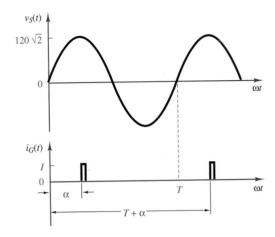

**Figure 11.28**

*Analysis:*

1. *Load voltage calculation.* As explained in the preceding section, the load voltage will have the appearance shown in Figure 11.29. The rms value of the load voltage as a function of the firing angle, $\alpha$, is therefore computed as follows:

$$\tilde{V}_L(\alpha) = \sqrt{\frac{(120\sqrt{2})^2}{2\pi} \int_\alpha^\pi \sin^2 \omega t' \, d(\omega t')}$$

$$= \frac{(120\sqrt{2})}{2} \sqrt{\frac{1}{\pi} \int_\alpha^\pi (1 - \cos(2\omega t')) d(\omega t')}$$

$$= \frac{(120\sqrt{2})}{2} \sqrt{1 - \frac{\alpha}{\pi} + \frac{1}{2} \sin(2\alpha)}$$

2. *Load power calculation.* We can now compute the load power for each of the three values of $\alpha$:

$$P_L = \frac{\tilde{V}^2}{R_L}$$

For $\alpha = 0$:

$$P_L = \frac{\tilde{V}^2}{R_L} = \frac{\left(\dfrac{120\sqrt{2}}{2}\right)^2}{240} = 30 \text{ W};$$

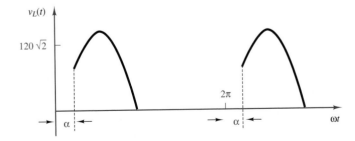

**Figure 11.29**

for $\alpha = \pi/2$:

$$P_L = \frac{\tilde{V}^2}{R_L} = \frac{\left(\frac{120\sqrt{2}}{2}\sqrt{1-\frac{1}{2}}\right)^2}{240} = 15 \text{ W}$$

for $\alpha = \pi$:

$$P_L = \frac{\tilde{V}^2}{R_L} = \frac{\left(\frac{120\sqrt{2}}{2}\sqrt{1-1}\right)^2}{240} = 0 \text{ W}$$

**Comments:** Note that no power is wasted when the firing angle is set for zero load voltage. This would not be the case if a resistive voltage divider were used to adjust the load voltage.

### EXAMPLE 11.7 Automotive Battery Charger

**Problem**

Qualitatively explain the operation of the **automotive battery charger** shown in Figure 11.30.

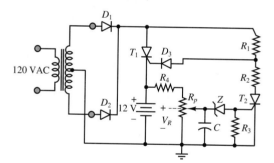

**Figure 11.30** Automotive battery charger

**Solution**

**Analysis:** The charging circuit is connected to a standard 110-V single-phase supply. Diodes $D_1$ and $D_2$ form a full-wave rectifier (see Figure 8.42); resistors $R_1$ and $R_2$ and thyristor $T_2$ form a variable voltage divider.

Assume that thyristor $T_2$ is not in the conducting state and that the anode voltage of $D_3$ is such that $D_3$ conducts. Then $T_1$ will be fired near the beginning of the positive half-cycle of the AC source voltage, and its period of conduction will be long, providing a substantial current to the battery (resistors $R_4$ and $R_p$ are sufficiently large that most of the current flowing through $T_1$ will go to the battery).

The potentiometer $R_p$ is set so that when the battery voltage is low, the voltage $V_R$ is not sufficient to turn on the Zener diode, $Z$. Thus, $Z$ is effectively an open circuit, and $T_2$ remains off (recall that we had initially assumed $T_2$ to be off—this confirms the correctness of the assumption). As the battery charges to a progressively higher value, $Z$

will eventually conduct; when $Z$ conducts, a gate current is injected into $T_2$, which is then turned on.

When $T_2$ conducts, the voltage across the $R_2$-$T_2$ series connection becomes significantly lower, because $T_2$ is now nearly a short circuit. Resistors $R_1$ and $R_2$ are selected so that when $T_2$ conducts, $D_3$ becomes reverse-biased. Once this condition occurs, $T_1$ is turned off and charging stops. You see that the circuit has built-in overcharging protection.

---

## EXAMPLE 11.8 Thyristor Circuit

### Problem

Determine the value of $R$ on the circuit of Figure 11.31 such that the average load current through the thyristor is 1 A.

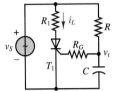

**Figure 11.31**

---

### Solution

**Known Quantities:** Resistances and source voltage.

**Find:** Resistor $R$ such that $\langle i_L \rangle = 1$ A.

**Schematics, Diagrams, Circuits, and Given Data:** $v_S = 200$ V rms, 250 Hz; $V_{AK \text{ on}} = 0$ V; $R_1 = 75\ \Omega$; $R_G = 1\ \text{k}\Omega$; $C = 1\ \mu\text{F}$.

**Assumptions:** The thyristor acts as an ideal diode when on ($V_{AK} > 0$).

**Analysis:** Figure 11.32 depicts the relative timing of the source voltage, $v_S(t)$, thyristor current, $i_L(t)$, and triggering voltage, $v_t(t)$. The expression for the source voltage is:

$$v_S(t) = \sqrt{2} \times 200 \sin(2\pi \times 250t)$$

The load current through the 74-$\Omega$ resistor is:

$$i_L(t) = \begin{cases} \dfrac{\sqrt{2} \times 200}{75} \sin(2\pi \times 250t) & \alpha \leq \omega t \leq \pi \\ 0 & \pi \leq \omega t \leq 2\pi \end{cases}$$

and the triggering voltage is:

$$v_t(t) = V_t \sin[(2\pi \times 250t) - \alpha]$$

The triggering voltage will go positive at the desired firing angle, $\alpha$, thus injecting a current into the gate of the thyristor, turning it on. Thus, the requirement that the average load current be equal to 1 A is equivalent to requiring that

$$\langle i_L(t) \rangle = \frac{1}{1\pi} \int_\alpha^\pi \frac{\sqrt{2} \times 200}{75} \sin(2\pi \omega t')\, d(\omega t') = 1 \text{ A}$$

Performing the integration, we determine that the requirement is

$$\frac{\sqrt{2} \times 200}{2\pi \times 75}(1 + \cos(\alpha)) = 1$$

Solving for $\alpha$, we find $\alpha = 48.23°$.

Now, to determine the value of $R$, we observe that the AC source voltage appears across the $RC$ circuit; thus, $v_t$ can be computed from an impedance voltage divider by

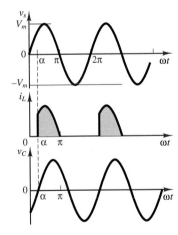

**Figure 11.32**

using phasor methods:

$$\mathbf{V}_t(j\omega) = \frac{\dfrac{1}{j\omega C}}{R + \dfrac{1}{j\omega C}} V_S(j\omega) = \frac{V_S}{\sqrt{1 + \omega^2 R^2 C^2}} \angle - \arctan(\omega RC).$$

We then observe that the phase of $\mathbf{V}_t(j\omega)$ is the firing angle $\alpha$, and we can therefore determine $\alpha$ by setting

$$-\arctan(\omega RC) = \alpha = 48.23°$$

$$R = \frac{\tan(\alpha)}{\omega C} = \frac{\tan(48.23)}{2\pi \times 250 \times 10^{-6}} = 713\ \Omega$$

**Focus on Computer-Aided Solutions:** An *Electronics Workbench*™ simulation of the circuit analyzed in this example is supplied in the accompanying CD-ROM. You may wish to experiment with changing the value of $R$ to see its effect on the average load current.

## Check Your Understanding

**11.2** Using the approximation given in equation 11.8, find the DC and AC load currents for the circuit of Figure 11.17 if $R = 10\ \Omega$, $L = 0.3$ H, $A = 170$ V, and $\omega = 377$ rad/s.

**11.3** Calculate the load voltage in Figure 11.26 for $A = 100$, $\alpha = \pi/3$.

**11.4** For the circuit in Example 11.6, the input AC voltage is 240 V. Find the rms value of the load voltage and the power at the firing angle $\alpha = \pi/4$.

## 11.6 ELECTRIC MOTOR DRIVES

The advent of high-power semiconductor devices has made it possible to design effective and relatively low-cost electronic supplies that take full advantage of the capabilities of the devices introduced in this chapter. Electronic power supplies for **DC and AC motors** have become one of the major fields of application of power electronic devices. The last section of this chapter is devoted to an introduction to two families of power supplies, or **electric drives: choppers,** or **DC-DC converters;** and **inverters,** or **DC-AC converters.** These circuits find widespread use in the control of AC and DC motors in a variety of applications and power ranges.

Before we delve into the discussion of the electronic supplies, it will be helpful to introduce the concept of quadrants of operation of a drive. Depending on the direction of current flow, and on the polarity of the voltage, an electronic drive can operate in one of four possible modes, as indicated in Figure 11.33.

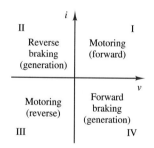

**Figure 11.33** The four quadrants of an electric drive

### Choppers (DC-DC Converters)

As the name suggests, a DC-DC converter is capable of converting a fixed DC supply to a variable DC supply. This feature is particularly useful in the control of the speed of a DC motor (described in greater detail in Chapter 17). In a DC motor, shown schematically in Figure 11.34, the developed torque, $T_m$, is proportional

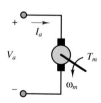

**Figure 11.34** DC motor

to the current supplied to the motor **armature,** $I_a$, while the **electromotive force (emf),** $E_a$, which is the voltage developed across the armature, is proportional to the speed of rotation of the motor, $\omega_m$. A DC motor is an electromechanical energy-conversion system; that is, it converts electrical to mechanical energy (or vice versa if it is used as a generator). If we recall that the product of torque and speed is equal to power in the mechanical domain, and that current times voltage is equal to power in the electrical domain, we conclude that in the ideal case of complete energy conversion, we have

$$E_a \times I_a = T_m \times \omega_m \tag{11.16}$$

Naturally, such ideal energy conversion cannot take place; however we can see that there is a correspondence between the four electrical quadrants of Figure 11.33 and the mechanical power output of the motor: namely, if the voltage and current are both positive or both negative, the electrical power will be positive, and so will the mechanical power. This corresponds to the **forward** ($i$, $v$ both positive) and **reverse** ($i$, $v$ both negative) **motoring** operation. Forward motoring corresponds to quadrant I, and reverse motoring to quadrant III in Figure 11.33. If the voltage and current are of opposite polarity (quadrants II and IV), electrical energy is flowing back to the electric drive; in mechanical terms this corresponds to a braking condition. Operation in the fourth quadrant can lead to **regenerative braking,** so called because power is regenerated by making current flow back to the source. This mode could be useful, for example, to recharge a battery supply, because the braking energy can be regenerated by returning it to the electric supply.

A simple circuit that can accomplish the task of providing a variable DC supply from a fixed DC source is the **step-down chopper (buck converter),** shown in Figure 11.35. The circuit consists of a "chopper" switch, denoted by the symbol $S$, and a free-wheeling diode, such as the one described in Section 11.5. The switch can be any of the power switches described in this chapter, for example, a power BJT or MOSFET, or a thyristor; see, for example, the BJT switch of Figure 11.4. The circuit to the right of the diode is a model of a DC motor, including the inductance and resistance of the armature windings, and the effect of the back emf $E_a$. When the switch is turned on (say, at $t = 0$), the supply $V_S$ is connected to the load and $v_o = V_S$. The load current, $i_o$, is determined by the motor parameters. When the switch is turned off, the load current continues to flow through the free-wheeling diode, but the output voltage is now $v_o = 0$. At time $T$, the switch is turned on again, and the cycle repeats.

Figure 11.36 depicts the $v_o$ and $i_o$ waveforms. The average value of the output voltage, $\langle v_o \rangle$, is given by the expression

$$\langle v_o \rangle = \frac{t_1}{T} V_S = \delta V_S \tag{11.17}$$

where $\delta$ is the **duty cycle** of the chopper. The step-down chopper has a useful range

$$0 \le \langle v_o \rangle \le V_S \tag{11.18}$$

It is also possible to increase the range of a DC-DC converter to above the supply voltage by making use of the energy-storage properties of an inductor; the resulting circuit is shown in Figure 11.37. When the chopper switch, $S$, is on, the supply current flows through the inductor and the closed switch, storing energy in the inductor; the output voltage, $v_o$, is zero, since the switch is a short

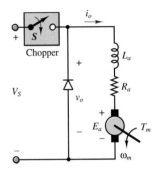

**Figure 11.35** Step-down chopper (buck converter)

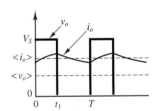

**Figure 11.36** Step-down chopper waveforms

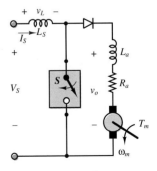

**Figure 11.37** Step-up chopper (boost converter)

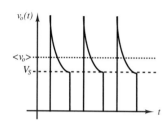

**Figure 11.38** Step-up chopper output voltage waveform (ideal)

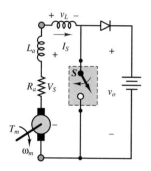

**Figure 11.39** Step-up chopper used for regenerative braking

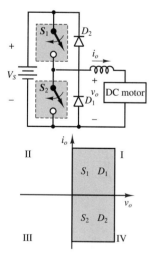

**Figure 11.40** Two-quadrant chopper

circuit. When the switch is open, the supply current will flow through the load via the diode; but the inductor voltage is negative during the transient following the opening of the switch and therefore adds to the source voltage: the energy stored in the inductor while the switch was closed is now released and transferred to the load. This stored energy makes it possible for the output voltage to be higher than the supply voltage for a finite period of time.

To maintain a constant average load current, the current increase between 0 and $t_1$ must equal the current decrease from $t_1$ to $T$. Therefore,

$$\frac{1}{L} \int_0^{t_1} V_s \, dt = \frac{1}{L} \int_{t_1}^{T} (\langle v_o \rangle - V_s) \, dt \qquad (11.19)$$

from which we can calculate

$$\langle v_o \rangle t_1 = (\langle v_o \rangle - V_s)(T - t_1). \qquad (11.20)$$

This results in an average output voltage given by the expression

$$\langle v_o \rangle = \frac{T}{1 - t_1} V_s = \frac{1}{1 - \frac{t_1}{T}} V_s = \frac{1}{1 - \delta} V_s \geq V_s. \qquad (11.21)$$

Since the duty cycle, $\delta$, is always less than 1, the theoretical range of the supply is

$$V_S \leq \langle v_o \rangle < \infty \qquad (11.22)$$

The waveforms for the boost converter are shown in Figure 11.38.

A step-up chopper can also be used to provide regenerative braking: if the "supply" voltage is the motor armature voltage and the output voltage is the fixed DC supply (battery) voltage, then power can be made to flow from the motor to the DC supply (i.e., recharging the battery). This configuration is shown in Figure 11.39.

Finally, the operation of the step-down and step-up choppers can be combined into a **two-quadrant chopper,** shown in Figure 11.40. The circuit shown schematically in Figure 11.40 can provide both regenerative braking and motoring operation in a DC motor. When switch $S_2$ is open, and switch $S_1$ serves as a chopper, the circuit operates as a step-down chopper, precisely as was described earlier in this section (convince yourself of this by redrawing the circuit with $S_2$ and $D_2$ replaced by open circuits). Thus, the drive and motor operate in the first quadrant (motoring operation). The output voltage, $v_o$, will switch between $V_S$ and zero, as shown in Figure 11.36, and the load current will flow in the direction indicated by the arrow in Figure 11.40; diode $D_1$ free-wheels whenever $S_1$ is open. Since both output voltage and current are positive, the system operates in the first quadrant.

When switch $S_1$ is open and switch $S_2$ serves as a chopper, the circuit resembles a step-up chopper. The source is the motor emf, $E_a$, and the load is the battery; this is the situation depicted in Figure 11.39. The current will now be negative, since the sum of the motor emf and the voltage across the inductor (corresponding to the energy stored during the "on" cycle of $S_2$) is greater than the battery voltage. Thus, the drive operates in the fourth quadrant.

Examples 11.9, 11.10 and 11.11 illustrate the operation of choppers as DC motor supplies.

## EXAMPLE  11.9  Operation of Step-Down Chopper (Buck Converter)

### Problem

Simulate the step-down chopper of Figure 11.35 and verify numerically that the average output voltage is given by equation 11.17.

### Solution

**Known Quantities:** Source voltage; load resistance and inductance; motor characteristics.

**Find:** $\langle v_o \rangle$.

**Schematics, Diagrams, Circuits, and Given Data:** $V_S = 220$ V; $R_a = 0.3\ \Omega$; $L_a = 15$ mH; $k_a\phi = 0.0167 \frac{V-s}{rev}$; $N = 0 - 2{,}000$ rev/min, $I_a = 25$ A.

## EXAMPLE  11.10  Operation of Step-Up Chopper (Boost Converter)

### Problem

Simulate the step-up chopper of Figure 11.37 and verify numerically that the average output voltage is given by equation 11.22.

### Solution

**Known Quantities:** Source voltage; source series inductance; load resistance and inductance; motor characteristics.

**Find:** $\langle v_o \rangle$.

**Schematics, Diagrams, Circuits, and Given Data:** $V_S = 220$; $L_S = 1$ H; $R_a = 0.3\ \Omega$; $L_a = 15$ mH; $k_a\phi = 0.0167 \frac{v-s}{rev}$; $N = 0 - 2{,}000$ rev/min.

## EXAMPLE  11.11  Two-Quadrant Chopper

### Problem

1. Determine the turn-on time of the chopper of Figure 11.40 in the motoring mode if $n = 500$ rev/min and $i_o = 90$ A. Also determine the power absorbed by the motor

armature winding; the power absorbed by the motor; and the power delivered by the source.

2. Determine the turn-on time of the chopper in the regenerative mode if $n = 380$ rev/min and $i_o = -90$ A. Also determine the power absorbed by the motor armature winding, the power absorbed by the motor, and the power delivered by the source.

**Solution**

**Known Quantities:** Supply voltage; motor parameters; chopping frequency armature resistance and inductance.

**Find:** For each of the two cases: $t_1$; $P_a$; $P_m$; $P_S$.

**Schematics, Diagrams, Circuits, and Given Data:**

1. $V_S = 120$ V; $E_a = 0.1n$; $R_a = 0.2\ \Omega$; $1/T = $ chopping frequency $= 300$ Hz.
2. $V_S = 120$ V; $E_a = 0.1n$; $R_a = 0.2\ \Omega$; $L_S \to \infty$; $1/T = $ chopping frequency $= 300$ Hz.

**Assumptions:** The switches in the chopper of Figure 11.40 act as ideal switches. Assume that the motor inductance is sufficiently small to be neglected in the calculations (i.e., assume a short circuit).

**Analysis:**

1. *Analysis of motoring operation.* To analyze motoring operation of the chopper, we refer to Figure 11.35 and apply KVL to the motor side:

$$\langle v_o \rangle = R_a I_a + E_a = R_a \langle i_o \rangle + 0.1n = 0.2 \times 90 + 0.1 \times 500 = 68 \text{ V}$$

From equation 11.17 we can then compute the duty cycle of the chopper, $\delta$:

$$\delta = \frac{t_1}{T} = \frac{\langle v_o \rangle}{V_S} = \frac{68}{120} = 0.567$$

Since the chopping frequency is 300 Hz, we can compute $t_1$:

$$t_1 = \frac{T}{\delta} = \frac{1}{300 \times 0.567} = 1.89 \text{ ms}$$

The power absorbed by the armature is:

$$P_a = R_a I_a^2 = R_a \langle i_o \rangle^2 = 0.2 \times 90^2 = 1.62 \text{ kW}$$

The power absorbed by the motor is:

$$P_m = E_a I_a = 0.1n \times \langle i_o \rangle = 0.1 \times 500 \times 90 = 4.5 \text{ kW}$$

The power delivered by the voltage supply is:

$$P_S = \delta V_S \langle i_o \rangle = 0.567 \times 120 \times 90 = 6.12 \text{ kW}$$

2. *Analysis of regenerative operation.* To analyze regenerative operation of the chopper, we refer to Figure 11.37 and apply KVL to the motor side, noting that now the current is flowing in the reverse direction:

$$\langle v_o \rangle = R_a I_a + E_a = R_a \langle i_o \rangle + E_a = -90 \times 0.2 + 0.1 \times 380 = 20 \text{ V}$$

We now turn to equation 11.22, and observe that in this equation the motor acts as the source, and the supply voltage as the load, thus:

$$V_S = \frac{1}{1 - \frac{t_1}{T}} \langle v_o \rangle \quad \text{or} \quad 120 = \frac{1}{1 - 300t_1} 20$$

We can then compute $t_1 = 2.8$ ms from the above equation. The duty cycle for the step-up chopper is now:

$$\delta = \frac{t_1}{T} = \frac{5}{6} = 0.833$$

The power absorbed by the armature is:

$$P_a = R_a I_a^2 = R_a \langle i_o \rangle^2 = 0.2 \times (-90)^2 = 1.62 \text{ kW}$$

The power absorbed by the motor will now be negative, since current is flowing in the reverse direction; the motor is in fact generating power:

$$P_m = E_a I_a = 0.1n \times \langle i_o \rangle = 0.1 \times 380 \times (-90) = -3.42 \text{ kW}$$

The power delivered by the voltage supply is:

$$P_S = \delta V_S \langle i_o \rangle = 0.567 \times 120 \times (-90) = -1.8 \text{ kW}$$

This power is negative because the supply is actually absorbing power, not delivering it.

**Comments:**

1. Note that the sum of the motor and armature power losses is equal to the power delivered by the source; this is to be expected, since we have assumed ideal (lossless) switches. In a practical chopper, the chopping circuit would actually absorb power; heat dissipation is therefore an important issue in the design of choppers. This chopper operates in quadrant I (see Figure 11.40).

2. Note that in the regenerative case the equivalent duty cycle is greater than 1. Note also that now the power absorbed by the motor is a negative quantity; that is, the motor delivers power to the rest of the circuit. However, the power absorbed by the armature resistance is still a positive quantity because the armature resistance dissipates power regardless of the direction of the current flow through it. Here $V_S$ might, for example, represent a battery pack in an electric vehicle, which would be recharged at the rate of 1.8 kW. The source of energy capable of producing this power is the inertial energy stored in the vehicle: when the vehicle decelerates, this mechanical energy causes the electric motor to act as a generator (see Chapter 17), producing the 90-ampere current in the reverse direction. This chopper operates in quadrant IV (see Figure 11.40).

## Inverters (DC-AC Converters)

As will be explained in Chapter 17, variable-speed drives for AC motors require a multiphase variable-frequency, variable-voltage supply. Such drives are called *DC-AC converters,* or *inverters.* Inverter circuits can be quite complex, so the objective of this section is to present a brief introduction to the subject, with the aim of illustrating the basic principles. A **voltage source inverter (VSI)** converts the output of a fixed DC supply (e.g., a battery) to a variable-frequency AC supply. Figure 11.41 depicts a **half-bridge VSI;** once again, the switches can be either bipolar or MOS transistors, or thyristors. The operation of this circuit is as follows. When switch $S_1$ is turned on, the output voltage is in the positive half-cycle, and $v_o = V_S/2$. To generate the negative half-cycle, switch $S_2$ is turned on, and

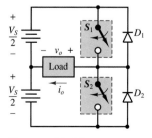

**Figure 11.41** Half-bridge voltage source inverter

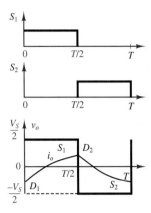

**Figure 11.42** Half-bridge voltage source inverter waveforms

$v_o = -V_S/2$. The switching sequence of $S_1$ and $S_2$ is shown in Figure 11.42. It is important that each switch be turned off before the other is turned on; otherwise, the DC supply would be short-circuited. Since the load is always going to be inductive in the case of a motor drive, it is important to observe that the load current, $i_o$, will lag the voltage waveform, as shown in Figure 11.42. As shown in this figure, there will be some portions of the cycle in which the voltage is positive but the current is negative. The function of diodes $D_1$ and $D_2$ is precisely to conduct the load current whenever it is of direction opposite to the polarity of the voltage. Without these diodes, there would be no load current in this case. Figure 11.42 also shows which element is conducting in each portion of the cycle.

A full-bridge version of the VSI can also be designed as shown in Figure 11.43; the associated output voltage waveform is shown in Figure 11.44. The operation of this circuit is analogous to that of the half-bridge VSI; switches $S_1$ and $S_2$ are fired during the first half-cycle, and switches $S_3$ and $S_4$ during the second half. Note that the full-bridge configuration allows the output voltage to swing from $V_S$ to $-V_S$. The diodes provide a path for the load current whenever the load voltage and current are of opposite polarity.

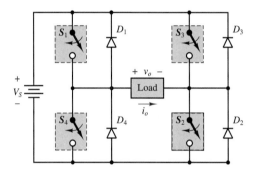

**Figure 11.43** Full-bridge voltage source inverter

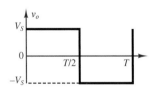

**Figure 11.44** Half-bridge voltage source inverter output voltage waveform

A three-phase version of the VSI is shown in Figure 11.45. Once again, the operation is analogous to that of the VSI circuits just presented. The related waveforms are shown in Figure 11.46. The top three waveforms depict the **pole voltages,** which are referenced to the DC supply neutral point, $o$. The pole voltages are obtained by firing the switches $S_1$ through $S_6$ at appropriate times. For example, if $S_1$ is fired at $\omega t = 0$, then pole $a$ is connected to the positive side of the DC supply, and $v_{ao} = V_S/2$; if $S_4$ is subsequently turned on at $\omega t = \pi$, then pole $a$ is connected to the negative side of the DC supply, and $v_{ao} = -V_S/2$. The other pairs of switches are then fired in an analogous sequence, shifted by 120 electrical degrees with respect to each other, to obtain the waveforms shown in the top three graphs of Figure 11.46. The **line voltages** are obtained from the pole voltages using the following relations:

$$v_{ab} = v_{ao} - v_{bo}$$
$$v_{bc} = v_{co} - v_{co}$$
$$v_{ca} = v_{co} - v_{ao}$$

(11.23)

and are shown in the second set of three diagrams in Figure 11.46. These are also phase-shifted by 120°. Now, we can also express the pole voltages in terms of the

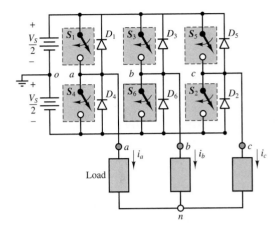

**Figure 11.45** Three-phase voltage source inverter

**load phase voltages,** $v_{an}$, $v_{bn}$, and $v_{cn}$:

$$v_{ao} = v_{an} - v_{no}$$

$$v_{bo} = v_{bn} - v_{no} \qquad\qquad\qquad\text{(11.24)}$$

$$v_{co} = v_{cn} - v_{no}$$

and since we must have $v_{an} + v_{bn} + v_{cn} = 0$ for balanced operation (see Chapter 7), we can derive the following relationship for the DC **supply neutral** ($o$) to **load neutral** ($n$) voltage:

$$v_{no} = \frac{v_{ao} + v_{bo} + v_{co}}{3} \qquad\qquad\qquad\text{(11.25)}$$

This voltage is also shown to be a square wave switching three times as fast as the inverter output voltage. Finally, to obtain the phase voltages, we make use of the relations

$$v_{an} = v_{ao} - v_{no} = \tfrac{2}{3}v_{ao} - \tfrac{1}{3}(v_{bo} + v_{co})$$

$$v_{bn} = v_{bo} - v_{no} = \tfrac{2}{3}v_{bo} - \tfrac{1}{3}(v_{ao} + v_{co}) \qquad\qquad\text{(11.26)}$$

$$v_{cn} = v_{co} - v_{no} = \tfrac{2}{3}v_{bo} - \tfrac{1}{3}(v_{ao} + v_{bo})$$

Only one phase voltage, $v_{an}$, is shown in the picture; however, it is straightforward to construct the other two phase voltages using equation 11.26. Note that the load phase voltage waveform shown in Figure 11.46 is a coarse stepwise approximation of a sinusoidal waveform; the corresponding load current, $i_a$, is a filtered version of the load voltage, since the load is inductive in nature, and is therefore somewhat smoothed with respect to the voltage waveform. The discontinuous nature of these waveforms creates a significant higher harmonic spectrum (see the box "Fourier Analysis" in Chapter 6), at frequencies that are integer multiples of the inverter output frequency; this is an unavoidable property of all inverters that employ switching circuits, but the problem can be reduced by using more complex switching schemes. Another major shortcoming of this AC supply is that if the DC supply is fixed, the amplitude of the inverter output is fixed.

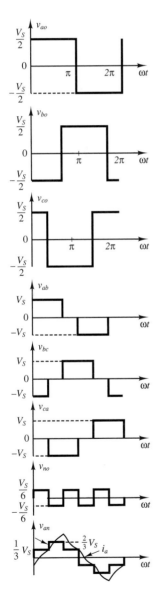

**Figure 11.46** Three-phase voltage source inverter waveforms

The VSI circuit described in the foregoing paragraphs can provide a variable-frequency supply provided that the commutation frequency of the electronic switches can be varied. Thus, in general, it is necessary to also provide the capability for timing circuits that can provide variable switching rates; this is often accomplished with a microprocessor (discussed in Chapter 14).

The limitations of the VSI of Figure 11.45 can be overcome with the use of more advanced switching schemes, such as *pulse-width modulation* (*PWM*) and *sinusoidal PWM*. The complexity of these schemes is beyond the scope of this book, and the interested reader is invited to explore a more advanced power electronics text to learn about advanced inverter circuits. We shall simply mention that it is possible to significantly reduce the harmonic content of the inverter waveforms and to provide variable-frequency, variable-amplitude, three-phase supplies for AC motors by means of power switching circuits under microprocessor control. These advances are finding a growing field of application in the electric vehicle arena. This subject is approached again in Chapter 17.

## Conclusion

- Power electronic devices can handle up to a few thousand volts and up to several hundred amperes and have a host of industrial applications. Various families of power electronic circuits and their application were discussed in this chapter.

- Voltage regulators are used in DC power supplies to provide a stable DC voltage output. The principal element of a voltage regulator is the Zener diode.

- Transistors find application both as power amplifiers and as switches; BJTs, MOSFETs, and IGBTs are all commonly employed, especially for switching functions. Each of these devices offers specific advantages, such as greater current capability, or faster response. Device technology is rapidly improving, especially among power MOSFETs.

- Power diodes and various types of thyristors find widespread application in rectifiers and controlled rectifiers, both for single- and three-phase circuits. Rectifiers are a necessary element of DC power supplies; controlled rectifiers also find application as DC motor drives and in many other variable-voltage applications.

- Electric motor drives based on power electronic devices allow for the implementation of sophisticated motor controls. DC motor drives include controlled regulators and choppers (DC-DC converters), while AC motor drives consist of inverter circuits (DC-AC converters). Both of the latter circuits make extensive use of high-power switching elements, such as MOSFETs, thyristors, BJTs, and IGBTs.

## CHECK YOUR UNDERSTANDING ANSWERS

| | |
|---|---|
| **CYU 11.1** | $P = 1.3$ W |
| **CYU 11.2** | $I_L = 5.4$ A; $I_{AC} = 0.75$ A; $\alpha = 84.95°$ |
| **CYU 11.3** | $V_L = 23.87$ V |
| **CYU 11.4** | 120 V, 60 W |

# HOMEWORK PROBLEMS

## Section 1: Regulators and Rectifiers

**11.1**   Repeat Example 11.1 for a 7-V Zener diode.

**11.2**   For the current regulator circuit shown in Figure P11.2, find the expression for $R_S$.

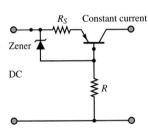

**Figure P11.2**

**11.3**   For the shunt-type voltage regulator shown in Figure P11.3, find the expression for the output voltage, $V_{out}$.

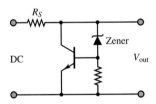

**Figure P11.3**

**11.4**   For the circuit shown in Figure 11.17, if the $LR$ load is replaced by a capacitor, draw the output waveform and label the values.

**11.5**   Draw $v_L(t)$ and label the values for the circuit in Figure 11.17 if the diode forward resistance is 50 Ω, the forward bias voltage is 0.7 V, and the load consists of a resistor $R = 10$ Ω and an inductor $L = 2$ H.

**11.6**   For the circuit shown in Figure P11.6, $v_{AC}$ is a sinusoid with 10-V peak amplitude, $R = 2$ kΩ, and the forward-conducting voltage of $D$ is 0.7 V.

  a.  Sketch the waveform of $v_L(t)$.

  b.  Find the average value of $v_L(t)$.

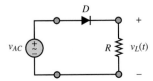

**Figure P11.6**

**11.7**   A vehicle battery charge circuit is shown in Figure P11.7. Describe the circuit, and draw the output waveform ($L_1$ and $L_2$ represent the inductances of the windings of the alternator).

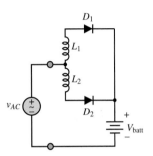

**Figure P11.7**

**11.8**   Repeat Example 11.2 for $\alpha = \pi/3$ and $\pi/6$.

**11.9**   The circuit shown in Figure P11.9 is a speed control system for a DC motor. Assume that the thyristors are fired at $\alpha = 60°$ and that the motor current is 20 A and is ripple free. The supply is 110 VAC (rms).

  a.  Sketch the output voltage waveform, $v_o$.

  b.  Compute the power absorbed by the motor.

  c.  Determine the volt-amperes generated by the supply.

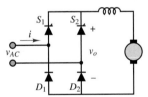

**Figure P11.9**

**11.10**   A full wave, single-phase controlled rectifier is used to control the speed of a DC motor. The circuit is similar to that of Figure 11.2, except for replacing the resistive load with a DC motor. The motor operates at 110 V and absorbs 4 kW of power. The AC supply is 80 V, 60 Hz. Assume that the motor inductance is very large (i.e., the motor current is ripple free), and that the motor constant is 0.055 V/rev/min. If the motor runs at 1,000 rev/min at rated current:

  a.  Determine the firing angle of the converter.

  b.  Determine the rms value of the supply current.

**11.11**  For the light dimmer circuit of Example 11.2, determine the load power at firing angles $\alpha = 0°$, $30°$, $60°$, $90°$, $120°$, $150°$, $180°$, and plot the load power as a function of $\alpha$.

**11.12**  In the circuit shown in Figure P11.12, if:

$V_L = 10$ V     $V_r = 10\% = 1$ V

$I_L = 650$ mA     $v_{line} = 170 \cos(\omega t)$

$\omega = 2{,}513$ rad/s

and if the diodes are fabricated from silicon, determine the conduction angle of the diodes.

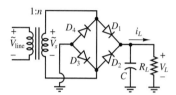

**Figure P11.12**

**11.13**  Assume that the conduction angle of the silicon diodes shown in the circuit of Figure P11.13 is:

$\phi = 23°$

$v_{s1}(t) = v_{s2}(t) = 8 \cos(\omega t)$ V

$\omega = 377$ rad/s     $R_L = 20$ k$\Omega$

$C = 0.5$ $\mu$F

Determine the rms value of the ripple voltage.

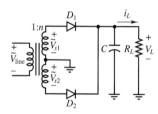

**Figure P11.13**

**11.14**  The diodes in the full-wave DC power supply shown in Figure P11.12 are silicon. If:

$I_L = 85$ mA     $V_L = 5.3$ V

$V_r = 0.6$ V     $\omega = 377$ rad/s

$v_{line} = 156 \cos(\omega t)$ V

$C = 1{,}023$ $\mu$F

$\phi =$ Conduction angle $= 23.90°$

determine the value of the average and peak current through each diode.

**11.15**  The diodes in the full-wave DC power supply shown in Figure P11.13 are silicon. If:

$I_L = 600$ mA     $V_L = 50$ V

$V_r = 8\% = 4$ V     $v_{line} = 170 \cos(\omega t)$ V

determine the value of the conduction angle for the diodes and the average and peak current through the diodes. The load voltage waveform is shown in Figure P11.15.

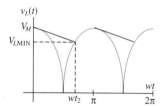

**Figure P11.15**

## Section 2: Choppers and DC Motor Supplies

**11.16**  The chopper of Figure 11.35 is used to control the speed of a DC motor. Let the supply voltage be 120 V and the armature resistance of the motor be 0.15 $\Omega$. The motor back emf constant is 0.05 V/rev/min and the chopper frequency is 250 Hz. Assume that the motor current is free of ripple and equal to 125 A at 120 rev/min.

a. Determine the duty cycle of the chopper, $\delta$, and the chopper on time, $t_1$.

b. Determine the power absorbed by the motor.

c. Determine the power generated by the supply.

**11.17**  The circuit of Figure 11.39 is used to provide regenerative braking in a traction motor. The motor constant is 0.3 V/rev/min and the supply voltage is 600 V. The armature resistance is $R_a = 0.2$ $\Omega$. If the motor speed is 800 rev/min and the motor current is 300 A:

a. Determine the duty cycle, $\delta$, of the chopper.

b. Determine the power fed back to the supply (battery).

**11.18**  For the two-quadrant chopper of Figure 10.40, assume that thyristors $S_1$ and $S_2$ are turned on for time $t_1$ and off for time $T - t_1$ ($T$ is the chopping period). Derive an expression for the average output voltage in terms of the supply voltage, $V_S$, and the duty cycle, $\delta$.

**11.19**  A step-up chopper is powered by an ideal 100-V battery pack. The load voltage waveform consists of rectangular pulses with "on" time $= 1$ ms and period equal to 2.5 ms. Calculate the average and rms value of the chopper supply voltage.

**11.20**  A buck converter connected to a 100-V battery pack supplies an $R$-$L$ load with $R = 0.5$ $\Omega$ and $L =$

1 mH. The switch (a thyristor) is switched "on" for 1 ms and the period of the switching waveform is 3 ms. Calculate the average value of the load voltage and the power supplied by the battery.

**11.21**  The converter of Problem 11.20 is used to supply a separately excited DC motor with $R_a = 0.2 \ \Omega$ and $L_a = 1$ mH. At the lowest speed of operation, the back emf, $E_a$, is equal to 10 V. Calculate the average value of the load current and voltage for this condition if the switching period is 3 ms and the duty cycle is 1/3.

**11.22**  A separately excited DC motor with $R_a = 0.33 \ \Omega$ and $L_a = 15$ mH is controlled by a DC chopper in the range 0–2,000 rev/min. The DC supply is 220 V. If the load torque is constant, and requires an average armature current of 25 A, calculate the range of duty cycles required if the motor armature constant is $K_a\phi = 0.00167$ V-s/rev.

**11.23**  A separately excited DC motor is rated at 10 kW, 240 V, 1,000 rev/min, and is supplied by a single-phase controlled bridge rectifier. The power supply is sinusoidal and rated at 240 V, 60 Hz. The motor armature resistance is 0.42 $\Omega$, and the motor constant is $K_a = 2$ V-s/rad. Calculate the speed, power factor, and efficiency for SCR firing angles $\alpha = 0°$ and $20°$ if the load torque is constant. Assume that additional inductance is present to ensure continuous conduction.

# CHAPTER

## 12

# Operational Amplifiers

I n this chapter we analyze the properties of the ideal amplifier and explore the features of a general-purpose amplifier circuit known as the *operational amplifier*. Understanding the gain and frequency response properties of the operational amplifier is essential for the user of electronic instrumentation. Fortunately, the availability of operational amplifiers in integrated circuit form has made the task of analyzing such circuits quite simple. The models presented in this chapter are based on concepts that have already been explored at length in earlier chapters, namely, Thévenin and Norton equivalent circuits and frequency response ideas.

Mastery of operational amplifier fundamentals is essential in any practical application of electronics. This chapter is aimed at developing your understanding of the fundamental properties of practical operational amplifiers. A number of useful applications are introduced in the examples and homework problems. Upon completion of the chapter, you should be able to:

- Analyze and design simple signal-conditioning circuits based on op-amps.
- Analyze and design simple active filters.
- Understand the operation of analog computers.
- Assess and understand the practical limitations of operational amplifiers.

## 12.1  AMPLIFIERS

One of the most important functions in electronic instrumentation is that of amplification. The need to amplify low-level electrical signals arises frequently in a number of applications. Perhaps the most familiar use of amplifiers arises in converting the low-voltage signal from a cassette tape player, a radio receiver, or a compact disk player to a level suitable for driving a pair of speakers. Figure 12.1 depicts a typical arrangement. Amplifiers have a number of applications of interest to the non–electrical engineer, such as the amplification of low-power signals from transducers (e.g., bioelectrodes, strain gauges, thermistors, and accelerometers) and other, less obvious functions that will be reviewed in this chapter—for example, filtering and impedance isolation. We turn first to the general features and characteristics of amplifiers, before delving into the analysis of the operational amplifier.

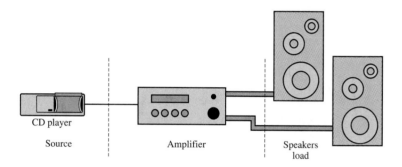

**Figure 12.1** Amplifier in audio system

### Ideal Amplifier Characteristics

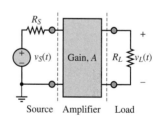

**Figure 12.2** A voltage amplifier

The simplest model for an amplifier is depicted in Figure 12.2, where a signal, $v_S(t)$, is shown being amplified by a constant factor $A$, called the *gain* of the amplifier. Ideally, the load voltage should be given by the expression

$$v_L(t) = Av_S(t) \tag{12.1}$$

Note that the source has been modeled as a Thévenin equivalent, and the load as an equivalent resistance. Thévenin's theorem guarantees that this picture can be representative of more complex circuits. Hence, the equivalent source circuit is the circuit the amplifier "sees" from its input port; and $R_L$, the load, is the equivalent resistance seen from the output port of the amplifier.

What would happen if the roles were reversed? That is, what does the source see when it "looks" into the input port of the amplifier, and what does the load see when it "looks" into the output port of the amplifier? While it is not clear at this point how one might characterize the internal circuitry of an amplifier (which is rather complex), it can be presumed that the amplifier will act as an equivalent load with respect to the source, and as an equivalent source with respect to the load. After all, this is a direct application of Thévenin's theorem. Figure 12.3 provides a pictorial representation of this simplified characterization of an amplifier. The "black box" of Figure 12.2 is now represented as an equivalent circuit with the following behavior. The input circuit has equivalent resistance $R_{in}$, so that the

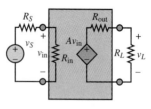

**Figure 12.3** Simple voltage amplifier model

input voltage, $v_{in}$, is given by

$$v_{in} = \frac{R_{in}}{R_S + R_{in}} v_S \tag{12.2}$$

The equivalent input voltage seen by the amplifier is then amplified by a constant factor, $A$. This is represented by the controlled voltage source $Av_{in}$. The controlled source appears in series with an internal resistor, $R_{out}$, denoting the internal (output) resistance of the amplifier. Thus, the voltage presented to the load is

$$v_L = Av_{in} \frac{R_L}{R_{out} + R_L} \tag{12.3}$$

or, substituting the equation for $v_{in}$,

$$v_L = \left( A \frac{R_{in}}{R_S + R_{in}} \frac{R_L}{R_{out} + R_L} \right) v_S \tag{12.4}$$

In other words, the load voltage is an amplified version of the source voltage.

Unfortunately, the amplification factor is now dependent on both the source and load impedances, and on the input and output resistance of the amplifier. Thus, a given amplifier would perform differently with different loads or sources. What are the desirable characteristics for a voltage amplifier that would make its performance relatively independent of source and load impedances? Consider, once again, the expression for $v_{in}$. If the input resistance of the amplifier, $R_{in}$, were very large, the source voltage, $v_S$, and the input voltage, $v_{in}$, would be approximately equal:

$$v_{in} \approx v_S \tag{12.5}$$

since

$$\lim_{R_{in} \to \infty} \left( \frac{R_{in}}{R_{in} + R_S} \right) = 1 \tag{12.6}$$

By an analogous argument, it can also be seen that the desired output resistance for the amplifier, $R_{out}$, should be very small, since for an amplifier with $R_{out} = 0$, the load voltage would be

$$v_L = Av_{in} \tag{12.7}$$

Combining these two results, we can see that as $R_{in}$ approaches infinity and $R_{out}$ approaches zero, the ideal amplifier magnifies the source voltage by a factor of $A$:

$$v_L = Av_S \tag{12.8}$$

just as was indicated in the "black box" amplifier of Figure 12.2.

Thus, two desirable characteristics for a general-purpose voltage amplifier are a very *large input impedance* and a very *small output impedance*. In the next sections it will be shown how operational amplifiers provide these desired characteristics.

## 12.2    THE OPERATIONAL AMPLIFIER

An **operational amplifier** is an **integrated circuit,** that is, a large collection of individual electrical and electronic circuits integrated on a single silicon wafer.

An operational amplifier—or op-amp—can perform a great number of operations, such as addition, filtering, or integration, which are all based on the properties of ideal amplifiers and of ideal circuit elements. The introduction of the operational amplifier in integrated circuit form marked the beginning of a new era in modern electronics. Since the introduction of the first IC op-amp, the trend in electronic instrumentation has been to move away from the discrete (individual-component) design of electronic circuits, toward the use of integrated circuits for a large number of applications. This statement is particularly true for applications of the type the non–electrical engineer is likely to encounter: op-amps are found in most measurement and instrumentation applications, serving as extremely versatile building blocks for any application that requires the processing of electrical signals.

In the following pages, simple circuit models of the op-amp will be introduced. The simplicity of the models will permit the use of the op-amp as a circuit element, or building block, without the need to describe its internal workings in detail. Integrated circuit technology has today reached such an advanced stage of development that it can be safely stated that for the purpose of many instrumentation applications, the op-amp can be treated as an ideal device. Following the introductory material presented in this chapter, more advanced instrumentation applications will be explored in Chapter 15.

### The Open-Loop Model

The ideal operational amplifier behaves very much as an ideal **difference amplifier,** that is, a device that amplifies the difference between two input voltages. Operational amplifiers are characterized by near-infinite input resistance and very small output resistance. As shown in Figure 12.4, the output of the op-amp is an amplified version of the difference between the voltages present at the two inputs:[1]

$$v_{out} = A_{V(OL)}(v^+ - v^-) \tag{12.9}$$

The input denoted by a positive sign is called the **noninverting input** (or terminal), while that represented with a negative sign is termed the **inverting input** (or terminal). The amplification factor, or gain, $A_{V(OL)}$, is called the **open-loop voltage gain** and is quite large by design, typically of the order of $10^5$ to $10^7$; it will soon become apparent why a large open-loop gain is a desirable characteristic. Together with the high input resistance and low output resistance, the effect of a large amplifier open-loop voltage gain, $A_{V(OL)}$, is such that op-amp circuits can be designed to perform very nearly as ideal voltage or current amplifiers. In effect, to analyze the performance of an op-amp circuit, only one assumption will be needed: that the current flowing into the input circuit of the amplifier is zero, or

$$i_{in} = 0 \tag{12.10}$$

This assumption is justified by the large input resistance and large open-loop gain of the **operational amplifier.** The model just introduced will be used to analyze three amplifier circuits in the next part of this section.

[1]The amplifier of Figure 12.4 is a *voltage amplifier;* another type of operational amplifier, called a *current* or *transconductance amplifier*, is described in the homework problems.

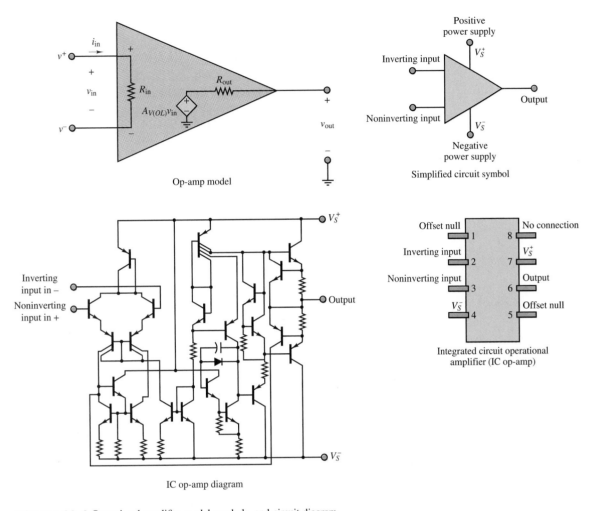

**Figure 12.4** Operational amplifier model symbols, and circuit diagram

## The Operational Amplifier in the Closed-Loop Mode

### *The Inverting Amplifier*

One of the more popular circuit configurations of the op-amp, because of its simplicity, is the so-called inverting amplifier, shown in Figure 12.5. The input signal to be amplified is connected to the inverting terminal, while the noninverting terminal is grounded. It will now be shown how it is possible to choose an (almost) arbitrary gain for this amplifier by selecting the ratio of two resistors. The analysis is begun by noting that at the inverting input node, KCL requires that

$$i_S + i_F = i_{\text{in}} \qquad (12.11)$$

The current $i_F$, which flows back to the inverting terminal from the output, is appropriately termed **feedback current,** because it represents an input to the amplifier that is "fed back" from the output. Applying Ohm's law, we may determine

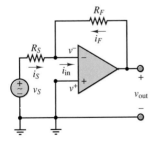

**Figure 12.5** Inverting amplifier

each of the three currents shown in Figure 12.5:

$$i_S = \frac{v_S - v^-}{R_S} \qquad i_F = \frac{v_{\text{out}} - v^-}{R_F} \qquad i_{\text{in}} = 0 \tag{12.12}$$

(the last by assumption, as stated earlier). The voltage at the noninverting input, $v^+$, is easily identified as zero, since it is directly connected to ground: $v^+ = 0$. Now, the **open-loop model for the op-amp** requires that

$$v_{\text{out}} = A_{V(\text{OL})}(v^+ - v^-) = -A_{V(\text{OL})}v^- \tag{12.13}$$

or

$$v^- = -\frac{v_{\text{out}}}{A_{V(\text{OL})}} \tag{12.14}$$

Having solved for the voltage present at the inverting input, $v^-$, in terms of $v_{\text{out}}$, we may now compute an expression for the amplifier gain, $v_{\text{out}}/v_S$. This quantity is called the **closed-loop gain,** because the presence of a feedback connection between the output and the input constitutes a closed loop.[2] Combining equations 12.11 and 12.12, we can write

$$i_S = -i_F \tag{12.15}$$

and

$$\frac{v_S}{R_S} + \frac{v_{\text{out}}}{A_{V(\text{OL})}R_S} = -\frac{v_{\text{out}}}{R_F} - \frac{v_{\text{out}}}{A_{V(\text{OL})}R_F} \tag{12.16}$$

leading to the expression

$$\frac{v_S}{R_S} = -\frac{v_{\text{out}}}{R_F} - \frac{v_{\text{out}}}{A_{V(\text{OL})}R_F} - \frac{v_{\text{out}}}{A_{V(\text{OL})}R_S} \tag{12.17}$$

or

$$v_S = -v_{\text{out}}\left(\frac{1}{R_F/R_S} + \frac{1}{A_{V(\text{OL})}R_F/R_S} + \frac{1}{A_{V(\text{OL})}}\right) \tag{12.18}$$

If the open-loop gain of the amplifier, $A_{V(\text{OL})}$, is sufficiently large, the terms $1/(A_{V(\text{OL})}R_F/R_S)$ and $1/A_{V(\text{OL})}$ are essentially negligible, compared with $1/(R_F/R_S)$. As stated earlier, typical values of $A_{V(\text{OL})}$ range from $10^5$ to $10^7$, and thus it is reasonable to conclude that, to a close approximation, the following expression describes the closed-loop gain of the inverting amplifier:

$$v_{\text{out}} = -\frac{R_F}{R_S}v_S \qquad \text{Inverting amplifier closed-loop gain} \tag{12.19}$$

That is, the closed-loop gain of an inverting amplifier may be selected simply by the appropriate choice of two externally connected resistors. The price for this extremely simple result is an inversion of the output with respect to the input—that is, a negative sign.

---

[2]This terminology is borrowed from the field of automatic controls, for which the theory of closed-loop feedback systems forms the foundation.

Next, we show that by making an additional assumption it is possible to simplify the analysis considerably. Consider that, as was shown for the inverting amplifier, the inverting terminal voltage is given by

$$v^- = -\frac{v_{\text{out}}}{A_{V(\text{OL})}} \tag{12.20}$$

Clearly, as $A_{V(\text{OL})}$ approaches infinity, the inverting-terminal voltage is going to be very small (practically, of the order of microvolts). It may then be assumed that *in the inverting amplifier*, $v^-$ is virtually zero:

$$v^- \approx 0 \tag{12.21}$$

This assumption prompts an interesting observation (which may not yet appear obvious at this point):

> The effect of the feedback connection from output to inverting input is to force the voltage at the inverting input to be equal to that at the noninverting input.

This is equivalent to stating that for an op-amp *with negative feedback*,

$$v^- \approx v^+ \tag{12.22}$$

The analysis of the operational amplifier can now be greatly simplified if the following two assumptions are made:

1. $i_{\text{in}} = 0$
2. $v^- = v^+$ $\tag{12.23}$

This technique will be tested in the next subsection by analyzing a noninverting amplifier configuration. Example 12.1 illustrates some simple design considerations.

---

## Why Feedback?

Why is such emphasis placed on the notion of an amplifier with a very large open-loop gain and with negative feedback? Why not just design an amplifier with a reasonable gain, say, $\times 10$, or $\times 100$, and just use it as such, without using feedback connections? In these paragraphs, we hope to answer these and other questions, introducing the concept of **negative feedback** in an intuitive fashion.

The fundamental reason for designing an amplifier with a very large open-loop gain is the flexibility it provides in the design of amplifiers with an (almost) arbitrary gain; it has already been shown that

the gain of the inverting amplifier is determined by the choice of two external resistors—undoubtedly a convenient feature! Negative feedback is the mechanism that enables us to enjoy such flexibility in the design of linear amplifiers.

To understand the role of feedback in the operational amplifier, consider the internal structure of the op-amp shown in Figure 12.4. The large open-loop gain causes any difference in voltage at the input terminals to appear greatly amplified at the output. When a negative feedback connection is provided, as shown, for example, in the inverting amplifier of

## Why Feedback? *(continued)*

Figure 12.5, the output voltage, $v_{\text{out}}$, causes a current, $i_F$, to flow through the feedback resistance so that KCL is satisfied at the inverting node. Assume, for a moment, that the differential voltage

$$\Delta v = v^+ - v^-$$

is identically zero. Then, the output voltage will continue to be such that KCL is satisfied at the inverting node, that is, such that the current $i_F$ is equal to the current $i_S$.

Suppose, now, that a small imbalance in voltage, $\Delta v$, is present at the input to the op-amp. Then the output voltage will be increased by an amount $A_{V(\text{OL})}\Delta v$. Thus, an incremental current approximately equal to $A_{V(\text{OL})}\Delta v/R_F$ will flow from output to input via the feedback resistor. The effect of this incremental current is to reduce the voltage difference $\Delta v$ to zero, so as to restore the original balance in the circuit. One way of viewing negative feedback, then, is to consider it a self-balancing mechanism, which allows the amplifier to preserve zero potential difference between its input terminals.

A practical example that illustrates a common application of negative feedback is the thermostat.

This simple temperature control system operates by comparing the desired ambient temperature and the temperature measured by a thermometer and turns a heat source on and off to maintain the difference between actual and desired temperature as close to zero as possible. An analogy may be made with the inverting amplifier if we consider that, in this case, negative feedback is used to keep the inverting-terminal voltage as close as possible to the noninverting-terminal voltage. The latter voltage is analogous to the desired ambient temperature in your home, while the former plays a role akin to that of the actual ambient temperature. The open-loop gain of the amplifier forces the two voltages to be close to each other, much the way the furnace raises the heat in the house to match the desired ambient temperature.

It is also possible to configure operational amplifiers in a **positive feedback** configuration if the output connection is tied to the noninverting input. We do not discuss this configuration in the present chapter, but present an example of it, the **voltage comparator,** in Chapter 15.

---

### EXAMPLE   12.1  Inverting Amplifier Circuit

#### Problem

Determine the voltage gain and output voltage for the inverting amplifier circuit of Figure 12.5. What will the uncertainty in the gain be if 5 and 10 percent tolerance resistors are used, respectively?

---

#### Solution

*Known Quantities:*  Feedback and source resistances, source voltage.

*Find:*  $A_V = v_{\text{out}}/v_{\text{in}}$; maximum percent change in $A_V$ for 5 and 10 percent tolerance resistors.

*Schematics, Diagrams, Circuits, and Given Data:*  $R_S = 1 \text{ k}\Omega$; $R_F = 10 \text{ k}\Omega$; $v_S(t) = A \cos(\omega t)$; $A = 0.015$ V; $\omega = 50$ rad/s.

*Assumptions:*  The amplifier behaves ideally; that is, the input current into the op-amp is zero and negative feedback forces $v^+ = v^-$.

**Analysis:** Using equation 12.19, we calculate the output voltage:

$$v_{\text{out}}(t) = A_V \times v_S(t) = -\frac{R_F}{R_S} \times v_S(t) = -10 \times 0.015 \; \cos(\omega t) = -0.15 \; \cos(\omega t)$$

The input and output waveforms are sketched in Figure 12.6.

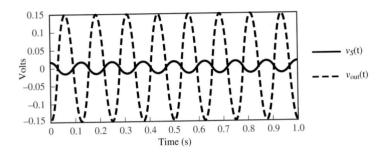

**Figure 12.6**

The nominal gain of the amplifier is $A_{V \text{ nom}} = -10$. If 5 percent tolerance resistors are employed, the worst-case error would occur at the extremes:

$$A_{V \min} = -\frac{R_{F \min}}{R_{S \max}} = -\frac{9,500}{1,050} = 9.05 \qquad A_{V \max} = -\frac{R_{F \max}}{R_{S \min}} = -\frac{10,500}{950} = 11.05$$

The percentage error is therefore computed as:

$$100 \times \frac{A_{V \text{ nom}} - A_{V \min}}{A_{V \text{ nom}}} = 100 \times \frac{10 - 9.05}{10} = 9.5\%$$

$$100 \times \frac{A_{V \text{ nom}} - A_{V \max}}{A_{V \text{ nom}}} = 100 \times \frac{10 - 11.05}{10} = -10.5\%$$

Thus, the amplifier gain could vary by as much as $\pm10$ percent (approximately) when 5 percent resistors are used. If 10 percent resistors were used, we would calculate a percent error of approximately $\pm 20$ percent, as shown below.

$$A_{V \min} = -\frac{R_{F \min}}{R_{S \max}} = -\frac{9,000}{1,100} = 8.18 \qquad A_{V \max} = -\frac{R_{F \max}}{R_{S \min}} = -\frac{11,000}{900} = 12.2$$

$$100 \times \frac{A_{V \text{ nom}} - A_{V \min}}{A_{V \text{ nom}}} = 100 \times \frac{10 - 8.18}{10} = 18.2\%$$

$$100 \times \frac{A_{V \text{ nom}} - A_{V \max}}{A_{V \text{ nom}}} = 100 \times \frac{10 - 12.2}{10} = -22.2\%$$

**Comments:** Note that the worst-case percent error in the amplifier gain is double the resistor tolerance.

**Focus on Computer-Aided Solutions:** An *Electronics Workbench*™ simulation of the circuit of Figure 12.5 can be found in the accompanying CD-ROM.

### The Summing Amplifier

A useful op-amp circuit that is based on the inverting amplifier is the **op-amp summer,** or **summing amplifier.** This circuit, shown in Figure 12.7, is used to

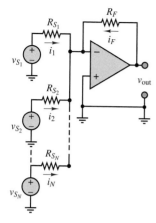

**Figure 12.7** Summing amplifier

add signal sources. The primary advantage of using the op-amp as a summer is that the summation occurs independently of load and source impedances, so that sources with different internal impedances will not interact with each other. The operation of the summing amplifier is best understood by application of KCL at the inverting node: the sum of the $N$ source currents and the feedback current must equal zero, so that

$$i_1 + i_2 + \cdots + i_N = -i_F \tag{12.24}$$

But each of the source currents is given by the following expression:

$$i_n = \frac{v_{S_n}}{R_{S_n}} \qquad n = 1, 2, \ldots, N \tag{12.25}$$

while the feedback current is

$$i_F = \frac{v_{\text{out}}}{R_F} \tag{12.26}$$

Combining equations 12.25 and 12.26, and using equation 12.15, we obtain the following result:

$$\sum_{n=1}^{N} \frac{v_{S_n}}{R_{S_n}} = -\frac{v_{\text{out}}}{R_F} \tag{12.27}$$

or

$$v_{\text{out}} = -\sum_{n=1}^{N} \frac{R_F}{R_{S_n}} v_{S_n} \tag{12.28}$$

That is, the output consists of the weighted sum of $N$ input signal sources, with the weighting factor for each source equal to the ratio of the feedback resistance to the source resistance.

### The Noninverting Amplifier

To avoid the negative gain (i.e., phase inversion) introduced by the inverting amplifier, a **noninverting amplifier** configuration is often employed. A typical noninverting amplifier is shown in Figure 12.8; note that the input signal is applied to the noninverting terminal this time.

The noninverting amplifier can be analyzed in much the same way as the inverting amplifier. Writing KCL at the inverting node yields

$$i_F = i_S + i_{\text{in}} \approx i_S \tag{12.29}$$

where

$$i_F = \frac{v_{\text{out}} - v^-}{R_F} \tag{12.30}$$

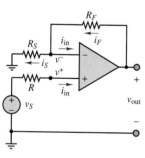

**Figure 12.8** Noninverting amplifier

$$i_S = \frac{v^-}{R_S} \tag{12.31}$$

Now, since $i_{\text{in}} = 0$, the voltage drop across the source resistance, $R$, is equal to zero. Thus,

$$v^+ = v_s \tag{12.32}$$

and, using equation 12.22,

$$v^- = v^+ = v_S \qquad\qquad (12.33)$$

Substituting this result in equations 12.29 and 12.30, we can easily show that

$$i_F = i_S \qquad\qquad (12.34)$$

or

$$\frac{v_{\text{out}} - v_S}{R_F} = \frac{v_S}{R_S} \qquad\qquad (12.35)$$

It is easy to manipulate equation 12.35 to obtain the result

| | |
|---|---|
| $\dfrac{v_{\text{out}}}{v_S} = 1 + \dfrac{R_F}{R_S}$    Noninverting amplifier closed-loop gain | **(12.36)** |

which is the closed-loop gain expression for a noninverting amplifier. Note that the gain of this type of amplifier is always positive and greater than (or equal to) 1.

The same result could have been obtained without making the assumption $v^+ = v^-$, at the expense of some additional work. The procedure one would follow in this latter case is analogous to the derivation carried out earlier for the inverting amplifier, and is left as an exercise.

In summary, in the preceding pages it has been shown that by constructing a nonideal amplifier with very large gain and near-infinite input resistance, it is possible to design amplifiers that have near-ideal performance and provide a variable range of gains, easily controlled by the selection of external resistors. The mechanism that allows this is negative feedback. From here on, unless otherwise noted, it will be reasonable and sufficient in analyzing new op-amp configurations to utilize the two assumptions

| | |
|---|---|
| 1. $i_{\text{in}} = 0$     Approximations used for ideal<br>2. $v^- = v^+$     op-amps with negative feedback | **(12.37)** |

## EXAMPLE  12.2  Voltage Follower

### Problem

Determine the closed-loop voltage gain and input resistance of the **voltage follower** circuit of Figure 12.9.

### Solution

***Known Quantities:***  Feedback and source resistances, source voltage.

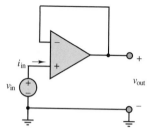

**Figure 12.9** Voltage follower

**Find:**

$$A_V = \frac{v_{out}}{v_{in}} \qquad r_i = \frac{v_{in}}{i_{in}}$$

**Assumptions:** The amplifier behaves ideally; that is, the input current into the op-amp is zero and negative feedback forces $v^+ = v^-$.

**Analysis:** From the ideal op-amp assumptions, $v^+ = v^-$. But $v^+ = v_{in}$ and $v^- = v_{out}$, thus:

$$v_{in} = v_{out}$$

The voltage follower's name derives from the ability of the output voltage to "follow" exactly the input voltage. To compute the input resistance of this amplifier, we observe that since the input current is zero,

$$r_i = \frac{v_{in}}{i_{in}} \to \infty$$

**Comments:** The input resistance of the voltage follower is the most important property of the amplifier: The extremely high input resistance of this amplifier (of the order of megohms to gigohms) permits virtually perfect isolation between source and load, and eliminates *loading* effects. Voltage followers, or *impedance buffers*, are commonly packaged in groups of four or more in integrated circuit (IC) form. The data sheets for one such IC are contained in the accompanying CD-ROM, and may also be found in the device templates for analog ICs in the *Electronics Workbench*™ libraries.

**Focus on Computer-Aided Solutions:** An *Electronics Workbench*™ simulation of the circuit of Figure 12.9 can be found in the accompanying CD-ROM.

### The Differential Amplifier

The third closed-loop model examined in this chapter is a combination of the inverting and noninverting amplifiers; it finds frequent use in situations where the difference between two signals needs to be amplified. The basic **differential amplifier** circuit is shown in Figure 12.10, where the two sources, $v_1$ and $v_2$, may be independent of each other, or may originate from the same process, as they do in "Focus on Measurements: Electrocardiogram (EKG) Amplifier."

The analysis of the differential amplifier may be approached by various methods; the one we select to use at this stage consists of:

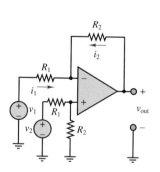

**Figure 12.10** Differential amplifier

1. Computing the noninverting- and inverting-terminal voltages, $v^+$ and $v^-$.
2. Equating the inverting and noninverting input voltages, $v^- = v^+$.
3. Applying KCL at the inverting node, where $i_2 = -i_1$.

Since it has been assumed that no current flows into the amplifier, the noninverting-terminal voltage is given by the following voltage divider:

$$v^+ = \frac{R_2}{R_1 + R_2} v_2 \qquad (12.38)$$

If the inverting-terminal voltage is assumed equal to $v^+$, then the currents $i_1$ and $i_2$ are found to be

$$i_1 = \frac{v_1 - v^+}{R_1} \tag{12.39}$$

and

$$i_2 = \frac{v_{\text{out}} - v^+}{R_2} \tag{12.40}$$

and since

$$i_2 = -i_1 \tag{12.41}$$

the following expression for the output voltage is obtained:

$$v_{\text{out}} = R_2 \left[ \frac{-v_1}{R_1} + \frac{1}{R_1 + R_2} v_2 + \frac{R_2}{R_1(R_1 + R_2)} v_2 \right] \tag{12.42}$$

$$v_{\text{out}} = \frac{R_2}{R_1}(v_2 - v_1) \qquad \text{Differential amplifier closed-loop gain}$$

Thus, the differential amplifier magnifies the difference between the two input signals by the closed-loop gain $R_2/R_1$.

In practice, it is often necessary to amplify the difference between two signals that are both corrupted by noise or some other form of interference. In such cases, the differential amplifier provides an invaluable tool in amplifying the desired signal while rejecting the noise. "Focus on Measurements: Electrocardiogram (EKG) Amplifier" provides a realistic look at a very common application of the differential amplifier.

## Electrocardiogram (EKG) Amplifier

**FOCUS ON MEASUREMENTS**

This example illustrates the principle behind a two-lead **electrocardiogram (EKG) measurement.** The desired cardiac waveform is given by the difference between the potentials measured by two electrodes suitably placed on the patient's chest, as shown in Figure 12.11. A healthy, noise-free EKG waveform, $v_1 - v_2$, is shown in Figure 12.12.

Unfortunately, the presence of electrical equipment powered by the 60-Hz, 110-VAC line current causes undesired interference at the electrode leads: the lead wires act as antennas and pick up some of the 60-Hz signal in addition to the desired EKG voltage. In effect, instead of recording the desired EKG signals, $v_1$ and $v_2$, the two electrodes provide the following inputs to the EKG amplifier, shown in Figure 12.13:

Lead 1:

$$v_1(t) + v_n(t) = v_1(t) + V_n \cos(377t + \phi_n)$$

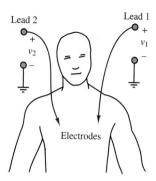

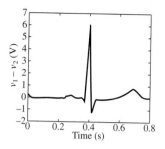

**Figure 12.12** EKG waveform

**Figure 12.11** Two-lead electrocardiogram

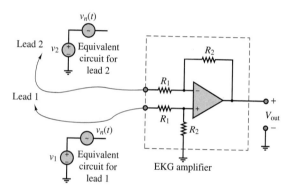

**Figure 12.13** EKG amplifier

Lead 2:

$$v_2(t) + v_n(t) = v_2(t) + V_n \cos(377t + \phi_n)$$

The interference signal, $V_n \cos(377t + \phi_n)$, is approximately the same at both leads, because the electrodes are chosen to be identical (e.g., they have the same lead lengths) and are in close proximity to each other. Further, the nature of the interference signal is such that it is common to both leads, since it is a property of the environment the EKG instrument is embedded in. On the basis of the analysis presented earlier, then,

$$v_{out} = \frac{R_2}{R_1}[(v_1 + v_n(t)) - (v_2 + v_n(t))]$$

or

$$v_{out} = \frac{R_2}{R_1}(v_1 - v_2)$$

Thus, the differential amplifier nullifies the effect of the 60-Hz interference, while amplifying the desired EKG waveform.

The preceding "Focus on Measurements" introduces the concept of so-called **common-mode** and **differential-mode signals.** The desired differential-mode EKG signal was amplified by the op-amp while the common-mode disturbance was canceled. Thus, the differential amplifier provides the ability to reject common-mode signal components (such as noise or undesired DC offsets) while amplifying the differential-mode components. This is a very desirable feature in instrumentation systems. In practice, rejection of the common-mode signal is not complete: some of the common-mode signal component will always appear in the output. This fact gives rise to a figure of merit called the *common-mode rejection ratio*, which is discussed in Section 12.6.

Often, to provide impedance isolation between bridge transducers and the differential amplifier stage, the signals $v_1$ and $v_2$ are amplified separately. This technique gives rise to the so-called **instrumentation amplifier (IA),** shown in Figure 12.14. Example 12.3 illustrates the calculation of the closed-loop gain for a typical instrumentation amplifier.

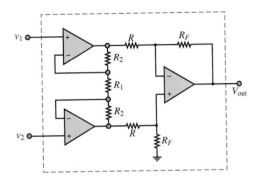

**Figure 12.14** Instrumentation amplifier

## EXAMPLE 12.3 Instrumentation Amplifier

### Problem

Determine the closed-loop voltage gain of the instrumentation amplifier circuit of Figure 12.14.

### Solution

***Known Quantities:*** Feedback and source resistances.

***Find:***

$$A_V = \frac{v_{\text{out}}}{v_1 - v_2} \tag{12.43}$$

***Assumptions:*** Assume ideal op-amps.

***Analysis:*** We consider the input circuit first. Thanks to the symmetry of the circuit, we can represent one half of the circuit as illustrated in Figure 12.15(a), depicting the lower half of the first *stage* of the instrumentation amplifier. We next recognize that the circuit of

Figure 12.15(a) is a noninverting amplifier (see Figure 12.8), and we can directly write the expression for the closed-loop voltage gain (equation 12.36):

$$A = 1 + \frac{R_2}{\frac{R_1}{2}} = 1 + \frac{2R_2}{R_1}$$

Each of the two inputs, $v_1$ and $v_2$, is therefore an input to the second *stage* of the instrumentation amplifier, shown in Figure 12.15(b). We recognize the second stage to be a differential amplifier (see Figure 12.10), and can therefore write the output voltage after equation 12.42:

$$v_{out} = \frac{R_F}{R}(Av_1 - Av_2) = \frac{R_F}{R}\left(1 + \frac{2R_2}{R_1}\right)(v_1 - v_2) \qquad \textbf{(12.44)}$$

from which we can compute the closed-loop voltage gain of the instrumentation amplifier:

$$A_V = \frac{v_{out}}{(v_1 - v_2)} = \frac{R_F}{R}\left(1 + \frac{2R_2}{R_1}\right)$$

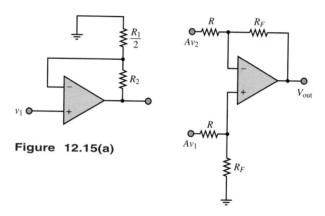

**Figure 12.15(a)**

**Figure 12.15(b)**

**Comments:** This circuit is analyzed in depth in Chapter 15.

**Focus on Computer-Aided Solutions:** An *Electronics Workbench*™ simulation of the circuit of Figure 12.14 can be found in the accompanying CD-ROM.

Because the instrumentation amplifier has widespread application—and in order to ensure the best possible match between resistors—the entire circuit of Figure 12.14 is often packaged as a single integrated circuit. The advantage of this configuration is that the resistors $R_1$ and $R_2$ can be matched much more precisely in an integrated circuit than would be possible using discrete components. A typical, commercially available integrated circuit package is the AD625. Data sheets for this device are provided in the accompanying CD-ROM.

Another simple op-amp circuit that finds widespread application in electronic instrumentation is the **level shifter.** Example 12.4 discusses its operation and its application. The following "Focus on Measurements" illustrates its use in a sensor calibration circuit.

## EXAMPLE 12.4  Level Shifter

### Problem

The level shifter of Figure 12.16 has the ability to add or subtract a DC offset to or from a signal. Analyze the circuit and design it so that it can remove a 1.8-VDC offset from a sensor output signal.

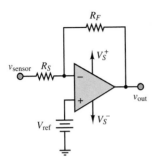

### Solution

**Known Quantities:**  Sensor (input) voltage; feedback and source resistors.

**Find:**  Value of $V_{ref}$ required to remove DC bias.

**Schematics, Diagrams, Circuits, and Given Data:**  $v_S(t) = 1.8 + 0.1 \cos(\omega t)$; $R_F = 220\ \text{k}\Omega$; $R_S = 10\ \text{k}\Omega$.

**Figure 12.16** Level shifter

**Assumptions:**  Assume ideal op-amp.

**Analysis:**  We first determine the closed-loop voltage gain of the circuit of Figure 12.16. The output voltage can be computed quite easily if we note that, upon applying the principle of superposition, the sensor voltage sees an inverting amplifier with gain $-R_F/R_S$, while the battery sees a noninverting amplifier with gain $(1 + R_F/R_S)$. Thus, we can write the output voltage as the sum of two outputs, due to each of the two sources:

$$v_{out} = -\frac{R_F}{R_S} v_{sensor} + \left(1 + \frac{R_F}{R_S}\right) V_{ref}$$

Substituting the expression for $v_{sensor}$ into the equation above, we find that:

$$v_{out} = -\frac{R_F}{R_S}(1.8 + 0.1\cos(\omega t)) + \left(1 + \frac{R_F}{R_S}\right) V_{ref}$$

$$= -\frac{R_F}{R_S}(0.1\ \cos(\omega t)) - \frac{R_F}{R_S}(1.8) + \left(1 + \frac{R_F}{R_S}\right) V_{ref}$$

Since the intent of the design is to remove the DC offset, we require that

$$-\frac{R_F}{R_S}(1.8) + \left(1 + \frac{R_F}{R_S}\right) V_{ref} = 0$$

or

$$V_{ref} = (1.8)\frac{\frac{R_F}{R_S}}{1 + \frac{R_F}{R_S}} = 1.714\ \text{V}$$

**Comments:**  The presence of a precision voltage source in the circuit is undesirable, because it may add considerable expense to the circuit design and, in the case of a battery, it is not adjustable. The circuit of Figure 12.17 illustrates how one can generate an adjustable voltage reference using the DC supplies already used by the op-amp, 2 resistors, $R$, and a potentiometer, $R_p$. The resistors $R$ are included in the circuit to prevent the potentiometer from being shorted to either supply voltage when the potentiometer is at the extreme positions. Using the voltage divider rule, we can write the following expression for the reference voltage generated by the resistive divider:

$$V_{ref} = \frac{(R + \Delta R)V_S^+ + (R + R_p - \Delta R)\,V_S^-}{2R + R_p}$$

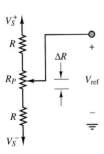

**Figure 12.17**

If the voltage supplies are symmetrical, as is almost always the case, one can further simplify the expression to:

$$V_{\text{ref}} = \pm \left( \frac{R + \Delta R}{2R + R_p} \right) V_S^+$$

Note that, by adjusting the potentiometer, $R_p$, one can obtain any value of $V_{\text{ref}}$ between the supply voltages.

---

## Sensor Calibration Circuit

In many practical instances, the output of a sensor is related to the physical variable we wish to measure in a form that requires some signal conditioning. The most desirable form of a sensor output is one in which the electrical output of the sensor (for example, voltage) is related to the physical variable by a constant factor. Such a relationship is depicted in Figure 12.18(a), where $k$ is the calibration constant relating voltage to temperature. Note that $k$ is a positive number, and that the *calibration curve* passes through the (0, 0) point. On the other hand, the sensor characteristic of Figure 12.18(b) is best described by the following equation:

$$v_{\text{sensor}} = -\beta T + V_0$$

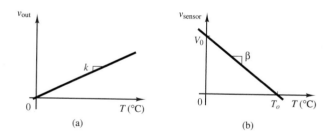

**Figure 12.18** Sensor calibration curves

It is possible to modify the sensor calibration curve of Figure 12.18(b) to the more desirable one of Figure 12.18(a) by means of the simple circuit displayed in Figure 12.19. This circuit provides the desired calibration constant $k$ by a simple gain adjustment, while the zero (or bias) offset is adjusted by means of a potentiometer connected to the voltage supplies. The detailed operation of the circuit is described in the following paragraphs.

As noted before, the nonideal characteristic can be described by the following equation:

$$v_{\text{sensor}} = -\beta T + V_0$$

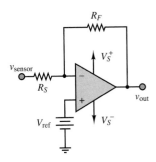

**Figure 12.19** Sensor calibration circuit

Then, the output of the op-amp circuit of Figure 12.19 may be determined by using the principle of superposition:

$$v_{\text{out}} = -\frac{R_F}{F_S} v_{\text{sensor}} + \left(1 + \frac{R_F}{R_S}\right) V_{\text{ref}}$$

$$= -\frac{R_F}{R_S}(-\beta T + V_0) + \left(1 + \frac{R_F}{R_S}\right) V_{\text{ref}}$$

After substituting the expression for the transducer voltage and after some manipulation, we see that by suitable choice of resistors, and of the reference voltage source, we can compensate for the nonideal transducer characteristic. We want the following expression to hold:

$$v_{\text{out}} = \frac{R_F}{R_S} \beta T + \left(1 + \frac{R_F}{R_S}\right) V_{\text{ref}} - \frac{R_F}{R_S} V_0 = kT$$

If we choose

$$\frac{R_F}{R_S}\beta = k$$

and

$$V_{\text{ref}} = \frac{R_F/R_S}{1 + R_F/R_S} V_0$$

then $v_{\text{out}} = kT$.

Note that

$$V_{\text{ref}} \approx V_0 \qquad \text{if} \qquad \frac{R_F}{R_S} \gg 1$$

and we can directly convert the characteristic of Figure 12.18(b) to that of Figure 12.18(a). Clearly, the effect of selecting the gain resistors is to change the magnitude of the slope of the calibration curve. The fact that the sign of the slope changes is purely a consequence of the inverting configuration of the amplifier. The reference voltage source simply shifts the DC level of the characteristic, so that the curve passes through the origin.

## Practical Op-Amp Design Considerations

The results presented in the preceding pages suggest that operational amplifiers permit the design of rather sophisticated circuit in a few very simple steps, simply by selecting appropriate resistor values. This is certainly true, provided that the circuit component selection satifies certain criteria. Here we summarize some important practical design criteria that the designer should keep in mind when selecting component values for op-amp circuits. Section 12.6 explores the practical limitations of op-amps in greater detail.

1. Use standard resistor values. While any arbitrary value of gain can in principle be achieved by selecting the appropriate combination of resistors, the designer is often constrained to the use of standard 5 percent resistor values (see Table 2.1). For example, if your design requires a gain of 25, you might be tempted to select, say, 100-k$\Omega$ and 4-k$\Omega$ resistors to achieve $R_F/R_S = 25$. However, inspection of Table 2.1 reveals that 4 k$\Omega$ is not a standard value; the closest 5 percent tolerance resistor value is 3.9 k$\Omega$, leading to a gain of 25.64. Can you find a combination of standard 5 percent resistors whose ratio is closer to 25?

2. Ensure that the load current is reasonable (do not select very small resistor values). Consider the same example given in 1. Suppose that the maximum output voltage is 10 V. The feedback current required by your design with $R_F = 100$ k$\Omega$ and $R_S = 4$ k$\Omega$ would be $I_F = 10/100,000 = 0.1$ mA. This is a very reasonable value for an op-amp, as you will see

in Section 12.6. If you tried to achieve the same gain using, say, a 10-$\Omega$ feedback resistor and a 0.39-$\Omega$ source resistor, the feedback current would become as large as 1 A. This is a value that is generally beyond the capabilities of a general-purpose op-amp, so the selection of exceedingly low resistor values is not acceptable. On the other hand, the selection of 10-k$\Omega$ and 390-$\Omega$ resistors would still lead to acceptable values of current, and would be equally good. As a general rule of thumb, you should avoid resistor values lower than 100 $\Omega$ in practical designs.

3. Avoid stray capacitance (do not select excessively large resistor values). The use of exceedingly large resistor values can cause unwanted signals to couple into the circuit through a mechanism known as *capacitive coupling*. This phenomenon is discussed in Chapter 15. Large resistance values can also cause other problems. As a general rule of thumb, you should avoid resistor values higher than 1 M$\Omega$ in practical designs.

4. Precision designs may be warranted. If a certain design requires that the amplifier gain be set to a very accurate value, it may be appropriate to use the (more expensive) option of precision resistors: for example, 1 percent tolerance resistors are commonly available, at a premium cost. Some of the examples and homework problems explore the variability in gain due to the use of higher and lower tolerance resistors.

# FOCUS ON METHODOLOGY

### Using Op-Amp Data Sheets

FIND IT
ON THE WEB

Here we illustrate use of **device data sheets** for two commonly used operational amplifiers. The first, the LM741, is a general-purpose (low-cost) amplifier; the second, the LMC6061 is a precision CMOS high-input-impedance single-supply amplifier. Excerpts from the data sheets are shown below, with some words of explanation. Later in this chapter we compare the electrical characteristics of these two op-amps in more detail. The complete data sheets can be found in the accompanying CD-ROM.

**LM 741 General Description and Connection Diagrams**—This table summarizes the general characteristics of the op-amp. The connection diagrams are shown. Note that the op-amp is available in various packages: a metal can package, a dual-in-line package (DIP), and two ceramic dual-in-line options. The dual-in-line (or SO) package is the one you are most likely to see in a laboratory. Note that in this configuration the integrated circuit has eight connections, or *pins*: two for the voltage supplies ($V^+$ and $V^-$); two inputs (inverting and noninverting); one output; two offset null connections (to be discussed layer in the chapter); and a no-connection pin (NC).

## LM741 Operational Amplifier

### General Description

The LM741 series are general purpose operational amplifier which feature improved performance over industry standards like the LM709. They are direct, plug-in replacements for the 709C, LM201, MC1439 and 748 in most applications.

The amplifiers offer many features which make their application nearly foolproof: overload protection on the input and output, no latch-up when the common mode range is exceeded, as well as freedom from oscillations.

The LM741C/LM741E are identical to the LM741/LM741A except that the LM741C/LM741E have their performance guaranteed over a 0°C to +70°C temperature range, instead of −55°C to +125°C.

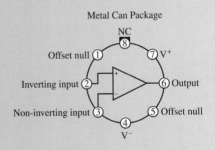

Metal Can Package

Order Number LM741H, LM741H/883*,
LM741AH/883 or LM41CH
See NS Package Number H08C

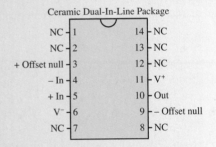

Ceramic Dual-In-Line Package

Order Number LM741J-14/883*, LM741AJ-14/883**
See NS Package Number J14A
*also available per JM38510/10101
**also available per JM38510/10102

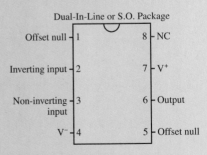

Dual-In-Line or S.O. Package

Order Number LM741J, LM741J/883,
LM741CM, LM741CN or LM741EN
See NS Package Number JO8A, MO8A or NO8E

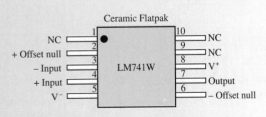

Ceramic Flatpak

Order Number LM741W/883
See NS Package Number W10A

*(Continued)*

*(Concluded)*

**LMC 6061 General Description and Connection Diagrams**—The description and diagram below reveal several similarities between the 741 and 6061 op-amp, but also some differences. The 6061 uses more advanced technology, and is characterized by some very desirable features (e.g., the very low power consumption of CMOS circuits results in typical supply currents of only 20 $\mu$A!). You can also see from the connection diagram that pins 1 and 5 (used for offset null connections in the 741) are not used in this IC. We return to this point later in the chapter. A further point of comparison between these two devices is their (1998) cost: the LM 741 (in quantities of 1,000) costs \$0.32/unit; the LMC 6061 sells for \$0.79/unit, also in quantities of 1,000 or more.

## LMC6061 Precision CMOS Single Micropower Operational Amplifier

### General Description

The LMC6061 is a precision single low offset voltage, micropower operational amplifier, capable of precision single supply operation. Performance characteristics include ultra low input bias current, high voltage gain, rail-to-rail output swing, and an input common mode voltage range that includes ground. These features, plus its low power consumption, make the LMC6061 ideally suited for battery powered applications. Other applications using the LMC6061 include precision full-wave rectifiers, integrators, references, sample-and-hold circuits, and true instrumentation amplifiers.

This device is built with National's advanced double-Poly Silicon-Gate CMOS process.

For designs that require higher speed, see the LMC6081 precision single operational amplifier.

For a dual or quad operational amplifier with similar features, see the LMC6062 or LMC6064 respectively.

**Features** (Typical Unless Otherwise Noted)
- Low offset voltage    100 $\mu$V
- Ultra low supply current    20 $\mu$A
- Operates from 4.5V to 15V single supply
- Ultra low input bias current 10 fA
- Output swing within 10 mV of supply rail, 100k load
- Input common-mode range includes $V^-$
- High voltage gain    140 dB
- Improved latchup immunity

**Applications**
- Instrumentation amplifier
- Photodiode and infrared detector preamplifier
- Transducer amplifiers
- Hand-held analytic instruments
- Medical instrumentation
- D/A converter
- Charge amplifier to piezoelectric transducers

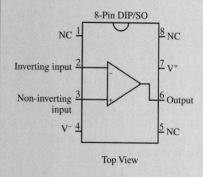

Top View

## Check Your Understanding

**12.1**  Consider an op-amp connected in the inverting configuration with a nominal closed-loop gain of $-R_F/R_S = -1,000$ (this would be the gain if the op-amp had an infinite open-loop gain). Determine the value of the closed-loop gain that includes the open-loop gain as a parameter, and compute the closed-loop gain for the following values of $A_{V(OL)}$: $10^7$, $10^6$, $10^5$, and $10^4$. How large do you think the open-loop gain should be for this op-amp, to achieve the desired closed-loop gain? [*Hint:* Do not assume that $A_{V(OL)}$ is negligible in Equation 12.18.]

**12.2**  Repeat Check Your Understanding Exercise 12.1 for $-R_F/R_S = -100$. What is the smallest value of $A_{V(OL)}$ you would recommend in this case?

**12.3**  Derive the result given for the differential amplifier by utilizing the principle of superposition. (Think of the differential amplifier as the combination of an inverting amplifier with input $= v_2$, plus a noninverting amplifier with input $= v_1$.)

**12.4**  For Example 12.4, find $\Delta R$ if the supply voltages are symmetrical at $\pm 15$ V and a 10-k$\Omega$ potentiometer is tied to two 10-k$\Omega$ resistors.

**12.5**  For the circuit of Example 12.4, find the range of values of $V_{ref}$ if the supply voltages are symmetrical at 15 V and a 1-k$\Omega$ potentiometer is tied to two 10-k$\Omega$ resistors.

**12.6**  Find the numerical values of $R_F/R_S$ and $V_{ref}$ if the temperature sensor of in "Focus on Measurements: Sensor Calibration Circuit" has $\beta = 0.235$ and $V_0 = 0.7$ V and the desired relationship is $v_{out} = 10T$.

## 12.3   ACTIVE FILTERS

The range of useful applications of an operational amplifier is greatly expanded if energy-storage elements are introduced into the design; the frequency-dependent properties of these elements, studied in Chapters 4 and 6, will prove useful in the design of various types of op-amp circuits. In particular, it will be shown that it is possible to shape the frequency response of an operational amplifier by appropriate use of complex impedances in the input and feedback circuits. The class of filters one can obtain by means of op-amp designs is called **active filters,** because op-amps can provide amplification (gain) in addition to the filtering effects already studied in Chapter 6 for passive circuits (i.e., circuits comprising exclusively resistors, capacitors, and inductors).

The easiest way to see how the frequency response of an op-amp can be shaped (almost) arbitrarily is to replace the resistors $R_F$ and $R_S$ in Figures 12.5 and 12.8 with impedances $Z_F$ and $Z_S$, as shown in Figure 12.20. It is a straightforward matter to show that in the case of the inverting amplifier, the expression for the closed loop gain is given by

$$\frac{\mathbf{V}_{out}}{\mathbf{V}_S}(j\omega) = -\frac{Z_F}{Z_S} \tag{12.45}$$

whereas for the noninverting case, the gain is

$$\frac{\mathbf{V}_{out}}{\mathbf{V}_S}(j\omega) = 1 + \frac{Z_F}{Z_S} \tag{12.46}$$

where $Z_F$ and $Z_S$ can be arbitrarily complex impedance functions and where $\mathbf{V}_S$, $\mathbf{V}_{out}$, $\mathbf{I}_F$, and $\mathbf{I}_S$ are all phasors. Thus, it is possible to shape the frequency response of an ideal op-amp filter simply by selecting suitable ratios of feedback impedance to source impedance. By connecting a circuit similar to the low-pass filters studied

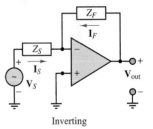

Inverting

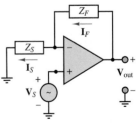

Noninverting

**Figure 12.20** Op-amp circuits employing complex impedances

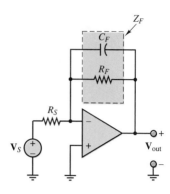

**Figure 12.21** Active low-pass filter

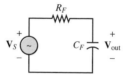

**Figure 12.22** Passive low-pass filter

in Chapter 6 in the feedback loop of an op-amp, the same filtering effect can be achieved and, in addition, the signal can be amplified.

The simplest op-amp low-pass filter is shown in Figure 12.21. Its analysis is quite simple if we take advantage of the fact that the closed-loop gain, as a function of frequency, is given by

$$A_{\text{LP}}(j\omega) = -\frac{Z_F}{Z_S} \tag{12.47}$$

where

$$Z_F = R_F \parallel \frac{1}{j\omega C_F} = \frac{R_F}{1 + j\omega C_F R_F} \tag{12.48}$$

and

$$Z_S = R_S \tag{12.49}$$

Note the similarity between $Z_F$ and the low-pass characteristic of the passive $RC$ circuit! The closed-loop gain $A_{\text{LP}}(j\omega)$ is then computed to be

$$A_{\text{LP}}(j\omega) = -\frac{Z_F}{Z_S} = -\frac{R_F/R_S}{1 + j\omega C_F R_F} \tag{12.50}$$

This expression can be factored into two terms. The first is an amplification factor analogous to the amplification that would be obtained with a simple inverting amplifier (i.e., the same circuit as that of Figure 12.21 with the capacitor removed); the second is a low-pass filter, with a cutoff frequency dictated by the parallel combination of $R_F$ and $C_F$ in the feedback loop. The filtering effect is completely analogous to that which would be attained by the passive circuit shown in Figure 12.22. However, the op-amp filter also provides amplification by a factor of $R_F/R_S$.

It should be apparent that the response of this op-amp filter is just an amplified version of that of the passive filter. Figure 12.23 depicts the amplitude response of the active low-pass filter (in the figure, $R_F/R_S = 10$ and $1/R_F C_F = 1$) in two different graphs; the first plots the amplitude ratio $\mathbf{V}_{\text{out}}(j\omega)$ versus radian frequency, $\omega$, on a logarithmic scale, while the second plots the amplitude ratio $20\log_{10} \mathbf{V}_S(j\omega)$ (in units of dB), also versus $\omega$ on a logarithmic scale. You will recall from Chapter 6 that dB frequency response plots are encountered very frequently. Note that in the dB plot, the slope of the filter response for frequencies significantly higher than the cutoff frequency,

$$\omega_0 = \frac{1}{R_F C_F} \tag{12.51}$$

is $-20$ dB/decade, while the slope for frequencies significantly lower than this cutoff frequency is equal to zero. The value of the response at the cutoff frequency is found to be, in units of dB,

$$|A_{\text{LP}}(j\omega_0)|_{\text{dB}} = 20\log_{10}\frac{R_F}{R_S} - 20\log_{10}\sqrt{2} \tag{12.52}$$

where

$$-20\log_{10}\sqrt{2} = -3 \text{ dB} \tag{12.53}$$

Thus, $\omega_0$ is also called the **3-dB frequency.**

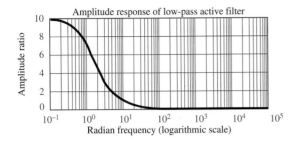

Figure 12.23 Normalized response of active low-pass filter

Among the advantages of such low-pass active filters is the ease with which the gain and the bandwidth can be adjusted by controlling the ratios $R_F/R_S$ and $1/R_FC_F$, respectively.

It is also possible to construct other types of filters by suitably connecting resistors and energy-storage elements to an op-amp. For example, a high-pass active filter can easily be obtained by using the circuit shown in Figure 12.24. Observe that the impedance of the input circuit is

$$Z_S = R_S + \frac{1}{j\omega C_S} \qquad (12.54)$$

and that of the feedback circuit is

$$Z_F = R_F \qquad (12.55)$$

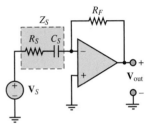

Figure 12.24 Active high-pass filter

Then, the following gain function for the op-amp circuit can be derived:

$$A_{\mathrm{HP}}(j\omega) = -\frac{Z_F}{Z_S} = -\frac{R_F}{R_S + 1/j\omega C_S} = -\frac{j\omega C_S R_F}{1 + j\omega R_S C_S} \qquad (12.56)$$

As $\omega$ approaches zero, so does the response of the filter, whereas as $\omega$ approaches infinity, according to the gain expression of equation 12.56, the gain of the amplifier approaches a constant:

$$\lim_{\omega \to \infty} A_{\mathrm{HP}}(j\omega) = -\frac{R_F}{R_S} \qquad (12.57)$$

That is, above a certain frequency range, the circuit acts as a linear amplifier. This is exactly the behavior one would expect of a high-pass filter. The high-pass response is depicted in Figure 12.25, in both linear and dB plots (in the

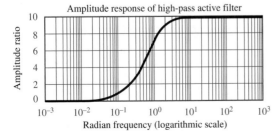
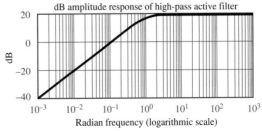

Figure 12.25 Normalized response of active high-pass filter

figure, $R_F/R_S = 10, 1/R_S C = 1$). Note that in the dB plot, the slope of the filter response for frequencies significantly lower than $\omega = 1/R_S C_S = 1$ is $+20$ dB/decade, while the slope for frequencies significantly higher than this cutoff (or 3-dB) frequency is equal to zero.

As a final example of active filters, let us look at a simple active band-pass filter configuration. This type of response may be realized simply by combining the high- and low-pass filters we examined earlier. The circuit is shown in Figure 12.26.

The analysis of the band-pass circuit follows the same structure used in previous examples. First, we evaluate the feedback and input impedances:

$$Z_F = R_F \parallel \frac{1}{j\omega C_F} = \frac{R_F}{1 + j\omega C_F R_F} \tag{12.58}$$

$$Z_S = R_S + \frac{1}{j\omega C_S} = \frac{1 + j\omega C_S R_S}{j\omega C_S} \tag{12.59}$$

Next, we compute the closed-loop frequency response of the op-amp, as follows:

$$A_{\mathrm{BP}}(j\omega) = -\frac{Z_F}{Z_S} = -\frac{j\omega C_S R_F}{(1 + j\omega C_F R_F)(1 + j\omega C_S R_S)} \tag{12.60}$$

The form of the op-amp response we just obtained should not appear as a surprise. It is very similar (although not identical) to the product of the low-pass and high-pass responses of equations 12.50 and 12.56. In particular, the denominator of $A_{\mathrm{BP}}(j\omega)$ is exactly the product of the denominators of $A_{\mathrm{LP}}(j\omega)$ and $A_{\mathrm{HP}}(j\omega)$. It is particularly enlightening to rewrite $A_{\mathrm{LP}}(j\omega)$ in a slightly different form, after making the observation that each $RC$ product corresponds to some "critical" frequency:

$$\omega_1 = \frac{1}{R_F C_S} \qquad \omega_{\mathrm{LP}} = \frac{1}{R_F C_F} \qquad \omega_{\mathrm{HP}} = \frac{1}{R_S C_S} \tag{12.61}$$

It is easy to verify that for the case where

$$\omega_{\mathrm{HP}} > \omega_{\mathrm{LP}} \tag{12.62}$$

the response of the op-amp filter may be represented as shown in Figure 12.27 in both linear and dB plots (in the figure, $\omega_1 = 1$, $\omega_{\mathrm{HP}} = 1{,}000$, $\omega_{\mathrm{LP}} = 10$). The dB plot is very revealing, for it shows that, in effect, the band-pass response is the graphical superposition of the low-pass and high-pass responses shown earlier. The two 3-dB (or cutoff) frequencies are the same as in $A_{\mathrm{LP}}(j\omega)$, $1/R_F C_F$; and in $A_{\mathrm{HP}}(j\omega)$, $1/R_S C_S$. The third frequency, $\omega_1 = 1/R_F C_S$, represents the point where the response of the filter crosses the 0-dB axis (rising slope). Since 0 dB corresponds to a gain of 1, this frequency is called the **unity gain frequency.**

The ideas developed thus far can be employed to construct more complex functions of frequency. In fact, most active filters one encounters in practical applications are based on circuits involving more than one or two energy-storage elements. By constructing suitable functions for $Z_F$ and $Z_S$, it is possible to realize filters with greater frequency selectivity (i.e., sharpness of cutoff), as well as flatter band-pass or band-rejection functions (that is, filters that either allow or reject signals in a limited band of frequencies). A few simple applications

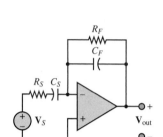

**Figure 12.26** Active band-pass filter

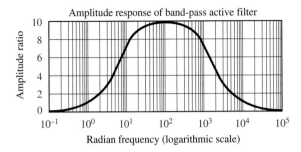

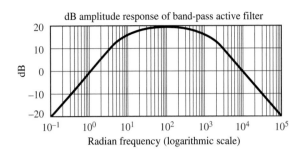

**Figure 12.27** Normalized amplitude response of active band-pass filter

are investigated in the homework problems. One remark that should be made in passing, though, pertains to the exclusive use of capacitors in the circuits analyzed thus far. One of the advantages of op-amp filters is that it is not necessary to use both capacitors and inductors to obtain a band-pass response. Suitable connections of capacitors can accomplish that task in an op-amp. This seemingly minor fact is of great importance in practice, because inductors are expensive to mass-produce to close tolerances and exact specifications and are often bulkier than capacitors with equivalent energy-storage capabilities. On the other hand, capacitors are easy to manufacture in a wide variety of tolerances and values, and in relatively compact packages, including in integrated circuit form.

Example 12.5 illustrates how it is possible to construct active filters with greater frequency selectivity by adding energy-storage elements to the design.

### EXAMPLE 12.5  Second-Order Low-Pass Filter

#### Problem

Determine the closed-loop voltage gain as a function of frequency for the op-amp circuit of Figure 12.28.

#### Solution

**Known Quantities:**  Feedback and source impedances.

**Find:**

$$A(j\omega) = \frac{\mathbf{V}_{\text{out}}(j\omega)}{\mathbf{V}_S(j\omega)}$$

**Schematics, Diagrams, Circuits, and Given Data:**  $R_2 C = L/R_1 = \omega_0$.

**Assumptions:**  Assume ideal op-amp.

**Analysis:**  The expression for the gain of the filter of Figure 12.28 can be determined by using equation 12.45:

$$A(j\omega) = \frac{\mathbf{V}_{\text{out}}(j\omega)}{\mathbf{V}_S(j\omega)} = -\frac{Z_F(j\omega)}{Z_S(j\omega)}$$

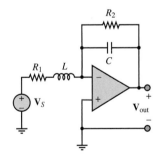

**Figure 12.28**

where

$$Z_F(j\omega) = R_2 || \frac{1}{j\omega C} = \frac{R_2}{1 + j\omega C R_2} = \frac{R_2}{1 + j\omega/\omega_0}$$

$$= R_1 + j\omega L = R_1 \left(1 + j\omega \frac{L}{R_1}\right) = R_1(1 + j\omega/\omega_0).$$

Thus, the gain of the filter is:

$$A(j\omega) = \frac{\frac{R_2}{1 + j\omega/\omega_0}}{R_1(1 + j\omega/\omega_0)} = \frac{R_2/R_1}{(1 + j\omega/\omega_0)^2}$$

**Comments:** Note the similarity between the expression for the gain of the filter of Figure 12.28 and that given in equation 12.50 for the gain of a (first-order) low-pass filter. Clearly, the circuit analyzed in this example is also a low-pass filter, of second order (as the quadratic denominator term suggests). Figure 12.29 compares the two responses in both linear and dB (Bode) magnitude plots. The slope of the dB plot for the second-order filter at higher frequencies is twice that of the first-order filter ($-40$ dB/decade versus $-20$ dB/decade). We should also remark that the use of an inductor in the filter design is not recommended in practice, as explained in the above section, and that we have used it in this example only because of the simplicity of the resulting gain expressions. Section 15.3 introduces design methods for practical high-order filters.

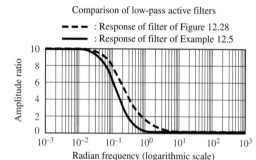

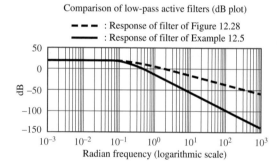

**Figure 12.29** Comparison of first- and second-order active filters

## Check Your Understanding

**12.7**  Design a low-pass filter with closed-loop gain of 100 and cutoff (3-dB) frequency equal to 800 Hz. Assume that only 0.01-$\mu$F capacitors are available. Find $R_F$ and $R_S$.

**12.8**  Repeat the design of Check Your Understanding Exercise 12.7 for a high-pass filter with cutoff frequency of 2,000 Hz. This time, however, assume that only standard values of resistors are available (see Table 2.1 for a table of standard values). Select the nearest component values, and calculate the percent error in gain and cutoff frequency with respect to the desired values.

**12.9**  Find the frequency corresponding to attenuation of 1 dB (with respect to the maximum value of the amplitude response) for the filter of Check Your Understanding Exercise 12.7.

**12.10**  What is the dB gain for the filter of Example 12.5 at the cutoff frequency, $\omega_0$? Find the 3-dB frequency for this filter in terms of the cutoff frequency, $\omega_0$, and note that the two are not the same.

## 12.4    INTEGRATOR AND DIFFERENTIATOR CIRCUITS

In the preceding sections, we examined the frequency response of op-amp circuits for sinusoidal inputs. However, certain op-amp circuits containing energy-storage elements reveal some of their more general properties if we analyze their response to inputs that are time-varying but not necessarily sinusoidal. Among such circuits are the commonly used integrator and differentiator; the analysis of these circuits is presented in the following paragraphs.

### The Ideal Integrator

Consider the circuit of Figure 12.30, where $v_S(t)$ is an arbitrary function of time (e.g., a pulse train, a triangular wave, or a square wave). The op-amp circuit shown provides an output that is proportional to the integral of $v_S(t)$. The analysis of the integrator circuit is, as always, based on the observation that

$$i_S(t) = -i_F(t) \tag{12.63}$$

where

$$i_S(t) = \frac{v_S(t)}{R_S} \tag{12.64}$$

It is also known that

$$i_F(t) = C_F \frac{dv_{\text{out}}(t)}{dt} \tag{12.65}$$

from the fundamental definition of the capacitor. The source voltage can then be expressed as a function of the derivative of the output voltage:

$$\frac{1}{R_S C_F} v_S(t) = -\frac{dv_{\text{out}}(t)}{dt} \tag{12.66}$$

By integrating both sides of equation 12.66, we obtain the following result:

$$v_{\text{out}}(t) = -\frac{1}{R_S C_F} \int_{-\infty}^{t} v_S(t') \, dt' \tag{12.67}$$

This equation states that the output voltage is the integral of the input voltage.

There are numerous applications of the op-amp integrator, most notably the **analog computer,** which will be discussed in Section 12.5. The following example illustrates the operation of the op-amp integrator.

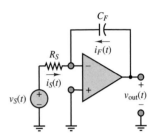

**Figure 12.30** Op-amp integrator

### EXAMPLE    12.6    Integrating a Square Wave

#### Problem

Determine the output voltage for the integrator circuit of Figure 12.31 if the input is a square wave of amplitude $\pm A$ and period $T$.

#### Solution

**Known Quantities:**  Feedback and source impedances; input waveform characteristics.

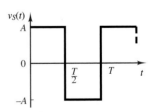

**Figure 12.31**

**Find:** $v_{out}(t)$.

**Schematics, Diagrams, Circuits, and Given Data:** $T = 10$ ms; $C_F = 1 \ \mu$F; $R_S = 10$ k$\Omega$.

**Assumptions:** Assume ideal op-amp. The square wave starts at $t = 0$ and therefore $v_{out}(0) = 0$.

**Analysis:** Following equation 12.67, we write the expression for the output of the integrator:

$$v_{out}(t) = -\frac{1}{R_F C_S} \int_{-\infty}^{t} v_S(t')dt' = -\frac{1}{R_F C_S} \left( \int_{-\infty}^{0} v_S(t')dt' + \int_{0}^{t} v_S(t')dt' \right)$$

$$= -\frac{1}{R_F C_S} \left( v_{out}(0) + \int_{0}^{t} v_S(t')dt' \right)$$

Next, we note that we can integrate the square wave in a piecewise fashion by observing that $v_S(t) = A$ for $0 \leq t < T/2$ and $v_S(t) = -A$ for $T/2 \leq t < T$. We consider the first half of the waveform:

$$v_{out}(t) = -\frac{1}{R_F C_S} \left( v_{out}(0) + \int_{0}^{t} v_S(t')dt' \right) = -100 \left( 0 + \int_{0}^{t} A dt' \right)$$

$$= -100 At \qquad 0 \leq t < \frac{T}{2}$$

$$v_{out}(t) = v_{out}\left(\frac{T}{2}\right) - \frac{1}{R_F C_S} \int_{T/2}^{t} v_S(t')dt' = -100A\frac{T}{2} - 100 \int_{T/2}^{t} (-A) \, dt'$$

$$= -100A\frac{T}{2} + 100A \left( t - \frac{T}{2} \right) = -100A \, (T - t) \qquad \frac{T}{2} \leq t < T$$

Since the waveform is periodic, the above result will repeat with period $T$, as shown in Figure 12.32.

**Comments:** The integral of a square wave is thus a triangular wave. This is a useful fact to remember. Note that the effect of the initial condition is very important, since it determines the starting point of the triangular wave.

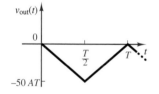

**Figure 12.32**

---

**FOCUS ON MEASUREMENTS**

### Charge Amplifiers

FIND IT

ON THE WEB

One of the most common families of transducers for the measurement of force, pressure, and acceleration is that of **piezoelectric transducers.** These transducers contain a piezoelectric crystal that generates an electric charge in response to deformation. Thus, if a force is applied to the crystal (leading to a displacement), a charge is generated within the crystal. If the external force generates a displacement $x_i$, then the transducer will generate a charge $q$ according to the expression

$$q = K_p x_i$$

Figure 12.33 depicts the basic structure of the piezoelectric transducer, and a simple circuit model. The model consists of a current source in parallel with a capacitor, where the current source represents the rate of change of the

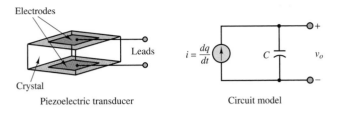

Piezoelectric transducer            Circuit model

**Figure 12.33** Piezoelectric transducer

charge generated in response to an external force and the capacitance is a
consequence of the structure of the transducer, which consists of a
piezoelectric crystal (e.g., quartz or Rochelle salt) sandwiched between
conducting electrodes (in effect, this is a parallel-plate capacitor).

Although it is possible, in principle, to employ a conventional voltage
amplifier to amplify the transducer output voltage, $v_o$, given by

$$v_o = \frac{1}{C} \int i \, dt = \frac{1}{C} \int \frac{dq}{dt} dt = \frac{q}{C} = \frac{K_p x_i}{C}$$

it is often advantageous to use a **charge amplifier.** The charge amplifier is
essentially an integrator circuit, as shown in Figure 12.34, characterized by
an extremely high input impedance.[3] The high impedance is essential;
otherwise, the charge generated by the transducer would leak to ground
through the input resistance of the amplifier.

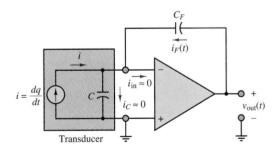

**Figure 12.34** Charge amplifier

Because of the high input impedance, the input current into the
amplifier is negligible; further, because of the high open-loop gain of the
amplifier, the inverting-terminal voltage is essentially at ground potential.
Thus, *the voltage across the transducer is effectively zero.* As a
consequence, to satisfy KCL, the feedback current, $i_F(t)$, must be equal and
opposite to the transducer current, $i$:

$$i_F(t) = -i$$

[3] Special op-amps are employed to achieve extremely high input impedance, through FET input
circuits.

and since

$$v_{\text{out}}(t) = \frac{1}{C_F} \int i_F(t)\, dt$$

it follows that the output voltage is proportional to the charge generated by the transducer, and therefore to the displacement:

$$v_{\text{out}}(t) = \frac{1}{C_F} \int -i\, dt = \frac{1}{C_F} \int -\frac{dq}{dt} dt = -\frac{q}{C_F} = -\frac{K_p x_i}{C_F}$$

Since the displacement is caused by an external force or pressure, this sensing principle is widely adopted in the measurement of force and pressure.

## The Ideal Differentiator

Using an argument similar to that employed for the integrator, we can derive a result for the ideal differentiator circuit of Figure 12.35. The relationship between input and output is obtained by observing that

$$i_S(t) = C_S \frac{dv_S(t)}{dt} \tag{12.68}$$

and

$$i_F(t) = \frac{v_{\text{out}}(t)}{R_F} \tag{12.69}$$

so that the output of the differentiator circuit is proportional to the derivative of the input:

$$v_{\text{out}}(t) = -R_F C_S \frac{dv_S(t)}{dt} \tag{12.70}$$

Although mathematically attractive, the differentiation property of this op-amp circuit is seldom used in practice, because differentiation tends to amplify any noise that may be present in a signal.

**Figure 12.35** Op-amp differentiator

## Check Your Understanding

**12.11**  Plot the frequency response of the ideal integrator in dB plot. Determine the slope of the curve in dB/decade. You may assume $R_S C_F = 10$.

**12.12**  Plot the frequency response of the ideal differentiator in a dB plot. Determine the slope of the curve in dB/decade. You may assume $R_F C_S = 100$.

**12.13**  Verify that if the triangular wave of Example 12.6 is the input to the ideal differentiator, the resulting output is the original square wave.

## 12.5    ANALOG COMPUTERS

Prior to the advent of digital computers, the solution of differential equations and the simulation of complex dynamic systems were conducted exclusively by means of analog computers. **Analog computers** still find application in engineering practice in the simulation of dynamic systems. The analog computer is a device that is based on three op-amp circuits introduced earlier in this chapter: the amplifier, the summer, and the integrator. These three building blocks permit the construction of circuits that can be used to solve differential equations and to simulate dynamic systems. Figure 12.36 depicts the three symbols that are typically employed to represent the principal functions of an analog computer.

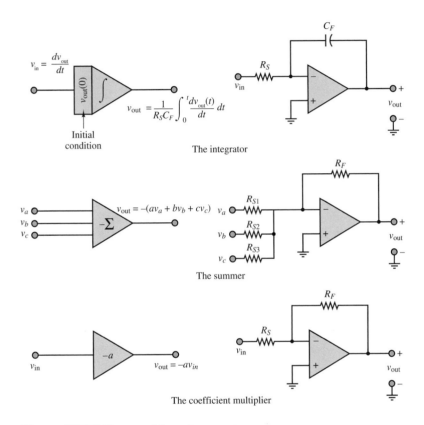

The integrator

The summer

The coefficient multiplier

**Figure 12.36** Elements of the analog computer

The simplest way to discuss the operation of the analog computer is to present an example. Consider the simple second-order mechanical system, shown in Figure 12.37, that represents, albeit in a greatly simplified fashion, one corner of an automobile suspension system. The mass, $M$, represents the mass of one quarter of the vehicle; the damper, $B$, represents the shock absorber; and the spring, $K$, represents the suspension spring (or strut). The differential equation of the system may be derived as follows:

$$M\frac{d^2 x_M}{dt^2} + B\left(\frac{dx_M}{dt} - \frac{dx_R}{dt}\right) + K(x_M - x_R) = 0 \qquad (12.71)$$

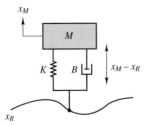

**Figure 12.37** Model of automobile suspension

Rearranging terms, we obtain the following equation, in which the terms related to the road displacement and velocity—$x_R$ and $dx_R/dt$, respectively—are the forcing functions:

$$M\frac{d^2 x_M}{dt^2} + B\frac{dx_M}{dt} + Kx_M = B\frac{dx_R}{dt} + Kx_R \qquad (12.72)$$

Assume that the car is traveling over a "washboard" surface on an unpaved road, such that the road profile is approximately described by the expression

$$x_R(t) = X\sin(\omega t) \qquad (12.73)$$

It follows, then, that the vertical velocity input to the suspension is given by the expression

$$\frac{dx_R}{dt} = \omega X\cos(\omega t) \qquad (12.74)$$

and we can write the equation for the suspension system in the form

$$M\frac{d^2 x_M}{dt^2} + B\frac{dx_M}{dt} + Kx_M = B\omega X\cos(\omega t) + KX\sin(\omega t) \qquad (12.75)$$

It would be desirable to solve the equation for the displacement, $x_M$, which represents the motion of the vehicle mass in response to the road excitation. The solution can be used as an aid in designing the suspension system that best absorbs the road vibration, providing a comfortable ride for the passengers. Equation 12.75 may be rearranged to obtain

$$\frac{d^2 x_M}{dt^2} = -\frac{B}{M}\frac{dx_M}{dt} - \frac{K}{M}x_M + \frac{B}{M}\omega X\cos(\omega t) + \frac{K}{M}X\sin(\omega t) \qquad (12.76)$$

This equation is now in a form appropriate for solution by repeated integration, since we have isolated the highest derivative term; thus, it will be sufficient to integrate the right-hand side twice to obtain the solution for the displacement of the vehicle mass, $x_M$.

Figure 12.38 depicts the three basic operations that need to be performed to integrate the differential equation describing the motion of the mass, $M$. Note that in each of the three blocks—the summer and the two integrators—the inversion due to the inverting amplifier configuration used for the integrator is already accounted for. Finally, the basic summing and integrating blocks together with three coefficient multipliers (inverting amplifiers) are connected in the configuration that corresponds to the preceding differential equation (equation 12.76). You can easily verify that the analog computer circuit of Figure 12.39 does indeed solve the differential equation in $x_M$ by repeated integration.

## Scaling in Analog Computers

One of the important issues in analog computing is that of scaling. Since the analog computer implements an electrical analog of a physical system, there is no guarantee that the voltages and currents in the analog computer circuits will be of the same order of magnitude as the physical variables (e.g., velocity, displacement, temperature, or flow) they simulate. Further, it is not necessary that the computer simulate the physical system with the same time scale; it may be desirable in practice to speed up or slow down the simulation. Thus, the interest in time scaling and magnitude scaling.

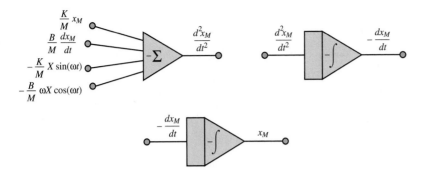

**Figure 12.38** Solution by repeated integration

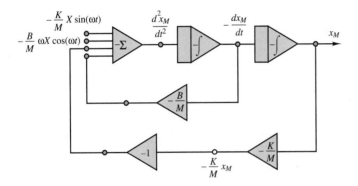

**Figure 12.39** Analog computer simulation of suspension system

Let Table 12.1 represent the physical and simulation variables. Considering time scaling first, let $t$ denote real time and $\tau$, the computer time variable. Then the time derivative of a physical variable can be expressed as

$$\frac{dx}{dt} = \frac{dx}{d\tau}\frac{d\tau}{dt} = \alpha\frac{dx}{d\tau} \tag{12.77}$$

where $\alpha$ is the scaling factor between real time and computer time:

$$\tau = \alpha t \tag{12.78}$$

For higher-order derivatives, the following relationship will hold:

$$\frac{d^n x}{dt^n} = \alpha^n \frac{d^n x}{d\tau^n} \tag{12.79}$$

**Table 12.1** Actual and simulated variables in analog computers

| Physical system | Analog simulation |
| --- | --- |
| Physical variable, $x$ | Voltage, $v$ |
| Time variable, $t$ | Simulated time, $\tau$ |

While time scaling is likely to be prompted by a desire to speed up or slow down a computation, magnitude scaling is motivated by several different factors:

1.  The relationship between physical variables and computer voltages (e.g., calibration constants).
2.  Overloading of the op-amp circuits (we shall see in Section 12.6 that one of the fundamental limitations of the operational amplifier is its voltage range).
3.  Loss of accuracy if voltages are too small (errors are usually expressed as a percentage of the full-scale range).

Thus, if the relationship between a physical variable and the computer voltage is $v = \beta x$, where $\beta$ is a magnitude scaling factor, the derivative terms will be affected according to the relation

$$\frac{dx}{dt} = \frac{1}{\beta}\frac{dv}{dt} \tag{12.80}$$

Note that different scaling factors may be introduced at each point in the analog computer simulation, and so there is no general rule with regard to magnitude scaling. For example, if $v = \beta_0 x$, it is entirely possible to have

$$\frac{dx}{dt} = \frac{1}{\beta_1}\frac{dv}{dt}$$

### EXAMPLE   12.7  Analog Computer Simulation of Automotive Suspension

**Problem**

Implement the analog computer simulation of the automotive suspension system of Figure 12.37.

---

**Solution**

**Known Quantities:**  Mass, spring rate, and damping parameters of automotive suspension.

**Find:**  Component values of analog computer circuit of Figure 12.39.

**Schematics, Diagrams, Circuits, and Given Data:**  $M = 400$ kg; $K = 1.6 \times 10^5$ N/m; $B = 20 \times 10^3$ N/m-s.

**Assumptions:**  Assume ideal op-amps. Express all resistors in M$\Omega$ and all capacitors in $\mu$F.

**Analysis:**  With reference to Figure 12.39, we observe that the analog computer simulation of the automotive suspension requires: a four-input summer, two integrators, two coefficient multipliers, and one sign inverter.
   Expressing all resistors in M$\Omega$ and all capacitors in $\mu$F is useful because each integrator has a multiplier of $-1/RC$. Using $R = 1$ M$\Omega$ and $C = 1$ $\mu$F results in $-1/RC = -1$. Figure 12.40 depicts the integrator configuration. The four-input summing amplifier uses 1-M$\Omega$ resistors throughout, so that the gain for each input is also equal to $-1$. On the other hand, the two coefficient multipliers are required to have gains $-K/M = -4{,}000$ and $-B/M = 50$, respectively. Thus we select 10-M$\Omega$ and 2.5-k$\Omega$ resistors for the first coefficient multiplier, and 1-M$\Omega$ and 20-k$\Omega$ resistors for the second.

Finally, the inverter can be realized with two 1-MΩ resistors. The various elements are depicted in Figure 12.40.

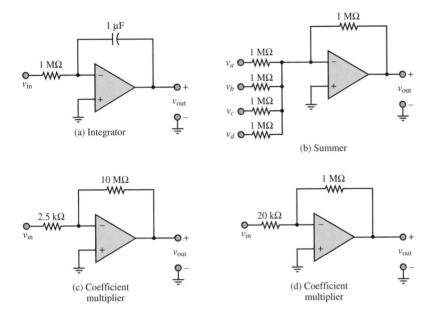

**Figure 12.40** Analog computer simulation of suspension system

**Comments:** Note that it is not necessary to employ five op-amps in this analog simulator. The summing amplifier function and the first integrator could, for example, be combined into a single op-amp, and one of the two coefficient multipliers could be eliminated. This idea is explored further in the homework problems.

## EXAMPLE 12.8 Deriving a Differential Equation from an Analog Computer Circuit

### Problem

Derive the differential equation corresponding to the analog computer simulator of Figure 12.41.

### Solution

**Known Quantities:**  Resistor and capacitor values.

**Find:**  Differential equation in $x(t)$.

**Schematics, Diagrams, Circuits, and Given Data:**  $R_1 = 0.4$ MΩ; $R_2 = R_3 = R_5 = 1$ MΩ; $R_4 = 2.5$ kΩ; $C_1 = C_2 = 1$ μF.

**Assumptions:**  Assume ideal op-amps.

**Analysis:**  We start the analysis from the right-hand side of the circuit, to determine the

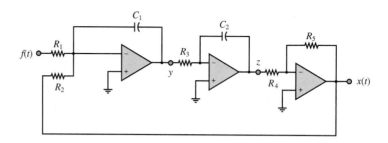

**Figure 12.41** Analog computer simulation of unknown system

intermediate variable $z$ as a function of $x$:

$$x = -\frac{R_5}{R_4}z = -400z$$

Next, we move to the left to determine the relationship between $y$ and $z$:

$$z = -\frac{1}{R_3 C_2}\int y(t')dt' \quad \text{or} \quad y = -\frac{dz}{dt}$$

Finally, we determine $y$ as a function of $x$ and $f$:

$$y = -\frac{1}{R_2 C_1}\int x(t')dt' - \frac{1}{R_1 C_1}\int f(t')dt' = -\int \left[x(t') + 2.5f(t')\right]dt'$$

or

$$\frac{dy}{dt} = -x - 2.5f$$

Substituting the expressions into one another and eliminating the variables $y$ and $z$, we obtain the differential equation in $x$:

$$x = -400z$$

$$\frac{dx}{dt} = -400\frac{dz}{dt} = 400y$$

$$\frac{d^2x}{dt^2} = 400\frac{dy}{dt} = 400\,(x - 2.5f)$$

and

$$\frac{d^2x}{dt^2} + 400x = -1,000f$$

**Comments:** Note that the summing and integrating functions have been combined into a single block in the first amplifier.

## Check Your Understanding

**12.14** Modify the gains of the coefficient multipliers in Example 12.7 if we wish to slow down the simulation by a factor of 10 (i.e., if $\alpha = 0.1$).

**12.15** For the simulation of Example 12.7, what will the largest magnitude of the voltage analog of $x_M$ be for a road displacement $x_R = 0.01 \sin(100t)$? [*Hint:* Use phasor techniques to compute the frequency response $x_M(\omega)/F(\omega)$, where $f(t) = B\, dx_R/dt +$

$Kx_R$, and evaluate the output voltage by multiplying the input by the magnitude of the frequency response at $\omega = 100$.]

**12.16**  Change the parameters of the analog computer simulation of Example 12.8 to simulate the differential equation $d^2x/dt^2 + 2x = -10f(t)$.

---

# 12.6    PHYSICAL LIMITATIONS OF OP-AMPS

Thus far, the operational amplifier has been treated as an ideal device, characterized by infinite input resistance, zero output resistance, and infinite open-loop voltage gain. Although this model is adequate to represent the behavior of the op-amp in a large number of applications, practical operational amplifiers are not ideal devices but exhibit a number of limitations that should be considered in the design of instrumentation. In particular, in dealing with relatively large voltages and currents, and in the presence of high-frequency signals, it is important to be aware of the nonideal properties of the op-amp. In the present section, we examine the principal limitations of the operational amplifier.

## Voltage Supply Limits

As indicated in Figure 11.4, operational amplifiers (and all amplifiers, in general) are powered by external DC voltage supplies, $V_S^+$ and $V_S^-$, which are usually symmetrical and of the order of $\pm 10$ V to $\pm 20$ V. Some op-amps are especially designed to operate from a single voltage supply, but for the sake of simplicity we shall from here on consider only symmetrical supplies. The effect of limiting supply voltages is that amplifiers are capable of amplifying signals *only within the range of their supply voltages;* it would be physically impossible for an amplifier to generate a voltage greater than $V_S^+$ or less than $V_S^-$. This limitation may be stated as follows:

$$V_S^- < v_{\text{out}} < V_S^+ \qquad\qquad (12.81)$$

For most op-amps, the limit is actually approximately 1.5 V less than the supply voltages. How does this practically affect the performance of an amplifier circuit? An example will best illustrate the idea.

---

### EXAMPLE  12.9  Voltage Supply Limits in an Inverting Amplifier

**Problem**

Compute and sketch the output voltage of the inverting amplifier of Figure 12.42.

**Solution**

**Known Quantities:**  Resistor and supply voltage values; input voltage.

**Find:**  $v_{\text{out}}(t)$.

**Schematics, Diagrams, Circuits, and Given Data:**  $R_S = 1$ k$\Omega$; $R_F = 10$ k$\Omega$; $R_L = 1$ k$\Omega$; $V_S^+ = 15$ V; $V_S^- = -15$ V; $v_S(t) = 2\ \sin(1{,}000t)$.

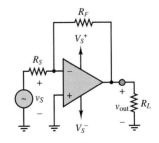

**Figure 12.42**

**Assumptions:** Assume supply voltage–limited op-amp.

**Analysis:** For an ideal op-amp the output would be:

$$v_{out}(t) = -\frac{R_F}{R_S}v_S(t) = -10 \times 2\sin(1,000t) = -20\sin(1,000t)$$

However, the supply voltage is limited to $\pm 15$ V, and the op-amp output voltage will therefore saturate before reaching the theoretical peak output value of $\pm 20$ V. Figure 12.43 depicts the output voltage waveform.

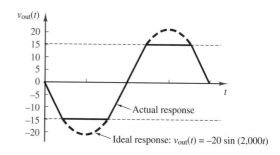

**Figure 12.43** Op-amp output with voltage supply limit

**Comments:** In a practical op-amp, saturation would be reached at 1.5 V below the supply voltages, or at approximately $\pm 13.5$ V.

**Focus on Computer-Aided Solutions:** An *Electronics Workbench*™ simulation of the circuit of Figure 12.42 can be found in the accompanying CD-ROM.

Note how the voltage supply limit actually causes the peaks of the sine wave to be clipped in an abrupt fashion. This type of hard nonlinearity changes the characteristics of the signal quite radically, and could lead to significant errors if not taken into account. Just to give an intuitive idea of how such clipping can affect a signal, have you ever wondered why rock guitar has a characteristic sound that is very different from the sound of classical or jazz guitar? The reason is that the "rock sound" is obtained by overamplifying the signal, attempting to exceed the voltage supply limits, and causing clipping similar in quality to the distortion introduced by voltage supply limits in an op-amp. This clipping broadens the spectral content of each tone and causes the sound to be distorted.

One of the circuits most directly affected by supply voltage limitations is the op-amp integrator. The following example illustrates how saturation of an integrator circuit can lead to severe signal distortion.

### EXAMPLE 12.10 Voltage Supply Limits in an Op-Amp Integrator

**Problem**

Compute and sketch the output voltage of the integrator of Figure 12.30.

**Solution**

**Known Quantities:**  Resistor, capacitor, and supply voltage values; input voltage.

**Find:**  $v_{\text{out}}(t)$.

**Schematics, Diagrams, Circuits, and Given Data:**  $R_S = 10\ \text{k}\Omega$; $C_F = 20\ \mu\text{F}$; $V_S^+ = 15\ \text{V}$; $V_S^- = -15\ \text{V}$; $v_S(t) = 0.5 + 0.3\ \cos(10t)$.

**Assumptions:**  Assume supply voltage–limited op-amp. The initial condition is $v_{\text{out}}(0) = 0$.

**Analysis:**  For an ideal op-amp integrator the output would be:

$$v_{\text{out}}(t) = -\frac{1}{R_S C_F} \int_{-\infty}^{t} v_S(t')dt' = -\frac{1}{0.2} \int_{-\infty}^{t} \left[0.5 + 0.3\cos(10t')\right]dt'$$

$$= -2.5t + 1.5\sin(10t)$$

However, the supply voltage is limited to $\pm 15$ V, and the integrator output voltage will therefore saturate at the lower supply voltage value of $-15$ V as the term $2.5t$ increases with time. Figure 12.44 depicts the output voltage waveform.

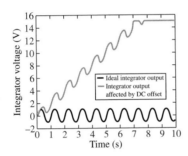

**Figure 12.44** Effect of DC offset on integrator

**Comments:**  Note that the DC offset in the waveform causes the integrator output voltage to increase linearly with time. The presence of even a very small DC offset will always cause integrator saturation. One solution to this problem is to include a large feedback resistor in parallel with the capacitor; this solution is explored in the homework problems.

**Focus on Computer-Aided Solutions:**  An *Electronics Workbench*™ simulation of a practical integrator (including the feedback resistance mentioned in the comments above) can be found in the accompanying CD-ROM. Try removing the feedback resistor to verify that saturation at the supply voltages is a real problem.

VIRTUAL LAB

## Frequency Response Limits

Another property of all amplifiers that may pose severe limitations to the op-amp is their finite bandwidth. We have so far assumed, in our ideal op-amp model, that the open-loop gain is a very large constant. In reality, $A_{V(\text{OL})}$ is a function of frequency and is characterized by a low-pass response. For a typical op-amp,

$$A_{V(OL)}(j\omega) = \frac{A_0}{1 + j\omega/\omega_0} \tag{12.82}$$

The cutoff frequency of the op-amp open-loop gain, $\omega_0$, represents approximately the point where the amplifier response starts to drop off as a function of frequency, and is analogous to the cutoff frequencies of the $RC$ and $RL$ circuits of Chapter 6. Figure 12.45 depicts $A_{V(OL)}(j\omega)$ in both linear and dB plots for the fairly typical values $A_0 = 10^6$ and $\omega_0 = 10\pi$. It should be apparent from this figure that the assumption of a very large open-loop gain becomes less and less accurate for increasing frequency. Recall the initial derivation of the closed-loop gain for the inverting amplifier: In obtaining the final result, $\mathbf{V}_{out}/\mathbf{V}_S = -R_F/R_S$, it was assumed that $A_{V(OL)} \to \infty$. This assumption is clearly inadequate at the higher frequencies.

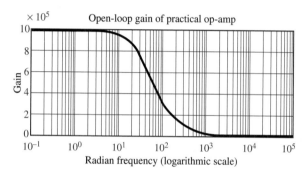
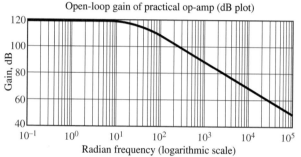

**Figure 12.45** Open-loop gain of practical op-amp

The finite bandwidth of the practical op-amp results in a fixed **gain-bandwidth product** for any given amplifier. The effect of a constant gain-bandwidth product is that as the closed-loop gain of the amplifier is increased, its 3-dB bandwidth is proportionally reduced, until, in the limit, if the amplifier were used in the open-loop mode, its gain would be equal to $A_0$ and its 3-dB bandwidth would be equal to $\omega_0$. The constant gain-bandwidth product is therefore equal to the product of the open-loop gain and the open-loop bandwidth of the amplifier: $A_0 \times \omega_0 = K$. When the amplifier is connected in a closed-loop configuration (e.g., as an inverting amplifier), its gain is typically much less than the open-loop gain and the 3-dB bandwidth of the amplifier is proportionally increased. To explain this further, Figure 12.46 depicts the case in which two different linear amplifiers (achieved through any two different negative feedback configurations) have been designed for the same op-amp. The first has closed-loop gain $A_1$, and the second has closed-loop gain $A_2$. The bold line in the figure indicates the open-loop frequency response, with gain $A_0$ and cutoff frequency $\omega_0$. As the gain decreases from the open-loop gain, $A_0$, to $A_1$, we see that the cutoff frequency increases from $\omega_0$ to $\omega_1$. If we further reduce the gain to $A_2$, we can expect the bandwidth to increase to $\omega_2$. Thus, *the product of gain and bandwidth in any given op-amp is constant.* That is,

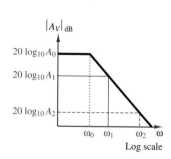

**Figure 12.46**

$$A_0 \times \omega_0 = A_1 \times \omega_1 = A_2 \times \omega_2 = K \tag{12.83}$$

# EXAMPLE 12.11 Gain-Bandwidth Product Limit in an Op-Amp

## Problem

Determine the maximum allowable closed-loop voltage gain of an op-amp, if the amplifier is required to have an audio-range bandwidth of 20 kHz.

## Solution

**Known Quantities:** Gain-bandwidth product.

**Find:** $A_{V \text{ max}}$.

**Schematics, Diagrams, Circuits, and Given Data:** $A_0 = 10^6$; $\omega_0 = 2\pi \times 5$ rad/s.

**Assumptions:** Assume gain-bandwidth product limited op-amp.

**Analysis:** The gain-bandwidth product of the op-amp is:

$$A_0 \times \omega_0 = K = 10^6 \times 2\pi \times 5 = \pi \times 10^7 \text{ rad/s.}$$

The desired bandwidth is $\omega_{\text{max}} = 2\pi \times 20,000$ rad/s, and the maximum allowable gain will therefore be:

$$A_{\text{max}} = \frac{K}{\omega_{\text{max}}} = \frac{\pi \times 10^7}{\pi \times 4 \times 10^4} = 250 \quad \frac{V}{V}$$

For any closed-loop voltage gain greater than 250, the amplifier would have reduced bandwidth.

**Comments:** If we desired to achieve gains greater than 250 and maintain the same bandwidth, two options would be available: (1) use a different op-amp with greater gain-bandwidth product or (2) connect two amplifiers in cascade, each with lower gain and greater bandwidth, such that the product of the gains would be greater than 250.

To further explore the first option, you may wish to look at the device data sheets for different op-amps (in the accompanying CD-ROM, or in the device templates in the *Electronics Workbench*™ libraries), and verify that op-amps can be designed (at a cost!) to have substantially greater gain-bandwidth product than the amplifier used in this example. The second option is examined in the next example.

# EXAMPLE 12.12 Increasing the Gain-Bandwidth Product by Means of Amplifiers in Cascade

## Problem

Determine the overall 3-dB bandwidth of the cascade amplifier of Figure 12.47.

## Solution

**Known Quantities:** Gain-bandwidth product and gain of each amplifier.

**Find:** $\omega_{3 \text{ dB}}$ of cascade amplifier.

**Schematics, Diagrams, Circuits, and Given Data:** $A_0 \times \omega_0 = K = 4\pi \times 10^6$ for each amplifier. $R_F/R_S = 100$ for each amplifier.

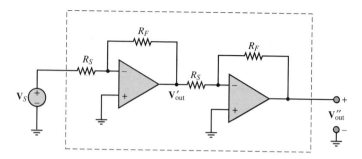

**Figure 12.47** Cascade amplifier

**Assumptions:** Assume gain-bandwidth product–limited (otherwise ideal) op-amps.

**Analysis:** Let $A_1$ and $\omega_1$ denote the gain and the 3-dB bandwidth of the first amplifier, respectively, and $A_2$ and $\omega_2$ those of the second amplifier.

The 3-dB bandwidth of the first amplifier is:

$$\omega_1 = \frac{K}{A_1} = \frac{4\pi \times 10^6}{10^2} = 4\pi \times 10^4 \quad \frac{\text{rad}}{s}$$

The second amplifier will also have:

$$\omega_2 = \frac{K}{A_2} = \frac{4\pi \times 10^6}{10^2} = 4\pi \times 10^4 \quad \frac{\text{rad}}{s}$$

Thus, the approximate bandwidth of the cascade amplifier is $4\pi \times 10^4$ and the gain of the cascade amplifier is $A_1 \times A_2 = 100 \times 100 = 10^4$.

Had we attempted to achieve the same gain with a single-stage amplifier having the same $K$, we would have achieved a bandwidth of only:

$$\omega_3 = \frac{K}{A_3} = \frac{4\pi \times 10^6}{10^4} = 4\pi \times 10^2 \quad \frac{\text{rad}}{s}$$

**Comments:** In practice, the actual 3-dB bandwidth of the cascade amplifier is not quite as large as that of each of the two stages, because the gain of each amplifier starts decreasing at frequencies somewhat lower than the nominal cutoff frequency. The calculation of the actual 3-dB bandwidth of the cascade amplifier is illustrated in Check Your Understanding Exercise 12.17.

## Input Offset Voltage

Another limitation of practical op-amps results because even in the absence of any external inputs, it is possible that an **offset voltage** will be present at the input of an op-amp. This voltage is usually denoted by $\pm V_{os}$ and it is caused by mismatches in the internal circuitry of the op-amp. The offset voltage appears as a differential input voltage between the inverting and noninverting input terminals. The presence of an additional input voltage will cause a DC bias error in the amplifier output, as illustrated in Example 12.13. Typical and maximum values of $V_{os}$ are quoted in manufacturers' data sheets. The worst-case effects due to the presence of offset voltages can therefore be predicted for any given application.

## EXAMPLE 12.13 Effect of Input Offset Voltage on an Amplifier

### Problem

Determine the effect of the input offset voltage $V_{os}$ on the output of the amplifier of Figure 12.48.

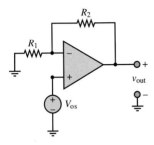

### Solution

**Known Quantities:** Nominal closed-loop voltage gain; input offset voltage.

**Find:** The offset voltage component in the output voltage, $v_{out,os}$.

**Schematics, Diagrams, Circuits, and Given Data:** $A_{nom} = 100$; $V_{os} = 1.5$ mV.

**Figure 12.48** Op-amp input offset voltage

**Assumptions:** Assume input-offset-voltage-limited (otherwise ideal) op-amp.

**Analysis:** The amplifier is connected in a noninverting configuration; thus its gain is:

$$A_{V\ nom} = 100 = 1 + \frac{R_F}{R_S}$$

The DC offset voltage, represented by an ideal voltage source, is represented as being directly applied to the noninverting input; thus:

$$V_{out,os} = A_{V\ nom} V_{os} = 100 V_{os} = 150 \quad \text{mV}$$

Thus, we should expect the output of the amplifier to be shifted upward by 150 mV.

**Comments:** The input offset voltage is not, of course, an external source, but it represents a voltage offset between the inputs of the op-amp. Figure 12.51 depicts how such an offset can be nulled.

The worst-case offset voltage is usually listed in the device data sheets (see the data sheets in the accompanying CD-ROM, or in the device templates in the *Electronics Workbench*™ libraries, for an illustration). Typical values are 2 mV for the 741c general-purpose op-amp and 5 mV for the FET-input TLO81.

## Input Bias Currents

Another nonideal characteristic of op-amps results from the presence of small input bias currents at the inverting and noninverting terminals. Once again, these are due to the internal construction of the input stage of an operational amplifier. Figure 12.49 illustrates the presence of nonzero input bias currents ($I_B$) going into an op-amp.

Typical values of $I_B$ depend on the semiconductor technology employed in the construction of the op-amp. Op-amps with bipolar transistor input stages may see input bias currents as large as 1 $\mu$A, while for FET input devices, the input bias currents are less than 1 nA. Since these currents depend on the internal design of the op-amp, they are not necessarily equal. One often designates the **input offset current** $I_{os}$ as

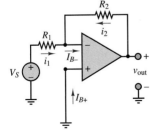

$$I_{os} = I_{B+} - I_{B-}$$

**(12.84)**    **Figure 12.49**

The latter parameter is sometimes more convenient from the standpoint of analysis. The following example illustrates the effect of the nonzero input bias current on a practical amplifier design.

## EXAMPLE 12.14 Effect of Input Offset Current on an Amplifier

### Problem

Determine the effect of the input offset current $I_{os}$ on the output of the amplifier of Figure 12.50.

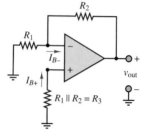

**Figure 12.50**

### Solution

**Known Quantities:** Resistor values; input offset current.

**Find:** The offset voltage component in the output voltage, $v_{out,os}$.

**Schematics, Diagrams, Circuits, and Given Data:** $I_{os} = 1 \ \mu A; \ R_2 = 10 \ k\Omega$.

**Assumptions:** Assume input-offset-current-limited (otherwise ideal) op-amp.

**Analysis:** We calculate the inverting and noninverting terminal voltages caused by the offset current in the absence of an external input:

$$v^+ = R_3 I_{B+} \qquad v^- = v^+ = R_3 I_{B+}$$

With these values we can apply KCL at the inverting node and write:

$$\frac{v_{out} - v^-}{R_2} - \frac{v^+}{R_1} = I_{B-}$$

$$\frac{v_{out}}{R_2} - \frac{-R_3 I_{B+}}{R_2} - \frac{-R_3 I_{B+}}{R_1} = I_{B-}$$

$$v_{out} = R_2 \left[ -I_{B+} R_3 \left( \frac{1}{R_2} + \frac{1}{R_1} \right) + I_{B-} \right] = -R_2 I_{os}$$

Thus, we should expect the output of the amplifier to be shifted downward by $R_2 I_{os}$, or $10^4 \times 10^{-6} = 10$ mV for the data given in this example.

**Comments:** Usually, the worst-case input offset current (or input bias currents) are listed in the device data sheets (see the data sheets in the accompanying CD-ROM, or in the device templates in the *Electronics Workbench*™ libraries, for an illustration). Values can range from 100 pA (for CMOS op-amps, e.g., LMC6061) to around 200 nA for a low-cost general-purpose amplifier (e.g., $\mu A$ 741c).

## Output Offset Adjustment

Both the offset voltage and the input offset current contribute to an output offset voltage $V_{out,os}$. Some op-amps provide a means for minimizing $V_{out,os}$. For example, the $\mu A741$ op-amp provides a connection for this procedure. Figure 12.51 shows a typical pin configuration for an op-amp in an eight-pin dual-in-line package (DIP) and the circuit used for nulling the output offset voltage. The variable resistor is adjusted until $v_{\text{out}}$ reaches a minimum (ideally, 0 volts). Nulling the output voltage in this manner removes the effect of both input offset voltage and current on the output.

## Slew Rate Limit

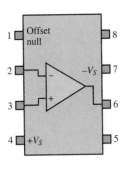

Another important restriction in the performance of a practical op-amp is associated with rapid changes in voltage. The op-amp can produce only a finite rate of change at its output. This limit rate is called the **slew rate.** Consider an ideal step input, where at $t = 0$ the input voltage is switched from zero to $V$ volts. Then we would expect the output to switch from 0 to $A \cdot V$ volts, where $A$ is the amplifier gain. However, $v_{\text{out}}(t)$ can change at only a finite rate; thus,

$$\left|\frac{dv_{\text{out}}(t)}{dt}\right|_{\text{max}} = S_0 = \text{Slew rate} \tag{12.85}$$

Figure 12.52 shows the response of an op-amp to an ideal step change in input voltage. Here, $S_0$, the slope of $v_{\text{out}}(t)$, represents the slew rate.

The slew rate limitation can affect sinusoidal signals, as well as signals that display abrupt changes, as does the step voltage of Figure 12.52. This may not be obvious until we examine the sinusoidal response more closely. It should be apparent that the maximum rate of change for a sinusoid occurs at the zero crossing, as shown by Figure 12.53. To evaluate the slope of the waveform at the zero crossing, let

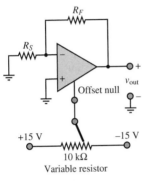

**Figure 12.51** Output off-set voltage adjustment

$$v(t) = A \sin \omega t \tag{12.86}$$

so that

$$\frac{dv(t)}{dt} = \omega A \cos \omega t \tag{12.87}$$

The maximum slope of the sinusoidal signal will therefore occur at $\omega t = 0, \pi, 2\pi, \ldots$, so that

$$\left|\frac{dv(t)}{dt}\right|_{\text{max}} = \omega \times A = S_0 \tag{12.88}$$

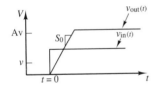

**Figure 12.52** Slew rate limit in op-amps

Thus, the maximum slope of a sinusoid is proportional to both the signal frequency and the amplitude. The curve shown by a dashed line in Figure 12.53 should indicate that as $\omega$ increases, so does the slope of $v(t)$ at the zero crossings. What is the direct consequence of this result, then? Example 12.15 gives an illustration of the effects of this slew rate limit.

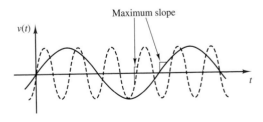

**Figure 12.53** The maximum slope of a sinusoidal signal varies with the signal frequency

## EXAMPLE  12.15  Effect of Slew Rate Limit on an Amplifier

### Problem

Determine the effect of the slew rate limit, $S_0$, on the output of an inverting amplifier for a sinusoidal input voltage of known amplitude and frequency.

### Solution

**Known Quantities:**  Slew rate limit, $S_0$; amplitude and frequency of sinusoidal input voltage; amplifier closed-loop gain.

**Find:**  Sketch the theoretically correct output and the actual output of the amplifier in the same graph.

**Schematics, Diagrams, Circuits, and Given Data:**  $S_0 = 1\ \text{V}/\mu\text{S}$; $v_S(t) = \sin(2\pi \times 10^5 t)$; $A_V = 10$.

**Assumptions:**  Assume slew-rate-limited (otherwise ideal) op-amp.

**Analysis:**  Given the closed-loop voltage gain of 10, we compute the theoretical output voltage to be:

$$v_{\text{out}}(t) = -10\ \sin(2\pi \times 10^5 t)$$

The maximum slope of the output voltage is then computed as follows:

$$\left| \frac{dv_{\text{out}}(t)}{dt} \right|_{\text{max}} = A\omega = 10 \times 2\pi \times 10^5 = 6.28\ \frac{V}{\mu s}$$

Clearly, the value calculated above far exceeds the slew rate limit. Figure 12.54 depicts the approximate appearance of the waveforms that one would measure in an experiment.

**Comments:**  Note that in this example the slew rate limit has been exceeded severely, and the output waveform is visibly distorted, to the point that it has effectively become a triangular wave. The effect of the slew rate limit is not always necessarily so dramatic and visible; thus one needs to pay attention to the specifications of a given op-amp. The slew rate limit is listed in the device data sheets (see the data sheets in the accompanying CD-ROM, or in the device templates in the *Electronics Workbench*™ libraries, for examples). Typical values can range from 13 V/$\mu$s, for the TLO81, to around 0.5 V/$\mu$s for a low-cost general-purpose amplifier (e.g., $\mu$A 741c).

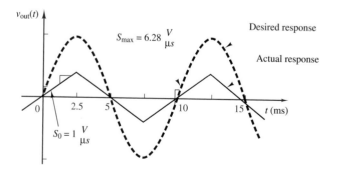

**Figure 12.54** Distortion introduced by slew rate limit

## Short-Circuit Output Current

Recall the model for the op-amp introduced in Section 12.2, which represented the internal circuits of the op-amp in terms of an equivalent input resistance, $R_{in}$, and a controlled voltage source, $A_V v_{in}$. In practice, the internal source is not ideal, because it cannot provide an infinite amount of current (either to the load, or to the feedback connection, or both). The immediate consequence of this nonideal op-amp characteristic is that the maximum output current of the amplifier is limited by the so-called short-circuit output current, $I_{SC}$:

$$|I_{out}| < I_{SC} \tag{12.89}$$

To further explain this point, consider that the op-amp needs to provide current to the feedback path (in order to "zero" the voltage differential at the input) and to whatever load resistance, $R_L$, may be connected to the output. Figure 12.55 illustrates this idea for the case of an inverting amplifier, where $I_{SC}$ is the load current that would be provided to a short-circuit load ($R_L = 0$).

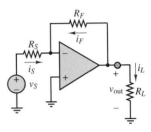

**Figure 12.55**

---

**EXAMPLE   12.16   Effect of Short-Circuit Current Limit
on an Amplifier**

### Problem

Determine the effect of the short-circuit limit, $I_{SC}$, on the output of an inverting amplifier for a sinusoidal input voltage of known amplitude.

---

### Solution

**Known Quantities:**  Short-circuit current limit, $I_{SC}$; amplitude of sinusoidal input voltage; amplifier closed-loop gain.

**Find:**  Compute the maximum allowable load resistance value, $R_{L\ min}$, and sketch the theoretical and actual output voltage waveforms for resistances smaller than $R_{L\ min}$.

**Schematics, Diagrams, Circuits, and Given Data:**  $I_{SC} = 50$ mA; $v_S(t)$ $= 0.05\ \sin(\omega t)$; $A_V = 100$.

**Assumptions:**  Assume short-circuit-current-limited (otherwise ideal) op-amp.

**Analysis:**  Given the closed-loop voltage gain of 100, we compute the theoretical output voltage to be:

$$v_{out}(t) = -A_V v_S(t) = -5\ \sin(\omega t)$$

To assess the effect of the short-circuit current limit, we calculate the peak value of the output voltage, since this is the condition that will require the maximum output current from the op-amp:

$$v_{out\ peak} = 5\ V$$

$$I_{SC} = 50\ mA$$

$$R_{L\ min} = \frac{v_{out\ peak}}{I_{SC}} = \frac{5\ V}{50\ mA} = 100\ \Omega$$

For any load resistance less than 100 $\Omega$, the required load current will be greater than $I_{SC}$.

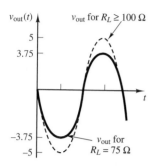

**Figure 12.56** Distortion introduced by short-circuit current limit

For example, if we chose a 75-$\Omega$ load resistor, we would find that

$$v_{\text{out peak}} = I_{\text{SC}} \times R_L = 3.75 \text{ V}$$

That is, the output voltage cannot reach the theoretically correct 5-V peak, and would be "compressed" to reach a peak voltage of only 3.75 V. This effect is depicted in Figure 12.56.

**Comments:** The short-circuit current limit is listed in the device data sheets (see the data sheets in the accompanying CD-ROM, or in the device templates in the *Electronics Workbench*™ libraries, for examples). Typical values for a low-cost general-purpose amplifier (e.g., 741c) are in the tens of milliamperes.

## Common-Mode Rejection Ratio (CMRR)

Early in this chapter, "Focus on Measurements: Electrocardiogram (EKG) Amplifier" introduces the notion of differential-mode and common-mode signals. If we define $A_{\text{dm}}$ as the **differential-mode gain** and $A_{\text{cm}}$ as the **common-mode gain** of the op-amp, the output of an op-amp can then be expressed as follows:

$$v_{\text{out}} = A_{\text{dm}}(v_2 - v_1) + A_{\text{cm}}\left(\frac{v_2 + v_1}{2}\right) \tag{12.90}$$

Under ideal conditions, $A_{\text{cm}}$ should be exactly zero, since the differential amplifier should completely reject common-mode signals. The departure from this ideal condition is a figure of merit for a differential amplifier and is measured by defining a quantity called the **common-mode rejection ratio (CMRR).** The CMRR is defined as the ratio of the differential-mode gain to the common-mode gain and should ideally be infinite:

$$\boxed{\text{CMRR} = \frac{A_{\text{dm}}}{A_{\text{cm}}}}$$

The CMRR is often expressed in units of decibels (dB). The common-mode rejection ratio idea is explored further in the problems at the end of the chapter.

## FOCUS ON METHODOLOGY

### Using Op-Amp Data Sheets—Comparison of LM741 and LMC6061

In this box we compare the LM741 and the LMC6061 op-amps that were introduced in an earlier methodology box. Excerpts from the data sheets are shown below, with some words of explanation. The complete data sheets can be found in the accompanying CD-ROM. You are encouraged to identify the information given below in the data sheets.

*LM741 Electrical Characteristics*— An abridged version of the electrical characteristics of the LM741 is shown below.

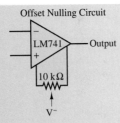

Offset Nulling Circuit

## Electrical Characteristics

| Parameter | Conditions | LM741A/LM741E | | | LM741 | | | LM741C | | | Units |
|---|---|---|---|---|---|---|---|---|---|---|---|
| | | Min | Typ | Max | Min | Typ | Max | Min | Typ | Max | |
| Input Offset Voltage | $T_A = 25°C$<br>$R_S \leq 10\,k\Omega$<br>$R_S \leq 50\Omega$ | | 0.8 | 3.0 | | 1.0 | 5.0 | | 2.0 | 6.0 | mV<br>mV |
| | $T_{AMIN} \leq T_A \leq T_{AMAX}$<br>$R_S \leq 50\Omega$<br>$R_S \leq 10\,k\Omega$ | | | 4.0 | | | 6.0 | | | 7.5 | mV<br>mV |
| Average Input Offset Voltage Drift | | | | 15 | | | | | | | $\mu V/°C$ |
| Input Offset Voltage Adjustment Range | $T_A = 25°C$, $V_S = \pm20V$ | ±10 | | | | ±15 | | | ±15 | | mV |
| Input Offset Current | $T_A = 25°C$ | | 3.0 | 30 | | 20 | 200 | | 20 | 200 | nA |
| | $T_{AMIN} \leq T_A \leq T_{AMAX}$ | | | 70 | | 85 | 500 | | | 300 | nA |
| Average Input Offset Current Drift | | | | 0.5 | | | | | | | $nA/°C$ |
| Input Bias Current | $T_A = 25°C$ | | 30 | 80 | | 80 | 500 | | 80 | 500 | nA |
| | $T_{AMIN} \leq T_A \leq T_{AMAX}$ | | | 0.210 | | | 1.5 | | | 0.8 | $\mu A$ |
| Input Resistance | $T_A = 25°C$, $V_S = \pm20V$ | 1.0 | 6.0 | | 0.3 | 2.0 | | 0.3 | 2.0 | | $M\Omega$ |
| | $T_{AMIN} \leq T_A \leq T_{AMAX}$,<br>$V_S = \pm20\,V$ | 0.5 | | | | | | | | | $M\Omega$ |
| Input Voltage Range | $T_A = 25°C$ | | | | | | | ±12 | ±13 | | V |
| | $T_{AMIN} \leq T_A \leq T_{AMAX}$ | | | | ±12 | ±13 | | | | | V |
| Large Signal Voltage Gain | $T_A = 25°C$, $R_L \geq 2\,k\Omega$<br>$V_S = \pm20V$, $V_O = \pm15V$<br>$V_S = \pm15V$, $V_O = \pm10V$ | 50 | | | | 50 | 200 | 20 | 200 | | V/mV<br>V/mV |
| | $T_{AMIN} \leq T_A \leq T_{AMAX}$,<br>$R_L \geq 2\,k\Omega$<br>$V_S = \pm20V$, $V_O = \pm15V$<br>$V_S = \pm15V$, $V_O = \pm10V$<br>$V_S = \pm5V$, $V_O = \pm2V$ | 32<br><br>10 | | | | 25 | | | 15 | | V/mV<br>V/mV<br>V/mV |
| Output Voltage Swing | $V_S = \pm20V$<br>$R_L \geq 10\,k\Omega$<br>$R_L \geq 2\,k\Omega$ | ±16<br>±15 | | | | | | | | | V<br>V |
| | $V_S = \pm15\,V$<br>$R_L \geq 10\,k\Omega$<br>$R_L \geq 2\,k\Omega$ | | | | ±12<br>±10 | ±14<br>±13 | | ±12<br>±10 | ±14<br>±13 | | V<br>V |

(*Continued*)

## Electrical Characteristics

| Parameter | Conditions | LM741A/LM741E | | | LM741 | | | LM741C | | | Units |
|---|---|---|---|---|---|---|---|---|---|---|---|
| | | Min | Typ | Max | Min | Typ | Max | Min | Typ | Max | |
| Output Short Circuit Current | $T_A = 25°C$<br>$T_{AMIN} \le T_A \le T_{AMAX}$ | 10<br>10 | 25 | 35<br>40 | | 25 | | | 25 | | mA<br>mA |
| Common-Mode Rejection Ratio | $T_{AMIN} \le T_A \le T_{AMAX}$<br>$R_S \le 10\,k\Omega,\ V_{CM} = \pm12V$<br>$R_S \le 50\Omega,\ V_{CM} = \pm12V$ | 80 | 95 | | 70 | 90 | | 70 | 90 | | dB<br>dB |
| Supply Voltage Rejection Ratio | $T_{AMIN} \le T_A \le T_{AMAX}$,<br>$V_S = \pm20V$ to $V_S = \pm5V$<br>$R_S \le 50\Omega$<br>$R_S \le 10\,k\Omega$ | 86 | 96 | 77 | 96 | | 77 | 96 | | | dB<br>dB |
| Transient Response<br>  Rise Time<br>  Overshoot | $T_A = 25°C$, Unity Gain | | 0.25<br>6.0 | 0.8<br>20 | | 0.3<br>5 | | | 0.3<br>5 | | $\mu$s<br>% |
| Bandwidth | $T_A = 25°C$ | 0.437 | 1.5 | | | | | | | | MHz |
| Slew Rate | $T_A = 25°C$, Unity Gain | 0.3 | 0.7 | | | 0.5 | | | 0.5 | | V/$\mu$s |
| Supply Current | $T_A = 25°C$ | | | | | 1.7 | 2.8 | | 1.7 | 2.8 | mA |
| Power Consumption | $T_A = 25°C$<br>$V_S = \pm20V$<br>$V_S = \pm15V$ | | 80 | 150<br>85 | | 50 | 85 | | 50 | 85 | mW<br>mW |
| LM741A | $V_S = \pm20V$<br>$T_A = T_{AMIN}$<br>$T_A = T_{AMAX}$ | | | 165<br>135 | | | | | | | mW<br>mW |
| LM741E | $V_S = \pm20V$<br>$T_A = T_{AMIN}$<br>$T_A = T_{AMAX}$ | | | 150<br>150 | | | | | | | mW<br>mW |
| LM741 | $V_S = \pm15V$<br>$T_A = T_{AMIN}$<br>$T_A = T_{AMAX}$ | | | | | 60<br>45 | 100<br>75 | | | | mW<br>mW |

*LMC6061 Electrical Characteristics—* An abridged version of the electrical characteristics of the LMC 6061 is shown below.

## DC Electrical Characteristics

Unless otherwise specified, all limits guaranteed for $T_J = 25°C$. **Boldface** limits apply at the temperature extremes. $V^+ = 5V$, $V^- = 0V$, $V_{CM} = 1.5V$, $V_O = 2.5V$ and $R_L > 1M$ unless otherwise specified.

| Symbol | Parameter | Conditions | Typ | LMC6061AM Limit | LMC6061AI Limit | LMC6061I Limit | Units |
|---|---|---|---|---|---|---|---|
| $V_{OS}$ | Input Offset Voltage | | 100 | 350<br>**1200** | 350<br>**900** | 800<br>**1300** | $\mu$V<br>Max |
| $TCV_{OS}$ | Input Offset Voltage Average Drift | | 1.0 | | | | $\mu$V/°C |

# DC Electrical Characteristics (Continued)

| Symbol | Parameter | Conditions | | Typ | LMC6061AM Limit | LMC6061AI Limit | LMC6061I Limit | Units |
|---|---|---|---|---|---|---|---|---|
| $I_B$ | Input Bias Current | | | 0.010 | **100** | **4** | **4** | pA Max |
| $I_{OS}$ | Input Offset Current | | | 0.005 | **100** | **2** | **2** | pA Max |
| $R_{IN}$ | Input Resistance | | | >10 | | | | Tera $\Omega$ |
| CMRR | Common Mode Rejection Ratio | $0V \leq V_{CM} \leq 12.0V$ $V^+ = 15V$ | | 85 | 75 **70** | 75 **72** | 66 **63** | dB Min |
| +PSRR | Positive Power Supply Rejection Ratio | $5V \leq V^+ \leq 15V$ $V_O = 2.5V$ | | 85 | 75 **70** | 75 **72** | 66 **63** | dB Min |
| −PSRR | Negative Power Supply Rejection Ratio | $0V \leq V^- \leq -10V$ | | 100 | 84 **70** | 84 **81** | 74 **71** | dB Min |
| $V_{CM}$ | Input Common-Mode Voltage Range | $V^+ = 5V$ and 15V for CMRR $\geq 60$ dB | | −0.4 | −0.1 **0** | −0.1 **0** | −0.1 **0** | V Max |
| | | | | $V^+ - 1.9$ | $V^+ - 2.3$ **$V^+$–2.6** | $V^+ - 2.3$ **$V^+$–2.5** | $V^+ - 2.3$ **$V^+$–2.5** | V Min |
| $A_V$ | Large Signal Voltage Gain | $R_L = 100$ k$\Omega$ (Note 7) | Sourcing | 4000 | 400 **200** | 400 **300** | 300 **200** | V/mV Min |
| | | | Sinking | 3000 | 180 **70** | 180 **100** | 90 **60** | V/mV Min |
| | | $R_L = 25$ k$\Omega$ (Note 7) | Sourcing | 3000 | 400 **150** | 400 **150** | 200 **80** | V/mV Min |
| | | | Sinking | 2000 | 100 **35** | 100 **50** | 70 **35** | V/mV Min |
| $V_O$ | Output Swing | $V^+ = 5V$ $R_L = 100$ k$\Omega$ to 2.5V | | 4.995 | 4.990 **4.970** | 4.990 **4.980** | 4.950 **4.925** | V Min |
| | | | | 0.005 | 0.010 **0.030** | 0.010 **0.020** | 0.050 **0.075** | V Max |
| | | $V^+ = 5V$ $R_L = 25$ k$\Omega$ to 2.5V | | 4.990 | 4.975 **4.955** | 4.975 **4.965** | 4.950 **4.850** | V Min |
| | | | | 0.010 | 0.020 **0.045** | 0.020 **0.035** | 0.050 **0.150** | V Max |
| | | $V^+ = 15V$ $R_L = 100$ k$\Omega$ to 7.5V | | 14.990 | 14.975 **14.955** | 14.975 **14.965** | 14.950 **14.925** | V Min |
| | | | | 0.010 | 0.025 **0.050** | 0.025 **0.035** | 0.050 **0.075** | V Max |
| | | $V^+ = 15V$ $R_L = 25$ k$\Omega$ to 7.5V | | 14.965 | 14.900 **14.800** | 14.900 **14.850** | 14.850 **14.800** | V Min |
| | | | | 0.025 | 0.050 **0.200** | 0.050 **0.150** | 0.100 **0.200** | V Max |

*(Continued)*

*(Concluded)*

## DC Electrical Characteristics (Continued)

| Symbol | Parameter | Conditions | Typ | LMC6061AM Limit | LMC6061AI Limit | LMC6061I Limit | Units |
|--------|-----------|-----------|-----|-----------------|-----------------|----------------|-------|
| $I_O$ | Output Current $V^+ = 5V$ | Sourcing, $V_O = 0V$ | 22 | 16 **8** | 16 **10** | 13 **8** | mA Min |
| | | Sinking, $V_O = 5V$ | 21 | 16 **7** | 16 **8** | 16 **8** | mA Min |
| $I_O$ | Output Current $V^+ = 15V$ | Sourcing, $V_O = 0V$ | 25 | 15 **9** | 15 **10** | 15 **10** | mA Min |
| | | Sinking, $V_O = 13V$ (Note 10) | 35 | 24 **7** | 24 **8** | 24 **8** | mA Min |
| $I_S$ | Supply Current | $V^+ = +5V$, $V_O = 1.5V$ | 20 | 24 **35** | 24 **32** | 32 **40** | mA Max |
| | | $V^+ = +15V$, $V_O = 7.5V$ | 24 | 30 **40** | 30 **38** | 40 **48** | $\mu$A Max |

## AC Electrical Characteristics

Unless otherwise specified, all limits guaranteed for $T_J = 25°C$. **Boldface** limits apply at the temperature extremes. $V^+ = 5V$, $V^- = 0V$, $V_{CM} = 1.5V$, $V_O = 2.5V$ and $R_L > 1M$ unless otherwise specified.

| Symbol | Parameter | Conditions | Typ | LMC6061AM Limit | LMC6061AI Limit | LMC6061I Limit | Units |
|--------|-----------|-----------|-----|-----------------|-----------------|----------------|-------|
| SR | Slew Rate | (Note 8) | 35 | 20 **8** | 20 **10** | 15 **7** | V/ms Min |
| GBW | Gain-Bandwidth Product | | 100 | | | | kHz |
| $\theta_m$ | Phase Margin | | 50 | | | | Deg |
| $e_n$ | Input-Referred Voltage Noise | $F = 1$ kHz | 83 | | | | nV/$\sqrt{Hz}$ |
| $i_n$ | Input-Referred Current Noise | $F = 1$ kHz | 0.0002 | | | pA/$\sqrt{Hz}$ | |
| T.H.D. | Total Harmonic Distortion | $F = 1$ kHz, $A_V = -5$ $R_L = 100$ k$\Omega$, $V_O = 2V_{PP}$ $\pm 5V$ Supply | 0.01 | | | | % % |

## Comparison:

*Input Offset Voltage*— Note that the typical input offset voltage in the 6061 is only 100 $\mu$V, versus 0.8 mV in the 741.

*Input Offset Voltage Adjustments*— The recommended circuit is shown for the 741, and a range of $\pm 15$ mV is given. The 6061 does not require offset voltage adjustment.

*Input Offset Current*— The 741 sheet reports typical value of 3 nA ($3 \times 10^{-9}$ A); the corresponding value for the 6061 is 0.005 pA ($5 \times 10^{-15}$ A)! This extremely low value is due to the MOS construction of the amplifier (see Chapter 9 for a discussion of MOS stage input impedance).

*Input Resistance*— The specifications related to input offset current are mirrored by the input resistance specifications. The 741 has a respectable typical input resistance of 6 M$\Omega$; the 6061 has an input resistance greater than 10 T$\Omega$ (1 teraohm = $10^{12}$ $\Omega$). Once again, this is the result of MOS construction.

*Large-Signal Voltage Gain*— The 741 lists a typical value of 50 V/mV (or $5 \times 10^4$) for its open-loop voltage gain; the 6061 lists values greater than or equal to 2,000 V/mV (or $2 \times 10^6$).

*CMRR*— The typical common-mode rejection ratio is 95 db for the 741 and 85 dB for the 6061.

*Slew rate*— 0.7 V/$\mu$s for the 741 and 35 V/ms for the 6061.

*Bandwidth*— The bandwidth for the 741 is listed as 1.5 MHz (this would be the unity gain bandwidth), while the 6061 lists a 100-kHz gain-bandwidth product.

*Output short circuit current*—25 mA for both devices.

Note that, while the LMC6061 is certainly superior to the LM741 op-amp in a number of categories, there are certain features (e.g., bandwidth and slew rate) that might cause a designer to prefer the 741 for a specific application.

## Check Your Understanding

**12.17**  In Example 12.12, we implicitly assumed that the gain of each amplifier was constant for frequencies up to the cutoff frequency. This is, in practice, not true, since the individual op-amp closed-loop gain starts dropping below the DC gain value according to the equation

$$A(j\omega) = \frac{A_1}{1 + j\omega/\omega_1}$$

Thus, the calculations carried out in the example are only approximate. Find an expression for the closed-loop gain of the cascade amplifier. [*Hint:* The combined gain is equal to the product of the individual closed-loop gains.]  What is the actual gain in dB at the cutoff frequency, $\omega_0$, for the cascade amplifier?

**12.18**  What is the 3-dB bandwidth of the cascade amplifier of Example 12.12? [*Hint:* The gain of the cascade amplifier is the product of the individual op-amp frequency responses. Compute the magnitude of this product and set the magnitude of the product of the individual frequency responses equal to $(1/\sqrt{2}) \times 10,000$, and then solve for $\omega$.]

Manufacturers generally supply values for the parameters discussed in this section in their device data specifications. Typical data sheets for common op-amps may be found in the accompanying CD-ROM.

## CONCLUSION

This chapter has described the fundamental properties and limitations of the operational amplifier.

- Ideal amplifiers represent fundamental building blocks of electronic instrumentation. With the concept of the ideal amplifiers in mind, one can design practical amplifiers, filters, integrators, and other useful signal-processing circuits.
- The operational amplifier closely approximates the characteristics of an ideal amplifier. The analysis of op-amp circuits may be carried out very easily, if it is assumed that the op-amp's input resistance and open-loop gain are very large. The inverting, noninverting, and differential amplifier configurations permit the design of useful electronic amplifiers simply by selecting a few external resistors.
- If energy-storage elements are used in the construction of op-amp circuits, it is possible to accomplish the functions of filtering, integration, and differentiation.

• The properties of summing amplifiers and integrators make it possible to build analog computers, which serve as an aid in the solution of differential equations, and in the simulation of dynamic systems.

• When op-amps are employed in more advanced applications, it is important to know that there are limitations on their performance that are not predicted by the simple op-amp model introduced at the beginning of the chapter. These include voltage supply limits, frequency response limits, offset voltages and currents, slew rate limits, and finite common-mode rejection ratio. In general, it is not difficult to compensate for these limitations in the design of op-amp circuits.

## CHECK YOUR UNDERSTANDING ANSWERS

| | |
|---|---|
| **CYU 12.1** | $A_{V(CL)} = 999.9, 999.0, 990.1, 909.1$   $A_{V(OL)\,min} = 10^6$ for 0.1% error in $A_{V(CL)}$. |
| **CYU 12.2** | $A_{V(CL)} = 99.99, 99.99, 99.90, 99.00$   $A_{V(OL)\,min} = 10^5$ for 0.1% error in $A_{V(CL)}$. |
| **CYU 12.4** | $\Delta R = 6{,}714\ \Omega$ |
| **CYU 12.5** | $V_{ref}$ can have values between $\pm 0.714$ V. |
| **CYU 12.6** | $R_F/R_S = 42.55$; $V_{ref} = 0.684$ V |
| **CYU 12.7** | $R_F = 19.9\ k\Omega$, $R_S = 199\ \Omega$ |
| **CYU 12.8** | $R_F = 820\ k\Omega$, $R_S = 8.2\ k\Omega$; error: gain = 0%, $\omega_{3\,dB} = 2.95\%$ |
| **CYU 12.9** | 407 Hz |
| **CYU 12.10** | $-6$ dB; $\omega_{3\,dB} = 0.642\omega_0$ |
| **CYU 12.11** | $-20$ dB/decade |
| **CYU 12.12** | $+20$ dB/decade |
| **CYU 12.14** | $B/M = 5$; $K/M = 40$ |
| **CYU 12.15** | $x_{M\ max} = 0.0082$ m |

**CYU 12.16**

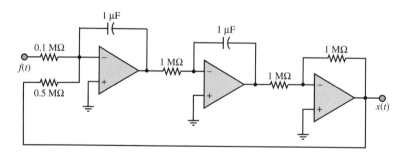

**CYU 12.17**      74 dB

**CYU 12.18**      $\omega_{3\,dB} = 2\pi \times 12{,}800$ rad/s

## HOMEWORK PROBLEMS

### Section 1: Ideal Amplifiers

**12.1**   The circuit shown in Figure P12.1 has a signal source, two stages of amplification, and a load. Determine, in dB, the power gain $G$ where:

$R_s = 0.6\ k\Omega$      $R_L = 0.6\ k\Omega$

$R_{i1} = 3\ k\Omega$      $R_{i2} = 3\ k\Omega$

$R_{o1} = 2\ k\Omega$      $R_{o2} = 2\ k\Omega$

$A_{VO1} = 100$      $G_{m2} = 350$ mΩ

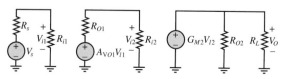

**Figure P12.1**

**12.2**  A temperature sensor in a production line under normal operating conditions produces a no-load (i.e., sensor current = 0) voltage:

$$v_s = V_{so} \cos(\omega t) \qquad R_s = 400 \ \Omega$$
$$V_{so} = 500 \ \text{mV} \qquad \omega = 6.28 \ \text{k rad/s}$$

The temperature is monitored on a display (the load) with a vertical line of light-emitting diodes. Normal conditions are indicated when a string of the bottommost diodes 2 cm in length is on. This requires that a voltage be supplied to the display input terminals where:

$$R_L = 12 \ \text{k}\Omega \qquad v_o = V_o \cos(\omega t) \qquad V_o = 6 \ \text{V}$$

The signal from the sensor must be amplified. Therefore, a voltage amplifier, shown in Figure P12.2, is connected between the sensor and CRT with:

$$R_i = 2 \ \text{k}\Omega \qquad R_o = 3 \ \text{k}\Omega$$

Determine the required no load gain of the amplifier.

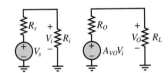

**Figure P12.2**

**12.3**  The circuit shown in Figure P12.3 has a signal source, two stages of amplification, and a load. Determine, in dB, the power gain $G = P_o/P_i$ where:

$$G = \frac{P_o}{P_i} \qquad P_i = \frac{V_{i1}^2}{R_{i1}}$$

$$R_s = 0.7 \ \text{k}\Omega \qquad R_L = 16 \ \Omega$$
$$R_{i1} = 1.1 \ \text{k}\Omega \qquad R_{i2} = 19 \ \text{k}\Omega$$
$$R_{o1} = 2.9 \ \text{k}\Omega \qquad R_{o2} = 22 \ \Omega$$
$$A_{VO1} = 65 \qquad G_{m2} = 130 \ \text{mS}$$

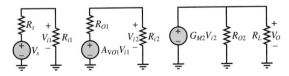

**Figure P12.3**

**12.4**  In the circuit of Figure 12.4,

$$R_s = 0.3 \ \text{k}\Omega \qquad R_L = 2 \ \text{k}\Omega$$
$$R_{i1} = R_{i2} = 7.7 \ \text{k}\Omega$$
$$R_{01} = R_{o2} = 1.3 \ \text{k}\Omega$$
$$A_{VO1} = A_{VO2} = 17$$

$$\frac{V_o}{V_{i1}} = 149.9$$

Determine:
a.  The power gain in dB.
b.  The overall voltage gain $v_o/v_s$.

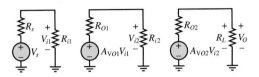

**Figure P12.4**

**12.5**  What approximations are usually made about the voltages and currents shown in Figure P12.5 for the ideal operational amplifier model?

**Figure P12.5**

**12.6**  What approximations are usually made about the circuit components and parameters shown in Figure P12.6 for the ideal op-amp model?

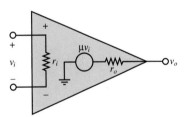

**Figure P12.6**

## Section 2: Op-Amp Circuits

**12.7**  Find $v_1$ in Figure P12.7(a) and (b). Note how the voltage follower holds $v_1$ in Figure P12.7(b) to $v_g/2$, while the 3-k$\Omega$ resistor "loads" the output in Figure P12.7(a).

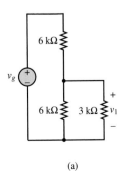

(a)

**Figure P12.7**

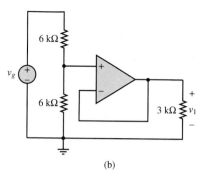

(b)

**Figure P12.7** Continued

**12.8** Determine an expression for the overall gain $A_V = v_o/v_i$ for the circuit of Figure P12.8. Assume the op-amp is ideal.

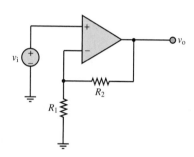

**Figure P12.8**

**12.9** In the circuit of Figure P12.9, find the current $i$.

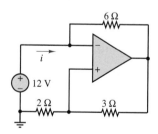

**Figure P12.9**

**12.10** Show that the circuit of Figure P12.10 is a noninverting summer.

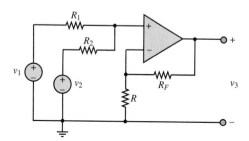

**Figure P12.10**

**12.11** Determine an expression for the overall gain $A_V = v_o/v_i$ for the circuit of Figure P12.11. Find the conductance $G = i_i/v_i$ seen by the voltage source, $v_i$. Assume the op-amps are ideal.

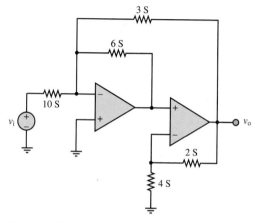

**Figure P12.11**

**12.12** Determine an expression for the overall gain $A_V = v_o/v_i$ for the circuit of Figure P12.12. Find the conductance $G = i_i/v_i$ seen by the voltage source, $v_i$. Assume the op-amp is ideal.

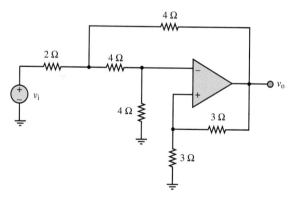

**Figure P12.12**

**12.13** In the circuit of Figure P12.13, it is critical that the gain remain within 2 percent of its nominal value, 16. Find the resistor, $R_S$, that will accomplish the nominal gain requirement, and state what the maximum and minimum values of $R_S$ can be. Will a *standard* 5 percent tolerance resistor be adequate to satisfy this requirement? (See Chapter 2 for resistor standard values.)

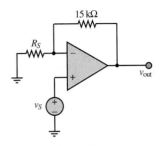

**Figure P12.13**

**12.14** An inverting amplifier uses two 10 percent tolerance resistors: $R_F = 33$ kΩ, and $R_S = 1.2$ kΩ.
a. What is the nominal gain of the amplifier?
b. What is the maximum value of $|A_V|$?
c. What is the minimum value of $|A_V|$?

**12.15** The circuit of Figure P12.15 will remove the DC portion of the input voltage, $v_1(t)$, while amplifying the AC portion. Let $v_1(t) = 10 + 10^{-3} \sin \omega t$ V, $R_F = 10$ kΩ, and $V_{batt} = 20$ V.
a. Find $R_S$ such that no DC voltage appears at the output.
b. What is $v_{out}(t)$, using $R_S$ from part a?

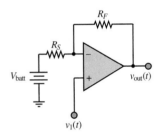

**Figure P12.15**

**12.16** Figure P12.16 shows a simple practical amplifier that uses the 741 op-amp. Pin numbers are as indicated. Assume the input resistance is $R = 2$ MΩ, the open-loop gain $K = 200,000$ and output resistance $R_o = 75$ Ω. Find the gain $A_V = v_o/v_i$ approximately.

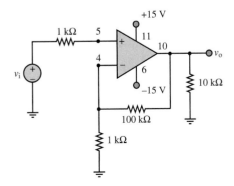

**Figure P12.16**

**12.17** Use an inverting summing amplifier to obtain the following weighted sum of four different signal sources:
$$v_{out} = -\left(\frac{1}{4} \sin \omega_1 t + 5 \sin \omega_2 t + 2 \sin \omega_3 t + 16 \sin \omega_4 t\right)$$
Assume that $R_F = 10$ kΩ, and determine the required source resistors.

**12.18** The amplifier shown in Figure P12.18 has a signal source, a load and one stage of amplification with:

$R_s = 11$ kΩ      $R_1 = 1$ kΩ
$R_F = 7$ kΩ      $R_L = 16$ Ω

Motorola MC1741C op amp:

$r_i = 2$ MΩ      $r_o = 25$ Ω
$\mu = 200,000$

In a first-approximation analysis, the op-amp parameters given above would be neglected and the op-amp modeled as an ideal device. In this problem, include their effects on the input resistance of the amplifier circuit.
a. Derive an expression for the input resistance $v_i/i_i$ including the effects of the op-amp.
b. Determine the value of the input resistance including the effects of the op-amp.
c. Determine the value of the input resistance assuming the op amp is ideal.

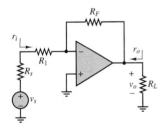

**Figure P12.18**

**12.19** In the circuit shown in Figure P12.19, if:
$R_1 = 50$ kΩ      $R_2 = 1.8$ kΩ
$R_F = 220$ kΩ
$v_s = 10^{-2} + 7 \times 10^{-6} \cos (\omega t)$ V

determine:
a. An expression for the output voltage.
b. The value of the output voltage.

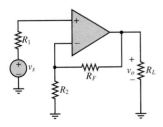

**Figure P12.19**

**12.20** If, in the circuit shown in Figure P12.20:

$$v_S = 17 \times 10^{-3} + 3 \times 10^{-3} \cos(\omega t)$$
$$R_s = 50 \ \Omega \qquad R_L = 200 \ \Omega$$

determine the output voltage.

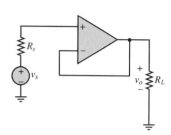

**Figure P12.20**

**12.21** In the circuit shown in Figure P12.21:

$$v_{S1} = 2.9 \times 10^{-3} \cos(\omega t)\text{V}$$
$$v_{S2} = 3.1 \times 10^{-3} \cos(\omega t)\text{V}$$
$$R_1 = 1 \ \text{k}\Omega \qquad R_2 = 3 \ \text{k}\Omega$$
$$R_3 = 13 \ \text{k}\Omega \qquad R_4 = 11 \ \text{k}\Omega$$

Determine the output voltage.

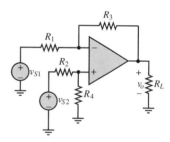

**Figure P12.21**

**12.22** In the circuit shown in Figure P12.21:

$$v_{S1} = 13 \ \text{mV} \qquad v_{S2} = 19 \ \text{mV}$$
$$R_1 = 1 \ \text{k}\Omega \qquad R_2 = 13 \ \text{k}\Omega$$
$$R_3 = 80 \ \text{k}\Omega \qquad R_4 = 68 \ \text{k}\Omega$$

Determine the output voltage.

**12.23** In the circuit shown in Figure P12.23, if:

$$v_{S1} = v_{S2} = 7 \ \text{mV}$$
$$R_F = 2.2 \ \text{k}\Omega \qquad R_1 = 850 \ \Omega$$
$$R_2 = 1.5 \ \text{k}\Omega$$

and the MC1741C op-amp has the following parameters:

$$r_i = 2 \ \text{M}\Omega \qquad \mu = 200{,}000$$
$$r_o = 25 \ \Omega$$

Determine the:

a. Output voltage.

b. Voltage gain for the two input signals.

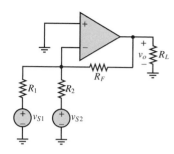

**Figure P12.23**

**12.24** In the circuit shown in Figure P12.21, the two voltage sources are temperature sensors with a response:

$$v_{S1} = kT_1 \qquad v_{S2} = kT_2$$

where:

$$k = 23 \ \text{mV/}^{\circ}\text{C}$$
$$R_1 = 11 \ \text{k}\Omega \qquad R_2 = 21 \ \text{k}\Omega$$
$$R_3 = 33 \ \text{k}\Omega \qquad R_4 = 56 \ \text{k}\Omega$$

$$T_1 = 35^{\circ}\text{C} \text{ and } T_2 = 100^{\circ}\text{C, determine:}$$

a. The output voltage.

b. The conditions required for the output voltage to depend oly on the difference between the two temperatures.

**12.25** In a differential amplifier, if:

$$A_{v1} = -20 \qquad A_{v2} = +22$$

derive expressions for and then determine the value of the common- and differential-mode gains.

**12.26** If, in the circuit shown in Figure P12.21:

$$v_{S1} = 1.3\text{V} \qquad v_{S2} = 1.9 \text{ V}$$
$$R_1 = R_2 = 5 \ \text{k}\Omega$$
$$R_3 = R_4 = 10 \ \text{k}\Omega \qquad R_L = 1.8 \ \text{k}\Omega$$

Determine:

a. The output voltage.

b. The common-mode component of the output voltage.

c. The differential-mode component of the output voltage.

**12.27** The two voltage sources shown in Figure P12.21 are pressure sensors where, for each source and with $P = $ pressure in $kPa$:

$$v_{S1,2} = A + BP_{1,2}$$
$$A = 0.3 \text{ V} \qquad B = 0.7 \ \tfrac{\text{V}}{\text{psi}}$$
$$R_1 = R_2 = 5 \ \text{k}\Omega$$
$$R_3 = R_4 = 10 \ \text{k}\Omega$$
$$R_L = 1.8 \ \text{k}\Omega$$

If $P_1 = 6 \ kPa$ and $P_2 = 5 \ kPa$, determine, using superposition, that part of the output voltage which is due to the:

a. Common mode input voltage.

b. Difference mode input voltage.

**12.28**  A linear potentiometer (variable resistor) $R_p$ is used to sense and give a signal voltage $v_y$ proportional to the current $y$ position of an $x$-$y$ plotter. A reference signal $v_R$ is supplied by the software controlling the plotter. The difference between these voltages must be amplified and supplied to a motor. The motor turns and changes the position of the pen and the position of the "pot" until the signal voltage is equal to the reference voltage (indicating the pen is in the desired position) and the motor voltage $= 0$. For proper operation the motor voltage must be 10 times the difference between the signal and reference voltage. For rotation in the proper direction, the motor voltage must be negative with respect to the signal voltage for the polarities shown. An additional requirement is that $i_P = 0$ to avoid loading the pot and causing an erroneous signal voltage.

a. Design an op-amp circuit which will achieve the specifications given. Redraw the circuit shown in Figure P12.28, replacing the box (drawn with dotted lines) with your circuit. Be sure to show how the signal voltage and output voltage are connected in your circuit.

b. Determine the value of each component in your circuit. The op-amp is a 741.

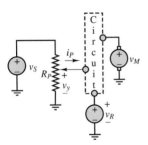

**Figure P12.28**

**12.29**  In the circuit shown in Figure P12.21:

$v_{S1} = 13 \text{ mV}$     $v_{S2} = 19 \text{ mV}$

$R_1 = 1 \text{ k}\Omega$     $R_2 = 13 \text{ k}\Omega$

$R_3 = 80 \text{ k}\Omega$     $R_4 = 68 \text{ k}\Omega$

Determine the output voltage.

**12.30**  Figure P12.30 shows a simple voltage-to-current converter. Show that the current $I_{out}$ through the light-emitting diode, and therefore its brightness, is proportional to the source voltage $V_s$ as long as $V_s > 0$.

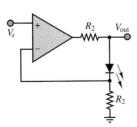

**Figure P12.30**

**12.31**  Figure P12.31 shows a simple current-to-voltage converter. Show that the voltage $V_{out}$ is proportional to the current generated by the cadmium sulfide cell. Also show that the transimpedance of the circuit $V_{out}/I_s$ is $-R$.

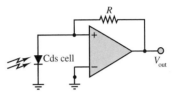

**Figure P12.31**

**12.32**  In some signal-processing applications, a *clamping* circuit is used to hold the output at a certain level even when the input continues to increase. One such circuit is shown in Figure P12.32. Assume the Zener diodes and op-amp are ideal. Determine the relationship between $v_o$ and $v_i$ and sketch it.

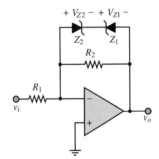

**Figure P12.32**

**12.33**  The circuit of Figure P12.33 serves as a voltage regulator whose output can be varied. Assume an ideal op-amp and that the Zener diode will hold its terminal voltage provided $i_Z \geq 0.1 I_Z$.

a. Find an expression for $v_o$ in terms of $V_Z$.

b. If $R_S$, $R_1$, $V_Z$, and $I_Z$ are known, specify the range of $V_S$ over which the circuit could regulate.

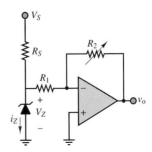

**Figure P12.33**

**12.34**  An op-amp voltmeter circuit as in Figure P12.34 is required to measure a maximum input of $E = 20$ mV. The op-amp input current is $I_B = 0.2\ \mu A$, and the meter circuit has $I_m = 100\ \mu A$ full-scale deflection and $r_m = 10\ k\Omega$. Determine suitable values for $R_3$ and $R_4$.

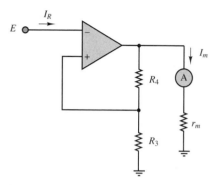

**Figure P12.34**

## Section 3: Filters, Integrators, and Differentiators

**12.35**  The circuit shown in Figure P12.35 is an active filter with:

$$C = 1\ \mu F \qquad R = 10\ k\Omega \qquad R_L = 1\ k\Omega$$

Determine:

a.  The gain (in dB) in the pass band.

b.  The cutoff frequency.

c.  If this is a low- or high-pass filter.

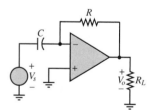

**Figure P12.35**

**12.36**  The op-amp circuit shown in Figure P12.36 is used as a filter.

$$C = 0.82\ \mu F \qquad R_L = 16\ \Omega$$
$$R_1 = 9.1\ k\Omega \qquad R_2 = 7.5\ k\Omega$$

Determine:

a.  If the circuit is a low- or high-pass filter.

b.  The gain, $V_o/V_s$, in dB in the pass-band, i.e., at the frequencies being passed by the filter.

c.  The cutoff frequency.

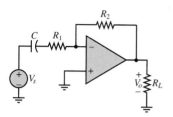

**Figure P12.36**

**12.37**  The op-amp circuit shown in Figure P12.36 is used as a filter.

$$C = 82\ pF \qquad R_L = 16\ \Omega$$
$$R_1 = 10\ k\Omega \qquad R_2 = 130\ k\Omega$$

Determine:

a.  If the circuit is a low- or high-pass filter.

b.  The gain, $V_o/V_s$, in dB in the pass-band, i.e., at the frequencies beign passed by the filter.

c.  The cutoff frequency.

**12.38**  The circuit shown in Figure 12.38 is an active filter with:

$$R_1 = 5\ k\Omega \qquad C = 8\ pF$$
$$R_2 = 68\ k\Omega \qquad R_L = 22\ k\Omega$$

Determine the cutoff frequencies and the magnitude of the voltage transfer function at very low and at very high frequencies.

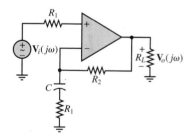

**Figure P12.38**

**12.39**  The circuit shown in Figure 12.39 is an active filter with:

$$R_1 = 1\ k\Omega \qquad R_2 = 5\ k\Omega$$
$$R_3 = 80\ k\Omega \qquad C = 5\ nF$$

Determine:

a.  An expression for the voltage transfer function in the standard form:

$$H_v(j\omega) = \frac{V_o(j\omega)}{V_i(j\omega)}$$

b.  The cutoff frequencies.

c.  The pass-band gain.

d.  The Bode plot.

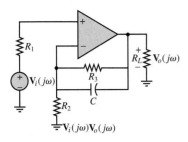

**Figure P12.39**

**12.40** The op-amp circuit shown in Figure P12.40 is used as a filter.

$R_1 = 9.1 \text{ k}\Omega$    $R_2 = 7.5 \text{ k}\Omega$
$C = 0.82 \,\mu\text{F}$    $R_L = 16 \,\Omega$

Determine:

a. If the circuit is a low- or high-pass filter.
b. An expression in standard form for the voltage transfer function.
c. The gain in dB in the pass-band, i.e., at the frequencies being passed by the filter, and the cutoff frequency.

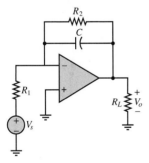

**Figure P12.40**

**12.41** The op-amp circuit shown in Figure P12.40 is a low-pass filter with:

$R_1 = 220 \,\Omega$    $R_2 = 68 \text{ k}\Omega$
$C = 0.47 \text{ nF}$    $R_L = 16 \,\Omega$

Determine:

a. An expression in standard form for the voltage transfer function.
b. The gain in dB in the pass-band, i.e., at the frequencies being passed by the filter, and the cutoff frequency.

**12.42** The circuit shown in Figure 12.42 is a band-pass filter. If

$R_1 = R_2 = 1 \text{ k}\Omega$
$C_1 = C_2 = 1 \,\mu\text{F}$

determine:

a. The pass-band gain.
b. The resonant frequency.
c. The cutoff frequencies.
d. The circuit $Q$.
e. The Bode plot.

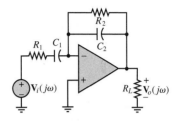

**Figure P12.42**

**12.43** The op-amp circuit shown in Figure P12.43 is a low-pass filter with:

$R_1 = 220 \,\Omega$    $R_2 = 68 \text{ k}\Omega$
$C = 0.47 \text{ nF}$    $R_L = 16 \,\Omega$

Determine:

a. An expression in standard form for the voltage transfer function.
b. The gain in dB in the pass-band, i.e., at the frequencies being passed by the filter, and the cutoff frequency.

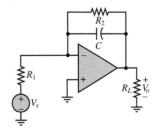

**Figure P12.43**

**12.44** The circuit shown in Figure P12.44 is a band-pass filter. If:

$R_1 = 220 \,\Omega$    $R_2 = 10 \text{ k}\Omega$
$C_1 = 2.2 \,\mu\text{F}$    $C_2 = 1 \text{ nF}$

determine the pass-band gain.

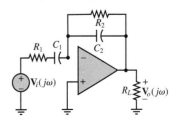

**Figure P12.44**

**12.45** Compute the frequency response of the circuit shown in Figure P12.45.

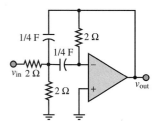

**Figure P12.45**

**12.46** The inverting amplifier shown in Figure P12.46 can be used as a low-pass filter.

a. Derive frequency response of the circuit.

b. If $R_1 = R_2 = 100$ k$\Omega$ and $C = 0.1$ $\mu$F, compute attenuation in dB at $\omega = 1,000$ rad/s.

c. Compute gain and phase at $\omega = 2,500$ rad/s.

d. Find the range of frequencies over which the attenuation is less than 1 dB.

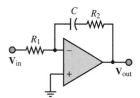

**Figure P12.46**

**12.47** Find an expression for the gain of the circuit of Figure P12.47.

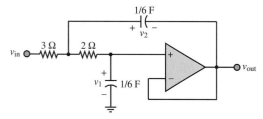

**Figure P12.47**

**12.48** For the circuit of Figure P12.48, sketch the amplitude response of $V_2/V_1$, indicating the half-power frequencies. Assume the op-amp is ideal.

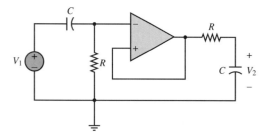

**Figure P12.48**

**12.49** Use the impedance concept to solve for the input impedance $Z_S = V_S/I_S$ of the circuit shown in Figure P12.49. Show that this input impedance has the equivalent form of an inductor $Z_L = j\omega L$. Find the simulated inductance in terms of $R_1$, $R_2$, $R_3$, $R_4$, and $C$.

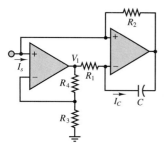

**Figure P12.49**

**12.50** The circuit shown in Figure P12.50(a) will give an output voltage which is either the integral or derivative of the source voltage shown in Figure P12.50(b) multiplied by some gain. If:

$$C = 1 \ \mu\text{F} \qquad R = 10 \ \text{k}\Omega \qquad R_L = 1 \ \text{k}\Omega$$

determine an expression for and plot the output voltage as a function of time.

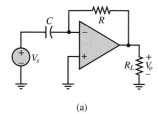

(a)

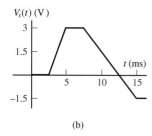

(b)

**Figure P12.50**

**12.51** The circuit shown in Figure P12.51(a) will give an output voltage which is either the integral or derivative of the supply voltage shown in Figure P12.51(b) multiplied by some gain.

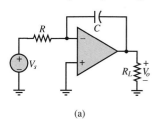

(a)

**Figure P12.51**

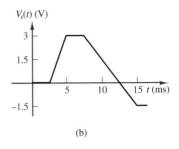

Figure P12.51 Continued

Determine:

a. An expression for the output voltage.
b. The value of the output voltage at $t = 5, 7.5, 12.5$, 15, and 20 ms and a plot of the output voltage as a function of time if:

$$C = 1\ \mu F \qquad R = 10\ k\Omega \qquad R_L = 1\ k\Omega$$

**12.52**   The circuit shown in Figure P12.52 is an integrator. The capacitor is initially uncharged, and the source voltage is

$$v_{in}(t) = 10 \times 10^{-3} + \sin(2{,}000\pi t)\ V$$

a. At $t = 0$, the switch, $S_1$, is closed. How long does it take before clipping occurs at the output if $R_s = 10\ k\Omega$ and $C_F = 0.008\ \mu F$?
b. At what times does the integration of the DC input cause the op-amp to saturate fully?

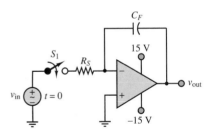

Figure P12.52

**12.53**   A practical integrator is shown in Figure 12.21 in the text. Note that the resistor in parallel with the feedback capacitor provides a path for the capacitor to discharge the DC voltage. Usually, the time constant $R_F C_F$ is chosen to be long enough not to interfere with the integration.

a. If $R_S = 10\ k\Omega$, $R_F = 2\ M\Omega$, $C_F = 0.008\ \mu F$, and $v_S(t) = 10\ V + \sin(2{,}000\pi t)\ V$, find $v_{out}(t)$ using phasor analysis.
b. Repeat part (a) if $R_F = 200\ k\Omega$, and if $R_F = 20\ k\Omega$.
c. Compare the time constants $R_F C_F$ with the period of the waveform for parts (a) and (b). What can you say about the time constant and the ability of the circuit to integrate?

**12.54**   The circuit of Figure 12.26 in the text is a practical differentiator. Assuming an ideal op-amp with $v_S(t) = 10 \times 10^{-3} \sin(2{,}000\pi t)$ V, $C_S = 100\ \mu F$, $C_F = 0.008\ \mu F$, $R_F = 2\ M\Omega$, and $R_S = 10\ k\Omega$,

a. Determine the frequency response, $V_o/V_S(\omega)$.
b. Use superposition to find the actual output voltage (remember that DC = 0 Hz).

**12.55**   Derive the differential equation corresponding to the analog computer simulation circuit of Figure P12.55.

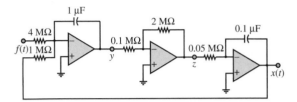

Figure P12.55

**12.56**   Construct the analog computer simulation corresponding to the following differential equation:

$$\frac{d^2x}{dt^2} + 100\frac{dx}{dt} + 10x = -5f(t)$$

## Section 4: Op-Amp Limitations

**12.57**   The ideal charge amplifier discussed in "Focus on Measurements: Charge Amplifiers" will saturate in the presence of any DC offsets, as discussed in Section 12.6. The circuit of Figure P12.57 represents a practical charge amplifier, in which the user is provided with a choice of three time constants—$\tau_{long} = R_L C_F$, $\tau_{medium} = R_M C_F$, $\tau_{short} = R_S C_F$—which can be selected by means of a switch. Assume that $R_L = 10\ M\Omega$, $R_M = 1\ M\Omega$, $R_S = 0.1\ M\Omega$, and $C_F = 0.1\ \mu F$. Analyze the frequency response of the practical charge amplifier for each case, and determine the lowest input signal frequency that can be amplified without excessive distortion for each case. Can this circuit amplify a DC signal?

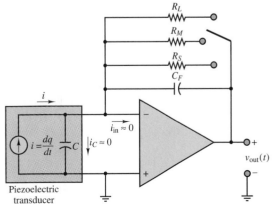

Figure P12.57

**12.58**  Consider a differential amplifier. We would desire the common-mode output to be less than 1 percent of the differential-mode output. Find the minimum dB common-mode rejection ratio (CMRR) to fulfill this requirement if the differential mode gain $A_{dm} = 1{,}000$. Let

$$v_1 = \sin(2{,}000\pi t) + 0.1\sin(120\pi t) \text{ V}$$

$$v_2 = \sin(2{,}000\pi t + 180°) + 0.1\sin(120\pi t) \text{ V}$$

$$v_o = A_{dm}(v_1 - v_2) + A_{cm}\left(\frac{v_1 + v_2}{2}\right)$$

**12.59**  Square wave testing can be used with operational amplifiers to estimate the *slew rate*, which is defined as the maximum rate at which the output can change (in $V/\mu s$). Input and output waveforms for a noninverting op-amp circuit are shown in Figure P12.59. As indicated, the rise time, $t_R$, of the output waveform is defined as the time it takes for that waveform to increase from 10 percent to 90 percent of its final value; i.e.,

$$t_R \overset{\triangle}{=} t_B - t_A = -\tau(\ln 0.1 - \ln 0.9) = 2.2\tau$$

where $\tau$ is the circuit time constant. Estimate the slew rate for the op-amp.

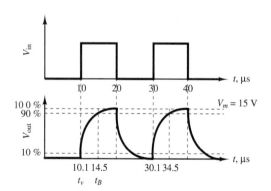

**Figure P12.59**

**12.60**  Consider an inverting amplifier with open-loop gain $10^5$. With reference to equation 12.18,

a. If $R_S = 10 \text{ k}\Omega$ and $R_F = 1 \text{ M}\Omega$, find the voltage gain $A_{V(CL)}$.

b. Repeat part a if $R_S = 10 \text{ k}\Omega$ and $R_F = 10 \text{ M}\Omega$.

c. Repeat part a if $R_S = 10 \text{ k}\Omega$ and $R_F = 100 \text{ M}\Omega$.

d. Using the resistor values of part c, find $A_{V(CL)}$ if $A_{V(OL)} \to \infty$.

**12.61**

a. If the op-amp of Figure P12.61 has an open-loop gain of $45 \times 10^5$, find the closed-loop gain for $R_F = R_S = 7.5 \text{ k}\Omega$, with reference to equation 12.18.

b. Repeat part a if $R_F = 5(R_S) = 37{,}500 \ \Omega$.

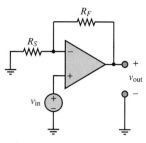

**Figure P12.61**

**12.62**  Given the unity-gain bandwidth for an ideal op-amp equal to 5.0 MHz, find the voltage gain at a frequency of $f = 500$ kHz.

**12.63**  The open-loop gain $A$ of real (nonideal) op-amps is very large at low frequencies but decreases markedly as frequency increases. As a result, the closed-loop gain of op-amp circuits can be strongly dependent on frequency. Determine the relationship between a finite and frequency-dependent open-loop gain $A_{V(OL)}(\omega)$ and the closed-loop gain $A_{V(CL)}(\omega)$ of an inverting amplifier as a function of frequency. Plot $A_{V(CL)}$ versus $\omega$. Notice that $-R_F/R_S$ is the low-frequency closed-loop gain.

**12.64**  A sinusoidal sound (pressure) wave $p(t)$ impinges upon a condenser microphone of sensitivity $S$ (mV/kPa). The voltage output of the microphone $v_s$ is amplified by two cascaded inverting amplifiers to produce an amplified signal $v_0$. Determine the peak amplitude of the sound wave (in dB) if $v_0 = 5 \ V_{RMS}$. Estimate the maximum peak magnitude of the sound wave in order that $v_0$ not contain any saturation effects of the op-amps.

**12.65**  If, in the circuit shown in Figure P12.65:

$$v_{S1} = 2.8 + 0.01\cos(\omega t) \text{ V}$$

$$v_{S2} = 3.5 - 0.007\cos(\omega t) \text{ V}$$

$$A_{v1} = -13 \qquad A_{v2} = 10 \qquad \omega = 4 \text{ krad/s}$$

determine the:

a. Common- and differential-mode input signals.

b. Common- and differential-mode gains.

c. Common- and differential-mode components of the output voltage.

d. Total output voltage.

e. Common-mode rejection ratio.

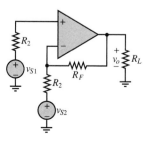

**Figure P12.65**

**12.66**  If, in the circuit shown in Figure P12.65:

$$v_{S1} = 3.5 + 0.01 \; \cos(\omega t) \; V$$
$$v_{S2} = 3.5 - 0.01 \; \cos(\omega t) \; V$$
$$A_{vc} = 10 \; dB \qquad A_{vd} = 20 \; dB$$
$$\omega = 4 \times 10^3 \; rad/s$$

determine the:

a.  Common- and differential-mode input voltages.
b.  The voltage gains for $v_{S1}$ and $v_{S2}$.
c.  Common-mode component and differential-mode component of the output voltage.
d.  The common-mode rejection ratio (CMRR) in dB.

**12.67**  If, in the circuit shown in Figure P12.67, the two voltage sources are temperature sensors with $T =$ temperature (Kelvin) and:

$$v_{S1} = kT_1 \qquad v_{S2} = kT_2$$

where:

$$k = 120 \; \mu V/K$$
$$R_1 = R_3 = R_4 = 5 \; k\Omega$$
$$R_2 = 3 \; k\Omega \qquad R_L = 600 \; \Omega$$

If:

$$T_1 = 310 \; K \qquad T_2 = 335 \; K$$

determine:

a.  The voltage gains for the two input voltages.
b.  The common-mode and differential-mode input voltage.
c.  The common-mode and differential-mode gains.

d.  The common-mode component and the differential-mode component of the output voltage.
e.  The common-mode rejection ratio (CMRR) in dB.

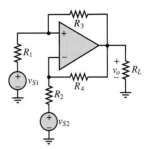

**Figure P12.67**

**12.68**  In the differential amplifier shown in Figure P12.67:

$$v_{S1} = 13 \; mV \qquad v_{S2} = 9 \; mV$$
$$v_o = v_{oc} + v_{od}$$
$$v_{oc} = 33 \; mV \qquad \text{(common-mode output voltage)}$$
$$v_{od} = 18 \; V \qquad \text{(differential-mode output voltage)}$$

Determine:

a.  The common-mode gain.
b.  The differential-mode gain.
c.  The common-mode rejection ratio in dB.

# C H A P T E R

# 13

# Digital Logic Circuits

igital computers have taken a prominent place in engineering and science over the last two decades, performing a number of essential functions such as numerical computations and data acquisition. It is not necessary to further stress the importance of these electronic systems in this book, since you are already familiar with personal computers and programming languages. The objective of the chapter is to discuss the essential features of digital logic circuits, which are at the heart of digital computers, by presenting an introduction to *combinational logic circuits.*

The chapter starts with a discussion of the binary number system, and continues with an introduction to Boolean algebra. The self-contained treatment of Boolean algebra will enable you to design simple logic functions using the techniques of combinational logic, and several practical examples are provided to demonstrate that even simple combinations of logic gates can serve to implement useful circuits in engineering practice. In a later section, we introduce a number of logic modules which can be described using simple logic gates but which provide more advanced functions. Among these, we discuss read-only memories, multiplexers, and decoders. Throughout the chapter, simple examples are given to demonstrate the usefulness of digital logic circuits in various engineering applications.

Chapter 13 provides the background needed to address the study of digital systems, which will be undertaken in Chapter 14. Upon completion of the chapter, you should be able to:

- Perform operations using the binary number system.
- Design simple combinational logic circuits using logic gates.
- Use Karnaugh maps to realize logical expressions.
- Interpret data sheets for multiplexers, decoders, and memory ICs.

## 13.1   ANALOG AND DIGITAL SIGNALS

One of the fundamental distinctions in the study of electronic circuits (and in the analysis of any signals derived from physical measurements) is that between analog and digital signals. As discussed in the preceding chapter, an **analog signal** is an electrical signal whose value varies in analogy with a physical quantity (e.g., temperature, force, or acceleration). For example, a voltage proportional to a measured variable pressure or to a vibration naturally varies in an analog fashion. Figure 13.1 depicts an analog function of time, $f(t)$. We note immediately that for each value of time, $t$, $f(t)$ can take one value among any of the values in a given range. For example, in the case of the output voltage of an op-amp, we expect the signal to take any value between $+V_{sat}$ and $-V_{sat}$, where $V_{sat}$ is the supply-imposed saturation voltage.

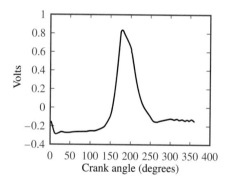

**Figure 13.1** Voltage analog of internal combustion engine in-cylinder pressure

A **digital signal,** on the other hand, can take only a *finite number of values.* This is an extremely important distinction, as will be shown shortly. An example of a digital signal is a signal that allows display of a temperature measurement on a digital readout. Let us hypothesize that the digital readout is three digits long and can display numbers from 0 to 100, and let us assume that the temperature sensor is correctly calibrated to measure temperatures from 0 to 100°F. Further, the output of the sensor ranges from 0 to 5 volts, where 0 V corresponds to 0°F and 5 V to 100°F. Therefore, the calibration constant of the sensor is

$$k_T = \frac{100° - 0°}{5 - 0} = 20° \text{ V}$$

Clearly, the output of the sensor is an analog signal; however, the display can show only a finite number of readouts (101, to be precise). Because the display itself can only take a value out of a discrete set of states—the integers from 0 to 100—we call it a digital display, indicating that the variable displayed is expressed in digital form.

Now, each temperature on the display corresponds to a *range of voltages:* each digit on the display represents one hundredth of the 5-volt range of the sensor, or 0.05 V = 50 mV. Thus, the display will read 0 if the sensor voltage is between 0 and 49 mV, 1 if it is between 50 and 99 mV, and so on. Figure 13.2 depicts the staircase function relationship between the analog voltage and the digital readout. This **quantization** of the sensor output voltage is in effect an approximation. If one wished to know the temperature with greater precision, a greater number of display digits could be employed.

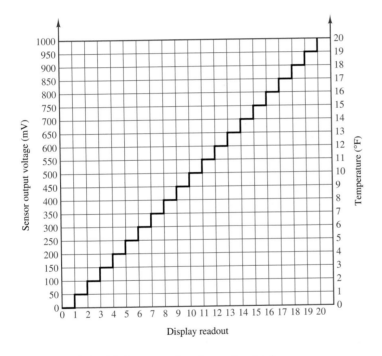

**Figure 13.2** Digital representation of an analog signal

The most common digital signals are binary signals. A **binary signal** is a signal that can take only one of two discrete values and is therefore characterized by transitions between two states. Figure 13.3 displays a typical binary signal. In binary arithmetic (which we discuss in the next section), the two discrete values $f_1$ and $f_0$ are represented by the numbers 1 and 0. In binary voltage waveforms, these values are represented by two voltage levels. For example, in the TTL convention (see Chapter 10), these values are (nominally) 5 V and 0 V, respectively; in CMOS circuits, these values can vary substantially. Other conventions are also used, including reversing the assignment—for example, by letting a 0-V level represent a logic 1 and a 5-V level represent a logic 0. Note that in a binary waveform, knowledge of the transition between one state and another (e.g., from $f_0$ to $f_1$ at

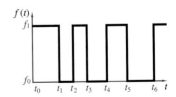

**Figure 13.3** A binary signal

$t = t_2$) is equivalent to knowledge of the state. Thus, digital logic circuits can operate by detecting transitions between voltage levels. The transitions are often called **edges** and can be positive ($f_0$ to $f_1$) or negative ($f_1$ to $f_0$). Virtually all of the signals handled by a computer are binary. From here on, whenever we speak of digital signals, you may assume that the text is referring to signals of the binary type, unless otherwise indicated.

## 13.2     THE BINARY NUMBER SYSTEM

The binary number system is a natural choice for representing the behavior of circuits that operate in one of two states (on or off, 1 or 0, or the like). The diode and transistor gates and switches studied in Chapter 10 fall in this category. Table 13.1 shows the correspondence between decimal and binary number systems for decimal numbers up to 16.

Binary numbers are based on powers of 2, whereas the decimal system is based on powers of 10. For example, the number 372 in the decimal system can be expressed as

$$372 = (3 \times 10^2) + (7 \times 10^1) + (2 \times 10^0)$$

while the binary number 10110 corresponds to the following combination of powers of 2:

$$10110 = (1 \times 2^4) + (0 \times 2^3) + (1 \times 2^2) + (1 \times 2^1) + (0 \times 2^0)$$

It is relatively simple to see the correspondence between the two number systems if we add the terms on the right-hand side of the previous expression. Let $n_2$ represent the number $n$ **base 2** (i.e., in the binary system) and $n_{10}$ the same number **base 10.** Then, our notation will be as follows:

$$10110_2 = 16 + 0 + 4 + 2 + 0 = 22_{10}$$

Note that a fractional number can also be similarly represented. For example, the number 3.25 in the decimal system may be represented as

$$3.25_{10} = 3 \times 10^0 + 2 \times 10^{-1} + 5 \times 10^{-2}$$

while in the binary system the number 10.011 corresponds to

$$10.011_2 = 1 \times 2^1 + 0 \times 2^0 + 0 \times 2^{-1} + 1 \times 2^{-2} + 1 \times 2^{-3}$$
$$= 2 + 0 + 0 + \tfrac{1}{4} + \tfrac{1}{8} = 2.375_{10}$$

Table 13.1 shows that it takes four binary digits, also called **bits,** to represent the decimal numbers up to 15. Usually, the rightmost bit is called the **least significant bit,** or **LSB,** and the leftmost bit is called the **most significant bit,** or **MSB.** Since binary numbers clearly require a larger number of digits than decimal numbers, the digits are usually grouped in sets of four, eight, or sixteen. Four bits are usually termed a **nibble,** eight bits are called a **byte,** and sixteen bits (or two bytes) form a **word.**

### Addition and Subtraction

The operations of addition and subtraction are based on the simple rules shown in Table 13.2. Note that, just as is done in the decimal system, a carry is generated

**Table 13.1** Conversion from decimal to binary

| Decimal number, $n_{10}$ | Binary number, $n_2$ |
|---|---|
| 0 | 0 |
| 1 | 1 |
| 2 | 10 |
| 3 | 11 |
| 4 | 100 |
| 5 | 101 |
| 6 | 110 |
| 7 | 111 |
| 8 | 1000 |
| 9 | 1001 |
| 10 | 1010 |
| 11 | 1011 |
| 12 | 1100 |
| 13 | 1101 |
| 14 | 1110 |
| 15 | 1111 |
| 16 | 10000 |

**Table 13.2** Rules for addition

$0 + 0 = 0$
$0 + 1 = 1$
$1 + 0 = 1$
$1 + 1 = 0$ (with a carry of 1)

whenever the sum of two digits exceeds the largest single-digit number in the given number system, which is 1 in the binary system. The carry is treated exactly as in the decimal system. A few examples of binary addition are shown in Figure 13.4, with their decimal counterparts.

| Decimal | Binary | Decimal | Binary | Decimal | Binary |
|---------|--------|---------|--------|---------|--------|
| 5 | 101 | 15 | 1111 | 3.25 | 11.01 |
| +6 | +110 | +20 | +10100 | +5.75 | +101.11 |
| 11 | 1011 | 35 | 100011 | 9.00 | 1001.00 |

(Note that in this example, $3.25 = 3\frac{1}{4}$ and $5.75 = 5\frac{3}{4}$.)

**Figure 13.4** Examples of binary addition

The procedure for subtracting binary numbers is based on the rules of Table 13.3. A few examples of binary subtraction are given in Figure 13.5, with their decimal counterparts.

| Decimal | Binary | Decimal | Binary | Decimal | Binary |
|---------|--------|---------|--------|---------|--------|
| 9 | 1001 | 16 | 10000 | 6.25 | 110.01 |
| −5 | −101 | −3 | −11 | −4.50 | −100.10 |
| 4 | 0100 | 13 | 01101 | 1.75 | 001.11 |

**Figure 13.5** Examples of binary subtraction

**Table 13.3** Rules for subtraction

$0 - 0 = 0$
$1 - 0 = 1$
$1 - 1 = 0$
$0 - 1 = 1$ (with a borrow of 1)

## Multiplication and Division

Whereas in the decimal system the multiplication table consists of $10^2 = 100$ entries, in the binary system we only have $2^2 = 4$ entries. Table 13.4 represents the complete multiplication table for the binary number system.

Division in the binary system is also based on rules analogous to those of the decimal system, with the two basic laws given in Table 13.5. Once again, we need be concerned with only two cases, and just as in the decimal system, division by zero is not contemplated.

**Table 13.4** Rules for multiplication

$0 \times 0 = 0$
$0 \times 1 = 0$
$1 \times 0 = 0$
$1 \times 1 = 1$

**Table 13.5** Rules for division

$0 \div 1 = 0$
$1 \div 1 = 1$

## Conversion from Decimal to Binary

The conversion of a decimal number to its binary equivalent is performed by successive division of the decimal number by 2, checking for the remainder each time. Figure 13.6 illustrates this idea with an example. The result obtained in Figure 13.6 may be easily verified by performing the opposite conversion, from binary to decimal:

$$110001 = 2^5 + 2^4 + 2^0 = 32 + 16 + 1 = 49$$

The same technique can be used for converting decimal fractional numbers to their binary form, provided that the whole number is separated from the fractional part and each is converted to binary form (separately), with the results added at the

| | Remainder |
|---|---|
| $49 \div 2 = 24$ | $+ 1$ |
| $24 \div 2 = 12$ | $+ 0$ |
| $12 \div 2 = 6$ | $+ 0$ |
| $6 \div 2 = 3$ | $+ 0$ |
| $3 \div 2 = 1$ | $+ 1$ |
| $1 \div 2 = 0$ | $+ 1$ |

$49_2 = 110001_2$

**Figure 13.6** Example of conversion from decimal to binary

| Remainder |
|---|
| $37 \div 2 = 18 + 1$ |
| $18 \div 2 = \phantom{0}9 + 0$ |
| $9 \div 2 = \phantom{0}4 + 1$ |
| $4 \div 2 = \phantom{0}2 + 0$ |
| $2 \div 2 = \phantom{0}1 + 0$ |
| $1 \div 2 = \phantom{0}0 + 1$ |

$$37_{10} = 100101_2$$

| |
|---|
| $2 \times 0.53 = 1.06 \rightarrow 1$ |
| $2 \times 0.06 = 0.12 \rightarrow 0$ |
| $2 \times 0.12 = 0.24 \rightarrow 0$ |
| $2 \times 0.24 = 0.48 \rightarrow 0$ |
| $2 \times 0.48 = 0.96 \rightarrow 0$ |
| $2 \times 0.96 = 1.92 \rightarrow 1$ |
| $2 \times 0.92 = 1.84 \rightarrow 1$ |
| $2 \times 0.84 = 1.68 \rightarrow 1$ |
| $2 \times 0.68 = 1.36 \rightarrow 1$ |
| $2 \times 0.36 = 0.72 \rightarrow 0$ |
| $2 \times 0.72 = 1.44 \rightarrow 1$ |

$$0.53_{10} = 0.10000111101$$

**Figure 13.7** Conversion from decimal to binary

end. Figure 13.7 outlines this procedure by converting the number 37.53 to binary form. The procedure is outlined in two steps. First, the integer part is converted; then, to convert the fractional part, one simple technique consists of multiplying the decimal fraction by 2 in successive stages. If the result exceeds 1, a 1 is needed to the right of the binary fraction being formed ($100101\ldots$, in our example). Otherwise, a 0 is added. This procedure is continued until no fractional terms are left. In this case, the decimal part is $0.53_{10}$, and Figure 13.7 illustrates the succession of calculations. Stopping the procedure outlined in Figure 13.7 after 11 digits results in the following approximation:

$$37.53_{10} = 100101.10000111101$$

Greater precision could be attained by continuing to add binary digits, at the expense of added complexity. Note that an infinite number of binary digits may be required to represent a decimal number *exactly*.

## Complements and Negative Numbers

To simplify the operation of subtraction in digital computers, **complements** are used almost exclusively. In practice, this corresponds to replacing the operation $X - Y$ with the operation $X + (-Y)$. This procedure results in considerable simplification, since the computer hardware need include only adding circuitry. Two types of complements are used with binary numbers: the **one's complement** and the **two's complement.**

The one's complement of an $n$-bit binary number is obtained by subtracting the number itself from $(2^n - 1)$. Two examples are as follows:

$$a = 0101$$
$$\text{One's complement of } a = (2^4 - 1) - a$$
$$= (1111) - (0101)$$
$$= 1010$$
$$b = 101101$$
$$\text{One's complement of } b = (2^6 - 1) - b$$
$$= (111111) - (101101)$$
$$= 010010$$

The two's complement of an $n$-bit binary number is obtained by subtracting the number itself from $2^n$. Two's complements of the same numbers $a$ and $b$ used in the preceding illustration are computed as follows:

$$a = 0101$$
$$\text{Two's complement of } a = 2^4 - a$$
$$= (10000) - (0101)$$
$$= 1011$$
$$b = 101101$$
$$\text{Two's complement of } b = 2^6 - b$$
$$= (1000000) - (101101)$$
$$= 010011$$

A simple rule that may be used to obtain the two's complement directly from a binary number is the following: Starting at the least significant (rightmost) bit, copy each bit *until the first 1 has been copied,* and then replace each successive 1 by a 0 and each 0 by a 1. You may wish to try this rule on the two previous examples to verify that it is much easier to use than the subtraction from $2^n$.

Different conventions exist in the binary system to represent whether a number is negative or positive. One convention, called the **sign-magnitude convention,** makes use of a *sign bit,* usually positioned at the beginning of the number, for which a value of 1 represents a minus sign and a value of 0, a plus sign. Thus, an eight-bit binary number would consist of a sign bit followed by seven *magnitude bits,* as shown in Figure 13.8(a). In a digital system that uses eight-bit signed integer words, we could represent integer numbers (decimal) in the range

$$-(2^7 - 1) \leq N \leq +(2^7 - 1)$$

or

$$-127 \leq N \leq +127$$

| Sign bit $b_7$ | $b_6$ | $b_5$ | $b_4$ | $b_3$ | $b_2$ | $b_1$ | $b_0$ |
|---|---|---|---|---|---|---|---|
| | ← Actual magnitude of binary number → | | | | | | |

(a)

| Sign bit $b_7$ | $b_6$ | $b_5$ | $b_4$ | $b_3$ | $b_2$ | $b_1$ | $b_0$ |
|---|---|---|---|---|---|---|---|
| | ← Actual magnitude of binary number (if $b_7 = 0$) →<br>← One's complement of binary number (if $b_7 = 1$) → | | | | | | |

(b)

| Sign bit $b_7$ | $b_6$ | $b_5$ | $b_4$ | $b_3$ | $b_2$ | $b_1$ | $b_0$ |
|---|---|---|---|---|---|---|---|
| | ← Actual magnitude of binary number (if $b_7 = 0$) →<br>← Two's complement of binary number (if $b_7 = 1$) → | | | | | | |

(c)

**Figure 13.8** (a) Eight-bit sign-magnitude binary number; (b) Eight-bit one's complement binary number; (c) Eight-bit two's complement binary number

A second convention uses the one's complement notation. In this convention, a sign bit is also used to indicate whether the number is positive (sign bit $= 0$) or negative (sign bit $= 1$). However, the magnitude of the binary number is represented by the true magnitude if the number is positive, and by its *one's complement* if the number is negative. Figure 13.8(b) illustrates the convention. For example, the number $(91)_{10}$ would be represented by the seven-bit binary number $(1011011)_2$ with a leading 0 (the sign bit): $(\mathbf{0}1011011)_2$. On the other hand, the number $(-91)_{10}$ would be represented by the seven-bit one's complement binary number $(0100100)_2$ with a leading 1 (the sign bit): $(\mathbf{1}0100100)_2$.

Most digital computers use the two's complement convention in performing integer arithmetic operations. The two's complement convention represents positive numbers by a sign bit of 0, followed by the true binary magnitude; negative numbers are represented by a sign bit of 1, *followed by the two's complement of the binary number,* as shown in Figure 13.8(c). The advantage of the two's complement convention is that the algebraic sum of two's complement binary numbers is carried out very simply by adding the two numbers *including the sign bit.* Example 13.1 illustrates two's complement addition.

## EXAMPLE 13.1 Two's Complement Operations

**Problem**

Perform the following subtractions using two's complement arithmetic:

1. $X - Y = 1011100 - 1110010$
2. $X - Y = 10101111 - 01110011$

**Solution**

*Analysis:* The two's complement subtractions are performed by replacing the operation $X - Y$ with the operation $X + (-Y)$. Thus, we first find the two's complement of $Y$ and add the result to $X$ in each of the two cases:

$$X - Y = 1011100 - 1110010 = 1011100 + (2^7 - 1110010)$$
$$= 1011100 + 0001110 = 1101010$$

Next, we add the *sign bit* (in boldface type) in front of each number (1 in first case since the difference $X - Y$ is a negative number):

$$X - Y = \mathbf{1}1101010$$

Repeating for the second subtraction gives:

$$X - Y = 10101111 - 01110011 = 10101111 + (2^8 - 01110011) = 10101111$$
$$+10001101 = 00111100$$
$$= \mathbf{0}00111100$$

where the first digit is a 0 because $X - Y$ is a positive number.

## The Hexadecimal System

Table 13.6  Hexa-decimal code

| | |
|---|---|
| 0 | 0000 |
| 1 | 0001 |
| 2 | 0010 |
| 3 | 0011 |
| 4 | 0100 |
| 5 | 0101 |
| 6 | 0110 |
| 7 | 0111 |
| 8 | 1000 |
| 9 | 1001 |
| A | 1010 |
| B | 1011 |
| C | 1100 |
| D | 1101 |
| E | 1110 |
| F | 1111 |

It should be apparent by now that representing numbers in base 2 and base 10 systems is purely a matter of convenience, given a specific application. Another base frequently used is the **hexadecimal system,** a direct derivation of the binary number system. In the hexadecimal (or hex) code, the bits in a binary number are subdivided into groups of four. Since there are 16 possible combinations for a four-bit number, the natural digits in the decimal system (0 through 9) are insufficient to represent a hex digit. To solve this problem, the first six letters of the alphabet are used, as shown in Table 13.6. Thus, in hex code, an eight-bit word corresponds to just two digits; for example:

$$1010\ 0111_2 = A7_{16}$$
$$0010\ 1001_2 = 29_{16}$$

## Binary Codes

In this subsection, we describe two common binary codes that are often used for practical reasons. The first is a method of representing decimal numbers in digital logic circuits that is referred to as **binary-coded decimal,** or **BCD, representation.** In effect, the simplest BCD representation is just a sequence of four-bit binary numbers that stops after the first 10 entries, as shown in Table 13.7. There are

Table 13.7  BCD code

| | |
|---|---|
| 0 | 0000 |
| 1 | 0001 |
| 2 | 0010 |
| 3 | 0011 |
| 4 | 0100 |
| 5 | 0101 |
| 6 | 0110 |
| 7 | 0111 |
| 8 | 1000 |
| 9 | 1001 |

Table 13.8  Three-bit Gray code

| Binary | Gray |
|---|---|
| 000 | 000 |
| 001 | 001 |
| 010 | 011 |
| 011 | 010 |
| 100 | 110 |
| 101 | 111 |
| 110 | 101 |
| 111 | 100 |

also other BCD codes, all reflecting the same principle: that each decimal digit is represented by a fixed-length binary word. One should realize that although this method is attractive because of its direct correspondence with the decimal system, it is not efficient. Consider, for example, the decimal number 68. Its binary representation by direct conversion is the seven-bit number 1000100. On the other hand, the corresponding BCD representation would require eight bits:

$$68_{10} = 01101000_{BCD}$$

Another code that finds many applications is the **Gray code.** This is simply a reshuffling of the binary code with the property that any two consecutive numbers differ only by one bit. Table 13.8 illustrates the three-bit Gray code. The Gray code can be very useful in practical applications, because in counting up or down according to this code, the binary representation of a number changes only one bit at a time. The next example illustrates an application of the Gray code to a practical engineering problem.

## Digital Position Encoders

FIND IT ON THE WEB

FOCUS ON MEASUREMENTS

**Position encoders** are devices that output a digital signal proportional to their (linear or angular) position. These devices are very useful in measuring instantaneous position in *motion control* applications. Motion control is a technique that is used when it is necessary to accurately control the motion of a moving object; examples are found in robotics, machine tools, and servomechanisms. For example, in positioning the arm of a robot to pick up an object, it is very important to know its exact position at all times. Since one is usually interested in both rotational and translational motion, two types of encoders are discussed in this example: *linear* and *angular* position encoders.

An optical position encoder consists of an *encoder pad,* which is either a strip (for translational motion) or a disk (for rotational motion) with alternating black and white areas. These areas are arranged to reproduce some binary code, as shown in Figure 13.9, where both the conventional binary and Gray codes are depicted for a four-bit linear encoder pad. A fixed array of photodiodes (see Chapter 8) senses the reflected light from each of the cells across a row of the encoder path; depending on the amount of light

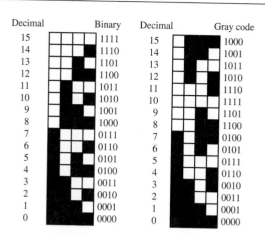

| Decimal | | Binary | Decimal | | Gray code |
|---|---|---|---|---|---|
| 15 | | 1111 | 15 | | 1000 |
| 14 | | 1110 | 14 | | 1001 |
| 13 | | 1101 | 13 | | 1011 |
| 12 | | 1100 | 12 | | 1010 |
| 11 | | 1011 | 11 | | 1110 |
| 10 | | 1010 | 10 | | 1111 |
| 9 | | 1001 | 9 | | 1101 |
| 8 | | 1000 | 8 | | 1100 |
| 7 | | 0111 | 7 | | 0100 |
| 6 | | 0110 | 6 | | 0101 |
| 5 | | 0101 | 5 | | 0111 |
| 4 | | 0100 | 4 | | 0110 |
| 3 | | 0011 | 3 | | 0010 |
| 2 | | 0010 | 2 | | 0011 |
| 1 | | 0001 | 1 | | 0001 |
| 0 | | 0000 | 0 | | 0000 |

**Figure 13.9** Binary and Gray code patterns for linear position encoders

reflected, each photodiode circuit will output a voltage corresponding to a binary 1 or 0. Thus, a different four-bit word is generated for each row of the encoder.

Suppose the encoder pad is 100 mm in length. Then its resolution can be computed as follows. The pad will be divided into $2^4 = 16$ segments, and each segment corresponds to an increment of $100/16$ mm $= 6.25$ mm. If greater resolution were necessary, more bits could be employed: an eight-bit pad of the same length would attain a resolution of $100/256$ mm $= 0.39$ mm.

A similar construction can be employed for the five-bit angular encoder of Figure 13.10. In this case, the angular resolution can be expressed in degrees of rotation, where $2^5 = 32$ sections correspond to $360°$. Thus, the resolution would be $360°/32 = 11.25°$. Once again, greater angular resolution could be obtained by employing a larger number of bits.

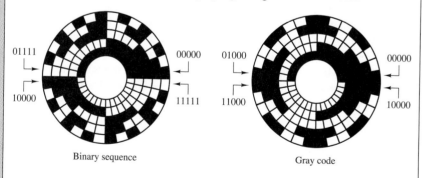

Binary sequence                      Gray code

**Figure 13.10** Binary and Gray code patterns for angular position encoders

## EXAMPLE   13.2  Conversion from Binary to Hexadecimal

### Problem

Convert the following binary numbers to hexadecimal form.

1.  100111
2.  1011101
3.  11001101
4.  101101111001
5.  100110110
6.  1101011011

---

## Solution

**Analysis:**  A simple method for binary to hexadecimal conversion consists of grouping each binary number into four-bit groups, and then performing the conversion for each four-bit word following Table 13.6:

1.  $100111_2 = 0010_2 0111_2 = 27_{16}$
2.  $1011101_2 = 0101_2 1101_2 = 5D_{16}$
3.  $11001101_2 = 1100_2 1101_2 = CD_{16}$
4.  $101101111001_2 = 1011_2 0111_2 1001_2 = B79_{16}$
5.  $100110110_2 = 0001_2 0011_2 0100_2 = 136_{16}$
6.  $1101011011_2 = 0011_2 0101_2 1011_2 = 35B_{16}$

**Comments:**  Note that we start grouping always from the right-hand side. The reverse process is equally easy: To convert from hexadecimal to binary, replace each hexadecimal number with the equivalent four-bit binary word.

---

# Check Your Understanding

**13.1**  Convert the following decimal numbers to binary form:

    a.   39           b.   59

    c.   512         d.   0.4475

    e.   $\frac{25}{32}$         f.   0.796875

    g.   256.75     h.   129.5625

    i.   4,096.90625

**13.2**  Convert the following binary numbers to decimal:

    a.   1101         b.   11011

    c.   10111       d.   0.1011

    e.   0.001101    f.   0.001101101

    g.   111011.1011   h.   1011011.001101

    i.   10110.0101011101

**13.3**  Perform the following additions and subtractions. Express the answer in decimal form for problems (a)–(d) and in binary form for problems (e)–(h).

    a.   $1001.1_2 + 1011.01_2$   b.   $100101_2 + 100101_2$

    c.   $0.1011_2 + 0.1101_2$   d.   $1011.01_2 + 1001.11_2$

    e.   $64_{10} - 32_{10}$     f.   $127_{10} - 63_{10}$

    g.   $93.5_{10} - 42.75_{10}$   h.   $(84\frac{9}{32})_{10} - (48\frac{5}{16})_{10}$

**13.4**  How many possible numbers can be represented in a 12-bit word?

**13.5**   If we use an eight-bit word with a sign bit (seven magnitude bits plus one sign bit) to represent voltages $-5$ V and $+5$ V, what is the smallest increment of voltage that can be represented?

**13.6**   Convert the following numbers from hex to binary or from binary to hex:

    a.   F83        b.   3C9

    c.   A6         d.   $110101110_2$

    e.   $10111001_2$   f.   $11011101101_2$

**13.7**   Find the two's complement of the following binary numbers:

    a. 11101001   b. 10010111   c. 1011110

**13.8**   Convert the following numbers from hex to binary, and find their two's complements:

    a. F43   b. 2B9   c. A6

## 13.3   BOOLEAN ALGEBRA

The mathematics associated with the binary number system (and with the more general field of logic) is called *Boolean,* in honor of the English mathematician George Boole, who published a treatise in 1854 entitled *An Investigation of the Laws of Thought, on Which Are Founded the Mathematical Theories of Logic and Probabilities.* The development of a *logical algebra,* as Boole called it, is one of the results of his investigations. The variables in a Boolean, or logic, expression can take only one of two values, usually represented by the numbers 0 and 1. These variables are sometimes referred to as true (1) and false (0). This convention is normally referred to as **positive logic.** There is also a **negative logic** convention in which the roles of logic 1 and logic 0 are reversed. In this book we shall employ only positive logic.

    Analysis of **logic functions,** that is, functions of logical (Boolean) variables, can be carried out in terms of truth tables. A truth table is a listing of all the possible values each of the Boolean variables can take, and of the corresponding value of the desired function. In the following paragraphs we shall define the basic logic functions upon which Boolean algebra is founded, and we shall describe each in terms of a set of rules and a truth table; in addition, we shall also introduce **logic gates.** Logic gates are physical devices (see Chapter 10) that can be used to implement logic functions.

### AND and OR Gates

The basis of **Boolean algebra** lies in the operations of **logical addition,** or the **OR** operation; and **logical multiplication,** or the **AND** operation. Both of these find a correspondence in simple logic gates, as we shall presently illustrate. Logical addition, although represented by the symbol $+$, differs from conventional algebraic addition, as shown in the last rule listed in Table 13.9. Note that this rule also differs from the last rule of binary addition studied in the previous section. Logical addition can be represented by the logic gate called an **OR gate,** whose symbol and whose inputs and outputs are shown in Figure 13.11. The OR gate represents the following logical statement:

    If either $X$ or $Y$ is true (1), then $Z$ is true(1).         **(13.1)**

**Table 13.9** Rules for logical addition (OR)

$0 + 0 = 0$
$0 + 1 = 1$
$1 + 0 = 1$
$1 + 1 = 1$

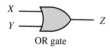

OR gate

| X | Y | Z |
|---|---|---|
| 0 | 0 | 0 |
| 0 | 1 | 1 |
| 1 | 0 | 1 |
| 1 | 1 | 1 |

Truth table

**Figure 13.11** Logical addition and the OR gate

This rule is embodied in the electronic gates discussed in Chapter 9, in which a logic 1 corresponds, say, to a 5-V signal and a logic 0 to a 0-V signal.

Logical multiplication is denoted by the center dot (·) and is defined by the rules of Table 13.10. Figure 13.12 depicts the **AND gate**, which corresponds to this operation. The AND gate corresponds to the following logical statement:

If both $X$ and $Y$ are true (1), then $Z$ is true (1). **(13.2)**

One can easily envision logic gates (AND and OR) with an arbitrary number of inputs; three- and four-input gates are not uncommon.

The rules that define a logic function are often represented in tabular form by means of a **truth table.** Truth tables for the AND and OR gates are shown in Figures 13.11 and 13.12. A truth table is nothing more than a tabular summary of all of the possible outputs of a logic gate, given all the possible input values. If the number of inputs is 3, the number of possible combinations grows from 4 to 8, but the basic idea is unchanged. Truth tables are very useful in defining logic functions. A typical logic design problem might specify requirements such as "the output $Z$ shall be logic 1 only when the condition ($X = 1$ AND $Y = 1$) OR ($W = 1$) occurs, and shall be logic 0 otherwise." The truth table for this particular logic function is shown in Figure 13.13 as an illustration. The design consists, then, of determining the combination of logic gates that exactly implements the required logic function. Truth tables can greatly simplify this procedure.

The AND and OR gates form the basis of all logic design in conjunction with the **NOT gate.** The NOT gate is essentially an inverter (which can be constructed using bipolar or field-effect transistors, as discussed in Chapter 10), and it provides the complement of the logic variable connected to its input. The complement of a logic variable $X$ is denoted by $\overline{X}$. The NOT gate has only one input, as shown in Figure 13.14.

To illustrate the use of the NOT gate, or inverter, we return to the design example of Figure 13.13, where we required that the output of a logic circuit be $Z = 1$ only if $X = 0$ AND $Y = 1$ OR if $W = 1$. We recognize that except for the requirement $X = 0$, this problem would be identical if we stated it as follows: "The output $Z$ shall be logic 1 only when the condition $(\overline{X} = 1$ AND $Y = 1)$ OR $(W = 1)$ occurs, and shall be logic 0 otherwise." If we use an inverter to convert $X$ to $\overline{X}$, we see that the required condition becomes $(\overline{X} = 1$ AND $Y = 1)$ OR $(W = 1)$. The formal solution to this elementary design exercise is illustrated in Figure 13.15.

In the course of the discussion of logic gates, extensive use will be made of truth tables to evaluate logic expressions. A set of basic rules will facilitate this task. Table 13.11 lists some of the rules of Boolean algebra; each of these can be proven by using a truth table, as will be shown in examples and exercises. An example proof for rule 16 is given in Figure 13.16 in the form of a truth table. This technique can be employed to prove any of the laws of Table 13.11. From the simple truth table in Figure 13.16, which was obtained step by step, we can clearly see that indeed $X \cdot (X + Y) = X$. This methodology for proving the validity of logical equations is called **proof by perfect induction.** The 19 rules of Table 13.11 can be used to simplify logic expressions.

To complete the introductory material on Boolean algebra, a few paragraphs need to be devoted to two very important theorems, called **De Morgan's theorems.**

**Table 13.10** Rules for logical multiplication (AND)

$0 \cdot 0 = 0$
$0 \cdot 1 = 0$
$1 \cdot 0 = 0$
$1 \cdot 1 = 1$

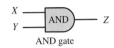

AND gate

| $X$ | $Y$ | $Z$ |
|---|---|---|
| 0 | 0 | 0 |
| 0 | 1 | 0 |
| 1 | 0 | 0 |
| 1 | 1 | 1 |

Truth table

**Figure 13.12** Logical multiplication and the AND gate

Logic gate realization of the statement "the output $Z$ shall be logic 1 only when the condition $(X = 1$ AND $Y = 1)$ OR $(W = 1)$ occurs, and shall be logic 0 otherwise."

| $X$ | $Y$ | $W$ | $Z$ |
|---|---|---|---|
| 0 | 0 | 0 | 0 |
| 0 | 0 | 1 | 1 |
| 0 | 1 | 0 | 0 |
| 0 | 1 | 1 | 1 |
| 1 | 0 | 0 | 0 |
| 1 | 0 | 1 | 1 |
| 1 | 1 | 0 | 1 |
| 1 | 1 | 1 | 1 |

Truth table

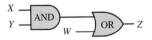

Solution using logic gates

**Figure 13.13** Example of logic function implementation with logic gates

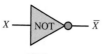

NOT gate

| $X$ | $\overline{X}$ |
|---|---|
| 1 | 0 |
| 0 | 1 |

Truth table for NOT gate

**Figure 13.14** Complements and the NOT gate

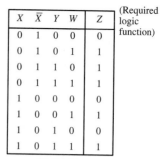

Truth table

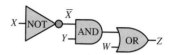

Solution using logic gates

**Figure 13.15** Solution of a logic problem using logic gates

Table 13.11   Rules of Boolean algebra

1. $0 + X = X$
2. $1 + X = 1$
3. $X + X = X$
4. $X + \overline{X} = 1$
5. $0 \cdot X = 0$
6. $1 \cdot X = X$
7. $X \cdot X = X$
8. $X \cdot \overline{X} = 0$
9. $\overline{\overline{X}} = X$
10. $X + Y = Y + X$     } Commutative law
11. $X \cdot Y = Y \cdot X$
12. $X + (Y + Z) = (X + Y) + Z$     } Associative law
13. $X \cdot (Y \cdot Z) = (X \cdot Y) \cdot Z$
14. $X \cdot (Y + Z) = X \cdot Y + X \cdot Z$      Distributive law
15. $X + X \cdot Z = X$      Absorption law
16. $X \cdot (X + Y) = X$
17. $(X + Y) \cdot (X + Z) = X + Y \cdot Z$
18. $X + \overline{X} \cdot Y = X + Y$
19. $X \cdot Y + Y \cdot Z + \overline{X} \cdot Z = X \cdot Y + \overline{X} \cdot Z$

| $X$ | $Y$ | $(X+Y)$ | $X \cdot (X+Y)$ |
|---|---|---|---|
| 0 | 0 | 0 | 0 |
| 0 | 1 | 1 | 0 |
| 1 | 0 | 1 | 1 |
| 1 | 1 | 1 | 1 |

**Figure 13.16** Proof of rule 16 by perfect induction

These are stated here in the form of logic functions:

$$\overline{(X + Y)} = \overline{X} \cdot \overline{Y} \tag{13.3}$$

$$\overline{(X \cdot Y)} = \overline{X} + \overline{Y} \tag{13.4}$$

These two laws state a very important property of logic functions:

Any logic function can be implemented using only OR and NOT gates, or using only AND and NOT gates.

De Morgan's laws can easily be visualized in terms of logic gates, as shown in Figure 13.17. The associated truth tables are proof of these theorems.

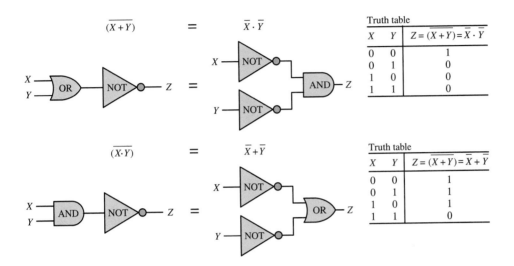

Truth table

| X | Y | $Z = \overline{(X+Y)} = \overline{X} \cdot \overline{Y}$ |
|---|---|---|
| 0 | 0 | 1 |
| 0 | 1 | 0 |
| 1 | 0 | 0 |
| 1 | 1 | 0 |

Truth table

| X | Y | $Z = \overline{(X+Y)} = \overline{X} + \overline{Y}$ |
|---|---|---|
| 0 | 0 | 1 |
| 0 | 1 | 1 |
| 1 | 0 | 1 |
| 1 | 1 | 0 |

**Figure 13.17** De Morgan's laws

The importance of De Morgan's laws is in the statement of the **duality** that exists between AND and OR operations: any function can be realized by just one of the two basic operations, plus the complement operation. This gives rise to two families of logic functions: **sums of products** and **product of sums,** as shown in Figure 13.18. Any logical expression can be reduced to either one of these two forms. Although the two forms are equivalent, it may well be true that one of the two has a simpler implementation (fewer gates). Example 13.3 illustrates this point.

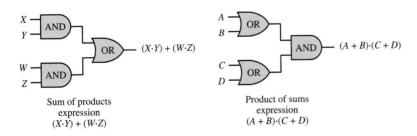

**Figure 13.18** Sum-of-products and product-of-sums logic functions

### EXAMPLE 13.3 Simplification of Logical Expression

**Problem**

Using the rules of Table 13.11, simplify the following function using the rules of Boolean algebra.

$$f(A, B, C, D) = \overline{A} \cdot \overline{B} \cdot D + \overline{A} \cdot B \cdot D + B \cdot C \cdot D + A \cdot C \cdot D$$

**Solution**

*Find:* Simplified expression for logical function of four variables.

*Analysis:*

$$
\begin{aligned}
f &= \overline{A} \cdot \overline{B} \cdot D + \overline{A} \cdot B \cdot D + B \cdot C \cdot D + A \cdot C \cdot D \\
&= \overline{A} \cdot D \cdot (\overline{B} + B) + B \cdot C \cdot D + A \cdot C \cdot D && \text{Rule 14} \\
&= \overline{A} \cdot D + B \cdot C \cdot D + A \cdot C \cdot D && \text{Rule 4} \\
&= (\overline{A} + A \cdot C) \cdot D + B \cdot C \cdot D && \text{Rule 14} \\
&= (\overline{A} + C) \cdot D + B \cdot C \cdot D && \text{Rule 18} \\
&= \overline{A} \cdot D + C \cdot D + B \cdot C \cdot D && \text{Rule 14} \\
&= \overline{A} \cdot D + C \cdot D \cdot (1 + B) && \text{Rule 14} \\
&= \overline{A} \cdot D + C \cdot D = (\overline{A} + C) \cdot D && \text{Rules 2 and 6}
\end{aligned}
$$

---

**FOCUS ON MEASUREMENTS**

### Fail-Safe Autopilot Logic

This example aims to illustrate the significance of De Morgan's laws and of the duality of the sum-of-products and product-of-sums forms. Suppose that a fail-safe autopilot system in a commercial aircraft requires that, prior to initiating a takeoff or landing maneuver, the following check must be passed: two of three possible pilots must be available. The three possibilities are the pilot, the co-pilot, and the autopilot. Imagine further that there exist switches in the pilot and co-pilot seats that are turned on by the weight of the crew, and that a self-check circuit exists to verify the proper operation of the autopilot system. Let the variable $X$ denote the pilot state (1 if the pilot is sitting at the controls), $Y$ denote the same condition for the co-pilot, and $Z$ denote the state of the autopilot, where $Z = 1$ indicates that the autopilot is functioning. Then, since we wish two of these conditions to be active before the maneuver can be initiated, the logic function corresponding to "system ready" is:

$$f = X \cdot Y + X \cdot Z + Y \cdot Z$$

This can also be verified by the truth table shown below.

| Pilot | Co-pilot | Autopilot | System ready |
|:-----:|:--------:|:---------:|:------------:|
| 0 | 0 | 0 | 0 |
| 0 | 0 | 1 | 0 |
| 0 | 1 | 0 | 0 |
| 0 | 1 | 1 | 1 |
| 1 | 0 | 0 | 0 |
| 1 | 0 | 1 | 1 |
| 1 | 1 | 0 | 1 |
| 1 | 1 | 1 | 1 |

The function $f$ defined above is based on the notion of a *positive check;* that is, it indicates when the system is ready. Let us now apply De Morgan's laws to the function $f$, which is in sum-of-products form:

$$\overline{f} = g = \overline{X \cdot Y + X \cdot Z + Y \cdot Z} = (\overline{X} + \overline{Y}) \cdot (\overline{X} + \overline{Z}) \cdot (\overline{Y} + \overline{Z})$$

The function $g$, in product-of-sums form, conveys exactly the same information as the function $f$, but it performs a negative check; in other words, $g$ verifies the *system not ready condition.* You see then that whether one chooses to implement the function in one form or another is simply a matter of choice; the two forms give exactly the same information.

---

## EXAMPLE 13.4 Realizing Logic Functions from Truth Tables

### Problem

Realize the logic function described by the truth table below.

| $A$ | $B$ | $C$ | $y$ |
|:---:|:---:|:---:|:---:|
| 0 | 0 | 0 | 0 |
| 0 | 0 | 1 | 1 |
| 0 | 1 | 0 | 0 |
| 0 | 1 | 1 | 1 |
| 1 | 0 | 0 | 1 |
| 1 | 0 | 1 | 1 |
| 1 | 1 | 0 | 1 |
| 1 | 1 | 1 | 1 |

---

### Solution

**Known Quantities:** Value of function $y(A, B, C)$ for each possible combination of logical variables $A, B, C$.

**Find:** Logical expression realizing the function $y$.

*Analysis:* To determine a logical expression for the function $y$, we first need to convert the truth table into a logical expression. We do so by expressing $y$ as the sum of the products of the three variables for each combination that yields $y = 1$. If the value of a variable is 1, we use the uncomplemented variable. If it's 0, we use the complemented variable. For example, the second row (first instance of $y = 1$) would yield the term $\overline{A} \cdot \overline{B} \cdot C$. Thus,

$$y = \overline{A} \cdot \overline{B} \cdot C + \overline{A} \cdot B \cdot C + A \cdot \overline{B} \cdot \overline{C} + A \cdot \overline{B} \cdot C + A \cdot B \cdot \overline{C} + A \cdot B \cdot C$$
$$= \overline{A} \cdot C(\overline{B} + B) + A \cdot \overline{B} \cdot (\overline{C} + C) + A \cdot B \cdot (\overline{C} + C)$$
$$= \overline{A} \cdot C + A \cdot \overline{B} + A \cdot B = \overline{A} \cdot C + A \cdot (\overline{B} + B) = \overline{A} \cdot C + A = A + C.$$

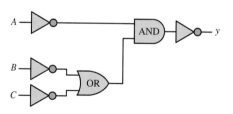

$A + C = y$　or

Thus, the function is a two-input OR gate, as shown in Figure 13.19.

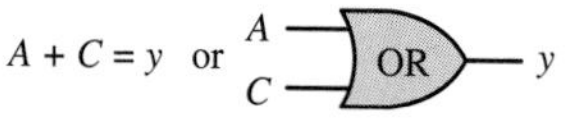

**Figure 13.19**

*Comments:* The derivation above has made use of two rules from Table 13.11: rules 4 and 18. Could you have predicted that the variable $B$ would not be used in the final realization? Why?

## EXAMPLE　13.5　DeMorgan's Theorem and Product-of-Sums Expressions

### Problem

Realize the logic function $y = A + B \cdot C$ in product-of-sums form. Implement the solution using AND, OR, and NOT gates.

### Solution

*Known Quantities:* Logical expression for the function $y(A, B, C)$.

*Find:* Physical realization using AND, OR, and NOT gates.

*Analysis:* We use the fact that $\overline{\overline{y}} = y$ and apply DeMorgan's theorem as follows:

$$\overline{y} = \overline{A + (B \cdot C)} = \overline{A} \cdot \overline{(B \cdot C)} = \overline{A} \cdot (\overline{B} + \overline{C})$$

$$\overline{\overline{y}} = y = \overline{\overline{A} \cdot (\overline{B} + \overline{C})}.$$

The above sum-of-products function is realized using complements of each variable (obtained using NOT gates) and is finally complemented as shown in Figure 13.20.

**Figure 13.20**

*Comments:* It should be evident that the original sum-of-products expression, which could be implemented with just one AND and one OR gate has a much more efficient realization. In the next section we show a systematic approach to function minimization.

**Focus on Computer-Aided Solutions:**  An *Electronics Workbench*™ simulation of the logic circuit of Figure 13.20 may be found in the accompanying CD-ROM.

VIRTUAL LAB

## NAND and NOR Gates

In addition to the AND and OR gates we have just analyzed, the complementary forms of these gates, called NAND and NOR, are very commonly used in practice. In fact, NAND and NOR gates form the basis of most practical logic circuits. Figure 13.21 depicts these two gates, and illustrates how they can be easily interpreted in terms of AND, OR, and NOT gates by virtue of De Morgan's laws. You can readily verify that the logic function implemented by the NAND and NOR gates corresponds, respectively, to AND and OR gates followed by an inverter. It is very important to note that, by De Morgan's laws, the NAND gate performs a *logical addition* on the *complements* of the inputs, while the NOR gate performs a *logical multiplication* on the *complements* of the inputs. Functionally, then, any logic function could be implemented with either NOR or NAND gates only.

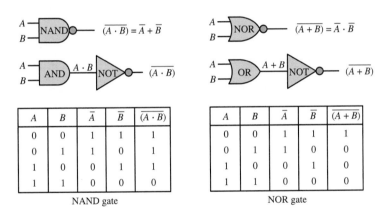

| $A$ | $B$ | $\bar{A}$ | $\bar{B}$ | $\overline{(A \cdot B)}$ |
|---|---|---|---|---|
| 0 | 0 | 1 | 1 | 1 |
| 0 | 1 | 1 | 0 | 1 |
| 1 | 0 | 0 | 1 | 1 |
| 1 | 1 | 0 | 0 | 0 |

NAND gate

| $A$ | $B$ | $\bar{A}$ | $\bar{B}$ | $\overline{(A + B)}$ |
|---|---|---|---|---|
| 0 | 0 | 1 | 1 | 1 |
| 0 | 1 | 1 | 0 | 0 |
| 1 | 0 | 0 | 1 | 0 |
| 1 | 1 | 0 | 0 | 0 |

NOR gate

**Figure 13.21** Equivalence of NAND and NOR gates with AND and OR gates

In the next section we shall learn how to systematically approach the design of logic functions. First, we provide a few examples to illustrate logic design with NAND and NOR gates.

## EXAMPLE   13.6  Realizing the AND Function with NAND Gates

### Problem

Use a truth table to show that the AND function can be realized using only NAND gates, and show the physical realization.

### Solution

***Known Quantities:***  AND and NAND truth tables.

| $A$ | $B(=A)$ | $A \cdot B$ | $\overline{(A \cdot B)}$ |
|---|---|---|---|
| 0 | 0 | 0 | 1 |
| 1 | 1 | 1 | 0 |

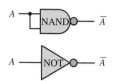

**Figure 13.22** NAND gate
as an inverters

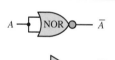

**Figure 13.23**

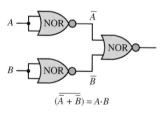

**Find:** AND realization using NAND gates.

**Assumptions:** Consider two-input functions and gates.

**Analysis:** The truth table below summarizes the two functions:

| $A$ | $B$ | NAND $\overline{A \cdot B}$ | AND $A \cdot B$ |
|---|---|---|---|
| 0 | 0 | 1 | 0 |
| 0 | 1 | 1 | 0 |
| 1 | 0 | 1 | 0 |
| 1 | 1 | 0 | 1 |

Clearly, to realize the AND function we need to simply invert the output of a NAND gate. This is easily accomplished if we observe that a NAND gate with its inputs tied together acts as an inverter; you can verify this in the above truth table by looking at the NAND output for the input combinations 0-0 and 1-1, or by referring to Figure 13.22. The final realization is shown in Figure 13.23.

**Comments:** NAND gates naturally implement functions that contain complemented products. Gates that employ negative logic are a natural consequence of the inverting characteristics of transistor switches (refer to Section 10.5). Thus, one should expect that NAND (and NOR) gates are very commonly employed in practice.

---

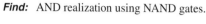

| $A$ | $B(=A)$ | $(A + B)$ | $\overline{(A + B)}$ |
|---|---|---|---|
| 0 | 0 | 0 | 1 |
| 1 | 1 | 1 | 0 |

**Figure 13.24** NOR gate as
an inverter

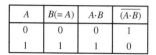

$(\overline{\overline{A} + \overline{B}}) = A \cdot B$

**Figure 13.25**

## EXAMPLE 13.7 Realizing the AND Function with NOR Gates

### Problem

Show analytically that the AND function can be realized using only NOR gates, and determine the physical realization.

---

### Solution

**Known Quantities:** AND and NOR functions.

**Find:** AND realization using NOR gates.

**Assumptions:** Consider two-input functions and gates.

**Analysis:** We can solve this problem using De Morgan's theorem. The output of an AND gate can be expressed as $f = A \cdot B$. Using De Morgan's theorem we write:

$$f = \overline{\overline{f}} = \overline{\overline{A \cdot B}} = \overline{\overline{A} + \overline{B}}$$

The above function is implemented very easily if we see that a NOR gate with its input tied together acts as a NOT gate (see Figure 13.24). Thus, the logic circuit of Figure 13.25 provides the desired answer.

**Comments:** NOR gates naturally implement functions that contain complemented sums. Gates that employ negative logic are a natural consequence of the inverting characteristics of transistor switches (refer to Section 10.5). Thus, one should expect that NOR (and NAND) gates are very commonly employed in practice.

## EXAMPLE 13.8 Realizing a Function with NAND and NOR Gates

### Problem

Realize the following function using only NAND and NOR gates:

$$y = \overline{(A \cdot B)} + C$$

### Solution

**Known Quantities:** Logical expression for $y$.

**Find:** Realization of $y$ using only NAND and NOR gates.

**Assumptions:** Consider two-input functions and gates.

**Analysis:** On the basis of the two preceding examples, we see that we can realize the term $Z = \overline{(A \cdot B)}$ using a two-input NAND gate, and the term $\overline{Z + C}$ using a two-input NOR gate. The solution is shown in Figure 13.26.

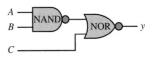

**Figure 13.26**

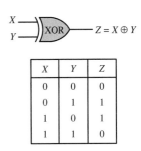

VIRTUAL LAB

## The XOR (Exclusive OR) Gate

It is rather common practice for a manufacturer of integrated circuits to provide common combinations of logic circuits in a single integrated circuit package. We review many of these common **logic modules** in Section 13.5. An example of this idea is provided by the **exclusive OR (XOR) gate,** which provides a logic function similar, but not identical, to the OR gate we have already studied. The XOR gate acts as an OR gate, except when its inputs are all logic 1s; in this case, the output is a logic 0 (thus the term *exclusive*). Figure 13.27 shows the logic circuit symbol adopted for this gate, and the corresponding truth table. The logic function implemented by the XOR gate is the following: "either $X$ or $Y$, but not both." This description can be extended to an arbitrary number of inputs.

The symbol adopted for the exclusive OR operation is $\oplus$, and so we shall write

$$Z = X \oplus Y$$

to denote this logic operation. The XOR gate can be obtained by a combination of the basic gates we are already familiar with. For example, if we observe that the XOR function corresponds to $Z = X \oplus Y = (X + Y) \cdot \overline{(X \cdot Y)}$, we can realize the XOR gate by means of the circuit shown in Figure 13.28.

Common IC logic gate configurations, device numbers, and data sheets are included in the CD-ROM that accompanies this book. These devices are typically available in both of the two more common device families, TTL and CMOS. The devices listed in the CD-ROM are available in CMOS technology under the numbers SN74AHXX. The same logic gate ICs are also available as TTL devices.

$Z = X \oplus Y$

| $X$ | $Y$ | $Z$ |
|-----|-----|-----|
| 0 | 0 | 0 |
| 0 | 1 | 1 |
| 1 | 0 | 1 |
| 1 | 1 | 0 |

Truth table

**Figure 13.27** XOR gate

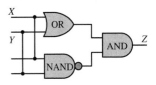

**Figure 13.28** Realization of an XOR gate

## Check Your Understanding

**13.9**  Show that one can obtain an OR gate using NAND gates only. [*Hint:* Use three NAND gates.]

**13.10**  Show that one can obtain an AND gate using NOR gates only. [*Hint:* Use three NOR gates.]

**13.11**  Prepare a step-by-step truth table for the following logic expressions:

   a. $\overline{(X + Y + Z)} + (X \cdot Y \cdot Z) \cdot \overline{X}$
   b. $\overline{X} \cdot Y \cdot Z + Y \cdot (Z + W)$
   c. $(X \cdot \overline{Y} + Z \cdot \overline{W}) \cdot (W \cdot X + \overline{Z} \cdot Y)$

[*Hint:* Your truth table must have $2^n$ entries, where $n$ is the number of logic variables.]

**13.12**  Implement the logic functions of Check Your Understanding Exercise 13.11 using NAND and NOR gates only. [*Hint:* Use De Morgan's theorems and the fact that $\overline{\overline{f}} = f$.]

**13.13**  Implement the logic functions of Check Your Understanding Exercise 13.11 using AND, OR, and NOT gates only.

**13.14**  Show that the XOR function can also be expressed as $Z = X \cdot \overline{Y} + Y \cdot \overline{X}$. Realize the corresponding function using NOT, AND, and OR gates. [*Hint:* Use truth tables for the logic function $Z$ ( as defined in the exercise) and for the XOR function.]

## 13.4  KARNAUGH MAPS AND LOGIC DESIGN

In examining the design of logic functions by means of logic gates, we have discovered that more than one solution is usually available for the implementation of a given logic expression. It should also be clear by now that some combinations of gates can implement a given function more efficiently than others. How can we be assured of having chosen the most efficient realization? Fortunately, there is a procedure that utilizes a map describing all possible combinations of the variables present in the logic function of interest. This map is called a **Karnaugh map,** after its inventor. Figure 13.29 depicts the appearance of Karnaugh maps for two-, three-, and four-variable expressions in two different forms. As can be seen, the row and column assignments for two or more variables are arranged so that all adjacent terms change by only one bit. For example, in the two-variable map, the columns next to column 01 are columns 00 and 11. Also note that each map consists of $2^N$ **cells,** where $N$ is the number of logic variables.

Each cell in a Karnaugh map contains a **minterm,** that is, a product of the $N$ variables that appear in our logic expression (in either uncomplemented or complemented form). For example, for the case of three variables ($N = 3$), there are $2^3 = 8$ such combinations, or minterms: $\overline{X} \cdot \overline{Y} \cdot \overline{Z}, \overline{X} \cdot \overline{Y} \cdot Z, \overline{X} \cdot Y \cdot \overline{Z}, \overline{X} \cdot Y \cdot Z,$ $X \cdot \overline{Y} \cdot \overline{Z}, X \cdot \overline{Y} \cdot Z, X \cdot Y \cdot \overline{Z},$ and $X \cdot Y \cdot Z$. The content of each cell—that is, the minterm—is the product of the variables appearing at the corresponding vertical and horizontal coordinates. For example, in the three-variable map, $X \cdot Y \cdot \overline{Z}$ appears at the intersection of $X \cdot Y$ and $\overline{Z}$. The map is filled by placing a value of 1 for any combination of variables for which the desired output is a 1. For example, consider the function of three variables for which we desire to have an output of 1 whenever the variables $X$, $Y$, and $Z$ have the following values:

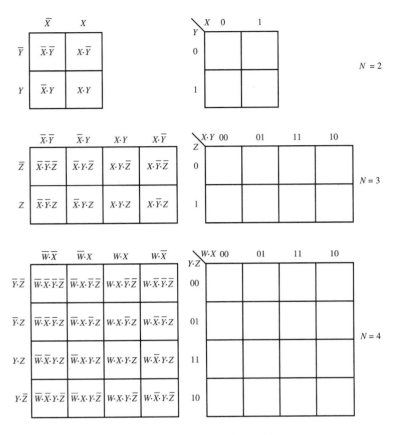

**Figure 13.29** Two-, three-, and four-variable Karnaugh maps

| $X = 0$ | $Y = 1$ | $Z = 0$ |
|---|---|---|
| $X = 0$ | $Y = 1$ | $Z = 1$ |
| $X = 1$ | $Y = 1$ | $Z = 0$ |
| $X = 1$ | $Y = 1$ | $Z = 1$ |

| | $\overline{X}\overline{Y}$ | $\overline{X}Y$ | $XY$ | $X\overline{Y}$ |
|---|---|---|---|---|
| $\overline{Z}$ | 0 | 1 | 1 | 0 |
| $Z$ | 0 | 1 | 1 | 0 |

Karnaugh map

| $X$ | $Y$ | $Z$ | Desired Function |
|---|---|---|---|
| 0 | 0 | 0 | 0 |
| 0 | 0 | 1 | 0 |
| 0 | 1 | 0 | 1 |
| 0 | 1 | 1 | 1 |
| 1 | 0 | 0 | 0 |
| 1 | 0 | 1 | 0 |
| 1 | 1 | 0 | 1 |
| 1 | 1 | 1 | 1 |

Truth table

**Figure 13.30** Truth table and Karnaugh map representations of a logic function

The same truth table is shown in Figure 13.30 together with the corresponding Karnaugh map.

The Karnaugh map provides an immediate view of the values of the function in graphical form. Further, the arrangement of the cells in the Karnaugh map is such that any two adjacent cells contain minterms that vary in only one variable. This property, as will be verified shortly, is quite useful in the design of logic functions by means of logic gates, especially if we consider the map to be continuously wrapping around itself, as if the top and bottom, and right and left, edges were touching each other. For the three-variable map given in Figure 13.29, for example, the cell $X \cdot \overline{Y} \cdot \overline{Z}$ is adjacent to $\overline{X} \cdot \overline{Y} \cdot \overline{Z}$ if we "roll" the map so that the right edge touches the left. Note that these two cells differ only in the variable $X$, a property we earlier claimed adjacent cells have.[1]

---

[1] A useful rule to remember is that in a two-variable map there are two minterms adjacent to any given minterm; in a three-variable map, three minterms are adjacent to any given minterm; in a four-variable map, the number is four, and so on.

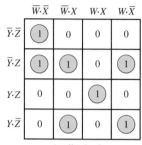

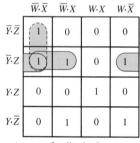

2 cell subcubes

**Figure 13.32** One- and two-cell subcubes for the Karnaugh map of Figure 13.31

Shown in Figure 13.31 is a more complex, four-variable logic function, which will serve as an example in explaining how Karnaugh maps can be used directly to implement a logic function. First, we define a subcube as a set of $2^m$ adjacent cells with logical value 1, for $m = 1, 2, 3, \ldots, N$. Thus, a subcube can consist of 1, 2, 4, 8, 16, 32,... cells. All possible subcubes for the four-variable map of Figure 13.31 are shown in Figure 13.32. Note that there are no four-cell subcubes in this particular case. Note also that there is some overlap between subcubes. Examples of four-cell and eight-cell subcubes are shown in Figure 13.33 for an arbitrary expression.

| X | Y | Y | Z | Desired Function |
|---|---|---|---|---|
| 0 | 0 | 0 | 0 | 1 |
| 0 | 0 | 0 | 1 | 1 |
| 0 | 0 | 1 | 0 | 0 |
| 0 | 0 | 1 | 1 | 0 |
| 0 | 1 | 0 | 0 | 0 |
| 0 | 1 | 0 | 1 | 1 |
| 0 | 1 | 1 | 0 | 1 |
| 0 | 1 | 1 | 1 | 0 |
| 1 | 0 | 0 | 0 | 0 |
| 1 | 0 | 0 | 1 | 1 |
| 1 | 0 | 1 | 0 | 1 |
| 1 | 0 | 1 | 1 | 0 |
| 1 | 1 | 0 | 0 | 0 |
| 1 | 1 | 0 | 1 | 0 |
| 1 | 1 | 1 | 0 | 0 |
| 1 | 1 | 1 | 1 | 1 |

Truth table for four-variable expression

| | $\overline{W}\cdot\overline{X}$ | $\overline{W}\cdot X$ | $W\cdot X$ | $W\cdot\overline{X}$ |
|---|---|---|---|---|
| $\overline{Y}\cdot\overline{Z}$ | 1 | 0 | 0 | 0 |
| $\overline{Y}\cdot Z$ | 1 | 1 | 0 | 1 |
| $Y\cdot Z$ | 0 | 0 | 1 | 0 |
| $Y\cdot\overline{Z}$ | 0 | 1 | 0 | 1 |

**Figure 13.31** Karnaugh map for a four-variable expression

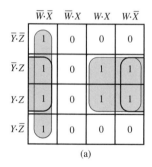

**Figure 13.33** Four- and eight-cell subcubes for an arbitrary logic function

In general, one tries to find the largest possible subcubes to cover all of the "1" entries in the map. How do maps and subcubes help in the realization of logic functions, then? The use of maps and subcubes in minimizing logic expressions is best explained by considering the following rule of Boolean algebra:

$$Y \cdot X + Y \cdot \overline{X} = Y$$

where the variable $Y$ could represent a product of logic variables (for example, we could similarly write $(Z \cdot W) \cdot X + (Z \cdot W) \cdot \overline{X} = Z \cdot W$ with $Y = Z \cdot W$). This rule is easily proven by factoring $Y$:

$$Y \cdot (X + \overline{X})$$

and observing that $X + \overline{X} = 1$, always. Then it should be clear that the variable $X$ need not appear in the expression at all. Let us apply this rule to a more complex logic expression, to verify that it can also apply to this case. Consider the logic expression

$$\overline{W} \cdot X \cdot \overline{Y} \cdot Z + \overline{W} \cdot \overline{X} \cdot \overline{Y} \cdot Z + W \cdot \overline{X} \cdot \overline{Y} \cdot Z + W \cdot X \cdot \overline{Y} \cdot Z$$

and factor it as follows:

$$\overline{W} \cdot Z \cdot \overline{Y} \cdot (X + \overline{X}) + W \cdot \overline{Y} \cdot Z \cdot (\overline{X} + X) = \overline{W} \cdot Z \cdot \overline{Y} + W \cdot \overline{Y} \cdot Z$$
$$= \overline{Y} \cdot Z \cdot (\overline{W} + W) = \overline{Y} \cdot Z$$

Quite a simplification! If we consider, now, a map in which we place a 1 in the cells corresponding to the minterms $\overline{W} \cdot X \cdot \overline{Y} \cdot Z$, $\overline{W} \cdot \overline{X} \cdot \overline{Y} \cdot Z$, $W \cdot \overline{X} \cdot \overline{Y} \cdot Z$, and $W \cdot X \cdot \overline{Y} \cdot Z$, forming the previous expression, we obtain the Karnaugh map of Figure 13.34. It can easily be verified that the map of Figure 13.34 shows a single four-cell subcube corresponding to the term $\overline{Y} \cdot Z$.

We have not established formal rules yet, but it definitely appears that the map method for simplifying Boolean expressions is a convenient tool. In effect, the map has performed the algebraic simplification automatically! We can see that in any subcube, one or more of the variables present will appear in both complemented *and* uncomplemented form in all their combinations with the other variables. These variables can be eliminated. As an illustration, in the *eight-cell* subcube case of Figure 13.35, the full-blown expression would be:

$$\overline{W} \cdot \overline{X} \cdot \overline{Y} \cdot \overline{Z} + \overline{W} \cdot X \cdot \overline{Y} \cdot \overline{Z} + W \cdot X \cdot \overline{Y} \cdot \overline{Z} + W \cdot \overline{X} \cdot \overline{Y} \cdot \overline{Z}$$

$$+\overline{W} \cdot \overline{X} \cdot Y \cdot \overline{Z} + \overline{W} \cdot X \cdot Y \cdot \overline{Z} + W \cdot X \cdot Y \cdot \overline{Z} + W \cdot \overline{X} \cdot Y \cdot \overline{Z}$$

However, if we consider the eight-cell subcube, we note that the three variables $X$, $W$, and $Z$ appear both in complemented and uncomplemented form in all their combinations with the other variables and thus can be removed from the expression. This reduces the seemingly unwieldy expression simply to $\overline{Y}$! In logic design terms, a simple inverter is sufficient to implement the expression.

The example just shown is a particularly simple one, but it illustrates how simple it can be to determine the minimal expression for a logic function. It should be apparent that the larger a subcube, the greater the simplification that will result. For subcubes that do not intersect, as in the previous example, the solution can be found easily, and is unique.

| | $\overline{W} \cdot \overline{X}$ | $\overline{W} \cdot X$ | $W \cdot X$ | $W \cdot \overline{X}$ |
|---|---|---|---|---|
| $\overline{Y} \cdot \overline{Z}$ | 0 | 0 | 0 | 0 |
| $\overline{Y} \cdot Z$ | 1 | 1 | 1 | 1 |
| $Y \cdot Z$ | 0 | 0 | 0 | 0 |
| $Y \cdot \overline{Z}$ | 0 | 0 | 0 | 0 |

**Figure 13.34** Karnaugh map for the function $\overline{W} \cdot X \cdot \overline{Y} \cdot Z + \overline{W} \cdot \overline{X} \cdot \overline{Y} \cdot Z + W \cdot \overline{X} \cdot \overline{Y} \cdot Z + W \cdot X \cdot \overline{Y} \cdot Z$

| | $\overline{W} \cdot \overline{X}$ | $\overline{W} \cdot X$ | $W \cdot X$ | $W \cdot \overline{X}$ |
|---|---|---|---|---|
| $\overline{Y} \cdot \overline{Z}$ | 1 | 1 | 1 | 1 |
| $\overline{Y} \cdot Z$ | 1 | 1 | 1 | 1 |
| $Y \cdot Z$ | 0 | 0 | 0 | 0 |
| $Y \cdot \overline{Z}$ | 0 | 0 | 0 | 0 |

**Figure 13.35**

## Sum-of-Products Realizations

Although not explicitly stated, the logic functions of the preceding section were all in sum-of-products form. As you know, it is also possible to realize logic functions in product-of-sums form. This section discusses the implementation of logic functions in sum-of-products form and gives a set of design rules. The next section will do the same for product-of-sums form logical expressions. The following rules are a useful aid in determining the minimal sum-of-products expression:

### FOCUS ON METHODOLOGY

**Sum-of-Products Realizations**

1. Begin with isolated cells. These must be used as they are, since no simplification is possible.
2. Find all cells that are adjacent to only one other cell, forming two-cell subcubes.
3. Find cells that form four-cell subcubes, eight-cell subcubes, and so forth.
4. The minimal expression is formed by the collection of the *smallest number of maximal subcubes*.

The following examples illustrate the application of these principles to a variety of problems.

| A | B | C | D | y |
|---|---|---|---|---|
| 0 | 0 | 0 | 0 | 1 |
| 0 | 0 | 0 | 1 | 1 |
| 0 | 0 | 1 | 0 | 1 |
| 0 | 0 | 1 | 1 | 0 |
| 0 | 1 | 0 | 0 | 0 |
| 0 | 1 | 0 | 1 | 1 |
| 0 | 1 | 1 | 0 | 0 |
| 0 | 1 | 1 | 1 | 0 |
| 1 | 0 | 0 | 0 | 1 |
| 1 | 0 | 0 | 1 | 1 |
| 1 | 0 | 1 | 0 | 0 |
| 1 | 0 | 1 | 1 | 1 |
| 1 | 1 | 0 | 0 | 0 |
| 1 | 1 | 0 | 1 | 1 |
| 1 | 1 | 1 | 0 | 0 |
| 1 | 1 | 1 | 1 | 1 |

**Figure 13.36**

### EXAMPLE 13.9 Logic Circuit Design Using Karnaugh Maps

**Problem**

Design a logic circuit that implements the truth table of Figure 13.36.

---

**Solution**

**Known Quantities:** Truth table for $y(A, B, C, D)$.

**Find:** Realization of $y$.

**Assumptions:** Two-, three-, and four-input gates are available.

**Analysis:** We use the Karnaugh map of Figure 13.37, which is shown with values of 1 and 0 already in place. We recognize four subcubes in the map; three are four-cell subcubes, and one is a two-cell subcube. The expressions for the subcubes are: $\overline{A} \cdot \overline{B} \cdot \overline{D}$ for the two-cell subcube; $\overline{B} \cdot \overline{C}$ for the subcube that wraps around the map; $\overline{C} \cdot D$ for the four-by-one subcube; and $A \cdot D$ for the square subcube at the bottom of the map. Thus, the expression for $y$ is:

$$y = \overline{A} \cdot \overline{B} \cdot \overline{D} + \overline{B} \cdot \overline{C} + \overline{C}D + AD.$$

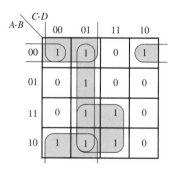

**Figure 13.37** Karnaugh map for Example 13.9

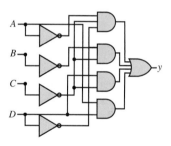

**Figure 13.38** Logic circuit realization of Karnaugh map of Figure 13.37

VIRTUAL LAB

The implementation of the above function with logic gates is shown in Figure 13.38.

**Comments:** The Karnaugh map covering of Figure 13.37 is a sum-of-products expression because we covered the map using the ones.

### EXAMPLE 13.10 Deriving a Sum-of-Products Expression from a Logic Circuit

**Problem**

Derive the truth table and minimum sum-of-products expression for the circuit of Figure 13.39.

## Solution

***Known Quantities:*** Logic circuit representing $f(x, y, z)$.

***Find:*** Expression for $f$ and corresponding truth table.

***Analysis:*** To determine the truth table, we write the expression corresponding to the logic circuit of Figure 13.39:

$$f = \overline{x} \cdot \overline{y} + y \cdot z$$

The truth table corresponding to this expression and the corresponding Karnaugh map with sum-of-products covering are shown in Figure 13.40.

**Figure 13.39**

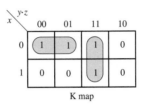

| $x$ | $y$ | $z$ | $f$ |
|---|---|---|---|
| 0 | 0 | 0 | 1 |
| 0 | 0 | 1 | 1 |
| 0 | 1 | 0 | 0 |
| 0 | 1 | 1 | 1 |
| 1 | 0 | 0 | 0 |
| 1 | 0 | 1 | 0 |
| 1 | 1 | 0 | 0 |
| 1 | 1 | 1 | 1 |

**Figure 13.40**

***Comments:*** If we used zeros in covering the Karnaugh map for this example, the resulting expression would be a product-of-sums. You may verify that, in the case of this example, the complexity of the circuit would be unchanged. Note also that there exists a third subcube $(x = 0, yz = 01, 11)$ that is not used because it does not help minimize the solution.

VIRTUAL LAB

## EXAMPLE   13.11   Realizing a Product-of-Sums Using Only NAND Gates

### Problem

Realize the following function in sum-of-products form, using only two-input NAND gates.

$$f = (\overline{x} + \overline{y}) \cdot (y + \overline{z})$$

### Solution

***Known quantities:*** $f(x, y, z)$.

***Find:*** Logic circuit for $f$ using only NAND gates.

***Analysis:*** The first step is to convert the expression for $f$ into an expression that can be easily implemented with NAND gates. We observe that direct application of De Morgan's theorem yields:

$$\overline{x} + \overline{y} = \overline{x \cdot y}$$

$$y + \overline{z} = \overline{z \cdot \overline{y}}$$

Thus, we can write the function as follows:

$$f = (\overline{x \cdot y}) \cdot (\overline{z \cdot \overline{y}})$$

and implement it with five NAND gates, as shown in Figure 13.41

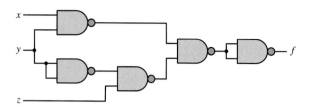

**Figure 13.41**

*Comments:* Note that we used two NAND gates as inverters—one to obtain $\overline{y}$, the other to invert the output of the fourth NAND gate, equal to $\overline{(\overline{x \cdot y}) \cdot (\overline{z \cdot \overline{y}})}$.

---

## EXAMPLE 13.12 Simplifying Expressions by Using Karnaugh Maps

### Problem

Simplify the following expression by using a Karnaugh map.

$$f = x \cdot y + \overline{x} \cdot z + y \cdot z$$

---

### Solution

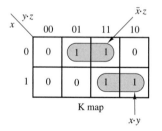

**Figure 13.42**

*Known Quantities:* $f(x, y, z)$.

*Find:* Minimal expression for $f$.

*Analysis:* We cover a three-term Karnaugh map to reflect the expression give above. The result is shown in Figure 13.42. It is clear that the Karnaugh map can be covered by using just two terms (subcubes): $f = x \cdot y + \overline{x} \cdot z$. Thus, the term $y \cdot z$ is redundant.

*Comments:* The Karnaugh map covering clearly shows that the term $y \cdot z$ corresponds to covering a third two-cell subcube vertically intersecting the two horizontal two-cell subcubes already shown. Clearly, the third subcube is redundant.

---

## EXAMPLE 13.13 Simplifying a Logic Circuit by Using the Karnaugh Map

### Problem

Derive the Karnaugh map corresponding to the circuit of Figure 13.43 and use the resulting map to simplify the expression.

---

### Solution

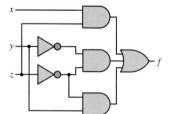

**Figure 13.43**

*Known Quantities:* Logic circuit.

*Find:* Simplified logic circuit.

**Analysis:**  We first determine the expression $f(x, y, z)$ from the logic circuit:

$$f = (x \cdot z) + (\overline{x} \cdot \overline{z}) + (y \cdot \overline{z})$$

This expression leads to the Karnaugh map shown in Figure 13.44. Inspection of the Karnaugh map reveals that the map could have been covered more efficiently by using four-cell subcubes. The improved map covering, corresponding to the simpler function $f = x + \overline{z}$, and the resulting logic circuit are shown in Figure 13.45.

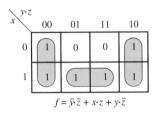

$$f = \overline{y} \cdot \overline{z} + x \cdot z + y \cdot \overline{z}$$

**Figure 13.44**

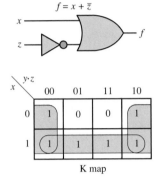

K map

**Figure 13.45**

VIRTUAL LAB

**Comments:**  In general, one wishes to cover the largest possible subcubes in a Karnaugh map.

## Product-of-Sums Realizations

Thus far, we have exclusively worked with sum-of-products expressions, that is, logic functions of the form $A \cdot B + C \cdot D$. We know, however, that De Morgan's laws state that there is an equivalent form that appears as a product of sums, for example, $(W + Y) \cdot (Y + Z)$. The two forms are completely equivalent, logically, but one of the two forms may lead to a realization involving a smaller number of gates. When using Karnaugh maps, we may obtain the product-of-sums form very simply by following these rules:

### FOCUS ON METHODOLOGY

**Product-of-Sums Realizations**

1. Solve for the 0s exactly as for the 1s in sum-of-products expressions.
2. Complement the resulting expression.

The same principles stated earlier apply in covering the map with subcubes and determining the minimal expression. The following examples illustrate how one form may result in a more efficient solution than the other.

### EXAMPLE 13.14  Comparison of Sum-of-Products and Product-of-Sums Designs

#### Problem

Realize the function $f$ described by the accompanying truth table using both 0 and 1 coverings in the Karnaugh map.

| x | y | z | f |
|---|---|---|---|
| 0 | 0 | 0 | 0 |
| 0 | 0 | 1 | 1 |
| 0 | 1 | 0 | 1 |
| 0 | 1 | 1 | 1 |
| 1 | 0 | 0 | 1 |
| 1 | 0 | 1 | 1 |
| 1 | 1 | 0 | 0 |
| 1 | 1 | 1 | 0 |

#### Solution

***Known Quantities:*** Truth table for logic function.

***Find:*** Realization in both sum-of-products and product-of-sums forms.

***Analysis:***

1. *Product-of-sums expression.* Product-of-sums expressions use zeros to determine the logical expression from a Karnaugh map. Figure 13.46 depicts the Karnaugh map covering with zeros, leading to the expression

$$f = (x + y + z) \cdot (\overline{x} + \overline{y})$$

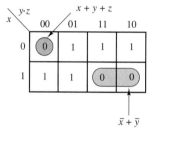

**Figure 13.46**

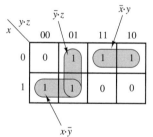

**Figure 13.47**

2. *Sum-of-products expression.* Sum-of-products expressions use ones to determine the logical expression from a Karnaugh map. Figure 13.47 depicts the Karnaugh map covering with ones, leading to the expression

$$f = (\overline{x} \cdot y) + (\overline{x} \cdot \overline{y}) + (\overline{y} \cdot z)$$

***Comments:*** The product-of-sums solution requires the use of five gates (two OR, two NOT, and one AND), while the sum-of-products solution will use six gates (one OR, two NOT, and three AND). Thus, solution 1 leads to the simpler design.

### EXAMPLE 13.15  Product-of-Sums Design

#### Problem

Realize the function $f$ described by the accompanying truth table in minimal product of sums form.

## Solution

**Known Quantities:** Truth table for logic function.

**Find:** Realization in minimal product-of-sums forms.

**Analysis:** We cover the Karnaugh map of Figure 13.48 using zeros, and obtain the following function:

$$f = \bar{z} \cdot (\bar{x} + \bar{y})$$

| $x$ | $y$ | $z$ | $f$ |
|---|---|---|---|
| 0 | 0 | 0 | 1 |
| 0 | 0 | 1 | 0 |
| 0 | 1 | 0 | 1 |
| 0 | 1 | 1 | 0 |
| 1 | 0 | 0 | 1 |
| 1 | 0 | 1 | 0 |
| 1 | 1 | 0 | 0 |
| 1 | 1 | 1 | 0 |

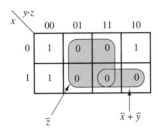

**Figure 13.48**

**Comments:** Is the sum-of-products solution simpler? Try it for yourself.

---

## Safety Circuit for Operation of a Stamping Press

In this example, the techniques illustrated in the preceding examples will be applied to a practical situation. To operate a stamping press, an operator must press two buttons ($b_1$ and $b_2$) one meter apart from each other and away from the press (this ensures that the operator's hands cannot be caught in the press). When the buttons are pressed, the logical variables $b_1$ and $b_2$ are equal to 1. Thus, we can define a new variable $A = b_1 \cdot b_2$; when $A = 1$, the operator's hands are safely away from the press. In addition to the safety requirement, however, other conditions must be satisfied before the operator can activate the press. The press is designed to operate on one of two workpieces, part I and part II, but not both. Thus, acceptable logic states for the press to be operated are "part I is in the press, but not part II" and "part II is in the press, but not part I." If we denote the presence of part I in the press by the logical variable $B = 1$ and the presence of part II by the logical variable $C = 1$, we can then impose additional requirements on the operation of the press. For example, a robot used to place either part in the press could activate a pair of switches (corresponding to logical variables $B$ and $C$) indicating which part, if any, is in the press. Finally, in order for the press to be operable, it must be "ready," meaning that it has to have completed any previous stamping operation. Let the logical variable $D = 1$

represent the ready condition. We have now represented the operation of the press in terms of four logical variables, summarized in the truth table of Table 13.12. Note that only two combinations of the logical variables will result in operation of the press: $ABCD = 1011$ and $ABCD = 1101$. You should verify that these two conditions correspond to the desired operation of the press. Using a Karnaugh map, realize the logic circuitry required to implement the truth table shown.

**Table 13.12**  Conditions for operation of stamping press

| (A) $b_1 \cdot b_2$ | (B) Part I is in press | (C) Part II is in press | (D) Press is operable | Press operation 1 = pressing; 0 = not pressing |
|---|---|---|---|---|
| 0 | 0 | 0 | 0 | 0 |
| 0 | 0 | 0 | 1 | 0 |
| 0 | 0 | 1 | 0 | 0 |
| 0 | 0 | 1 | 1 | 0 |
| 0 | 1 | 0 | 0 | 0 |
| 0 | 1 | 0 | 1 | 0 |
| 0 | 1 | 1 | 0 | 0 |
| 0 | 1 | 1 | 1 | 0 |
| 1 | 0 | 0 | 0 | 0 |
| 1 | 0 | 0 | 1 | 0 |
| 1 | 0 | 1 | 0 | 0 |
| 1 | 0 | 1 | 1 | 1 |
| 1 | 1 | 0 | 0 | 0 |
| 1 | 1 | 0 | 1 | 1 |
| 1 | 1 | 1 | 0 | 0 |
| 1 | 1 | 1 | 1 | 0 |

↑ Both buttons $(b_1, b_2)$ must be pressed for this to be a 1.

**Solution:**

Table 13.12 can be converted to a Karnaugh map, as shown in Figure 13.49. Since there are many more 0s than 1s in the table, the use of 0s in covering the map will lead to greater simplification. This will result in a product-of-sums expression. The four subcubes shown in Figure 13.49 yield the equation

$$A \cdot D \cdot (C + B) \cdot (\overline{C} + \overline{B})$$

By De Morgan's law, this equation is equivalent to

$$A \cdot D \cdot (C + B) \cdot \overline{(C \cdot B)}$$

which can be realized by the circuit of Figure 13.50.

For the purpose of comparison, the corresponding sum-of-products circuit is shown in Figure 13.51. Note that this circuit employs a greater number of gates and will therefore lead to a more expensive design.

**Focus on Computer-Aided Solutions—** An *Electronics Workbench*™ simulation of the logic circuit of Figure 13.50 may be found in the accompanying CD-ROM.

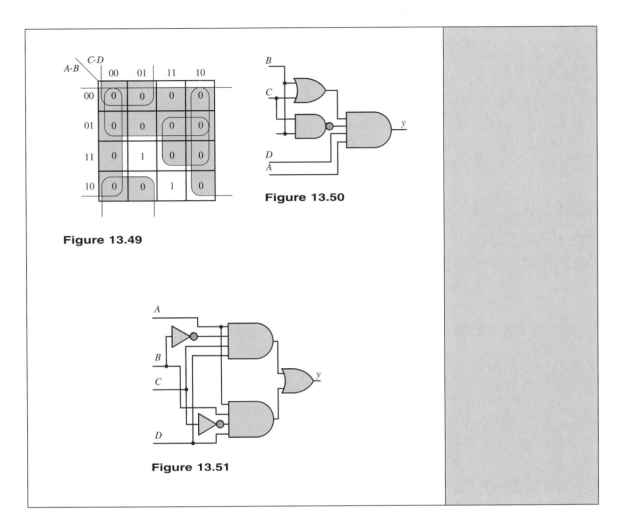

Figure 13.49

Figure 13.50

Figure 13.51

## Don't Care Conditions

Another simplification technique may be employed whenever the value of the logic function to be implemented can be either a 1 or a 0. This condition may result from the specification of the problem and is not uncommon. Whenever it does not matter whether a position in the map is filled by a 1 or a 0, we use a so-called **don't care** entry, denoted by an **x**. Then the don't care can be used as either a 1 or a 0, depending on which results in a greater simplification (i.e., helps in forming the smallest number of maximal subcubes). The following examples illustrate the use of don't cares.

---

**EXAMPLE   13.16   Using Don't Cares to Simplify Expressions—1**

### Problem

Use don't care entries to simplify the expression:

$$f(a, b, c, d) = \bar{a} \cdot \bar{b} \cdot \bar{c} \cdot d + \bar{a} \cdot \bar{b} \cdot c \cdot \bar{d} + \bar{a} \cdot \bar{b} \cdot c \cdot d + \bar{a} \cdot b \cdot \bar{c} \cdot d + a \cdot \bar{b} \cdot c \cdot d + a \cdot b \cdot \bar{c} \cdot \bar{d}$$

Note that the **x**'s never occur, and so they may be assigned a 1 or a 0, whichever will best simplify the expression.

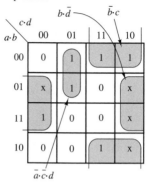

**Figure 13.52**

### Solution

**Known Quantities:**  Logical expression; don't care conditions.

**Find:**  Minimal realization.

**Schematics, Diagrams, Circuits, and Given Data:**  Don't care conditions: $f(a, b, c, d) = \{0100, 0110, 1010, 1110\}$.

**Analysis:**  We cover the Karnaugh map of Figure 13.52 using ones, and also using $x$ entries for each don't care condition. Treating all of the $x$ entries as ones, we complete the covering with two four-cell subcubes and one two-cell subcube, to obtain the following simplified expression:

$$f(a, b, c, d) = b \cdot \bar{d} + \bar{b} \cdot c + \bar{a} \cdot \bar{c} \cdot d$$

**Comments:**  Note that we could have also interpreted the don't care entries as zeros and tried to solve in product-of-sums form. Verify that the expression obtained above is indeed the minimal one.

---

### EXAMPLE   13.17   Using Don't Cares to Simplify Expressions—2

#### Problem

Find a minimum product-of-sums realization for the expression $f(a, b, c)$.

---

### Solution

**Known Quantities:**  Logical expression, don't care conditions.

**Find:**  Minimal realization.

**Schematics, Diagrams, Circuits, and Given Data:**

$$f(a, b, c) = 1 \text{ for } \{a, b, c\} = \{000, 010, 011\}$$

$$f(a, b, c) = \text{don't care for } \{a, b, c\} = \{100, 101, 110\}$$

**Analysis:**  We cover the Karnaugh map of Figure 13.53 using ones, and also using $x$ entries for each don't care condition. By appropriately selecting two of the three don't-care entries to be equal to 1, we complete the covering with one four-cell subcube and one two-cell subcube, to obtain the following minimal expression:

$$f(a, b, c) = \bar{a} \cdot b + \bar{c}$$

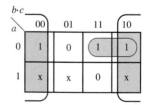

**Figure 13.53**

**Comments:**  Note that we have chosen to set one of the don't care entries equal to zero, since it would not lead to any further simplification.

---

### EXAMPLE   13.18   Using Don't Cares to Simplify Expressions—3

#### Problem

Find a minimum sum-of-products realization for the expression $f(a, b, c, d)$.

## Solution

**Known Quantities:**  Logical expression; don't care conditions.

**Find:**  Minimal realization.

**Schematics, Diagrams, Circuits, and Given Data**

$$f(a, b, c, d) = 1 \text{ for } \{a, b, c, d\} = \{0000, 0011, 0110, 1001\}$$

$$f(a, b, c, d) = \text{ don't care for } \{a, b, c, d\} = \{1010, 1011, 1101, 1110, 1111\}$$

**Analysis:**  We cover the Karnaugh map of Figure 13.54 using ones, and also using $x$ entries for each don't care condition. By appropriately selecting three of the four don't care entries to be equal to 1, we complete the covering with one four-cell subcube, two two-cell subcubes, and one one-cell subcube, to obtain the following expression:

$$f(a, b, c) = \overline{a} \cdot \overline{b} \cdot \overline{c} \cdot \overline{d} + b \cdot c \cdot \overline{d} + a \cdot d + \overline{b} \cdot c \cdot d$$

**Comments:**  Would the product-of-sums realization be simpler? Verify.

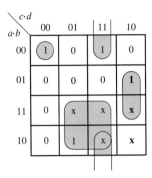

**Figure 13.54**

## Check Your Understanding

**13.15**  Simplify the following expression to show that it corresponds to the function $\overline{Z}$:

$$\overline{W} \cdot \overline{X} \cdot \overline{Y} \cdot \overline{Z} + \overline{W} \cdot X \cdot \overline{Y} \cdot \overline{Z} + W \cdot X \cdot \overline{Y} \cdot \overline{Z} + W \cdot \overline{X} \cdot \overline{Y} \cdot \overline{Z} + \overline{W} \cdot \overline{X} \cdot Y \cdot \overline{Z}$$
$$+ \overline{W} \cdot X \cdot Y \cdot \overline{Z} + W \cdot X \cdot Y \cdot \overline{Z} + W \cdot \overline{X} \cdot Y \cdot \overline{Z}$$

**13.16**  Simplify the following expression, using a Karnaugh map:

$$\overline{W} \cdot \overline{X} \cdot \overline{Y} \cdot \overline{Z} + \overline{W} \cdot \overline{X} \cdot Y \cdot \overline{Z} + W \cdot X \cdot \overline{Y} \cdot \overline{Z} + W \cdot \overline{X} \cdot \overline{Y} \cdot \overline{Z} + W \cdot \overline{X} \cdot Y \cdot \overline{Z}$$
$$+ W \cdot X \cdot Y \cdot \overline{Z}$$

**13.17**  Simplify the following expression, using a Karnaugh map:

$$\overline{W} \cdot \overline{X} \cdot \overline{Y} \cdot \overline{Z} + \overline{W} \cdot \overline{X} \cdot Y \cdot \overline{Z} + W \cdot X \cdot \overline{Y} \cdot \overline{Z} + W \cdot \overline{X} \cdot \overline{Y} \cdot \overline{Z}$$
$$+ W \cdot \overline{X} \cdot Y \cdot \overline{Z} + \overline{W} \cdot X \cdot \overline{Y} \cdot \overline{Z}$$

**13.18**  The function $y$ of Example 13.9 can be obtained with fewer gates if we use gates with three or four inputs. Find the minimum number of gates needed to obtain this function.

**13.19**  Verify that the product-of-sums expression for Example 13.14 can be realized with fewer gates.

**13.20**  Would a sum-of-products realization for Example 13.15 require fewer gates?

**13.21**  Prove that the circuit of Figure 13.51 can also be obtained from the sum of products.

**13.22**  In Example 13.16, assign a value of 0 to the don't care terms and derive the corresponding minimal expression. Is the new function simpler than the one obtained in Example 13.16?

**13.23**  In Example 13.17, assign a value of 0 to the don't care terms and derive the corresponding minimal expression. Is the new function simpler than the one obtained in Example 13.17?

**13.24**  In Example 13.17, assign a value of 1 to all don't care terms and derive the corresponding minimal expression. Is the new function simpler than the one obtained in Example 13.17?

**13.25**  In Example 13.18, assign a value of 0 to all don't care terms and derive the corresponding minimal expression.  Is the new function simpler than the one obtained in Example 13.18?

**13.26**  In Example 13.18, assign a value of 1 to all don't care terms and derive the corresponding minimal expression.  Is the new function simpler than the one obtained in Example 13.18?

## 13.5   COMBINATIONAL LOGIC MODULES

The basic logic gates described in the previous section are used to implement more advanced functions and are often combined to form logic modules, which, thanks to modern technology, are available in compact integrated circuit (IC) packages. In this section and the next, we discuss a few of the more common **combinational logic modules**, illustrating how these can be used to implement advanced logic functions.

### Multiplexers

**Multiplexers,** or **data selectors,** are combinational logic circuits that permit the selection of one of many inputs. A typical multiplexer (MUX) has $2^n$ **data lines,** $n$ **address** (or **data select**) **lines,** and one output. In addition, other control inputs (e.g., enables) may exist. Standard, commercially available MUXs allow for $n$ up to 4; however, two or more MUXs can be combined if a greater range is needed. The MUX allows for one of $2^n$ inputs to be selected as the data output; the selection of which input is to appear at the output is made by way of the address lines. Figure 13.55 depicts the block diagram of a four-input MUX. The input data lines are labeled $D_0$, $D_1$, $D_2$, and $D_3$; the **data select**, or address, **lines** are labeled $I_0$ and $I_1$; and the output is available in both complemented and uncomplemented form, and is thus labeled $F$, or $\overline{F}$. Finally, an **enable** input, labeled $E$, is also provided, as a means of enabling or disabling the MUX: if $E = 1$, the MUX is disabled; if $E = 0$, it is enabled. The negative logic (MUX off when $E = 1$ and on when $E = 0$) is represented by the small "bubble" at the enable input, which represents a complement operation (just as at the output of NAND and NOR gates). The enable input is useful whenever one is interested in a cascade of MUXs; this would be of interest if we needed to select a line from a large number, say $2^8 = 256$. Then two 4-input MUXs could be used to provide the data selection of 1 of 8.

The material described in the previous sections is quite adequate to describe the internal workings of a multiplexer. Figure 13.56 shows the internal construction of a 4-to-1 MUX using exclusively NAND gates (inverters are also used, but the reader will recall that a NAND gate can act as an inverter if properly connected).

In the design of digital systems (for example, microcomputers), a single line is often required to carry two or more different digital signals. However, only one signal at a time can be placed on the line. A MUX will allow us to select, at different instants, the signal we wish to place on this single line. This property is shown here for a 4-to-1 MUX. Figure 13.57 depicts the functional diagram of a 4-to-1 MUX, showing four data lines, $D_0$ through $D_3$, and two select lines, $I_0$ and $I_1$.

The data selector function of a MUX is best understood in terms of Table 13.13. In this truth table, the **x**'s represent don't care entries. As can be seen from the truth table, the output selects one of the data lines depending on the values of

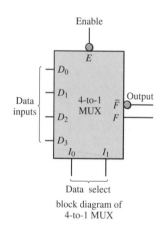

block diagram of
4-to-1 MUX

| $I_1$ | $I_0$ | $F$ |
|-------|-------|-----|
| 0 | 0 | $D_0$ |
| 0 | 1 | $D_1$ |
| 1 | 0 | $D_2$ |
| 1 | 1 | $D_3$ |

Truth table of
4-to-1 MUX

**Figure 13.55** 4-to-1 MUX

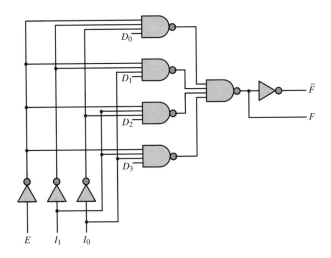

**Figure 13.56** Internal structure of the 4-to-1 MUX

**Table 13.13**

| $I_1$ | $I_0$ | $D_3$ | $D_2$ | $D_1$ | $D_0$ | $F$ |
|---|---|---|---|---|---|---|
| 0 | 0 | x | x | x | 0 | 0 |
| 0 | 0 | x | x | x | 1 | 1 |
| 0 | 1 | x | x | 0 | x | 0 |
| 0 | 1 | x | x | 1 | x | 1 |
| 1 | 0 | x | 0 | x | x | 0 |
| 1 | 0 | x | 1 | x | x | 1 |
| 1 | 1 | 0 | x | x | x | 0 |
| 1 | 1 | 1 | x | x | x | 1 |

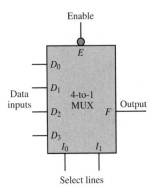

**Figure 13.57** Functional diagram of four-input MUX

$I_1$ and $I_0$, assuming that $I_0$ is the least significant bit. As an example, $I_1 I_0 = 10$ selects $D_2$, which means that the output, $F$, will select the value of the data line $D_2$. Therefore $F = 1$ if $D_2 = 1$ and $F = 0$ if $D_2 = 0$.

## Read-Only Memory (ROM)

Another common technique for implementing logic functions uses a **read-only memory**, or **ROM**. As the name implies, a ROM is a logic circuit that holds in storage ("memory") information—in the form of binary numbers—that cannot be altered but can be "read" by a logic circuit. A ROM is an array of memory cells, each of which can store either a 1 or a 0. The array consists of $2^m \times n$ cells, where $n$ is the number of bits in each word stored in ROM. To access the information stored in ROM, $m$ address lines are required. When an address is selected, in a fashion similar to the operation of the MUX, the binary word corresponding to the address selected appears at the output, which consists of $n$ bits, that is, the same number of bits as the stored words. In some sense, a ROM can be thought of as a MUX that has an output consisting of a word instead of a single bit.

Figure 13.58 depicts the conceptual arrangement of a ROM with $n = 4$ and $m = 2$. The ROM table has been filled with arbitrary 4-bit words, just for the

| ROM address | | ROM content (4-bit words) | | | | |
|---|---|---|---|---|---|---|
| $I_1$ | $I_0$ | $b_3$ | $b_2$ | $b_1$ | $b_0$ | |
| 0 | 0 | 0 | 1 | 1 | 0 | $W_0$ |
| 0 | 1 | 1 | 0 | 0 | 1 | $W_1$ |
| 1 | 0 | 0 | 1 | 1 | 0 | $W_2$ |
| 1 | 1 | 1 | 1 | 1 | 1 | $W_3$ |

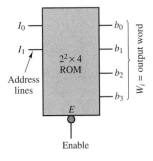

**Figure 13.58** Read-only memory

purpose of illustration. In Figure 13.58, if one were to select an enable input of 0 (i.e., on) and values for the address lines of $I_0 = 0$ and $I_1 = 1$, the output word would be $W_2 = 0110$, so that $b_0 = 0$, $b_1 = 1$, $b_2 = 1$, $b_3 = 0$. Depending on the content of the ROM and the number of address and output lines, one could implement an arbitrary logic function.

Unfortunately, the data stored in read-only memories must be entered during fabrication and cannot be altered later. A much more convenient type of read-only memory is the **erasable programmable read-only memory (EPROM),** the content of which can be easily programmed and stored and may be changed if needed. EPROMs find use in many practical applications, because of their flexibility in content and ease of programming. The following example illustrates the use of an EPROM to perform the linearization of a nonlinear function.

**FOCUS ON MEASUREMENTS**

### EPROM-Based Lookup Table for Automotive Fuel Injection System Control

One of the most common applications of EPROMs is the *arithmetic lookup table.* A lookup table is similar in concept to the familiar multiplication table and is used to store precomputed values of certain functions, eliminating the need for actually computing the function. A practical application of this concept is present in every automobile manufactured in the United States since the early 1980s, as part of the **exhaust emission control system.** In order for the catalytic converter to minimize the emissions of exhaust gases (especially hydrocarbons, oxides of nitrogen, and carbon monoxide), it is necessary to maintain the *air-to-fuel ratio* (A/F) as close as possible to the stoichiometric value, that is, 14.7 parts of air for each part of fuel. Most modern engines are equipped with fuel injection systems that are capable of delivering accurate amounts of fuel to each individual cylinder—thus, the task of maintaining an accurate A/F amounts to measuring the mass of air that is aspirated into each cylinder and computing the corresponding mass of fuel. Many automobiles are equipped with a *mass airflow sensor,* capable of measuring the mass of air drawn into each cylinder during each engine cycle. Let the output of the mass airflow sensor be denoted by the variable $M_A$, and let this variable represent the mass of air (in g) actually entering a cylinder during a particular stroke. It is then desired to compute the mass of fuel, $M_F$ (also expressed in g), required to achieve and A/F of 14.7. This computation is simply:

$$M_F = \frac{M_A}{14.7}$$

Although the above computation is a simple division, its actual calculation in a low-cost digital computer (such as would be used on an automobile) is rather complicated. It would be much simpler to tabulate a number of values of $M_A$, to precompute the variable $M_F$, and then to store the result of this computation into an EPROM. If the EPROM address were made to correspond to the tabulated values of air mass, and the content at each address to the corresponding fuel mass (according to the precomputed values of the expression $M_F = M_A/14.7$), it would not be necessary to perform the division by 14.7. For each measurement of air mass into one

**FIND IT ON THE WEB**

cylinder, an EPROM address is specified and the corresponding content is read. The content at the specific address is the mass of fuel required by that particular cylinder.

In practice, the fuel mass needs to be converted into a time interval corresponding to the duration of time during which the fuel injector is open. This final conversion factor can also be accounted for in the table. Suppose, for example, that the fuel injector is capable of injecting $K_F$ g of fuel per second; then the time duration, $T_F$, during which the injector should be open in order to inject $M_F$ g of fuel into the cylinder is given by:

$$T_F = \frac{M_F}{K_F} \text{ s}$$

Therefore, the complete expression to be precomputed and stored in the EPROM is:

$$T_F = \frac{M_A}{14.7 \times K_F} \text{ s}$$

Figure 13.59 illustrates this process graphically.

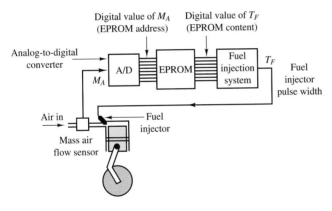

**Figure 13.59** Use of EPROM lookup table in automotive fuel injection system

To provide a numerical illustration, consider a hypothetical engine capable of aspirating air in the range $0 < M_A < 0.51$ g and equipped with fuel injectors capable of injecting at the rate of 1.36 g/s. Thus, the relationship between $T_F$ and $M_A$ is:

$$T_F = 50 \times M_A \text{ ms} = 0.05 M_A \text{ s}$$

If the digital value of $M_A$ is expressed in dg (decigrams, or tenths of g), the lookup table of Figure 13.60 can be implemented, illustrating the conversion capabilities provided by the EPROM. Note that in order to represent the quantities of interest in an appropriate binary format compatible with the 8-bit EPROM, the units of air mass and of time have been scaled.

| $M_A$ (g) $\times 10^{-2}$ | Address (digital value of $M_A$) | Content (digital value of $T_F$) | $T_F$(ms) $\times 10^{-1}$ |
|---|---|---|---|
| 0 | 00000000 | 00000000 | 0 |
| 1 | 00000001 | 00000101 | 5 |
| 2 | 00000010 | 00001010 | 10 |
| 3 | 00000011 | 00001111 | 15 |
| 4 | 00000100 | 00010100 | 20 |
| 5 | 00000101 | 00011001 | 25 |
| $\vdots$ | $\vdots$ | $\vdots$ | $\vdots$ |
| 51 | 00110011 | 11111111 | 255 |

**Figure 13.60** Lookup table for automotive fuel injection application

## Decoders and Read and Write Memory

**Decoders,** which are commonly used for applications such as address decoding or memory expansion, are combinational logic circuits as well. Our reason for introducing decoders is to show some of the internal organization of semiconductor memory devices. An important application of decoders in the organization of a memory system is discussed in Chapter 14.

Figure 13.61 shows the truth table for a 2-to-4 decoder. The decoder has an enable input, $\overline{G}$, and select inputs, $B$ and $A$. It also has four outputs, $Y_0$ through $Y_3$. When the enable input is logic 1, all decoder outputs are forced to logic 1 regardless of the select inputs.

This simple description of decoders permits a brief discussion of the internal organization of an **SRAM (static random-access or read and write memory).** SRAM is internally organized to provide memory with high speed (i.e., short access time), a large bit capacity, and low cost. The memory array in this memory device has a column length equal to the number of words, $W$, and a row length equal to the number of bits per word, $N$. To select a word, an $n$-to-$W$ decoder is needed. Since the address inputs to the decoder select only one of the decoder's outputs, the decoder selects one word in the memory array. Figure 13.62 shows the internal organization of a typical SRAM.

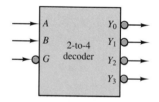

| Inputs | | | Outputs | | | |
|---|---|---|---|---|---|---|
| Enable | Select | | | | | |
| $\overline{G}$ | $A$ | $B$ | $Y_0$ | $Y_1$ | $Y_2$ | $Y_3$ |
| 1 | x | x | 1 | 1 | 1 | 1 |
| 0 | 0 | 0 | 0 | 1 | 1 | 1 |
| 0 | 0 | 1 | 1 | 0 | 1 | 1 |
| 0 | 1 | 0 | 1 | 1 | 0 | 1 |
| 0 | 1 | 1 | 1 | 1 | 1 | 0 |

**Figure 13.61** 2-to-4 decoder

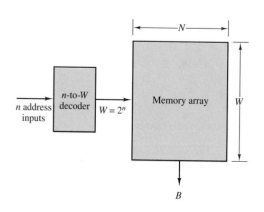

**Figure 13.62** Internal organization of SRAM

Thus, to choose the desired word from the memory array, the proper address inputs are required. As an example, if the number of words in the memory array is 8, a 3-to-8 decoder is needed. Data sheets for 2-to-4 and 3-to-8 decoders from a CMOS family data book are provided in the accompanying CD-ROM.

## Check Your Understanding

**13.27**  Which combination of the control lines will select the data line $D_3$ for a 4-to-1 MUX?

**13.28**  Show that an 8-to-1 MUX with eight data inputs ($D_0$ through $D_7$) and three control lines ($I_0$ through $I_2$) can be used as a data selector. Which combination of the control lines will select the data line $D_5$?

**13.29**  Which combination of the control lines will select the data line $D_4$ for an 8-to-1 MUX?

**13.30**  How many address inputs do you need if the number of words in a memory array is 16?

## CONCLUSION

- Digital logic circuits are at the basis of digital computers. Such circuits operate strictly on binary signals according to the laws of Boolean algebra.
- Combinational logic circuits can implement arbitrary Boolean logic functions.
- Combinational logic circuits include all of the logic gates—AND, OR, NAND, NOT, and XOR—as well as logic modules such as multiplexers and read-only memory.

## CHECK YOUR UNDERSTANDING ANSWERS

| | |
|---|---|
| **CYU 13.1** | (a) 100111; (b) 111011; (c) 100000000; (d) 0.011100; (e) 0.11001; (f) 0.110011; (g) 100000000.11; (h) 10000001.1001; (i) 1000000000000.11101 |
| **CYU 13.2** | (a) 13; (b) 27; (c) 23; (d) 0.6875; (e) 0.203125; (f) 0.2128906 0.2128906255; (g) 59.6875; (h) 91.203125; (i) 22.340820312 |
| **CYU 13.3** | (a) $20.75_{10}$; (b) $74_{10}$; (c) $1.5_{10}$; (d) $21_{10}$; (e) $100000_2$; (f) $1000000_2$; (g) $110010.11_2$; (h) $100011.11111_2$ |
| **CYU 13.4** | 4,096 |
| **CYU 13.5** | 39 mV |
| **CYU 13.6** | (a) 111110000011; (b) 00111001001; (c) 10100110; (d) 1AE; (e) B9; (f) 6ED |
| **CYU 13.7** | (a) 00010111; (b) 01101001; (c) 0100010 |
| **CYU 13.8** | (a) 0000 1011 1101; (b) 1101 0100 0111; (c) 0101 1010 |
| **CYU 13.16** | $W \cdot \overline{Z} + \overline{X} \cdot \overline{Z}$ |
| **CYU 13.17** | $\overline{Y} \cdot \overline{Z} + \overline{X} \cdot \overline{Z}$ |
| **CYU 13.18** | Nine gates |
| **CYU 13.20** | No |
| **CYU 13.22** | $f = a \cdot b \cdot \overline{c} \cdot \overline{d} + \overline{a} \cdot \overline{c} \cdot d + \overline{a} \cdot \overline{b} \cdot c + \overline{b} \cdot c \cdot d$ |
| **CYU 13.23** | $f = \overline{a} \cdot b + \overline{a} \cdot \overline{c}$ |
| **CYU 13.24** | $f = \overline{a} \cdot b + a \cdot \overline{b} + \overline{c}$ |
| **CYU 13.25** | $f = \overline{a} \cdot \overline{b} \cdot \overline{c} \cdot \overline{d} + \overline{a} \cdot \overline{b} \cdot c \cdot d + a \cdot \overline{b} \cdot \overline{c} \cdot d + \overline{a} \cdot b \cdot c \cdot \overline{d}$ |
| **CYU 13.26** | $f = \overline{a} \cdot \overline{b} \cdot \overline{c} \cdot \overline{d} + b \cdot c \cdot \overline{d} + a \cdot d + \overline{b} \cdot c \cdot d + a \cdot c$ |

**CYU 13.27**     $I_1 I_0 = 11$

**CYU 13.28**     For the first part, use the same method as in Check Your Understanding Exercise 13.27, but for an 8-to-1 MUX. For the second part, $I_2 I_1 I_0 = 101$.

**CYU 13.29**     $I_2 I_1 I_0 = 100$

**CYU 13.30**     4

# HOMEWORK PROBLEMS

## Section 1: Number Systems

**13.1**  Convert the following base 10 numbers to hex and binary:

a. 401    b. 273    c. 15    d. 38    e. 56

**13.2**  Convert the following hex numbers to base 10 and binary:

a. A    b. 66    c. 47    d. 21    e. 13

**13.3**  Convert the following base 10 numbers to binary:

a. 271.25    b. 53.375    c. 37.32    d. 54.27

**13.4**  Convert the following binary numbers to hex and base 10:

a. 1111    b. 1001101    c. 1100101    d. 1011100
e. 11101    f. 101000

**13.5**  Perform the following additions all in the binary system:

a.  11001011 + 101111
b.  10011001 + 1111011
c.  11101001 + 10011011

**13.6**  Perform the following subtractions all in the binary system:

a.  10001011 − 1101111
b.  10101001 − 111011
c.  11000011 − 10111011

**13.7**  Assuming that the most significant bit is the sign bit, find the decimal value of the following sign-magnitude form eight-bit binary numbers:

a. 11111000    b. 10011111    c. 01111001

**13.8**  Find the sign-magnitude form binary representation of the following decimal numbers:

a. 126    b. −126    c. 108    d. −98

**13.9**  Find the two's complement of the following binary numbers:

a. 1111    b. 1001101    c. 1011100    d. 11101

## Section 2: Combinational Logic

**13.10**  Use a truth table to prove that $B = AB + \overline{A}B$.

**13.11**  Use truth tables to prove that
$BC + B\overline{C} + \overline{B}A = A + B$.

**13.12**  Using the method of proof by perfect induction, show that

$$(X + Y) \cdot (\overline{X} + X \cdot Y) = Y$$

**13.13**  Using De Morgan's theorems and the rules of Boolean algebra, simplify the following logic function:

$$F(X, Y, Z) = \overline{X} \cdot \overline{Y} \cdot \overline{Z} + \overline{X} \cdot Y \cdot Z + X \cdot (\overline{Y + Z})$$

**13.14**  Simplify the expression
$f(A, B, C, D) = ABC + \overline{A}CD + \overline{B}CD$.

**13.15**  Simplify the logic function
$F(A, B, C) = \overline{A} \cdot B \cdot \overline{C} + \overline{A} \cdot B \cdot C + A \cdot B \cdot \overline{C} + A \cdot B \cdot C$
using Boolean algebra.

**13.16**  Find the logic function defined by the truth table given in Figure P13.16.

| A | B | C | F |
|---|---|---|---|
| 0 | 0 | 0 | 0 |
| 0 | 0 | 1 | 1 |
| 0 | 1 | 0 | 0 |
| 0 | 1 | 1 | 1 |
| 1 | 0 | 0 | 1 |
| 1 | 0 | 1 | 1 |
| 1 | 1 | 0 | 1 |
| 1 | 1 | 1 | 1 |

**Figure P13.16**

**13.17**  Determine the Boolean function describing the operation of the circuit shown in Figure P13.17.

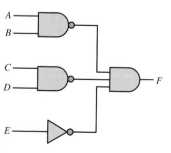

**Figure P13.17**

**13.18**  Use a truth table to show when the output of the circuit of Figure P13.18 is 1.

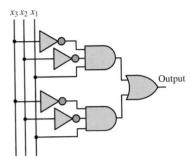

$x_3\, x_2\, x_1$

Output

**Figure P13.18**

**13.19**  Baseball is a complicated game and often the manager has a difficult time keeping track of all the rules of thumb that guide decisions. To assist your favorite baseball team you have been asked to design a logic circuit that will flash a light when the manager should give the steal sign. The rules have been laid out for you by a baseball fan with limited knowledge of the game as follows: Give the steal sign if there is a runner on first base and

a. There are no other runners, the pitcher is right-handed, and the runner is fast, or

b. There is one other runner on third-base, and one of the runners is fast, or

c. There is one other runner on second-base, the pitcher is left-handed, and both runners are fast.

Under no circumstances should the steal sign be given if all three bases have runners. Design a logic circuit that implements these rules to indicate when the steal sign should be given.

**13.20**  A small county board is composed of three commissioners. Each commissioner votes on measures presented to the board by pressing a button indicating whether the commissioner votes for or against a measure. If two or more commissioners vote for a measure it passes. Design a logic circuit that takes the three votes as inputs and lights either a green or red light to indicate whether or not a measure passed.

**13.21**  A water purification plant uses one tank for chemical sterilization and a second larger tank for settling and aeration. Each tank is equipped with two sensors that measure the height of water in each tank and the flow rate of water into each tank. When the height of water or flow rate is too high the sensors produce a logic high output. Design a logic circuit that sounds an alarm whenever the height of water in both tanks is too high and either of the flow rates is too high, or whenever both flow rates are too high and the height of water in either tank is also too high.

**13.22**  Many automobiles incorporate logic circuits to alert the driver of problems or potential problems. In one particular car, a buzzer is sounded whenever the

ignition key is turned and either a door is open or a seat belt is not fastened. The buzzer also sounds when the key is not turned but the lights are on. In addition, the car will not start unless the key is in the ignition, the car is in park, and all doors are closed and seat belts fastened. Design a logic circuit that takes all of the inputs listed and sounds the buzzer and starts the car when appropriate.

**13.23**  An on/off start-up signal governs the compressor motor of a large commercial air conditioning unit. In general, the start-up signal should be on whenever the output of a temperature sensor ($S$) exceeds a reference temperature. However, you are asked to limit the compressor start-ups to certain hours of the day and also enable service technicians to start up or shut down the compressor through a manual override. A time-of-day indicator ($D$) is available with on/off outputs as is a manual override switch ($M$). A separate timer ($T$) prohibits a compressor start-up within 10 minutes of a previous shutdown. Design a logic diagram that incorporates the state of all four devices ($S$, $D$, $M$, and $T$) and produces the correct on/off condition for the motor start-up.

## Section 3: Logic Design

**13.24**  Find the logic function corresponding to the truth table of Figure P13.24 in the simplest sum-of-products form.

| $A$ | $B$ | $C$ | $F$ |
|---|---|---|---|
| 0 | 0 | 0 | 1 |
| 0 | 0 | 1 | 0 |
| 0 | 1 | 0 | 0 |
| 0 | 1 | 1 | 0 |
| 1 | 0 | 0 | 1 |
| 1 | 0 | 1 | 0 |
| 1 | 1 | 0 | 1 |
| 1 | 1 | 1 | 1 |

**Figure P13.24**

**13.25**  Find the minimum expression for the output of the logic circuit shown in Figure P13.25.

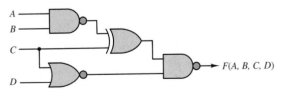

$A$

$B$

$C$

$D$

$F(A, B, C, D)$

**Figure P13.25**

**13.26**  Use a Karnaugh map to minimize the function
$f(A, B, C) = ABC + AB\overline{C} + A\overline{BC}.$

**13.27**

  a. Build the Karnaugh map for the logic function defined by the truth table of Figure P13.27.

  b. What is the minimum expression for this function?

  c. Realize $F$ using AND, OR, and NOT gates.

| A | B | C | D | F |
|---|---|---|---|---|
| 0 | 0 | 0 | 0 | 1 |
| 0 | 0 | 0 | 1 | 1 |
| 0 | 0 | 1 | 0 | 1 |
| 0 | 0 | 1 | 1 | 0 |
| 0 | 1 | 0 | 0 | 0 |
| 0 | 1 | 0 | 1 | 0 |
| 0 | 1 | 1 | 0 | 1 |
| 0 | 1 | 1 | 1 | 1 |
| 1 | 0 | 0 | 0 | 1 |
| 1 | 0 | 0 | 1 | 1 |
| 1 | 0 | 1 | 0 | 1 |
| 1 | 0 | 1 | 1 | 1 |
| 1 | 1 | 0 | 0 | 0 |
| 1 | 1 | 0 | 1 | 0 |
| 1 | 1 | 1 | 0 | 0 |
| 1 | 1 | 1 | 1 | 0 |

**Figure P13.27**

**13.28**  Fill in the Karnaugh map for the function defined by the truth table of Figure P13.28, and find the minimum expression for the function.

| A | B | C | f(A,B,C) |
|---|---|---|---|
| 0 | 0 | 0 | 0 |
| 0 | 0 | 1 | 1 |
| 0 | 1 | 0 | 1 |
| 0 | 1 | 1 | 0 |
| 1 | 0 | 0 | 1 |
| 1 | 0 | 1 | 0 |
| 1 | 1 | 0 | 0 |
| 1 | 1 | 1 | 1 |

**Figure P13.28**

**13.29**  A function, $F$, is defined such that it equals 1 when a 4-bit input code is equivalent to any of the decimal numbers 3, 6, 9, 12 or 15. $F$ is 0 for input codes 0, 2, 8 and 10. Other input values cannot occur. Use a Karnaugh map to determine a minimal expression for this function. Design and sketch a circuit to implement this function using only AND and NOT gates.

**13.30**  The function described in Figure P13.30 can be constructed using only two gates. Design the circuit.

| Input | | | Output |
|---|---|---|---|
| A | B | C | F |
| 0 | 0 | 0 | 0 |
| 0 | 0 | 1 | 0 |
| 0 | 1 | 0 | 0 |
| 0 | 1 | 1 | 1 |
| 1 | 0 | 0 | 0 |
| 1 | 0 | 1 | 0 |
| 1 | 1 | 0 | 1 |
| 1 | 1 | 1 | x |

**Figure P13.30**

**13.31**  Design a logic circuit which will produce the one's complement of an 8-bit signed binary number.

**13.32**  Construct the Karnaugh map for the logic function defined by the truth table of Figure P13.32, and find the minimum expression for the function.

| A | B | C | D | F |
|---|---|---|---|---|
| 0 | 0 | 0 | 0 | 1 |
| 0 | 0 | 0 | 1 | 0 |
| 0 | 0 | 1 | 0 | 1 |
| 0 | 0 | 1 | 1 | 0 |
| 0 | 1 | 0 | 0 | 0 |
| 0 | 1 | 0 | 1 | 0 |
| 0 | 1 | 1 | 0 | 0 |
| 0 | 1 | 1 | 1 | 1 |
| 1 | 0 | 0 | 0 | 1 |
| 1 | 0 | 0 | 1 | 0 |
| 1 | 0 | 1 | 0 | 1 |
| 1 | 0 | 1 | 1 | 0 |
| 1 | 1 | 0 | 0 | 1 |
| 1 | 1 | 0 | 1 | 1 |
| 1 | 1 | 1 | 0 | 1 |
| 1 | 1 | 1 | 1 | 0 |

**Figure P13.32**

**13.33**  Modify the circuit for Problem 13.31 so that it produces the two's complement of the 8-bit signed binary input.

**13.34**  Find the minimum output expression for the circuit of Figure P13.34.

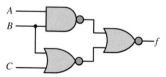

**Figure P13.34**

**13.35**  Design a combinational logic circuit which will add two 4-bit binary numbers.

**13.36**  Minimize the expression described in the truth table of Figure P13.36 and draw the circuit.

| A | B | C | F |
|---|---|---|---|
| 0 | 0 | 0 | 1 |
| 0 | 0 | 1 | 1 |
| 0 | 1 | 0 | 0 |
| 0 | 1 | 1 | 1 |
| 1 | 0 | 0 | 1 |
| 1 | 0 | 1 | 1 |
| 1 | 1 | 0 | 1 |
| 1 | 1 | 1 | 0 |

**Figure P13.36**

**13.37**  Find the minimum expression for the output of the logic circuit of Figure P13.37.

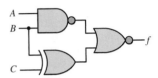

**Figure P13.37**

**13.38**  The objective of this problem is to design a combinational logic circuit which will aid in determination of the acceptability of emergency blood transfusions. It is known that human blood can be categorized into four types—A, B, AB, and O. Persons with type A blood can donate to both A and AB types, and can receive blood from both A and O types. Persons with type B blood can donate to both B and AB, and can receive from both B and O types. Persons with type AB blood can donate only to type AB, but can receive from any type. Persons with type O blood can donate to any type, but can receive only from type O. Make appropriate variable assignments and design a circuit which will approve or disapprove any particular transfusion based on these conditions.

**13.39**  Find the minimum expression for the logic function at the output of the logic circuit of Figure P13.39.

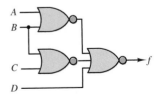

**Figure P13.39**

**13.40**  Design a combinational logic circuit which will accept a 4-bit binary number and:

If the number is even, divide it by $2_{10}$ and produce the binary result.
If the number is odd, multiply it by $2_{10}$ and produce the binary result.

**13.41**
  a. Fill in the Karnaugh map for the function defined in the truth table of Figure P13.41.
  b. What is the minimum expression for the function?
  c. Draw the circuit, using AND, OR, and NOT gates.

| A | B | C | f(A,B,C) |
|---|---|---|---|
| 0 | 0 | 0 | 1 |
| 0 | 0 | 1 | 1 |
| 0 | 1 | 0 | 0 |
| 0 | 1 | 1 | 1 |
| 1 | 0 | 0 | 1 |
| 1 | 0 | 1 | 1 |
| 1 | 1 | 0 | 1 |
| 1 | 1 | 1 | 0 |

**Figure P13.41**

**13.42**
  a. Fill in the Karnaugh map for the logic function defined by the truth table of Figure P13.42.
  b. What is the minimum expression for the function?

| A | B | C | D | F |
|---|---|---|---|---|
| 0 | 0 | 0 | 0 | 1 |
| 0 | 0 | 0 | 1 | 0 |
| 0 | 0 | 1 | 0 | 1 |
| 0 | 0 | 1 | 1 | 0 |
| 0 | 1 | 0 | 0 | 0 |
| 0 | 1 | 0 | 1 | 0 |
| 0 | 1 | 1 | 0 | 0 |
| 0 | 1 | 1 | 1 | 1 |
| 1 | 0 | 0 | 0 | 1 |
| 1 | 0 | 0 | 1 | 0 |
| 1 | 0 | 1 | 0 | 1 |
| 1 | 0 | 1 | 1 | 0 |
| 1 | 1 | 0 | 0 | 1 |
| 1 | 1 | 0 | 1 | 1 |
| 1 | 1 | 1 | 0 | 1 |
| 1 | 1 | 1 | 1 | 0 |

**Figure P13.42**

**13.43**
  a. Fill in the Karnaugh map for the logic function defined by the truth table of Figure P13.43.
  b. What is the minimum expression for the function?
  c. Realize the function, using only NAND gates.

| A | B | C | D | F |
|---|---|---|---|---|
| 0 | 0 | 0 | 0 | 1 |
| 0 | 0 | 0 | 1 | 1 |
| 0 | 0 | 1 | 0 | 1 |
| 0 | 0 | 1 | 1 | 1 |
| 0 | 1 | 0 | 0 | 0 |
| 0 | 1 | 0 | 1 | 1 |
| 0 | 1 | 1 | 0 | 0 |
| 0 | 1 | 1 | 1 | 1 |
| 1 | 0 | 0 | 0 | 1 |
| 1 | 0 | 0 | 1 | 1 |
| 1 | 0 | 1 | 0 | 0 |
| 1 | 0 | 1 | 1 | 0 |
| 1 | 1 | 0 | 0 | 0 |
| 1 | 1 | 0 | 1 | 1 |
| 1 | 1 | 1 | 0 | 1 |
| 1 | 1 | 1 | 1 | 0 |

**Figure P13.43**

**13.44** Design a circuit with a four-bit input representing the binary number $A_3A_2A_1A_0$. The output should be 1 if the input value is divisible by 3. Assume that the circuit is to be used only for the digits 0 through 9 (thus, values for 10 to 15 can be don't cares).

a. Draw the Karnaugh map and truth table for the function.

b. Determine the minimum expression for the function.

c. Draw the circuit, using only AND, OR, and NOT gates.

**13.45** Find the simplified sum-of-products representation of the function from the Karnaugh map shown in Figure P13.45. Note that x is the don't care term.

| C·D \ A·B | 00 | 01 | 11 | 10 |
|---|---|---|---|---|
| 00 | 0 | 1 | 0 | 0 |
| 01 | 1 | 1 | 0 | 0 |
| 11 | 0 | x | 1 | 0 |
| 10 | 0 | 0 | 1 | 0 |

**Figure P13.45**

**13.46** Can the circuit for Problem 13.40 be simplified if it is known that the input represents a BCD (binary-coded decimal) number, i.e., it can never be greater than $10_{10}$? If not, explain why not. Otherwise, design the simplified circuit.

**13.47** Find the simplified sum-of-products representation of the function from the Karnaugh map shown in Figure P13.47.

| C·D \ A·B | 00 | 01 | 11 | 10 |
|---|---|---|---|---|
| 00 | 0 | 1 | x | 0 |
| 01 | 0 | 1 | x | 0 |
| 11 | 0 | 1 | 0 | 1 |
| 10 | x | x | 1 | 0 |

**Figure P13.47**

**13.48** One method of ensuring reliability in data transmission systems is to transmit a parity bit along with every nibble, byte, or word of binary data transmitted. The parity bit confirms whether an even or odd number of 1's were transmitted in the data. In even-parity systems, the parity bit is set to 1 when the number of 1's in the transmitted data is odd. Odd-parity systems set the parity bit to 1 when the number of 1's in the transmitted data is even. Assume that a parity-bit is transmitted for every nibble of data. Design a logic circuit that checks the nibble of data and transmits the proper parity bit for both even- and odd-parity systems.

**13.49** Assume that a parity bit is transmitted for every nibble of data. Design two logic circuits that check a nibble of data and its parity bit to determine if there may have been an data transmission error. First assume an even-parity system, then an odd-parity system.

**13.50** Design a logic circuit that takes a 4-bit Gray code input from an optical encoder and translates it into two 4-bit nibbles of BCD code.

**13.51** Design a logic circuit that takes a 4-bit Gray code input from an optical encoder and determines if the input value is a multiple of 3.

**13.52** The 4221 code is a base 10–oriented code that assigns the weights 4221 to each of 4 bits in a nibble of data. Design a logic circuit that takes a BCD nibble as input and converts it to its 4221 equivalent. The logic circuit should also report an error in the BCD input if its value exceeds 1001.

**13.53** The 4-bit digital output of each of two sensors along an assembly line conveyor belt is proportional to the number of parts which pass by on the conveyor belt in a 30-second period. Design a logic circuit that reports an error if the outputs of the two sensors differ by more than one part per 30-second period.

## Section 4: Logic Modules

**13.54**

a. Fill in the Karnaugh map for the logic function defined by the truth table of Figure P13.54.

b. What is the minimum expression for the function?

c. Realize the function using a 1-of-8 multiplexer.

| A | B | C | D | $f(A,B,C,D)$ |
|---|---|---|---|---|
| 0 | 0 | 0 | 0 | 1 |
| 0 | 0 | 0 | 1 | 0 |
| 0 | 0 | 1 | 0 | 1 |
| 0 | 0 | 1 | 1 | 1 |
| 0 | 1 | 0 | 0 | 0 |
| 0 | 1 | 0 | 1 | 1 |
| 0 | 1 | 1 | 0 | 0 |
| 0 | 1 | 1 | 1 | 0 |
| 1 | 0 | 0 | 0 | 0 |
| 1 | 0 | 0 | 1 | 1 |
| 1 | 0 | 1 | 0 | 0 |
| 1 | 0 | 1 | 1 | 0 |
| 1 | 1 | 0 | 0 | 1 |
| 1 | 1 | 0 | 1 | 0 |
| 1 | 1 | 1 | 0 | 1 |
| 1 | 1 | 1 | 1 | 1 |

**Figure P13.54**

## 13.55

a. Fill in the truth table for the multiplexer circuit shown in Figure P13.55.

b. What binary function is performed by these multiplexers?

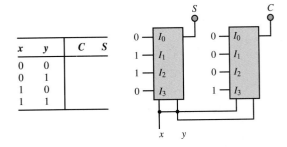

| x | y | C | S |
|---|---|---|---|
| 0 | 0 | | |
| 0 | 1 | | |
| 1 | 0 | | |
| 1 | 1 | | |

**Figure P13.55**

**13.56**  The circuit of Figure P13.56 can operate as a 4-to-16 decoder. Terminal EN denotes the enable input. Describe the operation of the 4-to-16 decoder. What is the role of logic variable $A$?

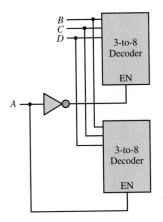

**Figure P13.56**

**13.57**  Show that the circuit given in Figure P13.57 converts 4-bit binary numbers to 4-bit Gray code.

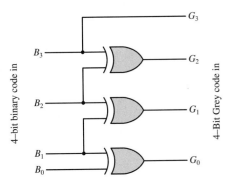

**Figure P13.57**

**13.58**  Suppose one of your classmates claims that the following Boolean expressions represent the conversion from 4-bit Gray code to 4-bit binary numbers:

$$B_3 = G_3$$
$$B_2 = G_3 \oplus G_2$$
$$B_1 = G_3 \oplus G_2 \oplus G_1$$
$$B_0 = G_3 \oplus G_2 \oplus G_1 \oplus G_0$$

a. Show that your classmate's claim is correct.

b. Draw the circuit which implements the conversion.

**13.59**  Select the proper inputs for a 4-input multiplexer to implement the function $f(A, B, C) = \overline{A}B\overline{C} + A\overline{B}C + AC$. Assume the inputs $I_0$, $I_1$, $I_2$, and $I_3$ correspond to $\overline{A}\,\overline{B}$, $\overline{A}B$, $A\overline{B}$, and $AB$, respectively, and that each input may be 0, 1, $\overline{C}$, or $C$.

**13.60**  Select the proper inputs for an 8-bit multiplexer to implement the function $f(A, B, C, D) = \sum(2, 5, 6, 8, 9, 10, 11, 13, 14)_{10}$. Assume the inputs $I_0$ through $I_7$ correspond to $\overline{A}\,\overline{B}\,\overline{C}$, $\overline{A}\,\overline{B}C$, $\overline{A}B\overline{C}$, $\overline{A}BC$, $A\overline{B}\,\overline{C}$, $A\overline{B}C$, $AB\overline{C}$, and $ABC$, respectively, and that each input may be 0, 1, $\overline{D}$, or $D$.

# C H A P T E R

## 14

# Digital Systems

The first half of Chapter 14 continues the analysis of digital circuits that was begun in Chapter 13 by focusing on sequential logic circuits, such as flip-flops, counters, and shift registers. The second half of the chapter is devoted to an overview of the basic functions of microcontrollers and microcomputers. During the last decade, microcomputers have become a standard tool in the analysis of engineering data, in the design of experiments, and in the control of plants and processes. No longer a specialized electronic device to be used only by appropriately trained computer engineers, today's microcomputer—perhaps more commonly represented by the ubiquitous *personal computer*—is a basic tool in the engineering profession. The common thread in its application in various engineering fields is its use in digital data acquisition instruments and digital controllers.

Modern microcomputers are relatively easy to program, have significant computing power and excellent memory storage capabilities, and can be readily interfaced with other instruments and electronic devices, such as transducers, printers, and other computers. The basic functions performed by the microcomputer in a typical digital data acquisition or control application are easily described: input signals (often analog, sometimes already in digital form) are acquired by the computer and processed by means of suitable software to produce the desired result (i.e., they undergo some kind of mathematical manipulation), which is then outputted to either a display or a storage device, or is used in controlling a process, a plant, or an experiment. The objective of this chapter is to describe these various processes, with the aim of giving the reader enough background information to understand the notation used in data books and instruction manuals.

Upon completing this chapter you should be able to:

- Analyze sequential circuits including *RS*, *D*, and *JK* flip-flops.
- Understand the operation of binary, decade, and ring counters.
- Design simple sequential circuits using state transition diagrams.
- Understand the basic architecture of microprocessors and microcomputers.

## 14.1 SEQUENTIAL LOGIC MODULES

The discussion of logic devices in Chapter 13 focuses on the general family of combinational logic devices. The feature that distinguishes combinational logic devices from the other major family—**sequential logic devices**—is that combinational logic circuits provide outputs that are based on a combination of present inputs only. On the other hand, sequential logic circuits depend on present and past input values. Because of this "memory" property, sequential circuits can store information; this capability opens a whole new area of application for digital logic circuits.

### Latches and Flip-Flops

The basic information-storage device in a digital circuit is called a **flip-flop.** There are many different varieties of flip-flops; however, all flip-flops share the following characteristics:

1. A flip-flop is a **bistable device;** that is, it can remain in one of two stable states (0 and 1) until appropriate conditions cause it to change state. Thus, a flip-flop can serve as a memory element.
2. A flip-flop has two outputs, one of which is the complement of the other.

#### RS Flip-Flop

It is customary to depict flip-flops by their block diagram and a name—such as $Q$ or $X$—representing the output variable. Figure 14.1 represents the so-called **RS flip-flop,** which has two inputs, denoted by $S$ and $R$, and two outputs, $Q$ and $\overline{Q}$. The value at $Q$ is called the *state* of the flip-flop. If $Q = 1$, we refer to the device as *being in the 1 state.* Thus, we need define only one of the two outputs of the flip-flop. The two inputs, $R$ and $S$, are used to change the state of the flip-flop, according to the following rules:

1. When $R = S = 0$, the flip-flop remains in its present state (whether 1 or 0).
2. When $S = 1$ and $R = 0$, the flip-flop is *set* to the 1 state (thus, the letter $S$, for **set**).
3. When $S = 0$ and $R = 1$, the flip-flop is *reset* to the 0 state (thus, the letter $R$, for **reset**).
4. It is not permitted for both $S$ and $R$ to be equal to 1. (This would correspond to requiring the flip-flop to set and reset at the same time.)

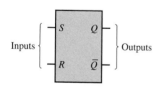

| S | R | Q |
|---|---|---|
| 0 | 0 | Present state |
| 0 | 1 | Reset |
| 1 | 0 | Set |
| 1 | 0 | Disallowed |

**Figure 14.1** *RS* flip-flop symbol and truth table

Inputs { S Q } Outputs { R $\overline{Q}$ }

The rules just described are easily remembered by noting that 1s on the $S$ and $R$ inputs correspond to the set and reset commands, respectively.

A convenient means of describing the series of transitions that occur as the signals sent to the flip-flop inputs change is the **timing diagram.** A timing diagram is a graph of the inputs and outputs of the $RS$ flip-flop (or any other logic device) depicting the transitions that occur over time. In effect, one could also represent these transitions in tabular form; however, the timing diagram provides a convenient visual representation of the evolution of the state of the flip-flop. Figure 14.2 depicts a table of transitions for an $RS$ flip-flop $Q$, as well as the corresponding timing diagram.

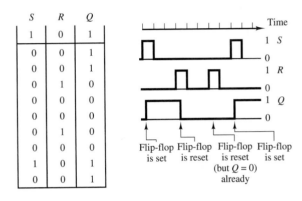

**Figure 14.2** Timing diagram for the $RS$ flip-flop

It is important to note that the $RS$ flip-flop is **level-sensitive.** This means that the set and reset operations are completed only after the $R$ and $S$ inputs have reached the appropriate levels. Thus, in Figure 14.2 we show the transitions in the $Q$ output as occurring with a small delay relative to the transitions in the $R$ and $S$ inputs.

It is instructive to illustrate how an $RS$ flip-flop can be constructed using simple logic gates. For example, Figure 14.3 depicts a realization of such a circuit consisting of four gates: two inverters and two NAND gates (actually, the same result could be achieved with four NAND gates). Consider the case in which the circuit is in the initial state $Q = 0$ (and therefore $\overline{Q} = 1$). If the input $S = 1$ is applied, the top NOT gate will see inputs $\overline{Q} = 1$ and $\overline{S} = 0$, so that $Q = (\overline{\overline{S} \cdot \overline{Q}}) = (\overline{0 \cdot 1}) = 1$—that is, the flip-flop is set. Note that when $Q$ is set to 1, $\overline{Q}$ becomes 0. This, however, does not affect the state of the $Q$ output, since replacing $\overline{Q}$ with 0 in the expression

$$Q = (\overline{\overline{S} \cdot \overline{Q}})$$

does not change the result:

$$Q = (\overline{0 \cdot 0}) = 1$$

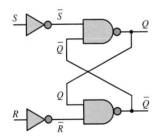

**Figure 14.3** Logic gate implementation of the $RS$ flip-flop

Thus, the cross-coupled feedback from outputs $Q$ and $\overline{Q}$ to the input of the NAND gates is such that the set condition sustains itself. It is straightforward to show (by symmetry) that a 1 input on the $R$ line causes the device to reset (i.e., causes $Q = 0$) and that this condition is also self-sustaining.

### EXAMPLE 14.1 *RS* Flip-Flop Timing Diagram

#### Problem

Determine the output of an *RS* flip-flop for the series of inputs given in the table below.

| R | 0 | 0 | 0 | 1 | 0 | 0 | 0 |
|---|---|---|---|---|---|---|---|
| S | 1 | 0 | 1 | 0 | 0 | 1 | 0 |

#### Solution

***Known Quantities:*** *RS* flip-flop truth table (Figure 14.1).

***Find:*** Output of *RS* flip-flop, *Q*.

***Analysis:*** We complete the timing diagram for the *RS* flip-flop following the rules stated earlier to determine the output of the device; the result is summarized below.

| R | 0 | 0 | 0 | 1 | 0 | 0 | 0 |
|---|---|---|---|---|---|---|---|
| S | 1 | 0 | 1 | 0 | 0 | 1 | 0 |
| Q | 1 | 1 | 1 | 0 | 0 | 1 | 1 |

A sketch of the waveforms, shown below, can also be generated to visualize the transitions.

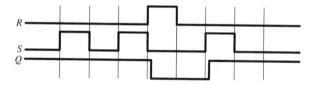

An extension of the *RS* flip-flop includes an additional enable input that is *gated* into each of the other two inputs. Figure 14.4 depicts an *RS* flip-flop consisting of two NOR gates. In addition, an enable input is connected through two AND gates to the *RS* flip-flop, so that an input to the *R* or *S* line will be effective only when the enable input is 1. Thus, any transitions will be controlled by the enable input, which acts as a synchronizing signal. The enable signal may consist of a **clock,** in which case the flip-flop is said to be **clocked** and its operation is said to be **synchronous.**

The same circuit of Figure 14.4 can be used to illustrate two additional features of flip-flops: the **preset** and **clear** functions, denoted by the inputs *P* and *C*, respectively. When *P* and *C* are 0, they do not affect the operation of the flip-flop. Setting *P* = 1 corresponds to setting *S* = 1, and therefore causes the flip-flop to go into the 1 state. Thus, the term *preset:* this function allows the user to preset the flip-flop to 1 at any time. When *C* is 1, the flip-flop is reset, or *cleared* (i.e., *Q* is made equal to 0). Note that these direct inputs are, in general,

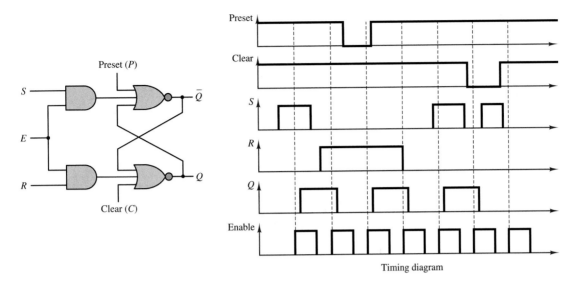

**Figure 14.4** *RS* flip-flop with enable, preset, and clear lines

asynchronous; therefore, they allow the user to preset or clear the flip-flop at any time. A set of timing waveforms illustrating the function of the enable, preset, and clear inputs is also shown in Figure 14.4. Note how transitions occur only when the enable input goes high (unless the preset or clear inputs are used to override the *RS* inputs).

Another extension of the *RS* flip-flop, called the **data latch,** or **delay element,** is shown in Figure 14.5. In this circuit, the *R* input is always equal to the inverted *S* input, so that whenever the enable input is high, the flip-flop is set. This device has the dual advantage of avoiding the potential conflict that might arise if both *R* and *S* were high and reducing the number of input connections by eliminating the reset input. This circuit is called a data latch or delay because once the enable input goes low, the flip-flop is latched to the previous value of the input. Thus, this device can serve as a basic memory element, delaying the output by one clock count with respect to the input.

### D Flip-Flop

The **D flip-flop** is an extension of the data latch that utilizes two *RS* flip-flops, as shown in Figure 14.6. In this circuit, a clock is connected to the enable input of each flip-flop. Since $Q_1$ sees an inverted clock signal, the latch is enabled when the clock waveform goes low. However, since $Q_2$ is disabled when the clock is low, the output of the *D* flip-flop will not switch to the 1 state until the clock goes high, enabling the second latch and transferring the state of $Q_1$ to $Q_2$. It is important to note that the *D* flip-flop changes state only on the positive edge of the clock waveform: $Q_1$ is set on the negative edge of the clock, and $Q_2$ (and therefore *Q*) is set on the positive edge of the clock, as shown in the timing diagram of Figure 14.6. This type of device is said to be **edge-triggered.** This feature is indicated by the "knife edge" drawn next to the CLK input in the

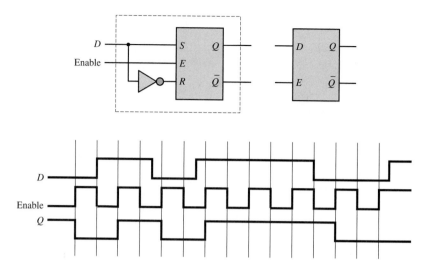

**Figure 14.5** Data latch and associated timing diagram

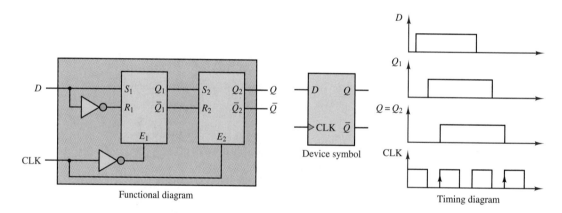

Functional diagram          Device symbol          Timing diagram

**Figure 14.6** $D$ flip-flop functional diagram, symbol, and timing waveforms

device symbol. The particular device described here is said to be positive edge–triggered, or **leading edge–triggered,** since the final output of the flip-flop is set on a positive-going clock transition.

On the basis of the rules stated in this section, the state of the $D$ flip-flop can be described by means of the following truth table:

| $D$ | **CLK** | $Q$ |
|-----|---------|-----|
| 0   | ↑       | 0   |
| 1   | ↑       | 1   |

where the symbol ↑ indicates the occurrence of a positive transition.

### JK Flip-Flop

Another very common type of flip-flop is the **JK flip-flop,** shown in Figure 14.7. The *JK* flip-flop operates according to the following rules:

- When *J* and *K* are both low, no change occurs in the state of the flip-flop.
- When $J = 0$ and $K = 1$, the flip-flop is reset to 0.
- When $J = 1$ and $K = 0$, the flip-flop is set to 1.
- When both *J* and *K* are high, the flip-flop will toggle between states at every negative transition of the clock input, denoted from here on by the symbol ↓.

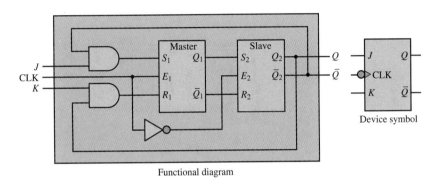

Functional diagram

**Figure 14.7** *JK* flip-flop functional diagram and device symbol

Note that, functionally, the operation of the *JK* flip-flop can also be explained in terms of two *RS* flip-flops. When the clock waveform goes high, the "master" flip-flop is enabled; the "slave" receives the state of the master upon a negative clock transition. The "bubble" at the clock input signifies that the device is negative or **trailing edge–triggered.** This behavior is similar to that of an *RS* flip-flop, except for the $J = 1$, $K = 1$ condition, which corresponds to a toggle mode rather than to a disallowed combination of inputs.

Figure 14.8 depicts the truth table for the *JK* flip-flop. It is important to note that when both inputs are 0 the flip-flop remains in its previous state at the occurrence of a clock transition; when either input is high and the other is low, the *JK* flip-flop behaves like the *RS* flip-flop, whereas if both inputs are high, the output "toggles" between states every time the clock waveform undergoes a negative transition.

Data sheets for various types of flip-flops may be found in the accompanying CD-ROM.

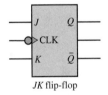

*JK* flip-flop

| $J_n$ | $K_n$ | $Q_{n+1}$ |
|---|---|---|
| 0 | 0 | $Q_n$ |
| 0 | 1 | 0 (reset) |
| 1 | 0 | 1 (set) |
| 1 | 1 | $\bar{Q}_n$ (toggle) |

**Figure 14.8** Truth table for the *JK* flip-flop

---

### EXAMPLE 14.2 The *T* Flip-Flop

**Problem**

Determine the truth table and timing diagram of the **T flip-flop** of Figure 14.9. Note that the *T* flip-flop is a *JK* flip-flop with its inputs tied together.

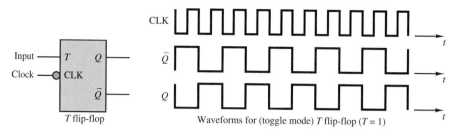

**Figure 14.9** $T$ flip-flop symbol and timing waveforms

### Solution

***Known Quantities:*** *JK* flip-flop rules of operation (Figure 14.8).

***Find:*** Truth table and timing diagram for $T$ flip-flop.

***Analysis:*** We recognize that the $T$ flip-flop is a *JK* flip-flop with its inputs tied together. Thus, the flip-flop will need only a two-element truth table to describe its operation, corresponding to the top and bottom entries in the *JK* flip-flop truth table of Figure 14.8. The truth table is shown below. A timing diagram is also included in Figure 14.9.

| $T$ | **CLK** | $Q_{k+1}$ |
|-----|---------|-----------|
| 0 | ↓ | $Q_k$ |
| 1 | ↓ | $\overline{Q_k}$ |

***Comments:*** The $T$ flip-flop takes its name from the fact that it *toggles* between the high and low state. Note that the toggling frequency is one half that of the clock. Thus the $T$ flip-flop also acts as a *divide-by-2* counter. Counters are explored in more detail in the next subsection.

### EXAMPLE   14.3  *JK* Flip-Flop Timing Diagram

### Problem

Determine the output of a *JK* flip-flop for the series of inputs given in the table below. The initial state of the flip-flop is $Q_0 = 1$.

| $J$ | 0 | 1 | 0 | 1 | 0 | 0 | 1 |
|-----|---|---|---|---|---|---|---|
| $K$ | 0 | 1 | 1 | 0 | 0 | 1 | 1 |

### Solution

***Known Quantities:*** *JK* flip-flop truth table (Figure 14.8).

***Find:*** Output of *RS* flip-flop, $Q$, as a function of the input transitions.

*Analysis:* We complete the timing diagram for the *JK* flip-flop following the rules of Figure 14.8; the result is summarized below.

| J | 0 | 0 | 0 | 1 | 0 | 0 | 0 |
|---|---|---|---|---|---|---|---|
| K | 1 | 0 | 1 | 0 | 0 | 1 | 0 |
| Q | 1 | 0 | 0 | 1 | 1 | 0 | 1 |

A sketch of the waveforms, shown below, can also be generated to visualize the transitions. Each vertical line corresponds to a clock transition.

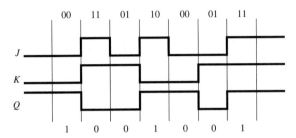

*Comments:* How would the timing diagram change if the initial state of the flip-flop were $Q_0 = 1$?

## Digital Counters

One of the more immediate applications of flip-flops is in the design of **counters.** A counter is a sequential logic device that can take one of $N$ possible states, stepping through these states in a sequential fashion. When the counter has reached its last state, it resets to zero and is ready to start counting again. For example, a three-bit **binary up counter** would have $2^3 = 8$ possible states, and might appear as shown in the functional block of Figure 14.10. The input clock waveform causes the counter to step through the eight states, making one transition for each clock pulse. We shall shortly see that a string of *JK* flip-flops can accomplish this task exactly. The device shown in Figure 14.10 also displays a reset input, which forces the counter output to equal 0: $b_2 b_1 b_0 = 000$.

Although binary counters are very useful in many applications, one is often interested in a **decade counter,** that is, a counter that counts from 0 to 9 and then resets. A four-bit binary counter can easily be configured in principle to provide this function by means of simple logic that resets the counter when it has reached the count $1001_2 = 9_{10}$. As shown in Figure 14.11, if we connect bits $b_3$ and $b_1$ to a four-input AND gate, along with $\bar{b}_2$ and $\bar{b}_0$, the output of the AND gate can be used to reset the counter after a count of 10. Additional logic can provide a "carry" bit whenever a reset condition is reached, which could be passed along to another decade counter, enabling counts up to 99. Decade counters can be cascaded so as to represent decimal digits in succession.

Although the decade counter of Figure 14.11 is attractive because of its simplicity, this configuration would never be used in practice, because of the

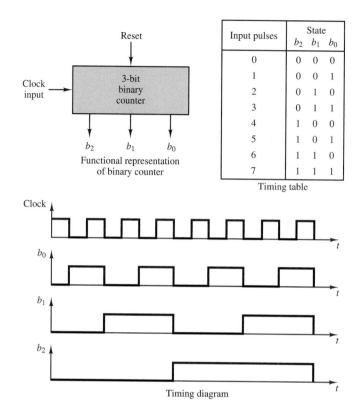

Figure 14.10 Binary up counter functional representation, state table, and timing waveforms

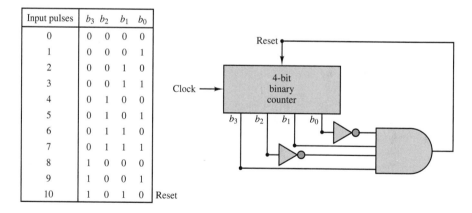

Figure 14.11 Decade counter

presence of **propagation delays.** These delays are caused by the finite response time of the individual transistors in each logic device and cannot be guaranteed to be identical for each gate and flip-flop. Thus, if the reset signal—which is presumed to be applied at exactly the same time to each of the four $JK$ flip-flops in the four-bit binary counter—does not cause the *JK* flip-flops to reset at exactly the same time on account of different propagation delays, then the binary word appearing at the output of the counter will change from 1001 to some other number, and the output of the four-input NAND gate will no longer be high.  In such a condition, the flip-flops that have not already reset will then not be able to reset, and the counting sequence will be irreparably compromised.

What can be done to obviate this problem?  The answer is to use a systematic approach to the design of sequential circuits making use of **state transition diagrams.**  This topic will be discussed in the next section.

A simple implementation of the binary counter we have described in terms of its functional behavior is shown in Figure 14.12.  The figure depicts a three-bit binary **ripple counter,** which is obtained from a cascade of three *JK* flip-flops. The transition table shown in the figure illustrates how the *Q* output of each stage becomes the clock input to the next stage, while each flip-flop is held in the toggle mode.  The output transitions assume that the clock, CLK, is a simple square wave (all *JK*s are negative edge–triggered).

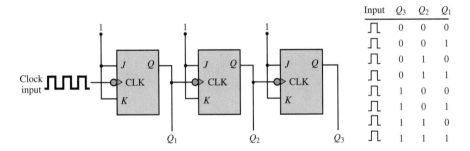

| Input | $Q_3$ | $Q_2$ | $Q_1$ |
|-------|-------|-------|-------|
| ⊓ | 0 | 0 | 0 |
| ⊓ | 0 | 0 | 1 |
| ⊓ | 0 | 1 | 0 |
| ⊓ | 0 | 1 | 1 |
| ⊓ | 1 | 0 | 0 |
| ⊓ | 1 | 0 | 1 |
| ⊓ | 1 | 1 | 0 |
| ⊓ | 1 | 1 | 1 |

**Figure 14.12** Ripple counter

This 3-bit ripple counter can easily be configured as a divide-by-8 mechanism, simply by adding an AND gate.  To divide the input clock rate by 8, one output pulse should be generated for every eight clock pulses.  If one were to output a pulse every time a binary 111 combination occurs, a simple AND gate would suffice to generate the required condition.  This solution is shown in Figure 14.13. Note that the square wave is also included as an input to the AND gate; this ensures that the output is only as wide as the input signal.  This application of ripple counters is further illustrated in the following example.

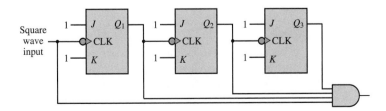

**Figure 14.13** Divide-by-8 circuit

### EXAMPLE 14.4 Divider Circuit

**Problem**

Draw the timing diagram for the clock input, $Q_0$ and $Q_1$, for the binary ripple counter of Figure 14.14.

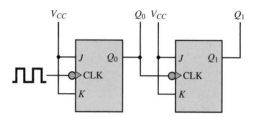

**Figure 14.14**

**Solution**

***Known Quantities:*** *JK* flip-flop truth table (Figure 14.8).

***Find:*** Output of each flip-flop, $Q$, as a function of the input clock transitions.

***Assumptions:*** Assume negative-edge–triggered devices.

***Analysis:*** Following the timing diagram of Figure 14.12, we see that $Q_0$ switches at half the frequency of the clock input, and that $Q_1$ switches at half the frequency of $Q_0$. Hence the timing diagram shown below.

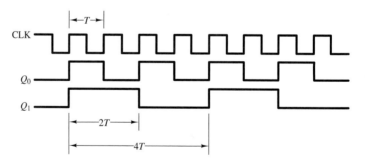

A slightly more complex version of the binary counter is the so-called **synchronous counter,** in which the input clock drives all of the flip-flops simultaneously. Figure 14.15 depicts a three-bit synchronous counter. In this figure, we have chosen to represent each flip-flop as a $T$ flip-flop. The clocks to all the flip-flops are incremented simultaneously. The reader should verify that $Q_0$ toggles to 1 first and then $Q_1$ toggles to 1, and that the AND gate ensures that $Q_2$ will toggle only after $Q_0$ and $Q_1$ have both reached the 1 state ($Q_0 \cdot Q_1 = 1$).

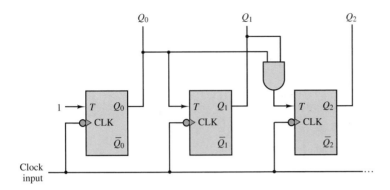

**Figure 14.15** Three-bit synchronous counter

Other common counters are the **ring counter,** illustrated in Example 14.5, and the **up-down counter,** which has an additional select input that determines whether the counter counts up or down. Data sheets for various counters may be found in the accompanying CD-ROM.

## EXAMPLE  14.5  Ring Counter

### Problem

Draw the timing diagram for the ring counter of Figure 14.16.

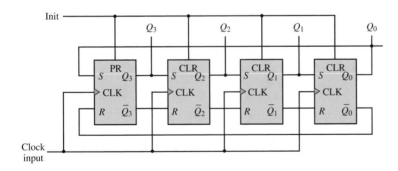

**Figure 14.16** Ring counter

### Solution

***Known Quantities:*** *JK* flip-flop truth table (Figure 14.8).

***Find:*** Output of each flip-flop, $Q$, as a function of the input clock transitions.

***Assumptions:*** Assume that prior to applying the clock input the Init line sees a positive transition (this initializes the counter by setting the state of the first flip-flop to 1 through a PR (preset) input, and all other states to zero through a CLR (clear) input).

***Analysis:*** With the initial state of $Q_3 = 0$, a clock transition will *set* $Q_3 = 1$. The clock also causes the other three flip-flops to see a *reset* input of 1, since

$Q_3 = Q_2 = Q_1 = Q_0 = 0$ at the time of the first clock pulse. Thus, $Q_2$, $Q_1$ and $Q_0$ remain in the zero state. At the second clock pulse, since $Q_3$ is now 1, the second flip-flop will see a *set* input of one, and its output will become $Q_2 = 1$. $Q_1$ and $Q_0$ remain in the zero state, and $Q_3$ is reset to 0. The pattern continues, causing the 1-state to ripple from left to right and back again. This rightward rotation gives the counter its name. The transition table is shown below.

| CLK | $Q_3$ | $Q_2$ | $Q_1$ | $Q_0$ |
|:---:|:---:|:---:|:---:|:---:|
| ↑ | 1 | 0 | 0 | 0 |
| ↑ | 0 | 0 | 1 | 0 |
| ↑ | 0 | 1 | 0 | 0 |
| ↑ | 0 | 0 | 0 | 1 |
| ↑ | 1 | 0 | 0 | 0 |
| ↑ | 0 | 1 | 0 | 0 |
| ↑ | 0 | 0 | 1 | 0 |

**VIRTUAL LAB**

***Comments:*** The shifting function implemented by the ring counter is used in the shift registers discussed in the following subsection.

***Focus on Computer-Aided Solutions:*** A ring counter simulation generated by Electronics Workbench™ may be found in the accompanying CD-ROM.

**FOCUS ON MEASUREMENTS**

## Digital Measurement of Angular Position and Velocity

Another type of angular position encoder, besides the angular encoder discussed in Chapter 13 in "Focus on Measurements: Position Encoders," is the slotted encoder shown in Figure 14.17. This encoder can be used in conjunction with a pair of counters and a high-frequency clock to determine the speed of rotation of the slotted wheel. As shown in Figure 14.18, a clock of known frequency is connected to a counter while another counter records the number of slot pulses detected by an optical slot detector as the wheel rotates. Dividing the counter values, one could obtain the speed of the rotating wheel in radians per second. For example, assume a clocking frequency of 1.2 kHz. If both counters are started at zero and at some instant the timer counter reads 2,850 and the encoder counter reads 3,050, then the speed of the rotating encoder is found to be:

$$1,200\frac{\text{cycles}}{\text{second}} \cdot \frac{2,850 \text{ slots}}{3,050 \text{ cycles}} = 1,121.3\frac{\text{slots}}{\text{second}}$$

and

$$1,121.3 \text{ slots per second} \times 1° \text{ per slot} \times 2\pi/360 \text{ rad/degree}$$
$$= 19.6 \text{ rad/s}$$

**FIND IT**

**ON THE WEB**

If this encoder is connected to a rotating shaft, it is possible to measure the angular position and velocity of the shaft. Such shaft encoders are used in **measuring the speed of rotation of electric motors, machine tools, engines,** and other rotating machinery.

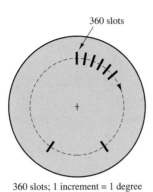

360 slots

360 slots; 1 increment = 1 degree

**Figure 14.17**

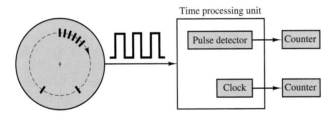

**Figure 14.18** Calculating the speed of rotation of the slotted wheel

A typical application of the slotted encoder is to compute the ignition and injection timing in an automotive engine. In an automotive engine, information related to speed is obtained from the camshaft and the flywheel, which have known reference points. The reference points determine the timing for the ignition firing points and fuel injection pulses, and are identified by special slot patterns on the camshaft and crankshaft. Two methods are used to detect the special slots (reference points): *period measurement with additional transition detection (PMA)*, and *period measurement with missing transition detection (PMM)*. In the PMA method, an additional slot (reference point) determines a known reference position on the crankshaft or camshaft. In the PMM method, the reference position is determined by the absence of a slot. Figure 14.19 illustrates a typical PMA pulse sequence, showing the presence of an additional pulse. The additional slot may be used to determine the timing for the ignition pulses relative to a known position of the crankshaft. Figure 14.20 depicts a typical PMM pulse sequence. Because the period of the pulses is known, the additional slot or the missing slot can be easily detected and used as a reference position. How would you implement these pulse sequences using ring counters?

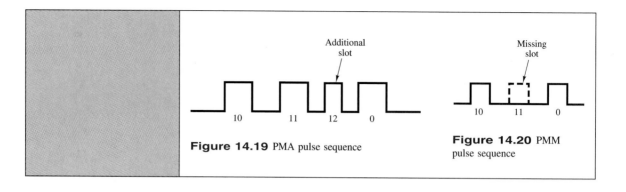

**Figure 14.19** PMA pulse sequence

**Figure 14.20** PMM pulse sequence

## Registers

A register consists of a cascade of flip-flops that can store binary data, one bit in each flip-flop. The simplest type of register is the parallel input–parallel output register shown in Figure 14.21. In this register, the "load" input pulse, which acts on all clocks simultaneously, causes the parallel inputs $b_0 b_1 b_2 b_3$ to be transferred to the respective flip-flops. The $D$ flip-flop employed in this register allows the transfer from $b_n$ to $Q_n$ to occur very directly. Thus, $D$ flip-flops are very commonly used in this type of application. The binary word $b_3 b_2 b_1 b_0$ is now "stored," each bit being represented by the state of a flip-flop. Until the "load" input is applied again and a new word appears at the parallel inputs, the register will preserve the stored word.

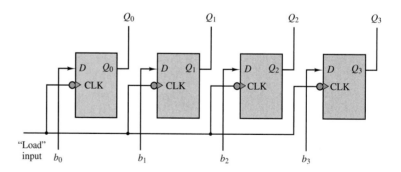

**Figure 14.21** Four-bit parallel register

The construction of the parallel register presumes that the $N$-bit word to be stored is available in parallel form. However, it is often true that a binary word will arrive in serial form, that is, one bit at a time. A register that can accommodate this type of logic signal is called a **shift register.** Figure 14.22 illustrates how the same basic structure of the parallel register applies to the shift register, except that the input is now applied to the first flip-flop and shifted along at each clock pulse. Note that this type of register provides both a serial and a parallel output.

Data sheets for some common registers are included in the accompanying CD-ROM.

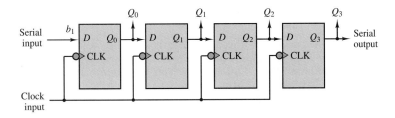

**Figure 14.22** Four-bit shift register

## Seven-Segment Display

A **seven-segment display** is a very convenient device for displaying digital data. The display is shown in Figure 14.23. Operation of a seven-segment display requires a decoder circuit to light the proper combinations of segments corresponding to the desired decimal digit.

This display, with the appropriate decoder driver, is capable of displaying values ranging from 0 to 9.

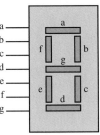

**Figure 14.23** Seven-segment display

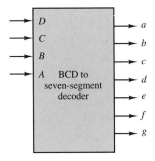

**Figure 14.24**

A typical BCD to seven-segment decoder function block is shown in Figure 14.24, where the lowercase letters correspond to the segments shown in Figure 14.23. The decoder features four data inputs ($A$, $B$, $C$, $D$), which are used to light the appropriate segment(s). The outputs of the decoder are connected to the seven-segment display. The decoder will light up the appropriate segments corresponding to the incoming value. A BCD to seven-segment decoder function is similar to the 2-to-4 decoder function described in Chapter 13 and shown in Figure 13.61. Data sheets for seven-segment display drivers may be found in the accompanying CD-ROM.

## Check Your Understanding

**14.1**   The circuit shown in Figure 14.25 also serves as an *RS* flip-flop and requires only two NOR gates. Analyze the circuit to prove that it operates as an *RS* flip-flop. [*Hint:* Use a truth table with two variables, *S* and *R*.]

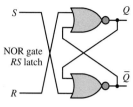

NOR gate
*RS* latch

**Figure 14.25**

**14.2**   Derive the detailed truth table and draw a timing diagram for the *JK* flip-flop, using the two–flip-flop model of Figure 14.7.

**14.3**   The speed of the rotating encoder of "Focus on Measurements: Digital Measurement of Angular Position and Velocity" is found to be 9,425 rad/s. The encoder timer reads 10 and the clock counter reads 300. Assuming that both the timer counter and the encoder counter started at zero, find the clock frequency.

## 14.2   SEQUENTIAL LOGIC DESIGN

The design of sequential circuits, just like the design of combinational circuits, can be carried out by means of a systematic procedure. You will recall how the Karnaugh map, introduced in Chapter 13, allowed us to formalize the design procedures for an arbitrary combinational circuit. The equivalent of a Karnaugh map for a sequential circuit is the **state diagram,** with its associated **state transition table.** To illustrate these concepts, it is best to proceed with an example. Consider the three-bit binary counter of Figure 14.26, which is made up of three $T$ flip-flops. You can easily verify that the input equations for this counter are $T_1 = 1$, $T_2 = q_1$, and $T_3 = q_1 q_2$. Knowing the inputs, we can determine the three outputs from these relationships at any time. The outputs $Q_1$, $Q_2$, and $Q_3$ form the **state** of the machine. It is straightforward to show that as the clock goes through a series of cycles, the counter will go through the transitions shown in Table 14.1, where we indicate the current state by lowercase $q$ and the next state by an uppercase $Q$. Note that the state diagram of Figure 14.26 provides information regarding the sequence of states assumed by the counter in graphical form. In a state diagram, each state is denoted by a circle called a **node,** and the transition from one state to another is indicated by a **directed edge,** that is, a line with a directional arrow. The analysis of sequential circuits consists of determining either their transition table or their state diagram.

The reverse of this analysis process is the design process. How can one systematically arrive at the design of a sequential circuit, such as a counter, by

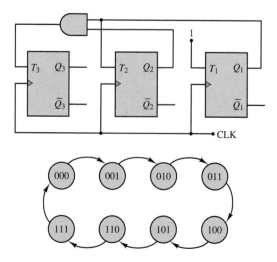

**Figure 14.26** Three-bit binary counter and state diagram

Table 14.1    State transition table for three-bit binary counter

| Current state | | | Input | | | Next state | | |
|---|---|---|---|---|---|---|---|---|
| $q_3$ | $q_2$ | $q_1$ | $T_3$ | $T_2$ | $T_1$ | $Q_3$ | $Q_2$ | $Q_1$ |
| 0 | 0 | 0 | 0 | 0 | 1 | 0 | 0 | 1 |
| 0 | 0 | 1 | 0 | 1 | 1 | 0 | 1 | 0 |
| 0 | 1 | 0 | 0 | 0 | 1 | 0 | 1 | 1 |
| 0 | 1 | 1 | 1 | 1 | 1 | 1 | 0 | 0 |
| 1 | 0 | 0 | 0 | 0 | 1 | 1 | 0 | 1 |
| 1 | 0 | 1 | 0 | 1 | 1 | 1 | 1 | 0 |
| 1 | 1 | 0 | 0 | 0 | 1 | 1 | 1 | 1 |
| 1 | 1 | 1 | 1 | 1 | 1 | 0 | 0 | 0 |

employing state transition tables and state diagrams? The design procedure will be explained in this section.

The initial specification for a logic circuit is usually in the form of either a transition table or a state diagram. The design will differ depending on the type of flip-flop used. Therefore one must first choose a flip-flop and define its behavior in the form of an excitation table. Truth tables and excitation tables for the $RS$, $D$, and $JK$ flip-flops are given in Tables 14.2, 14.3, and 14.4.

Table 14.2    Truth table and excitation table for $RS$ flip-flop

| Truth table for RS flip-flop | | | | Excitation table for RS flip-flop | | | |
|---|---|---|---|---|---|---|---|
| $S$ | $R$ | $Q_t$ | $Q_{t+1}$ | $Q_t$ | $Q_{t+1}$ | $S$ | $R$ |
| 0 | 0 | 0 | 0 | 0 | 0 | 0 | $d^b$ |
| 0 | 0 | 1 | 1 | 0 | 1 | 1 | 0 |
| 0 | 1 | 0 | 0 | 1 | 0 | 0 | 1 |
| 0 | 1 | 1 | 0 | 1 | 1 | d | 0 |
| 1 | 0 | 0 | 1 | | | | |
| 1 | 0 | 1 | 1 | | | | |
| 1 | 1 | $X^a$ | X | | | | |
| 1 | 1 | X | X | | | | |

$^a$ An X indicates that this combination of inputs is not allowed.
$^b$ A "d" denotes a don't care entry.

Table 14.3    Truth Table and excitation table for $D$ flip-flop

| Truth table for D flip-flop | | | Excitation table for D flip-flop | | |
|---|---|---|---|---|---|
| $D$ | $Q_t$ | $Q_{t+1}$ | $Q_t$ | $Q_{t+1}$ | $D$ |
| 0 | 0 | 0 | 0 | 0 | 0 |
| 0 | 1 | 0 | 0 | 1 | 1 |
| 1 | 0 | 1 | 1 | 0 | 0 |
| 1 | 1 | 1 | 1 | 1 | 1 |

The use of excitation tables will now be demonstrated through an example. Let us design a **modulo-4 binary up-down counter,** that is, a counter that can change state counting up or down in the binary sequence from 0 to 3. For example, if the current state of the counter is 2, an input of 1 will cause the counter to change state "up" to 3, while an input of 0 will cause the counter to count "down" to 1. The state diagram for this counter is given in Figure 14.27. We choose two $RS$ flip-flops for the implementation (the number of flip-flops must be sufficient to cover all the necessary states—two flip-flops are sufficient for a four-state machine) and begin constructing Table 14.5 by listing the possible inputs, denoted by the variable $x$, and their effect on the counter. Since the counter can have four states and there are two inputs, we must look at eight possible combinations. The first five columns of Table 14.5 describe the behavior of the counter for all possible inputs and present states; the behavior of the counter consists of determining the next state, denoted

**Table 14.4**  Truth table and excitation table for *JK* flip-flop

| *Truth table for JK flip-flop* | | | | *Excitation table for JK flip-flop* | | | |
|---|---|---|---|---|---|---|---|
| *J* | *K* | $Q_t$ | $Q_{t+1}$ | $Q_t$ | $Q_{t+1}$ | *J* | *K* |
| 0 | 0 | 0 | 0 | 0 | 0 | 0 | *d* |
| 0 | 0 | 1 | 1 | 0 | 1 | 1 | *d* |
| 0 | 1 | 0 | 0 | 1 | 0 | *d* | 1 |
| 0 | 1 | 1 | 0 | 1 | 1 | *d* | 0 |
| 1 | 0 | 0 | 1 | | | | |
| 1 | 0 | 1 | 1 | | | | |
| 1 | 1 | 0 | 1 | | | | |
| 1 | 1 | 1 | 0 | | | | |

**Table 14.5**  State transition table for modulo-4 up-down counter

| Input *x* | Current state $q_1$ | Current state $q_2$ | Next state $Q_1$ | Next state $Q_2$ | $S_1$ | $R_1$ | $S_2$ | $R_2$ | Output *y* |
|---|---|---|---|---|---|---|---|---|---|
| 0 | 0 | 0 | 1 | 1 | 1 | 0 | 1 | 0 | 1 |
| 0 | 0 | 1 | 0 | 0 | 0 | *d* | 0 | 1 | 0 |
| 0 | 1 | 0 | 0 | 1 | 0 | 1 | 1 | 0 | 1 |
| 0 | 1 | 1 | 1 | 0 | *d* | 0 | 0 | 1 | 0 |
| 1 | 0 | 0 | 0 | 1 | 0 | *d* | 1 | 0 | 1 |
| 1 | 0 | 1 | 1 | 0 | 1 | 0 | 0 | 1 | 0 |
| 1 | 1 | 0 | 1 | 1 | *d* | 0 | 1 | 0 | 1 |
| 1 | 1 | 1 | 0 | 0 | 0 | 1 | 0 | 1 | 0 |

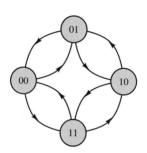

**Figure 14.27** State diagram of a modulo-4 up-down counter

by $Q_1Q_2$, given the input, $x$, and the current state, $q_1q_2$. Note that the first five columns of Table 14.5 contain exactly the same information that is given in the diagram of Figure 14.27. Now we can refer to the excitation table of the *RS* flip-flop to see what *R* and *S* inputs are required to obtain the desired counter function. For example, if $q_1 = 1$ and we wish to have $Q_1 = 0$, we must have $S_1 = 0$ and $R_1 = 1$ (we are resetting the first flip-flop). An entire state transition is handled by considering each flip-flop independently; for example, if we desire a transition from $q_1q_2 = 10$ to $Q_1Q_2 = 01$, we must have $S_1 = 0$ and $R_1 = 1$, as already stated, and $S_2 = 1$ and $R_2 = 0$. Repeating this analysis for each possible transition, one can then fill the next four columns of Table 14.5 with the values shown, where "d" represents a don't care condition.

So far, we have been able to determine the desired inputs for each flip-flop based on the counter input and on the desired state transition. Now we need to design a logic circuit that will cause the flip-flop inputs to be as stated in Table 14.5 in response to the input, $x$. This is a rather simple combinational logic problem, illustrated by the Karnaugh maps of Figure 14.28. From the Karnaugh maps we obtain the following expressions:

$$S_1 = \overline{x}\overline{q}_1\overline{q}_2 + x\overline{q}_1\overline{q}_2 = (\overline{x}\overline{q}_2 + xq_2)\overline{q}_1$$
$$R_1 = \overline{x}q_1\overline{q}_2 + xq_1q_2 = (\overline{x}\overline{q}_2 + xq_2)q_1$$
$$S_2 = \overline{q}_2$$
$$R_2 = q_2$$

which allow us to complete the design, as shown in Figure 14.29.

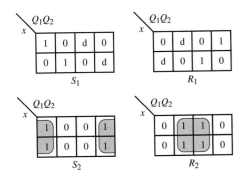

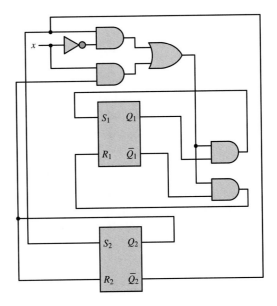

**Figure 14.28** Karnaugh maps for flip-flop inputs in modulo-4 counter

**Figure 14.29** Implementation of modulo-4 counter

The procedure outlined in this section can be applied to more complex sequential circuits using the same basic steps. More advanced problems are explored in the homework problems.

## 14.3  MICROCOMPUTERS

To bring the broad range of applicability of microcomputers in engineering into perspective, it will be useful to stop for a moment to consider the possible application of microcomputer systems to different fields. The following list—by no means exhaustive—provides a few suggestions; it would be a useful exercise to imagine other likely applications in your own discipline.

| | |
|---|---|
| Civil engineering | Measurement of stresses and vibration in structures |
| Chemical engineering | Process control |
| Industrial engineering | Control of manufacturing processes |
| Material and metallurgical engineering | Measurement of material properties |
| Marine engineering | Instrumentation to determine ship location, ship propulsion control |
| Aerospace engineering | Instrumentation for flight control and navigation |
| Mechanical engineering | Mechanical measurements, robotics, control of machine tools |
| Nuclear engineering | Radiation measurement, reactor instrumentation |
| Biomedical engineering | Measurement of physiological functions (e.g., electrocardiography and electroencephalography), control of experiments |

The massive presence of microcomputers in engineering laboratories and in plants and production facilities can be explained by considering the numerous advantages the computer can afford over more traditional instrumentation and control technologies. Consider, for example, the following points:

- A single microcomputer can perform computations and send signals from many different sensors measuring different parameters to many different display, storage, or control devices, under control of a single software program.

- The microcomputer is easily reprogrammed for any changes or adjustments to the measurement or control procedures, or in the computations.

- A permanent record of the activities performed by the microcomputer can be easily stored and retained.

It should be evident that the microcomputer can perform repetitive tasks, or tasks that require great accuracy and repeatability, far better than could be expected of human operators and analog instruments. What, then, does constitute a **digital data acquisition and control system?** Figure 14.30 depicts the basic blocks that form such a system. In the figure, the user of the microcomputer system is shown to interact with the microcomputer by means of software, often called **application software.** Application software is a collection of programs written either in **high-level languages,** such as C, C++, or Unix shell, or in **assembly language** (a programming language very close to the internal code used by the microcomputer). The particular application software used may be commercially available or may be provided by the user; a combination of these two cases is the norm.

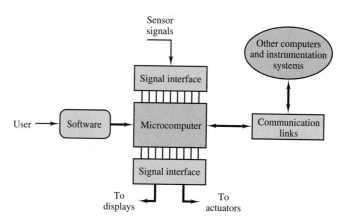

**Figure 14.30** Structure of a digital data acquisition and control system

The signals that originate from real-world sensors—signals related to temperatures, vibration, or flow, for example—are interfaced to the microcomputer by means of specialized circuitry that converts analog signals to digital form and times the flow of information into the microcomputer using a clock reference, which may be internal to the microcomputer or externally provided. The heart of the signal interface unit is the *analog-to-digital converter,* or *ADC,* which will be discussed in some detail in Chapter 15. Not all sensor signals are analog, though. For example, the position of a switch or an on-off valve might be of interest; signals of this type are binary in nature, and the signal interface unit can route such signals directly

# Mechatronics

Industry and the consumer market require engineering processes and products that are more reliable, more efficient, smaller, faster, and less expensive. The production and development of these devices require engineers who can understand and design systems from an integrated perspective. A **FIND IT** discipline that shows particular promise in this arena is ***mechatronic design,*** based on the integration of mechanical engineering, electrical engineering, and computer science (Figure 14.31). Most major programs in the United States don't emphasize mechatronics as a primary curriculum component, but there is an industry-motivated push to change this situation.

Mechatronics is an especially important and interesting domain for modern industry for a number of reasons. The automotive, aerospace, manufacturing, power systems, test and instrumentation, consumer and industrial electronics industries make use of and contribute to mechatronics. Mechatronic design has surfaced as a new philosophy of design, based on the integration of existing disciplines primarily mechanical, and electrical, electronic, and software engineering.[1-7] Design elements from these traditional disciplines don't simply exist side by side, but are deeply integrated in the design process. Whether a given functionality should be achieved electronically, by software, or by elements from electrical or mechanical engineering domains requires mastery of analysis and synthesis techniques from the different areas. Being a successful mechatronics design engineer requires an in-depth understanding of many, if not all, of its constituent disciplines.

One of the distinguishing features of the mechatronic approach to the design of products and processes is the use of *embedded microcontrollers*. These microcontrollers replace many mechanical functions with electronic ones, resulting in much greater flexibility, ease of redesign or reprogramming, the ability to implement distributed control in complex systems, and the ability to conduct automated data collection and reporting. Mechatronic design represents the fusion of traditional mechanical, electrical, and software engineering design methods with sensors and instrumentation technology, electric drive and actuator technology, and embedded real-time microprocessor systems and real-time software. Mechatronic systems range from heavy industrial machinery, to vehicle propulsion systems, to precision electromechanical motion control devices.

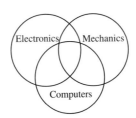

**Figure 14.31** Mechatronics as the intersection of three engineering disciplines

[1] S. Ashley, "Getting Hold on Mechatronics", *Mechanical Engineering,* Vol. 119, No. 5, 1997.

[2] D. Auslander, "What is Mechatronics?", *IEEE/ASME Trans. on Mechatronics,* Vol. 1, No. 1, 1996.

[3] F. Harashima, M. Tomizuka, and T. Fukuda, "Mechatronics—What Is It and How?", *IEEE/ASME Trans. on Mechatronics,* Vol. 1, No. 2, 1996.

[4] G. Rizzoni, A. Keyhani, "Design of Mechatronic Systems: An Integrated, Inter-Departmental Curriculum", *Mechatronics,* Vol. 5, No. 7, July 1995.

[5] G. Rizzoni, "Development of a Mechatronics Curriculum at the Ohio State University," *ASME International Mechanical Engineering Congress and Exposition,* San Francisco, 1996.

[6] D. Auslander, C. J. Kempf, *Mechatronics: Mechanical System Interfacing,* Prentice-Hall, Upper Saddle River, NJ, 1996, 242 pp.

[7] G. Rizzoni, A. Keyhani, G. Washington, G. Baumgartner, B. Chandrasekaran, "Education in Mechatronic Systems at the Ohio State University," *ASME International Mechanical Engineering Congress and Exposition, Proc. Dyn. Sys. and Control Division,* Anaheim, CA, November, 1998.

to the microcomputer. Once the sensor data has been acquired and converted to digital form, the microcomputer can perform computations on the data and either display or store the results, or generate command outputs to actuators through another signal interface. Actuators are devices that can generate a physical output (e.g., force, heat, flow, pressure) from an electrical input. Some actuators can be controlled directly by means of a digital signal (e.g., an on-off valve), but some require an analog input voltage or current, which can be obtained from the digital signal generated by the microcomputer by means of a *digital-to-analog converter,* or *DAC.*

In addition to the program control exercised by the user, the microcomputer may also respond to inputs originating from other computer and instrumentation systems through appropriate **communication links,** which also permit communication in the reverse direction. Thus, a microcomputer system dedicated to a complex task may consist of several microcomputers tied over a **communication network.**

The present chapter will describe the basic architecture and operation of a special class of microcomputers, called microcontrollers, while Chapter 14 will explore instrumentation-related issues.

## 14.4    MICROCOMPUTER ARCHITECTURE

Prior to delving into a description of how microcomputers interface with external devices (such as sensors and actuators) and communicate with the outside world, it will be useful to discuss the general architecture of a microcomputer, in order to establish a precise nomenclature and paint a clear picture of the major functions required in the operation of a typical microcomputer.

The general structure of a microcomputer is shown in Figure 14.32. It should be noted immediately that each of the blocks that are part of the microcomputer is connected with the **CPU bus,** which is the physical wire connection allowing each of the subsystems to communicate with the others. In effect, the CPU bus is simply a set of wires; note, however, that since only one set of signals can travel over the data bus at any one time, it is extremely important that the transmission of data between different parts of the microcomputer (e.g., from the A/D unit to memory) be managed properly, to prevent interference with other functions (e.g., the display of unwanted data on a video terminal). As will be explained shortly, the task of managing the operation of the CPU bus resides within the **central processing unit,** or **CPU.** The CPU has the task of managing the flow of data and coordinating the different functions of the microcomputer, in addition to performing the data processing—in effect, the CPU is the heart and brains of the microcomputer. Some of the major functions of the CPU will be discussed in more detail shortly.

One of the important features of the microcomputer is its ability to store data. This is made possible by different types of **memory** elements: **read-only memory, or ROM; read and write memory (random-access memory), or RAM;** and **mass storage memory,** such as hard drive or floppy disk, tape, or optical drives. ROM is *nonvolatile memory:* it will retain its data whether the operating power is on or off. ROM memory contains software programs that are used frequently by the microcomputer; one example is the *bootstrap program,* which is necessary to first start up the computer when power is turned on. RAM is memory that can be accessed very rapidly by the CPU; data can be either read from or written

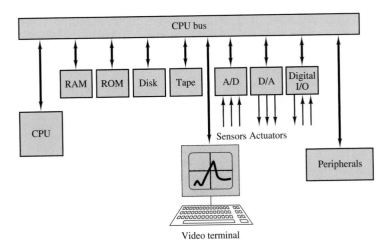

**Figure 14.32** Microcomputer architecture

into RAM very rapidly. RAM is therefore used primarily during the execution of programs to store partial or permanent results, as well as to store all of the software currently in use by the computer. The main difference between RAM and mass storage memory, such as a hard drive or a tape drive, is therefore in the speed of access: RAM can be accessed in tens of nanoseconds, whereas a hard drive requires access time on the order of microseconds, and tape drive, on the order of seconds. Another important distinction between RAM and mass memory is that the latter is far less expensive for an equivalent storage capability, the price typically being lower for longer access time.

A video terminal enables the user to enter programs and to display the data acquired by the microcomputer. The video terminal is one of many **peripherals** that enable the computer to communicate information to the outside world. Among these peripherals are printers, and devices that enable communication between computers, such as *modems* (a modem enables the computer to send and receive data over a telephone line). Finally, Figure 14.33 depicts real-time input/output (I/O) devices, such as analog-to-digital and digital-to-analog converters and digital I/O devices. These are the devices that allow the microcomputer to read signals from external sensors, to output signals to actuators, and to exchange data with other computers.

## 14.5    MICROCONTROLLERS

A **microcontroller** is a special-purpose microcomputer system, designed to perform the functions illustrated in Figure 14.28. Microcontrollers have become an essential part of many engineering products, processes and systems, and are often deeply embedded in the inner workings of many products and systems we use daily (for example, in automobile control systems, and in many consumer products and appliances, such as autofocus cameras and washing machines). This section introduces the operation of a general-purpose microcontroller using as an example the architecture of the Motorola 68HC05. Although much more powerful microcontrollers are available, the 8-bit, 2-MHz, HCMOS-technology (where "H" stands for "high speed") MC68HC05 contains all of the essential elements that make microcontrollers so useful. The material presented in this and the next sections is intended to serve as an overview. Supplemental material, including the details of the instruction set for the MC68HC05 and illustrative examples, may be found in the accompanying CD-ROM.

## Computer Architecture

Computer systems come in many different sizes, from the large mainframe computers that run entire companies and hospitals, to the powerful networked workstations that are the workhorse of computer-aided engineering design, to the ubiquitous personal computers, whether in desktop or laptop form, to the microcontrollers that are the subject of the remainder of this chapter. All computer systems are characterized by the same basic elements, although the details may differ significantly. Figure 14.32 depicts the overall structure of a computer system. A central processor unit (CPU) is timed by a **clock** to execute instructions contained in **memory** at a certain rate, determined by the frequency of the clock. The instructions contained in memory originate from computer **programs,** which are loaded into memory as needed. The flow of instruction execution is controlled by various external **inputs** and **outputs.** Inputs could consist of keyboard commands (as is often the case in personal computers), or of information provided by sensors, or of the position of switches (the last two inputs are very typical of microcontrollers). Typical outputs could be to a video display, a magnetic or optical storage device, a printer, or a plotter (all common with desktop PCs); microcontroller outputs are more likely to activate LED displays, relays, and actuators such as motors or valves.

We discuss the details of some typical microcontroller inputs and outputs later in this section and in the next section, where an application example is presented. Inputs and outputs can be either *analog* (that is, representing continuous values) or *digital* (representing discrete values). Digital inputs can be directly accepted by a CPU, while analog inputs require the use of an *analog-to-digital converter* (or ADC). ADCs are described in detail in the next chapter. Similarly, the CPU can directly generate digital outputs, while a *digital-to-analog converter* (DAC, also introduced in Chapter 15) is required to generate an analog output. Next, we outline the important properties of each of the elements of the block diagram of Figure 14.33.

### *The Central Processor Unit (CPU)*

The function of the CPU is to execute the program of instruction contained in memory. The CPU will therefore be required to **read** information from inputs and to **write** information to outputs. To accomplish these tasks, the CPU reads from and writes to memory. Microcontroller programs are usually much simpler than those that operate, say, in a desktop computer. This is because microcontrollers are usually dedicated to a few specific tasks. The **instruction set** of the M68HC05, which is the native programming language of this processor, is based on approximately 60 different instructions. We shall see in a later subsection ["Operation of the Central Processing Unit (CPU)"] how these instructions are used to execute desired functions.

### *The Clock*

The **clock** represents the heartbeat of the microcontroller. The clock function is typically implemented by a **crystal oscillator** that determines the basic clock cycle

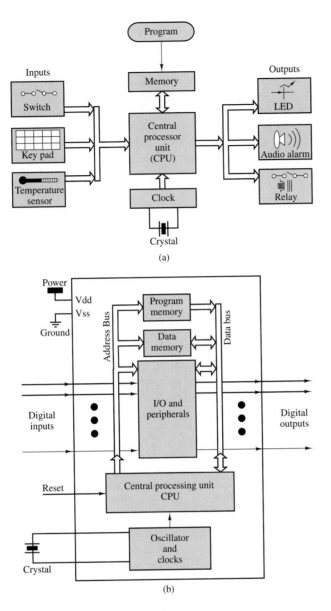

**Figure 14.33** (a) High-level block diagram of microcontroller; (b) Internal organization of microcontroller

for the execution of each instruction step. Each step takes one clock cycle to complete. We are all familiar with the rating of processor speeds in megahertz. Typical microcontrollers are capable of speeds in the megahertz to tens of megahertz range.

### Memory

The CPU needs to have access to different kinds of memory to execute programs. **Read-only memory (ROM)** is used for permanent programs and data that are necessary, for example, to boot and initialize the system. Information stored in ROM remains unchanged even when power to the computer is turned off. **Random**

**access read/write memory (RAM)** is used to temporarily store data and instructions. For example, the program that is executed by the CPU and the intermediate results of the calculations are stored in RAM. Many microcontrollers also employ **erasable programmable read-only memory (EPROM)** or **electrically erasable programmable read-only memory (EEPROM);** these types of memory are used like ROM, but can be reprogrammed relatively easily using special "EPROM burners." EPROM and EEPROM can be very useful if one wishes to make small but important changes to the functions executed by the microcontroller. For example, the microcontrollers used in automotive applications are usually the same from one model year to the next and across various vehicle platforms, but differences in control strategies and calibration data among vehicles, and changes and fixes required from one model year to the next are usually accommodated by means of EPROMs. Section 14.5 describes an automotive application.

Computer memory is arranged on the basis of **bits,** that is, a single digital variable with value of 0 or 1. Bits are grouped in **bytes,** consisting of 8 bits, and in **words,** consisting of 16 or 32 bits. While the size of a word can vary, a byte always consists of 8 bits. Small microcontrollers such as the 6805 have access to a relatively limited amount of memory (e.g., 64 Kbytes); more powerful microcontrollers may access as much as 1 Mbyte. Note that the memory capacity of a microcontroller is significantly smaller than that of the personal computers you are likely to use in your work.

Mass storage devices (magnetic and optical storage devices, such as hard drives and CD-ROMs) can also be used to increase a computer's access to data and information. Access to such external devices is much slower than access to ROM and RAM, and is therefore usually not practical in embedded microcontrollers.

### Computer Programs

A computer program is a listing of instructions to be executed by the CPU. The instructions are coded in a special **machine language** that consists of combinations of bytes. To assist the programmer, each CPU instruction is associated to a **mnemonic instruction code,** which associated a short word or abbreviated code to each instruction (for example, the instruction ASL in the M68HC05 stands for arithmetic shift left). More sophisticated software development systems allow the instructor to program in a higher-level language (often the C programming language); the high-level language program is then translated into machine code by a **compiler.** We shall devote one of the next subsections to programming issues.

## Number Systems and Number Codes in Digital Computers

### Number Systems

It should be already clear that computers operate on the basis of the binary number system. The binary and hexadecimal number systems were introduced in Chapter 13. The hexadecimal system is particularly well suited for use in computer codes, because it allows a much more compact notation than the binary code would. The hexadecimal code, as you will recall, permits expressing a four-bit binary number as a single-digit, using the numbers from 0 to 9 and the letters A to F. The range of possible combinations that can be expressed in a 16-bit word, for example, can be represented: in decimal numbers as the range from $010$ to $216 - 1 = 65,5351010$; as $0000\ 0000\ 0000\ 00002$ to $1111\ 1111\ 1111\ 11112$ in binary code; and as $000016$ to $FFFF16$ in hexadecimal code.

It's very common to precede a hexadecimal number with the $ symbol to differentiate it from a decimal representation; for example, $32 would be interpreted as a hexadecimal word, and 32 as a decimal number.

### Computer Codes

In addition to the codes that were described in Chapter 13 (binary, octal, hexadecimal, binary-coded decimal), a standard convention adopted by all computer manufacturers is the so-called **ASCII**[8] **code,** defined in Table 14.6. The ASCII code defines the alphanumeric characters that are typically associated with text used in programming.

Instructions to the CPU are coded as **operation codes,** or **opcodes.** Each opcode instructs the CPU to perform a sequence of steps that correspond to an operation (for example, an addition). Although all computers perform essentially the same basic tasks at the binary level, the manner in which these tasks are performed varies depending on the computer manufacturer, and therefore opcodes vary from manufacture to manufacturer. The **instruction set** of a specific computer is the set of all basic operations that the computer can perform. For example, the 6805 can execute 62 basic instructions, which are arranged into 210 unique opcodes. The difference between a basic instruction and an opcode is that the same basic instruction can be used in slightly different ways (in conjunction with other instructions) to perform a specific operation. Thus, opcodes are the basic building block of the programming language of a computer.

### Mnemonics and Assemblers

To assist the programmer in remembering and identifying the function of opcodes, **mnemonics** are used. A mnemonic is an alphabetic abbreviation that corresponds to a specific opcode. Thus, the programmer writes a program using mnemonics, and the program is translated into **machine code** (consisting of opcodes and data) by a computer program called an **assembler.** A more detailed discussion of programming issue follows in a later subsection.

## Memory Organization

Memory performs an essential function in microcontrollers. Different kinds of memory are used to store information of different types. The three basic types of memory are described in Section 13.5. ROM and EPROM are used to store the operating system and the programs used by the controller. RAM is used by the CPU to read and write instructions during the execution of a computer program.

Memory is usually organized in the form of a **memory map,** which is a graphical representation of the allocation of the memory used by a particular microcontroller. Example 14.6 describes the use of memory in a typical microcontroller.

---

[8]American Standard Code for Information Interchange.

Table 14.6  ASCII code

| Graphic or control | ASCII (hex) | Graphic or control | ASCII (hex) | Graphic or control | ASCII (hex) |
|---|---|---|---|---|---|
| NUL | 00 | + | 2B | V | 56 |
| SOH | 01 | , | 2C | W | 57 |
| STX | 02 | − | 2D | X | 58 |
| ETX | 03 | . | 2E | Y | 59 |
| EOT | 04 | / | 2F | Z | 5A |
| ENQ | 05 | 0 | 30 | [ | 5B |
| ACK | 06 | 1 | 31 | \ | 5C |
| BEL | 07 | 2 | 32 | ] | 5D |
| BS | 08 | 3 | 33 | ↑ | 5E |
| HT | 09 | 4 | 34 | ← | 5F |
| LF | 0A | 5 | 35 | ` | 60 |
| VT | 0B | 6 | 36 | a | 61 |
| FF | 0C | 7 | 37 | b | 62 |
| CR | 0D | 8 | 38 | c | 63 |
| SO | 0E | 9 | 39 | d | 64 |
| SI | 0F | : | 3A | e | 65 |
| DLE | 10 | ; | 3B | f | 66 |
| DC1 | 11 | < | 3C | g | 67 |
| DC2 | 12 | = | 3D | h | 68 |
| DC3 | 13 | > | 3E | i | 69 |
| DC4 | 14 | ? | 3F | j | 6A |
| NAK | 15 | @ | 40 | k | 6B |
| SYN | 16 | A | 41 | l | 6C |
| ETB | 17 | B | 42 | m | 6D |
| CAN | 18 | C | 43 | n | 6E |
| EM | 19 | D | 44 | o | 6F |
| SUB | 1A | E | 45 | p | 70 |
| ESC | 1B | F | 46 | q | 71 |
| FS | 1C | G | 47 | r | 72 |
| GS | 1D | H | 48 | s | 73 |
| RS | 1E | I | 49 | t | 74 |
| US | 1F | J | 4A | u | 75 |
| SP | 20 | K | 4B | v | 76 |
| ! | 21 | L | 4C | w | 77 |
| " | 22 | M | 4D | x | 78 |
| # | 23 | N | 4E | y | 79 |
| $ | 24 | O | 4F | z | 7A |
| % | 25 | P | 50 | { | 7B |
| & | 26 | Q | 51 | | | 7C |
| ' | 27 | R | 52 | } | 7D |
| ( | 28 | S | 53 | ~ | 7E |
| ) | 29 | T | 54 | DEL | 7F |
| * | 2A | U | 55 | | |

## EXAMPLE 14.6 Writing Data to and Reading Data from I/O Ports

### Problem

1. Write specified data to an I/O address.
2. Assume that an input device is connected to the address E6H. Write the code necessary to read a byte from this input device.

### Solution

*Known Quantities:* Desired data, I/O address.

*Find:* Write the appropriate sequence of commands using the MC68HC05 instruction set.

*Schematics, Diagrams, Circuits, and Given Data:* The data to be written to the I/O port is 36H (decimal number 36); the I/O address is A6H.

*Assumptions:* Data is written to the accumulator register first.

*Analysis:*

1. The command to write to an I/O port is *STA$⟨address⟩*. The command assumes that the data is in the accumulator register, thus we first must load the accumulator with the desired value:

       LDA #$36        ; load accumulator with 36H

       STA $00A6       ; write accumulator to I/O port A6H

2. The command to read from an I/O port is LDA⟨address⟩. The value read from the input port is stored in the accumulator. To store the byte into the accumulator register we use the command:

       LDA $00E6       ; load accumulator with E6H

*Comments:* The CD-ROM that accompanies this book contains the complete instruction set for the MC68HC05 microcontroller.

## Operation of the Central Processing Unit (CPU)

The M68HC05 is organized as follows. Five **CPU registers** can be directly accessed by the CPU (i.e., without the need to access memory); the memory map defines the names and types of the memory locations that are accessible to the CPU in addition to the registers.

The **accumulator,** or **A register,** is used to hold the results of arithmetic operations performed by the CPU.

The **index,** or **X register,** is used to point to an address in memory where the CPU will read or write information. This register is used to perform a function called *indexed addressing*, which is described in more detail in the M68HC05 instruction set found in the accompanying CD-ROM.

The **program counter (PC) register** keeps track of the address of the next instruction to be executed by the CPU.

The **condition code register (CCR)** holds information that reflects the status of prior CPU operations. For example, *branch instructions* look at the CCR to make either/or decisions.

The **stack pointer (SP) register** contains return address information and the previous content of all CPU registers, so that if the CPU is *interrupted* or a subroutine is initiated (we shall visit this concept soon), the status of the program prior to the **interrupt** or prior to branching to the subroutine is retained. Once the CPU has finished *servicing* the interrupt or has completed the subroutine, it can resume its previous operations by loading the contents of the SP register.

## Interrupts

**Interrupts** perform a very important function in microcontrollers by allowing the CPU to interrupt is normal flow of operations to respond to an external event. For example, an interrupt request may occur when an *analog-to-digital converter* (described in more detail in Chapter 15) has completed the conversion of an analog signal to digital form, so that the digital value of a sensor reading may be made available to the CPU for further processing. The following "Focus on Measurements" illustrates an automotive application of this concept.

**FOCUS ON MEASUREMENTS**

### Reading Sensor Data By Using Interrupts

In modern automotive instrumentation, a microprocessor performs all of the signal-processing operations for several measurements. A block diagram for such instrumentation is given in Figure 14.34. Depending on the technology used, the sensors' outputs can be either digital or analog. If the sensor signals are analog, they must be converted to digital format by means of an analog-to-digital converter (ADC), as shown in Figure 14.35. The analog-to-digital conversion process requires an amount of time that depends on the individual ADC, as will be explained in Chapter 15. After the conversion is completed, the ADC then signals the computer by changing the logic state on a separate line that sets its *interrupt request* flip-flop. This flip-flop stores the ADC's interrupt request until it is acknowledged (see Figure 14.36).

When an interrupt occurs, the processor automatically jumps to a designated program location and executes the interrupt service subroutine. For the ADC, this would be a subroutine to read the conversion results and store them in some appropriate location, or to perform an operation on them. When the processor responds to the interrupt, the interrupt request flip-flop is cleared by a direct signal from the processor. To resume the execution of the program at the proper point upon completion of the ADC service subroutine, the program counter content is automatically saved before control is transferred to the service subroutine. The service subroutine saves in a stack the content of any registers it uses, and restores the registers' content before returning.

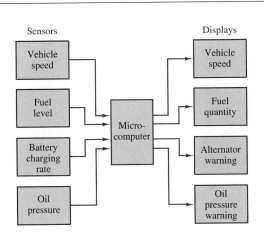

**Figure 14.34** Automotive instrumentation

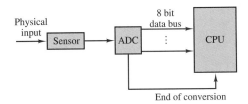

**Figure 14.35** Sensor interface

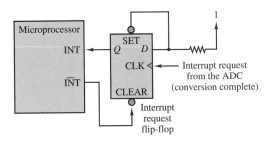

**Figure 14.36** Interrupt request in a microprocessor

The interrupt may occur at any point in a program's execution, independent of the internal clock; it is therefore referred to as an *asynchronous* event.

## Instruction Set for the MC68HC05 Microcontroller

The complete instruction set for the MC68HC05 microcontroller may be found in the accompanying CD-ROM.

## Programming and Application Development in a Microcontroller

Section 14.6 illustrates the use of a microcontroller in a very common application: the control of an automotive engine.

## 14.6    A TYPICAL AUTOMOTIVE ENGINE MICROCONTROLLER

This section gives the reader an insight to the functioning of a typical **automotive engine controller.** The system described is based on the features of commercially available 32-bit microprocessors.   Table 14.7 lists the microcontroller's major characteristics.

Table 14.7   Characteristics of automotive microcontroller

| Processor section | I/O section inputs | I/O section outputs |
|---|---|---|
| Microcoded timing channels (TPU—time processor unit) | Variable reluctance sensor interface[a] | Discrete low-side drivers |
| Discrete I/O channels | Hall sensor[b] interfaces (cam, crank) | PWM low-side drivers |
| PWM channels | Analog input | Low-side-driven fuel injectors[c] |
| 8-bit ADC channels | Exhaust gas oxygen sensors | Low-side-driven coil drivers |
| 10-bit ADC channels | Discrete pull-ups to ignition | High-side drivers |
| Boot memory (flash) | Discrete pull-downs to ground | High- and low-side current-controlled outputs |
| RAM | PWM/frequency inputs | Stepper motor driver |
| Serial communication (RS-232, CAN, Class II, UART) | Power and ground | H-bridge driver |

[a]See Chapter 16 "Focus on Measurements: Magnetic Reluctance Position Sensor" for an explanation of the operation of this sensor.
[b]See Chapter 7 "Focus on Measurements: How Hall-Effect Current Transducers Work" for an explanation of the operation of this sensor.
[c]See Chapter 11 for an example of a fuel injector driver.

## General Description

The controller consists of a processor section and an input/output (I/O) section, mounted in an enclosure. These two sections are usually combined onto one printed circuit board for cost considerations. A generic controller may be programmed with different software for a wide range of production engine applications, or specific hardware/software designs may be provided for each vehicle application. Through embedded software, the controller is able to communicate with a personal computer (PC) for software debugging and development. Serial bus–based instrumentation is also available to assist in the development process, before an application is released for vehicle production. A block diagram of a typical system configuration is given in Figure 14.37.

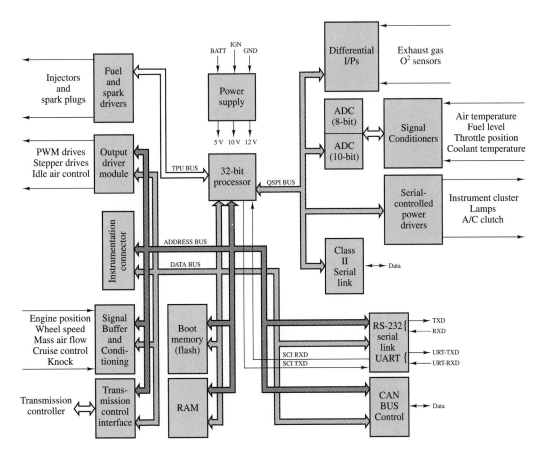

**Figure 14.37** System block diagram

## Processor Section

The processor comprises a 32-bit microprocessor, memory, analog and digital I/O, timing channels, serial communications, and all required supporting logic.

### *Microprocessor*

Typical of many systems, the controller printed circuit board is designed to accept a plug-in replacement microprocessor, in anticipation of updates for higher performance and/or additional features. Such modifications may require a new crystal, different value resistors, capacitors, etc. Typical crystal frequencies for automotive applications would be around 5 MHz, for a clock frequency of about 20 MHz.

A popular family of processors consists of a central processing unit and three integrated modules. Each of these modules can function on its own once it is initialized by the central processing unit.

### *Central Processing Unit*

The processor is equipped with a background mode, which can be used for program download or system diagnostics. No target software is required for operation of this system, provided proper hardware and software for a personal computer (desktop or laptop) are present.

### Timer Processor Unit (TPU)

The TPU has 16 channels of microcode-controlled pulse I/O. It comes with a predefined set of channel control programs called *primitives*. The TPU can be reprogrammed by using the on-chip RAM in the emulation mode. Programming for the TPU can be done by the end user.

### System Integration Module (SIM)

The SIM performs various functions, some of which include:

- External bus interface
- Interrupt detection
- Chip select activation
- Programmable timer interrupt
- Clock generation
- System protection
- System testing

The system is set up through the SIM setup registers.

### Queued Serial Module (QSM)

The QSM contains a serial communications interface (SCI) and a queued serial port interface (QSPI) module. The queuing feature of the QSPI bus, and its associated chip selects, enhances the overall performance of the processors.

## Memory

### Boot Memory

Single-chip processors contain their own flash boot memory. Alternatively, expanded mode processors support external boot memory, which may be a variety of technologies, including EPROM, EEPROM, or NOVRAM (non-volatile RAM). Because they are connected to the boot chip select on the processor, they must contain the boot program to set up the processor control registers. The remaining memory in the parts may be used for any of a number of purposes (program, parameter storage, data logging, etc.).

### RAM

Up to 1 Mbyte of RAM is supported by a typical processor. This should provide sufficient RAM to facilitate data logging and parameter storage.

### Memory Map

Processors control peripheral chips via programmable chip selects, which allow the user to change the memory map through control registers. The memory map is therefore not fixed until initialization has taken place. On power-up, the boot chip select is set to be active in the address range $000000 to $100000. After power-up, the chip selects can be set to create any desired memory map within the limits set by the hardware design.

### Chip Select

Examples of such chip selects are CSBOOT and CS0 to CS10. CSBOOT is always a chip select since it must be used before the registers are configured. CS0 to CS10 may be used for a variety of functions as they are programmable.

The hardware configuration of the board uses some of these chip selects for specific purposes. The Table 14.8 shows a typical use of the chip selects.

Table 14.8   Chip select allocation

| | |
|---|---|
| CSBOOT | Boot memory select. This chip select is combined with other signals in a PAL to produce chip selects for the boot memories. |
| CS0 | RAM select. |
| CS1 | Enable signal buffering and conditioning circuit. |
| CS2 | ADC enable. |
| CS3 | CAN serial interface IC. |
| CS4 | Output driver module enable. |
| CS5 | Transmission controller enable. |
| CS6 | Memory control. |
| CS7 | Spare. |
| CS8 | Spare. |
| CS9 | Spare. |
| CS910 | Spare. |

### Analog-to-Digital Converters

The system includes 10-bit and 8-bit ADCs. Both of these converters communicate with the processor via the QSPI bus. There is a low-pass RC filter on each of the ADC channels.

### Communications

Several different types of serial communications are available on microprocessor chips. Some examples are: SCI, QSPI, Class II, CAN, and RS-232.[9] All interfaces are made available to the user at the interface connector.

The SCI has both TXD (digital transmit) and RXD (digital receive) lines. The SCI lines are available at the interface connector directly and are routed to a level shifter circuit to provide the appropriate signal levels.

Class II is a single-wire serial bus for communication between microprocessor-controlled modules. The bus allows any module to communicate with any other module on the bus. A data link controller IC is used as an interface between the processor and the bus.

CAN is a European-standard, high-speed communications link using a two-wire interface. The controller communicates with the CAN controller over the parallel bus. Serial lines are found on the interface connectors.

---

[9]See Chapter 15 for an introduction to the RS-232 communication protocol.

### *Interrupts*

Many microprocessors have several interrupt input lines which, when set, will interrupt the software process being executed in favor of a new process. Some of these interrupt request lines may be software-maskable, such that their effect is ignored, while others may be nonmaskable, indicating that the processor will always service a request seen on these inputs. Each interrupt line will be prioritized to ensure an orderly process in the event that the processor receives more than one interrupt request. The nonmaskable interrupts will be ranked with the highest priority. Unused interrupt lines may be left disconnected, while a software fault routine may be serviced in the event the microprocessor experiences an input on one of these unused inputs.

### *Input/Output (I/O) Section*

For safety reasons, all outputs are disabled until such time as the processor passes its self-test routine and the system is able to establish proper control of all outputs. This is accomplished by use of the "computer not operating properly" (CNOP) signal, which is used by all significant devices within the controller. A custom circuit generates the CNOP signal.

A brief description of the features of the I/O section follows.

## Inputs

### *Discrete*

Discrete inputs may be "pulled up" to 12 V or "pulled down" to ground potential to ensure a recognizable default condition or to enable fault detection of failed inputs. All discrete inputs normally have some hardware filtering to provide noise immunity.

### *Analog*

Several 8-bit and 10-bit ADC inputs provide a means of feeding the processor digital signals from analog sensors. For maximum accuracy the ADCs should be powered and referenced to the same power supply that will be providing the voltage reference for the associated sensors that it handles. A 10-bit ADC provides a higher signal resolution (1024 bits) than the 8-bit ADC (256 bits). The ADCs communicate with the processor via the QSPI bus.

### *PWM/Frequency*

Pulse-width modulation frequency inputs are read by the TPU preprocessor which relieves the main processor from the burden of handling large amounts of time-dependent data (e.g., spark and fuel information). Microcode, embedded within the preprocessor, defines the way in which these types of signals are processed.

### *Knock*

Knock is a damaging, audible phenomenon that results from preignition of the fuel/air mixture in the combustion chamber. Piezoelectric sensors,[10] mounted

---

[10] See Chapter 11 "Focus on Measurements: Charge Amplifiers" for an introduction to piezoelectric sensors.

on the engine, interface with a custom integrated circuit to detect this damaging condition such that the processor can adjust spark timing to eliminate the knock.

## Outputs

### *Discrete*

All discrete outputs have self-shutdown and diagnostic capability.

### *PWM*

Outputs that are capable of being pulse-width–modulated can generally be used as a discrete output. These outputs also have self- shutdown and diagnostic capability. Alternatively, diagnostics can be implemented via a sense resistor fed back to an amplifier circuit and then on to an ADC.

### *Output Driver Module*

The output driver module is an application-specific integrated circuit (ASIC) designed for discrete I/O processing. It has programmable discrete I/O lines, programmable PWM lines, drives for external *pnp* switches, and set logic circuitry. It has a time-out line which is used as an active low "computer not operating properly" (CNOP) signal to other parts in the system and as a turn-off delay (TOD) for the power supply.

### *Fuel Injectors*

The fuel and spark driver IC controls the fuel injectors, under command of software, via the TPU bus. Diagnostic feedback capability is available by latching the feedback lines with a parallel to serial shift register and then shifting out the data via SPI control.

### *Spark Coil*

Ignition options include two different drive options. The first is driving external ignition coils. The second is high-side driving an external ignition (IGN) module.

Insulated-gate bipolar transistor (IGBT) coil drivers are available in the I/O section. These provide the ability to drive ignition coils directly. These IGBTs also have both analog and discrete feedback capabilities. A sense resistor from the IGBT is buffered and sent to two different circuits.

### *Exhaust Gas Recirculation (EGR) Valve Drivers*

A traditional discrete low-side drive or high-side drive is available for the linear EGR valve interface.

### *Current-Controlled Circuit*

A current-controlled circuit is provided to drive a force motor for direct application to a transmission control. The force motor circuit controls both the high and low side of the load. The force motor circuit is controlled by a PWM signal. Sampling the current and converting that into an analog value provides the feedback. An ADC can read this analog value.

### Stepper Motor

A stepper motor[11] driver is available for applications using a stepper-controlled idle air valve.

### Brushless Motor

A brushless motor[12] driver (BMD) has been provided for applications using this type of actuator. It is configurable for different drive modes.

### Power Supply

The power supply uses battery (BATT) and ignition (IGN) to provide different supply lines used by the controller circuitry. One example would be an independent, close-tolerance, tracking voltage reference, for all sensors that are being read by the ADC. The supply will operate reliably down to a battery voltage of 4–5 volts. The CNOP signal provides a simple means of allowing the controller to perform housekeeping tasks prior to shutdown after IGN goes low.

### Ground Structure

Care must be taken to avoid ground loops[13] within the controller that could result in interference with signal levels. Typical methods employed for this include separation of signal grounds, power grounds, and radio-frequency grounds. All three grounds have their own dedicated circuit board layer.

## 14.7   CONCLUSION

- Sequential logic circuits are digital logic circuits with memory capabilities; their operation is described by state transition tables and state diagrams. Counters and registers are the two principal classes of sequential circuits. Sequential circuits can be designed using a formal procedure analogous to the use of Karnaugh maps for combinational circuits.

- Digital systems play a prominent role in modern engineering. The microprocessor, in particular, has become an integral part of instrumentation and control systems. The microprocessor is very flexible in its application because it can be programmed to perform many different tasks. Depending on the computing power required, 8-, 16-, and 32-bit microprocessors are used for automating measurement and control functions in a variety of industrial applications.

## CHECK YOUR UNDERSTANDING ANSWERS

| **CYU 14.3** | 4.5 kHz | **CYU 14.4** | 2,048 | **CYU 14.5** | 32,768 |

---

[11] See Section 18.2 for an introduction to stepper motors.

[12] See Section 18.1 for an introduction to brushless DC motors.

[13] See Section 15.2 for an introduction to grounding and noise issues in circuit design.

# HOMEWORK PROBLEMS

**14.1**   The input to the circuit of Figure P14.1 is a square wave having a period of 2 s, maximum value of 5 V, and minimum value of 0 V. Assume all flip-flops are initially in the RESET state.

a. Explain what the circuit does.

b. Sketch the timing diagram, including the input and all four outputs.

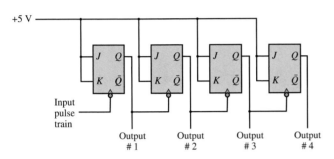

**Figure P14.1**

**14.2**   A binary pulse counter can be constructed by interconnecting $T$-type flip-flops in an appropriate manner. Assume it is desired to construct a counter which can count up to $100_{10}$.

a. How many flip-flops would be required?

b. Sketch the circuit needed to implement this counter.

**14.3**   Explain what the circuit of Figure P14.3 does and how it works. Hint: This circuit is called a 2-bit synchronous binary up-down counter.

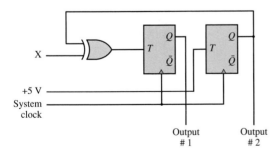

**Figure P14.3**

**14.4**   Suppose a circuit is constructed from 3 $D$-type flip-flops, with

$$D_0 = Q_2 \qquad D_1 = Q_2 \oplus Q_0 \qquad D_2 = Q_1$$

a. Draw the circuit diagram.

b. Assume the circuit starts with all flip-flops SET. Sketch a timing diagram which shows the outputs of all three flip-flops.

**14.5**   Suppose that you want to use a $D$ flip-flop for a laboratory experiment. However, you have only $T$ flip-flops. Assuming that you have all the logic gates available, make a $D$ flip-flop using a $T$ flip-flop and some logic gate(s).

**14.6**   Draw a timing diagram (four complete clock cycles) for $A_0$, $A_1$, and $A_2$ for the circuit of Figure P14.6. Assume that all initial values are 0. Note that all flip-flops are negative edge–triggered.

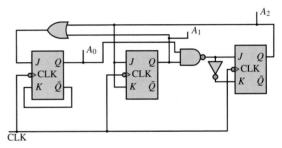

**Figure P14.6**

**14.7**   Assume that the slotted encoder shown in Figure P14.7 has a length of 1 meter and a total of 1,000 slots (i.e., there is one slot per millimeter). If a counter is incremented by 1 each time a slot goes past a sensor, design a digital counting system that determines the speed of the moving encoder (in meters per second).

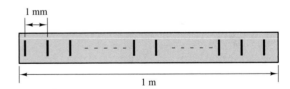

**Figure P14.7**

**14.8**   Find the output $Q$ for the circuit of Figure P14.8.

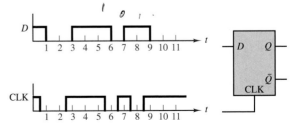

**Figure P14.8**

**14.9**   Describe how the ripple counter works. Why is it so named? What disadvantages can you think of for this counter?

**14.10**  Write the truth table for an *RS* flip-flop with enable (E), preset (P), and clear (C) lines.

**14.11**  A *JK* flip-flop is wired as shown in Figure P14.11 with a given input signal. Assuming that $Q$ is at logic 0 initially and the trailing edge triggering is effective, sketch the output $Q$.

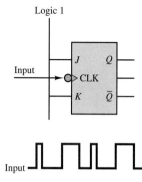

**Figure P14.11**

**14.12**  With reference to the *JK* flip-flop of Problem 14.11, assume that the output at the $Q$ terminal is made to serve as the input to a second *JK* flip-flop wired exactly as the first. Sketch the $Q$ output of the second flip-flop.

**14.13**  Assume that there is a flip-flop with the characteristic given in Figure P14.13, where $A$ and $B$ are the inputs to the flip-flop and $Q$ is the next state output. Using necessary logic gates, make a $T$ flip-flop from this flip-flop.

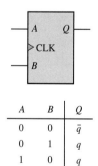

| A | B | Q |
|---|---|---|
| 0 | 0 | $\bar{q}$ |
| 0 | 1 | $q$ |
| 1 | 0 | $q$ |
| 1 | 1 | 0 |

**Figure P14.13**

**14.14**  A typical PC has 32 Mbytes of standard memory.
   a.  How many words is this?
   b.  How many nibbles is this?
   c.  How many bits is this?

**14.15**  Suppose a microprocessor has *n* registers.
   a.  How many control lines do you need to connect each register to all other registers?
   b.  How many control lines do you need if a bus is used?

**14.16**  Suppose it is desired to implement a 4-Kbyte 16-bit memory.
   a.  How many bits are required for the memory address register?
   b.  How many bits are required for the memory data register?

**14.17**  What is the distinction between volatile and nonvolatile memory?

**14.18**  Suppose a particular magnetic tape can be formatted with eight tracks per centimeter of tape width. The recording density is 200 bits/cm, and the transport mechanism moves the tape past the read heads at a velocity of 25 cm/s. How many bytes/s can be read from a 2-cm-wide tape?

**14.19**  Draw a block diagram of a circuit that will interface two interrupts, INT0 and INT1, to the INT input of a CPU so that INT1 has the higher priority and INT0 has the lower. In other words, a signal on INT1 is to be able to interrupt the CPU even when the CPU is currently handling an interrupt generated by INT0, but not vice versa.

# C H A P T E R

# 15

# Electronic Instrumentation and Measurements

This chapter introduces measurement and instrumentation systems and summarizes important concepts by building on the foundation provided in earlier chapters. The development of the chapter follows a logical thread, starting from the physical sensors and proceeding through wiring and grounding to signal conditioning and analog-to-digital conversion, and finally to digital data transmission.

The first section presents an overview of sensors commonly used in engineering measurements. Some sensing devices have already been covered in earlier chapters, and others will be discussed in later chapters; the main emphasis in this chapter will be on classifying physical sensors, and on providing additional detail on some sensors not presented elsewhere in this book—most notably, temperature transducers. The second section of the chapter describes the common signal connections and proper wiring and grounding techniques, with emphasis on noise sources and techniques for reducing undesired interference. Section 15.3 provides an essential introduction to digital signal conditioning, namely, a discussion of instrumentation amplifiers and active filters. The last three sections introduce analog-to-digital conversion, other integrated circuits used in instrumentation systems, and digital data transmission, respectively.

Upon completing this chapter, you should be able to:

- Recognize the principal classes of sensors.
- Design proper circuit connections to minimize noise in floating, grounded, and differential-source circuits.
- Understand the concepts of shielding and grounding.
- Specify, analyze, and design instrumentation amplifiers and simple active filters.
- Understand the processes of analog-to-digital and digital-to-analog conversion, and specify the requirements of a data acquisition system.
- Design simple instrumentation circuits using op-amps and integrated circuits.
- Understand the basic principles of digital data transmission.

## 15.1    MEASUREMENT SYSTEMS AND TRANSDUCERS

### Measurement Systems

In virtually every engineering application there is a need for measuring some physical quantities, such as forces, stresses, temperatures, pressures, flows, or displacements. These measurements are performed by physical devices called **sensors** or **transducers,** which are capable of converting a physical quantity to a more readily manipulated electrical quantity. Most sensors, therefore, convert the change of a physical quantity (e.g., humidity, temperature) to a corresponding (usually proportional) change in an electrical quantity (e.g., voltage or current). Often, the direct output of the sensor requires additional manipulation before the electrical output is available in a useful form. For example, the change in resistance resulting from a change in the surface stresses of a material—the quantity measured by the resistance strain gauges described in Chapter 2[1]—must first be converted to a change in voltage through a suitable circuit (the Wheatstone bridge) and then amplified from the millivolt to the volt level. The manipulations needed to produce the desired end result are referred to as *signal conditioning.* The wiring of the sensor to the signal conditioning circuitry requires significant attention to *grounding* and *shielding* procedures, to ensure that the resulting signal is as free from noise and interference as possible. Very often, the conditioned sensor signal is then converted to *digital* form and recorded in a computer for additional manipulation, or is displayed in some form. The apparatus used in manipulating a sensor output to produce a result that can be suitably displayed or stored is called a **measurement system.** Figure 15.1 depicts a typical computer-based measurement system in block diagram form.

### Sensor Classification

There is no standard and universally accepted classification of sensors. Depending on one's viewpoint, sensors may be grouped according to their physical characteristics (e.g., electronic sensors, resistive sensors), or by the physical variable or

---

[1] See "Focus on Measurements, Measurement of Force."

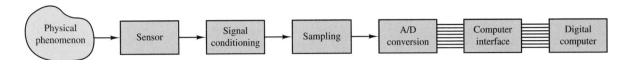

**Figure 15.1** Measurement system

quantity measured by the sensor (e.g., temperature, flow rate). Other classifications are also possible. Table 15.1 presents a partial classification of sensors grouped according to the variable sensed; we do not claim that the table is complete, but we can safely state that most of the engineering measurements of interest to the reader are likely to fall in the categories listed in Table 15.1. Also included in the table are references to the "Focus on Measurements" boxes that describe sensors elsewhere in this book.

A sensor is usually accompanied by a set of specifications that indicate its overall effectiveness in measuring the desired physical variable. The following definitions will help the reader understand sensor data sheets:

*Accuracy:* Conformity of the measurement to the true value, usually in percent of full-scale reading

*Error:* Difference between measurement and true value, usually in percent of full-scale reading

*Precision:* Number of significant figures of the measurement

*Resolution:* Smallest measurable increment

*Span:* Linear operating range

*Range:* The range of measurable values

*Linearity:* Conformity to an ideal linear calibration curve, usually in percent of reading or of full-scale reading (whichever is greater)

## Motion and Dimensional Measurements

The **measurement of motion and dimension** is perhaps the most commonly encountered engineering measurement. Measurements of interest include absolute position, relative position (displacement), velocity, acceleration, and jerk (the derivative of acceleration). These can be either translational or rotational measurements; usually, the same principle can be applied to obtain both kinds of measurements. These measurements are often based on changes in elementary properties, such as changes in the resistance of an element (e.g., strain gauges, potentiometers), in an electric field (e.g., capacitive sensors), or in a magnetic field (e.g., inductive, variable-reluctance, or eddy current sensors). Other mechanisms may be based on special materials (e.g., piezoelectric crystals), or on optical signals and imaging systems. Table 15.1 lists several examples of dimensional and motion measurement that can be found in this book.

## Force, Torque, and Pressure Measurements

Another very common class of **measurements** is that **of pressure and force,** and the related **measurement of torque.** Perhaps the single most common family of force and pressure transducers comprises those based on strain gauges (e.g.,

Table 15.1  Sensor classification

| Sensed variables | Sensors | *Focus on Measurements* boxes |
|---|---|---|
| Motion and dimensional variables | Resistive potentiometers | Resistive Throttle Position Sensor (Chapter 2) |
| | Strain gauges | Resistance Strain Gauges, The Wheatstone Bridge, and Force Measurements (Chapter 2) |
| | Differential transformers (LVDTs) | Linear Variable Differential Transformer (LVDT) (Chapter 16) |
| | Variable-reluctance sensors | Magnetic Reluctance Position Sensor (Chapter 16) |
| | Capacitive sensors | Capacitive Displacement Transducer and Microphone (Chapter 4); Peak Detector Circuit for Capacitive Displacement Transducer (Chapter 8) |
| | Piezoelectric sensors | Piezoelectric Sensor and Charge Amplifiers (Chapter 12) |
| | Electro-optical sensors | Digital Position Encoders; Digital Measurement of Angular Position and Velocity (Chapter 13) |
| | Moving-coil transducers | Seismic Transducer (Chapter 16) |
| | Seismic sensors | Seismic Transducer (Chapter 6) |
| Force, torque, and pressure | Strain gauges | Resistance Strain Gauges, The Wheatstone Bridge, and Force Measurements (Chapter 2) |
| | Piezoelectric sensors | Piezoelectric Sensor and Charge Amplifiers (Chapter 12) |
| | Capacitive sensors | Capacitive Displacement Transducer and Microphone (Chapter 4); Peak Detector Circuit for Capacitive Displacement Transducer (Chapter 8) |
| Flow | Pitot tube | |
| | Hot-wire anemometer | Hot-Wire Anemometer (Chapter 15) |
| | Differential pressure sensors | Differential Pressure Sensor (Chapter 15) |
| | Turbine meters | Turbine Meters (Chapter 15) |
| | Vortex shedding meters | |
| | Ultrasonic sensors | |
| | Electromagnetic sensors | |
| | Imaging systems | |
| Temperature | Thermocouples | Thermocouples (Chapter 15) |
| | Resistance thermometers (RTDs) | Resistance Thermometers (RTDs) (Chapter 15) |
| | Semiconductor thermometers | Diode Thermometer (Chapter 8) |
| | Radiation detectors | |
| Liquid level | Motion transducers | |
| | Force transducers | |
| | Differential-pressure measurement devices | |
| Humidity | Semiconductor sensors | |
| Chemical composition | Gas analysis equipment | |
| | Solid-state gas sensors | |

load cells, diaphragm pressure transducers). Also very common are piezoelectric transducers. Capacitive transducers again find application in the measurement of pressure. Table 15.1 indicates where the reader can find examples of these measurements in this book.

## Flow Measurements

In many engineering applications it is desirable to sense the flow rate of a fluid, whether compressible (gas) or incompressible (liquid). The **measurement of fluid flow rate** is a complex subject; in this section we simply summarize the concepts underlying some of the most common measurement techniques. Shown in Figure 15.2 are three different types of flow rate sensors. The sensor in Figure 15.2(a) is based on **differential pressure measurement** and on a **calibrated orifice:** the relationship between pressure across the orifice, $p_1 - p_2$, and flow rate through the orifice, $q$, is predetermined through the calibration; therefore, measuring the differential pressure is equivalent to measuring flow rate.

The sensor in Figure 15.2(b) is called a **hot-wire anemometer,** because it is based on a heated wire that is cooled by the flow of a gas. The resistance of the wire changes with temperature, and a Wheatstone bridge circuit converts this change in resistance to a change in voltage while the current is kept constant. Also commonly used are **hot-film anemometers,** where a heated film is used in place of the more delicate wire. A very common application of the latter type of sensor is in automotive engines, where control of the air-to-fuel ratio depends on measurement of the engine intake mass airflow rate.

Figure 15.2(c) depicts a **turbine flowmeter,** in which the fluid flow causes a turbine to rotate; the velocity of rotation of the turbine (which can be measured by a noncontact sensor—e.g., a magnetic pickup[2]) is related to the flow velocity.

Besides the techniques discussed in this chapter, many other techniques exist for measuring fluid flow, some of significant complexity.

## Temperature Measurements

One of the most frequently measured physical quantities is temperature. The need to measure temperature arises in just about every field of engineering. This subsection is devoted to summarizing two common **temperature sensors**—the **thermocouple** and the **resistance temperature detector (RTD)**—and their related signal-conditioning needs.

### *Thermocouples*

A thermocouple is formed by the junction of two dissimilar metals. This junction results on an open-circuit **thermoelectric voltage** due to the **Seebeck effect,** named after Thomas Seebeck, who discovered the phenomenon in 1821. Various types of thermocouples exist; they are usually classified according to the data of Table 15.2. The Seebeck coefficient shown in the table is specified at a given temperature because the output voltage of a thermocouple, $v$, has a nonlinear dependence on temperature. This dependence is typically expressed in terms of a polynomial of the following form:

$$T = a_0 + a_1 v + a_2 v^2 + a_3 v^3 + \cdots + a_n v^n \tag{15.1}$$

For example, the coefficients of the J thermocouple in the range $-100°C$ to $+1,000°C$ are as follows:

$a_0 = -0.048868252$    $a_1 = 19,873.14503$    $a_2 = -128,614.5353$

$a_3 = 11,569,199.78$    $a_4 = -264,917,531.4$    $a_5 = 2,018,441,314$

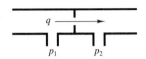

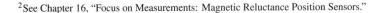

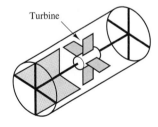

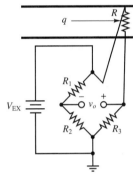

**Differential-pressure flowmeter:** A calibrated orifice and a pair of pressure transducers permit the measurement of flow rate.

(a)

**Hot-wire anemometer:** A heated wire is cooled by the gas flow. The resistance of the wire changes with temperature.

(b)

**Turbine flowmeter:** Fluid flow induces rotation of the turbine; measurement of turbine velocity provides an indication of flow rate.

(c)

**Figure 15.2** Devices for the measurement of flow

---

[2]See Chapter 16, "Focus on Measurements: Magnetic Reluctance Position Sensors."

Table 15.2 Thermocouple data

| Type | Elements +/– | Seebeck coefficient ($\mu$ V/°C) | Range (°C) | Range (mV) |
|------|--------------|----------------------------------|------------|------------|
| E | Chromel/constantan | 58.70 at 0°C | −270 to 1,000 | −9.835 to 76.358 |
| J | Iron/constantan | 50.37 at 0°C | −210 to 1,200 | −8.096 to 69.536 |
| K | Chromel/alumel | 39.48 at 0°C | −270 to 1,372 | −6.548 to 54.874 |
| R | Pt(10%)—Rh/Pt | 10.19 at 600°C | −50 to 1,768 | −0.236 to 18.698 |
| T | Copper/constantan | 38.74 at 0°C | −270 to 400 | −6.258 to 20.869 |
| S | Pt(13%)—Rh/Pt | 11.35 at 600°C | −50 to 1,768 | −0.226 to 21.108 |

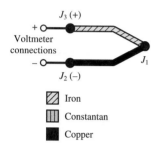

**Figure 15.3** J thermocouple circuit

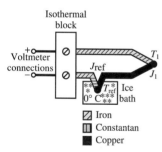

**Figure 15.4** Cold-junction–compensated thermocouple circuit

The use of a thermocouple requires special connections, because the junction of the thermocouple wires with other leads (such as voltmeter leads, for example) creates additional thermoelectric junctions that in effect act as additional thermocouples. For example, in the J thermocouple circuit of Figure 15.3, junction $J_1$ is exposed to the temperature to be measured, but junctions $J_2$ and $J_3$ also generate a thermoelectric voltage, which is dependent on the temperature at these junctions, that is, the temperature at the voltmeter connections. One would therefore have to know the voltages at these junctions, as well, in order to determine the actual thermoelectric voltage at $J_1$. To obviate this problem, a reference junction at known temperature can be employed; a traditional approach involves the use of a **cold junction,** so called because it often consists of an ice bath, one of the easiest means of obtaining a known reference temperature. Figure 15.4 depicts a thermocouple measurement using an ice bath. The voltage measured in Figure 15.4 is dependent on the temperature difference $T_1 - T_{ref}$, where $T_{ref} = 0°C$. The connections to the voltmeter are made at an *isothermal block,* kept at a constant temperature; note that the same metal is used in both of the connections to the isothermal block. Thus (still assuming a J thermocouple), there is no difference between the thermoelectric voltages at the two copper-iron junctions; these will add to zero at the voltmeter. The voltmeter will therefore read a voltage proportional to $T_1 - T_{ref}$.

An ice bath is not always a practical solution. Other cold junction temperature compensation techniques employ an additional temperature sensor to determine the actual temperature of the junctions $J_2$ and $J_3$ of Figure 15.3.

### Resistance Temperature Detectors (RTDs)

A resistance temperature detector (RTD) is a variable-resistance device whose resistance is a function of temperature. RTDs can be made with both positive and negative temperature coefficients and offer greater accuracy and stability than thermocouples. **Thermistors** are part of the RTD family. A characteristic of all RTDs is that they are *passive* devices, that is, they do not provide a useful output unless excited by an external source. The change in resistance in an RTD is usually converted to a change in voltage by forcing a current to flow through the device. An indirect result of this method is a **self-heating error,** caused by the $i^2 R$ heating of the device. Self-heating of an RTD is usually denoted by the amount of power that will raise the RTD temperature by 1°C. Reducing the excitation current can clearly help reduce self-heating, but it also reduces the output voltage.

The RTD resistance has a fairly linear dependence on temperature; a common definition of the **temperature coefficient** of an RTD is related to the change in resistance from $0°$ to $100°C$. Let $R_0$ be the resistance of the device at $0°C$ and $R_{100}$ the resistance at $100°C$. Then the temperature coefficient, $\alpha$, is defined to be

$$\alpha = \frac{R_{100} - R_0}{100 - 0} \frac{\Omega}{°C} \tag{15.2}$$

A more accurate representation of RTD temperature dependence can be obtained by using a nonlinear (cubic) equation and published tables of coefficients. As an example, a platinum RTD could be described either by the temperature coefficient $\alpha = 0.003911$, or by the equation

$$\begin{aligned}
R_T &= R_0(1 + AT - BT^2 - CT^3) \\
&= R_0(1 + 3.6962 \times 10^{-3}T - 5.8495 \times 10^{-7}T^2 \\
&\quad -4.2325 \times 10^{-12}T^3)
\end{aligned} \tag{15.3}$$

where the coefficient $C$ is equal to zero for temperatures above $0°C$.

Because RTDs have fairly low resistance, they are sensitive to error introduced by the added resistance of the lead wires connected to them; Figure 15.5 depicts the effect of the lead resistances, $r_L$, on the RTD measurement. Note that the measured voltage includes the resistance of the RTD as well as the resistance of the leads. If the leads used are long (greater than 3 m is a good rule of thumb), then the measurement will have to be adjusted for this error. Two possible solutions to the lead problems are the *four-wire* RTD measurement circuit and the *three-wire* Wheatstone bridge circuit, shown in Figure 15.6(a) and (b), respectively. In the circuit of Figure 15.6(a), the resistance of the lead wires from the excitation, $r_{L1}$ and $r_{L4}$, may be arbitrarily large, since the measurement is affected by the resistance of only the output lead wires, $r_{L2}$ and $r_{L3}$, which can be usually kept small by making these leads short. The circuit of Figure 15.6(b) takes advantage of the properties of the Wheatstone bridge to cancel out the unwanted effect of the lead wires while still producing an output dependent on the change in temperature.

## 15.2    WIRING, GROUNDING, AND NOISE

The importance of proper circuit connections cannot be overemphasized. Unfortunately, this is a subject that is rarely taught in introductory electrical engineering courses. The present section summarizes some important considerations regarding signal source connections, various types of input configurations, noise sources and coupling mechanisms, and means of minimizing the influence of noise on a measurement.

### Signal Sources and Measurement System Configurations

Before proper connection and wiring techniques can be presented, we must examine the difference between **grounded** and **floating signal sources.** Every sensor can be thought of as some kind of signal source; a general representation of the connection of a sensor to a measurement system is shown in Figure 15.7(a). The sensor is modeled as an ideal voltage source in series with a source resistance. Although this representation does not necessarily apply to all sensors, it will be adequate for the purposes of the present section. Figures 15.7(b) and (c) show two types of signal sources: grounded and floating. A grounded signal source is one

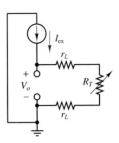

**Figure 15.5** Effect of connection leads on RTD temperature measurement

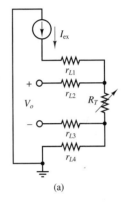

(a)

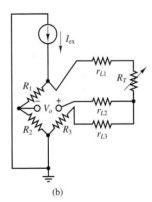

(b)

**Figure 15.6** Four-wire RTD circuit (a) and three-wire Wheatstone bridge RTD circuit (b)

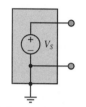

(a) Ideal signal source connected to measurement system

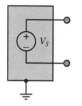

(b) Grounded signal source

(c) Floating signal source

**Figure 15.7** Measurement system and types of signal sources

in which a ground reference is established—for example, by connecting the *signal low* lead to a case or housing. A floating signal source is one in which neither signal lead is connected to ground; since ground potential is arbitrary, the signal source voltage levels (*signal low* and *signal high*) are at an unknown potential relative to the case ground. Thus, the signal is said to be *floating.* Whether a sensor can be characterized as a grounded or a floating signal source ultimately depends on the connection of the sensor to its case, but the choice of connection may depend on the nature of the source. For example, the thermocouple described in Section 15.1 is *intrinsically* a floating signal source, since the signal of interest is a difference between two voltages. The same thermocouple *could* become a grounded signal source if one of its two leads were directly connected to ground, but this is usually not a desirable arrangement for this particular sensor.

In analogy with a signal source, a measurement system can be either **ground-referenced** or **differential.** In a ground-referenced system, the signal low connection is tied to the instrument case ground; in a differential system, neither of the two signal connections is tied to ground. Thus, a differential measurement system is well suited to measuring the difference between two signal levels (such as the output of an ungrounded thermocouple).

One of the potential dangers in dealing with grounded signal sources is the introduction of **ground loops.** A ground loop is an undesired current path caused by the connection of two reference voltages to each other. This is illustrated in Figure 15.8, where a grounded signal source is shown connected to a ground-referenced measurement system. Notice that we have purposely denoted the signal source ground and the measurement system ground by two distinct symbols, to emphasize that these are not necessarily at the same potential—as also indicated by the voltage difference $\Delta V$. Now, one might be tempted to tie the two grounds to each other, but this would only result in a current flowing from one ground to the other, through the small (but nonzero) resistance of the wire connecting the two. The net effect of this ground loop would be that the voltage measured by the instrument would include the unknown ground voltage difference $\Delta V$, as shown in Figure 15.8. Since this latter voltage is unpredictable, you can see that ground loops can cause substantial errors in measuring systems. In addition, ground loops are the primary cause of conducted noise, as explained later in this section.

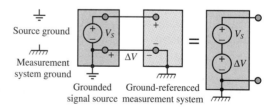

**Figure 15.8** Ground loop in ground-referenced measurement system

A differential measurement system is often a way to avoid ground loop problems, because the signal source and measurement system grounds are not connected to each other, and especially because the signal low input of the measuring instrument is not connected to either instrument case ground. The connection of a grounded signal source and a differential measurement system is depicted in Figure 15.9.

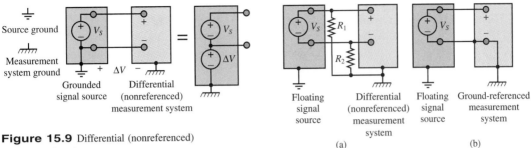

**Figure 15.9** Differential (nonreferenced) measurement system

**Figure 15.10** Measuring signals from a floating source: (a) differential input; (b) single-ended input

If the signal source connected to the differential measurement system is floating, as shown in Figure 15.10, it is often a recommended procedure to reference the signal to the instrument ground by means of two identical resistors that can provide a return path to ground for any currents present at the instrument. An example of such input currents would be the input bias currents inevitably present at the input of an operational or instrumentation amplifier.

The simple concepts illustrated in the preceding paragraphs and figures can assist the user and designer of instrumentation systems in making the best possible wiring connections for a given measurement.

## Noise Sources and Coupling Mechanisms

Noise—meaning any undesirable signal interfering with a measurement—is an unavoidable element of all measurements. Figure 15.11 depicts a block diagram of the three essential stages of a noisy measurement: a **noise source,** a **noise coupling mechanism,** and a sensor or associated signal-conditioning circuit. Noise sources are always present, and are often impossible to eliminate completely; typical sources of noise in practical measurements are the electromagnetic fields caused by fluorescent light fixtures, video monitors, power supplies, switching circuits, and high-voltage (or current) circuits. Many other sources exist, of course, but often the simple sources in our everyday environment are the most difficult to defeat.

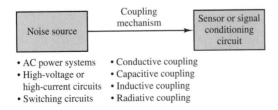

**Figure 15.11** Noise sources and coupling mechanisms

Figure 15.11 also indicates that various coupling mechanisms can exist between a noise source and an instrument. Noise coupling can be conductive; that

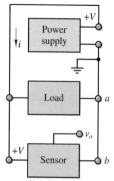

The ground loop created by the load circuit can cause a different ground potential between *a* and *b*.

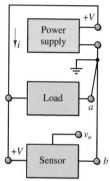

Separate ground returns for the load and the sensor circuit eliminate the ground loop.

**Figure 15.12** Conductive coupling: ground loop and separate ground returns

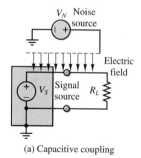

(a) Capacitive coupling

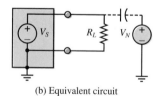

(b) Equivalent circuit

**Figure 15.13** Capacitive coupling and equivalent-circuit representation

is, noise currents may actually be conducted from the noise source to the instrument by physical wires. Noise can also be coupled capacitively, inductively, and radiatively.

Figure 15.12 illustrates how interference can be **conductively coupled** by way of a ground loop. In the figure, a power supply is connected to both a load and a sensor. We shall assume that the load may be switched on and off, and that it carries substantial currents. The top circuit contains a ground loop: the current *i* from the supply divides between the load and sensor; since the wire resistance is nonzero, a large current flowing through the load may cause the ground potential at point *a* to differ from the potential at point *b*. In this case, the measured sensor output is no longer $v_o$, but it is now equal to $v_o + v_{ba}$, where $v_{ba}$ is the potential difference from point *b* to point *a*. Now, if the load is switched on and off and its current is therefore subject to large, abrupt changes, these changes will be manifested in the voltage $v_{ba}$ and will appear as noise on the sensor output.

This problem can be cured simply and effectively by providing separate *ground returns* for the load and sensor, thus eliminating the ground loop.

The mechanism of **capacitive coupling** is rooted in electric fields that may be caused by sources of interference. The detailed electromagnetic analysis can be quite complex, but to understand the principle, refer to Figure 15.13(a), where a noise source is shown to generate an electric field. If a noise source conductor is sufficiently close to a conductor that is part of the measurement system, the two conductors (separated by air, a dielectric) will form a capacitor, through which any time-varying currents can flow. Figure 15.13(b) depicts an equivalent circuit in which the noise voltage $V_N$ couples to the measurement circuit through an imaginary capacitor, representing the actual capacitance of the noise path.

The dual of capacitive coupling is **inductive coupling.** This form of noise coupling is due to the magnetic field generated by current flowing through a conductor. If the current is large, the magnetic fields can be significant, and the **mutual inductance** (see Chapters 5 and 16) between the noise source and the measurement circuit causes the noise to couple to the measurement circuit. Thus, inductive coupling, as shown in Figure 15.14, results when undesired (unplanned) magnetic coupling ties the noise source to the measurement circuit.

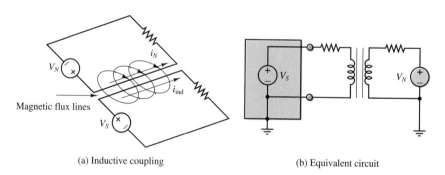

(a) Inductive coupling                    (b) Equivalent circuit

**Figure 15.14** Inductive coupling and equivalent-circuit representation

## Noise Reduction

Various techniques exist for minimizing the effect of undesired interference, in addition to proper wiring and grounding procedures. The two most common methods are **shielding** and the use of **twisted-pair wire.** A shielded cable is shown in Figure 15.15. The shield is made of a copper braid or of foil and is usually grounded at the source end *but not at the instrument end,* because this would result in a ground loop. The shield can protect the signal from a significant amount of electromagnetic interference, especially at lower frequencies. Shielded cables with various numbers of conductors are available commercially. However, shielding cannot prevent inductive coupling. The simplest method for minimizing inductive coupling is the use of twisted-pair wire; the reason for using twisted pair is that untwisted wire can offer large loops that can couple a substantial amount of electromagnetic radiation (see Section 16.1). Twisting drastically reduces the loop area, and with it the interference. Twisted pair is available commercially.

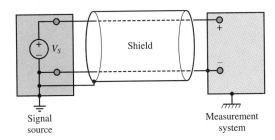

**Figure 15.15** Shielding

## 15.3    SIGNAL CONDITIONING

A properly wired, grounded, and shielded sensor connection is a necessary first stage of any well-designed measurement system. The next stage consists of any **signal conditioning** that may be required to manipulate the sensor output into a form appropriate for the intended use. Very often, the sensor output is meant to be fed into a digital computer, as illustrated in Figure 15.1. In this case, it is important to condition the signal so that it is compatible with the process of data acquisition. Two of the most important signal-conditioning functions are *amplification* and *filtering.* Both are discussed in the present section.

### Instrumentation Amplifiers

An **instrumentation amplifier (IA)** is a differential amplifier with very high input impedance, low bias current, and programmable gain that finds widespread application when low-level signals with large common-mode components are to be amplified in noisy environments. This situation occurs frequently when a low-level transducer signal needs to be preamplified, prior to further signal conditioning (e.g., filtering). Instrumentation amplifiers were briefly introduced in Chapter 12 (see Example 12.4), as an extension of the differential amplifier. You may recall that the IA introduced in Example 12.4 consisted of two stages, the first composed of two noninverting amplifiers, the second of a differential amplifier. Although the design in Chapter 12 is useful and is sometimes employed in practice, it suffers from a few drawbacks, most notably the requirement for very precisely matched resistors and source impedances to obtain the maximum possible cancellation of the common-mode signal. If the resistors are not matched exactly, the common-mode rejection ratio of the amplifier is significantly reduced, as the following will demonstrate.

The amplifier of Figure 15.16 has properly matched resistors ($R_2 = R'_2$, $R_F = R'_F$), except for resistors $R$ and $R'$, which differ by an amount $\Delta R$ such that

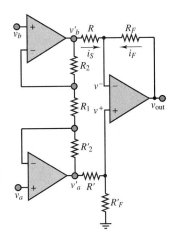

**Figure 15.16** Discrete op-amp instrumentation amplifier

$R' = R + \Delta R$. Let us compute the closed-loop gain for the amplifier. As shown in Example 12.4, the input-stage noninverting amplifiers have a closed-loop gain given by

$$A = \frac{v'_b}{v_b} = \frac{v'_a}{v_a} = 1 + \frac{2R_2}{R_1} \tag{15.4}$$

To compute the output voltage, we observe that the voltage at the noninverting terminal is

$$v^+ = \frac{R_F}{R_F + R + \Delta R} v'_a \tag{15.5}$$

and since the inverting-terminal voltage is $v^- = v^+$, the feedback current is given by

$$i_F = \frac{v_{\text{out}} - v^-}{R_F} = \frac{v_{\text{out}} - \frac{R_F}{R_F + R + \Delta R} v'_a}{R_F} \tag{15.6}$$

and the source current is

$$i_S = \frac{v'_b - v^-}{R} = \frac{v'_b - \frac{R_F}{R_F + R + \Delta R} v'_a}{R} \tag{15.7}$$

Applying KCL at the inverting node (under the usual assumption that the input current going into the op-amp is negligible), we set $i_F = -i_S$ and obtain the expression

$$\frac{v_{\text{out}}}{R_F} = \frac{v'_a}{R_F + R + \Delta R} - \frac{v'_b}{R} + \frac{R_F}{R} \frac{v'_a}{R_F + R + \Delta R}$$

$$= \left(1 + \frac{R_F}{R}\right) \frac{v'_a}{R_F + R + \Delta R} - \frac{v'_b}{R}$$

so that the output voltage may be computed to be

$$v_{\text{out}} = R_F \left(\frac{R + R_F}{R}\right) \frac{v'_a}{R_F + R + \Delta R} - \frac{R_F}{R} v'_b \tag{15.8}$$

Note that if the term $\Delta R$ in the denominator were zero, the same result would be obtained as in Example 12.4: $v_{\text{out}} = (R_F/R)(v'_a - v'_b)$; however, because of the resistor mismatch, there is a corresponding mismatch between the gains for the two differential signal components. Further—and more important—if the original signals, $v_a$ and $v_b$, contained both differential-mode and common-mode components:

$$v_a = v_{a\ \text{dif}} + v_{\text{com}} \qquad v_b = v_{b\ \text{dif}} + v_{\text{com}} \tag{15.9}$$

such that

$$v'_a = A(v_{a\ \text{dif}} + v_{\text{com}}) \qquad v'_b = A(v_{b\ \text{dif}} + v_{\text{com}}) \tag{15.10}$$

then the common-mode components would not cancel out in the output of the amplifier, because of the gain mismatch, and the output of the amplifier would be given by

$$v_{\text{out}} = R_F \left(\frac{R + R_F}{R}\right) \frac{A(v_{a\ \text{dif}} + v_{\text{com}})}{R_F + R + \Delta R} - \frac{R_F}{R} A(v_{b\ \text{dif}} + v_{\text{com}}) \tag{15.11}$$

resulting in the following output voltage:

$$v_{\text{out}} = v_{\text{out, dif}} + v_{\text{out, com}} \qquad \textbf{(15.12)}$$

with

$$v_{\text{out, dif}} = R_F \left( \frac{R + R_F}{R} \right) \frac{A v_{a \text{ dif}}}{R_F + R + \Delta R} - \frac{R_F}{R} A v_{b \text{ dif}} \qquad \textbf{(15.13)}$$

and

$$
\begin{aligned}
v_{\text{out, com}} &= R_F \left( \frac{R + R_F}{R} \right) \frac{A v_{\text{com}}}{R_F + R + \Delta R} - \frac{R_F}{R} A v_{\text{com}} \\
&= \frac{R_F}{R} \left( \frac{R + R_F}{R_F + R + \Delta R} - 1 \right) A v_{\text{com}}
\end{aligned}
\qquad \textbf{(15.14)}
$$

The common-mode rejection ratio (CMRR; see Section 12.6) is given in units of dB by

$$
\begin{aligned}
\text{CMRR}_{\text{dB}} &= \left| \frac{A_{\text{dif}}}{A_{\text{com}}} \right| = 20 \log_{10} \left| \frac{A_{\text{dif}}}{v_{\text{out, com}} / v_{\text{com}}} \right| \\[2mm]
&= 20 \log_{10} \left| \frac{A_{\text{dif}}}{\frac{R_F}{R} \left( \frac{R + R_F}{R_F + R + \Delta R} - 1 \right) A} \right|
\end{aligned}
\qquad \textbf{(15.15)}
$$

where $A_{\text{dif}}$ is the *differential gain* (which is usually assumed equal to the nominal design value). Since the common-mode gain, $v_{\text{out, com}} / v_{\text{com}}$, should ideally be zero, the theoretical CMRR for the instrumentation amplifier with perfectly matched resistors is infinite. In fact, even a small mismatch in the resistors used would dramatically reduce the CMRR, as the Check Your Understanding exercises at the end of this subsection illustrate. Even with resistors having 1 percent tolerance, the maximum CMRR that could be attained for typical values of resistors and an overall gain of 1,000 would be only 60 dB. In many practical applications, a requirement for a CMRR of 100 or 120 dB is not uncommon, and these would demand resistors of 0.01 percent tolerance (see Check Your Understanding Exercise 15.3). It should be evident, then, that the "discrete" design of the IA, employing three op-amps and discrete resistors, will not be adequate for the more demanding instrumentation applications.

## EXAMPLE  15.1  Common-Mode Gain and Rejection Ratio

### Problem

Compute the common-mode gain and common-mode rejection ratio (CMRR) for the amplifier of Figure 15.16.

### Solution

***Known Quantities:*** Amplifier nominal closed-loop gain; resistance values; resistor tolerance.

**Find:** $v_{\text{out, com}}/v_{\text{com}}$, $\text{CMRR}_{\text{dB}}$

**Analysis:** The common-mode gain is equal to the ratio of the common-mode output signal to the common-mode input:

$$\frac{v_{\text{out, com}}}{v_{\text{com}}} = 100 - \left(\frac{11}{11.02} - 1\right) = -0.1815$$

The CMRR (in units of dB) can be computed from equation 15.15 where

$$A_{\text{dif}} = A \times \frac{R_F}{R} = 100$$

and therefore,

$$\text{CMRR} = \left|\frac{A_{\text{dif}}}{A_{\text{com}}}\right|_{\text{dB}} = 20\log_{10}\left|\frac{A_{\text{dif}}}{v_{\text{out, com}}/v_{\text{com}}}\right| = 20\log_{10}\left|\frac{A_{\text{dif}}}{\frac{R_F}{R}\left(\frac{R+R_F}{R+R_F+\Delta R} - 1\right)A}\right|$$

$$= 20\log_{10}\left|\frac{100}{\frac{10}{1}\left(\frac{11}{11.02} - 1\right)10}\right| = 54.82 \text{ dB}$$

**Comments:** Note that, in general, it is difficult to determine exactly the level of resistor mismatch in an instrumentation amplifier. One usually makes reference to manufacturer data sheets, such as those you will find in the accompanying CD-ROM.

---

The general expression for the CMRR of the instrumentation amplifier of Figure 15.16, without assuming any of the resistors are matched, except for $R_2$ and $R_2'$, is

$$\text{CMRR} = \left|\frac{A_{\text{dif}}}{A_{\text{com}}}\right| = \left|\frac{(R_F/R)(1 + 2R_2/R_1)}{\frac{R_F}{R}\left[\frac{R_F'}{R_F}\left(\frac{R_F+R}{R_F'+R'}\right) - 1\right]}\right| \qquad \textbf{(15.16)}$$

and it can easily be shown that the CMRR is infinite if the resistors are perfectly matched.

Example 15.1 illustrated some of the problems that are encountered in the design of instrumentation amplifiers using discrete components. Many of these problems can be dealt with very effectively if the entire instrumentation amplifier is designed into a single *monolithic integrated circuit,* where the resistors can be carefully matched by appropriate fabrication techniques and many other problems can also be avoided. The functional structure of an IC instrumentation amplifier is depicted in Figure 15.17. Specifications for a common IC instrumentation amplifier (and a more accurate circuit description) are shown in Figure 15.18. Among the features worth mentioning here are the programmable gains, which the user can set by suitably connecting one or more of the resistors labeled $R_1$ to the appropriate connection. Note that the user may also choose to connect additional resistors to control the amplifier gain, without adversely affecting the amplifier's performance, since $R_1$ requires no matching. In addition to the pin connection that permits programmable gains, two additional pins are provided,

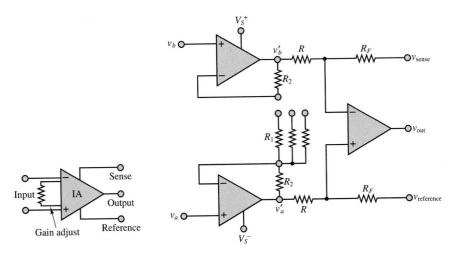

**Figure 15.17** IC instrumentation amplifier

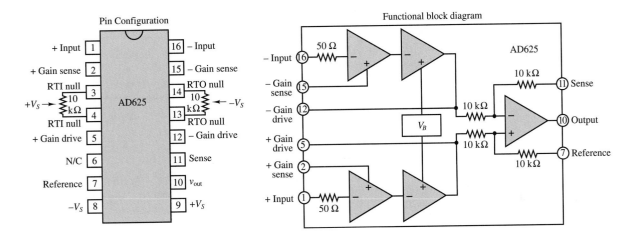

**Figure 15.18** AD625 instrumentation amplifier

called **sense** and **reference.** These additional connections are provided to the user
for the purpose of referencing the output voltage to a signal other than ground,
by means of the reference terminal, or of further amplifying the output current
(e.g., with a transistor stage), by connecting the sense terminal to the output of the
current amplifier.

---

## EXAMPLE  15.2  Instrumentation Amplifier Gain Configuration
### Using Internal Resistors

### Problem

Determine the possible input-stage gains that can be configured using the choice of
resistor values given for the instrumentation amplifier (IA) of Figure 15.17.

### Solution

**Known Quantities:** IA resistor values.

**Find:** $A$, for different resistor combinations.

**Schematics, Diagrams, Circuits, and Given Data:** $R_F = R = 10\text{ k}\Omega$; $R_2 = 20\text{ k}\Omega$; $R_1 = 80.2\ \Omega, 201\ \Omega, 404\ \Omega$.

**Analysis:** Recall that the gain of the input stage (for each of the differential inputs) can be calculated according to equation 15.4:

$$A = 1 + \frac{2R_2}{R_1}$$

Thus, by connecting each of the three resistors, we can obtain gains

$$A_1 = 1 + \frac{40,000}{80.2} = 500 \quad A_2 = 1 + \frac{40,000}{201} = 200 \quad A_3 = 1 + \frac{40,000}{404} = 100$$

It is also possible to obtain additional input-stage gains by connecting resistors in parallel:

$$80.2\|201 = 57.3\ \Omega\ (A_4 \approx 700) \qquad 80.2\|404 = 66.9\ \Omega\ (A_5 \approx 600)$$

$$404\|201 = 134.2\ \Omega\ (A_6 \approx 300)$$

**Comments:** The use of resistors supplied with the IA package is designed to reduce the uncertainty introduced by the use of external resistors, since the value of the internally supplied resistors can be controlled more precisely.

---

## Check Your Understanding

**15.1**   Use the definition of the common-mode rejection ratio (CMRR) given in equation 15.16 to compute the CMRR (in dB) of the amplifier of Example 15.1 if $R_F/R = 100$ and $A = 10$, and if $\Delta R = 5\%$ of $R$. Assume $R = 1\text{ k}\Omega$, $R_F = 100\text{ k}\Omega$.

**15.2**   Repeat Exercise 15.1 for a 1 percent variation in $R$.

**15.3**   Repeat Exercise 15.1 for a 0.01 percent variation in $R$.

**15.4**   Calculate the mismatch in gains for the differential components for the 5 percent resistance mismatch of Exercise 15.1.

**15.5**   Calculate the mismatch in gains for the differential components for the 1 percent resistance mismatch of Exercise 15.2.

**15.6**   What value of resistance $R_1$ would permit a gain of 1,000 for the IA of Example 15.2?

---

## Active Filters

The need to filter sensor signals that may be corrupted by noise or other interfering or undesired inputs has already been approached in two earlier chapters. In Chapter 6, simple passive filters made of resistors, capacitors, and inductors were analyzed. It was shown that three types of filter frequency response characteristics can be achieved with these simple circuits: low-pass, high-pass, and band-pass. In Chapter 12, the concept of active filters was introduced, to suggest that it may be desirable to exploit the properties of operational amplifiers to simplify filter design, to more easily match source and load impedances, and to eliminate the need for inductors. The aim of this section is to discuss more advanced active filter designs, which find widespread application in instrumentation circuits.

Figure 15.19 depicts the general characteristics of a low-pass active filter, indicating that within the pass-band of the filter, a certain deviation from the nominal filter gain, $A$, is accepted, as indicated by the minimum and maximum pass-band gains, $A + \varepsilon$ and $A - \varepsilon$. The width of the pass-band is indicated by the cutoff frequency, $\omega_C$. On the other hand, the stop-band, starting at the frequency $\omega_S$, does not allow a gain greater than $A_{\min}$. Different types of filter designs achieve different types of frequency responses, which are typically characterized either by having a particularly flat pass-band frequency response **(Butterworth filters),** or by a very rapid transition between pass-band and stop-band **(Chebyshev filters,** and **Cauer,** or **elliptical, filters),** or by some other characteristic, such as a linear phase response **(Bessel filters).** Achieving each of these properties usually involves trade-offs; for example, a very flat pass-band response will usually result in a relatively slow transition from pass-band to stop-band.

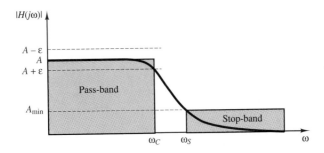

**Figure 15.19** Prototype low-pass filter response

In addition to selecting a filter from a certain family, it is also possible to select the *order* of the filter; this is equal to the order of the differential equation that describes the input-output relationship of a given filter. In general, the higher the order, the faster the transition from pass-band to stop-band (at the cost of greater phase shifts and amplitude distortion, however). Although the frequency response of Figure 15.19 pertains to a low-pass filter, similar definitions also apply to the other types of filters.

Butterworth filters are characterized by a *maximally flat* pass-band frequency response characteristic; their response is defined by a magnitude-squared function of frequency:

$$|H(j\omega)|^2 = \frac{H_0^2}{1 + \varepsilon^2 \omega^{2n}} \tag{15.17}$$

where $\varepsilon = 1$ for maximally flat response and $n$ is the order of the filter. Figure 15.20 depicts the frequency response (normalized to $\omega_C = 1$) of first-, second-, third-, and fourth-order Butterworth low-pass filters. The **Butterworth polynomials,** given in Table 15.3 in factored form, permit the design of the filter by specifying the denominator as a polynomial in $s$. For $s = j\omega$, one obtains the frequency

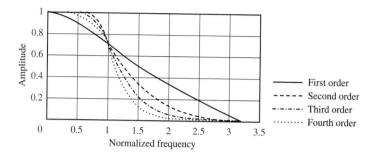

**Figure 15.20** Butterworth low-pass filter frequency response

Table 15.3  Butterworth polynomials in quadratic form

| Order $n$ | Quadratic factors |
|-----------|-------------------|
| 1 | $(s + 1)$ |
| 2 | $(s^2 + \sqrt{2}s + 1)$ |
| 3 | $(s + 1)(s^2 + s + 1)$ |
| 4 | $(s^2 + 0.7654s + 1)(s^2 + 1.8478s + 1)$ |
| 5 | $(s + 1)(s^2 + 0.6180s + 1)(s^2 + 1.6180s + 1)$ |

response of the filter.  Examples 15.4 and 15.5 illustrate filter design procedures that make use of these tables.

Figure 15.21 depicts the normalized frequency response of first- to fourth-order low-pass Chebyshev filters ($n = 1$ to 4), for $\varepsilon = 1.06$. Note that a certain amount of ripple is allowed in the pass-band; the amplitude of the ripple is defined by the parameter $\varepsilon$, and is constant throughout the pass-band. Thus, these filters are also called **equiripple filters.** Cauer, or elliptical, filters are similar to Chebyshev filters, except for being characterized by equiripple both in the pass-band and in the stop-band. Design tables exist to select the appropriate order of Butterworth, Chebyshev, or Cauer filter for a specific application.

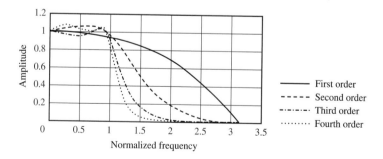

**Figure 15.21** Chebyshev low-pass filter frequency response

Three common configurations of second-order active filters, which can be used to implement **second-order** (or **quadratic) filter sections** using a single op-amp, are shown in Figure 15.22. These filters are called **constant-K**, or **Sallen and Key, filters** (after the names of the inventors). The analysis of these active filters, although somewhat more involved than that of the active filters presented in the preceding chapter, is based on the basic properties of the ideal operational amplifier discussed earlier. Consider, for example, the low-pass filter of Figure 15.22. The first unusual aspect of the filter is the presence of both negative and **positive feedback;** that is, feedback connections are provided to both the inverting and the noninverting terminals of the op-amp. The analysis method consists of finding expressions for the input terminal voltages of the op-amp, $v^+$ and $v^-$, and using these expressions to derive the input-output relationship for the filter. This analysis is left as a homework problem. The frequency response of the low-pass filter is given by

$$H(j\omega) = \frac{K(1/R_1 R_2 C_1 C_2)}{(j\omega)^2 + \left[\frac{1}{R_1 C_1} + \frac{1}{R_2 C_1} + \frac{1}{R_2 C_2}(K-1)\right] j\omega + \frac{1}{R_1 R_2 C_1 C_2}} \quad \textbf{(15.18)}$$

The above frequency response can be expressed in one of two more general forms:

$$H(j\omega) = \frac{K\omega_C^2}{(j\omega)^2 + \left(\frac{\omega_C}{Q}\right)(j\omega) + \omega_C^2}$$

and $\qquad\qquad\qquad\qquad\qquad\qquad\qquad\qquad\qquad\qquad\qquad\qquad$ **(15.19)**

$$H(j\omega) = \frac{K}{\frac{(j\omega)^2}{\omega_C^2} + \left(\frac{2\zeta}{\omega_C}\right)(j\omega) + 1}$$

The two forms are related to one another by the identity $2\zeta = Q^{-1}$. The parameter $\omega_C$ represents the cutoff frequency of the (low-pass) filter. The parameter $Q$ is called the **quality factor,** and represents the sharpness of the **resonant peak** in the frequency response of the filter (recall the material on **resonance** in Chapter 6). The parameter $\zeta$, which is called **damping ratio,** is proportional to the inverse of $Q$, and represents the degree of **damping** present in the filter: a filter with low damping (low $\zeta$), and therefore high $Q$, will have an **underdamped response** (see, again, Chapter 6), while a low-$Q$ filter will be highly damped.

The relationships between the three parameters of the second-order filter ($\omega_C$, $\zeta$, and $K$) and the resistors and capacitors is defined below for the low-pass Sallen and Key filter. A very desirable property of the Sallen and Key, or **constant-K, filter** is the fact that the low-frequency gain of the filter is independent of the cutoff frequency, and is determined simply by the ratio of resistors $R_A$ and $R_B$. The other four components define the cutoff frequency and damping ratio (or $Q$), as shown in equations 15.20.

$$K = 1 + \frac{R_A}{R_B}$$

$$\omega_C = \frac{1}{\sqrt{R_1 R_2 C_1 C_2}} \qquad\qquad \textbf{(15.20)}$$

$$\frac{1}{Q} = 2\zeta = \sqrt{\frac{R_2 C_2}{R_1 C_1}} + \sqrt{\frac{R_1 C_2}{R_2 C_1}} + (K-1)\sqrt{\frac{R_1 C_1}{R_2 C_2}}$$

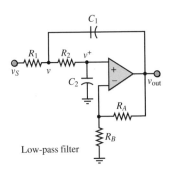

Low-pass filter

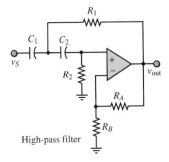

High-pass filter

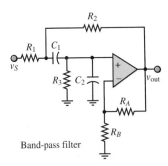

Band-pass filter

**Figure 15.22** Sallen and Key active filters

The quadratic filter sections of Figure 15.22 can be used to implement filters of arbitrary order and of different characteristics. For example, a fourth-order Butterworth filter can be realized by connecting two second-order Sallen and Key quadratic sections in cascade, and by observing that the component values of each section can be specified given the desired gain, cutoff frequency, and damping ratio (or quality factor). The following examples illustrate these procedures.

Data sheets for integrated-circuit filters may be found in the accompanying CD-ROM.

### EXAMPLE   15.3  Determining the Order of a Butterworth Filter

#### Problem

Determine the required order of a filter, given the filter specifications.

#### Solution

**Known Quantities:**  Filter gain at cutoff frequencies (pass-band and stop-band).

**Find:**  Order $n$ of filter.

**Schematics, Diagrams, Circuits, and Given Data:**  Pass-band gain: $-3$ dB at $\omega_C = 1$ rad/s; stop-band gain: $-40$ dB at $\omega_C = 4\omega_C$.

**Assumptions:**  Use a Butterworth filter response. Assume that the low-frequency gain $H_0 = 1$.

**Analysis:**  Using the magnitude-squared response for the Butterworth filter (equation 15.17),

$$|H(j\omega)|^2 = \frac{H_0^2}{1 + \varepsilon^2 \omega^{2n}}$$

with $\varepsilon = 1$, we obtain the following expression at the pass-band cutoff frequency, $\omega_C$:

$$|H(j\omega = j\omega_C)| = \frac{H_0}{\sqrt{1 + \omega_C^{2n}}} = \frac{H_0}{\sqrt{1 + 1^{2n}}} = \frac{H_0}{\sqrt{2}}.$$

This is already the desired value for the pass-band gain (3 dB below the low-frequency gain), since

$$20 \log_{10}\left(\frac{1}{\sqrt{2}}\right) = -3 \text{ dB}.$$

Note that the first requirement is automatically satisfied because of the nature of the Butterworth filter response.

The requirement for the stop-band gain imposes that the gain at frequencies at or above $\omega_S$ be less than $-40$ dB:

$$20 \log_{10} |H(j\omega = j\omega_S)| = 20 \log_{10} \frac{H_0}{\sqrt{1 + \omega_S^{2n}}} = 20 \log_{10} \frac{H_0}{\sqrt{1 + 4^{2n}}} \leq -40.$$

Thus:

$$20 \log_{10}(H_0) - 20 \log_{10}(\sqrt{1 + 4^{2n}}) \leq -40$$

or

$$\log_{10}(1 + 4^{2n}) \geq 4$$

$$(1 + 4^{2n}) \geq 10^4$$

$$2n \log_{10}(4) \geq \log_{10}(10^4 - 1)$$

Solving the above inequality we obtain: $n \geq 3.32$. Since $n$ must be an integer, we choose $n = 4$. Note that for $n = 4$, the actual gain at the stop-band frequency can be calculated to be:

$$|H(j\omega = j\omega_S)| = \frac{H_0}{\sqrt{1 + \omega_C^{2n}}} = \frac{1}{\sqrt{1 + 4^{2 \times 4}}} = -48.16 \text{ dB}$$

which is lower than the minimum desired gain of $-40$ dB, thus satisfying the specification.

**Comments:** Note that the $-3$ dB gain at the pass-band cutoff frequency is always satisfied in a Butterworth filter, since $\varepsilon = 1$.

## EXAMPLE 15.4  Design of Sallen and Key Filter

### Problem

Determine the cutoff frequency, DC gain, and quality factor for the Sallen and Key filter of Figure 15.22.

### Solution

**Known Quantities:**  Filter resistor and capacitor values.

**Find:**  $K$; $\omega_C$; $Q$.

**Schematics, Diagrams, Circuits, and Given Data:**  All resistors are 500 $\Omega$, all capacitors are 2 $\mu$F.

**Assumptions:**  None

**Analysis:**  Using the definitions given in equation 15.20, we compute:

$$K = 1 + \frac{R_A}{R_B} = 1 + \frac{500}{500} = 2$$

$$\omega_C = \frac{1}{\sqrt{R_1 R_2 C_1 C_2}} = \frac{1}{\sqrt{(500)^2 (2 \times 10^{-6})^2}} = 1,000 \text{ rad/s}$$

$$\frac{1}{Q} = 2\zeta = \sqrt{\frac{R_2 C_2}{R_1 C_1}} + \sqrt{\frac{R_1 C_2}{R_2 C_1}} + (1 - K)\sqrt{\frac{R_1 C_1}{R_2 C_2}} = 1$$

**Comments:**  What type of response does the filter analyzed above have? We can compare the filter response to that of a quadratic Butterworth filter (or other filter family) by determining the $Q$ of the filter. Once the gain and cutoff frequency have been defined, $Q$ is the parameter that distinguishes, say, a Butterworth from a Chebyshev filter. The Butterworth polynomial of order 2 is given in Table 15.3 as $(s^2 + \sqrt{2}s + 1)$. If we compare this expression to the denominator of equation 15.19, we obtain:

$$H(s) = \frac{K\omega_C^2}{s^2 + \left(\frac{\omega_C}{Q}\right)s + \omega_C^2} = \frac{1}{s^2 + \sqrt{2}s + 1}$$

Since the expressions for the quadratic polynomials of Table 15.3 are normalized to unity gain and cutoff frequency, we know that $K = 1$ and $\omega_C = 1$, and therefore we can solve for the value of $Q$ in a Butterworth filter by setting

$$\frac{1}{Q} = \sqrt{2} \quad \text{or} \quad Q = \frac{1}{\sqrt{2}} = 0.707$$

Thus, every second-order Butterworth filter is characterized by a $Q$ of 0.707; this corresponds to a damping ratio $\zeta = 0.5Q^{-1} = 0.707$, that is, to a lightly underdamped response. The next example considers the characteristics of a fourth-order Butterworth filter.

### EXAMPLE   15.5  Design of Fourth-Order Butterworth Filter

#### Problem

Design a fourth-order low-pass Butterworth filter using two quadratic Sallen and Key sections.

#### Solution

**Known Quantities:** Filter response; desired gain and cutoff frequency.

**Find:** Component values $R_1$, $R_2$, $C_1$, $C_2$, $R_A$, $R_B$, for each filter section.

**Schematics, Diagrams, Circuits, and Given Data:**  Gain $= 100$; cutoff frequency $= 400$ rad/s.

**Assumptions:** Use low-pass Sallen and Key filter prototype. In the sinusoidal steady state, $s \rightarrow j\omega$.

**Analysis:** Table 15.3 suggests that a Butterworth fourth-order filter is composed of the product of two quadratic responses. Our first objective is to determine the $Q$ of each of these two quadratic responses, so that we can design each of the two Sallen and Key quadratic sections. Comparing the standard quadratic low-pass filter response to the first of the two Butterworth polynomials for a normalized filter with $K = 1$ and $\omega_C = 1$, we have:

$$H(s) = \frac{K\omega_C^2}{s^2 + \left(\frac{\omega_C}{Q}\right)s + \omega_C^2} = \frac{1}{s^2 + 0.7654s + 1}$$

and, for a normalized filter with $K = 1$ and $\omega_C = 1$, we can solve for the value of $Q_1$, the $Q$ of the first section:

$$\frac{1}{Q_1} = 0.7654 \quad \text{or} \quad Q_1 = \frac{1}{0.7654} = 1.3065$$

Repeating the procedure for the second section we obtain:

$$\frac{1}{Q_2} = 1.8478 \quad \text{or} \quad Q_2 = \frac{1}{1.8478} = 0.5412$$

Having determined these values, we can now proceed to design two separate quadratic sections with the values of $Q$ computed above, and each with gain $K = 10$ (so that the product of the two sections yields a low-frequency gain of 100, as specified), and cutoff frequency $\omega_C = 400$ rad/s. The responses for the two sections are:

$$H(s) = \frac{K\omega_C^2}{s^2 + \left(\frac{\omega_C}{Q_1}\right)s + \omega_C^2} = \frac{1.6 \times 10^6}{s^2 + 306.16s + 1.6 \times 10^5}$$

$$= \frac{10}{6.25 \times 10^{-6}s^2 + 1.914 \times 10^{-3}s + 1}$$

and

$$H(s) = \frac{K\omega_C^2}{s^2 + \left(\frac{\omega_C}{Q_2}\right)s + \omega_C^2} = \frac{1.6 \times 10^6}{s^2 + 739.12s + 1.6 \times 10^5}$$

$$= \frac{10}{6.25 \times 10^{-6}s^2 + 4.62 \times 10^{-3}s + 1}$$

One of the important features of the Sallen and Key filter prototype is that we can choose the values for the resistors that set the circuit gain independently of those of the resistors that set the cutoff frequency (the converse is not true). Thus, we can separately select $K = 10$ for both stages by requiring that $1 + R_A/R_B = 10$, for example, $R_A = 100$ k$\Omega$, $R_A = 11.1$ k$\Omega$.

Next, we compute the component values using equations 15.20. Since we have only two equations and four unknowns, two values will have to be selected arbitrarily. We can write:

$$\omega_C = \sqrt{\frac{1}{R_1 R_2 C_1 C_2}} \Rightarrow \omega_C \sqrt{C_1 C_2} = \frac{1}{\sqrt{R_1 R_2}} \Rightarrow \sqrt{R_1 R_2} = \frac{1}{\omega_C \sqrt{C_1 C_2}}$$

and

$$\frac{1}{Q} = \sqrt{\frac{R_2}{R_1}} \cdot \sqrt{\frac{C_2}{C_1}} + \sqrt{\frac{R_1}{R_2}} \cdot \sqrt{\frac{C_2}{C_1}} + (1-K)\sqrt{\frac{R_1}{R_2}} \cdot \sqrt{\frac{C_1}{C_2}}.$$

Rearrange the last equation as:

$$\frac{1}{Q} = \sqrt{\frac{R_2}{R_1}} \cdot \sqrt{\frac{C_2}{C_1}} + \sqrt{\frac{R_1}{R_2}} \cdot \left[\sqrt{\frac{C_2}{C_1}} + (1-K) \cdot \sqrt{\frac{C_1}{C_2}}\right]$$

and operate a change of variables to obtain:

$$x = \sqrt{\frac{R_2}{R_1}} \qquad c = \left[\sqrt{\frac{C_2}{C_1}} + (1-K) \cdot \sqrt{\frac{C_1}{C_2}}\right] \qquad a = \sqrt{\frac{C_2}{C_1}} \qquad b = \frac{1}{Q}$$

$$b = ax + \frac{c}{x}$$

or

$$ax^2 - bx + c = 0$$

There is always a positive root in the preceding equation, as $a$ and $b$ are both positive. One can easily show that there is another positive root if

$$\sqrt{\frac{1}{4Q^2} + K - 1} < \sqrt{\frac{C_2}{C_1}} < \sqrt{K - 1}$$

and in that case there are two solutions for $R_1$, $R_2$; also, there are real solutions only if $C_2 < (K - 1)C_1$.

Solving the preceding equations gives:

$$x = \sqrt{\frac{R_2}{R_1}} = \frac{b + \sqrt{b^2 - 4 \cdot ac}}{2a}.$$

or

$$x = \sqrt{\frac{R_2}{R_1}} = \frac{\dfrac{1}{Q} + \sqrt{\dfrac{1}{Q^2} - 4 \cdot \dfrac{C_2}{C_1} + 4(K-1)}}{2\sqrt{\dfrac{C_2}{C_1}}}$$

Now, we have the new system:

$$\begin{cases} \sqrt{\dfrac{R_2}{R_1}} = \dfrac{\dfrac{1}{Q} + \sqrt{\dfrac{1}{Q^2} - 4 \cdot \dfrac{C_2}{C_1} + 4(K-1)}}{2\sqrt{\dfrac{C_2}{C_1}}} \\[4ex] \sqrt{R_1 R_2} = \dfrac{1}{w_C \sqrt{C_1 C_2}} \end{cases}$$

that can by easily be solved by substitution, as follows:

$$\sqrt{R_2} = \frac{\dfrac{1}{Q} + \sqrt{\dfrac{1}{Q^2} - 4 \cdot \dfrac{C_2}{C_1} + 4(K-1)}}{2\sqrt{\dfrac{C_2}{C_1}}} \sqrt{R_1}$$

$$R_1 \cdot \frac{\dfrac{1}{Q} + \sqrt{\dfrac{1}{Q^2} - 4 \cdot \dfrac{C_2}{C_1} + 4(K-1)}}{2\sqrt{\dfrac{C_2}{C_1}}} = \frac{1}{\omega_C \sqrt{C_1 C_2}}$$

That is,

$$R_1 = \frac{2\sqrt{\dfrac{C_2}{C_1}}}{\omega_C \sqrt{C_1 C_2} \left[ \dfrac{1}{Q} + \sqrt{\dfrac{1}{Q^2} - 4 \cdot \dfrac{C_2}{C_1} + 4(K-1)} \right]}$$

$$R_2 = \frac{\left[ \dfrac{1}{Q} + \sqrt{\dfrac{1}{Q^2} - 4 \cdot \dfrac{C_2}{C_1} + 4(K-1)} \right]}{2\sqrt{\dfrac{C_2}{C_1}} \cdot \omega_C \sqrt{C_1 C_2}}$$

If we assume 0.1 $\mu$F values for both $C_1$ and $C_2$ in each section, we can compute the value of the resistances required to complete the design:

*First section:*
$R_1 = 7{,}723\ \Omega$ (nearest standard 5% resistor value: 8.2 k$\Omega$)
$R_2 = 80{,}923\ \Omega$ (nearest standard 5% resistor value: 82 k$\Omega$)

*Second section:*
$R_1 = 6{,}411\ \Omega$ (nearest standard 5% resistor value: 6.8 k$\Omega$)
$R_2 = 97{,}484\ \Omega$ (nearest standard 5% resistor value: 100 k$\Omega$)

The designer may choose to employ high precision resistors or adjustable resistors, if desired.

**Comments:**  We have chosen to fix the values of the capacitors and compute the required values of the resistors because of the greater availability of resistor sizes.

**Focus on Computer-Aided Solutions:**  A Matlab file that performs the computations shown above for each quadratic section may be found in the accompanying CD-ROM. Note that the *m*-file can be easily modified to design a Butterworth filter of arbitrary order.

---

## Check Your Understanding

**15.7**    Determine the order of the filter required to satisfy the requirements of Example 15.3 if the stop-band frequency is moved to $\omega_S = 2\omega_C$.

**15.8**    What is the actual attenuation of the filter of Exercise 15.7 at the stop-band frequency, $\omega_S$?

**15.9**    Design a quadratic filter section with $Q = 1$ and a cutoff frequency of 10 rad/s. Note that there can be many solutions, depending on your design.

---

## 15.4   ANALOG-TO-DIGITAL AND DIGITAL-TO-ANALOG CONVERSION

To take advantage of the capabilities of a microcomputer, it is necessary to suitably interface signals to and from external devices with the microcomputer. Depending on the nature of the signal, either an analog or a digital interface circuit will be required. The advantages in memory storage, programming flexibility, and computational power afforded by today's digital computers are such that the instrumentation designer often chooses to convert an analog signal to an equivalent digital representation, to exploit the capabilities of a microprocessor in processing the signal. In many cases, the data converted from analog to digital form remain in digital form for ease of storage, or for further processing. In some instances it is necessary to convert the data back to analog form. The latter condition arises frequently in the context of control system design, where an analog measurement is converted to digital form and processed by a digital computer to generate a control action (e.g., raising or lowering the temperature of a process, or exerting a force or a torque); in such cases, the output of the digital computer is converted back to analog form, so that a continuous signal becomes available to the actuators. Figure 15.23 illustrates the general appearance of a digital measuring instrument and of a digital controller acting on a plant or process.

The objective of this section is to describe how the digital-to-analog (D/A) and analog-to-digital (A/D) conversion blocks of Figure 15.23 function. After illustrating discrete circuits that can implement simple A/D and D/A converters, we shall emphasize the use of ICs specially made for these tasks. Nowadays, it is uncommon (and impractical) to design such circuits using discrete components:

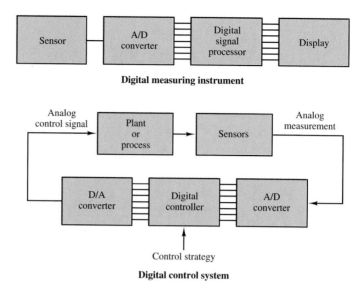

**Figure 15.23** Block diagrams of a digital measuring instrument and a digital control system

the performance and ease of use of IC packages make them the preferred choice in virtually all applications.

## Digital-to-Analog Converters

We discuss digital-to-analog conversion first because it is a necessary part of analog-to-digital conversion in many A/D conversion schemes. A **digital-to-analog converter (DAC)** will convert a binary word to an analog output voltage (or current). The binary word is represented in terms of 1s and 0s, where typically (but not necessarily), 1s correspond to a 5-volt level and 0s to a 0-volt signal. As an example, consider a four-bit binary word:

$$B = (b_3 b_2 b_1 b_0)_2 = (b_3 \cdot 2^3 + b_2 \cdot 2^2 + b_1 \cdot 2^1 + b_0 \cdot 2^0)_{10} \qquad \textbf{(15.21)}$$

The analog voltage corresponding to the digital word $B$ would be

$$v_a = (8b_3 + 4b_2 + 2b_1 + b_0)\,\delta v \qquad \textbf{(15.22)}$$

where $\delta v$ is the smallest *step size* by which $v_a$ can increment. This least step size will occur whenever the least significant bit (LSB), $b_0$, changes from 0 to 1, and is the smallest increment the digital number can make. We shall also shortly see that the analog voltage obtained by the D/A conversion process has a "staircase" appearance because of the discrete nature of the binary signal.

The step size is determined on the basis of each given application, and is usually determined on the basis of the number of bits in the digital word to be converted to an analog voltage. We can see that, by extending the previous example for an $n$-bit word, the maximum value $v_a$ can attain is

$$v_{a\,\max} = (2^{n-1} + 2^{n-2} + \cdots + 2^1 + 2^0)\,\delta v$$
$$= (2^n - 1)\,\delta v \qquad \textbf{(15.23)}$$

It is relatively simple to construct a DAC by taking advantage of the summing amplifier illustrated in Chapter 12. Consider the circuit shown in Figure 15.24, where each bit in the word to be converted is represented by means of a 5-V source and a switch. When the switch is closed, the bit takes a value of 1 (5 V); when the switch is open, the bit has value 0. Thus, the output of the DAC is proportional to the word $b_{n-1}b_{n-2}\ldots b_1b_0$.

You will recall that a property of the summing amplifier is that the sum of the currents at the inverting node is zero, yielding the relationship

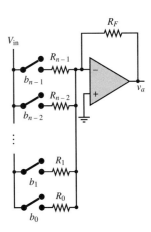

$$v_a = -\left(\frac{R_F}{R_i} \cdot b_i \cdot 5\right) \qquad i = 0, 1, \ldots, n-1 \qquad \textbf{(15.24)}$$

where $R_i$ is the resistor associated with each bit and $b_i$ is the decimal value of the $i$th bit (i.e., $b_0 = 2^0$, $b_1 = 2^1$, and so on). It is easy to verify that if we select

$$R_i = \frac{R_0}{2^i} \qquad \textbf{(15.25)}$$

**Figure 15.24** $n$-bit digital-to-analog converter (DAC)

we can obtain weighted gains for each bit so that

$$v_a = -\frac{R_F}{R_0}(2^{n-1}b_{n-1} + \cdots + 2^1 b_1 + 2^0 b_0) \cdot V_{in} \qquad \textbf{(15.26)}$$

and so that the analog output voltage is proportional to the decimal representation of the binary word. As an illustration, consider the case of a four-bit word; a reasonable choice for $R_0$ might be $R_0 = 10$ kΩ, yielding a resistor network consisting of 10-, 5-, 2.5-, and 1.25-kΩ resistors, as shown in Figure 15.25. The largest decimal value of a four-bit word is $2^4 - 1 = 15$, and so it is reasonable to divide this range into steps of 1 volt (i.e., $\delta v = 1$ V). Thus, the full-scale value of $v_a$ is 15 V:

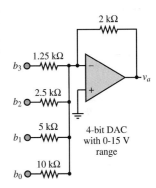

$$0 \le v_a \le 15 \text{ V}$$

and we select $R_F$ according to the following expression:

$$R_F = \frac{\delta v\, R_0}{V_{in}} = \frac{1 \cdot 10^4}{5} = 2 \text{ kΩ}$$

**Figure 15.25** Four-bit DAC

The corresponding four-bit DAC is shown in Figure 15.25.

The DAC transfer characteristic is such that the analog output voltage, $v_a$, has a steplike appearance, because of the discrete nature of the binary signal. The coarseness of the "staircase" can be adjusted by selecting the number of bits in the binary representation.

The practical design of a DAC is generally not carried out in terms of discrete components, because of problems such as the accuracy required of the resistor value. Many of the problems associated with this approach can be solved by designing the complete DAC circuit in integrated circuit (IC) form. The specifications stated by the IC manufacturer include the **resolution,** that is, the minimum nonzero voltage; the **full-scale accuracy;** the **output range;** the **output settling time;** the **power supply requirements;** and the **power dissipation.** The following examples illustrate the use of integrated circuit DACs.

## EXAMPLE 15.6 DAC Resolution

### Problem

Determine the smallest step size, or *resolution,* of an 8-bit DAC.

### Solution

***Known Quantities:*** Maximum analog voltage.

***Find:*** Resolution $\delta v$.

***Schematics, Diagrams, Circuits, and Given Data:*** $v_{a,\max} = 12$ V.

***Analysis:*** Using equation 15.23, we compute:

$$\delta v = \frac{v_{a,\max} - v_{a,\min}}{2^8 - 1} = \frac{12 - 0}{2^8 - 1} = 47.1 \text{ mV}$$

***Comments:*** Note that the resolution is dependent not only on the number of bits, but also on the analog voltage range (12 V in this case).

## EXAMPLE 15.7 Determining the Required Number of Bits in a DAC

### Problem

Find an expression for the required number of bits in a DAC using the definitions of *range* and *resolution.*

### Solution

***Known Quantities:*** Range and resolution of DAC. Voltage level corresponding to logic 1.

***Find:*** Number of DAC bits required.

***Schematics, Diagrams, Circuits, and Given Data:***
   *Range:* the analog voltage range of the DAC $= v_{a,\max} - v_{a,\min}$.
   *Resolution:* the minimum step size $\delta v$.
   $V_{\text{in}} =$ voltage level corresponding to logic 1.
   0 V $=$ voltage level corresponding to logic 0.

***Analysis:*** The maximum analog voltage output of the DAC is obtained when all bits are set to 1. Using equation 15.26 we can determine $v_{a,\max}$:

$$v_{a,\max} = V_{\text{in}} \frac{R_F}{R_0} (2^n - 1)$$

The minimum analog voltage output is realized when all bits are set to logic 0. In this case, since the voltage level associated with a logic 0 is 0 V, $v_{a,\min} = 0$. Thus, the range of this DAC is $v_{a,\max} - v_{a,\min} = v_{a,\max}$.

   The resolution was defined in the preceding example as:

$$\delta v = \frac{v_{a,\max} - v_{a,\min}}{2^n - 1}$$

Knowing both range and resolution, we can solve for the number of bits, $n$, as follows:

$$n = \frac{\log\left(\frac{v_{a,\,max} - v_{a,\,min}}{\delta v} + 1\right)}{\log 2} = \frac{\log\left(\frac{\text{range}}{\text{resolution}} + 1\right)}{\log 2}$$

Since $n$ must be an integer, the result of the above expression will be rounded up to the nearest integer. For example, if we require a 10-V range DAC with a resolution of 10 mV, we can compute the required number of bits to be:

$$n = \frac{\log\left(\frac{10}{10^{-2}} + 1\right)}{\log 2} = 9.97 \rightarrow 10 \text{ bits}$$

**Comments:** The result of this example is of direct use in the design of practical DAC circuit.

# EXAMPLE  15.8  Using DAC Device Data Sheets

## Problem

Using the data sheets for the AD7524 (supplied in the enclosed CD-ROM), answer the following questions:

1.  What is the best (smallest) resolution attainable for a range of 10 volts?
2.  What is the maximum allowable conversion frequency of this DAC?

## Solution

**Known Quantities:**  Desired range of DAC.

**Find:**  Resolution and maximum conversion frequency.

**Schematics, Diagrams, Circuits, and Given Data:**  Range $= 10$ V. DAC specifications found in device data sheet.

**Assumptions:**  The DAC is operated at full-scale range.

**Analysis:**

1.  From the data sheet we determine that the AD558 is an 8-bit converter. Thus, the best resolution that can be obtained is:

$$\delta v = \frac{v_{a,\,max} - v_{a,\,min}}{2^n - 1} = \frac{10}{2^8 - 1} = 39.2 \text{ mV}$$

2.  The maximum frequency of the DAC depends on the *settling time*. This is defined as the time required for the output to settle to within one half of the least significant bit (LSB) of its final value. Only one conversion can be performed during the settling time. The settling time is dependent on the voltage range, and for the 10-V range indicated in this problem it is equal to $T_S = 1 \ \mu$s. The corresponding maximum *sampling frequency* is $F_S = 1/T_S = 1$ MHz.

**Comments:**  The significance of the *sampling frequency* is discussed in the next subsection in connection with the *Nyquist sampling criterion*.

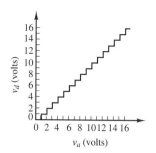

| $v_d$ | $b_3$ | $b_2$ | $b_1$ | $b_0$ |
|---|---|---|---|---|
| 0 | 0 | 0 | 0 | 0 |
| 1 | 0 | 0 | 0 | 1 |
| 2 | 0 | 0 | 1 | 0 |
| 3 | 0 | 0 | 1 | 1 |
| 4 | 0 | 1 | 0 | 0 |
| ⋮ | | ⋮ | | |
| 14 | 1 | 1 | 1 | 0 |
| 15 | 1 | 1 | 1 | 1 |

Quantized voltage — Binary representation

**Figure 15.26** A digital voltage representation of an analog voltage

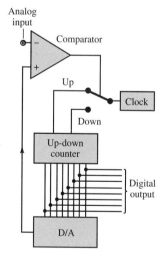

**Figure 15.27** Tracking ADC

## Analog-to-Digital Converters

The device that makes conversion of analog signals to digital form is the **analog-to-digital converter (ADC),** and, just like the DAC, it is also available as a single IC package. This section will illustrate the essential features of four types of ADCs: the tracking ADC, which utilizes a DAC to perform the conversion; the integrating ADC; the flash ADC; and the successive-approximation ADC. In addition to discussing analog-to-digital conversion, we shall also introduce the *sample-and-hold amplifier.*

### *Quantization*

The process of converting an analog voltage (or current) to digital form requires that the analog signal be quantized and encoded in binary form. The process of **quantization** consists of subdividing the range of the signal into a finite number of intervals; usually, one employs $2^n - 1$ intervals, where $n$ is the number of bits available for the corresponding binary word. Following this quantization, a binary word is assigned to each interval (i.e., to each range of voltages or currents); the binary word is then the digital representation of any voltage (current) that falls within that interval. You will note that the smaller the interval, the more accurate the digital representation is. However, some error is necessarily always present in the conversion process; this error is usually referred to as **quantization error.** Let $v_a$ represent the analog voltage and $v_d$ its quantized counterpart, as shown in Figure 15.26 for an analog voltage in the range 0–16 V. In the figure, the analog voltage $v_a$ takes on a value of $v_d = 0$ whenever it is in the range 0–1 V; for $1 \leq v_a < 2$, the corresponding value is $v_d = 1$; for $2 \leq v_a < 3$, $v_d = 2$; and so on, until, for $15 \leq v_a < 16$, we have $v_d = 15$. You see that if we now represent the quantized voltage $v_d$ by its binary counterpart, as shown in the table of Figure 15.26, each 1-volt analog interval corresponds to a unique binary word. In this example, a four-bit word is sufficient to represent the analog voltage, although the representation is not very accurate. As the number of bits increases, the quantized voltage is closer and closer to the original analog signal; however, the number of bits required to represent the quantized value increases.

### *Tracking ADC*

Although not the most efficient in all applications, the **tracking ADC** is an easy starting point to illustrate the operation of an ADC, in that it is based on the DAC presented in the previous section. The tracking ADC, shown in Figure 15.27, compares the analog input signal with the output of a DAC; the comparator output determines whether the DAC output is larger or smaller than the analog input to be converted to binary form. If the DAC output is smaller, then the comparator output will cause an up-down counter (see Chapter 14) to count up, until it reaches a level close to the analog signal; if the DAC output is larger than the analog signal, then the counter is forced to count down. Note that the rate at which the up-down counter is incremented is determined by the external clock, and that the binary counter output corresponds to the binary representation of the analog signal. A feature of the tracking ADC is that it follows ("tracks") the analog signal by changing one bit at a time.

*Integrating ADC*

The **integrating ADC** operates by charging and discharging a capacitor, according to the following principle: if one can ensure that the capacitor charges (discharges) linearly, then the time it will take for the capacitor to discharge is linearly related to the amplitude of the voltage that has charged the capacitor. In practice, to limit the time it takes to perform a conversion, the capacitor is not required to charge fully. Rather, a clock is used to allow the input (analog) voltage to charge the capacitor for a short period of time, determined by a fixed number of clock pulses. Then the capacitor is allowed to discharge through a known circuit, and the corresponding clock count is incremented until the capacitor is fully discharged. The latter condition is verified by a comparator, as shown in Figure 15.28. The clock count accumulated during the discharge time is proportional to the analog voltage.

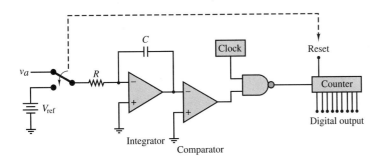

**Figure 15.28** Integrating ADC

In the figure, the switch causes the counter to reset when it is connected to the reference voltage, $V_{ref}$. The reference voltage is used to provide a known, linear discharge characteristic through the capacitor (see the material on the op-amp integrator in Chapter 12). When the comparator detects that the output of the integrator is equal to zero, it switches state and disables the NAND gate, thus stopping the count. The binary counter output is now the digital counterpart of the voltage $v_a$.

Other common types of ADC are the so-called **successive-approximation ADC** and the **flash ADC.**

*Flash ADC*

The **flash ADC** is fully parallel and is used for high-speed conversion. A resistive divider network of $2^n$ resistors divides the known voltage range into that many equal increments. A network of $2^n - 1$ comparators then compares the unknown voltage with that array of test voltages. All comparators with inputs exceeding the unknown are "on"; all others are "off." This comparator code can be converted to conventional binary by a digital priority encoder circuit. For example, assume that the three-bit flash ADC of Figure 15.29 is set up with $V_{ref} = 8$ V. An input of 6.2 V is provided. If we number the comparators from the top of Figure 15.29, the state of each of the seven comparators is as given in Table 15.4.

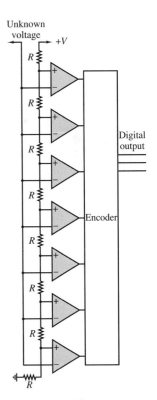

**Figure 15.29** A three-bit flash ADC

Table 15.4  State of comparators in a 3-bit flash ADC

| Comparator | Input on + line | Input on − line | Output |
|------------|-----------------|-----------------|--------|
| 1 | 7 V | 6.2 V | H |
| 2 | 6 V | 6.2 V | L |
| 3 | 5 V | 6.2 V | L |
| 4 | 4 V | 6.2 V | L |
| 5 | 3 V | 6.2 V | L |
| 6 | 2 V | 6.2 V | L |
| 7 | 1 V | 6.2 V | L |

## EXAMPLE  15.9  Flash ADC

**Problem**

How many comparators are needed in a 4-bit flash ADC?

**Solution**

***Known Quantities:*** ADC resolution.

***Find:*** Number of comparators required.

***Analysis:*** The number of comparators needed is $2^n - 1 = 15$.

***Comments:*** The flash ADC has the advantage of high speed because it can simultaneously determine the value of each bit thanks to the parallel comparators. However, because of the large number of comparators, flash ADCs tend to be expensive.

In the preceding discussion, we explored a few different techniques for converting an analog voltage to its digital counterpart; these methods—and any others—require a certain amount of time to perform the A/D conversion. This is the ADC **conversion time,** and is usually quoted as one of the main specifications of an ADC device. A natural question at this point would be: If the analog voltage changes during the analog-to-digital conversion and the conversion process itself takes a finite time, how fast can the analog input signal change while still allowing the ADC to provide a meaningful digital representation of the analog input? To resolve the uncertainty generated by the finite ADC conversion time of any practical converter, it is necessary to use a sample-and-hold amplifier. The objective of such an amplifier is to "freeze" the value of the analog waveform for a time sufficient for the ADC to complete its task.

A typical sample-and-hold amplifier is shown in Figure 15.30. It operates as follows. A MOSFET analog switch (see Chapter 10) is used to "sample" the analog waveform. Recall that when a voltage pulse is provided to the sample input of the MOSFET switch (the gate), the MOSFET enters the ohmic region and in

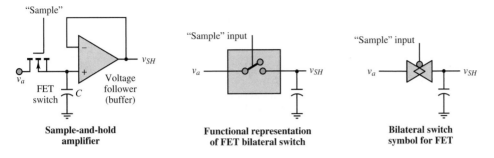

**Figure 15.30** Description of the sample-and-hold process

effect becomes nearly a short circuit for the duration of the sampling pulse. While the MOSFET conducts, the analog voltage, $v_a$, charges the "hold" capacitor, $C$, at a fast rate through the small "on" resistance of the MOSFET. The duration of the sampling pulse is sufficient to charge $C$ to the voltage $v_a$. Because the MOSFET is virtually a short circuit for the duration of the sampling pulse, the charging ($RC$) time constant is very small, and the capacitor charges very quickly. When the sampling pulse is over, the MOSFET returns to its nonconducting state, and the capacitor holds the sampled voltage without discharging, thanks to the extremely high input impedance of the voltage-follower (buffer) stage. Thus, $v_{SH}$ is the sampled-and-held value of $v_a$ at any given sampling time.

## EXAMPLE 15.10 Sample-and-Hold Amplifier

### Problem

Using the data sheets for the AD585 sample-and-hold amplifier (supplied in the enclosed CD-ROM), answer the following questions:

1.  What is the acquisition time of the AD582?
2.  How could acquisition time be reduced?

### Solution

**Known Quantities:** AD585 device data sheets.

**Find:** Acquisition time.

**Schematics, Diagrams, Circuits, and Given Data:** DAC specifications found in device data sheet. *Definition:* the acquisition time $T$ is the time required for the output of the sample-and-hold amplifier to reach its final value, within a specified error bound, after the amplifier has switched from the *sample mode* to the *hold mode*. The time $T$ includes the switch delay time, the slewing interval, and the amplifier settling time.

### Analysis:

1.  From the data sheets, the acquisition time for the AD585 is 3 $\mu$s.
2.  This acquisition time could be reduced by reducing the value of the holding capacitor, $C_H$.

*Comments:* The significance of the *sampling frequency* is discussed in the next subsection in connection with the *Nyquist sampling criterion.*

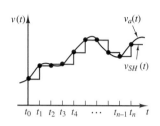

**Figure 15.31** Sampled data

The appearance of the output of a typical sample-and-hold circuit is shown in Figure 15.31, together with the analog signal to be sampled. The time interval between samples, or **sampling interval,** $t_n - t_{n-1}$, allows the ADC to perform the conversion and make the digital version of the sampled signal available, say, to a computer or to another data acquisition and storage system. The sampling interval needs to be at least as long as the A/D conversion time, of course, but it is reasonable to ask how frequently one needs to sample a signal to preserve its fundamental properties, that is, the basic shape of the waveform. One might instinctively be tempted to respond that it is best to sample as frequently as possible, within the limitations of the ADC, so as to capture all the features of the analog signal. In fact, this is not necessarily the best strategy. How should we select the appropriate sampling frequency for a given application? Fortunately, an entire body of knowledge exists with regard to sampling theory, which enables the practicing engineer to select the best sampling rate for any given application. Given the scope of this chapter, we have chosen not to delve into the details of sampling theory, but, rather, to provide the student with a statement of the fundamental result: the **Nyquist sampling criterion.**

> The Nyquist criterion states that to prevent aliasing[3] when sampling a signal, *the sample rate should be selected to be at least twice the highest-frequency component present in the signal.*

Thus, if we were sampling an audio signal (say, music), we would have to sample at a frequency of at least 40 kHz (twice the highest audible frequency, 20 kHz). In practice, it is advisable to select sampling frequencies substantially greater than the Nyquist rate; a good rule of thumb is 5 to 10 times greater. The following example illustrates how the designer might take the Nyquist criterion into account in designing a practical A/D conversion circuit.

### EXAMPLE 15.11 Performance Analysis of an Integrated-Circuit ADC

#### Problem

Using the data sheets for the AD574 (supplied in the enclosed CD-ROM), answer the following questions:

1. What is the accuracy (in volts) of the AD574?
2. What is the highest frequency signal that can be converted by this ADC without violating the Nyquist criterion?

---

[3] *Aliasing* is a form of signal distortion that occurs when an analog signal is sampled at an insufficient rate.

## Solution

**Known Quantities:**  ADC supply voltage; input voltage range.

**Find:**  ADC accuracy; maximum signal frequency for undistorted A/D conversion.

**Schematics, Diagrams, Circuits, and Given Data:**  $V_{CC} = 15$ V; $0 \leq V_{in} \leq 15$ V. ADC specifications found in device data sheet.

**Analysis:**

1.  From the data sheet we determine that the AD574 is a 12-bit converter. The accuracy is limited by the least-significant bit (LSB). For a 0–15-volt range, we can calculate the magnitude of the LSB to be:

$$\frac{V_{\text{in, max}} - V_{\text{in, min}}}{2^n - 1} = \frac{15}{2^{12} - 1} \times (\pm 1 \text{ bit}) = \pm 3.66 \text{ mV}$$

2.  The data sheet states that the maximum guaranteed conversion time of the ADC is 35 $\mu$s; therefore the highest conversion frequency for this ADC is:

$$f_{\text{max}} = \frac{1}{35 \times 10^{-6}} = 28.57 \text{ kHz}$$

Since the Nyquist criterion states that the maximum signal frequency that can be sampled wthout aliasing distortion is one half of the sampling frequency, we conclude that the maximum signal frequency that can be acquired by this ADC is approximatly 14 kHz.

**Comments:**  In practice, it is a good idea to *oversample* by a certain amount. A reasonable rule of thumb is to oversample by a factor of 2 to 5. Suppose we chose to oversample by a factor of 2; then we would not expect to have signal content above 7 kHz.

One way to ensure that the signal being sampled is limited to a 7-kHz bandwidth is to prefilter the signal with a low-pass filter having a cutoff frequency at or below 7 kHz. The active filters discussed in an earlier section of this chapter are often used for this purpose.

## Data Acquisition Systems

The structure of a data acquisition system, shown in Figure 15.32, can now be analyzed, at least qualitatively, since we have explored most of the basic building blocks. A typical data acquisition system often employs an *analog multiplexer,* to process several different input signals. A bank of bilateral analog MOSFET switches, such as the one we described together with the sample-and-hold amplifier, provides a simple and effective means of selecting which of the input signals should be sampled and converted to digital form. Control logic, employing standard gates and counters, is used to select the desired *channel* (input signal), and to trigger the sampling circuit and the ADC. When the A/D conversion is completed, the ADC sends an appropriate *end of conversion* signal to the control logic, thereby enabling the next channel to be sampled.

In the block diagram of Figure 15.32, four analog inputs are shown; if these were to be sampled at regular intervals, the sequence of events would appear as depicted in Figure 15.33. We notice, from a qualitative analysis of the figure, that the effective sampling rate for each channel is one fourth the actual external clock rate; thus, it is important to ensure that the sampling rate for each individual

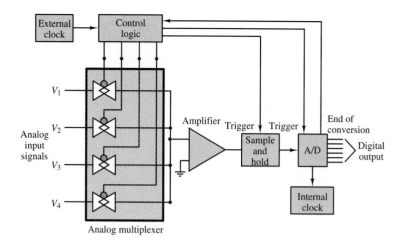

**Figure 15.32** Data acquisition system

channel satisfies the Nyquist criterion. Further, although each sample is held for four consecutive cycles of the external clock, we must notice that the ADC can use only one cycle of the external clock to complete the conversion, since its services will be required by the next channel during the next clock cycle. Thus, the internal clock that times the ADC must be sufficiently fast to allow for a complete conversion of any sample within the design range. These and several other issues are discussed in the next "Focus on Measurements" box.

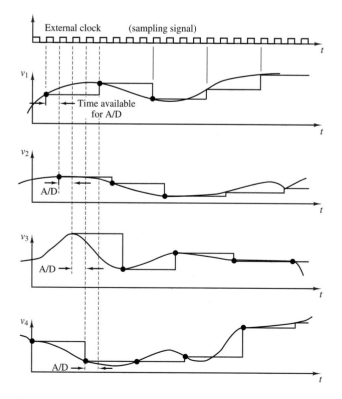

**Figure 15.33** Multiplexed sampled data

## Data Acquisition Card for Personal Computer

**FIND IT**
**ON THE WEB**

This example discusses the internal structure of a typical **data acquisition system,** such as might be used in process monitoring and control, instrumentation, and test applications. The data acquisition system discussed in this example is the AT-MIO-16 from National Instruments. The AT-MIO-16 is a high-performance, multifunction analog, digital, and timing input/output (I/O) board for the IBM PC/AT and compatibles. It contains a 12-bit ADC with up to 16 analog inputs, two 12-bit DACs with voltage outputs, eight lines of transistor-transistor-logic (TTL)–compatible digital I/O, and three 16-bit counter/timer channels for timing I/O. If additional analog inputs are required, the AMUX-64T analog multiplexer can be used. By cascading up to four AMUX-64T's, 256 single-ended or 128 differential inputs can be obtained. The AT-MIO-16 also uses the RTSI bus (real-time system interface bus) to synchronize multiboard analog, digital, and counter/timer operations by communicating system-level timing signals between boards.

Figure 15.34 is a block diagram of the AT-MIO-16 circuitry. Its major functions are described next.

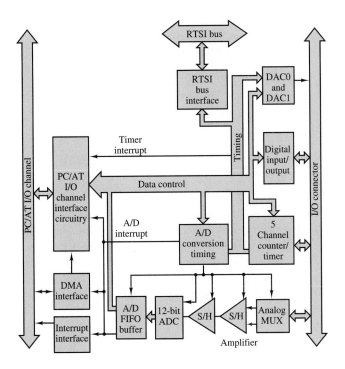

**Figure 15.34** AT-MIO-16 block diagram

### Analog Input:

The AT-MIO-16 has two CMOS analog input multiplexers connected to 16 analog input channels. This data acquisition board has a software-programmable gain amplifier that can be used with voltage gains of 1, 10, 100, or 500 (the AT-MIO-16L) to accommodate low-level analog input signals, or with gains of 1, 2, 4, or 8 (the AT-MIO-16H) for high-level analog input signals. The AT-MIO-16 has a 12-bit ADC that gives an analog signal resolution of 4.88 mV with gain of 1 and an input range of $\pm 10$ V. Finer resolutions up to 4.88 mV can be achieved by using gain and smaller input ranges. The board is available in three speeds: the AT-MIO-16(L/H)-9 contains a 9-$\mu$s ADC; the AT-MIO-16(L/H)-15 contains a 15-$\mu$s ADC; and the AT-MIO-16(L/H)-25 contains a 25-$\mu$s ADC. These conversions of the board have the following data acquisition sample rates on a single analog input channel:

| | Sampling rate | |
| Model number | Typical case | Worst case |
| --- | --- | --- |
| AT-MIO-16(L/H)-9 | 100 ksamples/s | 91 ksamples/s |
| AT-MIO-16(L/H)-15 | 71 ksamples/s | 59 ksamples/s |
| AT-MIO-16(L/H)-25 | 45 ksamples/s | 37 ksamples/s |

The timing of multiple A/D conversion is controlled either by the onboard counter/timer or by external timing signals. The onboard sample rate clock and sample counter control the onboard A/D timing. The AT-MIO-16 can generate both interrupts and DMA (direct memory access) requests on the PC/AT I/O channel. The interrupt can be generated when

1. An A/D conversion is available to be read from the A/D buffer.
2. The sample counter reaches its terminal count.
3. An error occurs.
4. One of the onboard timer clocks generates a pulse.

On the other hand, DMA requests can be generated whenever an A/D measurement is available from the A/D buffer.

**Analog Output:**

The AT-MIO-16 has two double-buffered multiplying 12-bit DACs that are connected to two analog output channels. The resolution of the 12-bit DACs is 2.44 mV in the unipolar mode or 4.88 mV in the bipolar mode with the onboard 10-V reference. Finer resolutions can be achieved by using smaller voltages on the external reference. The analog output channels have an accuracy of $\pm 0.5$ LSB and a differential linearity of $\pm 1$ LSB. Voltage offset and gain error can be trimmed to zero.

**Digital I/O:**

The AT-MIO-16 has eight digital I/O lines that are divided into two four-bit ports. The digital input circuitry has an eight-bit register that continuously reads the eight digital I/O lines, thus making read-back capability possible for the digital output ports, as well as reading incoming signals. The digital I/O lines are TTL-compatible.

**Counter/Timer:**

The AT-MIO-16 uses the AM9513A counter/timer for time-related functions. The AM9513A contains five independent 16-bit counter/timers.

A 1-MHz clock is the time baseline. Two of the AM9513A counter/timers are for multiple A/D conversion timing. The three remaining counters can be used for special data acquisition timing, such as expanding to a 32-bit sample counter or generating interrupts at user-programmable time intervals.

**RTSI Bus Interface:**
The AT-MIO-16 is interfaced to the RTSI bus. You can send or receive the external analog input control signal; the waveform-generation timing signals; the output of counters 1, 2, and 5; the gate of counter 1; and the source of counter 5. You can send the RTSI bus the frequency output of the AM9513A.

**PC/AT I/O Channel Interface:**
The PC/AT I/O channel interface circuitry includes address latches, address-decoding circuitry, data buffers, and interface timing and control signals.

**I/O Connector:**
The I/O connector is a 50-pin male ribbon cable connector.

**Software Support:**
The AT-MIO-16 also has software packages that control data acquisition functions on the PC-based data acquisition boards.

## Check Your Understanding

**15.10**  Apply KCL at the inverting node of the summing amplifier of Figure 15.25 to show that equation 15.24 holds whenever $R_i = R_0/2^i$.

**15.11**  If the maximum analog voltage ($V_{a\,max}$) of a 12-bit DAC is 15 volts, find the smallest step size ($\delta v$) by which $v_a$ can increment.

**15.12**  Repeat Example 15.6 for the case of an eight-bit word with $R_0 = 10$ k$\Omega$ and the same range of $v_a$. Find the value of $\delta v$ and $R_F$. Assume that ideal resistor values are available.

**15.13**  For Figure 15.25, find $V_{max}$ if $V_{in} = 4.5$ V.

**15.14**  For Figure 15.25, find the resolution if $V_{in} = 3.8$ V.

**15.15**  Find the minimum number of bits required in a DAC if the range of the DAC is from 0.5 to 15 V and the resolution of the DAC is 20 mV.

**15.16**  In Example 15.11, if the maximum conversion time available to you were 50 $\mu$s, what would be the highest-frequency signal you could expect to sample on the basis of the Nyquist criterion?

## 15.5    COMPARATOR AND TIMING CIRCUITS

Timing and comparator circuits find frequent application in instrumentation systems. The aim of this section is to introduce the foundations that will permit the student to understand the operation of op-amp comparators and multivibrators, and of an integrated circuit timer.

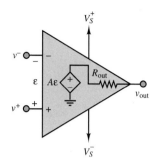

**Figure 15.35** Op-amp in open-loop mode

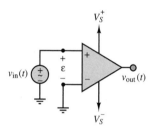

**Figure 15.36** Noninverting op-amp comparator

## The Op-Amp Comparator

The prototype of op-amp switching circuits is the op-amp comparator of Figure 15.35. This circuit, you will note, *does not employ feedback.* As a consequence of this,

$$v_{\text{out}} = A_{V(OL)}(v^+ - v^-) \tag{15.27}$$

Because of the large gain that characterizes the open-loop performance of the op-amp ($A_{V(OL)} > 10^5$), any small difference between input voltages, $\varepsilon$, will cause large outputs. In particular, for $\varepsilon$ of the order of a few tens of microvolts, the op-amp will go into saturation at either extreme, according to the voltage supply values and the polarity of the voltage difference (recall the discussion of the op-amp voltage supply limitations in Section 12.6). For example, if $\varepsilon$ were a 1-mV potential difference, the op-amp output would ideally be equal to 100 V, for an open-loop gain $A_{V(OL)} = 10^5$ (and in practice the op-amp would saturate at the voltage supply limits). Clearly, any difference between input voltages will cause the output to saturate toward either supply voltage, depending on the polarity of $\varepsilon$.

One can take advantage of this property to generate switching waveforms. Consider, for example, the circuit of Figure 15.36, in which a sinusoidal voltage source $v_{\text{in}}(t)$ of peak amplitude $V$ is connected to the noninverting input. In this circuit, in which the inverting terminal has been connected to ground, the differential input voltage is given by

$$\varepsilon = V \cos{(\omega t)} \tag{15.28}$$

and will be positive during the positive half-cycle of the sinusoid and negative during the negative half-cycle. Thus, the output will saturate toward $V_S^+$ or $V_S^-$, depending on the polarity of $\varepsilon$: the circuit is, in effect, *comparing* $v_{\text{in}}(t)$ and ground, producing a positive $v_{\text{out}}$ when $v_{\text{in}}(t)$ is positive, and a negative $v_{\text{out}}$ when $v_{\text{in}}(t)$ is negative, independent of the amplitude of $v_{\text{in}}(t)$ (provided, of course, that the peak amplitude of the sinusoidal input is at least 1 mV, or so). The circuit just described is therefore called a **comparator,** and in effect performs a binary decision, determining whether $v_{\text{in}}(t) > 0$ or $v_{\text{in}}(t) < 0$. The comparator is perhaps the simplest form of an *analog-to-digital converter,* that is, a circuit that converts a continuous waveform into discrete values. The comparator output consists of only two discrete levels: "greater than" and "less than" a reference voltage.

The input and output waveforms of the comparator are shown in Figure 15.37, where it is assumed that $V = 1$ V and that the saturation voltage corresponding to the $\pm 15$-V supplies is approximately $\pm 13.5$ V. This circuit will be termed a **noninverting comparator,** because a positive voltage differential, $\varepsilon$, gives rise to a positive output voltage. It should be evident that it is also possible to construct an inverting comparator by connecting the noninverting terminal to ground and connecting the input to the inverting terminal. Figure 15.38 depicts the waveforms for the **inverting comparator.** The analysis of any comparator circuit is greatly simplified if we observe that the output voltage is determined by the voltage difference present at the input terminals of the op-amp, according to the following relationship:

$$
\begin{array}{|c|}
\hline
\varepsilon > 0 \Rightarrow v_{\text{out}} = V_{\text{sat}}^+ \quad\quad \text{Operation of} \\
\varepsilon < 0 \Rightarrow v_{\text{out}} = V_{\text{sat}}^- \quad\quad \text{op-amp comparator} \\
\hline
\end{array}
\tag{15.29}
$$

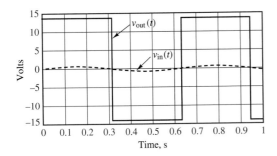

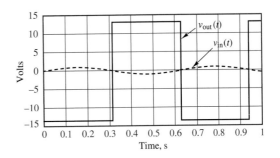

**Figure 15.37** Input and output of noninverting comparator

**Figure 15.38** Input and output of inverting comparator

where $V_{\text{sat}}$ is the saturation voltage for the op-amp (somewhat lower than the supply voltage, as discussed in Chapter 12). Typical values of supply voltages for practical op-amps are $\pm 5$ V to $\pm 24$ V.

A simple modification of the comparator circuit just described consists of connecting a fixed reference voltage to one of the input terminals; the effect of the reference voltage is to raise or lower the voltage level at which the comparator will switch from one extreme to the other. Example 15.12 describes one such circuit.

## EXAMPLE  15.12  Comparator with Offset

### Problem

Sketch the input and output waveforms of the comparator with offset shown in Figure 15.39.

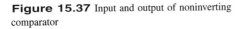

### Solution

**Known Quantities:** Input voltage, voltage offset.

**Find:** Output voltage, $v_{\text{out}}(t)$.

**Schematics, Diagrams, Circuits, and Given Data:** $v_{\text{in}}(t) = \sin(\omega t)$; $V_{\text{ref}} = 0.6$ V.

**Analysis:** We first compute the differential voltage across the inputs of the op-amp:

$$\varepsilon = v_{\text{in}} - V_{\text{ref}}$$

Then, using equation 15.29, we determine the switching conditions for the comparator:

$$v_{\text{in}} > V_{\text{ref}} \Rightarrow v_{\text{out}} = V_{\text{sat}}^{+}$$
$$v_{\text{in}} < V_{\text{ref}} \Rightarrow v_{\text{out}} = V_{\text{sat}}^{-}$$

Thus, the comparator will switch whenever the sinusoidal voltage rises above or falls below the reference voltage. Figure 15.40 depicts the appearance of the comparator output voltage. Note that comparator output waveform is no longer a symmetrical square wave.

**Comments:** Since it is not practical to use an additional external reference voltage source, one usually employs a potentiometer tied between the supply voltages to achieve

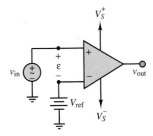

**Figure 15.39** Comparator with offset

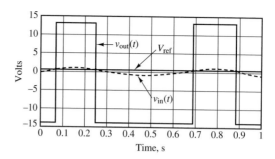

**Figure 15.40** Waveforms of comparator with offset

any value of $V_{ref}$ between the supply voltages by means of a resistive voltage divider. This circuit will be explored later in this chapter.

Another useful interpretation of the op-amp comparator can be obtained by considering its **input-output transfer characteristic.** Figure 15.41 displays a plot of $v_{out}$ versus $v_{in}$ for a noninverting zero-reference (no offset) comparator. This circuit is often called a **zero-crossing comparator,** because the output voltage goes through a transition ($V_{sat}$ to $-V_{sat}$, or vice versa) whenever the input voltage crosses the horizontal axis. You should be able to verify that Figure 15.42 displays the transfer characteristic for a comparator of the inverting type with a nonzero reference voltage.

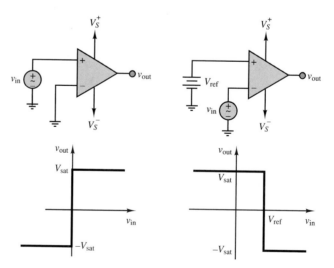

**Figure 15.41** Transfer characteristic of zero-crossing comparator

**Figure 15.42** Transfer characteristic of inverting comparator with offset

Very often, in converting an analog signal to a binary representation, one would like to use voltage levels other than $\pm V_{sat}$. Commonly used voltage levels in this type of switching circuit are 0 V and 5 V, respectively. This modified voltage transfer characteristic can be obtained by connecting a Zener diode between the output of the op-amp and the noninverting input, in the configuration sometimes called a **level** or **Zener clamp.** The circuit shown in Figure 15.43 is based on the fact that a reversed-biased Zener diode will hold a constant voltage, $V_Z$, as was shown in Chapter 8. When the diode is forward-biased, on the other hand, the output voltage becomes the negative of the offset voltage, $V_\gamma$. An additional advantage of the level clamp is that it reduces the switching time. Input and output waveforms for a Zener-clamped comparator are shown in Figure 15.44, for the case of a sinusoidal $v_{in}(t)$ of peak amplitude 1 V and Zener voltage equal to 5 V.

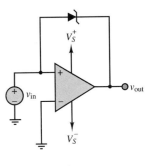

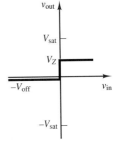

**Figure 15.43** Level-clamped comparator

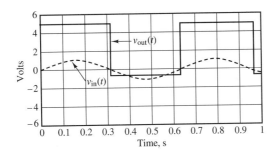

**Figure 15.44** Zener-clamped comparator waveforms

Although the Zener-clamped circuit illustrates a specific issue of interest in the design of comparator circuits, namely, the need to establish desired reference output voltages other than the supply saturation voltages, this type of circuit is rarely employed in practice. Special-purpose integrated circuit (IC) packages are available that are designed specifically to serve as comparators. These can typically accept relatively large inputs and have provision for selecting the desired reference voltage levels (or, sometimes, are internally clamped to a specified voltage range). A representative product is the LM311, which provides an open-collector output, as shown in Figure 15.45. The open-collector output allows the user to connect the output transistor to any supply voltage of choice by means of an external pull-up resistor, thus completing the output circuit. The actual value of the resistor is not critical, since the transistor is operated in the saturation mode; values between a few hundred and a few thousand ohms are typical. In the remainder of the chapter it will be assumed, unless otherwise noted, that the comparator output voltage will switch between 0 V and 5 V. Data sheets for integrated-circuit comparators may be found in the accompanying CD-ROM.

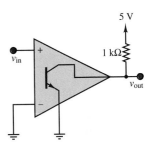

**Figure 15.45** Open-collector comparator output with representative external supply connection

## The Schmitt Trigger

One of the typical applications of the op-amp comparator is in detecting when an input voltage exceeds a present threshold level. The desired threshold is then represented by a DC reference, $V_{ref}$, connected to the noninverting input, and the input voltage source is connected to the inverting input, as in Figure 15.42. Under ideal conditions, for noise-free signals, and with an infinite slew rate for the op-amp, the operation of such a circuit would be as depicted in Figure 15.46. In practice, the presence of noise and the finite slew rate of practical op-amps will require special attention.

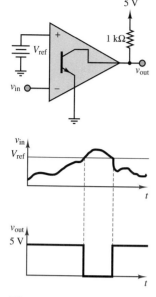

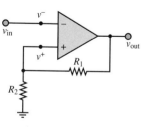

**Figure 15.46** Waveforms for inverting comparator with offset

Two improvements of this circuit will be discussed in this section: how to improve the switching speed of the comparator, and how to design a circuit that can operate correctly even in the presence of noisy signals. If the input to the comparator is changing slowly, the comparator will not switch instantaneously, since its open-loop gain is not infinite and, more important, its slew rate further limits the switching speed. In fact, commercially available comparators have slew rates that are typically much lower than those of conventional op-amps. In this case, the comparator output would not switch very quickly at all. Further, in the presence of noisy inputs, a conventional comparator is inadequate, because the input signal could cross the reference voltage level repeatedly and cause multiple triggering. Figure 15.47 depicts the latter occurrence.

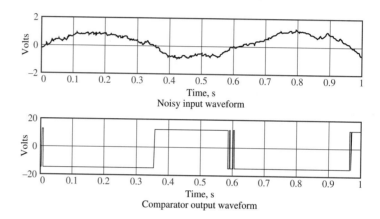

**Figure 15.47** Comparator response to noisy inputs

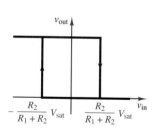

**Figure 15.48** Transfer characteristic of the Schmitt trigger

One very effective way of improving the performance of the comparator is by introducing positive feedback. As will be explained shortly, positive feedback can increase the switching speed of the comparator and provide noise immunity at the same time. Figure 15.48 depicts a comparator circuit in which the output has been tied back to the *noninverting* input (thus the terminology *positive* feedback) by means of a resistive voltage divider. The effect of this positive feedback connection is to provide a reference voltage at the noninverting input equal to a fraction of the comparator output voltage; since the comparator output is equal to either the positive or the negative saturation voltage, $\pm V_{\text{sat}}$, the reference voltage at the noninverting input can be either positive or negative.

Consider, first, the case when the comparator output is $v_{\text{out}} = +V_{\text{sat}}$. It follows that

$$v^+ = \frac{R_2}{R_2 + R_1} V_{\text{sat}} \tag{15.30}$$

and therefore the differential input voltage is

$$\varepsilon = v^+ - v^- = \frac{R_2}{R_2 + R_1} V_{\text{sat}} - v_{\text{in}} \tag{15.31}$$

For the comparator to switch from the positive to the negative saturation state, the differential voltage, $\varepsilon$, must then become negative; that is, the condition for the comparator to switch state becomes

$$v_{\text{in}} > \frac{R_2}{R_2 + R_1} V_{\text{sat}} \tag{15.32}$$

Since $[R_2/(R_2 + R_1)]V_{\text{sat}}$ is a positive voltage, the comparator will not switch when the input voltage crosses the zero level, but it will switch when the input voltage exceeds some positive voltage, which can be determined by appropriate choice of $R_1$ and $R_2$.

Consider, now, the case when the comparator output is $v_{\text{out}} = -V_{\text{sat}}$. Then

$$v^+ = -\frac{R_2}{R_2 + R_1} V_{\text{sat}} \tag{15.33}$$

and therefore

$$\varepsilon = v^+ - v^- = -\frac{R_2}{R_2 + R_1} V_{\text{sat}} - v_{\text{in}} \tag{15.34}$$

For the comparator to switch from the negative to the positive saturation state, the differential voltage, $\varepsilon$, must then become positive; the condition for the comparator to switch state is now

$$v_{\text{in}} < -\frac{R_2}{R_2 + R_1} V_{\text{sat}} \tag{15.35}$$

Thus, the comparator will not switch when the input voltage crosses the zero level (from the negative direction), but it will switch when the input voltage becomes more negative than a threshold voltage, determined by $R_1$ and $R_2$. Figure 15.48 depicts the effect of the different thresholds on the voltage transfer characteristic, showing the switching action by means of arrows.

The circuit just described finds frequent application and is called a **Schmitt trigger.**

If it is desired to switch about a voltage other than zero, a reference voltage can also be connected to the noninverting terminal, as shown in Figure 15.49. Now the expression for the noninverting terminal voltage is

$$v^+ = \frac{R_2}{R_2 + R_1} v_{\text{out}} + V_{\text{ref}} \frac{R_1}{R_2 + R_1} \tag{15.36}$$

and the switching levels for the Schmitt trigger are

$$v_{\text{in}} > \frac{R_2}{R_2 + R_1} V_{\text{sat}} + V_{\text{ref}} \frac{R_1}{R_2 + R_1} \tag{15.37}$$

for the positive-going transition, and

$$v_{\text{in}} < -\frac{R_2}{R_2 + R_1} V_{\text{sat}} + V_{\text{ref}} \frac{R_1}{R_2 + R_1} \tag{15.38}$$

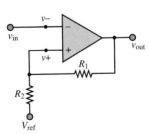

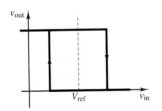

**Figure 15.49** Schmitt trigger (general circuit)

for the negative-going transition. In effect, the Schmitt trigger provides a noise-rejection range equal to $\pm[R_2/(R_2 + R_1)]V_{\text{sat}}$ within which the comparator cannot switch. Thus, if the noise amplitude is contained within this range, the Schmitt trigger will prevent multiple triggering. Figure 14.50 depicts the response of a Schmitt trigger with appropriate switching thresholds to a noisy waveform. Example 15.13 provides a numerical illustration of this process. Data sheets for Schmitt triggers may be found in the accompanying CD-ROM.

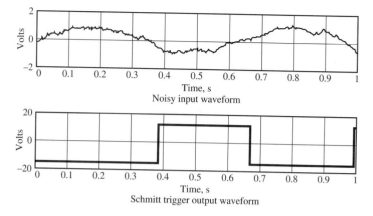

Figure 15.50 Schmitt trigger response to noisy waveforms

### EXAMPLE 15.13 Analysis and Design of Schmitt Trigger

**Problem**

Find the required resistor values for the Schmitt trigger circuit shown in Figure 15.51.

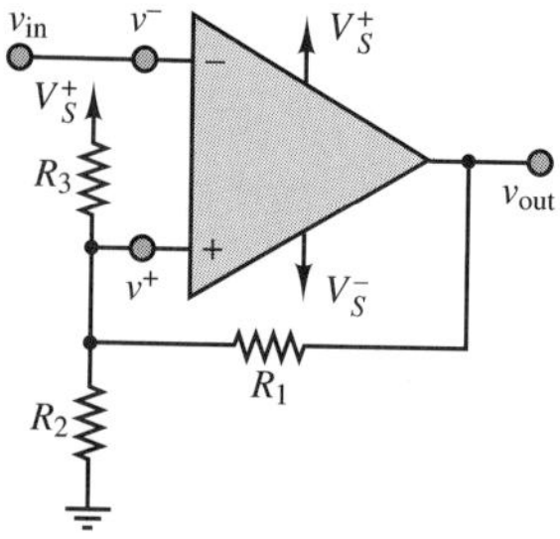

**Figure 15.51** Schmitt trigger

**Solution**

**Known Quantities:** Supply voltages and supply saturation voltages; reference voltage (offset); noise amplitude.

**Find:** $R_1$, $R_2$, $R_3$.

**Schematics, Diagrams, Circuits, and Given Data:** $|V_S| = 18$ V; $|V_{\text{sat}}| = 16.5$ V; $V_{\text{ref}} = 2$ V.

**Assumptions:** $|v_{\text{noise}}| = 100$ mV.

**Analysis:** We first observe that the offset voltage has been obtained by tying two resistors, forming a voltage divider, between the positive supply voltage and ground. This procedure avoids requiring a separate reference voltage source. From the circuit of Figure 15.51 we can calculate the noninverting voltage to be:

$$v^+ = \frac{R_2}{R_1 + R_2} v_{\text{out}} + \frac{R_2}{R_2 + R_3} V_S^+$$

Since the required noise protection level (the width of the transfer characteristic, symmetrically place about $V_{\text{ref}}$ in Figure 15.49) is $\Delta V = \pm 100$ mV, we can compute $R_1$ and $R_2$ from

$$\frac{\Delta v}{2} = \frac{R_2}{R_1 + R_2} V_{\text{sat}} = \frac{R_2}{R_1 + R_2} \times 16.5 = 0.1 \text{ V}$$

or

$$\frac{R_2}{R_1 + R_2} = \frac{0.1}{16.5}$$

The top half of Figure 15.52 depicts the $\pm 100$ mV noise protection band around the reference voltage. If we select a large value for one of the resistors, say, $R_1 = 100$ kΩ, we can calculate $R_2 \approx 610$ Ω.

To determine $R_3$, we note that

$$V_{\text{ref}} = \frac{R_2}{R_2 + R_3} V_S^+$$

or

$$2 = \frac{610}{610 + R_3} \times 18$$

and calculate $R_3 = 4.88 \text{ k}\Omega$. The design is complete. The transfer characteristic of the comparator and the associated waveforms are shown in Figure 15.52.

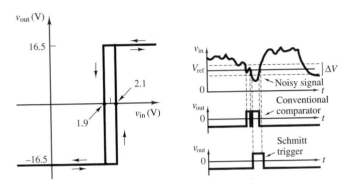

**Figure 15.52** Schmitt trigger waveforms and transfer characteristics

***Comments:*** In Figure 15.52 the Schmitt trigger output is compared to that of a comparator without noise protection. Note that a common comparator would be triggered twice in the presence of noise.

***Focus on Computer-Aided Tools:*** The Electronics Workbench solution of this example may be found in the accompanying CD-ROM.

## The Op-Amp Astable Multivibrator

This section describes an op-amp circuit useful in the generation of timing, or clock, waveforms. In the previous discussion, it was shown how it is possible to utilize positive feedback in a Schmitt trigger circuit. A small fraction of the large output saturation voltage was used to delay the switching of a comparator to make it less sensitive to random fluctuations in the input signal. A very similar circuit can be employed to generate a square-wave signal of fixed period and amplitude. This circuit is called an **astable multivibrator,** because it periodically switches between two states without ever reaching a stable state. The circuit is shown in Figure 15.53 in a somewhat simplified form.

The analysis of the circuit is similar to that of the Schmitt trigger, except for the presence of both negative and positive feedback connections. We start by postulating that the op-amp output is saturated at the positive supply saturation voltage. One can easily verify that, in this case, the noninverting terminal voltage is

$$v^+ = \frac{R_2}{R_2 + R_3} v_{\text{out}} \tag{15.39}$$

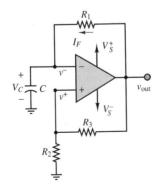

**Figure 15.53** Simplified op-amp astable multivibrator

and that the voltage at the inverting input is equal to the voltage across the capacitor, $v_C$. One can write an expression for the inverting terminal voltage, $v^-$, by noting that the capacitor charges at a rate determined by the feedback resistance and by the capacitance, according to the time constant:

$$\tau = R_1 C \tag{15.40}$$

Since the current flowing into the inverting terminal is negligible, the inverting input voltage is in effect given by

$$v^-(t) = v_C(t) = V_{\text{sat}}(1 - e^{-t/\tau}) \tag{15.41}$$

A sketch of the inverting input voltage as a function of time is shown in Figure 15.54.

The behavior of this op-amp multivibrator is best understood if the differential voltage, $\varepsilon = v^+ - v^-$, is considered. Assuming that the capacitor starts charging at $t = 0$ from zero initial charge and that $v_{\text{out}}$ is saturated at $+V_{\text{sat}}$, one can easily show that this condition will be sustained for as long as $\varepsilon$ remains positive. However, as the capacitor continues to charge, $v^-$ will eventually grow larger than $v^+$, and $\varepsilon$ will become negative. Since

$$v^+ = \frac{R_2}{R_2 + R_3} V_{\text{sat}} \tag{15.42}$$

$\varepsilon$ becomes negative as soon as the inverting-terminal voltage exceeds a threshold voltage:

$$v^- > \frac{R_2}{R_2 + R_3} V_{\text{sat}} \tag{15.43}$$

When the differential input voltage switches from a positive to a negative value, *the op-amp is forced to switch to the opposite extreme*, $-V_{\text{sat}}$. But when $v_{\text{out}}$ switches to the negative saturation voltage, we see a sudden sign reversal in $v^+$:

$$v^+ = -\frac{R_2}{R_2 + R_3} V_{\text{sat}} \tag{15.44}$$

Further, the capacitor, which has been charging toward $+V_{\text{sat}}$, now sees a negative voltage, $-V_{\text{sat}}$. This condition may be analyzed by resorting to the transient analysis methods of Chapter 5. The reversal of the output voltage causes the capacitor to discharge toward the new value of $v_{\text{out}}$, $-V_{\text{sat}}$, according to the function

$$v_C(t) = [v_C(t_0) + V_{\text{sat}}]e^{-(t-t_0)/\tau} - V_{\text{sat}} \tag{15.45}$$

where $t_0$ is the time at which the output voltage changes from $+V_{\text{sat}}$ to $-V_{\text{sat}}$. The discharging continues as long as the differential input voltage $\varepsilon$ remains negative, since this guarantees that $v_{\text{out}} = -V_{\text{sat}}$. But this condition cannot last indefinitely, because at some point, as the capacitor discharges, $v_C(t) = v^-(t)$ will become more negative than $v^+(t)$ and $\varepsilon$ will become positive again, causing the output of the op-amp to switch to $+V_{\text{sat}}$. This condition will occur when

$$v^- < -\frac{R_2}{R_2 + R_3} V_{\text{sat}} \tag{15.46}$$

Figure 15.55 graphically summarizes the operation of the astable multivibrator. The corresponding op-amp output is depicted in Figure 15.56. The period, $T$, of

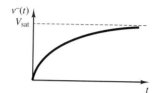

**Figure 15.54**

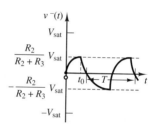

**Figure 15.55** Astable multivibrator inverting-terminal voltage

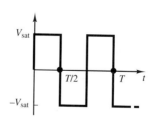

**Figure 15.56** Astable multivibrator output waveform

the waveform is determined by the charging and discharging cycle of the capacitor. It should be apparent that the introduction of a variable resistor $R_1$ gives rise to a circuit that can generate a variable-frequency square wave, since the frequency is directly determined by the time constant $\tau = R_1 C$. It can be shown that the period of the square-wave waveform resulting from the astable multivibrator is given by

$$T = 2R_1 C \log_e \left( \frac{2R_2}{R_3} + 1 \right) \tag{15.47}$$

## The Op-Amp Monostable Multivibrator (One-Shot)

An extension of the astable multivibrator is the **one-shot, or monostable multivibrator.** One-shots are available in integrated circuit form but can also be constructed from more general-purpose op-amps. In this section, we first look at the op-amp one-shot, taking special note of the analogy with the astable multivibrator. Figure 15.57 depicts an op-amp monostable multivibrator. Its operation is summarized in the associated timing diagram, showing that if a negative voltage pulse is applied at the input, $v_{in}$, the output of the op-amp will switch from a stable state, $+V_{sat}$, to the unstable state, $-V_{sat}$, for a period of time $T$, which is determined by a charging time constant, after which the output will resume its stable value, $+V_{sat}$. In the analysis of the one-shot circuit, it will be assumed that the diode is an offset diode, with offset voltage $V_{off}$, and that $v_{out} = +V_{sat}$ for $t < t_0$.

For $t < t_0$, before the negative input pulse is applied, the input terminal voltages are given by

$$v^+ = V_{sat} \frac{R_2}{R_2 + R_3} \tag{15.48}$$

and

$$v^- = V_\gamma \tag{15.49}$$

Provided that $v^+ > v^-$, the op-amp will remain in its positive saturated state, then, until some external condition causes $\varepsilon = v^+ - v^-$ to become negative. This condition is brought about at $t = t_0$ by the "trigger" pulse $v_{in}$, which briefly lowers the noninverting-terminal voltage by $\delta V$ volts. If $\delta V$ is sufficiently large, the following condition will be satisfied:

$$v^+ = V_{sat} \frac{R_2}{R_2 + R_3} - \delta V < V_\gamma \tag{15.50}$$

and the op-amp will switch to the negative saturation state, as indicated in Figure 15.57. The noninverting-terminal voltage may be expressed as a function of time for $t > t_0$, considering that $v_{out}$ has switched to $-V_{sat}$ at $t = t_0$ and the diode now acts as an open circuit, since $v^-$ is negative. With the diode out of the picture, then, the $R_1 C$ circuit will discharge, causing $v^-$ to drop from its initial voltage $(V_\gamma)$ toward $-V_{sat}$.

The circuit of Figure 15.58 depicts the equivalent switching circuit, illustrating that the capacitor discharges from the initial value of $v_C(t_0) = V_\gamma$ toward the final value, $-V_{sat}$. Recalling the analysis of transients in Chapter 5, we know that the capacitor voltage is given by the expression

$$v_C(t) = [v_C(t_0) + V_{sat}]e^{-(t-t_0)/R_1 C} - V_{sat} \tag{15.51}$$

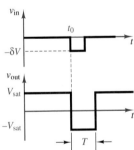

**Figure 15.57** Op-amp monostable multivibrator and typical waveforms

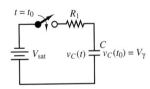

**Figure 15.58** Equivalent charging circuit for monostable multivibrator

As $v^-$ becomes more negative, it eventually becomes smaller than $v^+$. Thus, when the condition

$$v_C(t) = [v_C(0) + V_{sat}]e^{-(t-t_0)/R_1C} - V_{sat} = v^+ \tag{15.52}$$

is met, the comparator will switch back to the positive saturation state, $+V_{sat}$, and remain in that state until a new trigger pulse is provided. The value of the noninverting-terminal voltage during the time the output is in the negative saturated state is determined by the positive feedback circuit:

$$v^+ = -V_{sat}\frac{R_2}{R_2 + R_3} \tag{15.53}$$

Clearly, the duration of the output pulse provided by the one-shot is determined by the time constant of the $R_1C$ circuit, as well as by the value of the resistive voltage divider in the positive feedback network, as illustrated in Example 15.14.

## EXAMPLE 15.14 Analysis of Monostable Multivibrator (One-Shot)

### Problem

Calculate the duration of the output pulse for the monostable multivibrator of Figure 15.57.

### Solution

**Known Quantities:**  Component values; diode offset voltage; supply saturation voltage.

**Find:**  Duration of output pulse, $T$.

**Schematics, Diagrams, Circuits, and Given Data:**  $R_1 = 20$ k$\Omega$; $R_2 = 670$ $\Omega$; $R_3 = 100$ k$\Omega$; $C = 7$ $\mu$F; $V_\gamma = 0.6$ V; $V_{sat} = 16$ V.

**Assumptions:**  A small pulse of amplitude $\delta V$ is applied to the input of the monostable at $t_0 = 0$ (see Figure 15.57).

**Analysis:**  To compute the switching time of the circuit we need to determine when the capacitor will charge to a voltage such that $v^- = v_C$ is equal to $v^+$. We use equation 15.51 to compute the expression for the capacitor voltage:

$$v_C(t) = [v_C(0) + V_{sat}]e^{-t/R_1C} - V_{sat}$$

Note that $v_C(0) = 0.6$ V because of the diode connected in parallel to the capacitor. The noninverting terminal voltage is computed from:

$$v^+ = -V_{sat}\frac{R_2}{R_2 + R_3} = -0.1065 \text{ V}$$

Setting the two above expressions equal to one another we obtain:

$$-0.1065 = 16.6e^{-10.64\times10^{-3}T} - 16$$

or

$$T = -\frac{1}{10.64 \times 10^{-3}} \log_e \frac{15.8935}{16.6} = 4.09 \times 10^{-3} = 4.09 \text{ ms}$$

*Comments:*  In practice one rarely uses op-amps to build monostable multivibrators.
Integrated-circuit multivibrators are introduced next.

---

Monostable multivibrators are usually employed in IC package form.  An IC
one-shot can generate voltage pulses when triggered by a **rising** or a **falling edge,**
that is, by a transition in direction in the input voltage. Thus, a one-shot IC
offers the flexibility of external selection of the type of transition that will cause
a pulse to be generated:  a rising edge (from low voltage to high, typically zero
volts to some threshold level), or a falling edge (high-to-low transition).  Various
input connections are usually provided for selecting the preferred triggering mode,
and the time constant is usually set by selection of an external $RC$ circuit.  The
output pulse that may be generated by the one-shot can also occur as a positive or a
negative transition.  Figure 15.59 shows the response of a one-shot to a triggering
signal for the four conditions that may be attained with a typical one-shot.

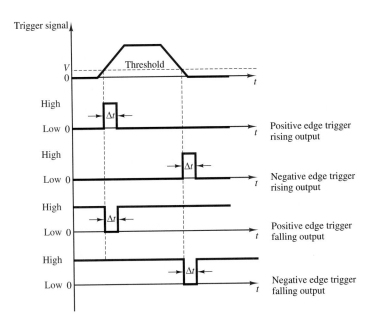

**Figure 15.59** IC monostable multivibrator waveforms

A typical IC one-shot circuit based on the 74123 (see the data sheet at the
end of the chapter) is displayed in Figure 15.60. The 74123 is a **dual one-shot,**
meaning that the package contains two monostable multivibrators, which can be
used independently.  The outputs of the one-shot are indicated by the symbols
$Q_1$, $\overline{Q}_1$, $Q_2$, and $\overline{Q}_2$, where the bar indicates the complement of the output.
For example, if $Q_1$ corresponds to a positive-going output pulse, $\overline{Q}_1$ indicates a
negative-going output pulse, of equal duration.

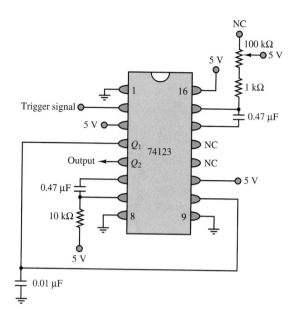

**Figure 15.60** Dual one-shot circuit

## Timer ICs: The NE555

The discussion of op-amp–based timing circuits presented in the previous sections served the purpose of introducing a large family of integrated circuits that can provide flexible timing waveforms. These fall—for our purposes—into one of two classes: pulse generators, and clock waveform generators. Chapters 13 and 14 delve into a more detailed analysis of digital timing circuits, a family to which the circuits of the previous sections belong. This section will now introduce a multipurpose integrated circuit that can perform both the monostable and astable multivibrator functions. The main advantage of the integrated circuit implementation of these circuits (as opposed to the discrete op-amp version previously discussed) lies in the greater accuracy and repeatability one can obtain with ICs, their ease of application, and the flexibility provided in the integrated circuit packages. The NE555 is a timer circuit capable of producing accurate time delays (pulses) or oscillation. In the time-delay, or monostable, mode, the time delay or pulse duration is controlled by an external $RC$ network. In the astable, or clock generator, mode, the frequency is controlled by two external resistors and one capacitor. Figure 15.61 depicts typical circuits for monostable and astable operation of the NE555. Note that the threshold level and the trigger level can also be externally controlled. For the monostable circuit, the pulse width can be computed from the following equation:

$$T = 1.1R_1C \tag{15.54}$$

For the astable circuit, the positive pulse width can be computed from the following equation:

$$T_+ = 0.69(R_1 + R_2)C \tag{15.55}$$

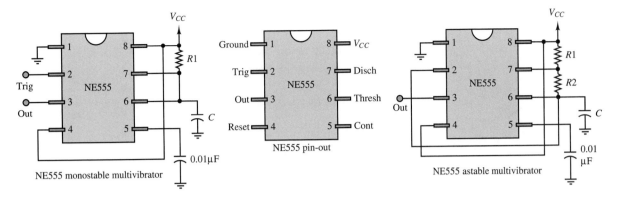

**Figure 15.61** NE555 timer

and the negative pulse width can be computed from

$$T_- = 0.69 R_2 C \tag{15.56}$$

The use of the NE555 timer is illustrated in Example 15.15.

## EXAMPLE 15.15  Analysis of 555 Timer

### Problem

Calculate the component values required to obtain a 0.421-ms pulse using the 555 timer monostable configuration of Figure 15.61.

### Solution

**Known Quantities:**  Desired pulse duration $T$.

**Find:**  Values of $R_1$, $C$.

**Schematics, Diagrams, Circuits, and Given Data:**  $T = 0.421$ ms.

**Assumptions:**  Assume a value for $C$.

**Analysis:**  Using equation 15.54,

$$T = 1.1 R_1 C$$

And assuming $C = 1$ $\mu$F, we calculate

$$0.421 \times 10^{-3} = 1.1 R_1 \times 10^{-6}$$

or

$$R_1 = 382.73 \ \Omega$$

**Comments:**  Any reasonable combination of $R_1$ and $C$ values can yield the desired design value of $T$. Thus, the component selection shown in this example is not unique.

**Focus on Computer-Aided Tools:**  The Electronics Workbench solution of this example may be found in the accompanying CD-ROM.

VIRTUAL LAB

## Check Your Understanding

**15.17**    Verify that the transfer characteristic of Figure 15.42 is correct.

**15.18**    For the comparator circuit shown in Figure 15.62, sketch the waveforms $v_{out}(t)$ and $v_S(t)$ if $v_S(t) = 0.1 \cos(\omega t)$ and $V_{ref} = 50$ mV.

**15.19**    Derive the expressions for the switching thresholds of the Schmitt trigger of Figure 15.48.

**15.20**    Explain why positive feedback increases the switching speed of a comparator.

**15.21**    Compute the period of the square wave generated by the multivibrator of Figure 15.53 if $C = 1$ μF, $R_1 = 10$ kΩ, $R_3 = 100$ kΩ, and $R_2 = 1$ kΩ.

**15.22**    Compute the value of $R_3$ that would increase the duration of the one-shot pulse in Example 15.14 to 10 ms.

**15.23**    Compute the value of $C$ that would increase the duration of the one-shot pulse in Example 15.14 to 10 ms.

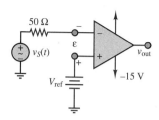

**Figure 15.62**

## 15.6    OTHER INSTRUMENTATION INTEGRATED CIRCUITS

The advent of low-cost integrated electronics and microprocessors has revolutionized instrumentation design. The transition from the use of analog, discrete circuits (e.g., discrete transistor amplifiers) for signal conditioning to (often digital) integrated circuits and to microprocessor-based instrumentation systems has taken place over a period of two decades, and is now nearly complete, with the exception of very specialized applications (e.g., very high frequency or low noise circuits). The aim of this section is to present a summary of some of the signal conditioning and processing functions that are readily available in low-cost integrated circuit form. The list is by no means exhaustive, and the reader is referred to the Web sites of the numerous integrated circuit manufacturers for more detailed information.

The nonelectrical engineer interested in the design of special-purpose instrumentation circuits can benefit from the wealth of information contained in the application notes available from IC manufacturers (often directly downloadable from the Web). You will find references to some of these resources in the accompanying CD-ROM.

In the preceding sections of this chapter we have already explored some IC instrumentation elements, namely, instrumentation amplifiers, op-amp active filters, digital-to-analog and analog-to-digital converters, sample-and-hold amplifiers, voltage comparators and timing ICs. Further, Chapter 12 delves into the basic operation of operational amplifiers, and Chapters 13 and 14 contain information on digital logic circuits and microprocessors and microcontrollers. In this section we briefly survey a number of common instrumentation applications not yet mentioned in this book, and provide references for some applications that have already been discussed.

### Amplifiers

A number of special-purpose IC amplifiers are available to perform a variety of functions. Although most of these amplifiers could be realized by using op-amps, these specially designed packages can save much design effort and provide better performance. The following list indicates some of the products available from one manufacturer (Analog Devices):

- Instrumentation amplifiers
- Logarithmic amplifiers
- RF amplifiers
- Sample-and-hold, track-and-hold amplifiers
- Variable-gain amplifiers
- (Audio) microphone preamplifiers
- (Audio) power amplifiers
- (Audio) voltage-controlled amplifiers

## DACs and ADCs

Digital-to-analog and analog-to-digital converters are also available in a variety of packages intended for general use or tailored to special applications:

> General-purpose ADCs ($\leq$ 1 Msample/s)
> High-speed ADCs ($>$ 1 Msample/s)
> DACs
> (Audio) Nyquist DAC
> (Audio) sigma-delta DAC
> (Audio) stereo ADC
> Digital radio ADCs

## Frequency-to-Voltage, Voltage-to-Frequency Converters and Phase-Locked Loops

The need for converting changes in instantaneous frequency to changes in an analog voltage (i.e., frequency demodulation) arises frequently in instrumentation applications. For example, *optical position encoders* (Chapter 13) and *magnetic position sensors* (Chapter 16) represent instantaneous velocity information as a frequency-modulated (FM) signal that can be demodulated by a frequency-to-voltage converter. Similarly, it is sometimes useful to encode analog voltage signals in FM form using a voltage-to-frequency converter.

**Phase-locked loops (PLLs)** can also be used for FM demodulation, as well as for a number of other related functions, such as tone decoding.

## Other Sensor and Signal Conditioning Circuits

Sensor and signal conditioning integrated circuit technology often permits the integration of sensing and signal conditioning on the same integrated circuit. In addition, specialized signal conditioning modules enable the design of rather complex instrumentation systems with relatively few IC building blocks. The enclosed CD-ROM contains device data sheets and application notes for a number of commercially available IC products, including **rms-to-DC converters,** Accelerometers and other **integrated sensors for industrial and automotive applications, and signal conditioning subsystems.** The following example, reprinted with permission of Analog Devices, consists of an application note discussing the use of integrated acceleration and tilt sensors in a car anti-theft alarm. This is only an example of the wealth of electronic information appended to this book.

## FOCUS ON MEASUREMENTS

## Using the ADXL202 Accelerometer as a Multifunction Sensor (Tilt, Vibration and Shock) in Car Alarms
### by Harvey Weinberg and Christophe Lemaire, Analog Devices

*By using an intelligent algorithm, the ADXL202 ($\pm 2$ g dual axis accelerometer) can serve as a low cost, multifunction sensor for vehicle security systems, capable of acting simultaneously as a shock/vibration detector as well as a tilt sensor (to detect towing or jacking up of the car). The accelerometer's output is passed through two parallel filters. A bandpass filter to extract shock/vibration information, and a low pass filter to extract tilt information. This application note describes the basics of such an implementation.*

### Introduction

The ADXL202 is a low cost, low power, complete dual axis accelerometer with a measurement range of $\pm 2$ g. The ADXL202 outputs analog and digital signals proportional to acceleration in each of the sensitive axes (see Figure 15.63).

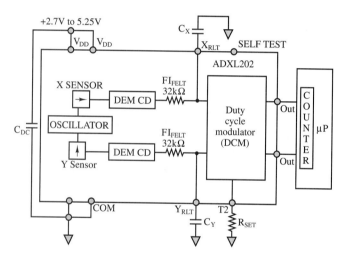

**Figure 15.63** ADXL202 block diagram

Currently automotive security systems use shock/vibration sensors to detect collision or forced intrusion into the car. Typically, these sensors are based on magneto-inductive sensing. Sensors of this type generally have adequate sensitivity, but fall short in other areas. Often a fair amount of signal conditioning and trimming is required between the shock sensor and microcontroller due to variations in magnetic material and Hall effect sensor sensitivity and their frequency response is fairly unpredictable due to inconsistency in mounting. In addition such sensors have no response to gravity-induced acceleration, so they are incapable of sensing inclination (a static acceleration). Tilt sensing is the most direct way of detecting if a vehicle is being jacked up, about to be towed, or being loaded onto a flatbed truck. These are some of the most common methods of car theft today.

The ADXL202 is a true accelerometer, easily capable of shock/vibration sensing with virtually no external signal conditioning circuitry. Since the ADXL202 is also sensitive to static (gravitational) acceleration, tilt sensing is also possible. Tilt sensing requires a very low noise floor which usually necessitates restricting the bandwidth of the accelerometer, while shock/vibration sensing requires wide bandwidth. These conflicting requirements may be met using clever design techniques.

**Principle of Operation**

The ADXL202 is set up to acquire acceleration from 0 to 200 Hz (the maximum frequency of interest). Figure 15.64 shows a block diagram of the system. The accelerometer's output is fed into two filters; a low pass filter with a corner frequency at 12.5 Hz used to lower the noise floor sufficiently for accurate tilt sensing, and a band pass filter to minimize the noise in the shock/vibration pass band of interest. The low pass filtered (tilt) output then goes to a differentiator (described in the Tilt Sensing section) where the determination is made as to whether the accelerometer actually sensed tilt or some other event such as noise or temperature drift. Then an auto-zero block performs further signal processing to reject temperature drift. The band pass filtered output goes to an integrator (described in the Shock Sensing section) that measures vibrational energy over a small period of time (40 ms). A decision as to whether or not to set off the alarm may then be made by the microcontroller. Most of these tasks are most easily implemented in the digital domain and require very little computational power.

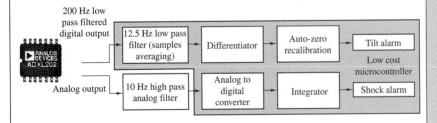

**Figure 15.64** Shock and tilt sensing using the ADXL202

Since the two measurements (shock/vibration and tilt) are basically exclusive and only share a common sensor, their respective signal processing tasks will be described separately.

**Tilt Sensing**

**Fundamentals**

The alarm system must detect a change in tilt slow enough to be the result of the vehicle being towed or jacked up, but must be immune to temperature changes and movement due to passing vehicles or wind. Note that the ADXL202 is most sensitive to tilt when its sensitive axes are perpendicular to the force of gravity, i.e., parallel to the earth's surface. Figure 15.65 shows that the change in projection of a 1 $g$ gravity-induced acceleration vector on the axis of sensitivity of the accelerometer will be more significant if the axis is tilted 10 degrees from the horizontal than if it is tilted by the same amount from the vertical.

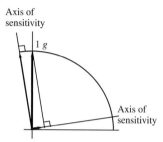

**Figure 15.65** Tilt sensitivity

However, the car may not always be level when the alarm is activated, and while the zero $g$ offset can be recalibrated for any initial inclination, effectively the farther from the horizontal the axes of sensitivity are, the less sensitive the system will be to tilt (see ADXL202 datasheet). In most cases, this should not be of great concern, since the sensitivity only declines by about 2.5 m$g$ per degree of tilt when inclination goes from zero (horizontal) to thirty degrees of tilt. Nevertheless, installation guidelines should recommend that the tilt sensing module containing the accelerometer be mounted such that the axes of sensitivity be as level as possible.

**Implementation**

In general we are interested in knowing if the inclination of the car has changed more than ±5 degrees from its inclination when initially parked. When the car is turned off, a measurement of the car's inclination is made. If the car's inclination is changed by more than ±5 degrees, an alarm is triggered. Alternatively, the rate of change of tilt may be evaluated and if its absolute value is above 0.2 degrees per second for several seconds the alarm may be triggered.

Each technique has certain advantages. The former algorithm is better at false alarm rejection due to jostling of the car, while the rate of change algorithm may be set up to react more quickly. Algorithms using a combination of both techniques may be used as well. It is left to the reader to decide which technique is best for their application. While all of the concepts presented here are valid for both algorithms, for consistency this application note will describe the former (absolute inclination) algorithm.

For the purpose of the following discussion, we will assume a less than perfect tilt sensitivity for the accelerometer of 15 m$g$ per degree of tilt, or 75 m$g$ for 5 degrees. The ADXL202 will be set up to have a bandwidth of 200 Hz so that vibration may be detected. A 200 Hz bandwidth will result in a noise floor of:

$$\text{Noise} = 500 \ \mu g \sqrt{\text{Hz}} \times (\sqrt{200 \times 1.5}) \ \text{rms}$$

$$\text{Noise} = 8.5 \ \text{m}g \ \text{rms}$$

or 34 m$g$ peak-to-peak of noise (using a peak-to-peak to rms ratio of 4:1), well within our 75 m$g$ requirement. For reliability purposes, we would like to have a noise floor about 10 times lower than this, or around 8 m$g$. Since towing a car takes at least a few seconds, we are free to narrow the bandwidth to lower the noise floor. An analog or digital low pass filter may be used, but since low pass filtering in the digital domain is very simple, it is the preferred method. By taking the average of 16 samples we reduce the effective bandwidth to 12.5 Hz (200 Hz/16 samples). The resulting noise performance is approximately 8.7 m$g$ peak-to-peak, close enough to our target.

Lowering the noise floor even further, by taking up to 128 samples for example, would result in about 3 m$g$ peak-to-peak of noise, which would allow us to easily detect the 15 m$g$ of static acceleration resulting from a change in tilt of less than a degree.

The typical zero $g$ drift due to temperature for the ADXL202 is 2 m$g$/°C. Since our trigger point for a tilt alarm could be as low as 15 m$g$, it is conceivable that temperature drift alone would cause a false alarm (a car parked overnight could easily experience more than 7.5 °C in ambient temperature change). Therefore we will include a differentiator to reject temperature drift.

In the event of the car being jacked up or lifted for towing, we would expect the rate of change in tilt to be faster than five degrees or 75 m$g$ per minute (or 1.25 m$g$ per second). Each time the acceleration is measured it is compared to the previous reading. If the change is less than 1.25 m$g$ per second we know that the change in accelerometer output is due to temperature drift. We can now add an auto-zero block that adjusts our "zero $g$" reference (that is the static acceleration sensed when the car was initially parked) to compensate for zero $g$ drift due to temperature.

### Shock Sensing

Generally for automotive shock/vibration sensing we are interested in signals between 10 and 200 Hz. Since the response of the ADXL202 extends from DC to 5 kHz, a band pass filter will have to be added to remove out of band signals. This band pass filter is most easily implemented in the analog domain (Figure 15.66 shows a simple 10 Hz high pass filter). When coupled with the 200 Hz low pass filter (from Xfilt and Yfilt on the ADXL202), a 10 to 200 Hz bandpass filter is realized.

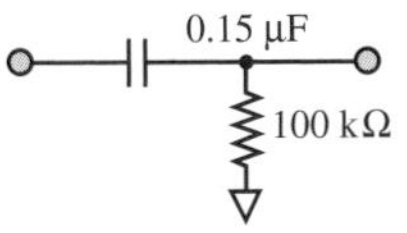

**Figure 15.66** 10 Hz
High Pass Filter

While analog bandpass filtering is very simple and requires no software overhead from the microcontroller, it does necessitate having an analog to digital converter. Today, even low cost microcontrollers can commonly be found, integrating an A to D converter on board. Bandpass filtering in the digital domain may be more effective, but may require a more powerful processor than one normally finds in automobile security systems. There are several methods for implementing band pass filters in the digital domain. Specific recommendations will not be given here since processor selection will influence what method will be most efficient.

Whether a digital or analog bandpass filter is used, the Nyquist criteria for signal sampling must be satisfied. That is that we must sample at least twice the maximum frequency of interest. Sampling at 400 Hz (for our 200 Hz pass band) gives us one sample every 2.5 ms. Our very simple software integrator will take the sum of the absolute value of 16 samples and evaluate if there is sufficient energy in that 40 ms period of time to warrant setting off the alarm (i.e., is the sum of 16 samples greater than some set point). It is assumed that no events will be missed in 40 ms.

**Design Trade-Offs**

The ADXL202 has digital (Pulse Width Modulated) as well as analog (312 mV/$g$) outputs. In theory, either output may be used. Using the PWM interface for tilt sensing is recommended for two reasons:

1.  We are interested in very small acceleration signals (on the order of 3 m$g$). This would correspond to approximately .94 mV. Probably not resolvable by the on board A to D converter of any microcontroller likely to be used in this application. The resolution of the pulse width modulator of the ADXL202 is around 14 bits, and is sufficient for resolution of 3 m$g$ acceleration signals.

2.  All signal processing will be done in the digital domain.

An analog interface for the shock/vibration sensor is recommended since, as previously mentioned, bandpass filtering in the digital domain may be beyond the capability of many microcontrollers. In addition using the PWM interface to acquire 200 Hz bandwidth requires that the PWM frequency be at least 4 kHz. 10 bit resolution implies that the microcontroller have a timer resolution of approximately 250 ns. Once again, probably beyond the capability of most microcontrollers.

## 15.7 DATA TRANSMISSION IN DIGITAL INSTRUMENTS

One of the necessary aspects of data acquisition and control systems is the ability to transmit and receive data. Often, a microcomputer-based data acquisition system is interfaced to other digital devices, such as digital instruments or other microcomputers. In these cases it is necessary to transfer data directly in digital form. In fact, it is usually preferable to transmit data that is already in digital form, rather than analog voltages or currents. Among the chief reasons for the choice of digital over analog is that digital data is less sensitive to noise and interference than analog signals: in receiving a binary signal transmitted over a data line, the only decision to be made is whether the value of a bit is 0 or 1. Compared with the difficulty in obtaining a precise measurement of an analog voltage or current, either of which could be corrupted by noise or interference in a variety of ways, the probability of making an error in discerning between binary 0s and 1s is very small. Further, as will be shown shortly, digital data is often coded in such a way that many transmission errors may be detected and corrected. Finally, storage and processing of digital data are much more readily accomplished than would be the case with analog signals. This section explores a few of the methods that are commonly employed in transmitting digital data; both parallel and serial interfaces are considered.

Digital signals in a microcomputer are carried by a bus, consisting of a set of parallel wires each carrying one bit of information. In addition to the signal-carrying wires, there are also control lines that determine under what conditions transmission may occur. A typical computer data bus consists of eight parallel wires and therefore enables the transmission of one byte; digital data is encoded in binary according to one of a few standard codes, such as the BCD code described in Chapter 13, or the ASCII code, introduced in Chapter 14 (see Table 14.6). This bus configuration is usually associated with **parallel transmission,** whereby all of the bits are transmitted simultaneously, along with some control bits.

Figure 15.67 depicts the general appearance of a parallel connection. Parallel data transmission can take place in one of two modes: **synchronous** or **asynchronous.** In synchronous transmission, a timing clock pulse is transmitted along with the data over a control line. The arrival of the clock pulse indicates that valid data has also arrived. While parallel synchronous transmission can be very fast, it requires the added complexity of a synchronizing clock, and is typically employed only for internal computer data transmission. Further, this type of communication can take place only over short distances (approximately 4 m). Asynchronous data transmission, on the other hand, does not take place at a fixed clock rate, but requires a **handshake protocol** between sending and receiving ends. The handshake protocol consists of the transmission of *data ready* and *acknowledge* signals over two separate control wires. Whenever the sending device is ready to transmit data, it sends a pulse over the *data ready* line. When this signal reaches the receiver, and if the receiver is ready to receive the data, an *acknowledge* pulse is sent back, indicating that the transmission may occur; at this point, the parallel data is transmitted.

Perhaps the most common parallel interface is based on the **IEEE 488 standard,** leading to the so-called IEEE 488 bus, also referred to as **GPIB** (for **general-purpose instrument bus**).

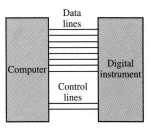

**Figure 15.67** Parallel data transmission

## The IEEE 488 Bus

The IEEE 488 bus, shown in Figure 15.68, is an eight-bit parallel asynchronous interface that has found common application in digital instrumentation applications. The physical bus consists of 16 lines, of which 8 are used to carry the data, 3 for the handshaking protocol, and the rest to control the data flow. The bus permits connection of up to 15 instruments and data rates of up to 1 Mbyte/s. There is a limitation, however, in the maximum total length of the bus cable, which is 20 m. The signals transmitted are TTL-compatible and employ negative logic (see Chapter 13), whereby a logic 0 corresponds to a TTL high state ($> 2$ V) and a logic 1 to a TTL low state ($< 0.8$ V). Often, the eight-bit word transmitted over an IEEE 488 bus is coded in ASCII format (see Table 14.6), as illustrated in Example 15.16.

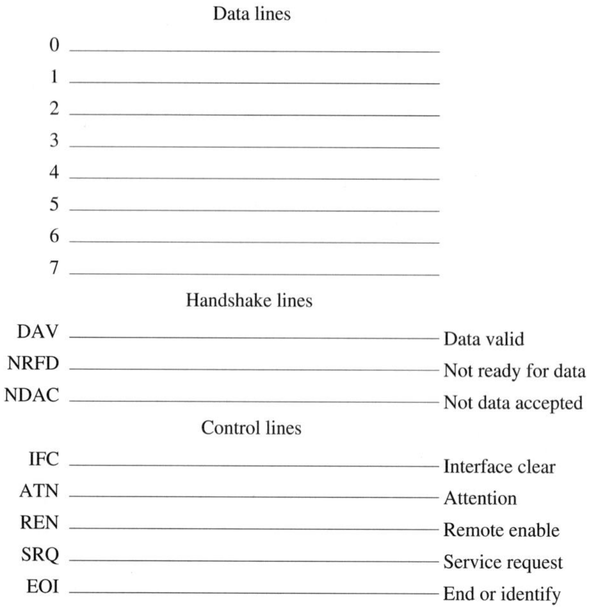

Data lines

**Figure 15.68** IEEE 488 bus

---

## EXAMPLE 15.16 ASCII to Binary Data Conversion Over IEEE 488 Bus

### Problem

Determine the actual binary data sent by a digital voltmeter over an IEEE 488 bus.

---

### Solution

**Known Quantities:** Digital voltmeter reading, $V$.

**Find:** Binary data sequence.

**Schematics, Diagrams, Circuits, and Given Data:** $V = 3.405$ V. ASCII conversion table (Table 14.6).

**Assumptions:** Data is encoded in ASCII format. Sequence is sent from most to least significant digit.

**Analysis:** Using Table 14.6, we can tabulate the conversion as follows:

| Control character | ASCII (hex) |
|:---:|:---|
| 3 | 33 |
| . | 2E |
| 4 | 34 |
| 0 | 30 |
| 5 | 35 |

The actual binary data sent can therefore be determined by converting the hex ASCII sequence into binary data (see Chapter 13):

33 2E 34 30 35 $\leftrightarrow$ 110011 101110 110100 110000 110101

**Comments:**  Note that the ASCII format is not very efficient: if you directly performed a base-10 to binary conversion (see Chapter 13), only eight bits (plus the decimal point) would be required.

___

In an IEEE 488 bus system, devices may play different roles and are typically classified as *controllers,* which manage the data flow; *talkers* (e.g., a digital voltmeter), which can only send data; *listeners* (e.g., a printer), which can only receive data; and *talkers/listeners* (e.g., a digital oscilloscope), which can receive as well as transmit data. The simplest system configuration might consist of just a talker and a listener. If more than two devices are present on the bus, a controller is necessary to determine when and how data transmission is to occur on the bus. For example, one of the key rules implemented by the controller is that only one talker can transmit at any one time; it is possible, however, for several listeners to be active on the bus simultaneously. If the data rates of the different listeners are different, the talker will have to transmit at the slowest rate, so that all of the listeners are assured of receiving the data correctly.

The set of rules by which the controller determines the order in which talking and listening are to take place is determined by a **protocol.** One aspect of the protocol is the handshake procedure, which enables the transmission of data. Since different devices (with different data rate capabilities) may be listening to the same talker, the handshake protocol must take into account these different capabilities. Let us discuss a typical handshake sequence that leads to transmission of data on an IEEE 488 bus. The three handshake lines used in the IEEE 488 have important characteristics that give the interface system wide flexibility, allowing interconnection of multiple devices that may operate at different speeds. The slowest active device controls the rate of data transfer, and more than one device can accept data simultaneously. The timing diagram of Figure 15.69 is used to illustrate the sequence in which the handshake and data transfer are performed:

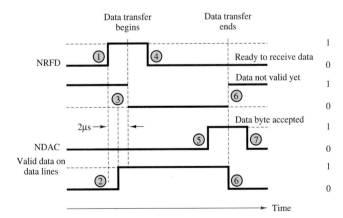

**Figure 15.69**  IEEE 488 data transmission protocol

1. All active listeners use the not ready for data (NRFD) line to indicate their state of readiness to accept a new piece of information. Nonreadiness to accept data is indicated if the NRFD line is held at zero volts. If even one active listener is not ready, the NRFD line of the entire bus is kept at zero volts and the active talker will not transmit the next byte. When all active listeners are ready and they have released the NRFD line, it now goes high.

2. The designated talker drives all eight data input/output lines, causing valid data to be placed on them.

3. Two microseconds after putting valid data on the data lines, the active talker pulls the data valid (DAV) line to zero volts and thereby signals the active listeners to read the information on the data bus. The 2-$\mu$s interval is required to allow the data put on the data lines to reach (settle to) valid logic levels.

4. After the DAV is asserted, the listeners respond by pulling the NRFD line back down to zero. This prevents any additional data transfers from being initiated. The listeners also begin accepting the data byte at their own rates.

5. When each listener has accepted the data, it releases the not data accepted (NDAC) line. Only when the last active listener has released its hold on the NDAC line will that line go to its high-voltage-level state.

6. (a) When the active talker sees that NDAC has come up to its high state, it stops driving the data line. (b) At the same time, the talker releases the DAV line, ending the data transfer. The talker may now put the next byte on the data bus.

7. The listeners pull down the NDAC line back to zero volts and put the byte "away."

Each of the instruments present on the data bus is distinguished by its own address, which is known to the controller; thus, the controller determines who the active talkers and listeners are on the bus by *addressing* them. To implement this and other functions, the controller uses the five control lines. Of these, ATN (attention) is used as a switch to indicate whether the controller is addressing or instructing the devices on the bus, or whether data transmission is taking place: when ATN is logic 1, the data lines contain either control information or addresses; with ATN $= 1$, only the controller is enabled to talk. When ATN $= 0$, only the devices that have been addressed can use the data lines. The IFC (interface clear) line is used to initialize the bus, or to clear it and reset it to a known condition in case of incorrect transmission. The REN (remote enable) line enables a remote instrument to be controlled by the bus; thus, any function that might normally be performed manually on the instrument (e.g., selecting a range or mode of operation) is now controlled by the bus via the data lines. The SRQ (service request) line is used by instruments on the bus whenever the instrument is ready to send or receive data; however, it is the controller who decides when to service the request. Finally, the EOI (end or identify) line can be used in two modes: when it is used by a talker, it signifies the end of a message; when it is used by the controller, it serves as a *polling* line, that is, a line used to interrogate the instrument about its data output.

Although it was mentioned earlier that the IEEE 488 bus can be used only over distances of up to 20 m, it is possible to extend its range of operation by connecting remote IEEE 488 bus systems over telephone communication lines. This can be accomplished by means of *bus extenders,* or by converting the parallel data to serial form (typically, in RS-232 format) and by transmitting the serial data over the phone lines by means of a modem. Serial communications and the RS-232 standard are discussed in the next section.

## The RS-232 Standard

The primary reason why parallel transmission of data is not used exclusively is the limited distance range over which it is possible to transmit data on a parallel bus. Although there are techniques which permit extending the range for parallel transmission, these are complex and costly. Therefore, **serial transmission** is frequently used, whenever data is to be transmitted over a significant distance. Since serial data travels along one single path and is transmitted one bit at a time, the cabling costs for long distances are relatively low; further, the transmitting and receiving units are also limited to processing just one signal, and are also much simpler and less expensive. Two modes of operation exist for serial transmission: **simplex,** which corresponds to transmission in one direction only; and **duplex,** which permits transmission in either direction. Simplex transmission requires only one receiver and one transmitter, at each end of the link; on the other hand, duplex transmission can occur in one of two manners: **half-duplex** and **full-duplex.** In the former, although transmission can take place in both directions, it cannot occur simultaneously in both directions; in the latter case, both ends can simultaneously transmit and receive. Full-duplex transmission is usually implemented by means of four wires.

The data rate of a serial transmission line is measured in bits per second, since the data is transmitted one bit at a time. The unit of 1 bit/s is called a **baud;** thus, reference is often made to the baud rate of a serial transmission. The baud rate can be translated into a parallel transmission rate in words per second if the structure of the word is known; for example, if a word consists of 10 bits (start and stop bits plus an 8-bit data word) and the transmission takes place at 1,200 baud, 120 words are being transmitted every second. Typical data rates for serial transmission are standardized; the most common rates (familiar to the users of personal computer modem connections) are 300, 600, 1,200, and 2,400 baud. Baud rates can be as low as 50 baud or as high as 19,200 baud.

Like parallel transmission, serial transmission can also occur either synchronously or asynchronously. In the serial case, it is also true that asynchronous transmission is less costly but not as fast. A handshake protocol is also required for asynchronous serial transmission, as explained in the following. The most popular data-coding scheme for serial transmission is, once again, the ASCII code, consisting of a 7-bit word plus a **parity bit,** for a total of 8 bits per character. The role of the parity bit is to permit error detection in the event of erroneous reception (or transmission) of a bit. To see this, let us discuss the sequence of handshake events for asynchronous serial transmission and the use of parity bits to correct for errors. In serial asynchronous systems, handshaking is performed by using start and stop bits at the beginning and end of each character that is transmitted. The beginning of the transmission of a serial asynchronous word is announced by the "start" bit, which is always a 0 state bit. For the next five to eight successive bit times (depending on the code and the number of bits that specify the word length in that code), the line is switched to the 1 and 0 states required to represent the character being sent. Following the last bit of the data and the parity bit (which will be explained next), there is one bit or more in the 1 state, indicating "idle." The time period associated with this transmission is called the "stop" bit interval.

If noise pulses affect the transmission line, it is possible that a bit in the transmission could be misread. Thus, following the 5 to 8 transmitted data bits, there is a parity bit that is used for error detection. Here is how the parity bit works. If the transmitter keeps track of the number of 1s in the word being sent, it can send a parity bit, a 1 or a 0, to ensure that the total number of 1s sent is always even (even parity) or odd (odd parity). Similarly, the receiver can keep track of the 1s received to see whether there was an error with the transmission. If an error is detected, retransmission of the word can be requested.

Serial data transmission occurs most frequently according to the **RS-232 standard.** The RS-232 standard is based on the transmission of voltage pulses at a preselected baud rate; the voltage pulses are in the range $-3$ to $-15$ V for a logic 0 and in the range $+3$ to $+15$ V for a logic 1. It is important to note that this amounts to a negative logic convention and that the signals are *not* TTL-compatible. The distance over which such transmission can take place is up to approximately 17 m (50 ft). The RS-232 standard was designed to make the transmission of digital data compatible with existing telephone lines; since phone lines were originally designed to carry analog voice signals, it became necessary to establish some standard procedures to make digital transmission possible over them. The resulting standard describes the mechanical and electrical characteristics of the interface between *data terminal equipment* (DTE) and *data communication equipment* (DCE). DTE consists of computers, terminals, digital instruments, and related peripherals; DCE includes all of those devices that are used to encode digital data in a format that permits their transmission over telephone lines. Thus, the standard specifies how data should be presented by the DTE to the DCE so that digital data can be transmitted over voice lines.

A typical example of DCE is the **modem.** A modem converts digital data to audio signals that are suitable for transmission over a telephone line and is also capable of performing the reverse function, by converting the audio signals back to digital form. The term *modem* stands for *mo*dulate-*dem*odulate, because a modem modulates a sinusoidal carrier using digital pulses (for transmission) and demodulates the modulated sinusoidal signal to recover the digital pulses (at reception). Three methods are commonly used for converting digital pulses to an audio signal: **amplitude-shift keying, frequency-shift keying,** and **phase-shift keying,** depending on whether the amplitude, phase, or frequency of the sinusoid is modulated by the digital pulses. Figure 15.70 depicts the essential block of a data transmission system based on the RS-232 standard, as well as examples of digital data encoded for transmission over a voice line.

In addition to the function just described, however, the RS-232 standard also provides a very useful set of specifications for the direct transmission of digital data between computers and instruments. In other words, communication between digital terminal instruments may occur directly in digital form (i.e., without digital communication devices encoding the digital data in a form compatible with analog voice lines). Thus, this standard is also frequently used for direct digital communication.

The RS-232 standard can be summarized as follows:

- Data signals are encoded according to a negative logic convention using voltage levels of $-3$ to $-15$ V for logic 1 and $+3$ to $+15$ V for logic 0.
- Control signals use a positive logic convention (opposite to that of data signals).

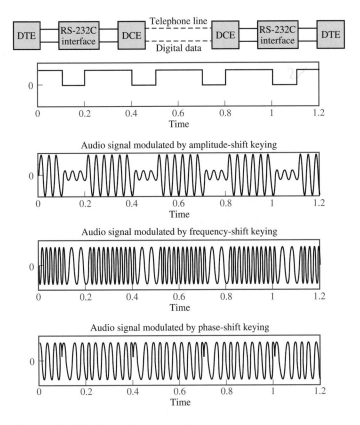

**Figure 15.70** Digital data encoded for analog transmission

· The maximum shunt capacitance of the load cannot exceed 2,500 pF; this, in effect, limits the maximum length of the cables used in the connection.

· The load resistance must be between 300 $\Omega$ and 3 k$\Omega$.

· Three wires are used for data transmission. One wire each is used for receiving and transmitting data; the third wire is a signal return line (signal ground). In addition, there are 22 wires that can be used for a variety of control purposes between the DTE and DCE.

· The male part of the connector is assigned to the DTE and the female part to the DCE. Figure 15.71 labels each of the wires in the 25-pin connector. Since each side of the connector has a *receive* and a *transmit* line, it has been decided by convention that the DCE transmits on the transmit line and receives on the receive line, while the DTE receives on the transmit line and transmits on the receive line.

· The baud rate is limited by the length of the cable; for a 17-m length, any rate from 50 baud to 19.2 kbaud is allowed. If a longer cable connection is desired, the maximum baud rate will decrease according to the length of the cable, and **line drivers** can be used to amplify the signals, which are transmitted over twisted-pair wires. Line drivers are simply signal amplifiers that are used directly on the digital signal, prior to encoding. For example, the signal generated by a DTE device (say a computer) may

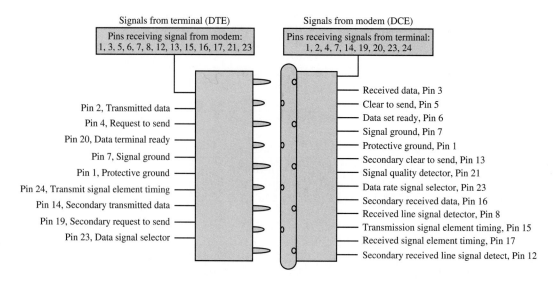

**Figure 15.71** RS-232 connections

be transmitted over a distance of up to 3,300 m (at a rate of 600 baud) prior to being encoded by the DCE.

· The serial data can be encoded according to any code, although the ASCII code is by far the most popular.

## Other Communication Network Standards

In addition to the popular RS-232 and IEEE bus standards, we should mention other communication standards that are commonplace or are rapidly becoming so. One is **Ethernet,** which operates at 10 Mb/s and is based on IEEE Standard 802.3. This is commonly used in office networks. Higher-speed networks include **fiber-distributed data interface** (FDDI), which specifies an optical fiber ring with a data rate of 100 Mb/s, and **asynchronous transfer mode** (ATM), a packet-oriented transfer mode moving data in fixed packets called *cells*. ATM does not operate at fixed speed. A typical speed is 155 Mb/s, but there are implementations running as fast as 2 Gb/s.

## Check Your Understanding

**15.24** Determine the actual binary data sent by a digital voltmeter reading 15.06 V over an IEEE 488 bus if the data is encoded in ASCII format (see Table 14.6). Assume that the sequence is from most to least significant digit.

## CONCLUSION

· Measurements and instrumentation are among the most important areas of electrical engineering because virtually all engineering disciplines require the ability to perform measurements of some kind.

- A measurement system consists of three essential elements: a sensor, signal-conditioning circuits, and recording or display devices. The last are often based on digital computers.

- Sensors are devices that convert a change in a physical variable into a corresponding change in an electrical variable—typically a voltage. A broad range of sensors exist to measure virtually all physical phenomena. Proper wiring, grounding, and shielding techniques are required to minimize undesired interference and noise.

- Often, sensor outputs need to be conditioned before further processing can take place. The most common signal-conditioning circuits are instrumentation amplifiers and active filters.

- If the conditioned sensor signals are to be recorded in digital form by a computer, it is necessary to perform an analog-to-digital conversion process; timing and comparator circuits are also often used in this context.

- Once the digital data corresponding to the measured quantity is available, the need for digital data transmission may arise. Standard transmission formats exist, of which the two most common are the IEEE 488 and the RS-232 standards.

# CHECK YOUR UNDERSTANDING ANSWERS

| | | | |
|---|---|---|---|
| **CYU 15.1** | 66 dB | **CYU 15.11** | 3.66 mV |
| **CYU 15.2** | 80 dB | **CYU 15.12** | $\delta v = 47.1$ mV; $R_F = 94.2\ \Omega$ |
| **CYU 15.3** | 120 dB | **CYU 15.13** | $-13.5$ V |
| **CYU 15.4** | $-6.1$ dB | **CYU 15.14** | 0.76 V |
| **CYU 15.5** | $-20.1$ dB | **CYU 15.15** | 10 |
| **CYU 15.6** | 40 $\Omega$ | **CYU 15.16** | $f_{max} = 10$ kHz |
| **CYU 15.7** | $n = 7$ | **CYU 15.21** | 63 $\mu$s |
| **CYU 15.8** | 42.1 dB | **CYU 15.22** | 9,300 $\Omega$ |
| **CYU 15.9** | $R_1 = R_2 = 1$ k$\Omega$; | **CYU 15.23** | 11.5 $\mu$F |
| | $C_1 = C_2 = 100\ \mu$F; $K = 2$ | **CYU 15.24** | 31 35 2E 30 36 |

# HOMEWORK PROBLEMS

## Section 1: Sensors and Measurements

**15.1**  Most motorcycles have engine speed tachometers, as well as speedometers, as part of their instrumentation. What differences, if any, are there between the two in terms of transducers?

**15.2**  Explain the differences between the engineering specifications you would write for a transducer to measure the frequency of an audible sound wave and a transducer to measure the frequency of a visible light wave.

**15.3**  A measurement of interest in the summer is the temperature-humidity index, consisting of the sum of the temperature and the percentage relative humidity. How would you measure this? Sketch a simple schematic diagram.

**15.4**  Consider a capacitive displacement transducer as shown in Figure P15.4. Its capacitance is determined by the equation

$$C = \frac{0.255A}{d}\ \text{F}$$

where $A$ = cross-sectional area of the transducer plate (in$^2$), and $d$ = air-gap length (in). Determine the change in voltage ($\Delta v_0$) when the air gap changes from 0.01 in to 0.015 in.

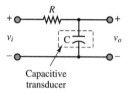

**Figure P15.4**

**15.5** The circuit of Figure P15.5 may be used for operation of a photodiode. The voltage $V_D$ is a reverse-bias voltage large enough to make diode current, $i_D$, proportional to the incident light intensity, $H$. Under this condition, $i_D/H = 0.5\ \mu\text{A-m}^2/\text{W}$.

a. Show that the output voltage, $V_{\text{out}}$, varies linearly with $H$.

b. If $H = 1,500$ W/m$^2$, $V_D = 7.5$ V, and an output voltage of 1 V is desired, determine an appropriate value for $R_L$.

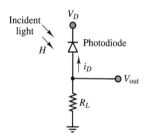

**Figure P15.5**

**15.6** G is a material constant equal to 0.055 V-m/N for quartz in compressive stress and 0.22 V-m/N for polyvinylidene fluoride in axial stress.

a. A force sensor uses a piezoelectric quartz crystal as the sensing element. The quartz element is 0.25 in thick and has a rectangular cross section of 0.09 in$^2$. The sensing element is compressed and the output voltage measured across the thickness. What is the output of the sensor in volts per newton?

b. A polyvinylidene fluoride film is used as a piezoelectric load sensor. The film is 30 μm thick, 1.5 cm wide, and 2.5 cm in the axial direction. It is stretched by the load in the axial direction, and the output voltage is measured across the thickness. What is the output of the sensor in volts per newton?

**15.7** Let $b$ be the damping constant of the mounting structure of a machine as pictured in Figure P15.7. It must be determined experimentally. First, the spring constant, $K$, is determined by measuring the resultant displacement under a static load. The mass, $m$, is directly measured. Finally, the damping ratio, $\xi$, is

measured using an impact test. The damping constant is given by $b = 2\xi\sqrt{Km}$. If the allowable levels of error in the measurements of $K$, $m$, and $\xi$ are $\pm5$ percent, $\pm2$ percent, and $\pm10$ percent respectively, estimate a percentage error limit for $b$.

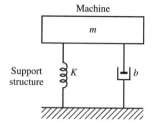

**Figure P15.7**

**15.8** The quality control system in a plant that makes acoustical ceiling tile uses a proximity sensor to measure the thickness of the wet pulp layer every 2 feet along the sheet, and the roller speed is adjusted based on the last 20 measurements. Briefly, the speed is adjusted unless the probability that the mean thickness lies within $\pm2\%$ of the sample mean exceeds 0.99. A typical set of measurements (in mm) is as follows:

8.2, 9.8, 9.92, 10.1, 9.98, 10.2, 10.2, 10.16, 10.0, 9.94,

9.9, 9.8, 10.1, 10.0, 10.2, 10.3, 9.94, 10.14, 10.22, 9.8

Would the speed of the rollers be adjusted based on these measurements?

**15.9** Discuss and contrast the following terms:

a. Measurement accuracy.

b. Instrument accuracy.

c. Measurement error.

d. Precision.

**15.10** Four sets of measurements were taken on the same response variable of a process using four different sensors. The true value of the response was known to be constant. The four sets of data are shown in Figure P15.10. Rank these data sets (and hence the sensors) with respect to:

a. Precision.

b. Accuracy.

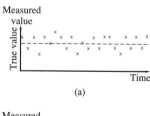

(a)

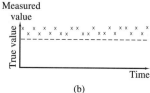

(b)

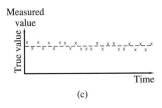

(c)

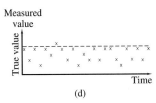

(d)

**Figure P15.10**

## Section 2: Instrumentation Amplifiers

**15.11** For the instrumentation amplifier of Figure P15.11, find the gain of the input stage if $R_1 = 1$ k$\Omega$ and $R_2 = 5$ k$\Omega$.

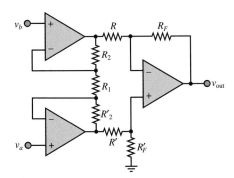

**Figure P15.11**

**15.12** Consider again the instrumentation amplifier of Figure P15.4. Let $R_1 = 1$ k$\Omega$. What value of $R_2$ should be used to make the gain of the input stage equal 50?

**15.13** Again consider the instrumentation amplifier of Figure P15.11. Let $R_2 = 10$ k$\Omega$. What value of $R_1$ will yield an input-stage gain of 16?

**15.14** For the IA of Figure 15.16, find the gain of the input stage if $R_1 = 1$ k$\Omega$ and $R_2 = 10$ k$\Omega$.

**15.15** For the IA of Figure 15.16, find the gain of the input stage if $R_1 = 1.5$ k$\Omega$ and $R_2 = 80$ k$\Omega$.

**15.16** Find the differential gain for the IA of Figure 15.16 if $R_2 = 5$ k$\Omega$, $R_1 = R' = R = 1$ k$\Omega$, and $R_F = 10$ k$\Omega$.

**15.17** Suppose, for the circuit of Figure P15.11, that $R_F = 200$ k$\Omega$, $R = 1$ k$\Omega$, and $\Delta R = 2\%$ of $R$. Calculate the common-mode rejection ratio (CMRR) of the instrumentation amplifier. Express your result in dB.

**15.18** Given the instrumentation amplifier of Figure P15.11, with the component values of Problem 15.17, calculate the mismatch in gains for the differential components. Express your result in dB.

**15.19** Given $R_F = 10$ k$\Omega$ and $R_1 = 2$ k$\Omega$ for the IA of Figure 15.16, find $R$ and $R_2$ so that a differential gain of 900 can be achieved.

## Section 3: Filters

**15.20** Replace the cutoff frequency specification of Example 15.3 with $\omega_C = 10$ rad/s and determine the order of the filter required to achieve 40 dB attenuation at $\omega_S = 24$ rad/s.

**15.21** The circuit of Figure P15.21 represents a low-pass filter with gain.
   a. Derive the relationship between output amplitude and input amplitude.
   b. Derive the relationship between output phase angle and input phase angle.

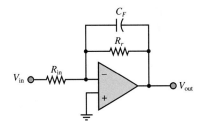

**Figure P15.21**

**15.22** Consider again the circuit of Figure P15.21. Let $R_{in} = 20$ k$\Omega$, $R_F = 100$ k$\Omega$, and $C_F = 100$ pF. Determine an expression for $v_{out}(t)$ if $v_{in}(t) = 2 \sin (2{,}000\pi t)$ V.

**15.23** Derive the frequency response of the low-pass filter of Figure 15.22.

**15.24** Derive the frequency response of the high-pass filter of Figure 15.22.

**15.25** Derive the frequency response of the band-pass filter of Figure 15.22.

**15.26** Consider again the circuit of Figure P15.21. Let $C_F = 100$ pF. Determine appropriate values for $R_{in}$ and $R_F$ if it is desired to construct a filter having a cutoff frequency of 20 kHz and a gain magnitude of 5.

**15.27** Design a second-order Butterworth high-pass filter with a 10-kHz cutoff frequency, a DC gain of 10, $Q = 5$, and $V_S = \pm 15$ V.

**15.28** Design a second-order Butterworth high-pass filter with a 25-kHz cutoff frequency, a DC gain of 15, $Q = 10$, and $V_S = \pm 15$ V.

**15.29** The circuit shown in Figure P15.29 is claimed to exhibit a second-order Butterworth low-pass voltage gain characteristic. Derive the characteristic and verify the claim.

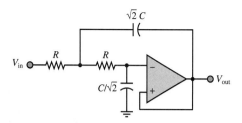

**Figure P15.29**

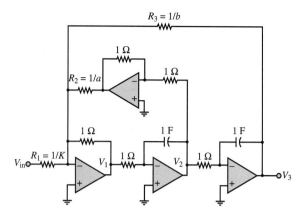

**Figure P15.34**

**15.30** Design a second-order Butterworth low-pass filter with a 15-kHz cutoff frequency, a DC gain of 15, $Q = 5$, and $V_S = \pm 15$ V.

**15.31** Design a band-pass filter with a low cutoff frequency of 200 Hz, a high cutoff frequency of 1 kHz, and a pass-band gain of 4. Calculate the value of $Q$ for the filter. Also, draw the approximate frequency response of this filter.

**15.32** Using the circuit of Figure P15.29, design a second-order low-pass Butterworth filter with a cutoff frequency of 10 Hz.

**15.33** A low-pass Sallen-Key filter is shown in Figure P15.33. Find the voltage gain $V_{out}/V_{in}$ as a function of frequency and generate its Bode magnitude plot. Show and observe that the cutoff frequency is $1/2\pi RC$ and that the low-frequency gain is $R_4/R_3$.

**15.35** The filter shown in Figure P15.35 is called an *infinite-gain multiple-feedback filter.* Derive the following expression for the filter's frequency response:

$$H(j\omega) = \frac{-(1/R_3R_2C_1C_2)R_3/R_1}{(j\omega)^2 + \left(\frac{1}{R_1C_1} + \frac{1}{R_2C_1} + \frac{1}{R_3C_1}\right)j\omega + \frac{1}{R_3R_2C_1C_2}}$$

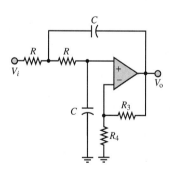

**Figure P15.33**

**15.34** The circuit shown in Figure P15.34 exhibits low-pass, high-pass, and band-pass voltage gain characteristics, depending on whether the output is taken at node 1, node 2, or node 3. Find the transfer functions relating each of these outputs to $V_{in}$, and determine which is which.

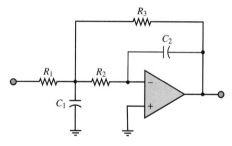

**Figure P15.35**

**15.36** The filter shown in Figure P15.36 is a Sallen and Key band-pass filter circuit, where $K$ is the DC gain of the filter. Derive the following expression for the filter's frequency response:

$$H(j\omega) = \frac{j\omega K/R_1C_1}{(j\omega)^2 + j\omega\left(\frac{1}{R_1C_1} + \frac{1}{R_3C_2} + \frac{1}{R_3C_1} + \frac{1-K}{R_2C_1}\right) + \frac{R_1+R_2}{R_1R_2R_3C_1C_2}}$$

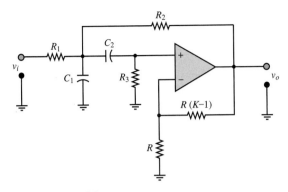

**Figure P15.36**

**15.37** Show that the expression for $Q$ in the filter of Problem 15.35 is given by

$$\frac{1}{Q} = \sqrt{R_2 R_3 \frac{C_2}{C_1}} \left( \frac{1}{R_1} + \frac{1}{R_2} + \frac{1}{R_3} \right)$$

## Section 4: Data Acquisition Systems

**15.38** List two advantages of digital signal processing over analog signal processing.

**15.39** Discuss the role of a multiplexer in a data acquisition system.

**15.40** The circuit shown in Figure P15.40 represents a sample-and-hold circuit, such as might be used in a successive-approximation ADC. Assume that the JFET is turned *on* when $V_G$ is high, and *off* when $V_G$ is low. Explain the operation of the circuit.

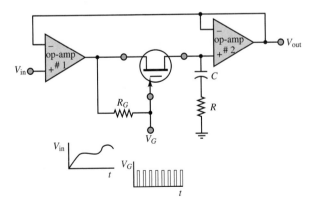

**Figure P15.40**

**15.41** For the circuit shown in Figure P15.40, let $V_{in}$ be a 1 kHz sinusoidal signal with $0°$ phase angle, 0 V DC offset, and 20 V peak-to-peak amplitude. Let $V_G$ be a rectangular pulse train, with pulse width 10 μs, and

period 100 μs, with the leading edge of the first pulse at $t = 0$.

   a. Sketch $V_{out}$ if the $RC$ circuit has a time constant equal to 20 μs.

   b. Sketch $V_{out}$ if the $RC$ circuit has a time constant equal to 1 ms.

**15.42** The unsigned decimal number $12_{10}$ is inputted to a four-bit DAC. Given that $R_F = R_0/15$, logic 0 corresponds to 0 V, and logic 1 corresponds to 4.5 V,

   a. What is the output of the DAC?

   b. What is the maximum voltage that can be outputted from the DAC?

   c. What is the resolution over the range 0 to 4.5 V?

   d. Find the number of bits required in the DAC if an improved resolution of 20 mV is desired.

**15.43** The unsigned decimal number $215_{10}$ is inputted to an eight-bit DAC. Given that $R_F = R_0/255$, logic 0 corresponds to 0 V, and logic 1 corresponds to 10 V,

   a. What is the output of the DAC?

   b. What is the maximum voltage that can be outputted from the DAC?

   c. What is the resolution over the range 0 to 10 V?

   d. Find the number of bits required in the DAC if an improved resolution of 3 mV is desired.

**15.44** The circuit shown in Figure P15.44 represents a simple 4-bit digital-to-analog converter. Each switch is controlled by the corresponding bit of the digital number—if the bit is 1 the switch is up; if the bit is 0 the switch is down. Let the digital number be represented by $b_3 b_2 b_1 b_0$. Determine an expression relating $v_o$ to the binary input bits.

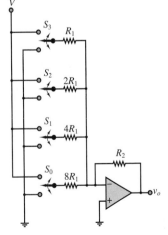

**Figure P15.44**

**15.45**  The unsigned decimal number $98_{10}$ is inputted to an eight-bit DAC. Given that $R_F = R_0/255$, logic 0 corresponds to 0 V and logic 1 corresponds to 4.5 V,

a. What is the output of the DAC?

b. What is the maximum voltage that can be outputted from the DAC?

c. What is the resolution over the range 0 to 4.5 V?

d. Find the number of bits required in the DAC if an improved resolution of 0.5 mV is desired.

**15.46**  For the DAC circuit shown in Figure P15.46 (using an ideal op-amp), what value of $R_F$ will give an output range of $-10 \leq V_0 \leq 0$ V? Assume that logic $0 = 0$ V and logic $1 = 5$ V.

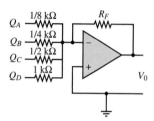

**Figure P15.46**

**15.47**  Explain how to redesign the circuit of Figure P15.44 so that the overall circuit is a "noninverting" device.

**15.48**  The circuit of Figure P15.48 has been suggested as a means of implementing the switches needed for the digital-to-analog converter of Figure P15.44. Explain how the circuit works.

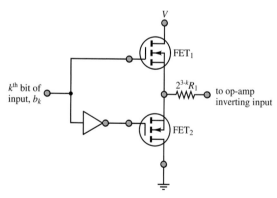

**Figure P15.48**

**15.49**  The unsigned decimal number $345_{10}$ is inputted to a 12-bit DAC. Given that $R_F = R_0/4,095$, logic 1 corresponds to 10 V, and logic 0 to 0 V,

a. What is the output of the DAC?

b. What is the maximum voltage that can be outputted from the DAC?

c. What is the resolution over the range 0 to 10 V?

d. Find the number of bits required in the DAC if an improved resolution of 0.5 mV is desired.

**15.50**  For the DAC circuit shown in Figure P15.46 (using an ideal op-amp), what value of $R_F$ will give an output range of $-15 \leq V_0 \leq 0$ V?

**15.51**  Using the model of Figure P15.44, design a 4-bit digital-to-analog converter whose output is given by

$$v_o = -\frac{1}{10}(8b_3 + 4b_2 + 2b_1 + b_0)V$$

**15.52**  A data acquisition system uses a DAC with a range of $\pm 15$ V and a resolution of 0.01 V. How many bits must be present in the DAC?

**15.53**  A data acquisition system uses a DAC with a range of $\pm 10$ V and a resolution of 0.04 V. How many bits must be present in the DAC?

**15.54**  A data acquisition system uses a DAC with a range of $-10$ to $+15$ V and a resolution of 0.004 V. How many bits must be present in the DAC?

**15.55**  A DAC is to be used to deliver velocity commands to a motor. The maximum velocity is to be 2,500 rev/min, and the minimum nonzero velocity is to be 1 rev/min. How many bits are required in the DAC? What will the resolution be?

**15.56**  Assume the full-scale value of the analog input voltage to a particular analog-to-digital converter is 10 V.

a. If this is a 3-bit device, what is the resolution of the output?

b. If this is an 8-bit device, what is its resolution?

c. Make a general comment about the relationship between the number of bits and the resolution of an ADC.

**15.57**  The voltage range of feedback signal from a process is $-5$ V to $+15$ V, and a resolution of 0.05 percent of the voltage range is required. How many bits are required for the DAC?

**15.58**  Eight channels of analog information are being used by a computer to close eight control loops. Assume that all analog signals have identical frequency content and are multiplexed into a single ADC. The ADC requires 100 μs per conversion. The closed-loop software requires 500 μs of computation and output time for four of the loops, and for the other four it requires 250 μs. What is the maximum frequency content that the analog signal can have according to the Nyquist criterion?

**15.59**  A rotary potentiometer is to be used as a remote rotational displacement sensor. The maximum displacement to be measured is 180°, and the potentiometer is rated for 10 V and 270° of rotation.

    a. What voltage increment must be resolved by an ADC to resolve an angular displacement of 0.5°? How many bits would be required in the ADC for full-range detection?

    b. The ADC requires a 10-V input voltage for full-scale binary output. If an amplifier is placed between the potentiometer and the ADC, what amplifier gain should be used to take advantage of the full range of the ADC?

**15.60**  Suppose it is desired to digitize a 250-kHz analog signal to 10 bits using a successive-approximation ADC. Estimate the maximum permissible conversion time for the ADC.

**15.61**  A torque sensor has been mounted on a farm tractor engine. The voltage produced by the torque sensor is to be sampled by an ADC. The rotational speed of the crankshaft is 800 rev/min. Because of speed fluctuation caused by the reciprocating action of the engine, frequency content is present in the torque signal at twice the shaft rotation frequency. What is the minimum sampling period that can be used to ensure that the Nyquist criterion is satisfied?

**15.62**  The output voltage of an aircraft altimeter is to be sampled using an ADC. The sensor outputs 0 V at 0 m altitude and outputs 10 V at 10,000 m altitude. If the allowable error in sensing ($\pm\frac{1}{2}$ LSB) is 10 m, find the minimum number of bits required for the ADC.

**15.63**  Consider a circuit that generates interrupts at fixed time intervals. Such a device is called a *real-time clock* and is used in control applications to establish the sample period as $T$ seconds for control algorithms. Show how this can be done with a square wave (clock) that has a period equal to the desired time interval between interrupts.

**15.64**  What is the minimum number of bits required to digitize an analog signal with a resolution of:

    a. 5%

    b. 2%

    c. 1%

## Section 5: Timing Circuits

**15.65**  A useful application that exploits the open-loop characteristics of op amps is known as a comparator. One particularly simple example known as a window comparator is shown in Figures P15.65(a) and (b). Show that $V_{out} = 0$ whenever $V_{low} < V_{in} < V_{high}$ and that $V_{out} = +V$ otherwise.

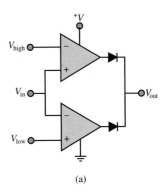

(a)

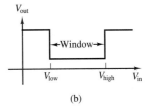

(b)

**Figure P15.65**

**15.66**  Design a Schmitt trigger to operate in the presence of noise with peak amplitude $= \pm150$ mV. The circuit is to switch around the reference value $-1$ V. Assume an op-amp with $\pm10$-V supplies ($V_{sat} = 8.5$ V).

**15.67**  In the circuit of Figure P15.67, $R_1 = 100\ \Omega$, $R_2 = 56\ \text{k}\Omega$, $R_i = R_1 \| R_2$, and $v_{in}$ is a 1-V peak-to-peak sine wave. Assuming that the supply voltages are $\pm15$ V, determine the threshold voltages (positive and negative $v^+$) and draw the output waveform.

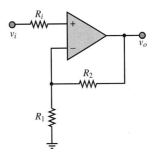

**Figure P15.67**

**15.68**  The circuit in Figure P15.68 shows how a Schmitt trigger might be constructed with an op-amp. Explain the operation of this circuit.

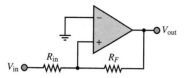

**Figure P15.68**

**15.69** Consider again the circuit of Figure P15.68. Let the op-amp be an LM741 with $\pm 15$ V bias supplies, and suppose $R_F$ is chosen to be 104 k$\Omega$. Assume $V_{in}$ is a 1-kHz sinusoidal signal with 1-V amplitude.
   a. Determine the appropriate value for $R_{in}$ if the output is to be high whenever $|V_{in}| \geq 0.25$ V.
   b. Sketch the input and output waveforms.

**15.70** For the circuit shown in Figure P15.70,
   a. Draw the output waveform for $v_{in}$ a 4-V peak-to-peak sine wave at 100 Hz and $V_{ref} = 2$ V.
   b. Draw the output waveform for $v_{in}$ a 4-V peak-to-peak sine wave at 100 Hz and $V_{ref} = -2$ V.

Note that the diodes placed at the input ensure that the differential voltage does not exceed the diode offset voltage.

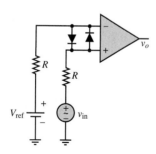

**Figure P15.70**

**15.71** Figure P15.71 shows a simple *go-no go* detector application of a comparator.
   a. Explain how the circuit works.
   b. Design a circuit (i.e., choose proper values for the resistors) such that the green LED will turn on when $V_{in}$ exceeds 5 V, and the red LED will be on whenever $V_{in}$ is less than 5 V. Assume only 15 V supplies are available.

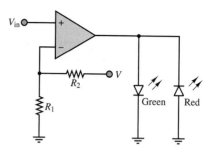

**Figure P15.71**

**15.72** For the circuit of Figure P15.72, $v_{in}$ is a 100-mV peak sine wave at 5 kHz, $R = 10$ k$\Omega$, and $D_1$ and $D_2$ are 6.2-V Zener diodes. Draw the output voltage waveform.

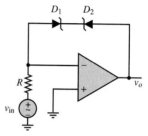

**Figure P15.72**

**15.73** Show that the period of oscillation of an op-amp astable multivibrator is given by the expression

$$T = 2R_1 C \log_e \left( \frac{2R_2}{R_3} + 1 \right)$$

**15.74** Use the data sheets for the 74123 monostable multivibrator to analyze the connection shown in Figure 15.60 in the text. Draw a timing diagram indicating the approximate duration of each pulse, assuming that the trigger signal consists of a positive-going transition.

**15.75** In the monostable multivibrator of Figure 15.61 in the text, $R_1 = 10$ k$\Omega$ and the output pulse width $T = 10$ ms. Determine the value of $C$.

## Section 6: Data Transmission

**15.76** An ASCII (hex) encoded message is given below. Decode the message.

41 53 43 49 49 20 64 65 63 6F 64 69 6E 67 20 69 73 20 65 61 73 79 21

**15.77** An ASCII (binary) encoded message is given below. Decode the messsage. Hint: Follow a line-by-line sequence, *not* column-by-column.

| | | | |
|---|---|---|---|
| 1010100 | 1101000 | 1101001 | 1110011 |
| 1101001 | 1101101 | 1100101 | 0101101 |
| 1101110 | 1100111 | 0100000 | 1110000 |
| | | | |
| 0100000 | 1101001 | 1110011 | 0100000 |
| 1100011 | 1101111 | 1101110 | 1110011 |
| 1110010 | 1101111 | 1100010 | 1101100 |
| | | | |
| 1100001 | 0100000 | 1110100 | |
| 1110101 | 1101101 | 1101001 | |
| 1100101 | 1101101 | 0101110 | |

**15.78** Express the following decimal numbers in ASCII form:

   a. 12

   b. 345.2

   c. 43.5

**15.79** Express the following words in ASCII form:

   a. Digital

   b. Computer

   c. Ascii

   d. ASCII

**15.80** Explain why data transmission over long distances is usually done via a serial scheme rather than parallel.

**15.81** A certain automated data-logging instrument has 16K-words of on-board memory. The device samples the variable of interest once every five minutes. How often must data be downloaded and the memory cleared in order to avoid losing any data?

**15.82** Explain why three wires are required for the handshaking technique employed by IEEE 488 bus systems.

**15.83** A CD-ROM can hold 650 Mbytes of information. Suppose the CD-ROMs are packed 50 per box. The manufacturer ships 100 boxes via commercial airliner from Los Angeles to New York. The distance between the two cities is 2,500 miles by air, and the airliner flies at a speed of 400 mi/h. What is the data transmission rate between the two cities in bits/s?

# PART III

# ELECTROMECHANICS

# C H A P T E R

# 16

# Principles of Electromechanics

The objective of this chapter is to introduce the fundamental notions of electromechanical energy conversion, leading to an understanding of the operation of various electromechanical transducers. The chapter also serves as an introduction to the material on electric machines to be presented in Chapters 17 and 18. The foundations for the material introduced in this chapter will be found in the circuit analysis chapters (1–7). In addition, the material on power electronics (Chapter 11) is also relevant, especially with reference to Chapters 17 and 18.

The subject of electromechanical energy conversion is one that should be of particular interest to the *non–electrical* engineer, because it forms one of the important points of contact between electrical engineering and other engineering disciplines. Electromechanical transducers are commonly used in the design of industrial and aerospace control systems and in biomedical applications, and they form the basis of many common appliances. In the course of our exploration of electromechanics, we shall illustrate the operation of practical devices, such as loudspeakers, relays, solenoids, sensors for the measurement of position and velocity, and other devices of practical interest.

Upon completion of the chapter, you should be able to:

- Analyze simple magnetic circuits, to determine electrical and mechanical performance and energy requirements.

- Size a relay or solenoid for a given application.
- Describe the energy-conversion process in electromechanical systems.
- Perform a simplified linear analysis of electromechanical transducers.

## 16.1    ELECTRICITY AND MAGNETISM

The notion that the phenomena of electricity and magnetism are interconnected was first proposed in the early 1800s by H. C. Oersted, a Danish physicist. Oersted showed that an electric current produces magnetic effects (more specifically, a magnetic field). Soon after, the French scientist André Marie Ampère expressed this relationship by means of a precise formulation, known as *Ampère's law*. A few years later, the English scientist Faraday illustrated how the converse of Ampère's law also holds true, that is, that a magnetic field can generate an electric field; in short, *Faraday's law* states that a changing magnetic field gives rise to a voltage. We shall undertake a more careful examination of both Ampère's and Faraday's laws in the course of this chapter.

As will be explained in the next few sections, the magnetic field forms a necessary connection between electrical and mechanical energy. Ampère's and Faraday's laws will formally illustrate the relationship between electric and magnetic fields, but it should already be evident from your own individual experience that the magnetic field can also convert magnetic energy to mechanical energy (for example, by lifting a piece of iron with a magnet). In effect, the devices we commonly refer to as *electromechanical* should more properly be referred to as electro*magneto*mechanical, since they almost invariably operate through a conversion from electrical to mechanical energy (or vice versa) by means of a magnetic field. Chapters 16 through 18 are concerned with the use of electricity and magnetic materials for the purpose of converting electrical energy to mechanical, and back.

### The Magnetic Field and Faraday's Law

The quantities used to quantify the strength of a magnetic field are the **magnetic flux,** $\phi$, in units of **webers** (Wb); and the **magnetic flux density, B,** in units of webers per square meter (Wb/m$^2$), or **teslas** (T). The latter quantity, as well as the associated **magnetic field intensity, H** (in units of amperes per meter, or A/m) are vectors.[1] Thus, the density of the magnetic flux and its intensity are in general described in vector form, in terms of the components present in each spatial direction (e.g., on the $x$, $y$, and $z$ axes). In discussing magnetic flux density and field intensity in this chapter and the next, we shall almost always assume that the field is a *scalar field*, that is, that it lies in a single spatial direction. This will simplify many explanations.

It is customary to represent the magnetic field by means of the familiar *lines of force* (a concept also due to Faraday); we visualize the strength of a magnetic field by observing the density of these lines in space. You probably know from a previous course in physics that such lines are closed in a magnetic field, that is, that they form continuous loops exiting at a magnetic north pole (by definition)

---

[1] We will use the boldface symbols **B** and **H** to denote the vector forms of $B$ and $H$; the standard typeface will represent the scalar flux density or field intensity in a given direction.

and entering at a magnetic south pole. The relative strengths of the magnetic fields generated by two magnets could be depicted as shown in Figure 16.1.

Magnetic fields are generated by electric charge in motion, and their effect is measured by the force they exert on a moving charge. As you may recall from previous physics courses, the vector force **f** exerted on a charge of $q$ moving at velocity **u** in the presence of a magnetic field with flux density **B** is given by the equation

$$\mathbf{f} = q\mathbf{u} \times \mathbf{B} \tag{16.1}$$

where the symbol $\times$ denotes the (vector) cross product. If the charge is moving at a velocity **u** in a direction that makes an angle $\theta$ with the magnetic field, then the magnitude of the force is given by

$$f = quB \sin \theta \tag{16.2}$$

and the direction of this force is at right angles with the plane formed by the vectors **B** and **u**. This relationship is depicted in Figure 16.2.

The magnetic flux, $\phi$, is then defined as the integral of the flux density over some surface area. For the simplified (but often useful) case of magnetic flux lines perpendicular to a cross-sectional area $A$, we can see that the flux is given by the following integral:

$$\phi = \int_A B \, dA \tag{16.3}$$

in webers (Wb), where the subscript $A$ indicates that the integral is evaluated over the surface $A$. Furthermore, if the flux were to be uniform over the cross-sectional area $A$ (a simplification that will be useful), the preceding integral could be approximated by the following expression:

$$\boxed{\phi = B \cdot A} \tag{16.4}$$

Figure 16.3 illustrates this idea, by showing hypothetical magnetic flux lines traversing a surface, delimited in the figure by a thin conducting wire.

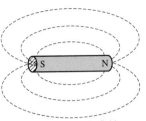

Weaker magnetic field

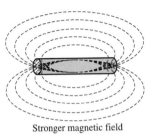

Stronger magnetic field

**Figure 16.1** Lines of force in a magnetic field

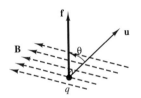

**Figure 16.2** Charge moving in a constant magnetic field

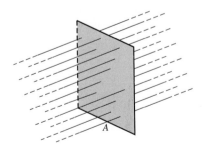

**Figure 16.3** Magnetic flux lines crossing a surface

**Faraday's law** states that if the imaginary surface $A$ were bounded by a conductor—for example, the thin wire of Figure 16.3—then a *changing* magnetic field would induce a voltage, and therefore a current, in the conductor. More

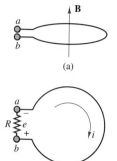

(a)

Current generating a magnetic flux
opposing the increase in flux due to **B**

(b)

**Figure 16.4** Flux direction

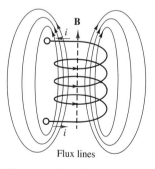

Right-hand rule

Flux lines

**Figure 16.5** Concept of
flux linkage

precisely, Faraday's law states that a time-varying flux causes an induced **electromotive force,** or **emf,** $e$, as follows:

$$e = -\frac{d\phi}{dt} \tag{16.5}$$

A little discussion is necessary at this point to explain the meaning of the minus sign in equation 16.5. Consider the one-turn coil of Figure 16.4, which forms a circular cross-sectional area, in the presence of a magnetic field with flux density **B** oriented in a direction perpendicular to the plane of the coil. If the magnetic field, and therefore the flux within the coil, is constant, no voltage will exist across terminals $a$ and $b$; if, however, the flux were increasing and terminals $a$ and $b$ were connected—for example, by means of a resistor, as indicated in Figure 16.4(b)—current would flow in the coil in such a way that *the magnetic flux generated by the current would oppose the increasing flux.* Thus, the flux induced by such a current would be in the direction opposite to that of the original flux density vector, **B.** This principle is known as **Lenz's law.** The reaction flux would then point downward in Figure 16.4(a), or into the page in Figure 16.4(b). Now, by virtue of the **right-hand rule,** this reaction flux would induce a current flowing clockwise in Figure 16.4(b), that is, a current that flows out of terminal $b$ and into terminal $a$. The resulting voltage across the hypothetical resistor $R$ would then be negative. If, on the other hand, the original flux were decreasing, current would be induced in the coil so as to reestablish the initial flux; but this would mean that the current would have to generate a flux in the upward direction in Figure 16.4(a) (or out of the page in Figure 16.4(b)). Thus, the resulting voltage would change sign.

The polarity of the induced voltage can usually be determined from physical considerations; therefore the minus sign in equation 16.5 is usually left out. We will use this convention throughout the chapter.

In practical applications, the size of the voltages induced by the changing magnetic field can be significantly increased if the conducting wire is coiled many times around, so as to multiply the area crossed by the magnetic flux lines many times over. For an $N$-turn coil with cross-sectional area $A$, for example, we have the emf

$$e = N\frac{d\phi}{dt} \tag{16.6}$$

Figure 16.5 shows an $N$-turn coil *linking* a certain amount of magnetic flux; you can see that if $N$ is very large and the coil is tightly wound (as is usually the case in the construction of practical devices), it is not unreasonable to presume that each turn of the coil links the same flux. It is convenient, in practice, to define the **flux linkage,** $\lambda$, as

$$\lambda = N\phi \tag{16.7}$$

so that

$$e = \frac{d\lambda}{dt} \tag{16.8}$$

Note that equation 16.8, relating the derivative of the flux linkage to the induced emf, is analogous to the equation describing current as the derivative of charge:

$$i = \frac{dq}{dt} \tag{16.9}$$

In other words, flux linkage can be viewed as the dual of charge in a circuit analysis sense, provided that we are aware of the simplifying assumptions just stated in the preceding paragraphs, namely, a uniform magnetic field perpendicular to the area delimited by a tightly wound coil. These assumptions are not at all unreasonable when applied to the inductor coils commonly employed in electric circuits.

What, then, are the physical mechanisms that can cause magnetic flux to change, and therefore to induce an electromotive force? Two such mechanisms are possible. The first consists of physically moving a permanent magnet in the vicinity of a coil—for example, so as to create a time-varying flux. The second requires that we first produce a magnetic field by means of an electric current (how this can be accomplished is discussed later in this section) and then vary the current, thus varying the associated magnetic field. The latter method is more practical in many circumstances, since it does not require the use of permanent magnets and allows variation of field strength by varying the applied current; however, the former method is conceptually simpler to visualize. The voltages induced by a moving magnetic field are called **motional voltages;** those generated by a time-varying magnetic field are termed **transformer voltages.** We shall be interested in both in this chapter, for different applications.

In the analysis of linear circuits in Chapter 4, we implicitly assumed that the relationship between flux linkage and current was a linear one:

$$\lambda = Li \qquad\qquad\qquad\qquad\qquad \textbf{(16.10)}$$

so that the effect of a time-varying current was to induce a transformer voltage across an inductor coil, according to the expression

$$v = L\frac{di}{dt} \qquad\qquad\qquad\qquad\qquad \textbf{(16.11)}$$

This is, in fact, the defining equation for the ideal **self-inductance,** $L$. In addition to self-inductance, however, it is also important to consider the **magnetic coupling** that can occur between neighboring circuits. Self-inductance measures the voltage induced in a circuit by the magnetic field generated by a current flowing in the same circuit. It is also possible that a second circuit in the vicinity of the first may experience an induced voltage as a consequence of the magnetic field generated in the first circuit. As we shall see in Section 16.4, this principle underlies the operation of all transformers.

## Self- and Mutual Inductance

Figure 16.6 depicts a pair of coils, one of which, $L_1$, is excited by a current, $i_1$, and therefore develops a magnetic field and a resulting induced voltage, $v_1$. The second coil, $L_2$, is not energized by a current, but links some of the flux generated by the current $i_1$ around $L_1$ because of its close proximity to the first coil. The magnetic coupling between the coils established by virtue of their proximity is described by a quantity called **mutual inductance** and defined by the symbol $M$. The mutual inductance is defined by the equation

$$v_2 = M\frac{di_1}{dt} \qquad\qquad\qquad\qquad\qquad \textbf{(16.12)}$$

The dots shown in the two figures indicate the polarity of the coupling between the coils. If the dots are at the same end of the coils, the voltage induced in coil 2 by a current in coil 1 has the same polarity as the voltage induced by the same current

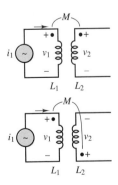

**Figure 16.6** Mutual inductance

in coil 1; otherwise, the voltages are in opposition, as shown in the lower part of Figure 16.6. Thus, the presence of such dots indicates that magnetic coupling is present between two coils. It should also be pointed out that if a current (and therefore a magnetic field) were present in the second coil, an additional voltage would be induced across coil 1. The voltage induced across a coil is, in general, equal to the sum of the voltages induced by self-inductance and mutual inductance.

**FOCUS ON
MEASUREMENTS**

### Linear Variable Differential Transformer (LVDT)

**FIND IT
ON THE WEB**

The **linear variable differential transformer** (LVDT) is a displacement transducer based on the mutual inductance concept just discussed. Figure 16.7 shows a simplified representation of an LVDT, which consists of a primary coil, subject to AC excitation ($v_{ex}$), and of a pair of identical secondary coils, which are connected so as to result in the output voltage

$$v_{out} = v_1 - v_2$$

The ferromagnetic core between the primary and secondary coils can be displaced in proportion to some external motion, $x$, and determines the magnetic coupling between primary and secondary coils. Intuitively, as the core is displaced upward, greater coupling will occur between the primary coil and the top secondary coil, thus inducing a greater voltage in the top secondary coil. Hence, $v_{out} > 0$ for positive displacements. The converse is true for negative displacements. More formally, if the primary coil has resistance $R_p$ and self-inductance $L_p$, we can write

$$i R_p + L_p \frac{di}{dt} = v_{ex}$$

and the voltages induced in the secondary coils are given by

$$v_1 = M_1 \frac{di}{dt}$$

$$v_2 = M_2 \frac{di}{dt}$$

so that

$$v_{out} = (M_1 - M_2) \frac{di}{dt}$$

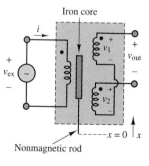

**Figure 16.7** Linear variable differential transformer

where $M_1$ and $M_2$ are the mutual inductances between the primary and the respective secondary coils. It should be apparent that each of the mutual inductances is dependent on the position of the iron core. For example, with the core at the *null position*, $M_1 = M_2$ and $v_{out} = 0$. The LVDT is typically designed so that $M_1 - M_2$ is linearly related to the displacement of the core, $x$.

Because the excitation is by necessity an AC signal (why?), the output voltage is actually given by the difference of two sinusoidal voltages at the same frequency, and is therefore itself a sinusoid, whose amplitude and phase depend on the displacement, $x$. Thus, $v_{out}$ is an *amplitude-modulated* (AM) signal, similar to the one discussed in "Focus on Measurements: Capacitive Displacement Transducer" in Chapter 4. To recover a signal proportional to the actual displacement, it is therefore necessary to use a demodulator circuit, such as the one discussed in "Focus on Measurements: Peak Detector for Capacitive Displacement Transducer" in Chapter 8.

In practical electromagnetic circuits, the self-inductance of a circuit is not necessarily constant; in particular, the inductance parameter, $L$, is not constant, in general, but depends on the strength of the magnetic field intensity, so that it will not be possible to use such a simple relationship as $v = L\,di/dt$, with $L$ constant. If we revisit the definition of the transformer voltage,

$$e = N\frac{d\phi}{dt} \qquad (16.13)$$

we see that in an inductor coil, the inductance is given by

$$L = \frac{N\phi}{i} = \frac{\lambda}{i} \qquad (16.14)$$

This expression implies that the relationship between current and flux in a magnetic structure is linear (the inductance being the slope of the line). In fact, the properties of ferromagnetic materials are such that the flux-current relationship is nonlinear, as we shall see in Section 16.3, so that the simple linear inductance parameter used in electric circuit analysis is not adequate to represent the behavior of the magnetic circuits of the present chapter. In any practical situation, the relationship between the flux linkage, $\lambda$, and the current is nonlinear, and might be described by a curve similar to that shown in Figure 16.8. Whenever the $i$-$\lambda$ curve is not a straight line, it is more convenient to analyze the magnetic system in terms of energy calculations, since the corresponding circuit equation would be nonlinear.

In a magnetic system, the energy stored in the magnetic field is equal to the integral of the instantaneous power, which is the product of voltage and current, just as in a conventional electrical circuit:

$$W_m = \int ei\,dt' \qquad (16.15)$$

However, in this case, the voltage corresponds to the induced emf, according to Faraday's law:

$$e = \frac{d\lambda}{dt} = N\frac{d\phi}{dt} \qquad (16.16)$$

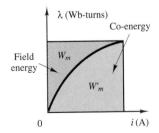

**Figure 16.8** Relationship between flux linkage, current, energy, and co-energy.

and is therefore related to the rate of change of the magnetic flux. The energy stored in the magnetic field could therefore be expressed in terms of the current by the integral

$$W_m = \int ei \, dt' = \int \frac{d\lambda}{dt} i \, dt' = \int i \, d\lambda' \qquad (16.17)$$

It should be straightforward to recognize that this energy is equal to the area above the $\lambda$-$i$ curve of Figure 16.8. From the same figure, it is also possible to define a fictitious (but sometimes useful) quantity called **co-energy,** equal to the area under the curve and identified by the symbol $W_m'$. From the figure, it is also possible to see that the co-energy can be expressed in terms of the stored energy by means of the following relationship:

$$W_m' = i\lambda - W_m \qquad (16.18)$$

Example 16.1 illustrates the calculation of energy, co-energy, and induced voltage using the concepts developed in these paragraphs.

The calculation of the energy stored in the magnetic field around a magnetic structure will be particularly useful later in the chapter, when the discussion turns to practical electromechanical transducers and it will be necessary to actually compute the forces generated in magnetic structures.

---

### EXAMPLE 16.1 Energy and Co-Energy Calculation for an Inductor

#### Problem

Compute the energy, co-energy, and incremental linear inductance for an iron core inductor with a given $\lambda$-$i$ relationship. Also compute the voltage across the terminals given the current through the coil.

---

#### Solution

***Known Quantities:*** $\lambda$-$i$ relationship; nominal value of $\lambda$; coil resistance; coil current.

***Find:*** $W_m$; $W_m'$; $L_\Delta$; $v$.

***Schematics, Diagrams, Circuits, and Given Data:*** $i = \lambda + 0.5\lambda^2$; $\lambda_0 = 0.5$ V · s; $R = 1 \, \Omega$; $i(t) = 0.625 + 0.01\sin(400t)$.

***Assumptions:*** Assume that the magnetic equation can be linearized and use the linear model in all circuit calculations.

***Analysis:***

1.  *Calculation of energy and co-energy.* From equation 16.17, we calculate the energy as follows.

$$W_m = \int_0^\lambda i(\lambda')d\lambda' = \frac{\lambda^2}{2} + \frac{\lambda^3}{6}$$

The above expression is valid in general; in our case, the inductor is operating at a nominal flux linkage $\lambda_0 = 0.5$ V-s and we can therefore evaluate the energy to be:

$$W_m(\lambda = \lambda_0) = \left. \left( \frac{\lambda^2}{2} + \frac{\lambda^3}{6} \right) \right|_{\lambda=0.5} = 0.1458 \text{ J}$$

Thus, after equation 16.18, the co-energy is given by:

$$W'_m = i\lambda - W_m$$

where

$$i = \lambda + 0.5\lambda^2 = 0.625 \text{ A}$$

and

$$W'_m = i\lambda - W_m = (0.625)(0.5) - (0.1458) = 0.1667 \text{ J}$$

2. *Calculation of incremental inductance.* If we know the nominal value of flux linkage (i.e., the operating point), we can calculate a linear inductance $L_\Delta$, valid around values of $\lambda$ close to the operating point $\lambda_0$:

$$L_\Delta = \left.\frac{d\lambda}{di}\right|_{\lambda=\lambda_0} = \left.\frac{1}{1+\lambda}\right|_{\lambda=0.5} = 0.667 \text{ H}$$

The above expressions can be used to analyze the circuit behavior of the inductor when the flux linkage is around $0.5$ V $\cdot$ s, or, equivalently, when the current through the inductor is around $0.625$ A.

3. *Circuit analysis using linearized model of inductor.* We can use the incremental linear inductance calculated above to compute the voltage across the inductor in the presence of a current $i(t) = 0.625 + 0.01 \sin(400t)$. Using the basic circuit definition of an inductor with series resistance $R$, the voltage across the inductor is given by:

$$v = iR + L_\Delta \frac{di}{dt} = [0.625 + 0.01\sin(400t)] \times 1 + 0.667 \times 4\cos(400t)$$

$$= 0.625 + 0.01\sin(400t) + 2.668\cos(400t) = 0.625 + 2.668\sin(400t + 89.8°)$$

**Comments:** The linear approximation in this case is not a bad one: the small sinusoidal current is oscillating around a much larger average current. In this type of situation, it is reasonable to assume that the inductor behaves linearly. This example explains why the linear inductor model introduced in Chapter 4 is an acceptable approximation in most circuit analysis problems.

## Ampère's Law

As explained in the previous section, Faraday's law is one of two fundamental laws relating electricity to magnetism. The second relationship, which forms a counterpart to Faraday's law, is **Ampère's law.** Qualitatively, Ampère's law states that the magnetic field intensity, **H,** in the vicinity of a conductor is related to the current carried by the conductor; thus Ampère's law establishes a dual relationship with Faraday's law.

In the previous section, we described the magnetic field in terms of its flux density, **B,** and flux $\phi$. To explain Ampère's law and the behavior of magnetic materials, we need to define a relationship between the magnetic field intensity, **H,** and the flux density, **B.** These quantities are related by:

$$\mathbf{B} = \mu\mathbf{H} = \mu_r\mu_0\mathbf{H} \qquad \text{Wb/m}^2 \text{ or T} \tag{16.19}$$

where the parameter $\mu$ is a scalar constant for a particular physical medium (at least, for the applications we consider here) and is called the **permeability** of the medium. The permeability of a material can be factored as the product of the permeability of free space, $\mu_0 = 4\pi \times 10^{-7}$ H/m, times the relative permeability, $\mu_r$, which varies greatly according to the medium. For example, for air and for most electrical conductors and insulators, $\mu_r$ is equal to 1. For ferromagnetic materials, the value of $\mu_r$ can take values in the hundreds or thousands. The size of $\mu_r$ represents a measure of the magnetic properties of the material. A consequence of Ampère's law is that, the larger the value of $\mu$, the smaller the current required to produce a large flux density in an electromagnetic structure. Consequently, many electromechanical devices make use of ferromagnetic materials, called iron cores, to enhance their magnetic properties. Table 16.1 gives approximate values of $\mu_r$ for some common materials.

**Table 16.1** Relative permeabilities for common materials

| Material | $\mu_r$ |
|---|---|
| Air | 1 |
| Permalloy | 100,000 |
| Cast steel | 1,000 |
| Sheet steel | 4,000 |
| Iron | 5,195 |

Conversely, the reason for introducing the magnetic field intensity is that it is independent of the properties of the materials employed in the construction of magnetic circuits. Thus, a given magnetic field intensity, **H,** will give rise to different flux densities in different materials. It will therefore be useful to define *sources* of magnetic energy in terms of the magnetic field intensity, so that different magnetic structures and materials can then be evaluated or compared for a given source. In analogy with electromotive force, this "source" will be termed **magnetomotive force (mmf).** As stated earlier, both the magnetic flux density and field intensity are vector quantities; however, for ease of analysis, scalar fields will be chosen by appropriately selecting the orientation of the fields, wherever possible.

Ampère's law states that the integral of the vector magnetic field intensity, **H,** around a closed path is equal to the total current linked by the closed path, $i$:

$$\oint \mathbf{H} \cdot d\mathbf{l} = \sum i \tag{16.20}$$

where $d\mathbf{l}$ is an increment in the direction of the closed path. If the path is in the same direction as the direction of the magnetic field, we can use scalar quantities to state that

$$\int H\,dl = \sum i \tag{16.21}$$

Figure 16.9 illustrates the case of a wire carrying a current $i$, and of a circular path of radius $r$ surrounding the wire. In this simple case, you can see that the magnetic field intensity, **H,** is determined by the familiar right-hand rule. This rule states that if the direction of the current $i$ points in the direction of the thumb of one's right hand, the resulting magnetic field encircles the conductor in the direction in which the other four fingers would encircle it. Thus, in the case of Figure 16.9, the closed-path integral becomes equal to $H \cdot (2\pi r)$, since the path and the magnetic field are in the same direction, and therefore the magnitude of the magnetic field intensity is given by

$$H = \frac{i}{2\pi r} \tag{16.22}$$

Now, the magnetic field intensity is unaffected by the material surrounding the conductor, but the flux density depends on the material properties, since $B = \mu H$. Thus, the density of flux lines around the conductor would be far greater in the presence of a magnetic material than if the conductor were surrounded by air. The field generated by a single conducting wire is not very strong; however, if we arrange the wire into a tightly wound coil with many turns, we can greatly increase

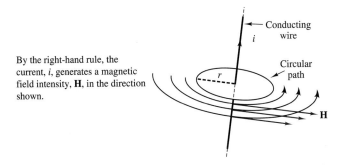

By the right-hand rule, the current, $i$, generates a magnetic field intensity, **H**, in the direction shown.

**Figure 16.9** Illustration of Ampère's law

the strength of the magnetic field. For such a coil, with $N$ turns, one can verify visually that the lines of force associated with the magnetic field link all of the turns of the conducting coil, so that we have effectively increased the current linked by the flux lines $N$-fold. The product $N \cdot i$ is a useful quantity in electromagnetic circuits, and is called the magnetomotive force,[2] $\mathcal{F}$ (often abbreviated mmf), in analogy with the electromotive force defined earlier:

$$\mathcal{F} = Ni \qquad \text{ampere-turns } (A \cdot t) \qquad \qquad \textbf{(16.23)}$$

Figure 16.10 illustrates the magnetic flux lines in the vicinity of a coil. The magnetic field generated by the coil can be made to generate a much greater flux density if the coil encloses a magnetic material. The most common ferromagnetic materials are steel and iron; in addition to these, many alloys and oxides of iron— as well as nickel—and some artificial ceramic materials called **ferrites** also exhibit magnetic properties. Winding a coil around a ferromagnetic material accomplishes two useful tasks at once: it forces the magnetic flux to be concentrated near the coil and—if the shape of the magnetic material is appropriate—completely confines the flux within the magnetic material, thus forcing the closed path for the flux lines to be almost entirely enclosed within the ferromagnetic material. Typical arrangements are the iron-core inductor and the toroid of Figure 16.11. The flux densities for these inductors are given by the expressions

$$B = \frac{\mu Ni}{l} \qquad \text{Flux density for tightly wound circular coil} \qquad \textbf{(16.24)}$$

$$B = \frac{\mu Ni}{2\pi r_2} \qquad \text{Flux density for toroidal coil} \qquad \textbf{(16.25)}$$

Intuitively, the presence of a high-permeability material near a source of magnetic flux causes the flux to preferentially concentrate in the high-$\mu$ material, rather than in air, much as a conducting path concentrates the current produced by an electric field in an electric circuit. In the course of this chapter, we shall

---

[2]Note that, although dimensionally equal to amperes, the units of magnetomotive force are ampere-turns.

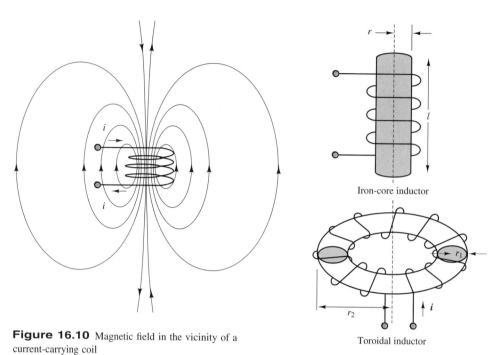

**Figure 16.10** Magnetic field in the vicinity of a current-carrying coil

Iron-core inductor

Toroidal inductor

**Figure 16.11** Practical inductors

continue to develop this analogy between electric circuits and magnetic circuits. Figure 16.12 depicts an example of a simple electromagnetic structure, which, as we shall see shortly, forms the basis of the practical transformer.

Table 16.2 summarizes the variables introduced thus far in the discussion of electricity and magnetism.

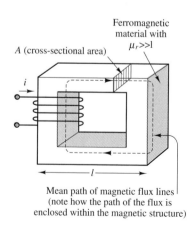

**Figure 16.12** A simple electromagnetic structure

Table 16.2  Magnetic variables and units

| Variable | Symbol | Units |
|---|---|---|
| Current | $I$ | A |
| Magnetic flux density | $B$ | $Wb/m^2 = T$ |
| Magnetic flux | $\phi$ | Wb |
| Magnetic field intensity | $H$ | A/m |
| Electromotive force | $e$ | V |
| Magnetomotive force | $\mathcal{F}$ | A · t |
| Flux linkage | $\lambda$ | Wb · t |

## Check Your Understanding

**16.1**  A coil having 100 turns is immersed in a magnetic field that is varying uniformly from 80 mWb to 30 mWb in 2 seconds. Find the induced voltage in the coil.

**16.2**  The magnitude of **H** at a radius of 0.5 m from a long linear conductor is $1\,A \cdot m^{-1}$. Find the current in the wire.

**16.3**  The relation between the flux linkages and the current for a magnetic material is given by $\lambda = 6i/(2i + 1)$ Wb · t. Determine the energy stored in the magnetic field for $\lambda = 2$ Wb · t.

**16.4**  Verify that for the linear case, where the flux is proportional to the mmf, the energy stored in the magnetic field is $\frac{1}{2}Li^2$.

## 16.2   MAGNETIC CIRCUITS

It is possible to analyze the operation of electromagnetic devices such as the one depicted in Figure 16.12 by means of magnetic equivalent circuits, similar in many respects to the equivalent electrical circuits of the earlier chapters. Before we can present this technique, however, we need to make a few simplifying approximations. The first of these approximations assumes that there exists a **mean path** for the magnetic flux, and that the corresponding mean flux density is approximately constant over the cross-sectional area of the magnetic structure. Thus, a coil wound around a core with cross-sectional area $A$ will have flux density

$$B = \frac{\phi}{A} \tag{16.26}$$

where $A$ is assumed to be perpendicular to the direction of the flux lines. Figure 16.12 illustrates such a mean path and the cross-sectional area, $A$. Knowing the flux density, we obtain the field intensity:

$$H = \frac{B}{\mu} = \frac{\phi}{A\mu} \tag{16.27}$$

But then, knowing the field intensity, we can relate the mmf of the coil, $\mathcal{F}$, to the product of the magnetic field intensity, $H$, and the length of the magnetic (mean) path, $l$, for one leg of the structure:

$$\mathcal{F} = N \cdot i = H \cdot l \tag{16.28}$$

In summary, the mmf is equal to the magnetic flux times the length of the magnetic path, divided by the permeability of the material times the cross-sectional area:

$$\mathcal{F} = \phi \frac{l}{\mu A} \tag{16.29}$$

A review of this formula reveals that the magnetomotive force, $\mathcal{F}$, may be viewed as being analogous to the voltage source in a series electrical circuit, and that the flux, $\phi$, is then equivalent to the electrical current in a series circuit and the term $l/\mu A$ to the *magnetic resistance* of one leg of the magnetic circuit. You will note that the term $l/\mu A$ is very similar to the term describing the resistance of a cylindrical conductor of length $l$ and cross-sectional area $A$, where the permeability, $\mu$, is analogous to the conductivity, $\sigma$. The term $l/\mu A$ occurs frequently enough to be assigned the name of **reluctance,** and the symbol $\mathcal{R}$. It is also important to recognize the relationship between the reluctance of a magnetic structure and its inductance. This can be derived easily starting from equation 16.14:

$$L = \frac{\lambda}{i} = \frac{N\phi}{i} = \frac{N}{i}\frac{Ni}{\mathcal{R}} = \frac{N^2}{\mathcal{R}} \quad \text{(H)} \tag{16.30}$$

In summary, when an $N$-turn coil carrying a current $i$ is wound around a magnetic core such as the one indicated in Figure 16.12, the mmf, $\mathcal{F}$, generated by the coil produces a flux, $\phi$, that is *mostly* concentrated within the core and is assumed to be uniform across the cross section. Within this simplified picture, then, the analysis of a magnetic circuit is analogous to that of resistive electrical circuits. This analogy is illustrated in Table 16.3 and in the examples in this section.

Table 16.3  Analogy between electric and magnetic circuits

| Electrical quantity | Magnetic quantity |
|---|---|
| Electrical field intensity, $E$, V/m | Magnetic field intensity, $H$, A·t/m |
| Voltage, $v$, V | Magnetomotive force, $\mathcal{F}$, A·t |
| Current, $i$, A | Magnetic flux, $\phi$, Wb |
| Current density, $J$, A/m$^2$ | Magnetic flux density, $B$, Wb/m$^2$ |
| Resistance, $R$, $\Omega$ | Reluctance, $\mathcal{R} = l/\mu A$, A·t/Wb |
| Conductivity, $\sigma$, $1/\Omega \cdot$ m | Permeability, $\mu$, Wb/A·m |

The usefulness of the magnetic circuit analogy can be emphasized by analyzing a magnetic core similar to that of Figure 16.12, but with a slightly modified geometry. Figure 16.13 depicts the magnetic structure and its equivalent circuit analogy. In the figure, we see that the mmf, $\mathcal{F} = Ni$, excites the magnetic circuit, which is composed of four legs: two of mean path length $l_1$ and cross-sectional area $A_1 = d_1 w$, and the other two of mean length $l_2$ and cross section $A_2 = d_2 w$. Thus, the reluctance encountered by the flux in its path around the magnetic core is given by the quantity $\mathcal{R}_{\text{series}}$, with

$$\mathcal{R}_{\text{series}} = 2\mathcal{R}_1 + 2\mathcal{R}_2$$

and

$$\mathcal{R}_1 = \frac{l_1}{\mu A_1} \qquad \mathcal{R}_2 = \frac{l_2}{\mu A_2}$$

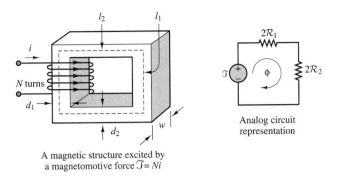

A magnetic structure excited by
a magnetomotive force $\mathcal{J} = Ni$

Analog circuit
representation

**Figure 16.13** Analogy between magnetic and electric circuits

It is important at this stage to review the assumptions and simplifications made in analyzing the magnetic structure of Figure 16.13:

1.   All of the magnetic flux is linked by all of the turns of the coil.
2.   The flux is confined exclusively within the magnetic core.
3.   The density of the flux is uniform across the cross-sectional area of the core.

You can probably see intuitively that the first of these assumptions might not hold true near the ends of the coil, but that it might be more reasonable if the coil is tightly wound. The second assumption is equivalent to stating that the relative permeability of the core is infinitely higher than that of air (presuming that this is the medium surrounding the core): if this were the case, the flux would indeed be confined within the core. It is worthwhile to note that we make a similar assumption when we treat wires in electric circuits as perfect conductors: the conductivity of copper is substantially greater than that of free space, by a factor of approximately $10^{15}$. In the case of magnetic materials, however, even for the best alloys, we have a relative permeability only on the order of $10^3$ to $10^4$. Thus, an approximation that is quite appropriate for electric circuits is not nearly as good in the case of magnetic circuits. Some of the flux in a structure such as those of Figures 16.12 and 16.13 would thus not be confined within the core (this is usually referred to as **leakage flux**). Finally, the assumption that the flux is uniform across the core cannot hold for a finite-permeability medium, but it is very helpful in giving an approximate *mean* behavior of the magnetic circuit.

The magnetic circuit analogy is therefore far from being exact. However, short of employing the tools of electromagnetic field theory and of vector calculus, or advanced numerical simulation software, it is the most convenient tool at the engineer's disposal for the analysis of magnetic structures. In the remainder of this chapter, the approximate analysis based on the electric circuit analogy will be used to obtain approximate solutions to problems involving a variety of useful magnetic circuits, many of which you are already familiar with. Among these will be the loudspeaker, solenoids, automotive fuel injectors, sensors for the measurement of linear and angular velocity and position, and other interesting applications.

## EXAMPLE 16.2 Analysis of Magnetic Structure and Equivalent Magnetic Circuit

### Problem

Calculate the flux, flux density, and field intensity on the magnetic structure of Figure 16.14.

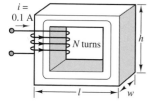

$i = 0.1$ A

$N$ turns

$h$

$l$

$w$

$l = 0.1$ m, $h = 0.1$ m, $w = 0.01$ m

**Figure 16.14**

### Solution

**Known Quantities:** Relative permeability; number of coil turns; coil current; structure geometry.

**Find:** $\phi$; $B$; $H$.

**Schematics, Diagrams, Circuits, and Given Data:** $\mu_r = 1,000$; $N = 500$ turns; $i = 0.1$ A. The magnetic circuit geometry is defined in Figures 16.14 and 16.15.

**Assumptions:** All magnetic flux is linked by the coil; the flux is confined to the magnetic core; the flux density is uniform.

**Analysis:**

1. *Calculation of magnetomotive force.* From equation 16.28, we calculate the magnetomotive force:

$$\mathcal{F} = \text{mmf} = Ni = (500 \text{ turns})(0.1 \text{ A}) = 50 \text{ A} \cdot \text{t}$$

2. *Calculation of mean path.* Next, we estimate the mean path of the magnetic flux. On the basis of the assumptions, we can calculate a mean path that runs through the geometric center of the magnetic structure, as shown in Figure 16.15. The path length is:

$$l_c = 4 \times 0.09 \text{ m} = 0.36 \text{ m}$$

The cross sectional area is $A = w^2 = (0.01)^2 = 0.0001 \text{ m}^2$.

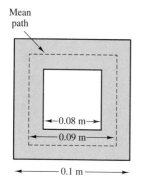

Mean path

←0.08 m→

←0.09 m→

←0.1 m→

**Figure 16.15**

3. *Calculation of reluctance.* Knowing the magnetic path length and cross sectional area we can calculate the reluctance of the circuit:

$$\mathcal{R} = \frac{l_c}{\mu A} = \frac{l_c}{\mu_r \mu_0 A} = \frac{0.36}{1,000 \times 4\pi \times 10^{-7} \times 0.0001} = 2.865 \times 10^6 \text{ A} \cdot \text{t/Wb}$$

The corresponding equivalent magnetic circuit is shown in Figure 16.16.

4. *Calculation of magnetic flux, flux density and field intensity.* On the basis of the assumptions, we can now calculate the magnetic flux:

$$\phi = \frac{\mathcal{F}}{\mathcal{R}} = \frac{50 \text{ A} \cdot \text{t}}{2.865 \times 10^6 \text{ A} \cdot \text{t/Wb}} = 1.75 \times 10^{-5} \text{ Wb}$$

the flux density:

$$B = \frac{\phi}{A} = \frac{\phi}{w^2} = \frac{1.75 \times 10^{-5} \text{ Wb}}{0.0001 \text{ m}^2} = 0.175 \text{ Wb/m}^2$$

and the magnetic field intensity:

$$H = \frac{B}{\mu} = \frac{B}{\mu_r \mu_0} = \frac{0.175 \text{ Wb/m}^2}{1,000 \times 4\pi \times 10^{-7} \text{ H/m}} = 139 \text{ A} \cdot \text{t/m}$$

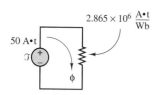

$2.865 \times 10^6 \ \frac{\text{A} \cdot \text{t}}{\text{Wb}}$

$50$ A·t

$\mathcal{F}$

$\phi$

**Figure 16.16**

**Comments:** This example has illustrated all the basic calculations that pertain to magnetic structures. Remember that the assumptions stated in this example (and earlier in the chapter) simplify the problem and make its approximate numerical solution possible in a few simple steps. In reality, flux leakage, fringing, and uneven distribution of flux across the structure would require the solution of three-dimensional equations using finite-element methods. These methods are not discussed in this book, but are necessary for practical engineering designs.

The usefulness of these approximate methods is that you can, for example, quickly calculate the approximate magnitude of the current required to generate a given magnetic flux or flux density. You shall soon see how these calculations can be used to determine electromagnetic energy and magnetic forces in practical structures.

The methodology described in this example is summarized in the following methodology box.

---

# FOCUS ON METHODOLOGY

**Magnetic Structures and Equivalent Magnetic Circuits**

**Direct Problem:**

*Given*—The structure geometry and the coil parameters (number of turns, current).

*Calculate*—The magnetic flux in the structure.

1. Compute the mmf.
2. Determine the length and cross section of the magnetic path for each continuous *leg* or section of the path.
3. Calculate the equivalent reluctance of the *leg*.
4. Generate the equivalent magnetic circuit diagram and calculate the total equivalent reluctance.
5. Calculate the flux, flux density, and magnetic field intensity, as needed.

**Inverse Problem:**

*Given*—The desired flux or flux density and structure geometry.

*Calculate*—The necessary coil current and number of turns.

1. Calculate the total equivalent reluctance of the structure from the desired flux.
2. Generate the equivalent magnetic circuit diagram.
3. Determine the mmf required to establish the required flux.
4. Choose the coil current and number of turns required to establish the desired mmf.

---

Consider the analysis of the same simple magnetic structure when an **air gap** is present. Air gaps are very common in magnetic structures; in rotating machines, for example, air gaps are necessary to allow for free rotation of the inner core of the machine. The magnetic circuit of Figure 16.17(a) differs from

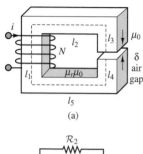

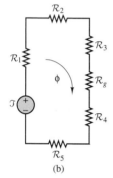

**Figure 16.17** (a) Magnetic circuit with air gap; (b) Equivalent representation of magnetic circuit with an air gap

the circuit analyzed in Example 16.2 simply because of the presence of an air gap; the effect of the gap is to break the continuity of the high-permeability path for the flux, adding a high-reluctance component to the equivalent circuit. The situation is analogous to adding a very large series resistance to a series electrical circuit. It should be evident from Figure 16.17(a) that the basic concept of reluctance still applies, although now two different permeabilities must be taken into account.

The equivalent circuit for the structure of Figure 16.17(a) may be drawn as shown in Figure 16.17(b), where $\mathcal{R}_n$ is the reluctance of path $l_n$, for $n = 1, 2, \ldots, 5$, and $\mathcal{R}_g$ is the reluctance of the air gap. The reluctances can be expressed as follows, if we assume that the magnetic structure has a uniform cross-sectional area, $A$:

$$\mathcal{R}_1 = \frac{l_1}{\mu_r \mu_0 A} \qquad \mathcal{R}_2 = \frac{l_2}{\mu_r \mu_0 A} \qquad \mathcal{R}_3 = \frac{l_3}{\mu_r \mu_0 A}$$

$$\mathcal{R}_4 = \frac{l_4}{\mu_r \mu_0 A} \qquad \mathcal{R}_5 = \frac{l_5}{\mu_r \mu_0 A} \qquad \mathcal{R}_g = \frac{\delta}{\mu_0 A_g} \tag{16.31}$$

Note that in computing $\mathcal{R}_g$, the length of the gap is given by $\delta$ and the permeability is given by $\mu_0$, as expected, but $A_g$ is different from the cross-sectional area, $A$, of the structure. The reason is that the flux lines exhibit a phenomenon known as **fringing** as they cross an air gap. The flux lines actually *bow out* of the gap defined by the cross section, $A$, not being contained by the high-permeability material any longer. Thus, it is customary to define an area $A_g$ that is greater than $A$, to account for this phenomenon. Example 16.3 describes in more detail the procedure for finding $A_g$ and also discusses the phenomenon of fringing.

## EXAMPLE 16.3 Magnetic Structure with Air Gaps

### Problem

Compute the equivalent reluctance of the magnetic circuit of Figure 16.18 and the flux density established in the bottom bar of the structure.

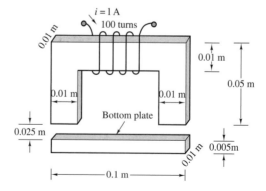

**Figure 16.18** Electromagnetic structure with air gaps

## Solution

**Known Quantities:** Relative permeability; number of coil turns; coil current; structure geometry.

**Find:** $\mathcal{R}_{eq}$; $B_{bar}$.

**Schematics, Diagrams, Circuits, and Given Data:** $\mu_r = 10,000$; $N = 100$ turns; $i = 1$ A.

**Assumptions:** All magnetic flux is linked by the coil; the flux is confined to the magnetic core; the flux density is uniform.

**Analysis:**

1. *Calculation of magnetomotive force.* From equation 16.28, we calculate the magnetomotive force:

$$\mathcal{F} = \text{mmf} = Ni = (100 \text{ turns})(1 \text{ A}) = 100 \text{ A} \cdot \text{t}$$

2. *Calculation of mean path.* Figure 16.19 depicts the geometry. The path length is:

$$l_c = l_1 + l_2 + l_3 + l_4 + l_5 + l_6 + l_g + l_g$$

However, the path must be broken into three legs: the upside-down U-shaped element, the air gaps, and the bar. We cannot treat these three parts as one because the relative permeability of the magnetic material is very different from that of the air gap. Thus, we define the following three paths, neglecting the very small (half bar thickness) lengths $l_5$ and $l_6$:

$$l_U = l_1 + l_2 + l_3 \qquad l_{bar} = l_4 + l_5 + l_6 \approx l_4 \qquad l_{gap} = l_g + l_g$$

where

$$l_U = 0.18 \text{ m} \qquad l_{bar} = 0.09 \text{ m} \qquad l_{gap} = 0.05 \text{ m}.$$

Next, we compute the cross-sectional area. For the magnetic structure, we calculate the square cross section to be: $A = w^2 = (0.01)^2 = 0.0001 \text{ m}^2$. For the air gap, we will make an empirical adjustment to account for the phenomenon of *fringing,* that is, to account for the tendency of the magnetic flux lines to bow out of the magnetic path, as illustrated in Figure 16.20. A rule of thumb used to account for fringing is to add the length of the gap to the actual cross-sectional area. Thus:

$$A_{gap} = (0.01 \text{ m} + l_g)^2 = (0.0125)^2 = 0.15625 \times 10^{-3} \text{ m}^2$$

3. *Calculation of reluctance.* Knowing the magnetic path length and cross sectional area we can calculate the reluctance of each of the legs of the circuit:

$$\mathcal{R}_U = \frac{l_U}{\mu_U A} = \frac{l_U}{\mu_r \mu_0 A} = \frac{0.18}{10,000 \times 4\pi \times 10^{-7} \times 0.0001}$$

$$= 1.43 \times 10^5 \text{ A} \cdot \text{t/Wb}$$

$$\mathcal{R}_{bar} = \frac{l_{bar}}{\mu_{bar} A} = \frac{l_{bar}}{\mu_r \mu_0 A} = \frac{0.09}{10,000 \times 4\pi \times 10^{-7} \times 0.0001}$$

$$= 0.715 \times 10^5 \text{ A} \cdot \text{t/Wb}$$

$$\mathcal{R}_{gap} = \frac{l_{gap}}{\mu_{gap} A_{gap}} = \frac{l_{gap}}{\mu_0 A_{gap}} = \frac{0.05}{4\pi \times 10^{-7} \times 0.0001} = 2.55 \times 10^7 \text{ A} \cdot \text{t/Wb}$$

Note that the reluctance of the air gap is dominant with respect to that of the magnetic

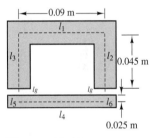

**Figure 16.19**

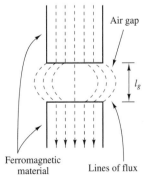

**Figure 16.20** Fringing effects in air gap

structure, in spite of the small dimension of the gap. This is because the relative permeability of the air gap is much smaller than that of the magnetic material.

The equivalent reluctance of the structure is:

$$\mathcal{R}_{eq} = \mathcal{R}_U + \mathcal{R}_{bar} + \mathcal{R}_{gap} = 1.43 \times 10^5 + 0.715 \times 10^5 + 2.55 \times 10^7$$

$$= 2.57 \times 10^7$$

Thus,

$$\mathcal{R}_{eq} \approx \mathcal{R}_{gap}$$

Since the gap reluctance is two orders of magnitude greater than the reluctance of the magnetic structure, it is reasonable to neglect the magnetic structure reluctance and work only with the gap reluctance in calculating the magnetic flux.

4.  *Calculation of magnetic flux and flux density in the bar.* From the result of the preceding sub-section, we calculate the flux

$$\phi = \frac{\mathcal{F}}{\mathcal{R}_{eq}} \approx \frac{\mathcal{F}}{\mathcal{R}_{gap}} = \frac{100 \text{ A} \cdot \text{t}}{2.55 \times 10^7 \text{ A} \cdot \text{t/Wb}} = 3.92 \times 10^{-6} \text{ Wb}$$

and the flux density in the bar:

$$B_{bar} = \frac{\phi}{A} = \frac{3.92 \times 10^{-6} \text{ Wb}}{0.0001 \text{ m}^2} = 39.2 \times 10^{-3} \text{ Wb/m}^2$$

**Comments:**  It is very common to neglect the reluctance of the magnetic material sections in these approximate calculations. We shall make this assumption very frequently in the remainder of the chapter.

---

### EXAMPLE   16.4  Magnetic Structure of Electric Motor

#### Problem

Figure 16.21 depicts the configuration of an electric motor. The electric motor consists of a *stator* and of a *rotor*. Compute the air gap flux and flux density.

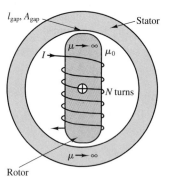

**Figure 16.21** Cross-sectional view of synchronous motor

#### Solution

**Known Quantities:**  Relative permeability; number of coil turns; coil current; structure geometry.

**Find:** $\phi_{gap}$; $B_{gap}$.

**Schematics, Diagrams, Circuits, and Given Data:** $\mu_r \to \infty$; $N = 1,000$ turns; $i = 10$ A; $l_{gap} = 0.01$ m; $A_{gap} = 0.1$ m$^2$. The magnetic circuit geometry is defined in Figure 16.21.

**Assumptions:** All magnetic flux is linked by the coil; the flux is confined to the magnetic core; the flux density is uniform. The reluctance of the magnetic structure is negligible.

**Analysis:**

1. *Calculation of magnetomotive force.* From equation 16.28, we calculate the magnetomotive force:

$$\mathcal{F} = \text{mmf} = Ni = (1,000 \text{ turns})(10 \text{ A}) = 10,000 \text{ A} \cdot \text{t}$$

2. *Calculation of reluctance.* Knowing the magnetic path length and cross sectional area, we can calculate the equivalent reluctance of the two gaps:

$$\mathcal{R}_{gap} = \frac{l_{gap}}{\mu_{gap} A_{gap}} = \frac{l_{gap}}{\mu_0 A_{gap}} = \frac{0.01}{4\pi \times 10^{-7} \times 0.2} = 3.97 \times 10^4 \text{ A} \cdot \text{t/Wb}$$

$$\mathcal{R}_{eq} = 2\mathcal{R}_{gap} = 7.94 \times 10^4 \text{ A} \cdot \text{t/Wb}$$

3. *Calculation of magnetic flux and flux density.* From the results of steps 1 and 2, we calculate the flux

$$\phi = \frac{\mathcal{F}}{\mathcal{R}_{eq}} = \frac{10,000 \text{ A} \cdot \text{t}}{7.94 \times 10^4 \text{ A} \cdot \text{t/Wb}} = 0.126 \text{ Wb}$$

and the flux density:

$$B_{bar} = \frac{\phi}{A} = \frac{0.126 \text{ Wb}}{0.1 \text{ m}^2} = 1.26 \text{ Wb/m}^2$$

**Comments:** Note that the flux and flux density in this structure are significantly larger than in the preceding example because of the larger mmf and larger gap area of this magnetic structure.

The subject of electric motors will be formally approached in Chapter 17.

---

## EXAMPLE   16.5   Equivalent Circuit of Magnetic Structure with Multiple Air Gaps

### Problem

Figure 16.23 depicts the configuration of a magnetic structure with two air gaps. Determine the equivalent circuit of the structure.

---

### Solution

**Known Quantities:** Structure geometry.

**Find:** Equivalent circuit diagram.

**Assumptions:** All magnetic flux is linked by the coil; the flux is confined to the magnetic core; the flux density is uniform. The reluctance of the magnetic structure is negligible.

**Analysis:**

1. *Calculation of magnetomotive force.*

$$\mathcal{F} = \text{mmf} = Ni$$

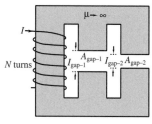

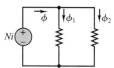

**Figure 16.22** Magnetic structure with two air gaps

2.  *Calculation of reluctance.* Knowing the magnetic path length and cross sectional area we can calculate the equivalent reluctance of the two gaps:

$$\mathcal{R}_{gap-1} = \frac{l_{gap-1}}{\mu_{gap-1} A_{gap-1}} = \frac{l_{gap-1}}{\mu_0 A_{gap-1}}$$

$$\mathcal{R}_{gap-1} = \frac{l_{gap-2}}{\mu_{gap-2} A_{gap-2}} = \frac{l_{gap-2}}{\mu_0 A_{gap-2}}$$

3.  *Calculation of magnetic flux and flux density.* Note that the flux must now divide between the two legs, and that a different air-gap flux will exist in each leg. Thus:

$$\phi_1 = \frac{Ni}{\mathcal{R}_{gap-1}} = \frac{Ni\mu_0 A_{gap-1}}{l_{gap-1}}$$

$$\phi_2 = \frac{Ni}{\mathcal{R}_{gap-2}} = \frac{Ni\mu_0 A_{gap-2}}{l_{gap-2}}$$

and the total flux generated by the coil is $\phi = \phi_1 + \phi_2$.
     The equivalent circuit is shown in the bottom half of Figure 16.22.

**Comments:** Note that the two legs of the structure act like resistors in a parallel circuit.

---

### EXAMPLE  16.6  Inductance, Stored Energy, and Induced Voltage

#### Problem

1.  Determine the inductance and the magnetic stored energy for the structure of Fig. 16.17(a). The structure is identical to that of Example 16.2 except for the air gap.
2.  Assume that the flux density in the air gap varies sinusoidally as $B(t) = B_0 \sin(\omega t)$. Determine the induced voltage across the coil, $e$.

---

#### Solution

**Known Quantities:** Relative permeability; number of coil turns; coil current; structure geometry; flux density in air gap.

**Find:** $L$; $W_m$; $e$.

**Schematics, Diagrams, Circuits, and Given Data:** $\mu_r \rightarrow \infty$; $N = 500$ turns; $i = 0.1$ A. The magnetic circuit geometry is defined in Figures 16.14 and 16.15. The air gap has $l_g = 0.002$ m. $B_0 = 0.6$ Wb/m$^2$.

**Assumptions:** All magnetic flux is linked by the coil; the flux is confined to the magnetic core; the flux density is uniform. The reluctance of the magnetic structure is negligible.

**Analysis:**

1.  *Part 1.* To calculate the inductance of this magnetic structure, we use equation 16.30:

$$L = \frac{N^2}{\mathcal{R}}$$

Thus, we need to first calculate the reluctance. Assuming that the reluctance of the structure is negligible, we have:

$$\mathcal{R}_{gap} = \frac{l_{gap}}{\mu_{gap} A_{gap}} = \frac{l_{gap}}{\mu_0 A_{gap}} = \frac{0.002}{4\pi \times 10^{-7} \times 0.0001} = 1.59 \times 10^7 \text{ A} \cdot \text{t/Wb}$$

and

$$L = \frac{N^2}{\mathcal{R}} = \frac{500^2}{1.59 \times 10^7} = 0.157 \text{ H}$$

Finally, we can calculate the stored magnetic energy as follows:

$$W_m = \frac{1}{2} L i^2 = \frac{1}{2} \times (0.157 \text{ H}) \times (0.1 \text{ A})^2 = 0.785 \times 10^{-3} \text{ J}$$

2.  *Part 2.* To calculate the induced voltage due to a time-varying magnetic flux, we use equation 16.16:

$$e = \frac{d\lambda}{dt} = N \frac{d\phi}{dt} = NA \frac{dB}{dt} = NAB_0 \omega \cos(\omega t)$$

$$= 500 \times 0.0001 \times 0.6 \times 377 \cos(377t) = 11.31 \cos(377t) \text{ V}$$

**Comments:**  The voltage induced across a coil in an electromagnetic transducer is a very important quantity called *back electromotive force,* or back emf.  We shall make use of this quantity in Sec. 16.5.

---

## Magnetic Reluctance Position Sensor

A simple magnetic structure, very similar to those examined in the previous examples, finds very common application in the so-called **variable-reluctance position sensor,** which, in turn, finds widespread application in a variety of configurations for the measurement of linear and angular velocity. Figure 16.23 depicts one particular configuration that is used in many applications. In this structure, a permanent magnet with a coil of wire wound around it forms the sensor; a steel disk (typically connected to a rotating shaft) has a number of tabs that pass between the pole pieces of the sensor. The area of the tab is assumed equal to the area of the cross section of the pole pieces and is equal to $a^2$. The reason for the name *variable-reluctance sensor* is that the reluctance of the magnetic structure is variable, depending on whether or not a ferromagnetic tab lies between the pole pieces of the magnet.

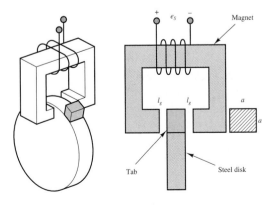

**Figure 16.23** Variable-reluctance position sensor

**FOCUS ON MEASUREMENTS**

The principle of operation of the sensor is that an electromotive force, $e_S$, is induced across the coil by the change in magnetic flux caused by the passage of the tab between the pole pieces when the disk is in motion. As the tab enters the volume between the pole pieces, the flux will increase, because of the lower reluctance of the configuration, until it reaches a maximum when the tab is centered between the poles of the magnet. Figure 16.24 depicts the approximate shape of the resulting voltage, which, according to Faraday's law, is given by

$$e_S = -\frac{d\phi}{dt}$$

The rate of change of flux is dictated by the geometry of the tab and of the pole pieces, and by the speed of rotation of the disk. It is important to note that, since the flux is changing only if the disk is rotating, this sensor cannot detect the static position of the disk.

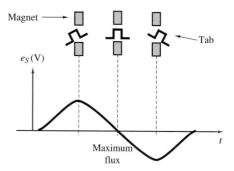

**Figure 16.24** Variable-reluctance position sensor waveform

One common application of this concept is in the measurement of the speed of rotation of rotating machines, including electric motors and internal combustion engines. In these applications, use is made of a *60-tooth wheel*, which permits the conversion of the speed rotation directly to units of revolutions per minute. The output of a variable-reluctance position sensor magnetically coupled to a rotating disk equipped with 60 tabs (teeth) is processed through a comparator or Schmitt trigger circuit (see Chapter 15). The voltage waveform generated by the sensor is nearly sinusoidal when the teeth are closely spaced, and it is characterized by one sinusoidal cycle for each tooth on the disk. If a negative zero-crossing detector (see Chapter 15) is employed, the trigger circuit will generate a pulse corresponding to the passage of each tooth, as shown in Figure 16.25. If the time between any two pulses is measured by means of a high-frequency clock, the speed of the engine can be directly determined in units of rev/min by means of a digital counter (see Chapter 14).

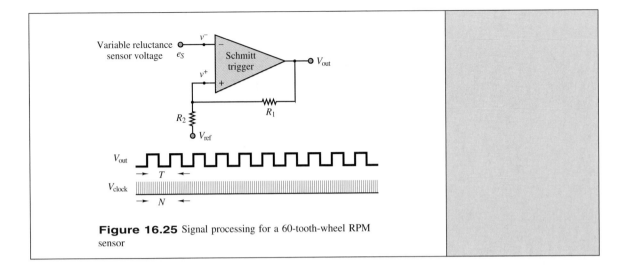

**Figure 16.25** Signal processing for a 60-tooth-wheel RPM sensor

## Voltage Calculation in Magnetic Reluctance Position Sensor

This example illustrates the calculation of the voltage induced in a magnetic reluctance sensor by a rotating toothed wheel. In particular, we will find an approximate expression for the reluctance and the induced voltage for the position sensor shown in Figure 16.26, and show that the induced voltage is speed-dependent. It will be assumed that the reluctance of the core and fringing at the air gaps are both negligible.

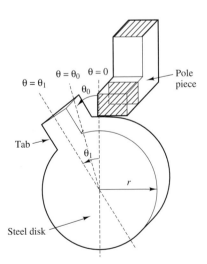

**Figure 16.26** Reluctance sensor for measurement of angular position

### Solution:

From the geometry shown in the preceding "Focus on Measurements," the equivalent reluctance of the magnetic structure is twice that of one gap, since the permeability of the tab and magnetic structure are assumed infinite (i.e., they have negligible reluctance). When the tab and the poles are aligned, the angle $\theta$ is zero, as shown in Figure 16.26, and the area of the air gap is maximum. For angles greater than $2\theta_0$, the magnetic length of the air gaps is so large that the magnetic field may reasonably be taken as zero.

To model the reluctance of the gaps, we assume the following simplified expression, where the area of overlap of the tab with the magnetic poles is assumed proportional to the angular displacement:

$$\mathcal{R} = \frac{2l_g}{\mu_0 A} = \frac{2l_g}{\mu_0 a r (\theta_1 - \theta)} \qquad \text{for} \qquad 0 < \theta < \theta_1$$

Naturally, this is an approximation; however, the approximation captures the essential idea of this transducer, namely, that the reluctance will decrease with increasing overlap area until it reaches a minimum, and then it will increase as the overlap area decreases. For $\theta = \theta_1$, that is, with the tab outside the magnetic pole pieces, we have $\mathcal{R}_{max} \to \infty$. For $\theta = 0$, that is, with the tab perfectly aligned with the pole pieces, we have $\mathcal{R}_{min} = 2l_g/\mu_0 a r \theta_1$. The flux $\phi$ may therefore be computed as follows:

$$\phi = \frac{Ni}{\mathcal{R}} = \frac{Ni\mu_0 a r (\theta_1 - \theta)}{2l_g}$$

The induced voltage $e_S$ is found by

$$e_S = \frac{d\phi}{dt} = -\frac{d\phi}{d\theta}\frac{d\theta}{dt} = \frac{Ni\mu_0 a r}{2l_g} \times \omega$$

where $\omega = d\theta/dt$ is the rotational speed of the steel disk. It should be evident that the induced voltage is speed-dependent. For $a = 1$ cm, $r = 10$ cm, $l_g = 0.1$ cm, $N = 100$ turns, $i = 10$ mA, $\theta_1 = 6° \approx 0.1$ rad, and $\omega = 400$ rad/s (approximately 3,800 rev/min), we have

$$\mathcal{R}_{max} = \frac{2 \times 0.1 \times 10^{-2}}{4\pi \times 10^{-7} \times 1 \times 10^{-2} \times 10 \times 10^{-2} \times 0.1}$$

$$= 1.59 \times 10^7 \text{ A} \cdot \text{t/Wb}$$

$$e_{S\,peak} = \frac{1{,}000 \times 10 \times 10^{-3} \times 4\pi \times 10^{-7} \times 1 \times 10^{-2} \times 10^{-1}}{2 \times 0.1 \times 10^{-2}} \times 400$$

$$= 2.5 \text{ mV}$$

That is, the peak amplitude of $e_S$ will be 2.5 mV.

## Check Your Understanding

**16.5**   If $\mathcal{R}_{eq} = 2\mathcal{R}_{gap}$ in Example 16.3, calculate $\phi$ and $B$.

**16.6**   Determine the equivalent reluctance of the structure of Figure 16.27 as seen by the "source" if $\mu_r$ for the structure is 1,000, $l = 5$ cm, and all of the legs are 1 cm on a side.

**16.7**   Find the equivalent reluctance of the magnetic circuit shown in Figure 16.28 if $\mu_r$ of the structure is infinite, $\delta = 2$ mm, and the physical cross section of the core is 1 cm$^2$. Do not neglect fringing.

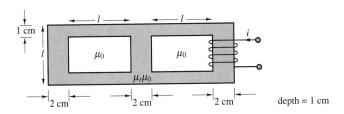

**Figure 16.27**

**16.8**   Find the equivalent magnetic circuit of the structure of Figure 16.29 if $\mu_r$ is infinite. Give expressions for each of the circuit values if the physical cross-sectional area of each of the legs is given by

$$A = l \times w$$

Do not neglect fringing.

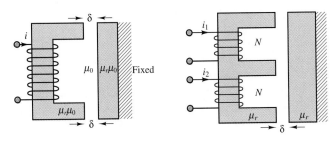

**Figure 16.28**              **Figure 16.29**

## 16.3   MAGNETIC MATERIALS AND *B-H* CURVES

In the analysis of magnetic circuits presented in the previous sections, the relative permeability, $\mu_r$, was treated as a constant. In fact, the relationship between the magnetic flux density, **B,** and the associated field intensity, **H,**

$$\mathbf{B} = \mu \mathbf{H} \qquad (16.32)$$

is characterized by the fact that the relative permeability of magnetic materials is not a constant, but is a function of the magnetic field intensity. In effect, all magnetic materials exhibit a phenomenon called **saturation,** whereby the flux density increases in proportion to the field intensity until it cannot do so any longer. Figure 16.30 illustrates the general behavior of all magnetic materials. You will note that since the *B-H* curve shown in the figure is nonlinear, the value of $\mu$ (which is the slope of the curve) depends on the intensity of the magnetic field.

To understand the reasons for the saturation of a magnetic material, we need to briefly review the mechanism of magnetization. The basic idea behind magnetic materials is that the spin of electrons constitutes motion of charge, and therefore

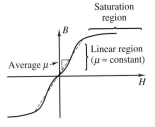

**Figure 16.30** Permeability and magnetic saturation effects

leads to magnetic effects, as explained in the introductory section of this chapter. In most materials, the electron spins cancel out, on the whole, and no net effect remains. In ferromagnetic materials, on the other hand, atoms can align so that the electron spins cause a net magnetic effect. In such materials, there exist small regions with strong magnetic properties (called **magnetic domains**), the effects of which are neutralized in unmagnetized material by other, similar regions that are oriented differently, in a random pattern. When the material is magnetized, the magnetic domains tend to align with each other, to a degree that is determined by the intensity of the applied magnetic field.

In effect, the large number of miniature magnets within the material are *polarized* by the external magnetic field. As the field increases, more and more domains become aligned. When all of the domains have become aligned, any further increase in magnetic field intensity does not yield an increase in flux density beyond the increase that would be caused in a nonmagnetic material. Thus, the relative permeability, $\mu_r$, approaches 1 in the saturation region. It should be apparent that an exact value of $\mu_r$ cannot be determined; the value of $\mu_r$ used in the earlier examples is to be interpreted as an average permeability, for intermediate values of flux density. As a point of reference, commercial magnetic steels saturate at flux densities around a few teslas. Figure 16.33, shown later in this section, will provide some actual $B$-$H$ curves for common ferromagnetic materials.

The phenomenon of saturation carries some interesting implications with regard to the operation of magnetic circuits: the results of the previous section would seem to imply that an increase in the mmf (that is, an increase in the current driving the coil) would lead to a proportional increase in the magnetic flux. This is true in the *linear region* of Figure 16.30; however, as the material reaches saturation, further increases in the driving current (or, equivalently, in the mmf) do not yield further increases in the magnetic flux.

There are two more features that cause magnetic materials to further deviate from the ideal model of the linear $B$-$H$ relationship: **eddy currents** and **hysteresis.** The first phenomenon consists of currents that are caused by any time-varying flux in the core material. As you know, a time-varying flux will induce a voltage, and therefore a current. When this happens inside the magnetic core, the induced voltage will cause "eddy" currents (the terminology should be self-explanatory) in the core, which depend on the resistivity of the core. Figure 16.31 illustrates the phenomenon of eddy currents. The effect of these currents is to dissipate energy in the form of heat. Eddy currents are reduced by selecting high-resistivity core materials, or by *laminating* the core, introducing tiny, discontinuous air gaps between core layers (see Figure 16.31). Lamination of the core reduces eddy currents greatly without affecting the magnetic properties of the core.

It is beyond the scope of this chapter to quantify the losses caused by induced eddy currents, but it will be important in Chapters 17 and 18 to be aware of this source of energy loss.

Hysteresis is another loss mechanism in magnetic materials; it displays a rather complex behavior, related to the magnetization properties of a material. The curve of Figure 16.32 reveals that the $B$-$H$ curve for a magnetic material during magnetization (as $H$ is increased) is displaced with respect to the curve that is measured when the material is demagnetized. To understand the hysteresis process, consider a core that has been energized for some time, with a field intensity of $H_1 A \cdot t/m$. As the current required to sustain the mmf corresponding to $H_1$ is decreased, we follow the hysteresis curve from the point $\alpha$ to the point $\beta$. When the mmf is exactly zero, the material displays the **remanent** (or **residual**)

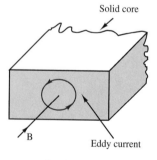

Solid core

B

Eddy current

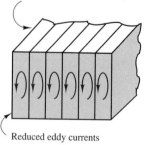

Laminated core
(the laminations are separated
by a thin layer of insulation)

Reduced eddy currents

**Figure 16.31** Eddy currents in magnetic structures

**magnetization** $B_r$. To bring the flux density to zero, we must further decrease the mmf (i.e., produce a negative current), until the field intensity reaches the value $-H_0$ (point $\gamma$ on the curve). As the mmf is made more negative, the curve eventually reaches the point $\alpha'$. If the excitation current to the coil is now increased, the magnetization curve will follow the path $\alpha' = \beta' = \gamma' = \alpha$, eventually returning to the original point in the $B$-$H$ plane, but via a different path.

The result of this process, by which an *excess magnetomotive force* is required to magnetize or demagnetize the material, is a net energy loss. It is difficult to evaluate this loss exactly; however, it can be shown that it is related to the area between the curves of Figure 16.32. There are experimental techniques that enable the approximate measurement of these losses.

Figures 16.33(a)–(c) depict magnetization curves for three very common ferromagnetic materials: cast iron, cast steel, and sheet steel. These curves will be useful in solving some of the homework problems.

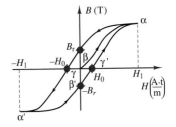

**Figure 16.32** Hysteresis in magnetization curves

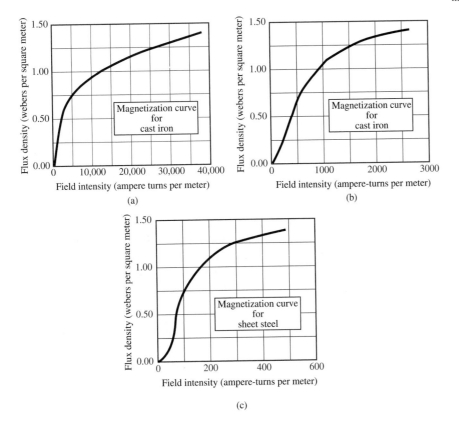

**Figure 16.33** (a) Magnetization curve for cast iron; (b) Magnetization curve for cast steel; (c) Magnetization curve for sheet steel

## 16.4    TRANSFORMERS

One of the more common magnetic structures in everyday applications is the **transformer.** The ideal transformer was introduced in Chapter 7 as a device that can step an AC voltage up or down by a fixed ratio, with a corresponding decrease or increase in current. The structure of a simple magnetic transformer is shown in

Figure 16.34, which illustrates that a transformer is very similar to the magnetic circuits described earlier in this chapter. Coil $L_1$ represents the input side of the transformer, while coil $L_2$ is the output coil; both coils are wound around the same magnetic structure, which we show here to be similar to the "square doughnut" of the earlier examples.

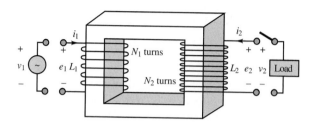

**Figure 16.34** Structure of a transformer

The ideal transformer operates on the basis of the same set of assumptions we made in earlier sections: the flux is confined to the core, the flux links all turns of both coils, and the permeability of the core is infinite. The last assumption is equivalent to stating that an arbitrarily small mmf is sufficient to establish a flux in the core. In addition, we assume that the ideal transformer coils offer negligible resistance to current flow.

The operation of a transformer requires a time-varying current; if a time-varying voltage is applied to the primary side of the transformer, a corresponding current will flow in $L_1$; this current acts as an mmf and causes a (time-varying) flux in the structure. But the existence of a changing flux will induce an emf across the secondary coil! Without the need for a direct electrical connection, the transformer can couple a source voltage at the primary to the load; the coupling occurs by means of the magnetic field acting on both coils. Thus, a transformer operates by converting electric energy to magnetic, and then back to electric. The following derivation illustrates this viewpoint in the ideal case (no loss of energy), and compares the result with the definition of the ideal transformer in Chapter 7.

If a time-varying voltage source is connected to the input side, then by virtue of Faraday's law, a corresponding time-varying flux $d\phi/dt$ is established in coil $L_1$:

$$e_1 = N_1 \frac{d\phi}{dt} = v_1 \tag{16.33}$$

But since the flux thus produced also links coil $L_2$, an emf is induced across the output coil as well:

$$e_2 = N_2 \frac{d\phi}{dt} = v_2 \tag{16.34}$$

This induced emf can be measured as the voltage $v_2$ at the output terminals, and one can readily see that the ratio of the open-circuit output voltage to input terminal voltage is

$$\frac{v_2}{v_1} = \frac{N_2}{N_1} \tag{16.35}$$

If a load current $i_2$ is now required by the connection of a load to the output circuit (by closing the switch in the figure), the corresponding mmf is $\mathcal{F}_2 = N_2 i_2$. This mmf, generated by the load current $i_2$, would cause the flux in the core to change; however, this is not possible, since a change in $\phi$ would cause a corresponding change in the voltage induced across the input coil. But this voltage is determined (fixed) by the source $v_1$ (and is therefore $d\phi/dt$), so that the input coil is forced to generate a **counter mmf** to oppose the mmf of the output coil; this is accomplished as the input coil draws a current $i_1$ from the source $v_1$ such that

$$i_1 N_1 = i_2 N_2 \tag{16.36}$$

or

$$\frac{i_2}{i_1} = \frac{N_1}{N_2} = \alpha \tag{16.37}$$

where $\alpha$ is the ratio of primary to secondary turns (the transformer ratio) and $N_1$ and $N_2$ are the primary and secondary turns, respectively. If there were any net difference between the input and output mmf, flux balance required by the input voltage source would not be satisfied. Thus, the two mmf's must be equal. As you can easily verify, these results are the same as in Chapter 7; in particular, the ideal transformer does not dissipate any power, since

$$v_1 i_1 = v_2 i_2 \tag{16.38}$$

Note the distinction we have made between the induced voltages (emf's), $e$, and the terminal voltages, $v$. In general, these are not the same.

The results obtained for the ideal case do not completely represent the physical nature of transformers. A number of loss mechanisms need to be included in a practical transformer model, to account for the effects of leakage flux, for various magnetic core losses (e.g., hysteresis), and for the unavoidable resistance of the wires that form the coils.

Commercial transformer ratings are usually given on the so-called **nameplate,** which indicates the normal operating conditions. The nameplate includes the following parameters:

- Primary-to-secondary voltage ratio
- Design frequency of operation
- (Apparent) rated output power

For example, a typical nameplate might read 480:240 V, 60 Hz, 2 kVA. The voltage ratio can be used to determine the turns ratio, while the rated output power represents the continuous power level that can be sustained without overheating. It is important that this power be rated as the apparent power in kVA, rather than real power in kW, since a load with low power factor would still draw current and therefore operate near rated power. Another important performance characteristic of a transformer is its **power efficiency,** defined by:

$$\text{Power efficiency} = \eta = \frac{\text{Output power}}{\text{Input power}} \tag{16.39}$$

The following examples illustrate the use of the nameplate ratings and the calculation of efficiency in a practical transformer, in addition to demonstrating the application of the circuit models.

### EXAMPLE 16.7 Transformer Nameplate

**Problem**

Determine the turns ratio and the rated currents of a transformer from nameplate data.

**Solution**

*Known Quantities:* Nameplate data.

*Find:* $\alpha = N_1/N_2$; $I_1$; $I_2$.

*Schematics, Diagrams, Circuits, and Given Data:* Nameplate data: 120 V/480 V; 48 kVA; 60 Hz.

*Assumptions:* Assume an ideal transformer.

*Analysis:* The first element in the nameplate data is a pair of voltages, indicating the primary and secondary voltages for which the transformer is rated. The ratio, $\alpha$, is found as follows:

$$\alpha = \frac{N_1}{N_2} = \frac{480}{120} = 4$$

To find the primary and secondary currents, we use the kVA rating (apparent power) of the transformer:

$$I_1 = \frac{|S|}{V_1} = \frac{48 \text{ kVA}}{480 \text{ V}} = 100 \text{ A} \qquad I_2 = \frac{|S|}{V_2} = \frac{48 \text{ kVA}}{120 \text{ V}} = 400 \text{ A}$$

*Comments:* In computing the rated currents, we have assumed that no losses take place in the transformer; in fact, there will be losses due to coil resistance and magnetic core effects. These losses result in heating of the transformer, and limit its rated performance.

### EXAMPLE 16.8 Impedance Transformer

**Problem**

Find the equivalent load impedance seen by the voltage source (i.e., reflected from secondary to primary) for the transformer of Figure 16.35.

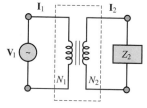

**Figure 16.35** Ideal transformer

**Solution**

*Known Quantities:* Transformer turns ratio, $\alpha$.

*Find:* Reflected impedance, $Z_2'$.

*Assumptions:* Assume an ideal transformer.

*Analysis:* By definition, the load impedance is equal to the ratio of secondary phasor voltage and current:

$$Z_2 = \frac{\mathbf{V}_2}{\mathbf{I}_2}$$

To find the reflected impedance we can express the above ratio in terms of primary voltage and current:

$$Z_2 = \frac{\mathbf{V}_2}{\mathbf{I}_2} = \frac{\frac{\mathbf{V}_1}{\alpha}}{\alpha \mathbf{I}_1} = \frac{1}{\alpha^2}\frac{\mathbf{V}_1}{\mathbf{I}_1}$$

where the ratio $\mathbf{V}_1/\mathbf{I}_1$ is the impedance seen by the source at the primary coil, that is, the *reflected load impedance* seen by the primary (source) side of the circuit. Thus, we can write the load impedance, $Z_2$, in terms of the primary circuit voltage and current; we call this the reflected impedance, $Z_2'$:

$$Z_2 = \frac{1}{\alpha^2}\frac{\mathbf{V}_1}{\mathbf{I}_1} = \frac{1}{\alpha^2}Z_1 = \frac{1}{\alpha^2}Z_2'.$$

Thus, $Z_2' = \alpha^2 Z_2$. Figure 16.36 depicts the equivalent circuit with the load impedance reflected back to the primary.

**Comments:** The equivalent reflected circuit calculations are convenient because all circuit elements can be referred to a single set of variable (i.e., only primary or secondary voltages and currents).

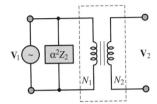

**Figure 16.36**

---

## Check Your Understanding

**16.9**   The high-voltage side of a transformer has 500 turns, and the low-voltage side has 100 turns. When the transformer is connected as a step-down transformer, the load current is 12 A. Calculate: (a) the turns ratio $\alpha$; (b) the primary current.

**16.10**   Calculate the turns ratio if the transformer in Check Your Understanding 16.9 is used as a step-up transformer.

**16.11**   The output of a transformer under certain conditions is 12 kW. The copper losses are 189 W and the core losses are 52 W. Calculate the efficiency of this transformer.

**16.12**   The output impedance of a servo amplifier is 250 Ω. The servomotor that the amplifier must drive has an impedance of 2.5 Ω. Calculate the turns ratio of the transformer required to match these impedances.

---

## 16.5    ELECTROMECHANICAL ENERGY CONVERSION

From the material developed thus far, it should be apparent that electromagnetomechanical devices are capable of converting mechanical forces and displacements to electromagnetic energy, and that the converse is also possible. The objective of this section is to formalize the basic principles of energy conversion in electromagnetomechanical systems, and to illustrate its usefulness and potential for application by presenting several examples of **energy transducers.** A transducer is a device that can convert electrical to mechanical energy (in this case, it is often called an **actuator**), or vice versa (in which case it is called a **sensor**).

Several physical mechanisms permit conversion of electrical to mechanical energy and back, the principal phenomena being the **piezoelectric effect,**[3] consisting of the generation of a change in electric field in the presence of strain in

---

[3]See "Focus on Measurements: Charge Amplifiers" in Chapter 12.

certain crystals (e.g., quartz), and **electrostriction** and **magnetostriction,** in which changes in the dimension of certain materials lead to a change in their electrical (or magnetic) properties. Although these effects lead to many interesting applications, this chapter is concerned only with transducers in which electrical energy is converted to mechanical energy through the coupling of a magnetic field. It is important to note that all rotating machines (motors and generators) fit the basic definition of electromechanical transducers we have just given.

## Forces in Magnetic Structures

Mechanical forces can be converted to electrical signals, and vice versa, by means of the coupling provided by energy stored in the magnetic field. In this subsection, we discuss the computation of mechanical forces and of the corresponding electromagnetic quantities of interest; these calculations are of great practical importance in the design and application of electromechanical actuators. For example, a problem of interest is the computation of the current required to generate a given force in an electromechanical structure. This is the kind of application that is likely to be encountered by the engineer in the selection of an electromechanical device for a given task.

As already seen in this chapter, an electromechanical system includes an electrical system and a mechanical system, in addition to means through which the two can interact. The principal focus of this chapter has been the coupling that occurs through an electromagnetic field common to both the electrical and the mechanical system; to understand electromechanical energy conversion, it will be important to understand the various energy storage and loss mechanisms in the electromagnetic field. Figure 16.37 illustrates the coupling between the electrical and mechanical systems. In the mechanical system, energy loss can occur because of the heat developed as a consequence of *friction*, while in the electrical system, analogous losses are incurred because of *resistance*. Loss mechanisms are also present in the magnetic coupling medium, since *eddy current losses* and *hysteresis losses* are unavoidable in ferromagnetic materials. Either system can supply energy, and either system can store energy. Thus, the figure depicts the flow of energy from the electrical to the mechanical system, accounting for these various losses. The same flow could be reversed if mechanical energy were converted to electrical form.

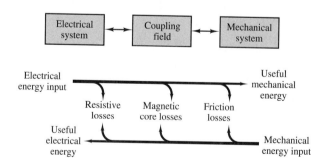

**Figure 16.37**

## Moving-Iron Transducers

One important class of electromagnetomechanical transducers is that of **moving-iron transducers.** The aim of this section is to derive an expression for the magnetic forces generated by such transducers and to illustrate the application of these calculations to simple, yet common devices such as electromagnets, solenoids, and relays. The simplest example of a moving-iron transducer is the **electromagnet** of Figure 16.38, in which the U-shaped element is fixed and the bar is movable. In the following paragraphs, we shall derive a relationship between the current applied to the coil, the displacement of the movable bar, and the magnetic force acting in the air gap.

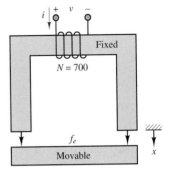

**Figure 16.38**

The principle that will be applied throughout the section is that in order for a mass to be displaced, some work needs to be done; this work corresponds to a change in the energy stored in the electromagnetic field, which causes the mass to be displaced. With reference to Figure 16.38, let $f_e$ represent the magnetic force acting on the bar and $x$ the displacement of the bar, in the direction shown. Then the net work into the electromagnetic field, $W_m$, is equal to the sum of the work done by the electrical circuit plus the work done by the mechanical system. Therefore, for an incremental amount of work, we can write

$$dW_m = ei\,dt - f_e\,dx \qquad (16.40)$$

where $e$ is the electromotive force across the coil and the negative sign is due to the sign convention indicated in Figure 16.38. Recalling that the emf $e$ is equal to the derivative of the flux linkage (equation 16.16), we can further expand equation 16.40 to obtain

$$dW_m = ei\,dt - f_e\,dx = i\frac{d\lambda}{dt}dt - f_e\,dx = i\,d\lambda - f_e\,dx \qquad (16.41)$$

or

$$f_e\,dx = i\,d\lambda - dW_m \qquad (16.42)$$

Now we must observe that the flux in the magnetic structure of Figure 16.38 depends on two variables, which are in effect independent: the current flowing through the coil, and the displacement of the bar. Each of these variables can cause the magnetic flux to change. Similarly, the energy stored in the electromagnetic field is also dependent on both current and displacement. Thus we can rewrite equation 16.42 as follows:

$$f_e = i\left(\frac{\partial\lambda}{\partial i}di + \frac{\partial\lambda}{\partial x}dx\right) - \left(\frac{\partial W_m}{\partial i}di + \frac{\partial W_m}{\partial x}dx\right) \qquad (16.43)$$

Since $i$ and $x$ are independent variables, we can write

$$f_e = i\frac{\partial\lambda}{\partial x} - \frac{\partial W_m}{\partial x} \qquad \text{and} \qquad 0 = i\frac{\partial\lambda}{\partial i} - \frac{\partial W_m}{\partial i} \qquad (16.44)$$

From the first of the expressions in equation 16.44, we obtain the relationship

$$f_e = \frac{\partial}{\partial x}(i\lambda - W_m) = \frac{\partial}{\partial x}(W_c) \qquad (16.45)$$

where the term $W_c$ corresponds to $W_m'$, defined as the co-energy in equation 16.18. Finally, we observe that the force acting to *pull* the bar toward the electromagnet structure, which we will call $f$, is of opposite sign relative to $f_e$, and therefore we can write

$$f = -f_e = -\frac{\partial}{\partial x}(W_c) = -\frac{\partial W_m}{\partial x} \qquad (16.46)$$

Equation 16.46 includes a very important assumption: that the energy is equal to the co-energy. If you make reference to Figure 16.8, you will realize that in general this is not true. Energy and co-energy are equal only if the $\lambda$-$i$ relationship is linear. Thus, the useful result of equation 16.46, stating that the magnetic force acting on the moving iron is proportional to the rate of change of stored energy with displacement, applies only for *linear magnetic structures*.

Thus, in order to determine the forces present in a magnetic structure, it will be necessary to compute the energy stored in the magnetic field. To simplify the analysis, it will be assumed hereafter that the structures analyzed are magnetically linear. This is, of course, only an approximation, in that it neglects a number of practical aspects of electromechanical systems (for example, the nonlinear $\lambda$-$i$ curves described earlier, and the core losses typical of magnetic materials), but it permits relatively simple analysis of many useful magnetic structures. Thus, although the analysis method presented in this section is only approximate, it will serve the purpose of providing a feeling for the direction and the magnitude of the forces and currents present in electromechanical devices. On the basis of a linear approximation, it can be shown that the stored energy in a magnetic structure is given by the expression

$$W_m = \frac{\phi \mathcal{F}}{2} \tag{16.47}$$

and since the flux and the mmf are related by the expression

$$\phi = \frac{Ni}{\mathcal{R}} = \frac{\mathcal{F}}{\mathcal{R}} \tag{16.48}$$

the stored energy can be related to the reluctance of the structure according to

$$W_m = \frac{\phi^2 \mathcal{R}(x)}{2} \tag{16.49}$$

where the reluctance has been explicitly shown to be a function of displacement, as is the case in a moving-iron transducer. Finally, then, we shall use the following approximate expression to compute the magnetic force acting on the moving iron:

$$\boxed{f = -\frac{dW_m}{dx} = -\frac{\phi^2}{2}\frac{d\mathcal{R}(x)}{dx}} \tag{16.50}$$

The following examples illustrate the application of this approximate technique for the computation of forces and currents (the two problems of practical engineering interest to the user of such electromechanical systems) in some common devices.

---

## EXAMPLE  16.9  An Electromagnet

### Problem

An electromagnet is used to support a solid piece of steel, as shown in Figure 16.38. Determine the minimum coil current required to support the weight for a given air gap.

---

### Solution

*Known Quantities:*  Force required to support weight; cross-sectional area of magnetic core; air gap dimension, number of coil turns.

**Find:** Coil current, $i$.

**Schematics, Diagrams, Circuits, and Given Data:** $F = 8{,}900$ N; $A = 0.01$ m$^2$; $x = 0.0015$ m.

**Assumptions:** Assume that the reluctance of the iron is negligible; neglect fringing.

**Analysis:** To compute the current we need to derive an expression for the force in the air gap. Using equation 16.50, we see that we need to compute the reluctance of the structure and the magnetic flux to derive an expression for the force.

Since we are neglecting the iron reluctance, we can write the expression for the reluctance as follows:

$$\mathcal{R}(x) = \frac{2x}{\mu_0 A} = \frac{2x}{4\pi \times 10^7 \times 0.01} = \frac{2x}{4\pi \times 10^{-7} \times 0.01} = 1.59 \times 10^8 x = \alpha x \; \text{A} \cdot \text{t/Wb}$$

Knowing the reluctance we can calculate the magnetic flux in the structure as a function of the coil current:

$$\phi = \frac{Ni}{\mathcal{R}(x)} = \frac{Ni}{\alpha x}$$

and the magnitude of the force in the air gap is given by the expression

$$|f| = \frac{\phi^2}{2} \frac{d\mathcal{R}(x)}{dx} = \frac{(Ni)^2}{2\alpha^2 x^2}\alpha = \frac{N^2 i^2}{2\alpha x^2}$$

Solving for the current, we calculate:

$$i^2 = \frac{2\alpha x^2 |f|}{N^2} = \frac{2 \times 1.59 \times 10^8 \times (0.0015)^2 \times 8{,}900}{700^2} = 13 \; \text{A}$$

$$i = \pm 3.6 \; \text{A}$$

**Comments:** As the air gap becomes smaller, the reluctance of the air gap decreases, to the point where the reluctance of the iron cannot be neglected. When the air gap is zero, the required, or *holding*, current is a minimum. Conversely, if the bar is initially positioned at a substantial distance from the electromagnet, the initial current required to exert the required force will be significantly larger than that computed in this example.

---

One of the more common practical applications of the concepts discussed in this section is the **solenoid.** Solenoids find application in a variety of electrically controlled valves. The action of a solenoid valve is such that when it is energized, the plunger moves in such a direction as to permit the flow of a fluid through a conduit, as shown schematically in Figure 16.39.

The following examples illustrate the calculations involved in the determination of forces and currents in a solenoid.

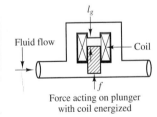

Force acting on plunger with coil energized

**Figure 16.39** Application of the solenoid as a valve

---

## EXAMPLE  16.10  A Solenoid

### Problem

Figure 16.40 depicts a simplified representation of a solenoid. The restoring force for the plunger is provided by a spring.

1.  Derive a general expression for the force exerted on the plunger as a function of the plunger position, $x$.

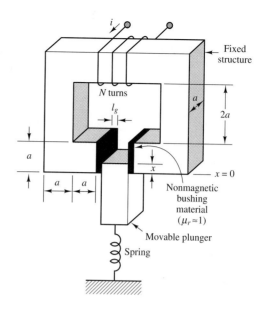

**Figure 16.40** A solenoid

2.  Determine the mmf required to pull the plunger to its end position $(x = a)$.

---

### Solution

**Known Quantities:** Geometry of magnetic structure; spring constant.

**Find:** $f$; mmf.

**Schematics, Diagrams, Circuits, and Given Data:** $a = 0.01$ m; $l_{gap} = 0.001$ m; $k = 1$ N/m.

**Assumptions:** Assume that the reluctance of the iron is negligible; neglect fringing. At $x = 0$ the plunger is in the gap by an infinitesimal displacement, $\varepsilon$.

### Analysis:

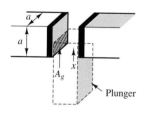

**Figure 16.41**

1.  *Force on the plunger.* To compute a general expression for the magnetic force exerted on the plunger, we need to derive an expression for the force in the air gap. Using equation 16.50, we see that we need to compute the reluctance of the structure and the magnetic flux to derive an expression for the force.

    Since we are neglecting the iron reluctance, we can write the expression for the reluctance as follows. Note that the area of the gap is variable, depending on the position of the plunger, as shown in Figure 16.41.

$$\mathcal{R}_{gap}(x) = 2 \times \frac{l_{gap}}{\mu_0 A_{gap}} = \frac{2l_{gap}}{\mu_0 a x}$$

The derivative of the reluctance with respect to the displacement of the plunger can then be computed to be:

$$\frac{d\mathcal{R}_{gap}(x)}{dx} = \frac{-2l_{gap}}{\mu_0 a x^2}$$

Knowing the reluctance, we can calculate the magnetic flux in the structure as a function of the coil current:

$$\phi = \frac{Ni}{\mathcal{R}(x)} = \frac{Ni\mu_0 ax}{2l_{gap}}$$

The force in the air gap is given by the expression

$$f_{gap} = \frac{\phi^2}{2}\frac{d\mathcal{R}(x)}{dx} = \frac{(Ni\mu_0 ax)^2}{8l_{gap}^2}\frac{(-2l_{gap})}{\mu_0 ax^2} = -\frac{\mu_0 a(Ni)^2}{4l_{gap}}$$

Thus, the force in the gap is proportional to the square of the current, and does not vary with plunger displacement.

2. *Calculation of magnetomotive force.* To determine the required magnetomotive force, we observe that the magnetic force must overcome the mechanical (restoring) force generated by the spring. Thus, $f_{gap} = kx = ka$. For the stated values, $f_{gap} = (10 \text{ N/m}) \times (0.01 \text{ m}) = 0.1$ N, and

$$Ni = \sqrt{\frac{4l_{gap}f_{gap}}{\mu_0 a}} = \sqrt{\frac{4 \times 0.001 \times 0.1}{4\pi \times 10^{-7} \times 0.01}} = 56.4 \text{ A} \cdot \text{t}$$

The required mmf can be most effectively realized by keeping the current value relatively low, and using a large number of turns.

**Comments:** The same mmf can be realized with an infinite number of combinations of current and number of turns; however, there are trade-offs involved. If the current is very large (and the number of turns small), the required wire diameter will be very large. Conversely, a small current will require a small wire diameter and a large number of turns. A homework problem explores this trade-off.

---

## EXAMPLE  16.11  Transient Response of a Solenoid

### Problem

Analyze the current response of the solenoid of Example 16.10 to a step change in excitation voltage. Plot the force and current as a function of time.

---

### Solution

**Known Quantities:** Coil inductance and resistance; applied current.

**Find:** Current and force response as a function of time.

**Schematics, Diagrams, Circuits, and Given Data:** See Example 16.10. $N = 1000$ turns. $V = 12$ V. $R_{coil} = 5 \; \Omega$.

**Assumptions:** The inductance of the solenoid is approximately constant, and is equal to the midrange value (plunger displacement equal to $a/2$).

**Analysis:** From Example 16.10, we have an expression for the reluctance of the solenoid:

$$\mathcal{R}_{gap}(x) = \frac{2l_{gap}}{\mu_0 ax}$$

Using equation 16.30, and assuming $x = a/2$, we calculate the inductance of the structure:

$$L \approx \frac{N^2}{\mathcal{R}_{gap}|_{x=a/2}} = \frac{N^2\mu_0 a^2}{4l_{gap}} = \frac{10^6 \times 4\pi \times 10^{-7} \times 10^{-4}}{4 \times 10^{-3}} = 31.4 \text{ mH}$$

The equivalent solenoid circuit is shown in Figure 16.42. When the switch is closed, the solenoid current rises exponentially with time constant $\tau = L/R = 6.3$ ms. As shown in Chapter 5, the response is of the form:

$$i(t) = \frac{V}{R}(1 - e^{-t/\tau}) = \frac{V}{R}(1 - e^{-Rt/L}) = \frac{12}{5}(1 - e^{-t/6.3 \times 10^{-3}}) \quad \text{A}$$

To determine how the magnetic force responds during the turn-on transient, we return to the expression for the force derived in Example 16.10:

$$f_{\text{gap}}(t) = \frac{\mu_0 a (Ni)^2}{4 l_{\text{gap}}} = \frac{4\pi \times 10^{-7} \times 10^{-2} \times 10^6}{4 \times 10^{-3}} i^2(t) = \pi i^2(t)$$

$$= \pi \left[ \frac{12}{5}(1 - e^{-t/6.3 \times 10^{-3}}) \right]^2$$

The two curves are plotted in Figure 16.42(b).

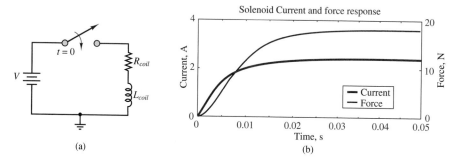

**Figure 16.42** Solenoid equivalent electrical circuit and step response

**Comments:** The assumption that the inductance is approximately constant is not quite accurate. The reluctance (and therefore the inductance) of the structure will change as the plunger moves into position. However, allowing for the inductance to be a function of plunger displacement causes the problem to become nonlinear, and requires numerical solution of the differential equation (i.e., the transient response results of Chapter 5 no longer apply). This issue is explored in a homework problem.

---

## Practical Facts About Solenoids

Solenoids can be used to produce linear or rotary motion, either in the *push* or *pull* mode. The most common solenoid types are listed below:

1. *Single-action linear* (push or pull). Linear stroke motion, with a restoring force (from a spring, for example) to return the solenoid to the neutral position.
2. *Double-acting linear.* Two solenoids back to back can act in either direction. Restoring force is provided by another mechanism (e.g., a spring).
3. *Mechanical latching solenoid* (bistable). An internal latching mechanism holds the solenoid in place against the load.
4. *Keep solenoid.* Fitted with a permanent magnet so that no power is needed to hold the load in the pulled-in position. Plunger is released by applying a current pulse of opposite polarity to that required to pull in the plunger.
5. *Rotary solenoid.* Constructed to permit rotary travel. Typical range is 25 to 95°. Return action via mechanical means (e.g., a spring).
6. *Reversing rotary solenoid.* Rotary motion is from one end to the other; when the solenoid is energized again it reverses direction.

Solenoid power ratings are dependent primarily on the current required by the coil, and on the coil resistance. $I^2R$ is the primary power sink, and solenoids are therefore limited by the heat they can dissipate. Solenoids can operated in continuous or pulsed mode. The power rating depends on the mode of operation, and can be increased by adding *hold-in resistors* to the circuit to reduce the *holding current* required for continuous operation. The hold resistor is switched into the circuit once the *pull-in* current required to pull the plunger has been applied, and the plunger has moved into place. The holding current can be significantly smaller than the pull-in current.

A common method to reduce the solenoid holding current employs a normally closed (NC) switch in parallel with a hold-in resistor. In Figure 16.43, when the push button (PB) closes the circuit, full voltage is applied to the solenoid coil, bypassing the resistor through the NC switch, connecting the resistor in series with the coil. The resistor will now limit the current to the value required to hold the solenoid in position.

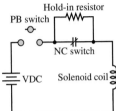

**Figure 16.43**

Another electromechanical device that finds common application in industrial practice is the **relay**. The relay is essentially an electromechanical switch that permits the opening and closing of electrical contacts by means of an electromagnetic structure similar to those discussed earlier in this section.

A relay such as would be used to start a high-voltage single-phase motor is shown in Figure 16.44. The magnetic structure has dimensions equal to 1 cm on all sides, and the transverse dimension is 8 cm. The relay works as follows. When the push button is pressed, an electrical current flows through the coil and generates a field in the magnetic structure. The resulting force draws the movable part toward the fixed part, causing an electrical contact to be made. The advantage of the relay is that a relatively low-level current can be used to control the opening and closing

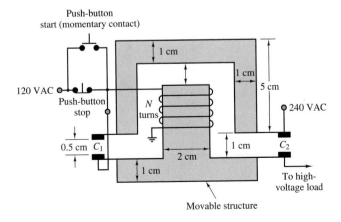

**Figure 16.44** A relay

of a circuit that can carry large currents. In this particular example, the relay is energized by a 120-VAC contact, establishing a connection in a 240-VAC circuit. Such relay circuits are commonly employed to remotely switch large industrial loads.

Circuit symbols for relays are shown in Figure 16.45. An example of the calculations that would typically be required in determining the mechanical and electrical characteristics of a simple relay are given in Example 16.12.

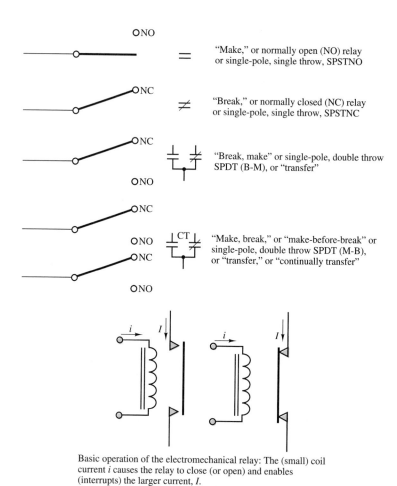

Basic operation of the electromechanical relay: The (small) coil current $i$ causes the relay to close (or open) and enables (interrupts) the larger current, $I$.
On the left: SPSTNO relay (magnetic field causes relay to close).
On the right: SPSTNC relay (magnetic field causes relay to open).

**Figure 16.45** Circuit symbols and basic operation of relays

## EXAMPLE 16.12 A Relay

### Problem

Figure 16.46 depicts a simplified representation of a relay. Determine the current required for the relay to make contact (i.e., pull in the ferromagnetic plate) from a distance $x$.

## Solution

**Known Quantities:** Relay geometry; restoring force to be overcome; distance between bar and relay contacts; number of coil turns.

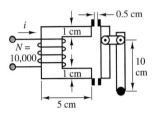

**Figure 16.46**

**Find:** $i$.

**Schematics, Diagrams, Circuits, and Given Data:** $A_{gap} = (0.01 \text{ m})^2$; $x = 0.05$ m; $f_{restore} = 5$ N; $N = 10,000$.

**Assumptions:** Assume that the reluctance of the iron is negligible; neglect fringing.

**Analysis:**

$$\mathcal{R}_{gap}(x) = \frac{2x}{\mu_0 A_{gap}}$$

The derivative of the reluctance with respect to the displacement of the plunger can then be computed to be:

$$\frac{d\mathcal{R}_{gap}(x)}{dx} = \frac{2}{\mu_0 A_{gap}}$$

Knowing the reluctance, we can calculate the magnetic flux in the structure as a function of the coil current:

$$\phi = \frac{Ni}{\mathcal{R}(x)} = \frac{Ni\mu_0 A_{gap}}{2}$$

and the force in the air gap is given by the expression

$$f_{gap} = \frac{\phi^2}{2}\frac{d\mathcal{R}(x)}{dx} = \frac{(Ni\mu_0 A_{gap})^2}{8}\frac{2}{\mu_0 A_{gap}} = \frac{\mu_0 A_{gap}(Ni)^2}{4}$$

The magnetic force must overcome a mechanical holding force of 5 N, thus,

$$f_{gap} = \frac{\mu_0 A_{gap}(Ni)^2}{4} = f_{restore} = 5 \text{ N}$$

or

$$i = \frac{1}{N}\sqrt{\frac{4f_{restore}}{\mu_0 A_{gap}}} = \frac{1}{10,000}\sqrt{\frac{20}{4\pi \times 10^{-7} \times 0.0001}} = 39.9 \text{ A}$$

**Comments:** The current required to close the relay is much larger than that required to hold the relay closed, because the reluctance of the structure is much smaller once the gap is reduced to zero.

## Moving-Coil Transducers

Another important class of electromagnetomechanical transducers is that of **moving-coil transducers.** This class of transducers includes a number of common devices, such as microphones, loudspeakers, and all electric motors and generators. The aim of this section is to explain the relationship between a fixed magnetic field, the emf across the moving coil, and the forces and motions of the moving element of the transducer.

The basic principle of operation of electromechanical transducers was presented in Section 16.1, where we stated that a magnetic field exerts a force on a charge moving through it. The equation describing this effect is

$$\mathbf{f} = q\mathbf{u} \times \mathbf{B} \tag{16.51}$$

which is a vector equation, as explained earlier. In order to correctly interpret equation 16.51, we must recall the right-hand rule and apply it to the transducer, illustrated in Figure 16.47, depicting a structure consisting of a sliding bar which makes contact with a fixed conducting frame. Although this structure does not represent a practical actuator, it will be a useful aid in explaining the operation of moving-coil transducers such as motors and generators. In Figure 16.47, and in all similar figures in this section, a small cross represents the "tail" of an arrow pointing into the page, while a dot represents an arrow pointing out of the page; this convention will be useful in visualizing three-dimensional pictures.

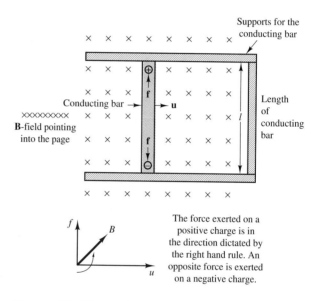

**Figure 16.47** A simple electromechanical motion transducer

### Motor Action

A moving-coil transducer can act as a motor when an externally supplied current flowing through the electrically conducting part of the transducer is converted into a force that can cause the moving part of the transducer to be displaced. Such a current would flow, for example, if the support of Figure 16.47 were made of conducting material, so that the conductor and the right-hand side of the support "rail" were to form a loop (in effect, a 1-turn coil). In order to understand the effects of this current flow in the conductor, one must consider the fact that a charge moving at a velocity $u'$ (along the conductor and perpendicular to the velocity of the conducting bar, as shown in Figure 16.48) corresponds to a current $i = dq/dt$ along the length $l$ of the conductor. This fact can be explained by considering the current $i$ along a differential element $dl$ and writing

$$i\,dl = \frac{dq}{dt} \cdot u'\,dt \tag{16.52}$$

since the differential element $dl$ would be traversed by the current in time $dt$ at a velocity $u'$. Thus we can write

$$i\,dl = dq\,u' \tag{16.53}$$

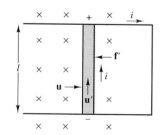

**Figure 16.48**

or

$$il = qu'$$                                                                        (**16.54**)

for the geometry of Figure 16.48. From Section 16.1, the force developed by a charge moving in a magnetic field is, in general, given by

$$\mathbf{f} = q\mathbf{u} \times \mathbf{B}$$                                      (**16.55**)

For the term $qu'$ we can substitute $i\mathbf{l}$, to obtain

$$\mathbf{f}' = i\mathbf{l} \times \mathbf{B}$$                                    (**16.56**)

Using the right-hand rule, we determine that the force $\mathbf{f}'$ generated by the current $i$ is in the direction that would push the conducting bar to the left. The magnitude of this force is $f' = Bli$ if the magnetic field and the direction of the current are perpendicular. If they are not, then we must consider the angle $\gamma$ formed by $\mathbf{B}$ and $\mathbf{l}$; in the more general case,

$$\boxed{f' = Bli \sin \gamma}$$                                                   (**16.57**)

The phenomenon we have just described is sometimes referred to as the "***Bli* law.**"

### *Generator Action*

The other mode of operation of a moving-coil transducer occurs when an external force causes the coil (i.e., the moving bar, in Figure 16.47) to be displaced. This external force is converted to an emf across the coil, as will be explained in the following paragraphs.

Since positive and negative charges are forced in opposite directions in the transducer of Figure 16.47, a potential difference will appear across the conducting bar; this potential difference is the electromotive force, or emf. The emf must be equal to the force exerted by the magnetic field. In short, the electric force per unit charge (or electric field ) $e/l$ must equal the magnetic force per unit charge $f/q = Bu$. Thus, the relationship

$$\boxed{e = Blu}$$                                                               (**16.58**)

which holds whenever $\mathbf{B}$, $\mathbf{l}$, and $\mathbf{u}$ are mutually perpendicular, as in Figure 16.49. If equation 16.58 is analyzed in greater depth, it can be seen that the product $lu$ (length times velocity) is the area crossed per unit time by the conductor. If one visualizes the conductor as "cutting" the flux lines into the base in Figure 16.48, it can be concluded that the electromotive force is equal to the *rate at which the conductor "cuts" the magnetic lines of flux*. It will be useful for you to carefully absorb this notion of conductors cutting lines of flux, since this will greatly simplify understanding the material in this section and in the next chapter.

In general, $\mathbf{B}$, $\mathbf{l}$, and $\mathbf{u}$ are not necessarily perpendicular. In this case one needs to consider the angles formed by the magnetic field with the normal to the plane containing $\mathbf{l}$ and $\mathbf{u}$, and the angle between $\mathbf{l}$ and $\mathbf{u}$.. The former is the angle $\alpha$ of Figure 16.49, the latter the angle $\beta$ in the same figure. It should be apparent that the optimum values of $\alpha$ and $\beta$ are $0°$ and $90°$, respectively. Thus, most practical

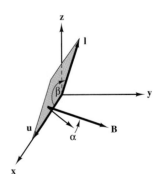

**Figure 16.49**

devices are constructed with these values of $\alpha$ and $\beta$. Unless otherwise noted, it will be tacitly assumed that this is the case. The "***Bli* law**" just illustrated explains how a moving conductor in a magnetic field can generate an electromotive force.

To summarize the electromechanical energy conversion that takes place in the simple device of Figure 16.47, we must note now that the presence of a current in the loop formed by the conductor and the rail requires that the conductor move to the right at a velocity $u$ (*Blu* law), thus cutting the lines of flux and generating the emf that gives rise to the current $i$. On the other hand, the same current causes a force $f'$ to be exerted on the conductor (*Bli* law) in the direction opposite to the movement of the conductor. Thus, it is necessary that an *externally applied force* $f_{ext}$ exist to cause the conductor to move to the right with a velocity $u$. The external force must overcome the force $f'$. This is the basis of electromechanical energy conversion.

An additional observation we must make at this point is that the current $i$ flowing around a closed loop generates a magnetic field, as explained in Section 16.1. Since this additional field is generated by a one-turn coil in our illustration, it is reasonable to assume that it is negligible with respect to the field already present (perhaps established by a permanent magnet). Finally, we must consider that this coil links a certain amount of flux, which changes as the conductor moves from left to right. The area crossed by the moving conductor in time $dt$ is

$$dA = lu\,dt \tag{16.59}$$

so that if the flux density, $B$, is uniform, the rate of change of the flux linked by the one-turn coil is

$$\frac{d\phi}{dt} = B\frac{dA}{dt} = Blu \tag{16.60}$$

In other words, *the rate of change* of the flux linked by the conducting loop is equal to the emf generated in the conductor. The student should realize that this statement simply confirms Faraday's law.

It was briefly mentioned that the *Blu* and *Bli* laws indicate that, thanks to the coupling action of the magnetic field, a conversion of mechanical to electrical energy—or the converse—is possible. The simple structures of Figures 16.47 and 16.48 can, again, serve as an illustration of this energy-conversion process, although we have not yet indicated how these idealized structures can be converted into a practical device. In this section we shall begin to introduce some physical considerations. Before we proceed any further, we should try to compute the power—electrical and mechanical—that is generated (or is required) by our ideal transducer. The electrical power is given by

$$P_E = ei = Blui \qquad \text{(W)} \tag{16.61}$$

while the mechanical power required, say, to move the conductor from left to right is given by the product of force and velocity:

$$P_M - f_{ext}u = Bliu \qquad \text{(W)} \tag{16.62}$$

The principle of conservation of energy thus states that in this ideal (lossless) transducer we can convert a given amount of electrical energy into mechanical energy, or vice versa. Once again we can utilize the same structure of Figure 16.47 to

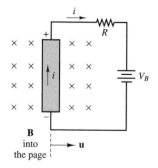

**B** into the page

**Figure 16.50** Motor and generator action in an ideal transducer

illustrate this reversible action. If the closed path containing the moving conductor is now formed from a closed circuit containing a resistance $R$ and a battery, $V_B$, as shown in Figure 16.50, the externally applied force, $f_{ext}$, generates a positive current $i$ into the battery provided that the emf is greater than $V_B$. When $e = Blu > V_B$, the ideal transducer acts as a *generator*. For any given set of values of $B, l, R$, and $V_B$, there will exist a velocity $u$ for which the current $i$ is positive. If the velocity is lower than this value—i.e., if $e = Blu < V_B$—then the current $i$ is negative, and the conductor is forced to move to the right. In this case the battery acts as a source of energy and the transducer acts as a *motor* (i.e., electrical energy drives the mechanical motion).

In practical transducers, we must be concerned with the inertia, friction, and elastic forces that are invariably present on the mechanical side of the transducer. Similarly, on the electrical side we must account for the inductance of the circuit, its resistance, and possibly some capacitance. Consider the structure of Figure 16.51. In the figure, the conducting bar has been placed on a surface with coefficient of sliding friction $d$; it has a mass $m$ and is attached to a fixed structure by means of a spring with spring constant $k$. The equivalent circuit representing the coil inductance and resistance is also shown.

If we recognize that $u = dx/dt$ in the figure, we can write the equation of motion for the conductor as:

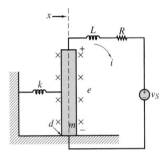

$$f + m\frac{du}{dt} + du + \frac{l}{k}\int u\, dt = f' = Bli \tag{16.63}$$

**Figure 16.51** A more realistic representation of the transducer of Figure 16.50

where the $Bli$ term represents the driving input that causes the mass to move. The driving input in this case is provided by the electrical energy source, $v_S$; thus the transducer acts as a motor, and $f$ is the net force acting on the mass of the conductor. On the electrical side, the circuit equation is:

$$v_S - L\frac{di}{dt} - Ri = e = Blu \tag{16.64}$$

Equations 16.63 and 16.64 could then be solved by knowing the excitation voltage, $v_S$, and the physical parameters of the mechanical and electrical circuits. For example, if the excitation voltage were sinusoidal, with

$$v_S(t) = V_S \cos \omega t \tag{16.65}$$

and the field density were constant:

$$B = B_0$$

we could postulate sinusoidal solutions for the transducer velocity, $u$, and current, $i$:

$$u = U \cos(\omega t + \theta_u) \qquad i = I \cos(\omega t + \theta_i) \tag{16.66}$$

and use phasor notation to solve for the unknowns $(U, I, \theta_u, \theta_i)$.

The results obtained in the present section apply directly to transducers that are based on translational (linear) motion. These basic principles of electromechanical energy conversion and the analysis methods developed in the section will be applied to practical transducers in a few examples.

The methods introduced in this section will later be applied in Chapters 17 and 18 to analyze rotating transducers, that is, electric motors and generators.

**FOCUS ON
MEASUREMENTS**

## Seismic Transducer

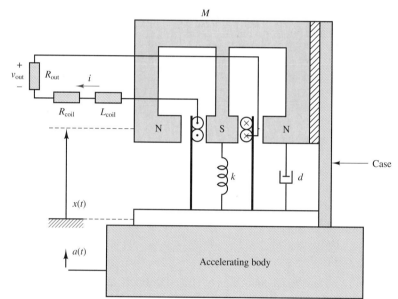

FIND IT
ON THE WEB

The device shown in Figure 16.52 is called a **seismic transducer** and can be used to measure the displacement, velocity, or acceleration of a body. The permanent magnet of mass $m$ is supported on the case by a spring, $k$, and there is some viscous damping, $d$, between the magnet and the case; the coil is fixed to the case. You may assume that the coil has length $l$ and resistance and inductance $R_{coil}$ and $L_{coil}$, respectively; the magnet exerts a magnetic field $B$. Find the transfer function between the output voltage, $v_{out}$, and the acceleration of the body, $a(t)$. Note that $x(t)$ is not equal to zero when the system is at rest. We shall ignore this offset displacement.

**Figure 16.52** An electromagnetomechanical seismic transducer

### Solution:

First we apply KVL around the electrical circuit to write the differential equation describing the electrical system:

$$L\frac{di}{dt} + (R_{coil} + R_{out})i + Bl\frac{dx}{dt} = 0$$

Also note that $v_{out} = -R_{out}i$. Next, we write the differential equation describing the mechanical system. The magnet experiences an inertial force due to the acceleration of the supporting body, $a(t)$, and to its own relative acceleration, $d^2x/dt^2$; thus, we can sketch a free-body diagram and apply Newton's second law to the permanent magnet, as shown in the sketch.

$$M\left(a + \frac{d^2x}{dt^2}\right) + d\frac{dx}{dt} + kx = Bli$$

Finally, using the Laplace transform, we determine the transfer function from $A(s)$ to $V_{\text{out}}(s)$. Let $R = R_{\text{coil}} + R_{\text{out}}$. Then

$$(Ls + R)I(s) + BlsX(s) = 0$$
$$BlI(s) - (Ms^2 + Ds + K)X(s) = MA(s)$$

Since we need the transfer function from $A$ to $V_{\text{out}}$, we use the expression

$$V_{\text{out}}(s) = -R_{\text{out}}I(s)$$

and, after some algebra, find that

$$I(s) = \frac{MBls\,A(s)}{(Ls + R)(Ms^2 + Ds + K) + B^2l^2s}$$

or

$$\frac{V_{\text{out}}(s)}{A(s)} = \frac{-MBs\,R_{\text{out}}}{(Ls + R)(Ms^2 + Ds + K) + B^2l^2s}$$

## EXAMPLE   16.13  A Loudspeaker

### Problem

A loudspeaker, shown in Figure 16.53, uses a permanent magnet and a moving coil to produce the vibrational motion that generates the pressure waves we perceive as sound. Vibration of the loudspeaker is caused by changes in the input current to a coil; the coil is, in turn, coupled to a magnetic structure that can produce time-varying forces on the speaker diaphragm. A simplified model for the mechanics of the speaker is also shown in Figure 16.53. The force exerted on the coil is also exerted on the mass of the speaker diaphragm, as shown in Figure 16.54, which depicts a free-body diagram of the forces acting on the loudspeaker diaphragm.

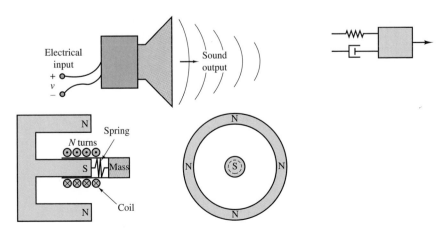

**Figure 16.53** Loudspeaker

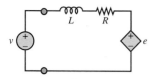

**Figure 16.55** Model of transducer electrical side

The force exerted on the mass, $f_i$, is the magnetic force due to current flow in the coil. The electrical circuit that describes the coil is shown in Figure 16.55, where $L$ represents the inductance of the coil, $R$ represents the resistance of the windlings, and $e$ is the emf induced by the coil moving through the magnetic field.

Determine the frequency response, $U(j\omega)/V(j\omega)$ of the speaker.

---

### Solution

**Known Quantities:** Circuit and mechanical parameters; magnetic flux density; number of coil turns; coil radius.

**Find:** Frequency response of loudspeaker, $U(j\omega)/V(j\omega)$.

**Schematics, Diagrams, Circuits, and Given Data:** Coil radius $= 0.05$ m; $L = 10$ mH; $R = 8$ $\Omega$; $m = 0.001$ kg; $d = 22.75$ N $\cdot$ s$^2$/m; $k = 5 \times 10^5$ N/m; $N = 47$; $B = 1$ T.

**Analysis:** To determine the frequency response of the loudspeaker, we write the differential equations that describe the electrical and mechanical subsystems. We apply KVL to the electrical circuit, using the circuit model of Figure 16.55, in which we have represented the $Blu$ term (motional voltage) in the form of a *back electromotive force*, $e$:

$$v - L\frac{di}{dt} - Ri - e = 0$$

or

$$L\frac{di}{dt} + Ri + Blu = v$$

Next, we apply Newton's second law to the mechanical system, consisting of: a lumped mass representing the mass of the moving diaphragm, $m$; an elastic (spring) term, which represents the elasticity of the diaphragm, $k$; and a damping coefficient, $d$, representing the frictional losses and aerodynamic damping affecting the moving diaphragm.

$$m\frac{du}{dt} = f_i - f_d - f_k = f_i - du - kx$$

where $f_i = Bli$ and therefore

$$-Bli + m\frac{du}{dt} + du + k\int_{-\infty}^{t} u(t')\,dt' = 0$$

Note that the two equations are *coupled*, that is, a mechanical variable appears in the electrical equation (the velocity $u$ in the $Blu$ term), and an electrical variable appears in the mechanical equation (the current $i$ in the $Bli$ term).

To derive the frequency response we Laplace-transform the two equations to obtain:

$$(sL + R)I(s) + BlU(s) = V(s)$$

$$-BlI(s) + \left(sm + d + \frac{k}{s}\right)U(s) = 0$$

We can write the above equations in matrix form and resort to Cramer's rule to solve for $U(s)$ as a function of $V(s)$:

$$\begin{bmatrix} (sL + R) & Bl \\ -Bl & \left(sm + d + \dfrac{k}{s}\right) \end{bmatrix} \begin{bmatrix} I(s) \\ U(s) \end{bmatrix} = \begin{bmatrix} V(s) \\ 0 \end{bmatrix}$$

with solution

$$U(s) = \frac{\det \begin{bmatrix} (sL + R) & V(s) \\ -Bl & 0 \end{bmatrix}}{\det \begin{bmatrix} (sL + R) & Bl \\ -Bl & \left(sm + d + \dfrac{k}{s}\right) \end{bmatrix}}$$

or

$$\frac{U(s)}{V(s)} = \frac{-Bl}{(sL + R)\left(sm + d + \dfrac{k}{s}\right) + (Bl)^2}$$

$$= \frac{-Bls}{(Lm)s^3 + (Rm + Ld)s^2 + (Rd + kL + (Bl)^2)s + (kR)}$$

To determine the frequency response of the loudspeaker, we let $s \rightarrow j\omega$ in the above expression:

$$\frac{U(j\omega)}{V(j\omega)} = \frac{-jBl\omega}{(kR) - (Rm + Ld)\omega^2 + j[(Rd + kL + (Bl)^2)\omega - (Lm)\omega^3]}$$

where $l = 2\pi Nr$, and substitute the appropriate numerical parameters:

$$\frac{U(j\omega)}{V(j\omega)} = \frac{-j14.8\omega}{(5 \times 10^5) - (0.008 + 0.2275)\omega^2 + j[(182 + 5,000 + 218)\omega - (10^{-5})\omega^3]}$$

$$= \frac{-j14.8\omega}{(5 \times 10^5) - (0.2355)\omega^2 + j[(5.4 \times 10^3)\omega - (10^{-5})\omega^3]}$$

The resulting frequency response is plotted in Figure 16.56.

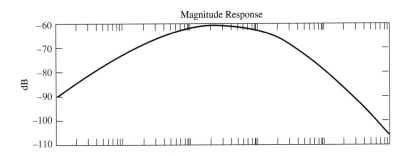

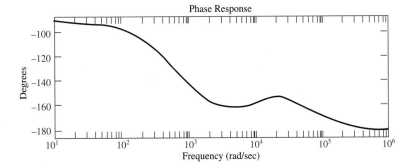

**Figure 16.56** Frequency response of loudspeaker

**Focus on Computer-Aided Tools:** A Matlab *m*-file containing the frequency response calculations leading to the frequency response (Bode) plot of Figure 16.56 may be found in the accompanying CD-ROM.

## Check Your Understanding

**16.13** The flux density of the earth's magnetic field is about $50\,\mu\text{T}$. Estimate the current required in a conductor of length 10 cm and mass 10 g to counteract the force of gravity if the wire is oriented in the optimum direction.

**16.14** In Example 16.13, we examined the frequency response of a loudspeaker. However, over a period of time, permanent magnets may become demagnetized. Find the frequency response of the same loudspeaker if the permanent magnet has lost its strength to a point where $B = 0.95$ T.

**16.15** In Example 16.10, a solenoid is used to exert force on a spring. Estimate the position of the plunger if the number of turns in the solenoid winding is 1,000 and the current going into the winding is 40 mA.

**16.16** For the circuit in Figure 16.47, the conducting bar is moving with a velocity of 6 m/s. The flux density is 0.5 Wb/m$^2$, and $l = 1.0$ m. Find the magnitude of the resulting induced voltage.

## CONCLUSION

- Magnetic fields form a coupling mechanism between electrical and mechanical systems, permitting the conversion of electrical energy to mechanical energy, and vice versa. The basic laws that govern such electromechanical energy conversion are Faraday's law, stating that a changing magnetic field can induce a voltage; and Ampère's law, stating that a current flowing through a conductor generates a magnetic field.

- The two fundamental variables in the analysis of magnetic structures are the magnetomotive force and the magnetic flux; if some simplifying approximations are made, these quantities are linearly related through the reluctance parameter, much in the same way as voltage and current are related through resistance according to Ohm's law. This simplified analysis permits approximate calculations of required forces and currents to be conducted with relative ease in magnetic structures.

- Magnetic materials are characterized by a number of nonideal properties, which should be considered in the detailed analysis of a magnetic structure. The most important phenomena are saturation, eddy currents, and hysteresis.

- Electromechanical transducers, which convert electrical signals to mechanical forces, or mechanical motion to electrical signals, can be analyzed according to the techniques presented in this chapter. Examples of such transducers are electromagnets, position and velocity sensors, relays, solenoids, and loudspeakers.

## CHECK YOUR UNDERSTANDING ANSWERS

| | |
|---|---|
| **CYU 16.1** | $e = -2.5$ V |
| **CYU 16.2** | $I = \pi$ A |
| **CYU 16.3** | $W_m = 0.648$ J |
| **CYU 16.5** | $\phi = 3.94 \times 10^{-6}$ Wb; $B = 0.0788$ Wb/m$^2$ |
| **CYU 16.6** | $\mathcal{R}_{\text{eq}} = 1.41 \times 10^6$ A · t/Wb |

**CYU 16.7**    $\mathcal{R}_{eq} = 22 \times 10^6 \; A \cdot t/Wb$

**CYU 16.8**    $\mathcal{R}_g = \mathcal{R}_1 = \mathcal{R}_2 = \mathcal{R}_3 = \delta/\mu_0(l + \delta)(w + \delta); \; \mathcal{F}_1 = Ni_1; \; \mathcal{F}_2 = Ni_2$

**CYU 16.9**    $\alpha = 5; \; I_1 = I_2/\alpha = 2.4 \; A$

**CYU 16.10**   $\alpha = 0.2$

**CYU 16.11**   $\eta = 98\%$

**CYU 16.12**   $\alpha = 10$

**CYU 16.13**   $i = 196 \times 10^2 \; A$

**CYU 16.14**   $U(j\omega)/V(j\omega) = 0.056(j\omega/15{,}950)/(1 + j\omega/15{,}950)(1 + j\omega/31{,}347)$

**CYU 16.15**   $x = 0.5 \; cm$

**CYU 16.16**   3 V

# HOMEWORK PROBLEMS

## Section 1: Electricity and Magnetism

**16.1**    An iron-core inductor has the following characteristic:

$$i = \lambda + 0.5\lambda^2$$

a.  Determine the energy, co-energy, and incremental inductance for $\lambda = 0.5 \; V \cdot s$.

b.  Given that the coil resistance is 1 $\Omega$ and that

$$i(t) = 0.625 + 0.01 \sin 400t \; A$$

determine the voltage across the terminals on the inductor.

**16.2**    For the electromagnet of Figure P16.2:

a.  Find the flux density in the core.

b.  Sketch the magnetic flux lines and indicate their direction.

c.  Indicate the north and south poles of the magnet.

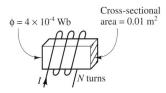

**Figure P16.2**

**16.3**    An iron-core inductor has the characteristic shown in Figure P16.3:

a.  Determine the energy and the incremental inductance for $i = 1.0 \; A$.

b.  Given that the coil resistance is 2 $\Omega$ and that $i(t) = 0.5 \sin 2\pi t$, determine the voltage across the terminals of the inductor.

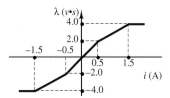

**Figure P16.3**

**16.4**    A single loop of wire carrying current $I_2$ is placed near the end of a solenoid having $N$ turns and carrying current $I_1$, as shown in Figure P16.4. The solenoid is fastened to a horizontal surface, but the single coil is free to move. With the currents directed as shown, is there a resultant force on the single coil? If so, in what direction? Why?

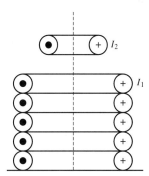

**Figure P16.4**

**16.5**    The electromagnet of Figure P16.5 has reluctance given by $\mathcal{R}(x) = 7 \times 10^8(0.002 + x) \; H^{-1}$, where $x$ is the length of the variable gap in meters. The coil has 980 turns and 30 $\Omega$ resistance. For an applied voltage of 120 VDC, find:

a.  The energy stored in the magnetic field for $x = 0.005 \; m$.

b. The magnetic force for $x = 0.005$ m.

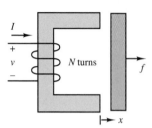

**Figure P16.5**

**16.6**  A practical LVDT is typically connected to a resistive load. Derive the LVDT equations in the presence of a resistive load, $R_L$, connected across the output terminals, using the results of "Focus on Measurements: Linear Variable Differential Transformer."

**16.7**  On the basis of the equations of "Focus on Measurements: Linear Variable Differential Transformer," and of the results of Problem 16.6, derive the frequency response of the LVDT, and determine the range of frequencies for which the device will have maximum sensitivity for a given excitation. [*Hint:* Compute $dv_{out}/dv_{ex}$, and set the derivative equal to zero to determine the maximum sensitivity.]

**16.8**  A wire of length 20 cm vibrates in one direction in a constant magnetic field with a flux density of 0.1 T; see Figure P16.8. The position of the wire as a function of time is given by $x(t) = 0.1 \sin 10t$ m. Find the induced emf across the length of the wire as a function of time.

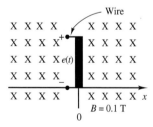

**Figure P16.8**

**16.9**  The wire of Problem 16.8 induces a time-varying emf of

$$e_1(t) = 0.02 \cos 10t$$

A second wire is placed in the same magnetic field but has a length of 0.1 m, as shown in Figure P16.9. The position of this wire is given by $x(t) = 1 - 0.1 \sin 10t$. Find the induced emf $e(t)$ defined by the difference in emf's $e_1(t)$ and $e_2(t)$.

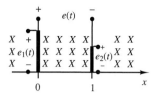

**Figure P16.9**

**16.10**  A conducting bar shown in Figure 16.48 in the text, is carrying 4 A of current in the presence of a magnetic field; $B = 0.3$ Wb/m$^2$. Find the magnitude and direction of the force induced on the conducting bar.

**16.11**  A wire, shown in Figure P16.11, is moving in the presence of a magnetic field, with $B = 0.4$ Wb/m$^2$. Find the magnitude and direction of the induced voltage in the wire.

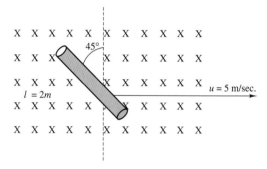

**Figure P16.11**

## Section 2: Magnetic Circuits

**16.12**
  a. Find the reluctance of a magnetic circuit if a magnetic flux $\phi = 4.2 \times 10^{-4}$ Wb is established by an impressed mmf of 400 A $\cdot$ t.
  b. Find the magnetizing force, H, in SI units if the magnetic circuit is 6 inches in length.

**16.13**  For the circuit shown in Figure P16.13:
  a. Determine the reluctance values and show the magnetic circuit, assuming that $\mu = 3,000\mu_0$.
  b. Determine the inductance of the device.
  c. The inductance of the device can be modified by cutting an air gap in the magnetic structure. If a gap of 0.1 mm is cut in the arm of length $l_3$, what is the new value of inductance?
  d. As the gap is increased in size (length), what is the limiting value of inductance? Neglect leakage flux and fringing effects.

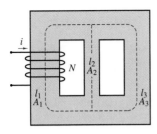

N = 100 turns    $A_2$ = 25 cm²

$l_1$ = 30 cm    $l_3$ = 30 cm

$A_1$ = 100 cm²    $A_3$ = 100 cm²

$l_2$ = 10 cm

**Figure P16.13**

**16.14** The magnetic circuit shown in Figure P16.14 has two parallel paths. Find the flux and flux density in each of the legs of the magnetic circuit. Neglect fringing at the air gaps and any leakage fields. $N$ = 1,000 turns, $i$ = 0.2 A, $l_{g1}$ = 0.02 cm, and $l_{g2}$ = 0.04 cm. Assume the reluctance of the magnetic core to be negligible.

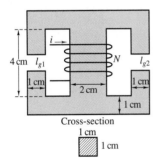

**Figure P16.14**

**16.15** Find the current necessary to establish a flux of $\phi = 3 \times 10^{-4}$ Wb in the series magnetic circuit of Figure P16.15. Here, $l_{iron} = l_{steel} = 0.3$ m, Area (throughout) = $5 \times 10^{-4}$ m², and $N$ = 100 turns.

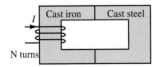

**Figure P16.15**

**16.16**

a. Find the current, $I$, required to establish a flux $\phi = 2.4 \times 10^{-4}$ Wb in the magnetic circuit of Figure P16.16. Here, Area(throughout) = $2 \times 10^{-4}$ m², $l_{ab} = l_{ef} = 0.05$ m, $l_{af} = l_{be} = 0.02$ m, $l_{bc} = l_{dc}$, and the material is sheet steel.

b. Compare the mmf drop across the air gap to that across the rest of the magnetic circuit. Discuss your results using the value of $\mu$ for each material.

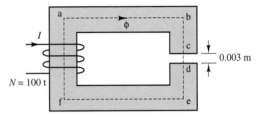

**Figure P16.16**

**16.17** Find the magnetic flux, $\phi$, established in the series magnetic circuit of Figure P16.17.

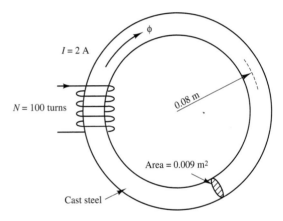

**Figure P16.17**

**16.18** For the series-parallel magnetic circuit of Figure P16.18, find the value of $I$ required to establish a flux in the gap of $\phi = 2 \times 10^{-4}$ Wb. Here, $l_{ab} = l_{bg} = l_{gh} = l_{ha} = 0.2$ m, $l_{bc} = l_{fg} = 0.1$ m, $l_{cd} = l_{ef} = 0.099$ m, and the material is sheet steel.

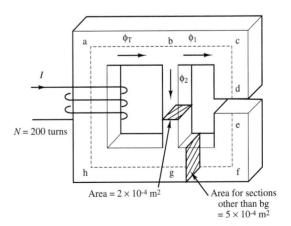

**Figure P16.18**

**16.19**  Refer to the actuator of Figure P16.19. The entire device is made of sheet steel. The coil has 2,000 turns. The armature is stationary so that the length of the air gaps, g = 10 mm, is fixed. A direct current passing through the coil produces a flux density of 1.2 T in the gaps. Determine:

a.  The coil current.

b.  The energy stored in the air gaps.

c.  The energy stored in the steel.

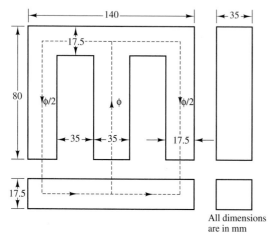

**Figure P16.19**

**16.20**  A core is shown in Figure P16.20, with $\mu_r = 2,000$ and $N = 100$. Find:

a.  The current needed to produce a flux density of 0.4 Wb/m² in the center leg.

b.  The current needed to produce a flux density of 0.8 Wb/m² in the center leg.

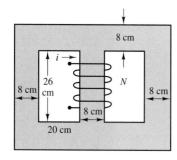

Cross-section
8 cm

**Figure P16.20**

## Section 3:  **Transformers**

**16.21**  For the transformer shown in Figure P16.21, $N = 1,000$ turns, $l_1 = 16$ cm, $A_1 = 4$ cm², $l_2 = 22$ cm, $A_2 = 4$ cm², $l_3 = 5$ cm, and $A_3 = 2$ cm². The relative permeability of the material is $\mu_r = 1,500$.

a.  Construct the equivalent magnetic circuit, and find the reluctance associated with each part of the circuit.

b.  Determine the self-inductance and mutual inductance for the pair of coils (i.e., $L_{11}$, $L_{22}$, and $M = L_{12} = L_{21}$).

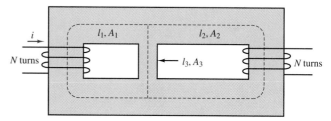

**Figure P16.21**

**16.22**  A transformer is delivering power to a 300-$\Omega$ resistive load. To achieve the desired power transfer, the turns ratio is chosen so that the resistive load referred to the primary is 7,500 $\Omega$. The parameter values, *referred to the secondary winding,* are:

$$r_1 = 20 \ \Omega \qquad L_1 = 1.0 \ \text{mH} \qquad L_m = 25 \ \text{mH}$$
$$r_2 = 20 \ \Omega \qquad L_2 = 1.0 \ \text{mH}$$

Core losses are negligible.

a.  Determine the turns ratio.

b.  Determine the input voltage, current, and power and the efficiency when this transformer is delivering 12 W to the 300-$\Omega$ load at a frequency $f = 10,000/2\pi$ Hz.

**16.23**  A 220/20-V transformer has 50 turns on its low-voltage side. Calculate

a.  The number of turns on its high side.

b.  The turns ratio $\alpha$ when it is used as a step-down transformer.

c.  The turns ratio $\alpha$ when it is used as a step-up transformer.

**16.24**  The high-voltage side of a transformer has 750 turns, and the low-voltage side 50 turns. When the high side is connected to a rated voltage of 120 V, 60 Hz, a rated load of 40 A is connected to the low side. Calculate

a.  The turns ratio.

b.  The secondary voltage (assuming no internal transformer impedance voltage drops).

c.  The resistance of the load.

**16.25**   A transformer is to be used to match an 8-Ω loudspeaker to a 500-Ω audio line. What is the turns ratio of the transformer, and what are the voltages at the primary and secondary terminals when 10 W of audio power is delivered to the speaker? Assume that the speaker is a resistive load and the transformer is ideal.

**16.26**   The high-voltage side of a step-down transformer has 800 turns, and the low-voltage side has 100 turns. A voltage of 240 VAC is applied to the high side, and the load impedance is 3 Ω (low side). Find

a. The secondary voltage and current.

b. The primary current.

c. The primary input impedance from the ratio of primary voltage and current.

d. The primary input impedance.

**16.27**   Calculate the transformer ratio of the transformer in Problem 16.26 when it is used as a step-up transformer.

**16.28**   A 2,300/240-V, 60-Hz, 4.6-kVA transformer is designed to have an induced emf of 2.5 V/turn. Assuming an ideal transformer, find

a. The number of high-side turns, $N_h$, and low-side turns, $N_l$.

b. The rated current of the high-voltage side, $I_h$.

c. The transformer ratio when the device is used as a step-up transformer.

## Section 4:
## Electromechanical Transducers

**16.29**   For the electromagnet of Example 16.9:

a. Calculate the current required to keep the bar in place. (*Hint:* The air gap becomes zero and the iron reluctance cannot be neglected.)

b. If the bar is initially 0.1 m away from the electromagnet, what initial current would be required to lift the magnet?

**16.30**   With reference to Example 16.10, determine the best combination of current magnitude and wire diameter to reduce the volume of the solenoid coil to a minimum. Will this minimum volume result in the lowest possible resistance? How does the power dissipation of the coil change with the wire gauge and current value? To solve this problem you will need to find a table of wire gauge diameter, resistance, and current ratings. Table 2.2 in this book contains some information. The solution can only be found numerically.

**16.31**   Derive the same result obtained in Example 16.10 using equation 16.46 and the definition of inductance given in equation 16.30. You will first compute the inductance of the magnetic circuit as a function of the reluctance, then compute the stored magnetic energy, and finally write the expression for the magnetic force given in equation 16.46.

**16.32**   Derive the same result obtained in Example 16.11 using equation 16.46 and the definition of inductance given in equation 16.30. You will first compute the inductance of the magnetic circuit as a function of the reluctance, then compute the stored magnetic energy, and finally write the expression for the magnetic force given in equation 16.46.

**16.33**   With reference to Example 16.11, generate a simulation program (e.g., using Simulink™) that accounts for the fact that the solenoid inductance is not constant, but is a function of plunger position. Compare graphically the current and force step responses of the constant-$L$ simplified solenoid model to the step responses obtained in Example 16.11.

**16.34**   With reference to Example 16.12, calculate the required holding current to keep the relay closed.

**16.35**   The relay circuit shown in Figure P16.35 has the following parameters: $A_{gap} = 0.001$ m²; $N = 500$ turns; $L = 0.02$ m; $\mu = \mu_0 = 4\pi \times 10^{-7}$ (neglect the iron reluctance); $k = 1000$ N/m, $R = 18$ Ω. What is the minimum DC supply voltage, $v$, for which the relay will make contact when the electrical switch is closed?

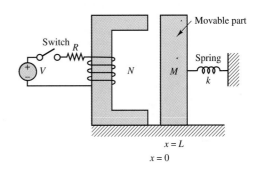

**Figure P16.35**

**16.36**   The magnetic circuit shown in Figure P16.36 is a very simplified representation of devices used as *surface roughness sensors*. The stylus is in contact with the surface and causes the plunger to move along with the surface. Assume that the flux $\phi$ in the gap is given by the expression $\phi = \beta/\mathcal{R}(x)$, where $\beta$ is a known constant and $\mathcal{R}(x)$ is the reluctance of the gap. The emf $e$ is measured to determine the surface profile. Derive an expression for the displacement $x$ as a function of the various parameters of the magnetic circuit and of the measured emf. (Assume a frictionless contact between the moving plunger and the magnetic structure and that the plunger is restrained to vertical motion only. The cross-sectional area of the plunger is $A$.)

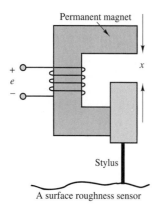

**Figure P16.36** A surface
roughness sensor

**16.37** The electrodynamic shaker shown in
Figure P16.37 is commonly used as a vibration tester.
A constant current is used to generate a magnetic field
in which the armature coil of length $l$ is immersed.
The shaker platform with mass $m$ is mounted in the
fixed structure by way of a spring with stiffness $k$. The
platform is rigidly attached to the armature coil, which
slides on the fixed structure thanks to frictionless
bearings.

a. Neglecting iron reluctance, determine the
   reluctance of the fixed structure, and hence
   compute the strength of the magnetic flux density,
   $B$, in which the armature coil is immersed.

b. Knowing $B$, determine the dynamic equations of
   motion of the shaker, assuming that the moving coil
   has resistance $R$ and inductance $L$.

c. Derive the transfer function and frequency response
   function of the shaker mass *velocity* in response to
   the input voltage $V_S$.

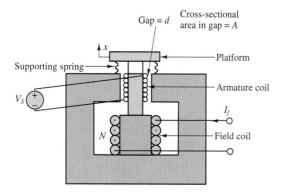

**Figure P16.37** Electrodynamic shaker

**16.38** A cylindrical solenoid is shown in Figure P16.38.
The plunger may move freely along its axis. The air gap
between the shell and the plunger is uniform and equal
to 1 mm, and the diameter, $d$, is 25 mm. If the exciting
coil carries a current of 7.5 A, find the force acting
on the plunger when $x = 2$ mm. Assume $N = 200$
turns, and neglect the reluctance of the steel shell.

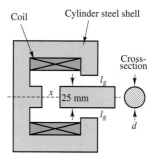

**Figure P16.38**

**16.39** The double-excited electromechanical system
shown in Figure P16.39 moves horizontally. Assuming
that resistance, magnetic leakage, and fringing are
negligible, the permeability of the core is very large,
and the cross section of the structure is $w \times w$, find

a. The reluctance of the magnetic circuit.

b. The magnetic energy stored in the air gap.

c. The force on the movable part as a function of its
   position.

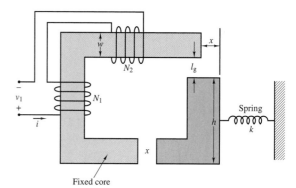

**Figure P16.39**

**16.40** Determine the force, $F$, between the faces of the
poles (stationary coil and plunger) of the solenoid
pictured in Figure P16.40 when it is energized. When
energized, the plunger is drawn into the coil and comes
to rest with only a negligible air gap separating the
two. The flux density in the cast steel pathway is 1.1 T.
The diameter of the plunger is 10 mm.

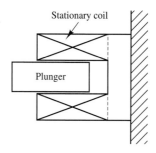

**Figure P16.40**

**16.41**  An electromagnet is used to support a solid piece of steel as shown in Example 15.10. A force of 10,000 N is required to support the weight. The cross-sectional area of the magnetic core (the fixed part) is 0.01 m². The coil has 1,000 turns. Determine the minimum current that can keep the weight from falling for $x = 1.0$ mm. Assume negligible reluctance for the steel parts and negligible fringing in the air gaps.

**16.42**  The armature, frame, and core of a 12-VDC control relay are made of sheet steel. The average length of the magnetic circuit is 12 cm when the relay is energized, and the average cross section of the magnetic circuit is 0.60 cm². The coil is wound with 250 turns and carries 50 mA. Determine:

a. The flux density, $B$, in the magnetic circuit of the relay when the coil is energized.

b. The force, $\mathcal{F}$, exerted on the armature to close it when the coil is energized.

**16.43**  Derive and sketch the frequency response of the loudspeaker of Example 16.13 for (1) $k = 50,000$ N/m and (2) $k = 5 \times 10^6$ N/m. Describe qualitatively how the loudspeaker frequency response changes as the spring stiffness, $k$, increases and decreases. What will the frequency response be in the limit as $k$ approaches zero? What kind of speaker would this condition correspond to?

**16.44**  A relay is shown in Figure P16.44. Find the differential equations describing the system.

**16.45**  A solenoid having a cross section of 5 cm² is shown in Figure P16.45.

a. Calculate the force exerted on the plunger when the distance $x$ is 2 cm and the current in the coil (where $N = 100$ turns) is 5 A. Assume that the fringing and leakage effects are negligible. The relative permeabilities of the magnetic material and the nonmagnetic sleeve are 2,000 and 1.

b. Develop a set of defferential equations governing the behavior of the solenoid.

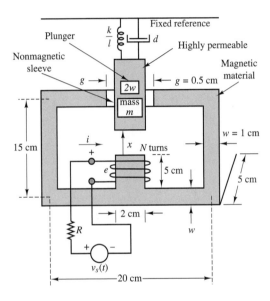

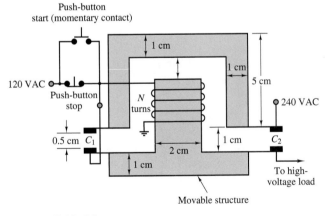

**Figure P16.44**

# C H A P T E R

# 17

# Introduction to Electric Machines

The objective of this chapter is to introduce the basic operation of rotating electric machines. The operation of the three major classes of electric machines—DC, synchronous, and induction—will first be described as intuitively as possible, building on the material presented in Chapter 16. The second part of the chapter will be devoted to a discussion of the applications and selection criteria for the different classes of machines.

The emphasis of this chapter will be on explaining the properties of each type of machine, with its advantages and disadvantages with regard to other types; and on classifying these machines in terms of their performance characteristics and preferred field of application. Chapter 18 will be devoted to a survey of special-purpose electric machines—many of which find common application in industry—such as stepper motors, brushless DC motors, switched reluctance motors, and single-phase induction motors. Selected examples and application notes will discuss some current issues of interest.

By the end of this chapter, you should be able to:

• Describe the principles of operation of DC and AC motors and generators.

• Interpret the nameplate data of an electric machine.

• Interpret the torque-speed characteristic of an electric machine.

• Specify the requirements of a machine given an application.

## 17.1 ROTATING ELECTRIC MACHINES

The range of sizes and power ratings and the different physical features of rotating machines are such that the task of explaining the operation of rotating machines in a single chapter may appear formidable at first. Some features of rotating machines, however, are common to all such devices. This introductory section is aimed at explaining the common properties of all rotating electric machines. We begin our discussion with reference to Figure 17.1, in which a hypothetical rotating machine is depicted in a cross-sectional view. In the figure, a box with a cross inscribed in it indicates current flowing into the page, while a dot represents current out of the plane of the page.

In Figure 17.1, we identify a **stator,** of cylindrical shape, and a **rotor,** which, as the name indicates, rotates inside the stator, separated from the latter by means of an air gap. The rotor and stator each consist of a magnetic core, some electrical insulation, and the windings necessary to establish a magnetic flux (unless this is created by a permanent magnet). The rotor is mounted on a bearing-supported shaft, which can be connected to *mechanical loads* (if the machine is a motor) or to a *prime mover* (if the machine is a generator) by means of belts, pulleys, chains, or other mechanical couplings. The windings carry the electric currents that generate the magnetic fields and flow to the electrical loads, and also provide the closed loops in which voltages will be induced (by virtue of Faraday's law, as discussed in the previous chapter).

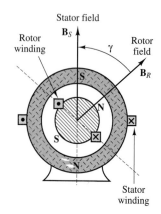

**Figure 17.1** A rotating electric machine

### Basic Classification of Electric Machines

An immediate distinction can be made between different types of windings characterized by the nature of the current they carry. If the current serves the sole purpose of providing a magnetic field and is independent of the load, it is called·a *magnetizing,* or excitation, current, and the winding is termed a **field winding.** Field currents are nearly always DC and are of relatively low power, since their only purpose is to magnetize the core (recall the important role of high-permeability cores in generating large magnetic fluxes from relatively small currents). On the other hand, if the winding carries only the load current, it is called an **armature.** In DC and AC synchronous machines, separate windings exist to carry field and armature currents. In the induction motor, the magnetizing and load currents flow in the same winding, called the *input winding,* or *primary;* the output winding is then called the *secondary.* As we shall see, this terminology, which is reminiscent of transformers, is particularly appropriate for induction motors, which bear a significant analogy to the operation of the transformers studied in Chapters 7 and 16. Table 17.1 characterizes the principal machines in terms of their field and armature configuration.

It is also useful to classify electric machines in terms of their energy-conversion characteristics. A machine acts as a **generator** if it converts mechanical energy from a prime mover—e.g., an internal combustion engine—to electrical form. Examples of generators are the large machines used in power-generating plants, or the common automotive alternator. A machine is classified as a **motor** if it converts electrical energy to mechanical form. The latter class of machines is probably of more direct interest to you, because of its widespread application in engineering practice. Electric motors are used to provide forces and torques to generate motion in countless industrial applications. Machine tools, robots, punches, presses, mills, and propulsion systems for electric vehicles are but a few examples of the application of electric machines in engineering.

Table 17.1   Configurations of the three types of electric machines

| Machine type | Winding | Winding type | Location | Current |
|---|---|---|---|---|
| DC | Input and output | Armature | Rotor | AC (winding) |
| | | | | DC (at brushes) |
| | Magnetizing | Field | Stator | DC |
| Synchronous | Input and output | Armature | Stator | AC |
| | Magnetizing | Field | Rotor | DC |
| Induction | Input | Primary | Stator | AC |
| | Output | Secondary | Rotor | AC |

Note that in Figure 17.1 we have explicitly shown the direction of two magnetic fields: that of the rotor, $\mathbf{B}_R$, and that of the stator, $\mathbf{B}_S$. Although these fields are generated by different means in different machines (e.g., permanent magnets, AC currents, DC currents), the presence of these fields is what causes a rotating machine to turn and enables the generation of electric power. In particular, we see that in Figure 17.1 the north pole of the rotor field will seek to align itself with the south pole of the stator field. It is this magnetic attraction force that permits the generation of torque in an electric motor; conversely, a generator exploits the laws of electromagnetic induction to convert a changing magnetic field to an electric current.

To simplify the discussion in later sections, we shall presently introduce some basic concepts that apply to all rotating electric machines. Referring to Figure 17.2, we note that for all machines the force on a wire is given by the expression

$$\mathbf{f} = i_w \mathbf{l} \times \mathbf{B} \tag{17.1}$$

where $i_w$ is the current in the wire, $\mathbf{l}$ is a vector along the direction of the wire, and $\times$ denotes the cross product of two vectors. Then the torque for a multiturn coil

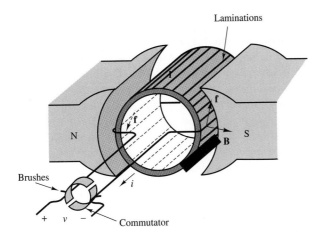

**Figure 17.2** Stator and rotor fields and the force acting on a rotating machine

becomes

$$T = K B i_w \sin \alpha \tag{17.2}$$

where

$B$ = magnetic flux density caused by the stator field

$K$ = constant depending on coil geometry

$\alpha$ = angle between **B** and the normal to the plane of the coil

In the hypothetical machine of Figure 17.2, there are two magnetic fields: one generated within the stator, the other within the rotor windings. Either (but not both) of these fields could be generated either by a current or by a permanent magnet. Thus, we could replace the permanent-magnet stator of Figure 17.2 with a suitably arranged winding to generate a stator field in the same direction. If the stator were made of a toroidal coil of radius $R$ (see Chapter 16), then the magnetic field of the stator would generate a flux density $B$, where

$$B = \mu H = \mu \frac{Ni}{2\pi R} \tag{17.3}$$

and where $N$ is the number of turns and $i$ is the coil current. The direction of the torque is always the direction determined by the rotor and stator fields as they seek to align to each other (i.e., counterclockwise in the diagram of Figure 17.1).

It is important to note that Figure 17.2 is merely a general indication of the major features and characteristics of rotating machines. A variety of configurations exist, depending on whether each of the fields is generated by a current in a coil or by a permanent magnet, and on whether the load and magnetizing currents are direct or alternating. The type of excitation (AC or DC) provided to the windings permits a first classification of electric machines (see Table 17.1). According to this classification, one can define the following types of machines:

- *Direct-current machines:* DC current in both stator and rotor
- *Synchronous machines:* AC current in one winding, DC in the other
- *Induction machines:* AC current in both

In most industrial applications, the induction machine is the preferred choice, because of the simplicity of its construction. However, the analysis of the performance of an induction machine is rather complex. On the other hand, DC machines are quite complex in their construction but can be analyzed relatively simply with the analytical tools we have already acquired. Therefore, the progression of this chapter will be as follows. We start with a section that discusses the physical construction of DC machines, both motors and generators. Then we continue with a discussion of synchronous machines, in which one of the currents is now alternating, since these can easily be understood as an extension of DC machines. Finally, we consider the case where both rotor and stator currents are alternating, and analyze the induction machine.

## Performance Characteristics of Electric Machines

As already stated earlier in this chapter, electric machines are **energy-conversion devices,** and we are therefore interested in their energy-conversion **efficiency.** Typical applications of electric machines as motors or generators must take into consideration the energy losses associated with these devices. Figure 17.3(a) and (b) represent the various loss mechanisms you must consider in analyzing the efficiency of an electric machine for the case of direct-current machines. It is important for you to keep in mind this conceptual flow of energy when analyzing electric machines. The sources of loss in a rotating machine can be separated into three fundamental groups: electrical ($I^2R$) losses, core losses, and mechanical losses.

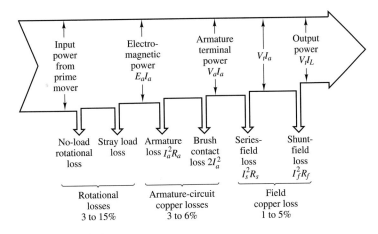

**Figure 17.3(a)** Generator losses, direct current

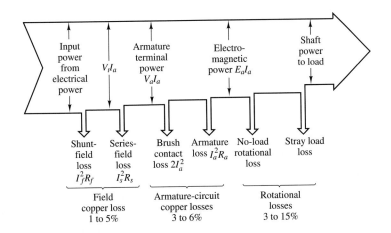

**Figure 17.3(b)** Motor losses, direct current

$I^2R$ losses are usually computed on the basis of the DC resistance of the windings at 75°C; in practice, these losses vary with operating conditions. The difference between the nominal and actual $I^2R$ loss is usually lumped under the category of *stray-load loss*. In direct-current machines, it is also necessary to account for the *brush contact loss* associated with slip rings and commutators.

Mechanical losses are due to *friction* (mostly in the bearings) and *windage,* that is, the air drag force that opposes the motion of the rotor. In addition, if external devices (e.g., blowers) are required to circulate air through the machine for cooling purposes, the energy expended by these devices is also included in the mechanical losses.

Open-circuit core losses consist of *hysteresis* and *eddy current* losses, with only the excitation winding energized (see Chapter 16 for a discussion of hysteresis and eddy currents). Often these losses are summed with friction and windage losses to give rise to the *no-load rotational loss*. The latter quantity is useful if one simply wishes to compute efficiency. Since open-circuit core losses do not account for the changes in flux density caused by the presence of load currents, an additional magnetic loss is incurred that is not accounted for in this term. *Stray-load losses* are used to lump the effects of nonideal current distribution in the windings and of the additional core losses just mentioned. Stray-load losses are difficult to determine exactly and are often assumed to be equal to 1.0 percent of the output power for DC machines; these losses can be determined by experiment in synchronous and induction machines.

The performance of an electric machine can be quantified in a number of ways. In the case of an electric motor, it is usually portrayed in the form of a graphical **torque-speed characteristic** and **efficiency map.** The torque-speed characteristic of a motor describes how the torque supplied by the machine varies as a function of the speed of rotation of the motor for steady speeds. As we shall see in later sections, the torque-speed curves vary in shape with the type of motor (DC, induction, synchronous) and are very useful in determining the performance of the motor when connected to a mechanical load. Figure 17.4(a) depicts the torque-speed curve of a hypothetical motor. Figure 17.4(b) depicts a typical efficiency map for a DC machine. It is quite likely that in most engineering applications, the engineer is required to make a decision regarding the performance characteristics of the motor best suited to a specified task. In this context, the torque-speed curve of a machine is a very useful piece of information.

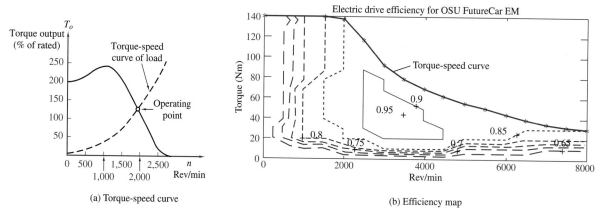

**Figure 17.4** Torque-speed and efficiency curves for an electric motor

The first feature we note of the torque-speed characteristic is that it bears a strong resemblance to the $i$-$v$ characteristics used in earlier chapters to represent the behavior of electrical sources. It should be clear that, according to this torque-speed curve, the motor is not an ideal source of torque (if it were, the curve would appear as a horizontal line across the speed range). One can readily see, for example, that the hypothetical motor represented by the curves of Figure 17.4(a) would produce maximum torque in the range of speeds between approximately 800 and 1,400 rev/min. What determines the actual speed of the motor (and therefore its output torque and power) is the torque-speed characteristic of the load connected to it, much as a resistive load determines the current drawn from a voltage source. In the figure, we display the torque-speed curve of a load, represented by the dashed line; the operating point of the motor-load pair is determined by the intersection of the two curves.

Another important observation pertains to the fact that the motor of Figure 17.4(a) produces a nonzero torque at zero speed. This fact implies that as soon as electric power is connected to the motor, the latter is capable of sup-plying a certain amount of torque; this zero-speed torque is called the **starting torque.** If the load the motor is connected to requires less than the starting torque the motor can provide, then the motor can accelerate the load, until the motor speed and torque settle to a stable value, at the operating point. The motor-load pair of Figure 17.4(a) would behave in the manner just described. However, there may well be circumstances in which a motor might not be able to provide a sufficient starting torque to overcome the static load torque that opposes its motion. Thus, we see that a torque-speed characteristic can offer valuable insight into the operation of a motor. As we proceed to discuss each type of machine in greater detail, we shall devote some time to the discussion of its torque-speed curve.

The efficiency of an electric machine is also an important design and per-formance characteristic. The **1995 Department of Energy Energy Policy Act,** also known as EPACT, has required electric motor manufacturers to guarantee a minimum efficiency. The efficiency of an electric motor is usually described using a contour plot of the efficiency value (a number between 0 and 1) in the torque-speed plane. This representation permits a determination of the motor efficiency as a function of its performance and operating conditions. Figure 17.4(b) depicts the efficiency map of an electric drive used in a hybrid-electric vehicle—a 20-kW permanent magnet AC (or brushless DC) machine. We shall discuss this type of machine in Chapter 18. Note that the peak efficiency can be as high as 0.95 (95 percent), but that the efficiency decreases significantly away from the optimum point (around 3500 rev/min and 45 N-m), to values as low as 0.65.

The most common means of conveying information regarding electric ma-chines is the *nameplate*. Typical information conveyed by the nameplate is:

1.  Type of device (e.g., DC motor, alternator)
2.  Manufacturer
3.  Rated voltage and frequency
4.  Rated current and volt-amperes
5.  Rated speed and horsepower

The **rated voltage** is the terminal voltage for which the machine was designed, and which will provide the desired magnetic flux. Operation at higher voltages will increase magnetic core losses, because of excessive core saturation. The **rated current** and **rated volt-amperes** are an indication of the typical current and power levels at the terminal that will not cause undue overheating due to copper losses ($I^2R$ losses) in the windings. These ratings are not absolutely precise, but they give an indication of the range of excitations for which the motor will perform without overheating. Peak power operation in a motor may exceed rated torque, power, or currents by a substantial factor (up to as much as 6 or 7 times the rated value); however, continuous operation of the motor above the rated performance will cause the machine to overheat, and eventually to sustain damage. Thus, it is important to consider both peak and continuous power requirements when selecting a motor for a specific application. An analogous discussion is valid for the speed rating: While an electric machine may operate above rated speed for limited periods of time, the large centrifugal forces generated at high rotational speeds will eventually cause undesirable mechanical stresses, especially in the rotor windings, leading eventually even to self-destruction.

Another important feature of electric machines is the **regulation** of the machine speed or voltage, depending on whether it is used as a motor or as a generator, respectively. Regulation is the ability to maintain speed or voltage constant in the face of load variations. The ability to closely regulate speed in a motor or voltage in a generator is an important feature of electric machines; regulation is often improved by means of feedback control mechanisms, some of which will be briefly introduced in this chapter. We shall take the following definitions as being adequate for the intended purpose of this chapter:

$$\text{Speed regulation} = \frac{\text{Speed at no load} - \text{Speed at rated load}}{\text{Speed at rated load}} \qquad \textbf{(17.4)}$$

$$\text{Voltage regulation} = \frac{\text{Voltage at no load} - \text{Voltage at rated load}}{\text{Voltage at rated load}} \qquad \textbf{(17.5)}$$

Please note that the rated value is usually taken to be the nameplate value, and that the meaning of *load* changes depending on whether the machine is a motor, in which case the load is mechanical, or a generator, in which case the load is electrical.

---

### EXAMPLE 17.1 Regulation

**Problem**

Find the percent speed regulation of a shunt DC motor.

---

**Solution**

**Known Quantities:** No-load speed, speed at rated load.

**Find:** Percent speed regulation, SR%.

**Schematics, Diagrams, Circuits, and Given Data:**
  $n_{nl}$ = no-load speed = 1,800 rev/min
  $n_{nr}$ = rated-load speed = 1,760 rev/min

**Analysis:**

$$SR\% = \frac{n_{nl} - n_{rl}}{n_{rl}} \times 100 = \frac{1,800 - 1,760}{1,800} \times 100 = 2.27\%$$

*Comments:*  Speed regulation is an intrinsic property of a motor; however, external speed controls can be used to regulate the speed of a motor to any (physically achievable) desired value. Some motor control concepts are discussed later in this chapter.

## EXAMPLE 17.2 Nameplate Data

### Problem

Discuss the nameplate data, shown below, of a typical induction motor.

### Solution

*Known Quantities:*  Nameplate data.

*Find:*  Motor characteristics.

*Schematics, Diagrams, Circuits, and Given Data:*  The nameplate appears below.

| MODEL | 19308 J-X | | |
|---|---|---|---|
| TYPE | CJ4B | FRAME | 324TS |
| VOLTS | 230/460 | °C AMB. | 40 |
| | | INS. CL. | B |
| FRT. BRG | 210SF | EXT. BRG | 312SF |
| SERV FACT | 1.0 | OPER INSTR | C-517 |
| PHASE \| 3 | Hz \| 60 | CODE \| G | WDGS \| 1 |
| H.P. | 40 | | |
| R.P.M. | 3,565 | | |
| AMPS | 106/53 | | |
| NEMA NOM. | EFF | | |
| NOM. P.F. | | | |
| DUTY | CONT. | NEMA DESIGN | B |

*Analysis:*  The nameplate of a typical induction motor is shown in the table above. The model number (sometimes abbreviated as MOD) uniquely identifies the motor to the manufacturer. It may be a style number, a model number, an identification number, or an instruction sheet reference number.

The term *frame* (sometimes abbreviated as FR) refers principally to the physical size of the machine, as well as to certain construction features.

Ambient temperature (abbreviated as AMB, or MAX. AMB) refers to the maximum ambient temperature in which the motor is capable of operating. Operation of the motor in a higher ambient temperature may result in shortened motor life and reduced torque.

Insulation class (abbreviated as INS. CL.) refers to the type of insulation used in the motor. Most often used are class A (105°C) and class B (130°C).

The duty (DUTY), or time rating, denotes the length of time the motor is expected to be able to carry the rated load under usual service conditions. "CONT." means that the machine can be operated continuously.

The "CODE" letter sets the limits of starting kVA per horsepower for the machine. There are 19 levels, denoted by the letters A through V, excluding I, O, and Q.

Service factor (abbreviated as SERV FACT) is a term defined by NEMA (the National Electrical Manufacturers Association) as follows: "The service factor of a general-purpose alternating-current motor is a multiplier which, when applied to the rated horsepower, indicates a permissible horsepower loading which may be carried under the conditions specified for the service factor."

The voltage figure given on the nameplate refers to the voltage of the supply circuit to which the motor should be connected. Sometimes two voltages are given, for example, 230/460. In this case, the machine is intended for use on either a 230-V or a 460-V circuit. Special instructions will be provided for connecting the motor for each of the voltages.

The term "BRG" indicates the nature of the bearings supporting the motor shaft.

---

## EXAMPLE  17.3  Torque-Speed Curves

### Problem

Discuss the significance of the torque-speed curve of an electric motor.

---

### Solution

A variable-torque variable-speed motor has a torque output that varies directly with speed; hence, the horsepower output varies directly with the speed. Motors with this characteristic are commonly used with fans, blowers, and centrifugal pumps. Figure 17.5 shows typical torque-speed curves for this type of motor. Superimposed on the motor torque-speed curve is the torque-speed curve for a typical fan where the input power to the fan varies as the cube of the fan speed. Point $A$ is the desired operating point, which could be determined graphically by plotting the load line and the motor torque-speed curve on the same graph, as illustrated in Figure 17.5.

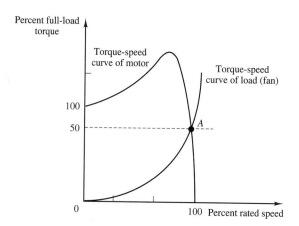

**Figure 17.5**

## Basic Operation of All Rotating Machines

We have already seen in Chapter 16 how the magnetic field in electromechanical devices provides a form of coupling between electrical and mechanical systems. Intuitively, one can identify two aspects of this coupling, both of which play a role in the operation of electric machines:

1.  Magnetic attraction and repulsion forces generate mechanical torque.
2.  The magnetic field can induce a voltage in the machine windings (coils) by virtue of Faraday's law.

Thus, we may think of the operation of an electric machine in terms of either a motor or a generator, depending on whether the input power is electrical and mechanical power is produced (motor action), or the input power is mechanical and the output power is electrical (generator action). Figure 17.6 illustrates the two cases graphically.

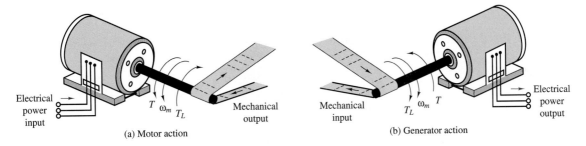

**Figure 17.6** Generator and motor action in an electric machine

The coupling magnetic field performs a dual role, which may be explained as follows. When a current $i$ flows through conductors placed in a magnetic field, a force is produced on each conductor, according to equation 17.1. If these conductors are attached to a cylindrical structure, a torque is generated, and if the structure is free to rotate, then it will rotate at an angular velocity $\omega_m$. As the conductors rotate, however, they move through a magnetic field and cut through flux lines, thus generating an electromotive force in opposition to the excitation. This emf is also called "counter" emf, as it opposes the source of the current $i$. If, on the other hand, the rotating element of the machine is driven by a prime mover (for example, an internal combustion engine), then an emf is generated across the coil that is rotating in the magnetic field (the armature). If a load is connected to the armature, a current $i$ will flow to the load, and this current flow will in turn cause a reaction torque on the armature that opposes the torque imposed by the prime mover.

You see, then, that for energy conversion to take place, two elements are required:

1.  A coupling field, **B**, usually generated in the field winding.
2.  An armature winding that supports the load current, $i$, and the emf, $e$.

## Magnetic Poles in Electric Machines

Before discussing the actual construction of a rotating machine, we should spend a few paragraphs to illustrate the significance of **magnetic poles** in an electric machine. In an electric machine, torque is developed as a consequence of magnetic forces of attraction and repulsion between magnetic poles on the stator and on the rotor; these poles produce a torque that accelerates the rotor and a reaction torque on the stator. Naturally, we would like a construction such that the torque generated as a consequence of the magnetic forces is continuous and in a constant direction. This can be accomplished if the number of rotor poles is equal to the number of stator poles. It is also important to observe that the number of poles must be even, since there have to be equal numbers of north and south poles.

The motion and associated electromagnetic torque of an electric machine are the result of two magnetic fields that are trying to align with each other so that the south pole of one field attracts the north pole of the other. Figure 17.7 illustrates this action by analogy with two permanent magnets, one of which is allowed to rotate about its center of mass.

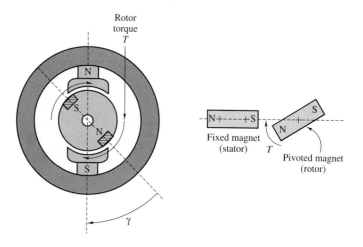

**Figure 17.7** Alignment action of poles

Figure 17.8 depicts a two-pole machine in which the stator poles are constructed in such a way as to project closer to the rotor than to the stator structure. This type of construction is rather common, and poles constructed in this fashion are called **salient poles.** Note that the rotor could also be constructed to have salient poles.

To understand magnetic polarity, we need to consider the direction of the magnetic field in a coil carrying current. Figure 17.9 shows how the *right-hand rule* can be employed to determine the direction of the magnetic flux. If one were to grasp the coil with the right hand, with the fingers curling in the direction of current flow, then the thumb would be pointing in the direction of the magnetic flux. Magnetic flux is by convention viewed as entering the south pole and exiting from the north pole. Thus, to determine whether a magnetic pole is north or south, we must consider the direction of the flux. Figure 17.10 shows a cross section of a coil wound around a pair of salient rotor poles. In this case, one can readily identify the direction of the magnetic flux and therefore the magnetic polarity of the poles by applying the right-hand rule, as illustrated in the figure.

Often, however, the coil windings are not arranged as simply as in the case of salient poles. In many machines, the windings are embedded in slots cut into the stator or rotor, so that the situation is similar to that of the stator depicted in

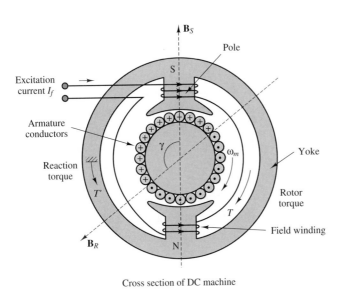

Cross section of DC machine

**Figure 17.8** A two-pole machine with salient stator poles

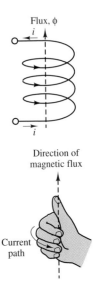

**Figure 17.9** Right-hand rule

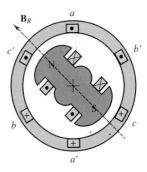

**Figure 17.10** Magnetic field in a salient rotor winding

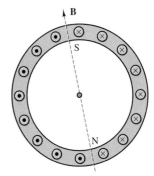

**Figure 17.11** Magnetic field of stator

Figure 17.11. This figure is a cross section in which the wire connections between "crosses" and "dots" have been cut away. In Figure 17.11, the dashed line indicates the axis of the stator flux according to the right-hand rule, indicating that the slotted stator in effect behaves like a pole pair. The north and south poles indicated in the figure are a consequence of the fact that the flux exits the bottom part of the structure (thus, the north pole indicated in the figure) and enters the top half of the structure (thus, the south pole). In particular, if you consider that the windings are arranged so that the current entering the right-hand side of the stator (to the right of the dashed line) flows through the back end of the stator and then flows outward from the left-hand side of the stator slots (left of the dashed line), you can visualize the windings in the slots as behaving in a manner similar to the coils of Figure 17.10, where the flux axis of Figure 17.11 corresponds to the flux axis of each of the coils of Figure 17.10. The actual circuit that permits current flow is completed by the front and back ends of the stator, where the wires are connected according to the pattern $a$-$a'$, $b$-$b'$, $c$-$c'$, as depicted in the figure.

Another important consideration that facilitates understanding the operation of electric machines pertains to the use of AC currents. It should be apparent by now that if the current flowing into the slotted stator is alternating, the direction of the flux will also alternate, so that in effect the two poles will reverse polarity every time the current reverses direction, that is, every half-cycle of the sinusoidal current. Further—since the magnetic flux is approximately proportional to the current in the coil—as the amplitude of the current oscillates in a sinusoidal fashion, so will the flux density in the structure. Thus, *the magnetic field developed in the stator changes both spatially and in time.*

This property is typical of AC machines, where a *rotating magnetic field* is established by energizing the coil with an alternating current. As we shall see in the next section, the principles underlying the operation of DC and AC machines are quite different: in a direct-current machine, there is no rotating field, but a mechanical switching arrangement (the *commutator*) makes it possible for the rotor and stator magnetic fields to always align at right angles to each other.

The accompanying CD-ROM includes 2-D "movies" of the most common types of electric machines. You might wish to explore these animations to better understand the basic concepts described in this section.

## Check Your Understanding

**17.1**  The percent speed regulation of a motor is 10 percent. If the full-load speed is $50\pi$ rad/s, find (a) the no-load speed in rad/s, and (b) the no-load speed in rev/min.

**17.2**  The percent voltage regulation for a 250-V generator is 10 percent. Find the no-load voltage of the generator.

**17.3**  The nameplate of a three-phase induction motor indicates the following values:

$$\begin{aligned}
\text{H.P.} &= 10 & \text{Volt} &= 220 \text{ V} \\
\text{R.P.M.} &= 1{,}750 & \text{Service factor} &= 1.15 \\
\text{Temperature rise} &= 60°\text{C} & \text{Amp} &= 30\text{A}
\end{aligned}$$

Find the rated torque, rated volt-amperes, and maximum continuous output power.

**17.4**  A motor having the characteristics shown in Figure 17.4 is to drive a load; the load has a linear torque-speed curve and requires 150 percent of rated torque at 1,500 rev/min. Find the operating point for this motor-load pair.

## 17.2     DIRECT-CURRENT MACHINES

As explained in the introductory section, direct-current (DC) machines are easier to analyze than their AC counterparts, although their actual construction is made rather complex by the need to have a commutator, which reverses the direction of currents and fluxes to produce a net torque. The objective of this section is to describe the major construction features and the operation of direct-current machines, as well as to develop simple circuit models that are useful in analyzing the performance of this class of machines.

### Physical Structure of DC Machines

A representative DC machine was depicted in Figure 17.8, with the magnetic poles clearly identified, for both the stator and the rotor. Figure 17.12 is a photograph of the same type of machine. Note the salient pole construction of the stator and the slotted rotor. As previously stated, the torque developed by the machine is a consequence of the magnetic forces between stator and rotor poles. This torque is maximum when the angle $\gamma$ between the rotor and stator poles is $90°$. Also, as you can see from the figure, in a DC machine the armature is usually on the rotor, and the field winding is on the stator.

To keep this torque angle constant as the rotor spins on its shaft, a mechanical switch, called a **commutator,** is configured so the current distribution in the rotor winding remains constant and therefore the rotor poles are consistently at $90°$ with respect to the fixed stator poles. In a DC machine, the magnetizing current is DC, so that there is no spatial alternation of the stator poles due to time-varying currents. To understand the operation of the commutator, consider the simplified diagram of Figure 17.13. In the figure, the brushes are fixed, and the rotor revolves at an angular velocity $\omega_m$; the instantaneous position of the rotor is given by the expression $\theta = \omega_m t - \gamma$.

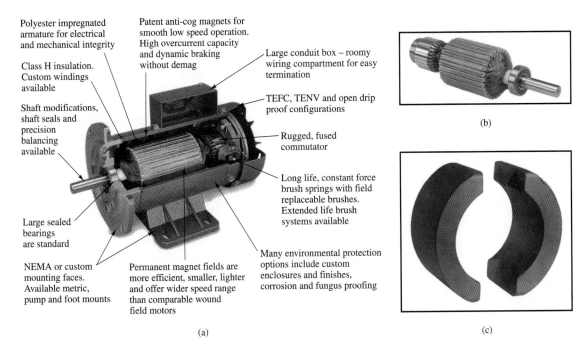

Polyester impregnated armature for electrical and mechanical integrity

Class H insulation. Custom windings available

Shaft modifications, shaft seals and precision balancing available

Patent anti-cog magnets for smooth low speed operation. High overcurrent capacity and dynamic braking without demag

Large conduit box – roomy wiring compartment for easy termination

TEFC, TENV and open drip proof configurations

Rugged, fused commutator

Long life, constant force brush springs with field replaceable brushes. Extended life brush systems available

Large sealed bearings are standard

NEMA or custom mounting faces. Available metric, pump and foot mounts

Permanent magnet fields are more efficient, smaller, lighter and offer wider speed range than comparable wound field motors

Many environmental protection options include custom enclosures and finishes, corrosion and fungus proofing

(a)

(b)

(c)

**Figure 17.12** (a) DC machine; (b) rotor; (c) permanent magnet stator

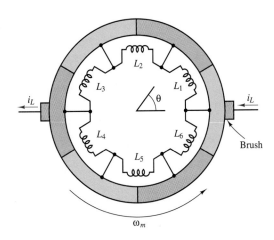

**Figure 17.13** Rotor winding and commutator

The commutator is fixed to the rotor and is made up in this example of six segments that are made of electrically conducting material but are insulated from each other. Further, the rotor windings are configured so that they form six coils, connected to the commutator segments as shown in Figure 17.13.

As the commutator rotates counterclockwise, the rotor magnetic field rotates with it up to $\theta = 30°$. At that point, the direction of the current changes in coils $L_3$ and $L_6$ as the brushes make contact with the next segment. Now the direction of the magnetic field is $-30°$. As the commutator continues to rotate, the direction of the rotor field will again change from $-30°$ to $+30°$, and it will switch again when the brushes switch to the next pair of segments. In this machine, then, the torque angle, $\gamma$, is not always 90°, but can vary by as much as ±30°; the actual torque produced by the machine would fluctuate by as much as ±14 percent, since the torque is proportional to $\sin\gamma$. As the number of segments increases, the torque fluctuation produced by the commutation is greatly reduced. In a practical machine, for example, one might have as many as 60 segments, and the variation of $\gamma$ from 90° would be only ±3°, with a torque fluctuation of less than 1 percent. Thus, the DC machine can produce a nearly constant torque (as a motor) or voltage (as a generator).

## Configuration of DC Machines

In DC machines, the field excitation that provides the magnetizing current is occasionally provided by an external source, in which case the machine is said to be **separately excited** (Figure 17.14(a)). More often, the field excitation is derived from the armature voltage and the machine is said to be **self-excited.** The latter configuration does not require the use of a separate source for the field excitation and is therefore frequently preferred. If a machine is in the separately excited configuration, an additional source, $V_f$, is required. In the self-excited case, one method used to provide the field excitation is to connect the field in parallel with the armature; since the field winding typically has significantly higher resistance than the armature circuit (remember that it is the armature that carries the load current), this will not draw excessive current from the armature. Further, a series resistor can be added to the field circuit to provide the means for adjusting the field current independent of the armature voltage. This configuration is called a **shunt-connected** machine and is depicted in Figure 17.14(b). Another method for self-exciting a DC machine consists of connecting the field in series with the armature, leading to the **series-connected** machine, depicted in Figure 17.14(c); in this case, the field winding will support the entire armature current, and thus the field coil must have low resistance (and therefore relatively few turns). This configuration is rarely used for generators, since the generated voltage and the load voltage must always differ by the voltage drop across the field coil, which varies with the load current. Thus, a series generator would have poor (large) regulation. However, series-connected motors are commonly used in certain applications, as will be discussed in a later section.

The third type of DC machine is the **compound-connected** machine, which consists of a combination of the shunt and series configurations. Figures 17.14(d) and (e) show the two types of connections, called the **short shunt** and the **long shunt,** respectively. Each of these configurations may be connected so that the series part of the field adds to the shunt part (**cumulative compounding**) or so that it subtracts (**differential compounding**).

## DC Machine Models

As stated earlier, it is relatively easy to develop a simple model of a DC machine, which is well suited to performance analysis, without the need to resort to the details of the construction of the machine itself. This section will illustrate the development of such models in two steps. First, steady-state models relating field and armature currents and voltages to speed and torque are introduced; second, the differential equations describing the dynamic behavior of DC machines are derived.

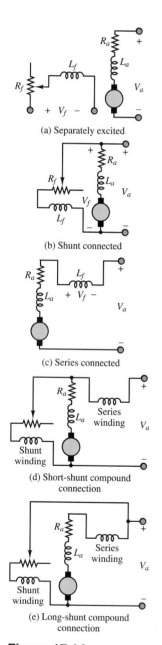

(a) Separately excited

(b) Shunt connected

(c) Series connected

(d) Short-shunt compound connection

(e) Long-shunt compound connection

**Figure 17.14**

When a field excitation is established, a magnetic flux, $\phi$, is generated by the field current, $I_f$. From equation 17.2, we know that the torque acting on the rotor is proportional to the product of the magnetic field and the current in the load-carrying wire; the latter current is the armature current, $I_a$ ($i_w$, in equation 16.2). Assuming that, by virtue of the commutator, the torque angle, $\gamma$, is kept very close to 90°, and therefore $\sin \gamma = 1$, we obtain the following expression for the torque (in units of N-m) in a DC machine:

$$T = k_T \phi I_a \qquad \text{for } \gamma = 90° \tag{17.6}$$

You may recall that this is simply a consequence of the $Bli$ law of Chapter 16. The mechanical power generated (or absorbed) is equal to the product of the machine torque and the mechanical speed of rotation, $\omega_m$ (in rad/s), and is therefore given by

$$P_m = \omega_m T = \omega_m k_T \phi I_a \tag{17.7}$$

Recall now that the rotation of the armature conductors in the field generated by the field excitation causes a **back emf**, $E_b$, in a direction that opposes the rotation of the armature. According to the $Blu$ law (see Chapter 16), then, this back emf is given by the expression

$$E_b = k_a \phi \omega_m \tag{17.8}$$

where $k_a$ is called the **armature constant** and is related to the geometry and magnetic properties of the structure. The voltage $E_b$ represents a countervoltage (opposing the DC excitation) in the case of a motor, and the generated voltage in the case of a generator. Thus, the electric power dissipated (or generated) by the machine is given by the product of the back emf and the armature current:

$$P_e = E_b I_a \tag{17.9}$$

The constants $k_T$ and $k_a$ in equations 17.6 and 17.8 are related to geometry factors, such as the dimension of the rotor and the number of turns in the armature winding; and to properties of materials, such as the permeability of the magnetic materials. Note that in the ideal energy-conversion case, $P_m = P_e$, and therefore $k_a = k_T$. We shall in general assume such ideal conversion of electrical to mechanical energy (or vice versa) and will therefore treat the two constants as being identical: $k_a = k_T$. The constant $k_a$ is given by

$$k_a = \frac{pN}{2\pi M} \tag{17.10}$$

where

$\qquad p$ = number of magnetic poles

$\qquad N$ = number of conductors per coil

$\qquad M$ = number of parallel paths in armature winding

An important observation concerning the units of angular speed must be made at this point. The equality (under the no-loss assumption) between the constants $k_a$ and $k_T$ in equations 17.6 and 17.8 results from the choice of consistent units, namely, volts and amperes for the electrical quantities, and newton-meters and radians per second for the mechanical quantities. You should be aware that it is fairly common practice to refer to the speed of rotation of an electric machine in units of revolutions per minute (rev/min).[1] In this book, we shall uniformly use the symbol $n$ to denote angular speed in rev/min; the following relationship should be committed to memory:

$$n \text{ (rev/min)} = \frac{60}{2\pi}\omega_m \text{ (rad/s)} \qquad \textbf{(17.11)}$$

If the speed is expressed in rev/min, the armature constant changes as follows:

$$E_b = k_a'\phi n \qquad \textbf{(17.12)}$$

where

$$k_a' = \frac{pN}{60M} \qquad \textbf{(17.13)}$$

Having introduced the basic equations relating torque, speed, voltages, and currents in electric machines, we may now consider the interaction of these quantities in a DC machine at steady state, that is, operating at constant speed and field excitation. Figure 17.15 depicts the electrical circuit model of a separately excited DC machine, illustrating both motor and generator action. It is very important to note the reference direction of armature current flow, and of the developed torque, in order to make a distinction between the two modes of operation. The field excitation is shown as a voltage, $V_f$, generating the field current, $I_f$, that flows through a variable resistor, $R_f$, and through the field coil, $L_f$. The variable resistor permits adjustment of the field excitation. The armature circuit, on the other hand, consists of a voltage source representing the back emf, $E_b$, the armature resistance, $R_a$, and the armature voltage, $V_a$. This model is appropriate both for motor and for generator action. When $V_a < E_b$, the machine acts as a generator ($I_a$ flows out of the machine). When $V_a > E_b$, the machine acts as a motor ($I_a$ flows into the machine). Thus, according to the circuit model of Figure 17.15, the operation of a DC machine at steady state (i.e., with the inductors in the circuit replaced by short circuits) is described by the following equations:

$$-I_f + \frac{V_f}{R_f} = 0 \quad \text{and} \quad V_a - R_a I_a - E_b = 0 \quad \text{(motor action)}$$

$$\textbf{(17.14)}$$

$$-I_f + \frac{V_f}{R_f} = 0 \quad \text{and} \quad V_a + R_a I_a - E_b = 0 \quad \text{(generator action)}$$

Equation pair 17.14 together with equations 17.6 and 17.8 may be used to determine the steady-state operating condition of a DC machine.

The circuit model of Figure 17.15 permits the derivation of a simple set of differential equations that describe the *dynamic* analysis of a DC machine. The dynamic equations describing the behavior of a separately excited DC machine are as follows:

$$V_a(t) - I_a(t)R_a - L_a\frac{dI_a(t)}{dt} - E_b(t) = 0 \quad \text{(armature circuit)} \qquad \textbf{(17.15a)}$$

$$V_f(t) - I_f(t)R_f - L_f\frac{dI_f(t)}{dt} = 0 \qquad \text{(field circuit)} \qquad \textbf{(17.15b)}$$

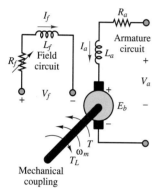

(a) Motor reference direction

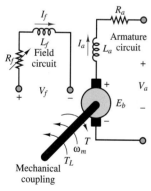

(b) Generator reference direction

**Figure 17.15** Electrical circuit model of a separately excited DC machine

---

[1]Note that the abbreviation RPM, although certainly familiar to the reader, is not a standard unit, and its use should be discouraged.

These equations can be related to the operation of the machine in the presence of a load. If we assume that the motor is rigidly connected to an inertial load with moment of inertia $J$ and that the friction losses in the load are represented by a viscous friction coefficient, $b$, then the torque developed by the machine (in the motor mode of operation) can be written as follows:

$$T(t) = T_L + b\omega_m(t) + J\frac{d\omega_m(t)}{dt} \qquad \textbf{(17.16)}$$

where $T_L$ is the load torque. $T_L$ is typically either constant or some function of speed, $\omega_m$, in a motor. In the case of a generator, the load torque is replaced by the torque supplied by a prime mover, and the machine torque, $T(t)$, opposes the motion of the prime mover, as shown in Figure 17.15. Since the machine torque is related to the armature and field currents by equation 17.6, equations 17.16 and 17.17 are coupled to each other; this coupling may be expressed as follows:

$$T(t) = k_a\phi I_a(t) \qquad \textbf{(17.17)}$$

or

$$k_a\phi I_a(t) = T_L + b\omega_m(t) + J\frac{d\omega_m(t)}{dt} \qquad \textbf{(17.18)}$$

The dynamic equations described in this section apply to any DC machine. In the case of a *separately excited* machine, a further simplification is possible, since the flux is established by virtue of a separate field excitation, and therefore

$$\phi = \frac{N_f}{\mathcal{R}}I_f = k_f I_f \qquad \textbf{(17.19)}$$

where $N_f$ is the number of turns in the field coil, $\mathcal{R}$ is the reluctance of the structure, and $I_f$ is the field current.

## 17.3  DIRECT-CURRENT GENERATORS

To analyze the performance of a DC generator, it would be useful to obtain an open-circuit characteristic capable of predicting the voltage generated when the machine is driven at a constant speed $\omega_m$ by a prime mover. The common arrangement is to drive the machine at rated speed by means of a prime mover (or an electric motor). Then, with no load connected to the armature terminals, the armature voltage is recorded as the field current is increased from zero to some value sufficient to produce an armature voltage greater than the rated voltage. Since the load terminals are open-circuited, $I_a = 0$ and $E_b = V_a$; and since $k_a\phi = E_b/\omega_m$, the magnetization curve makes it possible to determine the value of $k_a\phi$ corresponding to a given field current, $I_f$, for the rated speed.

Figure 17.16 depicts a typical magnetization curve. Note that the armature voltage is nonzero even when no field current is present. This phenomenon is due to the *residual magnetization* of the iron core. The dashed lines in Figure 17.16 are called **field resistance curves** and are a plot of the voltage that appears across the field winding plus rheostat (variable resistor; see Figure 17.15) versus the field current, for various values of field winding plus rheostat resistance. Thus, the slope of the line is equal to the total field circuit resistance, $R_f$.

The operation of a DC generator may be readily understood with reference to the magnetization curve of Figure 17.16. As soon as the armature is connected

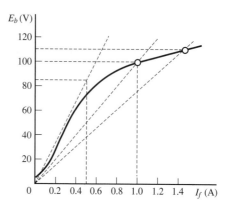

**Figure 17.16** DC machine magnetization curve

across the shunt circuit consisting of the field winding and the rheostat, a current will flow through the winding, and this will in turn act to increase the emf across the armature. This **buildup** process continues until the two curves meet, that is, until the current flowing through the field winding is exactly that required to induce the emf. By changing the rheostat setting, the operating point at the intersection of the two curves can be displaced, as shown in Figure 17.16, and the generator can therefore be made to supply different voltages. The following examples illustrate the operation of the separately excited DC generator.

### EXAMPLE 17.4 Separately Excited DC Generator

#### Problem

A separately excited DC generator is characterized by the magnetization curve of Figure 17.16.

1. If the prime mover is driving the generator at 800 rev/min, what is the no-load terminal voltage, $V_a$?
2. If a 1-$\Omega$ load is connected to the generator, what is the generated voltage?

#### Solution

**Known Quantities:** Generator magnetization curve and ratings.

**Find:** Terminal voltage with no load and 1-$\Omega$ load.

**Schematics, Diagrams, Circuits, and Given Data:**
Generator ratings: 100 V, 100 A, 1,000 rev/min.
Circuit parameters: $R_a = 0.14\ \Omega$; $V_f = 100$ V; $R_f = 100\ \Omega$.

**Analysis:**

1. The field current in the machine is

$$I_f = \frac{V_f}{R_f} = \frac{100\ \text{V}}{100\ \Omega} = 1\ \text{A}$$

From the magnetization curve, it can be seen that this field current will produce

100 V at a speed of 1,000 rev/min. Since this generator is actually running at 800 rev/min, the induced emf may be found by assuming a linear relationship between speed and emf. This approximation is reasonable, provided that the departure from the nominal operating condition is small. Let $n_0$ and $E_{b0}$ be the nominal speed and emf, respectively (i.e., 1,000 rev/min and 100 V); then,

$$\frac{E_b}{E_{b0}} = \frac{n}{n_0}$$

and therefore

$$E_b = \frac{n}{n_0} E_{b0} = \frac{800 \text{ rev/min}}{1,000 \text{ rev/min}} \times 100 \text{ V} = 80 \text{ V}$$

The open-circuit (output) terminal voltage of the generator is equal to the emf from the circuit model of Figure 17.15; therefore:

$$V_a = E_b = 80 \text{ V}$$

2.  When a load resistance is connected to the circuit (the practical situation), the terminal (or load) voltage is no longer equal to $E_b$, since there will be a voltage drop across the armature winding resistance. The armature (or load) current may be determined from the expression

$$I_a = I_L = \frac{E_b}{R_a + R_L} = \frac{80 \text{ V}}{(0.14 + 1)\Omega} = 70.2 \text{ A}$$

where $R_L = 1\ \Omega$ is the load resistance. The terminal (load) voltage is therefore given by

$$V_L = I_L R_L = 70.2 \times 1 = 70.2 \text{ V}$$

## EXAMPLE  17.5  Separately Excited DC Generator

### Problem

Determine the following quantities for a separately excited DC:

1.  Induced voltage
2.  Machine constant
3.  Torque developed at rated conditions

### Solution

**Known Quantities:**  Generator ratings and machine parameters.

**Find:**  $E_b, k_a, T$.

**Schematics, Diagrams, Circuits, and Given Data:**
Generator ratings: 1,000 kW; 2,000 V; 3,600 rev/min
Circuit parameters: $R_a = 0.1\ \Omega$; flux per pole $= \phi = 0.5$ Wb

**Analysis:**

1.  The armature current may be found by observing that the rated power is equal to the product of the terminal (load) voltage and the armature (load) current; thus,

$$I_a = \frac{P_{\text{rated}}}{V_L} = \frac{1,000 \times 10^3}{2,000} = 500 \text{ A}$$

The generated voltage is equal to the sum of the terminal voltage and the voltage drop across the armature resistance (see Figure 16.14):

$$E_b = V_a + I_a R_a = 2,000 + 500 \times 0.1 = 2,050 \text{ V}$$

2. The speed of rotation of the machine in units of rad/s is

$$\omega_m = \frac{2\pi n}{60} = \frac{2\pi \times 3,600 \text{ rev/min}}{60 \text{ s/min}} = 377 \text{ rad/s}$$

Thus, the machine constant is found to be

$$k_a = \frac{E_b}{\phi \omega_m} = \frac{2,050 \text{ V}}{0.5 \text{ Wb} \times 377 \text{ rad/s}} = 10.876 \frac{\text{V-s}}{\text{Wb-rad}}$$

3. The torque developed is found from equation 16.6:

$$T = k_a \phi I_a = 10.875 \text{ V-s/Wb-rad} \times 0.5 \text{ Wb} \times 500 \text{ A} = 2,718.9 \text{ N-m}$$

**Comments:** In many practical cases, it is not actually necessary to know the armature constant and the flux separately, but it is sufficient to know the value of the product $k_a \phi$. For example, suppose that the armature resistance of a DC machine is known and that, given a known field excitation, the armature current, load voltage, and speed of the machine can be measured. Then, the product $k_a \phi$ may be determined from equation 16.20, as follows:

$$k_a \phi = \frac{E_b}{\omega_m} = \frac{V_L + I_a(R_a + R_S)}{\omega_m}$$

where $V_L$, $I_a$, and $\omega_m$ are measured quantities for given operating conditions.

---

Since the compound-connected generator contains both a shunt and a series field winding, it is the most general configuration, and the most useful for developing a circuit model that is as general as possible. In the following discussion, we shall consider the so-called short-shunt, compound-connected generator, in which the flux produced by the series winding adds to that of the shunt winding. Figure 17.17 depicts the equivalent circuit for the compound generator; circuit models for the shunt generator and for the rarely used series generator can be obtained by removing the shunt or series field winding element, respectively. In the circuit of Figure 17.17, the generator armature has been replaced by a voltage source corresponding to the induced emf and a series resistance, $R_a$, corresponding to the resistance of the armature windings. The equations describing the DC generator at steady state (i.e., with the inductors acting as short circuits) are:

**DC Generator Steady-State Equations**

$$E_b = k_a \phi \omega_m \text{ V} \tag{17.20}$$

$$T = \frac{P}{\omega_m} = \frac{E_b I_a}{\omega_m} = k_a \phi I_a \text{ N-m} \tag{17.21}$$

$$V_L = E_b - I_a R_a - I_S R_S \tag{17.22}$$

$$I_a = I_S + I_f \tag{17.23}$$

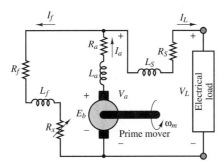

**Figure 17.17** Compound generator circuit
model

Note that in the circuit of Figure 17.17, the load and armature voltages are not equal, in general, because of the presence of a series field winding, represented by the resistor $R_S$ and by the inductor $L_S$ where the subscript "$S$" stands for "series." The expression for the armature emf is dependent on the air-gap flux, $\phi$, to which the series and shunt windings in the compound generator both contribute, according to the expression

$$\phi = \phi_{\text{sh}} \pm \phi_S = \phi_{\text{sh}} \pm k_S I_a \qquad (17.24)$$

## Check Your Understanding

**17.5**   A 24-coil, 2-pole DC generator has 16 turns per coil in its armature winding. The field excitation is 0.05 Wb per pole, and the armature angular velocity is 180 rad/s. Find the machine constant and the total induced voltage.

**17.6**   A 1,000-kW, 1,000-V, 2,400-rev/min separately excited DC generator has an armature circuit resistance of 0.04 $\Omega$. The flux per pole is 0.4 Wb. Find: (a) the induced voltage; (b) the machine constant; and (c) the torque developed at the rated conditions.

**17.7**   A 100-kW, 250-V shunt generator has a field circuit resistance of 50 $\Omega$ and an armature circuit resistance of 0.05 $\Omega$. Find: (a) the full-load line current flowing to the load; (b) the field current; (c) the armature current; and (d) the full-load generator voltage.

## 17.4    DIRECT-CURRENT MOTORS

DC motors are widely used in applications requiring accurate speed control—for example, in servo systems. Having developed a circuit model and analysis methods for the DC generator, we can extend these results to DC motors, since these are in effect DC generators with the roles of input and output reversed. Once again, we shall analyze the motor by means of both its magnetization curve and a circuit model. It will be useful to begin our discussion by referring to the schematic diagram of a cumulatively compounded motor, as shown in Figure 17.18. The choice of the compound-connected motor is the most convenient, since its model can be used to represent either a series or a shunt motor with minor modifications.

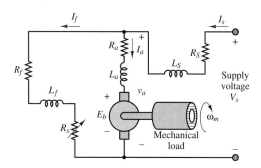

**Figure 17.18** Equivalent circuit of a cumulatively compounded motor

The equations that govern the behavior of the DC motor follow and are similar to those used for the generator. Note that the only differences between these equations and those that describe the DC generator appear in the last two equations in the group, where the source voltage is equal to the *sum* of the emf and the voltage drop across the series field resistance and armature resistance, and where the source current now equals the *sum* of the field shunt and armature series currents.

---

**DC Motor Steady-State Equations**

$$E_b = k_a \phi \omega_m \tag{17.25}$$

$$T = k_a \phi I_a \tag{17.26}$$

$$V_s = E_b + I_a R_a + I_s R_S \tag{17.27}$$

$$I_s = I_f + I_a \tag{17.28}$$

---

Note that in these equations we have replaced the symbols $V_L$ and $I_L$, used in the generator circuit model to represent the generator load current and voltage, with the symbols $V_s$ and $I_s$, indicating the presence of an external source.

## Speed-Torque and Dynamic Characteristics of DC Motors

### The Shunt Motor

In a shunt motor (similar to the configuration of Figure 17.18, but with the series field short-circuited), the armature current is found by dividing the net voltage across the armature circuit (source voltage minus back emf) by the armature resistance:

$$I_a = \frac{V_s - k_a \phi \omega_m}{R_a} \tag{17.29}$$

An expression for the armature current may also be obtained from equation 16.26, as follows:

$$I_a = \frac{T}{k_a \phi} \tag{17.30}$$

It is then possible to relate the torque requirements to the speed of the motor by substituting equation 17.29 in equation 17.30:

$$\frac{T}{k_a \phi} = \frac{V_s - k_a \phi \omega_m}{R_a} \tag{17.31}$$

Equation 17.31 describes the steady-state torque-speed characteristic of the shunt motor. To understand this performance equation, we observe that if $V_s$, $k_a$, $\phi$, and $R_a$ are fixed in equation 17.31 (the flux is essentially constant in the shunt motor for a fixed $V_s$), then the speed of the motor is directly related to the armature current. Now consider the case where the load applied to the motor is suddenly increased, causing the speed of the motor to drop. As the speed decreases, the armature current increases, according to equation 17.29. The excess armature current causes the motor to develop additional torque, according to equation 17.30, until a new equilibrium is reached between the higher armature current and developed torque and the lower speed of rotation. The equilibrium point is dictated by the balance of mechanical and electrical power, in accordance with the relation

$$E_b I_a = T \omega_m \tag{17.32}$$

Thus, the shunt DC motor will adjust to variations in load by changing its speed to preserve this power balance. The torque-speed curves for the shunt motor may be obtained by rewriting the equation relating the speed to the armature current:

$$\omega_m = \frac{V_s - I_a R_a}{k_a \phi} = \frac{V_s}{k_a \phi} - \frac{R_a T}{(k_a \phi)^2} \tag{17.33}$$

To interpret equation 17.33, one can start by considering the motor operating at rated speed and torque. As the load torque is reduced, the armature current will also decrease, causing the speed to increase in accordance with equation 17.33. The increase in speed depends on the extent of the voltage drop across the armature resistance, $I_a R_a$. The change in speed will be on the same order of magnitude as this drop; it typically takes values around 10 percent. This corresponds to a relatively good speed regulation, which is an attractive feature of the shunt DC motor (recall the discussion of regulation in Section 17.1). Normalized torque and speed vs. power curves for the shunt motor are shown in Figure 17.19. Note that, over a reasonably broad range of powers, up to rated value, the curve is relatively flat, indicating that the DC shunt motor acts as a reasonably constant-speed motor.

The dynamic behavior of the shunt motor is described by equations 17.15 through 17.18, with the additional relation

$$I_a(t) = I_s(t) - I_f(t) \tag{17.34}$$

### Compound Motors

It is interesting to compare the performance of the shunt motor with that of the compound-connected motor; the comparison is easily made if we recall that a series field resistance appears in series with the armature resistance and that the

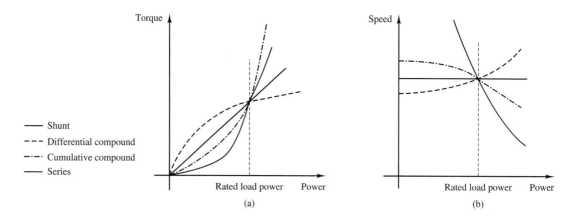

**Figure 17.19** DC motor operating characteristics

flux is due to the contributions of both series and shunt fields. Thus, the speed equation becomes

$$\omega_m = \frac{V_s - I_a(R_a + R_S)}{k_a(\phi_{\text{sh}} \pm \phi_S)} \tag{17.35}$$

where

$+$ in the denominator is for a cumulatively compounded motor.

$-$ in the denominator is for a differentially compounded motor.

$\phi_{\text{sh}}$ is the flux set up by the shunt field winding, assuming that it is constant.

$\phi_S$ is the flux set up by the series field winding, $\phi_S = k_S I_a$.

For the cumulatively compound motor, two effects are apparent: the flux is increased by the presence of a series component, $\phi_S$; and the voltage drop due to $I_a$ in the numerator term is increased by an amount proportional to the resistance of the series field winding, $R_S$. As a consequence, when the load to the motor is reduced, the numerator increases more dramatically than in the case of the shunt motor, because of the corresponding decrease in armature current, while at the same time the series flux decreases. Each of these effects causes the speed to increase; therefore, it stands to reason that the speed regulation of the compound-connected motor is poorer than that of the shunt motor. Normalized torque and speed vs. power curves for the compound motor (both differential and cumulative connections) are shown in Figure 17.19.

The differential equation describing the behavior of a compound motor differs from that for the shunt motor in having additional terms due to the series field component:

$$V_s = E_b(t) + I_a(t)R_a + L_a\frac{dI_a(t)}{dt} + I_s(t)R_S + L_S\frac{dI_s(t)}{dt}$$
$$= V_a(t) + I_s(t)R_S + L_S\frac{dI_s(t)}{dt} \tag{17.36}$$

The differential equation for the field circuit can be written as

$$V_a = I_f(t)(R_f + R_x) + L_f\frac{dI_f(t)}{dt} \tag{17.37}$$

We also have the following basic relations:

$$I_a(t) = I_s(t) - I_f(t) \tag{17.38}$$

and

$$E_b(t) = k_a I_a(t)\omega_m(t) \quad \text{and} \quad T(t) = k_a\phi I_a(t) \tag{17.39}$$

### Series Motors

The series motor [see Figure 17.14(c)] behaves somewhat differently from the shunt and compound motors because the flux is established solely by virtue of the series current flowing through the armature. It is relatively simple to derive an expression for the emf and torque equations for the series motor if we approximate the relationship between flux and armature current by assuming that the motor operates in the linear region of its magnetization curve. Then we can write

$$\phi = k_S I_a \tag{17.40}$$

and the emf and torque equations become

$$E_b = k_a\omega_m\phi = k_a\omega_m k_S I_a \tag{17.41}$$

$$T = k_a\phi I_a = k_a k_S I_a^2 \tag{17.42}$$

The circuit equation for the series motor becomes

$$V_s = E_b + I_a(R_a + R_S) = (k_a\omega_m k_S + R_T)I_a \tag{17.43}$$

where $R_a$ is the armature resistance, $R_S$ is the series field winding resistance, and $R_T$ is the total series resistance. From equation 17.43, we can solve for $I_a$ and substitute in the torque expression (equation 17.42) to obtain the following torque-speed relationship:

$$T = k_a k_S \frac{V_s^2}{(k_a\omega_m k_S + R_T)^2} \tag{17.44}$$

which indicates the inverse squared relationship between torque and speed in the series motor. This expression describes a behavior that can, under certain conditions, become unstable. Since the speed increases when the load torque is reduced, one can readily see that if one were to disconnect the load altogether, the speed would tend to increase to dangerous values. To prevent excessive speeds, series motors are always mechanically coupled to the load. This feature is not necessarily a drawback, though, because series motors can develop very high torque at low speeds, and therefore can serve very well for traction-type loads (e.g., conveyor belts or vehicle propulsion systems). Torque and speed vs. power curves for the series motor are also shown in Figure 17.19.

The differential equation for the armature circuit of the motor can be given as

$$\begin{aligned} V_s &= I_a(t)(R_a + R_S) + L_a\frac{dI_a(t)}{dt} + L_S\frac{dI_a(t)}{dt} + E_b \\ &= I_a(t)(R_a + R_S) + L_a\frac{dI_a(t)}{dt} + L_S\frac{dI_a(t)}{dt} + k_a k_S I_a\omega_m \end{aligned} \tag{17.45}$$

### Permanent-Magnet DC Motors

Permanent-magnet (PM) DC motors have become increasingly common in applications requiring relatively low torques and efficient use of space. The construction of PM direct-current motors differs from that of the motors considered thus far in that the magnetic field of the stator is produced by suitably located poles made of magnetic materials. Thus, the basic principle of operation, including the idea of commutation, is unchanged with respect to the wound-stator DC motor. What changes is that there is no need to provide a field excitation, whether separately or by means of the self-excitation techniques discussed in the preceding sections. Therefore, the PM motor is intrinsically simpler than its wound-stator counterpart.

The equations that describe the operation of the PM motor follow. The torque produced is related to the armature current by a torque constant, $k_{PM}$, which is determined by the geometry of the motor:

$$T = k_{TPM} I_a \tag{17.46}$$

As in the conventional DC motor, the rotation of the rotor produces the usual counter or back emf, $E_b$, which is linearly related to speed by a voltage constant, $k_{aPM}$:

$$E_b = k_{aPM} \omega_m \tag{17.47}$$

The equivalent circuit of the PM motor is particularly simple, since we need not model the effects of a field winding. Figure 17.20 shows the circuit model and the torque-speed curve of a PM motor.

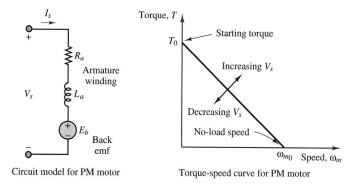

Circuit model for PM motor          Torque-speed curve for PM motor

**Figure 17.20** Circuit model and torque-speed curve of PM motor

We can use the circuit model of Figure 17.20 to predict the torque-speed curve shown in the same figure, as follows. From the circuit model, for a constant speed (and therefore constant current), we may consider the inductor a short circuit and write the equation

$$V_s = I_a R_a + E_b = I_a R_a + k_{aPM} \omega_m$$

$$= \frac{T}{k_{TPM}} R_a + k_{aPM} \omega_m \tag{17.48}$$

thus obtaining the equations relating speed and torque:

$$\omega_m = \frac{V_s}{k_{aPM}} - \frac{T R_a}{k_{aPM} k_{TPM}} \tag{17.49}$$

and

$$T = \frac{V_s}{R_a}k_{TPM} - \frac{\omega_m}{R_a}k_{aPM}k_{TPM} \qquad (17.50)$$

From these equations, one can extract the stall torque, $T_0$, that is, the zero-speed torque:

$$T_0 = \frac{V_s}{R_a}k_{TPM} \qquad (17.51)$$

and the no-load speed, $\omega_{m0}$:

$$\omega_{m0} = \frac{V_s}{k_{aPM}} \qquad (17.52)$$

Under dynamic conditions, assuming an inertia plus viscous friction load, the torque produced by the motor can be expressed as

$$T = k_{TPM}I_a(t) = T_{\text{load}}(t) + d\omega_m(t) + J\frac{d\omega_m(t)}{dt} \qquad (17.53)$$

The differential equation for the armature circuit of the motor is therefore given by:

$$\begin{aligned} V_s &= I_a(t)R_a + L_a\frac{dI_a(t)}{dt} + E_b \\ &= I_a(t)R_a + L_a\frac{dI_a(t)}{dt} + k_{aPM}\omega_m(t) \end{aligned} \qquad (17.54)$$

The fact that the air-gap flux is constant in a permanent-magnet DC motor makes its characteristics somewhat different from those of the wound DC motor. A direct comparison of PM and wound-field DC motors reveals the following advantages and disadvantages of each configuration.

---

**Comparison of Wound-Field and PM DC Motors**

1.  PM motors are smaller and lighter than wound motors for a given power rating. Further, their efficiency is greater because there are no field winding losses.

2.  An additional advantage of PM motors is their essentially linear speed-torque characteristic, which makes analysis (and control) much easier. Reversal of rotation is also accomplished easily, by reversing the polarity of the source.

3.  A major disadvantage of PM motors is that they can become demagnetized by exposure to excessive magnetic fields, application of excessive voltage, or operation at excessively high or low temperatures.

4.  A less obvious drawback of PM motors is that their performance is subject to greater variability from motor to motor than is the case for wound motors, because of variations in the magnetic materials.

---

In summary, four basic types of **DC motors** are commonly used. Their principal operating characteristics are summarized as follows, and their general torque and speed versus power characteristics are depicted in Figure 17.19, assuming motors with identical voltage, power, and speed ratings.

*Shunt wound motor:* Field connected in parallel with the armature. With constant armature voltage and field excitation, the motor has good speed regulation (flat speed-torque characteristic).

*Compound wound motor:* Field winding has both series and shunt components. This motor offers better starting torque than the shunt motor, but worse speed regulation.

*Series wound motor:* The field winding is in series with the armature. The motor has very high starting torque and poor speed regulation. It is useful for low-speed, high-torque applications.

*Permanent-magnet motor:* Field windings are replaced by permanent magnets. The motor has adequate starting torque, with speed regulation somewhat worse than that of the compound wound motor.

---

### EXAMPLE  17.6  DC Shunt Motor Analysis

**Problem**

Find the speed and torque generated by a four-pole DC shunt motor.

---

**Solution**

***Known Quantities:***  Motor ratings; circuit and magnetic parameters.

***Find:***  $\omega_m$, $T$.

***Schematics, Diagrams, Circuits, and Given Data:***
Motor ratings: 3 hp; 240 V; 120 rev/min.
Circuit and magnetic parameters: $I_S = 30$ A; $I_f = 1.4$ A; $R_a = 0.6\ \Omega$; $\phi = 20$ mWb; $N = 1,000$; $M = 4$ (see equation 17.10).

***Analysis:***  We convert the power to SI units:

$$P_{\text{RATED}} = 3\text{ hp} \times 746\frac{\text{W}}{\text{hp}} = 2.238\text{ W}$$

Next, we compute the armature current as the difference between source and field current (equation 17.34):

$$I_a = I_s - I_f = 30 - 1.4 = 28.6\text{ A}$$

The no-load armature voltage, $E_b$, is given by:

$$E_b = V_s - I_a R_a = 240 - 28.6 \times 0.6 = 222.84\text{ V}$$

and equation 17.10 can be used to determine the armature constant:

$$k_a = \frac{pN}{2\pi M} = \frac{4 \times 1000}{2\pi \times 4} = 159.15\frac{\text{V-s}}{\text{Wb-rad}}$$

Knowing the motor constant, we can calculate the speed, after equation 17.25:

$$\omega_m = \frac{E_a}{k_a\phi} = \frac{222.84\text{ V}}{159.15\dfrac{\text{V-s}}{\text{Wb-rad}} \times 0.002\text{ Wb}} = 70\frac{\text{rad}}{\text{s}}$$

Finally, the torque developed by the motor can be found as the ratio of the power to the angular velocity:

$$T = \frac{P}{\omega_m} = \frac{2{,}238 \text{ W}}{70\dfrac{\text{rad}}{\text{s}}} = 32 \text{ N-m}$$

## EXAMPLE 17.7 DC Shunt Motor Analysis

### Problem

Determine the following quantities for the DC shunt motor, connected as shown in the circuit Figure 17.21:

1.  Field current required for full-load operation
2.  No-load speed
3.  Plot the speed torque curve of the machine in the range from no-load torque to rated torque
4.  Power output at rated torque.

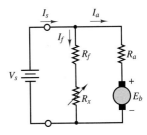

**Figure 17.21** Shunt motor configuration

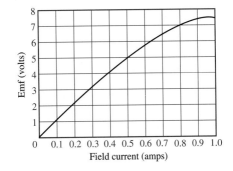

**Figure 17.22** Magnetization curve for a small DC motor

### Solution

***Known Quantities:*** Magnetization curve, rated current, rated speed, circuit parameters.

***Find:*** $I_f$; $n_{\text{no-load}}$; $T$-$n$ curve, $P_{\text{rated}}$.

***Schematics, Diagrams, Circuits, and Given Data:*** Figure 17.22 (magnetization curve)
Motor ratings: 8 A, 120 rev/min
Circuit parameters: $R_a = 0.2 \ \Omega$; $V_s = 7.2$ V; $N$ = number of coil turns in winding = 200

***Analysis:***

1.  To find the field current, we must find the generated emf since $R_f$ is not known. Writing KVL around the armature circuit, we obtain

$$V_s = E_b + I_a R_a$$
$$E_b = V_s - I_a R_a = 7.2 - 8(0.2) = 5.6 \text{ V}$$

Having found the back emf, we can find the field current from the magnetization curve. At $E_b = 5.6$ V, we find that the field current and field resistance are

$$I_f = 0.6 \text{ A} \qquad \text{and} \qquad R_f = \frac{7.2}{0.6} = 12 \ \Omega$$

2.  To obtain the no-load speed, we use the equations

$$E_b = k_a \phi \frac{2\pi n}{60} \qquad T = k_a \phi I_a$$

leading to

$$V_s = I_a R_a + E_b = I_a R_a + k_a \phi \frac{2\pi}{60} n$$

or

$$n = \frac{V_s - I_a R_a}{k_a \phi (2\pi/60)}$$

At no load, and assuming no mechanical losses, the torque is zero, and we see that the current $I_a$ must also be zero in the torque equation ($T = k_a \phi I_a$). Thus, the motor speed at no load is given by

$$n_{\text{no-load}} = \frac{V_s}{k_a \phi (2\pi/60)}$$

We can obtain an expression for $k_a \phi$ knowing that, at full load,

$$E_b = 5.6 \text{ V} = k_a \phi \frac{2\pi n}{60}$$

so that, for constant field excitation,

$$k_a \phi = E_b \left( \frac{60}{2\pi n} \right) = 5.6 \left( \frac{60}{2\pi (120)} \right) = 0.44563 \frac{\text{V} \cdot \text{s}}{\text{rad}}$$

Finally, we may solve for the no-load speed, in rev/min:

$$n_{\text{no-load}} = \frac{V_s}{k_a \phi (2\pi/60)} = \frac{7.2}{(0.44563)(2\pi/60)}$$

$$= 154.3 \text{ rev/min}$$

3.  The torque at rated speed and load may be found as follows:

$$T_{\text{rated load}} = k_a \phi I_a = (0.44563)(8) = 3.565 \text{ N-m}$$

Now we have the two points necessary to construct the torque-speed curve for this motor, which is shown in Figure 17.23.

4.  The power is related to the torque by the frequency of the shaft:

$$P_{\text{rated}} = T \omega_m = (3.565) \left( \frac{120}{60} \right) 2\pi = 44.8 \text{ W}$$

or, equivalently,

$$P = 44.8 \text{ W} \times \frac{1}{746} \frac{\text{hp}}{\text{W}} = 0.06 \text{ hp}$$

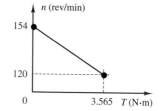

**Figure 17.23**

## EXAMPLE 17.8 DC Series Motor Analysis

### Problem

Determine the torque developed by a DC series motor when the current supplied to the motor is 60 A.

## Solution

**Known Quantities:**  Motor ratings; operating conditions.

**Find:**  $T_{60}$, torque delivered at 60-A series current.

**Schematics, Diagrams, Circuits, and Given Data:**  Motor ratings: 10 hp; 115 V; full load speed $= 1,800$ rev/min
Operating conditions: motor draws 40 A

**Assumptions:**  The motor operates in the linear region of the magnetization curve.

**Analysis:**  Within the linear region of operation, the flux per pole is directly proportional to the current in the field winding. That is,

$$\phi = k_S I_a$$

The full-load speed is

$$n = 1,800 \text{ rev/min}$$

or

$$\omega_m = \frac{2\pi n}{60} = 60\pi \text{ rad/s}$$

Rated output power is

$$P_{\text{rated}} = 10 \text{ hp} \times 746 \text{ W/hp} = 7,460 \text{ W}$$

and full-load torque is

$$T_{40\text{A}} = \frac{P_{\text{rated}}}{\omega_m} = \frac{7,460}{60\pi} = 39.58 \text{ N-m}$$

Thus, the machine constant may be computed from the torque equation for the series motor:

$$T = k_a k_S I_a^2 = K I_a^2$$

At full load,

$$K = k_a k_S = \frac{39.58 \text{ N-m}}{40^2 \text{ A}^2} = 0.0247 \frac{\text{N-m}}{\text{A}^2}$$

and we can compute the torque developed for a 60-A supply current to be

$$T_{60\text{A}} = K I_a^2 = 0.0247 \times 60^2 = 88.92 \text{ N-m}$$

## EXAMPLE  17.9  Dynamic Response of PM DC Motor

### Problem

Develop a set of differential equations and a transfer function describing the dynamic response of the motor angular velocity of a PM DC motor connected to a mechanical load.

## Solution

**Known Quantities:**  PM DC motor circuit model; mechanical load model.

**Find:**  Differential equations and transfer functions of electromechanical system.

**Analysis:** The dynamic response of the electromechanical system can be determined by applying KVL to the electrical circuit (Figure 17.20), and Newton's second law to the mechanical system. These equations will be coupled to one another, as you shall see, because of the nature of the motor back emf and torque equations.

Applying KVL and equation 17.47 to the electrical circuit we obtain:

$$V_L(t) - R_a I_a(t) - L_a \frac{dI_a(t)}{dt} - E_b(t) = 0$$

or

$$L_a \frac{dI_a(t)}{dt} + R_a I_a(t) + K_{a\text{PM}}\omega_m(t) = V_L(t)$$

Applying Newton's second law and equation 17.46 to the load inertia, we obtain:

$$J \frac{d\omega(t)}{dt} = T(t) - T_{\text{load}}(t) - b\omega$$

or

$$-K_{T\,\text{PM}} I_a(t) + J \frac{d\omega(t)}{dt} + b\omega(t) = T_{\text{load}}(t)$$

These two differential equations are coupled because the first depends on $\omega_m$ and the second on $I_a$. Thus, they need to be solved simultaneously.

To derive the transfer function, we Laplace-transform the two equations to obtain:

$$(sL_a + R_a)I_a(s) + K_{a\text{PM}}\Omega(s) = V_L(s)$$

$$-K_{T\,\text{PM}} I_a(s) + (sJ + b)\Omega(s) = T_{\text{load}}(s)$$

We can write the above equations in matrix form and resort to Cramer's rule to solve for $\Omega_m(s)$ as a function of $V_L(s)$ and $T_{\text{load}}(s)$.

$$\begin{bmatrix} (sL_a + R_a) & K_{a\text{PM}} \\ -K_{T\,\text{PM}} & (sJ + b) \end{bmatrix} \begin{bmatrix} I_a(s) \\ \Omega_m(s) \end{bmatrix} = \begin{bmatrix} V_L(s) \\ T_{\text{load}}(s) \end{bmatrix}$$

with solution

$$\Omega_m(s) = \frac{\det \begin{bmatrix} (sL_a + R_a) & V_L(s) \\ K_{T\,\text{PM}} & T_{\text{load}}(s) \end{bmatrix}}{\det \begin{bmatrix} (sL_a + R_a) & K_{a\,\text{PM}} \\ -K_{T\,\text{PM}} & (sJ + b) \end{bmatrix}}$$

or

$$\Omega_m(s) = \frac{(sL_a + R_a)}{(sL_a + R_a)(sJ + b) + K_{a\,\text{PM}}K_{T\,\text{PM}}} T_{\text{load}}(s)$$

$$+ \frac{K_{T\,\text{PM}}}{(sL_a + R_a)(sJ + b) + K_{a\,\text{PM}}K_{T\,\text{PM}}} V_L(s)$$

**Comments:** Note that the dynamic response of the motor angular velocity depends on both the input voltage and on the load torque. This problem is explored further in the homework problems.

## DC Drives and DC Motor Speed Control

The advances made in power semiconductors have made it possible to realize low-cost **speed control systems for DC motors.** The basic operation of *controlled rectifier* and *chopper* drives for DC motors was described in Chapter 11. In the present section we describe some of the considerations that are behind the choice of a specific drive type, and of some of the loads that are likely to be encountered.

*Constant-torque loads* are quite common, and are characterized by a need for constant torque over the entire speed range. This need is usually due to friction; the load will demand increasing horsepower at higher speeds, since power is the product of speed and torque. Thus, the power required will increase linearly with speed. This type of loading is characteristic of conveyors, extruders, and surface winders.

Another type of load is one that requires *constant horsepower* over the speed range of the motor. Since torque is inversely proportional to speed with constant horsepower, this type of load will require higher torque at low speeds. Examples of constant-horsepower loads are machine tool spindles (e.g., lathes). This type of application requires very high starting torques.

*Variable-torque loads* are also common. In this case, the load torque is related to the speed in some fashion, either linearly or geometrically. For some loads, for example, torque is proportional to the speed (and thus horsepower is proportional to speed squared); examples of loads of this type are positive displacement pumps. More common than the linear relationship is the squared-speed dependence of inertial loads such as centrifugal pumps, some fans, and all loads in which a flywheel is used for energy storage.

To select the appropriate motor and adjustable speed drive for a given application, we need to examine how each method for speed adjustment operates on a DC motor. Armature voltage control serves to smoothly adjust speed from 0 to 100 percent of the nameplate rated value (i.e., base speed), provided that the field excitation is also equal to the rated value. Within this range, it is possible to fully control motor speed for a constant-torque load, thus providing a linear increase in horsepower, as shown in Figure 17.24. Field weakening allows for increases in speed of up to several times the base speed; however, field control changes the characteristics of the DC motor from constant torque to constant horsepower, and therefore the torque output drops with speed, as shown in Figure 17.24. Operation above base speed requires special provision for field control, in addition to the circuitry required for armature voltage control, and is therefore more complex and costly.

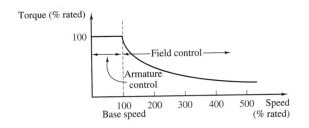 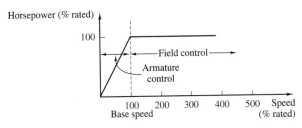

**Figure 17.24** Speed control in DC motors

## Check Your Understanding

**17.8**  A series motor draws a current of 25 A and develops a torque of 100 N-m Find: (a) the torque when the current rises to 30 A if the field is unsaturated; and (b) the torque when the current rises to 30 A and the increase in current produces a 10 percent increase in flux.

**17.9**  A 200-V DC shunt motor draws 10 A at 1,800 rev/min. The armature circuit resistance is 0.15 $\Omega$ and the field winding resistance is 350 $\Omega$. What is the torque developed by the motor?

**17.10**   Describe the cause-and-effect behavior of the speed control method of changing armature voltage for a shunt DC motor.

---

## 17.5   AC MACHINES

From the previous sections, it should be apparent that it is possible to obtain a wide range of performance characteristics from DC machines, as both motors and generators. A logical question at this point should be, Would it not be more convenient in some cases to take advantage of the single- or multiphase AC power that is available virtually everywhere than to expend energy and use additional hardware to rectify and regulate the DC supplies required by direct-current motors? The answer to this very obvious question is certainly a resounding yes. In fact, the AC induction motor is the workhorse of many industrial applications, and synchronous generators are used almost exclusively for the generation of electric power worldwide. Thus, it is appropriate to devote a significant portion of this chapter to the study of AC machines, and of induction motors in particular. The objective of this section is to explain the basic operation of both synchronous and induction machines, and to outline their performance characteristics. In doing so, we shall also point out the relative advantages and disadvantages of these machines in comparison with direct-current machines. The motor "movies" included in the CD-ROM may help you visualize the operation of AC machines.

### Rotating Magnetic Fields

As mentioned in Section 17.1, the fundamental principle of operation of AC machines is the generation of a rotating magnetic field, which causes the rotor to turn at a speed that depends on the speed of rotation of the magnetic field. We shall now explain how a rotating magnetic field can be generated in the stator and air gap of an AC machine by means of AC currents.

Consider the stator shown in Figure 17.25, which supports windings $a\text{-}a'$, $b\text{-}b'$ and $c\text{-}c'$. The coils are geometrically spaced 120° apart, and a three-phase voltage is applied to the coils. As you may recall from the discussion of AC power in Chapter 7, the currents generated by a three-phase source are also spaced by 120°, as illustrated in Figure 17.26. The phase voltages referenced the neutral

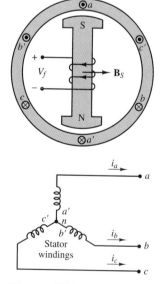

**Figure 17.25** Two-pole three-phase stator

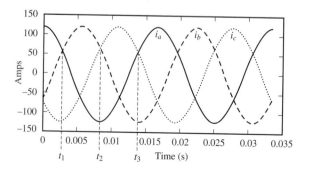

**Figure 17.26** Three-phase stator winding currents

terminal, would then be given by the expressions

$$v_a = A \cos(\omega_e t)$$

$$v_b = A \cos\left(\omega_e t - \frac{2\pi}{3}\right)$$

$$v_c = A \cos\left(\omega_e t + \frac{2\pi}{3}\right)$$

where $\omega_e$ is the frequency of the AC supply, or line frequency. The coils in each winding are arranged in such a way that the flux distribution generated by any one winding is approximately sinusoidal. Such a flux distribution may be obtained by appropriately arranging groups of coils for each winding over the stator surface. Since the coils are spaced $120°$ apart, the flux distribution resulting from the sum of the contributions of the three windings is the sum of the fluxes due to the separate windings, as shown in Figure 17.27. Thus, the flux in a three-phase machine rotates in space according to the vector diagram of Figure 17.28, and is constant in amplitude. A stationary observer on the machine's stator would see a sinusoidally varying flux distribution as shown in Figure 17.27.

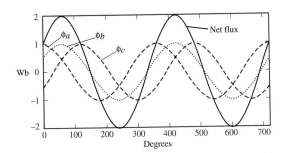

**Figure 17.27** Flux distribution in a three-phase stator winding as a function of angle of rotation

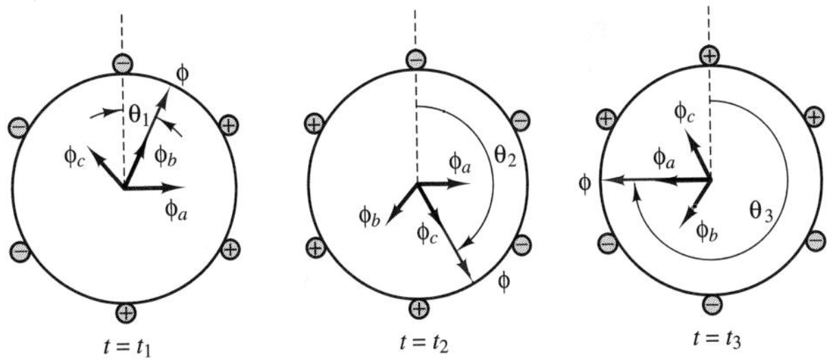

**Figure 17.28** Rotating flux in a three-phase machine

Since the resultant flux of Figure 17.27 is generated by the currents of Figure 17.26, the speed of rotation of the flux must be related to the frequency of the sinusoidal phase currents. In the case of the stator of Figure 17.25, the number of magnetic poles resulting from the winding configuration is two; however, it is also possible to configure the windings so that they have more poles. For example, Figure 17.29 depicts a simplified view of a four-pole stator.

In general, the speed of the rotating magnetic field is determined by the frequency of the excitation current, $f$, and by the number of poles present in the stator, $p$, according to the equation

$$n_s = \frac{120f}{p}\text{rev/min}$$

or

$$\omega_s = \frac{2\pi n_s}{60} = \frac{2\pi \times 2f}{p}$$

where $n_s$ (or $\omega_s$) is usually called the **synchronous speed.**

**(17.55)**

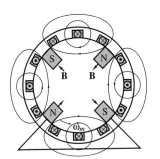

**Figure 17.29** Four-pole stator

Now, the structure of the windings in the preceding discussion is the same whether the AC machine is a motor or a generator; the distinction between the two depends on the direction of power flow. In a generator, the electromagnetic torque is a reaction torque that opposes rotation of the machine; this is the torque against which the prime mover does work. In a motor, on the other hand, the rotational (motional) voltage generated in the armature opposes the applied voltage; this voltage is the counter (or back) emf. Thus, the description of the rotating magnetic field given thus far applies to both motor and generator action in AC machines.

As described a few paragraphs earlier, the stator magnetic field rotates in an AC machine, and therefore the rotor cannot "catch up" with the stator field and is in constant pursuit of it. The speed of rotation of the rotor will therefore depend on the number of magnetic poles present in the stator and in the rotor. The magnitude of the torque produced in the machine is a function of the angle $\gamma$ between the stator and rotor magnetic fields; precise expressions for this torque depend on how the magnetic fields are generated and will be given separately for the two cases of synchronous and induction machines. What is common to all rotating machines is that the number of stator and rotor poles must be identical if any torque is to be generated. Further, the number of poles must be even, since for each north pole there must be a corresponding south pole.

One important desired feature in an electric machine is an ability to generate a constant electromagnetic torque. With a constant-torque machine, one can avoid torque pulsations that could lead to undesired mechanical vibration in the motor itself and in other mechanical components attached to the motor (e.g., mechanical loads, such as spindles or belt drives). A constant torque may not always be achieved, although it will be shown that it is possible to accomplish this goal when the excitation currents are multiphase. A general rule of thumb, in this respect, is that it is desirable, insofar as possible, to produce a constant flux per pole.

## 17.6     THE ALTERNATOR (SYNCHRONOUS GENERATOR)

One of the most common AC machines is the **synchronous generator,** or **alternator.** In this machine, the field winding is on the rotor, and the connection is made by means of brushes, in an arrangement similar to that of the DC machines studied earlier. The rotor field is obtained by means of a DC current provided to the rotor winding, or by permanent magnets. The rotor is then connected to a mechanical source of power and rotates at a speed that we will consider constant to simplify the analysis.

Figure 17.30 depicts a two-pole three-phase synchronous machine. Figure 17.31 depicts a four-pole three-phase alternator, in which the rotor poles are generated by means of a wound salient pole configuration and the stator poles are the result of windings embedded in the stator according to the simplified arrangement shown in the figure, where each of the pairs $a/a'$, $b/b'$, and so on, contributes to the generation of the magnetic poles, as follows. The group $a/a'$, $b/b'$, $c/c'$ produces a sinusoidally distributed flux (see Figure 17.27) corresponding to one of the pole pairs, while the group $-a/-a'$, $-b/-b'$, $-c/-c'$ contributes the other pole pair. The connections of the coils making up the windings are also shown in Figure 17.31. Note that the coils form a wye connection (see Chapter 7). The resulting flux distribution is such that the flux completes two sinusoidal cycles around the circumference of the air gap. Note also that each arm of the

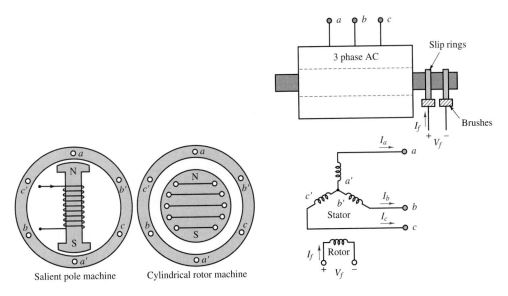

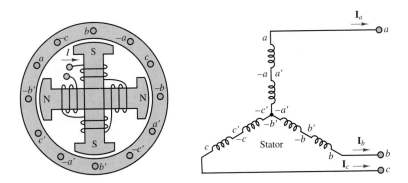

Salient pole machine    Cylindrical rotor machine

**Figure 17.30** Two-pole synchronous machine

**Figure 17.31** Four-pole three-phase alternator

three-phase wye connection has been divided into two coils, wound in different locations, according to the schematic stator diagram of Figure 17.31. One could then envision analogous configurations with greater numbers of poles, obtained in the same fashion, that is, by dividing each arm of a wye connection into more windings.

The arrangement shown in Figure 17.31 requires that a further distinction be made between mechanical degrees, $\theta_m$, and electrical degrees, $\theta_e$. In the four-pole alternator, the flux will see two complete cycles during one rotation of the rotor, and therefore the voltage that is generated in the coils will also oscillate at twice the frequency of rotation. In general, the electrical degrees (or radians) are related to the mechanical degrees by the expression

$$\theta_e = \frac{p}{2}\theta_m \qquad \qquad \textbf{(17.56)}$$

where $p$ is the number of poles. In effect, the voltage across a coil of the machine goes through one cycle every time a pair of poles moves past the coil. Thus, the frequency of the voltage generated by a synchronous generator is

$$f = \frac{p}{2} \frac{n}{60} \text{ Hz} \tag{17.57}$$

where $n$ is the mechanical speed in rev/min. Alternatively, if the speed is expressed in rad/s, we have

$$\omega_e = \frac{p}{2} \omega_m \tag{17.58}$$

where $\omega_m$ is the mechanical speed of rotation in rad/s. The number of poles employed in a synchronous generator is then determined by two factors: the frequency desired of the generated voltage (e.g., 60 Hz, if the generator is used to produce AC power), and the speed of rotation of the prime mover. In the latter respect, there is a significant difference, for example, between the speed of rotation of a steam turbine generator and that of a hydroelectric generator, the former being much greater.

A common application of the alternator is in automotive battery-charging systems—in which, however, the generated AC voltage is rectified to provide the DC current required for charging the battery. Figure 17.32 depicts an automotive alternator.

**Figure 17.32** Automotive alternator (Courtesy: Delphi Automotive Systems)

## 17.7  THE SYNCHRONOUS MOTOR

Synchronous motors are virtually identical to synchronous generators with regard to their construction, except for an additional winding for helping start the motor and minimizing motor speed over- and undershoots. The principle of operation is, of course, the opposite: an AC excitation provided to the armature generates a magnetic field in the air gap between stator and rotor, resulting in a mechanical torque. To generate the rotor magnetic field, some DC current must be provided to the field windings; this is often accomplished by means of an **exciter,** which consists of a small DC generator propelled by the motor itself, and therefore mechanically connected to it. It was mentioned earlier that to obtain a constant torque in an electric motor, it is necessary to keep the rotor and stator magnetic fields constant relative to each other. This means that the electromagnetically rotating field in the stator and the mechanically rotating rotor field should be aligned at all times. The only condition for which this can occur is if both fields are rotating at the synchronous speed, $n_s = 120f/p$. Thus, synchronous motors are by their very nature constant-speed motors.

For a non–salient pole (cylindrical-rotor) synchronous machine, the torque can be written in terms of the AC stator current, $i_S(t)$, and of the DC rotor current, $I_f$:

$$T = k i_S(t) I_f \sin(\gamma) \tag{17.59}$$

where $\gamma$ is the angle between the stator and rotor fields (see Figure 17.7). Let the angular speed of rotation be

$$\omega_m = \frac{d\theta_m}{dt} \text{ rad/s} \tag{17.60}$$

where $\omega_m = 2\pi n/60$, and let $\omega_e$ be the electrical frequency of $i_S(t)$, where $i_S(t) = \sqrt{2}I_S \sin(\omega_e t)$. Then the torque may be expressed as follows:

$$T = k\sqrt{2}I_S \sin(\omega_e t) I_f \sin(\gamma) \tag{17.61}$$

where $k$ is a machine constant, $I_S$ is the rms value of the stator current, and $I_f$ is the DC rotor current. Now, the rotor angle $\gamma$ can be expressed as a function of time by

$$\gamma = \gamma_0 + \omega_m t \tag{17.62}$$

where $\gamma_0$ is the angular position of the rotor at $t = 0$; the torque expression then becomes

$$T = k\sqrt{2} I_S I_f \sin(\omega_e t) \sin(\omega_m t + \gamma_0)$$
$$= k\frac{\sqrt{2}}{2} I_S I_f \cos[(\omega_m - \omega_e)t - \gamma_0] - \cos[(\omega_m + \omega_e)t + \gamma_0] \tag{17.63}$$

It is a straightforward matter to show that the average value of this torque, $\langle T \rangle$, is different from zero only if $\omega_m = \pm\omega_e$, that is, only if the motor is turning at the synchronous speed. The resulting average torque is then given by

$$\langle T \rangle = k\sqrt{2} I_S I_f \cos(\gamma_0) \tag{17.64}$$

Note that equation 17.63 corresponds to the sum of an average torque plus a fluctuating component at twice the original electrical (or mechanical) frequency. The fluctuating component results because, in the foregoing derivation, a single-phase current was assumed. The use of multiphase currents reduces the torque fluctuation to zero and permits the generation of a constant torque.

A per-phase circuit model describing the synchronous motor is shown in Figure 17.33, where the rotor circuit is represented by a field winding equivalent resistance and inductance, $R_f$ and $L_f$, respectively, and the stator circuit is represented by equivalent stator winding inductance and resistance, $L_S$ and $R_S$, respectively, and by the induced emf, $E_b$. From the exact equivalent circuit as given in Figure 17.33, we have

$$V_S = E_b + I_S(R_S + jX_S) \tag{17.65}$$

where $X_S$ is known as the synchronous reactance and includes magnetizing reactance.

The motor power is

$$P_{\text{out}} = \omega_S T = |V_S||I_S|\cos(\theta) \tag{17.66}$$

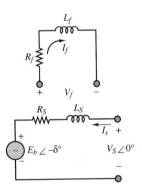

**Figure 17.33** Per-phase circuit model

for each phase, where $T$ is the developed torque and $\theta$ is the angle between $V_S$ and $I_S$.

When the phase winding resistance $R_S$ is neglected, the circuit model of a synchronous machine can be redrawn as shown in Figure 17.34. The input power (per phase) is equal to the output power in this circuit, since no power is dissipated in the circuit; that is:

$$P_\phi = P_{\text{in}} = P_{\text{out}} = |V_S||I_S|\cos(\theta) \tag{17.67}$$

Also by inspection of Figure 17.34, we have

$$d = |E_b| \sin(\delta) = |I_S| X_S \cos(\theta) \tag{17.68}$$

Then

$$|E_b||V_S| \sin(\delta) = |V_S||I_S| X_S \cos(\theta) = X_S P_\phi \tag{17.69}$$

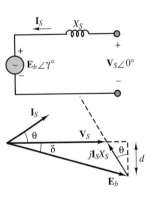

**Figure 17.34**

The total power of a three-phase synchronous machine is then given by

$$P = (3)\frac{|\mathbf{V}_S||\mathbf{E}_b|}{X_S}\sin(\delta) \qquad (17.70)$$

Because of the dependence of the power upon the angle $\delta$, this angle has come to be called the **power angle.** If $\delta$ is zero, the synchronous machine cannot develop useful power. The developed power has its maximum value at $\delta$ equal to 90°. If we assume that $|\mathbf{E}_b|$ and $|\mathbf{V}_S|$ are constant, we can draw the curve shown in Figure 17.35, relating the power and power angle in a synchronous machine.

A synchronous generator is usually operated at a power angle varying from 15° to 25°. For synchronous motors and small loads, $\delta$ is close to 0°, and the motor torque is just sufficient to overcome its own windage and friction losses; as the load increases, the rotor field falls further out of phase with the stator field (although the two are still rotating at the same speed), until $\delta$ reaches a maximum at 90°. If the load torque exceeds the maximum torque, which is produced for $\delta = 90°$, the motor is forced to slow down below synchronous speed. This condition is undesirable, and provisions are usually made to shut the motor down automatically whenever synchronism is lost. The maximum torque is called the **pull-out torque** and is an important measure of the performance of the synchronous motor.

Accounting for each of the phases, the total torque is given by

$$T = \frac{m}{\omega_s}|\mathbf{V}_S||\mathbf{I}_S|\cos(\theta) \qquad (17.71)$$

where $m$ is the number of phases. From Figure 17.34, we have $E_b\sin(\delta) = X_S I_S\cos(\theta)$. Therefore, for a three-phase machine, the developed torque is

$$T = \frac{P}{\omega_s} = \frac{3}{\omega_s}\frac{|\mathbf{V}_S||\mathbf{E}_b|}{X_S}\sin(\delta) \qquad \text{N-m} \qquad (17.72)$$

Typically, analysis of multiphase motors is performed on a per-phase basis, as illustrated in the examples that follow.

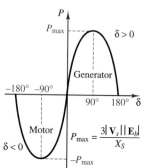

**Figure 17.35** Power versus power angle for a synchronous machine

### EXAMPLE   17.10  Synchronous Motor Analysis

**Problem**

Find the kVA rating, the induced voltage and the power angle of the rotor for a fully loaded synchronous motor.

**Solution**

**Known Quantities:**  Motor ratings; motor synchronous impedance.

**Find:**  $S$; $\mathbf{E}_b$; $\delta$.

**Schematics, Diagrams, Circuits, and Given Data:**  Motor ratings: 460 V; 3 $\phi$; pf = 0.707 lagging; full-load stator current: 12.5 A. $Z_S = 1 + j12\ \Omega$.

**Assumptions:**  Use per-phase analysis.

**Analysis:**  The circuit model for the motor is shown in Figure 17.36. The per-phase current in the wye-connected stator winding is

$$I_S = |\mathbf{I}_S| = 12.5\ \text{A}$$

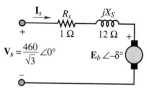

**Figure 17.36**

The per-phase voltage is

$$V_S = |\mathbf{V}_S| = \frac{460 \text{ V}}{\sqrt{3}} = 265.58 \text{ V}$$

The kVA rating of the motor is expressed in terms of the apparent power, $S$ (see Chapter 7):

$$S = 3V_S I_S = 3 \times 265.58 \text{ V} \times 12.5 \text{ A} = 9{,}959 \text{ W}$$

From the equivalent circuit, we have

$$\mathbf{E}_b = \mathbf{V}_S - \mathbf{I}_S(R_S + jX_S)$$
$$= 265.58 - (12.5\angle - 45° \text{ A}) \times (1 + j12 \; \Omega) = 179.31\angle - 32.83° \text{ V}$$

The induced line voltage is defined to be

$$V_{\text{line}} = \sqrt{3}E_b = \sqrt{3} \times 179.31 \text{ V} = 310.57 \text{ V}$$

From the expression for $\mathbf{E}_b$, we can find the power angle:

$$\delta = -32.83°$$

**Comments:** The minus sign indicates that the machine is in the motor mode.

---

## EXAMPLE  17.11  Synchronous Motor Analysis

### Problem

Find the stator current, the line current and the induced voltage for a synchronous motor.

---

### Solution

**Known Quantities:** Motor ratings; motor synchronous impedance.

**Find:** $\mathbf{I}_S$; $\mathbf{I}_{\text{line}}$; $\mathbf{E}_b$.

**Schematics, Diagrams, Circuits, and Given Data:** Motor ratings: 208 V; 3 $\phi$; 45 kVA; 60 Hz; pf = 0.8 leading; $Z_S = 0 + j2.5 \; \Omega$. Friction and windage losses: 1.5 kW; core losses: 1.0 kW. Load power: 15 hp.

**Assumptions:** Use per-phase analysis.

**Analysis:** The output power of the motor is 15 hp; that is:

$$P_{\text{out}} = 15 \text{ hp} \times 0.746 \text{ kW/hp} = 11.19 \text{ kW}$$

The electric power supplied to the machine is

$$P_{\text{in}} = P_{\text{out}} + P_{\text{mech}} + P_{\text{core-loss}} + P_{\text{elec-loss}}$$
$$= 11.19 \text{ kW} + 1.5 \text{ kW} + 1.0 \text{ kW} + 0 \text{ kW} = 13.69 \text{kW}$$

As discussed in Chapter 7, the resulting line current is

$$I_{\text{line}} = \frac{P_{\text{in}}}{\sqrt{3}V \cos\theta} = \frac{13{,}690 \text{ W}}{\sqrt{3} \times 208 \text{ V} \times 0.8} = 47.5 \text{ A}$$

Because of the $\Delta$ connection, the armature current is

$$\mathbf{I}_S = \frac{1}{\sqrt{3}}\mathbf{I}_{\text{line}} = 27.4\angle 36.87° \text{ A}$$

The emf may be found from the equivalent circuit and KVL:

$$\mathbf{E}_b = \mathbf{V}_S - jX_S\mathbf{I}_S$$

$$= 208\angle 0° - j2.5\ \Omega(27.4\angle 36.87°\ \text{A}) = 255\angle - 12.4°\ \text{V}$$

The power angle is

$$\delta = -12.4°$$

Synchronous motors are not very commonly used in practice, for various reasons, among which are that they are essentially required to operate at constant speed (unless a variable-frequency AC supply is available) and that they are not self-starting. Further, separate AC and DC supplies are required. It will be seen shortly that the induction motor overcomes most of these drawbacks.

## Check Your Understanding

**17.11**   A synchronous generator has a multipolar construction that permits changing its synchronous speed. If only two poles are energized, at 50 Hz, the speed is 3,000 rev/min. If the number of poles is progressively increased to 4, 6, 8, 10, and 12, find the synchronous speed for each configuration.

**17.12**   Draw the complete equivalent circuit of a synchronous generator and its phasor diagram.

**17.13**   Find an expression for the maximum pull-out torque of the synchronous motor.

## 17.8    THE INDUCTION MOTOR

The induction motor is the most widely used electric machine, because of its relative simplicity of construction. The stator winding of an induction machine is similar to that of a synchronous machine; thus, the description of the three-phase winding of Figure 17.25 also applies to induction machines. The primary advantage of the induction machine, which is almost exclusively used as a motor (its performance as a generator is not very good), is that no separate excitation is required for the rotor. The rotor typically consists of one of two arrangements: a **squirrel cage,** or a **wound rotor.** The former contains conducting bars short-circuited at the end and embedded within it; the latter consists of a multiphase winding similar to that used for the stator, but electrically short-circuited.

In either case, the induction motor operates by virtue of currents induced from the stator field in the rotor. In this respect, its operation is similar to that of a transformer, in that currents in the stator (which acts as a primary coil) induce currents in the rotor (acting as a secondary coil). In most induction motors, no external electrical connection is required for the rotor, thus permitting a simple, rugged construction, without the need for slip rings or brushes. Unlike the synchronous motor, the induction motor does not operate at synchronous speed, but at a somewhat lower speed, which is dependent on the load. Figure 17.37 illustrates the appearance of a squirrel-cage induction motor. The following discussion will focus mainly on this very common configuration.

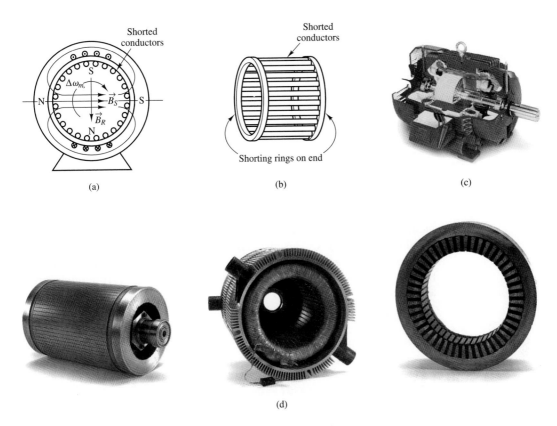

**Figure 17.37** (a) Squirrel-cage induction motor; (b) conductors in rotor; (c) photo of squirrel-cage induction motor; (d) views of Smokin' Buckey motor: rotor, stator, and cross section of stator (Courtesy: David H. Koether Photography.)

You are by now acquainted with the notion of a rotating stator magnetic field. Imagine now that a squirrel-cage rotor is inserted in a stator in which such a rotating magnetic field is present. The stator field will induce voltages in the cage conductors, and if the stator field is generated by a three-phase source, the resulting rotor currents—which circulate in the bars of the squirrel cage, with the conducting path completed by the shorting rings at the end of the cage—are also three-phase, and are determined by the magnitude of the induced voltages and by the impedance of the rotor. Since the rotor currents are induced by the stator field, the number of poles and the speed of rotation of the induced magnetic field are the same as those of the stator field, *if the rotor is at rest*. Thus, when a stator field is initially applied, the rotor field is synchronous with it, and the fields are stationary with respect to each other. Thus, according to the earlier discussion, a *starting torque* is generated.

If the starting torque is sufficient to cause the rotor to start spinning, the rotor will accelerate up to its operating speed. However, an induction motor can never reach synchronous speed; if it did, the rotor would appear to be stationary with respect to the rotating stator field, since it would be rotating at the same speed. But in the absence of relative motion between the stator and rotor fields, no voltage would be induced in the rotor. Thus, an induction motor is limited to speeds somewhere below the synchronous speed, $n_s$. Let the speed of rotation of the rotor be $n$; then, the rotor is losing ground with respect to the rotation of the stator field at a speed $(n_s - n)$. In effect, this is equivalent to backward motion of the rotor at the **slip speed,** defined by $(n_s - n)$. The **slip,** $s$, is usually defined as a fraction of $n_s$:

$$s = \frac{n_s - n}{n_s} \tag{17.73}$$

which leads to the following expression for the rotor speed:

$$n = n_s(1 - s) \tag{17.74}$$

The slip, $s$, is a function of the load, and the amount of slip in a given motor is dependent on its construction and rotor type (squirrel cage or wound rotor). Since there is a relative motion between the stator and rotor fields, voltages will be induced in the rotor at a frequency called the **slip frequency,** related to the relative speed of the two fields. This gives rise to an interesting phenomenon: the rotor field travels relative to the rotor at the slip speed $sn_s$, but the rotor is mechanically traveling at the speed $(1 - s)n_s$, so that the net effect is that the rotor field travels at the speed

$$sn_s + (1 - s)n_s = n_s \tag{17.75}$$

that is, at synchronous speed. The fact that the rotor field rotates at synchronous speed—although the rotor itself does not—is extremely important, because it means that the stator and rotor fields will continue to be stationary with respect to each other, and therefore a net torque can be produced.

As in the case of DC and synchronous motors, important characteristics of induction motors are the starting torque, the maximum torque, and the torque-speed curve. These will be discussed shortly, after some analysis of the induction motor is performed in the next few paragraphs.

### EXAMPLE 17.12 Induction Motor Analysis

**Problem**

Find the full load rotor slip and frequency of the induced voltage at rated speed in a four-pole induction motor.

**Solution**

**Known Quantities:** Motor ratings.

**Find:** $s$; $f_R$.

**Schematics, Diagrams, Circuits, and Given Data:** Motor ratings: 230 V; 60 Hz; full-load speed: 1,725 rev/min.

**Analysis:** The synchronous speed of the motor is

$$n_s = \frac{120f}{p} = \frac{60f}{p/2} = \frac{60 \text{ s/min} \times 60 \text{ rev/s}}{4/2} = 1{,}800 \text{ rev/min}$$

The slip is

$$s = \frac{n_s - n}{n_s} = \frac{1{,}800 \text{ rev/min} - 1{,}725 \text{ rev/min}}{1{,}800 \text{ rev/min}} = 0.0417$$

The rotor frequency, $f_R$, is

$$f_R = sf = 0.0417 \times 60 \text{ Hz} = 2.5 \text{ Hz}$$

---

The induction motor can be described by means of an equivalent circuit, which is essentially that of a rotating transformer. (See Chapter 16 for a circuit model of the transformer.) Figure 17.38 depicts such a circuit model, where:

$R_S$ = stator resistance per phase, $R_R$ = rotor resistance per phase

$X_S$ = stator reactance per phase, $X_R$ = rotor reactance per phase

$X_m$ = magnetizing (mutual) reactance

$R_C$ = equivalent core-loss resistance

$E_S$ = per-phase induced voltage in stator windings

$E_R$ = per-phase induced voltage in rotor windings

The primary internal stator voltage, $\mathbf{E}_S$, is coupled to the secondary rotor voltage, $\mathbf{E}_R$, by an ideal transformer with an effective turns ratio $\alpha$. For the rotor circuit, the induced voltage at any slip will be

$$\mathbf{E}_R = s\mathbf{E}_{R0} \tag{17.76}$$

where $\mathbf{E}_{R0}$ is the induced rotor voltage at the condition in which the rotor is stationary. Also, $X_R = \omega_R L_R = 2\pi f_R L_R = 2\pi s f L_R = s X_{R0}$, where $X_{R0} = 2\pi f L_R$ is the reactance when the rotor is stationary. The rotor current is given by the expression

$$\mathbf{I}_R = \frac{\mathbf{E}_R}{R_R + jX_R} = \frac{s\mathbf{E}_{R0}}{R_R + jsX_{R0}} = \frac{\mathbf{E}_{R0}}{R_R/s + jX_{R0}} \tag{17.77}$$

The resulting rotor equivalent circuit is shown in Figure 17.39.

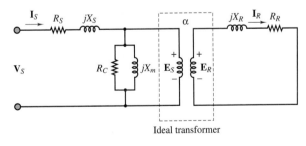

Figure 17.38 Circuit model for induction machine

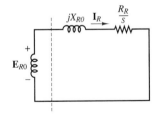

**Figure 17.39** Rotor circuit

The voltages, currents, and impedances on the secondary (rotor) side can be reflected to the primary (stator) by means of the effective turns ratio. When this transformation is effected, the transformed rotor voltage is given by

$$\mathbf{E}_2 = \mathbf{E}'_R = \alpha\mathbf{E}_{R0} \tag{17.78}$$

The transformed (reflected) rotor current is

$$\mathbf{I}_2 = \frac{\mathbf{I}_R}{\alpha} \tag{17.79}$$

The transformed rotor resistance can be defined as

$$R_2 = \alpha^2 R_R \tag{17.80}$$

and the transformed rotor reactance can be defined by

$$X_2 = \alpha^2 X_{R0} \tag{17.81}$$

The final per-phase equivalent circuit of the induction motor is shown in Figure 17.40.

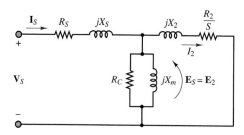

**Figure 17.40** Equivalent circuit of an induction machine

The following examples illustrate the use of the circuit model in determining the performance of the induction motor.

### EXAMPLE 17.13 Induction Motor Analysis

#### Problem

Determine the following quantities for an induction motor using the circuit model of Figures 17.38 to 17.40.

1. Speed
2. Stator current
3. Power factor
4. Output torque

#### Solution

**Known Quantities:**  Motor ratings; circuit parameters.

**Find:**  $n$; $\omega_m$; $\mathbf{I}_S$; pf; $T$.

**Schematics, Diagrams, Circuits, and Given Data:**  Motor ratings: 460 V; 60 Hz; four poles; $s = 0.022$; $P = 14$ hp
$R_S = 0.641\ \Omega$; $R_2 = 0.332\ \Omega$; $X_S = 1.106\ \Omega$; $X_S = 0.464\ \Omega$; $X_m = 26.3\ \Omega$

**Assumptions:**  Use per-phase analysis. Neglect core losses ($R_C = 0$).

## Analysis:

1. The per-phase equivalent circuit is shown in Figure 17.40. The synchronous speed is found to be

$$n_s = \frac{120f}{p} = \frac{60 \text{ s/min} \times 60 \text{ rev/s}}{4/2} = 1{,}800 \text{ rev/min}$$

or

$$\omega_s = 1{,}800 \text{ rev/min} \times \frac{2\pi \text{ rad}}{60 \text{ s/min}} = 188.5 \text{ rad/s}$$

The rotor mechanical speed is

$$n = (1-s)n_s \text{ rev/min} = 1{,}760 \text{ rev/min}$$

or

$$\omega_m = (1-s)\omega_s \text{ rad/s} = 184.4 \text{ rad/s}$$

2. The reflected rotor impedance is found from the parameters of the per-phase circuit to be

$$Z_2 = \frac{R_2}{s} + jX_2 = \frac{0.332}{0.022} + j0.464 \ \Omega$$

$$= 15.09 + j0.464 \ \Omega$$

The combined magnetization plus rotor impedance is therefore equal to

$$Z = \frac{1}{1/jX_m + 1/Z_2} = \frac{1}{-j0.038 + 0.0662\angle - 1.76°} = 12.94\angle 31.1° \ \Omega$$

and the total impedance is

$$Z_{\text{total}} = Z_S + Z = 0.641 + j1.106 + 11.08 + j6.68$$

$$= 11.72 + j7.79 = 14.07\angle 33.6° \ \Omega$$

Finally, the stator current is given by

$$\mathbf{I}_S = \frac{\mathbf{V}_S}{Z_{\text{total}}} = \frac{460/\sqrt{3}\angle 0° \text{ V}}{14.07\angle 33.6° \ \Omega} = 18.88\angle - 33.6° \text{ A}$$

3. The power factor is

$$\text{pf} = \cos 33.6° = 0.883 \qquad \text{lagging}$$

4. The output power, $P_{\text{out}}$, is

$$P_{\text{out}} = 14 \text{ hp} \times 746 \text{ W/hp} = 10.444 \text{ kW}$$

and the output torque is

$$T = \frac{P_{\text{out}}}{\omega_m} = \frac{10{,}444 \text{ W}}{184.4 \text{ rad/s}} = 56.64 \text{ N-m}$$

---

## EXAMPLE  17.14  Induction Motor Analysis

### Problem

Determine the following quantities for a three-phase induction motor using the circuit model of Figures 17.39 to 17.41.

1. Stator current

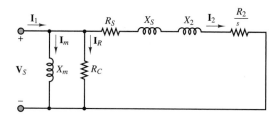

**Figure 17.41**

2. Power factor
3. Full-load electromagnetic torque

---

## Solution

**Known Quantities:** Motor ratings; circuit parameters.

**Find:** $\mathbf{I}_S$; pf; $T$.

**Schematics, Diagrams, Circuits, and Given Data:** Motor ratings: 500 V; 3 $\phi$; 50 Hz; $p = 8$; $s = 0.05$; $P = 14$ hp.
Circuit parameters: $R_S = 0.13\ \Omega$; $R'_R = 0.32\ \Omega$; $X_S = 0.6\ \Omega$; $X'_R = 1.48\ \Omega$; $Y_m = G_C + j B_m$ = magnetic branch admittance describing core loss and mutual inductance $= 0.004 - j0.05\ \Omega^{-1}$; stator to rotor turns ratio $= 1 : \alpha = 1 : 1.57$.

**Assumptions:** Use per-phase analysis. Neglect mechanical losses.

**Analysis:** The approximate equivalent circuit of the three-phase induction motor on a per-phase basis is shown in Figure 17.41. The parameters of the model are calculated as follows:

$$R_2 = R'_R \times \left(\frac{1}{\alpha}\right)^2 = 0.32 \times \left(\frac{1}{1.57}\right)^2 = 0.13\ \Omega$$

$$X_2 = X'_R \times \left(\frac{1}{\alpha}\right)^2 = 1.48 \times \left(\frac{1}{1.57}\right)^2 = 0.6\ \Omega$$

$$Z = R_S + \frac{R_2}{S} + j(X_S + X_2)$$

$$= 0.13 + \frac{0.13}{0.05} + j(0.6 + 0.6) = 2.73 + j1.2\ \Omega$$

Using the approximate circuit,

$$\mathbf{I}_2 = \frac{\mathbf{V}_S}{Z} = \frac{(500/\sqrt{3})\angle 0°\ \text{V}}{2.73 + j1.2\ \Omega} = 88.8 - j39\ \text{A}$$

$$\mathbf{I}_R = \mathbf{V}_S G_S = 288.7\ \text{V} \times 0.004\ \Omega^{-1} = 1.15\ \text{A}$$

$$\mathbf{I}_m = -j\mathbf{V}_S B_m = 288.7\ \text{V} \times (-j0.05)\Omega = -j14.4\ \text{A}$$

$$\mathbf{I}_1 = \mathbf{I}_2 + \mathbf{I}_R + \mathbf{I}_m = 89.95 - j53.4\ \text{A}$$

$$\text{Input power factor} = \frac{\text{Re}[\mathbf{I}_1]}{|\mathbf{I}_1|} = \frac{89.95}{104.6} = 0.86\ \text{lagging}$$

$$\text{Torque} = \frac{3P}{\omega_S} = \frac{3I_2^2 R_2/s}{4\pi f/p} = 935\ \text{N-m}$$

## Performance of Induction Motors

The performance of induction motors can be described by torque-speed curves similar to those already used for DC motors. Figure 17.42 depicts an induction motor torque-speed curve, with five torque ratings marked *a* through *e*. Point *a* is the *starting torque,* also called **breakaway torque,** and is the torque available with the rotor "locked," that is, in a stationary position. At this condition, the frequency of the voltage induced in the rotor is highest, since it is equal to the frequency of rotation of the stator field; consequently, the inductive reactance of the rotor is greatest. As the rotor accelerates, the torque drops off, reaching a maximum value called the **pull-up torque** (point *b*); this typically occurs somewhere between 25 and 40 percent of synchronous speed. As the rotor speed continues to increase, the rotor reactance decreases further (since the frequency of the induced voltage is determined by the relative speed of rotation of the rotor with respect to the stator field). The torque becomes a maximum when the rotor inductive reactance is equal to the rotor resistance; maximum torque is also called **breakdown torque** (point *c*). Beyond this point, the torque drops off, until it is zero at synchronous speed, as discussed earlier. Also marked on the curve are the *150 percent torque* (point *d*), and the *rated torque* (point *e*).

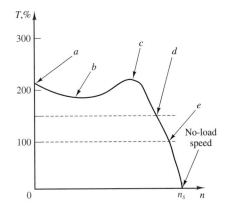

**Figure 17.42** Performance curve for induction motor

A general formula for the computation of the induction motor steady-state torque-speed characteristic is

$$T = \frac{1}{\omega_e} \frac{m V_S^2 R_R / s}{[(R_S + \frac{R_R}{s})^2 + (X_S + X_R)^2]} \qquad \textbf{(17.82)}$$

where *m* is the number of phases.

Different construction arrangements permit the design of **induction motors** with different torque-speed curves, thus permitting the user to select the motor that best suits a given application. Figure 17.43 depicts the four basic classifications, classes A, B, C, and D, as defined by NEMA. The determining features in the classification are the locked-rotor torque and current, the breakdown torque, the pull-up torque, and the percent slip. Class A motors have a higher breakdown torque than class B motors, and a slip of 5 percent or less. Motors in this class are often designed for a specific application. Class B motors are general-purpose motors; this is the

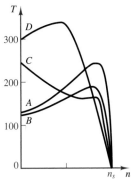

**Figure 17.43** Induction motor classification

most commonly used type of induction motor, with typical values of slip of 3 to 5 percent. Class C motors have a high starting torque for a given starting current, and a low slip. These motors are typically used in applications demanding high starting torque but having relatively normal running loads, once running speed has been reached. Class D motors are characterized by high starting torque, high slip, low starting current, and low full-load speed. A typical value of slip is around 13 percent.

Factors that should be considered in the selection of an AC motor for a given application are the *speed range,* both minimum and maximum, and the speed variation. For example, it is important to determine whether constant speed is required; what variation might be allowed, either in speed or in torque; or whether variable-speed operation is required, in which case a variable-speed drive will be needed. The torque requirements are obviously important as well. The starting and running torque should be considered; they depend on the type of load. Starting torque can vary from a small percentage of full-load to several times full-load torque. Furthermore, the excess torque available at start-up determines the *acceleration characteristics* of the motor. Similarly, *deceleration characteristics* should be considered, to determine whether external braking might be required.

Another factor to be considered is the *duty cycle* of the motor. The duty cycle, which depends on the nature of the application, is an important consideration when the motor is used in repetitive, noncontinuous operation, such as is encountered in some types of machine tools. If the motor operates at zero or reduced load for periods of time, the duty cycle—that is, the percentage of the time the motor is loaded—is an important selection criterion. Last, but by no means least, are the *heating properties* of a motor. Motor temperature is determined by internal losses and by ventilation; motors operating at a reduced speed may not generate sufficient cooling, and forced ventilation may be required.

Thus far, we have not considered the dynamic characteristics of induction motors. Among the integral-horsepower induction motors (i.e., motors with horsepower rating greater than one), the most common dynamic problems are associated with starting and stopping and with the ability of the motor to continue operation during supply system transient disturbances. Dynamic analysis methods for induction motors depend to a considerable extent on the nature and complexity of the problem and the associated precision requirements. When the electrical transients in the motor are to be included as well as the motional transients, and especially when the motor is an important element in a large network, the simple transient equivalent circuit of Figure 17.44 provides a good starting approximation. In the circuit model of Figure 17.44, $X'_S$ is called the *transient reactance.* The voltage $E'_S$ is called the *voltage behind the transient reactance* and is assumed to be equal to the initial value of the induced voltage, at the start of the transient. $R_S$ is the stator resistance. The dynamic analysis problem consists of selecting a sufficiently simple but reasonably realistic representation that will not unduly complicate the dynamic analysis, particularly through the introduction of nonlinearities.

It should be remarked that the basic equations of the induction machine, as derived from first principles, are quite nonlinear. Thus, an accurate dynamic analysis of the induction motor, without any linearizing approximations, requires the use of computer simulation.

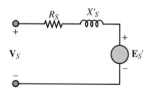

**Figure 17.44** Simplified induction motor dynamic model

# AC Motor Speed and Torque Control

As explained in an earlier section, AC machines are constrained to fixed-speed or near fixed-speed operation when supplied by a constant-frequency source. Several simple methods exist to provide limited **speed control in an AC induction machines;** more complex methods, involving the use of advanced power electronics circuits can be used if the intended application requires wide-bandwidth control of motor speed or torque. In this subsection we provide a general overview of available solutions.

### Pole Number Control

The (conceptually) easiest method to implement speed control in an induction machine is by *varying the number of poles*. Equation 17.55 explains the dependence of synchronous speed in an AC machine on the supply frequency and on the number of poles. For machines operated at 60 Hz, the following speeds can be achieved by varying the number of magnetic poles in the stator winding:

| Number of poles | 2 | 4 | 6 | 8 | 12 |
|---|---|---|---|---|---|
| $n$, rev/min | 3,600 | 1,800 | 1,200 | 800 | 600 |

Motor stators can be wound so that the number of pole pairs in the stators can be varied by switching between possible winding connections. Such switching requires that care be taken in timing it to avoid damage to the machine.

### Slip Control

Since the rotor speed is inherently dependent on the slip, *slip control* is a valid means of achieving some speed variation in an induction machine. Since motor torque falls with the square of the voltage (see equation 17.82), it is possible to change the slip by changing the motor torque through a reduction in motor voltage. This procedure allows for speed control over the range of speeds that allow for stable motor operation. With reference to Figure 17.42, this is possible only above point $c$, that is, above the *breakdown torque*.

### Rotor Control

For motors with wound rotors, it is possible to connect the rotor slip rings to resistors; adding resistance to the rotor increases the losses in the rotor, and therefore causes the rotor speed to decrease. This method is also limited to operation above the *breakdown torque* though it should be noted that the shape of the motor torque-speed characteristic changes when the rotor resistance is changed.

### Frequency Regulation

The last two methods cause additional losses to be introduced in the machine. If a variable-frequency supply is used, motor speed can be controlled without any additional losses. As seen in equation 17.55, the motor speed is directly dependent on the supply frequency, as the supply frequency determines the speed of the rotating magnetic field. However, to maintain the same motor torque characteristics over a range of speeds, the motor voltage must change with frequency, to maintain a constant torque. Thus, generally, the V/Hz ratio should be held constant.

This condition is difficult to achieve at start-up and at very low frequencies, in which cases the voltage must be raised above the constant V/Hz ratio that will be appropriate at higher frequency.

## Adjustable-Frequency Drives

The advances made in the last two decades in power electronics and microcontrollers (see Chapters 11 and 14) have made AC machines employing **adjustable-frequency drives** well-suited to many common engineering applications that until recently required the use of the more easily speed-controlled DC drives. An adjustable-frequency drive consists of four major subsystems, as shown in Figure 17.45.

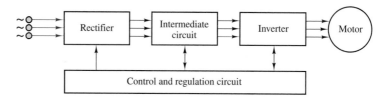

**Figure 17.45** General configuration of adjustable-frequency drive

The diagram of Figure 17.45 assumes that a three-phase AC supply is available; the three-phase AC voltage is rectified using a controlled or uncontrolled **rectifier** (see Chapter 8 for a description of uncontrolled rectifiers and Chapter 11 for a description of controlled rectifiers). An **intermediate circuit** is sometimes necessary to further condition the rectified voltage and current. An **inverter** is then used to convert the fixed DC voltage to a variable frequency and variable amplitude AC voltage. This is accomplished via **pulse-amplitude modulation** (PAM) or increasingly, via **pulse-width modulation** (PWM) techniques. Figure 17.46

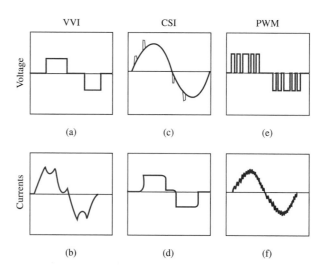

**Figure 17.46** Typical adjustable-frequency controller voltage and current waveforms. (Courtesy: Rockwell Automation, Reliance Electric)

illustrates how approximately sinusoidal currents and voltages of variable frequency can be obtained by suitable shaping a train of pulses. It is important to understand that the technology used to generate such wave shapes is based on the simple power switching concepts underlying the voltage-source inverter (VSI) drive described in Chapter 11. DC-AC inverters come in many different configurations; the interested reader will find additional information and resources in the accompanying CD-ROM.

## Check Your Understanding

**17.14**  A three-phase induction motor has six poles.  (a) If the line frequency is 60 Hz, calculate the speed of the magnetic field in rev/min.  (b) Repeat the calculation if the frequency is changed to 50 Hz.

**17.15**  A four-pole induction motor operating at a frequency of 60 Hz has a full-load slip of 4 percent. Find the frequency of the voltage induced in the rotor (a) at the instant of starting and (b) at full load.

**17.16**  A four-pole, 1,746-rev/min, 220-V, 3-phase, 60-Hz, 10-hp, Y-connected induction machine has the following parameters: $R_S = 0.4$ $\Omega$, $R_2 = 0.14$ $\Omega$, $X_m = 16$ $\Omega$, $X_S = 0.35$ $\Omega$, $X_2 = 0.35$ $\Omega$, $R_C = 0$. Using Figure 16.39, find: (a) the stator current; (b) the rotor current; (c) the motor power factor; and (d) the total stator power input.

## CONCLUSION

The principles developed in Chapter 17 can be applied to rotating electric machines, to explain how mechanical energy can be converted to electrical energy, and vice versa. The former function is performed by electric generators, while the latter is provided by electric motors.

- Electric machines are described in terms of their mechanical characteristics, their torque-speed curves, and their electrical characteristics, including current and voltage requirements. Losses and efficiency are an important part of the operation of electric machines, and it should be recognized that there will be electrical losses (due to the resistance of the windings), mechanical losses (friction and windage), and magnetic core losses (eddy currents, hysteresis). The main mechanical components of an electric machine are the stator, rotor, and air gap. Electrically, the important parameters are the armature (load current-carrying) circuit, and the field (magnetizing) circuit. Magnetic fields establish the coupling between the electrical and mechanical systems.

- Electric machines are broadly classified into DC and AC machines; the former use DC excitation for both the field and armature circuits, while the latter may be further subdivided into two classes: synchronous machines, and induction motors. AC synchronous machines are characterized by DC field and AC armature excitation. Induction machines (of the squirrel-cage type), on the other hand, do not require a field excitation, since this is provided by electromagnetic induction. Typically, DC machines have the armature winding on the rotor, while AC machines have it on the stator.

- The performance of electric machines can be approximately predicted with the use of circuit models, or of performance curves. The selection of a particular machine for a given application is driven by many factors, including the availability of suitable electrical supplies (or prime movers), the type of load, and various other concerns, of which heat dissipation and thermal characteristics are probably the most important.

# CHECK YOUR UNDERSTANDING ANSWERS

| | |
|---|---|
| **CYU 17.1** | (a) $\omega = 55\pi$ rad/s;    (b) $n = 1,650$ rev/min |
| **CYU 17.2** | 275 V |
| **CYU 17.3** | $T = 40.7$ N-m; volt-amperes $= 11,431$ VA; $P_{\max} = 11.5$ hp |
| **CYU 17.4** | 170% of rated torque; 1,700 rev/min |
| **CYU 17.5** | $k_a = 5.1$; $E_b = 45.9$ V |
| **CYU 17.6** | (a) $E_b = 1,040$ V;    (b) $k_a = 10.34$; (c) $T = 4,138$ N-m |
| **CYU 17.7** | (a) 400 A;    (b) 5 A;    (c) 405 A;    (d) 270.25 V |
| **CYU 17.8** | (a) 144 N-m;    (b) 132 N-m |
| **CYU 17.9** | $T = \dfrac{P}{\omega_m} = 9.93$ N-m |
| **CYU 17.10** | Increasing the armature voltage leads to an increase in armature current. Consequently, the motor torque increases until it exceeds the load torque, causing the speed to increase as well. The corresponding increase in back emf, however, causes the armature current to drop and the motor torque to decrease until a balance condition is reached between motor and load torque and the motor runs at constant speed. |
| **CYU 17.11** | 1,500 rev/min; 1,000 rev/min; 750 rev/min; 600 rev/min; 500 rev/min |
| **CYU 17.13** | $T_{\max} = \dfrac{3V_S E_b}{\omega_m X_S}$ |
| **CYU 17.14** | (a) $n = 1,200$ rev/min; (b) $n = 1,000$ rev/min |
| **CYU 17.15** | (a) $f_R = 60$ Hz; (b) $f_R = 2.4$ Hz |
| **CYU 17.16** | (a) $25.92\angle -22.45°$ A; (b) $24.39\angle -6.51°$ A; (c) 0.9243; (d) 8,476 W |

# HOMEWORK PROBLEMS

## Section 1: Electric Machine Fundamentals

**17.1**  Calculate the force exerted by each conductor, 6 in. long, on the armature of a DC motor when it carries a current of 90 A and lies in a field the density of which is $5.2 \times 10^{-4}$ Wb per square in.

**17.2**  In a DC machine, the air-gap flux density is 4 Wb/m². The area of the pole face is 2 cm $\times$ 4 cm. Find the flux per pole in the machine.

**17.3**  The power rating of a motor can be modified to account for different ambient temperature, according to the following table:

| Ambient temperature | 30°C | 35°C | 40°C |
|---|---|---|---|
| Variation of rated power | +8% | +5% | 0 |

| Ambient temperature | 45°C | 50°C | 55°C |
|---|---|---|---|
| Variation of rated power | −5% | −12.5% | −25% |

A motor with $P_e = 10$ kW is rated up to 85°C. Find the actual power for each of the following conditions:

    a.  Ambient temperature is 50°C.

    b.  Ambient temperature is 25°C.

**17.4**  The speed-torque characteristic of an induction motor has been empirically determined as follows:

| Speed | 1,470 | 1,440 | 1,410 | 1,300 | 1,100 |
|---|---|---|---|---|---|
| Torque | 3 | 6 | 9 | 13 | 15 |
| **Speed** | 900 | 750 | 350 | 0 | rev/min |
| Torque | 13 | 11 | 7 | 5 | N-m |

The motor will drive a load requiring a starting torque of 4 N-m and increase linearly with speed to 8 N-m at 1,500 rev/min.

    a.  Find the steady-state operating point of the motor.

    b.  Equation 17.82 predicts that the motor speed can be regulated in the face of changes in load torque by adjusting the stator voltage. Find the change in voltage required to maintain the speed at the operating point of part a if the load torque increases to 10 N-m.

## Section 2: Direct-Current Machines

**17.5**  A 220-V shunt motor has an armature resistance of 0.32 Ω and a field resistance of 110 ohms. At no load the armature current is 6 A and the speed is 1,800 rpm.

Assume that the flux does not vary with load and calculate:

a. The speed of the motor when the line current is 62 A (assume a 2-volt brush drop).

b. The speed regulation of the motor.

**17.6**  A 120-V, 10-A shunt generator has an armature resistance of 0.6 Ω. The shunt field current is 2 A. Determine the voltage regulation of the generator.

**17.7**  A 50-hp, 550-volt shunt motor has an armature resistance, including brushes, of 0.36 ohm. When operating at rated load and speed, the armature takes 75 amp. What resistance should be inserted in the armature circuit to obtain a 20 percent speed reduction when the motor is developing 70 percent of rated torque? Assume that there is no flux change.

**17.8**  A 20-kW, 230-V separately excited generator has an armature resistance of 0.2 Ω and a load current of 100 A. Find:

a. The generated voltage when the terminal voltage is 230 V.

b. The output power.

**17.9**  A 10-kW, 120-VDC series generator has an armature resistance of 0.1 Ω and a series field resistance of 0.05 Ω. Assuming that it is delivering rated current at rated speed, find (a) the armature current and (b) the generated voltage.

**17.10**  The armature resistance of a 30-kW, 440-V shunt generator is 0.1 Ω. Its shunt field resistance is 200 Ω. Find

a. The power developed at rated load.

b. The load, field, and armature currents.

c. The electrical power loss.

**17.11**  A four-pole, 450-kW, 4.6-kV shunt generator has armature and field resistances of 2 and 333 Ω. The generator is operating at the rated speed of 3,600 rev/min. Find the no-load voltage of the generator and terminal voltage at half load.

**17.12**  A shunt DC motor has a shunt field resistance of 400 Ω and an armature resistance of 0.2 Ω. The motor nameplate rating values are 440 V, 1,200 rev/min, 100 hp, and full-load efficiency of 90 percent. Find:

a. The motor line current.

b. The field and armature currents.

c. The counter emf at rated speed.

d. The output torque.

**17.13**  A 30-kW, 240-V generator is running at half load at 1,800 rev/min with efficiency of 85 percent. Find the total losses and input power.

**17.14**  A 240-volt series motor has an armature resistance of 0.42 Ω and a series-field resistance of

0.18 Ω. If the speed is 500 rev/min when the current is 36 A, what will be the motor speed when the load reduces the line current to 21 A? (Assume a 3-volt brush drop and that the flux is proportional to the current.)

**17.15**  A 220-VDC shunt motor has an armature resistance of 0.2 Ω and a rated armature current of 50 A. Find

a. The voltage generated in the armature.

b. The power developed.

**17.16**  A 550-volt series motor takes 112 A and operates at 820 rev/min when the load is 75 hp. If the effective armature-circuit resistance is 0.15 Ω, calculate the horsepower output of the motor when the current drops to 84 A, assuming that the flux is reduced by 15 percent.

**17.17**  A 200-VDC shunt motor has the following parameters:

$$R_a = 0.1 \ \Omega \qquad R_f = 100 \ \Omega$$

When running at 1,100 rev/min with no load connected to the shaft, the motor draws 4 A from the line. Find $E$ and the rotational losses at 1,100 rev/min (assuming that the stray-load losses can be neglected).

**17.18**  A 230-VDC shunt motor has the following parameters:

$$R_a = 0.5 \ \Omega \qquad R_f = 75 \ \Omega$$
$$P_{\text{rot}} = 500 \ \text{W} \ (\text{at } 1{,}120 \ \text{rev/min})$$

When loaded, the motor draws 46 A from the line. Find

a. The speed, $P_{\text{dev}}$, and $T_{\text{sh}}$.

b. If $L_f = 25$ H, $L_a = 0.008$ H, and the terminal voltage has a 115-V change, find $i_a(t)$ and $\omega_m(t)$.

**17.19**  A 200-VDC shunt motor with an armature resistance of 0.1 Ω and a field resistance of 100 Ω draws a line current of 5 A when running with no load at 955 rev/min. Determine the motor speed, the motor efficiency, the total losses (i.e., rotational and $I^2 R$ losses), and the load torque (i.e., $T_{\text{sh}}$) that will result when the motor draws 40 A from the line. Assume rotational power losses are proportional to the square of shaft speed.

**17.20**  A self-excited DC shunt generator is delivering 20 A to a 100-V line when it is driven at 200 rad/s. The magnetization characteristic is shown in Figure P17.20. It is known that $R_a = 1.0 \ \Omega$ and $R_f = 100 \ \Omega$. When the generator is disconnected from the line, the drive motor speeds up to 220 rad/s. What is the terminal voltage?

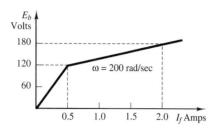

**Figure P17.20**

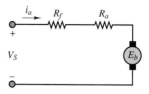

**Figure P17.23**

**17.21** A 50-hp, 230-volt shunt motor has a field resistance of 17.7 $\Omega$ and operates at full load when the line current is 181 A at 1,350 rev/min. To increase the speed of the motor to 1,600 rev/min, a resistance of 5.3 $\Omega$ is "cut in" via the field rheostat; the line current then increases to 190 A. Calculate:

a. The power loss in the field and its percentage of the total power input for the 1,350 rev/min speed.

b. The power losses in the field and the field rheostat for the 1,600 rev/min speed.

c. The percent losses in the field and in the field rheostat at 1,600 rev/min.

**17.22** A 10-hp, 230-volt shunt-wound motor has rated speed of 1,000 rev/min and full-load efficiency of 86 percent. Armature circuit resistance is 0.26 $\Omega$; field-circuit resistance is 225 $\Omega$. If this motor is operating under rated load and the field flux is very quickly reduced to 50 percent of its normal value, what will be the effect upon counter emf, armature current and torque? What effect will this change have upon the operation of the motor, and what will be its speed when stable operating conditions have been regained?

**17.23** The machine of example 17.7 is being used in a series connection. That is, the field coil is connected in series with the armature, as shown in Figure P17.23. The machine is to be operated under the same conditions as in the previous example, that is, $n = 120$ rev/min, $I_a = 8$ A. In the operating region, $\phi = kI_f$, and $k = 200$. The armature resistance is 0.2 $\Omega$, and the resistance of the field winding is negligible.

a. Find the number of field winding turns necessary for full-load operation.

b. Find the torque output for the following speeds:

1. $n' = 2n$    3. $n' = n/2$
2. $n' = 3n$    4. $n' = n/4$

c. Plot the speed-torque characteristic for the conditions of part b.

**17.24** With reference to Example 17.9, assume that the load torque applied to the PM DC motor is zero. Determine the response of the motor speed to a step change in input voltage. Derive expressions for the natural frequency and damping ratio of the second order system. What determines whether the system is over- or underdamped?

**17.25** A motor with polar moment of inertia $J$ develops torque according to the relationship $T = a\omega + b$. The motor drives a load defined by the torque-speed relationship $T_L = c\omega^2 + d$. If the four coefficients are all positive constants, determine the equilibrium speeds of the motor-load pair, and whether these speeds are stable.

**17.26** Assume that a motor has known friction and windage losses described by the equation $T_{FW} = b\omega$. Sketch the $T$-$\omega$ characteristic of the motor if the load torque, $T_L$, is constant, and the $T_L$-$\omega$ characteristic if the motor torque is constant. Assume that $T_{FW}$ at full speed is equal to 30 percent of the load torque.

**17.27** A PM DC motor is rated at 6 V, 3350 rev/min, and has the following parameters: $r_a = 7 \Omega$, $L_a = 120$ mH, $k_T = 7 \times 10^{-3}$ N-m/A, $J = 1 \times 10^{-6}$ kg-m$^2$. The no-load armature current is 0.15 A.

a. In the steady-state no-load condition, the magnetic torque must be balanced by an internal damping torque; find the damping coefficient, $b$. Now sketch a model of the motor, write the dynamic equations, and determine the transfer function from armature voltage to motor speed. What is the approximate 3-dB bandwidth of the motor?

b. Now let the motor be connected to a pump with inertia $J_L = 1 \times 10^{-4}$ kg-m$^2$, damping coefficient $b_L = 5 \times 10^{-3}$ N-m-s, and load torque $T_L = 3.5 \times 1-^{-3}$ N-m. Sketch the model describing the motor-load configuration, and write the dynamic equations for this system; determine the new transfer function from armature voltage to motor speed. What is the approximate 3-dB bandwidth of the motor/pump system?

**17.28** A PM DC motor with torque constant $k_{PM}$ is used to power a hydraulic pump; the pump is a positive displacement type and generates a flow proportional to the pump velocity: $q_p = k_p\omega$. The fluid travels through a conduit of negligible resistance; an accumulator is included to smooth out the pulsations of the pump. A hydraulic load (modelled by a fluid resistance, $R$) is connected between the pipe and a reservoir (assumed at zero pressure). Sketch the motor-pump circuit. Derive the dynamic equations for the system and determine the transfer function between motor voltage and the pressure across the load.

**17.29**   A shunt motor in Figure P17.29 is characterized by a field coefficient $k_f = 0.12$ V-s/A-rad, such that the back emf is given by the expression $E_b = k_f I_f \omega$, and the motor torque by the expression $T = k_f I_f I_a$. The motor drives an inertia/viscous friction load with parameters $J = 0.8$ kg-m$^2$ and $b = 0.6$ N-m-s/rad. The field equation may be approximated by $V_S = R_f I_f$. The armature resistance is $R_a = 0.75$ Ω, and the field resistance is $R_f = 60$ Ω. The system is perturbed around the nominal operating point $V_{S0} = 150$ V, $\omega_0 = 200$ rad/s, $I_{a0} = 186.67$ A, respectively.

a. Derive the dynamic system equations in *symbolic form*.

b. Linearize the equations you obtained in part a.

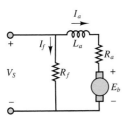

**Figure P17.29**

**17.30**   A PM DC motor is rigidly coupled to a fan; the fan load torque is described by the expression $T_L = 5 + 0.05\omega + 0.001\omega^2$ where torque is in N-m and speed in rad/s. The motor has $k_a\phi = k_T\phi = 2.42$. $R_a = 0.2$ Ω, and the inductance is negligible. If the motor voltage is 50 V, what will be the speed of rotation of the motor and fan?

**17.31**   A separately excited DC motor has the following parameters:

$R_a = 0.1$ Ω     $R_f = 100$ Ω     $L_a = 0.2$ H

$L_f = 0.02$ H     $K_a = 0.8$     $K_f = 0.9$

The motor load is an inertia with $J = 0.5$ kg-m$^2$, $b = 2$ N-m-rad/s. No external load torque is applied.

a. Sketch a diagram of the system and derive the (three) differential equations.

b. Sketch a simulation block diagram of the system (you should have three integrators).

c. Code the diagram using Simulink.

d. Run the following simulations:

*Armature control.* Assume a constant field with $V_f = 100$ V; now simulate the response of the system when the armature voltage changes in step fashion from 50 V to 75 V. Save and plot the current and angular speed responses.
*Field control.* Assume a constant armature voltage with $V_a = 100$ V; now simulate the response of the system when the field voltage changes in step fashion from 75 V to 50 V. This procedure is called *field weakening*. Save and plot the current and angular speed responses.

**17.32**   Determine the transfer functions from *input voltage* to *angular velocity* and from *load torque* to *angular velocity* for a PM DC motor rigidly connected to an inertial load. Assume resistance and inductance parameters $R_a$, $L_a$ let the armature constant be $k_a$. Assume ideal energy conversion, so that $k_a = k_T$. The motor has inertia $J_m$ and damping coefficient $b_m$, and is rigidly connected to an inertial load with inertia $J$ and damping coefficient $b$. The load torque $T_L$ acts on the load inertia to oppose the magnetic torque.

**17.33**   Assume that the coupling between the motor and the inertial load of the proceding problem is flexible (e.g., a long shaft). This can be modeled by adding a torsional spring between the motor inertia and the load inertia. Now we can no longer lump the two inertias and damping coefficients as if they were one, but we need to write separate equations for the two inertias. In total, there will be three equations in this system:
1. the motor electrical equation
2. the motor mechanical equation ($J_m$ and $B_m$)
3. the load mechanical equation ($J$ and $B$)

a. Sketch a diagram of the system.

b. Use free-body diagrams to write each of the two mechanical equations. Set up the equations in matrix form.

c. Compute the transfer function from input voltage to load inertia speed using the method of determinants.

**17.34**   A wound DC motor is connected in both a shunt and series configuration. Assume generic resistance and inductance parameters $R_a$, $R_f$, $L_a$, $L_f$; let the field magnetization constant be $k_f$ and the armature constant be $k_a$. Assume ideal energy conversion, so that $k_a = k_T$. The motor has inertia $J_m$ and damping coefficient $b_m$, and is rigidly connected to an inertial load with inertia $J$ and damping coefficient $b$.

a. Sketch a system-level diagram of the two configurations that illustrates both the mechanical and electrical systems.

b. Write an expression for the torque-speed curve of the motor in each configuration.

c. Write the differential equations of the motor-load system in each configuration.

d. Determine whether the differential equations of each system are linear; if one (or both) are nonlinear, could they be made linear with some simple assumption? Explain clearly under what conditions this would be the case.

## Section 3: AC Synchronous Machines

**17.35**  A non-salient pole, Y-connected, three-phase, two-pole synchronous machine has a synchronous reactance of 7 Ω and negligible resistance and rotational losses. One point on the open-circuit characteristic is given by $V_o = 400$ V (phase voltage) for a field current of 3.32 A. The machine is to be operated as a motor, with a terminal voltage of 400 V (phase voltage). The armature current is 50 A, with power factor 0.85, leading. Determine $E_b$, field current, torque developed, and power angle $\delta$.

**17.36**  A factory load of 900 kW at 0.6 power factor lagging is to be increased by the addition of a synchronous motor that takes 450 kW. At what power factor must this motor operate, and what must be its KVA input if the overall power factor is to be 0.9 lagging?

**17.37**  A non-salient pole, Y-connected, three-phase, two-pole synchronous generator is connected to a 400-V (line to line), 60-Hz, three-phase line. The stator impedance is $0.5 + j1.6$ Ω (per phase). The generator is delivering rated current (36 A) at unity power factor to the line. Determine the power angle for this load and the value of $E_b$ for this condition. Sketch the phasor diagram, showing $\mathbf{E}_b$, $\mathbf{I}_S$, and $\mathbf{V}_S$.

**17.38**  A non-salient pole, three-phase, two-pole synchronous motor is connected in parallel with a three-phase, Y-connected load so that the per-phase equivalent circuit is as shown in Figure P17.38. The parallel combination is connected to a 220-V (line to line), 60-Hz, three-phase line. The load current $\mathbf{I}_L$ is 25 A at a power factor of 0.866 inductive. The motor has $X_S = 2$ Ω and is operating with $I_f = 1$ A and $T = 50$ N-m at a power angle of $-30°$. (Neglect all losses for the motor.) Find $\mathbf{I}_S$, $P_{in}$ (to the motor), the overall power factor (i.e., angle between $\mathbf{I}_1$ and $\mathbf{V}_S$), and the total power drawn from the line.

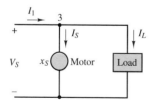

**Figure P17.38**

**17.39**  An automotive alternator is rated 500 V-A and 20 V. It delivers its rated V-A at a power factor of 0.85. The resistance per phase is 0.05 Ω, and the field takes 2 A at 12 V. If friction and windage loss is 25 W and core loss is 30 W, calculate the percent efficiency under rated conditions.

**17.40**  A four-pole, three-phase, Y-connected, non–salient pole synchronous motor has a synchronous reactance of 10 Ω. This motor is connected to a $230\sqrt{3}$ V (line to line), 60-Hz, three-phase line and is driving a load such that $T_{shaft} = 30$ N-m. The line current is 15 A, leading the phase voltage. Assuming that all losses can be neglected, determine the power angle $\delta$ and $E$ for this condition. If the load is removed, what is the line current, and is it leading or lagging the voltage?

**17.41**  A 10-hp, 230-V, 60 Hz, three-phase wye-connected synchronous motor delivers full load at a power factor of 0.8 leading. The synchronous reactance is 6 Ω, the rotational loss is 230 W, and the field loss is 50 W. Find

a.  The armature current.

b.  The motor efficiency.

c.  The power angle.

Neglect the stator winding resistance.

**17.42**  The circuit of Figure P17.42 represents a voltage regulator for a car alternator. Briefly, explain the function of $Q$, $D$, $Z$, and SCR. Note that unlike other alternators, a car alternator is *not* driven at constant speed.

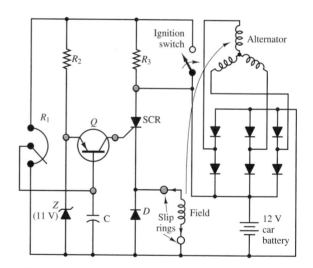

**Figure P17.42**

**17.43**  It has been determined by test that the synchronous reactance, $X_s$, and armature resistance, $r_a$, of a 2,300-V, 500-KVA, three-phase synchronous generator are 8.0 Ω and 0.1 Ω, respectively. If the machine is operating at rated load and voltage at a power factor of 0.867 lagging, find the generated voltage per phase and the torque angle.

**17.44**  A 2,000-hp, unity power factor, three-phase, Y-connected, 2,300-V, 30-pole, 60-Hz synchronous motor has a synchronous reactance of 1.95 $\Omega$ per phase. Neglect all losses. Find the maximum power and torque.

**17.45**   A 1,200-V, three-phase, wye-connected synchronous motor takes 110 kW (exclusive of field winding loss) when operated under a certain load at 1,200 rev/min. The back emf of the motor is 2,000 V. The synchronous reactance is 10 $\Omega$ per phase, with negligible winding resistance. Find the line current and the torque developed by the motor.

**17.46**   The per-phase impedance of a 600-V, three-phase, Y-connected synchronous motor is $5 + j50\ \Omega$. The motor takes 24 kW at a leading power factor of 0.707. Determine the induced voltage and the power angle of the motor.

## Section 4: AC Induction Machines

**17.47**   A 74.6-kW, three-phase, 440-V (line to line), four-pole, 60-Hz induction motor has the following (per-phase) parameters referred to the stator circuit:

$$R_S = 0.06\ \Omega \qquad X_S = 0.3\ \Omega \qquad X_m = 5\ \Omega$$
$$R_R = 0.08\ \Omega \qquad X_R = 0.3\ \Omega$$

The no-load power input is 3,240 W at a current of 45 A. Determine the line current, the input power, the developed torque, the shaft torque, and the efficiency at $s = 0.02$.

**17.48**   A 60-Hz, four-pole, Y-connected induction motor is connected to a 400-V (line to line), three-phase, 60-Hz line. The equivalent circuit parameters are:

$$R_S = 0.2\ \Omega \qquad R_R = 0.1\ \Omega$$
$$X_S = 0.5\ \Omega \qquad X_R = 0.2\ \Omega$$
$$X_m = 20\ \Omega$$

When the machine is running at 1,755 rev/min, the total rotational and stray-load losses are 800 W. Determine the slip, input current, total input power, mechanical power developed, shaft torque, and efficiency.

**17.49**   A three-phase, 60-Hz induction motor has eight poles and operates with a slip of 0.05 for a certain load. Determine

a. The speed of the rotor with respect to the stator.

b. The speed of the rotor with respect to the stator magnetic field.

c. The speed of the rotor magnetic field with respect to the rotor.

d. The speed of the rotor magnetic field with respect to the stator magnetic field.

**17.50**   A three-phase, two-pole, 400-V (per phase), 60-Hz induction motor develops 37 kW (total) of mechanical power ($P_m$) at a certain speed. The rotational loss at this speed is 800 W (total). (Stray-load loss is negligible.)

a. If the total power transferred to the rotor is 40 kW, determine the slip and the output torque.

b. If the total power into the motor ($P_{in}$) is 45 kW and $R_S$ is 0.5 $\Omega$ find $I_S$ and the power factor.

**17.51**   The nameplate speed of a 25-Hz induction motor is 720 rev/min. If the speed at no load is 745 rev/min, calculate:

a. The slip

b. The percent regulation

**17.52**   The name plate of a squirrel-cage four-pole induction motor has the following information: 25 hp, 220 volts, three-phase, 60 Hz, 830 rev/min, 64A line current. If the motor draws 20,800 watts when operating at full load, calculate:

a. Slip

b. Percent regulation if the no-load speed is 895 rpm

c. Power factor

d. Torque

e. Efficiency

**17.53**   A 60-Hz, four-pole, Y-connected induction motor is connected to a 200-V (line to line), three-phase, 60 Hz line. The equivalent circuit parameters are:

$$R_S = 0.48\ \Omega \qquad \text{Rotational loss torque} = 3.5\ \text{N-m}$$
$$X_S = 0.8\ \Omega \qquad R_R = 0.42\ \Omega\ (\text{referred to the stator})$$
$$X_m = 30\ \Omega \qquad X_R = 0.8\ \Omega\ (\text{referred to the stator})$$

The motor is operating at slip $s = 0.04$. Determine the input current, input power, mechanical power, and shaft torque (assuming that stray-load losses are negligible).

**17.54**

a. A three-phase, 220-V, 60-Hz induction motor runs at 1,140 rev/min. Determine the number of poles (for minimum slip), the slip, and the frequency of the rotor currents.

b. To reduce the starting current, a three-phase squirrel-cage induction motor is started by reducing the line voltage to $V_s/2$. By what factor are the starting torque and the starting current reduced?

**17.55**   A six-pole induction motor for vehicle traction has a 50-kW rating and is 85 percent efficient. If the supply is 220 V at 60 Hz, compute the motor speed and torque at a slip of 0.04.

**17.56**   An AC induction machine has six poles and is designed for 60-Hz, 240-V (rms) operation. When the machine operates with 10 percent slip, it produces 60 N-m of torque.

a. The machine is now used in conjunction with a friction load which opposes a torque of 50 N-m. Determine the speed and slip of the machine when used with the above mentioned load.

b. If the machine has an efficiency of 92 percent, what minimum rms current is required for operation with the load of part a)?

[*Hint:* you may assume that the speed torque curve is approximately linear in the region of interest.]

**17.57**  A blocked-rotor test was performed on a 5-hp, 220-V, four-pole, 60 Hz, three-phase induction motor. The following data were obtained: $V = 48$ V, $I = 18$ A, $P = 610$ W. Calculate:

a. The equivalent stator resistance per phase, $R_S$
b. The equivalent rotor resistance per phase, $R_R$
c. The equivalent blocked-rotor reactance per phase, $X_R$

**17.58**  Calculate the starting torque of the motor of Problem 17.58 when it is started at:

a. 220 V
b. 110 V

The starting torque equation is:

$$T = \frac{m}{\omega_e} \cdot V_S^2 \cdot \frac{R_R}{(R_R + R_S)^2 + (X_R + X_S)^2}$$

**17.59**  Find the speed of the rotating field of a six-pole, three-phase motor connected to (a) a 60-Hz line and (b) a 50-Hz line, in rev/min and rad/s.

**17.60**  A six-pole, three-phase, 440-V, 60-Hz induction motor has the following model impedances:

$$R_S = 0.8 \ \Omega \qquad X_S = 0.7 \ \Omega$$
$$R_R = 0.3 \ \Omega \qquad X_R = 0.7 \ \Omega$$
$$X_m = 35 \ \Omega$$

Calculate the input current and power factor of the motor for a speed of 1,200 rev/min.

**17.61**  An eight-pole, three-phase, 220-V, 60-Hz induction motor has the following model impedances:

$$R_S = 0.78 \ \Omega \qquad X_S = 0.56 \ \Omega \qquad X_m = 32 \ \Omega$$
$$R_R = 0.28 \ \Omega \qquad X_R = 0.84 \ \Omega$$

Find the input current and power factor of this motor for $s = 0.02$.

**17.62**  A nameplate is given in Example 17.2. Find the rated torque, rated volt amperes, and maximum continuous output power for this motor.

**17.63**  A 3-phase induction motor, at rated voltage and frequency, has a starting torque of 140 percent and a maximum torque of 210 percent of full-load torque. Neglect stator resistance and rotational losses and assume constant rotor resistance. Determine:

a. The slip at full load.
b. The slip at maximum torque.
c. The rotor current at starting as a percent of full-load rotor current.

# C H A P T E R

# 18

# Special-Purpose Electric Machines

The objective of this chapter is to introduce the operating principles and performance characteristics of a number of special-purpose electric machines that find widespread engineering application in a variety of fields, ranging from robotics to vehicle propulsion, aerospace, and automotive control. In Chapters 16 and 17, you were introduced to the operating principles of the major classes of electric machines: DC machines, synchronous machines, and induction motors. The machines discussed in Chapter 18 operate according to the essential principles described earlier, but are also characterized by unique features that set them apart from the machines described in Chapter 17. The first of these special-purpose machines is the brushless DC motor. Next, we discuss stepping motors, illustrating a very natural match between electromechanical devices and digital logic. The switched reluctance motor is presented next. A discussion of universal motors and single-phase induction motors follows, with a brief description of the types of electronic drives used to supply power to these machines. The discussion of the electronic drives ties the electromechanics material with the subject of power electronics introduced in Chapter 11.

The last section of this chapter covers design and performance specifications related to the application of electric machines.

The machines introduced in Chapter 18 are used in many applications requiring fractional horsepower, or the ability to accurately control position, velocity, or torque. By the end of this chapter, you should be able to:

- Describe the principles of operation of brushless DC motors, stepping motors, switched reluctance motors, universal motors, and single-phase motors.
- Select an appropriate motor and determine the required performance characteristics for a given application.
- Describe the characteristics of the electronic drives required for each of the important motors discussed in the chapter.

## 18.1    BRUSHLESS DC MOTORS

FIND IT
ON THE WEB

In spite of its name, the **brushless DC motor** is actually not a DC motor, but (typically) a permanent-magnet synchronous machine; the name is actually due not to the construction of the machine, but to the fact that its operating characteristics resemble those of a shunt DC motor with constant field current. This characteristic can be obtained by providing the motor with a power supply whose electrical frequency is always identical to the mechanical frequency of rotation of the rotor. To generate a source of variable frequency, use is made of DC-to-AC converters (inverters), consisting of banks of transistors that are switched on and off at a frequency corresponding to the rotor speed; thus, the inverter converts a DC source to an AC source of variable frequency. As far as the user is concerned, then, the source of excitation of a brushless DC motor is DC, although the current that actually flows through the motor windings is AC. (The operation of the inverters will be explained shortly.) In effect, the brushless DC motor is a synchronous motor in which the torque angle, $\delta$, is kept constant by an appropriate excitation current.

Brushless DC motors also require measurement of the position of the rotor to determine its speed of rotation, and to generate a supply current at the same frequency. This function is accomplished by means of a position-sensing arrangement that usually consists either of a magnetic Hall-effect position sensor, which senses the passage of each pole in the rotor, or of an optical encoder similar to the encoders discussed in Chapter 13.

Figure 18.1(a) depicts the appearance of a brushless DC motor. Note how the multiphase winding is similar to that of the synchronous motor of Chapter 17. Figure 18.1(b) depicts the construction of a typical brushless DC servomotor. The brushless motor consists of a stator with a multiphase winding, usually three-phase; a permanent-magnet rotor; and a rotor position sensor. It is interesting to observe that since the commutation is performed electronically by switching the current to the motor—rather than by brushes, as in DC motors—the brushless motor can be produced in many different configurations, including, for example, very flat ("pancake") motors. Figure 18.1 shows the classical configuration of inside rotor, outside stator. For simple machines, it is also possible to resort to an outside rotor, with greater ease of magnet attachment and inherently smoother rotation, but with inferior thermal characteristics, since a stator encased within the rotor structure cannot be cooled efficiently.

In conventional DC motors, the supply voltage is limited by brush wear and sparking that can occur at the commutator, often resulting in the need for transformers to step down the supply voltage. In brushless DC motors, on the other hand, such a concern does not exist, because the commutation is performed electronically without the need for brushes. Further, since, in general, the armature

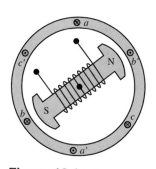

**Figure 18.1a** Two-pole brushless DC motor with three-phase stator winding

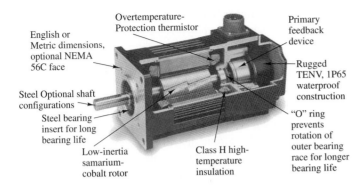

**Figure 18.1b** A typical brushless DC servomotor

(load-carrying winding) is on the stator and thus the losses are concentrated in the stator, liquid cooling (if required) is feasible and does not involve excessive complexity. You will recall that in a conventional DC motor the armature is on the rotor, and therefore auxiliary liquid cooling is very difficult to implement.

Another important advantage of brushless DC motors is that by sealing the stator, submersible units can be built. In addition to these operational advantages, it should be noted that these motors are also characterized by easier construction: the construction of the stator in a brushless DC motor is similar to that in traditional induction motors, and is therefore suitable for automated production. The windings may also be fitted with temperature sensors, providing the possibility of additional thermal protection.

The permanent-magnet rotor is typically made either of rare earth magnets (Sm-Co) or of ceramic magnets (ferrites). Rare earth magnets have outstanding magnetic properties, but they are expensive and in limited supply, and therefore the more commonly employed materials are ceramic magnets. Rare earth magnet motors can be a cost-effective solution—since they allow much greater fluxes to be generated by a given supply current—in applications where high speed, high efficiency, and small size are important. Brushless DC motors can be rated up to 250 kW at 50,000 rev/min. The rotor position sensor must be designed for operation inside the motor, and must withstand the backlash, vibrations, and temperature range typical of motor operation.

Brushless DC motors do require a position-sensing device, though, to permit proper switching of the supply current. Recall that the brushless DC motor replaces the cumbersome mechanical commutation arrangement with electronic switching of the supply current. The most commonly used position-sensing devices are *position encoders* and **resolvers.** The resolver, shown in Figure 18.2, is a rotating machine that is mechanically coupled to the rotor of the brushless motor and consists of two stator and two rotor windings; the stator windings are excited by an AC signal, and the resulting rotor voltages are proportional to the sine and cosine of the angle of rotation of the rotor, thus providing a signal that can be directly related to the instantaneous position of the rotor. The resolver has two major disadvantages: First, it requires a separate AC supply; second, the resolver output must be appropriately decoded to obtain a usable position signal. For these reasons, *angular position encoders* (see Chapter 13) are often used. You will

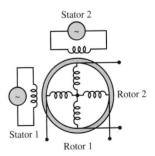

**Figure 18.2** Resolver

recall that such encoders provide a digital signal directly related to the position of a rotating shaft. Their output can therefore be directly used to drive the current supply for a brushless motor.

To understand the operation of the brushless DC motor, it will be useful to make an analogy with the operation of a permanent-magnet (PM) DC motor. As discussed in Chapter 17, in a permanent-magnet DC motor, a fixed magnetic field generated by the permanent magnets interacts with the perpendicular field induced by the currents in the rotor windings, thus creating a mechanical torque. As the rotor turns in response to this torque, however, the angle between the stator and rotor fields is reduced, so that the torque would be nullified within a rotation of 90 electrical degrees. To sustain the torque acting on the rotor, permanent-magnet DC motors incorporate a commutator, fixed to the rotor shaft. The commutator switches the supply current to the stator so as to maintain a constant angle, $\delta = 90°$, between interacting fields. Because the current is continually switched between windings as the rotor turns, the current in each stator winding is actually alternating, at a frequency proportional to the number of motor magnetic poles and the speed.

The basic principle of operation of the brushless DC motor is essentially the same, with the important difference that the supply current switching takes place electronically, instead of mechanically. Figure 18.3 depicts a transistor switching circuit capable of switching a DC supply so as to provide the appropriate currents to a three-phase rotor winding. The electronic switching device consists of a rotor position sensor, fixed on the motor shaft, and an electronic switching module that can supply each stator winding. Diagrams of the phase-to-phase back emf's and the switching sequence of the inverter are shown in Figure 18.4. The back emf waveforms shown in Figure 18.4 are called *trapezoidal*; the total back emf of the inverter is obtained by piecewise addition of the motor phase voltages and is a constant voltage, proportional to motor speed. You should visually verify that the addition of the three phase voltages of Figure 18.4 leads to a constant voltage. The brushless DC motor is therefore similar to a standard permanent-magnet DC motor, and can be described by the following simplified equations:

$$V = k_a \omega_m + R_w I \tag{18.1}$$

$$T = k_T I \tag{18.2}$$

where

$$k_a = k_T$$

and where

$$V = \text{motor voltage}$$
$$k_a = \text{armature constant}$$
$$\omega_m = \text{mechanical speed}$$
$$R_w = \text{winding resistance}$$
$$T = \text{motor torque}$$
$$k_T = \text{torque constant}$$
$$I = \text{motor (armature) current}$$

The speed and torque of a brushless DC motor can therefore be controlled with any variable-speed DC supply, such as one of the supplies briefly discussed in

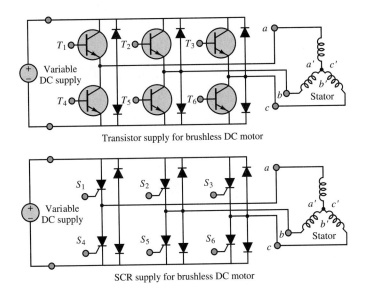

Figure 18.3 Transistor and SCR drives for a brushless DC motor

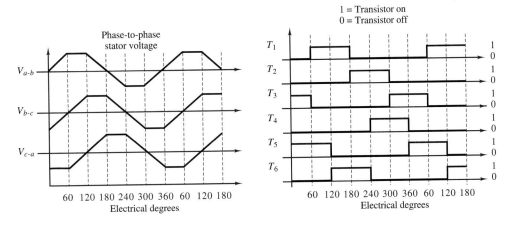

**Figure 18.4** Phase voltages and transistor (SCR) switching sequence for the brushless DC motor drive of Figure 18.3

Chapter 11. Further, since the brushless motor has intrinsically higher torque and lower inertia than its DC counterpart, its response speed is superior to that obtained from traditional DC motors. Figure 18.5 depicts the torque-speed (a) and efficiency (b) curves of a commercially produced brushless DC motor.

One important difference between the conventional DC and the brushless motor, however, is due to the coarseness of the electronic switching compared with the mechanical switching of the brush-type DC motor (recall the discussion of torque ripple due to the commutation effect in DC motors in Chapter 17). In practice, one cannot obtain the exact trapezoidal emf of Figure 18.4 by means of the transistor switching circuit of Figure 18.3, and a voltage ripple results as a consequence, leading to a torque ripple in the motor. Additional phase windings on the stator could solve the problem, at the expense of further complexity in the

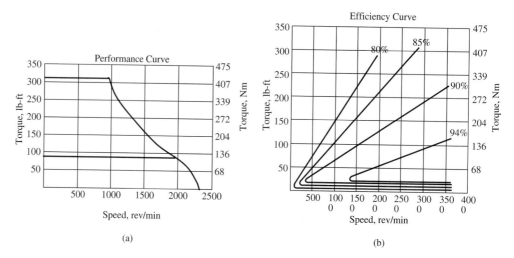

**Figure 18.5** Performance and efficiency characteristics of brushless DC motor *(Courtesy Pacific Scientific)*

drive electronics, since the switching sequence would be more complex. Thus, brushless motors suffer from an inherent trade-off between torque ripple and drive complexity.

Among other applications, brushless DC motors find use in the design of servo loops in control systems—for example, in computer disk drives, and in propulsion systems for electric vehicles. The comparisons between the conventional DC motor and the brushless DC motor are summarized in the following table:

Conventional DC motors

Advantages
  1. Controllability over a wide range of speeds
  2. Capability of rapid acceleration and deceleration
  3. Convenient control of shaft speed and position by servo amplifiers
Disadvantages
  1. Commutation (through brushes) causes wear, electrical noise, and sparking

Brushless DC motors

Advantages
  1. Controllability over a wide range of speeds
  2. Capability of rapid acceleration and deceleration
  3. Convenient control of shaft speed and position
  4. No mechanical wear or sparking problem due to commutation
  5. Better heat dissipation capabilities
Disadvantages
  1. Need for more complex power electronics than the brush-type DC motor for equivalent power rating and control range

# EXAMPLE   18.1   Sinusoidal Torque Generation in Brushless DC Motors

## Problem

Show that the use of sinusoidal currents in a brushless DC motor can result in a ripple-free torque.

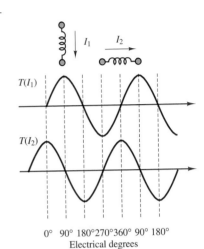

## Solution

**Known Quantities:**  Coil (phase) currents.

**Find:**  Total output torque, $T$.

**Schematics, Diagrams, Circuits, and Given Data:**  $I_{m1} = I_m \sin\theta$; $I_{m2} = I_m \cos\theta$.

**Assumptions:**  The field coil is wound in a two-phase circuit; each winding is sinusoidally spaced. Sinusoidal currents can be generated by suitable power electronics circuits.

**Analysis:**  Using equation 18.2, we determine that the torques generated by the currents in each of the two coils of the two-phase stator are:

$$T_1 = k_T I_{m1} \sin\theta$$

$$T_2 = k_T I_{m2} \cos\theta$$

The sinusoidal form of the torques is due to the sinusoidal distribution of the stator windings in each phase, which are spaced 90 degrees out of phase with one another so as to produce sine-cosine components, and is shown in Figure 18.6.

The net torque produced by the motor is the sum of the two phase torques:

$$T = T_1 + T_2 = k_T I_{m1} \sin\theta + k_T I_{m2} \cos\theta = k_T[(I_m \sin\theta)\sin\theta + (I_m \cos\theta)\cos\theta]$$

$$= k_T I_m[\sin^2\theta + \cos^2\theta] = k_T I_m$$

Thus, the torque generated by the motor is constant, or ripple-free.

**Comments:**  Note that this scheme requires two features:  sinusoidally spaced two-phase windings and sinusoidal phase currents. It is also very important that both the windings and the currents be exactly 90° out of phase.

**Figure 18.6** Sinusoidal torque-generation circuit and current waveforms for a brushless DC motor

# EXAMPLE   18.2   Selecting a Trapezoidal Speed Profile to Match a Desired Motion Profile

## Problem

Determine the trapezoidal speed profile required to move a load 0.5 m in 5 s. Analyze the motion of the motor.

## Solution

**Known Quantities:**  Desired load motion profile.

**Find:**  Required trapezoidal speed profile.

**Schematics, Diagrams, Circuits, and Given Data:**  The motor covers 0.5 m in 100 revolutions. Trapezoidal profile characteristics as shown in Figure 18.7.

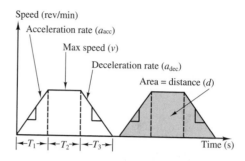

**Figure 18.7** Trapezoidal profile

**Assumptions:**  Assume a trapezoidal speed profile, and that the motor will accelerate for 1 s and decelerate for 1 s.

**Analysis:**  Define the following quantities:

$d$ = motor travel (rev); $v$ = motor speed (rev/s)

$T_1$ = acceleration time (s); $T_2$ = time at maximum speed (s); $T_3$ = deceleration time (s)

$a$ = acceleration or deceleration rate (rev/s$^2$)

From the above definitions, we can calculate the maximum rotational velocity of the motor as follows. For constant acceleration, the expressions for the motor displacement and velocity are:

$$d = \frac{1}{2}at^2 \qquad \text{and} \qquad v = d' = at$$

From the above expressions, we can relate the maximum velocity to the acceleration and deceleration rates:

$$a_{\text{acc}} = \frac{v}{T_1} \qquad a_{\text{dec}} = \frac{v}{T_3}$$

Now we can write an expression for the total motor travel (100 revolutions):

$$d = \frac{1}{2}a_{\text{acc}}T_1^2 + vT_2 + \frac{1}{2}a_{\text{dec}}T_3^2 = \frac{1}{2}\frac{v}{T_1}T_1^2 + vT_2 + \frac{1}{2}\frac{v}{T_3}T_3^2$$

$$= \frac{1}{2}vT_1 + vT_2 + \frac{1}{2}vT_3 = v\left(\frac{1}{2}T_1 + T_2 + \frac{1}{2}T_3\right)$$

Note that the above expression is quite general, and could be used also for asymmetrical profiles. Using the given numbers, we calculate the maximum velocity to be:

$$v = \frac{d}{\left(\frac{1}{2}T_1 + T_2 + \frac{1}{2}T_3\right)} = \frac{100 \text{ rev}}{(0.5 + 3 + 0.5)\text{s}} = 25 \text{ rev/s}$$

which corresponds to $25 \times 60 = 1{,}500$ rev/min.

**Comments:**  The results derived in this example are very useful—trapezoidal speed profiles are very common in servomotors.

## Check Your Understanding

**18.1**   List four features of the brushless DC motor that differ from a conventional shunt-type DC motor.

**18.2**   Repeat Example 18.2 for $I = 7.2$ A.

**18.3**   Convert the result of Exercise 18.2 to oz-in.

**18.4**   Verify that the transistor and SCR switching circuits of Figure 18.3 perform the same function.

## 18.2   STEPPING MOTORS

**Stepping,** or **stepper, motors** are motors that convert digital information to mechanical motion. The principles of operation of stepping motors have been known since the 1920s; however, their application has seen a dramatic rise with the increased use of digital computers. Stepping motors, as the name suggests, rotate in distinct steps, and their position can be controlled by means of logic signals. Typical applications of stepping motors are line printers, positioning of heads in magnetic disk drives, and any other situation where continuous or stepwise displacements are required.

Stepping motors can generally be classified in one of three categories: variable-reluctance, permanent-magnet, and hybrid types. It will soon be shown that the principles of operation of each of these devices bear a definite resemblance to those of devices already encountered in this book. Stepping motors have a number of special features that make them particularly useful in practical applications. Perhaps the most important feature of a stepping motor is that the angle of rotation of the motor is directly proportional to the number of input pulses; further, the angle error per step is very small and does not accumulate. Stepping motors are also capable of rapid responses: starting, stopping, and reversing commands, and can be driven directly by digital signals. Another important feature is a self-holding capability that makes it possible for the rotor to be held in the stopped position without the use of brakes. Finally, a wide range of rotating speeds—proportional to the frequency of the pulse signal—may be attained in these motors.

Figure 18.8 depicts the general appearance of three types of stepping motors. The **permanent-magnet-rotor stepping motor,** Figure 18.8(a), permits a nonzero holding torque when the motor is not energized. Depending on the construction of the motor, it is typically possible to obtain step angles of 7.5, 11.25, 15, 18, 45, or 90°. The angle of rotation is determined by the number of stator poles, as will be illustrated shortly in an example. The **variable-reluctance stepping motor,** Figure 18.8(b), has an iron multipole rotor and a laminated wound stator, and rotates when the teeth on the rotor are attracted to the electromagnetically energized stator teeth. The rotor inertia of a variable-reluctance stepping motor is low, and the response is very quick, but the allowable load inertia is small. When the windings are not energized, the static torque of this type of motor is zero. Generally, the step angle of the variable-reluctance stepping motor is 15°.

The **hybrid stepping motor,** Figure 18.8(c), is characterized by multitoothed stator and rotor, the rotor having an axially magnetized concentric magnet around its shaft. It can be seen that this configuration is a mixture of the variable-reluctance

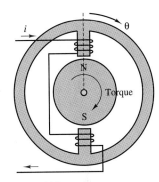

(a) Permanent-magnet
stepping motor

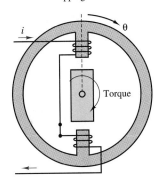

(b) Variable-reluctance
stepping motor

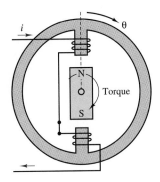

(c) Hybrid stepping motor

**Figure 18.8** Stepping motor configurations

and permanent-magnet types. This type of motor generally has high accuracy and high torque and can be configured to provide a step angle as small as 1.8°. Figure 18.9(a)–(e) depict the construction of a VR step motor.

For any of these configurations, the principle of operation is essentially the same: when the coils are energized, magnetic poles are generated in the stator, and

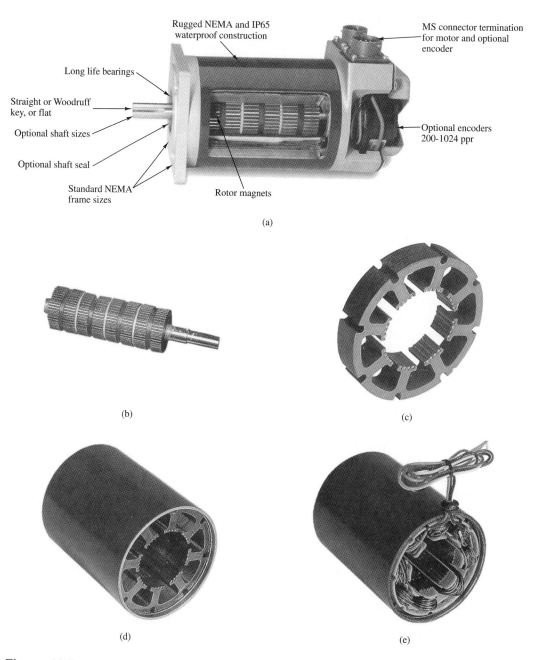

**Figure 18.9** VR Stepper Motor (Courtesy: Pacific Scientific Motor Products Division.) (a) Complete motor assembly. (b) PM rotor. (c) Stator cross section. (d) Fully assembled stator. (e) Stator with windings.

the rotor will align in accordance with the direction of the magnetic field developed in the stator. By reversing the phase of the currents in the coils, or by energizing only some of the coils (this is possible in motors with more than two stator poles), the alignment of the stator magnetic field can take one of a discrete number of positions; if the currents in the coils are pulsed in the appropriate sequence, the rotor will advance in a step-by-step fashion. Thus, this type of motor can be very useful whenever precise incremental motion must be attained. As mentioned earlier, typical applications are printer wheels, computer disk drives, and plotters. Other applications are found in the control of the position of valves (e.g., control of the throttle valve in an engine, or of a hydraulic valve in a fluid power system), and in drug-dispensing apparatus for clinical applications.

The following examples illustrate the operation of a four-pole, two-phase permanent-magnet stepping motor, and of a similar motor of the variable-reluctance type. The operation of these motors is representative of all stepping motors.

## EXAMPLE 18.3 Analysis of Two-Phase, Four-Pole Step Motor

### Problem

Determine the full-step single-phase, full-step two-phase, and half-step current excitation sequences for the PM step motor of Figure 18.10.

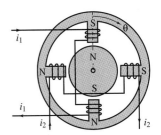

### Solution

**Known Quantities:** Phase currents.

**Find:** Full-step sequence for the motor.

**Assumptions:** The motor currents at the start of the sequence are $i_1 > 0$ and $i_2 = 0$.

**Figure 18.10** Two-phase four-pole PM stepper motor

**Analysis:** With the initial currents assumed (phase 1 energized), the motor will be at rest if the rotor is in the position shown in Figure 18.10. A single-phase sequence consists of turning on each of the two coils in sequence, reversing the polarity of the currents every other time. Then, the PM rotor will align with the stator poles according to the polarity of the magnetic field generated by each coil's pole pair. For example, if coil 1 is turned off and coil 2 is turned on with a positive current polarity, the rotor will rotate clockwise by 90°. Table 18.1 depicts the (bipolar) sequence of coil currents, and the corresponding motor position.

Table 18.1  Full-step, single-phase sequence

| $i_1$ | $i_2$ | $\theta$ |
| --- | --- | --- |
| + | 0 | 0° |
| 0 | + | 90° |
| − | 0 | 180° |
| 0 | − | 270° |
| + | 0 | 0° |

Table 18.2  Full-step, two-phase sequence

| $i_1$ | $i_2$ | $\theta$ |
| --- | --- | --- |
| + | + | 45° |
| − | + | 135° |
| − | − | 225° |
| + | − | 315° |
| + | + | 45° |

Table 18.3  Half-step sequence

| $i_1$ | $i_2$ | $\theta$ |
| --- | --- | --- |
| + | 0 | 0° |
| + | + | 45° |
| 0 | + | 90° |
| − | + | 135° |
| − | 0 | 180° |
| − | − | 225° |
| 0 | − | 270° |
| + | − | 315° |
| + | 0 | 0° |

If both coils are activated, it is possible to cause the rotor to align between stator poles, also in increments of 90°, but shifted in phase by 45° with respect to the single-phase stepping sequence. Table 18.2 illustrates this stepping sequence.

Finally, if one combines the two sequences (easily accomplished, since the current commands for the two sequences are distinct), it is possible to obtain increments of 45°. Table 18.3 depicts the half-step sequence. Any finer resolution would require increasing the number of windings and teeth in the stator.

**Comments:**  The simplicity of the electronic controls required by this type of machine is one of the very attractive features of step motors.

---

## EXAMPLE   18.4  Analysis of Variable-Reluctance Step Motor

### Problem

Determine the current excitation sequences required to achieve 45° steps in the VR step motor of Figure 18.11.

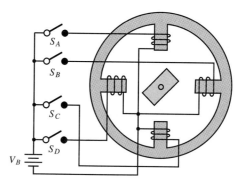

**Figure 18.11** Two-phase four-pole VR stepping motor

---

### Solution

**Known Quantities:**  Phase currents.

**Find:**  Current excitation sequence for 45° steps.

**Assumptions:**  The motor currents at the start of the sequence are $i_1 > 0$ and $i_2 = 0$.

**Analysis:**  The operation of the variable-reluctance (VR) step motor (with a salient-pole rotor) is simpler than that of the PM type, because the rotor is not magnetically polarized, and therefore it is not necessary to have bipolar currents to achieve the desired rotor motion. The stator of Figure 18.10 is excited by DC currents supplied by a single (unipolar) voltage supply. The switches shown in the figure could be controlled by a logic circuit similar to the ones described in Chapters 13 and 14. Note that four separate coils are used.

Figure 18.12 depicts how the first three steps of the sequence could be achieved. These are summarized in Table 18.4.

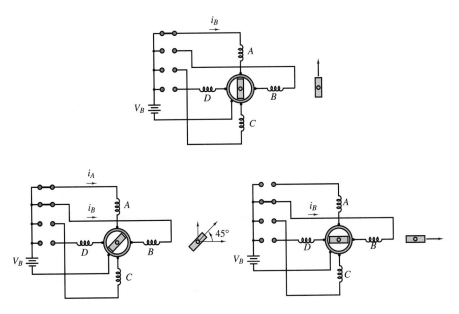

**Figure 18.12** Two-phase four-pole VR motor positioning sequence

**Table 18.4** Current excitation sequence for VR step motor

| $S_A$ | $S_B$ | $S_C$ | $S_D$ | **Rotor position** |
|-------|-------|-------|-------|--------------------|
| 1 | 0 | 0 | 0 | 0° |
| 1 | 1 | 0 | 0 | 45° |
| 0 | 1 | 0 | 0 | 90° |
| 0 | 1 | 1 | 0 | 135° |
| 0 | 0 | 1 | 0 | 180° |
| 0 | 0 | 1 | 1 | 225° |
| 0 | 0 | 0 | 1 | 270° |
| 1 | 0 | 0 | 1 | 315° |
| 1 | 0 | 0 | 0 | 360° |

**Comments:** Note that the circuit required to drive this circuit is even simpler than the one required by the PM step motor.

### EXAMPLE 18.5  Step Angle Determination of VR Step Motor

#### Problem

Determine an expression for the step angle of a VR step motor based on the number of teeth on the rotor and stator and on the number of phases.

#### Solution

**Known Quantities:** Number of rotor and stator teeth; number of phases.

*Find:* Step angle.

*Schematics, Diagrams, Circuits, and Given Data:* $t$ = number of teeth = 4; $m$ = number of phases = 3.

*Analysis:* The number of steps in a revolution, $N$, is given by the product of the number of teeth and the number of phases (e.g., in the preceding example it is equal to 2 teeth × 4 phases = 8 steps). Thus, $N = tm$.

The step angle increment, or *resolution*, is equal to $\Delta\theta = 360°/N$. For the motor described in this example,

$$\Delta\theta = \frac{360°}{N} = \frac{360°}{tm} = \frac{360°}{12 \times 3} = 10°$$

## EXAMPLE 18.6 Torque Equation of Step Motor

### Problem

Calculate the torque generated by a step motor.

### Solution

*Known Quantities:* $t$ = number of teeth per phase; $L$ = axial length of rotor; $g$ = rotor-to-stator radial air gap; $r$ = rotor radius; $\mathcal{F}$ = mmf developed across the two air gaps (in series) through which a line of flux must pass in one phase. Expression for the motor torque.

*Find:* Torque developed by the motor.

*Schematics, Diagrams, Circuits, and Given Data:* $t = 16$ (48 steps, 3-phase excitation); $L = 6.35 \times 10^{-3}$ m; $g = 6.35 \times 10^{-5}$ m; $r = 1.29 \times 10^{-2}$ m; $\mathcal{F} = 720$ A-t.

$$T = 0.314 \times 10^{-6} \frac{tL(r + g/2)\mathcal{F}^2}{g} \quad \text{N-m}$$

*Analysis:* Using the expression given above gives

$$T = 0.314 \times 10^{-6} \frac{tL(r + g/2)\mathcal{F}^2}{g}$$

$$= 0.314 \times 10^{-6} \frac{16 \times 6.35 \times 10^{-3}(1.29 \times 10^{-2} + 3.175 \times 10^{-5})720^2}{6.35 \times 10^{-5}}$$

$$= 3.37 \text{ N-m}$$

From the preceding examples, you should now have a feeling for the operation of variable-reluctance and PM stepping motors. The hybrid configuration is characterized by multitooth rotors that are made of magnetic materials, thus providing a variable-reluctance geometry in conjunction with a permanent-magnet rotor.

An ideal torque-speed characteristic for a stepper motor is shown in Figure 18.13. Two distinct modes of operation are marked on the curve: the **locked-step mode,** and the **slewing mode.** In the first mode, the rotor comes to

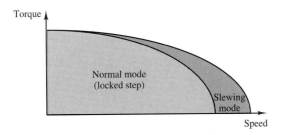

**Figure 18.13** Ideal torque-speed characteristic of a stepping motor

rest (or at least decelerates) between steps; this is the mode commonly used to achieve a given rotor position. In the locked-step mode, the rotor can be started, stopped, and reversed. The slewing mode, on the other hand, does not allow stopping or reversal of the rotor, although the rotor still advances in synchronism with the stepping sequence, as described in the preceding examples. This second mode could be used, for example, in rewinding or fast-forwarding a tape drive.

The power supply, or driver, required by a stepping motor is shown in block diagram form in Figure 18.14; it includes a DC power supply, to provide the required current to drive the motor, in addition to logic and switching circuits to provide the appropriate inputs at the right time. One of the important considerations in driving a stepping motor is the excitation mode, which can be one-phase or two-phase. The driver is the circuit that arranges, distributes, and amplifies pulse trains from the logic circuit determining the stepping sequence; the driver excites each winding of the stepping motor at specified times. In the **one-phase excitation mode,** current is supplied to one phase at a time, with the advantages of low power consumption and good step-angle accuracy. Input signal pulses and the change in the condition of each phase excitation are shown in Figure 18.15. In the **two-phase excitation mode,** current is simultaneously provided to two phases. Input signal pulses and the change in the condition of each phase excitation are also shown in Figure 18.16.

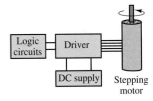

**Figure 18.14** Power supply for stepping motor

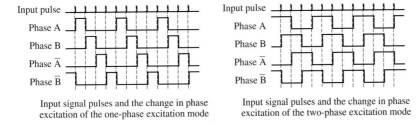

Input signal pulses and the change in phase excitation of the one-phase excitation mode

Input signal pulses and the change in phase excitation of the two-phase excitation mode

**Figure 18.15** One- and two-phase excitation waveforms for stepper motors

In addition to the classification of the excitation by phase, stepping motor drives are also classified according to whether the drive supplies are unipolar or bipolar, that is, whether they can supply current in one or both directions. Unipolar excitation is clearly simpler, although in the case of the two-phase excitation mode,

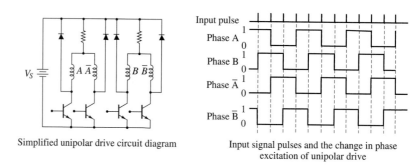

Simplified unipolar drive circuit diagram

Input signal pulses and the change in phase
excitation of unipolar drive

**Figure 18.16** Unipolar drive for stepper motor

only half of the motor windings are used, with an obvious decrease in performance.
Figure 18.17 shows a circuit diagram of the unipolar drive and the sequence of
phase excitation.

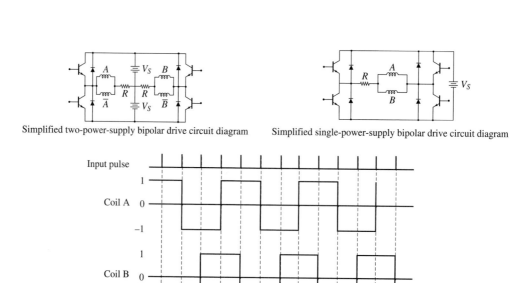

Simplified two-power-supply bipolar drive circuit diagram          Simplified single-power-supply bipolar drive circuit diagram

Input pulse signals and the change in phase excitation of bipolar drive

**Figure 18.17** Bipolar drive for stepper motors

When a bipolar drive is used, motor windings are used effectively, because
of the bidirectional exciting current; when operated in this mode, a stepping motor
can generate a large output torque at low speed compared with the unipolar drive.
Figure 18.17 shows two versions of the bipolar drive. The first requires two power
supplies, one for each polarity, while the second requires only one power supply
but needs four switching transistors per phase to reverse the polarity.

## Check Your Understanding

**18.5**   Explain why the term *variable-reluctance* is used to describe the class of stepping motors so named.

**18.6**   Determine the smallest increment in angular position that can be obtained with a PM stepper motor with six stator teeth and three-phase current excitation.

**18.7**   Express the stepping sequence of the variable-reluctance stepping motor of Example 18.4 as a four-digit binary sequence.

**18.8**   Derive the excitation waveforms corresponding to the direction of rotation opposite to that caused by the stepping sequence shown in Figure 18.14.

**18.9**   For Example 18.6, express the torque in units of lb-in.

## 18.3    SWITCHED RELUCTANCE MOTORS

The **switched reluctance (SR) machine** is the simplest electric machine that permits variable-speed operation. Today, this machine finds increasingly common application in variable-speed drives for industrial applications and in traction drives for automotive propulsion. It is a widely held belief that the SR motor forms the basis of an ideal electric and hybrid-electric vehicle traction drive because of its low cost.

Figure 18.18 depicts the simplest configuration of a reluctance machine and illustrates how the reluctance and inductance of the machine change as a function of position. Note that the magnetic circuit consists only of iron and air—no permanent magnets are required! Note also that the rotor is a salient-pole iron element, which is the lowest-cost rotor that can be manufactured. When a current is supplied to the coil, the rotor will experience a torque seeking to align it with the magnetic poles of the stator; when $\theta = 0$, the torque is zero and the rotor will no longer move, having reached its minimum reluctance position. Note that minimum reluctance corresponds to minimum stored energy in the system. Thus,

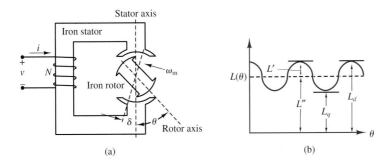

**Figure 18.18** (a) Basic reluctance machine and (b) inductance variation as a function of position

the torque in the motor is developed because of the change in reluctance with rotor position. This principle makes the reluctance machine different from all other (AC or DC) machines discussed so far. Note also that this machine is one of a few machines, along with the induction motor and VR step motor, to be *singly excited*, that is, to have a single source of magnetic field (whether generated by a coil or by a permanent magnet). One can think of the basic reluctance machine as a salient-pole synchronous machine without any field excitation.

The *switched* reluctance machine is a special variation of the simple reluctance machine shown in Figure 18.18 that relies on continuous switching of currents in the stator to guarantee motion of the rotor. It is also a true reluctance machine in that it has *salient poles* both in the rotor and in the stator. The configuration of a typical SR machine is shown in Figure 18.19. Note that the configuration of the SR machine is very similar to that of a VR step motor, discussed in the preceding section. The primary difference between the two is that the SR machine is designed for continuous and not stepped (discrete) motion. The advent of low-cost power semiconductors, especially GTOs, IGBTs and power MOSFETs (see Chapter 11) has made it possible to reliably control SR machines. With reference to Figure 18.19, you can see that the stator of a SR machine is wound through slots, with simple solenoid-type windings, and is similar to that of an induction or synchronous AC machine. This stator can be excited by any multiphase source, such as the three-phase sources described in Chapter 17. The SR machine is excited by discrete current pulses that must be timed with respect to the position of the rotor poles with respect to the stator poles, thus requiring **position feedback.** The speed of the rotor is determine by the switching frequency of the stator coil currents.

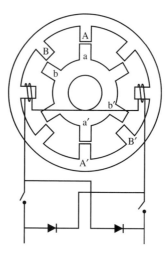

**Figure 18.19** Configuration of switched reluctance machine

## Operating Principles of SR Machine

Torque production in an SR machine depends on the variation in stored magnetic energy as a function of position. Consider the simple reluctance machine of

Figure 18.18, and assume that the variation in winding inductance with rotor position is sinusoidal. The inductance will vary at twice the excitation frequency, because of the number of poles:

$$L(\theta) = L'' + L' \cos 2\theta \tag{18.3}$$

We shall determine the torque generated by the machine given the excitation current

$$i(t) = I_m \sin(\omega t) \tag{18.4}$$

The magnetic stored energy (see Chapter 16) is given by the expression

$$W_m = \frac{1}{2}L(\theta)i^2(t) \tag{18.5}$$

and the flux linkage is

$$\lambda(\theta) = L(\theta)i(t) \tag{18.6}$$

From Chapter 16, we know that the torque can be written as follows:

$$T_m = -\frac{\partial W_m}{\partial \theta} + i\frac{\partial \lambda}{\partial \theta} = -\frac{1}{2}i^2\frac{\partial L}{\partial \theta} + i^2\frac{\partial L}{\partial \theta} = \frac{1}{2}i^2\frac{\partial L}{\partial \theta}. \tag{18.7}$$

Given the known sinusoidal current and inductance variations, we can write the torque expression as:

$$T_m = -I_m^2 L' \sin(2\theta) \sin^2(\omega t). \tag{18.8}$$

It can be shown, with the use of trigonometric identities, that if the rotor rotates at angular velocity $\omega_m$, such that $\theta = \omega_m t - \theta_0$ (with $\theta_0$ equal to the rotor position at $t = 0$), the torque of the SR machine will be nonzero only if the frequency of the sinusoidal stator current is $\omega = \omega_m$. If the electrical frequency is synchronous with the mechanical frequency, then the average torque will be given by the expression

$$\langle T_m \rangle = \frac{1}{4}I_m^2 L' \sin(2\theta_0) = \frac{1}{8}I_m^2(L_d - L_q) \sin(2\theta_0) \tag{18.9}$$

We can draw some conclusions from this simplified analysis of the SR machine:

1. The reluctance machine develops an average torque only at one particular (synchronous) speed, $\omega = \omega_m$. Thus, the reluctance machine is a synchronous machine.
2. The torque developed by the machine is proportional to $(L_d - L_q)$, and is therefore dependent on the amplitude of the variation in inductance (or reluctance); thus, this torque is called **reluctance torque.** The values $L_d$ and $L_q$ are called **direct axis inductance,** and **quadrature axis inductance,** respectively.
3. The torque varies with the angle $\theta_0$, which is therefore equivalent to the *torque angle* $\delta$ defined in Chapter 17 for synchronous machines. The maximum torque occurs at $\theta_0 = \pi/4$, and is called the **pull-out torque.**

The above equations have been derived for a continuous reluctance machine; a switched reluctance machine has discontinuous currents, and will therefore have nonsinusoidal reluctance (inductance) variations and a discontinuous torque. Figure 18.20 depicts the typical appearance of the $L$ and $T_m$ curves for an SR

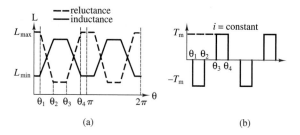

**Figure 18.20** Inductance and torque variation in switched reluctance machine

machine. It can be shown that the magnetic torque generated by an SR machine may be expressed in the form

$$\langle T_m \rangle = \frac{1}{4\pi}(KmP)\tilde{I}^2(L_{max} - L_{min}) \tag{18.10}$$

where $\tilde{I}$ is the rms current, $P$ is the number of pulses per revolution, $m$ is the number of phases, $L_{max}$ and $L_{min}$ are the maximum and minimum inductances seen by the exciting coils, and $K$ is a physical constant. Note that the rms value in equation 18.10 includes also the higher harmonics.

## 18.4    SINGLE-PHASE AC MOTORS

In Chapter 17, two types of AC machines were discussed: synchronous and induction. In the discussion of these devices, especially in their motor applications, three-phase excitation was assumed; however, in many practical applications—and especially in small household appliances and small industrial motors—three-phase sources are not readily available, and it would be desirable to use single-phase excitation. Unfortunately, single-phase power does not lend itself to the generation of a rotating magnetic field: single-phase currents in the winding of an AC machine lead to a magnetic field that pulsates in amplitude but does not rotate in space. Thus, it would not be possible to use the AC machines described in Chapter 17 if only single-phase power were available. The aim of this section is to discuss the construction and the operating and performance characteristics of single-phase AC motors. The discussion will focus mainly on the **universal motor** and on **single-phase induction motors.**

**Fractional-horsepower** (as opposed to **integral-horsepower**) **motors** represent by far the major share of the industrial production of electric motors. Many fractional-horsepower motors are designed for single-phase use, since single-phase AC power is readily available practically anywhere. Many applications are related to household appliances: refrigerator compressors, air conditioners, fans, electric tools, washer and dryer motors, and others. For the rest of this chapter, we shall examine qualitatively the principle of operation of single-phase motors and look at a few applications. The variety of designs for practical single-phase motors is such that it would not be possible to present the detailed principles of operation for all common types. However, it is hoped that the introduction provided in this chapter will help you in decoding the manufacturer's specifications for a given motor, and in making a preliminary selection for a given application.

## Fractional Horsepower Motors

A small motor, as defined by the American Standards Association (ASA) and the National Electrical Manufacturers Association (NEMA; see Chapter 17) is a "motor built in a frame smaller than that having a continuous rating of 1 hp, open type, at 1700 to 1800 rev/min." Small motors are generally considered *fractional-horsepower motors*. However, since the determination is based on frame size and on a given speed range, the classification of a motor is not always obvious. Let us give two examples.

1.  Consider a $\frac{3}{4}$-hp, 1,200-rev/min motor. This motor is not considered a fractional-horsepower motor, because of its frame size. If the same frame size were used for an 1,800-rev/min motor, it would produce a rating of more than 1 hp. Thus, it is considered an integral-horsepower motor of

$$0.75 \text{ hp} \times \frac{1,800}{1,200} = 1.125 \text{ hp}$$

In other words, since the motor is capable of integral-horsepower performance at speeds of 1,700 to 1,800 rev/min, it is classified as an integral-horsepower motor.

2.  Consider now a 1.25-hp, 3,600-rev/min motor. This motor is classified as a fractional-horsepower motor, in spite of the fact that its power output is actually greater than 1 hp. If the same motor were used at a speed of 1,800 rev/min, it would produce a rating of less than 1 hp:

$$1.25 \times \frac{1,800}{3,600} = 0.625 \text{ hp}$$

Thus, we see once again that some attention must be paid to the speed of operation of the motor in determining its classification. The term *fractional horsepower* relates more to the physical size of the machine than to the actual power output rating.

## The Universal Motor

If it were possible to operate a DC motor from a single-phase AC supply, a wide range of simple applications would become readily available. Recall that the direction of the torque produced by a DC machine is determined by the direction of current flow in the armature conductors and by the polarity of the field; torque is developed in a DC machine because the commutator arrangement permits the field and armature currents to remain in phase, thus producing torque in a constant direction. A similar result can be obtained by using an AC supply, and by connecting the armature and field windings in series, as shown in Figure 18.21. A series DC motor connected in this configuration can therefore operate on a single-phase AC supply, and is referred to as a **universal motor.** An additional consideration is that, because of the AC excitation, it is necessary to reduce AC core losses by laminating the stator; thus, the universal motor differs from the series DC motor discussed in Chapter 17 in its construction features. Typical torque-speed curves for AC and DC operation of a universal motor are shown in Figure 18.22. As shown in Figure 18.21, the load current is sinusoidal and therefore reverses direction each half-cycle; however, the torque generated by the motor is always in the same direction, resulting in a pulsating torque, with nonzero average value.

As in the case of a DC series motor, the best method for controlling the speed of a universal motor is to change its (rms) input voltage. The higher the rms input voltage, the greater the resulting speed of the motor. Approximate torque-

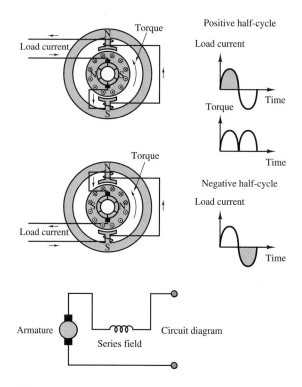

**Figure 18.21** Operation and circuit diagram of a universal motor

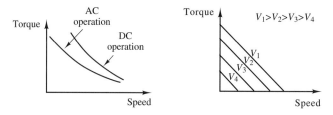

**Figure 18.22** Torque-speed curve of a universal motor

**Figure 18.23** Torque-speed characteristics of a universal motor

speed characteristics of a universal motor as a function of voltage are shown in Figure 18.23.

## EXAMPLE   18.7  Analysis of Universal Motor

### Problem

Find the following quantities for a universal motor:

1. Back emf
2. Power output
3. Shaft torque
4. Motor efficiency

## Solution

**Known Quantities:**  Motor operating data and circuit parameters.

**Find:**  $\mathbf{E}_b$; $P_{\text{out}}$; $T_{\text{out}}$; $\eta$ (efficiency).

**Schematics, Diagrams, Circuits, and Given Data:**  *Motor operating data:* 120 V; 60 Hz; 2 poles; 800 rev/min; 17.85 A (full load); pf $= 0.912$ (lagging).
*Circuit parameters:* $R_f = 0.65\ \Omega$; $X_f = 1.2\ \Omega$; $R_a = 1.36\ \Omega$; $X_{\text{fa}} = 1.6\ \Omega$.

**Assumptions:**  Use the circuit model for the series motor described in Chapter 17. The rotational losses amount to 80 W.

**Analysis:**  The circuit model for the series machine is shown in Figure 18.24. We shall use this model with the understanding that all currents and voltages are now phasors.

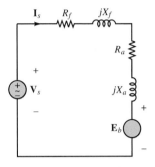

**Figure 18.24** Equivalent circuit of a universal motor

1. *Back emf computation.* To determine the back emf, we need to calculate the voltage across the armature coil and subtract it from the supply voltage:

$$\mathbf{E}_b = \mathbf{V}_S - \mathbf{I}_S(R_f + jX_f + R_a + jX_a)$$
$$= \mathbf{V}_S - (I_S\angle\theta)(R_f + jX_f + R_a + jX_a)$$

The impedance angle is the only unknown quantity and it may be found from the power factor:

$$\text{pf} = \cos(\theta) = 0.912 \text{ (lagging)} \qquad \theta = \arccos(0.912) = -24.22°$$

Thus,

$$\mathbf{E}_b = \mathbf{V}_S - (I_S\angle\theta)(R_f + jX_f + R_a + jX_a)$$
$$= 120 - (17.85\angle 22.24°)(0.65 + j1.2 + 1.36 + j1.6)$$
$$= 73.56\angle - 24.8°\ \text{V}$$

2. *Output power calculation.* The total power developed by the motor is equal to the product of the back emf times the series current:

$$P_{\text{total}} = E_b I_S = 73.56 \times 17.85 = 1{,}313.15\ \text{W}$$

The mechanical (output) power of the motor is the difference between the total power and the rotational losses:

$$P_{\text{out}} = P_{\text{total}} - P_{\text{rot}} = 1{,}313.15 - 80 = 1{,}233.15\ \text{W}$$

3. *Output (shaft) torque calculation.* The output torque is equal to the ratio of output power to shaft speed:

$$T_{\text{out}} = \frac{P_{\text{out}}}{\omega} = \frac{1{,}233.15\ \text{W}}{\dfrac{2\pi \times 800}{60}\ \text{rad/s}} = 14.72\ \text{N-m}$$

4. *Efficiency calculation.* The efficiency of the motor is defined as the ratio of output power to input power:

$$\eta = \frac{P_{\text{out}}}{P_{\text{in}}} = \frac{P_{\text{out}}}{V_S I_S \cos\theta} = \frac{1{,}233.15}{1{,}953.5} = 63.12\%$$

**Comments:**  Note that the analysis of this machine is very similar to that of the series DC motor, except for the use of phasors. It is very important to notice that in calculating the input power, one has to consider the power factor of the motor to obtain the *real power.*

## EXAMPLE   18.8  Universal Motor Torque Expression

### Problem

Compute an expression for the average torque generated by a universal motor, based on the circuit diagram of Figure 18.24.

### Solution

**Known Quantities:**  Circuit model of motor.

**Find:**  Expression for average torque, $T_{ave}$.

**Assumptions:**  The motor operates in the linear region of the magnetization curve.

**Analysis:**  With reference to Chapter 17, we know that the flux produced in a series motor by the series current $i_S(t)$ is $\phi = k_S i_S(t)$. The instantaneous torque produced by the machine is given by the expression

$$T(t) = k_T \phi(t) i_s(t)$$

If the source waveform has period $\tau = 2\tau/\omega$, we can calculate the average power by integrating the instantaneous torque over one period:

$$T_{ave}(t) = \frac{\omega}{2\pi} \int_0^{2\pi/\omega} k_T k_S i_S^2(t') dt' = \frac{\omega}{2\pi} \int_0^{2\pi/\omega} k_T k_S I_S^2 (\sin^2 \omega t') dt' = \frac{1}{2} k_T k_S I_S^2$$

where $I_S$ is the rms value of the (series) armature current

**Comments:**  The series motor can produce a nonzero average torque when excited by an AC current because of the quadratic nature of the instantaneous torque. A PM DC machine, which has a linear torque-current relationship, would generate zero average torque if driven from an AC supply.

## Single-Phase Induction Motors

A typical single-phase induction motor bears close resemblance to the polyphase squirrel-cage induction motor discussed in Chapter 17, the major difference being in the configuration of the stator winding. A simplified schematic diagram of such a motor, with a single winding, is shown in Figure 18.25; the winding is typically distributed around the stator so as to produce an approximately sinusoidal mmf.

Assume that the mmf for a practical motor can be generated so as to approximate the following function:

$$\mathcal{F} = F_{max} \cos(\omega t) \cos(\theta_m)$$

This function can be written as the sum of two components, as follows:

$$\mathcal{F}^+ = \tfrac{1}{2} F_{max} \cos(\theta_m - \omega t)$$
$$\mathcal{F}^- = \tfrac{1}{2} F_{max} \cos(\theta_m + \omega t)$$

These two components may be interpreted as representing two mmf waves traveling in opposite directions around the stator. Each of these mmf's produces torque

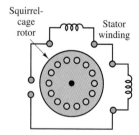

**Figure 18.25** Single-phase induction motor

Squirrel-cage rotor

Stator winding

according to the induction principles described in Chapter 17; however, the two components are equal and opposite, and no net torque results if the rotor is at rest. The resulting mmf is pulsating (i.e., changing in amplitude), but not rotating in space, as it would be in a polyphase stator. If the rotor is made to turn in either direction, however, the two mmf's will not be equal any longer, because the motion of the rotor will induce an additional mmf, which will add to one of the two mmf's and subtract from the other. Thus, a net torque will be established, causing the motor to continue its rotation in the same direction in which it was started. In particular, if the rotor is started in the forward direction, the forward mmf, $\mathcal{F}^+$, will be greater than the backward mmf, and the motor will continue to rotate in the forward direction.

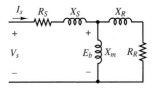

Figure 18.26 depicts an equivalent circuit for the single-phase induction motor *with stationary rotor*, where the parameters in the circuit are defined as follows:

**Figure 18.26** Circuit model for single-phase induction motor with rotor at standstill

$V_s$ = supply voltage

$R_S$ = resistance of stator winding

$X_S$ = leakage reactance of stator winding

$X_m$ = magnetizing reactance of stator winding

$X_R$ = leakage reactance of rotor referred to stator at standstill

$R_R$ = leakage resistance of rotor referred to stator at standstill

$E_b$ = voltage induced in the stator winding by the (stationary) pulsating flux in air gap

Figure 18.27 depicts the equivalent circuit for the same motor with the rotor rotating with slip $s$. Note that the circuit is asymmetrical, because of the different air-gap flux forward and backward components, $E_f$ and $E_b$, respectively. The factors of 0.5 come from the resolution of the pulsating stator mmf into forward and backward components. Note further that the reflected rotor impedance is asymmetrical because of the presence of the slip parameter in the expression for the reflected rotor resistance. Further, the circuit model also confirms that the forward induced voltage, $E_f$, must be greater than the backward voltage, $E_b$, since the slip is always less than 1.

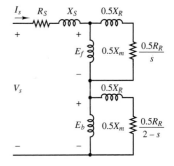

**Figure 18.27** Circuit model for single-phase induction motor with rotor in motion

It can be shown that the torque components in the forward and backward directions are given by the expressions

$$T_f = \frac{P_f}{\omega_s} \qquad (18.11)$$

and

$$T_b = \frac{P_b}{\omega_s} \qquad (18.12)$$

where $\omega_s$ is the synchronous speed and

$$P_f = I_s^2 R_f \qquad (18.13)$$

Here, $R_f$ is the resistive component of the forward field impedance; also,

$$P_b = I_s^2 R_b \qquad (18.14)$$

where $R_b$ is the resistive component of the backward field impedance. Since the torque produced by the backward field is in the opposite direction to that produced by the forward field, the net torque will consist of the difference between the two:

$$T = T_f - T_b = \frac{I_s^2(R_f - R_b)}{\omega_s} \tag{18.15a}$$

The mechanical power developed by the motor is

$$P_{\text{mech}} = T\omega_m = T\omega_s(1 - s) = (P_f - P_b)(1 - s)$$
$$= I_s^2(R_f - R_b)(1 - s) \tag{18.15b}$$

---

### EXAMPLE   18.9   Slip in a Single-Phase Induction Motor

**Problem**

Find the slip of the field in the forward and backward directions for a single-phase induction machine.

---

**Solution**

**Known Quantities:**  Motor operating characteristics.

**Find:**  Forward slip, $s_f$; backward slip, $s_b$.

**Schematics, Diagrams, Circuits, and Given Data:**  Motor operating characteristics: 115 V; 60 Hz; 4 poles; 1710 rev/min.

**Analysis:**  We first determine the synchronous speed of the motor:

$$n_s = \frac{120f}{p} = \frac{120 \times 60}{4} = 1,800 \text{ rev/min}$$

The slip in the forward direction (direction of rotation of the motor) can now be computed:

$$s_f = \frac{n_s - n}{n_s} = \frac{1,800 - 1,710}{1,800} = 0.05 = 5\%$$

The slip in the backward direction can be computed as follows, with reference to Figure 18.26:

$$s_b = 2 - s_f = 2 - 0.05 = 1.95$$

---

### EXAMPLE   18.10   Analysis of Single-Phase Induction Motor

**Problem**

Find the input current and generated torque for a single-phase induction motor.

---

**Solution**

**Known Quantities:**  Motor operating characteristics and circuit parameters.

**Find:** Motor input (stator) current, $I_S$; motor torque, $T$.

**Schematics, Diagrams, Circuits, and Given Data:** *Motor operating data:* $\frac{1}{4}$ hp; 110-V; 60 Hz; 4 poles
*Circuit parameters:* $R_S = 1.5\ \Omega$; $X_S = 2\ \Omega$; $R_R = 3\ \Omega$; $X_R = 2\ \Omega$; $X_m = 50\ \Omega$; $s = 0.05$.

**Assumptions:** The motor is operated at rated voltage and frequency.

**Analysis:** With reference to the equivalent circuit of Figure 18.26, you can easily show that the impedance seen by the backward emf, $E_b$, is much smaller than that seen by the forward emf, $E_f$. This corresponds to stating that the backward component of the magnetizing impedance (which is in parallel with the backward component of the rotor impedance) is much larger than the backward component of the rotor impedance, and can therefore be neglected. This approximation is generally true for values of slip less than 0.15, and corresponds to stating that

$$0.5Z_b = 0.5\frac{jX_m\left(\dfrac{R_R}{2-s} + jX_R\right)}{\dfrac{R_R}{2-s} + j(X_m + X_R)} \approx 0.5\left(\frac{R_R}{2-s} + jX_R\right)$$

The approximate circuit based on this simplification is shown in Figure 18.28.

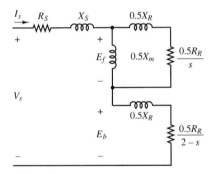

**Figure 18.28** Approximate circuit model for single-phase induction motor

Using the approximate circuit, we find that the impedance seen by the backward emf is given by the expression

$$0.5Z_b = 0.5\left(\frac{R_R}{2-s} + jX_R\right) = 0.5(1.538 + j2) = 0.5(R_b + jX_b)$$

The impedance seen by the forward emf is, on the other hand, given by the exact expression:

$$0.5Z_f = 0.5\frac{jX_m\left(\dfrac{R_R}{s} + jX_R\right)}{\dfrac{R_R}{s} + j(X_m + X_R)}$$

$$= 0.5\frac{j50(60 + j2)}{60 + j(50 + 2)} = 11.9 + j14.69 = 0.5R_f + j0.5X_f$$

If we let $Z_S = R_S + jX_S = 1.5 + j2\ \Omega$, we can write an expression for the total impedance of the motor as follows:

$$Z = Z_S + 0.5Z_f + 0.5Z_b = 14.169 + j17.69 = 22.66\angle 51.3°\ \Omega$$

Knowing the total series impedance we can calculate the stator current:

$$\mathbf{I}_S = \frac{\mathbf{V}_S}{Z} = \frac{110\ \text{V}}{22.66\angle 51.3°\ \Omega} = 4.85\angle -51.3°\ \text{A}$$

We can now calculate the power absorbed by the motor by separately computing the real power absorbed in the forward and backward fields:

$$P_f = I_S^2 \times 0.5R_f = (4.85)^2 \times 11.9 = 279.9\ \text{W}$$
$$P_b = I_S^2 \times 0.5R_b = (4.85)^2 \times 0.769 = 18.1\ \text{W}$$

The net power is the difference between the two components, thus, $P = P_f - P_b = 261.8$ W, and the torque developed by the motor is equal to the ratio of the power to the motor speed. The synchronous speed can be computed to be:

$$\omega_s = \frac{4\pi f}{p} = 188.5\ \text{rad/s}$$

and, if we assume negligible rotational losses, we have:

$$T = \frac{P}{\omega} = \frac{P}{(1-s) \times \omega_s} = \frac{261.8\ \text{W}}{0.95 \times 188.5\ \text{rad/s}} = 1.46\ \text{N-m}$$

**Comments:** Note that the power factor of the motor is pf $= \cos(-51.3°) = 0.625$. Such low power factors are typical of single-phase motors.

---

## EXAMPLE   18.11  Analysis of Single-Phase Induction Motor

### Problem

Find the following quantities for the single-phase machine of Example 18.10:

1. Output torque
2. Output power
3. Efficiency

---

### Solution

**Known Quantities:**  Motor operating characteristics.

**Find:**  Motor torque, $T$; output power, $P_{\text{out}}$; efficiency, $\eta$.

**Schematics, Diagrams, Circuits, and Given Data:**  Motor operating data: $\frac{1}{4}$ hp; 110 V; 60 Hz; 4 poles; $s = 0.05$.

**Assumptions:**  The motor is operated at rated voltage and frequency. The combined rotational and core losses are $P_{\text{rot}} + P_{\text{core}} = 30$ W.

**Analysis:**

1. *Output power calculation.* The motor generated power is the difference between the forward and backward components, as explained in Example 18.10. Thus,

$$P = P_f - P_b = 261.8\ \text{W}$$

The motor power is the difference between the generated power and the losses:

$$P_{\text{out}} = P - P_{\text{loss}} = 261.8 - 30 \text{ W} = 231.8 \text{ W}$$

2. *Shaft torque calculation.* The shaft speed is:

$$\omega = (1 - s) \times \omega_s = (1 - s) \times \frac{4\pi f}{p} = 179 \text{ rad/s}$$

and the torque is:

$$T = \frac{P_{\text{out}}}{\omega} = \frac{231.8 \text{ W}}{179 \text{ rad/s}} = 1.29 \text{ N-m}$$

3. *Efficiency calculation.* To calculate the overall efficiency of the motor we must account for three loss mechanisms: mechanical losses, core losses, and electrical losses. The first two are given as a lumped number; the electrical losses can be computed by calculating the $I^2R$ losses in the stator and forward and backward circuits:

$$P_{S\text{ loss}} = I_S^2 R_s = (4.85)^2 \times 1.5 = 35.3 \text{ W}$$

$$P_{R_f\text{ loss}} = sP_f = 0.05 \times 279.9 = 14 \text{ W}$$

$$P_{R_b\text{ loss}} = (1 - s)P_b = 1.95 \times 18.1 = 35.3 \text{ W}$$

$$P_{\text{elec}} = P_{S\text{ loss}} + P_{R_f\text{ loss}} + P_{R_b\text{ loss}} = 114.6 \text{ W}$$

The efficiency can be calculated according to the following expression:

$$\eta = 1 - \frac{\sum \text{losses}}{P_{\text{out}} + \sum \text{losses}} = 1 - \frac{P_{\text{rot}} + P_{\text{core}} + P_{\text{elec}}}{P_{\text{out}} + P_{\text{rot}} + P_{\text{core}} + P_{\text{elec}}}$$

$$= 1 - \frac{30 + 114.16 \text{ W}}{2131.8 + 30 + 114.16 \text{ W}} = 0.617 = 61.7\%$$

**Comments:**  Note that the overall efficiency of this machine is fairly low. Multiphase AC machines can achieve significantly higher efficiencies.

The equations and circuit models in the preceding examples suggest that a single-phase induction motor is capable of sustaining a torque, and of reaching its operating speed, once it is started by external means. However, because the magnetic field in a single-phase winding is stationary, a single-phase motor is not self-starting. The speed-torque characteristic of a typical single-phase induction motor shown in Figure 18.29 clearly shows that the starting torque for this motor is zero. The curve also shows that the motor can operate in either direction, depending on the direction of the initial starting torque, which must be provided by separate means.

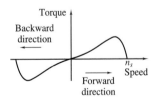

**Figure 18.29** Torque-speed curve of a single-phase induction motor

## Classification of Single-Phase Induction Motors

Thus far, we have not mentioned how the initial starting torque can be provided to a single-phase motor. In practice, single-phase motors are classified by their starting and running characteristics, and several methods exist to provide nonzero starting torque. The aim of this section is to classify single-phase motors by describing their configuration on the basis of the method of starting. For each class of motor, a torque-speed characteristic will also be described.

### Split-Phase Motors

**Split-phase motors** are constructed with two separate stator windings, called **main** and **auxiliary windings;** the axes of the two windings are actually at 90° with respect to each other, as shown in Figure 18.30. The auxiliary winding current is designed to be out of phase with the main winding current, as a result of different reactances of the two windings. Different winding reactances can be attained by having a different ratio of resistance to inductance—for example, by increasing the resistance of the auxiliary winding. In particular, the auxiliary winding current, $I_{aux}$, leads the main winding current, $I_{main}$. The net effect is that the motor sees a two-phase (unbalanced) current that results in a rotating magnetic field, as in any polyphase stator arrangement. Thus, the motor has a nonzero starting torque, as shown in Figure 18.31. Once the motor has started, a centrifugal switch is used to disconnect the auxiliary winding, since a single winding is sufficient to sustain the motion of the rotor. The switching action permits the use of relatively high-resistance windings, since these are not used during normal operation and therefore one need not be concerned with the losses associated with a higher-resistance winding. Figure 18.31 also depicts the combined effect of the two modes of operation of the split-phase motor.

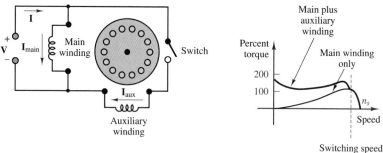

**Figure 18.30** Split-phase motor

**Figure 18.31** Torque-speed curve of split-phase motor

Split-phase motors have appropriate characteristics (at very low cost) for fans, blowers, centrifugal pumps, and other applications in the range of $\frac{1}{20}$ to $\frac{1}{2}$ hp.

### Capacitor-Type Motors

Another method for obtaining a phase difference between two currents that will give rise to a rotating magnetic field is by the addition of a capacitor. Motors that use this arrangement are termed **capacitor-type motors.** These motors make different use of capacitors to provide starting or running capabilities, or a combination of the two. The **capacitor-start motor** is essentially identical to the split-phase motor, except for the addition of a capacitor in series with the auxiliary winding, as shown in Figure 18.32. The addition of the capacitor changes the reactance of the auxiliary circuit in such a way as to cause the auxiliary current to lead the main current. The advantage of using the capacitor as a means for achieving a phase split is that greater starting torque may be obtained than with the split-phase arrangement. A centrifugal switching arrangement is used to disconnect the auxiliary winding above a certain speed, in the neighborhood of 75 percent of synchronous speed.

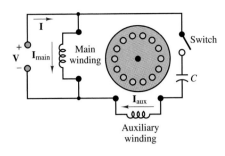

**Figure 18.32** Capacitor-start motor

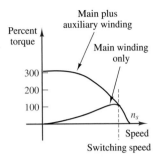

**Figure 18.33** Torque-speed curve for a capacitor-start motor

Figure 18.33 depicts the torque-speed characteristic of a capacitor-start motor. Because of their higher starting torque, these motors are very useful in connection with loads that present a high static torque. Examples of such loads are compressors, pumps, and refrigeration and air-conditioning equipment.

It is also possible to use the capacitor-start motor without the centrifugal switch, leading to a simpler design. Motors with this design are called **permanent split-capacitor motors;** they offer a compromise between running and starting characteristics. A typical torque-speed curve is shown in Figure 18.34.

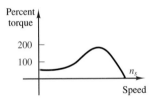

**Figure 18.34** Torque-speed curve for a permanent split-capacitor motor

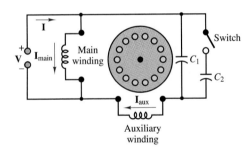

**Figure 18.35** Capacitor-start capacitor-run motor

A further compromise can be achieved by using two capacitors, one to obtain a permanent phase split and the resulting improvement in running characteristics, the other to improve the starting torque. A small capacitance is sufficient to improve the running performance, while a much larger capacitor provides the temporary improvement in starting torque. A motor with this design is called a **capacitor-start capacitor-run motor;** its schematic diagram is shown in Figure 18.35. Its torque-speed characteristic is similar to that of a capacitor-start motor.

---

## EXAMPLE    18.12  Analysis of Capacitor-Start Motor

**Problem**

With reference to Figure 18.32, find the required starting capacitance.

**Solution**

***Known Quantities:*** Motor operating characteristics; motor circuit parameters.

***Find:*** Starting capacitance, $C$.

***Schematics, Diagrams, Circuits, and Given Data:***
*Motor operating data:* $\frac{1}{3}$ hp; 120 V; 60 Hz
*Circuit parameters:* $R_m = 4.5\ \Omega$; $C_m = 3.7\ \Omega$; $R_a = 9.5\ \Omega$; $X_a = 3.5\ \Omega$

***Analysis:*** The purpose of the starting capacitor is to cause the auxiliary winding current, $\mathbf{I}_{aux}$, at standstill to lead the main winding current, $\mathbf{I}_{main}$, by 90°. The 90° phase lead will provide the maximum starting torque. Figure 18.36 shows the phasor diagram for these two currents and the voltage. The impedance angle of the main winding is:

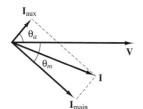

**Figure 18.36** Starting phasor diagram for capacitor-start motor

$$\theta_m = \arctan\left(\frac{X_m}{R_m}\right) = \arctan\left(\frac{3.7\ \Omega}{4.5\ \Omega}\right) = 39.4°$$

Knowing that the desired phase shift between the main and auxiliary impedance angles is −90° (see Figure 18.36), we compute the impedance angle of the auxiliary winding:

$$\theta_a = 39.4° - 90° = -50.6°$$

The minus sign indicates that $\mathbf{I}_{aux}$ leads the terminal voltage. The required capacitance can now be calculated from the relationship

$$\arctan\left(\frac{X_a - X_C}{R_a}\right) = -50.6°$$

$$X_C = -R_a \times \tan(-50.6°) + X_a = -9.5 \times (-1.21) + 3.5 = 15.07\ \Omega$$

and we can compute the desired capacitance to be:

$$C = \frac{1}{\omega X_C} = \frac{1}{377 \times 15.07} = 176 \times 10^{-6}\ \text{F} = 176\mu\text{F}$$

---

### *Shaded-Pole Motors*

The last type of single-phase induction motor discussed in this chapter is the **shaded-pole motor.** This type of motor operates on a different principle from the motors discussed thus far. The stator of a shaded-pole motor has a salient pole construction, as shown in Figure 18.37, that includes a shading coil consisting of a copper band wound around part of each pole. The flux in the shaded portion of the pole lags behind the flux in the unshaded part, achieving an effect similar to a rotation of the flux in the direction of the shaded part of the pole. This flux rotation in effect produces a rotating field that enables the motor to have a starting torque. This construction technique is rather inexpensive and is used in motors up to about $\frac{1}{20}$ hp.

A typical torque-speed characteristic for a shaded-pole motor is given in Figure 18.38.

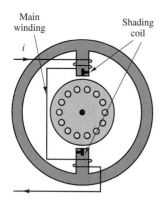

**Figure 18.37** Shaded-pole motor

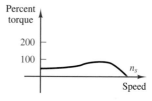

**Figure 18.38** Torque-speed curve of a shaded-pole motor

## EXAMPLE  18.13  Split-Phase Motor Nameplate Analysis

### Problem

The table below depicts a split-phase motor nameplate. Determine the following quantities using nameplate data:

1. Rated slip
2. Synchronous speed
3. Rated torque

### Solution

**Known Quantities:** Nameplate data.

**Find:** $s$; $\omega_S$; $T$.

**Schematics, Diagrams, Circuits, and Given Data:**

| Thermal Protected | | | | Split-Phase Fan & Blower Motor | | | |
|---|---|---|---|---|---|---|---|
| MOD | 4k800 | HP | 1/3 | RPM | 1,725 | KVA CODE | N |
| V | 115 | | | A. | 5.5 | | |
| FR | 48Y | Hz | 60 | PH | 1 | DUTY | CONT. |
| INS.CL. | B | MAX. AMB | 40 C | S. F. | 1.35 | BRG | SLEEVE |

**Analysis:** An explanation of the nameplate for a typical electric motor was given in Chapter 17. This example focuses on a few specific items of interest in the case of a split-phase motor. As you can see, the nameplate directly indicates the split-phase motor classification. Following the Hz designation is the phase information. AC systems may have one, two, or three phases. Single-phase and three-phase systems are the most common.

The code letter following "KVA CODE" indicates the locked-rotor kilovolt-amperes per horsepower, as explained in *NEMA Motor and Generator Standards,* NEMA Publication Number MG1-10.37. The symbol "N" means that this motor has a maximum locked-rotor kilovolt-amperes per horsepower of 12.5. Since the motor is rated at $\frac{1}{3}$ hp, the maximum locked-rotor kilovolt-amperes is $12.5/3 = 4.167$. The maximum locked-rotor amperes at 115 V will be 4.167 kVA/115 V = 36.23 A.

A large percentage of fractional-horsepower motors are now provided with built-in thermal protection. The use of such protection will also be indicated in the motor nameplate—here, for example, by "THERMAL PROTECTED."

Bearing is abbreviated as "BRG." Fractional-horsepower motors normally use one of two types of bearings: sleeve or ball.

A variety of additional information may appear on the nameplate. This may include instructions for connecting the motor to a source of supply, reversing the direction of rotation, lubricating the motor, or operating it safely.

For the machine in this example, the synchronous speed is

$$n_s = 1,800 \text{ rev/min}$$

The slip at rated speed is

$$s = \frac{n_s - n}{n_s} = \frac{1,800 - 1,725}{1,800} = 0.042$$

The power is

$$P = \tfrac{1}{3}\,\text{hp} = \tfrac{1}{3} \times 746\,\frac{\text{W}}{\text{hp}} = 248.7\,\text{W}$$

The rated torque is

$$T = \frac{K \times P}{n}$$

where the constant $K$ is given by

$K = 0.97376$ when $T$ is expressed in meter-kilograms

$K = 9.549$ when $T$ is expressed in newton-meters

$$T = 9.549 \times \frac{248.7}{1{,}725} = 1.377\,\text{N-m}$$

## Summary of Single-Phase Motor Characteristics

Four basic classes of single-phase motors are commonly used:

1. Single-phase induction motors are used for the larger home and small business tasks, such as furnace oil burner pumps, or hot water or hot air circulators. Refrigerator compressors, lathes, and bench-mounted circular saws are also powered with induction motors.

2. Shaded-pole motors are used in the smaller sizes for quiet, low-cost applications. The size range is from $\tfrac{1}{30}$ hp (24.9 W) to $\tfrac{1}{2}$ hp (373 W), particularly for fans and similar drives in which the starting torque is low.

3. Universal motors will operate on any household AC frequency or on DC without modification or adjustment. They can develop very high speed while loaded, and very high power for their size. Vacuum cleaners, sewing machines, kitchen food mixers, portable electric drills, portable circular saws, and home motion-picture projectors are examples of applications of universal motors.

4. The capacitor-type motor finds its widest field of application at low speeds (below 900 rev/min) and in ratings from $\tfrac{3}{4}$ hp (0.5595 kW) to 3 hp (2.238 kW) at all speeds, especially in fan drives.

## Check Your Understanding

**18.10**  Prove that the mmf for the single-phase induction motor can be expressed as the sum of two components traveling in opposite directions.

**18.11**  For the circuit shown in Figure 18.36, draw the starting phasor diagram relating $\mathbf{V}, \mathbf{I}, \mathbf{I}_{main}$, and $\mathbf{I}_{aux}$.

**18.12**

    a. What is the zero-speed torque for a single-phase induction motor?

    b. If external torque is applied to the machine, what is its terminal speed?

## 18.5    MOTOR SELECTION AND APPLICATION

The objective of this section is to outline the selection process of a motor for application to an electrical drive, and to summarize the characteristics of the most common drive motors, with emphasis on fractional-horsepower applications. An electrical motor should satisfy a set of precise requirements to be considered for a specific application. These include:

1.  Starting characteristics (torque and current)
2.  Acceleration characteristics (dependent on the load)
3.  Efficiency at rated load
4.  Overload capability
5.  Electrical and thermal safety
6.  Cost

These requirements suggest that the specific details of the application should be clear in the designer's mind. For example, the nature of the load, of the available electrical supplies, and of the ambient conditions should be carefully investigated before the motor selection process is initiated. Once the application environment is known, it is usually possible to narrow the selection of a drive motor to a few choices. In this section we shall provide some insight into the motor selection process.

### Motor Performance Calculations

To better understand the **motor selection process,** it is important to review some of the basic ideas underlying the motion of the motor and of the load. For rotational systems, the relationships of Table 18.5 hold. Figure 18.39 summarizes the various types of load profiles that are likely to be encountered in practical applications. These include constant-torque loads; viscous friction-type loads, where torque is proportional to speed; loads in which the torque is proportional to a power of speed (e.g., fans, pumps); and constant-power loads, where torque is inversely proportional to speed.

FIND IT
ON THE WEB

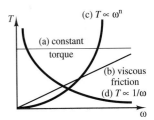

**Figure 18.39** Typical load torque-speed curves

**Table 18.5**  Equations of motion and definitions of variables

| Equations of motion | Definition of terms |
|---|---|
| $\omega_2 = \omega_1 + \alpha t$ | $\omega_1$ = initial velocity (rad-s$^{-1}$) |
| $\theta = \omega_1 t + \frac{1}{2}(\alpha t^2)$ | $\omega_2$ = final velocity (rad-s$^{-1}$) |
| $\omega_2^2 = \omega_1^2 + 2\alpha\theta$ | $\alpha$ = acceleration (rad-s$^{-2}$) |
| $P = T\omega$ | $\theta$ = angular displacement (rad) |
| $T = Fr = J\alpha$ | $n$ = speed of rotation (rev/min) |
| $W = T\theta$ | $P$ = output power (W) |
|  | $W$ = work (J) |
| $\omega = \dfrac{2\pi n}{60} = 0.105n$ | $F$ = force (N) |
|  | $T$ = torque (N-m) |
| $J = mk^2$ | $J$ = polar moment of inertia (kg-m$^2$) |
|  | $r$ = radius of arm (m) |
|  | $k$ = radius of gyration (m) |
|  | $m$ = mass, kg |

### Reflected Load Inertia Calculations

To calculate the motor requirements, one must compute the required torque referenced to the motor output shaft. Since gearing systems are often employed, the inertias of all rotating components must be referred back to the motor shaft. Using the terminology of Table 18.5, we then conclude that the reflected load torque at the motor shaft, $T_r$, is related to the load torque, $T_L$, by the relationship

$$T_r = \frac{\omega_L}{\omega_r} T_L \tag{18.16}$$

where $\omega_r = \omega_m$ is the motor shaft speed, and the ratio of load speed to motor speed is equal to the gear ratio:

$$\frac{\omega_L}{\omega_r} = \frac{n_L}{n_r} \tag{18.17}$$

If we equate the kinetic energy on the motor side to that on the load side, we can also derive an expression for the reflected load inertia:

$$J_L \omega_L^2 = J_r \omega_r^2 \tag{18.18}$$

or

$$J_r = J_L \left( \frac{n_L}{n_r} \right)^2 \tag{18.19}$$

Thus, the reflected inertia seen by the motor at the shaft is equal to the load inertia times the square of the gear ratio. Note that this is a mechanical impedance transformation similar to that used in the case of transformers. For all practical purposes, one can think of a gearing system as a mechanical transformer that, in the ideal case, conserves power. Under this ideal gearing assumption, it can be shown that the acceleration of the load is given by the following expression:

$$\alpha = \frac{T_{m\ \text{peak}}}{\dfrac{\omega_r}{\omega_L} \left[ J_m + \dfrac{J_L}{\left( \dfrac{\omega_r}{\omega_L} \right)^2} \right]} \tag{18.20}$$

where the numerator on the right-hand side is the peak torque the motor can produce and $J_m$ is the motor inertia. If one wished to determine what gear ratio were required to obtain maximum acceleration of the load, equation 18.20 would be differentiated and set equal to zero, to obtain the result

$$\frac{\omega_r}{\omega_L} = \sqrt{\frac{J_L}{J_m}} \tag{18.21}$$

This expression implies that the load inertia should be made equivalent to the motor inertia by appropriate gearing, in order to obtain the best acceleration. Substituting equation 18.21 in 18.20, we can show that the maximum acceleration is given by

$$\alpha_{\text{max}} = \frac{1}{2} \frac{T_{m\ \text{peak}} \omega_L}{J_m \omega_r} \tag{18.22}$$

Equations 18.16 through 18.22 are very useful in sizing a motor and in determining whether any gearing will be necessary to achieve the desired performance.

### Torque Definitions

Although the definition of various torques was introduced in Chapter 17, it will be useful to briefly review these definitions in light of the preceding subsection. In sizing a motor, it is important to ensure that the motor is capable of overcoming the static friction at start-up, to accelerate to the desired speed in an acceptable fashion, and to handle any overloads that may occur. The following definitions will help in the analysis:

1.  **Locked-rotor,** or **static, torque:** The minimum torque the motor will develop at rest for all angular positions under rated conditions.
2.  **Breakdown torque:** The maximum torque a motor will develop under rated conditions without an abrupt drop in speed.
3.  **Full-load torque:** The torque necessary to produce rated power output at full-load speed.
4.  **Acceleration torque:** At any specified speed, acceleration torque, i.e., the torque available for acceleration, is $T_{acc} = T_m - T_L - T_F$, where $T_m$ is the motor torque, $T_L$ is the load torque, and $T_F$ is the frictional load torque.

Clearly, in order for the motor to accelerate to full-speed operation, the motor torque at standstill must exceed the total static-load torque. When the motor torque-speed curve intersects the load torque-speed curve, then a balanced operating condition has been reached.

### Acceleration Calculations

The equation that defines the acceleration characteristics of a motor-load pair is

$$T_m - T_L = J \frac{d\omega}{dt} \tag{18.23}$$

where $T_L$ is the total load torque. From this equation we can calculate the time required to accelerate the load from a speed $\omega_1$ to a speed $\omega_2$ as follows:

$$t = J \int_{\omega_1}^{\omega_2} \frac{d\omega}{T_m - T_L} \; (s) \tag{18.24}$$

or, in units of rev/min,

$$t = \frac{2\pi J_T}{60T} (n_1 - n_2) \; (s) \tag{18.25}$$

where $T$ is the net torque (motor torque minus load torque) and $J_T$ is the total system inertia (motor inertia plus reflected load inertia).

### Efficiency Calculations

The efficiency of a motor is the ratio of the mechanical power output to the electrical power input, that is, the effectiveness of the electromechanical energy conversion. We have already discussed the sources of loss in Chapter 17 and classified them as electrical losses, magnetic losses, and mechanical losses; refer to Section 17.1 for these definitions. The efficiency of a motor, $\eta$, is defined by:

$$\eta = 1 - \left(\frac{P_{\text{loss}}}{P_{\text{input}}}\right) = 1 - \left(\frac{P_{\text{loss}}}{VI}\right) \tag{18.26}$$

### Thermal Calculations

The calculation of the temperature rise and thermal dissipation in a motor can be quite complex and depends very much on the motor construction. For the purpose of illustration, we briefly discuss only the thermal characteristics of a DC motor and perform some thermal calculations for this type of machine.

Thermal dissipation is one of the most important limiting factors in the operation of DC machines. We assume that most power losses take place in the armature (a reasonable assumption, since most of the electrical power flows through the armature circuit), and we use the thermal-electrical system analogy where the thermal difference, $\Delta\theta°$, is given by

$$\Delta\theta° = I_a^2 R_a \times R_T \tag{18.27}$$

and where $I_a$ is the armature current, $R_a$ the armature resistance, and $R_T$ the thermal resistance of the rotor. The thermal time constant of the motor is then defined to be the time (in seconds) taken by the armature to reach 63 percent of the temperature rise corresponding to a given constant power dissipation. Now, the maximum continuous torque the motor can develop is related to the power dissipation, because the motor torque is proportional to the armature current:

$$T_{\text{max}} = K_T I_{a\,\text{max}} = K_T \sqrt{\frac{P_{\text{diss}}}{R_{\text{max}}}} = K_T \sqrt{\frac{\Delta\theta°}{R_T R_{\text{max}}}} \tag{18.28}$$

where $P_{\text{diss}}$ is the dissipated power and $R_{\text{max}}$ the rotor resistance at the maximum temperature, $R_T$ is the rotor thermal resistance at ambient temperature, and $\Delta\theta$ is the temperature rise. The temperature rise of copper can be determined from the known resistance of the wound rotor by computing the maximum temperature as follows:

$$\theta_{\text{max}}° = \left[\frac{R_{\text{max}}}{R_T}\right] (\theta_{\text{ambient}}° + 235) - 235 \tag{18.29}$$

and by computing $\Delta\theta° = \theta_{\text{max}}° - \theta_{\text{ambient}}°$, it is possible to use equation 18.28 to determine the maximum acceptable torque.

Conversely, to ensure that a given motor can operate within its thermal limits, one can calculate an average rms current requirement, $I_{\text{rms}}$, consisting of the acceleration, deceleration, and running current, and use it to compute the temperature as follows:

$$\Delta\theta° = I_{\text{rms}}^2 R_a R_T \tag{18.30}$$

## Motor Selection

The range of electric motor applications is so broad that it is difficult to establish precise rules for motor selection. The differences between applications such as vehicle traction, robot motion, micromotors, disk drives, manufacturing machines, and pump systems, for example, are so many that it is virtually impossible to specify what the best motor would be, unless the application and its environment are clearly specified. The aim of this subsection is simply to outline a procedure that can help in narrowing down the choice of a suitable drive motor to a few most likely candidates.

The first step in selecting a motor is the analysis of the requirements imposed by the application; these can be divided into three groups: (1) motor requirements, (2) load requirements, and (3) control requirements. Table 18.6 summarizes the important considerations for each of these.

Table 18.6  Motor selection requirements

| Motor requirements | Load requirements | Control requirements |
| --- | --- | --- |
| Operating speed | Determine worst-case operating conditions | Available power (AC, DC) |
| Life span and maintenance | Dynamic acceleration, full-load and overload conditions | Motor operating voltage and current |
| Torque characteristics | Starting conditions | Open- or closed-loop |
| Mechanical aspects (size, weight, noise level, environment) | Transients | Forward/reverse operation |
| Applicable standards (e.g., radio frequency interference) | Need for gearing, selection of optimum gear ratio | Motoring and/or braking |
| Overload characteristics | Frictional characteristics | Torque, position, or speed control |
| Thermal dissipation characteristics | | Accuracy of speed or position control |
| | | Controller complexity and cost |

On the basis of the requirements listed in Table 18.6, one can undertake the task of selecting a motor for a specific application.

### Motion Requirements

The first step in the drive selection process is to understand the application-driven specifications, that is, issues such as the type of motion, duty cycle, the required acceleration and gearing system, and the type of control that may be required (position, velocity, torque).

### Motor Sizing

The second step in the drive selection process concerns the sizing of the motor itself. This is done by first calculating the maximum speed; next, the reflected inertia of the load and drive components is calculated, as discussed earlier in this section. From the inertia calculations, the peak torque required by the application can be calculated. The maximum speed and torque requirements thus obtained will narrow the field significantly. Next, one should determine the appropriate constants for each of the candidate motors; these include, in general, inertias, resistances (electrical and thermal), and torque and back emf constants. With these constants it becomes possible to determine that the motor can operate within its thermal specifications by calculating the **temperature rise** of the machine in operation. This, of course, can be a greater limitation during certain portions of the motion cycle—for example, during a hard acceleration.

*Defining the Power Requirements*

This step involves calculating peak voltage and current, to determine the supply requirements.

*Choosing a Transmission*

Although we have been assuming that the mechanical drive system was known beforehand, so that the reflected inertia and peak torque could be calculated, there are many issues that need to be investigated in establishing the drive system—for example, the effect that elastic couplings might have in creating mechanical resonances; noise and vibration characteristics; and backlash due to gearing system imperfections.

It should be apparent from this brief discussion that the process of selecting an electromechanical drive is quite complex, and that it requires a good understanding of many aspects of engineering, including heat transfer, kinematics, dynamics, electronics, systems, and, of course, electromechanics. We hope that this brief introduction will provide the motivation to pursue further studies in this exciting area of engineering.

## CONCLUSION

- The most common engineering applications of electric motors make use of a number of special-purpose motors, with operating characteristics that may be derived from the fundamental principles presented in Chapters 16 and 17.
- The brushless DC motor is a PM synchronous motor in which the mechanical commutation of conventional DC motors is replaced by electronic commutation. Brushless DC motors can be made quite compact and find application in electric vehicle propulsion and motion control.
- Stepping motors—of the variable-reluctance, PM, or hybrid type—permit fine angular displacement control by moving in fixed, discrete steps. Typical applications are in robotics and control systems. Switched reluctance machines are gaining more widespread acceptance because of their simplicity, since they require no permanent magnets and very simple stator windings. Candidate applications for SR machines are low-cost industrial uses and vehicular propulsion.
- The universal motor is very similar in construction to a series DC motor but can operate on AC supplies; its speed can be controlled by means of electronic circuitry of modest complexity. Thus, the universal motor finds common application in both variable- and fixed-speed appliances, such as power drills and vacuum cleaners, respectively.
- Squirrel-cage induction motors can be operated on a single-phase AC supply if a means is provided for establishing a starting torque. Various techniques are commonly employed, such as split-phase, capacitor-start, and shaded-pole construction. The different types are characterized by differing torque-speed characteristics that make the single-phase induction motor a very versatile device. This is the most commonly employed electric machine.

## ANSWERS TO CHECK YOUR UNDERSTANDING

**CYU 18.2**     $T = 1.07$ N-m

**CYU 18.3**     $T = 150.9$ oz-in

**CYU 18.6**     $\Delta\theta = 20°$

**CYU 18.7**

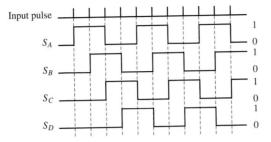

**CYU 18.8**

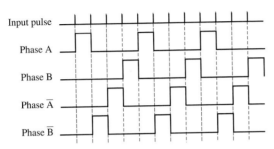

**CYU 18.9**     29.875 lb-in

**CYU 18.11**

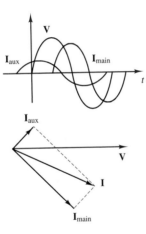

**CYU 18.12**     (a) Zero, without a start winding; (b) $n = (1 - s)n_s$, where $n_s = 120 f/p$ and $s = $ slip (determined by shaft load).

---

# HOMEWORK PROBLEMS

## Section 1: Brushless DC Motors

**18.1**   It is found that $\lambda_m = 0.1$ V-s for a permanent magnet six-pole two-phase synchronous machine. Calculate the amplitude (peak value) of the open-circuit phase voltage measured when the rotor is turned at 60 rev/sec.

**18.2**   A four-pole two-phase brushless dc motor is driven by a mechanical source at $n = 3600$ rev/min. The

open-circuit voltage across one of the phases is 50 V rms.

a.  Calculate $\lambda$.

b.  The mechanical source is removed and the following voltages are applied: $V_a = \sqrt{2}\,25\cos\theta$, $V_b = \sqrt{2}\,25\sin\theta$ where $\theta = \omega_e t$. Calculate the no-load rotor speed $\omega$ in rad/s.

**18.3** With reference to Example 18.2, we wish to shorten the trapezoidal speed profile cycle time by accelerating the motor to a maximum speed of 1,800 rev/min. If we still allow 1 s for acceleration and deceleration, how long will the cycle time be?

**18.4** With reference to the triangular speed profile of Figure P18.4, determine the speed profile required to move a load 0.5 m or 100 revolutions in 3 s.

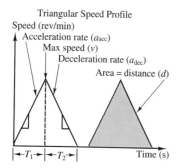

**Figure P18.4**

## Section 2: Step Motors

**18.5** With reference to Example 18.4, design a logic circuit that uses the logic design principles of Chapters 13 and 14 to achieve the step sequence given in Table 18.4. [*Hint:* use a counter and logic gates.]

**18.6** A PM stepper motor has six poles and a bipolar supply (i.e., the current into each coil pair can be either positive or negative). Figure 18.10 depicts a four-pole stepper motor as an example; the motor described in this problem has two additional poles. The spacing between the poles is uniform. Determine the size of the smallest achievable step in degrees.

**18.7** Derive the dynamic equation for a stepping motor coupled to a load. The motor moment of inertia is $J_m$, the load moment of inertia is $J_L$, the viscous damping coefficient is $D$, and motor friction torque is $T_f$.

**18.8** Sketch the rotor-stator configuration of a hybrid stepper motor capable of 18° steps. [*Hint:* The rotor will have five teeth.]

**18.9** Use a binary counter and logic gates to implement the stepping motor binary sequence of Check Your Understanding Exercise 18.7.

**18.10** A two-phase permanent-magnet stepper motor has 50 rotor teeth. When the rotor is driven by an external mechanical source at $\omega = 100$ rad/s, the measured open circuit phase voltage is 25 V, peak-to-peak. Calculate $\lambda$. If $i_a = 1$A and $i_b = 0$, express the developed torque. Assume the winding resistance is 0.1 $\Omega$.

**18.11** The schematic diagram of a four-phase, two-pole PM stepper motor is shown in Figure P18.11. The phase coils are excited in sequence by means of a logic circuit. Find

a. The logic schedule for full-stepping of this motor.

b. The displacement angle of the full step.

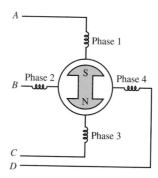

**Figure P18.11**

**18.12** A PM stepper motor is designed to provide a full-step angle of 15°. Find the number of stator and rotor poles.

**18.13** A bridge driver scheme for a two-phase stepping motor is shown in Figure P18.13. Find the excitation sequences of the bridge operation (fill in the blanks of the table).

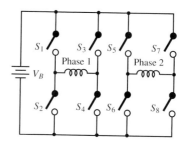

| Clock state | Reset | 1 | 2 | 3 | 4 | 5 | 6 | 7 | 8 |
|---|---|---|---|---|---|---|---|---|---|
| $S_1$ | | | | | | | | | |
| $S_2$ | | | | | | | | | |
| $S_3$ | | | | | | | | | |
| $S_4$ | | | | | | | | | |
| $S_5$ | | | | | | | | | |
| $S_6$ | | | | | | | | | |
| $S_7$ | | | | | | | | | |
| $S_8$ | | | | | | | | | |

**Figure P18.13**

**18.14** A permanent-magnet stepper motor with a 15° step angle is used to directly drive a 0.100 in. lead screw. Determine:

a. The resolution of the stepper motor in steps/revolution,

b. The distance the lead screw travels (in inches) for each 15° step of the stepper motor,

c. The number of full 15° steps required to move the lead screw and the stepper motor shaft through 17.5 revolutions, and

d. The shaft speed (in rev/min) when the stepping frequency is 220 pps.

## Section 3: Single-Phase AC Motors

**18.15**  Determine whether the following motors are integral- or fractional-horse power motors:

a. $\frac{3}{4}$ hp, 900 rev/min

b. $1\frac{1}{2}$ hp, 3,600 rev/min

c. $\frac{3}{4}$ hp, 1,800 rev/min

d. $1\frac{1}{2}$ hp, 6,000 rev/min

**18.16**  The spatial fluctuation of the stator mmf $\mathcal{F}_1$ is expressed as

$$\mathcal{F}_1 = F_{1(peak)} \cos \theta$$

where $\theta$ is the electrical angle measured from the stator coil axis and $F_{1(peak)}$ is the instantaneous value of the mmf wave at the coil axis and is proportional to the instantaneous stator current. If the stator current is a cosine function of time, the instantaneous value of the spatial peak of the pulsating mmf wave is

$$F_{1(peak)} = F_{1(max)} \cos \omega t$$

where $F_{1(max)}$ is the peak value corresponding to maximum instantaneous current. Derive the expression for $\mathcal{F}_1$, and verify that for a single-phase winding, both forward and backward components are present.

**18.17**  A 200-V, 60-Hz, 10-hp single-phase induction motor operates at an efficiency of 0.86 and a power factor of 0.9. What capacitor should be placed in parallel with the motor so that the feeder supplying the motor will operate at unity power factor?

**18.18**  A 230-V, 50-Hz, two-pole single-phase induction motor is designed to run at 3 percent slip, Find the slip in the opposite direction of rotation. What is the speed of the motor in the normal direction of rotation?

**18.19**  Determine the amount of time (in seconds) it will take for a stepper motor with a 15° step angle, operating in one-phase excitation mode, to rotate through 28 rev when the pulse rate is 180 pps. *Note:* $t = \theta/\omega$.

**18.20**  A $\frac{1}{4}$-hp, 110-V, 60-Hz, four-pole capacitor-start motor has the following parameters:

$$R_S = 2.02 \ \Omega \qquad X_S = 2.8 \ \Omega$$
$$R_R = 4.12 \ \Omega \qquad X_R = 2.12 \ \Omega$$
$$X_m = 66.8 \ \Omega \qquad s = 0.05$$

Find

a. The stator current.

b. The mechanical power.

c. The rotor speed.

**18.21**  A $\frac{1}{4}$-hp, four-pole, 110-V, 60-Hz single-phase induction motor has the following data:

$$R_S = 1.86 \ \Omega \qquad X_S = 2.56 \ \Omega$$
$$R_R = 3.56 \ \Omega \qquad X_R = 2.56 \ \Omega$$
$$X_m = 53.5 \ \Omega \qquad s = 0.05$$

Find the mechanical power output.

**18.22**  A one-phase, 115-V, 60-Hz, four-pole induction motor has the following parameters:

$$R_S = 0.5 \ \Omega \qquad X_S = 0.4 \ \Omega$$
$$R_R = 0.25 \ \Omega \qquad X_R = 0.4 \ \Omega$$
$$X_m = 35 \ \Omega$$

Find the input current and developed torque when the motor speed is 1,730 rev/min.

**18.23**  The no-load test of a single-phase induction motor is made by running the motor without load at rated voltage and rated frequency. Derive the equivalent circuit of a single-phase induction motor for the no-load test. [*Hint:* The no-load slip is very small.]

**18.24**  Derive the equivalent circuit of a single-phase induction motor for the locked-rotor test. Neglect the magnetizing current.

**18.25**  The design for a $\frac{1}{8}$-hp, two-pole, 115-V universal motor gives the effective resistances of the armature and series field as 4 $\Omega$ and 6 $\Omega$, respectively. The output torque is 0.17 N-m when the motor is drawing rated current of 1.5 A (rms) at a power factor of 0.88 at rated speed. Find:

a. The full-load efficiency.

b. The rated speed.

c. The full-load copper losses.

d. The combined windage, friction, and iron losses.

e. The motor speed when the rms current is 0.5 A, neglecting phase differences and saturation.

**18.26**  A 240-V, 60-Hz, two-pole universal motor operates at a speed of 12,000 rev/min on full load and draws a current of 6.5 A at 0.94 power factor lagging. The series field-winding impedance is $4.55 + j3.2$ ohms, and the armature circuit impedance is $6.15 + j9.4$ ohms. Find

a. The back emf of the motor.

b. The mechanical power developed by the motor.

c. The power output if the rotational loss is 65 W.

d. The efficiency of the motor.

**18.27**  A single-phase motor is drawing 20 A from a 400-V, 50-Hz supply. The power factor is 0.8 lagging.

What value of capacitor connected across the circuit will be necessary to raise the power factor to unity?

**18.28**   A three-phase induction motor is required to operate from a single-phase source. One possible connection is shown in Figure P18.28. Will the motor work? Explain why or why not.

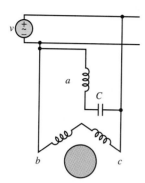

**Figure P18.28**

**18.29**   In performing a brake-load test upon a 1/4-hp capacitor-start motor with its output adjusted to rated value, the following data were obtained: $E = 115$ volts; $I = 3.8$ amp; $P = 310$ W; rev/min = 1725. Calculate:

a. Efficiency

b. Power factor

c. Torque in pound-inches

## Section 4: Motor Performance and Selection

**18.30**   What type of motor would you select to perform the following tasks? Justify your selection.

a. Vacuum cleaner

b. Refrigerator

c. Air conditioner compressor

d. Air conditioner fan

e. Variable-speed sewing machine

f. Clock

g. Electric drill

h. Tape drive

i. X-Y plotter

**18.31**   A 5-hp, 1,150-rev/min shunt motor has its speed controlled by means of a tapped field resistor as shown in Figure P18.31. With the tap at position 3, determine

the speed of the motor and the torque available at the maximum permissible load.

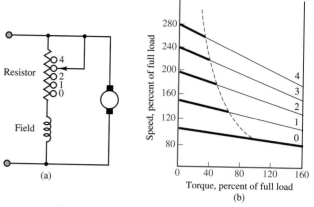

**Figure P18.31**

**18.32**   Which single-phase motor would you choose for the following applications?

a. Inexpensive analog electric clock.

b. Bathroom ventilator fan.

c. Escalator which must start under all load conditions.

d. Kitchen blender.

e. Table model circular saw operating at about 3,500 rev/min.

f. Hand-held circular saw operating at 15,000 rev/min.

g. Water pump.

**18.33**   The power required to drive a fan varies as the cube of the speed. If a motor driving a shaft-mounted fan is loaded to 100 percent of its horsepower rating on the top speed connection, what is the horsepower output in percent of rating:

a. At a speed reduction of 20 percent?

b. At a speed reduction of 30 percent?

c. At a speed reduction of 50 percent?

**18.34**   An industrial plant has a load of 800 kW at a power factor of 0.8 lagging. It is desired to purchase a synchronous motor of sufficient capacity to deliver a load of 200 kW and also serve to correct the overall plant power factor to 0.92. Assuming that the synchronous motor has an efficiency of 91 percent, determine its KVA input rating and the power factor at which it will operate.

# APPENDIX

# A

# Linear Algebra and Complex Numbers

## A.1 SOLVING SIMULTANEOUS LINEAR EQUATIONS, CRAMER'S RULE AND MATRIX EQUATION

The solution of simultaneous equations, such as those that are often seen in circuit theory, may be obtained relatively easily by using Cramer's rule. This method applies to $2 \times 2$ or larger systems of equations. Cramer's rule requires the use of the concept of determinant. The method of determinants is valuable because it is systematic, general, and useful in solving complicated problems. A determinant is a scalar defined on a square array of numbers, or matrix, such as

$$det(A) = |A| = \begin{vmatrix} a_{11} & a_{12} \\ a_{21} & a_{22} \end{vmatrix} \tag{A.1}$$

In this case the matrix is a $2 \times 2$ array, with two rows and two columns and its determinant is defined as

$$det = a_{11}a_{22} - a_{12}a_{21} \tag{A.2}$$

A third-order, or $3 \times 3$, determinant such as

$$det(A) = \begin{vmatrix} a_{11} & a_{12} & a_{13} \\ a_{21} & a_{22} & a_{23} \\ a_{31} & a_{32} & a_{33} \end{vmatrix} \tag{A.3}$$

is given by

$$\begin{aligned} det = a_{11}(a_{22}a_{33} - a_{23}a_{32}) &-a_{12}(a_{21}a_{33} - a_{23}a_{31}) \\ &+a_{13}(a_{21}a_{32} - a_{22}a_{31}) \end{aligned} \tag{A.4}$$

For higher-order determinants, you may refer to a linear algebra book. To illustrate Cramer's method, a set of two equations in general form will be solved here. A set of two linear simultaneous algebraic equations in two unknowns can be written in the form

$$\begin{aligned} a_{11}x_1 + a_{12}x_2 &= b_1 \\ a_{21}x_1 + a_{22}x_2 &= b_2 \end{aligned} \tag{A.5}$$

where $x_1$ and $x_2$ are the two unknowns to be solved for. The coefficients $a_{11}$, $a_{12}$, $a_{21}$ and $a_{22}$ are known quantities. The two quantities on the right-hand side, $b_1$ and $b_2$, are also known (these are typically the source currents and voltages in a circuit problem). The set of equations can be arranged in matrix form, as shown in equation A.6.

$$\begin{bmatrix} a_{11} & a_{12} \\ a_{21} & a_{22} \end{bmatrix} \begin{bmatrix} x_1 \\ x_2 \end{bmatrix} = \begin{bmatrix} b_1 \\ b_2 \end{bmatrix} \tag{A.6}$$

In equation A.6, a coefficient matrix multiplied by a vector of unknown variables is equated to a right-hand-side vector. Cramer's rule can then be applied to find $x_1$ and $x_2$ using the following formulas:

$$x_1 = \frac{\begin{vmatrix} b_1 & a_{12} \\ b_2 & a_{22} \end{vmatrix}}{\begin{vmatrix} a_{11} & a_{12} \\ a_{21} & a_{22} \end{vmatrix}}$$

$$x_2 = \frac{\begin{vmatrix} a_{11} & b_1 \\ a_{21} & b_2 \end{vmatrix}}{\begin{vmatrix} a_{11} & a_{12} \\ a_{21} & a_{22} \end{vmatrix}} \tag{A.7}$$

Thus, the solution is given by the ratio of two determinants: the denominator is the determinant of the matrix of coefficients, while the numerator is the determinant of the same matrix with the right-hand-side vector ($\begin{bmatrix} b_1 \\ b_2 \end{bmatrix}$ in this case) substituted in place of the column of the coefficient matrix corresponding to the desired variable (i.e., first column for $x_1$, second column for $x_2$, etc.). In a circuit analysis problem, the coefficient matrix is formed by the resistance (or conductance) values, the vector of unknowns is composed of the mesh currents (or node voltages), and the right-hand-side vector contains the source currents or voltages.

In practice, many calculations involve solving higher-order systems of linear equations. Therefore, a variety of computer software packages are often used to solve higher-order systems of linear equations.

## Check Your Understanding

**A.1**  Use Cramer's rule to solve the system

$$5v_1 + 4v_2 = 6$$
$$3v_1 + 2v_2 = 4$$

**A.2**  Use Cramer's rule to solve the system

$$i_1 + 2i_2 + i_3 = 6$$
$$i_1 + i_2 - 2i_3 = 1$$
$$i_1 - i_2 + i_3 = 0$$

**A.3**  Convert the following system of linear equations into a matrix equation as shown in equation A.6 and find matrices $A$ and $b$.

$$2i_1 - 2i_2 + 3i_3 = -10$$
$$-3i_1 + 3i_2 - 2i_3 + i_4 = -2$$
$$5i_1 - i_2 + 4i_3 - 4i_4 = 4$$
$$i_1 - 4i_2 + i_3 + 2i_4 = 0$$

## A.2    INTRODUCTION TO COMPLEX ALGEBRA

From your earliest training in arithmetic, you have dealt with real numbers such as 4, $-2$, $\frac{5}{9}$, $\pi$, $e$, etc., which may be used to measure distances in one direction or another from a fixed point. However, a number that satisfies the equation

$$x^2 + 9 = 0 \tag{A.8}$$

is not a real number. Imaginary numbers were introduced to solve equations such as equation A.8. Imaginary numbers add a new dimension to our number system. To deal with imaginary numbers, a new element, $j$, is added to the number system having the property

$$j^2 = -1$$

or                                                                                      **(A.9)**

$$j = \sqrt{-1}$$

Thus, we have $j^3 = -j$, $j^4 = 1$, $j^5 = j$, etc. Using equation A.9, you can see that the solutions to equation A.8 are $\pm j3$. In mathematics, the symbol $i$ is used for the imaginary unit, but this might be confused with current in electrical engineering. Therefore, the symbol $j$ is used in this book.

A complex number (indicated in boldface notation) is an expression of the form

$$A = a + jb \tag{A.10}$$

where $a$ and $b$ are real numbers. The complex number $A$ has a real part, $a$, and an imaginary part, $b$, which can be expressed as

$$a = \operatorname{Re} A$$
$$b = \operatorname{Im} A \tag{A.11}$$

It is important to note that $a$ and $b$ are both real numbers. The complex number $a + jb$ can be represented on a rectangular coordinate plane, called the *complex plane,* by interpreting it as a point $(a, b)$. That is, the horizontal coordinate is $a$ in real axis and the vertical coordinate is $b$ in imaginary axis, as shown in Figure A.1. The complex number $A = a + jb$ can also be uniquely located in the complex plane by specifying the distance $r$ along a straight line from the origin and the angle $\theta$, which this line makes with the real axis, as shown in Figure A.1. From the right triangle of Figure A.1, we can see that:

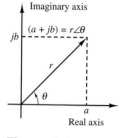

**Figure A.1** Polar form representation of complex numbers

$$r = \sqrt{a^2 + b^2}$$
$$\theta = \tan^{-1}\left(\frac{b}{a}\right)$$
$$a = r \cos \theta$$
$$b = r \sin \theta \tag{A.12}$$

Then, we can represent a complex number by the expression:

$$A = re^{j\theta} = r\angle\theta \tag{A.13}$$

which is called the polar form of the complex number. The number $r$ is called the magnitude (or amplitude) and the number $\theta$ is called the angle (or argument). The two numbers are usually denoted by: $r = |A|$ and $\theta = \arg A = \angle A$.

Given a complex number $A = a + jb$, the complex conjugate of $A$, denoted by the symbol $A^*$, is defined by the following equalities:

$$\operatorname{Re} A^* = \operatorname{Re} A$$
$$\operatorname{Im} A^* = -\operatorname{Im} A \tag{A.14}$$

That is, the sign of the imaginary part is reversed in the complex conjugate.

Finally, we should remark that two complex numbers are equal *if and only if* the real parts are equal and the imaginary parts are equal. This is equivalent to stating that two complex numbers are equal only if their magnitudes are equal and their arguments are equal.

The following examples and exercises should help clarify these explanations.

### EXAMPLE  A.1

Convert the complex number $A = 3 + j4$ to its polar form.

#### Solution

$$r = \sqrt{3^2 + 4^2} = 5 \quad \theta = \tan^{-1}\left(\frac{4}{3}\right) = 53.13°$$
$$A = 5\angle53.13°$$

## EXAMPLE   A.2

Convert the number $A = 4\angle -60°$ to its complex form.

### Solution

$$a = 4\cos(-60°) = 4\cos(60°) = 2$$
$$b = 4\sin(-60°) = -4\sin(60°) = -2\sqrt{3}$$

Thus, $A = 2 - j2\sqrt{3}$.

*Addition* and *subtraction* of complex numbers take place according to the following rules:

$$(a_1 + jb_1) + (a_2 + jb_2) = (a_1 + a_2) + j(b_1 + b_2)$$
$$(a_1 + jb_1) - (a_2 + jb_2) = (a_1 - a_2) + j(b_1 - b_2)$$

(A.15)

*Multiplication* of complex numbers in polar form follows the law of exponents. That is, the magnitude of the product is the product of the individual magnitudes, and the angle of the product is the sum of the individual angles, as shown below.

$$(A)(B) = (Ae^{j\theta})(Be^{j\phi}) = ABe^{j(\theta+\phi)} = AB\angle(\theta + \phi)$$

(A.16)

If the numbers are given in rectangular form and the product is desired in rectangular form, it may be more convenient to perform the multiplication directly, using the rule that $j^2 = -1$, as illustrated in equation A.17.

$$(a_1 + jb_1)(a_2 + jb_2) = a_1a_2 + ja_1b_2 + ja_2b_1 + j^2b_1b_2$$
$$= (a_1a_2 + j^2b_1b_2) + j(a_1b_2 + a_2b_1)$$
$$= (a_1a_2 - b_1b_2) + j(a_1b_2 + a_2b_1)$$

(A.17)

*Division* of complex numbers in polar form follows the law of exponents. That is, the magnitude of the quotient is the quotient of the magnitudes, and the angle of the quotient is the difference of the angles, as shown in equation A.18.

$$\frac{A}{B} = \frac{Ae^{j\theta}}{Be^{j\phi}} = \frac{A\angle\theta}{B\angle\phi} = \frac{A}{B}\angle(\theta - \phi)$$

(A.18)

Division in the rectangular form can be accomplished by multiplying the numerator and denominator by the complex conjugate of the denominator. Multiplying the denominator by its complex conjugate converts the denominator to a real number and simplifies division. This is shown in Example A.4. Powers and roots of a complex number in polar form follow the laws of exponents, as shown in equations A.19 and A.20.

$$A^n = (Ae^{j\theta})^n = A^ne^{jn\theta} = A^n\angle n\theta$$

(A.19)

$$A^{1/n} = (Ae^{j\theta})^{1/n} = A^{1/n}e^{j1/n\theta}$$

$$= \sqrt[n]{A}\angle\left(\frac{\theta + k2\pi}{n}\right) \quad k = 0, \pm1, \pm2, \ldots \quad \textbf{(A.20)}$$

## EXAMPLE  A.3

Perform the following operations given that $A = 2 + j3$ and $B = 5 - j4$.
 (a) $A + B$ (b) $A - B$ (c) $2A + 3B$

### Solution

$$A + B = (2 + 5) + j(3 + (-4)) = 7 - j$$
$$A - B = (2 - 5) + j(3 - (-4)) = -3 + j7$$

For part c, $2A = 4 + j6$ and $3B = 15 - j12$. Thus,
$2A + 3B = (4 + 15) + j(6 + (-12)) = 19 - j6$

## EXAMPLE  A.4

Perform the following operations both in rectangular and polar form, given that
$A = 3 + j3$ and $B = 1 + j\sqrt{3}$.
 (a) $AB$ (b) $A \div B$

### Solution

(a)  In rectangular form:

$$AB = (3 + j3)(1 + j\sqrt{3}) = 3 + j3\sqrt{3} + j3 + j^2 3\sqrt{3}$$
$$= (3 + j^2 3\sqrt{3}) + j(3 + j3\sqrt{3})$$
$$= (3 - 3\sqrt{3}) + j(3 + j3\sqrt{3})$$

To obtain the answer in polar form, we need to convert $A$ and $B$ to their polar forms:

$$A = 3\sqrt{2}e^{j45°} = 3\sqrt{2}\angle 45°$$
$$B = \sqrt{4}e^{j60°} = 2\angle 60°$$

Then,

$$AB = (3\sqrt{2}e^{j45°})(\sqrt{4}e^{j60°}) = 6\sqrt{2}\angle 105°$$

(b)  To find $A \div B$ in rectangular form, we can multiply $A$ and $B$ by $B^*$.

$$\frac{A}{B} = \frac{3 + j3}{1 + j\sqrt{3}}\frac{1 - j\sqrt{3}}{1 - j\sqrt{3}}$$

Then,

$$\frac{A}{B} = \frac{(3 + 3\sqrt{3}) + j(3 - 3\sqrt{3})}{4}$$

In polar form, the same operation may be performed as follows:

$$A \div B = \frac{3\sqrt{2}\angle 45°}{2\angle 60°} = \frac{3\sqrt{2}}{2}\angle(45° - 60°) = \frac{3\sqrt{2}}{2}\angle -15°$$

## Euler's Identity

This formula extends the usual definition of the exponential function to allow for complex numbers as arguments. Euler's identity states that

$$e^{j\theta} = \cos\theta + j\sin\theta \qquad (A.21)$$

All the standard trigonometry formulas in the complex plane are direct consequences of Euler's identity. The two important formulas are:

$$\cos\theta = \frac{e^{j\theta} + e^{-j\theta}}{2} \qquad \sin\theta = \frac{e^{j\theta} - e^{-j\theta}}{2j} \qquad (A.22)$$

## EXAMPLE  A.5

Using Euler's formula, show that

$$\cos\theta = \frac{e^{j\theta} + e^{-j\theta}}{2}$$

## Solution

Using Euler's formula

$$e^{j\theta} = \cos\theta + j\sin\theta$$

Extending the above formula, we can obtain

$$e^{-j\theta} = \cos(-\theta) + j\sin(-\theta) = \cos\theta - j\sin\theta$$

Thus,

$$\cos\theta = \frac{e^{j\theta} + e^{-j\theta}}{2}$$

## Check Your Understanding

**A.4**  In a certain AC circuit, $V = ZI$ where $Z = 7.75\angle 90°$ and $I = 2\angle -45°$. Find $V$.

**A.5**  In a certain AC circuit, $V = ZI$ where $Z = 5\angle 82°$ and $V = 30\angle 45°$. Find $I$.

**A.6**  Show that the polar form of $AB$ in Example A.4 is equivalent to its rectangular form.

**A.7**  Show that the polar form of $A \div B$ in Example A.4 is equivalent to its rectangular form.

**A.8**  Using Euler's formula, show that $\sin\theta = (e^{j\theta} - e^{-j\theta})/2j$.

# A P P E N D I X

# B

# Fundamentals of Engineering (FE) Examination

## B.1  INTRODUCTION

The *Fundamentals of Engineering* (FE) examination[1] is one of four steps to be completed toward registering as a Professional Engineer (PE). Each of the 50 states in the United States has laws that regulate the practice of engineering; these laws are designed to ensure that registered professional engineers have demonstrated sufficient competence and experience. The same exam is administered at designated times throughout the country, but each state's Board of Registration administers the exam and supplies information and registration forms. You will find a Web reference to the **phone numbers of the Boards of Registration in the 50 states** in the accompanying CD-ROM.

The four steps required to become a Professional Engineer are:

1. *Education.* Usually satisfied by completing a B.S. degree in engineering from an accredited college or university.

2. *Fundamentals of Engineering Examination.* An 8-hour examination described in the next section of this appendix.

3. *Experience.* Following successful completion of the Fundamentals of Engineering Examination, 2 to 4 years of engineering experience are required.

[1] This exam used to be called "Engineer in Training."

4. *Principles and Practices of Engineering Examination.* A second 8-hour examination, also known as the Professional Engineer (PE) Examination, which requires in-depth knowledge of one particular branch of engineering.

This appendix provides a review of the background material in electrical engineering required in the Electrical Engineering part of the FE exam. This exam is prepared by the National Council of Examiners for Engineering and Surveying[2] (NCEES).

## B.2   EXAM FORMAT AND CONTENT

The FE Examination is given twice a year (April and October) and is divided into two 4-hour sessions. The morning (AM) session consists of 120 multiple-choice questions (five possible answers). Each question is assigned 1 point. The topics covered in the AM session are listed below.

Mathematics

Statics

Dynamics

Mechanics of Materials

Fluid Mechanics

Thermodynamics

Chemistry

Materials Science/Structure of Matter

Electrical Circuits

Engineering Economics

The topics covered in the morning session are intended to be general and are drawn from the material covered in the first 2 years of an ABET curriculum. In particular, the Electrical Circuits section of the exam is intended to cover the material presented in the first electrical engineering core requirement of an undergraduate program.

The afternoon (PM) session lasts 4 hours, and consists of 60 multiple-choice questions, each worth 2 points. The PM session is devoted to specific disciplines, based on material drawn from the last 2 years of an ABET engineering curriculum. The six disciplines are:

General

Chemical

Civil

Electrical

Industrial

Mechanical

The candidate is required to select one discipline out of the six listed above.

---

[2]P.O. Box 1686 (1826 Seneca Road), Clemson, SC 29633-1686

## B.3    THE ELECTRICAL ENGINEERING SECTION OF THE FE EXAM

The following is a listing of Electric Circuits subject areas, which may be tested. Note that the AM session contains material that is required of all candidates, while the PM session topics are of interest to those candidates who choose the Electrical discipline for the PM session. The chapter covering each of these topics in this book is shown in parentheses next to the each topic. Note that the topics in the AM and PM Electrical Circuits sessions are the same; however, the PM session questions are more in-depth, and are intended for electrical engineering majors. This appendix is aimed primarily at the AM session.

AM    Electrical Circuits
- Three-Phase Circuits
  - Line and phase currents and voltages for delta and wye (Chapter 7)
  - Computation of power (Chapter 7)
- AC Circuits (Chapters 4 and 7)
  - Algebra of complex numbers (Appendix A)
  - Watts, Vars, apparent power and power factor (Chapter 7)
  - Phasor representation (Chapter 4)
  - Resonance (Chapter 6)
  - Reactance and impedance, susceptance and admittance (Chapter 4)
  - Concepts of time domain and frequency domain (Chapters 4 and 6)
  - Ideal Transformers (Chapter 7)
- Diode Applications (Chapter 8)
- DC Circuits (Chapter 3)
  - Relationship between current and charge
  - Current law and voltage law
  - Power and energy
  - Equivalent series and parallel elements
  - Theorems
    - Maximum power
    - Norton
    - Thévenin
- Electric and Magnetic fields (Chapter 16)
  - Work done in moving a charge in an electric field
  - Basic relationships; i.e., force versus distance
  - Faraday's law
  - Magnetic field of a current carrying conductor
  - Relationship between voltage and work (Chapter 2)
- Capacitance and inductance
  - Energy storage (Chapter 4)
  - *RL* and *RC* transients (Chapter 5)
    - Final value
    - Initial value
    - Time constant
- Operational amplifiers (ideal) (Chapter 12)
- Transients (Chapter 5)

PM    Electrical Circuits
- Three-phase circuits
- AC circuits

Diode applications
DC circuits
Electric and magnetic fields
Capacitance and inductance
Operational amplifiers
Transients

## B.4    REVIEW FOR THE ELECTRICAL ENGINEERING SECTION OF THE FE EXAMINATION

The remainder of this appendix is devoted to the presentation of background material and sample questions on each of the major topics of the Electrical Engineering section of the FE Examination. You will note that the topics have been rearranged in the more natural order in which they appear in this book.

### DC Circuits

The analysis of DC circuits forms the foundation of electrical engineering. Chapters 2 and 3 cover this material with a wealth of examples. The following exercises illustrate the general type of questions that might be encountered in the FE exam.

---

### Check Your Understanding

**B.1**    Assuming the connecting wires and the battery have negligible resistance, the voltage across the 25-$\Omega$ resistance in Figure B.1 is

a. 25 V        b. 60 V        c. 50 V        d. 15 V        e. 12.5 V

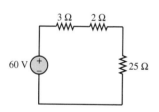

**Figure B.1**

**Solution:**
This problem calls for application of the voltage divider rule, discussed in Section 2.6. Applying the voltage divider rule to the circuit of Figure B.2, we have

$$v_{25\,\Omega} = 60\left(\frac{25}{3+2+25}\right) = 50 \text{ V}$$

Thus, the answer is c.

**B.2**    Assuming the connecting wires and the battery have negligible resistance, the voltage across the 6-$\Omega$ resistor in Figure B.2 is

a. 6 V        b. 3.5 V        c. 12 V        d. 4 V        e. 3 V

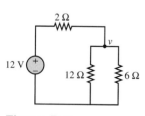

**Figure B.2**

**Solution:**
This problem can be solved most readily by applying nodal analysis (Section 3.1), since one of the node voltages is already known. Applying KCL at the node $v$, we obtain

$$\frac{12 - v}{2} = \frac{v}{6} + \frac{v}{12}$$

This equation can be solved to show that $v = 4$ V. Note that it is also possible to solve this problem by mesh analysis (Section 3.2). You are encouraged to try this method as well.

**B.3**    A 125-V battery charger is used to charge a 75-V battery with internal resistance of 1.5 $\Omega$. If the charging current is not to exceed 5 A, the minimum resistance in series with the charger must be

a. 10 Ω    b. 5 Ω    c. 38.5 Ω    d. 41.5 Ω    e. 8.5 Ω

**Solution:**
The circuit of Figure B.3 describes the charging arrangement. Applying KVL to the circuit of Figure B.3, we obtain

$$i_{max} R + 1.5 i_{max} - 125 + 75 = 0$$

and using $i = i_{max} = 5$ A, we can find $R$ from the following equation:

$$5R + 7.5 - 125 + 75 = 0$$
$$R = 8.5 \ \Omega$$

Thus, e is the correct answer.

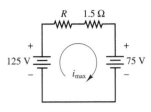

**Figure B.3**

## Capacitance and Inductance

The material on capacitance and inductance pertains to two basic areas: energy storage in these elements, and transient response of the circuits containing these elements. The examples below deal with the former part (covered in Chapter 4); the latter is covered under the heading "Transients" in this appendix, and in Chapter 5 of the book.

## Check Your Understanding

**B.4**   A coil with inductance of 1 H and negligible resistance carries the current shown in Figure B.4. The maximum energy stored in the inductor is:

a. 2 J    b. 0.5 J    c. 0.25 J    d. 1 J    e. 0.2 J

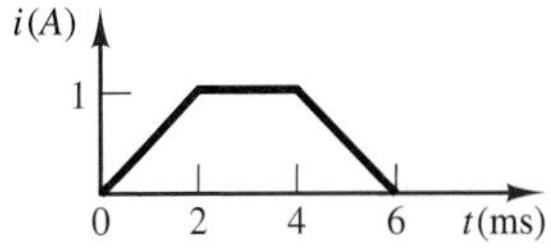

**Solution:**
The energy stored in an inductor is $W = \frac{1}{2} L i^2$ (see Section 4.1). Since the maximum current is 1 A, the maximum energy will be $W_{max} = \frac{1}{2} L i_{max}^2 = \frac{1}{2}$ J. Thus, b is the correct answer.

**Figure B.4**

**B.5**   The maximum voltage that will appear across the coil is:

a. 5 V    b. 100 V    c. 250 V    d. 500 V    e. 5,000 V

**Solution:**
Since the voltage across an inductor is given by $v = L(di/dt)$, we need to find the maximum (positive) value of $di/dt$. This will occur anywhere between $t = 0$ and $t = 2$ ms. The corresponding slope is:

$$\left. \frac{di}{dt} \right|_{max} = \frac{1}{2 \times 10^{-3}} = 500$$

Therefore $v_{max} = 1 \times 500 = 500$ V, and the correct answer is d.

## AC Circuits

AC circuit analysis emphasis proficiency in the use of complex algebra (see Appendix A for a review of this subject, including sample exercises), and is primarily concerned with the use of AC circuits in electrical power systems. The following exercises illustrate typical problems.

## Check Your Understanding

**B.6** A voltage sine wave of peak value 100 V is in phase with a current sine wave of peak value 4 A. When the phase angle is $60°$ later than a time at which the voltage and the current are both zero, the instantaneous power is most nearly

a. 250 W　　b. 200 W　　c. 400 W　　d. 150 W　　e. 100 W

**Solution:**

As discussed in Section 7.1, the instantaneous AC power $p(t)$ is

$$p(t) = \frac{VI}{2} \cos\theta + \frac{VI}{2} \cos(2\omega t + \theta_V + \theta_I)$$

In this problem, when the phase angle is $60°$ later than a "zero crossing," we have $\theta_V = \theta_I = 0$, $\theta = \theta_V - \theta_I = 0$, $2\omega t = 120°$. Thus, we can compute the power at this instant as

$$p = \frac{\left(\frac{100}{\sqrt{2}}\frac{4}{\sqrt{2}}\right)}{2} + \frac{\left(\frac{100}{\sqrt{2}}\frac{4}{\sqrt{2}}\right)}{2} \cos 120° = 250 \text{ W}$$

The correct answer is a.

**B.7** A sinusoidal voltage whose amplitude is $20\sqrt{2}$ V is applied to a 5-$\Omega$ resistor. The root-mean-square value of the current is

a. 5.66 A　　b. 4 A　　c. 7.07 A　　d. 8 A　　e. 10 A

**Solution:**

From Section 4.2, we know that

$$V_{\text{rms}} = \frac{V}{\sqrt{2}} = \frac{20\sqrt{2}}{\sqrt{2}} = 20 \text{ V}$$

Thus, $I_{\text{rms}} = 20/5 = 4$ A. Therefore, b is the correct answer.

**B.8** The magnitude of the steady-state root-mean-square voltage across the capacitor in the circuit of Figure B.5 is

a. 30 V　　b. 15 V　　c. 10 V　　d. 45 V　　e. 60 V

**Solution:**

This problem requires the use of impedances (Section 4.4). Using the voltage divider rule for impedances, we write the voltage across the capacitor as

$$\mathbf{V} = 30\angle 0° \times \frac{-j10}{10 - j10 + j10}$$

$$= 30\angle 0° \times (-j1) = 30\angle 0° \times 1\angle -90° = 30\angle 90°$$

Thus, the rms amplitude of the voltage across the capacitor is 30 V, and a is the correct answer. Note the importance of the phase angle in this kind of problem.

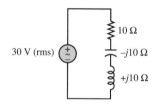

30 V (rms)

10 $\Omega$

$-j10$ $\Omega$

$+j10$ $\Omega$

**Figure B.5**

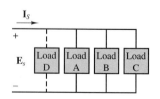

$\mathbf{I}_S$

+

$\mathbf{E}_s$

Load D　Load A　Load B　Load C

–

**Figure B.6**

The next set of questions (Exercises B.9 to B.28) pertain to single-phase AC power calculations and refer to the single-phase electrical network shown in Figure B.6. In this figure, $\mathbf{E}_S = 480\angle 0°$ V; $\mathbf{I}_S = 100\angle -15°$ A; $\omega = 120\pi$ rad/s. Further, load A is a bank of single-phase induction machines. The bank has an efficiency ($\eta$) of 80 percent, a power factor of 0.70 lagging, and a load of 20 hp. Load B is a bank of overexcited single-phase synchronous machines. The machines draw 15 kVA and the load current leads the line voltage by 30 degrees. Load C is a lighting (resistive) load and absorbs 10 kW. Load D is a proposed single-phase capacitor that will correct the source power factor to unity. This material is covered in Sections 7.1 and 7.2.

# Check Your Understanding

**B.9**   The root-mean-square magnitude of load A current, $I_A$, is most nearly

a. 44.4 A    b. 31.08 A    c. 60 A    d. 38.85 A    e. 55.5 A

**Solution:**

The output power $P_O$ of the single-phase induction motor is: $P_O = 20 \times 746 = 14{,}920$ W. The input electric power $P_{in}$ is:

$$P_{in} = \frac{P_O}{\eta} = \frac{14{,}920}{0.80} = 18{,}650 \text{ W}$$

$P_{in}$ can be expressed as:

$$P_{in} = E_S I_A \cos \theta_A$$

Therefore, the rms magnitude of the current $\mathbf{I}_A$ is found as

$$I_A = \frac{P_{in}}{E_S \cos \theta_A} = \frac{18{,}650}{480 \times 0.70} = 55.5015 \approx 55.5 \text{ A}$$

Thus, the correct answer is e.

**B.10**   The phase angle of $\mathbf{I}_A$ with respect to the line voltage $\mathbf{E}_S$ is most nearly

a. 36.87°    b. 60°    c. 45.6°    d. 30°    e. 48°

**Solution:**

The phase angle between $\mathbf{I}_A$ and $\mathbf{E}_S$ is:

$$\theta = \cos^{-1} 0.70 = 45.57° \approx 45.6°$$

The correct answer is c.

**B.11**   The power absorbed by synchronous machines is most nearly

a. 20,000 W    b. 7,500 W    c. 13,000 W    d. 12,990 W    e. 15,000 W

**Solution:**

The apparent power, $S$, is known to be 15 kVA, and $\theta$ is 30°. From the power triangle, we have

$$P = S \cos \theta$$

Therefore, the power drawn by the bank of synchronous motors is:

$$P = 15{,}000 \times \cos 30° = 12{,}990.38 \approx 12.99 \text{ kW}$$

The answer is d.

**B.12**   The power factor of the system before load $D$ is installed is most nearly

a. 0.70 lagging    b. 0.866 leading    c. 0.866 lagging    d. 0.966 leading    e. 0.966 lagging

**Solution:**

From the expression for the current $\mathbf{I}_S$, we have

$$pf = \cos \theta = \cos(0° - (-15°)) = \cos 15° = 0.966 \text{ lagging}$$

The correct answer is e.

**B.13**   The capacitance of the capacitor that will give a unity power factor of the system is most nearly

a. 219 $\mu$F    b. 187 $\mu$F    c. 132.7 $\mu$F    d. 240 $\mu$F    e. 132.7 pF

**Solution:**

The reactive power $Q_A$ in load $A$ is:

$$Q_A = P_A \times \tan \theta_A$$
$$\theta_A = \cos^{-1} 0.70 = 45.57°$$

Therefore,

$$Q_A = 18,650 \times \tan 45.57° = 19,025 \text{ VAR}$$

The total reactive power $Q_B$ in load $B$ is:

$$Q_B = S \times \sin \theta_B = 15,000 \times \sin(-30°) = -7,500 \text{ VAR}$$

The total reactive power $Q$ is:

$$Q = Q_A + Q_B = 19,025 - 7,500 = 11,525 \text{ VAR}$$

To cancel this reactive power, we set

$$Q_C = -Q = -11,525 \text{ VAR}$$

and

$$Q_C = -\frac{E_S^2}{X_C} \text{ and } X_C = -\frac{1}{\omega C}$$

Therefore, the capacitance required to obtain a power factor of unity is:

$$C = -\frac{Q_C}{\omega E_S^2} = \frac{11,525}{120\pi \times 480^2} = 132.7 \ \mu\text{F}$$

The correct answer is c.

## Transients

The FE Exam is primarily focused on first-order transients, as covered in Chapter 5. The concepts of initial and final value and of time constant are essential to the solution of problems of this nature.

## Check Your Understanding

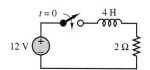

**Figure B.7**

**B.14**   The expression for the current in the 2-$\Omega$ resistor in Figure B.7 for time greater than zero is:

a. $-3e^{-0.5t} + 3$ A    b. $3e^{-0.5t} + 3$ A    c. $-3e^{0.5t} + 3$ A
d. $-6e^{0.5t} + 6$ A    e. $6e^{-0.5t} + 6$ A

**Solution:**
Applying KVL to the circuit when the switch is closed ($t \geq 0$), we have

$$4\frac{di}{dt} + 2i = 12$$

Solving the differential equation:

$$i(t) = Ke^{-0.5t} + 6 \text{ A} \qquad t \geq 0$$
$$i(0^-) = i(0^+) = 0$$

Therefore,

$$i(t) = -6e^{-0.5t} + 6 \text{ A} \qquad t \geq 0$$

Therefore, the correct answer is d.

## Three-Phase Circuits

Three-phase circuits are covered in Chapter 7. The FE exam requires proficiency in the analysis of simple delta and wye circuits, as illustrated in the following examples.

---

## Check Your Understanding

**B.15**  A 3-phase circuit is shown in Figure B.8. Load resistors (66 $\Omega$) are connected in delta and supplied by a 220-volt balanced three-phase source through three lines of 2-ohm resistance. The magnitude of the root-mean-square, line-to-line voltage across each 66-ohm resistor is most nearly

a. 198 V     b. 110 V     c. 201 V     d. 220 V     e. 120 V

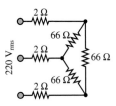

**Figure B.8**

**Solution:**
Since the load in this problem is $\Delta$-connected, it must first be converted to an equivalent Y-form. The phase impedance of the $\Delta$-connected load is 66 $\Omega$, so the equivalent phase impedance of the corresponding Y-form is:

$$Z_Y = \frac{Z_\Delta}{3} = \frac{66}{3} = 22 \ \Omega$$

The phase voltage is:

$$\mathbf{V}_\phi = \frac{208}{\sqrt{3}} \angle 0° \ \text{V}$$

The per-phase equivalent circuit of this problem is shown in Figure B.9. The load voltage is obtained by the voltage divider rule:

$$\mathbf{V}_L = \frac{22}{2 + 22} \times \mathbf{V}_\phi = \frac{208 \times 22}{\sqrt{3} \times 24} = \frac{201.67}{\sqrt{3}} \angle 0° \ \text{V}$$

The rms line-to-line voltage across the 66 $\Omega$ resistor therefore is:

$$\mathbf{V}_{66\Omega} = \sqrt{3}\mathbf{V}_L = 201.67 \angle 0° \ \text{V}$$

The correct answer is c.

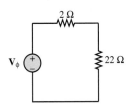

**Figure B.9**

**B.16**  A three-phase load is composed of three impedance of $9.0 + j9.0$ ohms and connected in wye. The balanced three-phase source is 208 volt (line to line). The current in each line is most nearly

a. 40 A     b. 16.3 A     c. 13.3 A     d. 9 A     e. 6 A

**Solution:**
The phase voltage is $208/\sqrt{3}$. Therefore, the magnitude of phase current $I_{an}$ is

$$I_{an} = \frac{\left(\frac{208}{\sqrt{3}}\right)}{9 + j9} = \frac{208}{\sqrt{3}} \frac{1}{12.73} = 9.43 \ \text{A}$$

The line current is equal to the phase current in a Y-connected system. The answer is d.

**Figure B.10**

The next four questions refer to a three-phase system with line-to-line voltage of 220 V rms, with *ABC* phase sequence, and with phase reference $V_{AB}$ shown in the phase diagram of Figure B.10. The load is a balanced delta connection, shown in Figure B.11 with branch impedances $Z = 30 - j40$ ohms, $j = \sqrt{-1}$.

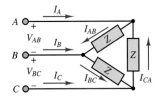

**Figure B.11**

## Check Your Understanding

**B.17**  The phase current is most nearly

a. $4.4\angle 53.13°$ A    b. $2.4\angle 53.13°$ A    c. $4.4\angle 0°$ A
d. $4.4\angle -53.13°$ A    e. $2.4\angle -53.13°$ A

**Solution:**

The load current $\mathbf{I}_{AB}$ is given by

$$\mathbf{I}_{AB} = \frac{\mathbf{V}_{AB}}{Z} = \frac{220\angle 0°}{30 - j40} = \frac{220\angle 0°}{50\angle -53.13°} = 4.4\angle 53.13° \text{A}$$

The correct answer is a.

**B.18**  The line current $I_A$ (in amperes) is most nearly

a. $4.4\angle -186.87°$    b. $4.4\angle 23°$    c. 7    d. $7.6\angle 23°$    e. $7\angle -186.87°$

**Solution:**

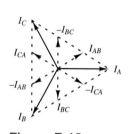

**Figure B.12**

The current phasor diagram for this $\Delta$-connected system is shown in Figure B.12. From the relationship among three-phase currents, we have

$$\mathbf{I}_{CA} = \mathbf{I}_{AB}\angle -240° = 4.4\angle(53.13° - 240°) = 4.4\angle -186.87° \text{A}$$

From the phasor diagram, we have

$$\mathbf{I}_A = \mathbf{I}_{AB} - \mathbf{I}_{CA} = 4.4\angle 53.13° - 4.4\angle -186.87°$$
$$= 2.64 + j3.52 - (-4.37 + j0.53) = 7.62\angle 23.1° \text{A}$$

Therefore, the correct answer is d.

**B.19**  The power factor is most nearly

a. 1.0    b. 0.6 leading    c. 0.866 leading    d. 0    e. 0.8 lagging

**Solution:**

The impedance angle $\theta$ is:

$$\theta = \tan^{-1}\frac{40}{30} = 53.13°$$

Therefore, the power factor is:

$$\text{pf} = \cos\theta = 0.6, \text{ leading}$$

We can also get the answer directly from the expression for the current $\mathbf{I}_{AB}$. The correct answer is b.

**B.20**  The total real power delivered from the source to the load is most nearly

a. 1496 W    b. 580 W    c. 1742 W    d. 2904 W    e. 850 W

**Solution:**

The total power $P$ delivered to the balanced load is:

$$P = 3V_{AB}I_{AB}\cos\theta = 3 \times 220 \times 4.4 \times 0.6 = 1742.4 \approx 1742 \text{ W}$$

The answer is c.

## Diode Applications

Chapter 8 covers all of the diode material that is relevant to the FE exam. The following examples illustrate typical questions.

## Check Your Understanding

**B.21**  The circuit of Figure B.13 is a

a. Peak detector.    b. Half-wave rectifier.    c. Bridge rectifier.
d. Voltage doubler.    e. Full-wave rectifier.

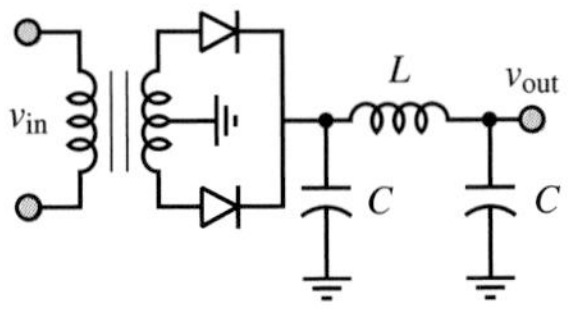

**Solution:**

The correct answer is e.

**B.22**  The inductor $L$ and the capacitor $C$ serve the function of

a. Converting the AC input to DC output.
b. Increasing the peak value of the output voltage.
c. Protecting the diodes.
d. A high-pass filter.
e. Reducing the ripple component of the output voltage.

**Solution:**

The correct answer is e.

**B.23**  The ideal diode $D$ in Figure B.14 will always conduct if:

a. $V_1$ is greater than $V_2$.    b. $V_2$ is greater than $V_1$.
c. $V_1$ is greater than 1 V.    d. $R_2$ is an open circuit.
e. $R_1$ is an open circuit.

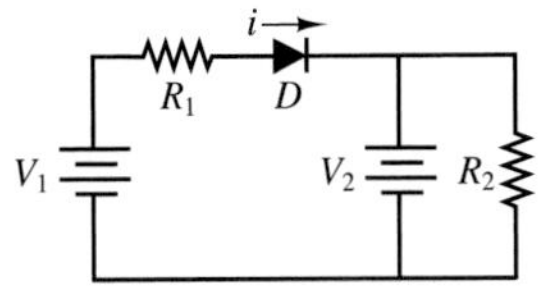

**Solution:**

Using the methodology developed in Chapter 8, we assume that $D$ conducts, resulting in the following expression for the diode current.

$$i = \frac{V_1 - V_2}{R_1}$$

This expression will result in a positive current only if $V_1$ is greater than $V_2$. When this condition applies, the assumption that the diode conducts is correct; thus, a is the correct answer.

**Figure B.13**

**Figure B.14**

## Operational Amplifiers

The coverage of operational amplifiers in this book is well beyond the scope of the FE Exam. Sections 12.1 to 12.3 cover all of the required material on ideal op-amps.

## Check Your Understanding

**B.24**  In the circuit of Figure B.15, which value is closest to $v_3$ if $R_1 = 2.2\,\text{k}\Omega$, $R_2 = 1.5\,\text{k}\Omega$, $R_3 = 18\,\text{k}\Omega$, $v_1 = 120\,\text{mV}$ and $v_2 = -40\,\text{mV}$?

a. $-250\,\text{mV}$    b. $500\,\text{mV}$    c. $-500\,\text{mV}$    d. $1.46\,\text{V}$    e. $-1.46\,\text{V}$

**Solution:**

Using the summing amplifier equation in Chapter 12, we calculate:

$$v_3 = -\frac{R_3}{R_1}v_1 - \frac{R_3}{R_2}v_2 = -\frac{18}{2.2}(120) - \frac{18}{1.5}(-40) = -501.8\,\text{mV}$$

Thus, the nearest answer is c.

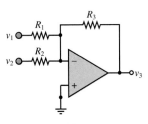

**Figure B.15**

**B.25**  In Figure B.15, if $R_1 = 2.2\,\text{k}\Omega$, $R_3 = 18\text{k}\Omega$, $v_1 = 120\,\text{mV}$ and $v_2 = -40\,\text{mV}$, choose the value of $R_2$ such that $v_3 = 0$.

a. $1.2\,\text{k}\Omega$    b. $5\,\text{k}\Omega$    c. $7.33\,\text{k}\Omega$    d. $0.733\,\text{k}\Omega$    e. $0.5\,\text{k}\Omega$

**Solution:**

Using the summing amplifier equation in Chapter 12, we calculate:

$$v_3 = -\frac{R_3}{R_1}v_1 - \frac{R_3}{R_2}v_2 \text{ or } 0 = -\frac{18}{2.2}(120) - \frac{18}{R_2}(-40)$$

$$\frac{120}{2.2} = \frac{40}{R_2}$$

$$R_2 = \frac{2.2 \times 40}{120} = 0.733 \text{ k}\Omega$$

Thus, the nearest answer is d.

## Electric and Magnetic Fields

Some of the basic ideas on electric fields, voltage, charge, and work are covered in Chapter 2. Faraday's law and other magnetic field concepts are covered in Chapter 16. Two examples are given below.

## Check Your Understanding

**B.26**   Which of the following is a true characteristic of magnetic flux lines?

a. They cross each other.
b. They begin and end on electric charges.
c. They are parabolic.
d. They are continuous.
e. None of the above.

**Solution:**

As discussed in Chapter 15, magnetic flux lines are continuous. Thus, d is the correct answer.

**B.27**   For the circuit of Figure B.16, where $i = 2A$, $\phi = 1 \times 10^{-3}$ Wb, cross-sectional area $= 5$ in$^2$, and the mean flux path length $= 2$ in, the total reluctance $\mathcal{R}$ of the magnetic circuit in $(A \cdot t \cdot \text{in}^2)/$Wb is

a. $1 \times 10^5$   b. $2 \times 10^5$   c. $1.5 \times 10^5$   d. $3.5 \times 10^4$   e. $2 \times 10^5$

**Solution:**

From Chapter 16, the relationship between magnetomotive force and flux is

$$\mathcal{F} = \phi \mathcal{R}$$

Also, the magnetomotive force is related to the current $i$ by

$$\mathcal{F} = Ni = 100 \times 2 = 200 \text{ A} \cdot \text{t}$$

which means that the reluctance is

$$\mathcal{R} = \frac{\mathcal{F}}{\phi} = \frac{200}{1 \times 10^{-3}} = 200,000 \frac{A \cdot t \cdot \text{in}^2}{Wb}$$

Therefore, the answer is b.

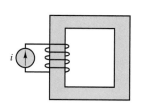

**Figure B.16**

The nonelectrical engineer will be frequently exposed to electrical machinery. This subject is covered in Chapters 17 and 18. The most popular electric machines in engineering applications are the DC motor and the AC induction motor. The former is discussed in the first half of Chapter 17, while the latter is discussed in Chapter 17 (three-phase motors) and in Chapter 18 (single-phase motors).

## Check Your Understanding

**B.28**  A four-pole synchronous motor operating from a 60-Hz supply will have a synchronous speed, in rev/min, of

a. 900     b. 1,800     c. 1,200     d. 3,600     e. 4,800

**Solution:**

$$\text{Frequency} = (\text{no. of poles}/2) \times \text{rev/min}/60)$$

Therefore,

$$60 = (4/2) \times (\text{rev/min}/60)$$

or rev/min = 1,800. Thus, b is the answer.

**B.29**  The armature resistance of a 55-hp, 525-V, DC shunt wound motor is 0.4 $\Omega$. The full-load armature current of this motor is 80 A. If the initial starting current is 175 percent of the full-load value, the resistance of the starting coil in $\Omega$ should be nearest to

a. 6.6125     b. 3.75     c. 3.35     d. 4.15     e. 2.75

**Solution:**

$$I_{\text{start}}(\text{starting current}) = 1.75 \times 80 = 140 \text{ A}$$

Therefore, the total resistance is

$$R = \frac{V}{I_S} = \frac{525}{140} = 3.75 \ \Omega$$

Thus, the resistance of the starter is $3.75 - 0.4 = 3.35 \ \Omega$. The correct answer is c.

**B.30**  The speed of an AC electric motor

a. Is independent of the frequency.
b. Is directly proportional to the square of the frequency.
c. Varies directly with the number of poles.
d. Varies inversely with the number of poles.
e. None of the above.

**Solution:**

The speed varies inversely with the number of poles and directly with the frequency. Thus, d is the correct answer.

# APPENDIX
## C

# Answers to Selected Problems

## Chapter 2

**2.4**  a. 360,000 C
b. $224.7 \times 10^{22}$

**2.12**  Element A: 300 W (dissipating)
Element B: 375 W (dissipating)
Element C: 675 W (supplying)

**2.15**  2.88 Ω, 1.92 Ω; 1.152 Ω.

**2.19**  $R = 18$ kΩ, $v = 16$ V, $v_1 = 2$ V, $I = 0.5$ mA.

**2.28**  a.  12 Ω    b.  0.5 A    c.  3 W    d.  −2 V    e.  1 W

**2.49**  a.  $v_{out}(x) = \dfrac{e^x}{2203}$    b. $x = 9.08$ cm

**2.54**  a.  $r_B = 0.061$ Ω    b.  $r_B = 84.1$ Ω

**2.56**  a.  $i \approx 1$ mA    b.  $r_a = 9.28$ Ω

**2.60**

|   | With meter in circuit | Without meter in circuit |
|---|---|---|
| a | 8.61 mA | 8.92 mA |
| b | 39.6 mA | 47.2 mA |
| c | 61.9 mA | 82.6 mA |
| d | 65.6 mA | 89.3 mA |

# Chapter 3

**3.3**  $i_1 = -0.143A$   $i_2 = -0.856A$

**3.13**  $v = 0.66V$

**3.17**  $i = 8.29A$

**3.20**  $A_V = \dfrac{v_2}{v_1} = -0.04$

**3.32**  $R_{TH} = 4\ \Omega$   $v_{TH} = -2.14$ V

**3.36**  $I_N = i_{SC} = 12.5$ A   $R_N = R_{TH} = 0.57\Omega$

**3.39**  $I_N = -3.05$ A

**3.46**  $i_o = 2i = 0 \Rightarrow R_N = \infty$

**3.58**  a.  $I = 52.2$ mA; $V = 4.57$ V      b. $R_{inc} = 43.8\ \Omega$
         c.  $I = 73$ mA; $V = 5.40$ V; $R_{inc} = 37\ \Omega$

# Chapter 4

**4.1**  $v_L(t) = 377\ \sin\left(377t - \dfrac{5\pi}{6}\right)$ V

**4.7**  $w_{1F} = 72$ J   $w_{1H} = 8$ J   $w_{2H} = 4$ J   $w_{2F} = 36$ J

**4.18**  $x_{rms} = 2.87$ V

**4.20**  $v_{rms} = 6.40$ V

**4.23**  $i_{rms} = 1.1547$ A

**4.38**  $i(t) = 2.12\cos\left(\omega t - \dfrac{\pi}{8}\right)$ A

**4.43**  $v_{out}(t) = 90\cos(\omega t)$ V

**4.48**  a.  $Z_T = 500 + j10.01\ \Omega$, $\mathbf{V}_T = 10\angle 0°$ V      b.  $Z_T = 500 - j10.01\ \Omega$, $\mathbf{V}_T = 10\angle 0°$ V

**4.50**  $i(t) = 0.2357\cos(2t + 45°)$ A

**4.54**  $Z_T = 2\ \Omega$

**4.57**  $V_T = 1.414\angle 45°$ V

# Chapter 5

**5.21**  a.  $v_c(0^-) = v_c(0^+) = 0$ V      b.  $\tau = 48$ s   c.  $v_c(t) = -8e^{-1/48t} + 8$ V   $t > 0$
         d.  $v_c(0) = 0$ V; $v_c(\tau) = 5.06$ V;   $v_c(2\tau) = 6.9$ V; $v_c(5\tau) = 7.95$ V;   $v_c(10\tau) = 8.0$ V

**5.23**  a.  $v_c(0^-) = 11.67$ V      b.  $V_c(t) = -11.09e^{-3/70t} + 11.67$ V   $t > 0$

**5.26**  $\tau = 0.923$ ms; $\tau' = 1.333$ ms.

**5.28**  a.  $\tau = 5.005$ ms      b.  $\tau = 5\ \mu s$

**5.39**  For $t > 5s$, the inductor current is:   $i(t) = 2 + e^{-0.041t}\{-3.641\cos[0.220(t-5)] + 1.77\sin[0.220(t-5)]\}$ A

**5.42**  $L = 1.6$ mH, $R = 560\ \Omega$

**5.45**  $v(t) = -12e^{-2t} - 12te^{-2t} + 12$ V for $t > 0$

**5.48**  $v(t) = 18e^{-t} - 3e^{-6t}$ V   $t > 0$

# Chapter 6

**6.2**  a.  $\left|\dfrac{\mathbf{V}_{out}}{\mathbf{V}_{in}}(j\omega)\right| = \dfrac{1}{\sqrt{4 + 0.01\omega^2}}$   $\phi(j\omega) = -\arctan(0.05\omega)$

**6.5**  a.  $\omega_{CO} = \sqrt{\dfrac{B}{C}}$
         b.  At high frequencies the slope is zero.
         c.  At low frequencies the Bode plot is sloping up at 20 dB/decade.
         d.  At high frequencies, $|V| \rightarrow \dfrac{A}{\sqrt{C}}$

**6.9**   a.   $Z_{ab} = \dfrac{j\omega L}{1 - \omega^2 LC}$     b.   $\omega = \dfrac{1}{\sqrt{LC}} = 10^6$ rad/s

**6.35**  a.   $Z_f(\omega = 2\pi 60) = 0.581 \times 10^6 - j30.8 \times 10^3$     b.   $|V_o(\omega = 2\pi 60)| = 258 \, \mu$ V

**6.38**  $\dfrac{V_o(j\omega)}{V_S(j\omega)} = \dfrac{-j\omega^3 K - \omega^2 K^2}{-j\omega^3 2K - \omega^2 \left(2K^2 + \dfrac{1}{K^2}\right)} + j\omega 2K + 1$ where $K = 2000\pi$

# Chapter 7

**7.2**   5.76 kW

**7.3**   a.   800 W     b.   800 W     c.   478.6 W     d.   1700 W

**7.6**   a.   0.56     b.   56.25°     c.   $5.56 \pm j8.31 \, \Omega$     d.   5.56 $\Omega$

**7.10**  a.   0.848 leading     b.   0.9925 lagging     c.   0.08716 leading     d.   0.9487 lagging

**7.13**  $P_1 = 112.5$ W, $Q_1 = 48.4$ VAR     $P_2 = -7.81$ W, $Q_2 = -33.2$ VAR

**7.17**  $S = 2,070.3\angle 28.97°$ VA

**7.20**  $\mathbf{I}_{OLD} = 10.4\angle -0.644$ A $\mathbf{I}_{NEW} = 8.77\angle -0.318$ A $C = 77.6 \, \mu$F

**7.30**  a.   869.57 A     b.   400 kW     c.   320 kW     d.   280 kW     e.   0.75

**7.34**  $\mathbf{V}_S = 245.1\angle -3.136$ V

**7.38**  $11.448\angle 2.36$ A

**7.41**  $\mathbf{I}_R = 22$ A; $\mathbf{I}_W = 22\angle(2\pi/3)$ A;   $\mathbf{I}_B = 22\angle\left(\dfrac{4\pi}{3}\right)$ A; $\mathbf{I}_N = 0$ A

**7.51**  $\mathbf{I}_A = 5.3\angle(0.244)$ A; $\mathbf{I}_B = 5.3\angle(-1.85)$ A $\mathbf{I}_C = 5.3\angle(-1.85)$ A; $\mathbf{I}_N = 2.56\angle\left(\dfrac{-\pi}{6}\right)$ A $P = 2262.8$ W

# Chapter 8

**8.9**   a.   $R = 860\Omega$     b.   $E_{min} = 1.56$ V

**8.11**  a.   Reverse-biased   b.   Forward-biased   c.   Reverse-biased   d.   Forward-biased   e.   Forward-biased

**8.12**  $V_{in} > 0$

**8.13**  a.   $D_2$ and $D_4$ are forward-biased; $D_1$ and $D_3$ are reverse-biased. $v_{out} = -4.3$ V.
         b.   $D_1$ and $D_2$ are reverse-biased; $D_3$ is forward-biased. $v_{out} = -9.3$ V.
         c.   $D_1$ is reverse-biased; $D_2$ is forward-biased. $v_{out} = 4.3$ V.

**8.20**  a.   Source voltage is a 16.97-V peak sinusoid. Load voltage is a 15.77-V half-wave rectified sinusoid. D1 and D4 are on during the positive half cycle of the source voltage; D2 and D3 are on during the negative half cycle of the source voltage.
         b.   The load voltage waveform is a periodic exponential charging-discharging waveform going from 1.96 to 15.77 V and back.
         c.   Now the amplitude of the "ripple" is greatly reduced, going from 13.34 V to 15.77 V.

**8.36**  $i = 14$ mA and $v = 0.6$ V

**8.38**  $R_{S_{min}} = 46.2 \, \Omega$; $R_{S_{max}} = 133.3 \, \Omega$

**8.51**  $\langle v_{out}\rangle = 4.3$ V

**8.53**  a.   $I_{SW} = 0$; $I_S = I_B = 0.31$ A
         b.   $I_S = 13$ A, $I_B = -0.96$ A; $I_{SW} = 13.96$ A
         c.   The battery will be drained of its charge because of the small resistance.
         a'.  $I_{SW} = 0$; $I_S = I_B = 0.25$ A
         b'.  $I_S = 13$ A, $I_B = 0$ A; $I_{SW} = 13$ A.
         c'.  The battery will not be drained, because of the large reverse resistance of the diode.

# Chapter 9

**9.2**   a.  Saturation    b.  Active    c.  Saturation    d.  Cutoff

**9.4**   a.  $V_{CE} = 6.65$ V    b.  $I_C = 5.9$ mA    c.  $P \approx 39$ mW

**9.13**  a.  $V_B = 0.3$ V   b.  $I_B = 15 \, \mu A$   c.  $I_E = 800 \, \mu A$   d.  $I_C = 785 \, \mu A$   e.  $\beta = 52.333$   f.  $\alpha = 0.981$

**9.23**  Active (triode) region.

**9.26**  $V_D = 2.2$ V; $R = 44.5$ kΩ

**9.31**  $i_D = 0.25$ mA

**9.33**  a.  $r_{DS} = 250 \, \Omega$    b.  $v_{GS} = 0$ V

# Chapter 10

**10.1**  a.  $I_B = 0.142$ mA; $I_C = 21.3$ mA; $V_{CE} = 5.78$ V    b.  $h_{ie} = 4.93$ kΩ; $h_{fe} \approx \beta = 150$

**10.5**  a.  $R_E = 1.81$ kΩ    b.  $R_C = 8.27$ kΩ    c.  $A_V = -4.15$

**10.8**  a.  $P_{max} = 0.02694 + 0.5 = 527$ mW    b.  3.69 ms

**10.10** a.  $A_i = 9,300$    b.  $R_{in} = 27.7$ kΩ

**10.48** $5,833 \, \Omega \le R_B \le 18,333 \, \Omega$

**10.52** This circuit performs the function of a 2-input NAND gate.

**10.55** The two transistors at the top are cut off and the two at the bottom are on.

# Chapter 11

**11.2**  $R_S = \dfrac{V_Z - V_{BE}}{I} = \dfrac{V_Z - 0.6}{I}$

**11.3**  $V_{out} = V_Z + V_\gamma$

**11.6**  a.  $v_L(t) = (10 - 0.7) \sin \omega t$ (positive half cycle)    b.  $\langle v_L \rangle \approx 2.96$ V

# Chapter 12

**12.7**  a.  $v_1 = \dfrac{v_g}{4}$    b.  $v_1 = \dfrac{v_g}{2}$

**12.11** $A_v = -2; G = 10$ S

**12.15** a.  $R_S = 10$ kΩ    b.  $v_o(t) = 0.001\,2 \times 10^{-3} \sin \omega t$ V

**12.17** $R_{S1} = 40$ kΩ; $R_{S2} = 2$ kΩ; $R_{S3} = 5$ kΩ; $R_{S4} = 0.625$ kΩ

**12.34** $R_3 \approx 102$ kΩ, $R_4 \approx 5$ MΩ

**12.46** a.  $\dfrac{V_{out}}{V_{in}} = -\dfrac{1 + j\omega R_2 C}{j\omega R_1 C}$    b.  0.04 dB    c.  Gain $= -1.0008$; phase $= 3.1$ rad    d.  $\omega > 196.5$ rad/s

**12.52** a.  $t = 104$ ms    b.  $t = 120$ ms

**12.55** $\dfrac{d^2 x}{dt^2} - 4,000x(t) = 16,000 f(t)$

**12.58** $CMRR_{min} = 74$ dB

**12.61** a.  $A_{CL} = 1.9999$    b.  $A_{CL} = 5.9999$

# Chapter 13

**13.2**  a.  10, 1010    b.  102, 1100110    c.  71, 1000111    d.  33, 100001    e.  19, 10011

**13.6**  a.  11100    b.  1101110    c.  1000

**13.13** $F(X, Y, Z) = \overline{YZ} + XYZ$

**13.16** $F = A + C$

**13.25**    $F(A, B, C, D) = (\overline{CD})((\overline{A} + \overline{B})\overline{C} + ABC)$

**13.28**    $f = \overline{BC} + A\overline{B} + \overline{A}BC$

**13.34**    $\overline{A} + \overline{B}$

**13.38**    $f = B\overline{D} + AC\overline{D}$

**13.45**    $F = \overline{AC}D + \overline{A}B\overline{C} + ABC$

**13.56**    The circuit operates as a 4 of 16 decoder.

## Chapter 14

**14.7**    A 10-kHz clock increments a 16-bit binary counter. The count is held by a latch, and then converted to BCD for use with seven-segment displays.

**14.7**    This is briefly discussed in the digital counters section, in Chapter 14.

**14.15**    a.  $n(n-1)$    b.  $2n$

**14.17**    "Static" means that memory contents do not have to be refreshed. "Nonvolatile" means that the information in the memory is not lost when the power is off.

## Chapter 15

**15.14**    $A = 21$

**15.16**    $A_{\text{diff}} = 110$

**15.20**    $n = 2$

**15.24**    $\dfrac{v_{\text{out}}(j\omega)}{v_s(j\omega)} = \dfrac{(j\omega)^2/K}{(j\omega)^2 + j\omega K_1 - K_2}$

$K = \dfrac{R_B}{R_A + R_B}$

$K_1 = \dfrac{1}{KR_1C_1} - \dfrac{1}{R_2C_2} - \dfrac{1}{R_1C_1} - \dfrac{1}{R_2C_1}$

$K_2 = \dfrac{1}{R_1R_2C_1C_2}$

**15.28**    $C_1 = C_2 = 1\ \mu\text{F};\ R_1 = 1.8\ \Omega,\ R_2 = 23\ \Omega$

**15.39**    1. Digital signals are less subject to noise.
2. Digital signals are directly compatible with digital computers, and can therefore be easily stored on a disk, or exchanged between computers.

**15.43**    a.  $v_a = -3.6\ \text{V}$    b.  $(v_a)_{\text{max}} = -4.5\ \text{V}$    c.  $\delta v = 0.3\ \text{V}$    d.  $n = 8$

**15.50**    a.  $v_a = -0.8425\ \text{V}$    b.  $(v_a)_{\text{max}} = -10\ \text{V}$    c.  $\delta v = 2.44\ \text{mV}$    d.  Choose a 15-bit ADC.

**15.55**    $n = 13$

**15.59**    $f_{\text{max}} = 104.15\ \text{Hz}$

**15.63**    $n = 9$

**15.67**    Since the required noise protection level is $\pm150\ \text{mV}$, $R_1 = 100\ \text{k}\Omega$, $R_2 = 1.8\ \text{k}\Omega$. $R_3 = 16.2\ \text{k}\Omega$

**15.76**    $C = 0.91\ \mu\text{F}$

## Chapter 16

**16.3**    a.  $W_m = 1.25\ \text{J}$    $L_\Delta = 2\ \text{H}$    b.  $v_L(t) = \sin(2\pi t) + 4\pi\cos(2\pi t)$

**16.22**    a.  $N = 5$    b.  $\mathbf{V}_1 = 353.6\angle -0.015\ \text{V}$    $\mathbf{I}_1 = 0.066\angle -0.89\ \text{A}$    $P_{\text{in}} = 14.98\ \text{W}$    efficiency $= \eta = 80.1\%$

**16.27**    $\alpha = 1/8$

**16.38**    $f = 173.5\ \text{N}$

**16.39**  a.  $R = \dfrac{x}{\mu_0 w^2} + \dfrac{l_g}{\mu_0 w(w-x)}$     b.  $W_m = \dfrac{(N_1 + N_2)^2 i^2}{2R}$     c.  $f = \dfrac{i^2}{2}\dfrac{(N_1 + N_2)^2 \mu_0 w^2}{x^2}$

**16.41**  $i = 1.784$ A

**16.45**  a.  $f = -1.55$ N

## Chapter 17

**17.2**   $\phi = 0.32$ mWb

**17.9**   a.  $I_a = 83.33$ A     b.  $V_a = 124.17$ V

**17.14**  $P_{loss} = 2.647$ kW; $P_{in} = 17.647$ kW.

**17.19**  a.  $i_a = 42.93$ A, $\omega_m = 106.3$ rad/sec    $P_{dev} = 8.952$ kW    $P_o = 8,452$ W    $T_{sh} = 72.1$ N-m

**17.24**  a.  $N_{series} = \dfrac{120}{8} = 15$ turns

b.  (1) $T_X = 1.13$ NM    (2) $T_X = 0.55$ N-m    (3) $T_X = 9.54$ N-m    (4) $T_X = 20.53$ N-m

**17.39**  $\mathbf{I}_s = 54.23\angle(0.42)$ A   $P_{in\,motor} = 18.85$ kW   $P_{in\,total} = 27.10$ kW   pf $= 0.991$ leading

**17.45**  $P_{max} = 3.096$ MW   $T_{max} = 123.2$ kN-m

**17.49**  $s = 0.025$; $\mathbf{I}_s = 54.6\angle(0.35)$ A;   $P_{in} = 35.6$ kW; $P_{sh} = 32.17$ kW;   $T_{sh} = 175$ N-m; efficiency $= 0.904$

**17.54**  $I_s = 11.02\angle -26.2°$ A; $P_{in} = 3,426$ W; $P_m = 3,121$ W; $T_{sh} = 17.24$ N-m

**17.62**  $I_s = 9.39\angle -(0.474)$ A   pf $= 0.8898$ lagging

## Chapter 18

**18.7**   $v = Ri + L\dfrac{di}{dt} + K_E\omega \quad T = K_T i = (J_m + J_L)\dfrac{d\omega}{dt} + D\omega + T_F + T_L$

**18.12**  The motor will require 24 stator teeth and 2 rotor teeth.

**18.16**   $F_1 = F_{1\,max}\cos(\omega t)\cos\theta$

$$= \frac{1}{2}F_{1\,max}\cos\theta\cos(\omega t) - \frac{1}{2}F_{1\,max}\cos\theta\sin(\omega t)$$

$$+ \frac{1}{2}F_{1\,max}\cos\theta\cos(\omega t) + \frac{1}{2}F_{1\,max}\cos\theta\sin(\omega t)$$

$$= F_{CW} + F_{CCW}$$

**18.21**  $P_{mech} = 201.68$ W

**18.25**  a.  efficiency $= 61.43\%$    b.  5,238.1 rev/min    c.  22.5 W    d.  36.05 W    e.  6,091.9 rev/min

**18.28**  It will work. The $b$ and $c$ windings will produce a magnetic field similar to a single phase machine, that is, two components rotating in opposite directions and the $a$ winding would act as a starting winding. The phase shift provided by the capacitor is needed to provide a nonzero starting torque.

# Index